TURING 图灵原创

U0277751

iOS开发指南

从Hello World到App Store上架

（第5版）

关东升 著

人民邮电出版社
北京

图书在版编目（CIP）数据

iOS开发指南 ：从Hello World到App Store上架 / 关东升著. -- 5版. -- 北京 ：人民邮电出版社，2017.6（2024.5重印）
（图灵原创）
ISBN 978-7-115-45063-0

Ⅰ. ①i… Ⅱ. ①关… Ⅲ. ①移动终端－应用程序－程序设计－教材 Ⅳ. ①TN929.53

中国版本图书馆CIP数据核字(2017)第040974号

内容提要

本书是 iOS 开发权威教程，以 Swift 和 Objective-C 两种语言进行讲解。书中分 5 部分介绍如何从零起步编写并上线 iOS 应用：第一部分介绍 iOS 开发的基础知识，包括界面构建技术、基本控件、协议、表视图、界面布局、屏幕适配、导航、iPad 应用开发、iOS 设备手势、Quartz 2D、iOS 动画等；第二部分介绍本地数据持久化、数据交换格式和 Web Service，第三部分为进阶篇，介绍了定位服务、苹果地图、Contacts 框架、应用扩展、用户通知等内容；第四部分介绍测试、调试和优化等相关知识；第五部分为实战篇，涵盖代码版本管理、项目依赖管理、App Store 发布流程，以及一个真实 iOS 应用的设计、编程、测试与发布过程。

与上一版相比，本书不仅基于 iOS 10 进行了全面更新，还增加并修订了设备手势、Quartz 2D、动画、用户扩展、用户通知、Core Data 等内容。

本书适合 iOS 开发人员阅读。

◆ 著　　　关东升
责任编辑　王军花
责任印制　彭志环

◆ 人民邮电出版社出版发行　　北京市丰台区成寿寺路 11 号
邮编　100164　　电子邮件　315@ptpress.com.cn
网址　https://www.ptpress.com.cn
固安县铭成印刷有限公司印刷

◆ 开本：880×1230　1/16
印张：46　　　　2017 年 6 月第 5 版
字数：1523千字　　　2024 年 5 月河北第 21 次印刷

定价：119.00元

读者服务热线：(010)84084456-6009　印装质量热线：(010)81055316

反盗版热线：(010)81055315

广告经营许可证：京东市监广登字 20170147 号

前　言

北京时间2016年10月13日凌晨，苹果公司正式推出了iOS 10系统。此次，Swift 3.0也正式推出，Swift经过两年的发展已经非常健壮了，很多项目也转而使用Swift语言开发，但是很多老项目还使用Objective-C语言开发。在这个大背景下，我们原来编写的《iOS开发指南：从零基础到App Store上架》得到了广大读者的认可，很多读者希望我们将其升级为iOS 10版本，并且能够使用Swift和Objective-C两种语言进行讲解，并比较它们的不同之处。几个月过去了，我们终于在2016年年底将书稿提交给出版社。

内容和组织结构

本书是我们团队编写的iOS系列丛书中的一本，目的是使有Swift或Objective-C基础的程序员通过学习本书，从零基础学习如何在App Store上发布一款应用。全书共5部分。

第一部分为基础篇，共13章内容，介绍了iOS的一些基础知识。

第1章介绍了iOS的开发背景以及本书约定。

第2章使用故事板技术创建了HelloWorld，同时讨论了iOS工程模板、应用的运行机制和生命周期、视图的生命周期等，最后介绍了如何使用API帮助文档和官方案例。

第3章重点介绍了Cocoa Touch框架中构建界面的相关类，重点介绍了构建界面的三种技术：故事板、XIB和纯代码。

第4章重点介绍了标签、按钮、文本框、文本视图、开关、滑块、分段控件、网页控件、活动指示器、进度条、警告框、操作表、工具栏、导航栏等基本控件，而且每个示例都采用故事板和纯代码两种方式实现。

第5章首先介绍了数据源协议和委托协议，然后介绍了高级视图：选择器和集合视图。

第6章探讨了表视图的组成、表视图类的构成和表视图的分类，使我们对表视图有了一个整体上的认识。接下来，介绍了如何实现简单表视图和分节表视图，以及表视图中索引、搜索栏和分组的使用，然后学习了如何对表视图单元格进行删除、插入、移动等操作，最后介绍了表视图UI设计模式方面的内容。

第7章讲述了界面布局，其中介绍了iOS应用布局模式，并比较了传统界面布局和Auto Layout布局技术的区别。此外，还介绍了静态表布局和堆视图StackView等布局相关的内容。

第8章介绍了iOS多分辨率屏幕适配方法，其中主要介绍了Size Class技术。

第9章讨论了如何判断应用是否需要一个导航功能，并且知道在什么情况下选择平铺导航、标签导航、树形结构导航，或者同时综合使用这3种导航模式。

第10章首先介绍了iPhone和iPad设备使用场景上的差异，然后介绍了iPad树形结构导航、iPad模态视图和Popover视图，最后介绍了iOS分屏多任务。

第11章介绍了iOS设备手势识别，这些手势包括Tap（点击）、Long Press（长按）、Pan（拖动）、Swipe（滑动）、Rotation（旋转）、Pinch（手指的合拢和张开）和Screen Edge Pan（屏幕边缘平移）。

第12章主要介绍Quartz 2D绘图技术，其中包括UIKit绘图技术、绘制视图的路径、绘制图像和文本、坐标、Quartz坐标和坐标变换。

第13章介绍了iOS动画技术，其中包括视图动画和Core Animation框架等。

第二部分为数据与网络通信篇，共3章，介绍了iOS本地数据持久化、数据交换以及如何进行网络通信，实现数据传输。

第14章讨论了iOS本地数据持久化的问题。首先分析了数据存取的几种方式以及每种数据存取方式适合什么样的场景，然后分别举例介绍了每种存取方式的实现。

第15章重点介绍了数据交换格式，其中XML和JSON是主要的方式。

第16章介绍了Web Service，并重点讨论了REST风格Web Service，其中具体访问Web Service的API包括：NSURLSession、AFNetworking框架和Alamofire框架。

第三部分为进阶篇，共5章，介绍了iOS的一些高级知识。

第17章讨论了iOS中的定位服务技术，包括地理信息编码和反编码查询。此外，还介绍了苹果的微定位技术iBeacon。

第18章讨论了iOS苹果地图的使用，包括显示地图、添加标注以及跟踪用户位置变化等。最后，我们介绍了程序外地图的使用。

第19章首先介绍了如何在iOS中访问通讯录框架——Contacts框架，然后介绍了如何使用Contacts框架访问联系人信息、写入联系人信息等。最后，还介绍了如何使用ContactsUI框架提供系统界面实现选择联系人、显示和修改联系人以及创建联系人的操作。

第20章讨论了iOS 10应用扩展，首先介绍了应用扩展的概念，然后介绍了Today扩展、开发表情包和Message框架。

第21章介绍了iOS 10用户通知，其中首先介绍了用户通知的概念，接着介绍了如何开发本地通知和推送通知。

第四部分为测试、调试和优化篇，共4章，介绍了iOS高级内容，包括测试、调试和优化等相关知识。

第22章首先介绍了iOS中都有哪些调试工具并重点介绍了几个常用的工具，具体包括日志与断言的输出、异常栈报告分析，接下来讲解了如何在设备上调试应用，最后分析了Xcode设备管理工具的使用。

第23章讨论了测试驱动的iOS开发，学习了测试驱动开发流程，单元测试框架XCTest，以及如何基于分层架构进行单元测试。

第24章介绍了UI测试框架。在这一章最后，我们还介绍了基于分层架构进行UI测试。

第25章分析了iOS中的性能优化方法，其中包括内存优化、资源文件优化、延迟加载、持久化优化、使用可重用对象、并发处理与多核CPU等。

第五部分为实战篇，共4章，介绍了iOS项目开发过程中相关的技术，包括：代码版本管理、项目依赖管理，以及App Store发布流程，最后还从无到有地介绍了一个真实的iOS应用：MyNotes应用和2020东京奥运会应用。

第26章介绍了如何使用Git进行代码版本控制，其中包括Git服务器的搭建、Git常用命令和协同开发。此外，还介绍了在Xcode中如何配置和使用Git工具。

第27章讨论了iOS和macOS项目依赖管理工具，其中包括：CocoaPods和Carthage。

第28章探讨了如何在App Store上发布应用，介绍了应用的发布流程以及应用审核不通过的一些常见原因。

第29章介绍了完整的iOS应用分析设计、编程、测试和发布过程，其中采用了敏捷开发方法。此外，该项目采用分层架构设计，这对于学习iOS架构是非常重要的。

书中并没有包括多媒体等知识，我们会在另外一本iOS开发书中介绍，具体进展请读者关注智捷iOS课堂官方网站http://www.51work6.com。

本书服务网址

为了更好地为广大读者提供服务，我们专门为本书建立了一个服务网址www.51work6.com/book/ios15.php，大家可以查看相关出版进度，对书中内容发表评论，并提出宝贵意见。

源代码

书中包括了100多个完整的案例项目源代码，大家可以到本书网站www.51work6.com/book/ios15.php下载或者到图灵社区本书主页（www.ituring.com.cn/book/1932）免费注册下载。

勘误与支持

我们在网站www.51work6.com/book/ios15.php中建立了一个勘误专区，可以及时地把书中的问题、失误和纠正反馈给广大读者。如果你发现了任何问题，均可以在网上留言，也可以发送电子邮件到eorient@sina.com，我们会在第一时间回复你。此外，你也可以通过新浪微博与我们联系，我的微博为@tony_关东升 。

致谢

在此感谢图灵的王军花编辑给我们提供的宝贵意见，感谢智捷iOS课堂团队的赵志荣参与内容讨论和审核，感谢赵大羽老师手绘了书中全部草图，并从专业的角度修改书中图片，力求更加真实完美地奉献给广大读者。此外，还要感谢我的家人容忍我的忙碌，以及对我的关心和照顾，使我能抽出这么多时间，投入全部精力专心编写本书。

由于时间仓促，书中难免存在不妥之处，请读者原谅。

关东升

2016年11月于北京

目　　录

第一部分　基　础　篇

第 1 章　开篇综述……2
1.1　iOS 概述……2
1.1.1　iOS 介绍……2
1.1.2　iOS 10 新特性……2
1.2　开发环境及开发工具……3
1.3　本书约定……4
1.3.1　案例代码约定……4
1.3.2　图示的约定……5
1.3.3　方法命名约定……6
1.3.4　构造函数命名约定……7
1.3.5　错误处理约定……7
第 2 章　第一个 iOS 应用程序……8
2.1　创建 HelloWorld 工程……8
2.1.1　通过 Xcode 创建工程……9
2.1.2　添加标签……11
2.1.3　运行应用……13
2.1.4　Xcode 中的 iOS 工程模板……13
2.1.5　应用剖析……14
2.2　应用生命周期……16
2.2.1　非运行状态——应用启动场景……18
2.2.2　点击 Home 键——应用退出场景……18
2.2.3　挂起重新运行场景……21
2.2.4　内存清除：应用终止场景……21
2.3　设置产品属性……22
2.3.1　Xcode 中的工程和目标……22
2.3.2　设置常用的产品属性……24
2.4　iOS API 简介……26
2.4.1　API 概述……26
2.4.2　如何使用 API 帮助文档……26
2.5　小结……28
第 3 章　Cocoa Touch 框架与构建应用界面……29
3.1　视图控制器……29
3.1.1　视图控制器种类……29
3.1.2　视图的生命周期……29
3.2　视图……30
3.2.1　UIView 继承层次结构……30
3.2.2　视图分类……32
3.2.3　应用界面构建层次……33
3.3　使用故事板构建界面……34
3.3.1　什么是故事板……34
3.3.2　场景和过渡……37
3.4　使用 XIB 文件构建界面……38
3.4.1　重构 HelloWorld……38
3.4.2　XIB 与故事板比较……42
3.5　使用纯代码构建界面……42
3.5.1　重构 HelloWorld……42
3.5.2　视图的几个重要属性……44
3.6　三种构建界面技术讨论……46
3.6.1　所见即所得……46
3.6.2　原型驱动开发……46
3.6.3　团队协同开发……47
3.7　小结……47
第 4 章　UIView 与视图……48
4.1　标签与按钮……48
4.1.1　Interface Builder 实现……48
4.1.2　代码实现……51
4.2　事件处理……53
4.2.1　Interface Builder 实现……53
4.2.2　代码实现……55
4.3　访问视图……57
4.3.1　Interface Builder 实现……57
4.3.2　代码实现……58
4.4　TextField 和 TextView……60
4.4.1　Interface Builder 实现……60
4.4.2　代码实现……62
4.4.3　键盘的打开和关闭……64
4.4.4　关闭和打开键盘的通知……64
4.4.5　键盘的种类……65
4.5　开关控件、分段控件和滑块控件……67

4.5.1　开关控件……67
4.5.2　分段控件……69
4.5.3　滑块控件……70
4.6　Web视图：WKWebView类……72
4.7　警告框和操作表……77
4.7.1　使用UIAlertController实现警告框……77
4.7.2　使用UIAlertController实现操作表……79
4.8　等待相关的控件与进度条……81
4.8.1　活动指示器ActivityIndicatorView……81
4.8.2　进度条ProgressView……83
4.9　工具栏和导航栏……86
4.9.1　工具栏……86
4.9.2　导航栏……90
4.10　小结……95
第5章　委托协议、数据源协议与高级视图……96
5.1　视图中的委托协议和数据源协议……96
5.2　选择器……96
5.2.1　日期选择器……96
5.2.2　普通选择器……101
5.2.3　数据源协议与委托协议……106
5.3　集合视图……108
5.3.1　集合视图的组成……108
5.3.2　集合视图数据源协议与委托协议……109
5.4　案例：奥运会比赛项目……109
5.4.1　创建工程……110
5.4.2　自定义集合视图单元格……110
5.4.3　添加集合视图……112
5.4.4　实现数据源协议……114
5.4.5　实现委托协议……115
5.5　小结……115
第6章　表视图……116
6.1　概述……116
6.1.1　表视图的组成……116
6.1.2　表视图的相关类……117
6.1.3　表视图分类……117
6.1.4　单元格的组成和样式……119
6.1.5　数据源协议与委托协议……121
6.2　简单表视图……122
6.2.1　实现协议方法……122
6.2.2　UIViewController根视图控制器……123
6.2.3　UITableViewController根视图控制器……129
6.3　自定义表视图单元格……133
6.3.1　Interface Builder实现……133
6.3.2　代码实现……137
6.4　添加搜索栏……138
6.5　分节表视图……143
6.5.1　添加索引……143
6.5.2　分组……146
6.6　插入和删除单元格……147
6.6.1　Interface Builder实现……149
6.6.2　代码实现……154
6.7　移动单元格……155
6.8　表视图UI设计模式……157
6.8.1　分页模式……157
6.8.2　下拉刷新模式……158
6.8.3　下拉刷新控件……159
6.9　小结……160
第7章　界面布局……161
7.1　界面布局概述……161
7.1.1　表单布局模式……161
7.1.2　列表布局模式……161
7.1.3　网格布局模式……162
7.2　iOS中各种“栏”……162
7.3　传统界面布局问题……163
7.4　Auto Layout布局技术……164
7.4.1　在Interface Builder中管理Auto Layout约束……164
7.4.2　案例：Auto Layout布局……164
7.5　静态表布局……169
7.5.1　什么是静态表……169
7.5.2　案例：iMessage应用登录界面……170
7.6　使用堆视图StackView……174
7.6.1　堆视图与布局……174
7.6.2　案例：堆视图布局……175
7.7　小结……180
第8章　屏幕适配……181
8.1　iOS屏幕的多样性……181
8.1.1　iOS屏幕介绍……181
8.1.2　iOS的3种分辨率……182
8.1.3　获得iOS设备的屏幕信息……183
8.2　Size Class与iOS多屏幕适配……184
8.2.1　在Xcode 6和Xcode 7中使用Size Class技术……184
8.2.2　Size Class的九宫格……185
8.2.3　Size Class的四个象限……186
8.2.4　在Xcode 8中使用Size Class……187
8.2.5　案例：使用Size Class……190
8.3　资源目录与图片资源适配……192

8.4 小结 ······ 195

第 9 章 视图控制器与导航模式 ······ 196

9.1 概述 ······ 196
9.1.1 视图控制器的种类 ······ 196
9.1.2 导航模式 ······ 196
9.2 模态视图 ······ 197
9.2.1 Interface Builder 实现 ······ 199
9.2.2 代码实现 ······ 205
9.3 平铺导航 ······ 206
9.3.1 应用场景 ······ 206
9.3.2 基于分屏导航的实现 ······ 208
9.3.3 基于电子书导航的实现 ······ 213
9.4 标签导航 ······ 219
9.4.1 应用场景 ······ 219
9.4.2 Interface Builder 实现 ······ 220
9.4.3 代码实现 ······ 223
9.5 树形结构导航 ······ 225
9.5.1 应用场景 ······ 225
9.5.2 Interface Builder 实现 ······ 226
9.5.3 代码实现 ······ 233
9.6 组合使用导航模式 ······ 235
9.6.1 应用场景 ······ 235
9.6.2 Interface Builder 实现 ······ 236
9.6.3 代码实现 ······ 243
9.7 小结 ······ 246

第 10 章 iPad 应用开发 ······ 247

10.1 iPad 与 iPhone 应用开发的差异 ······ 247
10.1.1 应用场景不同 ······ 247
10.1.2 导航模式不同 ······ 247
10.1.3 API 不同 ······ 247
10.2 iPad 树形结构导航 ······ 248
10.2.1 “邮件”应用中的树形结构导航 ······ 248
10.2.2 Master-Detail 应用程序模板 ······ 249
10.2.3 使用 Interface Builder 实现 SplitViewSample 案例 ······ 254
10.2.4 使用代码实现 SplitViewSample 案例 ······ 258
10.3 iPad 模态视图 ······ 261
10.3.1 “邮件”应用中的模态导航 ······ 261
10.3.2 iPad 模态导航相关 API ······ 262
10.3.3 使用 Interface Builder 实现 ModalViewSample 案例 ······ 264
10.3.4 使用代码实现 ModalViewSample 案例 ······ 268
10.4 Popover 视图 ······ 271
10.4.1 Popover 相关 API ······ 272
10.4.2 PopoverViewSample 案例 ······ 272
10.5 分屏多任务 ······ 276
10.5.1 Slide Over 多任务 ······ 276
10.5.2 分屏视图多任务 ······ 278
10.5.3 画中画多任务 ······ 280
10.6 iPad 分屏多任务适配开发 ······ 280
10.6.1 分屏多任务前提条件 ······ 280
10.6.2 分屏多任务适配 ······ 281
10.7 小结 ······ 282

第 11 章 手势识别 ······ 283

11.1 手势种类 ······ 283
11.2 手势识别器 ······ 284
11.2.1 视图对象与手势识别 ······ 284
11.2.2 手势识别状态 ······ 285
11.2.3 实例：识别 Tap 手势 ······ 285
11.2.4 实例：识别 Long Press 手势 ······ 290
11.2.5 实例：识别 Pan 手势 ······ 291
11.2.6 实例：Swipe 手势 ······ 293
11.2.7 实例：Rotation 手势 ······ 295
11.2.8 实例：Pinch 手势 ······ 297
11.2.9 实例：Screen Edge Pan 手势 ······ 299
11.3 小结 ······ 301

第 12 章 Quartz 2D 绘图技术 ······ 302

12.1 绘制技术基础 ······ 302
12.1.1 视图绘制周期 ······ 302
12.1.2 实例：填充屏幕 ······ 302
12.1.3 填充与描边 ······ 305
12.1.4 绘制图像和文本 ······ 306
12.2 Quartz 图形上下文 ······ 308
12.3 Quartz 路径 ······ 309
12.3.1 Quartz 路径概述 ······ 309
12.3.2 实例：使用贝塞尔曲线 ······ 311
12.4 Quartz 坐标变换 ······ 312
12.4.1 坐标系 ······ 312
12.4.2 2D 图形的基本变换 ······ 315
12.4.3 CTM 变换 ······ 317
12.4.4 仿射变换 ······ 322
12.5 小结 ······ 322

第 13 章 动画技术 ······ 323

13.1 视图动画 ······ 323
13.1.1 动画块 ······ 323

13.1.2 动画结束的处理……325
13.1.3 过渡动画……326
13.2 Core Animation 框架……329
13.2.1 图层……329
13.2.2 隐式动画……331
13.2.3 显式动画……333
13.2.4 关键帧动画……335
13.2.5 使用路径……336
13.3 小结……338

第二部分 数据与网络通信篇

第 14 章 数据持久化……340
14.1 概述……340
14.1.1 沙箱目录……340
14.1.2 持久化方式……341
14.2 实例：MyNotes 应用……341
14.3 属性列表……343
14.4 使用 SQLite 数据库……349
14.4.1 SQLite 数据类型……350
14.4.2 添加 SQLite3 库……350
14.4.3 配置 Swift 环境……351
14.4.4 创建数据库……351
14.4.5 查询数据……353
14.4.6 修改数据……355
14.5 iOS 10 中的 Core Data 技术……357
14.5.1 对象关系映射技术……357
14.5.2 添加 Core Data 支持……358
14.5.3 Core Data 栈……359
14.6 案例：采用 Core Data 重构 MyNotes 应用……360
14.6.1 建模和生成实体……360
14.6.2 Core Data 栈 DAO……366
14.6.3 查询数据……367
14.6.4 修改数据……369
14.7 小结……370
第 15 章 数据交换格式……371
15.1 XML 数据交换格式……372
15.1.1 XML 文档结构……372
15.1.2 解析 XML 文档……373
15.2 案例：MyNotes 应用读取 XML 数据……374
15.2.1 使用 NSXML 解析……376
15.2.2 使用 TBXML 解析……379
15.3 JSON 数据交换格式……385
15.3.1 JSON 文档结构……385
15.3.2 JSON 数据编码/解码……386
15.4 案例：MyNotes 应用 JSON 解码……387
15.5 小结……388
第 16 章 REST Web Service……389
16.1 概述……389
16.1.1 REST Web Service 概念……389
16.1.2 HTTP 协议……389
16.1.3 HTTPS 协议……390
16.1.4 苹果 ATS 限制……390
16.2 使用 NSURLSession……391
16.2.1 NSURLSession API……392
16.2.2 简单会话实现 GET 请求……393
16.2.3 默认会话实现 GET 请求……396
16.2.4 实现 POST 请求……397
16.2.5 下载数据……398
16.3 实例：使用 NSURLSession 重构 MyNotes 案例……400
16.3.1 插入方法……401
16.3.2 修改方法……402
16.3.3 删除方法……402
16.4 使用 AFNetworking 框架……405
16.4.1 比较 ASIHTTPRequest、AFNetworking 和 MKNetworkKit……405
16.4.2 安装和配置 AFNetworking 框架……405
16.4.3 实现 GET 请求……407
16.4.4 实现 POST 请求……408
16.4.5 下载数据……408
16.4.6 上传数据……410
16.5 使用为 Swift 设计的网络框架：Alamofire……412
16.5.1 安装和配置 Alamofire 框架……412
16.5.2 实现 GET 请求……413
16.5.3 实现 POST 请求……414
16.5.4 下载数据……414
16.5.5 上传数据……415
16.6 反馈网络信息改善用户体验……416
16.6.1 使用下拉刷新控件改善用户体验……416
16.6.2 使用活动指示器控件……419
16.6.3 使用网络活动指示器……420
16.7 小结……421

第三部分 进 阶 篇

第 17 章 定位服务……424
17.1 定位服务概述……424
17.1.1 定位服务编程……425
17.1.2 测试定位服务……428

17.2 管理定位服务 ······ 432
17.2.1 应用启动与停止下的定位服务管理 ······ 432
17.2.2 视图切换下的定位服务管理 ······ 432
17.2.3 应用前后台切换下的定位服务管理 ······ 433
17.2.4 设置自动暂停位置服务 ······ 436
17.2.5 后台位置服务管理 ······ 436
17.3 地理信息编码与反编码 ······ 437
17.3.1 地理信息反编码 ······ 437
17.3.2 实例：地理信息反编码 ······ 437
17.3.3 地理信息编码查询 ······ 438
17.3.4 实例：地理信息编码查询 ······ 439
17.4 小结 ······ 440
第 18 章 苹果地图应用 ······ 441
18.1 使用 iOS 苹果地图 ······ 441
18.1.1 显示地图 ······ 441
18.1.2 显示 3D 地图 ······ 445
18.2 添加标注 ······ 446
18.2.1 实现查询 ······ 447
18.2.2 在地图上添加标注 ······ 449
18.3 跟踪用户位置变化 ······ 450
18.4 使用程序外地图 ······ 451
18.5 小结 ······ 454
第 19 章 访问通讯录 ······ 455
19.1 通讯录的安全访问设置 ······ 455
19.2 使用 Contacts 框架读取联系人信息 ······ 456
19.2.1 查询联系人 ······ 457
19.2.2 读取单值属性 ······ 460
19.2.3 读取多值属性 ······ 461
19.2.4 读取图片属性 ······ 464
19.3 使用 Contacts 框架写入联系人信息 ······ 464
19.3.1 创建联系人 ······ 465
19.3.2 修改联系人 ······ 467
19.3.3 删除联系人 ······ 468
19.4 使用系统提供的界面 ······ 469
19.4.1 选择联系人 ······ 470
19.4.2 显示和修改联系人 ······ 472
19.5 小结 ······ 475
第 20 章 iOS 10 应用扩展 ······ 476
20.1 应用扩展概述 ······ 476
20.1.1 iOS 10 应用扩展种类 ······ 476
20.1.2 应用扩展工作原理 ······ 477
20.1.3 应用扩展的生命周期 ······ 478
20.2 Today 应用扩展 ······ 478
20.2.1 使用 Today 应用扩展 ······ 479
20.2.2 实例：奥运会倒计时牌 ······ 480
20.3 开发表情包 ······ 483
20.3.1 iMessage 应用 ······ 483
20.3.2 表情包 ······ 484
20.3.3 实例：开发表情包 ······ 484
20.4 Message 框架 ······ 488
20.4.1 Message 框架的主要 API ······ 488
20.4.2 消息布局 ······ 488
20.4.3 消息扩展界面的收缩和展开 ······ 489
20.4.4 消息应用的生命周期 ······ 490
20.4.5 消息会话 ······ 490
20.4.6 实例：高斯模糊滤镜 ······ 491
20.5 小结 ······ 494
第 21 章 重装上阵的 iOS 10 用户通知 ······ 495
21.1 用户通知概述 ······ 495
21.1.1 通知种类 ······ 495
21.1.2 通知界面 ······ 495
21.1.3 设置通知 ······ 497
21.2 开发本地通知 ······ 498
21.2.1 开发本地通知案例 ······ 498
21.2.2 请求授权 ······ 500
21.2.3 通知的创建与发送 ······ 501
21.2.4 通知接收后的处理 ······ 503
21.3 开发推送通知 ······ 503
21.3.1 推送通知机理 ······ 504
21.3.2 生成 SSL 证书 ······ 504
21.3.3 iOS 客户端编程 ······ 513
21.3.4 在 iOS 设备上运行客户端 ······ 515
21.3.5 内容提供者推送通知 ······ 518
21.4 小结 ······ 519
第四部分 测试、调试和优化篇
第 22 章 找出程序中的 bug——调试 ······ 522
22.1 Xcode 调试工具 ······ 522
22.1.1 定位编译错误 ······ 522
22.1.2 查看和显示日志 ······ 523
22.1.3 设置和查看断点 ······ 524
22.1.4 调试工具 ······ 530
22.1.5 输出窗口 ······ 532
22.1.6 变量查看窗口 ······ 532
22.1.7 查看线程 ······ 533
22.2 LLDB 调试工具 ······ 535
22.2.1 断点命令 ······ 535

22.2.2 观察点命令……537
22.2.3 查看变量和计算表达式命令……538
22.3 日志与断言输出……541
22.3.1 使用 NSLog 函数……541
22.3.2 使用断言……542
22.4 异常栈报告分析……543
22.4.1 跟踪异常栈……543
22.4.2 分析栈报告……545
22.5 在 iOS 设备上调试……546
22.5.1 Xcode 设置……546
22.5.2 设备设置……546
22.6 Xcode 设备管理工具……548
22.6.1 查看设备上的应用程序……548
22.6.2 设备日志……550
22.7 小结……551
第 23 章 iOS 测试驱动与单元测试……552
23.1 测试驱动的软件开发概述……552
23.1.1 测试驱动的软件开发流程……552
23.1.2 测试驱动的软件开发案例……553
23.1.3 iOS 单元测试框架……555
23.2 使用 XCTest 测试框架……555
23.2.1 添加 XCTest 到工程……555
23.2.2 编写 XCTest 测试方法……558
23.2.3 运行测试用例目标……561
23.2.4 分析测试报告……562
23.3 异步单元测试……564
23.4 性能测试……567
23.4.1 测试用例代码……568
23.4.2 分析测试结果……568
23.5 小结……571
第 24 章 iOS 应用 UI 测试……572
24.1 UI 测试概述……572
24.2 添加 UI 测试到工程……572
24.2.1 创建工程时添加 UI 测试框架……572
24.2.2 在现有工程中添加 UI 测试用例目标…573
24.3 录制脚本……575
24.3.1 录制之前的准备……575
24.3.2 录制过程……575
24.3.3 修改录制脚本……577
24.4 访问 UI 元素……577
24.4.1 UI 元素的层次结构树……577
24.4.2 UI 测试中相关 API……579
24.5 表示逻辑组件测试最佳实践……579
24.5.1 备忘录查询操作……580
24.5.2 增加备忘录操作……580
24.5.3 删除备忘录操作……581
24.5.4 显示备忘录详细信息操作……582
24.6 小结……582
第 25 章 让你的程序“飞”起来——性能优化……583
25.1 内存优化……583
25.1.1 内存管理……583
25.1.2 使用 Analyze 工具检查内存泄漏……583
25.1.3 使用 Instruments 工具检查内存泄漏…587
25.1.4 使用 Instruments 工具检查僵尸对象…592
25.1.5 autorelease 的使用问题……594
25.1.6 响应内存警告……595
25.2 优化资源文件……596
25.2.1 图片文件优化……597
25.2.2 音频文件优化……598
25.3 延迟加载……599
25.3.1 资源文件的延迟加载……599
25.3.2 故事板文件的延迟加载……603
25.3.3 XIB 文件的延迟加载……605
25.4 数据持久化的优化……607
25.4.1 使用文件……607
25.4.2 使用 SQLite 数据库……611
25.4.3 使用 Core Data……612
25.5 可重用对象的使用……613
25.5.1 表视图中的可重用对象……614
25.5.2 集合视图中的可重用对象……615
25.5.3 地图视图中的可重用对象……617
25.6 并发处理……618
25.6.1 一些概念……618
25.6.2 主线程阻塞问题……618
25.6.3 选择 NSThread、NSOperation 还是 GCD……619
25.6.4 GCD 技术……619
25.7 小结……620

第五部分 实 战 篇

第 26 章 管理好你的程序代码——代码版本控制……622
26.1 概述……622
26.1.1 版本控制历史……622
26.1.2 基本概念……623
26.2 Git 代码版本控制……623

26.2.1 服务器搭建……623
26.2.2 Gitolite服务器管理……625
26.2.3 Git常用命令……627
26.2.4 Git分支……628
26.2.5 Git协同开发……632
26.2.6 Xcode中Git的配置与使用……634
26.3 GitHub代码托管服务……642
26.3.1 创建和配置GitHub账号……642
26.3.2 创建代码库……645
26.3.3 删除代码库……646
26.3.4 派生代码库……647
26.3.5 管理组织……650
26.4 小结……653
第27章 项目依赖管理……654
27.1 使用CocoaPods工具管理依赖……654
27.1.1 安装CocoaPods……654
27.1.2 搜索库……655
27.1.3 项目与第三方库搭配形式……656
27.1.4 实例：静态链接库形式管理依赖……657
27.1.5 实例：框架形式管理依赖……659
27.2 使用Carthage工具管理依赖……660
27.2.1 安装Carthage……660
27.2.2 项目与第三方库搭配形式……661
27.2.3 Cartfile文件……661
27.2.4 实例：重构MyNotes依赖关系……662
27.3 小结……664
第28章 把应用放到App Store上……665
28.1 收官……665
28.1.1 在Xcode中添加图标……665
28.1.2 在Xcode中添加启动界面……668
28.1.3 调整Identity和Deployment Info属性……671
28.2 为发布进行编译……672
28.2.1 创建开发者证书……672
28.2.2 创建App ID……675
28.2.3 创建描述文件……676
28.2.4 发布编译……679
28.3 发布上架……680
28.3.1 创建应用……681
28.3.2 应用定价……683
28.3.3 基本信息输入……684
28.3.4 上传应用……687
28.3.5 提交审核……689
28.4 常见审核不通过的原因……691
28.4.1 功能问题……691
28.4.2 用户界面问题……691
28.4.3 商业问题……691
28.4.4 不当内容……691
28.4.5 其他问题……692
28.5 小结……692
第29章 iOS开发项目实战——2020东京奥运会应用开发及App Store发布……693
29.1 应用分析与设计……693
29.1.1 应用概述……693
29.1.2 需求分析……693
29.1.3 原型设计……694
29.1.4 数据库设计……694
29.2 任务1：创建应用工程……695
29.2.1 迭代1.1：创建工程……696
29.2.2 迭代1.2：发布到GitHub……696
29.3 任务2：数据库与数据持久化逻辑组件开发……696
29.3.1 迭代2.1：编写数据库DDL脚本……696
29.3.2 迭代2.2：插入初始数据到数据库……697
29.3.3 迭代2.3：数据库版本控制……697
29.3.4 迭代2.4：配置数据持久化逻辑组件……698
29.3.5 迭代2.5：编写实体类……698
29.3.6 迭代2.6：编写DAO类……700
29.3.7 迭代2.7：数据库帮助类DBHelper……704
29.3.8 迭代2.8：发布到GitHub……706
29.4 任务3：表示逻辑组件开发……706
29.4.1 迭代3.1：使用资源目录管理图片和图标资源……707
29.4.2 迭代3.2：根据原型设计初步设计故事板……708
29.4.3 迭代3.3："首页"模块……709
29.4.4 迭代3.4："比赛项目"模块……710
29.4.5 迭代3.5："比赛日程"模块……714
29.4.6 迭代3.6："倒计时"模块……718
29.4.7 迭代3.7："关于我们"模块……719
29.4.8 迭代3.8：发布到GitHub……720
29.5 任务4：收工……720
29.5.1 迭代4.1：添加图标……720
29.5.2 迭代4.2：设计和添加启动界面……720
29.5.3 迭代4.3：性能测试与改善……721
29.5.4 迭代4.4：发布到GitHub……722
29.5.5 迭代4.5：在App Store上发布应用……722
29.6 小结……722

Part 1 第一部分

基 础 篇

本部分内容

- 第 1 章 开篇综述
- 第 2 章 第一个 iOS 应用程序
- 第 3 章 Cocoa Touch 框架与构建应用界面
- 第 4 章 UIView 与视图
- 第 5 章 委托协议、数据源协议与高级视图
- 第 6 章 表视图
- 第 7 章 界面布局
- 第 8 章 屏幕适配
- 第 9 章 视图控制器与导航模式
- 第 10 章 iPad 应用开发
- 第 11 章 手势识别
- 第 12 章 Quartz 2D 绘图技术
- 第 13 章 动画技术

第 1 章

开篇综述

App Store自上线以来，创造了很多神话，给我们这些程序员提供了展示自己的舞台，给了我们创意的空间，给了我们创业的机会。下面就让我们从这里开始iOS开发之旅吧。

1.1 iOS 概述

在本节中，我们将了解什么是iOS以及iOS 10有哪些新特性。

1.1.1 iOS 介绍

iOS是由苹果公司开发的移动设备操作系统，这些移动设备包括iPhone、iPod touch和iPad等，目前最新的操作系统是iOS 10。

苹果公司最早于2007年1月9日的Macworld大会上公布了这个系统，最初是设计给iPhone使用的，后来陆续被应用到iPod touch和iPad等产品上。iOS与苹果的macOS 操作系统一样，都属于类Unix的商业操作系统。

原本这个系统的名字为iPhone OS，因为主要应用于iPhone和iPod touch设备，后来在2010 WWDC大会上宣布改名为iOS。

1.1.2 iOS 10 新特性

iOS的最新版本为iOS 10。苹果公司于北京时间2016年10月13日开放其正式版的下载，它支持iPhone 5、iPhone5c、iPhone 5s、iPhone 6、iPhone 6 Plus、iPhone 6s、iPhone 6s Plus、iPhone 7、iPhone 7 Plus、iPod touch 6、iPad 4、iPad Air、iPad Air 2、iPad mini 2、iPad mini 3、iPad mini 4和iPad Pro等设备。

现在我们先简要介绍一下iOS 10几个重要的变化。

- **用户通知增强**。iOS 10增强了用户通知功能，通知栏可以展示图片和媒体内容，开发人员还可以自定义用户通知界面。
- **增强Siri功能**。Siri会更加智能，具有更强的语言识别能力。Siri将成为一个人工智能机器人，具备深度学习的功能。
- **增强iMesseage功能**。为了增强用户之间的即时聊天功能，iMesseage在聊天时可以发送图片、视频、链接和表情贴纸等。
- **开放Siri API**。从iOS 10开始，苹果开放Siri API给第三方开发者。
- **开发iMesseage**。从iOS 10开始，第三方开发者可以开发iMesseage和表情包应用或扩展。
- **增强HomeKit功能**。HomeKit可以与Siri相连接。
- **增强电话功能**。更新了电话功能，增加来电时识别骚扰电话。允许用户使用360手机卫士、腾讯手机管家拦截欺诈电话。

1.2 开发环境及开发工具

苹果公司于2008年3月6日发布了iPhone和iPod touch的应用程序开发包，其中包括Xcode开发工具、iPhone SDK和iPhone手机模拟器。第一个Beta版本是iPhone SDK 1.2b1（build 5A147p），发布后立即就能使用，但是同时推出的App Store所需要的固件更新直到2008年7月11日才发布。编写本书时，iOS SDK 10.1版本已经发布。

iOS开发工具主要是Xcode。自从Xcode 3.1发布以后，Xcode就成为iPhone软件开发工具包的开发环境。Xcode可以用于开发Mac OS X和iOS应用程序，其版本与SDK相对应。例如，Xcode 7与iOS SDK 9对应、Xcode 8与iOS SDK 10对应。

在Xcode 4.1之前，还有一个配套使用的工具Interface Builder，它是Xcode套件的一部分，用来设计窗体和视图，可用于“所见即所得”地拖曳控件并定义事件等，其数据以XML的形式存储在XIB文件中。自Xcode 4.1起，Interface Builder成为了Xcode的一部分，与Xcode集成在一起。

打开Xcode 8工具，看到的主界面如图1-1所示。该界面主要分成5个区域：①号区域是菜单栏，其中的菜单几乎包括Xcode的所有功能；②号区域是工具栏，其中的按钮可以完成大部分工作；③号区域是导航面板，主要是对工作空间中的内容进行导航；④号区域是代码编辑区，我们的编码工作就是在这里完成的；⑤号区域是检查器，包括：文件检查器、属性检查器、标识检查器和尺寸检查器等。

导航面板上面还有一排按钮，如图1-2所示，默认选中的是“文件”导航面板。关于各个按钮的具体用法，我们会在以后用到的时候详细介绍。

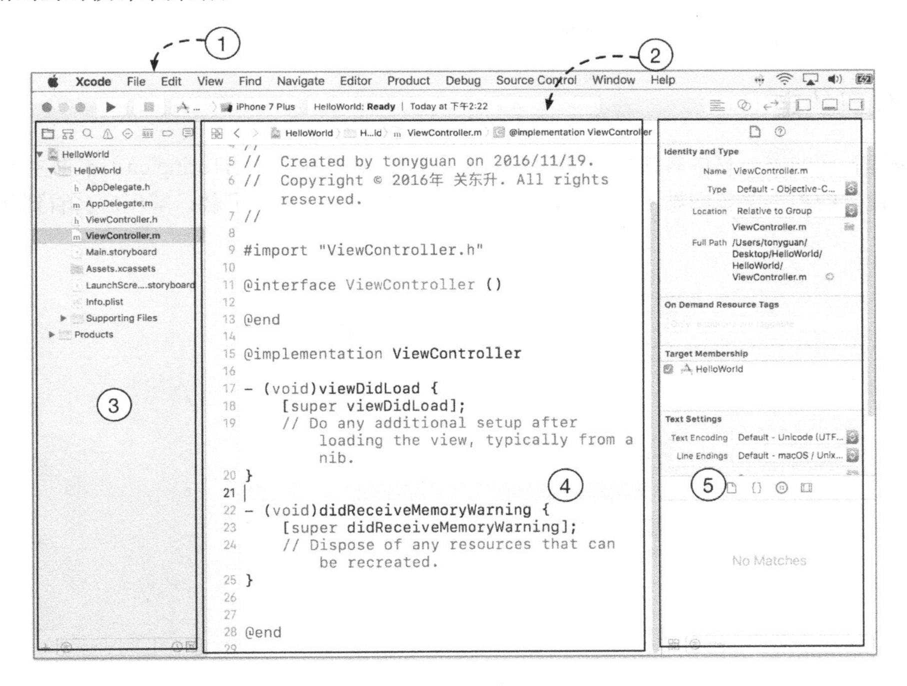

图1-1 Xcode主界面

图1-2 Xcode导航面板

选中导航面板时，导航栏下面也有一排按钮，如图1-3所示。这是辅助按钮，它们的功能都与该导航面板的内容相关。对于不同的导航面板，这些按钮也是不同的。

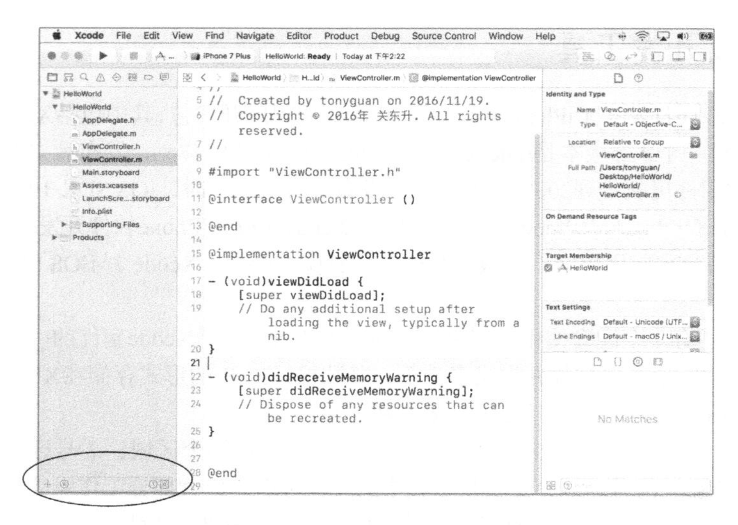

图1-3　导航面板的辅助按钮

1.3　本书约定

为了方便大家阅读本书，本节介绍一下书中案例代码和图示的相关约定。

1.3.1　案例代码约定

作为一本编程方面的书，书中有很多案例代码，我们可以从图灵网站（iTuring.cn）本书主页免费注册下载，或者从智捷课堂提供的本书服务网站（www.51work6.com/book/ios15.php）下载，解压后会看到如图1-4所示的目录结构。

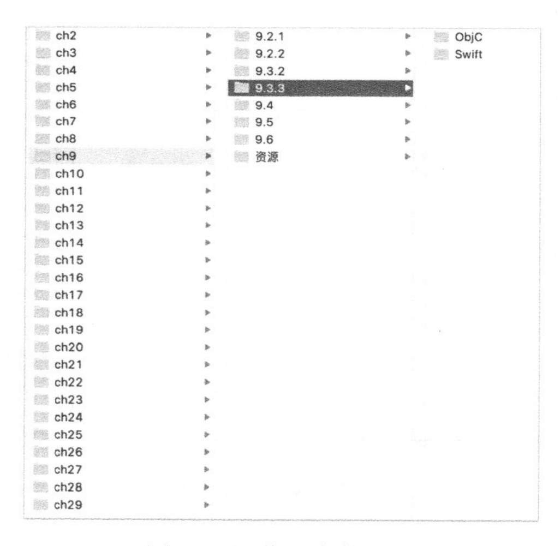

图1-4　源代码文件目录

ch2 ~ ch29代表第2章到第29章的案例代码或一些资源文件，其中工程或工作空间的命名有如下几种形式。

- 二级目录标号，如“3.4”说明是3.4节中使用的工程（或工作空间）案例。
- 三级目录标号，如“3.8.3”说明是3.8.3节中使用的工程（或工作空间）案例。
- 有 ~ 的情况，如“3.8.1~3.8.2”说明是3.8.1节到3.8.2节共同使用的工程（或工作空间）案例。
- 对于没有标号的情况，由其所在的父目录说明是哪个章节的案例工程（或工作空间），如“TokyoOlympics”

说明是在第29章中使用的。

此外，一般情况下，会有ObjC和Swift两个目录，ObjC目录中是Objective-C版本的工程，Swift目录中是Swift版本的工程。

书中出现的代码如果有Objective-C和Swift两个版本，会以左右两栏的形式显示，左栏为Swift代码，右栏为Objective-C代码。

1.3.2 图示的约定

为了更形象有效地说明知识点或描述操作，本书添加了很多图示，下面简要说明图示中一些符号的含义。

- **图中的圈框**。有时读者会看到如图1-5所示的圈框，其中是选中或要重点说明的内容。

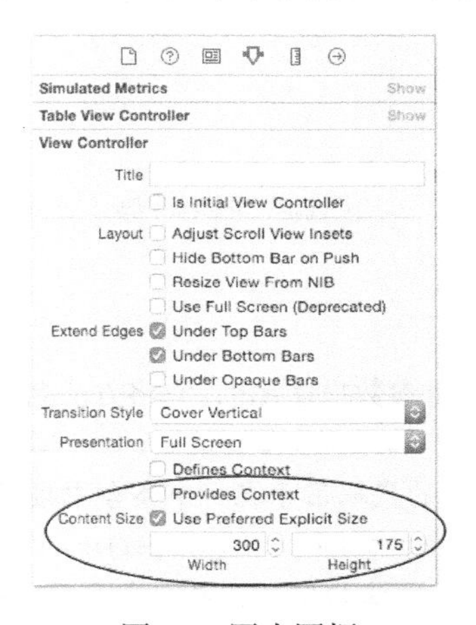

图1-5 图中圈框

- **图中的箭头**。如图1-6和图1-7所示，箭头用于说明用户的动作，一般箭尾是动作开始的地方，箭头指向动作结束的地方。图1-7所示的虚线箭头在书中用得比较多，常用来描述设置控件的属性等操作，箭头指向代表打开*XXX*检查器。

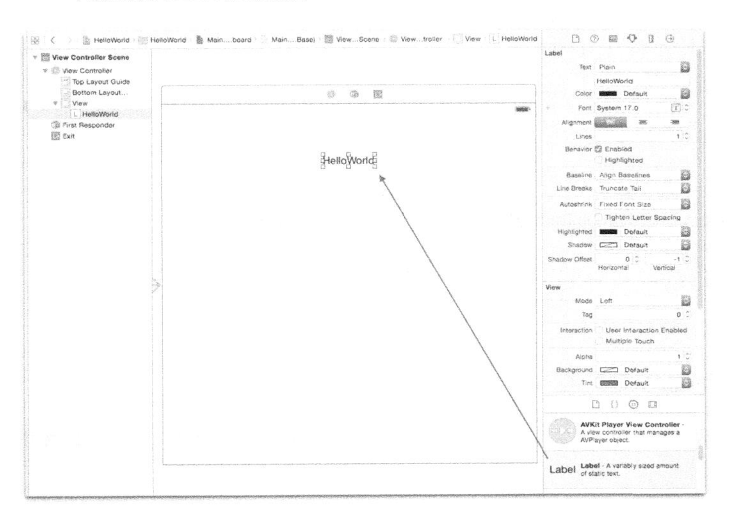

图1-6 图中箭头1

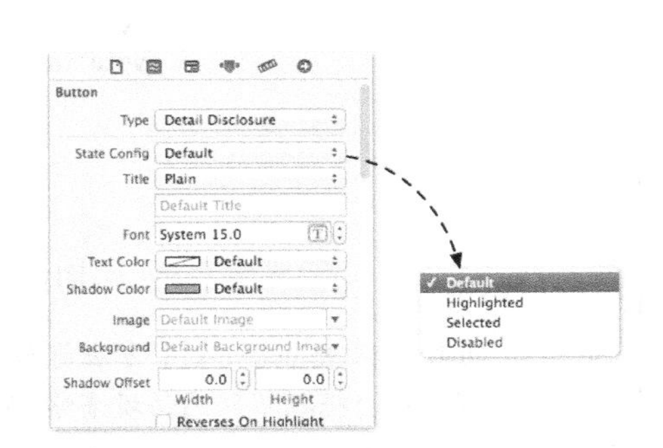

图1-7 图中箭头2

- **图中手势**。为了描述操作，我们在图中放置了👆等手势符号，这说明点击了该处的按钮。如图1-8所示，屏幕下方的"更多..."按钮上面就有这个手势，说明用户点击了"更多..."按钮。

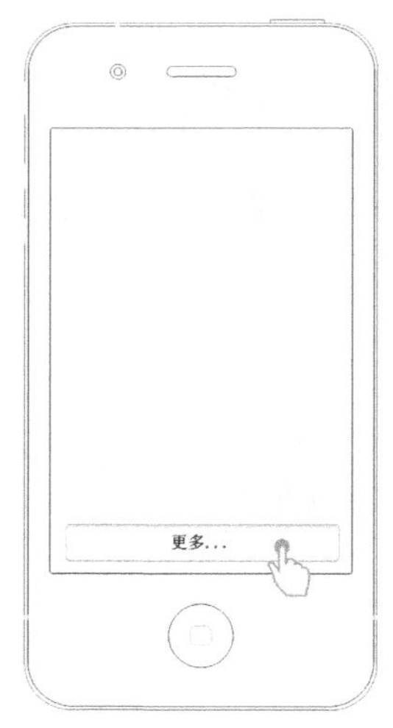

图1-8　图中手势

1.3.3　方法命名约定

苹果在官方文档中采用Objective-C多重参数描述API，它将方法名按照参数的个数分成几个部分。

提示

关于Objective-C多重参数，下面的代码用于在一个集合中按照索引插入元素：

```
- (void)insertObject:(id)anObject atIndex:(NSInteger)index
```

图1-9说明了Objective-C多重参数方法的定义，第一个参数是anObject，参数类型是id类型；第二个参数是index，参数类型是NSUInteger，这叫作多重参数。它的返回类型是void，方法签名是insertObject:atIndex:。方法类型标识符中的-代表方法是实例方法，+代表方法是类方法。关于实例方法和类方法，我们将在后面的内容中讨论。如果上面的方法变成C或C++形式，则是下面的样子：

```
void insertObjectAtIndex(id anObject, NSUInteger index)
```

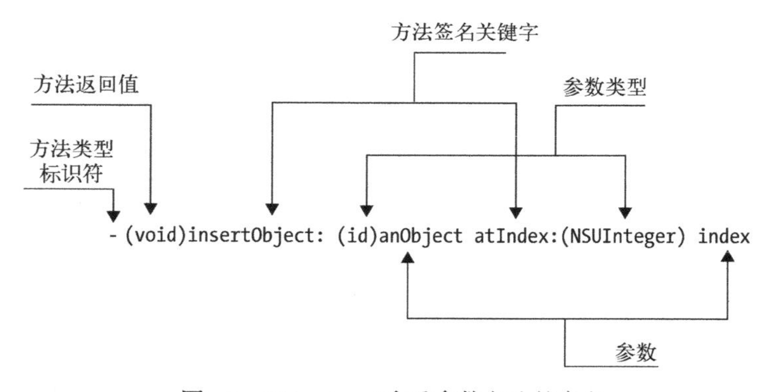

图1-9　Objective-C多重参数方法的定义

另外，在Objective-C中方法前面-表示实例方法，+表示静态方法，方法中的:表示有参数。在本书中有时也会根据不同的语境省略-和+。

苹果公司在推出Swift语言后，仍然采用多重参数描述API。在Swift 3之前，描述同一个API时，Objective-C语言与Swift语言基本上是一样的。例如：UITableView中的numberOfRowsInSection:方法，Objective-C的描述方式如下：

```
- (NSInteger)numberOfRowsInSection:(NSInteger)section
```

Swift 2的描述方式如下：

```
func numberOfRowsInSection(_ section: Int) -> Int
```

其中方法名都是numberOfRowsInSection，参数都是section。

但是从Swift 3开始，苹果主张“去Objective-C语言化”，使得Objective-C语言与Swift语言描述同一个API时差别越来越大。上述方法使用Swift 3的描述方式如下：

```
func numberOfRows(inSection section: Int) -> Int
```

其方法名与Swift 2比较有很大的变化，它将numberOfRowsInSection拆分成为多个部分，其中numberOfRows部分作为参数名，inSection部分作为参数名，并且首字符小写。

由于本书采用双语言介绍，为了统一命名，我们默认采用Objective-C的表述方式，如numberOfRowsInSection:，而且默认情况下实例方法会省略-，特殊情况下会加以说明。

1.3.4 构造函数命名约定

构造函数是特殊的方法，Objective-C和Swift也采用多重参数描述API，但是更为特殊。例如UITableView的构造函数是- initWithFrame:style:，该构造函数表示成Swift语言的形式如下：

```
init(frame frame: CGRect, style style: UITableViewStyle)
```

表示成Objective-C语言的形式如下：

```
- (instancetype)initWithFrame:(CGRect)frame style:(UITableViewStyle)style
```

为了统一命名，我们也采用苹果官方的提法，即在本书中提到Swift构造函数时也采用-initWithFrame:style:多重参数形式，特殊情况下则会加以说明。

1.3.5 错误处理约定

从Swift 2开始，错误处理机制与Objective-C有了很大差别。下面的代码通过NSFileManager的-copyItemAtPath:toPath:error:方法实现文件复制：

```
let defaultDBPath = frameworkBundlePath!.appendingPathComponent("NotesList.plist")

do {
    try fileManager.copyItem(atPath: defaultDBPath, toPath: self.plistFilePath)
} catch {
    let nserror = error as NSError
    NSLog("数据保存错误: %@", nserror.localizedDescription)
    assert(false, "错误写入文件")
}
```

```
NSString *defaultDBPath = [frameworkBundlePath
stringByAppendingPathComponent:@"NotesList.plist"];

NSError *error;
BOOL success = [fileManager copyItemAtPath:defaultDBPath
toPath:self.plistFilePath error:&error];

if (error) {
    NSAssert(success, @"错误写入文件");
}
```

其中Swift 2采用do-try-catch错误处理模式，catch语句是捕获错误。而Objective-C则是给方法传递一个&error（NSError地址）。当方法调用完成之后，若发生错误，则error不为nil，若没有发生错误，则error为nil。

对于copyItemAtPath:toPath:error:方法，Objective-C的声明如下：

```
- (BOOL)copyItemAtPath:(NSString *)srcPath toPath:(NSString *)dstPath error:(NSError * _Nullable *)error;
```

Swift 3的声明如下：

```
func copyItem(atPath srcPath: String, toPath dstPath: String) throws
```

可见，两种语言的描述方式有很大的差别，Swift会比Objective-C少一个参数error，但是Swift会在后面多一个throws来表示抛出错误。要详细了解有关Swift的错误处理模式，可以参考我写的《从零开始学Swift（第2版）》一书。

第 2 章

第一个iOS应用程序

从控制台输出HelloWorld是我学习C语言的第一步，也是我人生中非常重要的一步。多年后的今天，我希望仍以HelloWorld作为第一步，为大家开启一个神奇、瑰丽的世界——iOS。

本章以HelloWorld作为切入点，向大家系统介绍什么是iOS应用以及如何使用Xcode创建iOS应用。

2.1 创建 HelloWorld 工程

在学习之初，我们有必要对使用Xcode创建iOS工程做一个整体概览，这里通过创建一个基于故事板的HelloWorld iPhone工程来详述其中涉及的知识点。

实现HelloWorld应用后，界面上会展示字符串HelloWorld（如图2-1所示），其中主要包含Label（标签）。

图2-1　HelloWorld的iPhone界面

实现HelloWorld应用可以分为如下3个步骤：

(1) 通过Xcode创建工程

(2) 添加Label标签

(3) 运行应用

2.1.1 通过 Xcode 创建工程

启动Xcode，然后点击File→New→Project菜单，在打开的Choose a template for your new project界面中选择Single View Application工程模板（如图2-2所示）。

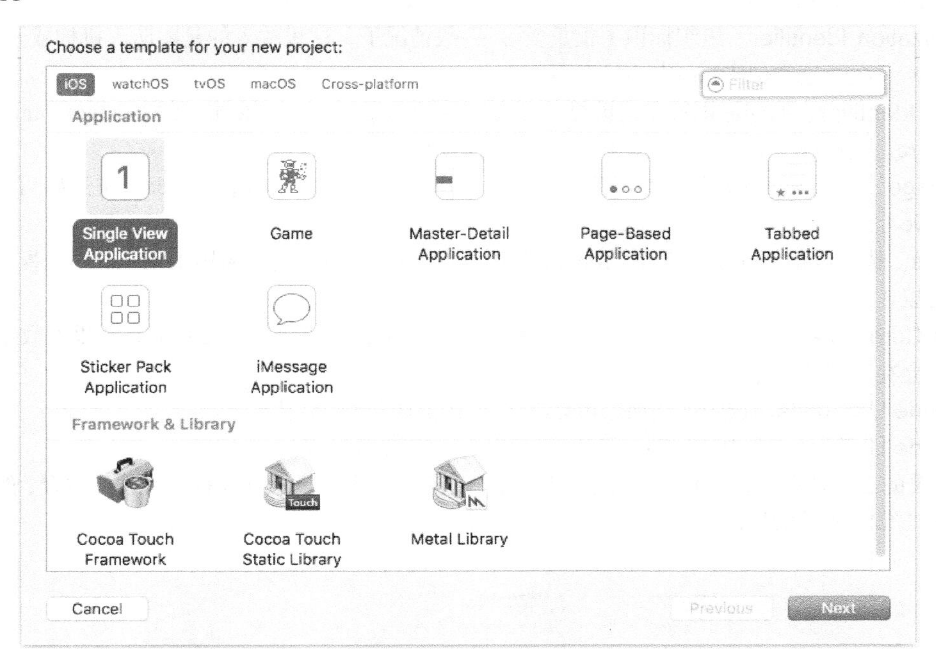

图2-2 选择工程模板

接着点击Next按钮，随即出现如图2-3所示的界面。

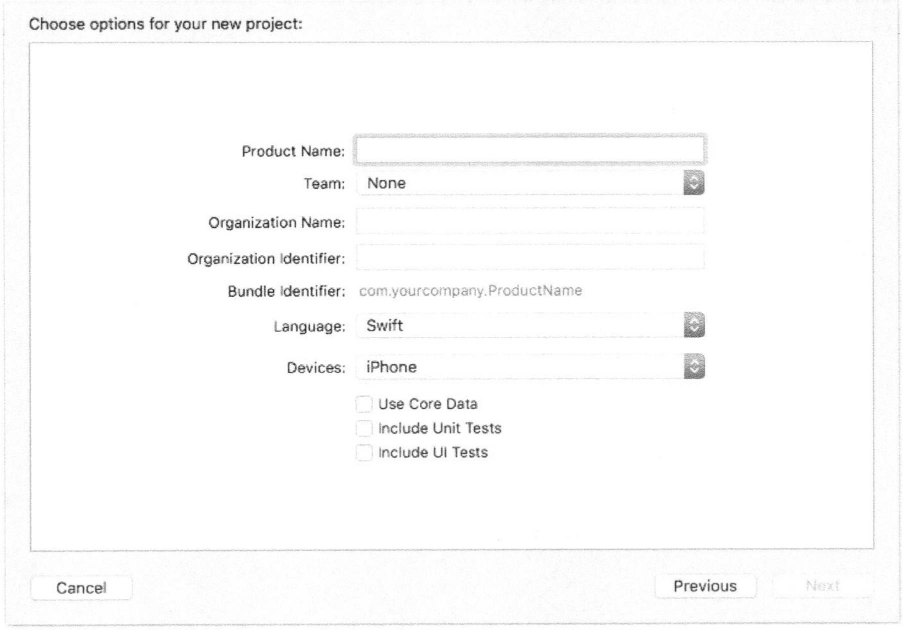

图2-3 新工程中的选项

这里我们可以按照提示并结合自己的实际情况和需要输入相关内容。下面简要说明图2-3中的选项。

- Product Name。工程名字。
- Team。App Store上开发者的名字，没有时可以设为None。
- Organization Name。组织名称，可以是团队、机构或开发者名字。
- Organization Identifier。组织标识（很重要）。一般情况下，这里输入的是团队、机构或开发的域名（如com.51work6），这类似于Java中的包命名。
- Bundle Identifier。捆绑标识符（很重要）。该标识符由Organization Identifier + Product Name构成。因为在App Store上发布应用时会用到它，所以它的命名不可重复。
- Language。开发语言选择。这里可以选择开发应用所使用的语言，从Xcode 6开始可以选择Swift或Objective-C。
- Devices。选择设备。可以构建基于iPhone或iPad的工程，也可以构建通用工程。通用工程在iPhone和iPad上都可以正常运行。
    - Use Core Data。选中该选项，可以在工程中添加Core Data代码。Core Data是苹果的数据持久化技术，我们会在第13章中介绍该技术。
    - Include Unit Tests。选中该选项，可以在工程中添加单元测试代码。
    - Include UI Tests。选中该选项，可以在工程中添加UI测试代码。

设置完相关的工程选项后，点击Next按钮，进入下一级界面。根据提示选择存放文件的位置，然后点击Create按钮，将出现如图2-4所示的界面。

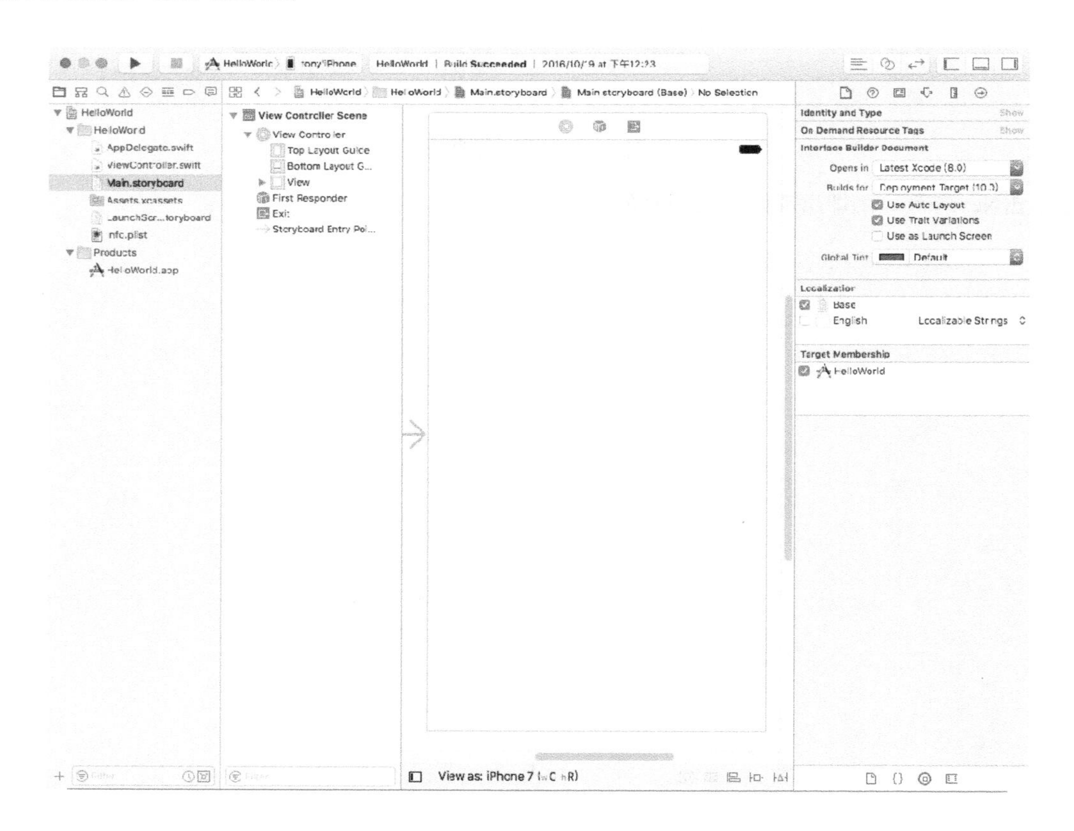

(a)

图2-4　新创建的工程（图a为Swift版，图b为Objective-C版）

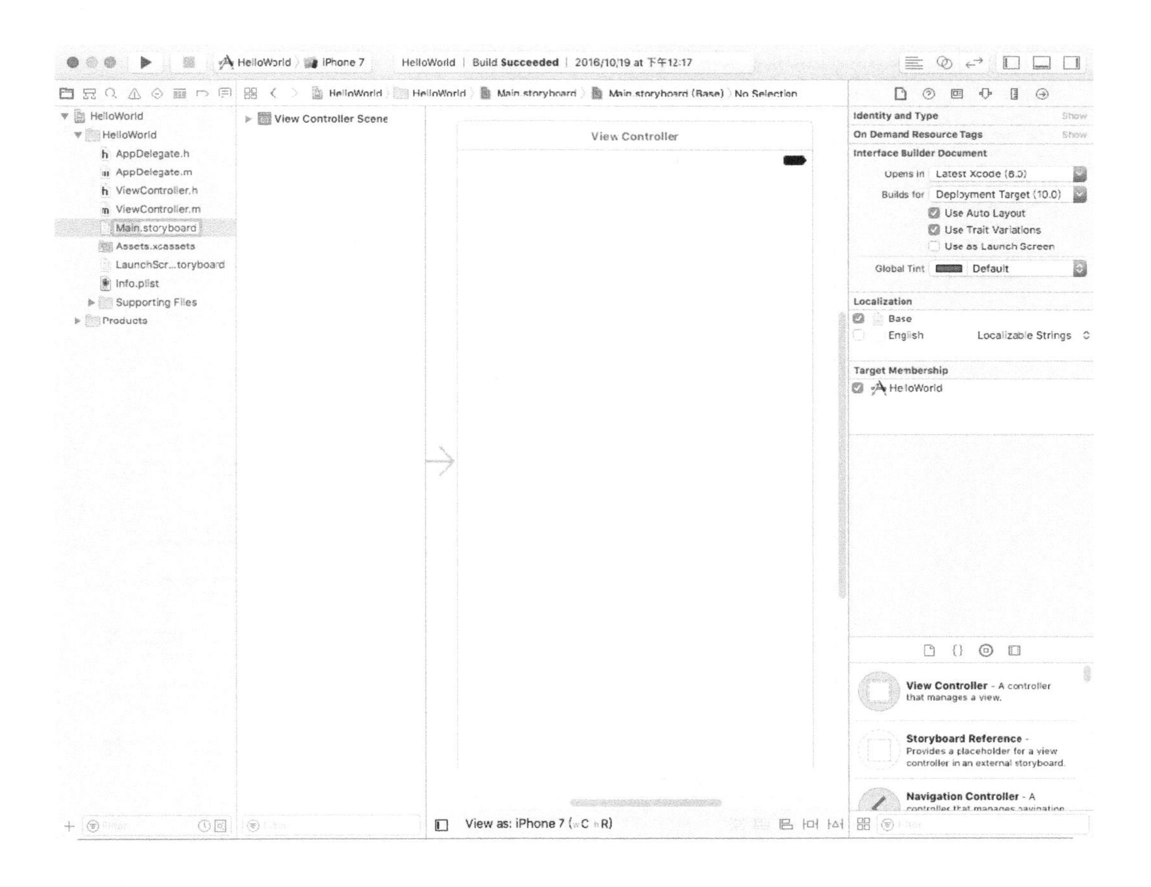

(b)

图2-4 （续）

2.1.2 添加标签

在图2-5中，右边的“显示对象库”按钮◎用于显示对象库，拖动滚动条找到Label，将其拖曳到View设计界面上并调整其位置。双击Label，使其处于编辑状态（也可以通过控件的属性来设置），在其中输入HelloWorld。

添加HelloWorld标签后，需要设置标签的位置。拖曳标签，此时会出现蓝色虚线，如图2-6所示，这说明该标签现在处于居中位置。

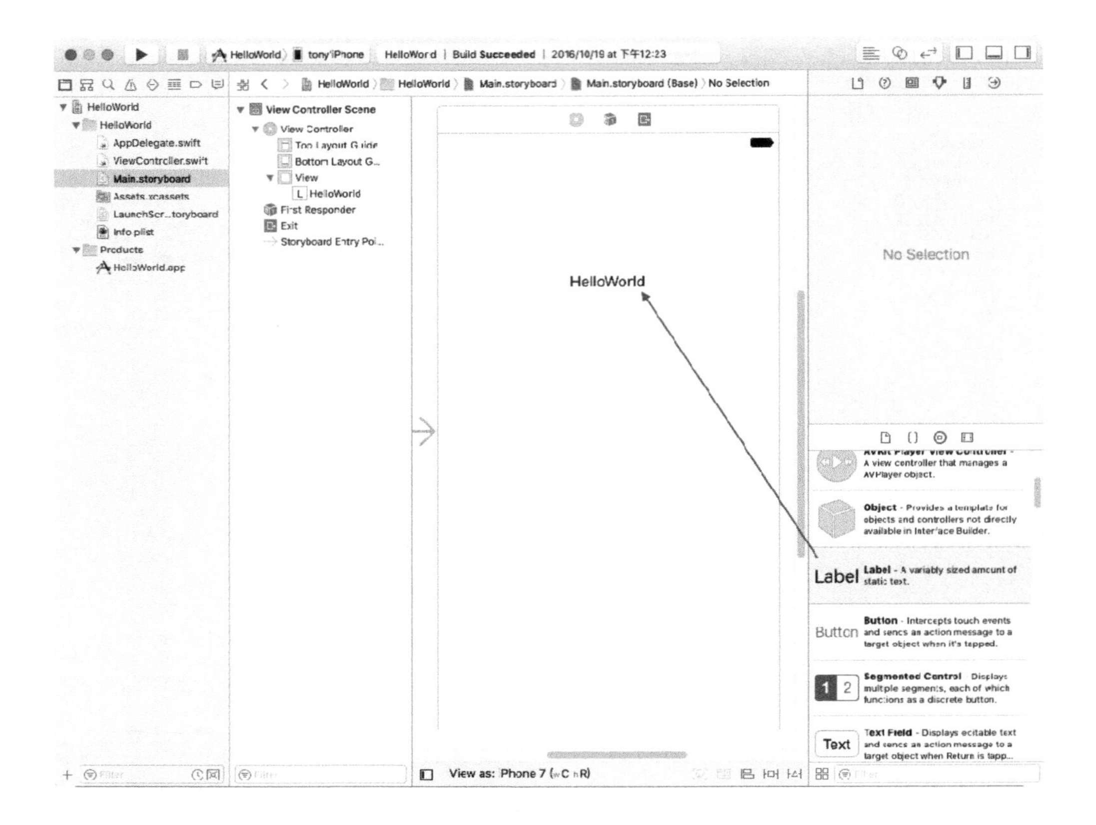

图2-5　添加标签视图

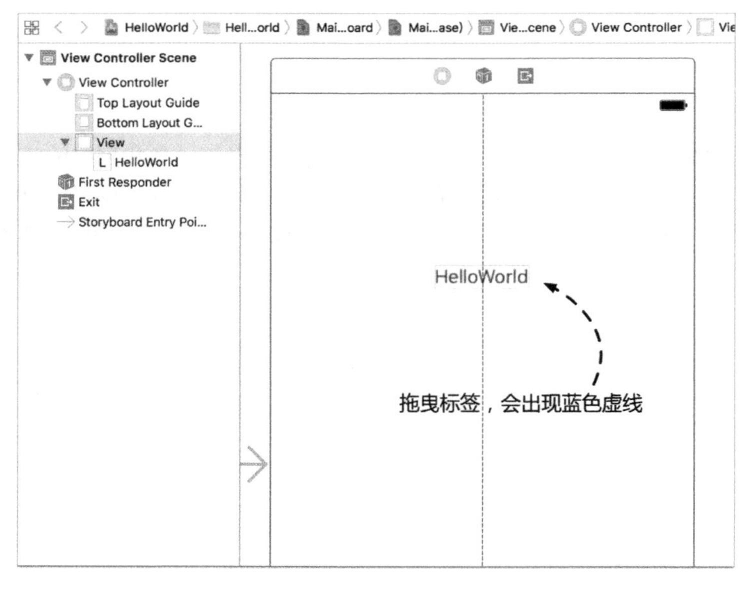

图2-6　拖曳标签使其居中

2.1.3 运行应用

至此，整个工程创建完毕。如图2-7所示，选择运行的模拟器或设备，然后点击左上角的“运行”按钮▶，即可看到运行结果。

图2-7 运行应用

在没有输入任何代码的情况下，就已经利用Xcode工具的Single View Application模板创建了一个工程并成功运行，Xcode之强大可见一斑。

注意

本案例中运行时选择的模拟器要与设计界面中视图大小（如图2-5所示，设计界面底部View as: iPhone 7可以选择不同的视图大小）匹配，否则会发现标签并非居中。例如，设计界面视图选择iPhone 7，则模拟器也应该选择iPhone 6、iPhone 6s或iPhone 7，因为这三种设备的屏幕宽度相同。而图2-6所进行的操作只是保证在iPhone 7等相同宽度屏幕上的居中，如果其他设备上也能居中，需要使用Auto Layout约束，甚至还需要使用Size Class技术，这些会在第8章中介绍。在学习Auto Layout约束和Size Class技术之前，暂时不考虑在其他设备上的布局问题。

2.1.4 Xcode中的iOS工程模板

从图2-2中可以看出，iOS工程模板分为两类——Application和Framework & Library，下面将分别介绍这两类模板。

1. Application类型

我们大部分的开发工作都是从使用Application类型模板创建iOS程序开始的。该类型共包含7个模板，具体如下所示。

- Single View Application。可以构建简单的单个视图应用。
- Game。可以构建基于iOS的游戏应用。
- Master-Detail Application。可以构建树形结构导航模式应用，生成的代码中包含了导航控制器和表视图控制器等。

- ❑ Page-Based Application。可以构建类似于电子书效果的应用，这是一种平铺导航。
- ❑ Tabbed Application。可以构建标签导航模式的应用，生成的代码中包含了标签控制器和标签栏等。
- ❑ Sticker Pack Application。可以构建表情包（Sticker Pack）应用。
- ❑ iMessage Application。可以构建聊天应用。

2. Framework & Library类型

- ❑ Cocoa Touch Framework。可以让我们创建自己的iOS框架。
- ❑ Cocoa Touch Static Library。可以让我们创建自己的静态库。出于代码安全和多个工程重用代码的考虑，我们可以将一些类或者函数编写成静态库。静态库不能独立运行，编译成功时会生成名为lib*XXX*.a的文件（例如libHelloWorld.a）。
- ❑ Metal Library。可以让我们创建自己的Metal①库。

2.1.5　应用剖析

在创建HelloWorld的过程中，我们生成了很多文件（展开Xcode左边的项目导航视图可以看到，如图2-8所示），它们各自的作用是什么？彼此间又是怎样的一种关系呢？

如图2-8所示，导航视图下有HelloWorld和Products两个组，其中HelloWorld组中放置HelloWorld工程的主要代码，Products组中放置了编译后的工程。下面我们重点介绍HelloWorld组中的内容。

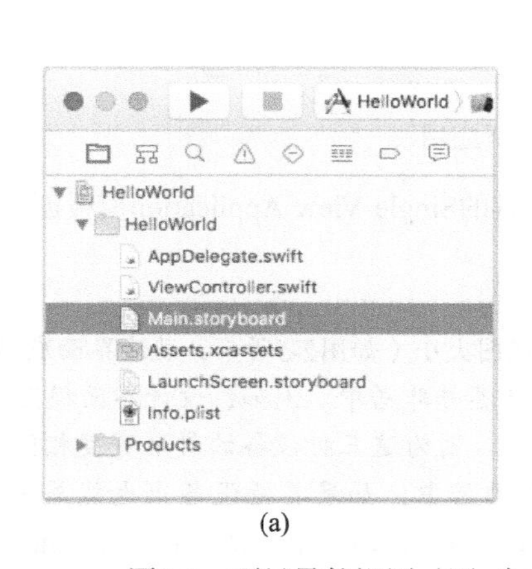

(a)

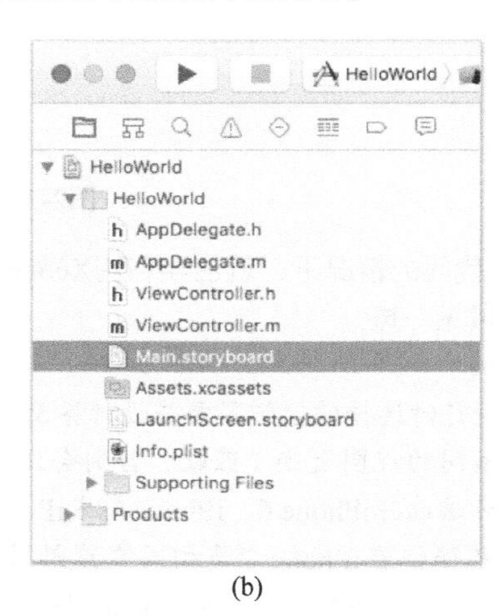

(b)

图2-8　项目导航视图（图a为Swift版，图b为Objective-C版）

在HelloWorld组中，有两个类`AppDelegate`和`ViewController`，两个界面布局文件Main.storyboard和LaunchScreen.storyboard，其中Main.storyboard文件是故事板文件（稍后我们会详细介绍），LaunchScreen.storyboard是应用启动界面故事板文件。Assets.xcassets文件夹是资源目录（asset catalog），可以用来管理图片。Info.plist是工程属性文件。

> **说明**　在访问资源文件时，文件夹和组是有区别的。访问文件夹中的资源时，需要将文件夹作为路径。如果icon.png文件放在image文件夹下，则访问它的路径是image/icon.png；如果image是组，则访问它的路径是icon.png。

① Metal是一个兼顾图形与计算功能的，面向底层、低开销的硬件加速应用程序接口（API），类似于将OpenGL与OpenCL的功能集成到了同一个API上，最初支持它的系统是iOS 8。——引自于维基百科：https://zh.wikipedia.org/wiki/Metal_(API)

我们主要的编码工作就是在AppDelegate和ViewController这两个类中进行的，它们的类图如图2-9所示。

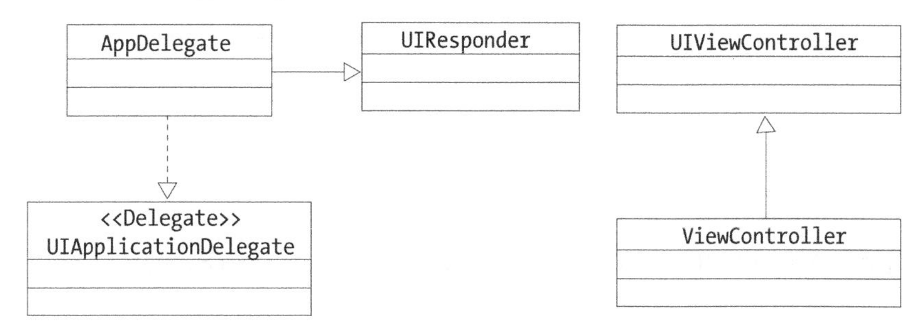

图2-9　HelloWorld工程中的类图

AppDelegate是应用程序委托对象，它继承了UIResponder类，并实现了UIApplicationDelegate委托协议。UIResponder类可以使子类AppDelegate具有处理响应事件的能力，而UIApplicationDelegate委托协议使AppDelegate能够成为应用程序委托对象，这种对象能够响应应用程序的生命周期。相应地，AppDelegate的子类也可以实现这两个功能。

ViewController类继承自UIViewController类，它是视图控制器类，在工程中扮演着根视图和用户事件控制类的角色。

AppDelegate类是应用程序委托对象，这个类中继承的一系列方法在应用生命周期的不同阶段会被回调，其定义如下：

```
//
//AppDelegate.swift
//
import UIKit

@UIApplicationMain
class AppDelegate: UIResponder, UIApplicationDelegate {

    var window: UIWindow?

    func application(_ application: UIApplication,
    ➥didFinishLaunchingWithOptions launchOptions:
    ➥[UIApplicationLaunchOptionsKey: Any]?) -> Bool {          ①
        return true
    }

    func applicationWillResignActive(_ application: UIApplication) {
    }

    func applicationDidEnterBackground(_ application: UIApplication) {
    }

    func applicationWillEnterForeground(_ application: UIApplication) {
    }

    func applicationDidBecomeActive(_ application: UIApplication) {
    }

    func applicationWillTerminate(_ application: UIApplication) {
    }

}
```

```
//
//AppDelegate.h
//

#import <UIKit/UIKit.h>

@interface AppDelegate : UIResponder <UIApplicationDelegate>

@property (strong, nonatomic) UIWindow *window;

@end

//
//AppDelegate.m
//

#import "AppDelegate.h"
@interface AppDelegate ()
@end

@implementation AppDelegate


- (BOOL)application:(UIApplication *)application
➥didFinishLaunchingWithOptions:(NSDictionary *)launchOptions {          ①
    return YES;
}

- (void)applicationWillResignActive:(UIApplication *)application {
}

- (void)applicationDidEnterBackground:(UIApplication *)application {
}

- (void)applicationWillEnterForeground:(UIApplication *)application {
```

```
}

- (void)applicationDidBecomeActive:(UIApplication *)application {
}

- (void)applicationWillTerminate:(UIApplication *)application {
}

@end
```

启动HelloWorld时，首先会调用第①行的application:didFinishLaunchingWithOptions:方法，其他方法稍后再详细介绍。

2.2　应用生命周期

作为应用程序的委托对象，AppDelegate类在应用生命周期的不同阶段会回调不同的方法。首先，让我们先了解一下iOS应用的不同状态及其彼此间的关系，如图2-10所示。

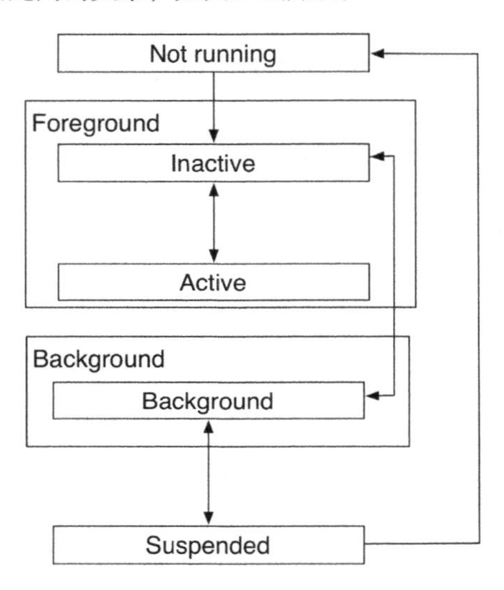

图2-10　iOS应用状态图

下面简要介绍一下iOS应用的5种状态。

- Not running（**非运行状态**）。应用没有运行或被系统终止。
- Inactive（**前台非活动状态**）。应用正在进入前台状态，但是还不能接受事件处理。
- Active（**前台活动状态**）。应用进入前台状态，能接受事件处理。
- Background（**后台状态**）。应用进入后台后，依然能够执行代码。如果有可执行的代码，就会执行代码，如果没有可执行的代码或者将可执行的代码执行完毕，应用会马上进入挂起状态。
- Suspended（**挂起状态**）。被挂起的应用进入一种“冷冻”状态，不能执行代码。如果系统内存不够，应用会被终止。

在应用状态跃迁的过程中，iOS系统会回调AppDelegate中的一些方法，并且发送一些通知。实际上，在应用的生命周期中用到的方法和通知很多，我们选取了几个主要的方法和通知进行详细介绍，具体如表2-1所示。

表2-1 状态跃迁过程中应用回调的方法和本地通知

方　法	本地通知	说　明
application:didFinishLaunching➥WithOptions:	UIApplicationDidFinishLaunchingNotification	应用启动并进行初始化时会调用该方法并发出通知。这个阶段会实例化根视图控制器
applicationDidBecomeActive:	UIApplicationDidBecomeActiveNotification	应用进入前台并处于活动状态时调用该方法并发出通知。这个阶段可以恢复UI的状态（例如游戏状态等）
applicationWillResignActive:	UIApplicationWillResignActiveNotification	应用从活动状态进入到非活动状态时调用该方法并发出通知。这个阶段可以保存UI的状态（例如游戏状态等）
applicationDidEnterBackground:	UIApplicationDidEnterBackgroundNotification	应用进入后台时调用该方法并发出通知。这个阶段可以保存用户数据，释放一些资源（例如释放数据库资源等）
applicationWillEnterForeground:	UIApplicationWillEnterForegroundNotification	应用进入到前台，但是还没有处于活动状态时调用该方法并发出通知。这个阶段可以恢复用户数据
applicationWillTerminate:	UIApplicationWillTerminateNotification	应用被终止时调用该方法并发出通知，但内存清除时除外。这个阶段释放一些资源，也可以保存用户数据

为了便于观察应用程序的运行状态，我们在 AppDelegate 类中的方法内添加一些日志输出，具体代码如下：

```swift
//
//AppDelegate.swift
//

import UIKit

@UIApplicationMain
class AppDelegate: UIResponder, UIApplicationDelegate {

    var window: UIWindow?

    func application(_ application: UIApplication, didFinishLaunchingWithOptions
➥launchOptions: [UIApplicationLaunchOptionsKey: Any]?) -> Bool {
        print("application:didFinishLaunchingWithOptions:")
        return true
    }

    func applicationWillResignActive(_ application: UIApplication) {
        print("applicationWillResignActive:")
    }

    func applicationDidEnterBackground(_ application: UIApplication) {
        print("applicationDidEnterBackground:")
    }

    func applicationWillEnterForeground(_ application: UIApplication) {
        print("applicationWillEnterForeground:")
    }

    func applicationDidBecomeActive(_ application: UIApplication) {
        print("applicationDidBecomeActive:")
    }

    func applicationWillTerminate(_ application: UIApplication) {
        print("applicationWillTerminate:")
```

```objc
//
//AppDelegate.m
//

#import "AppDelegate.h"

@interface AppDelegate ()

@end

@implementation AppDelegate


- (BOOL)application:(UIApplication *)application
➥didFinishLaunchingWithOptions:(NSDictionary *)launchOptions {
    NSLog(@"%@", @"application:didFinishLaunchingWithOptions:");
    return YES;
}

- (void)applicationWillResignActive:(UIApplication *)application {
    NSLog(@"%@", @"applicationWillResignActive:");
}

- (void)applicationDidEnterBackground:(UIApplication *)application {
    NSLog(@"%@", @"applicationDidEnterBackground:");
}

- (void)applicationWillEnterForeground:(UIApplication *)application {
    NSLog(@"%@", @"applicationWillEnterForeground:");
}

- (void)applicationDidBecomeActive:(UIApplication *)application {
    NSLog(@"%@", @"applicationDidBecomeActive:");
}
```

```
    }
}

- (void)applicationWillTerminate:(UIApplication *)application {
    NSLog(@"%@", @"applicationWillTerminate:");
}

@end
```

为了让大家更直观地了解各状态与其相应的方法、通知间的关系，下面我们以几个应用场景为切入点进行系统分析。

2.2.1　非运行状态——应用启动场景

场景描述：用户点击应用图标的时候，可能是第一次启动这个应用，也可能是应用终止后再次启动。该场景的状态跃迁过程见图2-11，共经历两个阶段3个状态：Not running→Inactive→Active。

- **Not running→Inactive阶段**：调用`application:didFinishLaunchingWithOptions:`方法，发出`UIApplicationDidFinishLaunchingNotification`通知。
- **Inactive→Active阶段**：调用`applicationDidBecomeActive:`方法，发出`UIApplicationDidBecomeActiveNotification`通知。

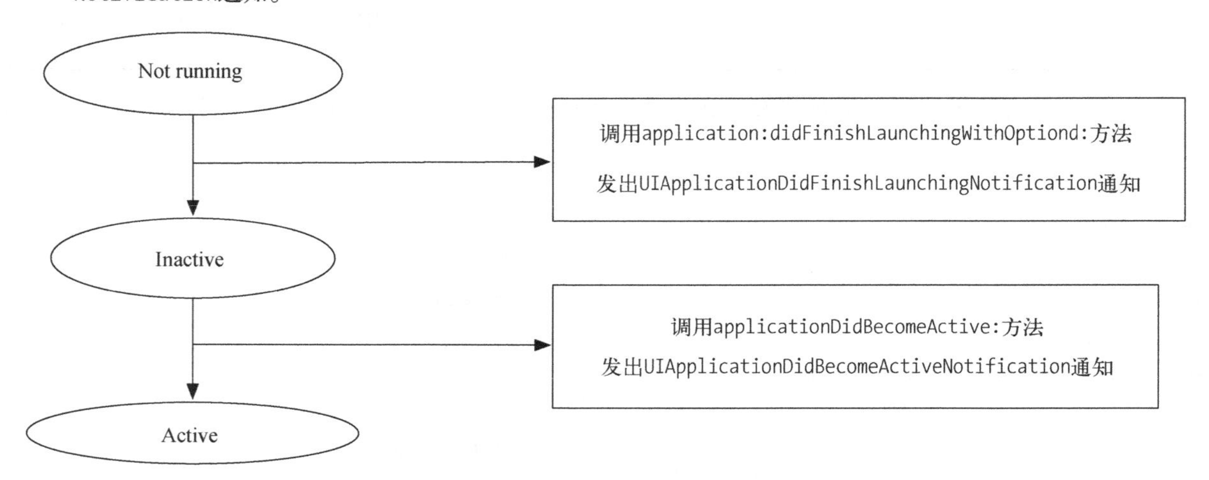

图2-11　应用启动场景的状态跃迁过程

2.2.2　点击 Home 键——应用退出场景

场景描述：应用处于运行状态（即Active状态）时，点击Home键应用会退出。该场景的状态跃迁过程可以分成两种情况：可以在后台运行或者挂起，不可以在后台运行或者挂起。根据工程属性文件（如Info.plist）中的相关属性Application does not run in background（如图2-12所示）的设置，我们可以控制这两种状态。如果采用文本编辑器打开Info.plist文件，该设置项对应的键是`UIApplicationExitsOnSuspend`。

状态跃迁的第一种情况：应用可以在后台运行或者挂起。该场景的状态跃迁过程见图2-13，共经历3个阶段4个状态：Active→Inactive→Background→Suspended。

- **Active→Inactive阶段**：调用`applicationWillResignActive:`方法，发出`UIApplicationWillResignActiveNotification`通知。
- **Inactive→Background阶段**：应用从非活动状态进入到后台（不涉及我们要重点说明的方法和通知）。
- **Background→Suspended阶段**：调用`applicationDidEnterBackground:`方法，发出`UIApplicationDidEnterBackgroundNotification`通知。

HelloWorld › HelloWorld › Supporting Files › Info.plist › No Selection

Key	Type	Value
▼Information Property List	Dictionary	(15 items)
Localization native development region	String	en
Executable file	String	$(EXECUTABLE_NAME)
Bundle identifier	String	com.51work6.$(PRODUCT_NAME:rfc1034identifier)
InfoDictionary version	String	6.0
Bundle name	String	$(PRODUCT_NAME)
Bundle OS Type code	String	APPL
Bundle versions string, short	String	1.0
Bundle creator OS Type code	String	????
Bundle version	String	1
Application requires iPhone environment	Boolean	YES
Launch screen interface file base name	String	LaunchScreen
Main storyboard file base name	String	Main
▶Required device capabilities	Array	(1 item)
Application does not run in background	Boolean	YES
▶Supported interface orientations	Array	(3 items)

图2-12 属性设置

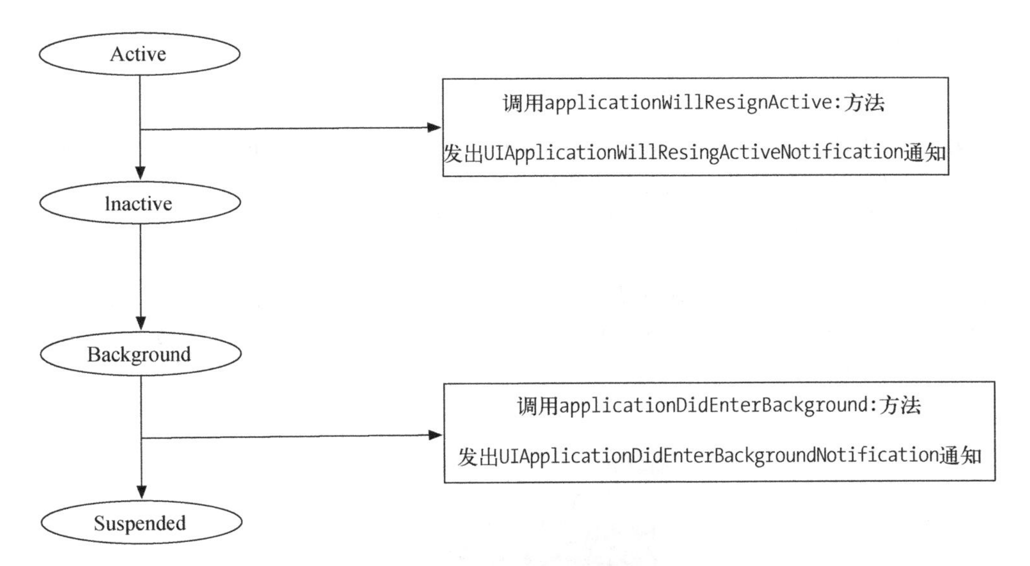

图2-13 应用在后台运行或者挂起

状态跃迁的第二种情况：应用不可以在后台运行或者挂起。其状态跃迁情况见图2-14，共经历4个阶段5个状态：Active→Inactive→Background→Suspended→Not running 。

- Active→Inactive阶段：应用由活动状态转为非活动状态（不涉及我们要重点说明的方法和通知）。
- Inactive→Background阶段：应用从非活动状态进入到后台（不涉及我们要重点说明的方法和通知）。
- Background→Suspended阶段：调用applicationDidEnterBackground:方法，发出UIApplicationDidEnterBackgroundNotification通知。
- Suspended→Not running阶段：调用applicationWillTerminate:方法，发出UIApplicationWillTerminateNotification通知。

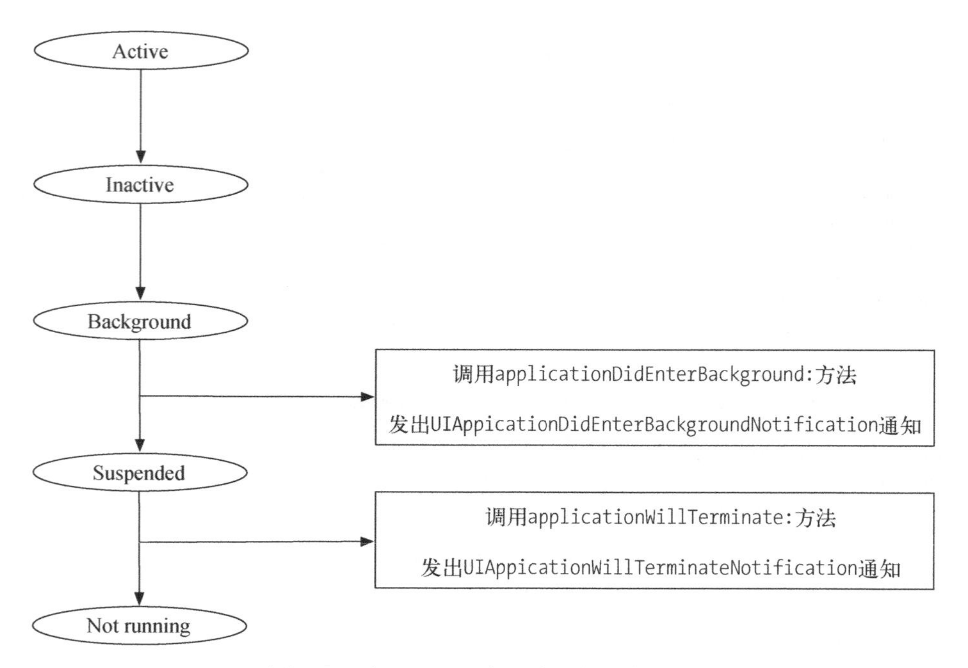

图2-14 应用不可以在后台运行或者挂起

iOS在iOS 4之前不支持多任务，点击Home键时，应用会退出并中断；而在iOS 4之后（包括iOS 4），操作系统能够支持多任务处理，轻按一下Home键时应用会进入后台但不会中断（内存不够的情况除外）。

应用在后台也可以进行部分处理工作，处理完成后进入挂起状态。

说明 轻按两下Home键可以快速进入iOS多任务栏，图2-15是iOS 9及其后续版本的多任务栏，此时可以看到处于后台运行或挂起状态的应用，也可能有处于终止状态的应用。向上滑动应用界面，可以删除这些应用并释放内存。

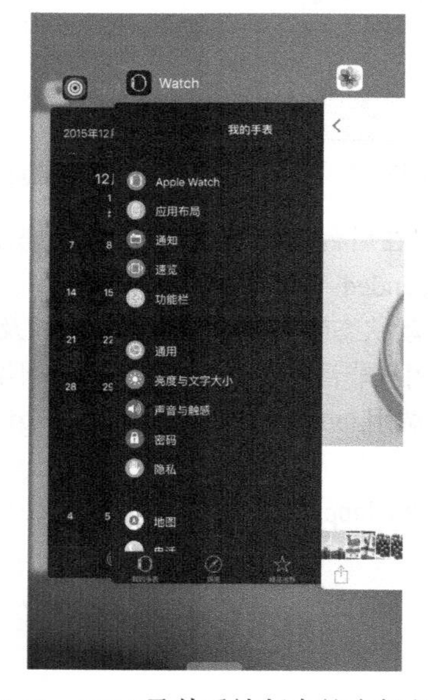

图2-15 iOS 9及其后续版本的多任务栏

2.2.3　挂起重新运行场景

场景描述：挂起状态的应用重新运行。该场景的状态跃迁过程如图2-16所示，共经历3个阶段4个状态：Suspended→Background→Inactive→Active。

- **Suspended→Background阶段：**应用从挂起状态进入后台（不涉及我们讲述的几个方法和通知）。
- **Background→Inactive阶段：**调用applicationWillEnterForeground:方法，发出UIApplicationWillEnter-ForegroundNotification通知。
- **Inactive→Active阶段：**调用applicationDidBecomeActive:方法，发出UIApplicationDidBecomeActive-Notification通知。

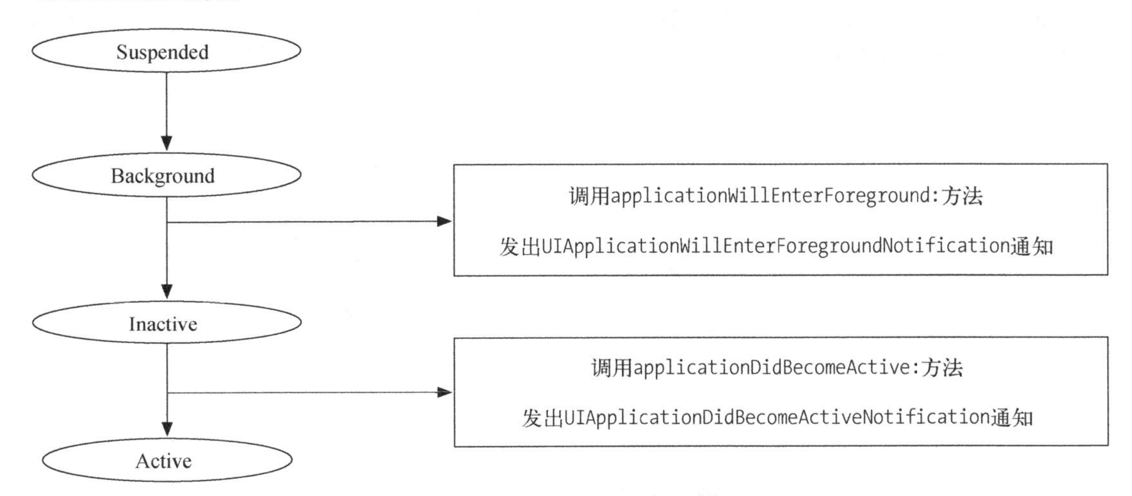

图2-16　挂起重新运行场景的状态跃迁过程

2.2.4　内存清除：应用终止场景

场景描述：应用在后台处理完成时进入挂起状态（这是一种休眠状态），如果这时发出低内存警告，为了满足其他应用对内存的需要，该应用就会从内存中清除从而终止运行。该场景的状态跃迁过程如图2-17所示。

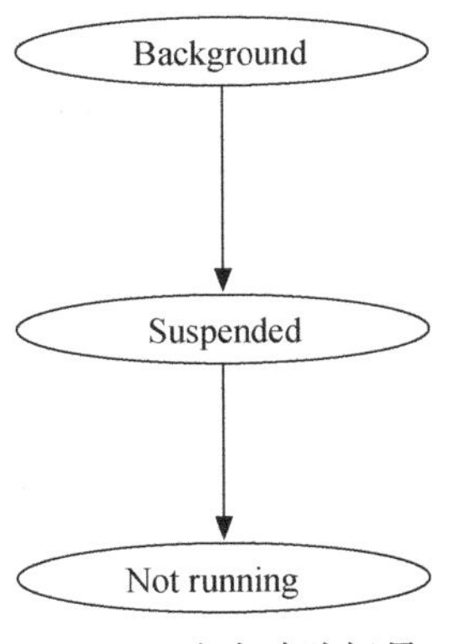

图2-17　内存清除场景

内存清除的时候应用终止运行。内存清除有两种情况，可能是系统强制清除内存，也可能是由使用者从任务栏中手动清除（即删掉应用）。内存清除后如果应用再次运行，上一次的运行状态不会被保存，相当于应用第一次运行。

在内存清除场景下，应用不会调用任何方法，也不会发出任何通知。

2.3　设置产品属性

在前面讲解应用生命周期时，为了禁止应用在后台运行，我们将工程属性文件Info.plist中的Application does not run in background属性修改为YES（即`UIApplicationExitsOnSuspend = YES`），这项操作就属于产品属性的设置。在Xcode中，产品与目标直接相关，而目标（Target）与工程（Project）直接相关。

2.3.1　Xcode 中的工程和目标

我们首先解释一下前面提到的目标概念，一个目标就是一个编译后的产品。图2-18所示的界面是使用Xcode创建的HelloWorld工程。一个工程中可以包含多个目标，一个目标包含了一些源程序文件、资源文件和编译说明文件等内容，其中编译说明文件通过“编译参数设置”（Build Settings）和“编译阶段”（Build Phases）设置。

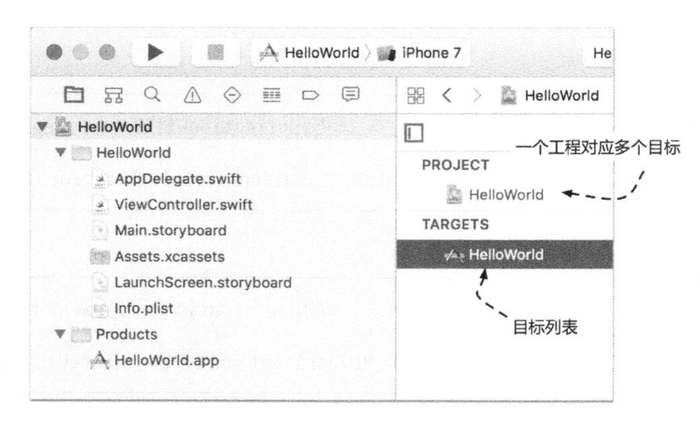

图2-18　Xcode的工程和目标

在目标列表上面还有一个工程，其中也包含一些“编译参数设置”和“编译阶段”设置项目。目标继承了工程的设置，而且还可以覆盖工程的设置。

我们可以在Xcode工程中添加更多的目标，下面就为之前使用故事板实现的HelloWorld工程增加一个目标。

首先，依次选择File→New→Target菜单项，此时会弹出一个选择模板对话框，从中选择iOS→Application→Single View Application模板，如图2-19所示。

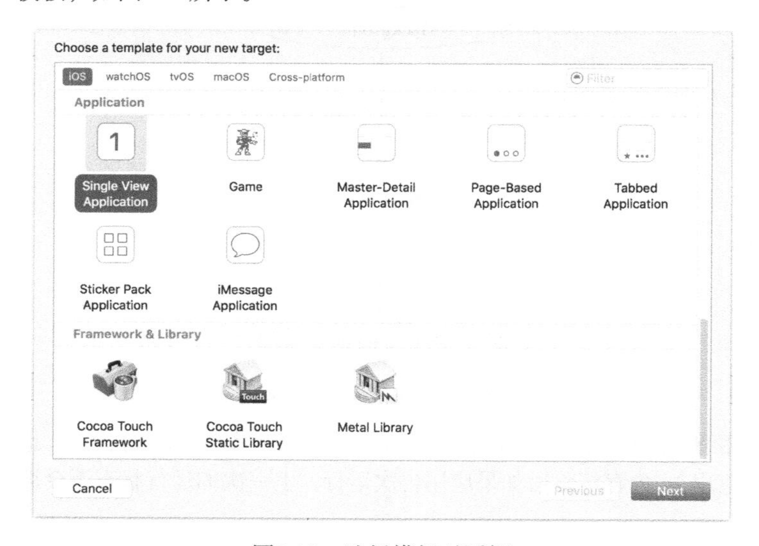

图2-19　选择模板对话框

然后点击Next按钮，此时会出现如图2-20所示的对话框，该对话框与新工程对话框非常类似。

Choose options for your new target:

Product Name:
Team: None
Organization Name: 关东升
Organization Identifier: com.51work6
Bundle Identifier: com.51work6.ProductName
Language: Swift
Devices: iPhone
Use Core Data
Include Unit Tests
Include UI Tests
Project: HelloWorld

Cancel
Previous
Finish

图2-20 目标的一些选项设定

根据自己的需要逐一设定后，点击Finish按钮，我们就成功为HelloWorld新增了一个目标（此时可以发现目标列表中有两个目标了），并同时生成一套完整的AppDelegate、ViewController、Main.storyboard和LaunchScreen.storyboard等文件，它们独立于原来的HelloWorld目标而存在，如图2-21所示。

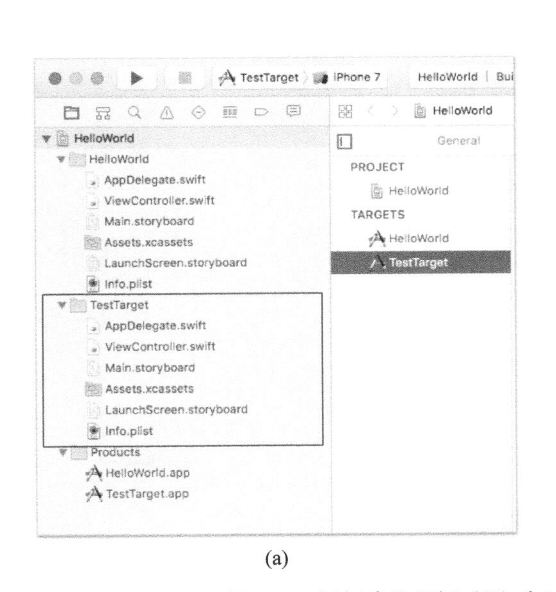

(a)

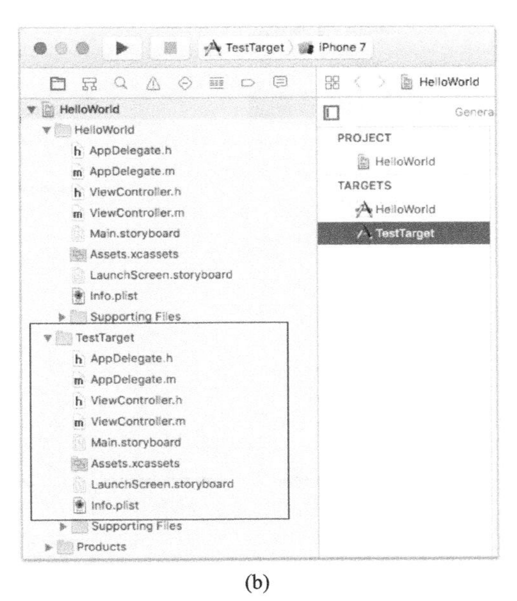

(b)

图2-21 新创建的目标（图a为Swift版，图b为Objective-C版）

要指定运行哪一个目标，可以通过选择不同的方案（Scheme）来实现。如图2-22所示，在Xcode的左上角选择TestTarget→iPhone 7，就可以在iPhone 7模拟器上运行TestTarget了。

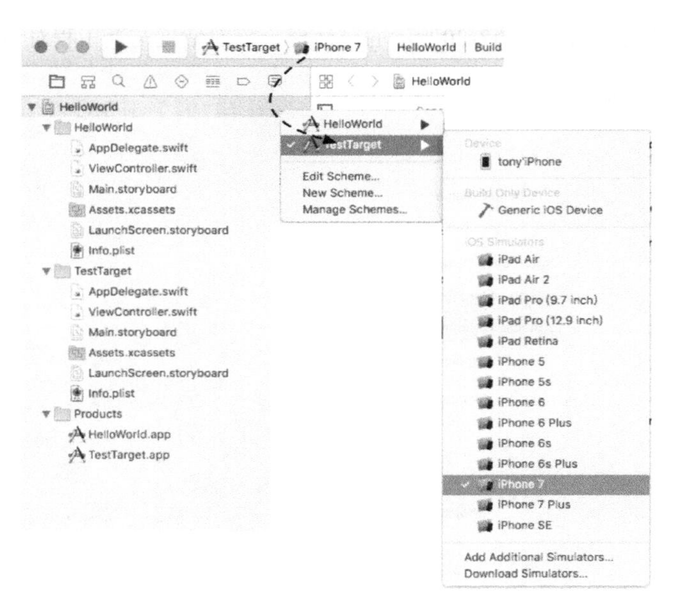

图2-22　选择方案

说明　什么是方案（Scheme）？在Xcode中，方案是指一个要编译执行的目标，其中包括编译这个目标的配置信息，以及运行目标的测试方案。一个Xcode工程中可以包含多个方案，但每次只能有一个是活动方案。要修改方案，可以在图2-22所示的下拉菜单中选择Edit Scheme，此时会弹出如图2-23所示的对话框，在其中进行设置即可。

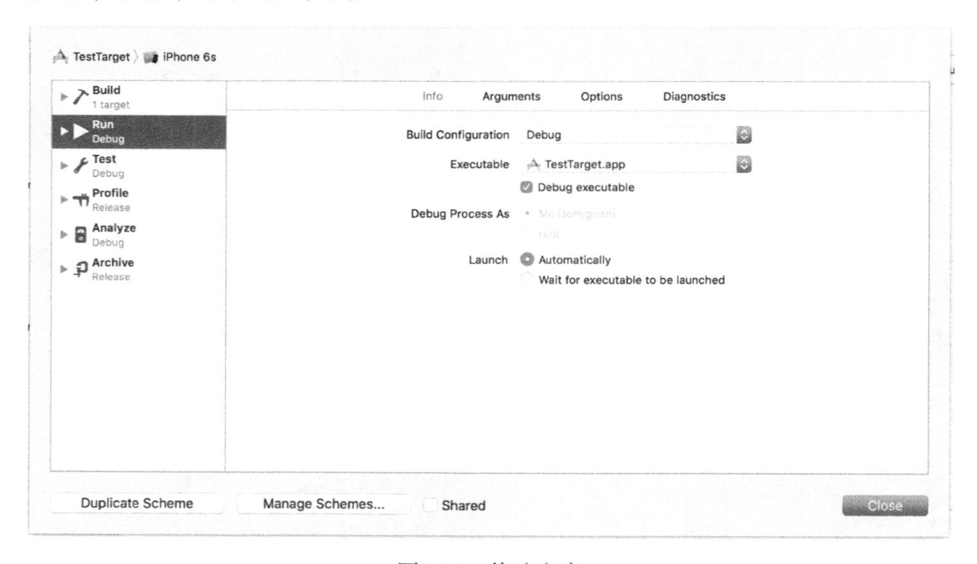

图2-23　修改方案

2.3.2　设置常用的产品属性

Xcode中的目标设置继承自工程设置。对于目标和工程下都有的一些设置项，我们可根据需要对目标进行再设置，此设置可覆盖工程的设置。

工程中的属性设置相对比较简单，大家可以参考官方的相关资料。这里介绍目标中两个常用的产品属性。

1. 设定屏幕方向

如图2-24所示，在导航面板中选择TestTarget，然后在右侧选择General选项卡，此时可以发现下面的Device Orientation区域中有4个复选框，它们代表设备支持的4个方向，选中则代表支持该指定方向。

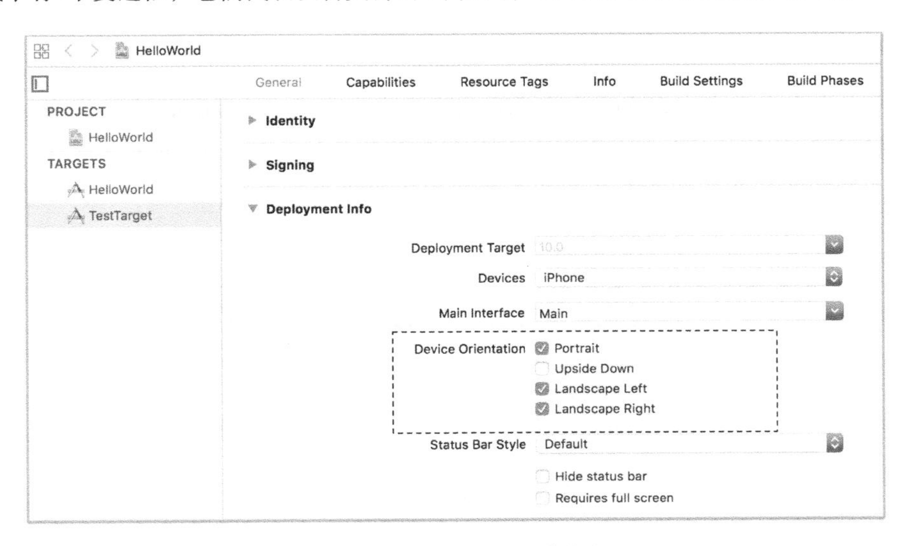

图2-24　设定支持的屏幕方向

提示　如何在模拟器上旋转屏幕方向？我们可以在模拟器运行后，通过快捷键“command + 方向键”实现屏幕方向旋转。

2. 设置设备支持情况

我们可以让应用支持iPhone或iPad设备，或者同时支持iPhone和iPad设备。如图2-25所示，在Deployment Info区域中找到Devices下拉列表，从中选择iPhone、iPad或者Universal选项，其中Universal表示同时支持iPhone和iPad设备。

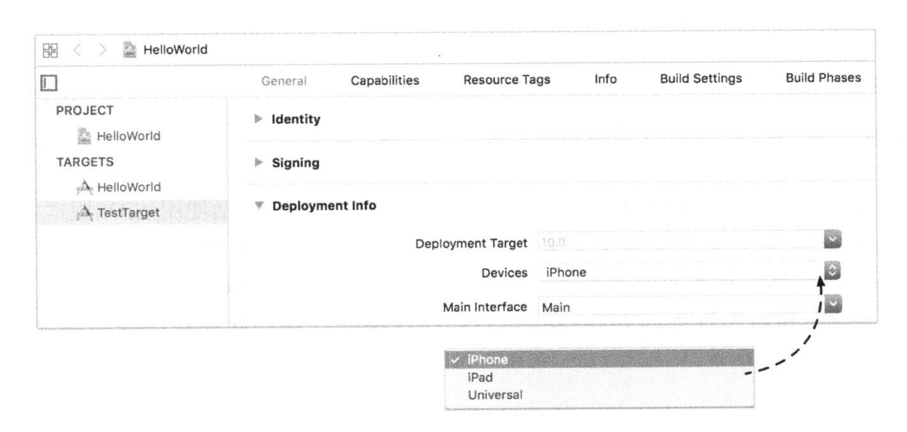

图2-25　设置设备支持情况

事实上，产品的相关属性还有很多，我们会在后面继续了解。

2.4 iOS API 简介

苹果的iOS API在不同版本间有很多变化，本书采用的是iOS 10。本节中，我们会介绍iOS 10有哪些API，说明如何使用这些API的帮助文档。

2.4.1 API 概述

iOS的整体架构图参见图2-26，分为4层——Cocoa Touch层、Media层、Core Services层和Core OS层。下面概要介绍一下这4层。

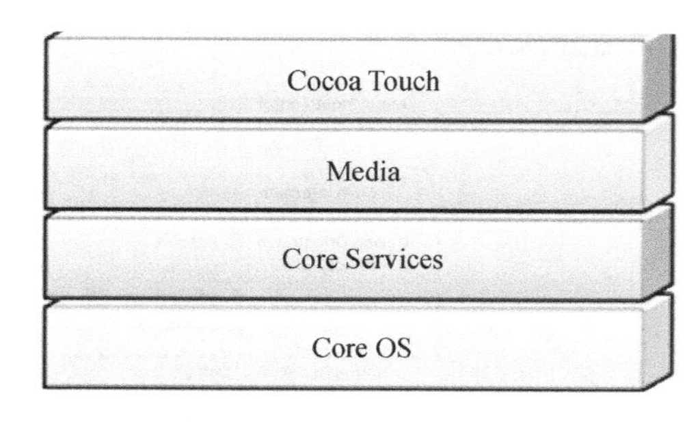

图2-26　iOS整体架构图

- Cocoa Touch层。该层提供了构建iOS应用的一些基本系统服务（如多任务、触摸输入和推送通知等）和关键框架，例如UIKit、WatchKit、GameKit和MapKit等。
- Media层。Media层提供了图形、音频、视频和AirPlay技术，例如Core Audio、Core Graphics、Core Text和Core Video等。
- Core Services层。该层为应用提供一些基本的服务，一般不提供界面，例如CloudKit、HealthKit、HomeKit、应用内购买、SQLite数据库和XML支持等技术。
- Core OS层。该层提供了一些与硬件和网络相关的低级服务，例如应用程序沙箱机制、代码数字签名、安全相关的服务、访问低能耗蓝牙设备和访问重力加速计。

2.4.2 如何使用 API 帮助文档

对于初学者来说，学会在Xcode中使用API帮助文档是非常重要的，下面通过一个例子来介绍API帮助文档的用法。

在编写HelloWorld程序时，可以看到ViewController的代码，如果我们对`didReceiveMemoryWarning`方法感到困惑，可以查找帮助文档。如果只是简单查看帮助信息，可以选中该方法，然后选择右边的快捷帮助检查器Ⓘ，如图2-27所示。

在打开的Xcode快捷帮助检查器窗口中可以看到该方法的描述，其中包括使用的iOS版本、相关主题以及一些相关示例。这里需要说明的是，如果需要查看官方的示例，直接从这里下载即可。

如果想查询比较完整的、全面的帮助文档，可以按住Alt键双击`didReceiveMemoryWarning`方法名，此时会打开一个API帮助搜索结果窗口（如图2-28所示），然后选择感兴趣的主题，进入API帮助界面（如图2-29所示）。

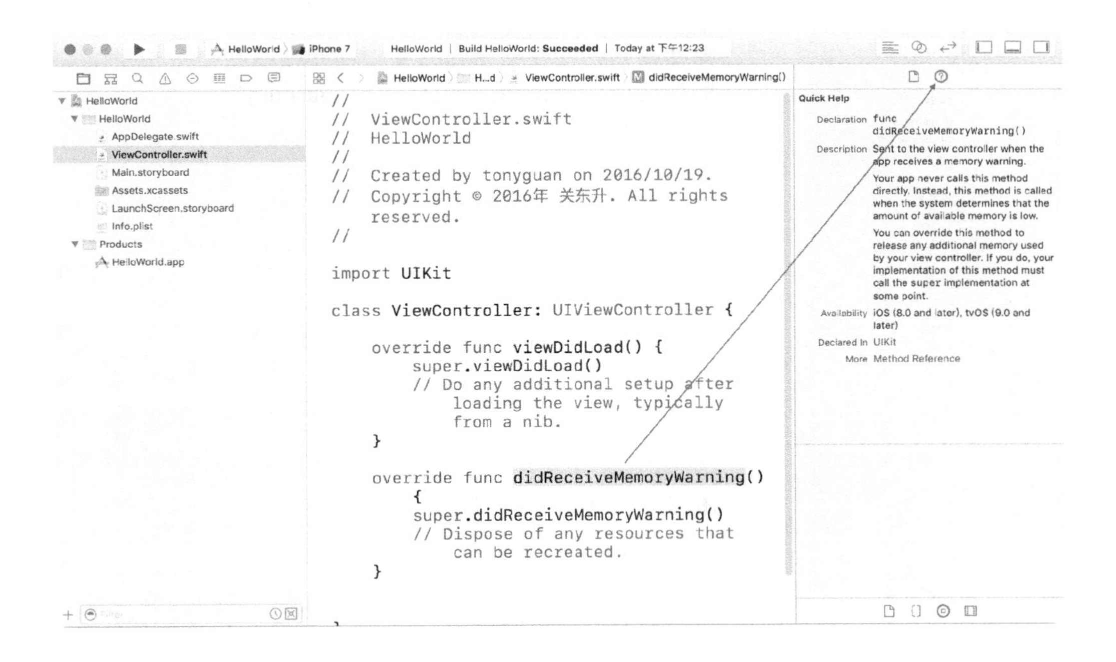

(a)

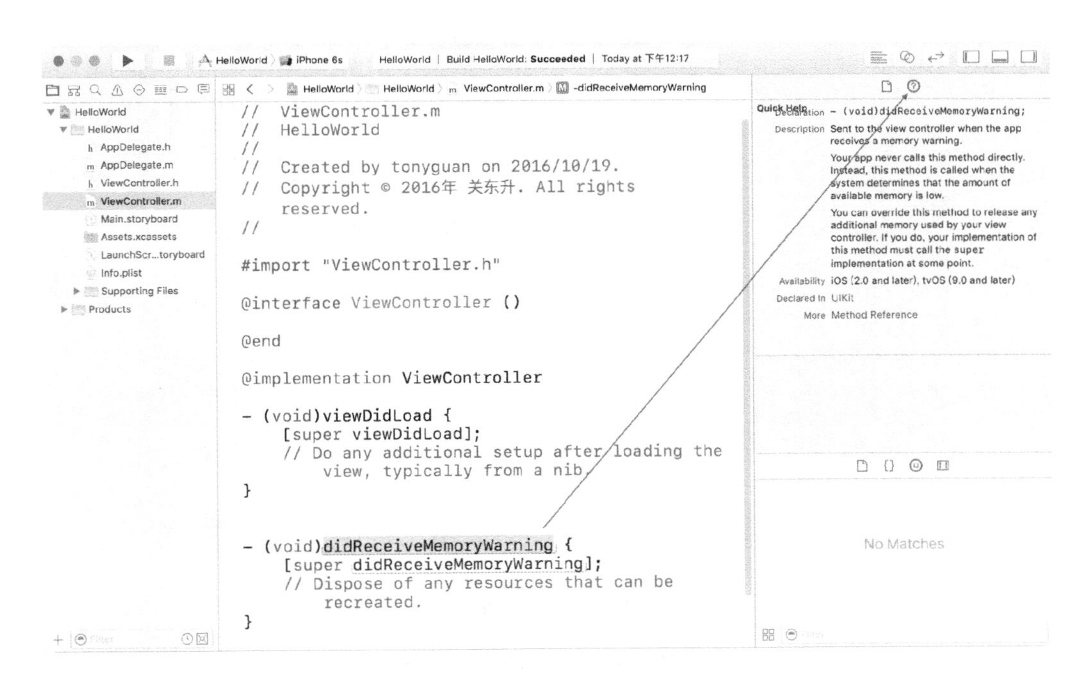

(b)

图2-27 Xcode快捷帮助检查器（图a为Swift版，图b为Objective-C版）

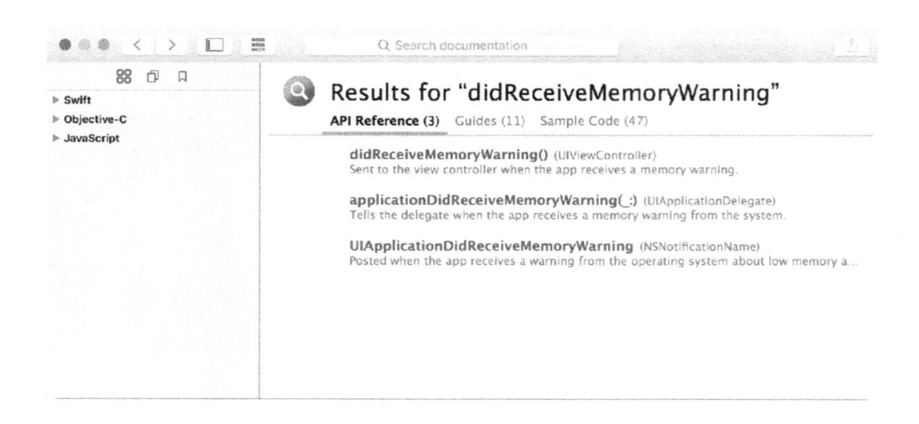

图2-28　API帮助搜索结果窗口

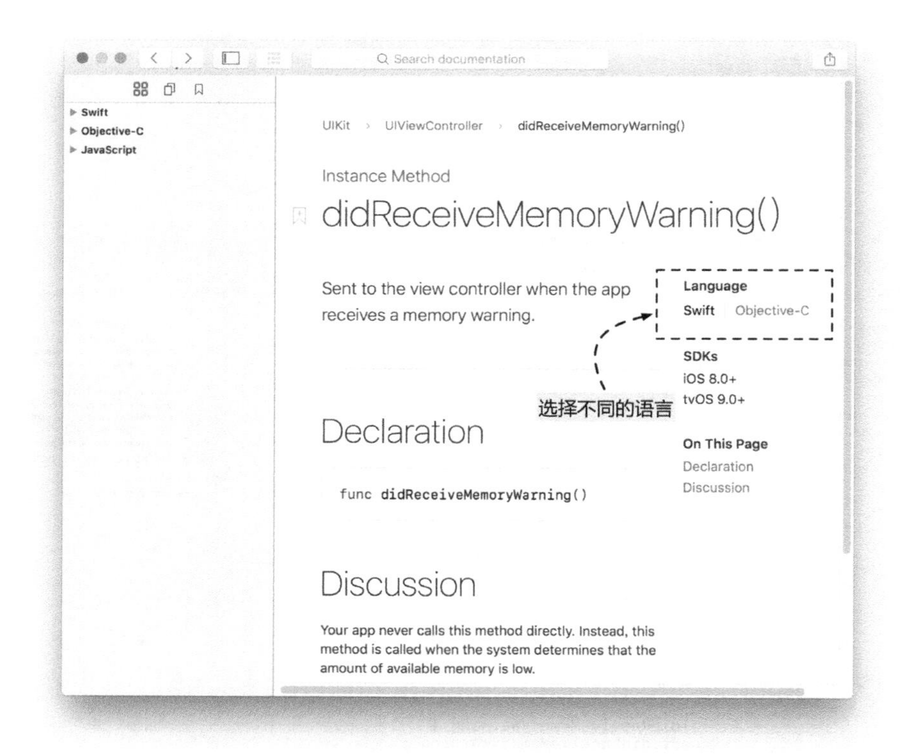

图2-29　API帮助界面

在图2-29所示的API帮助界面中，可以切换Swift和Objective-C两种语言。

2.5　小结

在本章中，我们首先通过HelloWorld工程讨论了iOS工程模板、应用的运行机制和生命周期，然后介绍了几项常用产品属性的设置。最后，我们向大家介绍了API帮助文档的用法。

第3章

Cocoa Touch框架与构建应用界面

苹果应用程序的界面几乎都是由Cocoa和Cocoa Touch框架中的类和协议等元素构建的，这些元素通过MVC（Model-View-Controller，模型–视图–控制器）模式有效地组织起来。

macOS应用的界面主要使用Cocoa框架开发，Cocoa是一种为应用程序提供丰富用户体验的框架，其核心是：Foundation和Application Kit（AppKit）框架。iOS应用界面主要使用Cocoa Touch框架开发，核心是：Foundation和UIKit框架。Cocoa Touch框架有一套自己的MVC模式，其中视图和控制器是UIKit中的`UIView`（及其子类）和`UIViewController`（及其子类）。

本章重点介绍iOS应用界面的构建基础。

3.1 视图控制器

在Cocoa Touch中，`UIViewController`是所有控制器的根类，视图控制器有很多种。下面先介绍一下iOS中视图控制器的种类和视图的生命周期。

3.1.1 视图控制器种类

在Cocoa Touch中，视图控制器有很多，有些用于显示视图，有些起到导航（界面跳转）的作用，有些还有其他用途。与导航相关的视图控制器整理如下。

- **`UIViewController`**。用于自定义视图控制器的导航。例如，对于两个界面的跳转，可以用一个`UIViewController`来控制另外两个`UIViewController`。
- **`UINavigationController`**。导航控制器，它与`UITableViewController`结合使用，能够构建树形结构导航模式。
- **`UITabBarController`**。标签栏控制器，用于构建树形标签导航模式。
- **`UIPageViewController`**。呈现电子书导航风格的控制器。
- **`UISplitViewController`**。可以把屏幕分割成几块的视图控制器，主要为iPad屏幕设计。
- **`UIPopoverController`**。呈现“气泡”风格视图的控制器，主要为iPad屏幕设计。

视图控制器随着iOS版本的变化而变化，例如`UISplitViewController`和`UIPopoverController`是随着iPad的出现而推出的；`UIPageViewController`则用于构建电子书和电子杂志应用。

3.1.2 视图的生命周期

在应用运行过程中视图会显示不同的状态，这就是视图的生命周期。视图生命周期的不同阶段会回调视图控制器的不同方法，如图3-1所示。

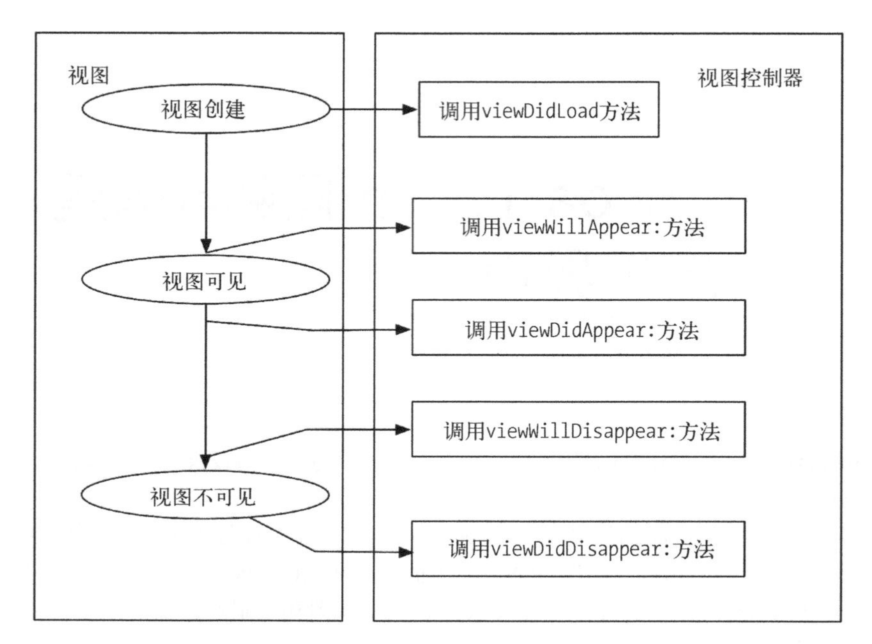

图3-1　视图控制器生命周期

视图创建并加载到内存中时，会调用viewDidLoad方法，这时视图并未出现。该方法中通常会对所控制的视图进行初始化处理。

视图可见前后会调用viewWillAppear:方法和viewDidAppear:方法，视图不可见前后会调用viewWillDisappear:方法和viewDidDisappear:方法。这4个方法调用父类相应的方法以实现其功能，编码时调用父类方法的位置可根据实际情况做调整，参见如下代码：

```
override func viewWillAppear(_ animated: Bool) {
    super.viewWillAppear(animated)
}
```

```
- (void)viewWillAppear:(BOOL)animated {
    [super viewWillAppear:animated];
}
```

viewDidLoad方法在应用运行的时候只调用一次，而上述这4个方法可以被反复调用多次，它们的使用很广泛，但同时也具有很强的技巧性。例如，有的应用会使用重力加速计，重力加速计会不断轮询设备以实时获得设备在Z轴、X轴和Y轴方向的重力加速度。不断地轮询必然会耗费大量电能，进而影响电池的使用寿命，我们利用这4个方法适时地打开或者关闭重力加速计，来达到节约电能的目的。然而，怎么使用这4个方法才能做到“适时”是一个值得思考的问题。

除了上述5种方法，还有很多其他方法可用。随着学习的深入，我们会逐一向大家介绍。

3.2　视图

在Cocoa Touch框架中，提供了视图的“根”类——UIView。

3.2.1　UIView继承层次结构

从继承关系上看，UIView是所有视图的“根”，这就构成了如图3-2所示的UIView类的继承层次。

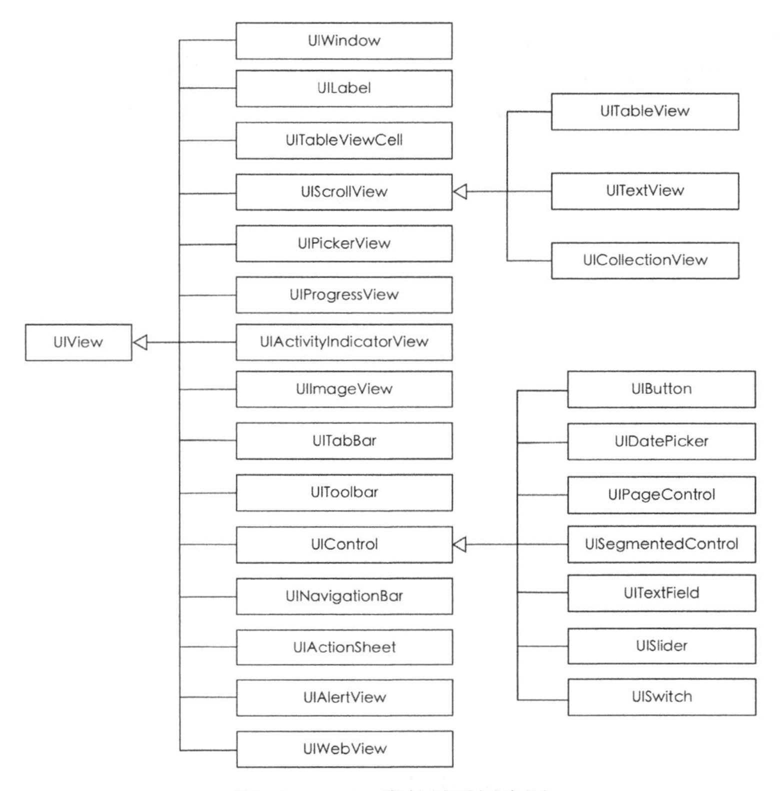

图3-2 UIView类的继承层次图

在UIView类的继承层次图中可见特殊的视图——UIControl类，该类是控件类，其子类有UIButton、UITextField和UISlider等。之所以称它们为“控件类”，是因为它们都有能力响应一些高级事件。为了查看这些事件，我们可以在Interface Builder中拖曳一个UIButton控件到设计界面，然后选中这个控件，点击右上角的 ⊙ 按钮（如图3-3所示），打开连接检查器。

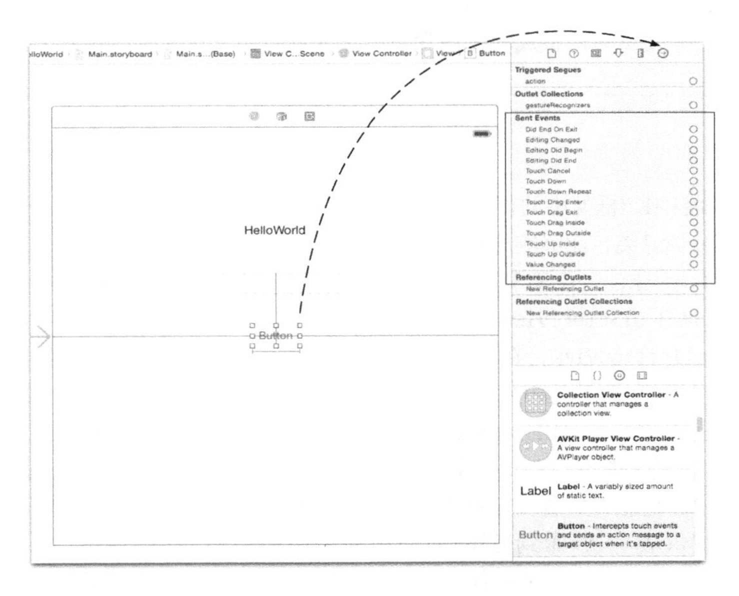

图3-3 UIButton的事件

其中Sent Events栏中的内容就是UIButton对应的高级事件。UIControl类以外的视图没有这些高级事件，这可以借助HelloWorld工程中的Label控件验证一下。选中UILabel控件，打开连接检查器，如图3-4所示。可以发现，UILabel的连接检查器中没有Sent Events栏，即没有高级事件，不可以响应高级事件。

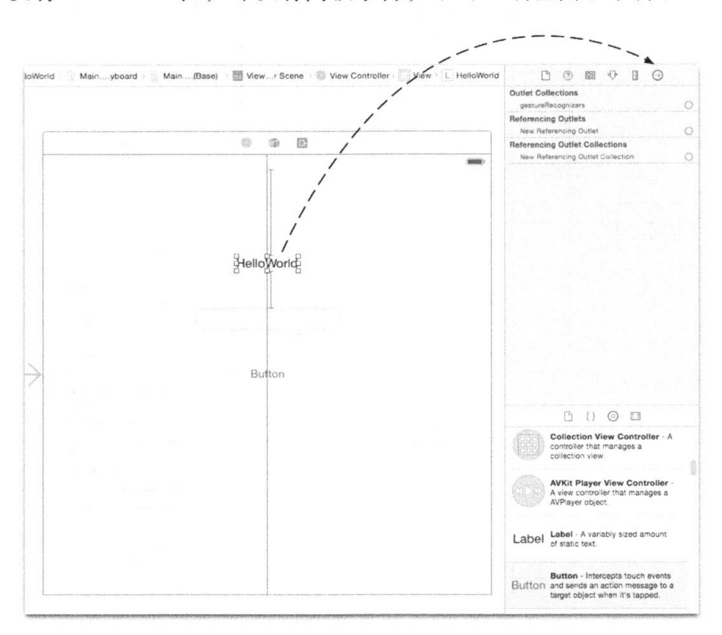

图3-4　UILabel没有高级事件

事实上，视图也可以响应事件，但这些事件比较低级，需要开发人员自己处理。很多手势的开发都以这些低级事件为基础。

注意　在后面章节中，很多视图（例如UILabel、文本视图和进度条等）并未继承UIControl类，但也习惯上称为控件，这是开发中约定俗成的一种常用归类方式，与严格意义上的概念性分类有差别。

3.2.2　视图分类

为了便于开发，苹果将UIKit框架中的视图分成以下几个类别。

- **控件**。继承自UIControl类，能够响应用户高级事件。
- **窗口**。它是UIWindow对象。一个iOS应用只有一个UIWindow对象，它是所有子视图的“根”容器。
- **容器视图**。它包括了UIScrollView、UIToolbar及它们的子类。UIScrollView的子类有UITextView、UITableView和UICollectionView，在内容超出屏幕时，它们可以提供水平或垂直滚动条。UIToolbar是非常特殊的容器，它能够包含其他控件，一般置于屏幕底部，特殊情况下也可以置于屏幕顶部。
- **显示视图**。用于显示信息，包括UIImageView、UILabel、UIProgressView和UIActivityIndicatorView等。
- **文本和Web视图**。提供了能够显示多行文本的视图，包括UITextView和UIWebView，其中UITextView也属于容器视图，UIWebView是能够加载和显示HTML代码的视图。
- **导航视图**。为用户提供从一个屏幕到另外一个屏幕的导航（或跳转）视图，它包括UITabBar和UINavigationBar。
- **警告框和操作表**。用于给用户提供一种反馈或者与用户进行交互。警告框是以动画形式弹出来的视图；而操作表是给用户提供可选操作的视图，在iPhone中它会从屏幕底部滑出，在iPad中则会出现在屏幕中央。

3.2.3 应用界面构建层次

iOS应用界面是由若干个视图构建而成的，这些视图对象采用树形构建。图3-5是一个应用界面的构建层次图，该应用有一个UIWindow，其中包含一个UIView根视图。根视图下又有3个子视图——Button1、Label2和UIView（View2），其中子视图UIView（View2）中存在一个按钮Button3。

一般情况下，应用中只包含一个UIWindow。从UI构建层次上讲，UIWindow包含了一个根视图UIView。根视图一般也只有一个，放于UIWindow中。根视图的类型决定了应用程序的类型。图3-5中各视图对象间的关系如图3-6所示。

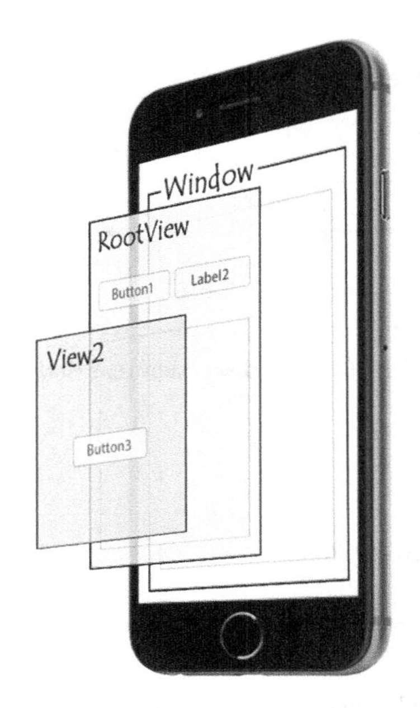

图3-5 应用界面的构建层次图

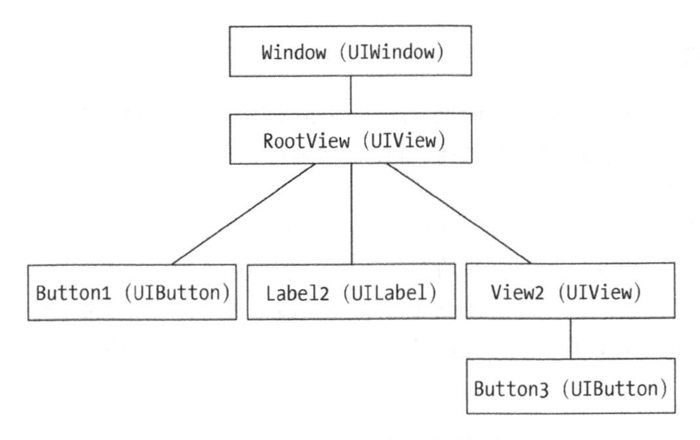

图3-6 各视图对象间的关系

应用界面的构建层次是一种树形结构，"树根"是Window，RootView根视图是"树干"，其他视图对象为"树冠"。在层次结构中，上下两个视图是"父子关系"。除了Window，每个视图的父视图有且只有一个，子视图可以有多个。它们间的关系涉及3个属性，如图3-7所示。

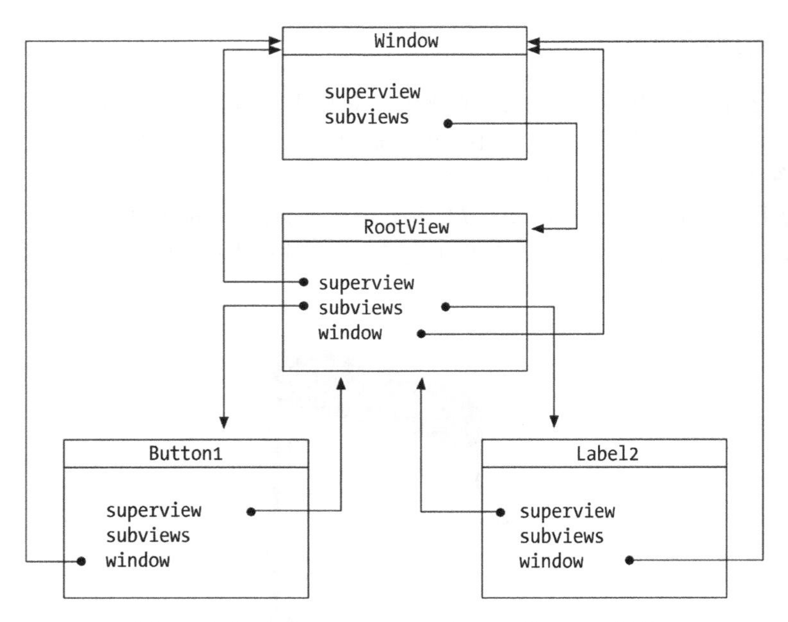

图3-7　视图中的superview、subviews和window属性

下面简要介绍这3个属性的含义。

- **superview**。获得父视图对象。
- **subviews**。获得子视图对象集合。
- **window**。获得视图所在的界面Window对象。

3.3　使用故事板构建界面

在iOS应用开发的过程中，构建一个界面可以采用3种方式：故事板文件、XIB文件和纯代码实现。接下来分别介绍一下，本节我们先介绍故事板。

3.3.1　什么是故事板

在上一章介绍的HelloWorld工程中有一个Main.storyboard文件，称为“故事板”（storyboard）文件。那么，究竟什么是故事板[①]？“故事板”的概念来源于电影行业和动画行业，也称为“分镜头”。

图3-8是一个视频广告的故事板。

在iOS和macOS应用开发中，故事板技术可以用来构建界面。它本质上是一个XML文件，可以用来描述应用中有哪些界面、界面中有哪些视图元素，它们的布局、事件处理，以及界面之间是如何导航（或跳转）的。

下面举例说明故事板的用法。在如图3-9所示的标签应用中，有两个不同的界面，两个标签分别与其对应，点击标签可实现两个界面的互相切换，这就是标签栏导航模式。有关导航模式的知识，我们会在第9章中详细介绍。

① “在电影电视中，故事板的作用是安排剧情中的重要镜头，它们相当于一个可视化的剧本。故事板展示了各个镜头之间的关系，以及它们是如何串联起来，给观众一个完整的体验。”——引自于百度百科：http://baike.baidu. com/view/189750.htm

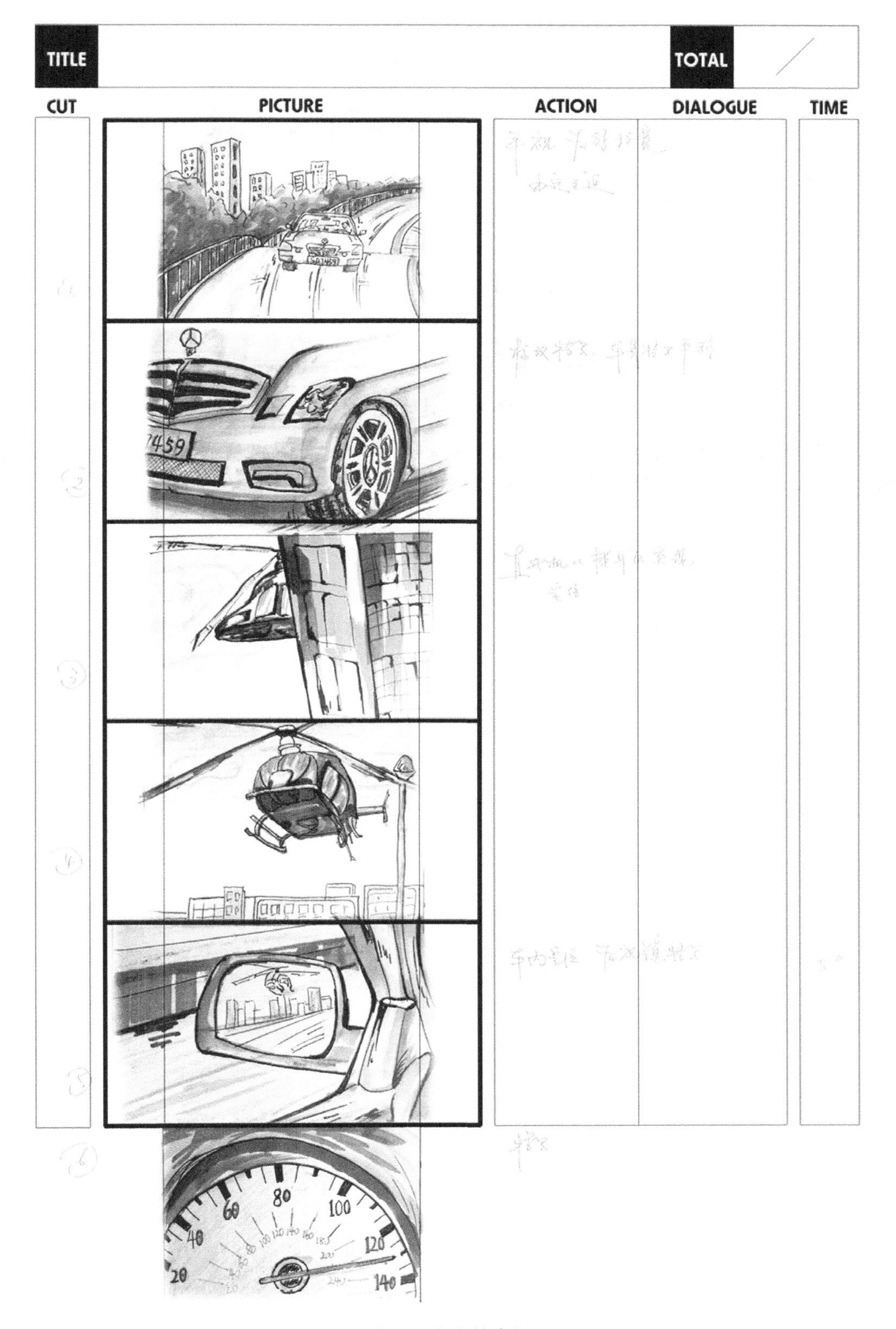

图3-8　广告故事板

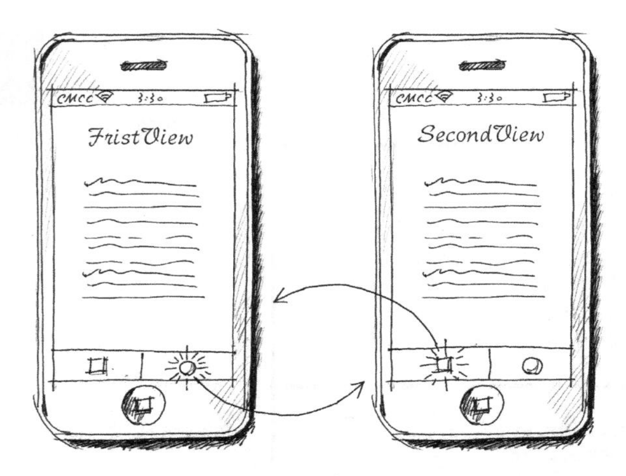

图3-9　设计原型图

要创建标签栏导航模式应用，可以选择Tabbed Application模板（如图2-2所示）创建工程，在生成的工程中打开Main.storyboard文件。此时，你会看到如图3-10所示的设计视图。

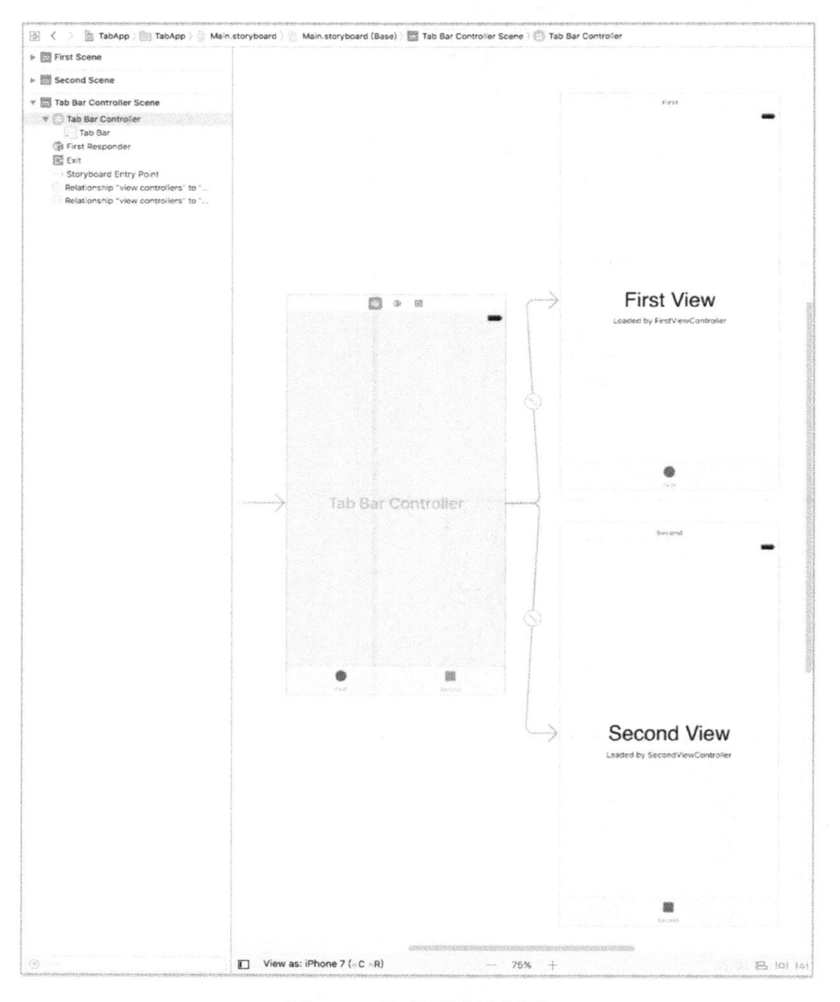

图3-10　故事板设计视图

从图3-10中可以看到，该应用包含两个视图，并且两个视图存在切换关系。

3.3.2　场景和过渡

图3-11演示了故事板中两个非常重要的概念——场景（scene）和过渡（segue）。一个场景中包含一个视图控制器，视图控制器通过管理视图来显示界面；视图控制器有一个view属性，该属性可用于获得它所管理的视图。多个场景之间通过“过渡”连接，过渡定义了场景之间的导航（或跳转）方式。

讨论　故事板中的segue是一个“只可意会，不可言传”的词，有人将segue翻译为“联线”，这种翻译方式过于直接，与故事板和场景这些词格格不入。我推荐将segue翻译为“过渡”，这是因为故事板和场景等概念来自于电影行业，segue是指两个电影场景之间的“转场”，这种转场的专业术语称为“过渡”。

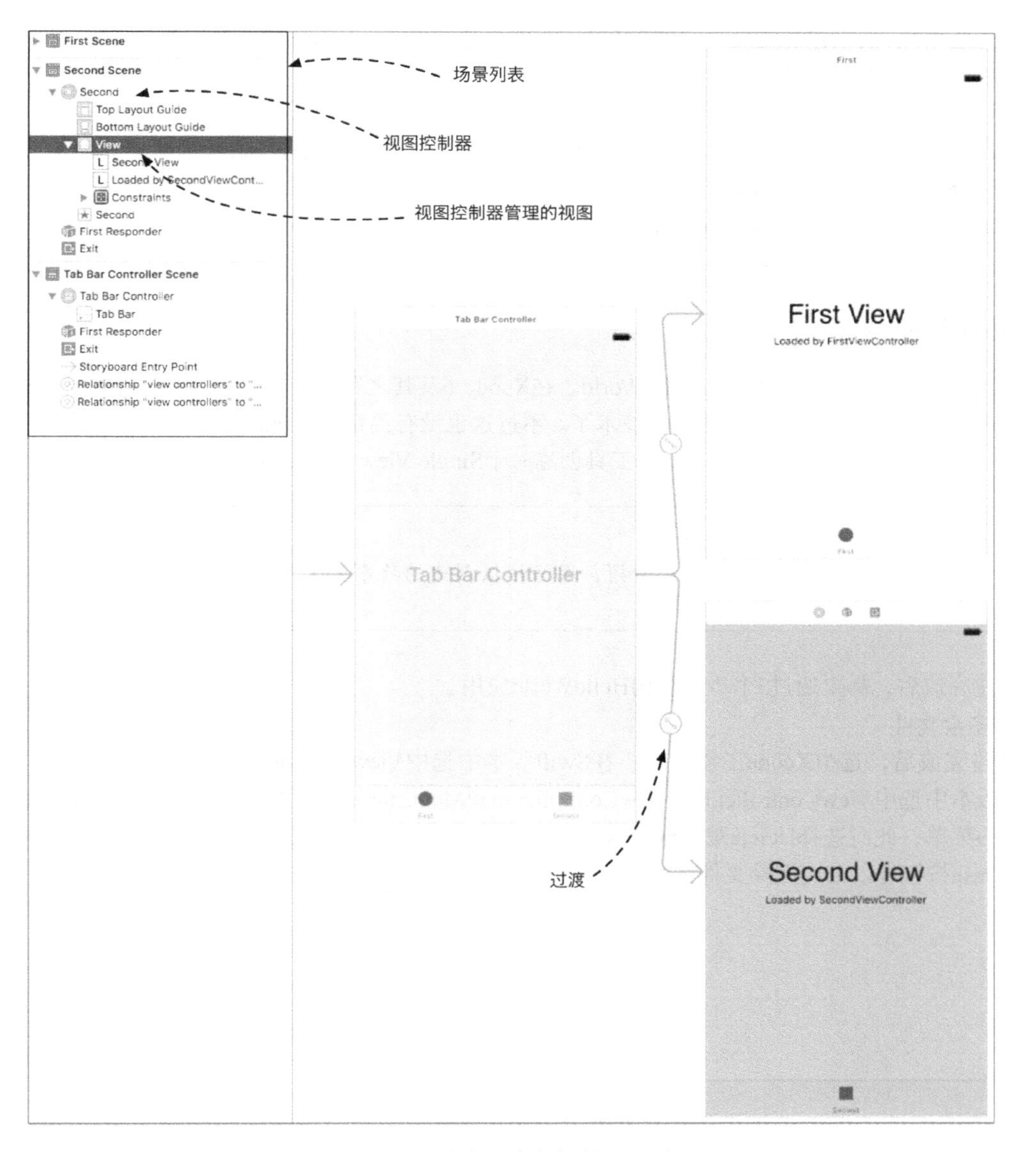

图3-11　故事板中的场景和过渡

场景之间采用什么样的过渡类型，与具体的导航模式有关。下面我们介绍一下过渡类型。

- Show Segue。在当前视图上展示视图，如果当前控制器是导航控制器，那么视图的显示会采用树形结构导航模式。如果不是导航控制器，那么视图显示会采用模态（Modal）导航模式。
- Show Detail Segue。与Show Segue非常类似，区别在于当采用UISplitViewController控制器时，在Detail视图中显示视图。如果在其他导航控制器中使用这个过渡，那么它和Show Segue的效果是一样的。
- Modal Segue。视图显示会采用模态导航模式。
- Popover Segue。用于在iPad设备上呈现浮动窗口，这些导航模式会在后面介绍。

有关树形结构导航模式和模态导航模式等的详细内容，我们将在后面详细介绍。另外，关于过渡的使用细节，也将在后面逐一介绍。

3.4　使用XIB文件构建界面

在一些老版本Xcode创建的工程中，经常会看到XIB文件，事实上XIB与故事板是非常相似的技术。本节中，我们使用XIB文件来构建界面。

提示　在苹果的官方资料和API命名中会看到NIB，那么NIB与XIB是怎样一种关系呢？最初苹果的界面是使用NIB文件构建的，后来由于文件格式采用了XML格式，于是更名为XIB，但很多人还一直沿袭NIB这个叫法，所以目前为止NIB等同于XIB。

3.4.1　重构HelloWorld

下面详细介绍一下通过XIB实现的HelloWorld。在Xcode 6及其之后的版本中，已经没有能够创建XIB的工程模板了，这也可见苹果公司重点支持故事板技术了。不过这也没有关系，创建故事板的办法还是有的。先参考上一章创建HelloWorld工程的方法，通过Xcode工具创建一个Single View Application工程。

提示　事实上，选择哪个工程模板都无所谓，因为模板创建的故事板文件是要删除的。

工程创建完成后，需要通过3个步骤重构HelloWorld应用。

1. 删除多余文件

工程创建完成后，选中Xcode工程文件：在Swift版本中选中ViewController.swift和Main.storyboard文件；在Objective-C版本中选中ViewController.h、ViewController.m和Main.storyboard文件。然后点击右键，此时会弹出如图3-12所示的菜单，此时选择Delete菜单删除这两个文件，这时会弹出如图3-13所示的删除确认对话框，此时点击Move to Trash按钮可以彻底删除文件，而Remove References按钮只是从工程中删除文件。

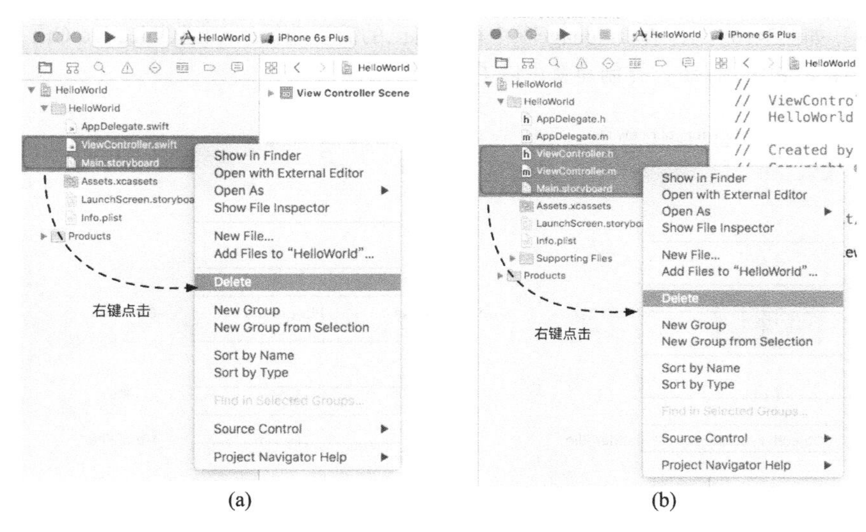

图3-12　删除工程中文件（图a为Swift版，图b为Objective-C版）

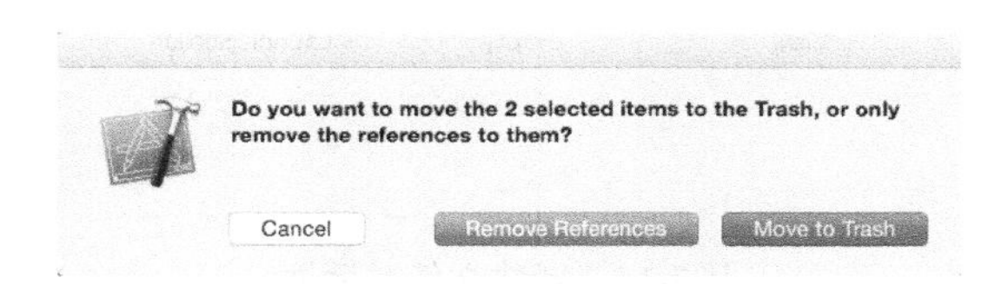

图3-13　删除确认对话框（Swift版是删除2个文件，Objective-C版是删除3个文件）

另外，由于删除了Main.storyboard文件，但是工程默认还会加载Main.storyboard文件，所以还需要设置工程属性。如图3-14所示，选择TARGETS→HelloWorld→Deployment Info→Main Interface。Main Interface是应用加载的故事板文件名，默认为Main，我们删除默认的Main，使其内容为空。

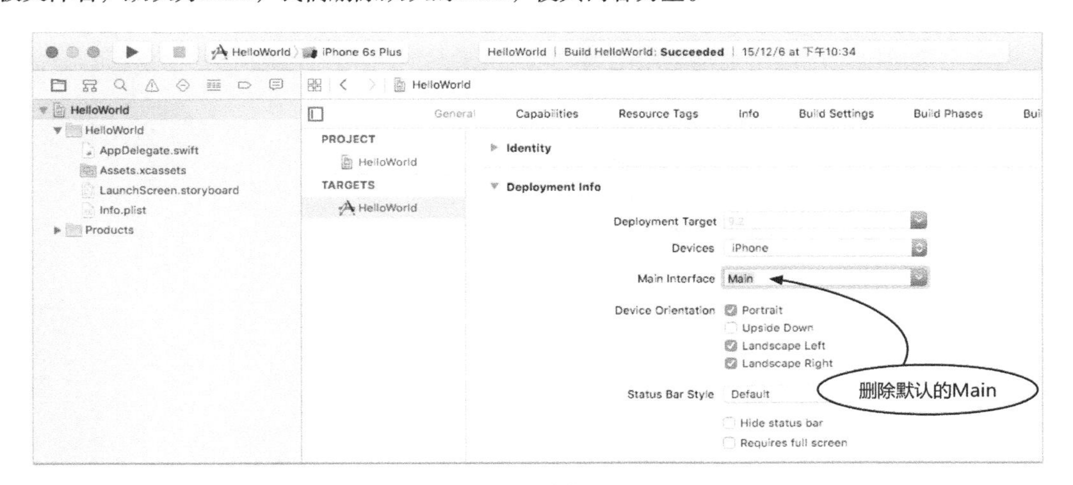

图3-14　删除Main

2. 添加视图控制器

在Xcode工程中选择菜单File→New→File...，此时会弹出如图3-15所示的新建文件对话框，从中选择iOS→Source→Cocoa Touch Class，然后点击Next按钮。此时会弹出一个对话框，在Class中输入RootViewController，在Subclass of中输入UIViewController，选中Also create XIB file（选中它，可以在创建视图控制器类的时候同时创建对应的XIB文件），如图3-16所示。在Language项中根据自己的需要选择Swift或Objective-C语言。选择完成之后，点击Next按钮，选择文件保存目录，并创建就可以了。

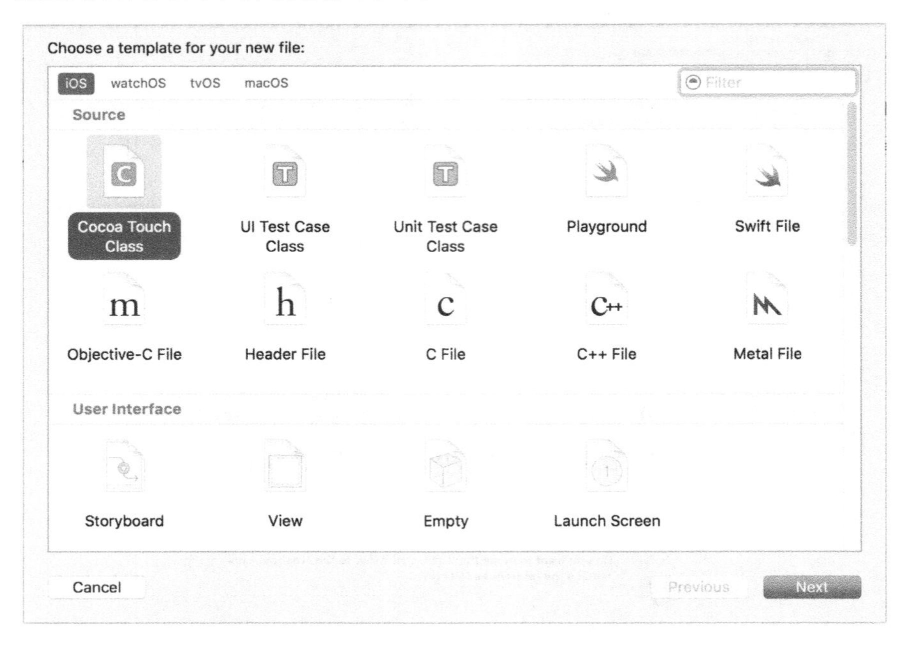

图3-15　新建文件对话框

图3-16　新建视图控制器对话框

要进行界面设计，可以打开RootViewController.xib文件，具体可参考故事板实现的HelloWorld设计界面。具体步骤不再赘述，完成之后的界面如图3-17所示。从图中可见没有场景和过渡等内容，其中File's Owner（文件所有者）在故事板中是没有的，它表示当前XIB文件的所有者是谁，本例中是RootViewController视图控制器。

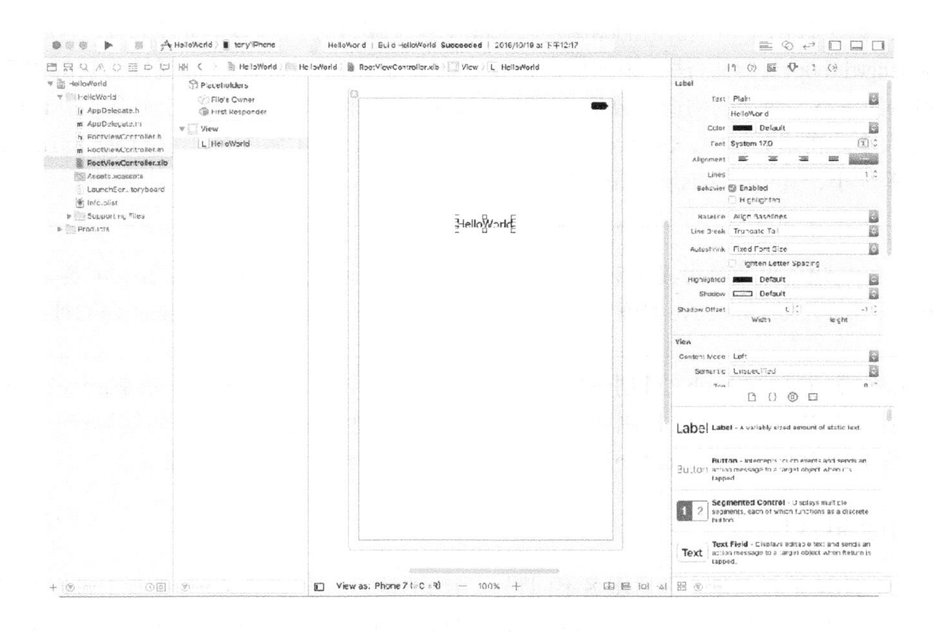

图3-17 打开XIB文件

3. 修改AppDelegate类

下面看看代码部分。打开AppDelegate类，其代码如下：

```
//AppDelegate.swift文件
@UIApplicationMain
class AppDelegate: UIResponder, UIApplicationDelegate {

    var window: UIWindow?                                                    ①


    func application(_ application: UIApplication, didFinishLaunchingWithOptions
    ➥launchOptions: [UIApplicationLaunchOptionsKey: Any]?) -> Bool {

        self.window = UIWindow(frame: UIScreen.main.bounds)                  ②
        self.window?.rootViewController =
        ➥RootViewController(nibName: "RootViewController", bundle:nil)      ③
        self.window?.makeKeyAndVisible()                                     ④

        return true
    }
    ......
}
```

```
//AppDelegate.h文件
#import <UIKit/UIKit.h>
@interface AppDelegate : UIResponder <UIApplicationDelegate>

@property (strong, nonatomic) UIWindow *window;                              ①

@end

//AppDelegate.m文件
#import "AppDelegate.h"
#import "RootViewController.h"

@interface AppDelegate ()

@end

@implementation AppDelegate


- (BOOL)application:(UIApplication *)application
➥didFinishLaunchingWithOptions:(NSDictionary *)launchOptions {

    self.window = [[UIWindow alloc] initWithFrame:[[UIScreen mainScreen]
    ➥bounds]];                                                              ②
    self.window.rootViewController =
    ➥[[RootViewController alloc] initWithNibName:@"RootView
    ➥Controller" bundle:nil];                                                ③
    [self.window makeKeyAndVisible];                                         ④

    return YES;
}
......

@end
```

上述代码中，第①行用于声明UIWindow属性window（这个属性在3.2.3节中介绍过）。然后在 application:didFinishLaunchingWithOptions:方法中添加代码，其中第②行用于实例化window属性，它的类型是UIWindow。UIWindow是UIView的子类，使用的构造函数是initWithFrame:，用来初始化视图的frame属性，该属性能够描述一个视图的位置和大小。在第②行中，UIScreen.main.bounds（Objective-C版中是[[UIScreen mainScreen] bounds]）用于获得当前屏幕大小，其中bounds属性是屏幕的边界。

提示　第②行在Swift 2中的语句是UIScreen.mainScreen().bounds，类似于Objective-C类名和方法名。而Swift 3中主张“去Objective-C语言化”，它的很多类和方法命名不再与Objective-C对应。

第③行通过RootViewController.xib文件创建RootViewController根控制器对象，构造函数是initWithNibName:bundle:，其中第一个参数指定XIB文件名。根控制器对象创建完成后，把它赋值给window的属性rootViewController，UIWindow有一个根视图属性rootViewController。最后，第④行用于显示应用程序窗口。

3.4.2　XIB与故事板比较

那么，故事板与XIB是否只是文件后缀名不同呢？当然不是，一般而言，一个工程中可以有多个XIB文件，一个XIB文件对应着一个视图控制器。图3-18是分别采用XIB和故事板构建的标签栏导航应用程序，其中包含两个视图控制器：FirstViewController和SecondViewController。图3-18a是使用XIB实现的工程，每个视图控制都有一个XIB文件与之对应，这个XIB文件用来设计该视图控制器的管理界面，不能设计其他界面，更无法描述界面之间如何导航。图3-18b是使用故事板实现的工程，这个应用程序在3.3节中已经介绍了，工程中包含一个故事板文件，两个视图控制器没有对应的XIB，视图控制器管理的界面是在故事板中设计的。

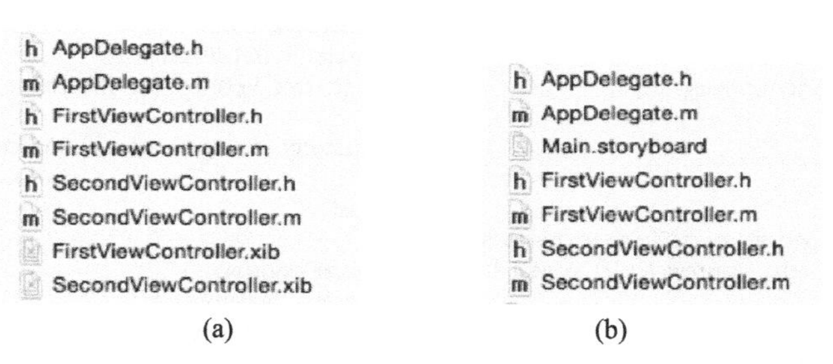

图3-18　XIB和故事板比较

故事板不仅可以描述应用的整体界面结构，还可以描述单个界面，而且能够描述界面之间的导航关系。XIB文件只能描述应用的单个界面，不能描述界面之间如何导航。

3.5　使用纯代码构建界面

代码是万能的，通过代码完全可以构建应用界面，但是调试起来非常麻烦。界面每次的修改结果，只能重新运行才能看到，不是“所见即所得”的，这是最大的问题。

3.5.1　重构HelloWorld

如果使用代码构建HelloWorld工程，则可以参考上一节的XIB实现，参考3.4.1节创建工程并删除ViewController和Main.storyboard等文件。然后再创建一个根视图控制器，但是创建时不需要选中Also create XIB file（见图3-16）。

下面看看代码部分。AppDelegate类的代码如下：

```
//AppDelegate.swift文件
@UIApplicationMain
class AppDelegate: UIResponder, UIApplicationDelegate {

    var window: UIWindow?


    func application(_ application: UIApplication, didFinishLaunchingWithOptions
    ➥launchOptions: [UIApplicationLaunchOptionsKey: Any]?) -> Bool {

        self.window = UIWindow(frame: UIScreen.main.bounds)                    ①
        self.window?.rootViewController = RootViewController()
        self.window?.backgroundColor = UIColor.white                           ②
        self.window?.makeKeyAndVisible()

        return true
    }
    ......
}
```

```
//AppDelegate.m文件
#import "AppDelegate.h"
#import "RootViewController.h"

@interface AppDelegate ()

@end

@implementation AppDelegate


- (BOOL)application:(UIApplication *)application
➥didFinishLaunchingWithOptions:(NSDictionary *)launchOptions {

    self.window = [[UIWindow alloc] initWithFrame:[[UIScreen mainScreen]
    ➥bounds]];                                                                ①
    self.window.rootViewController = [[RootViewController alloc] init];
    self.window.backgroundColor = [UIColor whiteColor];                       ②
    [self.window makeKeyAndVisible];

    return YES;
}
......

@end
```

上述代码与XIB实现非常相似，不同的地方在于创建视图控制器RootViewController时（详见第①行代码），没有采用基于XIB的构造函数initWithNibName:bundle:，而是采用init默认构造函数。第②行用于设置window的背景颜色，背景颜色属性是backgroundColor。

由于没有设计界面，添加Label控件到根视图的过程，需要在根视图控制器中通过代码实现。根视图控制器RootViewController类的主要代码如下：

```
//RootViewController.swift文件
class RootViewController: UIViewController {

    override func viewDidLoad() {
        super.viewDidLoad()

        let screen = UIScreen.main.bounds                                      ①
        let labelWidth:CGFloat = 90
        let labelHeight:CGFloat = 20
        let labelTopView:CGFloat = 150
        let frame = CGRect(x: (screen.size.width - labelWidth)/2,
        ➥y: labelTopView, width: labelWidth, height: labelHeight)             ②
        let label = UILabel(frame: frame)                                      ③
        label.text = "HelloWorld"

        //字体左右居中
        label.textAlignment = NSTextAlignment.center
        self.view.addSubview(label)                                            ④
    }
    ......
}
```

```
//RootViewController.m文件
@implementation RootViewController

- (void)viewDidLoad {
    [super viewDidLoad];

    CGRect screen = [[UIScreen mainScreen] bounds];                            ①
    CGFloat labelWidth = 90;
    CGFloat labelHeight = 20;
    CGFloat labelTopView = 150;
    CGRect frame = CGRectMake((screen.size.width - labelWidth)/2 ,
    ➥labelTopView, labelWidth, labelHeight);                                  ②
    UILabel* label = [[UILabel alloc] initWithFrame:frame];                    ③

    label.text = @"HelloWorld";
    //字体左右居中
    label.textAlignment = NSTextAlignmentCenter;
    [self.view addSubview:label];                                              ④
}
......
@end
```

上述代码中，第①行用于获得屏幕的边界，其返回值是CGRect类型。CGRect是描述视图对象位置和大小的结构体。第③行创建UILabel对象，构造函数中的frame参数也是CGRect实例。很多视图对象都可以通过frame参数创建。

创建完视图对象后，请一定不要忘记要通过- addSubview:方法把它添加到父视图中。第④行用于将Label对象添加到根视图上。

提示 CGRect类型事实上是一种结构体类型。在Objective-C中创建CGRect结构体实例，需要使用CGRectMake函数，该函数的定义为CGRectMake(CGFloat x, CGFloat y, CGFloat width, CGFloat height)，见Objective-C代码第②行。而在Swift中创建CGRect结构体实例时，需要使用构造函数，函数定义为init(x: CGFloat, y: CGFloat, width: CGFloat, height: CGFloat)，见Swift代码第②行。另外，CGFloat事实上是float类型的别名。

3.5.2　视图的几个重要属性

在纯代码编写界面时，常常遇到一些重要视图属性，其中frame和bounds这两个属性容易混淆。

- frame属性表示该视图在父视图坐标系统（相对于父视图）中的位置和大小。
- bounds属性表示该视图在本地坐标系统（相对于自己）中的位置和大小。

frame属性用得很多，bounds属性用得比较少，这主要是因为bounds属性的坐标系是本地坐标。

下面我们通过一个实例解释一下frame和bounds属性的区别。修改3.5.1节的HelloWorld实例代码，根视图控制器RootViewController类的主要代码如下：

```
//RootViewController.swift文件
class RootViewController: UIViewController {

    override func viewDidLoad() {
        super.viewDidLoad()

        //创建视图viewA
        let viewA = UIView()                                                    ①
        viewA.backgroundColor = UIColor.gray
        //设置viewA的frame属性
        viewA.frame = CGRect(x: 0, y: 0, width: 300, height: 400)
        //将viewA添加到根视图中
        self.view.addSubview(viewA)

        //创建视图viewB
        let viewB = UIView()                                                    ②
        viewB.backgroundColor = UIColor.green
        //设置viewB的frame属性
        viewB.frame = CGRect(x: 50, y: 100, width: 100, height: 200)
        //将viewB添加到viewA视图中
        viewA.addSubview(viewB)                                                 ③

        NSLog("frame_x: %.2f, frame_y: %.2f",
        ➥viewB.frame.origin.x, viewB.frame.origin.y)                           ④
        NSLog("frame_w: %.2f, frame_h: %.2f",
        ➥viewB.frame.size.width, viewB.frame.size.height)                      ⑤

        NSLog("bounds_x: %.2f, bounds_y: %.2f",
        ➥viewB.bounds.origin.x, viewB.bounds.origin.y)                         ⑥
        NSLog("bounds_w: %.2f, bounds_h: %.2f",
        ➥viewB.bounds.size.width, viewB.bounds.size.height)                    ⑦

    }

    ......
}
```

```
//RootViewController.m文件
@implementation RootViewController

- (void)viewDidLoad {
    [super viewDidLoad];

    //创建视图viewA
    UIView* viewA = [[UIView alloc] init];                                      ①
    viewA.backgroundColor = [UIColor grayColor];
    //设置viewA的frame属性
    viewA.frame = CGRectMake(0, 0, 300, 400);
    //将viewA添加到根视图中
    [self.view addSubview: viewA];

    //创建视图viewB
    UIView* viewB = [[UIView alloc] init];                                      ②
    viewB.backgroundColor = [UIColor greenColor];
    //设置viewB的frame属性
    viewB.frame = CGRectMake(50, 100, 100, 200);
    //将viewB添加到viewA视图中
    [viewA addSubview: viewB];                                                  ③

    NSLog(@"frame_x: %.2f, frame_y: %.2f", viewB.frame.origin.x,
    ➥viewB.frame.origin.y);                                                    ④
    NSLog(@"frame_w: %.2f, frame_h: %.2f",
    ➥viewB.frame.size.width, viewB.frame.size.height);                         ⑤

    NSLog(@"bounds_x: %.2f, bounds_y: %.2f",
    ➥viewB.bounds.origin.x, viewB.bounds.origin.y);                            ⑥
    NSLog(@"bounds_w: %.2f, bounds_h: %.2f",
    ➥viewB.bounds.size.width, viewB.bounds.size.height);                       ⑦
}
......

@end
```

上述代码创建了两个视图对象viewA和viewB，见第①行和第②行代码。因为使用UIView的无参数构造函数init，而不是initWithFrame:，所以我们需要重新设置frame属性。第③行将viewB视图添加到viewA中，即viewA

是viewB的父视图。

第④行用于输出viewB视图的frame属性中的坐标点，第⑤行用于输出viewB视图的frame属性中的高和宽。frame属性的类型是CGRect结构体，该结构体如图3-19所示。CGRect结构体有两个成员：origin和size。origin坐标点类型，是CGPoint结构体，size是高和宽类型，是CGSize结构体。CGPoint结构体有两个成员x和y，CGSize结构体有width和height两个成员。

第⑥行用于输出viewB视图的bounds属性中的坐标点，第⑦行用于输出viewB视图的bounds属性中的高和宽。

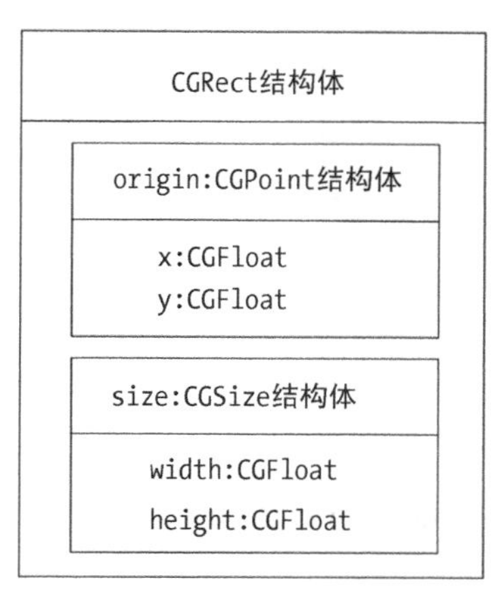

图3-19　CGRect结构体

运行实例，输出日志如下：

```
frame_x: 50.00, frame_y: 100.00
frame_ w: 100.00, frame_h: 200.00
bounds_x: 0.00, bounds_y: 0.00
bounds_w: 100.00, bounds_h: 200.00
```

实例的运行界面如图3-20所示。

为了帮助理解frame和bounds属性，我绘制了如图3-21所示的示意图，从图中可见viewA和viewB的frame属性以及bounds属性。

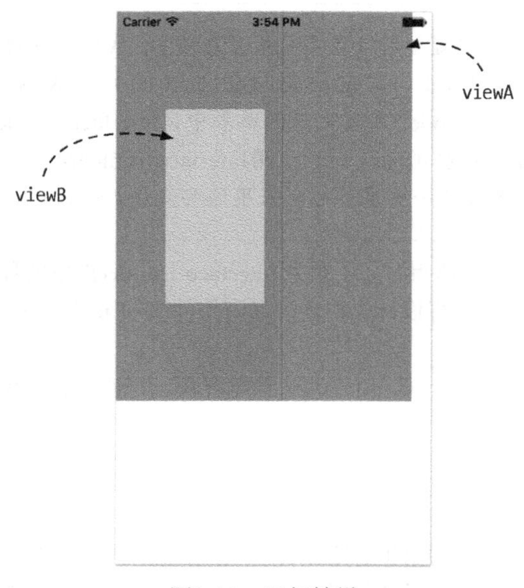

图3-20　运行结果

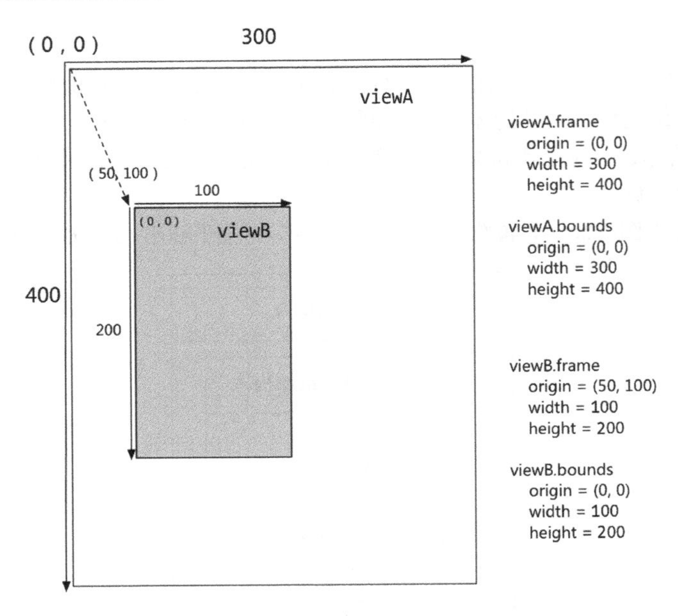

图3-21　viewA和viewB示意图

3.6 三种构建界面技术讨论

经过前面的学习，广大读者已经对三种界面构建技术有了一定程度的了解。一直以来，也有很多开发人员问我一个问题：三种技术哪个好，哪个不好？这里先不给出结论，而是从几个方面讨论一下，然后再做总结。

3.6.1 所见即所得

所见即所得（What You See Is What You Get，WYSIWYG）是目前主流的界面设计思路，通过提供一个界面设计工具，能够实时地进行设计和预览界面。

提示

默认情况下，我们在Xcode中直接打开故事板或XIB文件，然后会看到一个设计界面。事实上，Xcode工具中集成了一个界面设计工具——Interface Builder（IB）。在Xcode早期版本中，Interface Builder与Xcode是相对独立的，在Xcode 5及其之后的版本中，Interface Builder被嵌入到Xcode工具中，所以在一些资料中会有这样的一些说法，如“使用Interface Builder打开故事板文件”或“使用Interface Builder打开XIB文件”。本书的一些章节也会采用这些说法，而且有时也会将Interface Builder称为IB。

故事板和XIB技术当然是“所见即所得”了，使用Interface Builder提供的界面设计工具，能够使开发人员从对象库中拖曳视图和控件到设计窗口进行设计，并且可以为控件添加事件处理代码。在故事板中，界面设计工具还可以拖曳视图控制器到故事板场景中。

纯代码构建界面不是“所见即所得”的，由于无法预览界面，每次在代码中修改视图的参数后，都要重新编译、运行，开发起来工作量非常大。

3.6.2 原型驱动开发

所谓“原型驱动”，就是开发一个应用的可交互原型，然后用户测试这个原型，从而找出设计上的缺陷，以及设计方案的可行性，还可以帮助用户明确他的需求。原型驱动开发通过不断修改原型，快速迭代设计，进而驱动用户需求的确定，以及设计方案的改进。

能够制作原型的工具有很多，甚至使用笔和纸就可以绘制原型了，但是制作可交互原型的工具很少，有些设计人员使用Flash来制作可交互原型，然而Flash工具制作起原型来非常耗时。事实上，Xcode中的故事板是最好的可交互原型制作工具，不懂技术的美工或产品经理可以在Xcode中设计故事板，然后将这个Xcode工程发布到iOS设备上，这样不写一行代码就可以制作可交互原型。这个故事板原型经过多次迭代后，会获得一个成熟的Xcode工程，然后将这个Xcode工程移交给开发人员进行开发，这样就大大提高了开发速度，缩短了开发周期，提高了开发效率。

提示 如果你对通过在Xcode故事板中制作可交互原型感兴趣，可以阅读我写的《交互设计的艺术：iOS 7拟物化到扁平化革命》一书。

如果通过XIB和纯代码制作可交互原型，就必须编写一定数量的代码。编写代码对于开发人员不是问题，而对于不懂技术的美工或产品经理却是不现实的事情。

3.6.3　团队协同开发

很多情况下，一个应用是由一个团队多人协同开发的，但是由于一个工程中只有一个故事板文件，多人同时修改故事板就会发生版本冲突，这是故事板技术的一个缺点。而采用XIB和纯代码构建的界面，文件是多个、分散的，不会发生版本冲突。在团队协同开发代码版本管理方面，XIB和纯代码方式要比故事板方式有优势。

提示 事实上，这个问题是可以解决的，就是故事板文件统一由一个人负责修改，不允许其他人员修改，或者错时修改，这样通过项目管理手段规避版本冲突的问题。客观地说，一个工程中只要有类似的共享文件，就存在这个问题，完全依赖于版本管理软件是不现实的，通过项目管理手段解决这些问题是务实的做法。

说明 另外，苹果在 iOS 9 及其之后的版本中引入故事板引用（Storyboard References）技术。故事板引用技术允许我们将主故事板文件分割成几个小故事板文件，这可以在一定程度上解决团队协同开发的问题。故事板引用技术超出了本书范围，读者可以参考其他资料。

经过前面的解释说明，可见这三种技术各有千秋。我们可以根据自己的喜好选择技术，但是过于偏激的做法不是一个好的技术方案，我更喜欢以故事板为主，以XIB和纯代码为辅的技术方案。

3.7　小结

本章重点介绍了Cocoa Touch 框架中构建界面的相关类，重点介绍了构建界面的三种技术：故事板、XIB和纯代码。

第 4 章

UIView与视图

4

视图和控件是应用的基本元素。在学习iOS之初，我们要掌握一些常用视图和控件的特点及其使用方式。

4.1 标签与按钮

标签和按钮是两个常用的控件，下面我们通过一个LabelButton案例介绍一下它们的用法。

该案例的设计原型草图如图4-1所示，其中包含一个标签和一个按钮。当点击按钮的时候，标签文本会从初始的Label变为HelloWorld。

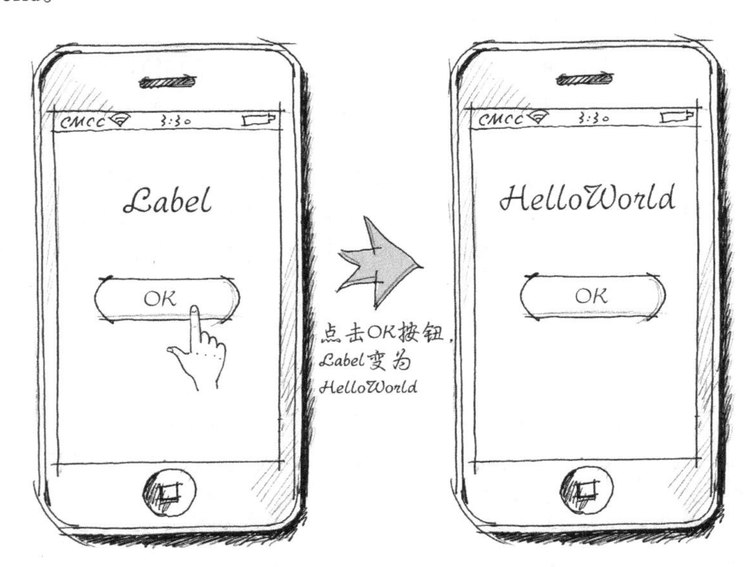

图4-1　LabelButton案例设计原型草图

4.1.1 Interface Builder 实现

使用Single View Application模板创建一个名为LabelButton的工程（具体创建过程参见2.1.1节）。

1. 添加Label

打开Main.storyboard文件，从对象库中拖曳一个Label（其属性检查器如图4-2所示），并将该Label摆放在设计视图的中间位置。

由图4-2可以看出，标签的属性检查器包括Label和View两个组。Label组主要是文本相关的属性，而View组主要是从视图的角度对视图进行设置，所有的视图都具有View组。

前面提到过，通过双击Label可以设置Label显示的内容，这个属性就是Label下的Text属性。当然，我们也可以用代码来设置Text属性。

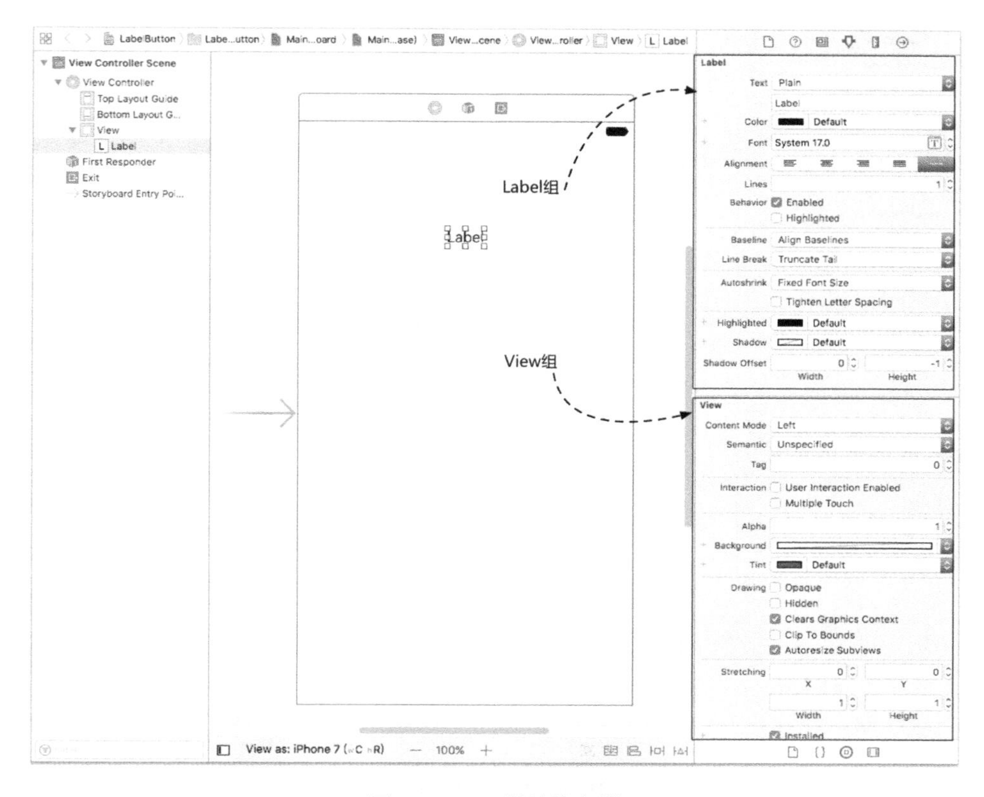

图4-2　Label属性检查器

提示　对象库中包含了控制器、视图、控件和手势等很多对象。随着版本的升级，对象库还在不断扩充和完善，短时间内可能无法找到指定的对象，此时我们可以借助对象库下方的搜索栏来查找，如图4-3所示。

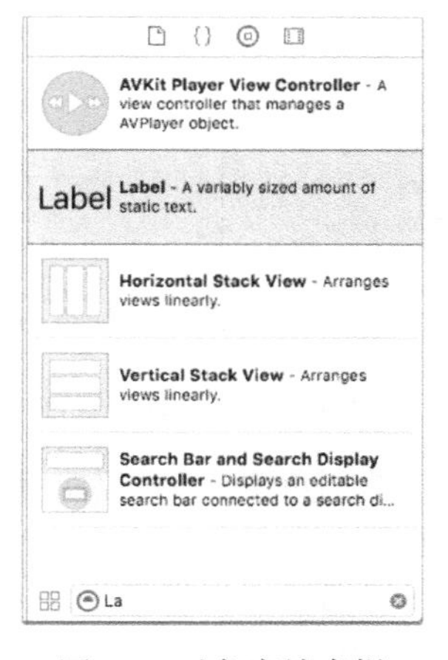

图4-3　对象库搜索栏

2. 添加Button

从对象库中拖曳一个Button并将其摆放到Label的正下方，如图4-4所示。

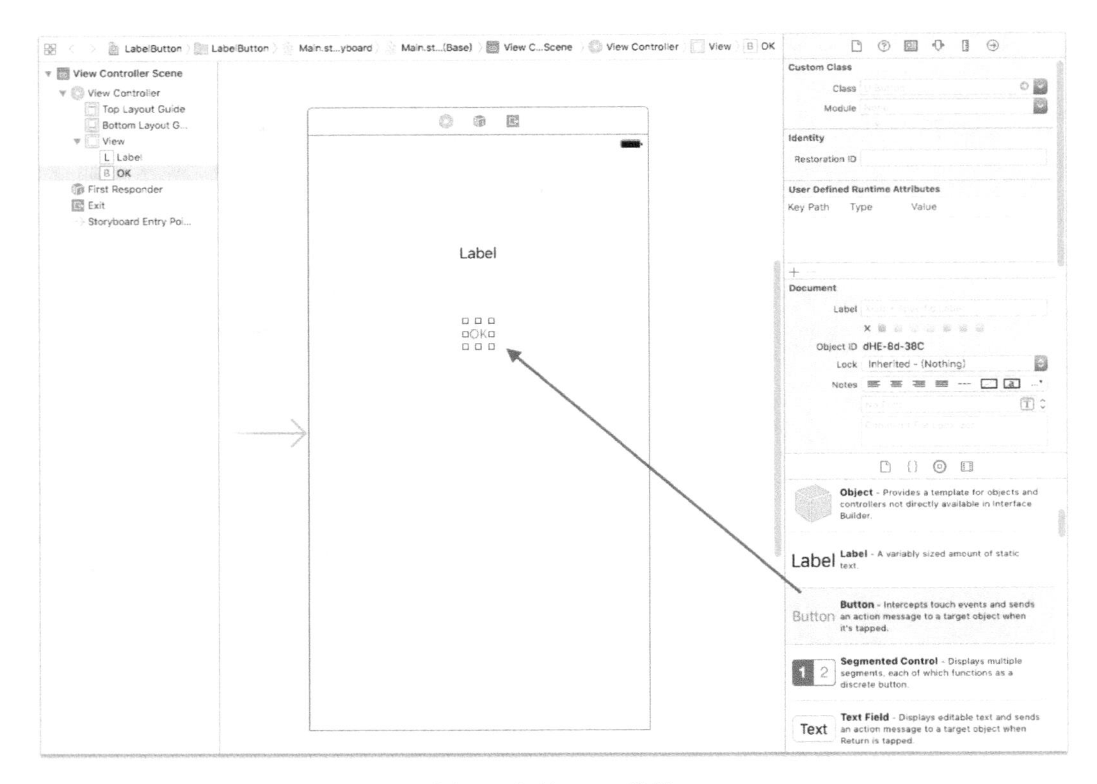

图4-4　摆放Button控件

双击该Button，输入文本OK。现在Button的状态是默认状态，我们可以运行一下看看效果。

为了美观，往往还要通过属性检查器优化该按钮。打开其属性检查器，如图4-5左图所示，打开Type下拉列表（如图4-5右图所示），其中各选项的含义如下所示。

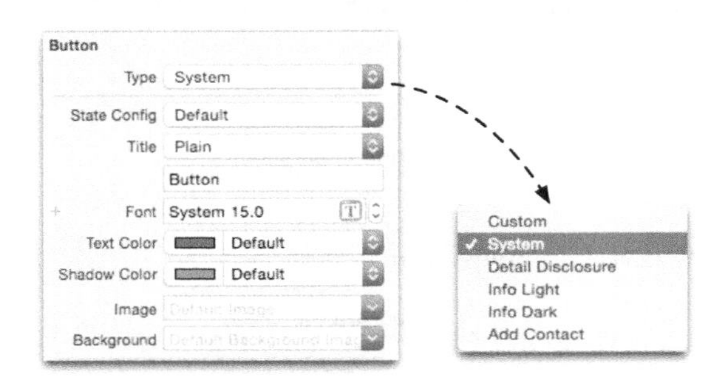

图4-5　按钮的Type属性

- Custom。自定义类型。如果我们不喜欢圆角按钮，可以使用该类型。
- System。系统默认属性，表示该按钮没有边框。在iOS 7之前，按钮默认为圆角矩形。
- Detail Disclosure。细节展示按钮ⓘ，主要用于表视图中的细节展示。
- Info Light和Info Dark。这两个是信息按钮ⓘ（样式上与细节展示按钮一样），表示有一些信息需要展示，或有可以设置的内容。
- Add Contact。添加联系人按钮⊕。

State Config下拉列表中有5种状态，分别是Default（默认）、Highlighted（高亮）、Focused（获得焦点）、Selected（选择）和Disabled（不可用），如图4-6所示。选择不同的State Config选项，可以设置不同状态下的属性。

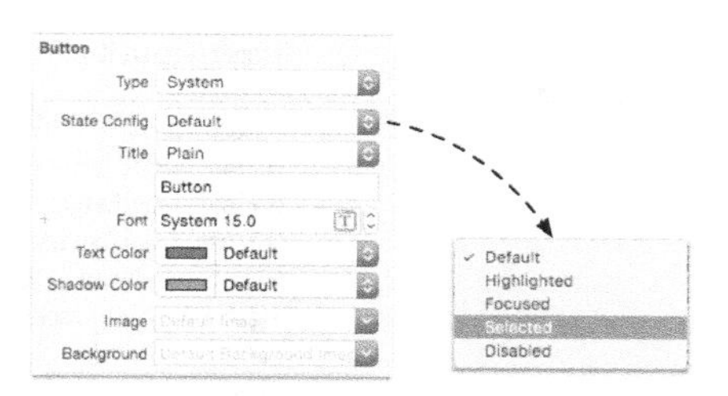

图4-6　按钮的State Config属性

如果希望点击按钮时按钮中央高亮显示，可以按照图4-7所示勾选Drawing中的Shows Touch On Highlight复选框。

图4-7　高亮状态的设置

为了能突出高亮效果，我们可以把按钮背景设置为深色，这可以到属性检查器的View→Background中设置。是点击OK按钮时的高亮效果，其中按钮中央会出现一个光圈。

提示　UIKit至少有两种按钮：一种是UIButton类型的普通按钮，该按钮可以有文字，也可以有图片；另一种是放置于工具栏或导航栏中的UIBarButtonItem，它虽然可以当按钮用，但是从类的继承关系上看，它不是UIView的子类。

4.1.2　代码实现

上一节我们介绍使用Interface Builder实现LabelButton案例，这一节我们介绍代码实现的LabelButton案例。

首先使用Single View Application模板创建一个名为LabelButton的工程（具体创建过程参见2.1.1节），然后参考3.5.1节将LabelButton修改为纯代码工程。

AppDelegate类与3.5.1节的案例完全一样，这里不再赘述了。我们重点介绍一下视图控制器类ViewController，其主要代码如下：

```
//ViewController.swift文件
class ViewController: UIViewController {

    override func viewDidLoad() {
        super.viewDidLoad()

        let screen = UIScreen.main.bounds

        let labelWidth:CGFloat = 90
        let labelHeight:CGFloat = 20
```

```
//ViewController.m文件
#import "ViewController.h"

@interface ViewController ()

@end

@implementation ViewController

- (void)viewDidLoad {
```

```swift
        let labelTopView:CGFloat = 150
        let labelFrame = CGRect(x: (screen.size.width - labelWidth)/2 ,
        ➥y: labelTopView, width: labelWidth, height: labelHeight)

        let label = UILabel(frame: labelFrame)

        label.text = "Label"
        //字体左右居中
        label.textAlignment = NSTextAlignment.center
        self.view.addSubview(label)

        let button = UIButton(type: UIButtonType.system)                ①
        button.setTitle("OK", for: UIControlState.normal)               ②

        let buttonWidth:CGFloat = 60
        let buttonHeight:CGFloat = 20
        let buttonTopView:CGFloat = 240

        button.frame = CGRect(x: (screen.size.width - buttonWidth)/2 ,
        ➥y: buttonTopView, width: buttonWidth, height: buttonHeight)


        self.view.addSubview(button)
    }

    ......

}
```

```objc
    [super viewDidLoad];

    CGRect screen = [[UIScreen mainScreen] bounds];
    CGFloat labelWidth = 90;
    CGFloat labelHeight = 20;
    CGFloat labelTopView = 150;
    CGRect labelFrame = CGRectMake((screen.size.width - labelWidth)/2 ,
    ➥labelTopView, labelWidth, labelHeight);

    UILabel* label = [[UILabel alloc] initWithFrame:labelFrame];

    label.text = @"Label";
    //字体左右居中
    label.textAlignment = NSTextAlignmentCenter;
    [self.view addSubview:label];

    UIButton* button = [UIButton buttonWithType:UIButtonTypeSystem];      ①
    [button setTitle:@"OK" forState:UIControlStateNormal];                ②

    CGFloat buttonWidth = 60;
    CGFloat buttonHeight = 20;
    CGFloat buttonTopView = 240;

    button.frame = CGRectMake((screen.size.width - buttonWidth)/2 ,
    ➥buttonTopView, buttonWidth, buttonHeight);

    [self.view addSubview:button];
}

......

@end
```

上述代码中，第①行通过按钮的Type属性创建UIButton对象，相当于通过属性检查器（如图4-5所示）设置Type属性。

Objective-C版中使用静态工厂方法+ (instancetype)buttonWithType:(UIButtonType)buttonType获得UIButton对象，Swift版中创建UIButton对象使用的是构造函数init(type buttonType: UIButtonType)，其中参数buttonType用于指定按钮的Type属性。UIButtonType枚举成员的说明见表4-1。

表4-1　UIButtonType枚举成员

Objective-C枚举成员	Swift枚举成员	说　明
UIButtonTypeCustom	custom	自定义类型按钮
UIButtonTypeSystem	system	系统默认类型按钮
UIButtonTypeDetailDisclosure	detailDisclosure	细节展示按钮ⓘ
UIButtonTypeInfoLight	infoLight	Light风格的信息按钮ⓘ
UIButtonTypeInfoDark	infoDark	Dark风格的信息按钮ⓘ
UIButtonTypeContactAdd	contactAdd	添加联系人按钮⊕

第②行代码用于设置UIButton某种状态下的title。setTitle:forState:（Swift版是setTitle(_:for:)）方法的第一个参数是title，第二个参数是UIControlState。UIControlState枚举成员的说明见表4-2。

表4-2　UIControlState枚举成员

Objective-C枚举成员	Swift枚举成员	说　明
UIControlStateNormal	normal	默认状态
UIControlStateHighlighted	highlighted	高亮状态
UIControlStateDisabled	disabled	不可用状态
UIControlStateSelected	selected	选择状态

4.2 事件处理

上一节并没有完全实现图4-1所示的LabelButton案例，这一节我们先为按钮添加事件处理，这也可以通过Interface Builder和代码两种方式实现。

4.2.1 Interface Builder 实现

Interface Builder实现就是在XIB或故事板文件中通过连线实现的，如图4-8所示，需要在按钮（XIB或故事板文件中）与onClick:方法（ViewController代码中）之间建立连线，这种连接通过Interface Builder可视化地拖曳完成。

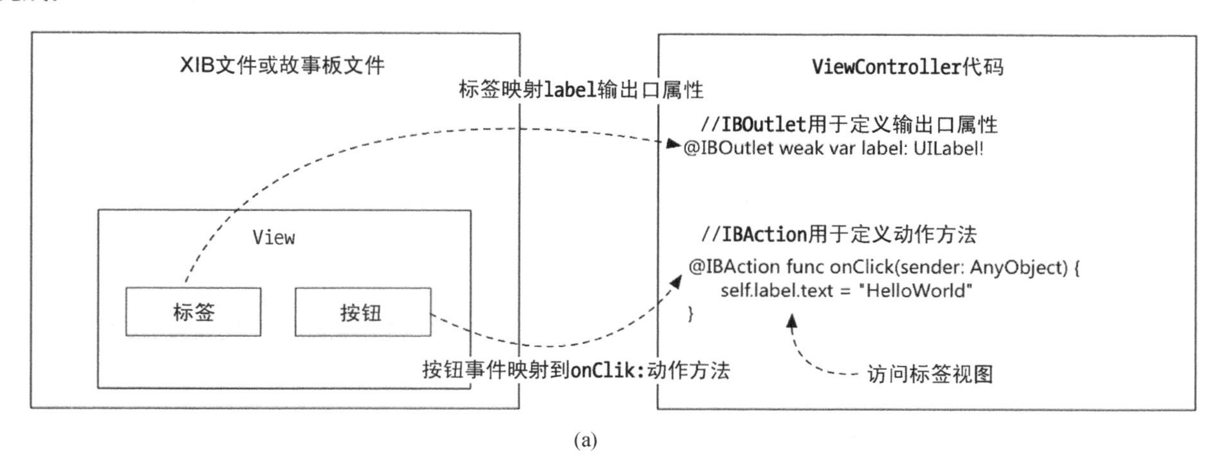

(a)

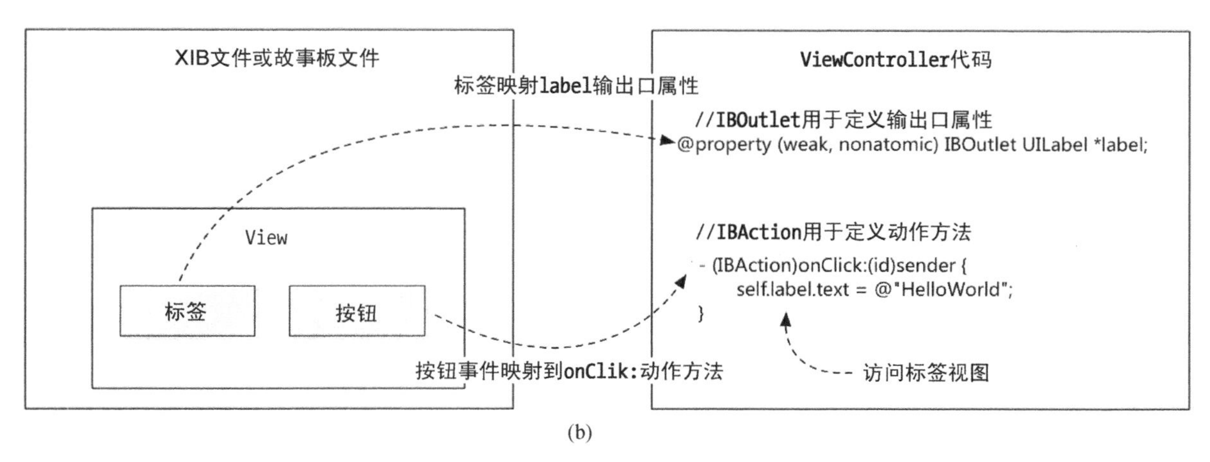

(b)

图4-8 动作方法和输出口属性的作用原理（图a为Swift版，图b为Objective-C版）

在Interface Builder中按钮事件与动作方法实现连线的具体步骤如下。

打开故事板或XIB文件，点击左上角第一组按钮中的“打开辅助编辑器”按钮，此时会打开如图4-9所示的界面。然后，选中Button，按住control键，同时拖曳鼠标到辅助编辑器窗口，如图4-10所示。

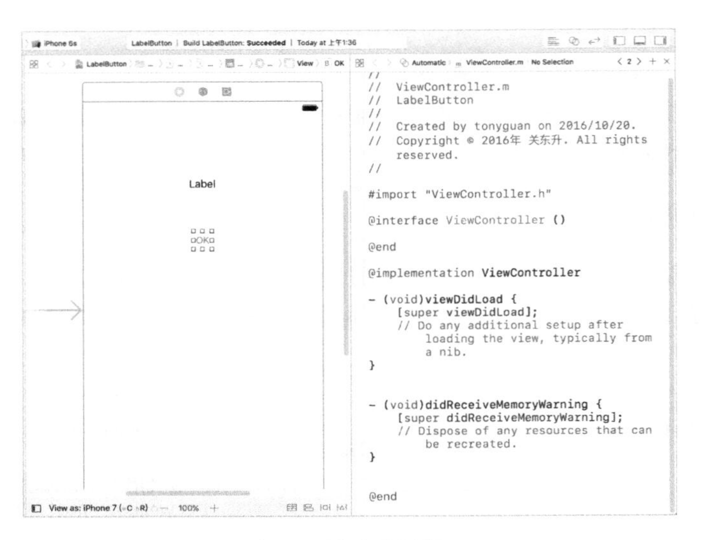

图4-9　辅助编辑器

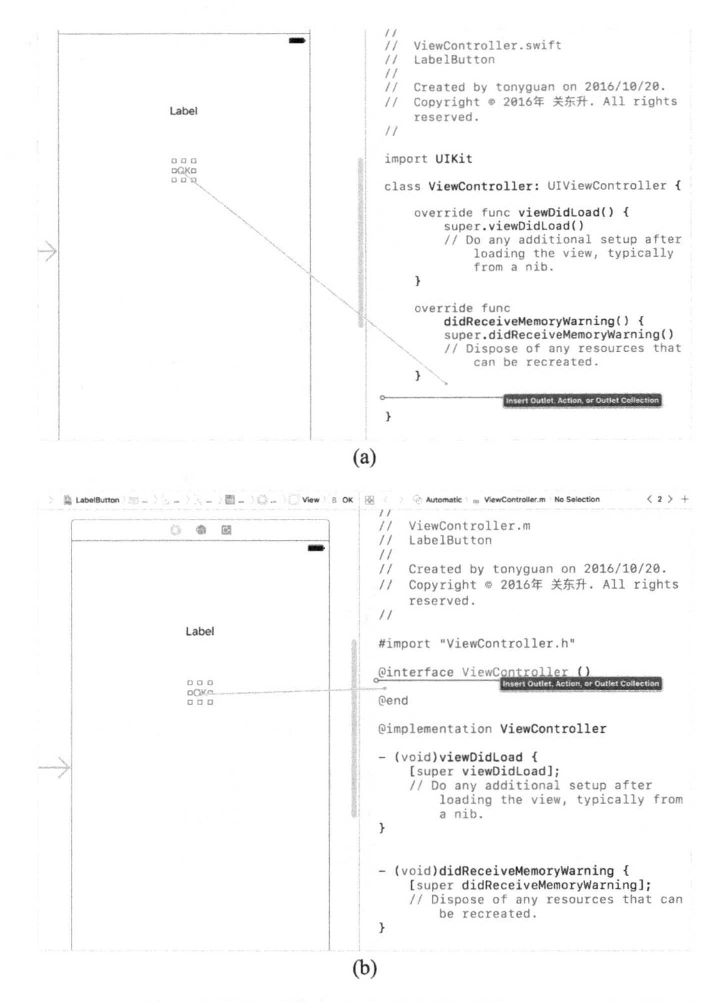

图4-10　Interface Builder中按钮事件与动作方法实现连线（图a为Swift版，图b为Objective-C版）

说明

Swift版本则在ViewController类内部松开鼠标。在Objective-C版本中，如果辅助编辑器打开的是ViewController.m文件，则可以在`@interface ViewController ()...@end`之间松开鼠标；如果打开的是ViewController.h文件，则可以在`@interface ViewController...@end`之间松开鼠标。

这时松开鼠标，则弹出如图4-11所示的对话框，其中我们将Connection选择为Action；在Name中输入onClick，这里Name是事件处理方法，命名是由用户自己确定的；Event下拉框中可以选择按钮事件，按钮的默认事件是Touch Up Inside（手指触摸抬起），这是按钮的默认事件，每一个控件都有一个默认事件。设置完成后，点击Connect按钮，会生成如下代码：

```swift
@IBAction func onClick(_ sender: AnyObject) {
}
```

```objectivec
- (IBAction)onClick:(id)sender {
}
```

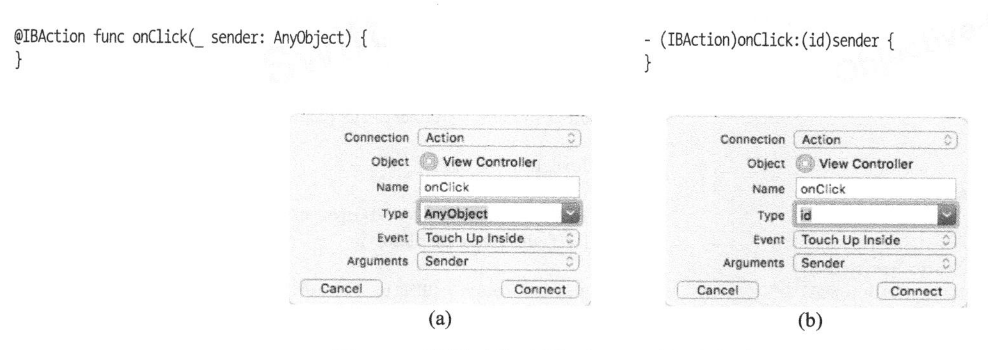

图4-11 设置动作（图a为Swift版，图b为Objective-C版）

方法返回类型为`IBAction`，这种方法被称为"动作方法"，`sender`参数是事件源。打开图4-11中的Arguments下拉列表，此时会弹出如图4-12所示的下拉框，从中可以选择参数的个数，其中选项None表示不会带参数的方法，生成的代码如下：

```swift
@IBAction func onClick() {
}
```

```objectivec
- (IBAction)onClick {
}
```

图4-12 选择参数

选项Sender and Event表示方法带有两个参数，第一个参数是事件源sender，第二个参数是事件对象，类型是`UIEvent`，生成的代码如下：

```swift
@IBAction func onClick(_ sender: AnyObject, forEvent event: UIEvent) {
}
```

```objectivec
- (IBAction) onClick:(id)sender forEvent:(UIEvent *)event {
}
```

代码编写至此，动作方法还不能访问故事板或XIB中的标签等视图。为了便于测试，我们可以先在`onClick`方法中输出日志信息，代码如下：

```swift
@IBAction func onClick(_ sender: AnyObject) {
    print("OK Button onClick.")
}
```

```objectivec
- (IBAction)onClick:(id)sender {
    NSLog(@"OK Button onClick.");
}
```

4.2.2 代码实现

这是通过`UIControl`类的`addTarget:action:forControlEvents:`（Swift版本是`addTarget(_:action:for:)`）方法

实现的，主要代码如下：

```
//ViewController.swift文件
class ViewController: UIViewController {

    override func viewDidLoad() {
        super.viewDidLoad()

        let screen = UIScreen.main.bounds

        let labelWidth:CGFloat = 90
        let labelHeight:CGFloat = 20
        let labelTopView:CGFloat = 150
        let labelFrame = CGRect(x: (screen.size.width - labelWidth)/2 ,
        ➥y: labelTopView, width: labelWidth, height: labelHeight)

        let label = UILabel(frame: labelFrame)

        label.text = "Label"
        //字体左右居中
        label.textAlignment = NSTextAlignment.center
        self.view.addSubview(label)

        let button = UIButton(type: UIButtonType.system)
        button.setTitle("OK", for: UIControlState.normal)

        let buttonWidth:CGFloat = 60
        let buttonHeight:CGFloat = 20
        let buttonTopView:CGFloat = 240

        button.frame = CGRect(x: (screen.size.width - buttonWidth)/2 ,
        ➥y: buttonTopView, width: buttonWidth, height: buttonHeight)
        button.addTarget(self, action: #selector(onClick(_:)),
        ➥for: UIControlEvents.touchUpInside)                              ①

        self.view.addSubview(button)
    }

    ......
    func onClick(_ sender: AnyObject) {                                     ②
        print("OK Button onClick.")
    }

}
```

```
//ViewController.m文件
#import "ViewController.h"
......
@implementation ViewController

- (void)viewDidLoad {
    [super viewDidLoad];

    CGRect screen = [[UIScreen mainScreen] bounds];
    CGFloat labelWidth = 90;
    CGFloat labelHeight = 20;
    CGFloat labelTopView = 150;
    CGRect labelFrame = CGRectMake((screen.size.width - labelWidth)/2 ,
    ➥labelTopView, labelWidth, labelHeight);

    UILabel* label = [[UILabel alloc] initWithFrame:labelFrame];

    label.text = @"Label";
    //字体左右居中
    label.textAlignment = NSTextAlignmentCenter;
    [self.view addSubview:label];

    UIButton* button = [UIButton buttonWithType:UIButtonTypeSystem];
    [button setTitle:@"OK" forState:UIControlStateNormal];

    CGFloat buttonWidth = 60;
    CGFloat buttonHeight = 20;
    CGFloat buttonTopView = 240;

    button.frame = CGRectMake((screen.size.width - buttonWidth)/2 ,
    ➥buttonTopView, buttonWidth, buttonHeight);

    [button addTarget:self action:@selector(onClick:)
    ➥forControlEvents:UIControlEventTouchUpInside];                          ①

    [self.view addSubview:button];
}
......
- (void)onClick:(id)sender {                                                ②
    NSLog(@"OK Button onClick.");
}

@end
```

上述代码中，第①行调用了addTarget:action:forControlEvents:方法实现，该方法的第一个参数是target，即事件处理者，本例中是self。第二个参数是action，是选择器（Selector）类型，它指向事件处理方法，在Objective-C选择器的表示方式为@selector(onClick:)；在Swift中，选择器的表示方法是#selector(onClick(_:))，onClick:是在第②行代码中定义的方法。第三个参数是事件，UIControlEventTouchUpInside（Swift版是touchUpInside）是按钮的触摸点击事件，它属于枚举类型UIControlEvents中的成员。

提示

在代码实现的事件处理方法中，返回值不必是IBAction，可以声明为void。

如果调用如下的无参数事件处理方法：

```
func onClick() {
}
```

```
- (void)onClick:(id)sender {
}
```

那么调用代码如下：

```
button.addTarget(self, action: #selector(onClick),
➥forControlEvents: UIControlEvents.TouchUpInside)
```

```
[button addTarget:self action:@selector(onClick)
➥forControlEvents:UIControlEventTouchUpInside];
```

如果调用如下两个参数的事件处理方法：

```
func onClick(_ sender: AnyObject, forEvent event: UIEvent) {
}
```

```
- (void) onClick:(id)sender forEvent:(UIEvent *)event {
}
```

那么调用代码如下：

```
button.addTarget(self, action: #selector(onClick(_:forEvent:)),
➥for: UIControlEvents.touchUpInside)
```

```
[button addTarget:self action:@selector(onClick:forEvent:)
➥forControlEvents:UIControlEventTouchUpInside];
```

4.3 访问视图

上一节中我们已经实现了如图4-1所示的LabelButton案例的事件处理，但是还需要在事件处理方法中访问标签视图，将它的Text属性改为“HelloWorld”。访问视图也可以通过Interface Builder和代码两种方式实现。

4.3.1 Interface Builder 实现

这是在XIB或故事板文件中通过连线实现，具体步骤如下。

打开故事板或XIB文件，参考4.2.1节辅助编辑器界面，选中标签Label，同时按住control键，将其拖曳到右边的辅助编辑器窗口后松开鼠标，如图4-13所示。

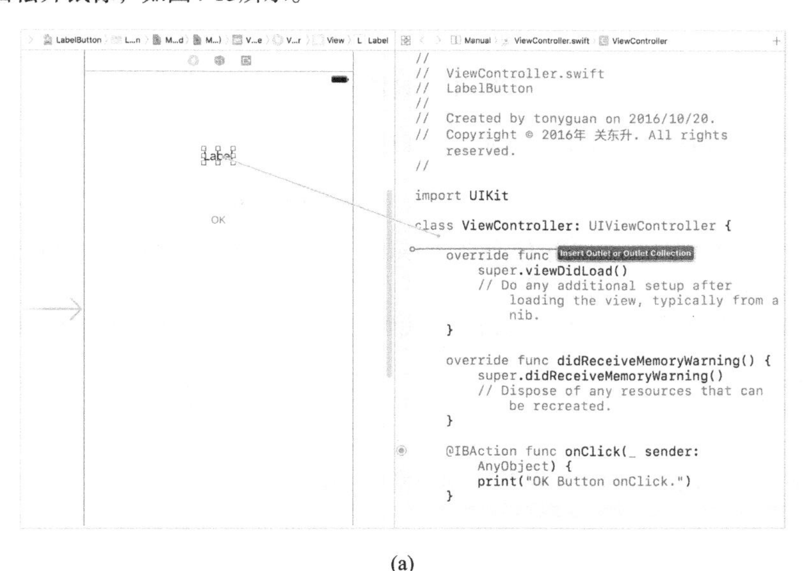

(a)

图4-13 在Interface Builder中实现标签与label属性连线（图a为Swift版，图b为Objective-C版）

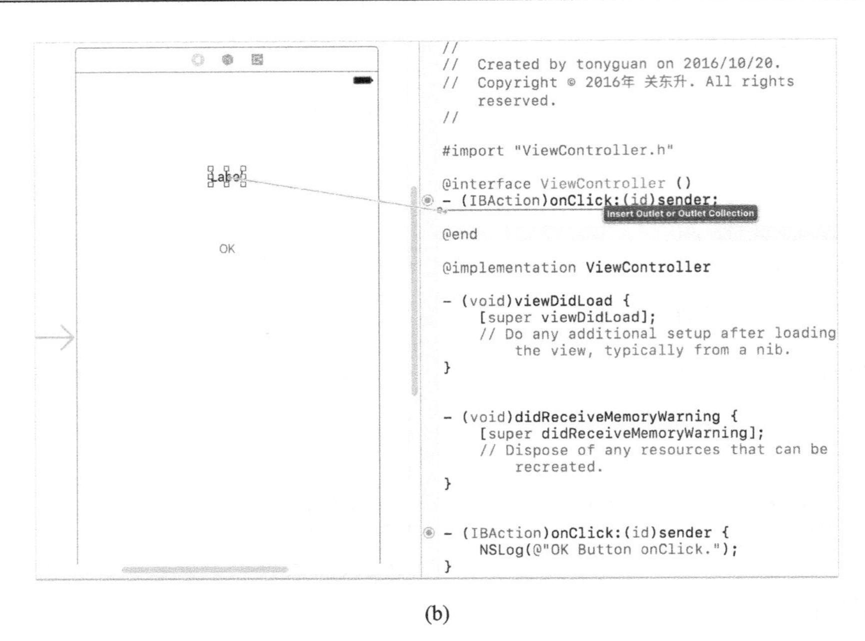

(b)

图4-13　（续）

松开鼠标后会弹出一个对话框，如图4-14所示。我们在Connection栏中选择Outlet，将输出口命名为label。

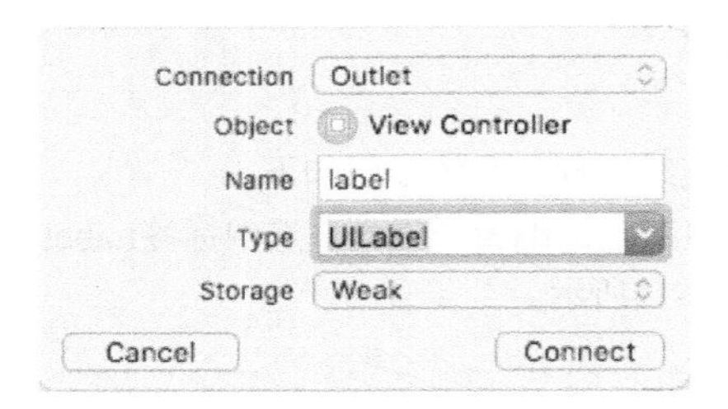

图4-14　设置输出口

点击Connect按钮，右边的编辑界面将自动添加下面这行代码：

```
@IBOutlet weak var label: UILabel!
```

```
@property (weak, nonatomic) IBOutlet UILabel *label;
```

label属性定义时，是使用Outlet关键字修饰的，label属性被称为输出口属性。在Objective-C版本中，Outlet关键字也可以修饰成员变量，称为输出口变量。

这样标签与label属性连线成功后，可以修改程序代码。在onClick:方法中添加如下代码：

```
@IBAction func onClick(_ sender: AnyObject) {
    self.label.text = "HelloWorld"
}
```

```
- (IBAction)onClick:(id)sender {
    self.label.text = @"HelloWorld";
}
```

此时点击OK按钮，标签的文本内容从原来的Label成功切换为HelloWorld。

4.3.2　代码实现

在代码中要想访问视图，需要将这些视图定义为属性，Objective-C版本中也可以将视图定义为成员变量，不需要使用Outlet关键字修饰。

主要代码如下：

```
//ViewController.swift文件
class ViewController: UIViewController {
    var label: UILabel!                                                          ①
    override func viewDidLoad() {
        super.viewDidLoad()

        let screen = UIScreen.main.bounds

        let labelWidth:CGFloat = 90
        let labelHeight:CGFloat = 20
        let labelTopView:CGFloat = 150
        self.label = UILabel(frame: CGRect(x: (screen.size.width - labelWidth)/2 ,
        ➥y: labelTopView, width: labelWidth, height: labelHeight))               ②

        self.label.text = "Label"
        //字体左右居中
        self.label.textAlignment = .center
        self.view.addSubview(self.label)

        let button = UIButton(type: UIButtonType.system)
        button.setTitle("OK", for: UIControlState())

        let buttonWidth:CGFloat = 60
        let buttonHeight:CGFloat = 20
        let buttonTopView:CGFloat = 240

        button.frame = CGRect(x: (screen.size.width - buttonWidth)/2 ,
        ➥y: buttonTopView, width: buttonWidth, height: buttonHeight)

        button.addTarget(self, action: #selector(ViewController.onClick(_:)),
        ➥for: UIControlEvents.touchUpInside)

        self.view.addSubview(button)
    }
    ......
    func onClick(_ sender: AnyObject) {
        self.label.text = "HelloWorld"
    }
}
```

```
//ViewController.m文件
#import "ViewController.h"
@interface ViewController ()
@property (strong, nonatomic) UILabel *label;                                    ①
@end

@implementation ViewController
- (void)viewDidLoad {
    [super viewDidLoad];

    CGRect screen = [[UIScreen mainScreen] bounds];
    CGFloat labelWidth = 90;
    CGFloat labelHeight = 20;
    CGFloat labelTopView = 150;

    self.label = [[UILabel alloc] initWithFrame:CGRectMake((
    ➥screen.size.width - labelWidth)/2 ,
    ➥labelTopView, labelWidth, labelHeight)];                                   ②

    self.label.text = @"Label";
    //字体左右居中
    self.label.textAlignment = NSTextAlignmentCenter;
    [self.view addSubview:self.label];

    UIButton* button = [UIButton buttonWithType:UIButtonTypeSystem];
    [button setTitle:@"OK" forState:UIControlStateNormal];

    CGFloat buttonWidth = 60;
    CGFloat buttonHeight = 20;
    CGFloat buttonTopView = 240;

    button.frame = CGRectMake((screen.size.width - buttonWidth)/2 ,
    ➥buttonTopView, buttonWidth, buttonHeight);

    [button addTarget:self action:@selector(onClick:)
    ➥forControlEvents:UIControlEventTouchUpInside];

    [self.view addSubview:button];

}

- (void)onClick:(id)sender {
    self.label.text = @"HelloWorld";
}
......
@end
```

上述代码中，第①行用于定义label属性，注意这个属性是strong的（Swift属性默认是strong的）。如果label属性是weak的，那么执行代码第②行创建label属性对象，但label属性对象的内存马上就会被释放。strong属性可以保持刚刚创建的对象内存不释放，视图控制器拥有label属性对象的所有权。

说明 为什么Interface Builder实现中label属性是weak的？这是因为Interface Builder实现时label等视图是在故事板或XIB文件中定义的，当应用程序启动时会根据故事板或XIB文件的描述创建label等视图对象，对象所有权在故事板或XIB，它们对label等视图对象是强引用的。由于对象所有权不是视图控制器，所以在视图控制器中使用它时不能定义为strong，只能是定义为weak。

4.4 TextField 和 TextView

与标签一样，TextField和TextView也是文本类视图，是可以编辑文本内容的。

在文本内容编辑方面，三者都可以通过代码、双击控件和设置属性检查器中的Text属性来实现，但是TextField和TextView比标签多了一个键盘设置。另外，TextField和TextView还各有一个委托协议。考虑到这些，我们将TextField和TextView这二者单列在一节。

下面我们通过图4-15所示的TextFieldTextView案例向大家展示TextField控件和TextView控件的用法。其中包括两个标签（Name:和Abstract:）、一个TextField和一个TextView，当TextField和TextView获得焦点进入编辑状态时，键盘会从屏幕下方滑出来，此时点击return键可以关闭键盘。

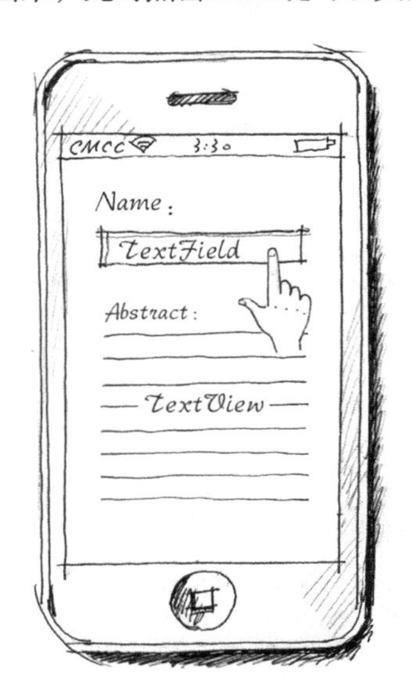

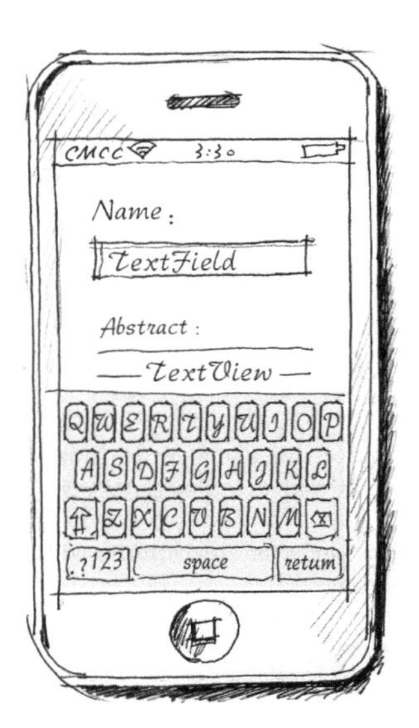

图4-15 TextFieldTextView案例设计原型草图

4.4.1 Interface Builder 实现

在UIKit框架中，TextField由`UITextField`类创建。此外，它还有对应的`UITextFieldDelegate`委托协议。委托可以帮助响应事件处理。`UITextField`继承了`UIControl`，隶属于真正的“控件”，而`UITextView`继承了`UIScrollView`，并不属于“控件”。

我们首先介绍一下使用Interface Builder实现的TextFieldTextView案例。首先使用Single View Application模板创建一个名为TextFieldTextView的工程。打开Main.storyboard设计界面，从对象库中拖曳两个标签控件到设计界面，分别将其命名为Name:和Abstract:，在Name:标签下摆放一个TextField。打开TextField属性检查器，在Placeholder属性中输入enter your book name作为提示，运行时该文本是浅灰色，当有输入动作时文本消失。我们可以利用TextField后面的清除按钮清除TextField的内容，如图4-16所示。

现在我们就为TextField添加清除按钮：打开TextField的属性检查器，进入Clear Button下拉列表，从中选择Is always visible，如图4-17所示。

Tony Guan

图4-16 TextField的清除按钮

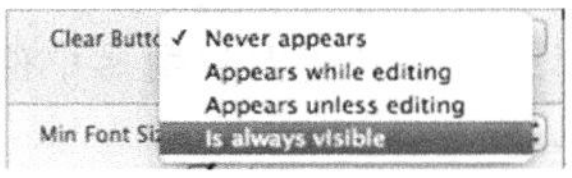

Clear Butt ✓ Never appears
Appears while editing
Appears unless editing
Min Font Si Is always visible

图4-17 选择清除按钮属性

现在来添加TextView。TextView是一个可展示和编辑多行文本的控件，由UITextView类创建。TextView控件有对应的UITextViewDelegate委托协议，我们可以借助委托来响应事件。

回到Interface Builder设计界面，在第二个标签Abstract:下面放置一个TextView控件。

在视图控制器ViewController的代码中需要实现UITextFieldDelegate和UITextViewDelegate委托协议，相关代码如下：

```
//ViewController.swift文件
import UIKit
class ViewController: UIViewController, UITextFieldDelegate,
➥UITextViewDelegate {                                                ①

    override func viewDidLoad() {
        super.viewDidLoad()
    }

    override func didReceiveMemoryWarning() {
        super.didReceiveMemoryWarning()
    }

    //MARK: -- 实现UITextFieldDelegate委托协议方法
    func textFieldShouldReturn(_ textField: UITextField) -> Bool {    ②
        print("TextField获得焦点，点击return键")
        return true
    }

    //MARK: -- 实现UITextViewDelegate委托协议方法
    func textView(_ textView: UITextView, shouldChangeTextInRange
➥range: NSRange, replacementText text: String) -> Bool {             ③
        if (text == "\n") {
            print("TextView获得焦点，点击return键")
            return false
        }
        return true
    }
}
```

```
//在ViewController.m中实现委托协议
#import "ViewController.h"
@interface ViewController () <UITextFieldDelegate, UITextViewDelegate>   ①
@end
@implementation ViewController

- (void)viewDidLoad {
    [super viewDidLoad];
}

- (void)didReceiveMemoryWarning {
    [super didReceiveMemoryWarning];
}

#pragma mark -- 实现UITextFieldDelegate委托协议方法
- (BOOL)textFieldShouldReturn:(UITextField *)textField {                 ②
    NSLog(@"TextField获得焦点，点击return键");
    return YES;
}

#pragma mark --实现UITextViewDelegate委托协议方法
-(BOOL)textView:(UITextView *)textView shouldChangeTextInRange:
➥(NSRange)range replacementText:(NSString *) text {                     ③
    if([text isEqualToString:@"\n"]) {
        NSLog(@"TextView获得焦点，点击return键");
        return NO;
    }
    return YES;
}

@end
```

上述代码中，第①行实现了UITextFieldDelegate和UITextViewDelegate委托协议。在Objective-C版本中，委托协议的实现有两种方法：

- 在ViewController.m中的ViewController ()扩展中实现，本例采用这种实现方式；
- 在ViewController.h中实现委托协议。

示例代码如下：

```
//ViewController.h中实现委托协议
@interface ViewController : UIViewController<UITextFieldDelegate, UITextViewDelegate>

@end
```

代码第②行的方法是UITextFieldDelegate委托协议中的方法，当TextField获得焦点时，点击return键会调用该方法；代码第③行的方法是UITextViewDelegate委托协议中要求实现的方法，当修改TextView内的文本时，会调用该方法。判断语句if (text == "\n"){...}（Objective-C版本中是if([text isEqualToString:@"\n"]) {...}），用于判断键盘输入的字符是否为\n，\n表示return键。

如果ViewController只是实现UITextFieldDelegate和UITextViewDelegate委托协议，并不意味着ViewController真正成为TextField和TextView的委托对象，我们还需要将ViewController当前对象赋值给TextView和TextField控件的delegate委托属性，否则代码第②行和第③行的委托方法不会被调用。

这个过程的实现可以通过Interface Builder或代码完成。使用Interface Builder的具体实现过程如下。

在Interface Builder中打开故事板（或XIB）文件，右击TextField控件，此时弹出的快捷菜单如图4-18所示，此时用鼠标拖曳Outlets→delegate后面的小圆点到左边的View Controller上。然后，我们以同样的方式将TextView控件Outlets→delegate后面的小圆点拖曳到左边的View Controller上。

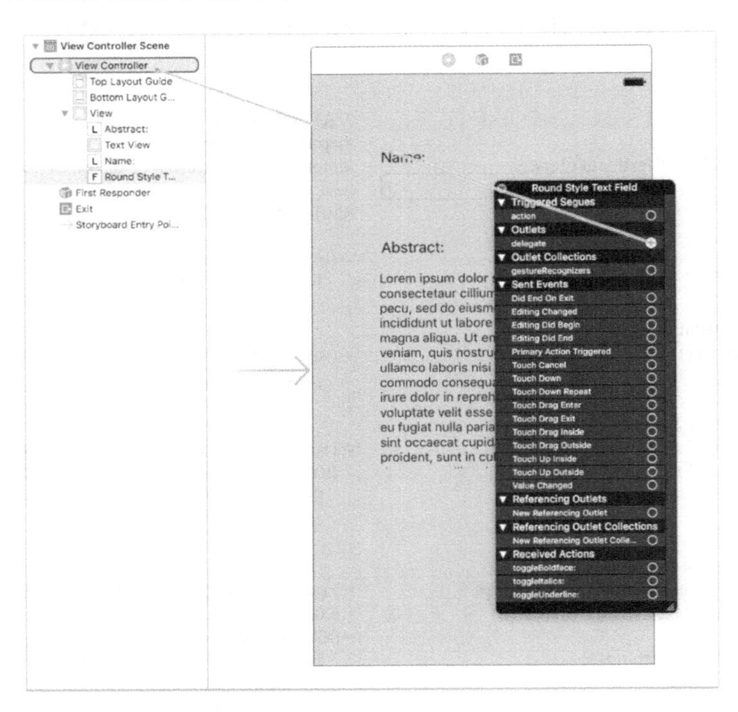

图4-18　在Interface Builder中分配委托

4.4.2　代码实现

上一节中，我们介绍了采用Interface Builder实现的TextFieldTextView案例，这一节就来介绍代码实现的TextFieldTextView案例。

首先，使用Single View Application模板创建一个名为TextFieldTextView的工程，然后参考3.5.1节将TextFieldTextView修改为纯代码工程。

其中AppDelegate类与3.5.1节的案例完全一样，这里不再赘述了。我们重点介绍一下视图控制器类ViewController，其主要代码如下：

```
//ViewController.swift文件
import UIKit
class ViewController: UIViewController, UITextFieldDelegate, UITextViewDelegate {

    override func viewDidLoad() {
        super.viewDidLoad()

        let screen = UIScreen.main.bounds
        let textFieldWidth:CGFloat = 223
        let textFieldHeight:CGFloat = 30
        let textFieldTopView:CGFloat = 150
        let textFieldFrame = CGRect(x: (screen.size.width - textFieldWidth)/2,
        ➥y: textFieldTopView, width: textFieldWidth, height:
        ➥textFieldHeight)
        let textField = UITextField(frame: textFieldFrame)                ①
```

```
//ViewController.m文件
#import "ViewController.h"
@interface ViewController () <UITextFieldDelegate, UITextViewDelegate>
@end
@implementation ViewController

- (void)viewDidLoad {
    [super viewDidLoad];

    CGRect screen = [[UIScreen mainScreen] bounds];
    CGFloat textFieldWidth = 223;
    CGFloat textFieldHeight = 30;
    CGFloat textFieldTopView = 150;
    UITextField* textField = [[UITextField alloc] initWithFrame:
    ➥CGRectMake ((screen.size.width - textFieldWidth)/2 , textFieldTopView,
    ➥textFieldWidth, textFieldHeight)];                                  ①
```

```swift
        textField.borderStyle = UITextBorderStyle.roundedRect                    ②
        textField.delegate = self                                                ③
        self.view.addSubview(textField)

        //labelName标签与textField之间的距离
        let labelNameTextFieldSpace: CGFloat = 30
        let labelNameFrame = CGRect(x: textField.frame.origin.x,
➥y: textField.frame.origin.y - labelNameTextFieldSpace, width: 51,
➥height: 21)
        let labelName = UILabel(frame: labelNameFrame)                           ④
        labelName.text = "Name:"
        self.view.addSubview(labelName)

        let textViewWidth:CGFloat = 236
        let textViewHeight: CGFloat = 198
        let textViewTopView: CGFloat = 240
        let textViewFrame = CGRect(x: (screen.size.width - textViewWidth)/2,
➥y: textViewTopView, width: textViewWidth, height: textViewHeight)
        let textView = UITextView(frame: textViewFrame)                          ⑤

        textView.text = "Lorem ipsum dolor sit er elit lamet, ..."

        textView.delegate = self                                                 ⑥
        self.view.addSubview(textView)

        //labelAbstract标签与textView之间的距离
        let labelAbstractTextViewSpace: CGFloat = 30
        let labelAbstractFrame = CGRect(x: textView.frame.origin.x,
➥y: textView.frame.origin.y - labelAbstractTextViewSpace, width: 103,
➥height: 21)
        let labelAbstract = UILabel(frame: labelAbstractFrame)                   ⑦

        labelAbstract.text = "Abstract:"
        self.view.addSubview(labelAbstract)
    }
    ......
    //MARK: -- 实现UITextFieldDelegate委托协议方法
    func textFieldShouldReturn(textField: UITextField) -> Bool {
        print("TextField获得焦点，点击return键")
        return true
    }

    //MARK: -- 实现UITextViewDelegate委托协议方法
    func textView(textView: UITextView, shouldChangeTextInRange range: NSRange,
➥replacementText text: String) -> Bool {
        if (text == "\n") {
            print("TextView获得焦点，点击return键")
            return false
        }
        return true
    }
}
```

```objectivec
    textField.borderStyle = UITextBorderStyleRoundedRect;                        ②
    textField.delegate = self;                                                   ③

    [self.view addSubview:textField];

    //labelName标签与textField之间的距离
    CGFloat labelNameTextFieldSpace = 30;
    UILabel* labelName = [[UILabel alloc] initWithFrame:CGRectMake(
➥textField.frame.origin.x, textField.frame.origin.y -
➥labelNameTextFieldSpace, 51, 21)];                                              ④

    labelName.text = @"Name:";
    [self.view addSubview:labelName];

    CGFloat textViewWidth = 236;
    CGFloat textViewHeight = 198;
    CGFloat textViewTopView = 240;
    UITextView* textView = [[UITextView alloc] initWithFrame:CGRectMake((
➥screen.size.width - textViewWidth)/2 , textViewTopView,
➥textViewWidth, textViewHeight)];                                                ⑤

    textView.text = @"Lorem ipsum dolor sit er elit lamet, ...";

    textView.delegate = self;                                                    ⑥

    [self.view addSubview:textView];

    //labelAbstract标签与textView之间的距离
    CGFloat labelAbstractTextViewSpace = 30;
    UILabel* labelAbstract = [[UILabel alloc]
➥initWithFrame:CGRectMake(textView.frame.origin.x, textView.frame.
➥origin.y - labelAbstractTextViewSpace, 103, 21)];                              ⑦
    labelAbstract.text = @"Abstract:";
    [self.view addSubview:labelAbstract];
}
......
#pragma mark -- 实现UITextFieldDelegate委托协议方法
- (BOOL)textFieldShouldReturn:(UITextField *)textField {
    NSLog(@"TextField获得焦点，点击return键");
    return YES;
}

#pragma mark -- 实现UITextViewDelegate委托协议方法
-(BOOL)textView:(UITextView *)textView shouldChangeTextInRange:(NSRange)range
➥replacementText:(NSString *)text {
    if([text isEqualToString:@"\n"]) {
        NSLog(@"TextView获得焦点，点击return键");
        return NO;
    }
    return YES;
}
@end
```

上述代码中，第①行用于实例化TextField控件对象，第②行用于设置TextField样式，样式属性是borderStyle。TextField样式有4种，对应UITextBorderStyle枚举类型中定义的4个成员，详见表4-3。

表4-3 UITextBorderStyle枚举成员

Objective-C枚举成员	Swift枚举成员	说　明
UITextBorderStyleNone	none	默认样式
UITextBorderStyleLine	line	直线样式
UITextBorderStyleBezel	bezel	带阴影样式
UITextBorderStyleRoundedRect	roundedRect	圆角样式

第③行用于将当前视图控制器self赋值给TextField控件的delegate委托属性。第④行用于创建标签对象，用来在界面中显示Name:文字，这个标签与TextField左对齐。标签的X轴坐标是textField.frame.origin.x，与TextField的X轴坐标相同；标签的Y轴坐标是textField.frame.origin.y - labelNameTextFieldSpace，即标签在TextField之上的labelNameTextFieldSpace位置。

Name:文字标签的位置是相对于TextField的位置计算出来的，这就是相对布局。

第⑤行用于实例化TextView视图对象，第⑥行将当前视图控制器self赋值给TextView的delegate委托属性。第⑦行用于设置界面中Abstract:文字标签位置，它也是相对布局（相对于TextView）。

4.4.3　键盘的打开和关闭

一旦TextField和TextView等视图处于编辑状态，系统就会智能地弹出键盘，而不需要我们做任何额外的操作。但是，关闭键盘就不像打开键盘这样顺利了，我们需要用代码去实现。

首先，我们要了解键盘不能自动关闭的原因。当TextField或TextView处于编辑状态时，这些控件变成了“第一响应者”，此时若要关闭键盘，就要放弃“第一响应者”的身份。在iOS中，事件沿着响应者链从一个响应者传到下一个响应者，如果其中一个响应者没有对事件做出响应，那么该事件会继续向下传递。

顾名思义，“第一响应者”是响应者链中的第一个，不同视图成为“第一响应者”之后的“表现”不太一致。TextField和TextView等输入类型的控件会导致弹出键盘，而我们只有让这些视图放弃它们的“第一响应者”身份，键盘才会关闭。

要想放弃“第一响应者”身份，需要调用UIResponder类中的resignFirstResponder方法，此方法一般在点击键盘的return键或者是背景视图时触发，本例采用点击return键关闭键盘的方式。要实现这个操作，我们可以利用TextField和TextView的委托协议实现。相关的实现代码在ViewController文件中，具体如下所示：

```swift
//ViewController.swift文件
class ViewController: UIViewController, UITextFieldDelegate, UITextViewDelegate {
    ......
    //MARK: -- 实现UITextFieldDelegate委托协议方法
    func textFieldShouldReturn(_ textField: UITextField) -> Bool {
        textField.resignFirstResponder()                                    ①
        return true
    }

    //MARK: -- 实现UITextViewDelegate委托协议方法
    func textView(_ textView: UITextView, shouldChangeTextInRange range:
➥NSRange, replacementText text: String) -> Bool {
        if (text == "\n") {
            textView.resignFirstResponder()                                 ②
            return false
        }
        return true
    }
}
```

```objc
//ViewController.m文件
@implementation ViewController

......
#pragma mark -- 实现UITextFieldDelegate委托协议方法
- (BOOL)textFieldShouldReturn:(UITextField *)textField  {
    [textField resignFirstResponder];                                       ①
    return YES;
}
#pragma mark --实现UITextViewDelegate委托协议方法
-(BOOL)textView:(UITextView *)textView shouldChangeTextInRange:
➥(NSRange)range replacementText:(NSString *)text  {
    if([text isEqualToString:@"\n"]) {
        [textView resignFirstResponder];                                    ②
        return NO;
    }
    return YES;
}

@end
```

在上述代码中，第①行和第②行将关闭键盘。

4.4.4　关闭和打开键盘的通知

在关闭和打开键盘时，iOS系统会分别发出如下广播通知。

- **UIKeyboardDidHideNotification**。键盘隐藏通知，Swift版本中用Notification.Name.UIKeyboardDidHide表示，Notification.Name是结构体，UIKeyboardDidHide是结构体的静态属性。
- **UIKeyboardDidShowNotification**。键盘出现通知，Swift版本中用Notification.Name.UIKeyboardDidHide表示，Notification.Name是结构体，UIKeyboardDidShow是结构体的静态属性。

使用广播通知的时候，我们需要注意在合适的时机注册和注销通知，而ViewController中的有关代码如下：

```
//ViewController.swift文件
override func viewWillAppear(_ animated: Bool) {
    super.viewWillAppear (animated)
    //注册键盘出现通知
    NotificationCenter.default.addObserver(self, selector: #selector
    ➥(keyboardDidShow(_:)),
    ➥name: Notification.Name.UIKeyboardDidShow, object: nil)                ①
    //注册键盘隐藏通知
    NotificationCenter.default.addObserver(self, selector: #selector
    ➥(keyboardDidHide(_:)),
    ➥name: Notification.Name.UIKeyboardDidHide, object: nil)                ②
}

override func viewWillDisappear(_ animated: Bool) {
    super.viewWillDisappear(animated)
    //注销键盘出现通知
    NotificationCenter.default.removeObserver(self,
    ➥name: Notification.Name.UIKeyboardDidShow, object: nil)                ③
    //注销键盘隐藏通知
    NotificationCenter.default.removeObserver(self,
    ➥name: Notification.Name.UIKeyboardDidHide, object: nil)                ④
}

func keyboardDidShow(_ notification: Notification) {
    print("键盘打开")
}

func keyboardDidHide(_ notification: Notification) {
    print("键盘关闭")
}
```

```
//ViewController.m文件
-(void) viewWillAppear:(BOOL)animated {
    [super viewWillAppear:animated];
    //注册键盘出现通知
    [[NSNotificationCenter defaultCenter] addObserver:self
    ➥selector:@selector (keyboardDidShow:)
    ➥name: UIKeyboardDidShowNotification object:nil];                      ①
    //注册键盘隐藏通知
    [[NSNotificationCenter defaultCenter] addObserver:self
    ➥selector:@selector (keyboardDidHide:)
    ➥name: UIKeyboardDidHideNotification object:nil];                      ②
}

-(void) viewWillDisappear:(BOOL)animated {
    [super viewWillDisappear:animated];
    //注销键盘出现通知
    [[NSNotificationCenter defaultCenter] removeObserver:self
    ➥name: UIKeyboardDidShowNotification object:nil];                      ③
    //注销键盘隐藏通知
    [[NSNotificationCenter defaultCenter] removeObserver:self
    ➥name: UIKeyboardDidHideNotification object:nil];                      ④
}

-(void) keyboardDidShow: (NSNotification *)notification {
    NSLog(@"键盘打开");
}

-(void) keyboardDidHide: (NSNotification *)notification {
    NSLog(@"键盘关闭");
}
```

注册通知在viewWillAppear:方法中进行，见代码第①行和第②行；注销通知在viewWillDisappear:方法中进行，见代码第③行和第④行。keyboardDidShow:消息是在键盘打开时发出的，keyboardDidHide:消息是在键盘关闭时发出的。

注意 由于Swift 3采用了去Objective-C语法，Objective-C的NSNotificationCenter类在Swift 3中变为NotificationCenter，去掉了NS前缀。在Objective-C中需要调用defaultCenter方法，而在Swift 3中则是调用defaultCenter属性。

4.4.5 键盘的种类

我们之前所看到的键盘都是系统默认的类型。在iOS中，打开有输入动作的控件的属性检查器，可以发现Keyboard Type的下拉选项有11个，分别表示11种类型的键盘。如图4-19所示，我们可以根据需要进行选择。

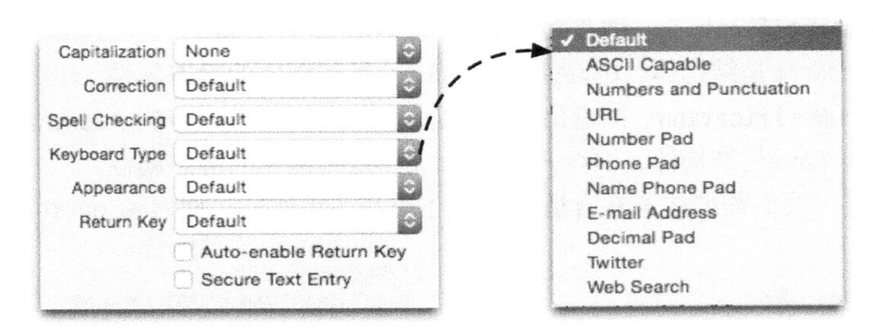

图4-19　选择键盘类型

选择不同的键盘类型，iOS上则会弹出不同的键盘，这些键盘的样式如图4-20至图4-23所示。

图4-20　ASCII键盘

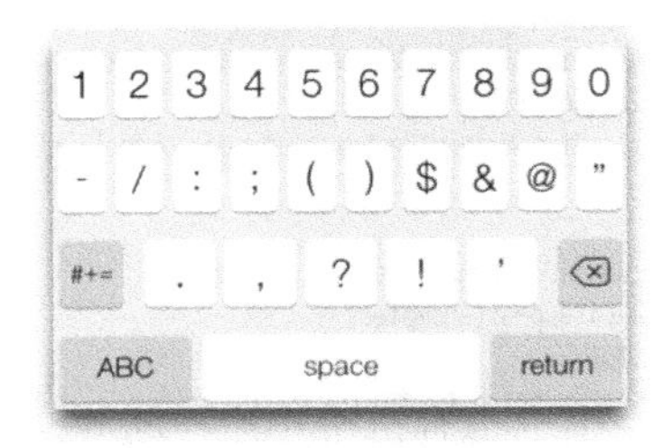

图4-21　数字和标点符号键盘

图4-22　邮箱键盘

图4-23　电话拨号键盘

除了可以为控件选择合适的键盘类型外，我们还可以自定义return键的文本，而文本的内容根据有输入动作的控件而定。如果控件内输入的是查询条件，我们可以将return键的文本设置为Go或者Search，示意接下来进行的就是查找动作。return键的文本设置如图4-24所示。

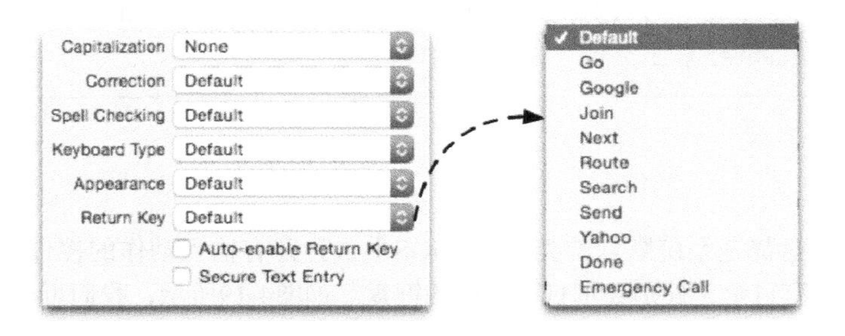

图4-24　选择return键

如果通过代码来设置键盘类型（Keyboard Type）和return键，代码如下：

```
textField.returnKeyType = UIReturnKeyType.next                        ①
textField.keyboardType = UIKeyboardType.numbersAndPunctuation         ②
```

```
textField.returnKeyType = UIReturnKeyNext;                            ①
textField.keyboardType = UIKeyboardTypeNumbersAndPunctuation;         ②
```

```
textView.returnKeyType = UIReturnKeyType.go                    ③
textView.keyboardType = UIKeyboardType.default                 ④
```

```
textView.returnKeyType = UIReturnKeyGo;                        ③
textView.keyboardType = UIKeyboardTypeDefault;                 ④
```

上述代码中，第①行用于设置textField的return键为Next，代码第③行用于设置textView的return键为Go。代码第②行用于设置textField的键盘类型为NumbersAndPunctuation，代码第④行用于设置textView键盘为默认类型。

4.5 开关控件、分段控件和滑块控件

开关控件、滑块控件和分段控件都是UIControl的子类，下面我们将通过一个示例为大家讲解这3个控件。

如图4-25所示，该案例包括两个开关控件、一个分段控件、两个标签控件和一个滑块控件。两个开关控件的值保持一致，点击其中一个，令其值为ON，另一个也会随之改变；一个有两段的分段控件，左侧和右侧的段分别命名为Left和Right，点击Right时两个开关控件消失，点击Left时两个开关控件显示；后面的滑块控件可以改变标签SliderValue:的内容，把滑块变化的数值显示在后面。

图4-25 SwitchSliderSegmentedControl案例原型设计图

4.5.1 开关控件

开关控件类是UISwitch，它的功能类似于Windows系统中的复选框，且只有两种状态——true和false，两种状态的切换方法是setOn:animated:。下面我们通过Interface Builder和代码两种方式介绍SwitchSliderSegmentedControl案例如何添加开关控件。

1. Interface Builder实现

使用Single View Application模板创建一个名为SwitchSliderSegmentedControl的工程（具体创建过程请参见2.1.1节）。打开Main.storyboard文件，从对象库中拖曳两个开关控件到设计界面，参考4.3.1节将两个开关控件连线到ViewController输出口属性，输出口属性分别命名为leftSwitch和rightSwitch。参考4.2.1节将两个开关控件的默认事件连线到ViewController的switchValueChanged:动作方法；该方法的作用是同时设置两个开关的值，使它们的状态保持一致，其实现代码如下：

```swift
@IBAction func switchValueChanged (_ sender: AnyObject) {
    var witchSwitch = sender as UISwitch
    var setting = witchSwitch.on
    self.leftSwitch.setOn(setting, animated: true)
    self.rightSwitch.setOn(setting, animated: true)
}
```

```objc
- (IBAction)switchValueChanged:(id)sender {
    UISwitch *witchSwitch = (UISwitch *)sender;
    BOOL setting = witchSwitch.isOn;
    [self.leftSwitch setOn:setting animated:TRUE];
    [self.rightSwitch setOn:setting animated:TRUE];
}
```

2. 代码实现

上一节中，我们介绍了使用Interface Builder在SwitchSliderSegmentedControl案例中添加leftSwitch和rightSwitch控件，这一节介绍如何通过代码添加leftSwitch和rightSwitch控件。

其他步骤不再赘述，我们重点介绍一下视图控制器类ViewController，主要代码如下：

```swift
//ViewController.swift文件
class ViewController: UIViewController {

    var leftSwitch: UISwitch!
    var rightSwitch: UISwitch!
    var sliderValue: UILabel!

    override func viewDidLoad() {
        super.viewDidLoad()

        let screen = UIScreen.mainScreen().bounds

        ///1.添加rightSwitch控件
        //rightSwitch与屏幕的左边距
        //leftSwitch与屏幕的右边距
        let switchScreenSpace:CGFloat = 39                                    ①

        self.rightSwitch = UISwitch()
        var frame = self.rightSwitch.frame
        frame.origin = CGPoint(x: switchScreenSpace, y: 98)                   ②
        //重新设置控件的位置
        self.rightSwitch.frame = frame
        //设置控件状态
        self.rightSwitch.isOn = true
        //指定事件处理方法
        self.rightSwitch.addTarget(self,
        ➥action: #selector(switchValueChanged(_:)), for: .valueChanged)
        self.view.addSubview(self.rightSwitch)                                ③

        ///2.添加leftSwitch控件
        self.leftSwitch = UISwitch()                                          ④
        frame = self.leftSwitch.frame
        frame.origin = CGPoint(x: screen.size.width -
        ➥(frame.size.width + switchScreenSpace), y: 98)
        //重新设置控件的位置
        self.leftSwitch.frame = frame
        //设置控件状态
        self.leftSwitch.on = true
        //指定事件处理方法
        self.leftSwitch.addTarget(self,
        ➥action: #selector(switchValueChanged(_:)), for: .valueChanged)
        self.view.addSubview(self.leftSwitch)                                 ⑤
        ......
    }
    //使两个开关的值保持一致
    @IBAction func switchValueChanged(_ sender: AnyObject) {
        let witchSwitch = sender as! UISwitch
        let setting = witchSwitch.on
        self.leftSwitch.setOn(setting, animated: true)
        self.rightSwitch.setOn(setting, animated: true)
```

```objc
//ViewController.m文件
#import "ViewController.h"
@interface ViewController ()

@property (strong, nonatomic) UISwitch *rightSwitch;
@property (strong, nonatomic) UISwitch *leftSwitch;
@property (strong, nonatomic) UILabel *sliderValue;
@end

@implementation ViewController
- (void)viewDidLoad {
    [super viewDidLoad];

    CGRect screen = [[UIScreen mainScreen] bounds];

    ///1.添加rightSwitch控件
    //rightSwitch与屏幕的左边距
    //leftSwitch与屏幕的右边距
    CGFloat switchScreenSpace = 39;                                           ①

    self.rightSwitch = [[UISwitch alloc] init];
    CGRect frame = self.rightSwitch.frame;
    frame.origin = CGPointMake(switchScreenSpace, 98);                        ②
    //重新设置控件的位置
    self.rightSwitch.frame = frame;
    //设置控件状态
    self.rightSwitch.on = TRUE;
    //指定事件处理方法
    [self.rightSwitch addTarget:self action:@selector(switchValueChanged:)
    ➥forControlEvents:UIControlEventValueChanged];
    [self.view addSubview:self.rightSwitch];                                  ③

    ///2.添加leftSwitch控件
    self.leftSwitch = [[UISwitch alloc] init];                                ④
    frame = self.leftSwitch.frame;
    frame.origin = CGPointMake(screen.size.width - (frame.size.width +
    ➥switchScreenSpace), 98);
    //重新设置控件的位置
    self.leftSwitch.frame = frame;
    //设置控件状态
    self.leftSwitch.on = TRUE;
    //指定事件处理方法
    [self.leftSwitch addTarget:self action:@selector(switchValueChanged:)
    ➥forControlEvents:UIControlEventValueChanged];
    [self.view addSubview:self.leftSwitch];                                   ⑤
    ......
}
......
//使两个开关的值保持一致
- (void)switchValueChanged:(id)sender {
    UISwitch *witchSwitch = (UISwitch *)sender;
    BOOL setting = witchSwitch.isOn;
    [self.leftSwitch setOn:setting animated:TRUE];
```

```
    }
    ......
}
```

```
        [self.rightSwitch setOn:setting animated:TRUE];
    }
@end
```

上述代码中，第①行~第③行用于创建rightSwitch控件，第④行~第⑤行用于创建leftSwitch控件。在代码中创建Switch控件时使用的是默认构造函数init，所以frame属性设置不是在实例化时设置的。第②行用于设置frame中的origin，origin是控件的坐标原点。

4.5.2　分段控件

分段控件也是一种选择控件，其功能类似于Windows中的单选按钮。它由两段或更多段构成，每段相当于一个独立的按钮。它有3种样式——Plain、Bordered和Bar样式，但是在iOS 7及其之后的版本中，这3种样式没有什么区别。它们的样式如图4-26所示。

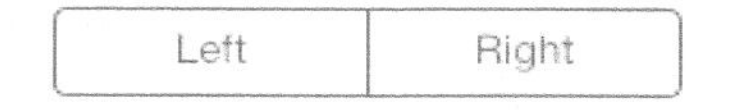

图4-26　分段控件样式

如果想要设置多段，可以在如图4-27所示的属性检查器中修改Segments属性。如果想要设置每一个段的标题和图标，可以按照图4-27所示，首先从Segment下拉框选择具体的段，然后修改下面的Title（段标题）属性或Image属性（段图标）。

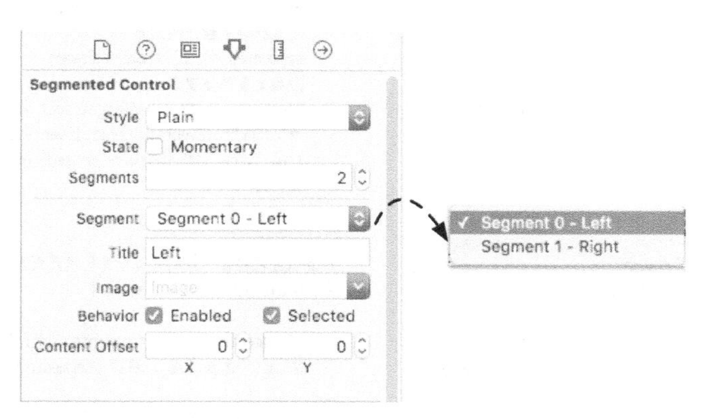

图4-27　分段控件属性

下面我们通过Interface Builder和代码两种方式在SwitchSliderSegmentedControl案例中添加分段控件。

1. Interface Builder实现

在开关控件下方拖曳一个分段控件，双击使其处于编辑状态，依次输入文本Left和Right。

下面我们看看实现代码，视图控制器ViewController中实现touchDown:的代码如下：

```swift
@IBAction func touchDown(_ sender: AnyObject) {

    let segmentedControl = sender as! UISegmentedControl
    print("选择的段 : ", segmentedControl.selectedSegmentIndex)             ①

    if (self.leftSwitch.hidden == true) {
        self.rightSwitch.hidden = false
        self.leftSwitch.hidden  = false
    }else {
        self.rightSwitch.hidden = true
        self.leftSwitch.hidden  = true
    }
}
```

```objc
- (IBAction)touchDown:(id)sender {

    UISegmentedControl *segmentedControl = (UISegmentedControl *)sender;
    NSLog(@"选择的段 : %li", segmentedControl.selectedSegmentIndex);       ①

    if (self.leftSwitch.hidden) {
        self.rightSwitch.hidden = FALSE;
        self.leftSwitch.hidden  = FALSE;
    }else{
        self.leftSwitch.hidden  = TRUE;
        self.rightSwitch.hidden = TRUE;
    }
}
```

分段控件中有多个段，如何判断选择了哪个段？第①行的selectedSegmentIndex属性可以获得选择的段索引，段索引是从0开始的。

2. 代码实现

上一节中，我们介绍了使用Interface Builder在SwitchSliderSegmentedControl案例中添加了分段控件，这一节介绍如何通过代码添加分段控件。

其他步骤不再赘述，这里重点介绍一下视图控制器类ViewController，主要代码如下：

```
//ViewController.swift文件
class ViewController: UIViewController {
    ......
    override func viewDidLoad() {
        super.viewDidLoad()
        ......
        var frame = self.rightSwitch.frame
        ......
        ///3.添加segmentedControl控件
        let segments = ["Right", "Left"]
        let segmentedControl = UISegmentedControl(items: segments)

        let scWidth:CGFloat = 220
        let scHeight:CGFloat = 29 //29为默认高度
        let scTopView:CGFloat = 186
        frame = CGRect(x: (screen.size.width - scWidth)/2 ,
        ➥y: scTopView, width: scWidth, height: scHeight)
        //重新设置控件的位置
        segmentedControl.frame = frame
        //指定事件处理方法
        segmentedControl.addTarget(self, action: #selector(touchDown(_:)),
        ➥for: .valueChanged)
        self.view.addSubview(segmentedControl)
        ......
    }
    ......
    //点击分段控件，控制开关控件的隐藏或显示
    @IBAction func touchDown(_ sender: AnyObject) {

        let segmentedControl = sender as! UISegmentedControl
        print("选择的段 : ", segmentedControl.selectedSegmentIndex)

        if (self.leftSwitch.hidden == true) {
            self.rightSwitch.hidden = false
            self.leftSwitch.hidden  = false
        }else {
            self.rightSwitch.hidden = true
            self.leftSwitch.hidden  = true
        }
    }
    ......
}
```

```
//ViewController.m文件
@implementation ViewController

- (void)viewDidLoad {
    [super viewDidLoad];
    ......

    ///3.添加segmentedControl控件
    NSArray* segments = @[@"Right", @"Left"];
    UISegmentedControl *segmentedControl =
    ➥[[UISegmentedControl alloc] initWithItems:segments];

    CGFloat scWidth = 220;
    CGFloat scHeight = 29; //29为默认高度
    CGFloat scTopView = 186;
    frame = CGRectMake((screen.size.width - scWidth)/2 ,
    ➥scTopView, scWidth, scHeight);
    //重新设置控件的位置
    segmentedControl.frame = frame;
    //指定事件处理方法
    [segmentedControl addTarget:self action:@selector(touchDown:)
    ➥forControlEvents:UIControlEventValueChanged];
    [self.view addSubview:segmentedControl];
    ......
}
......
//点击分段控件，控制开关控件的隐藏或显示
- (void)touchDown:(id)sender {

    UISegmentedControl *segmentedControl = (UISegmentedControl *)sender;
    NSLog(@"选择的段 : %li", segmentedControl.selectedSegmentIndex);

    if (self.leftSwitch.hidden) {
        self.rightSwitch.hidden = FALSE;
        self.leftSwitch.hidden  = FALSE;
    }else{
        self.leftSwitch.hidden  = TRUE;
        self.rightSwitch.hidden = TRUE;
    }
}

......

@end
```

上述代码前面基本上都介绍了，这里不再赘述。

4.5.3　滑块控件

下面我们通过Interface Builder和代码两种方式在SwitchSliderSegmentedControl案例中添加滑块控件。

1. Interface Builder实现

在视图上拖曳一个滑块控件，然后将其水平放置。打开它的属性检查器，如图4-28所示，将其最小值

（Minimum）、最大值（Maximum）、当前值（Current）依次设定为0、100、50。另外，我们可以为滑块控件两边添加最小值图片（Min Image）和最大值图片（Max Image）。

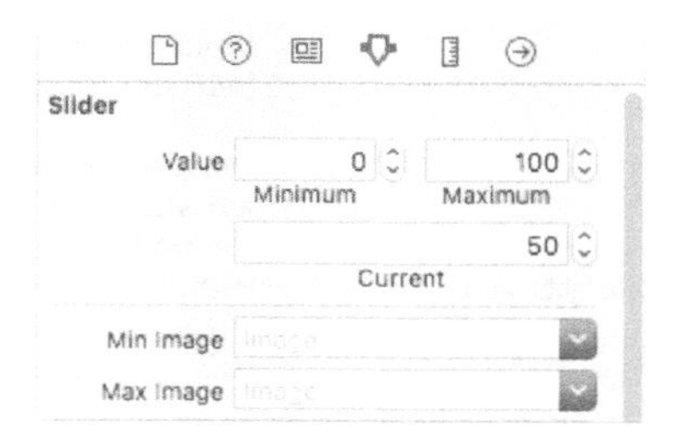

图4-28 设置滑块控件属性

在滑块上方拖曳两个标签，将左侧标签的文本改为SliderValue:，将右侧标签的文本清除，并为其实现输出口，命名为SliderValue。右侧的标签用于显示滑块的值，也就是滑块控制着标签的值，这里我们为滑块实现一个动作，将其命名为sliderValueChange。实现sliderValueChange:方法的代码如下：

```
@IBAction func sliderValueChange(_ sender: AnyObject) {
    var slider = sender as UISlider
    let progressAsInt = Int(slider.value)                        ①
    let newText = NSString(format: "%d", progressAsInt)          ②
    self.sliderValue.text = newText
}
```

```
- (IBAction)sliderValueChange:(id)sender {
    UISlider *slider = (UISlider *)sender;
    int progressAsInt = (int)(slider.value);                     ①
    NSString *newText = [[NSString alloc]
    ➥initWithFormat:@"%d",progressAsInt];                       ②
    self.sliderValue.text = newText;
}
```

第①行用于设置滑块控件的值，由于取出的值是浮点类型，我们需要将其强制转换为整数类型。第②行用于格式化字符串。

2. 代码实现

上一节中，我们在使用Interface Builder实现的SwitchSliderSegmentedControl案例中添加了滑块控件，这一节介绍如何通过代码添加滑块控件。

其他步骤不再赘述，我们重点介绍一下视图控制器类ViewController，其主要代码如下：

```
//ViewController.swift文件
class ViewController: UIViewController {
    ......
    override func viewDidLoad() {
        super.viewDidLoad()
        ......
        ///4.添加slider控件
        let sliderWidth:CGFloat  = 300
        let sliderHeight:CGFloat  = 31 //31为默认高度
        let sliderdTopView:CGFloat  = 298
        let slider = UISlider(frame: CGRect(x: (
        ➥screen.size.width - sliderWidth)/2 ,
        ➥y: sliderdTopView, width: sliderWidth, height: sliderHeight))

        slider.minimumValue = 0.0
        slider.maximumValue = 100.0
        slider.value = 50.00

        //指定事件处理方法
        slider.addTarget(self, action: #selector(sliderValueChange(_:)),
        ➥for: .valueChanged)
        self.view.addSubview(slider)

        ///5.添加SliderValue:标签
        //SliderValue:标签与Slider之间的距离
        let labelSliderValueSliderSpace:CGFloat = 30
```

```
//ViewController.m文件
@implementation ViewController

- (void)viewDidLoad {
    [super viewDidLoad];
    ......
    ///4.添加slider控件
    CGFloat sliderWidth = 300;
    CGFloat sliderHeight = 31; //31为默认高度
    CGFloat sliderdTopView = 298;
    UISlider *slider = [[UISlider alloc]
    ➥initWithFrame:CGRectMake((screen.size.width- sliderWidth)/2 ,
    ➥sliderdTopView, sliderWidth, sliderHeight)];

    slider.minimumValue = 0.0f;
    slider.maximumValue = 100.0f;
    slider.value = 50.00f;

    //指定事件处理方法
    [slider addTarget:self action:@selector(sliderValueChange:)
    ➥forControlEvents:UIControlEventValueChanged];

    [self.view addSubview:slider];

    ///5.添加SliderValue:标签
    //SliderValue:标签与Slider之间的距离
```

```
        let labelSliderValue = UILabel(frame: CGRect(x: slider.frame.origin.x,
        ➥y: slider.frame.origin.y - labelSliderValueSliderSpace, width: 103,
        ➥height: 21))

        labelSliderValue.text = "SliderValue: "
        self.view.addSubview(labelSliderValue)

        ///6.添加sliderValue标签
        self.sliderValue = UILabel(frame: CGRect(x:
        ➥labelSliderValue.frame.origin.x + 120,
        ➥y: labelSliderValue.frame.origin.y, width: 50, height: 21))

        self.sliderValue.text = "50"
        self.view.addSubview(self.sliderValue)
        ......
    }
    ......
    //用标签显示滑块的值
    @IBAction func sliderValueChange(_ sender: AnyObject) {
        let slider = sender as! UISlider
        let progressAsInt = Int(slider.value)
        let newText = String(format: "%d", progressAsInt)
        print("滑块的值 : ", newText)
        self.sliderValue.text = newText
    }
    ......
}
```

```
    CGFloat labelSliderValueSliderSpace = 30;
    UILabel* labelSliderValue = [[UILabel alloc]
    ➥initWithFrame:CGRectMake(slider.frame.origin.x, slider.frame.origin.y -
    ➥labelSliderValueSliderSpace, 103, 21)];
    labelSliderValue.text = @"SliderValue: ";
    [self.view addSubview:labelSliderValue];

    ///6.添加sliderValue标签
    self.sliderValue = [[UILabel alloc]
    ➥initWithFrame:CGRectMake(labelSliderValue.frame.origin.x + 120,
    ➥labelSliderValue.frame.origin.y, 50, 21)];
    self.sliderValue.text = @"50";
    [self.view addSubview:self.sliderValue];
}
......

//用标签显示滑块的值
- (void)sliderValueChange:(id)sender {
    UISlider *slider = (UISlider *)sender;
    int progressAsInt = (int)(slider.value);
    NSString *newText = [[NSString alloc]initWithFormat:@"%d",progressAsInt];
    NSLog(@"滑块的值 : %@", newText);
    self.sliderValue.text = newText;
}
@end
```

上述代码前面基本都介绍了，这里不再赘述。

4.6　Web 视图：WKWebView 类

Web技术可以应用于iOS开发，苹果公司允许发布本地+Web的混合应用。很多情况下，使用Web技术构建界面很有优势，例如：提供丰富的界面布局、显示多行不同风格的文本、显示图片、播放音频和视频等。Web视图能够完成显示HTML、解析CSS和执行JavaScript等操作。

WKWebView是苹果在iOS 8中发布的新Web视图，旨在替换iOS中的UIWebView和macOS中的WebView。WKWebView很好地解决了UIWebView的内存占用大和加载速度慢等问题。

WKWebView可以加载本地HTML代码或者网络资源。

本地资源的加载一般采用同步方式，数据可以来源于本地文件或者是硬编码的HTML字符串，相关方法如下。

- **- loadHTMLString:baseURL:**。设定主页文件的基本路径，通过一个HTML字符串加载主页数据。
- **- loadData:MIMEType:characterEncodingName:baseURL:**。指定MIME类型、编码集和NSData对象加载一个主页数据，并设定主页文件的基本路径。

使用这两个方法时，我们需要注意字符集问题，而采用什么样的字符集取决于HTML文件。

加载网络资源时，我们一般采用的是异步加载方式，使用的方法是WKWebView的loadRequest:方法，该方法的参数是NSURLRequest对象，该对象在构建的时候必须严格遵守某种协议格式，例如：

- http://www.sina.com.cn，HTTP协议；
- file://localhost/Users/tonyguan/.../index.html，文件传输协议。

其中http://和file://是协议名，不能省略。上网的时候我们常常将http://省略，一般的浏览器仍然可以解析输入的URL，但是在WKWebView的loadRequest:方法中，该字符串一定不能省略！

由于我们采用异步请求加载网络资源，所以还要实现相应的WKNavigationDelegate委托协议。请求加载网络资源的不同阶段会触发WKNavigationDelegate委托对象的不同方法。

下面我们通过一个案例（如图4-29所示）来了解WKWebView这3个方法的用法。该案例有3个按钮，分别为loadHTMLString、loadData和loadRequest，点击这3个按钮会分别触发WKWebView的3个加载方法。

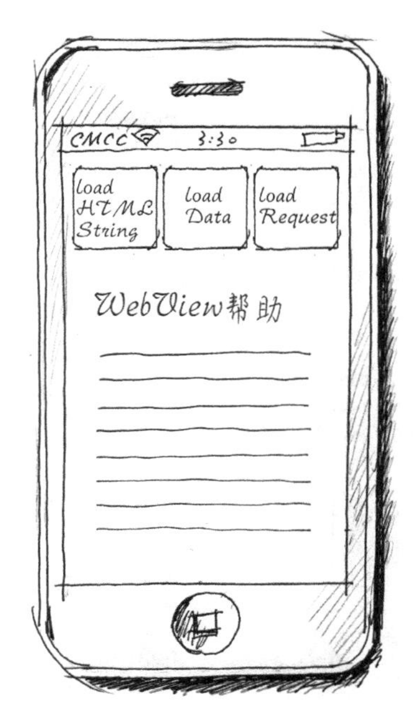

图4-29 WebViewSample案例原型设计图

> **提示** 由于在Interface Builder控件库中没有可拖曳的WKWebView视图对象，所以要想添加WKWebView对象到UI界面，只能通过代码实现。

使用Single View Application模板创建一个名为WebViewSample的工程，然后参考3.5.1节将WebViewSample修改为纯代码工程。

其中AppDelegate类与3.5.1节的案例完全一样，这里不再赘述。我们重点介绍一下视图控制器类ViewController，其主要代码如下：

```
//ViewController.swift文件
import UIKit
import WebKit                                                    ①

class ViewController: UIViewController, WKNavigationDelegate {   ②

    var webView: WKWebView!                                      ③

    override func viewDidLoad() {
        super.viewDidLoad()

        let screen = UIScreen.main.bounds
        ///按钮栏
        //按钮栏宽
        let buttonBarWidth: CGFloat = 316
        let buttonBar = UIView(frame: CGRect(x: (screen.size.width -
        ➥buttonBarWidth) / 2, y: 20, width: buttonBarWidth, height: 30)) ④
        self.view.addSubview(buttonBar)

        ///1.添加LoadHTMLString按钮
        let buttonLoadHTMLString = UIButton(type: UIButtonType.system)
        buttonLoadHTMLString.setTitle("LoadHTMLString", for: UIControlState())
        buttonLoadHTMLString.frame = CGRect(x: 0, y: 0, width: 117, height: 30)
        //指定事件处理方法
        buttonLoadHTMLString.addTarget(self,
```

```
//ViewController.m文件
#import "ViewController.h"
#import <WebKit/WebKit.h>                                        ①
@interface ViewController () <WKNavigationDelegate>              ②
@property(nonatomic, strong) WKWebView* webView;                 ③
@end

@implementation ViewController

- (void)viewDidLoad {
    [super viewDidLoad];

    CGRect screen = [[UIScreen mainScreen] bounds];

    ///按钮栏
    //按钮栏宽
    CGFloat buttonBarWidth = 316;
    UIView* buttonBar = [[UIView alloc]
    ➥initWithFrame: CGRectMake((screen.size.width - buttonBarWidth) / 2, 20,
    ➥buttonBarWidth, 30)];                                       ④

    [self.view addSubview:buttonBar];

    ///1.添加LoadHTMLString按钮
```

```
    ➥action: #selector(testLoadHTMLString(_:)), for: .touchUpInside)
    buttonBar.addSubview(buttonLoadHTMLString)                          ⑤

    ///2.添加LoadData按钮
    let buttonLoadData = UIButton(type: UIButtonType.system)
    buttonLoadData.setTitle("LoadData", for: UIControlState())
    buttonLoadData.frame = CGRect(x: 137, y: 0, width: 67, height: 30)
    //指定事件处理方法
    buttonLoadData.addTarget(self,
    ➥action: #selector(testLoadData(_:)), for: .touchUpInside)
    buttonBar.addSubview(buttonLoadData)

    ///3.添加LoadRequest按钮
    let buttonLoadRequest = UIButton(type: UIButtonType.system)
    buttonLoadRequest.setTitle("LoadRequest", for: UIControlState())
    buttonLoadRequest.frame = CGRect(x: 224, y: 0, width: 92, height: 30)
    //指定事件处理方法
    buttonLoadRequest.addTarget(self,
    ➥action: #selector(testLoadRequest(_:)), for: .touchUpInside)
    buttonBar.addSubview(buttonLoadRequest)

    ///4.添加WKWebView
    self.webView = WKWebView(frame: CGRect(x: 0, y: 60,
    ➥width: screen.size.width, height: screen.size.height - 80))
    self.view.addSubview(self.webView)                                  ⑥
}
......
func testLoadHTMLString(_ sender: AnyObject) {
    let htmlPath = Bundle.main.path(forResource: "index", ofType: "html")
    let bundleUrl = URL(fileURLWithPath: Bundle.main.bundlePath)
    do {
        let html = try NSString(contentsOfFile: htmlPath!,
        ➥encoding: String.Encoding.utf8.rawValue)                       ⑦
        self.webView.loadHTMLString(html as String, baseURL: bundleUrl)
    } catch let err as NSError {
        NSLog("加载失败 error :  %@", err.localizedDescription)
    }
}

func testLoadData(_ sender: AnyObject) {

    let htmlPath = Bundle.main.path(forResource: "index", ofType: "html")
    let bundleUrl = URL(fileURLWithPath: Bundle.main.bundlePath)
    let htmlData = try? Data(contentsOf: URL(fileURLWithPath: htmlPath!))

    self.webView.load(htmlData!, mimeType: "text/html",
    ➥ characterEncodingName: "UTF-8", baseURL: bundleUrl)
}

func testLoadRequest(_ sender: AnyObject) {
    let url = URL(string: "http://51work6.com")
    let request = URLRequest(url: url!)
    self.webView.load(request)⑧
    self.webView.navigationDelegate = self                              ⑨
}

//MARK: --实现WKNavigationDelegate委托协议
//开始加载时调用
func webView(_ webView: WKWebView,
➥didStartProvisionalNavigation navigation: WKNavigation!) {
    print("开始加载")
}

//当内容开始返回时调用
func webView(_ webView: WKWebView,
➥didCommitNavigation navigation: WKNavigation!) {
    print("内容开始返回")
}
```

```
    UIButton* buttonLoadHTMLString = [UIButton buttonWithType:
    ➥UIButtonTypeSystem];
    [buttonLoadHTMLString setTitle:@"LoadHTMLString" forState:
    ➥UIControlStateNormal];
    buttonLoadHTMLString.frame = CGRectMake(0, 0, 117, 30);
    //指定事件处理方法
    [buttonLoadHTMLString addTarget:self action:@selector(
    ➥testLoadHTML String:) forControlEvents: UIControlEventTouchUpInside];
    [buttonBar addSubview: buttonLoadHTMLString];                       ⑤

    ///2.添加LoadData按钮
    UIButton* buttonLoadData = [UIButton buttonWithType:UIButtonTypeSystem];
    [buttonLoadData setTitle:@"LoadData" forState:UIControlStateNormal];
    buttonLoadData.frame = CGRectMake(137, 0, 67, 30);
    //指定事件处理方法
    [buttonLoadData addTarget:self action:@selector(testLoadData:) forControl
    ➥Events:  UIControlEventTouchUpInside];
    [buttonBar addSubview: buttonLoadData];

    ///3.添加LoadRequest按钮
    UIButton* buttonLoadRequest = [UIButton buttonWithType:
    ➥UIButtonTypeSystem];

    [buttonLoadRequest setTitle:@"LoadRequest" forState:
    ➥UIControlState Normal];
    buttonLoadRequest.frame = CGRectMake(224, 0, 92, 30);
    //指定事件处理方法
    [buttonLoadRequest addTarget:self action:@selector(testLoadRequest:)
    ➥forControlEvents: UIControlEventTouchUpInside];
    [buttonBar addSubview: buttonLoadRequest];

    ///4.添加WKWebView
    self.webView = [[WKWebView alloc] initWithFrame: CGRectMake(0, 60,
    ➥screen.size.width, screen.size.height - 80)];

    [self.view addSubview: self.webView];                               ⑥
}
......
- (void)testLoadHTMLString:(id)sender {
    NSString *htmlPath = [[NSBundle mainBundle] pathForResource:@"index"
    ➥ofType:@"html"];
    NSURL *bundleUrl = [NSURL fileURLWithPath:[[NSBundle mainBundle]
    ➥bundlePath]];
    NSError *error = nil;

    NSString *html = [[NSString alloc] initWithContentsOfFile:htmlPath
    ➥encoding: NSUTF8StringEncoding error:&error];                     ⑦

    if (error == nil) {//数据加载没有错误的情况下
        [self.webView loadHTMLString:html baseURL:bundleUrl];
    }
}

- (void)testLoadData:(id)sender {

    NSString *htmlPath = [[NSBundle mainBundle] pathForResource:@"index"
    ➥ofType:@"html"];
    NSURL *bundleUrl = [NSURL fileURLWithPath:[[NSBundle mainBundle]
    ➥bundlePath]];
    NSData *htmlData = [[NSData alloc]  initWithContentsOfFile: htmlPath];

    [self.webView loadData:htmlData MIMEType:@"text/html"
    ➥characterEncodingName:@"UTF-8" baseURL:bundleUrl];

}
```

```
    //加载完成之后调用
    func webView(_ webView: WKWebView,
    ➥didFinishNavigation navigation: WKNavigation!) {
        print("加载完成")
    }

    //加载失败时调用
    func webView(_ webView: WKWebView,
    ➥didFailProvisionalNavigation navigation: WKNavigation!,
    ➥withError error: NSError) {
        print("加载失败 error :  ", error.localizedDescription)
    }
}
```

```
- (void)testLoadRequest:(id)sender {

    NSURL * url = [NSURL URLWithString: @"http://www.51work6.com"];
    NSURLRequest * request = [NSURLRequest requestWithURL:url];
    [self.webView loadRequest:request];                                   ⑧
    self.webView.navigationDelegate = self;                               ⑨
}

#pragma mark -- 实现WKNavigationDelegate委托协议
//开始加载时调用
-(void)webView:(WKWebView *)webView
➥didStartProvisionalNavigation:(WKNavigation *)navigation {
    NSLog(@"开始加载");
}
//当内容开始返回时调用
-(void)webView:(WKWebView *)webView didCommitNavigation:(
➥WKNavigation *)navigation {
    NSLog(@"内容开始返回");
}

//加载完成之后调用
-(void)webView:(WKWebView *)webView didFinishNavigation:(
➥WKNavigation *)navigation {
    NSLog(@"加载完成");
}

//加载失败时调用
-(void)webView:(WKWebView *)webView
➥didFailProvisionalNavigation:(WKNavigation *)navigation
➥withError: (NSError *)error {
    NSLog(@"加载失败 error :  %@", error.description);
}

@end
```

上述代码中，因为需要使用WKWebView、WKNavigationDelegate和WKNavigation，所以需要在第①行中引入WebKit模块，在Objective-C版中需要引入<WebKit/WebKit.h>头文件。

第②行用于在视图控制器ViewController中声明实现WKNavigationDelegate委托协议。与WKWebView相关的协议有：WKNavigationDelegate和WKUIDelegate。WKNavigationDelegate主要与Web视图界面加载过程有关，WKUIDelegate主要与Web视图界面显示和提示框相关。本例使用WKNavigationDelegate委托协议，它的主要方法如下。

- **- webView:didStartProvisionalNavigation:**。该方法在Web视图开始加载界面时调用。
- **- webView:didCommitNavigation:**。该方法是当内容开始返回时调用。
- **- webView:didFinishNavigation:**。该方法在Web视图完成加载之后调用。
- **- webView:didFailProvisionalNavigation:withError:**。该方法在Web视图加载失败时调用。

第③行用于定义WKWebView的属性webView。

第④行~第⑥行用于构建UI界面。我们在代码第④行创建按钮栏，该按钮栏是UIView对象，包含了LoadHTMLString按钮、LoadData按钮和LoadRequest按钮，本案例界面的构建层次如图4-30所示。将按钮添加到按钮栏类似于第⑤行代码。

提示 为什么将按钮添加到按钮栏，然后再将按钮栏添加到根视图，而不是直接将按钮添加到根视图呢？这是因为我想将3个按钮作为一个整体视图来对待，让这个整体视图在屏幕中居中。

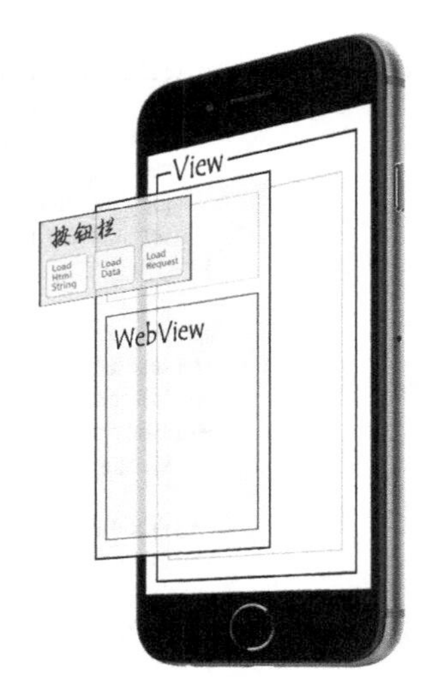

图4-30　WebViewSample案例界面构建层次图

提示　在使用Xcode 7及其之后的版本运行应用时，点击“LoadRequest按钮”请求网络资源，则会输出如下错误：The resource could not be loaded because the ATS App Transport Security policy requires the use of a secure connection。这是苹果引入的新特性ATS App Transport Security，ATS要求应用程序访问网络资源时必须使用HTTPS协议，但是很多情况下使用HTTP。如果使用HTTP协议，可以修改Xcode工程属性文件Info.plist，按照图4-31所示添加App Transport Security Settings键，其中在Xcode 7中，是在该键下再添加Allow Arbitrary Loads键（Xcode 7之后是Allow Arbitrary Loads in Web Content），将其值修改为YES。

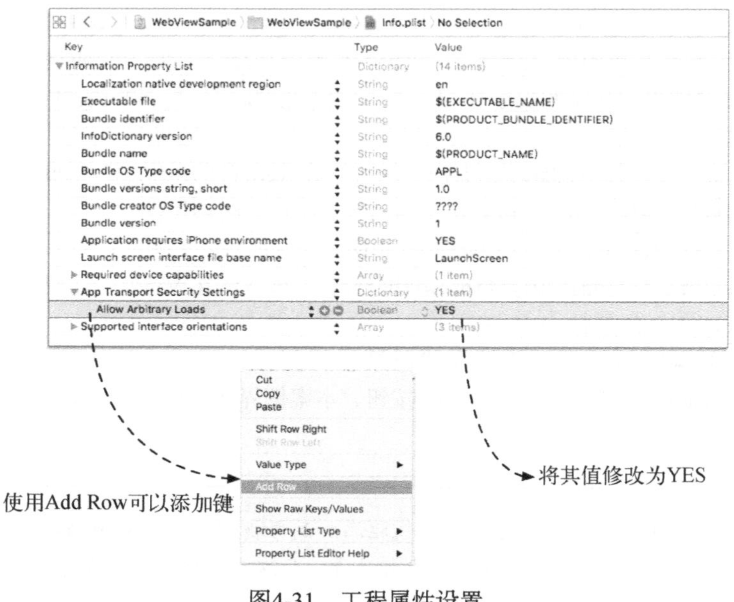

图4-31　工程属性设置

4.7 警告框和操作表

应用如何与用户交流呢？警告框（AlertView）和操作表（ActionSheet）就是为此而设计的。

本节案例的原型草图如图4-32所示，其中有两个按钮——“Test警告框”和“Test操作表”：点击“Test警告框”按钮时弹出警告框，其中有两个按钮No和Yes；点击“Test操作表”按钮时，屏幕下方将滑出操作表。

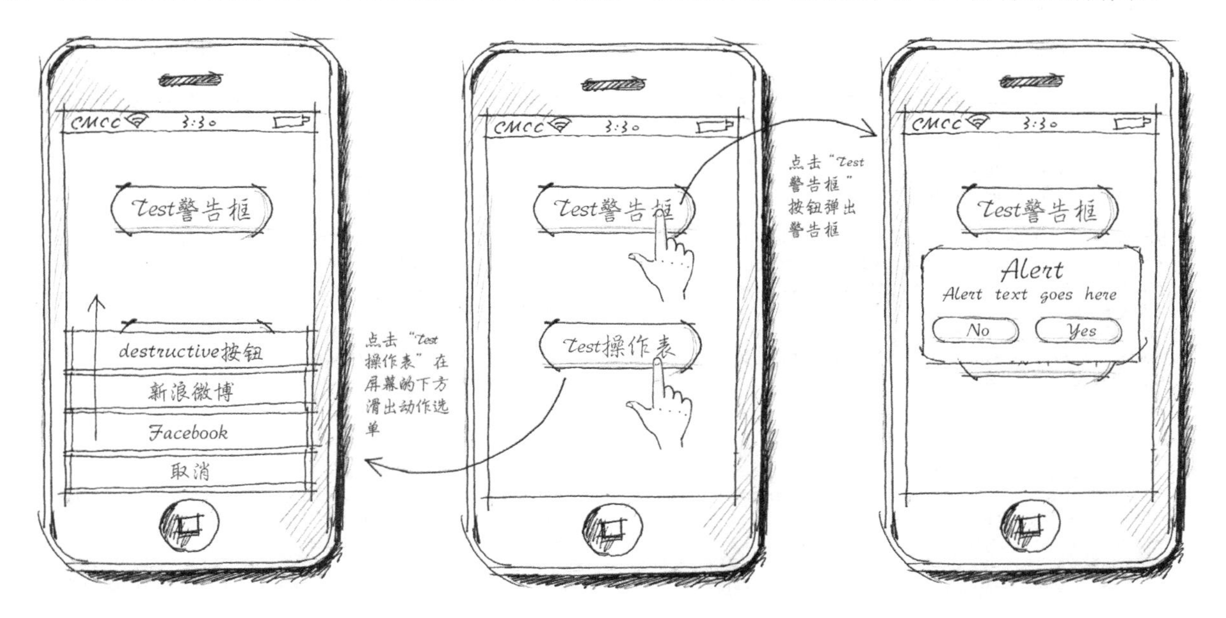

图4-32 AlertViewActionSheet案例原型草图

4.7.1 使用 UIAlertController 实现警告框

iOS中的警告框用于给用户以警告或提示，最多有两个按钮，超过两个就应该使用操作表。由于在iOS中警告框是“模态”的①，因此不应该随意使用。一般情况下，警告框的使用场景有如下几个。

- **应用不能继续运行。** 例如，无法获得网络数据或者功能不能完成的时候，给用户一个警告，这种警告框只需一个按钮。
- **询问另外的解决方案。** 好多应用在不能继续运行时，会给出另外的解决方案，让用户去选择。例如，Wi-Fi网络无法连接时，是否可以使用4G网络。
- **询问对操作的授权。** 当应用访问用户的一些隐私信息时，需要用户授权，例如用户当前的位置、通讯录或日程表等。

警告框在iOS 8之前使用UIAlertView视图，而在iOS 8及其之后的版本中使用UIAlertController控制器，可以实现警告框和操作表。UIAlertController控制器中不仅可以添加按钮，还可以添加文本框和自定义视图到警告框和操作表，响应事件可以通过闭包（Objective-C版为代码块）实现，而不用委托协议实现。

下面看看AlertViewActionSheet案例中警告框的实现过程。由于UIAlertController在Interface Builder控件库中没有可拖曳对象，所以AlertViewActionSheet案例采用纯代码实现。

首先，我们使用Single View Application模板创建一个名为AlertViewActionSheet的工程，然后参考3.5.1节将AlertViewActionSheet修改为纯代码工程。

这一节我们介绍如何通过代码方式添加警告框，ViewController的主要代码如下：

① “模态”表示不关闭它就不能做别的事情。

```swift
//ViewController.swift文件
import UIKit

class ViewController: UIViewController {

    override func viewDidLoad() {
        super.viewDidLoad()

        let screen = UIScreen.mainScreen().bounds
        let buttonAlertView = UIButton(type: UIButtonType.system)           ①
        buttonAlertView.setTitle("Test警告框", for: UIControlState())

        let buttonAlertViewWidth: CGFloat = 100
        let buttonAlertViewHeight: CGFloat = 30
        let buttonAlertViewTopView: CGFloat = 130

        buttonAlertView.frame = CGRect(x: (screen.size.width -
        ➥buttonAlertViewWidth)/2 , y: buttonAlertViewTopView,
        ➥width: buttonAlertViewWidth, height: buttonAlertViewHeight)
        //指定事件处理方法
        buttonAlertView.addTarget(self,
        ➥action: #selector(testAlertView(_:)), for: .touchUpInside)
        self.view.addSubview(buttonAlertView)                               ②
        ......
    }
    ......
    func testAlertView(_ sender: AnyObject) {
        let alertController: UIAlertController = UIAlertController(title:
        ➥"Alert", message: "Alert text goes here",
        ➥preferredStyle: UIAlertControllerStyle.alert)                     ③

        let noAction = UIAlertAction(title: "No", style: .cancel)
        ➥{ (alertAction) -> Void in
            print("Tap No Button")
        }                                                                   ④

        let yesAction = UIAlertAction(title: "Yes", style: .default)
        ➥{ (alertAction) -> Void in
            print("Tap Yes Button")
        }                                                                   ⑤
        alertController.addAction(noAction)                                 ⑥
        alertController.addAction(yesAction)                                ⑦

        //显示
        self.presentViewController(alertController, animated: true,
        ➥completion: nil)                                                  ⑧
    }
    ......
}
```

```objectivec
//ViewController.m文件
@implementation ViewController

- (void)viewDidLoad {
    [super viewDidLoad];

    CGRect screen = [[UIScreen mainScreen] bounds];

    UIButton* buttonAlertView = [UIButton buttonWithType:
    ➥UIButtonTypeSystem];                                                  ①
    [buttonAlertView setTitle:@"Test警告框" forState:UIControlStateNormal];

    CGFloat buttonAlertViewWidth = 100;
    CGFloat buttonAlertViewHeight = 30;
    CGFloat buttonAlertViewTopView = 130;

    buttonAlertView.frame = CGRectMake((screen.size.width -
    ➥buttonAlertViewWidth)/2 , buttonAlertViewTopView,
    ➥buttonAlertViewWidth,buttonAlertViewHeight);
    //指定事件处理方法
    [buttonAlertView addTarget:self action:@selector(testAlertView:)
     ➥forControlEvents: UIControlEventTouchUpInside];
    [self.view addSubview:buttonAlertView];                                 ②
    .....
}
......
- (void)testAlertView:(id)sender {
    UIAlertController* alertController  = [UIAlertController alertController
    ➥WithTitle:@"Alert" message: @"Alert text goes here"
    ➥preferredStyle:UIAlertControllerStyleAlert];                          ③

    UIAlertAction* noAction = [UIAlertAction actionWithTitle:@"No"
    ➥style:UIAlertActionStyleCancel handler:^(UIAlertAction *action) {
        NSLog(@"Tap No Button");
    }];                                                                     ④

    UIAlertAction* yesAction = [UIAlertAction actionWithTitle:@"Yes"
    ➥style:UIAlertActionStyleDefault handler:^(UIAlertAction *action) {
        NSLog(@"Tap Yes Button");
    }];                                                                     ⑤

    [alertController addAction:noAction];                                   ⑥
    [alertController addAction:yesAction];                                  ⑦

    //显示
    [self presentViewController:alertController animated:true
    ➥completion: nil];                                                     ⑧

}
......
@end
```

上述代码中，第①行~第②行用于创建并设置“Test警告框”按钮控件，具体代码前面已经解释了，这里不再赘述。

在testAlertView:方法中，第③行创建并初始化UIAlertController对象，构造函数中的第一个参数是警告框标题，第二个参数message是警告框内容，第三个参数preferredStyle是对话框类型。而对话框类型是在UIAlertControllerStyle枚举类型中定义的。UIAlertControllerStyle枚举成员说明见表4-4。

表4-4　UIAlertControllerStyle枚举成员

Objective-C枚举成员	Swift枚举成员	说　明
UIAlertControllerStyleActionSheet	actionSheet	操作表（默认）
UIAlertControllerStyleAlert	alert	警告框

另外，第④行和第⑤行都创建了UIAlertAction对象，每个UIAlertAction对象对应一个按钮动作。UIAlertAction构造函数的第一个参数是按钮标题，第二个参数style是按钮样式，第三个参数是与按钮动作相关的闭包（Objective-C版本是代码块）。

按钮样式是在UIAlertActionStyle枚举类型中定义的。UIAlertActionStyle枚举成员说明见表4-5。

表4-5 UIAlertActionStyle枚举成员

Objective-C枚举成员	Swift枚举成员	说　明
UIAlertActionStyleDefault	default	默认样式，粗体显示标题
UIAlertActionStyleCancel	cancel	取消按钮样式
UIAlertActionStyleDestructive	destructive	破坏性按钮样式，红色显示标题

第⑥行和第⑦行用于将两个UIAlertAction对象添加到UIAlertController对象中。最后，第⑧行显示UIAlertController对象。

4.7.2 使用 UIAlertController 实现操作表

如果想给用户提供多于两个的选择，比如想把应用中的某个图片发给新浪微博或者Facebook等平台，就应该使用操作表。在iPhone下运行操作表，它会从屏幕下方滑出来（如图4-33所示），其布局中最下面是“取消”按钮，该按钮离用户的大拇指最近，最容易被点击到。如果选项中有一个破坏性的操作，将会被放在最上面，是大拇指最不容易碰到的位置，并且其颜色是红色的。

在iPad中，操作表的布局与iPhone有所不同，如图4-34所示。在iPad中，操作表不是在底部滑出来的，而是以气泡的形式出现在触发它的按钮的周围，气泡的箭头一般指向弹出它的按钮或视图。此外，它还没有“取消”按钮，即便是在程序代码中定义了“取消”按钮，也不会显示它。

提示 图4-34出现的气泡的样式视图被称为“浮动层”视图，气泡的箭头是浮动层的Anchor Point（锚点）。浮动层视图主要应用于iPad和iPhone Plus等大屏幕设备。

图4-33 iPhone中的操作表

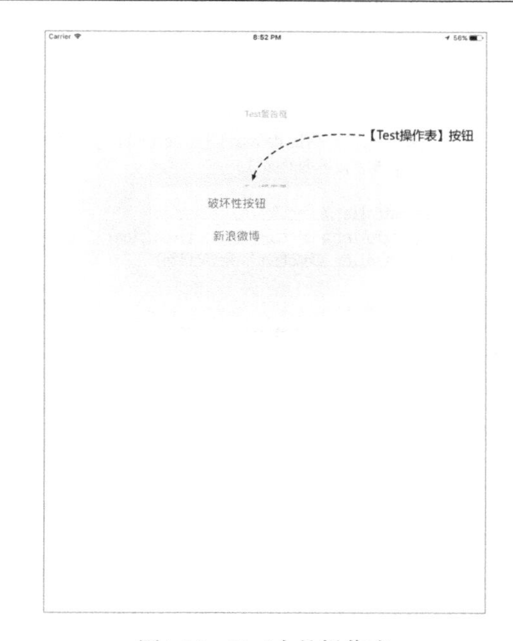

图4-34 iPad中的操作表

上一节中，我们在AlertViewActionSheet案例中添加了警告框，下面来添加操作表。ViewController的主要代码如下：

```
//ViewController.swift文件
import UIKit
class ViewController: UIViewController {

    override func viewDidLoad() {
        super.viewDidLoad()

        let screen = UIScreen.main.bounds
        ......
        let buttonActionSheet = UIButton(type: UIButtonType.system)          ①
        buttonActionSheet.setTitle("Test操作表", for: UIControlState())

        let buttonActionSheetWidth: CGFloat = 100
        let buttonActionSheetHeight: CGFloat = 30
        let buttonActionSheetTopView: CGFloat = 260

        buttonActionSheet.frame = CGRect(x: (screen.size.width -
        ➥buttonActionSheetWidth)/2 ,
        ➥y: buttonActionSheetTopView, width: buttonActionSheetWidth,
        ➥height: buttonActionSheetHeight)
        //指定事件处理方法
        buttonActionSheet.addTarget(self,
        ➥action: #selector(testActionSheet(_:)), for: .touchUpInside)

        self.view.addSubview(buttonActionSheet)                               ②
    }
    ......
    func testActionSheet(_ sender: AnyObject) {

        let actionSheetController = UIAlertController()                       ③
        let cancelAction = UIAlertAction(title: "取消", style:
        ➥UIAlertActionStyle.cancel) { (alertAction) -> Void in
            print("Tap 取消 Button")
        }

        let destructiveAction = UIAlertAction(title: "破坏性按钮",
        ➥style: UIAlertActionStyle.destructive) { (alertAction) -> Void in
            print("Tap 破坏性按钮 Button")
        }

        let otherAction = UIAlertAction(title: "新浪微博",
        ➥style: UIAlertActionStyle.default) { (alertAction) -> Void in
            print("Tap 新浪微博 Button")
        }
        actionSheetController.addAction(cancelAction)
        actionSheetController.addAction(destructiveAction)
        actionSheetController.addAction(otherAction)

        //为iPad设备设置锚点
        actionSheetController.popoverPresentationController?.sourceView =
        ➥sender as ? UIView                                                   ④
        //显示
        self.presentViewController(actionSheetController, animated: true,
        ➥completion: nil)
    }
    ......
}
```

```
//ViewController.m文件
@implementation ViewController

- (void)viewDidLoad {
    [super viewDidLoad];

    CGRect screen = [[UIScreen mainScreen] bounds];
    ......
    UIButton* buttonActionSheet = [UIButton
    ➥buttonWithType:UIButtonTypeSystem];                                      ①
    [buttonActionSheet setTitle:@"Test操作表" forState:UIControlStateNormal];

    CGFloat buttonActionSheetWidth = 100;
    CGFloat buttonActionSheetHeight = 30;
    CGFloat buttonActionSheetTopView = 260;

    buttonActionSheet.frame = CGRectMake((screen.size.width -
    ➥buttonActionSheetWidth)/2 ,
    ➥buttonActionSheetTopView, buttonActionSheetWidth,
    ➥buttonActionSheetHeight);
    //指定事件处理方法
    [buttonActionSheet addTarget:self action:@selector(testActionSheet:)
    ➥forControlEvents: UIControlEventTouchUpInside];
    [self.view addSubview:buttonActionSheet];                                 ②
    ......

}
......
- (void)testActionSheet:(id)sender {

    UIAlertController* actionSheetController =
    ➥[[UIAlertController alloc] init];                                        ③

    UIAlertAction* cancelAction = [UIAlertAction actionWithTitle:@"取消"
    ➥style:UIAlertActionStyleCancel handler:^(UIAlertAction *action) {
        NSLog(@"Tap 取消 Button");
    }];

    UIAlertAction* destructiveAction = [UIAlertAction actionWithTitle:@"
    ➥破坏性按钮" style:UIAlertActionStyleDestructive
    ➥handler:^(UIAlertAction *action) {
        NSLog(@"Tap 破坏性按钮 Button");
    }];

    UIAlertAction* otherAction = [UIAlertAction actionWithTitle:@"新浪微博"
    ➥style:UIAlertActionStyleDefault handler:^(UIAlertAction *action) {
        NSLog(@"Tap 新浪微博 Button");
    }];

    [actionSheetController addAction:cancelAction];
    [actionSheetController addAction:destructiveAction];
    [actionSheetController addAction:otherAction];

    //为iPad设备浮动层设置锚点
    actionSheetController.popoverPresentationController.sourceView = sender; ④
    //显示
    [self presentViewController:actionSheetController animated:true
    ➥completion:nil];
}
......
@end
```

上述代码中，第①行~第②行创建并设置“Test操作表”按钮控件，具体代码前面已经解释了，这里不再赘述。

第③行创建并初始化UIAlertController对象，构造函数是默认的init，UIAlertController的默认样式是操作表。

第④行用于iPad设备浮动层设置锚点。在iPad设备下运行操作表时会以浮动层视图方式显示。我们需要设置actionSheetController的popoverPresentationController属性，该属性是UIPopoverPresentationController类型。UIPopoverPresentationController是浮动层视图控制器。要设置浮动层锚点，可以通过UIPopoverPresentationController的如下3个属性之一设置即可。

- **barButtonItem**。设置导航栏按钮（UIBarButtonItem）作为锚点。关于UIBarButtonItem的内容，我们会在4.9节中介绍。
- **sourceView**。设置一个视图（UIView）作为锚点，本例中我们设置了该属性，见第④行的sender，它是事件源，即我们点击的按钮。按钮也是一个视图。
- **sourceRect**。指定一个区域（CGRect）作为锚点。

4.8 等待相关的控件与进度条

在请求完成之前，我们经常会用到活动指示器ActivityIndicatorView和进度条ProgressView，其中活动指示器可以消除用户的心理等待时间，而进度条可以指示请求的进度。

下面我们通过一个案例讲解这两个控件，其原型设计草图如图4-35所示，其中有两个按钮——Upload和Download，分别对应于活动指示器和进度条。点击Upload按钮，活动指示器开始旋转，再次点击该按钮，它会停止旋转；点击Download按钮，进度条开始前进，完成时弹出一个对话框。

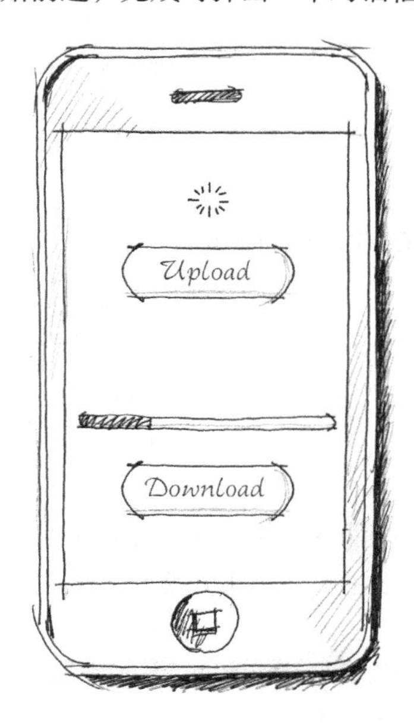

图4-35 UploadDownload案例原型设计草图

4.8.1 活动指示器 ActivityIndicatorView

如果不知道什么时候结束任务，我们可以使用活动指示器，活动指示器类是UIActivityIndicatorView。

1. Interface Builder实现

使用Single View Application模板创建一个名为UploadDownload的工程（具体创建过程请参见2.1.1节）。打开Interface Builder设计界面，在视图上拖曳一个ActivityIndicatorView控件和一个按钮，将按钮命名为Upload。为了与白色的活动指示器区分，我们将后面的视图背景设置为黑色。在如图4-36所示的ActivityIndicatorView控件的属性检查器中，选择Style为Large White。Style有3个：Large White、White和Gray。另外，Behavior属性中的Animating被选中后，运行时控件会处于活动状态。Behavior属性中的Hides When Stopped被选中后，当控件处于非活动状态时，控件会隐藏。

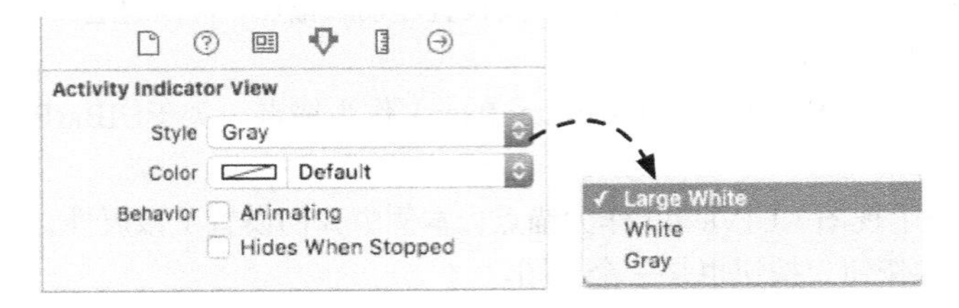

图4-36　ActivityIndicatorView控件的属性检测器

打开Interface Builder，实现按钮的动作方法和活动指示器的输出口属性。ViewController中的相关代码如下：

```
//ViewController.swift文件
class ViewController: UIViewController {

    @IBOutlet weak var activityIndicatorView: UIActivityIndicatorView!

    @IBAction func startToMove(_ sender: AnyObject) {

    }
}
```

```
//ViewController.m文件
@interface ViewController ()

@property (weak, nonatomic) IBOutlet UIActivityIndicatorView
➥*activityIndicatorView;

@end

@implementation ViewController

- (IBAction)startToMove:(id)sender {

}

@end
```

在ViewController中，点击Upload按钮的实现代码如下：

```
@IBAction func startToMove(_ sender: AnyObject) {
    if (self.activityIndicatorView.isAnimating()) {
        self.activityIndicatorView.stopAnimating()
    }else{
        self.activityIndicatorView.startAnimating()
    }
}
```

```
- (IBAction)startToMove:(id)sender {
    if ([self.activityIndicatorView isAnimating]) {
        [self.activityIndicatorView stopAnimating];
    }else{
        [self.activityIndicatorView startAnimating];
    }
}
```

在上述代码中，isAnimating方法用于判断ActivityIndicatorView是否处于活动状态，stopAnimating方法用于停止旋转（进入非活动状态），startAnimating方法用于开始旋转（进入活动状态）。

2. 代码实现

上一节中，我们使用Interface Builder在UploadDownload案例中添加了ActivityIndicatorView控件，这一节我们介绍如何通过代码添加ActivityIndicatorView控件。

其他步骤不再赘述，这里我们重点介绍一下视图控制器类ViewController。ViewController的主要代码如下：

```
//ViewController.swift文件
class ViewController: UIViewController {
    var activityIndicatorView: UIActivityIndicatorView!

    override func viewDidLoad() {
        super.viewDidLoad()

        self.view.backgroundColor = UIColor.black

        let screen = UIScreen.main.bounds
```

```
//ViewController.m文件
#import "ViewController.h"
@interface ViewController ()
@property (strong, nonatomic) UIActivityIndicatorView *activityIndicatorView;
@end

@implementation ViewController

- (void)viewDidLoad {
    [super viewDidLoad];
```

```
        ///1.获得指示器
        self.activityIndicatorView =
        ➥UIActivityIndicatorView(activityIndicatorStyle: .whiteLarge)      ①
        var frame = self.activityIndicatorView.frame
        frame.origin = CGPoint(x: (screen.size.width - frame.size.width) / 2,
        ➥y: 84)
        //重新设置控件的位置
        self.activityIndicatorView.frame = frame

        self.activityIndicatorView.hidesWhenStopped = false                  ②
        self.view.addSubview(self.activityIndicatorView)

        ///2.Upload按钮
        let buttonUpload = UIButton(type: UIButtonType.system)
        buttonUpload.setTitle("Upload", for: UIControlState())

        let buttonUploadWidth:CGFloat = 50
        let buttonUploadHeight:CGFloat = 30
        let buttonUploadTopView:CGFloat = 190

        buttonUpload.frame = CGRect(x: (screen.size.width - buttonUploadWidth)/2 ,
        ➥y: buttonUploadTopView, width: buttonUploadWidth, height:
        ➥buttonUploadHeight)
        //指定事件处理方法
        buttonUpload.addTarget(self, action: #selector(startToMove(_:)),
        ➥for: .touchUpInside)
        self.view.addSubview(buttonUpload)
        ......
    }

    func startToMove(sender: AnyObject) {
        if (self.activityIndicatorView.isAnimating()) {
            self.activityIndicatorView.stopAnimating()
        }else{
            self.activityIndicatorView.startAnimating()
        }
    }
    ......
}
```

```
    self.view.backgroundColor = [UIColor blackColor];

    CGRect screen = [[UIScreen mainScreen] bounds];

    ///1.获得指示器
    self.activityIndicatorView = [[UIActivityIndicatorView alloc] initWith
    ➥ActivityIndicatorStyle: UIActivityIndicatorViewStyleWhiteLarge];     ①

    CGRect frame = self.activityIndicatorView.frame;
    frame.origin = CGPointMake((screen.size.width - frame.size.width) / 2, 84);
    //重新设置控件的位置
    self.activityIndicatorView.frame = frame;

    self.activityIndicatorView.hidesWhenStopped = false;                  ②
    [self.view addSubview:self.activityIndicatorView];

    ///2.Upload按钮
    UIButton* buttonUpload = [UIButton buttonWithType:UIButtonTypeSystem];
    [buttonUpload setTitle:@"Upload" forState:UIControlStateNormal];

    CGFloat buttonUploadWidth = 50;
    CGFloat buttonUploadHeight = 30;
    CGFloat buttonUploadTopView = 190;

    buttonUpload.frame = CGRectMake((screen.size.width - buttonUpload
    ➥Width)/2 , buttonUploadTopView, buttonUploadWidth,
    ➥buttonUploadHeight);

    //指定事件处理方法
    [buttonUpload addTarget:self action:@selector(startToMove:)
    ➥forControlEvents:UIControlEventTouchUpInside];
    [self.view addSubview:buttonUpload];
    ......
}
- (void)startToMove:(id)sender {
    if ([self.activityIndicatorView isAnimating]) {
        [self.activityIndicatorView stopAnimating];
    }else{
        [self.activityIndicatorView startAnimating];
    }
}
......
@end
```

上述代码中，第①行实例化活动指示器UIActivityIndicatorView，使用的构造函数是initWithActivity-IndicatorStyle:，参数是设置活动指示器的样式，参数类型是UIActivityIndicatorViewStyle枚举类型，其枚举成员说明见表4-6。

表4-6 UIActivityIndicatorViewStyle枚举成员

Objective-C枚举成员	Swift枚举成员	说　明
UIActivityIndicatorViewStyleWhiteLarge	whiteLarge	白色且大样式
UIActivityIndicatorViewStyleWhite	white	白色样式
UIActivityIndicatorViewStyleGray	gray	灰色样式

第②行用于设置ActivityIndicatorView的hidesWhenStopped属性，该属性是图4-36所示Behavior中的Hides When Stopped属性。若hidesWhenStopped属性设置为true，则当控件处于非活动状态时，控件会隐藏。

4.8.2 进度条 ProgressView

进度条体现了任务执行的进度，同活动指示器一样，也有消除用户心理等待时间的作用。

为了模拟真实的任务进度变化，我们在案例中引入了定时器，可以在特定的时间间隔后向某对象发出消息。

1. Interface Builder实现

打开Interface Builder，实现按钮的动作和进度条的输出口，ViewController中的相关代码如下：

```
//ViewController.swift文件
class ViewController: UIViewController {

    @IBOutlet weak var progressView: UIProgressView!
    var timer: Timer!

    @IBAction func downloadProgress(_ sender: AnyObject) {

    }

}
```

```
//ViewController.m文件
@interface ViewController ()

@property (weak, nonatomic) IBOutlet UIProgressView *progressView;
@property(nonatomic,strong) NSTimer *timer;

@end

@implementation ViewController

- (IBAction)downloadProgress:(id)sender {

}

@end
```

其中Download按钮的实现代码如下：

```
//ViewController.swift文件
class ViewController: UIViewController {
    ......
    @IBAction func downloadProgress(_ sender: AnyObject) {
        self.timer = Timer.scheduledTimer(timeInterval: 1.0, target: self,
                                          selector: #selector(download),
                                          userInfo: nil, repeats: true)   ①
    }

    func download() {                                                     ②
        self.progressView.progress = self.progressView.progress + 0.1
        if (self.progressView.progress == 1.0) {
            self.timer.invalidate()

            let alertController: UIAlertController =
            ➥UIAlertController(title: "download completed! ",
            ➥message: "", preferredStyle: .alert)                          ③
            let okAction = UIAlertAction(title: "OK", style: .default,
            ➥handler:nil)
            alertController.addAction(okAction)

            //显示
            self.present(alertController, animated: true, completion: nil) ④
        }
    }
}
```

```
//ViewController.m文件
- (IBAction)downloadProgress:(id)sender {

    self.timer = [NSTimer scheduledTimerWithTimeInterval:1.0
                                                  target:self
                                                selector:@selector(download)
                                                userInfo:nil repeats:YES];    ①
}

-(void)download {                                                            ②
    self.progressView.progress = self.progressView.progress+0.1;
    if (self.progressView.progress == 1.0) {
        [self.timer invalidate];

        UIAlertController* alertController  = [UIAlertController
        ➥alertControllerWithTitle:@"download completed! "
        ➥message: @"" preferredStyle:UIAlertControllerStyleAlert];           ③

        UIAlertAction* okAction = [UIAlertAction actionWithTitle:@"Ok"
        ➥style:UIAlertActionStyleCancel handler:nil];

        [alertController addAction:okAction];

        //显示
        [self presentViewController:alertController animated:true
        ➥completion:nil];                                                    ④
    }

}
```

上述代码中，第①行中使用了NSTimer类，Swift版本是Timer，它是定时器类，它的静态方法是+ scheduled-TimerWithTimeInterval:target:selector:userInfo:repeats:，在Swift版本中使用的静态方法scheduledTimer(timeInterval: target:selector:userInfo:repeats:)。该方法可以在给定的时间间隔调用指定的方法。其中，第一个参数用于设定间隔时间；第二个参数target用于指定发送消息给哪个对象；第三个参数selector指定要调用的方法名，相当于一个函数指针；第四个参数userInfo可以给消息发送参数；第五个参数repeats表示是否重复。

第②行的download方法是定时器调用的方法，在定时器完成任务后一定要停止它，这可以通过调用定时器的invalidate方法来实现。第③行用于创建UIAlertController对象。第④行用于显示警告框视图，这在4.7节已经详细介绍过。

2. 代码实现

上一节中，我们使用Interface Builder在UploadDownload案例中添加了ProgressView控件，这一节介绍如何通过代码添加ProgressView控件。

其他步骤不再赘述，这里重点介绍一下视图控制器类ViewController，其主要代码如下：

```
//ViewController.swift文件
class ViewController: UIViewController {

    var progressView: UIProgressView!
    var timer: Timer!

    override func viewDidLoad() {
        super.viewDidLoad()
        ......

        ///3.进度条
        let progressViewWidth:CGFloat = 200
        let progressViewHeight:CGFloat = 2
        let progressViewTopView:CGFloat = 283

        self.progressView = UIProgressView(frame:
        ➥CGRectMake((screen.size.width - progressViewWidth)/2 ,
        ➥progressViewTopView, progressViewWidth, progressViewHeight))

        self.view.addSubview(self.progressView)

        ///4.Download按钮
        let buttonDownload = UIButton(type: UIButtonType.system)
        buttonUpload.setTitle("Upload", for: UIControlState())

        let buttonDownloadWidth:CGFloat = 69
        let buttonDownloadHeight:CGFloat = 30
        let buttonDownloadTopView:CGFloat = 384

        buttonDownload.frame = CGRect(x: (screen.size.width -
        ➥buttonUploadWidth)/2 , y: buttonUploadTopView,
        ➥width: buttonUploadWidth, height: buttonUploadHeight)
        //指定事件处理方法
        buttonUpload.addTarget(self, action: #selector(startToMove(_:)),
        ➥for: .touchUpInside)

        self.view.addSubview(buttonDownload)

    }
    ......

    func downloadProgress(_ sender: AnyObject) {
        self.timer = Timer.scheduledTimer(timeInterval: 1.0, target: self,
                                    selector: #selector(download),
                                    userInfo: nil, repeats: true)
    }

    func download() {
        self.progressView.progress = self.progressView.progress + 0.1
        if (self.progressView.progress == 1.0) {
            self.timer.invalidate()

      let alertController: UIAlertController = UIAlertController(title:
      ➥"download completed! ", message: "", preferredStyle: .alert)
            let okAction = UIAlertAction(title: "OK", style: .default,
            ➥handler: nil)
            alertController.addAction(okAction)
```

```
//ViewController.m文件

#import "ViewController.h"

@interface ViewController ()

@property (strong, nonatomic) UIProgressView *progressView;
@property(nonatomic, strong) NSTimer *timer;

@end

@implementation ViewController

- (void)viewDidLoad {
    [super viewDidLoad];
    ......

    ///3.进度条
    CGFloat progressViewWidth = 200;
    CGFloat progressViewHeight = 2;
    CGFloat progressViewTopView = 283;

    self.progressView = [[UIProgressView alloc] initWithFrame:CGRectMake
    ➥((screen.size.width -
    ➥progressViewWidth)/2 , progressViewTopView,
    ➥progressViewWidth, progressViewHeight)];

    [self.view addSubview: self.progressView];

    ///4.Download按钮
    UIButton* buttonDownload = [UIButton buttonWithType:UIButtonTypeSystem];
    [buttonDownload setTitle:@"Download" forState:UIControlStateNormal];

    CGFloat buttonDownloadWidth = 69;
    CGFloat buttonDownloadHeight = 30;
    CGFloat buttonDownloadTopView = 384;

    buttonDownload.frame = CGRectMake((screen.size.width -
    ➥buttonDownloadWidth)/2 , buttonDownloadTopView, buttonDownloadWidth,
    ➥buttonDownloadHeight);
    //指定事件处理方法
    [buttonDownload addTarget:self action:@selector(downloadProgress:)
    ➥forControlEvents: UIControlEventTouchUpInside];
    [self.view addSubview:buttonDownload];


}
......

- (void)downloadProgress:(id)sender {
    self.timer = [NSTimer scheduledTimerWithTimeInterval:1.0
                            target:self
                            selector:@selector(download)
                            userInfo:nil
                            repeats:TRUE];
}
```

```
            //显示
            self.present(alertController, animated: true, completion: nil)
        }
    }

}
```

```
-(void)download{
    self.progressView.progress = self.progressView.progress + 0.1;
    if (self.progressView.progress == 1.0) {
        [self.timer invalidate];

        UIAlertController* alertController  = [UIAlertController
        ➥alertControllerWithTitle:@"download completed! " message: @""
        ➥preferredStyle:UIAlertControllerStyleAlert];

        UIAlertAction* okAction = [UIAlertAction actionWithTitle:@"Ok"
        ➥style:UIAlertActionStyleCancel handler:nil];

        [alertController addAction:okAction];

        //显示
        [self presentViewController:alertController animated:true
        ➥completion:nil];
    }

}

@end
```

上述代码前面基本都已经解释了，这里不再赘述。

4.9　工具栏和导航栏

工具栏和导航栏的应用有很大差别，但是有一个共同的特性，那就是其中都可以放置UIBarButtonItem。UIBarButtonItem是工具栏和导航栏中的按钮，在事件响应方面与UIButton类似。

4.9.1　工具栏

工具栏类为UIToolbar。在iPhone中，工具栏位于屏幕底部，如果是竖屏布局工具栏，按钮数不能超过5个，否则第5个按钮（即屏幕中显示的最后一个）是“更多”按钮，如图4-37所示。在iPad中，工具栏位于屏幕顶部，按钮的数量没有限制。

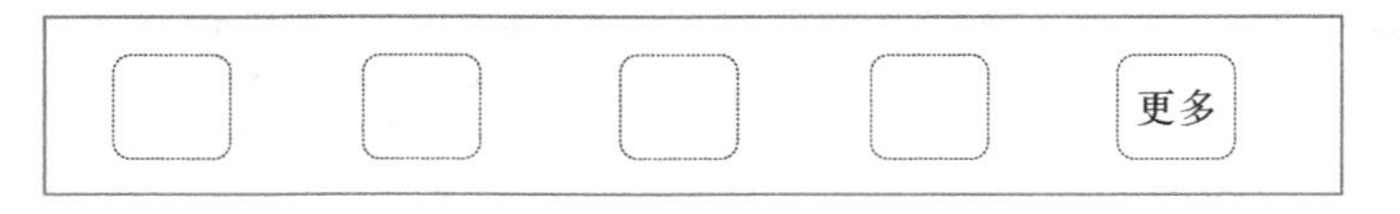

图4-37　iPhone工具栏中的按钮

工具栏是工具栏按钮（UIBarButtonItem）的容器。在UIBarButtonItem中，除了我们看到的按钮，还有“固定空格”和“可变空格”，它们的作用是在各个按钮之间插入一定的空间，如图4-38所示。这样处理以后，工具栏给用户的视觉效果会更好。

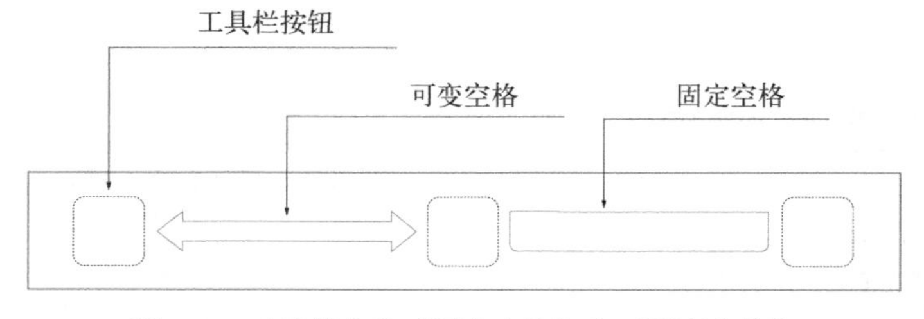

图4-38　工具栏中的“固定空格”和“可变空格”

在工具栏中，我们除了可以放置UIBarButtonItem，还可以放置其他自定义视图，但这种操作只在特殊情况下使用。下面我们用一个案例（其原型设计图如图4-39所示）来介绍一下工具栏的用法。其中，工具栏中有两个按钮Save和Open，界面中央有一个标签，点击Save和Open按钮均会改变标签的内容。

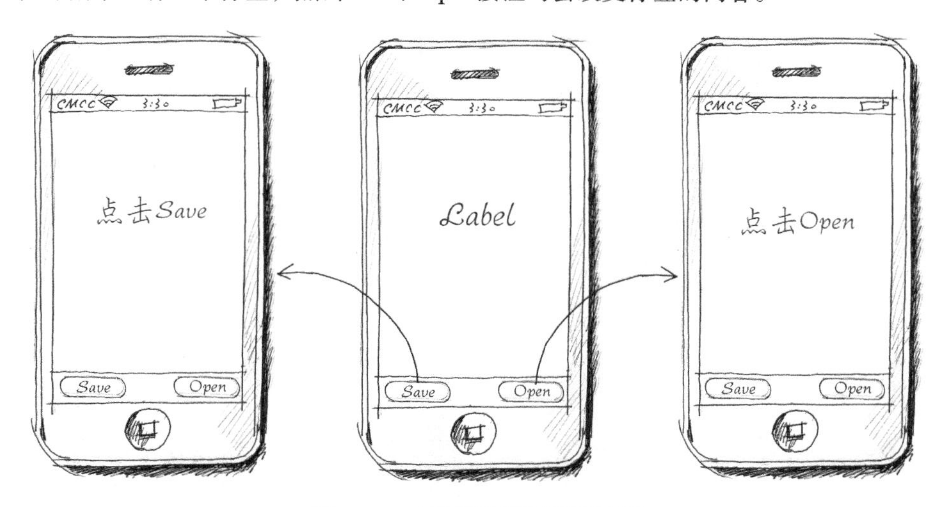

图4-39　工具栏案例原型设计图

1. Interface Builder实现

使用Single View Application模板创建一个名为ToolbarSample的工程。打开Interface Builder设计界面，摆放两个按钮控件，如图4-40所示，从对象库中拖曳一个Toolbar到设计界面底部并将其摆放到合适的位置。

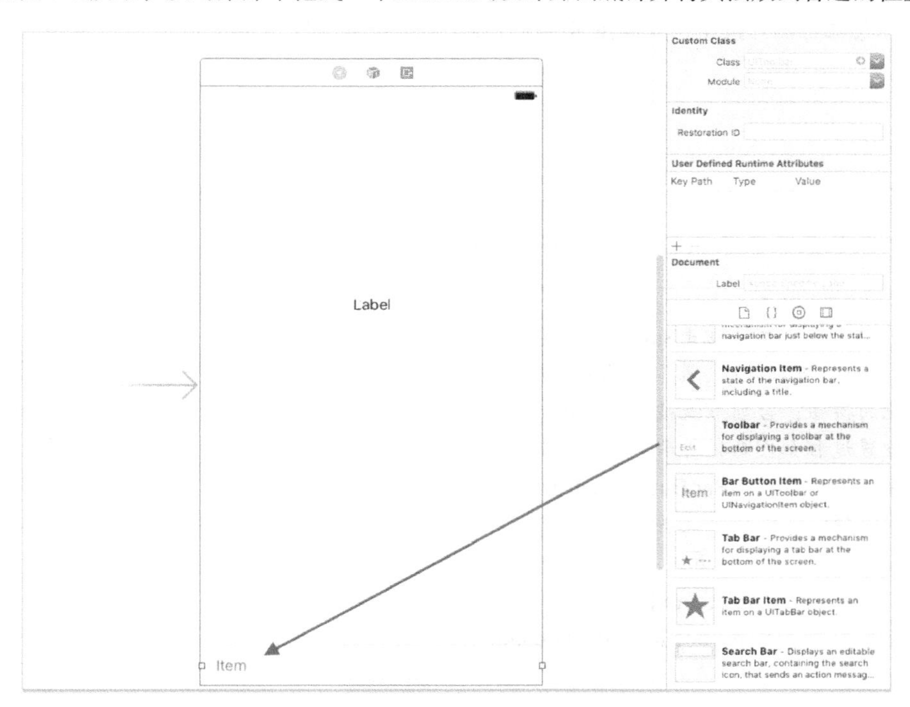

图4-40　在Interface Builder中添加工具栏

再拖曳一个工具栏按钮到工具栏，然后拖曳一个“可变空格”到两个按钮之间，如图4-41所示。

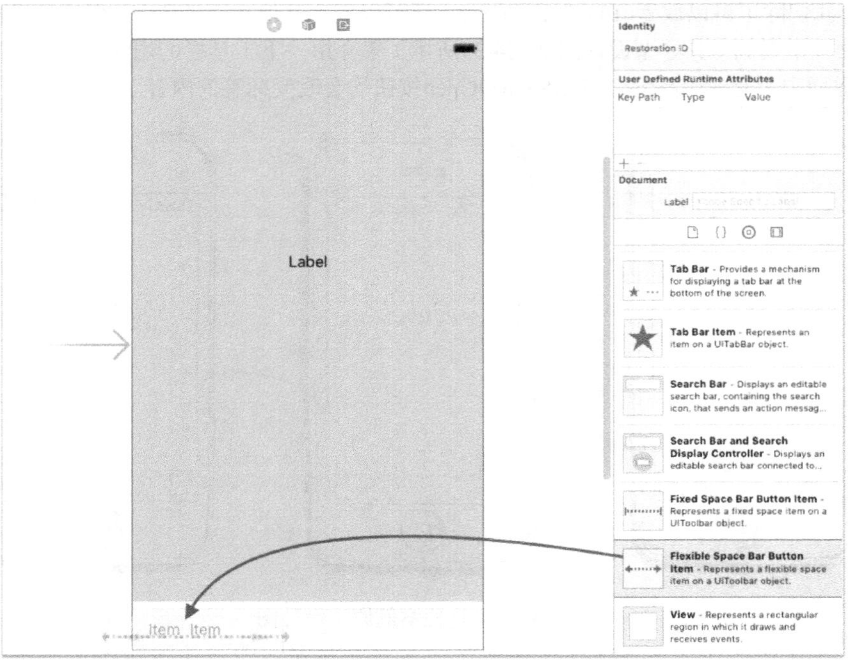

图4-41　在工具栏中添加按钮和“可变空格”

双击选中按钮，修改按钮上的标题。当然，你也可以打开如图4-42所示的属性检查器，直接编辑Bar Item下的Title属性。如果想添加图片按钮，直接在属性检查器中修改Image属性即可。

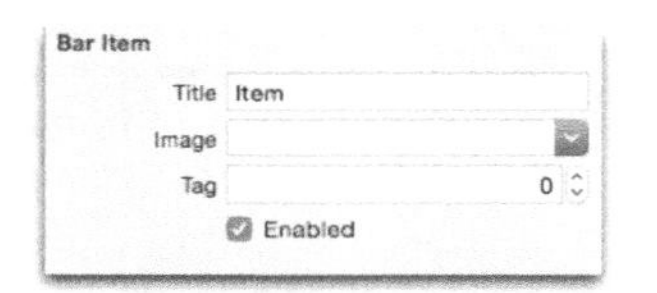

图4-42　工具栏按钮属性检查器

提示

本案例中工具栏左按钮是Save，它比较特殊。它是iOS系统的标准按钮，属于苹果公司规定必须优先处理的按钮，否则可能被拒绝在App Store上发布。设置过程如图4-43所示，选择按钮，然后打开其属性检查器，在Bar Button Item下的Identifier属性中选择Save选项。这些按钮的用途可以在苹果HIG（iOS人机交互开发指南）文档中找到：https://developer.apple.com/ios/human-interface-guidelines/#//apple_ ref/doc/uid/TP40006556。

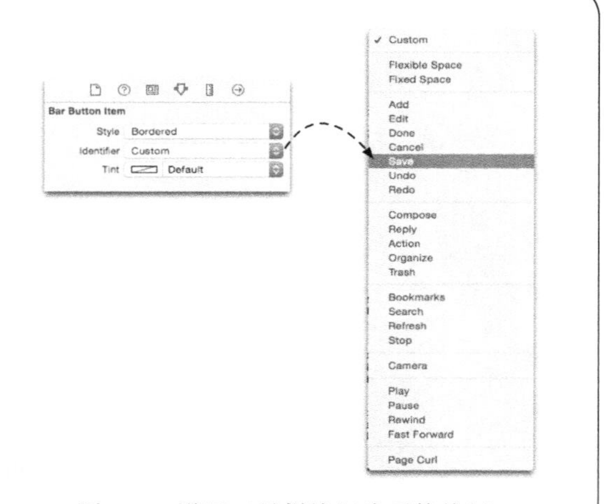

图4-43　设置工具栏按钮为系统按钮

下面我们看看ViewController中的相关代码：

```
//ViewController.swift文件
class ViewController: UIViewController {

    @IBOutlet weak var label: UILabel!

    override func viewDidLoad() {
        super.viewDidLoad()
    }

    override func didReceiveMemoryWarning() {
        super.didReceiveMemoryWarning()
    }

    @IBAction func save(_ sender: AnyObject) {
        self.label.text = "点击Save"
    }

    @IBAction func open(_ sender: AnyObject) {
        self.label.text = "点击Open"
    }
}
```

```
//ViewController.m文件
@interface ViewController ()

@property (weak, nonatomic) IBOutlet UILabel *label;

@end

@implementation ViewController

- (IBAction)save:(id)sender {
    self.label.text = @"点击Save";
}

- (IBAction)open:(id)sender {
    self.label.text = @"点击Open";
}
@end
```

在上述代码中，我们定义了输出口类型的UILabel属性label、用于响应Save按钮点击动作事件的save:方法，以及用于响应Open按钮点击动作事件的open:方法。编写完ViewController代码后，我们还需要通过Interface Builder为输出口和动作事件连线。

2. 代码实现

上一节中，我们使用Interface Builder在ToolbarSample案例中添加了Toolbar，这一节介绍如何通过代码添加Toolbar。

其他步骤不再赘述，我们重点介绍一下视图控制器类ViewController，其主要代码如下：

```
//ViewController.swift文件
class ViewController: UIViewController {

    var label: UILabel!

    override func viewDidLoad() {
        super.viewDidLoad()

        let screen = UIScreen.mainScreen().bounds

        let toolbarHeight:CGFloat = 44 //44是默认高度

        ///1.添加Toolbar
        let toolbar = UIToolbar(frame: CGRect(x: 0, y: screen.size.height -
        ➥toolbarHeight,
        ➥width: screen.size.width, height: toolbarHeight))                    ①

        let saveButtonItem = UIBarButtonItem(barButtonSystemItem: .save,
        ➥target: self,
        ➥action: #selector(ViewController.save(_:)))                          ②

        let openButtonItem = UIBarButtonItem(title: "Open", style: .plain,
        ➥target: self,
        ➥action: #selector(ViewController.open(_:)))                          ③

        let flexibleButtonItem = UIBarButtonItem(barButtonSystem
```

```
//ViewController.m文件
#import "ViewController.h"

@interface ViewController ()

@property (strong, nonatomic) UILabel *label;

@end

@implementation ViewController

- (void)viewDidLoad {
    [super viewDidLoad];

    CGRect screen = [[UIScreen mainScreen] bounds];

    CGFloat toolbarHeight = 44; //44是默认高度

    UIToolbar* toolbar = [[UIToolbar alloc] initWithFrame:CGRectMake(0,
    ➥screen.size.height - toolbarHeight, screen.size.width, toolbarHeight)];
                                                                            ①

    UIBarButtonItem *saveButtonItem = [[UIBarButtonItem alloc]
    ➥initWithBarButtonSystemItem: UIBarButtonSystemItemSave
    ➥target:self action:@selector(save:)];                                  ②

    UIBarButtonItem *openButtonItem = [[UIBarButtonItem alloc]
    ➥initWithTitle:@"Open" style:UIBarButtonItemStylePlain
    ➥target:self action:@selector(open:)];                                  ③
```

```
        ➥Item: .flexibleSpace,
        ➥target: nil, action: nil)                                         ④

        toolbar.items = [saveButtonItem, flexibleButtonItem,
        ➥openButtonItem]                                                  ⑤
        self.view.addSubview(toolbar)

        ///2.添加标签
        let labelWidth:CGFloat = 84
        let labelHeight:CGFloat = 21
        let labelTopView:CGFloat = 250
        self.label = UILabel(frame: CGRect(x: (screen.size.width - labelWidth)/2 ,
        ➥y: labelTopView, width: labelWidth, height: labelHeight))

        self.label.text = "Label"
        //字体左右居中
        self.label.textAlignment = .center
        self.view.addSubview(self.label)

    }

    func save(_ sender: AnyObject) {
        self.label.text = "点击Save"
    }

    func open(_ sender: AnyObject) {
        self.label.text = "点击Open"
    }

}
```

```
    UIBarButtonItem *flexibleButtonItem = [[UIBarButtonItem alloc]
    ➥initWithBarButtonSystemItem: UIBarButtonSystemItemFlexibleSpace
    ➥target:nil action:nil];                                              ④

    toolbar.items = @[saveButtonItem, flexibleButtonItem, openButtonItem]; ⑤
    [self.view addSubview:toolbar];

    CGFloat labelWidth = 84;
    CGFloat labelHeight = 21;
    CGFloat labelTopView = 250;

    self.label = [[UILabel alloc] initWithFrame:CGRectMake((
    ➥screen.size.width - labelWidth)/2 , labelTopView, labelWidth,
    ➥labelHeight)];

    self.label.text = @"Label";
    //字体左右居中
    self.label.textAlignment = NSTextAlignmentCenter;
    [self.view addSubview:self.label];

}

- (void)save:(id)sender {
    self.label.text = @"点击Save";
}

- (void)open:(id)sender {
    self.label.text = @"点击Open";
}

@end
```

上述代码中，第①行用于创建UIToolbar对象。第②行用于创建Save工具栏按钮，类型是UIBarButtonItem，使用了构造函数initWithBarButtonSystemItem:target:action:，其中第一个参数指定系统按钮样式，样式参考图4-43，这些样式是在UIBarButtonSystemItem枚举类型中定义的。本例中Save工具栏按钮是UIBarButtonSystemItem枚举类型中的save成员（Objective-C版本是UIBarButtonSystemItemSave）。

第③行用于创建Open工具栏按钮，它不是系统按钮，因此使用的构造函数是- initWithTitle:style:target:action:，第一个参数是按钮标题。

第④行用于创建如图4-38所示的“可变空格”，它属于系统按钮，样式使用UIBarButtonSystemItem中的flexibleSpace成员（Objective-C版本中是UIBarButtonSystemItemFlexibleSpace）。类似地，“固定空格”样式使用fixedSpace（Objective-C版本中是UIBarButtonSystemItemFixedSpace）。

第⑤行用于设置UIToolbar对象的items属性，它是工具栏中按钮的集合。

4.9.2　导航栏

导航栏主要用于导航，考虑的是整个应用，而工具栏应用于当前界面，考虑的是局部界面。相关类和概念如下所示。

- **UINavigationController**。导航控制器，可以构建树形导航模式应用的根控制器，这将在第9章中介绍。
- **UINavigationBar**。导航栏，它与导航控制器是一对一的关系，管理一个视图控制器的栈，用来显示树形结构中的视图。
- **UINavigationItem**。导航栏项目，在每个界面中都会看到。它分为左、中、右3个区域：左侧区域一般放置一个返回按钮（设定属性是backBarButtonItem）或左按钮（设定属性是leftBarButtonItem）；右侧区域一般放置一个右按钮（设定属性是rightBarButtonItem）；中间区域是标题（属性是title）或者提示信息（属性是prompt）。导航栏与导航栏项目是一对多的关系，如图4-44所示。

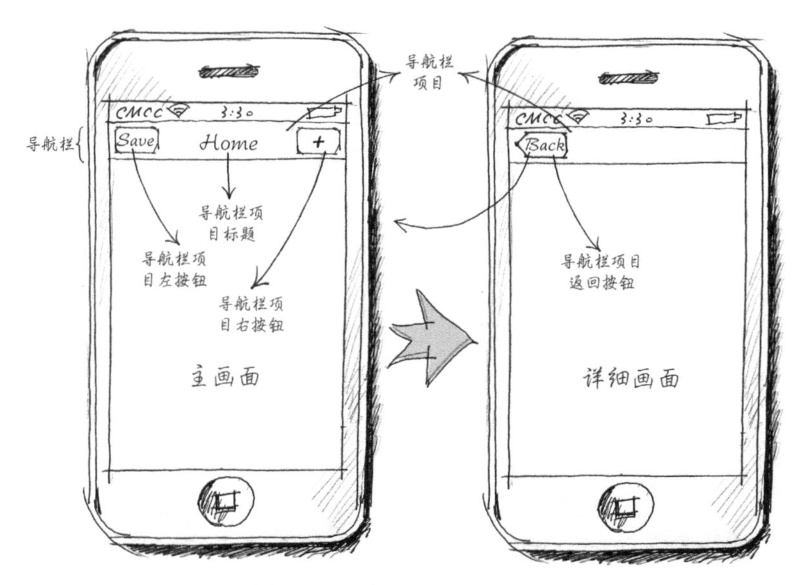

图4-44 导航栏和导航栏项目

- **UIBarButtonItem**。与工具栏中的按钮一样，它是导航栏中的左右按钮。

下面我们用一个案例介绍一下导航栏的用法，该案例的原型设计图如图4-45所示。

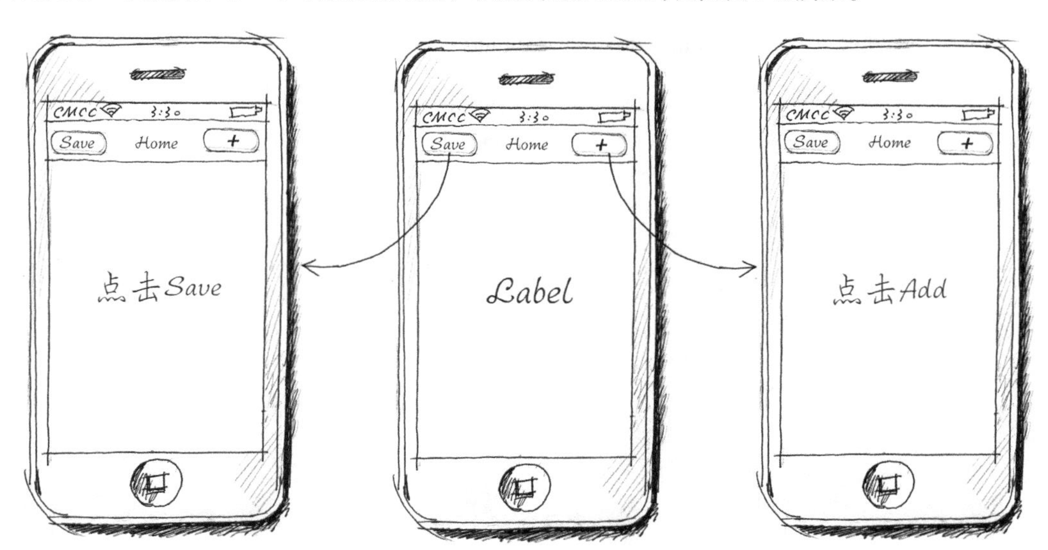

图4-45 导航栏案例原型设计图

该导航栏中共有两个按钮Save和+，点击Save按钮，将标签的内容改为“点击Save”，点击+按钮，将标签的内容改为“点击Add”。需要说明的是，这里的Save和+按钮也是iOS系统标准按钮。

1. Interface Builder实现

使用Single View Application模板创建工程名为NavigationBarSample的应用，然后打开Interface Builder设计界面，并从对象库中拖曳一个Navigation Bar到设计界面顶部（距离视图顶部20点，这样不会遮挡状态栏），并将其摆放到合适的位置，如图4-46所示。

然后在导航栏项目中的左右两个区域分别拖曳一个Bar Button Item，为导航栏项目添加左右按钮，如图4-47所示。

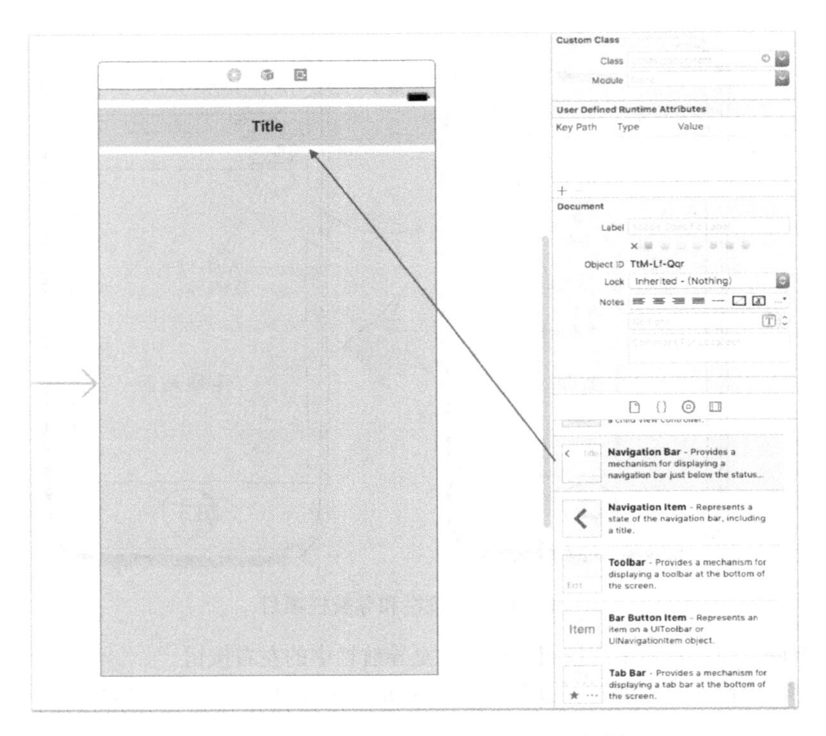

图4-46　在Interface Builder中添加导航栏

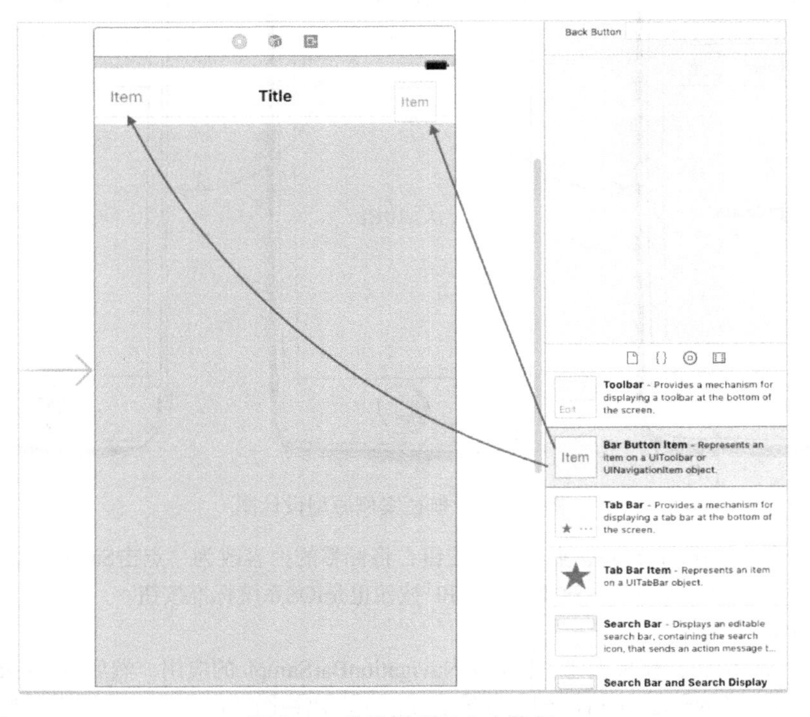

图4-47　导航栏项目左右按钮

左右按钮（即：Save按钮和+按钮）的identifier属性设置，可以参考上一节的Save按钮设置，只是要注意+按钮的identifier属性是Add，具体步骤不再赘述。

选择导航栏项目，打开其属性检查器，将Title属性修改为Home，如图4-48所示。

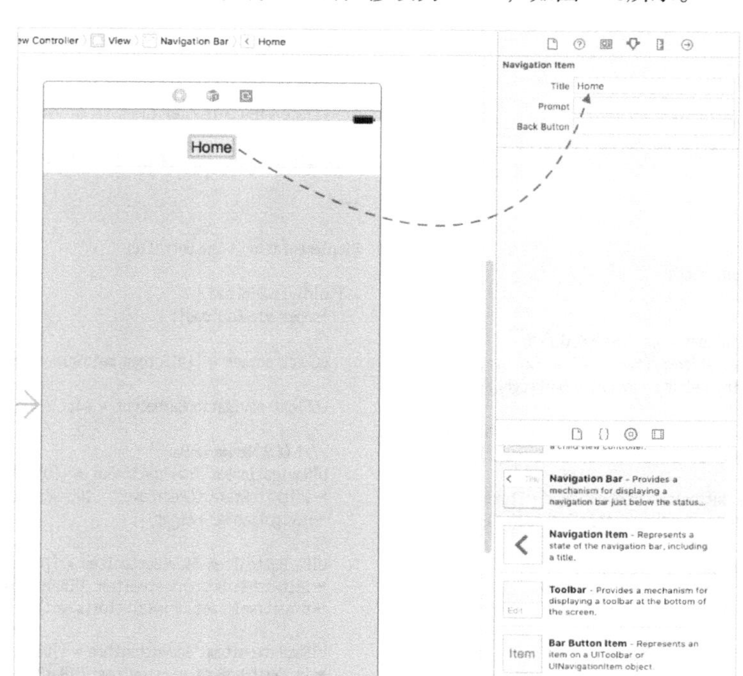

图4-48 修改导航栏项目标题

案例实现代码ViewController的内容如下：

```
//ViewController.swift文件
import UIKit

class ViewController: UIViewController {

    @IBOutlet weak var label: UILabel!

    ......

    @IBAction func save(_ sender: AnyObject) {
        self.label.text = "点击Save"
    }

    @IBAction func add(_ sender: AnyObject) {
        self.label.text = "点击Add"
    }
}
```

```
//ViewController.m文件
@interface ViewController ()

@property (weak, nonatomic) IBOutlet UILabel *label;

@end

@implementation ViewController

......

- (IBAction)save:(id)sender {
    self.label.text = @"点击Save";
}

- (IBAction)add:(id)sender {
    self.label.text = @"点击Add";
}
@end
```

上述代码定义了输出口类型的UILabel属性label、用于响应Save按钮的点击动作事件的save:方法、用于响应+按钮的点击动作事件add:方法。编写完代码后，我们还需要通过Interface Builder为输出口和动作事件连线。

一般情况下，如果涉及导航栏，都是多界面的应用，这是因为导航栏的用途就是导航，而单界面不需要导航。但是本案例中只有一个界面，主要关注导航栏和导航栏项目的用法。关于导航模式的相关内容，我们以后再介绍。

2. 代码实现

上一节中，我们使用Interface Builder在NavigationBarSample案例中添加了NavigationBar，这一节介绍如何通过代码添加NavigationBar。

其他步骤不再赘述，这里重点介绍一下视图控制器类ViewController，其主要代码如下：

```
//ViewController.swift文件
import UIKit

class ViewController: UIViewController {

    var label: UILabel!

    override func viewDidLoad() {
        super.viewDidLoad()

        let screen = UIScreen.main.bounds

        ///1.创建NavigationBar
        let navigationBarHeight:CGFloat = 44 //44是默认高度
        let navigationBar = UINavigationBar(frame: CGRect(x: 0, y: 20,
        ➥width: screen.size.width, height: navigationBarHeight))              ①

        let saveButtonItem = UIBarButtonItem(barButtonSystemItem: .save,
        ➥target: self, action: #selector(ViewController.save(_:)))

        let addButtonItem = UIBarButtonItem(barButtonSystemItem: .add,
        ➥target: self, action: #selector(ViewController.add(_:)))

        let navigationItem = UINavigationItem(title: "")                       ②
        navigationItem.leftBarButtonItem = saveButtonItem                      ③
        navigationItem.rightBarButtonItem = addButtonItem                      ④

        navigationBar.items = [navigationItem]                                 ⑤
        self.view.addSubview(navigationBar)

        ///2.添加标签
        let labelWidth:CGFloat = 84
        let labelHeight:CGFloat = 21
        let labelTopView:CGFloat = 198
        self.label = UILabel(frame: CGRect(x: (screen.size.width -
        ➥labelWidth)/2 ,
        ➥y: labelTopView, width: labelWidth, height: labelHeight))

        self.label.text = "Label"
        //字体左右居中
        self.label.textAlignment = .center
        self.view.addSubview(self.label)

    }


    func save(_ sender: AnyObject) {
        self.label.text = "点击Save"
    }

    func add(_ sender: AnyObject) {
        self.label.text = "点击Add"
    }
}
```

```
//ViewController.m文件
#import "ViewController.h"

@interface ViewController ()

@property (strong, nonatomic) UILabel *label;

@end

@implementation ViewController

- (void)viewDidLoad {
    [super viewDidLoad];

    CGRect screen = [[UIScreen mainScreen] bounds];

    CGFloat navigationBarHeight = 44; //44是默认高度

    ///1.创建NavigationBar
    UINavigationBar *navigationBar = [[UINavigationBar alloc]
    ➥initWithFrame:CGRectMake(0, 20, screen.size.width,
    ➥navigationBar Height)];                                                  ①

    UIBarButtonItem *saveButtonItem = [[UIBarButtonItem alloc]
    ➥initWithBarButtonSystemItem: UIBarButtonSystemItemSave
    ➥target:self action:@selector(save:)];

    UIBarButtonItem *addButtonItem = [[UIBarButtonItem alloc]
    ➥initWithBarButtonSystemItem: UIBarButtonSystemItemAdd
    ➥target:self action:@selector(add:)];

    UINavigationItem *navigationItem = [[UINavigationItem alloc]
    ➥initWithTitle:@""];                                                      ②
    navigationItem.leftBarButtonItem = saveButtonItem;                         ③
    navigationItem.rightBarButtonItem = addButtonItem;                         ④

    navigationBar.items = @[navigationItem];                                   ⑤

    [self.view addSubview:navigationBar];

    ///2.添加标签
    CGFloat labelWidth = 84;
    CGFloat labelHeight = 21;
    CGFloat labelTopView = 198;

    self.label = [[UILabel alloc] initWithFrame:CGRectMake((screen.size.width -
    ➥labelWidth)/2 , labelTopView, labelWidth, labelHeight)];

    self.label.text = @"Label";
    //字体左右居中
    self.label.textAlignment = NSTextAlignmentCenter;
    [self.view addSubview:self.label];

}

- (void)didReceiveMemoryWarning {
    [super didReceiveMemoryWarning];
}

- (void)save:(id)sender {
    self.label.text = @"点击Save";
}

- (void)add:(id)sender {
    self.label.text = @"点击Add";
}

@end
```

上述代码中，第①行用于创建UINavigationBar导航栏对象，该对象中可以放置多个UINavigationItem导航栏项目对象。第②行用于创建导航栏项目对象，使用的构造函数是initWithTitle:，参数是显示在导航栏项目中间的文字。导航栏项目对象还可以放置左右两个按钮，导航栏按钮与工具栏按钮一样都是UIBarButtonItem类型，见第③行和第④行。

第⑤行用于将多个导航栏项目放置在数组中，并赋值导航栏对象属性items。

4.10 小结

本章首先向大家介绍了视图和控件之间的关系以及应用界面的构建层次，然后介绍了标签、按钮、文本框、文本视图、开关、滑块、分段控件、网页控件、活动指示器、进度条、警告框、操作表、工具栏、导航栏等基本控件。

第5章 委托协议、数据源协议与高级视图

在macOS和iOS应用开发中，一些高级视图会用到数据源协议和委托协议。本章中，我们将介绍数据源协议、委托协议以及一些高级视图（选择器和集合视图）。

5.1 视图中的委托协议和数据源协议

一些高级视图中的功能比较复杂，我们需要将这些复杂功能从视图本身剥离开来，由另外一些类完成。当然，这些类需要遵守视图指定的协议。相对简单一点儿的视图只指定了委托协议，这些视图我们在第4章中介绍过，包括UITextField和UITextView，对应的委托协议命名规则是“视图名 + Delegate”，这两个视图对应的委托协议是UITextFieldDelegate和UITextViewDelegate。

对于复杂的高级视图，我们同时指定委托协议和数据源协议，这些视图在iOS中不是很多，其中主要有选择器（UIPickerView）、集合视图（UICollectionView）和表视图（UITableView），对应的数据源协议的命名规则是“视图名 + DataSource”。

委托协议是macOS和iOS应用开发中非常重要的设计模式之一。关于委托设计模式的原理，我们将在第11章中详细介绍。视图中的数据源协议与委托协议一样，都是委托设计模式的具体实现，只不过它们的角色不同。委托对象负责控制控件外观，如选择器的宽度、行高等信息。此外，委托对象还负责对控件的事件和状态变化作出反应。数据源对象是控件与应用数据（模型）的桥梁，如选择器的行数、拨轮数等信息。委托中的方法在实现时是可选的，而数据源中的方法一般是必须实现的。

本章中，我们重点介绍选择器和集合视图，而表视图将在下一章中详细介绍。

5.2 选择器

玩过老虎机吗？选择器的外形很像一台老虎机，当你拨动选择器时，它还会像老虎机一样发出“咔咔”的声音，还有真实的拨盘旋转的感觉。虽然它看起来像老虎机，但是它并不是用来娱乐的，而是iOS中的标准控件，主要用于为用户提供选择。在软件领域，有句话很经典：“有输入的地方，就要验证。”当界面是用户注册界面时，其中有一个“出生日期”字段，你是给用户一个文本框吗？如果是，他可能会输入类似“2018-1-18”这样不合法的日期，而我们需要确保输入内容的合法性。为了更方便操作，我们希望用户以选择的方式完成信息输入，此时选择器便应运而生了。

5.2.1 日期选择器

日期是最复杂的，为此iOS推出了UIDatePicker日期选择器，用于实现对日期的选择。日期选择器有4种模式：日期、日期时间、时间和倒计时定时器，如图5-1至图5-4所示。

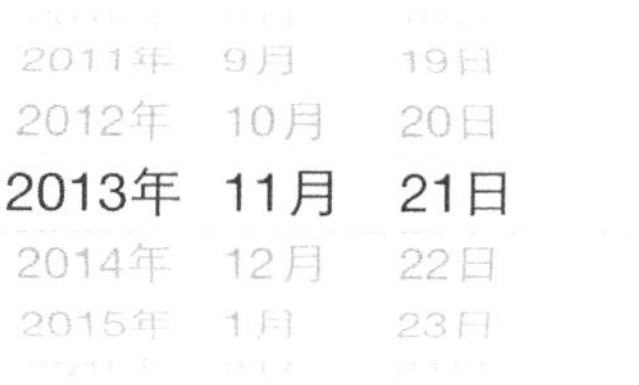

2011年 9月 19日
2012年 10月 20日
2013年 11月 21日
2014年 12月 22日
2015年 1月 23日

图5-1 日期模式

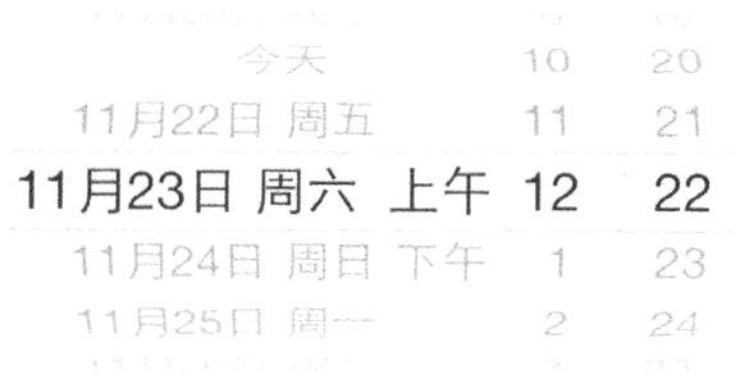

今天 10 20
11月22日 周五 11 21
11月23日 周六 上午 12 22
11月24日 周日 下午 1 23
11月25日 周一 2 24

图5-2 日期时间模式

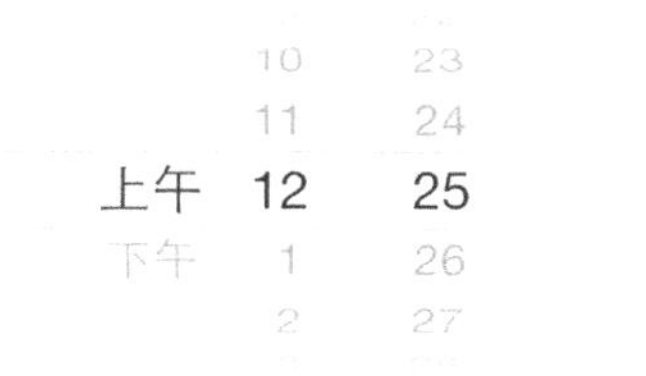

10 23
11 24
上午 12 25
下午 1 26
2 27

图5-3 时间模式

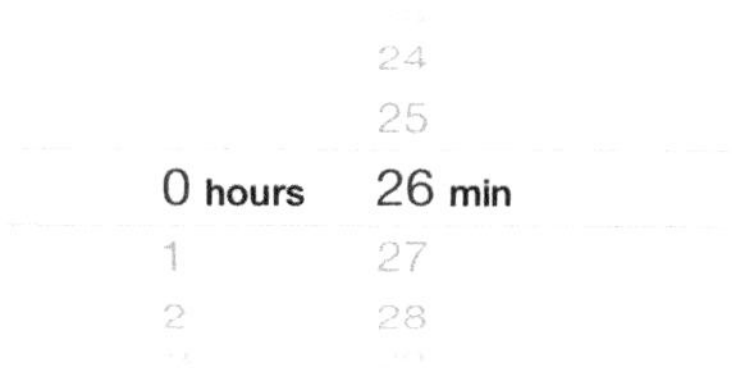

24
25
0 hours 26 min
1 27
2 28

图5-4 倒计时定时器模式

下面我们通过图5-5所示的案例来学习日期选择器，其中有日期选择器、一个标签和一个按钮。点击该按钮，选中的日期将显示在标签上。

CMCC
3:30
January 23 2008
February 24 2009
March 25 2010
April 26 2011
May 27 2012
Label
Button

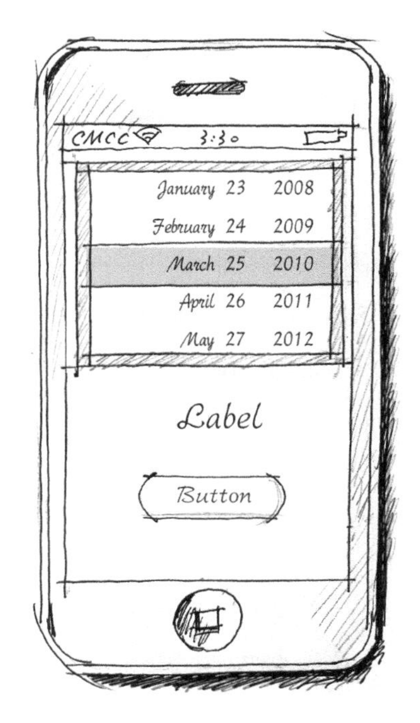

图5-5 DatePickerSample案例原型草图

下面我们具体介绍案例实现过程。

1. Interface Builder实现

使用Single View Application模板创建一个名为DatePickerSample的工程。打开Main.storyboard设计界面，从对象库中拖曳控件到设计界面，如图5-6所示。

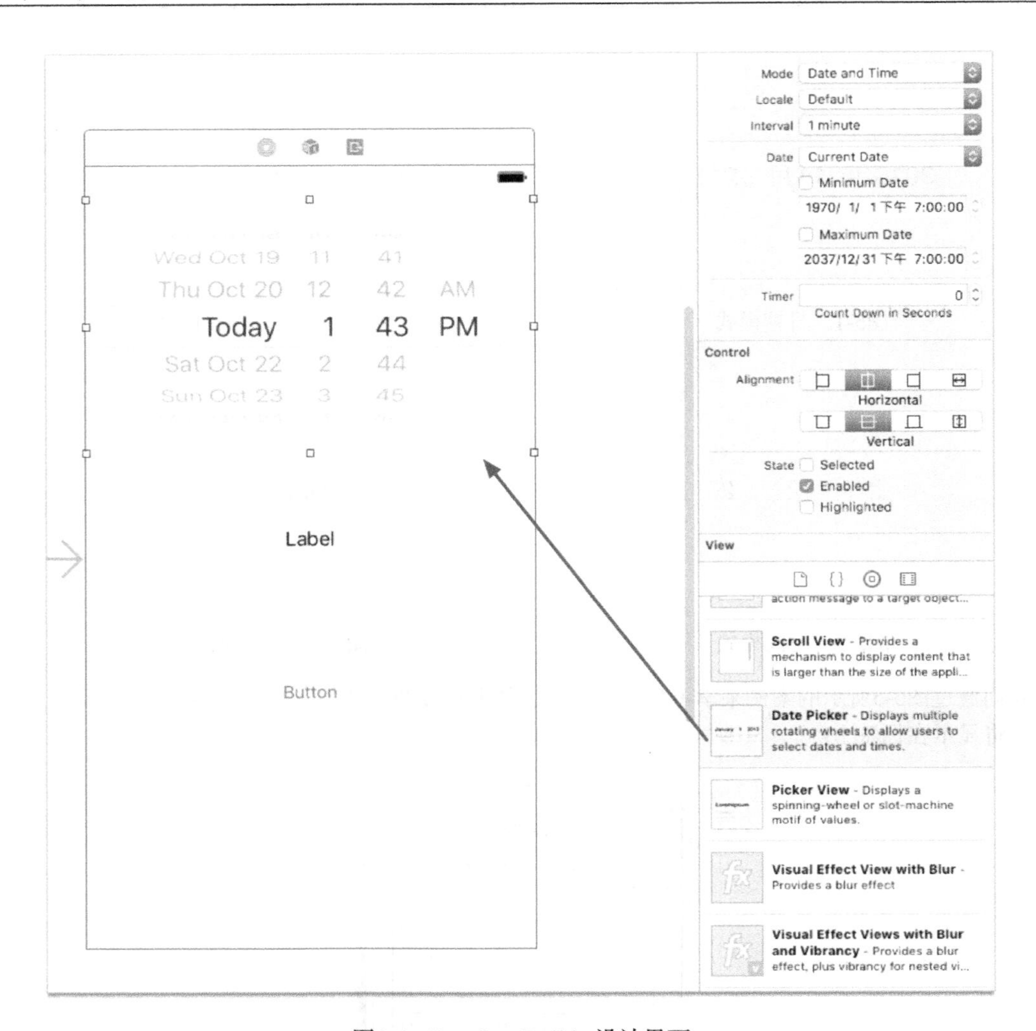

图5-6　Interface Builder设计界面

选择Date Picker，打开其属性检查器，如图5-7所示。

图5-7　Date Picker属性检查器

这些属性项的含义如下所示。

- Mode。设定日期选择器的模式。
- Locale。设定本地化，日期选择器会按照本地习惯和文字显示日期。

- Interval。设定间隔时间，单位为分钟。
- Date。设定开始时间。
- Constraints。设定能显示的最大和最小日期。
- Timer。在倒计时定时器模式下倒计时的秒数。

下面我们看看视图控制器ViewController的相关代码，具体如下：

```
//ViewController.swift文件
class ViewController: UIViewController {

    @IBOutlet weak var datePicker: UIDatePicker!                           ①
    @IBOutlet weak var label: UILabel!                                     ②

    @IBAction func onclick(_ sender: AnyObject) {                          ③

        var theDate : NSDate = self.datePicker.date                        ④
        let desc = theDate.description(with: Locale.current)               ⑤
        print("the date picked is: ", desc)

        var dateFormatter : NSDateFormatter = NSDateFormatter()            ⑥
        dateFormatter.dateFormat = "YYYY-MM-dd HH:mm:ss"                   ⑦
        print("the date formate is: ",
        ➥dateFormatter.stringFromDate(the Date))                          ⑧

        self.label.text = dateFormatter.stringFromDate(theDate)

    }
    ......
}
```

```
//ViewController.m文件
#import "ViewController.h"

@interface ViewController ()

@property (weak, nonatomic) IBOutlet UIDatePicker *datePicker;             ①
@property (weak, nonatomic) IBOutlet UILabel *label;                       ②

@end

@implementation ViewController

- (IBAction)onclick:(id)sender {                                           ③

    NSDate * theDate = self.datePicker.date;                               ④
    NSLog(@"the date picked is: %@", [theDate descriptionWithLocale:
    ➥[NSLocale currentLocale]]);                                          ⑤
    NSDateFormatter * dateFormatter = [[NSDateFormatter alloc] init] ;     ⑥
    dateFormatter.dateFormat=@"YYYY-MM-dd HH:mm:ss";                       ⑦
    NSLog(@"the date formate is: %@",
    ➥[dateFormatter stringFromDate:theDate]);                             ⑧
    self.label.text = [dateFormatter stringFromDate:theDate];

}
@end
```

在上述代码中，我们在第①行和第②行中定义了输出口属性UIDatePicker和UILabel，第③行定义按钮点击事件onclick。

第④行中的self.datePicker.date是通过UIDatePicker的date属性返回NSDate数据，该属性返回选中的时间。第⑤行中，NSDate的descriptionWithLocale:方法返回本地化的日期信息，其中NSLocale的静态方法currentLocale返回当前NSLocale对象（而在Swift 3版本中，NSLocale类变换为Locale，通过current属性返回本地化的日期信息）。

第⑥行用于创建NSDateFormatter对象，它是日期格式对象。第⑦行中，dateFormatter的dateFormat属性用于设置日期格式为YYYY-MM-dd HH:mm:ss。第⑧行中dateFormatter的stringFromDate:方法用于从日期对象返回日期字符串。

2. 代码实现

上一节中，我们使用Interface Builder在DatePickerSample案例中添加了日期选择器，这一节介绍如何通过代码添加日期选择器。

其他步骤不再赘述，这里重点介绍一下视图控制器类ViewController，其主要代码如下：

```
//ViewController.swift文件
class ViewController: UIViewController {

    var datePicker: UIDatePicker!
    var label: UILabel!

    override func viewDidLoad() {
        super.viewDidLoad()

        let screen = UIScreen.mainScreen().bounds
```

```
//ViewController.m文件
#import "ViewController.h"

@interface ViewController ()

@property (strong, nonatomic) UIDatePicker *datePicker;
@property (strong, nonatomic) UILabel *label;

@end

@implementation ViewController
```

```
        ///1.日期选择器
        let datePickerWidth:CGFloat = 320
        let datePickerHeight:CGFloat = 167

        self.datePicker = UIDatePicker(frame: CGRectMake(0, 0,
        ➥datePickerWidth, datePickerHeight))                              ①
        //zh-Hans简体中文
        self.datePicker.locale = NSLocale(localeIdentifier: "zh-Hans")     ②
        //设置日期时间模式
        self.datePicker.datePickerMode = .DateAndTime                      ③

        self.view.addSubview(self.datePicker)

        ///2.添加标签
        let labelWidth:CGFloat = 200
        let labelHeight:CGFloat = 21
        let labelTopView:CGFloat = 281
        self.label = UILabel(frame: CGRectMake((screen.size.width -
        ➥labelWidth)/2 ,labelTopView, labelWidth, labelHeight))

        self.label.text = "Label"
        //字体左右居中
        self.label.textAlignment = .Center
        self.view.addSubview(self.label)

        ///3.Button按钮
        let button = UIButton(type: UIButtonType.System)
        button.setTitle("Button", forState: UIControlState.Normal)

        let buttonWidth:CGFloat = 46
        let buttonHeight:CGFloat = 30
        let buttonTopView:CGFloat = 379

        button.frame = CGRectMake((screen.size.width - buttonWidth)/2 ,
        ➥buttonTopView, buttonWidth, buttonHeight)
        //指定事件处理方法
        button.addTarget(self, action: Selector("onclick:"),
        ➥forControlEvents: .TouchUpInside)

        self.view.addSubview(button)

    }
    ......

    func onclick(sender: AnyObject) {

        let theDate : NSDate = self.datePicker.date
        let desc = theDate.descriptionWithLocale(NSLocale.currentLocale())
        NSLog("the date picked is: %@", desc)

        let dateFormatter : NSDateFormatter = NSDateFormatter()
        dateFormatter.dateFormat = "YYYY-MM-dd HH:mm:ss"
        NSLog("the date formate is: %@", dateFormatter.stringFromDate
        ➥(theDate))

        self.label.text = dateFormatter.stringFromDate(theDate)

    }
}
```

```
- (void)viewDidLoad {
    [super viewDidLoad];

    CGRect screen = [[UIScreen mainScreen] bounds];

    ///1.日期选择器
    CGFloat datePickerWidth = 320;
    CGFloat datePickerHeight = 167;

    self.datePicker = [[UIDatePicker alloc] initWithFrame:CGRectMake(0, 0,
    ➥datePickerWidth, datePickerHeight)];                                  ①
    //zh-Hans简体中文
    self.datePicker.locale = [NSLocale localeWithLocaleIdentifier:
    ➥@"zh-Hans"];                                                          ②
    //设置日期时间模式
    self.datePicker.datePickerMode = UIDatePickerModeDateAndTime;           ③

    [self.view addSubview:self.datePicker];

    ///2.添加标签
    CGFloat labelWidth = 200;
    CGFloat labelHeight = 21;
    CGFloat labelTopView = 281;

    self.label = [[UILabel alloc] initWithFrame:CGRectMake((screen.size.width-
    ➥labelWidth)/2 , labelTopView, labelWidth, labelHeight)];

    self.label.text = @"Label";
    //字体左右居中
    self.label.textAlignment = NSTextAlignmentCenter;
    [self.view addSubview:self.label];

    ///3.Button按钮
    UIButton* button= [UIButton buttonWithType:UIButtonTypeSystem];
    [button setTitle:@"Button" forState:UIControlStateNormal];

    CGFloat buttonWidth = 46;
    CGFloat buttonHeight = 30;
    CGFloat buttonTopView = 379;

    button.frame = CGRectMake((screen.size.width - buttonWidth)/2 ,
    ➥buttonTopView, buttonWidth, buttonHeight);
    //指定事件处理方法
    [button addTarget:self action:@selector(onclick:) forControlEvents:
    ➥UIControlEventTouchUpInside];
    [self.view addSubview:button];

}
......

- (void)onclick:(id)sender {

    NSDate * theDate = self.datePicker.date;
    NSLog(@"the date picked is: %@", [theDate
    ➥descriptionWithLocale:[NSLocale currentLocale]]);
    NSDateFormatter * dateFormatter = [[NSDateFormatter alloc] init] ;
    dateFormatter.dateFormat = @"YYYY-MM-dd HH:mm:ss";
    NSLog(@"the date formate is: %@", [dateFormatter stringFromDate:theDate]);
    self.label.text = [dateFormatter stringFromDate:theDate];

}

@end
```

上述代码中，第①行用于实例化日期选择器UIDatePicker，第②行用于设置日期选择器的locale属性，可以参考图5-7所示的Locale属性，zh-Hans表示设置本地化属性为简体中文。第③行用于设置日期时间模式（即datePickerMode属性），可以参考图5-7所示的Mode属性，日期时间模式是在UIDatePickerMode枚举类型中定义的成员。UIDatePickerMode枚举成员说明见表5-1。

表5-1 UIDatePickerMode枚举成员

Swift枚举成员	Objective-C枚举成员	说　明
time	UIDatePickerModeTime	时间模式见图5-3
date	UIDatePickerModeDate	日期模式见图5-1
dateAndTime	UIDatePickerModeDateAndTime	日期时间模式见图5-2
countDownTimer	UIDatePickerModeCountDownTimer	倒计时定时器模式见图5-4

5.2.2 普通选择器

有时候，我们可能还需要输入日期除外的其他内容，比如籍贯。籍贯要选择省，省下面还要有市等信息，普通选择器UIPickerView就能够满足用户的这些需求。UIPickerView并非UIDatePicker的父类，它非常灵活，拨盘的个数可以设定，每一个拨盘的内容也可以设定。与UIDatePicker不同的是，UIPickerView需要两个非常重要的协议——UIPickerViewDataSource和UIPickerViewDelegate。

下面我们通过一个案例学习一下普通选择器的用法。图5-8是"选择籍贯"的界面，其中有一个选择器、一个标签和一个按钮，第一个拨轮是所在的省，第二个拨轮是这个省下面可以选择的市。点击按钮，选择器中选中的两个拨轮内容将显示在标签上，具体的实现过程如下所示。

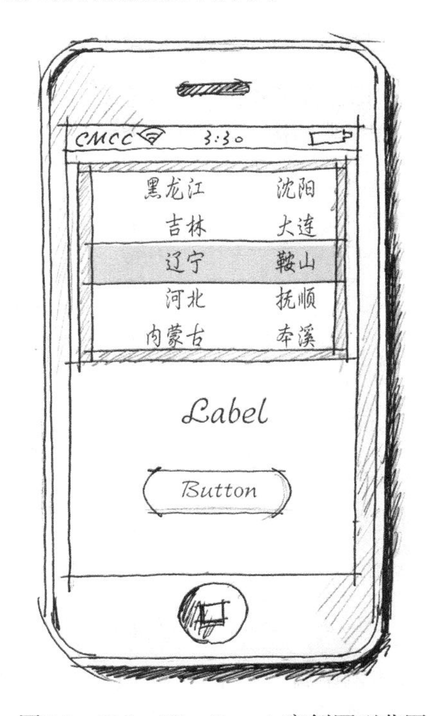

图5-8 PickerViewSample案例原型草图

1. Interface Builder实现

使用Single View Application模板创建一个名为PickerViewSample的工程。打开Main.storyboard设计界面，从对象库中拖曳相关控件到设计界面，如图5-9所示。

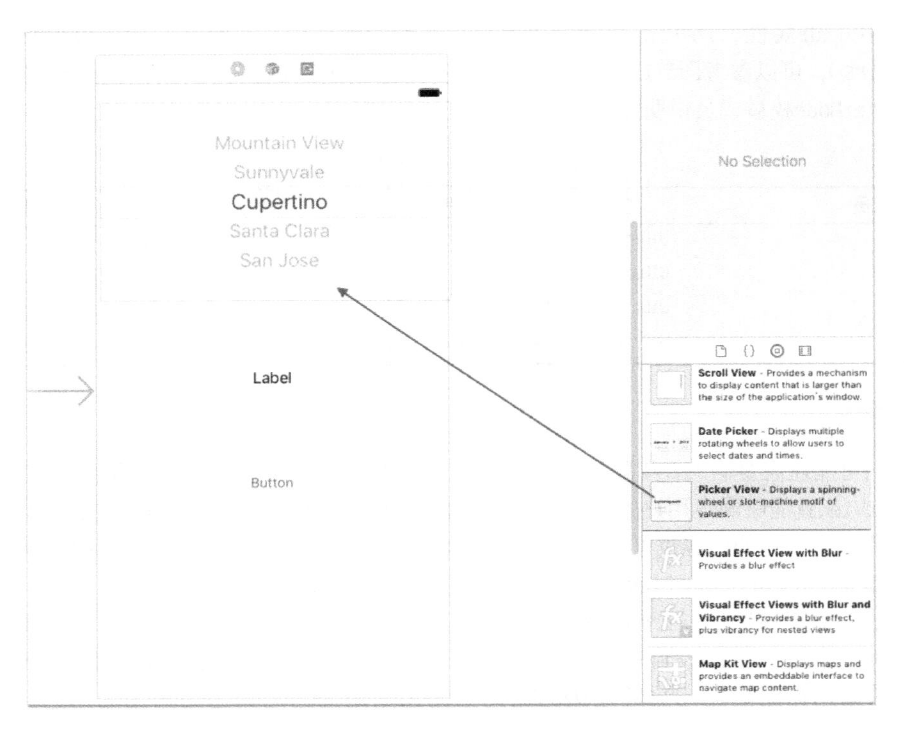

图5-9　Interface Builder设计界面

该案例中省份和市的数据是联动的，即选择了省份后，与它对应的市也会跟着一起变化，省市的信息放在provinces_cities.plist文件中。这个文件采用字典结构，如图5-10所示。

Key	Type	Value
▼ Root	Dictionary	(3 items)
▼ 吉林省	Array	(8 items)
Item 0	String	长春
Item 1	String	吉林
Item 2	String	四平
Item 3	String	辽源
Item 4	String	通化
Item 5	String	白山
Item 6	String	松原
Item 7	String	白城
▶ 黑龙江省	Array	(12 items)
▶ 辽宁省	Array	(14 items)

图5-10　provinces_cities.plist文件

属性列表文件provinces_cities.plist在本例中属于资源文件，资源文件与源程序文件一起被编译打包。我们需要将资源文件添加到Xcode工程中，具体步骤是，右击Picker View Sample组，从弹出的菜单中选择Add Files to "PickerViewSample"...菜单项（如图5-11所示），此时将弹出选择文件对话框。这里选择本章代码中resource文件夹中的provinces_cities.plist文件，然后点击Options按钮，如图5-12所示。此时会打开详细设置选项，在Destination中选中Copy items if needed复选框，这样可以将文件复制到我们的工程目录中。然后，我们在Add to targets中选择PickerViewSample，接着点击Add按钮将provinces_cities.plist文件添加到Xcode工程中。

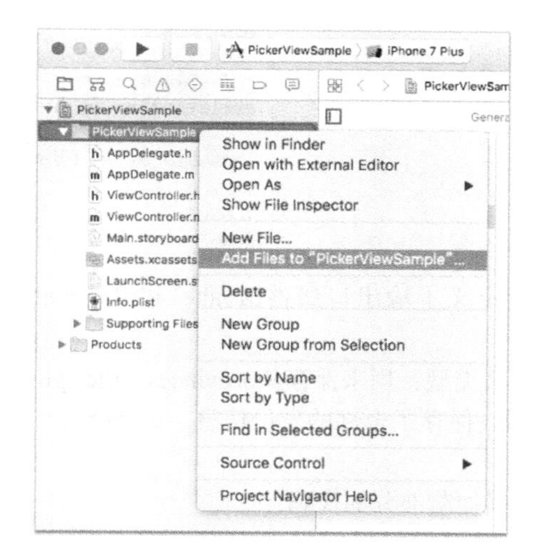

图5-11 添加资源文件

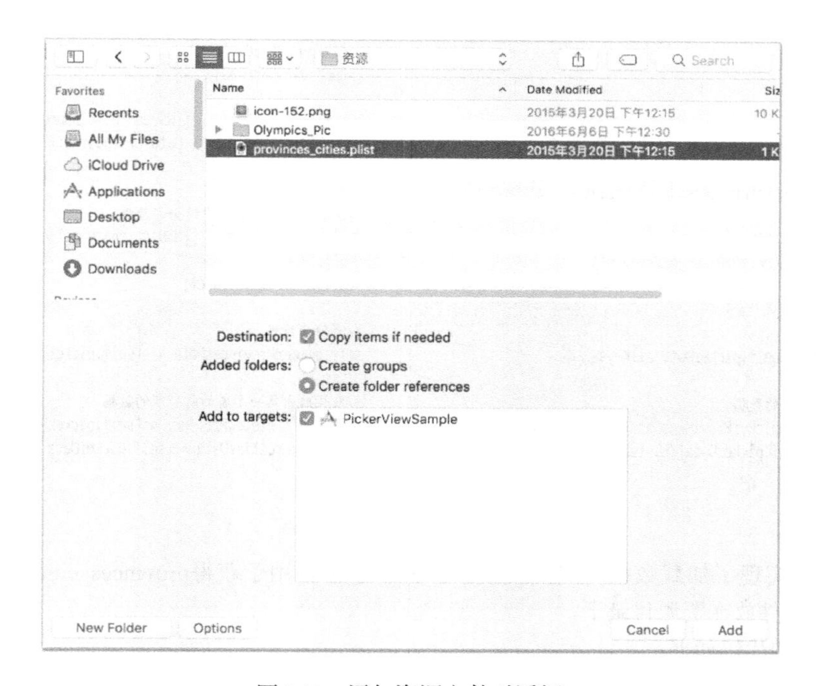

图5-12 添加资源文件对话框

然后再看看ViewController中属性、输出口和动作相关的代码：

```
//ViewController.swift文件
class ViewController: UIViewController  {

    @IBOutlet weak var label: UILabel!                          ①
    @IBOutlet weak var pickerView: UIPickerView!                ②

    var pickerData : NSDictionary!          //保存全部数据
    var pickerProvincesData: NSArray!       //当前的省数据
    var pickerCitiesData : NSArray!         //当前省下面的市数据

    @IBAction func onclick(_ sender: AnyObject) {               ③
```

```
//ViewController.h文件

@interface ViewController : UIViewController

@property (weak, nonatomic) IBOutlet UILabel *label;                     ①
@property (weak, nonatomic) IBOutlet UIPickerView *pickerView;           ②

@property (nonatomic, strong) NSDictionary *pickerData;   //保存全部数据
@property (nonatomic, strong) NSArray *pickerProvincesData;//当前的省数据
@property (nonatomic, strong) NSArray *pickerCitiesData;//当前省下面的市数据
```

```
    }
    ......
}
```

```
@end

//ViewController.m文件
@implementation ViewController

- (IBAction)onclick:(id)sender {                                    ③
}
......
@end
```

上述代码中，第①行和第②行定义了输出口属性UILabel和UIPickerView，第③行定义了一个动作事件onclick，用于响应按钮点击事件。

装载数据的属性pickerData是字典类型，用来保存从provinces_cities.plist文件中读取的全部内容。

pickerProvincesData是数组类型，保存了全部的省份信息。pickerCitiesData也是数组类型，保存了当前选中省份下的市信息。

我们再看看ViewController中数据加载部分的代码：

```
//ViewController.swift文件
override func viewDidLoad() {
    super.viewDidLoad()

    let plistPath = Bundle.main.path(forResource: "provinces_cities",
    ➥ofType: "plist")                                                  ①
    //获取属性列表文件中的全部数据
    let dict = NSDictionary(contentsOfFile: plistPath!)
    self.pickerData = dict

    //省份名数据
    self.pickerProvincesData = self.pickerData.allKeys

    //默认取出第一个省的所有市的数据
    let seletedProvince = self.pickerProvincesData[0] as! String
    self.pickerCitiesData = self.pickerData[seletedProvince] as! NSArray

}
```

```
//ViewController.m文件

@implementation ViewController

- (void)viewDidLoad  {
    [super viewDidLoad];

    NSString *plistPath = [[NSBundle mainBundle]
                            pathForResource:@"provinces_cities"
                                     ofType:@"plist"];                 ①

    //获取属性列表文件中的全部数据
    NSDictionary *dict = [[NSDictionary alloc] initWithContentsOfFile:
    ➥plistPath];
    self.pickerData = dict;

    //省份名数据
    self.pickerProvincesData =  [self.pickerData allKeys];

    //默认取出第一个省的所有市的数据
    NSString *seletedProvince = [self.pickerProvincesData objectAtIndex:0];
    self.pickerCitiesData = [self.pickerData objectForKey:seletedProvince];

}
@end
```

viewDidLoad方法实现了加载数据到成员变量中，其中第①行用于获得provinces_cities.plist文件的全路径。provinces_cities.plist文件放在资源目录下。

用户点击按钮时的代码如下：

```
@IBAction func onclick(_ sender: AnyObject) {

    let row1 = self.pickerView.selectedRowInComponent(0)
    let row2 = self.pickerView.selectedRowInComponent(1)
    let selected1 = self.pickerProvincesData[row1] as! String
    let selected2 = self.pickerCitiesData[row2] as! String

    let title = String(format: "%@, %@市", selected1,selected2)

    self.label.text = title;
}
```

```
- (IBAction)onclick:(id)sender {

    NSInteger row1 = [self.pickerView selectedRowInComponent:0];
    NSInteger row2 = [self.pickerView selectedRowInComponent:1];
    NSString *selected1 = [self.pickerProvincesData objectAtIndex:row1];
    NSString *selected2 = [self.pickerCitiesData objectAtIndex:row2];

    NSString *title = [[NSString alloc] initWithFormat:@"%@, %@市",
    ➥selected1,selected2];

    self.label.text = title;

}
```

UIPickerView的Component属性就是指拨盘，selectedRowInComponent:方法返回拨盘中被选定行的索引，索引是从0开始的。

2. 代码实现

上一节中，我们使用Interface Builder在PickerViewSample案例中添加了普通选择器，这一节介绍如何通过代码实现普通选择器。

其他步骤不再赘述，我们重点介绍一下视图控制器类ViewController，其主要代码如下：

```
//ViewController.swift文件
class ViewController: UIViewController {

    var label: UILabel!
    var pickerView: UIPickerView!

    <参考Interface Builder实现部分>

    override func viewDidLoad() {
        super.viewDidLoad()

        <参考Interface Builder实现部分>

        let screen = UIScreen.main.bounds

        ///1.选择器
        let pickerViewWidth: CGFloat = 320
        let pickerViewHeight: CGFloat = 162
        self.pickerView = UIPickerView(frame: CGRect(x: 0, y: 0,
        ➥width: pickerViewWidth, height: pickerViewHeight))

        self.view.addSubview(self.pickerView)

        ///2.添加标签
        let labelWidth:CGFloat = 200
        let labelHeight:CGFloat = 21
        let labelTopView:CGFloat = 273
        self.label = UILabel(frame: CGRect(x: (screen.size.width -
        ➥labelWidth)/2 ,
        ➥y: labelTopView, width: labelWidth, height: labelHeight))

        self.label.text = "Label"
        //字体左右居中
        self.label.textAlignment = .center
        self.view.addSubview(self.label)

        ///3.Button按钮
        let button = UIButton(type: UIButtonType.system)
        button.setTitle("Button", for: UIControlState())

        let buttonWidth:CGFloat = 46
        let buttonHeight:CGFloat = 30
        let buttonTopView:CGFloat = 374

        button.frame = CGRect(x: (screen.size.width - buttonWidth)/2 ,
        ➥y: buttonTopView, width: buttonWidth, height: buttonHeight)
        //指定事件处理方法
        button.addTarget(self, action: #selector(onclick(_:)),
        ➥for: .touchUpInside)

        self.view.addSubview(button)

    }

    @IBAction func onclick(_ sender: AnyObject) {
        <参考Interface Builder实现部分>
    }
```

```
//ViewController.m文件
#import "ViewController.h"

@interface ViewController ()

@property (strong, nonatomic) UIPickerView *pickerView;
@property (strong, nonatomic) UILabel *label;

<参考Interface Builder实现部分>

@end

@implementation ViewController

- (void)viewDidLoad {
    [super viewDidLoad];

    <参考Interface Builder实现部分>

    CGRect screen = [[UIScreen mainScreen] bounds];

    ///1.选择器
    CGFloat pickerViewWidth = 320;
    CGFloat pickerViewHeight = 162;
    self.pickerView = [[UIPickerView alloc] initWithFrame:CGRectMake(0, 0,
    ➥pickerViewWidth, pickerViewHeight)];

    [self.view addSubview:self.pickerView];

    ///2.添加标签
    CGFloat labelWidth = 200;
    CGFloat labelHeight = 21;
    CGFloat labelTopView = 273;
    self.label = [[UILabel alloc] initWithFrame:CGRectMake((screen.size.width-
    ➥labelWidth)/2 , labelTopView, labelWidth, labelHeight)];

    self.label.text = @"Label";
    //字体左右居中
    self.label.textAlignment = NSTextAlignmentCenter;
    [self.view addSubview:self.label];

    ///3.Button按钮
    UIButton* button= [UIButton buttonWithType:UIButtonTypeSystem];
    [button setTitle:@"Button" forState:UIControlStateNormal];

    CGFloat buttonWidth = 46;
    CGFloat buttonHeight = 30;
    CGFloat buttonTopView = 374;

    button.frame = CGRectMake((screen.size.width - buttonWidth)/2 ,
    ➥buttonTop View, buttonWidth, buttonHeight);
    //指定事件处理方法
    [button addTarget:self action:@selector(onclick:) forControlEvents:
    ➥UIControlEventTouchUpInside];
    [self.view addSubview:button];

}
```

```
    ......
}
```

```
- (void)onclick:(id)sender {

    <参考Interface Builder实现部分>
}
......

@end
```

上述代码前面基本上已经介绍过了，这里不再赘述。

5.2.3　数据源协议与委托协议

UIPickerView的委托协议是UIPickerViewDelegate，数据源是UIPickerViewDataSource。我们需要在ViewController视图控制器中声明实现UIPickerViewDelegate和UIPickerViewDataSource协议。ViewController的相关代码如下：

```
//ViewController.swift文件
class ViewController: UIViewController, UIPickerViewDelegate,
➥UIPickerViewDataSource {                                          ①
    ......
}
```

```
//ViewController.m文件
#import "ViewController.h"

@interface ViewController () <UIPickerViewDelegate, UIPickerViewDataSource> ①

......

@end
```

上述代码中，第①行是声明实现协议，对应的Objective-C版本可以在ViewController.h或者ViewController.m文件中声明实现协议。

下面我们具体看看这两个协议中的方法。UIPickerViewDataSource中的方法有如下两个。

- **numberOfComponentsInPickerView:**。为选择器中拨轮的数目，Swift语言的表示方式为numberOfComponents(in:)。
- **pickerView:numberOfRowsInComponent:**。为选择器中某个拨轮的行数，Swift语言的表示方式为pickerView(_:numberOfRowsInComponent:)。

在ViewController中，UIPickerViewDataSource的实现代码如下：

```
//MARK: -- 实现UIPickerViewDataSource协议
func numberOfComponents(in pickerView: UIPickerView) -> Int {
    return 2
}

func pickerView(_ pickerView: UIPickerView,
➥numberOfRowsInComponent component: Int) -> Int {
    if (component == 0) {//省份个数
        return self.pickerProvincesData.count
    } else {            //市的个数
        return self.pickerCitiesData.count
    }
}
```

```
#pragma mark 实现UIPickerViewDataSource协议
- (NSInteger)numberOfComponentsInPickerView:(UIPickerView *)pickerView {
    return 2;
}

- (NSInteger)pickerView:(UIPickerView *)pickerView
➥numberOfRowsInComponent:(NSInteger)component {
    if (component == 0) {    //省份个数
        return [self.pickerProvincesData count];
    } else {              //市的个数
        return [self.pickerCitiesData count];
    }
}
```

UIPickerViewDelegate中的常用方法有如下两个。

- **pickerView:titleForRow:forComponent:**。为选择器中某个拨轮的行提供显示数据，Swift语言的表示方式为pickerView(_:titleForRow:forComponent:)。
- **pickerView:didSelectRow:inComponent:**。选中选择器的某个拨轮中的某行时调用，Swift语言的表示方式为pickerView(_:didSelectRow:inComponent:)。

在ViewController中实现UIPickerViewDelegate的代码如下：

```
//MARK: -- 实现UIPickerViewDelegate协议
func pickerView(_ pickerView: UIPickerView, titleForRow row: Int,
➥forComponent component: Int) -> String? {
    if (component == 0) {//选择省份名
        return self.pickerProvincesData[row] as? String
    } else {//选择市名
        return self.pickerCitiesData[row] as? String
    }
}
func pickerView(_ pickerView: UIPickerView, didSelectRow row: Int,
➥inComponent component: Int) {
    if (component == 0) {
        let seletedProvince = self.pickerProvincesData[row] as! String
        self.pickerCitiesData = self.pickerData[seletedProvince] as! NSArray
        self.pickerView.reloadComponent(1)
    }
}
```

```
#pragma mark 实现UIPickerViewDelegate协议
-(NSString *)pickerView:(UIPickerView *)pickerView
➥titleForRow:(NSInteger)row forComponent:(NSInteger)component {
    if (component == 0) {//选择省份名
        return [self.pickerProvincesData objectAtIndex:row];
    } else {//选择市名
        return [self.pickerCitiesData objectAtIndex:row];
    }
}

- (void)pickerView:(UIPickerView *)pickerView
    didSelectRow:(NSInteger)row inComponent:(NSInteger)component {
    if (component == 0) {
        NSString *seletedProvince = [self.pickerProvincesData
        ➥objectAtIndex:row];
        NSArray *array = [self.pickerData objectForKey:seletedProvince];
        self.pickerCitiesData = array;
        [self.pickerView reloadComponent:1];
    }
}
```

最后，不要忘记将委托和数据源的实现对象ViewController分配给UIPickerView的委托属性delegate和数据源属性dataSource，这可以通过代码或Interface Builder进行分配。下面的代码就是用来实现分配的：

```
override func viewDidLoad() {
    ......
    self.pickerView.dataSource = self
    self.pickerView.delegate = self
}
```

```
- (void)viewDidLoad {
    ......
    self.pickerView.dataSource = self;
    self.pickerView.delegate = self;
}
```

在Interface Builder中分配的过程是：打开故事板文件，右击选择器，此时会弹出右键菜单，如图5-13所示，将Outlets→dataSource后面的小圆点拖曳到左边的View Controller上，然后以同样的方式将Outlets→delegate后面的小圆点拖曳到左边的View Controller上。

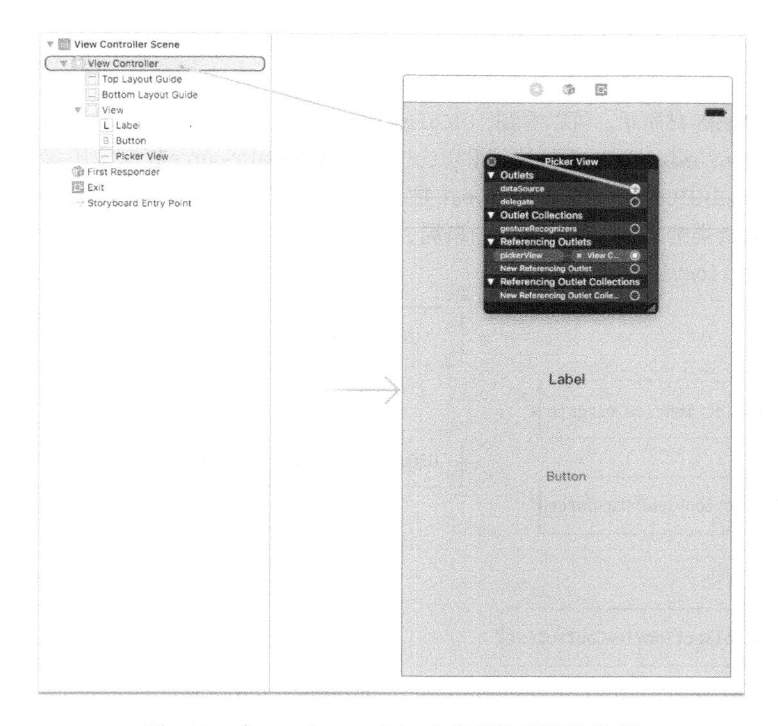

图5-13 在Interface Builder中分配数据源和委托

5.3　集合视图

为了增强网格视图开发，iOS 6中开放了集合视图API。这种网格视图的开源代码在开源社区中很早就有，但是都比较麻烦，而iOS 6的集合视图API使用起来却非常方便。

5.3.1　集合视图的组成

图5-14显示了集合视图的组成，它有4个重要的组成部分。

- **单元格**。它是集合视图中的一个单元格。
- **节**。它是集合视图中的一个行数据，由多个单元格构成。
- **补充视图**。它是节的头和脚。
- **装饰视图**。集合视图中的背景视图。

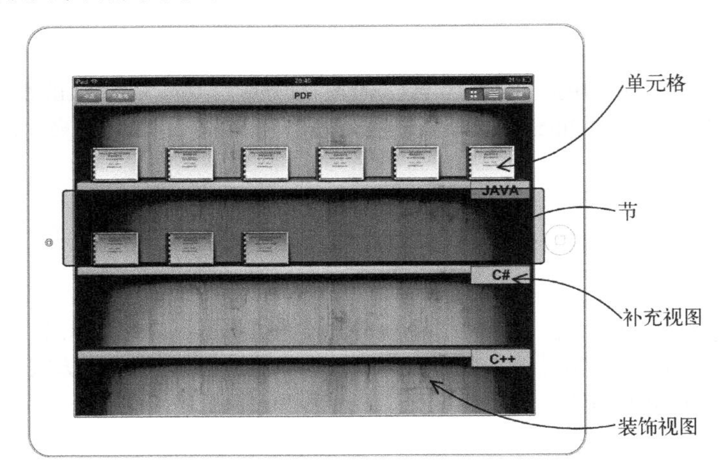

图5-14　集合视图组成

集合视图类的构成如图5-15所示。可以看到，UICollectionView继承自UIScrollView。与选择器类似，集合视图也有两个协议：UICollectionViewDelegate委托协议和UICollectionViewDataSource数据源协议。UICollectionViewCell是单元格类，它的布局是由UICollectionViewLayout类定义的，它是一个抽象类。UICollectionViewFlowLayout类是UICollectionViewLayout类的子类。对于复杂的布局，可以自定义UICollectionViewLayout类。UICollectionView对应的控制器是UICollectionViewController类。

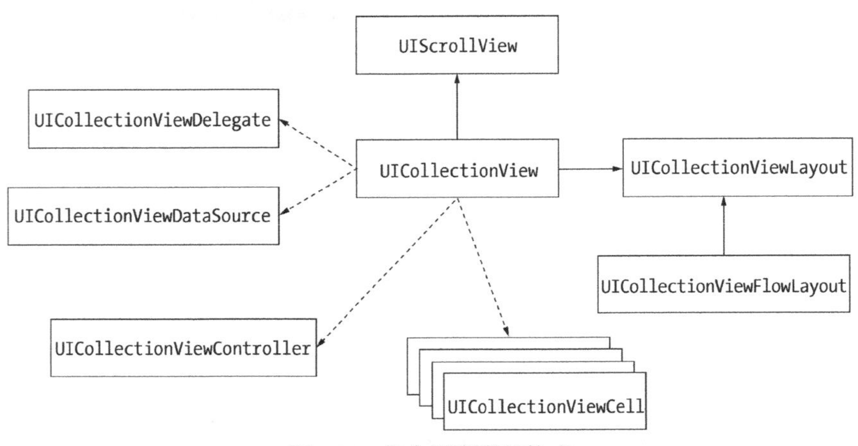

图5-15　集合视图类的构成

5.3.2 集合视图数据源协议与委托协议

集合视图的委托协议是UICollectionViewDelegate，数据源协议是UICollectionViewDataSource。

UICollectionViewDataSource中的方法有如下4个。

- **collectionView:numberOfItemsInSection:**。提供某个节中的列数，Swift语言的表示方式为collectionView(_:numberOfItemsInSection:)。
- **numberOfSectionsInCollectionView:**。提供视图中节的个数，Swift语言的表示方式为numberOfSections(in:)。
- **collectionView:cellForItemAtIndexPath:**。为某个单元格提供显示数据，Swift语言的表示方式为collectionView(_:cellForItemAt:)。
- **collectionView:viewForSupplementaryElementOfKind:atIndexPath:**。为补充视图提供显示数据，Swift语言的表示方式为collectionView(_:viewForSupplementaryElementOfKind:at:)。

UICollectionViewDelegate中的方法很多，其中较为重要的方法如下。

- **collectionView:didSelectItemAtIndexPath:**。选择单元格之后触发，Swift语言的表示方式为collectionView(_:didSelectItemAt:)。
- **collectionView:didDeselectItemAtIndexPath:**。取消选择单元格之后触发，Swift语言的表示方式为collectionView(_:didDeselectItemAt:)。

5.4 案例：奥运会比赛项目

图5-16是使用集合视图展示奥运会比赛项目的案例，其中有8个比赛项目，点击其中一个，会输出一些日志信息。该案例的具体实现过程如下所示。

图5-16 奥运会比赛项目案例原型草图

CollectionViewSample案例可以通过Interface Builder和代码两种方式实现。Interface Builder在布局时有很多参数需要设置，比较麻烦。我推荐本案例（可以使用所有集合视图应用）采用代码实现，具体步骤如下。

5.4.1　创建工程

首先，使用Single View Application模板创建一个名为CollectionViewSample的工程。然后，需要添加资源图片和属性列表文件到工程，读者可以从本节工程中找到Olympics_Pic文件夹并将其添加到工程中，具体可参考5.2.2节操作弹出对话框。如图5-17所示，选择本章代码中resource文件夹中的Olympics_Pic文件夹，并在Destination中选中Copy items if needed，这样可以将Olympics_Pic文件夹复制到我们的工程目录中，在Add folders中选中Create groups选项。Create groups表示添加文件夹到Xcode工程时作为一个组，而Create folder references选项表示添加文件夹到Xcode工程时作为一个文件夹。

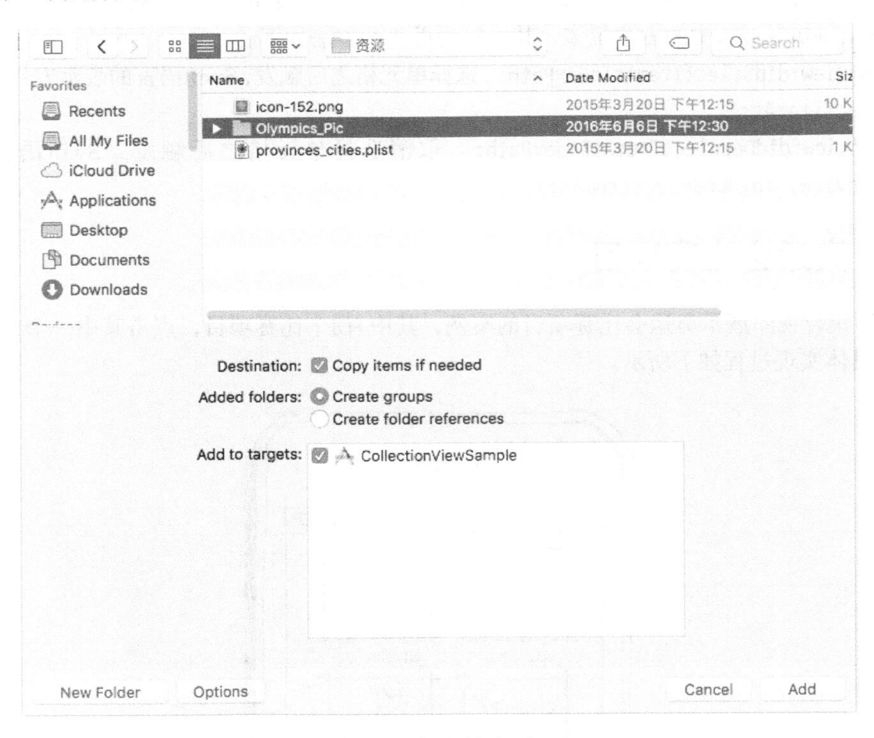

图5-17　添加文件夹到工程

5.4.2　自定义集合视图单元格

集合视图单元格是集合视图中最为重要的组成部分，没有样式和风格定义，它可以在故事板中设计，也可以通过代码来设定。单元格就是一个视图，可以在内部放置其他视图或控件。

首先，需要添加一个自定义单元格类，它继承自`UICollectionViewCell`。我们可以在Xcode中创建单元格类，具体步骤是：选择CollectionViewSample组，再选择菜单File→New→File，此时会弹出如图5-18所示的对话框。此时选择iOS→Source→Cocoa Touch Class，然后点击Next按钮，弹出如图5-19所示的对话框。然后，在Class项目中输入EventCollectionViewCell，在Subclass of项目中选择UICollectionViewCell，在Language中选择Swift或Objective-C。选择完成之后，点击Next按钮创建单元格类`EventCollectionViewCell`。

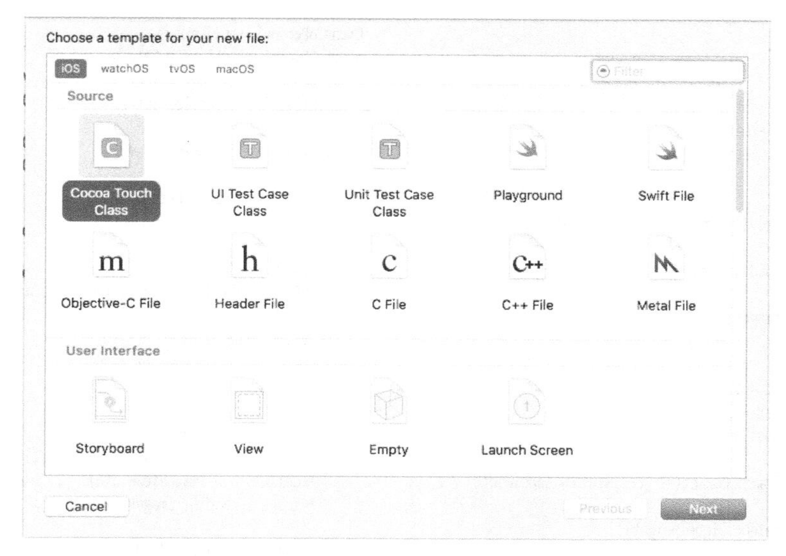

图5-18　选择文件模板

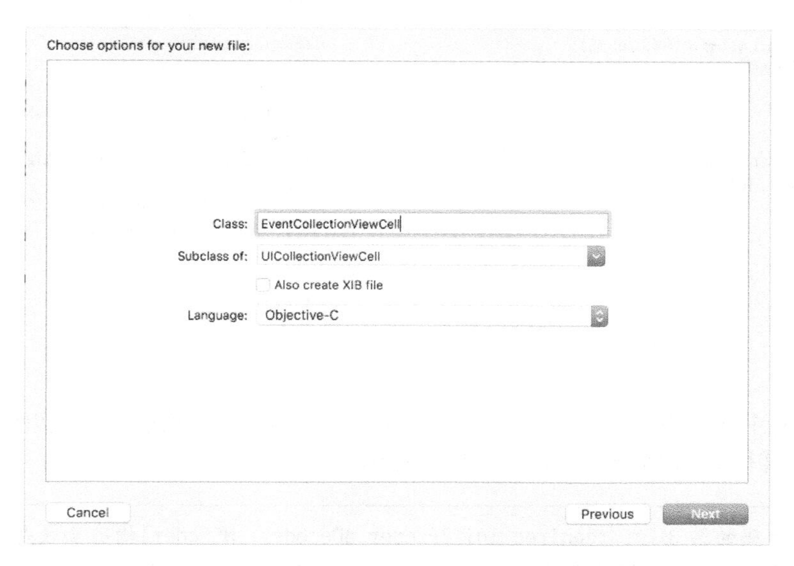

图5-19　添加自定义单元格类

我们还需要在单元格中添加ImageView和Label视图属性，还要在适当的方法中实例化这些属性对象。之前我们通常都是在视图控制器的viewDidLoad方法中实例化当前UI界面中的视图，而集合视图单元格本质上是一个视图，不是视图控制器，没有viewDidLoad方法，这种情况下可以在视图的构造函数中实例化这些属性对象。

EventCollectionViewCell的代码如下：

```
//EventCollectionViewCell.swift文件
class EventCollectionViewCell: UICollectionViewCell {

    var imageView: UIImageView!
    var label: UILabel!

    override init(frame: CGRect) {                                   ①
        super.init(frame: frame)
```

```
//EventCollectionViewCell.h文件

@interface EventCollectionViewCell : UICollectionViewCell

@property (strong, nonatomic) UIImageView *imageView;
@property (strong, nonatomic) UILabel *label;

@end
```

```
        //单元格的宽度
        let cellWidth: CGFloat = self.frame.size.width

        let imageViewWidth: CGFloat = 101
        let imageViewHeight: CGFloat = 101
        let imageViewTopView: CGFloat = 15

        ///1.添加ImageView
        self.imageView = UIImageView(frame: CGRect(x: (cellWidth -
        ➥imageViewWidth) / 2,
        ➥y: imageViewTopView, width: imageViewWidth, height: imageViewHeight))

        self.addSubview(self.imageView)                               ②

        ///2.添加标签
        let labelWidth: CGFloat = 101
        let labelHeight: CGFloat = 16
        let labelTopView: CGFloat = 120

        self.label = UILabel(frame: CGRect(x: (cellWidth - labelWidth) / 2, y:
        ➥labelTopView,
        ➥width: labelWidth, height: labelHeight))

        //字体左右居中
        self.label.textAlignment = .center
        //设置字体
        self.label.font = UIFont.systemFont(ofSize: 13)
        self.addSubview(self.label)

    }

    required init?(coder aDecoder: NSCoder) {                         ③
        fatalError("init(coder:) has not been implemented")
    }

}
```

```
//EventCollectionViewCell.m文件
#import "EventCollectionViewCell.h"

@implementation EventCollectionViewCell

- (id)initWithFrame:(CGRect)frame                                     ①
{
    self = [super initWithFrame:frame];
    if (self) {

        //单元格的宽度
        CGFloat cellWidth = self.frame.size.width;

        CGFloat imageViewWidth = 101;
        CGFloat imageViewHeight = 101;
        CGFloat imageViewTopView = 15;

        ///1.添加ImageView
        self.imageView = [[UIImageView alloc] initWithFrame: CGRectMake((
        ➥cellWidth - imageViewWidth) / 2, imageViewTopView,
        ➥imageViewWidth, imageViewHeight)];

        [self addSubview:self.imageView];                             ②

        ///2.添加标签
        CGFloat labelWidth = 101;
        CGFloat labelHeight = 16;
        CGFloat labelTopView = 120;
        self.label = [[UILabel alloc] initWithFrame:CGRectMake((cellWidth -
        ➥labelWidth) / 2, labelTopView, labelWidth, labelHeight)];
        //字体左右居中
        self.label.textAlignment = NSTextAlignmentCenter;
        //设置字体
        self.label.font = [UIFont systemFontOfSize:13];
        [self addSubview:self.label];

    }
    return self;
}

@end
```

上述代码中，第①行重写构造函数。系统会默认调用该构造函数，在该构造函数中实例化单元格包含的各个子视图属性对象。第②行将这两个子视图添加到单元格视图中。

提示

在Swift版代码的第③行，`required init?(coder aDecoder: NSCoder)`构造函数是单元格视图父类要求必须实现的。单元格视图父类中声明实现NSCoding协议，子类中要求实现该构造函数，但本例中并未用到该构造函数。

5.4.3 添加集合视图

视图控制器类ViewController中添加集合视图的相关代码如下：

```
//ViewController.swift文件
//集合视图列数，即：每一行有几个单元格
let COL_NUM = 3
class ViewController: UIViewController,
➥UICollectionViewDataSource, UICollectionViewDelegate {

    var events : NSArray!
    var collectionView: UICollectionView!
```

```
//ViewController.m文件
#import "ViewController.h"
#import "EventCollectionViewCell.h"

@interface ViewController () <UICollectionViewDataSource,
UICollectionViewDelegate>

@property (strong, nonatomic) NSArray * events;
```

```swift
    override func viewDidLoad() {
        super.viewDidLoad()

        let plistPath = Bundle.main.path(forResource: "events", ofType:
        ➥"plist")
        //获取属性列表文件中的全部数据
        self.events = NSArray(contentsOfFile: plistPath!)

        self.setupCollectionView()
    }

    func setupCollectionView() {

        //1.创建流式布局
        let layout = UICollectionViewFlowLayout()                        ①
        //2.设置每个单元格的尺寸
        layout.itemSize = CGSize(width: 80, height: 80)                  ②
        //3.设置整个collectionView的内边距
        layout.sectionInset = UIEdgeInsetsMake(15, 15, 30, 15)           ③

        let screenSize  = UIScreen.main.bounds.size;
        //重新设置iPhone 6/6s/7/7s/Plus
        if (screenSize.height > 568) {                                   ④
            layout.itemSize = CGSize(width: 100, height: 100)
            layout.sectionInset = UIEdgeInsetsMake(15, 15, 20, 15)
        }

        //4.设置单元格之间的间距
        layout.minimumInteritemSpacing = 5                               ⑤

        self.collectionView = UICollectionView(frame: self.view.frame,
        ➥collectionViewLayout: layout)                                  ⑥

        //设置可重用单元格标识与单元格类型
        self.collectionView.registerClass(EventCollectionViewCell.self,
        ➥forCellWithReuseIdentifier: "cellIdentifier" )                 ⑦

        self.collectionView.backgroundColor = UIColor.white

        self.collectionView.delegate = self                              ⑧
        self.collectionView.dataSource = self                            ⑨

        self.view.addSubview(self.collectionView)
    }

    <实现数据源协议的代码参考Interface Builder部分>
}
```

```objc
@property (strong, nonatomic) UICollectionView* collectionView;

@end

@implementation ViewController

- (void)viewDidLoad {
    [super viewDidLoad];

    NSString *plistPath = [[NSBundle mainBundle] pathForResource:@"events"
                                                          ofType:@"plist"];
    //获取属性列表文件中的全部数据
    self.events = [[NSArray alloc] initWithContentsOfFile:plistPath];

    [self setupCollectionView];

}

- (void) setupCollectionView {

    //1.创建流式布局
    UICollectionViewFlowLayout *layout = [[UICollectionViewFlowLayout alloc]
    ➥init];                                                             ①
    //2.设置每个单元格的尺寸
    layout.itemSize = CGSizeMake(80, 80);                                ②
    //3.设置整个collectionView的内边距
    layout.sectionInset = UIEdgeInsetsMake(15, 15, 30, 15);              ③

    CGSize screenSize  = [UIScreen mainScreen].bounds.size;
    //重新设置iPhone 6/6s/7/7s/Plus
    if (screenSize.height > 568) {                                       ④
        layout.itemSize = CGSizeMake(100, 100);
        layout.sectionInset = UIEdgeInsetsMake(15, 15, 20, 15);
    }
    //4.设置单元格之间的间距
    layout.minimumInteritemSpacing = 10;                                 ⑤

    self.collectionView = [[UICollectionView alloc] initWithFrame:self.view.
    ➥framecollectionViewLayout:layout];                                 ⑥

    //设置可重用单元格标识与单元格类型
    [self.collectionView registerClass:[EventCollectionViewCell class]
    ➥forCellWithReuseIdentifier:@"cellIdentifier" ];                    ⑦

    self.collectionView.backgroundColor = [UIColor whiteColor];

    self.collectionView.delegate = self;                                 ⑧
    self.collectionView.dataSource = self;                               ⑨

    [self.view addSubview:self.collectionView];
}

<实现数据源协议的代码参考Interface Builder部分>

@end
```

上述代码中，第①行用于创建布局管理器对象UICollectionViewFlowLayout。UICollectionViewFlowLayout是一种流式布局管理器，即从左到右从上到下布局。第②行~第⑤行用于设置流式布局管理器的属性，为了帮助读者熟悉这些属性，我绘制了图5-20，说明如下。

- itemSize属性（见代码第②行）描述了每个单元格的尺寸。
- sectionInset属性（见代码第③行）描述了collectionView的内边距，类型是UIEdgeInsets结构体。UIEdgeInsets包括：top（上边界）、left（左边界）、bottom（底边界）和right（右边界）4个成员。UIEdgeInsetsMake函数可以创建UIEdgeInsets结构体实例，它的4个参数依次是top、left、bottom和right。

- minimumInteritemSpacing属性（见代码第⑤行）描述了单元格之间的最小间距。

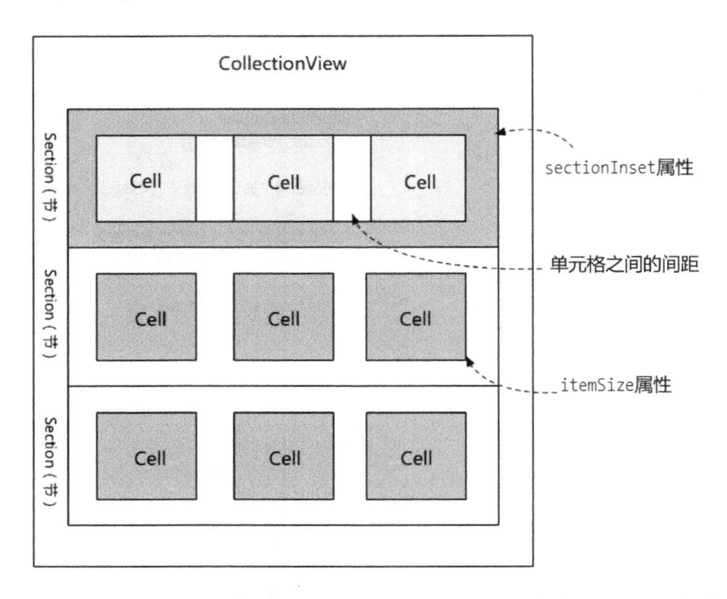

图5-20　流式布局管理器属性与Interface Builder中各属性的对应关系

第④行用于判断是否是iPhone 6/6s/7/7s，以及iPhone Plus设备。由于这些设备比较宽，需要重新设置itemSize和sectionInset属性。

第⑥行创建集合视图UICollectionView对象的构造函数，其中第一个参数是设置集合视图frame属性，第二个参数是设置布局管理器对象。

第⑦行用于设置可重用单元格标识。在Swift代码中，EventCollectionViewCell.self表达式用于获取EventCollectionViewCell类的类型，对应的Objective-C代码为[EventCollectionViewCell class]。

第⑧行和第⑨行用于设置当前视图控制器self为集合视图委托对象和数据源对象。

5.4.4　实现数据源协议

在视图控制器ViewController中实现数据源协议UICollectionViewDataSource的代码如下：

```
//ViewController.swift文件

//集合视图列数，即每一行有几个单元格
let COL_NUM = 3

class ViewController: UIViewController,
➥UICollectionViewDataSource, UICollectionViewDelegate {
    ......

    //MARK: -- UICollectionViewDataSource
    func numberOfSections(in collectionView: UICollectionView) -> Int {      ①

        let num = self.events.count % COL_NUM
        if (num == 0) {//偶数
            return self.events.count / COL_NUM
        } else {        //奇数
            return self.events.count / COL_NUM + 1
        }
    }

    func collectionView(_ collectionView: UICollectionView,
```

```
//ViewController.m文件
#import "EventCollectionViewCell.h"

//集合视图列数，即每一行有几个单元格
#define COL_NUM 3

@interface ViewController ()
➥<UICollectionViewDataSource, UICollectionViewDelegate>
......
- (NSInteger)numberOfSectionsInCollectionView:(
➥UICollectionView *)collectionView {                                     ①

    int num = [self.events count] % COL_NUM;
    if (num == 0) {//偶数
        return [self.events count] / COL_NUM;
    } else {        //奇数
        return [self.events count] / COL_NUM + 1;
    }
}

- (NSInteger)collectionView:(UICollectionView *)collectionView
```

```swift
➥numberOfItemsInSection section: Int) -> Int {
    return COL_NUM
}

func collectionView(_ collectionView: UICollectionView,
➥cellForItemAt indexPath: IndexPath) -> UICollectionViewCell {

    let cell = collectionView.dequeueReusableCell(withReuseIdentifier:
    ➥"cellIdentifier",
    ➥for: indexPath) as! EventCollectionViewCell                    ②

    //计算events集合下标索引
    let idx = indexPath.section * COL_NUM + indexPath.row;           ③

    if (self.events.count <= idx) {//防止下标越界
        return cell;
    }

    let event = self.events[idx] as! NSDictionary

    cell.label.text = event["name"] as? String
    cell.imageView.image = UIImage(named: event["image"] as! String)

    return cell
}
......
}
```

```objc
➥numberOfItemsInSection:(NSInteger)section {
    return COL_NUM;
}

- (UICollectionViewCell *)collectionView:(UICollectionView *)collectionView
➥cellForItemAtIndexPath:(NSIndexPath *)indexPath {

    EventsCollectionViewCell *cell = [collectionView
    ➥dequeueReusableCellWithReuseIdentifier:
    ➥@"cellIdentifier" forIndexPath:indexPath];                      ②

    //计算events集合下标索引
    NSInteger idx = indexPath.section * COL_NUM + indexPath.row;      ③

    if (self.events.count <= idx) {//防止下标越界
        return cell;
    }

    NSDictionary *event = self.events[idx];

    cell.label.text = [event objectForKey:@"name"];
    cell.imageView.image = [UIImage imageNamed:event[@"image"]];

    return cell;
}
......

@end
```

上述代码中，第①行实现的方法会返回视图中节的个数，这里需要注意数据的行数是否能与COL_NUM（每一行有几个单元格）整除，不能整除时要加多一行。

第②行实现的方法返回集合视图的单元格。单元格的获得采用可重用设计，其中第一个参数是可重用单元格标识符，它是在setupCollectionView方法中设置的，见5.4.3节setupCollectionView方法的第①行；第二个参数是indexPath，它是NSIndexPath类型（Swift版本是IndexPath）。NSIndexPath是一种数据结构，是一种复杂多维数组结构，其中我们常用的属性有section和row。第③行使用了这两个属性，section是集合视图节索引，row是集合视图当前节中列（单元格）的索引。

5.4.5 实现委托协议

在视图控制器ViewController中实现委托协议UICollectionViewDelegate的代码如下：

```swift
func collectionView(_ collectionView: UICollectionView,
➥didSelectItemAt indexPath: IndexPath) {
    let event = self.events[(indexPath as NSIndexPath).section * COL_NUM +
    ➥(indexPath as NSIndexPath).row] as! NSDictionary
    print("select event name : ", event["name"]!)
}
```

```objc
- (void)collectionView:(UICollectionView *)collectionView
➥didSelectItemAtIndexPath:(NSIndexPath *)indexPath {
    NSDictionary *event = self.events[indexPath.section * COL_NUM +
    ➥indexPath.row];
    NSLog(@"select event name : %@", event[@"name"]);
}
```

运行上述代码，得到的输出结果如下：

```
select event name : basketball
select event name : athletics
select event name : archery
```

5.5 小结

本章向大家介绍了数据源协议和委托协议，并且介绍了高级视图——选择器和集合视图。

第 6 章

表视图

表视图是iOS开发中使用最频繁的视图。一般情况下，我们都会选择以表的形式来展现数据，比如通讯录和频道列表等。在表视图中，分节、分组和索引等功能使我们所展示的数据看起来更规整、更有条理。更令人兴奋的是，表视图还可以利用细节展示等功能多层次地展示数据。但与其他控件相比，表视图的使用相对比较复杂。

6.1 概述

在本节中，我们将带大家了解表视图中的一些概念、相关类、表视图的分类、单元格的组成和样式，以及表视图的两个协议——`UITableViewDelegate`委托和`UITableViewDataSource`数据源。

6.1.1 表视图的组成

在iOS中，表视图是最重要的视图，它有很多概念，这些概念之间的关系如图6-1所示。

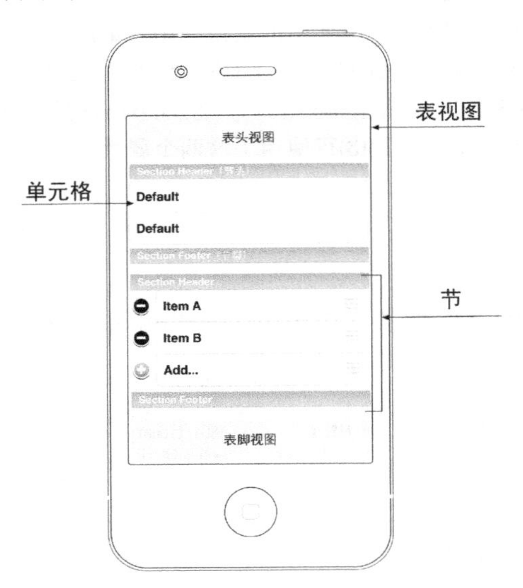

图6-1　表视图组成图

下面我们简要介绍一下这些概念。

- **表头视图**（table header view）。表视图最上边的视图，用于展示表视图的信息。如图6-2a所示，表头视图放置了一个搜索栏。
- **表脚视图**（table footer view）。表视图最下边的视图，用于展示表视图的信息，例如表视图分页时显示“更多”等信息。如图6-2b所示，可以在表视图分页时显示“加载中”等信息。
- **单元格**（cell）。它是组成表视图每一行的单位视图。

- 节（section）。它由多个单元格组成，有节头（section header）和节脚（section footer）。
  - **节头**。节的头，描述节的信息，如图6-3所示，文字左对齐。
  - **节脚**。节的脚，也可描述节的信息和声明，如图6-3所示，文字左对齐。

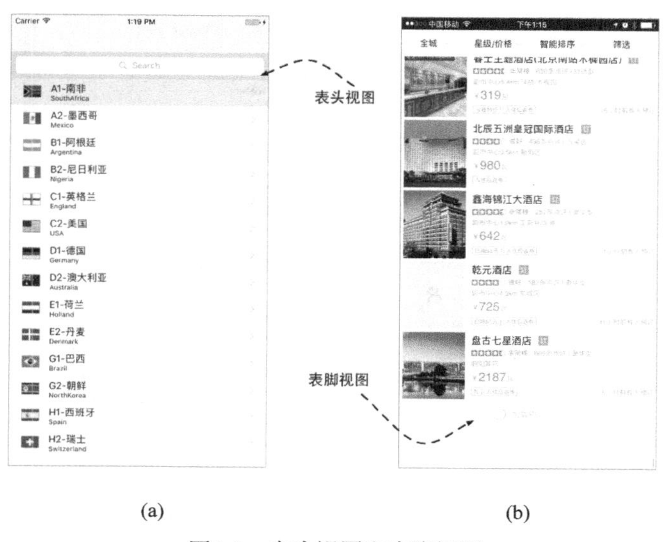

图6-2 表头视图和表脚视图

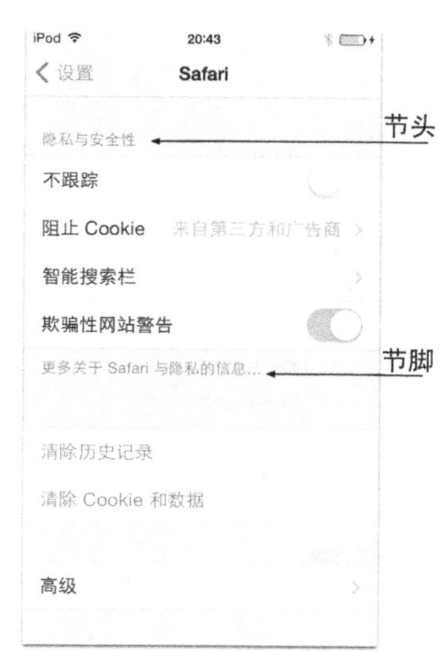

图6-3 节头和节脚

6.1.2 表视图的相关类

表视图（UITableView）继承自UIScrollView，且有两个协议：UITableViewDelegate委托协议和UITableViewDataSource数据源协议。此外，表视图还包含很多其他类，其中UITableViewCell类是单元格类，UITableViewController类是UITableView的控制器，UITableViewHeaderFooterView类（它是iOS 6之后才有的新类）用于为节头和节脚提供视图，这些类的构成如图6-4所示。

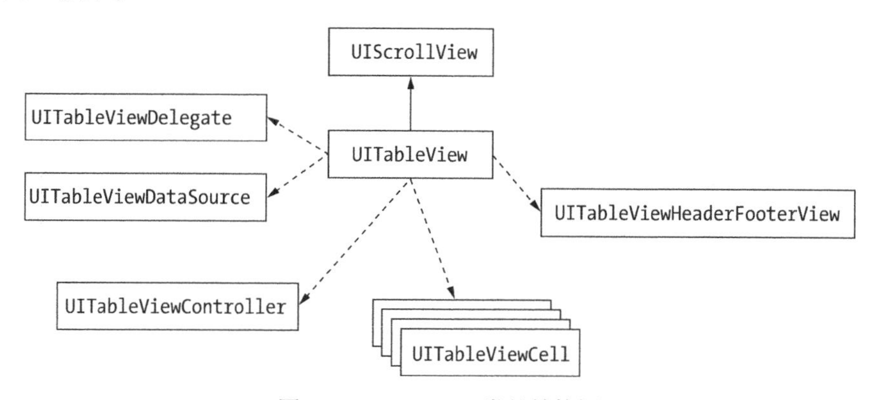

图6-4 UITableView类的结构图

图6-4所示的类只是表视图使用过程中涉及的几个主要类，其他的类和常量我们将在使用过程中逐一介绍。

6.1.3 表视图分类

iOS中的表视图主要分为普通表视图（如图6-5所示）和分组表视图（如图6-6所示），下面简要介绍一下这两种视图。

- **普通表视图。** 主要用于动态表[1]，动态表一般在单元格数目未知的情况下使用。
- **分组表视图。** 可以用于动态表和静态表[2]。动态表分组时，单元格分成不同部分，而每一部分中单元格中的数据是类似的，如图6-6a所示。静态表分组时，会将功能类似的视图放置在一起，如图6-6b所示。

图6-5　普通表视图

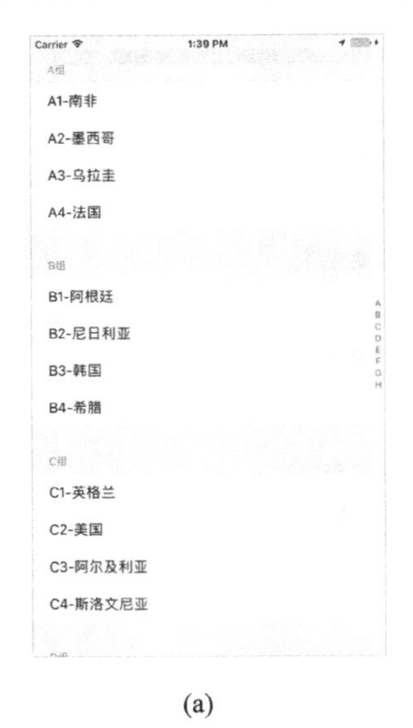

(a)

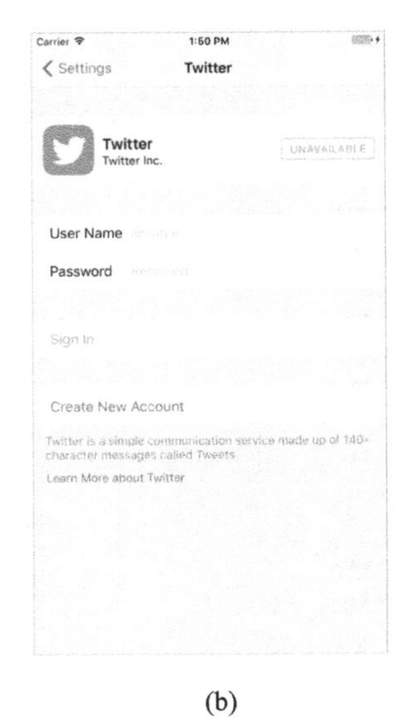

(b)

图6-6　分组表视图

此外，在表视图中还可以带有索引列、选择列和搜索栏等，下面介绍一下具有这种特征的表视图。

图6-7所示的是索引表视图。一般情况下，在表视图超过一屏时应该添加索引列。图6-8所示的是选择表视图，用于给用户提供一个选择列表。由于iOS标准控件没有复选框控件，所以一般使用选择表视图来替代其他平台的复选框控件。

图6-7　索引表视图

图6-8　选择表视图

图6-9所示的是带有搜索栏的表视图。由于单元格很多，所以我们需要借助搜索栏进行过滤。搜索栏一般放在表头，也就是说，只有表视图翻到最顶端时才会看到搜索栏。图6-10所示的是分页表视图。一般情况下，Twitter、

---

① 动态表是展示动态数据的表。

② 静态表一般用于控件的界面布局。

微博等需要网络请求的列表会使用分页表视图。分页表视图的表头中有刷新和加载等待标识，表脚中会有“更多”按钮或“加载更多”标识。对于此功能，iOS 6之后提供了下拉刷新控件。

图6-9 搜索栏表视图

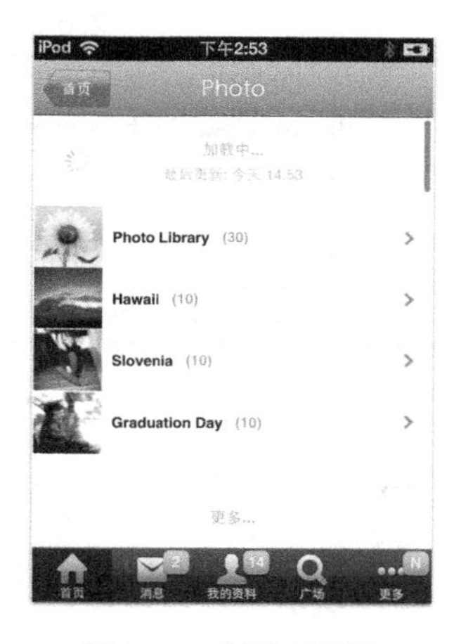

图6-10 分页表视图

表视图的分类不是绝对的。苹果提供了一些表视图的使用模式，使用时我们应首先考虑这些使用模式。当然，必要的话，我们还要根据业务需要进行合理创新。

6.1.4 单元格的组成和样式

如图6-11所示，单元格由图标、标题和扩展视图等组成。

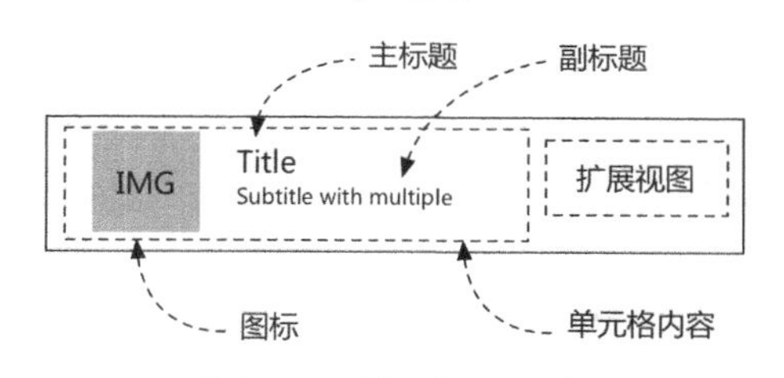

图6-11 单元格的组成

当然，单元格可以有很多样式，我们可以根据需要进行选择。图标、主标题和副标题可以有选择地设置，扩展视图可以内置或者自定义，其中内置的扩展视图是在枚举类型UITableViewCellAccessoryType中定义的，枚举类型UITableViewCellAccessoryType中定义的成员如表6-1所示。

表6-1 UITableViewCellAccessoryType枚举成员

Swift枚举成员	Objective-C枚举成员	说明
none	UITableViewCellAccessoryNone	没有扩展图标
disclosureIndicator	UITableViewCellAccessoryDisclosureIndicator	扩展指示器，触摸该图标将切换到下一级表视图，图标为 ›
detailDisclosureButton	UITableViewCellAccessoryDetailDisclosureButton	细节展示按钮，触摸该单元格的时候，表视图会以视图的方式显示当前单元格的更多详细信息，图标为 ⓘ ›
checkmark	UITableViewCellAccessoryCheckmark	选中标志，表示该行被选中，图标为✓
detailButton	UITableViewCellAccessoryDetailButton	细节按钮，触摸该单元格的时候，会显示当前单元格附加信息，图标为ⓘ

在开发过程中，我们应该首先考虑苹果公司提供的一些固有的单元格样式。iOS API提供的单元格样式是在枚举类型UITableViewCellStyle中定义的。枚举类型UITableViewCellStyle中定义的成员如表6-2所示。

表6-2　UITableViewCellStyle枚举成员

Swift枚举成员	Objective-C枚举成员	说　明
default	UITableViewCellStyleDefault	默认样式，如图6-12所示，只有图标和主标题
subtitle	UITableViewCellStyleSubtitle	Subtitle样式，如图6-13所示，有图标、主标题和副标题，副标题在主标题的下面
value1	UITableViewCellStyleValue1	Value1样式，如图6-14所示，有主标题和副标题，可以有图标
value2	UITableViewCellStyleValue2	Value2样式，如图6-15所示，有主标题和副标题，无图标

图6-12　默认样式

图6-13　Subtitle样式

图6-14　Value1样式

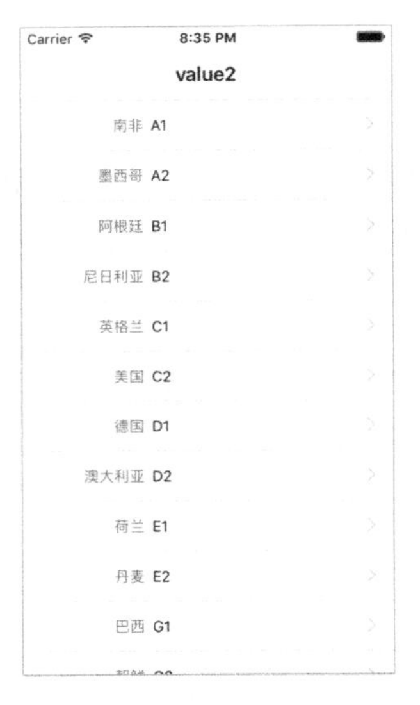

图6-15　Value2标题样式

如果以上单元格样式都不能满足业务需求，可以考虑自定义单元格。

6.1.5 数据源协议与委托协议

与UIPickerView等复杂控件类似，表视图在开发过程中也会使用委托协议和数据源协议，而表视图UITableView的数据源协议是UITableViewDataSource，委托协议是UITableViewDelegate。UITableViewDataSource协议中的主要方法如表6-3所示，其中必须要实现的方法有tableView:numberOfRowsInSection:[Swift版是tableView(_:numberOfRowsInSection:)]和tableView:cellForRowAtIndex-Path:[Swift版是tableView(_:cellForRowAt:)]。

表6-3 UITableViewDataSource协议的主要方法

方　法	返回类型	说　明
tableView:cellForRowAtIndexPath: [Swift版是tableView(_:cellForRowAt:)]	UITableViewCell	为表视图单元格提供数据，该方法是必须实现的方法
tableView:numberOfRowsInSection: [Swift版是tableView(_:numberOfRowsInSection:)]	Int	返回某个节中的行数
tableView:titleForHeaderInSection: [Swift版是tableView(_:titleForHeaderInSection:)]	String	返回节头的标题
tableView:titleForFooterInSection: [Swift版是tableView(_:titleForFooterInSection:)]	String	返回节脚的标题
numberOfSectionsInTableView: [Swift版是numberOfSections(in:)]	Int	返回节的个数
sectionIndexTitlesForTableView: [Swift版是sectionIndexTitles(for:)]	[AnyObject] (Objective-C版本为NSArray)	提供表视图节索引标题
tableView:commitEditingStyle:forRowAtIndexPath: [Swift版是tableView(_:commit:forRowAt:)]	无	为删除或修改提供数据

UITableViewDelegate协议主要用来设定表视图中节头和节脚的标题，并响应一些动作事件，主要的方法见表6-4，它们都是可选的。

表6-4 UITableViewDelegate协议的主要方法

方　法	返回类型	说　明
tableView:viewForHeaderInSection: [Swift版是tableView(_:viewForHeaderInSection:)]	UIView	为节头准备自定义视图，iOS 6之后可以使用UITableViewHeaderFooterView
tableView:viewForFooterInSection: [Swift版是tableView(_:viewForFooterInSection:)]	UIView	为节脚准备自定义视图，iOS 6之后可以使用UITableViewHeaderFooterView
tableView:didEndDisplayingHeaderView:forSection: [Swift版是tableView(_:didEndDisplayingHeaderView:forSection:)]	无	该方法在节头从屏幕中消失时触发
tableView:didEndDisplayingFooterView:forSection: [Swift版是tableView(_:didEndDisplayingFooterView:forSection:)]	无	当节脚从屏幕中消失时触发
tableView:didEndDisplayingCell:forRowAtIndexPath: [Swift版是tableView(_:didEndDisplaying:forRowAt:)]	无	当单元格从屏幕中消失时触发
tableView:didSelectRowAtIndexPath: [Swift版是tableView:didSelectRowAtIndexPath:]	无	响应选择表视图单元格时调用的方法
tableView:editActionsForRowAtIndexPath: [Swift版是tableView(_:editActionsForRowAt:)]	[AnyObject] (Objective-C版本为NSArray)	响应沿单元格水平滑动事件（iOS 8及其之后版本提供的方法）

此外，相关的方法还有很多，随着学习的深入，我们会在一些案例和项目中进一步接触。

6.2 简单表视图

表视图的形式灵活多变，本着由浅入深的原则，我们先从简单表视图开始学习。

下面我们创建一个如图6-16所示的简单表视图，其中单元格使用默认样式，有图标和主标题，具体创建步骤如下所示。

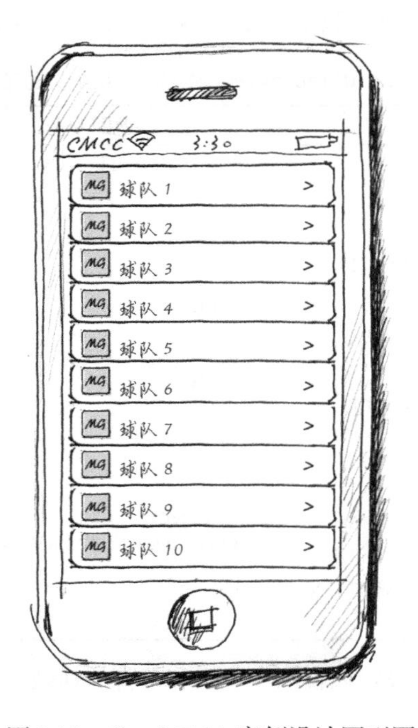

图6-16　SimpleTable案例设计原型图

6.2.1 实现协议方法

鉴于要创建的是一个最基本的表，我们只需实现UITableViewDataSource协议中必须要实现的方法即可，详见6.1.5节。简单表视图的时序图如图6-17所示，其中构造函数initWithFrame:style:在实例化表视图时调用。

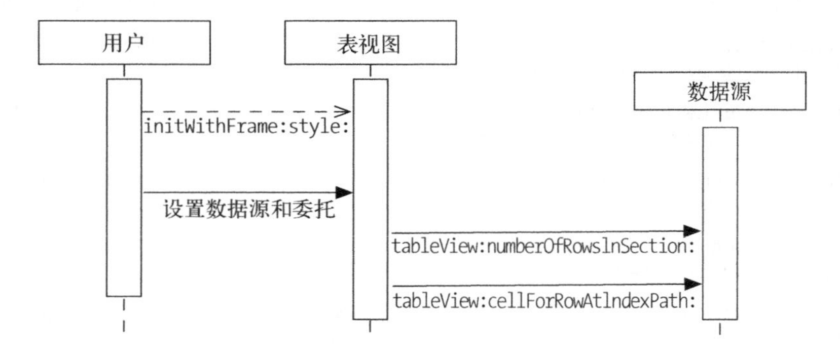

图6-17　简单表视图的时序图

表视图创建是在视图控制器加载时完成的，表视图显示的时候会调用表视图数据源对象的tableView:numberOfRowsInSection:方法，询问当前节中的行数。表视图单元格显示的时候会调用表视图数据源对象的tableView:cellForRowAtIndexPath:方法为单元格提供显示数据。

6.2.2 UIViewController 根视图控制器

SimpleTable案例的整个界面是一个表视图，这样的应用界面在构建时，可以使用如下两个方案。

- 方案一，采用默认的UIView作为根视图，然后把表视图（UITableVIew）作为UIView的子视图，如图6-18a所示。这种情况下，根视图控制器是普通的视图控制器UIViewController。
- 方案二，采用表视图（UITableVIew）作为根视图，如图6-18b所示。这种情况下，根视图控制器是表视图控制器UITableViewController。

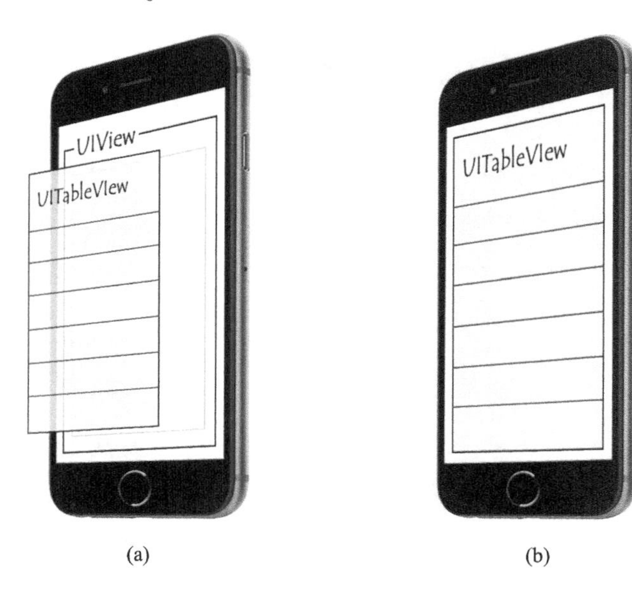

图6-18 构建根视图表视图应用界面

提示 这两种界面构建方案适合于构建整个界面完全是一个视图的情况，例如集合视图应用。我们在第5章的集合视图案例中采用的是第一种方案，事实上也可以采用第二种方案，但需要将根视图控制器由UITableViewController换成UICollectionViewController。

上述两种界面构建方法，都可以通过故事板或XIB中的Interface Builder实现，也可以通过代码实现。我们先介绍使用Interface Builder实现的UIViewController根视图控制器构建界面的过程。

1. Interface Builder实现

使用Single View Application模板创建一个工程，将工程命名为SimpleTable。

然后，我们需要添加资源图片和属性列表文件到工程，读者可以从本章的“资源”文件夹中找到“球队图片”文件夹（这是本案例需要的资源图片），以及描述球队信息的属性列表文件team.plist。参考5.4节，将它们添加到工程中。

打开Main.storyboard故事板文件，我们需要从对象库中拖曳一个Table View（表视图）到设计界面中，如图6-19所示，将整个表视图覆盖整个View。

如图6-19所示，默认情况下表视图中没有任何单元格，我们需要选择表视图来打开表视图属性检查器。如图6-20所示，设置Table View→Prototype Cells为1，注意不要添加多个，否则会发生错误。这时我们会发现，Table View下面添加了一个Table View Cell。

> **提示**　Table View→Content中有两个选项——Dynamic Prototypes和Static Cells，这两个选项只有在故事板中才有。Dynamic Prototypes用于构建“动态表”，而Static Cells的相关内容我们会在“静态表”中详细介绍。

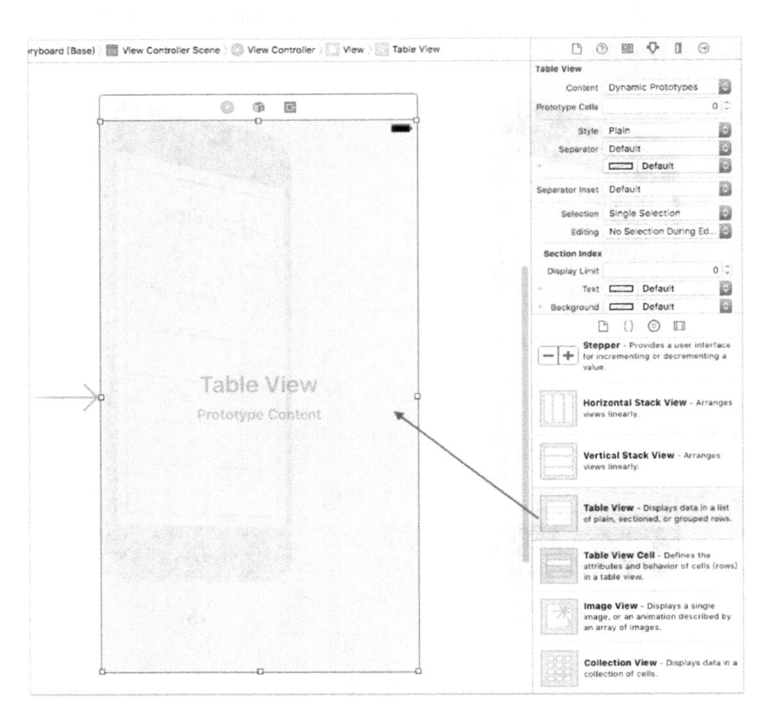

图6-19　拖曳表视图到设计界面

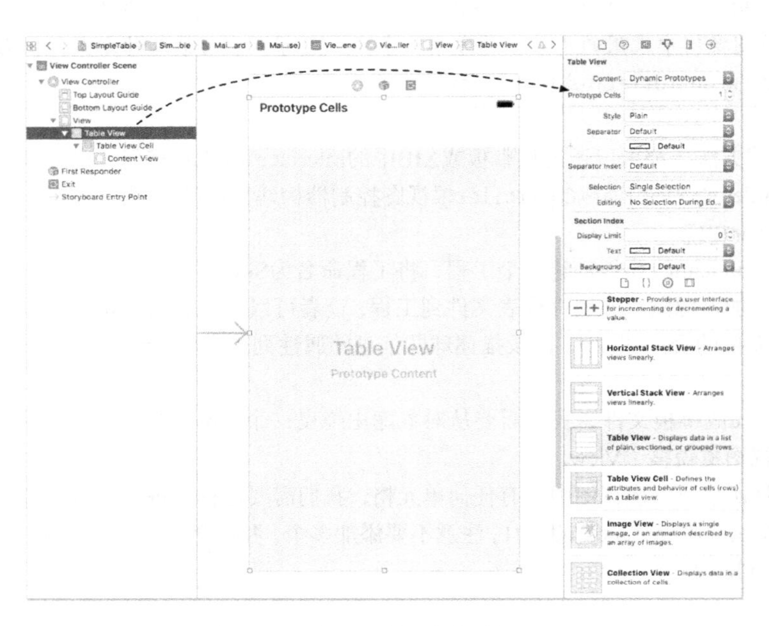

图6-20　在表视图中添加单元格

此外，表视图的单元格也是需要重用的。表视图中的可重用单元格与集合视图中的可重用单元格的概念一样。创建和获得可重用单元格有两种方式：Interface Builder和代码方式。我们先看看Interface Builder方式：选择View Controller Scene中的Table View Cell（表视图单元格），打开其属性检查器。如图6-21所示，其中Style属性用于设置单元格样式，其选项与6.1节中描述的表视图单元格的样式一致，而Identifier属性指可重用单元格的标识符，本例中我们设置为CellIdentifier。

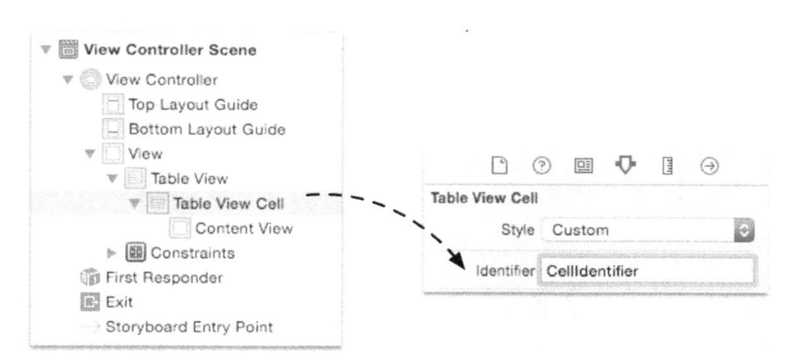

图6-21 设置可重用单元格的标识

在Interface Builder中设置完成后，就可以在视图控制器中编写代码了。视图控制器ViewController的主要代码如下：

```
//ViewController.swift文件
class ViewController: UIViewController,
➥UITableViewDataSource, UITableViewDelegate {                                   ①

    var listTeams : NSArray!

    override func viewDidLoad() {
        super.viewDidLoad()
        let plistPath = Bundle.main.path(forResource: "team", ofType: "plist")
        //获取属性列表文件中的全部数据
        self.listTeams = NSArray(contentsOfFile: plistPath!)                    ②
    }

    //MARK: -- UITableViewDataSource 协议方法
    func tableView(_ tableView: UITableView, numberOfRowsInSection section:
    ➥Int)-> Int {
        return self.listTeams.count                                             ③
    }

    func tableView(_ tableView: UITableView, cellForRowAt indexPath: IndexPath)
    ➥-> UITableViewCell {

        let cell = tableView.dequeueReusableCell(withIdentifier:
        ➥"CellIdentifier", for: indexPath)                                     ④

        let row = (indexPath as NSIndexPath).row

        let rowDict = self.listTeams[row] as! NSDictionary
        cell.textLabel?.text = rowDict["name"] as? String

        let imagePath = String(format: "%@.png", rowDict["image"] as! String)
        cell.imageView?.image = UIImage(named: imagePath)

        cell.accessoryType = .disclosureIndicator                               ⑤

        return cell
    }
}
```

```
//ViewController.h文件
#import <UIKit/UIKit.h>

@interface ViewController : UIViewController
➥<UITableViewDataSource, UITableViewDelegate>                                  ①

@end

//ViewController.m文件
#import "ViewController.h"

@interface ViewController ()

@property (nonatomic, strong) NSArray *listTeams;

@end

@implementation ViewController

- (void)viewDidLoad {
    [super viewDidLoad];
    NSString *plistPath = [[NSBundle mainBundle]pathForResource:
    ➥@"team" ofType:@"plist"];
    //获取属性列表文件中的全部数据
    self.listTeams = [[NSArray alloc] initWithContentsOfFile:plistPath];      ②

}

#pragma mark -- UITableViewDataSource 协议方法
- (NSInteger)tableView:(UITableView *)tableView
➥numberOfRowsInSection:(NSInteger)section {
    return [self.listTeams count];                                              ③
}

- (UITableViewCell *)tableView:(UITableView *)tableView
➥cellForRowAtIndexPath:(NSIndexPath *)indexPath {

    UITableViewCell *cell = [tableView
```

```
➥dequeueReusableCellWithIdentifier:@"CellIdentifier"
➥forIndexPath:indexPath];                                              ④

    NSUInteger row = [indexPath row];

    NSDictionary *rowDict = self.listTeams[row];
    cell.textLabel.text = rowDict[@"name"];

    NSString *imagePath = [[NSString alloc] initWithFormat: @"%@.png",
➥rowDict[@"image"]];
    cell.imageView.image = [UIImage imageNamed:imagePath];

    cell.accessoryType = UITableViewCellAccessoryDisclosureIndicator;      ⑤

    return cell;
}

@end
```

上述代码中，第①行用于声明根视图控制器类ViewController，可见它继承自UIViewController。另外，第①行同时声明了ViewController实现UITableViewDataSource和UITableViewDelegate协议。对于本例而言，UITableViewDelegate协议不需要实现。

第②行读取属性列表文件team.plist到集合属性listTeams中。team.plist文件结构如图6-22所示。

Key	Type	Value
▼Root	Array	(14 items)
▼Item 0	Dictionary	(2 items)
name	String	A1-南非
image	String	SouthAfrica
▼Item 1	Dictionary	(2 items)
name	String	A2-墨西哥
image	String	Mexico
▼Item 2	Dictionary	(2 items)
name	String	B1-阿根廷
image	String	Argentina
▶Item 3	Dictionary	(2 items)
▶Item 4	Dictionary	(2 items)
▶Item 5	Dictionary	(2 items)
▶Item 6	Dictionary	(2 items)
▶Item 7	Dictionary	(2 items)
▶Item 8	Dictionary	(2 items)
▶Item 9	Dictionary	(2 items)
▶Item 10	Dictionary	(2 items)
▶Item 11	Dictionary	(2 items)
▶Item 12	Dictionary	(2 items)
▶Item 13	Dictionary	(2 items)

图6-22　属性列表文件team.plist结构

第③行是表视图数据协议要求实现的方法，这里返回集合listTeams中的元素个数。

第④行通过可重用标识符获得可重用单元格对象，其中CellIdentifier是可重用单元格标识符，是我们在图6-21中设置的。

第⑤行用于设置扩展视图，单元格的accessoryType属性是扩展视图样式属性，扩展视图样式的定义见表6-1。

最后，不要忘记将委托和数据源的实现对象ViewController分配给表视图的委托属性delegate和数据源属性dataSource。通过Interface Builder进行分配的过程是：打开故事板文件，右击选择器，弹出如图6-23所示的右键菜单，将Outlets→dataSource后面的小圆点拖曳到左边的View Controller上，然后以同样的方式将Outlets→delegate后面的小圆点拖曳到左边的View Controller上。

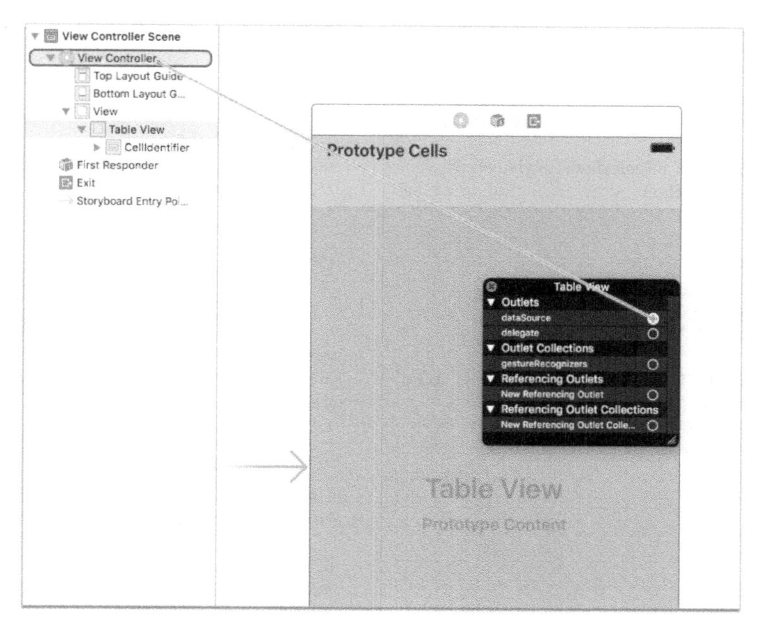

图6-23　在Interface Builder中分配数据源和委托

完成上述代码和操作过程后，就可以测试运行一下了。前面还只是通过Interface Builder实现的SimpleTable案例，下面我们通过代码实现该案例。

2. 代码实现

首先使用Single View Application模板创建SimpleTable工程，然后参考3.5.1节将SimpleTable工程修改为纯代码工程。

其中AppDelegate类与3.5.1节的案例完全一样，这里不再赘述，我们重点介绍一下视图控制器类ViewController，其代码如下：

```
//ViewController.swift文件
let CellIdentifier = "CellIdentifier"

class ViewController: UIViewController, UITableViewDataSource,
➥UITableViewDelegate {

    var listTeams : NSArray!
    var tableView : UITableView!

    override func viewDidLoad() {
        super.viewDidLoad()
        let plistPath = Bundle.main.path(forResource: "team", ofType: "plist")
        //获取属性列表文件中的全部数据
        self.listTeams = NSArray(contentsOfFile: plistPath!)

        self.tableView = UITableView(frame: self.view.frame, style: .plain) ①

        //设置表视图委托对象为self
        self.tableView.delegate = self
        //设置表视图数据对象为self
        self.tableView.dataSource = self

        self.view.addSubview(self.tableView)

    }
    ......
```

```
//ViewController.m文件
#import <UIKit/UIKit.h>

@interface ViewController : UIViewController
➥<UITableViewDataSource, UITableViewDelegate>

@end

#import "ViewController.h"

#define CellIdentifier @"CellIdentifier"

@interface ViewController ()

@property (nonatomic, strong) NSArray *listTeams;
@property (nonatomic, strong) UITableView *tableView;

@end

@implementation ViewController

- (void)viewDidLoad {
    [super viewDidLoad];

    NSString *plistPath = [[NSBundle mainBundle] pathForResource:@"team"
                                                           ofType:@"plist"];
    //获取属性列表文件中的全部数据
```

```
    func tableView(_ tableView: UITableView, cellForRowAt indexPath: IndexPath)
    ➥-> UITableViewCell {

        var cell: UITableViewCell! =
        ➥tableView.dequeueReusableCell(withIdentifier: CellIdentifier)    ②
        if (cell == nil) {                                                 ③
            cell = UITableViewCell(style: UITableViewCellStyle.default,
            ➥reuseIdentifier:CellIdentifier)                              ④
        }

        let row = (indexPath as NSIndexPath).row

        let rowDict = self.listTeams[row] as! NSDictionary
        cell.textLabel?.text = rowDict["name"] as? String

        let imagePath = String(format: "%@.png", rowDict["image"] as! String)
        cell.imageView?.image = UIImage(named: imagePath)

        cell.accessoryType = .disclosureIndicator

        return cell
    }
}
```

```
    self.listTeams = [[NSArray alloc] initWithContentsOfFile:plistPath];

    self.tableView = [[UITableView alloc] initWithFrame:self.view.frame
    ➥style:UITableViewStylePlain];                                        ①

    //设置表视图委托对象为self
    self.tableView.delegate = self;
    //设置表视图数据对象为self
    self.tableView.dataSource = self;

    [self.view addSubview:self.tableView];

}

#pragma mark -- UITableViewDataSource 协议方法
- (NSInteger)tableView:(UITableView *)tableView
➥numberOfRowsInSection:(NSInteger)section {
    return [self.listTeams count];
}

- (UITableViewCell *)tableView:(UITableView *)tableView
➥cellForRowAtIndexPath:(NSIndexPath *)indexPath {

    UITableViewCell *cell = [tableView
    ➥dequeueReusableCellWithIdentifier:CellIdentifier];                   ②
    if (cell == nil) {                                                     ③
        cell = [[UITableViewCell alloc]
        ➥initWithStyle:UITableViewCellStyleDefault
        ➥reuseIdentifier:CellIdentifier];                                 ④
    }

    NSUInteger row = [indexPath row];

    NSDictionary *rowDict = self.listTeams[row];
    cell.textLabel.text = rowDict[@"name"];

    NSString *imagePath = [[NSString alloc] initWithFormat: @"%@.png",
    ➥rowDict[@"image"]];
    cell.imageView.image = [UIImage imageNamed:imagePath];

    cell.accessoryType = UITableViewCellAccessoryDisclosureIndicator;

    return cell;
}

@end
```

上述代码中，第①行用于实例化表视图对象，使用的构造函数是`initWithFrame:style:`（Swift版是`init(frame:style:)`）。其中第一个参数是表视图的frame属性，由于需要覆盖整个界面，这里取值是根视图frame。第二个参数是表视图的样式，表视图样式有两种，是在枚举类型UITableViewStyle中定义的。枚举类型UITableViewCellStyle中定义的成员如表6-5所示。

表6-5　UITableViewStyle枚举成员

Swift枚举成员	Objective-C枚举成员	说　明
plain	UITableViewStylePlain	普通表视图
grouped	UITableViewStyleGrouped	分组表视图

第②行通过可重用标识符获得可重用单元格对象，类似于图6-21中的设置。

提示

获得可重用单元格对象时，Interface Builder实现采用的是dequeueReusableCellWithIdentifier:-forIndexPath:（Swift版是dequeueReusableCell(withIdentifier:for:)）方法，而代码实现采用的是dequeueReusableCellWithIdentifier:（Swift版是dequeueReusableCell(withIdentifier:)）方法。dequeueReusableCellWithIdentifier:方法要配合第③行和第④行使用，即先判断是否找到可以重用的单元格，如果没有，则通过单元格initWithStyle:reuseIdentifier:（Swift版是init(frame:style:)）构造函数创建单元格对象。

6.2.3 UITableViewController 根视图控制器

上一节我们介绍了根视图控制器为UIViewController实现方式的SimpleTable案例，本节介绍根视图控制器为UITableViewController实现方式的SimpleTable案例。

我们先介绍使用Interface Builder实现的UITableViewController根视图控制器构建界面的过程。

1. Interface Builder实现

参考6.2.2节创建工程SimpleTable。打开Interface Builder设计界面，由于模板生成的视图控制器不是表视图控制器，所以需要在View Controller Scene中删除View Controller，然后再从控件库中拖曳一个Table View Controller到设计界面，如图6-24所示。

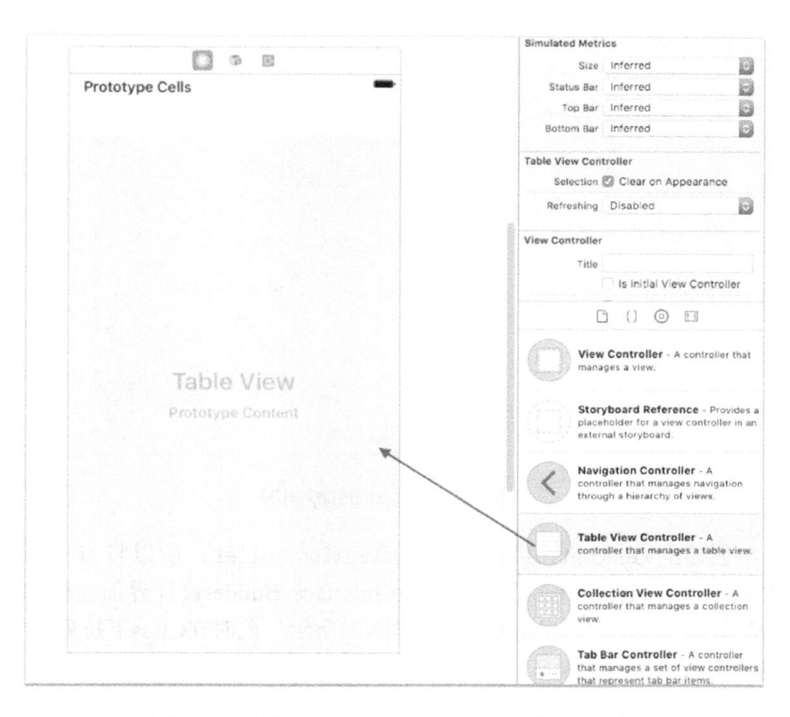

图6-24 拖曳Table View Controller到设计界面

由于工程初始视图控制器被删除了，重新添加一个表视图控制器后，我们需要设置表视图控制器为初始视图控制器（如图6-25所示），否则应用启动后是"黑屏"。这个设置过程是：选择场景中的Table View Controller，然后选择右边的属性检查器，选中View Controller→Is Initial View Controller复选框，之后的设计界面如图6-26所示。此时视图控制器上出现一个箭头，这个箭头表示该视图控制器是初始视图控制器，初始视图控制器就是根视图控制器。

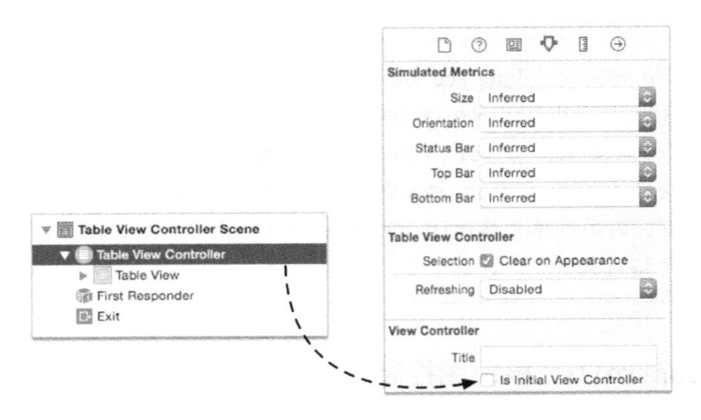

图6-25　设置表视图控制器为初始视图控制器

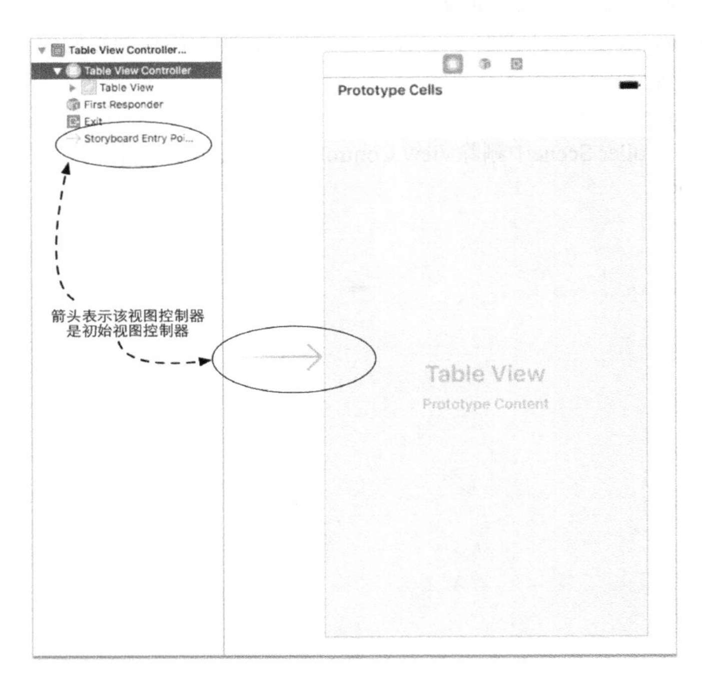

图6-26　初始视图控制器

由于采用Xcode版本创建的`ViewController`父类是`UIViewController`，所以将`ViewController`父类从原来的`UIViewController`修改为`UITableViewController`。然后在Interface Builder设计界面左侧的Scene列表中选择Table View Controller，打开表视图控制器的标识检查器，如图6-27所示。此时在Class下拉列表中选择ViewController，这是我们自己编写的视图控制器。

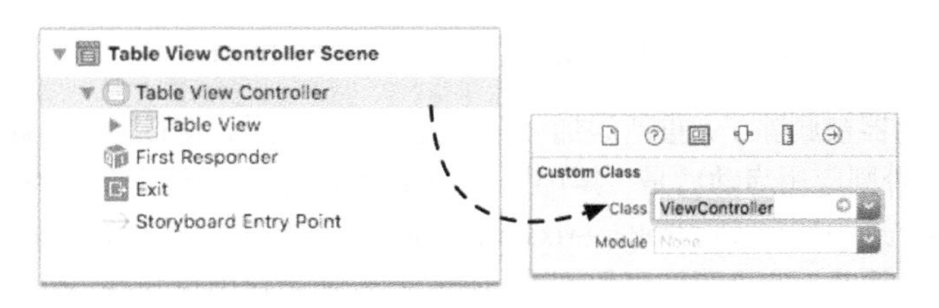

图6-27　表视图控制器的标识检查器

然后参考6.2.2节设置可重用单元格标识符。

代码部分几乎与6.2.2 节的Interface Builder实现完全一样，ViewController中的主要代码如下：

```
//ViewController.swift文件
class ViewController: UITableViewController {                              ①

    //MARK: -- UITableViewDataSource 协议方法
    override func tableView(_ tableView: UITableView,
    ➥numberOfRowsInSection section: Int) -> Int {                          ②
        ......
    }

    override func tableView(_ tableView: UITableView,
    ➥cellForRowAtIndexPath indexPath: NSIndexPath) -> UITableViewCell {    ③
        ......
    }
    ......
}
```

```
//ViewController.h文件
#import <UIKit/UIKit.h>

@interface ViewController : UITableViewController                          ①

@end
```

上述代码中，第①行用于声明视图控制器ViewController。

提示 ViewController父类是UITableViewController类型有很多好处，首先不需要声明实现UITableView-DataSource和UITableViewDelegate协议，其次不需要把ViewController分配给表视图的委托属性delegate和数据源属性dataSource，也不需要代码实现分配。这是因为UITableViewController已经实现了UITableViewDataSource和UITableViewDelegate协议，并且分配表视图的委托属性delegate和数据源属性dataSource。

在Swift版中，UITableViewDataSource和UITableViewDelegate协议方法前面加override关键字，见第②行和第③行。这是因为这些方法是重写父类UITableViewController的方法，而不是直接实现UITableViewDataSource和UITableViewDelegate协议的方法。

2. 代码实现

参考6.2.2节创建工程SimpleTable。代码实现不需要Interface Builder，我们直接看看有什么不同。

代码部分几乎与6.2.2 节的代码实现完全一样，ViewController中的主要代码如下：

```
//ViewController.swift文件
class ViewController: UITableViewController {                              ①

    var listTeams : NSArray!
    var tableView : UITableView!                                           ②

    override func viewDidLoad() {
        super.viewDidLoad()
        let plistPath = Bundle.main.path(forResource: "team", ofType: "plist")
        //获取属性列表文件中的全部数据
        self.listTeams = NSArray(contentsOfFile: plistPath!)

        self.tableView = UITableView(frame: self.view.frame, style: .plain) ③

        self.tableView.delegate = self                                     ④
        self.tableView.dataSource = self                                   ⑤

        self.view.addSubview(self.tableView)                               ⑥
    }

    //MARK: -- UITableViewDataSource 协议方法
    override func tableView(_ tableView: UITableView,
    ➥numberOfRowsInSection section: Int) -> Int {
```

```
//ViewController.h文件
#import <UIKit/UIKit.h>

@interface ViewController : UITableViewController                          ①

@end

//ViewController.m文件
#import "ViewController.h"

#define CellIdentifier @"CellIdentifier"

@interface ViewController ()

@property (nonatomic, strong) NSArray *listTeams;
@property (nonatomic, strong) UITableView *tableView;                      ②

@end

@implementation ViewController

- (void)viewDidLoad {
    [super viewDidLoad];
```

```
        ......
    }

    override func tableView(_ tableView: UITableView,
    ➥cellForRowAtIndexPath indexPath: NSIndexPath) -> UITableViewCell {
        ......
    }
    ......
}
```

```
NSString *plistPath = [[NSBundle mainBundle]pathForResource:@"team"
                                                     ofType:@"plist"];
//获取属性列表文件中的全部数据
self.listTeams = [[NSArray alloc] initWithContentsOfFile:plistPath];

self.tableView = [[UITableView alloc] initWithFrame:self.view.frame
➥ style:UITableViewStylePlain];                                        ③

//设置表视图委托对象为self
self.tableView.delegate = self;                                         ④
//设置表视图数据对象为self
self.tableView.dataSource = self;                                       ⑤

[self.view addSubview:self.tableView];                                  ⑥
}
```

上述代码中，第①行用于声明视图控制器ViewController，父类是UITableViewController，不需要再声明实现UITableViewDataSource和UITableViewDelegate协议了。第②行~第⑥行是不再需要的代码。

提示 综上所述，广大读者会发现，在表视图和集合视图构建界面，采用UITableViewController作为根视图控制器，要比采用UIViewController作为根视图控制器简单一些。不足之处是使用Interface Builder实现时，Xcode没有一个合适的工程模板，我们需要自己重新做。而在使用代码实现时，这就不是问题了。所以我推荐使用UITableViewController作为根视图控制器方案，如果不特殊说明，本章自此之后，表视图案例都采用UITableViewController作为根视图控制器方案。

运行之后的效果如图6-28所示。注意，运行模拟器时，基于Retina显示屏3.5英寸和4英寸的效果最好。这里我们可以将单元格的样式Default替换为其他3种，来体验一下其他3种单元格样式的效果。

图6-28　简单表案例运行结果

> **提示** 如果我们在iOS 7之后的系统中运行，会发现表视图顶部与状态栏重叠了，这是因为iOS 7之后的状态栏是透明的。事实上，这个问题不需要担心。往往使用表视图的时候，表视图的顶部还会有一个导航栏，添加导航栏之后表视图顶部不会与状态栏重叠。

6.3　自定义表视图单元格

当苹果公司提供的单元格样式不能满足业务需求时，我们可以自定义单元格。在iOS 5之前，自定义单元格有两种实现方式：通过代码实现和用XIB技术实现。用XIB技术实现相对比较简单：创建一个.xib文件，然后再自定义一个继承`UITableViewCell`的单元格类即可。在iOS 5之后，我们又有了新的选择——用故事板实现，这种方式比XIB方式更简单一些。

这里我们把6.2节所示的SimpleTable案例的原型图修改一下，改后的原型图如图6-29所示。下面我们分别采用Interface Builder和代码实现案例CustomCell。

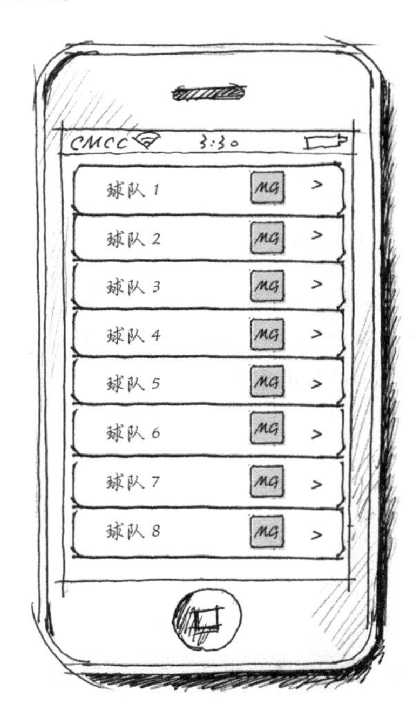

图6-29　使用CustomCell自定义单元格设计原型图

6.3.1　Interface Builder 实现

采用Single View Application工程模板创建一个名为CustomCell的表视图工程，操作过程参考6.2.3节中的Interface Builder实现部分，修改根视图控制器为表视图控制器`UITableViewController`。

打开故事板文件，选中单元格，从对象库中拖曳一个Label和Image View控件到单元格内部，如图6-30所示，调整好它们的位置。

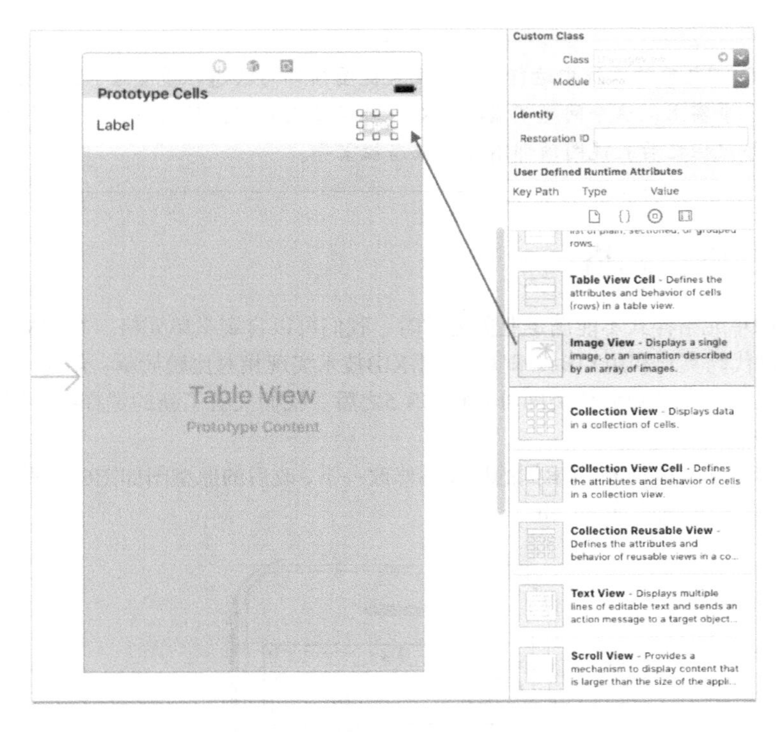

图6-30　设计表视图单元格

创建自定义单元格类`CustomCell`，具体操作方法为：右击工程名，在弹出的快捷菜单中选择New File...，然后在打开的Choose a template for your new file对话框中选择Cocoa Touch Class文件模板，如图6-31所示；点击Next按钮，此时会弹出如图6-32所示的对话框，在Subclass of中选择`UITableViewCell`为其父类，在Class项目中输入CustomCell，然后点击Next按钮创建文件。

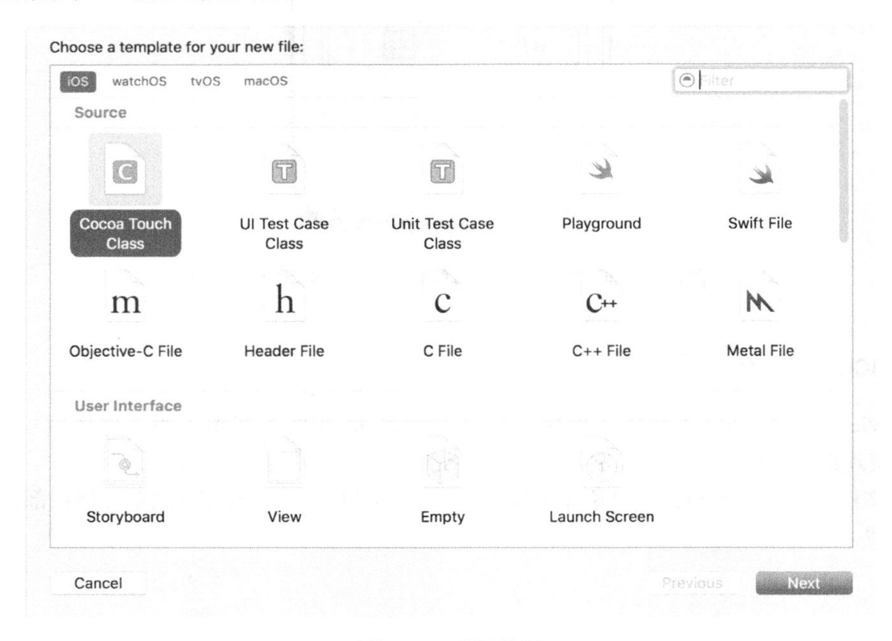

图6-31　选择模板

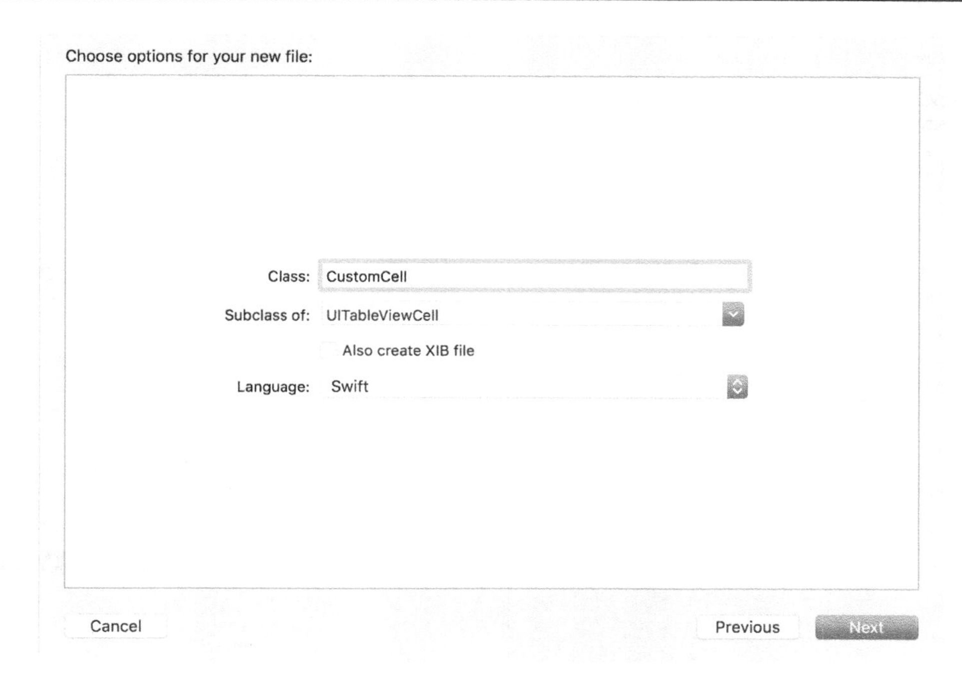

图6-32　创建自定义单元格类CustomCell

在Interface Builder设计界面中选择View Controller Scene中的Table View Cell，然后打开单元格的标识检查器，如图6-33所示，在Class下拉列表中选择CustomCell类。

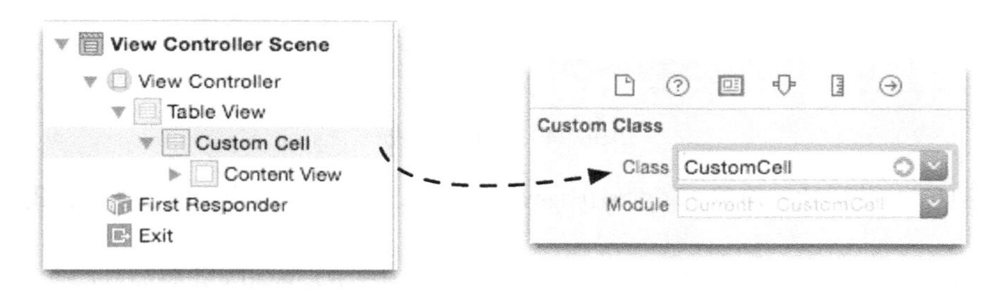

图6-33　选择CustomCell类

接着，为Label和Image View控件连接输出口。打开辅助编辑器，找到CustomCell.swift（Objective-C版本中是CustomCell.h）文件，选中单元格中的Image View视图，同时按住control键，将Image View拖曳到如图6-34所示的位置，此时释放鼠标，会弹出一个对话框，此时参考4.3.1节在弹出的对话框中设置输出口，并将其命名为myImageView。然后，请使用同样的方法将Label控件与输出口属性myLabel连接好。

设置完成后，CustomCell类的代码如下：

```
//CustomCell.swift文件
import UIKit

class CustomCell: UITableViewCell {

    @IBOutlet weak var myImageView: UIImageView!

    @IBOutlet weak var myLabel: UILabel!

    override func awakeFromNib() {
        super.awakeFromNib()
```

```
//CustomCell.h文件
#import <UIKit/UIKit.h>

@interface CustomCell : UITableViewCell
@property (weak, nonatomic) IBOutlet UILabel *name;
@property (weak, nonatomic) IBOutlet UIImageView *image;

@end
```

```
    }

    override func setSelected(selected: Bool, animated: Bool) {
        super.setSelected(selected, animated: animated)
    }

}
```

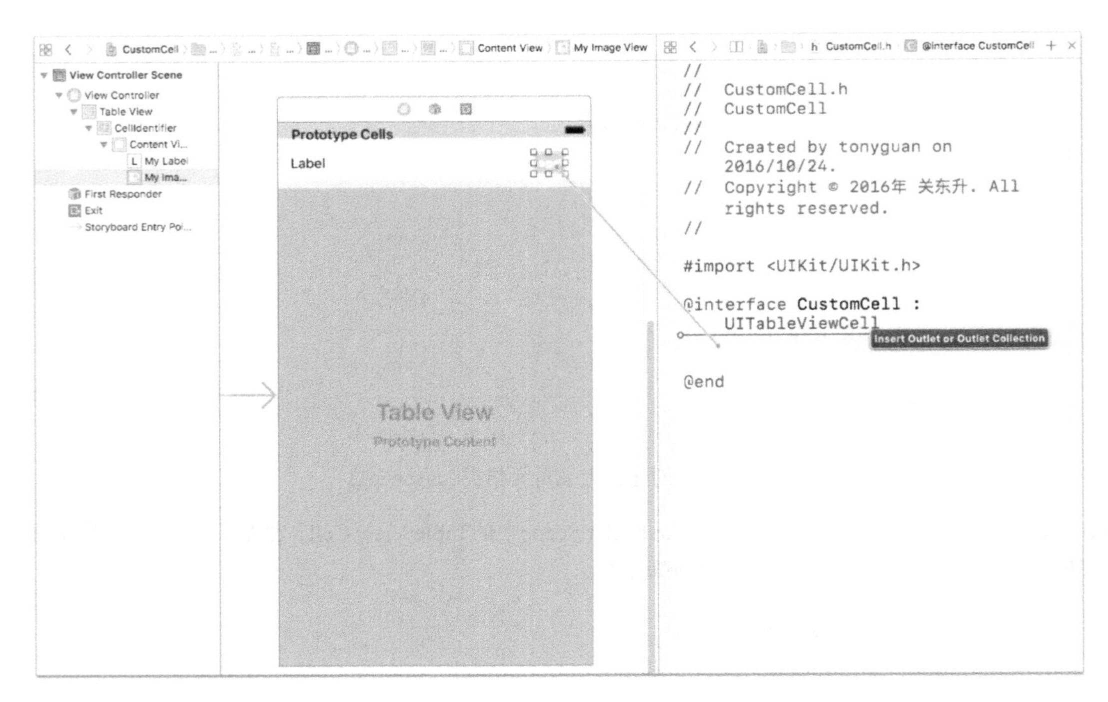

图6-34　输出口连线

其中CustomCell类的代码比较简单，属性myImageView和myLabel是定义输出口时添加的，我们不需要修改其他代码。

打开ViewController类，其中tableView:cellForRowAtIndexPath:方法的相关代码如下：

```
override func tableView(_ tableView: UITableView,
➥cellForRowAtindexPath: IndexPath) -> UITableViewCell {

    let cell:CustomCell! = tableView.dequeueReusableCell(withIdentifier:
    ➥cellIdentifier, for:indexPath) as? CustomCell                          ①

    let row = (indexPath as NSIndexPath).row

    let rowDict = self.listTeams[row] as! NSDictionary
    cell.myLabel.text = rowDict["name"] as? String                           ②

    let imageFile = rowDict["image"] as? String
    let imagePath = String(format: "%@.png", imageFile!)

    cell.myImageView.image = UIImage(named: imagePath)                       ③

    cell.accessoryType = .disclosureIndicator

    return cell
}
```

```
- (UITableViewCell *)tableView:(UITableView *)tableView
➥cellForRowAtIndexPath:(NSIndexPath *)indexPath {

    CustomCell *cell = [tableView dequeueReusableCellWithIdentifier:
    ➥cellIdentifier forIndexPath:indexPath];                                ①

    NSUInteger row = [indexPath row];

    NSDictionary *rowDict = self.listTeams[row];
    cell.myLabel.text = rowDict[@"name"]                                     ②

    NSString *imageFile = rowDict[@"image"];
    NSString *imagePath = [[NSString alloc] initWithFormat:@"%@.png",
    ➥imageFile];

    cell.myImageView.image = [UIImage imageNamed:imagePath];                 ③

    cell.accessoryType = UITableViewCellAccessoryDisclosureIndicator;

    return cell;
}
```

上述代码中，第①行获得可重用单元格，单元格类型是我们自定义的CustomCell类型。第②行和第③行用于设置自定义单元格的Label和Image View控件的内容。

运行之后的效果如图6-35所示。注意，运行模拟器时，基于Retina显示屏4.7英寸的效果最好。

图6-35　CustomCell案例的运行结果

6.3.2　代码实现

上一节中我们在故事板中使用Interface Builder实现了CustomCell案例，这一节介绍如何通过代码实现CustomCell案例。

首先，使用Single View Application工程模板创建一个名为CustomCell的表视图工程，操作过程参考6.2.3节中的代码实现部分，修改根视图控制器为表视图控制器UITableViewController。

我们重点介绍自定义单元格类CustomCell和视图控制器类ViewController。CustomCell中的主要代码如下：

```
//CustomCell.swift文件
import UIKit
class CustomCell: UITableViewCell {

    var myLabel: UILabel!
    var myImageView: UIImageView!

    override init(style: UITableViewCellStyle, reuseIdentifier: String?) {  ①

        super.init(style: style, reuseIdentifier: reuseIdentifier)          ②

        //单元格的高度
        let cellHeight: CGFloat = self.frame.size.height

        let imageViewWidth: CGFloat = 39
```

```
//CustomCell.h文件
#import <UIKit/UIKit.h>

@interface CustomCell : UITableViewCell

@property (strong, nonatomic) UILabel *myLabel;
@property (strong, nonatomic) UIImageView *myImageView;

@end

//CustomCell.m文件
#import "CustomCell.h"

@implementation CustomCell
```

```swift
        let imageViewHeight: CGFloat = 28
        let imageViewLeftView: CGFloat = 300

        ///1.添加ImageView
        self.myImageView = UIImageView(frame: CGRect(x: imageViewLeftView,
        ➥y: (cellHeight - imageViewHeight) / 2, width: imageViewWidth,
        ➥height: imageViewHeight))                                    ③

        self.addSubview(self.myImageView)

        ///2.添加标签
        let labelWidth: CGFloat = 120
        let labelHeight: CGFloat = 21
        let labelLeftView: CGFloat = 15

        self.myLabel = UILabel(frame: CGRect(x: labelLeftView,
        ➥y: (cellHeight - labelHeight) / 2, width: labelWidth, height:
        ➥labelHeight))

        self.addSubview(self.myLabel)

    }

    required init?(coder aDecoder: NSCoder) {                          ④
        fatalError("init(coder:) has not been implemented")
    }

}
```

```objectivec
-(instancetype)initWithStyle:(UITableViewCellStyle)style
➥reuseIdentifier:(NSString *)reuseIdentifier {                          ①

    self = [super initWithStyle:style reuseIdentifier:reuseIdentifier];  ②

    if (self) {

        //单元格的高度
        CGFloat cellHeight = self.frame.size.height;

        CGFloat imageViewWidth = 39;
        CGFloat imageViewHeight = 28;
        CGFloat imageViewLeftView= 300;

        ///1.添加ImageView
        self.myImageView = [[UIImageView alloc] initWithFrame:
        ➥CGRectMake(imageViewLeftView,(cellHeight - imageViewHeight) / 2,
        ➥imageViewWidth, imageViewHeight)];                              ③

        [self addSubview:self.myImageView];

        ///2.添加标签
        CGFloat labelWidth = 120;
        CGFloat labelHeight = 21;
        CGFloat labelLeftView =15;
        self.myLabel = [[UILabel alloc]
        ➥initWithFrame:CGRectMake(labelLeftView,
        ➥(cellHeight - labelHeight) / 2, labelWidth, labelHeight)];

        [self addSubview:self.myLabel];
    }

    return self;
}

@end
```

用代码编写自定义表视图单元格时，需要初始化其中的子视图，这个过程是在initWithStyle: reuseIdentifier:构造函数中实现的，见第①行。第②行用于调用父类构造函数。

第③行用于创建并初始化ImageView对象，表达式(cellHeight - imageViewHeight) / 2是ImageView的X轴坐标，可以使ImageView在单元格中垂直居中。

另外，在Swift版第④行中，required init?(coder aDecoder: NSCoder)构造函数是单元格视图父类要求必须实现的。单元格视图父类中声明实现NSCoding协议，子类中要求实现该构造函数，但本例中并未用到该构造函数。

6.4　添加搜索栏

当表视图中的数据量比较大的时候，要找到指定的数据并不是件轻而易举的事情，幸好iOS给我们提供了一个搜索栏控件（UISearchBar）。一般情况下，搜索栏置于表视图的表头，只有翻到顶部搜索栏才会出现。但是，也有很多设计师会把搜索栏固定放置于屏幕顶部，使其不随表视图的翻动而移动。将搜索栏一直放在屏幕顶部必然导致屏幕的部分空间一直被占用，而iPhone的屏幕本来就很小，因此这样设计不会带来太好的用户体验。

搜索栏有多种样式，如表6-6所示。

表6-6　搜索栏样式说明

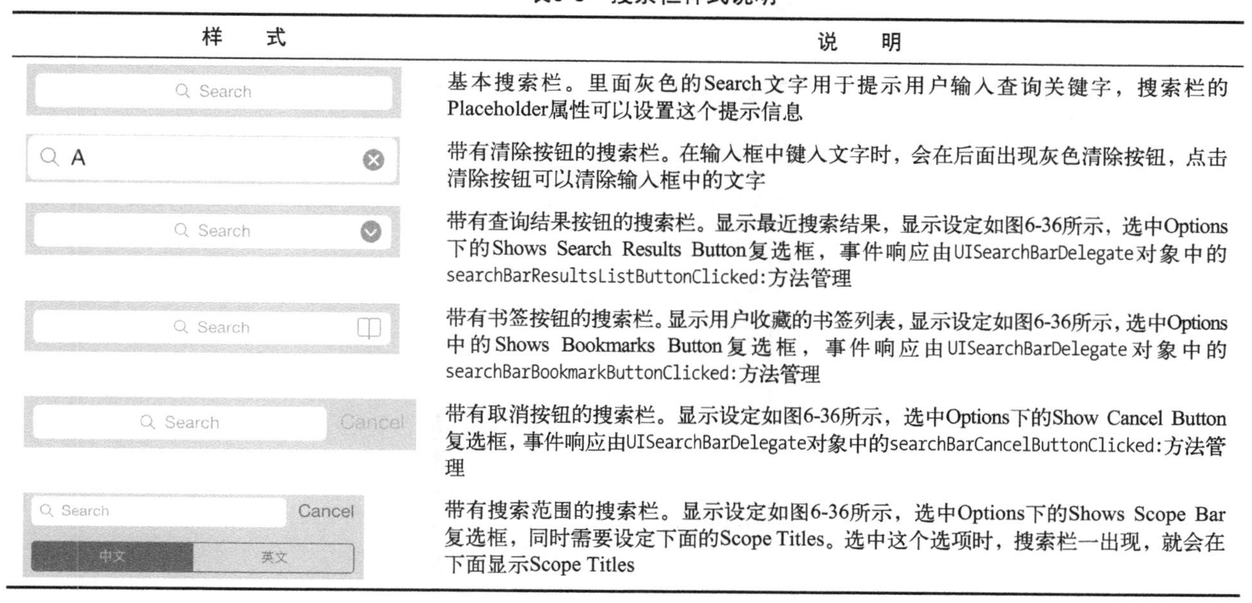

样　式	说　明
Search	基本搜索栏。里面灰色的Search文字用于提示用户输入查询关键字，搜索栏的Placeholder属性可以设置这个提示信息
A	带有清除按钮的搜索栏。在输入框中键入文字时，会在后面出现灰色清除按钮，点击清除按钮可以清除输入框中的文字
Search	带有查询结果按钮的搜索栏。显示最近搜索结果，显示设定如图6-36所示，选中Options下的Shows Search Results Button复选框，事件响应由`UISearchBarDelegate`对象中的`searchBarResultsListButtonClicked:`方法管理
Search	带有书签按钮的搜索栏。显示用户收藏的书签列表，显示设定如图6-36所示，选中Options中的Shows Bookmarks Button复选框，事件响应由`UISearchBarDelegate`对象中的`searchBarBookmarkButtonClicked:`方法管理
Search　Cancel	带有取消按钮的搜索栏。显示设定如图6-36所示，选中Options下的Show Cancel Button复选框，事件响应由`UISearchBarDelegate`对象中的`searchBarCancelButtonClicked:`方法管理
Search　Cancel 中文　英文	带有搜索范围的搜索栏。显示设定如图6-36所示，选中Options下的Shows Scope Bar复选框，同时需要设定下面的Scope Titles。选中这个选项时，搜索栏一出现，就会在下面显示Scope Titles

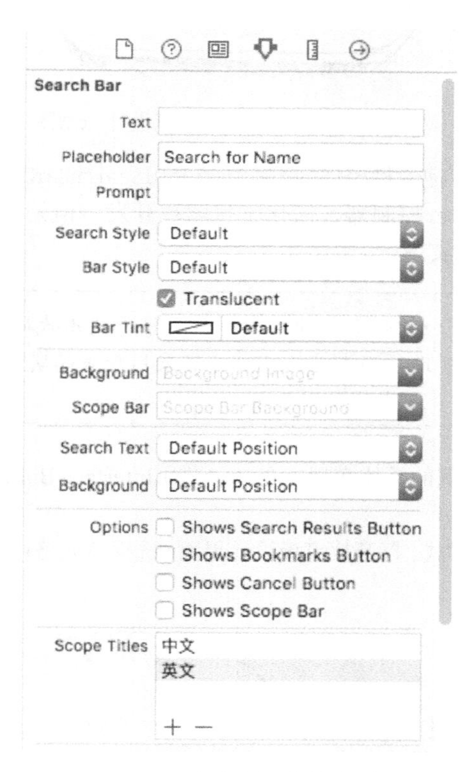

图6-36　搜索栏属性检查器

下面我们通过一个名为SearchbarTable的工程来介绍如何在表视图中添加搜索栏，案例原型图见图6-37。我们在6.2节简单表案例的基础上添加搜索栏，工程的创建过程参照上一节。有别于简单表视图的是，本案例中的单元格样式采用了有副标题的样式，其中副标题用于展示球队的英文名称，主标题是该球队的中文名称。在输入查询内容时，搜索栏下面会出现搜索范围栏，它有按中文查询和按英文查询两种查询方式。

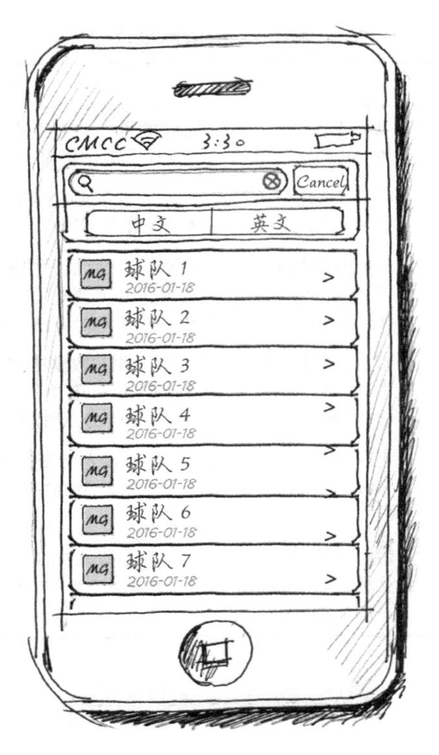

图6-37　SearchbarTable案例设计原型图

搜索栏是一个比较复杂的控件，搜索栏控件的委托协议是UISearchBarDelegate，不需要数据源协议。iOS 8及之后版本提供了UISearchController控制器，简化了搜索栏开发，但是UISearchController控制器不能通过Interface Builder实现，只能通过代码实现。

提示　考虑使用UISearchController控制器管理搜索栏，我推荐使用代码实现SearchbarTable案例。我们可以在搜索栏部分使用代码实现，而在其他UI部分还采用原来的实现方式，例如表视图部分可以还采用Interface Builder实现。

下面我们介绍SearchbarTable案例的具体实现，在6.2.3节用Interface Builder实现简单表案例的基础上添加搜索栏。

为了在单元格中显示副标题的样式，打开故事板设计界面，选择单元格样式，选择单元格属性Style为Subtitle，如图6-38所示。

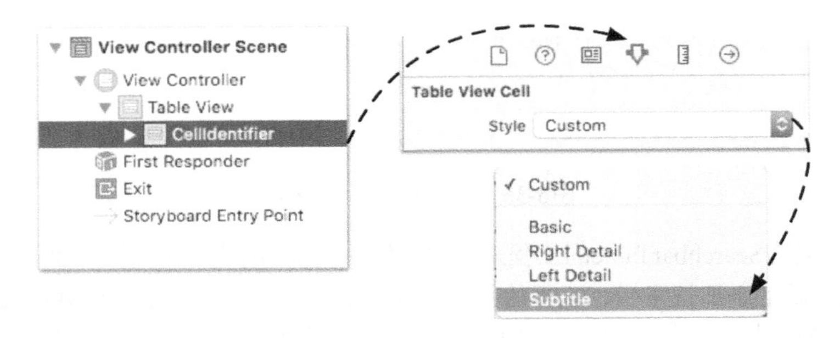

图6-38　选择单元格属性Style为Subtitle

下面我们看看视图控制器ViewController的定义和属性，以及视图加载方法viewDidLoad：

```
//ViewController.swift文件
import UIKit

class ViewController: UITableViewController,
➥UISearchBarDelegate, UISearchResultsUpdating {

    var searchController : UISearchController!

    //全部数据
    var listTeams : NSArray!
    //过滤后的数据
    var listFilterTeams : NSArray!

    override func viewDidLoad() {
        super.viewDidLoad()

        let plistPath = Bundle.main.path(forResource: "team", ofType: "plist")
        //获取属性列表文件中的全部数据
        self.listTeams = NSArray(contentsOfFile: plistPath!)
        //查询所有数据
        self.filterContentForSearchText("", scope:-1)

        //实例化UISearchController
        self.searchController = UISearchController(searchResults
        ➥Controller:nil)
        //设置self为更新搜索结果对象
        self.searchController.searchResultsUpdater = self
        //在搜索时，设置背景为灰色
        self.searchController.dimsBackgroundDuringPresentation = false

        //设置搜索范围栏中的按钮
        self.searchController.searchBar.scopeButtonTitles = ["中文", "英文"]
        self.searchController.searchBar.delegate = self

        //将搜索栏放到表视图的表头中
        self.tableView.tableHeaderView = self.searchController.searchBar

        self.searchController.searchBar.sizeToFit()
    }
    ......
}
```

```
//ViewController.m文件
#import "ViewController.h"

@interface ViewController () <UISearchBarDelegate, UISearchResultsUpdating>

@property (strong, nonatomic) UISearchController *searchController;

//全部数据
@property (nonatomic, strong) NSArray *listTeams;
//过滤后的数据
@property (nonatomic, strong) NSMutableArray *listFilterTeams;

//内容过滤方法
- (void)filterContentForSearchText:(NSString*)
➥searchText scope:(NSUInteger)scope;

@end

@implementation ViewController

- (void)viewDidLoad {
    [super viewDidLoad];

    NSString *plistPath = [[NSBundle mainBundle] pathForResource:@"team"
                                                          ofType:@"plist"];
    //获取属性列表文件中的全部数据
    self.listTeams = [[NSArray alloc] initWithContentsOfFile:plistPath];

    //查询所有数据
    [self filterContentForSearchText:@"" scope:-1];

    //实例化UISearchController
    self.searchController = [[UISearchController alloc]
    ➥initWithSearchResultsController:nil];
    //设置self为更新搜索结果对象
    self.searchController.searchResultsUpdater = self;
    //在搜索时，设置背景为灰色
    self.searchController.dimsBackgroundDuringPresentation = FALSE;

    //设置搜索范围栏中的按钮
    self.searchController.searchBar.scopeButtonTitles = @[@"中文", @"英文"];
    self.searchController.searchBar.delegate = self;

    //将搜索栏放到表视图的表头中
    self.tableView.tableHeaderView = self.searchController.searchBar;

    [self.searchController.searchBar sizeToFit];
}
......
@end
```

在定义ViewController类时，指定继承UITableViewController，并声明实现UISearchBarDelegate和UISearch-ResultsUpdating协议。属性listTeams用于装载全部球队的信息，是数组类型NSArray。listFilterTeams是查询之后的球队信息，它是可变数组类型NSMutableArray，是listTeams的子集。

为了查询方便，我们自定义了过滤结果集方法，该方法在ViewController中的实现代码如下：

```
func filterContentForSearchText(_ searchText: NSString, scope: Int) {
    if(searchText.length == 0) {
        //查询所有
        self.listFilterTeams = NSMutableArray(array:self.listTeams)
        return
    }
    var tempArray : NSArray!
```

```
- (void)filterContentForSearchText:(NSString*)searchText
➥scope:(NSUInteger)scope {

    if([searchText length]==0) {
        //查询所有
        self.listFilterTeams = [NSMutableArray arrayWithArray:
        ➥self.listTeams];
```

```
    if (scope == 0) {       //中文, name字段是中文名
        let scopePredicate = NSPredicate(format:"SELF.name contains[c] %@",
        ➥searchText)                                                    ①
        tempArray = self.listTeams.filtered(using: scopePredicate) as NSArray!
                                                                         ②
        self.listFilterTeams = NSMutableArray(array: tempArray)
    } else if (scope == 1) { //英文, image字段保存英文名
        let scopePredicate = NSPredicate(format:"SELF.image contains[c] %@",
        ➥searchText)
        tempArray = self.listTeams.filtered(using: scopePredicate) as NSArray!
        self.listFilterTeams = NSMutableArray(array: tempArray)
    } else {      //查询所有
        self.listFilterTeams = NSMutableArray(array: self.listTeams)
    }
}
```

```
        return;
    }

    NSPredicate *scopePredicate;
    NSArray *tempArray ;

    switch (scope) {
        case 0://中文, name字段是中文名
            scopePredicate = [NSPredicate
            ➥predicateWithFormat:@"SELF.name contains[c] %@",searchText]; ①
            tempArray =[self.listTeams filteredArrayUsingPredicate:scope
            ➥Predicate];                                                 ②
            self.listFilterTeams = [NSMutableArray arrayWithArray:tempArray];
            break;
        case 1: //英文, image字段保存英文名
            scopePredicate = [NSPredicate
            ➥predicateWithFormat:@"SELF.image contains[c] %@",searchText];
            tempArray =[self.listTeams filteredArrayUsingPredicate:
            ➥scope Predicate];
            self.listFilterTeams = [NSMutableArray arrayWithArray:tempArray];
            break;
        default:
            //查询所有
            self.listFilterTeams = [NSMutableArray arrayWithArray:
            ➥self.list Teams];
            break;
    }
}
```

上述代码用于定义filterContentForSearchText方法，其中参数searchText是要过滤结果的条件，参数scope是搜索范围栏中选择按钮的索引。本例中有两个按钮，我们将它们的值分别设置为0、1。

第①行用于进行中文查询（匹配字典中的name键），其中NSPredicate是谓词，可以定义一个查询条件，用来在内存中过滤集合对象，NSPredicate构造器中的format参数用于设置Predicate字符串格式。本例中的@"SELF.name contains[c] %@"是Predicate字符串，它有点像SQL语句或是HQL（Hibernate Query Language），其中SELF代表要查询的对象，SELF.name是查询对象的name字段（字典对象的键或实体对象的属性），contains[c]是包含字符的意思，其中小写c表示不区分大小写。

提示 关于Predicate字符串的语法，大家可以参考https://developer.apple.com/library/mac/#documentation/Cocoa/Conceptual/Predicates/Articles/pSyntax.html，也可以参考我写的另一本书《从零开始学Swift（第2版）》第22章。

在第②行中，NSArray的-filteredArrayUsingPredicate:方法是按照前面的条件进行过滤，结果返回的还是NSArray对象。我们需要重新构建一个NSMutableArray对象，才可以将结果放到属性listFilterTeams中。

下面还有两个NSPredicate的例子：

```
var array = NSMutableArray(array : ["Bill", "Ben", "Chris", "Melissa"])
let bPredicate = NSPredicate(format: "SELF beginswith[c] '@%'", "b")
let beginWithB = array.filtered(using: bPredicate)
//beginWithB 包含 { @"Bill", @"Ben" }.
let sPredicate = NSPredicate(format: "SELF contains[c] '@%'", "s")
array.filtered(using: sPredicate)
//数组包含 { @"Chris", @"Melissa" }
```

```
NSMutableArray *array = [NSMutableArray arrayWithObjects:@"Bill",
➥@"Ben", @"Chris", @"Melissa",  nil];
NSPredicate *bPredicate = [NSPredicate predicateWithFormat:
➥@"SELF beginswith[c] 'b'"];
NSArray *beginWithB = [array filteredArrayUsingPredicate:bPredicate];
//beginWithB 包含 { @"Bill", @"Ben" }.

NSPredicate *sPredicate = [NSPredicate predicateWithFormat:
➥@"SELF contains[c]  's'"];
[array filterUsingPredicate:sPredicate];
//数组包含 { @"Chris", @"Melissa" }
```

下面我们看看如何实现UISearchBarDelegate和UISearchResultsUpdating协议方法：

```swift
//MARK: -- 实现UISearchBarDelegate协议方法
func searchBar(_ searchBar: UISearchBar,
➥selectedScopeButtonIndexDidChange selectedScope: Int) {
    self.updateSearchResults(for: self.searchController)            ①
}

//MARK: -- 实现UISearchResultsUpdating协议方法
func updateSearchResults(for searchController: UISearchController) {
    let searchString = searchController.searchBar.text
    self.filterContentForSearchText(searchString! as NSString,
    ➥scope: searchController.searchBar.selectedScopeButtonIndex)   ②
    self.tableView.reloadData()                                     ③
}
```

```objectivec
#pragma mark -- 实现UISearchBarDelegate协议方法
- (void)searchBar:(UISearchBar *)searchBar
➥selectedScopeButtonIndexDidChange:(NSInteger)selectedScope {
    [self updateSearchResultsForSearchController:self.searchController];  ①
}

#pragma mark --  实现UISearchResultsUpdating协议方法
- (void)updateSearchResultsForSearchController:(UISearchController *)
➥searchController {
    NSString *searchString = searchController.searchBar.text;
    //查询
    [self filterContentForSearchText:searchString
    ➥scope:searchController.searchBar.selectedScopeButtonIndex];        ②
    [self.tableView reloadData];                                         ③
}
```

上述代码中，第①行实现UISearchBarDelegate委托协议的searchBar:selectedScopeButtonIndexDidChange:（Swift版是searchBar(_:selectedScopeButtonIndexDidChange:)）方法，点击搜索范围栏按钮时调用该方法。

第②行用于实现UISearchResultsUpdating委托协议的 updateSearchResultsForSearchController:（Swift版是updateSearchResults(for:)）方法，当搜索栏成为第一响应者，并且内容被改变时调用该方法。重新搜索完成后，一定要重新加载表视图，见第③行。

6.5 分节表视图

上一节中的简单表视图只有一节（section），它实际上是分节表视图的一个特例。一个表可以有多个节，节也有头有脚，分节是添加索引和分组的前提。

在简单表视图的例子中，我们省略了如下代码：

```swift
func numberOfSections(in tableView: UITableView) -> Int {
    return 1
}
```

```objectivec
- (NSInteger)numberOfSectionsInTableView:(UITableView *)tableView {
    return 1;
}
```

上述方法的返回值是表视图中节的个数，一旦返回值大于1，其他很多方法都要相应地有所变化。另外，我们还可能会用 tableView:titleForHeaderInSection:和tableView:titleForFooterInSection:方法来设置节头和节脚的标题。

6.5.1 添加索引

当表视图中有大量数据集合时，除了添加搜索栏，还可以通过添加索引来辅助查询。

为一个表视图建立索引的规则与在数据库表中建立索引的规则类似，但也有一定差别。对于图6-39所示的表，索引列中的索引标题几乎与显示的标题完全一样，这种情况下我们还需要索引吗？该表的另一个问题就是索引列与扩展视图发生了冲突，当你点击索引列时，往往会点击到扩展视图的图标。索引列表的正确使用方式应该像英文字典的索引一样，A字母代表A开头的所有单词，如图6-40所示。

索引的正确使用原则如下所示。

- **索引标题不能与显示的标题完全一样。**如果与要显示的标题一致，索引就变得毫无意义，如图6-39所示。
- **索引标题应具有代表性，能代表一个数据集合。**如图6-40所示，索引标题A下有一系列符合要求的数据。
- **如果采用了索引列表视图，一般情况下就不再使用扩展视图。**索引列表视图与扩展视图并存的时候，两者会存在冲突，点击索引标题时，很容易点击到扩展视图。

接下来，我们通过一个案例来演示正确使用索引的方式。修改属性列表文件（文件名为team_dictionary.plist），该文件的数据结构与上一节有所区别，如图6-41所示。

图6-39　错误使用索引

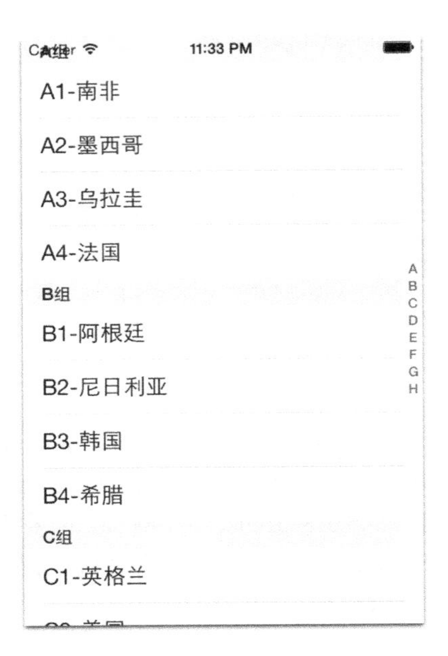

图6-40　正确使用索引

Key	Type	Value
▼Root	Dictionary	(8 items)
▼A组	Array	(4 items)
Item 0	String	A1-南非
Item 1	String	A2-墨西哥
Item 2	String	A3-乌拉圭
Item 3	String	A4-法国
▼B组	Array	(4 items)
Item 0	String	B1-阿根廷
Item 1	String	B2-尼日利亚
Item 2	String	B3-韩国
Item 3	String	B4-希腊
▶C组	Array	(4 items)
▶D组	Array	(4 items)
▶E组	Array	(4 items)
▶F组	Array	(4 items)
▶G组	Array	(4 items)
▶H组	Array	(4 items)

图6-41　属性列表文件team_dictionary.plist

使用Single View Application模板创建一个名为IndexTable的工程。下面我们看看视图控制器ViewController的定义和属性，以及视图加载方法viewDidLoad：

```
//ViewController.swift文件
class ViewController: UITableViewController {

    //从team_dictionary.plist文件中读取出来的数据
    var dictData : NSDictionary!
    //小组名集合
    var listGroupname : NSArray!

    override func viewDidLoad() {
```

```
//ViewController.m文件
#import "ViewController.h"

@interface ViewController ()

//从team_dictionary.plist文件中读取出来的数据
@property (nonatomic, strong) NSDictionary *dictData;
//小组名集合
@property (nonatomic, strong) NSArray *listGroupname;
```

```
        super.viewDidLoad()
        let plistPath = Bundle.main.path(forResource: "team_dictionary",
        ➥ofType: "plist")
        //获取属性列表文件中的全部数据
        self.dictData = NSDictionary(contentsOfFile: plistPath!)

        let tempList  = self.dictData.allKeys as! [String]             ①
        //对key进行排序
        self.listGroupname = tempList.sorted(by: <) as NSArray!         ②

    }
    ......
}
```

```
@end

@implementation ViewController

- (void)viewDidLoad {
    [super viewDidLoad];

    NSBundle *bundle = [NSBundle mainBundle];
    NSString *plistPath = [bundle pathForResource:@"team_dictionary"
    ➥ofType:@"plist"];
    //获取属性列表文件中的全部数据
    self.dictData = [[NSDictionary alloc] initWithContentsOfFile:
    ➥plistPath];

    NSArray* tempList  = [self.dictData allKeys];                       ①
    //对key进行排序
    self.listGroupname = [tempList sortedArrayUsingSelector:
    ➥@selector(compare:)];                                             ②
}
......
```

属性dictData用于从属性列表文件team_dictionary.plist中读取字典类型数据。属性listGroupname保存了小组名的集合，是从dictData属性中取出的，listGroupname属性是dictData的键的集合。

第①行用于从字典中取出来所有键，它的顺序是混乱状态（D组,C组,B组,H组,A组,G组,F组,E组），这是因为它是散列结构，内部结构是无序的。我们需要使用第②行代码重新对其排序：在Objective-C版中调用NSArray的sortedArrayUsingSelector:方法进行排序，其参数是Selector（选择器）类型，选择器指向compare:方法；在Swift版中通过调用字符串数组sorted(by: <)方法，其中<表示升序（如果是>，则表示降序）。

此外，我们还需要修改数据源方法tableView:numberOfRowsInSection:和tableView:cellForRowAtIndexPath:，具体代码如下所示：

```
//实现数据源协议方法
override func tableView(_ tableView: UITableView,
➥numberOfRowsInSection section: Int) -> Int {
    //按照节索引从小组名数组中获得组名
    var groupName = self.listGroupname[section] as String
    //将组名作为key，从字典中取出球队数组集合
    var listTeams = self.dictData[groupName] as NSArray
    return listTeams.count
}

override func tableView(_ tableView: UITableView,
➥cellForRowAtindexPath: IndexPath) -> UITableViewCell

    let cellIdentifier = "CellIdentifier"
    let cell:UITableViewCell! = tableView.dequeueReusableCell(withIdentifier:
    ➥cellIdentifier, for: indexPath)

    //获得选择的节
    let section = (indexPath as NSIndexPath).section
    //获得选择节中选中的行索引
    let row = (indexPath as NSIndexPath).row
    //按照节索引从小组名数组中获得组名
    let groupName = self.listGroupname[section] as! String
    //将组名作为key，从字典中取出球队数组集合
    let listTeams = self.dictData[groupName] as! NSArray

    cell.textLabel?.text = listTeams[row] as? String

    return cell
}
```

```
#pragma mark -- 实现数据源协议方法
- (NSInteger)tableView:(UITableView *)tableView
➥numberOfRowsInSection:(NSInteger)section {
    //按照节索引从小组名数组中获得组名
    NSString *groupName = self.listGroupname[section];
    //将组名作为key，从字典中取出球队数组集合
    NSArray *listTeams = self.dictData[groupName];
    return [listTeams count];
}


- (UITableViewCell *)tableView:(UITableView *)tableView
➥cellForRowAtIndexPath:(NSIndexPath *)indexPath {

    static NSString *CellIdentifier = @"CellIdentifier";
    UITableViewCell *cell = [tableView dequeueReusableCellWithIdentifier:
    ➥CellIdentifier forIndexPath:indexPath];

    //获得选择的节
    NSUInteger section = [indexPath section];
    //获得选择节中选中的行索引
    NSUInteger row = [indexPath row];
    //按照节索引从小组名数组中获得组名
    NSString *groupName = self.listGroupname[section];
    //将组名作为key，从字典中取出球队数组集合
    NSArray *listTeams = self.dictData[groupName];

    cell.textLabel.text = listTeams[row];

    return cell;
}
```

在表视图分节时，需要实现数据源中的numberOfSectionsInTableView:和tableView:titleForHeaderInSection:方法，具体实现代码如下：

```swift
override func numberOfSections(in tableView: UITableView) -> Int {
    return self.listGroupname.count
}

override func tableView(_ tableView: UITableView,
➥titleForHeaderInSection section: Int) -> String? {
    let groupName = self.listGroupname[section] as! String
    return groupName
}
```

```objc
- (NSInteger)numberOfSectionsInTableView:(UITableView *)tableView {
    return [self.listGroupname count];
}

- (NSString *)tableView:(UITableView *)tableView
➥titleForHeaderInSection:(NSInteger)section {
    NSString *groupName = self.listGroupname[section];
    return groupName;
}
```

上面这几个方法已实现了分节。分节只是添加索引的前提，数据源的sectionIndexTitlesForTableView:方法才与索引直接相关。我们在该方法的listGroupname集合中存放的数据是“A组,B组,C组,D组,E组,F组,G组,H组”，这些数据在索引列中显示的结果是A,B,C,D,E,F,G,H，将后面的“组”字符截取掉：

```swift
override func sectionIndexTitles(for tableView: UITableView) -> [String]? {
    var listTitles = [String]()
    //把“A组”改为“A”
    for item in self.listGroupname {
        let title = (item as AnyObject).substring(to: 1) as String
        listTitles.append(title)
    }
    return listTitles
}
```

```objc
-(NSArray *) sectionIndexTitlesForTableView: (UITableView *) tableView {
    NSMutableArray *listTitles = [[NSMutableArray alloc] init];
    //把“A组”改为“A”
    for (NSString *item in self.listGroupname) {
        NSString *title = [item  substringToIndex:1];
        [listTitles addObject:title];
    }
    return listTitles;
}
```

此时再看看运行结果。

6.5.2　分组

在Interface Builder设计器中选择表视图，打开其属性检查器，从Style属性下拉列表中选择Grouped选项，如图6-42所示。

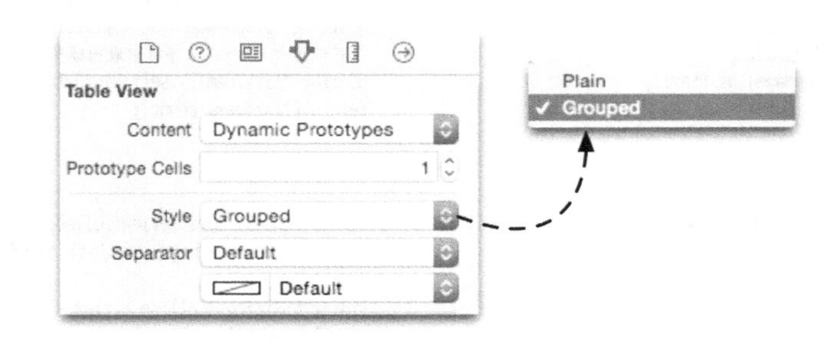

图6-42　表视图属性检查器

实现分组的代码如下：

```swift
let tableView = UITableView(frame: self.view.frame, style: .grouped)
```

```objc
UITableView *tableView = [[UITableView alloc] initWithFrame:self.view.frame
➥style:UITableViewStyleGrouped];
```

运行一下，得到的结果如图6-43所示，这个结果你是否满意呢？分析一下：我们分组的目的是让相关单元格放在“组”上（这个功能已经实现），但是界面中“组”的间距比较大，并不适合大量数据集的展示。需要说明的是，在数据量较小的情况下，没必要使用索引。

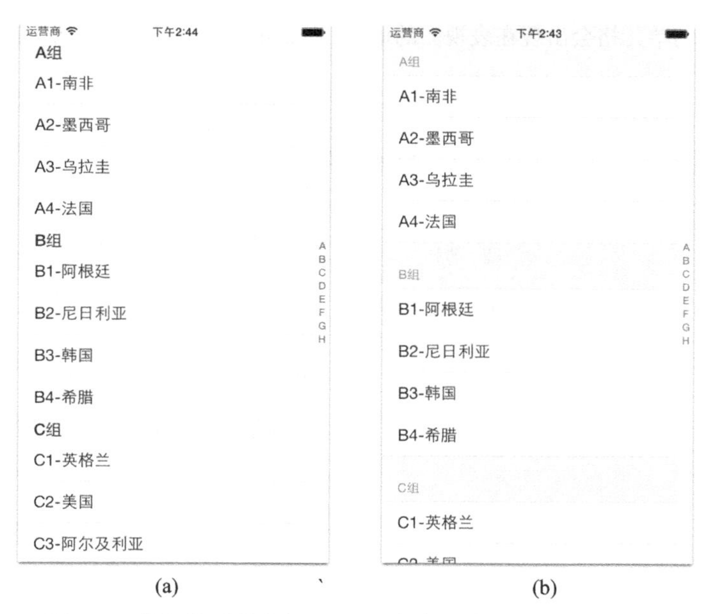

图6-43　分组前后的表视图（图a为分组前，图b为分组后）

6.6　插入和删除单元格

对于表视图，我们不仅需要浏览数据，有时还需要删除、插入和移动单元格等。本节先介绍单元格的插入和删除。

表视图一旦进入插入和删除状态，单元格的左边就会出现一个“编辑控件”，如图6-44所示。这个区域会显示删除控件或插入控件，具体显示哪个图标在表视图委托协议的`tableView:editingStyleForRowAtIndexPath:`方法中设定。

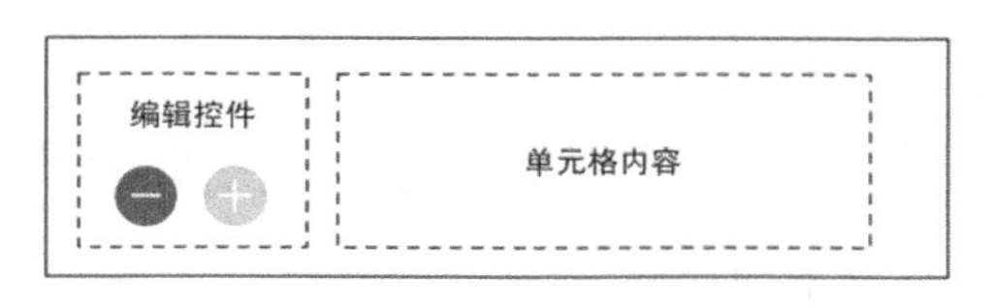

图6-44　单元格编辑控件

为了防止用户误操作，删除过程需要确认。删除控件时，删除控件从图6-45变成图6-46所示的样式，同时右侧会出现一个Delete按钮，点击该按钮才会成功删除数据。

图6-45　单元格删除控件

图6-46　单元格删除确认控件

提示　在iOS中，还有一个删除表视图单元格手势，那就是在单元格中从右往左滑动手势，此时也会在单元格右边出现Delete按钮。

插入数据时，新插入的单元格会出现在表视图的最后，如图6-47所示。当点击插入控件时，会增加一行数据，此操作可重复进行。

图6-47　插入和删除单元格的DeleteAddCell案例

单元格操作的核心是如下两个方法：表视图委托对象的tableView:editingStyleForRowAtIndexPath:方法和表视图数据源对象的tableView:commitEditingStyle:forRowAtIndexPath:方法。插入和删除单元格的时序图如图6-48所示。

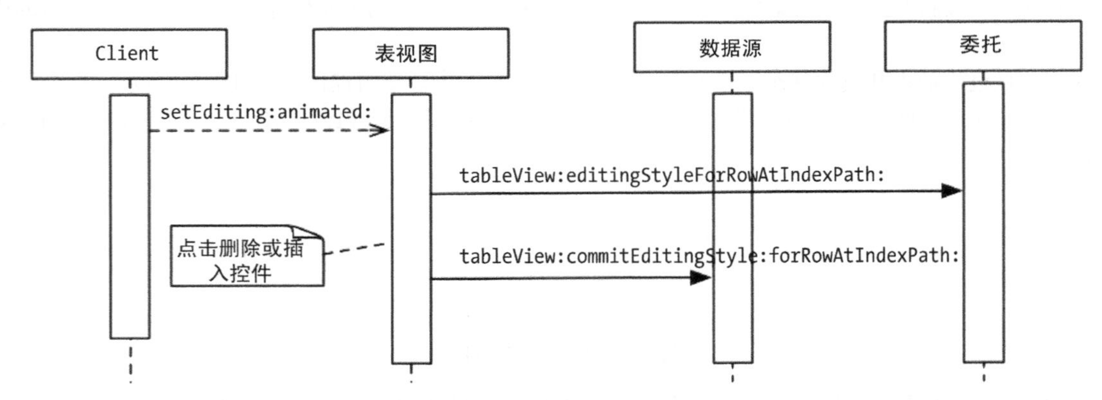

图6-48　插入和删除单元格的时序图

setEditing:animated:方法设定视图能否进入编辑状态，然后调用委托协议中的tableView:editingStyleForRowAtIndexPath:方法进行单元格编辑图标的设置。当用户删除或插入控件时，委托方法向数据源发出tableView:commitEditingStyle:forRowAtIndexPath:消息实现删除或插入的处理。

下面我们介绍如何通过Interface Builder和代码实现如图6-49所示的DeleteAddCell案例。

提示　在图6-49所示的DeleteAddCell案例中，界面上方有一个导航栏，用来显示界面标题和放置操作按钮。在表视图界面中添加导航栏视图的内容可以参考4.9.2节，但是在这种方式运行的结果中，导航栏与状态栏不能有机地融合起来（如图6-49a所示）。我们可以将表视图控制器嵌入到导航控制器中，这种处理方式可以获得如图6-49b所示的效果。

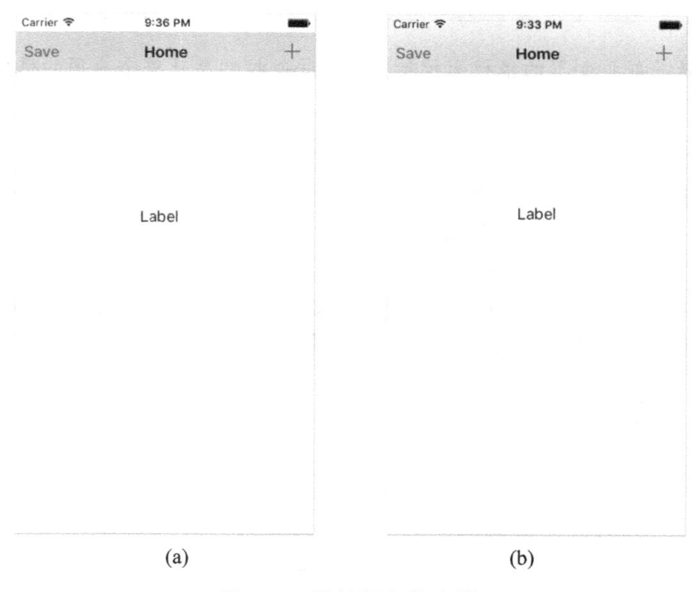

(a) (b)

图6-49 导航栏与状态栏

6.6.1 Interface Builder 实现

使用Single View Application模板创建一个名为 DeleteAddCell的工程。然后，打开Interface Builder设计界面，删除View Controller，然后从对象库中拖曳一个Navigation Controller到设计界面，如图6-50所示。添加Navigation Controller的同时也会添加一个，它是Root View Controller表视图控制器，这个表视图控制器是嵌入到导航控制器（Navigation Controller）中的。

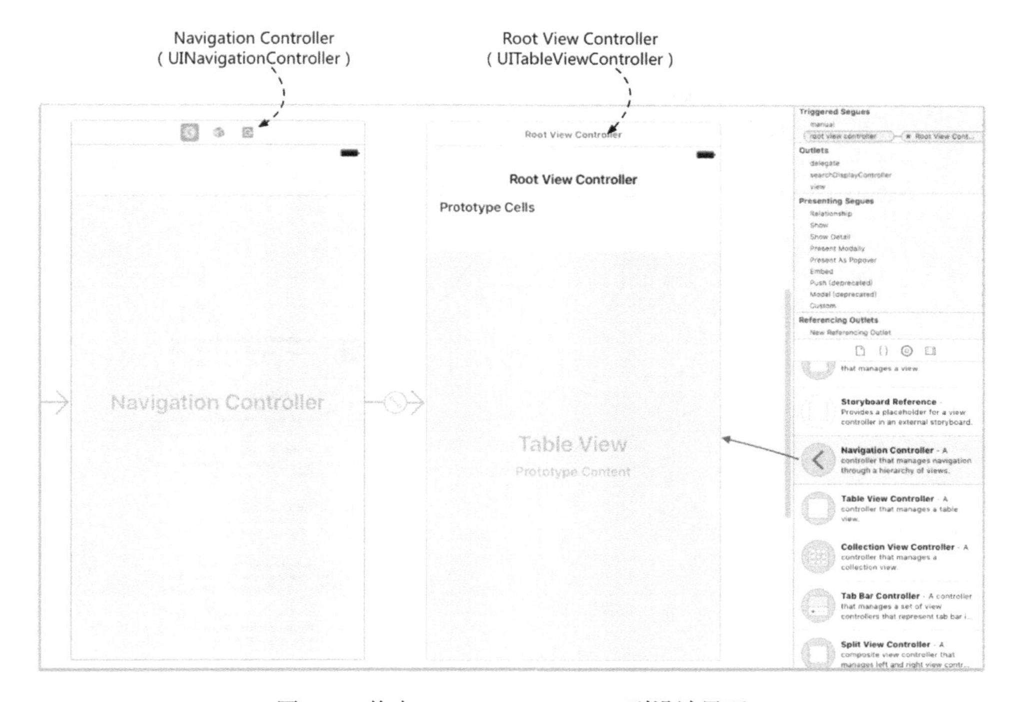

图6-50 拖曳Navigation Controller到设计界面

注意　表视图控制器是导航控制器的根视图控制器，而导航控制器是整个应用的根视图控制器。

由于工程的初始视图控制器被删除了，重新添加一个导航控制器时，需要设置导航控制器为初始视图控制器，否则屏幕会出现“黑屏”，具体设置过程参考6.2.3节。

然后修改`ViewController`的父类，从原来的`UIViewController`修改为`UITableViewController`。

在Interface Builder设计界面左侧的Scene列表中选择Root View Controller，打开表视图控制器的标识检查器，在Class下拉列表中选择`ViewController`（这是我们自己编写的视图控制器），如图6-51所示。

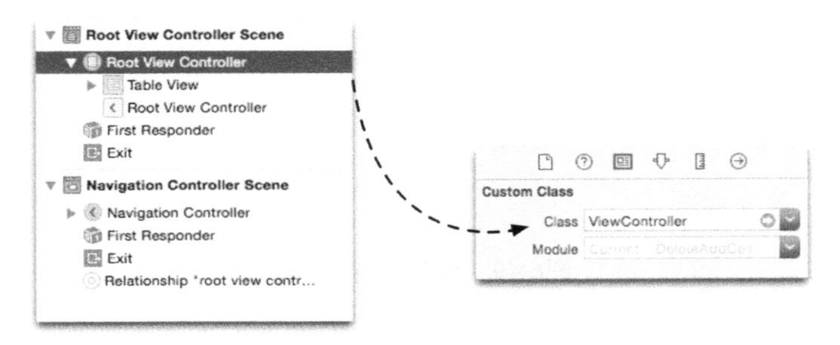

图6-51　设置根视图控制器的标识检查器

当插入单元格时，应该有一个控件能够接收用户输入的信息，这个控件应该是TextField文本输入框，所以我们在插入的单元格里放置了一个文本框。但是在Interface Builder中，把文本框放入到单元格中是比较困难的，我们可以将Text Field文本输入框添加到View Controller Scene中，如图6-52所示。然后打开其属性检查器，将Font属性设置为System 20.0，将Placeholder属性设置为Add...。

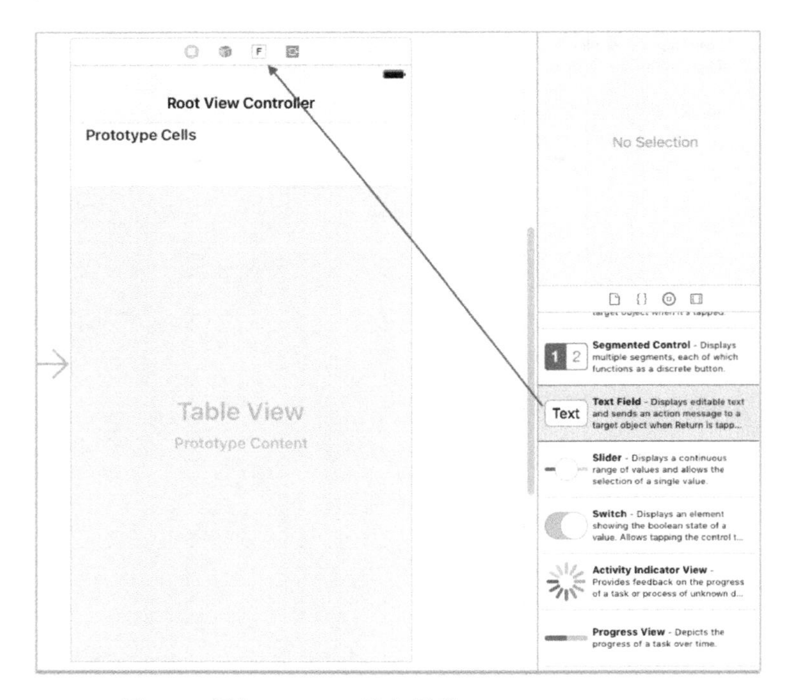

图6-52　添加Text Field输入框到View Controller Scene

然后我们需要为刚才拖曳的Text Field控件定义输出口，并将其与视图控制器连线，如图6-53所示。我们将Text Field控件拖曳到右边的编辑辅助窗口，松开鼠标，在弹出对话框的name中输入txtField，然后点击Connect按钮定义输出口。

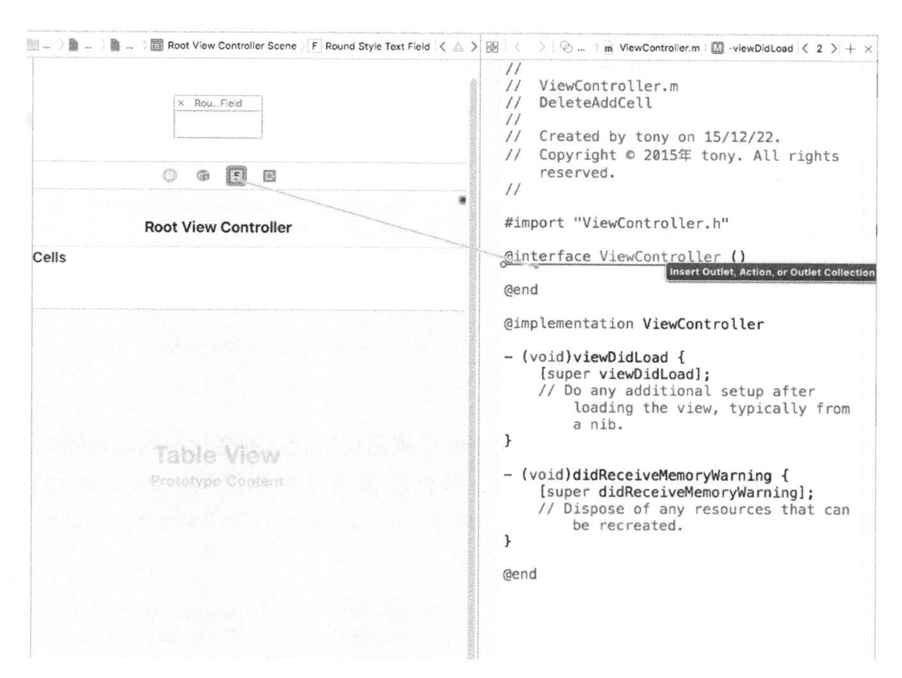

图6-53 为Text Field控件定义输出口

下面我们看看视图控制器ViewController的定义和属性，以及视图加载方法viewDidLoad：

```
//ViewController.swift文件
class ViewController: UITableViewController, UITextFieldDelegate {

    @IBOutlet  var txtField: UITextField!
    var listTeams : NSMutableArray!

    override func viewDidLoad() {
        super.viewDidLoad()

        //设置导航栏
        self.navigationItem.rightBarButtonItem = self.editButtonItem()      ①
        self.navigationItem.title = "单元格插入和删除"

        //设置单元格文本框
        self.txtField.hidden = true
        self.txtField.delegate = self

        self.listTeams = NSMutableArray(array: ["黑龙江", "吉林", "辽宁"])

    }

}
```

```
//ViewController.m文件
#import "ViewController.h"

@interface ViewController () <UITextFieldDelegate>

@property (strong, nonatomic) IBOutlet UITextField *txtField;
@property (nonatomic, strong) NSMutableArray *listTeams;

@end

@implementation ViewController

- (void)viewDidLoad {
    [super viewDidLoad];

    //设置导航栏
    self.navigationItem.rightBarButtonItem = self.editButtonItem;      ①
    self.navigationItem.title = @"单元格插入和删除";

    //设置单元格文本框
    self.txtField.hidden = TRUE;
    self.txtField.delegate = self;

    self.listTeams = [[NSMutableArray alloc] initWithObjects:@"黑龙江",
    ➥@"吉林", @"辽宁", nil];
}
```

在定义ViewController类的时候，需要实现UITextFieldDelegate协议，它是TextField文本框所需要的。

listTeams属性是可变数组集合，用于装载表视图的数据。这里将其声明为可变的，是为了可以对它进行删除或修改。

在视图加载方法viewDidLoad中，第①行用于将编辑按钮设置为导航栏右边的按钮。编辑按钮是视图控制器中已经定义好的按钮，用self.editButtonItem()方法可以取得编辑按钮的对象（Objective-C版是self.editButtonItem属性）。编辑按钮的样式可以在Edit和Done之间切换，如何切换取决于当前视图是否处于编辑状态。点击编辑按钮时，会调用setEditing:animated:方法，其代码如下：

```
//UIViewController生命周期方法，用于响应视图编辑状态变化
override func setEditing(_ editing: Bool, animated: Bool) {
    super.setEditing(editing, animated: animated)
    self.tableView.setEditing(editing, animated: true)
    if editing {
        self.txtField.hidden = false
    } else {
        self.txtField.hidden = true
    }
}
```

```
#pragma mark -- UIViewController生命周期方法，用于响应视图编辑状态变化
- (void)setEditing:(BOOL)editing animated:(BOOL)animated {
    [super setEditing:editing animated:animated];

    [self.tableView setEditing:editing animated: TRUE];
    if (editing) {
        self.txtField.hidden = FALSE;
    } else {
        self.txtField.hidden = TRUE;
    }
}
```

该方法是UIViewController生命周期的方法，用于响应视图编辑状态的变化。当表视图处于编辑状态时，文本框需要显示出来；当表视图处于非编辑状态时，我们应该将文本框隐藏。ViewController中还需要实现UITableViewDataSource协议中的numberOfRowsInSection:和tableView:cellForRowAtIndexPath:方法，它们的代码如下：

```
//MARK: -- 实现数据源方法
override func tableView(_ tableView: UITableView,
➥numberOfRowsInSection section: Int) -> Int {
    return self.listTeams.count + 1
}

override func tableView(_ tableView: UITableView,
➥cellForRowAt indexPath: IndexPath) -> UITableViewCell {

    let cellIdentifier = "CellIdentifier"

    let b_addCell = ((indexPath as NSIndexPath).row == self.listTeams.
    ➥count)

    let cell:UITableViewCell! = tableView.dequeueReusableCell
    ➥(withIdentifier: cellIdentifier, for: indexPath)

    if (!b_addCell) {
        cell.accessoryType = .disclosureIndicator
        cell.textLabel?.text = self.listTeams[(indexPath as NSIndexPath).
        ➥row] as? String
    } else {
        self.txtField.frame = CGRect(x: 40,y: 0,width: 300,height: cell.
        ➥frame.size.height)
        self.txtField.borderStyle = .none
        self.txtField.placeholder = "Add..."
        self.txtField.text = ""
        cell.addSubview(self.txtField)
    }
    return cell
}
```

```
#pragma mark -- UITableViewDataSource 协议方法
- (NSInteger)tableView:(UITableView *)tableView
➥numberOfRowsInSection:(NSInteger)section {
    return [self.listTeams count] + 1;
}

- (UITableViewCell *)tableView:(UITableView *)tableView
➥cellForRowAtIndexPath:(NSIndexPath *)indexPath {

    static NSString *cellIdentifier = @"CellIdentifier";

    BOOL b_addCell = (indexPath.row == self.listTeams.count);

    UITableViewCell *cell = [tableView dequeueReusableCellWithIdentifier:
    ➥cellIdentifier forIndexPath:indexPath];

    if (!b_addCell) {
        cell.accessoryType = UITableViewCellAccessoryDisclosureIndicator;
        cell.textLabel.text = self.listTeams[indexPath.row];
    } else {
        self.txtField.frame = CGRectMake(40,0,300,cell.frame.size.height);
        self.txtField.borderStyle = UITextBorderStyleNone;
        self.txtField.placeholder = @"Add...";
        self.txtField.text = @"";
        [cell addSubview:self.txtField];
    }

    return cell;
}
```

numberOfRowsInSection:方法返回的不是listTeams集合的长度，而是“listTeams集合的长度+1”，这是因为我们需要为插入准备一个空的单元格，必须在此处预先指定。

在tableView:cellForRowAtIndexPath:方法中要注意的是，单元格要分两种情况来处理：一种是普通单元格，

另一种是要插入的那个单元格，在插入的单元格中需要在其内容视图中添加文本框。

tableView:editingStyleForRowAtIndexPath:方法用于单元格编辑图标的设定，其代码如下：

```
override func tableView(_ tableView: UITableView,
➥editingStyleForRowAt indexPath:IndexPath) -> UITableViewCellEditing
➥Style {
    if ((indexPath as NSIndexPath).row == self.listTeams.count) {
        return .insert
    } else {
        return .delete
    }
}
```

```
- (UITableViewCellEditingStyle)tableView:(UITableView *)tableView
➥editingStyleForRowAtIndexPath:(NSIndexPath *)indexPath {
    if (indexPath.row == [self.listTeams count]) {
        return UITableViewCellEditingStyleInsert;
    } else {
        return UITableViewCellEditingStyleDelete;
    }
}
```

上述代码的返回类型是枚举类型UITableViewCellEditingStyle，该枚举类型中的成员如表6-7所述。

表6-7 UITableViewCellEditingStyle枚举成员

Swift枚举成员	Objective-C枚举成员	说明
none	UITableViewCellEditingStyleNone	默认样式，没有图标
delete	UITableViewCellEditingStyleDelete	删除样式，图标是⊖
insert	UITableViewCellEditingStyleInsert	插入样式，图标是⊕

tableView:commitEditingStyle:forRowAtIndexPath:方法用于实现删除或插入处理，其代码如下：

```
override func tableView(_ tableView: UITableView,
➥commit editingStyle: UITableViewCellEditingStyle,
➥forRowAt indexPath: IndexPath) {

    var indexPaths =[indexPath]                                          ①

    if (editingStyle == .delete) {                                       ②
        self.listTeams.removeObject(at: (indexPath as NSIndexPath).row)
        self.tableView.deleteRows(at: indexPaths, with: .fade)           ③
    } else if (editingStyle == .insert) {                                ④
        self.listTeams.insert(self.txtField.text! , at: self.listTeams.count)
        self.tableView.insertRows(at: indexPaths, with: .fade)           ⑤
    }
    self.tableView.reloadData()                                          ⑥
}
```

```
- (void)tableView:(UITableView *)tableView
➥commitEditingStyle:(UITableViewCellEditingStyle)editingStyle
➥forRowAtIndexPath:(NSIndexPath *)indexPath {
    NSArray* indexPaths = [NSArray arrayWithObject:indexPath];           ①

    if (editingStyle == UITableViewCellEditingStyleDelete) {             ②
        [self.listTeams removeObjectAtIndex: indexPath.row];
        [self.tableView deleteRowsAtIndexPaths:indexPaths
        ➥withRowAnimation:UITableViewRowAnimationFade];                  ③
    } else if (editingStyle == UITableViewCellEditingStyleInsert) {      ④

        [self.listTeams insertObject:self.txtField.text atIndex:[
        ➥self.listTeams count]];
        [self.tableView insertRowsAtIndexPaths:indexPaths
        ➥withRowAnimation:UITableViewRowAnimationFade];                  ⑤

    }
    [self.tableView reloadData];                                         ⑥

}
```

在删除单元格数据时（见第②行），本例中删除的是内存对象listTeams集合中的数据，但是如果数据来源于数据库，则应该删除的是数据库里的数据。第③行是删除表视图单元格的方法，其中indexPaths参数是NSIndexPath对象集合，该集合是要删除单元格的索引。第①行用于创建NSIndexPath对象集合。

插入单元格数据时（见第④行），本例中插入的是内存对象listTeams集合中的数据，但是如果数据来源于数据库，则应该插入数据库里的数据。第⑤行是插入表视图单元格的方法，其中参数indexPaths是要插入单元格的索引，而withRowAnimation参数（Swift版本是with）可以设置插入时的动画效果。

最后，第⑥行用于重新加载表视图数据。插入和删除单元格都需要重新加载数据。

上述代码足以完成单元格的插入和删除。为了更加友好，我们还添加了其他一些方法：UITableViewDelegate协议中的tableView:shouldHighlightRowAtIndexPath:方法，其代码如下：

```
override func tableView(_ tableView: UITableView,
➥shouldHighlightRowAtindexPath: IndexPath) -> Bool  {
```

```
- (BOOL)tableView:(UITableView *)tableView
➥shouldHighlightRowAtIndexPath:(NSIndexPath *)indexPath {
```

```swift
    if (indexPath.row == self.listTeams.count) {
        return false
    } else {
        return true
    }
}
```

```objc
    if (indexPath.row == [self.listTeams count]) {
        return FALSE;
    } else {
        return TRUE;
    }
}
```

一般情况下，我们不希望用户能够选择表视图的最后一个单元格，因为它没有内容，如图6-54所示。如果tableView:shouldHighlightRowAtIndexPath:方法返回FALSE，用户就能够选择最后一个单元格，但这样用户就发觉不到它的存在了。

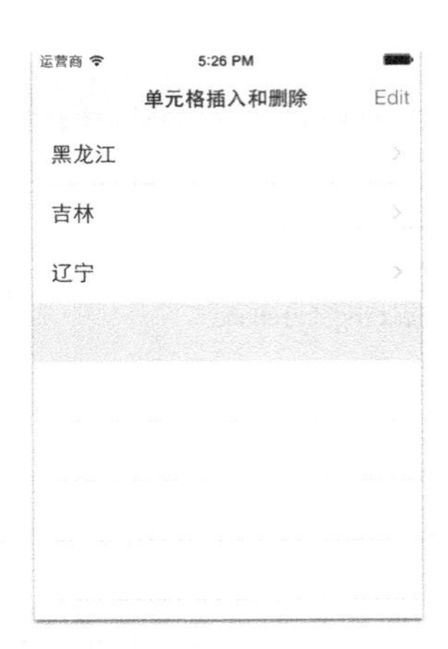

图6-54　单元格选择

6.6.2　代码实现

上一节我们使用Interface Builder实现了DeleteAddCell案例，这一节介绍如何通过代码实现这个案例。

首先，使用Single View Application工程模板创建一个名为DeleteAddCell的表视图工程，具体操作过程可参考6.2.3节，修改ViewController的父视图控制器为表视图控制器UITableViewController。

下面看看代码部分，其中AppDelegate与其他的代码实现工程有些不同，这是因为应用的根视图控制器是UINavigationController（导航控制器）。AppDelegate的主要代码如下：

```swift
//AppDelegate.swift文件
@UIApplicationMain
class AppDelegate: UIResponder, UIApplicationDelegate {

    var window: UIWindow?

    func application(_ application: UIApplication,
    ➥didFinishLaunchingWithOptions
    ➥launchOptions: [UIApplicationLaunchOptionsKey: Any]?) -> Bool {

        self.window = UIWindow(frame: UIScreen.main.bounds)
        self.window?.backgroundColor = UIColor.white
        let navigationController = UINavigationController(rootViewController:
        ➥ViewController())                                                ①
        self.window?.rootViewController = navigationController             ②
        self.window?.makeKeyAndVisible()

        return true
    }
    ......
}
```

```objc
//AppDelegate.m文件
......
- (BOOL)application:(UIApplication *)application
➥didFinishLaunchingWithOptions:(NSDictionary *)launchOptions {

    self.window = [[UIWindow alloc] initWithFrame:[[UIScreen mainScreen]
    ➥bounds]];
    self.window.backgroundColor = [UIColor whiteColor];

    ViewController* viewController = [[ViewController alloc] init];
    UINavigationController* navigationController = [[UINavigationController
    ➥alloc] initWithRootViewController:viewController];                  ①

    self.window.rootViewController = navigationController;                ②
    [self.window makeKeyAndVisible];

    return TRUE;
}
......
```

上述代码中，第①行用于实例化UINavigationController，构造函数是initWithRootViewController:，其中参数viewController是UINavigationController的根视图控制器。第②行将根控制器对象（即UINavigationController实例化对象）创建完成，把它赋值给window的属性rootViewController。

修改ViewController的代码：

```
//ViewController.swift文件
override func viewDidLoad() {
    super.viewDidLoad()

    //设置导航栏
    self.navigationItem.rightBarButtonItem =  self.editButtonItem
    self.navigationItem.title = "单元格插入和删除"

    //设置单元格文本框
    self.txtField = UITextField()                                    ①
    self.txtField.isHidden = true
    self.txtField.delegate = self
    ......
}

override func tableView(_ tableView: UITableView,
➥cellForRowAt indexPath: IndexPath) -> UITableViewCell
    let cellIdentifier = "CellIdentifier"

    let b_addCell = ((indexPath as NSIndexPath).row == self.listTeams.count)

    var cell: UITableViewCell! = tableView
    ➥.dequeueReusableCell(withIdentifier: cellIdentifier)            ②
    if (cell == nil) {
        cell = UITableViewCell(style: UITableViewCellStyle.default,
            ➥reuseIdentifier:cellIdentifier)                          ③
    }
    ......
    return cell
}
```

```
//ViewController.m文件
- (void)viewDidLoad {
    [super viewDidLoad];

    //设置导航栏
    self.navigationItem.rightBarButtonItem =  self.editButtonItem;
    self.navigationItem.title = @"单元格插入和删除";

    //设置单元格文本框
    self.txtField = [[UITextField alloc] init];                      ①
    self.txtField.hidden = TRUE;
    self.txtField.delegate = self;
    ......
}

- (UITableViewCell *)tableView:(UITableView *)tableView
➥cellForRowAtIndexPath:(NSIndexPath *)indexPath {

    static NSString *cellIdentifier = @"CellIdentifier";

    BOOL b_addCell = (indexPath.row == self.listTeams.count);

    UITableViewCell *cell = [tableView dequeueReusableCellWithIdentifier:
    ➥cellIdentifier];                                                ②


    if (cell == nil) {
        cell = [[UITableViewCell alloc] initWithStyle:
        ➥UITableViewCellStyleDefault reuseIdentifier:cellIdentifier];  ③
    }
    ......
    return cell;
}
```

上述代码中，第①行、第②行和第③行是修改过的代码：第①行用于实例化UITextField对象，第②行通过可重用标识符获得单元格对象，第③行用于在单元格对象为nil的情况下，实例化单元格对象。这里获得可重用单元格对象的方法与Interface Builder实现不同，这个问题我们在6.2.2节的代码实现部分介绍过，这里不再赘述。

6.7　移动单元格

在表视图中，单元格的顺序可以重新排列，本书中将其称为移动单元格。移动单元格与插入和删除单元格类似，单元格的后面会有重排序控件，如图6-55所示。

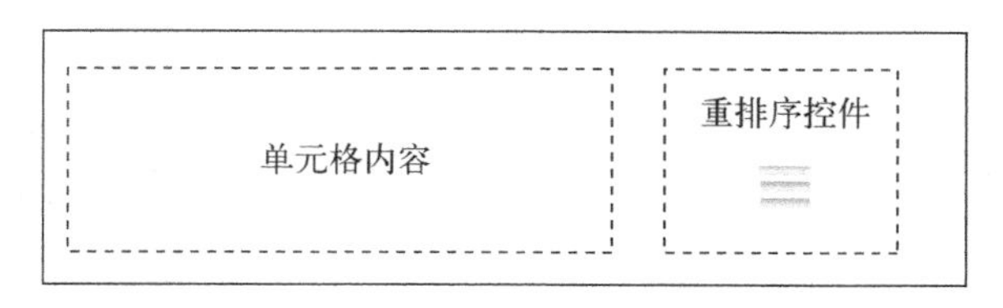

图6-55　重排序控件

图6-56是处于编辑状态的单元格，图6-57是处于移动状态的单元格。

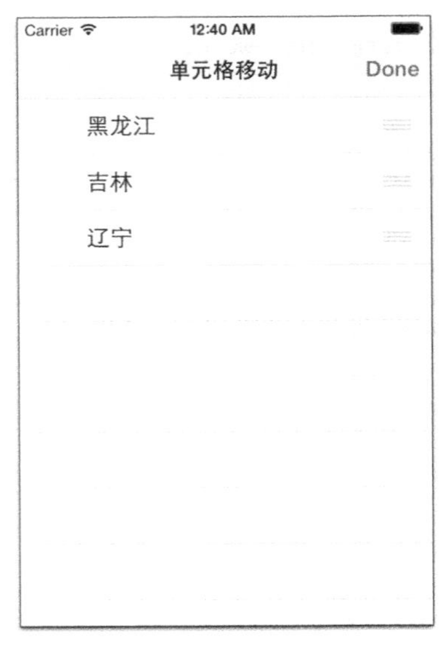

图6-56　处于编辑状态的单元格

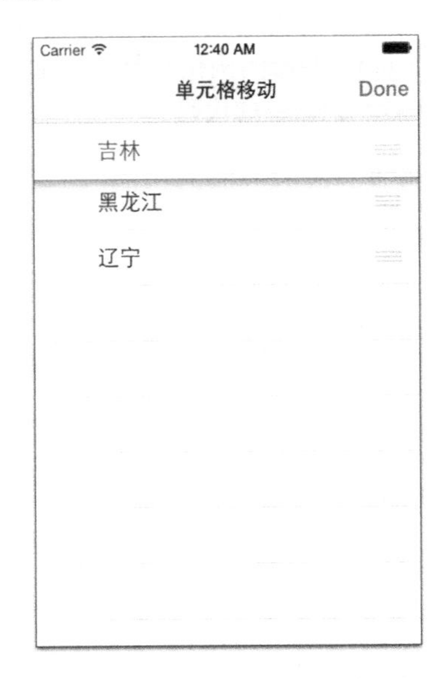

图6-57　处于移动状态的单元格

移动单元格时需要实现数据源的tableView:canMoveRowAtIndexPath:和tableView:moveRowAtIndexPath:toIndexPath:方法，其时序图如图6-58所示。

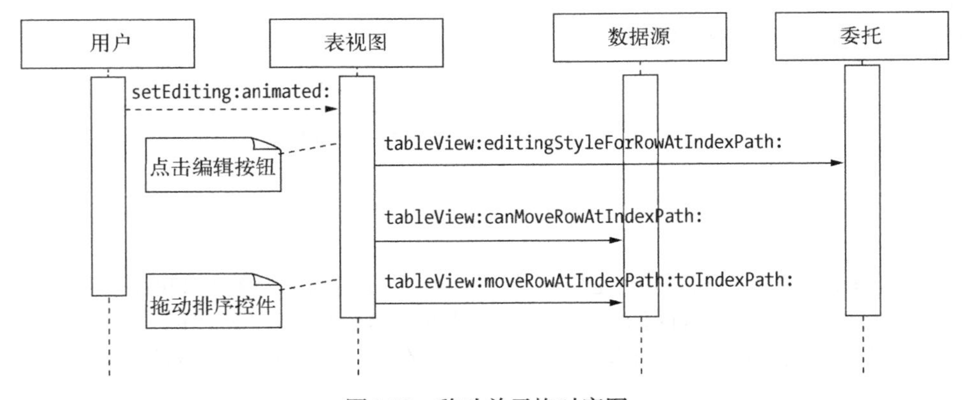

图6-58　移动单元格时序图

时序图反映的是当用户点击编辑按钮时，系统通过setEditing:animated:方法设定视图能否进入编辑状态。

然后调用协议方法tableView:editingStyleForRowAtIndexPath:进行单元格编辑图标的设定，这个方法不是必须实现的。如果不设定它，默认情况下删除和重排序图标同时存在，如图6-59所示。tableView:editingStyleForRowAtIndexPath:方法的代码如下：

```
override func tableView(_ tableView: UITableView,
➥editingStyleForRowAt indexPath: IndexPath) -> UITableViewCellEditingStyle {
    return .none
}
```

```
- (UITableViewCellEditingStyle)tableView:(UITableView *)tableView
➥editingStyleForRowAtIndexPath:(NSIndexPath *)indexPath {
    return  UITableViewCellEditingStyleNone;
}
```

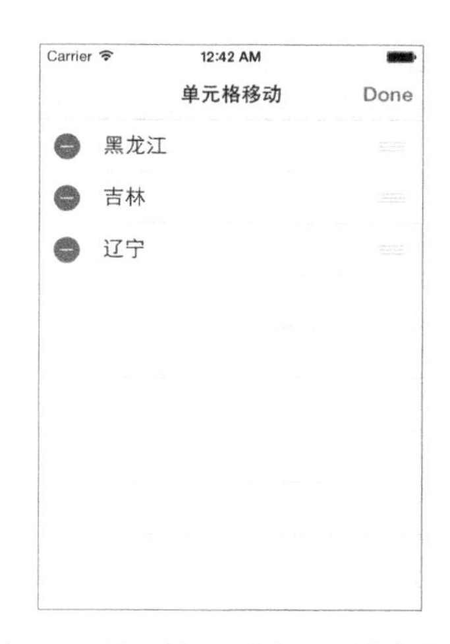

图6-59 单元格有删除和重排序控件

当tableView:canMoveRowAtIndexPath:方法返回true时，表示可以移动单元格；返回false时，表示不能移动单元格。其代码如下：

```
override func tableView(_ tableView: UITableView,
➥canMoveRowAtindexPath: IndexPath) -> Bool {
    return true
}
```

```
- (BOOL)tableView:(UITableView *)tableView
➥canMoveRowAtIndexPath:(NSIndexPath*)indexPath {
    return TRUE;
}
```

当用户拖动排序控件时，会触发tableView:moveRowAtIndexPath:toIndexPath:方法[Swift版本是tableView(_:moveRowAt:to:)]，该方法会对listTeams数据重新进行排序：

```
override func tableView(_ tableView: UITableView,
➥moveRowAt sourceIndexPath: IndexPath,
➥to destinationIndexPath: IndexPath) {

    let stringToMove = self.listTeams[(sourceIndexPath as NSIndexPath).row]
    ➥as! String

    self.listTeams.removeObject(at: (sourceIndexPath as NSIndexPath).row)
    self.listTeams.insert(stringToMove, at: (destinationIndexPath as
    ➥NSIndexPath).row)
}
```

```
- (void)tableView:(UITableView *)tableView
➥moveRowAtIndexPath:(NSIndexPath*)sourceIndexPath
➥toIndexPath:(NSIndexPath *)destinationIndexPath {

    NSString *stringToMove = self.listTeams[sourceIndexPath.row];

    [self.listTeams removeObjectAtIndex:sourceIndexPath.row];
    [self.listTeams insertObject:stringToMove atIndex:
    ➥destinationIndexPath.row];
}
```

上述代码是实现移动单元格的主要代码，其他代码参考6.6节，更详细的代码可以参考本书实例代码。

6.8 表视图 UI 设计模式

在iOS中，表视图应用极其广泛，本节将向大家介绍表视图中的两个UI设计模式——分页模式和下拉刷新（Pull-to-Refresh）模式，这两种模式已经成为移动平台开发的标准。

6.8.1 分页模式

想一想，你的新浪微博是否可以一次将所有的微博信息返回到你的设备屏幕里？当数据量很大时，一次返回所有信息这种方式会严重影响应用的性能，造成网络堵塞。通常，我们利用分页模式来解决请求大量数据的问题。

分页模式是先请求少量数据，例如一次50条，当翻动屏幕已显示50条数据之后，应用会再次请求50条。

根据触发方式的不同，请求分为主动请求和被动请求。图6-60为主动请求模式，即当条件满足时，再次请求下50条数据是自动发出的，并且一般会在表视图的表脚出现活动指示器，请求结束后活动指示器会隐藏起来。图

6-61为被动请求模式，当条件满足时，表视图的表脚中会显示出一个响应点击事件的控件。这个控件一般是一个按钮，按钮标签一般会设为“更多”。当我们点击“更多”按钮时，应用会向服务器发出请求，请求结束后，“更多”按钮会隐藏起来。

图6-60　主动请求

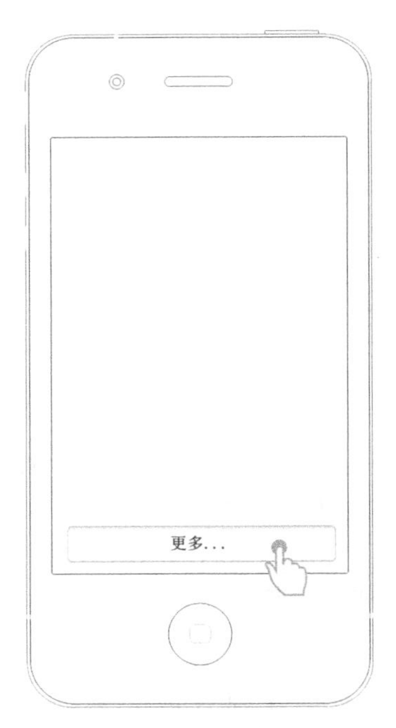

图6-61　被动请求

6.8.2　下拉刷新模式

下拉刷新是重新刷新表视图或列表，以便重新加载数据，这种模式广泛用于移动平台。下拉刷新与分页相反，当屏幕翻到顶部时，再往下拉屏幕，程序就开始重新请求数据。此时表视图的表头部分会出现活动指示器，请求结束后表视图的表头会消失，如图6-62所示。可以看到，下拉刷新模式带有箭头动画效果。

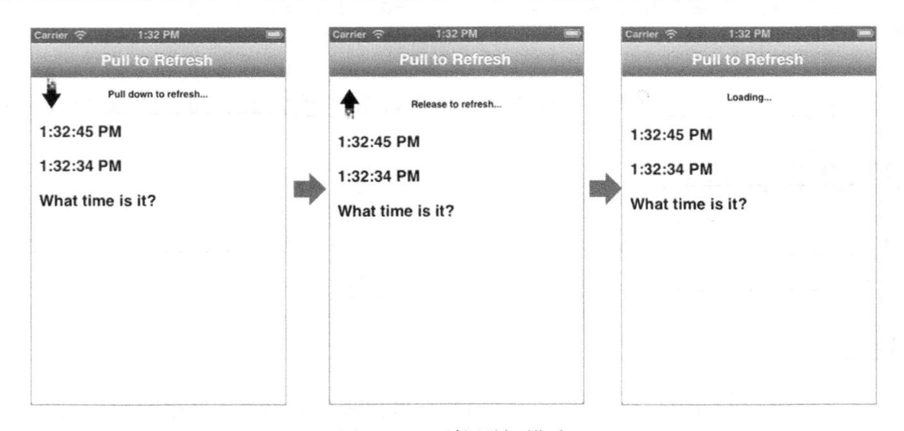

图6-62　下拉刷新模式

很多开源社区中都有下拉刷新的实现代码供大家参考，比如GitHub上的https://github.com/leah/PullToRefresh.git。

6.8.3 下拉刷新控件

苹果在iOS 6之后推出了下拉刷新控件UIRefreshControl。图6-63是iOS中的下拉刷新效果。

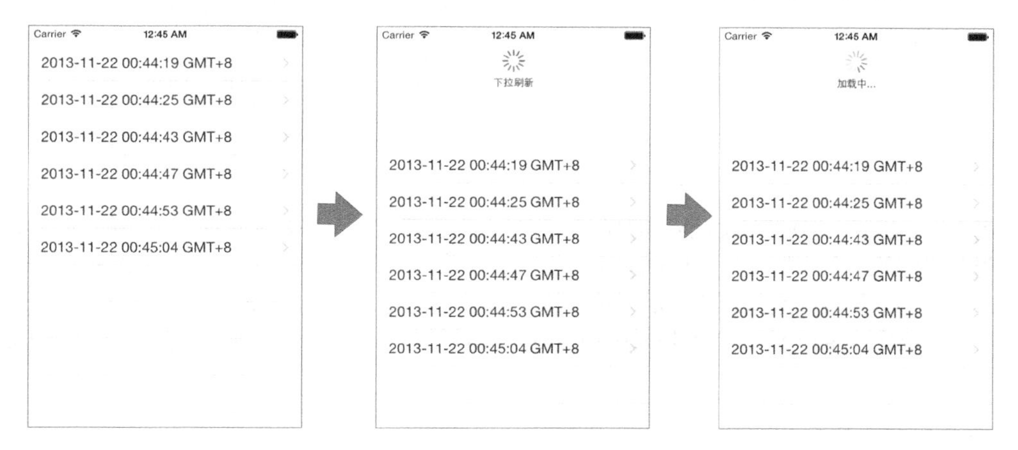

图6-63 iOS的下拉刷新

下拉刷新控件不能通过Interface Builder设计，而需要通过代码构建。UITableViewController提供了一个refreshControl属性，这个属性是UIRefreshControl类型，通过设置该属性，可以为表视图添加下拉刷新控件。下拉刷新控件的布局等问题不必考虑，UITableViewController会将其自动放置于表视图顶部，我们只需要设置它的状态。

提示 下拉刷新控件UIRefreshControl只能应用于表视图和滚动视图界面，目前不能是其他视图。

下面通过一个RefreshControlSample案例介绍下拉刷新控件的用法，这里参考6.2.3节创建案例工程。

提示 下拉刷新控件在Interface Builder控件库中没有可拖曳对象，创建下拉刷新控件只能通过代码实现，但是RefreshControlSample案例的其他UI部分可以通过Interface Builder或代码实现。

下面我们看看视图控制器ViewController的定义和属性，以及视图加载方法viewDidLoad：

```
//ViewController.swift文件
class ViewController: UITableViewController {

    var Logs : NSMutableArray!

    override func viewDidLoad() {
        super.viewDidLoad()
        //初始化变量和时间
        self.Logs = NSMutableArray()
        var date = NSDate()
        self.Logs.addObject(date)

        //初始化UIRefreshControl
        var rc = UIRefreshControl()                                              ①
        rc.attributedTitle = NSAttributedString(string: "下拉刷新")              ②
        rc.addTarget(self, action: #selector(refreshTableView),
```

```
//ViewController.m文件
#import "ViewController.h"

@interface ViewController ()

@property (nonatomic,strong) NSMutableArray* Logs;

@end

@implementation ViewController

- (void)viewDidLoad {
    [super viewDidLoad];

    //初始化变量和时间
    self.Logs = [[NSMutableArray alloc] init];
```

```swift
        ➥for: UIControlEvents.valueChanged)                                    ③
        self.refreshControl = rc
    }
}
```

```objc
    NSDate *date = [[NSDate alloc] init];
    [self.Logs addObject:date];

    //初始化UIRefreshControl
    UIRefreshControl *rc = [[UIRefreshControl alloc] init];                      ①
    rc.attributedTitle = [[NSAttributedString alloc]initWithString:@"下拉
    ➥刷新"];                                                                   ②
    [rc addTarget:self action:@selector(refreshTableView)
    ➥forControlEvents:UIControlEventValueChanged];                              ③
    self.refreshControl = rc;

}
```

上述代码中，Logs属性存放了NSDate日期列表，用于在表视图中显示需要的数据。在viewDidLoad方法中，我们初始化了当前时间的一条模拟数据。第①行用于创建UIRefreshControl对象。第②行用于设置UIRefreshControl对象的attributedTitle属性，它是用于显示下拉控件的标题。第③行通过编程方式为UIRefreshControl控件添加UIControlEventValueChanged（Swift版本是valueChanged）事件。refreshTableView是该事件的处理方法，相关代码如下：

```swift
func refreshTableView() {

    if (self.refreshControl?.refreshing == true) {
        self.refreshControl?.attributedTitle = NSAttributedString(string:
        ➥"加载中...")                                                          ①
        //添加新的模拟数据
        var date = NSDate()
        self.Logs.addObject(date)

        self.refreshControl?.endRefreshing()                                   ②
        self.refreshControl?.attributedTitle = NSAttributedString(string:
        ➥"下拉刷新")                                                           ③

        self.tableView.reloadData()                                            ④
    }
}
```

```objc
-(void) refreshTableView {

    if (self.refreshControl.refreshing) {
        self.refreshControl.attributedTitle =
        ➥[[NSAttributedString alloc]initWithString:@ "加载中..."];              ①
        //添加新的模拟数据
        NSDate *date = [[NSDate alloc] init];
        [self.Logs addObject:date];

        [self.refreshControl endRefreshing];                                   ②
        self.refreshControl.attributedTitle =
        ➥[[NSAttributedString alloc]initWithString:@"下拉刷新"];               ③

        [self.tableView reloadData];                                           ④

    }
}
```

UIRefreshControl的refreshing属性可以判断控件是否处于刷新状态，刷新状态的图标是我们常见的活动指示器。第①行用于将显示标题设置为“加载中...”。

在刷新操作完成的时候，endRefreshing方法可以停止下拉刷新控件，回到初始状态，显示的标题文本为“下拉刷新”，见第②行和第③行代码。第③行用于重新设置下拉刷新控件的标题，第④行再重新加载表视图。

上述代码是实现下拉刷新控件的主要代码，其他代码参考6.2.3节，更详细的代码可以参考本书实例代码。

请运行一下，看看效果。在应用能够实现用户需求的前提下，良好的用户体检是我们更高的追求，因为我们的产品不仅仅是具备某种功能的工具，更是一件艺术品，是我们心、智、力的结晶。

6.9　小结

在本章中，首先我们带大家建立对表视图的整体认识，了解表视图的组成、表视图类的构成、表视图的分类以及表视图的两个重要协议（委托协议和数据源协议）。接着，我们讨论了如何实现简单表视图和分节表视图，以及表视图中索引、搜索栏、分组的用法，然后介绍了如何对表视图单元格进行删除、插入、移动等操作，最后阐述了表视图UI设计模式的内容。

第7章 界面布局

掌握控件后，就需要将这些控件摆放到屏幕上，这个过程要考虑到苹果公司的iOS人机界面设计规范，还要考虑不同设备屏幕的适配问题。本章将讨论界面布局问题。

7.1 界面布局概述

界面布局就是控件在界面中的摆放过程，这个过程需要参考苹果公司的iOS人机界面设计规范。虽然App Store上不遵守这一规范的个性化应用也很多，但是本书重点还是介绍一下界面布局。

这一规范下的界面布局可以归纳出3种主要的界面布局UI设计模式，如图7-1至图7-3所示。

7.1.1 表单布局模式

表单布局（如图7-1所示）提供一种与用户交互的界面，例如：登录界面和注册界面。表单布局可以采用静态表实现，稍后会在7.5节中详细介绍。

图7-1　表单布局

7.1.2 列表布局模式

列表布局（如图7-2所示）是指当需要展示大量数据的时候，可以通过列表或网格布局实现。列表布局使用动态表视图，这在第6章中已经介绍过。动态视图需要实现表视图的委托协议和数据源协议相关方法。

图7-2　列表布局

7.1.3　网格布局模式

网格布局（如图7-3所示）与列表布局类似，但列表布局只有一列，网格布局却可以有多列，这种布局使用集合视图实现。

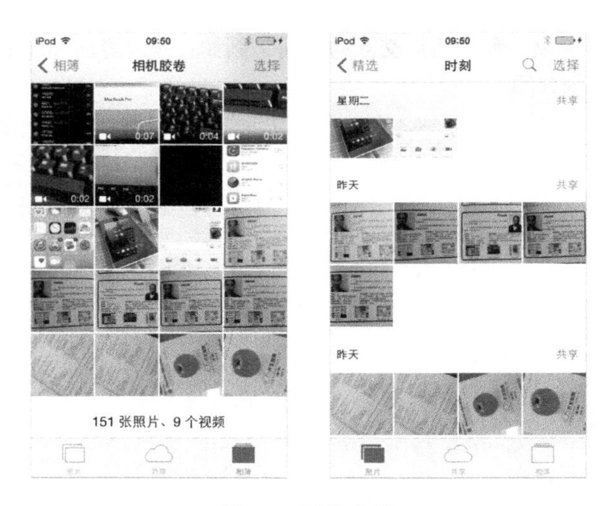

图7-3　网格布局

另外，自iOS 9起，推出了堆视图StackView（详情可参见7.6节），可以实现管理垂直方向和水平方向排列的布局。

7.2　iOS 中各种“栏”

在iOS屏幕布局中会用到各种“栏”，一般会有状态栏、标签栏、导航栏、工具栏和搜索栏等，这些栏的尺寸是固定的。如图7-4所示，在iPhone竖屏幕中，状态栏占用20点[①]，导航栏占用44点，标签栏占用49点。实际上，这些尺寸在iPhone横屏幕和iPad上也保持不变。

---

① “点”是一个与设备无关的单位，Interface Builder设计器和程序代码中的单位都是“点”。

提示 在iOS界面中，44是个很神奇的数字，它是经过人体工程学计算而得的高度数值，不仅是导航栏，工具栏、搜索栏、搜索范围栏和表视图单元格等的高度都是44点。

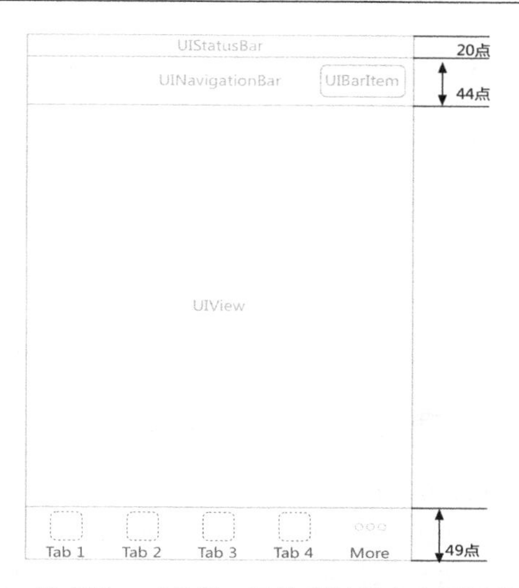

图7-4　状态栏、工具栏、导航栏以及内容视图的尺寸

7.3　传统界面布局问题

图7-5是iPhone 6/6s/7设备竖屏界面，其中有3个按钮。但是如果旋转屏幕（竖屏到横屏），我们会发现界面如图7-6所示，其中只有Button1，其他两个按钮不见了，事实上是超出屏幕了。但是如果该案例运行在iPhone 6 Plus/6s Plus/7 Plus等设备上，那么界面布局会如图7-7所示，其中Button1不会刚好居中，其他两个按钮也会出现布局问题。

图7-5　iPhone 6/6s/7等设备竖屏

图7-6　iPhone 6/6s/7等设备横屏

图7-7　iPhone 6 Plus/6s Plus/7 Plus等设备竖屏

这些布局问题都是我们在实际开发时经常遇到的，主要原因在于iOS设备的多样化。苹果的Auto Layout和Size Class技术可以帮助我们解决这些问题，Auto Layout偏重于解决布局问题，而Size Class侧重于解决屏幕适配问题。本章中，我们将介绍Auto Layout技术。

提示　在模拟器中旋转屏幕的快捷键是：command + 左右箭头，command是苹果键⌘。

7.4 Auto Layout 布局技术

Auto Layout布局技术在iOS 6之后被引入到iOS系统，它可以帮助我们解决复杂多样的iOS设备屏幕问题。Auto Layout为空间布局定义了一套约束（constraint），这套约束定义了视图之间的关系。

7.4.1 在 Interface Builder 中管理 Auto Layout 约束

Auto Layout约束的管理可以使用Interface Builder和代码这两种方式。当然也可以混合使用这两种方式。代码方式使用起来非常灵活，但是需要掌握很多API，而Interface Builder很容易学习，我推荐使用该方式。

针对Auto Layout布局，Interface Builder设计器提供了一些操作按钮，如图7-8所示，下面简要介绍这些按钮的作用。

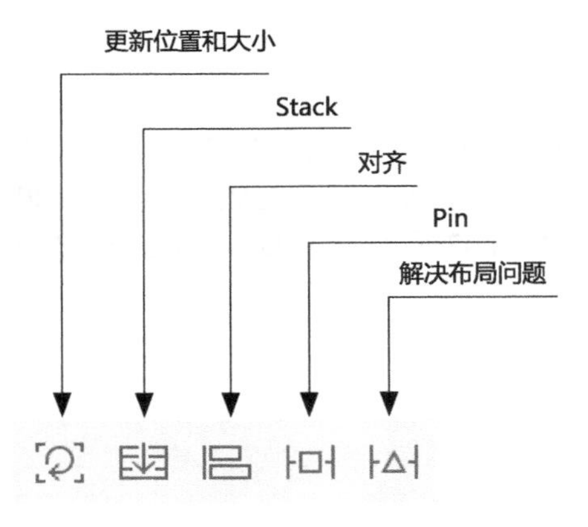

图7-8　布局按钮

- **更新位置和大小**：更新视图的Frame属性，该属性用于描述视图的位置和大小。
- **Stack**：将视图对象添加到堆视图StackView中。
- **对齐**：创建对齐约束，例如使视图在容器中居中。
- **Pin**：创建距离和位置的相关约束，例如视图的高度，或指定与其他视图的水平距离。
- **解决布局问题**：顾名思义，就是用于解决布局中的问题。

7.4.2 案例：Auto Layout 布局

下面我们通过一个案例介绍传统布局技术的用法。如图7-9左图所示，界面中有3个按钮。将屏幕向右旋转至横屏时，效果如图7-9右图所示，这3个按钮仍然需要能够很好地摆放在屏幕中。

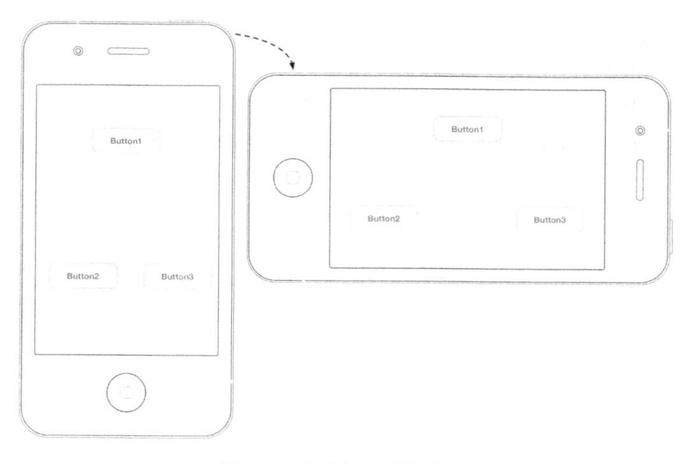

图7-9　案例原型设计图

下面我们采用Auto Layout布局技术实现该案例。创建案例的具体过程为：使用Single View Application模板创建一个名为AutolayoutSample的工程。使用Interface Builder打开Main.storyboard，从对象库中拖曳3个按钮到设计界面。为了保证竖屏和横屏两种情况下都能正常显示这3个按钮，需要为按钮添加一些约束条件。

- ❑ Button1。水平居中，与屏幕的上边距为绝对距离。
- ❑ Button2。左边距和下边距为绝对距离。
- ❑ Button3。右边距和下边距为绝对距离，与Button2顶对齐。

下面看看如何添加它们。

1. Button1水平居中

Button1水平居中的添加步骤如图7-10所示。首先，点击"对齐"按钮 ，从弹出的对齐菜单中选中Horizontally in Container，后面的数字是与中间轴的偏移量，设置为0表示刚好居中。设置完成后，请点击Add 1 Constraint按钮，添加约束并关闭菜单。

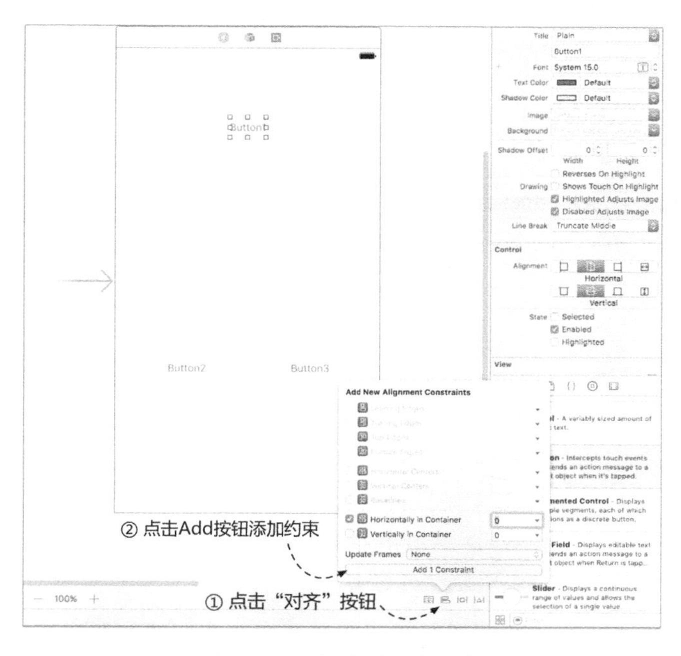

图7-10　添加水平居中约束

2. Button1上边距绝对约束

添加该约束的具体步骤如图7-11所示。点击Pin按钮 ⊢□⊣ ，从弹出的Pin菜单中可以看到上、下、左、右4个方向的线段，虚线线段代表相对距离，实线线段代表绝对距离，点击可以相互切换。由于Button1与屏幕的上边距是绝对距离，我们需要将上边距设置为实线。设置完成后，请点击Add 1 Constraint按钮，添加约束并关闭菜单。

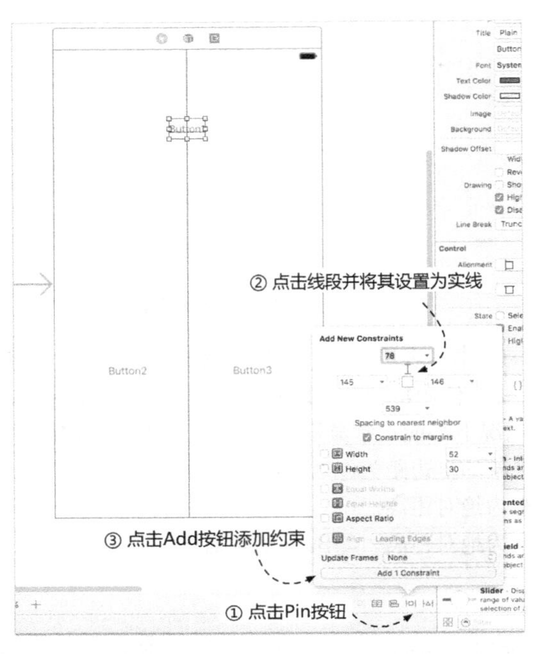

图7-11　添加上边距绝对约束

Button1的约束添加完成之后，如图7-12所示，选择其中一个约束，则在设计界面中可以看到该约束被加粗和添加阴影显示。这里也可以选中某个约束，通过Delete键删除它。

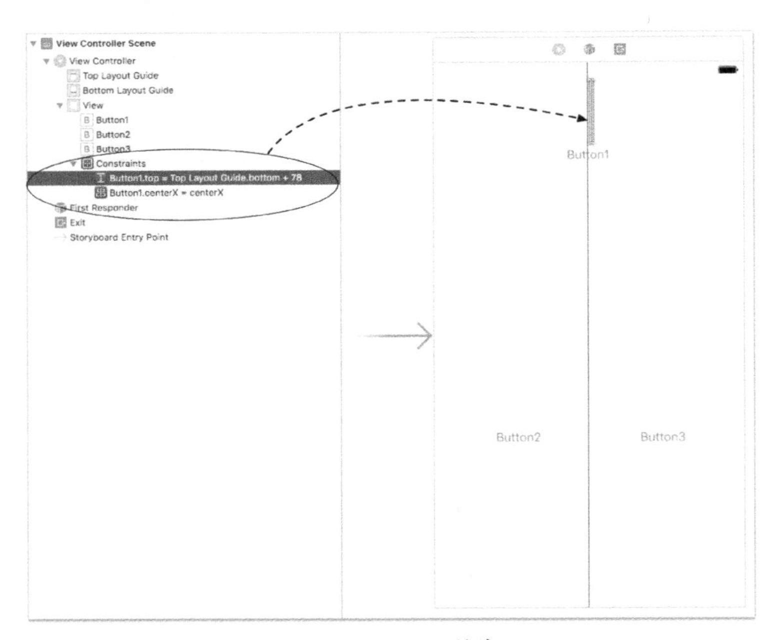

图7-12　Button1约束

3. Button2左边距和下边距绝对约束

点击Pin按钮 ，弹出Pin菜单，添加约束（如图7-13所示），将左边距和下边距设置为实线。设置完成后，请点击Add 2 Constraints按钮，添加约束并关闭菜单。

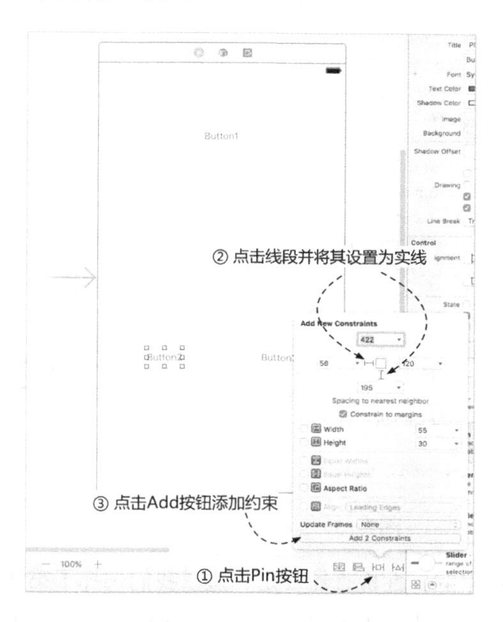

图7-13　添加左边距和下边距绝对约束

4. Button3右边距和下边距绝对约束

参考添加Button2左边距和下边距绝对约束，来添加Button3右边距和下边距绝对约束，具体步骤不再赘述。

5. Button3与Button2顶对齐约束

添加Button3与Button2顶对齐约束，添加步骤如图7-14所示。首先选中Button3和Button2两个按钮，然后点击对齐按钮 ，从弹出的对齐菜单中选中Top Edges（Top Edges表示顶边对齐），设置完成后点击Add 1 Constraint按钮，添加约束并关闭菜单。

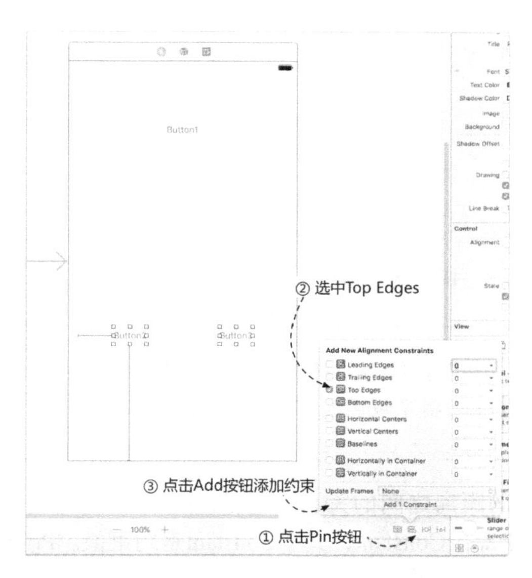

图7-14　添加对齐约束

添加完约束后，打开View Controller Scene视图中View下面的Constraints项，我们发现其中有7个约束，如图7-15所示。

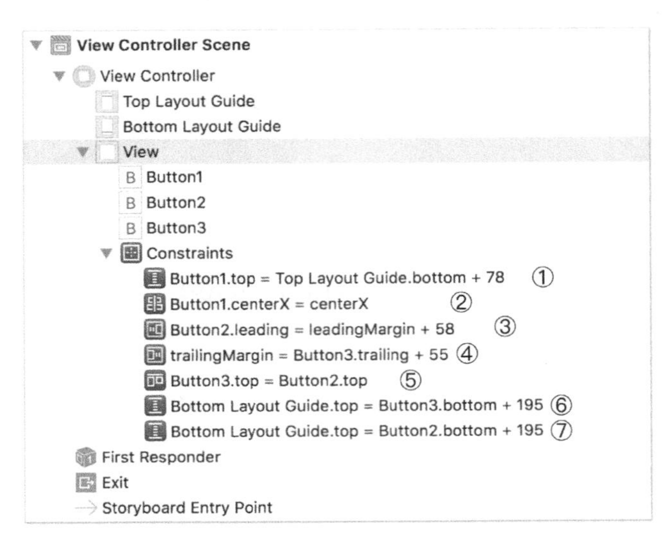

图7-15　Auto Layout视图约束

如果想更加精确，可以通过约束的属性检查器设定，这里我们选择了第②个约束。如图7-16所示，其中First Item和Second Item是约束相关的两个项目。First Item是Button1.Top，表示到Button1的顶部。如图7-17所示，选中Button1.Top，会看到相关的视图是Button1。Second Item是Top Layout Guide.Bottom，表示到Top Layout Guide的底边距。如图7-18所示，选中Top Layout Guide.Bottom，会看到相关的视图是Top Layout Guide。

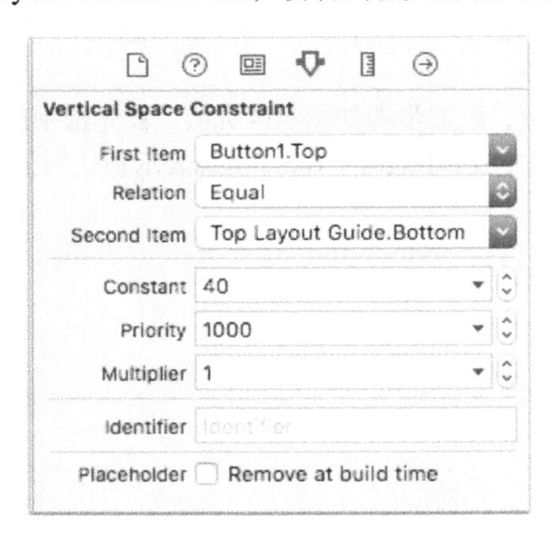

图7-16　Auto Layout约束属性检查器

提示　Top Layout Guide可以在如图7-18所示的视图控制器中看到，它限制了视图能够放置的最高位置。类似地，Bottom Layout Guide限制了视图能够放置的最低位置。

图7-16中的Relation是指设定的距离之间的关系，包括3个选项——等于、大于等于和小于等于，即等于Constant、大于等于Constant和小于等于Constant，其中Constant是约束值。Priority是约束等级，当有相同的约束作用于两个视图之间时，等级高的约束优先。

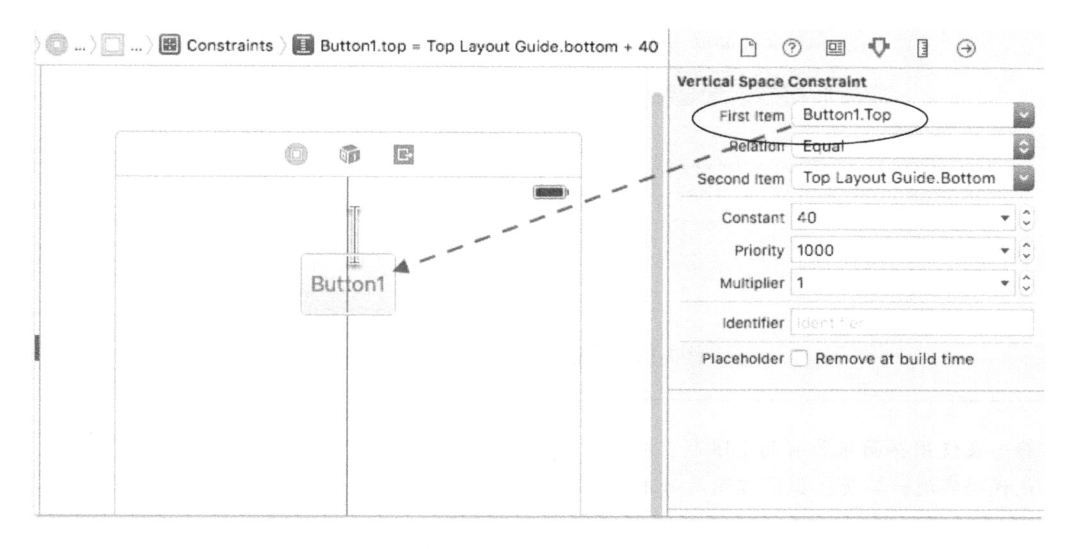

图7-17　选中Button1.Top

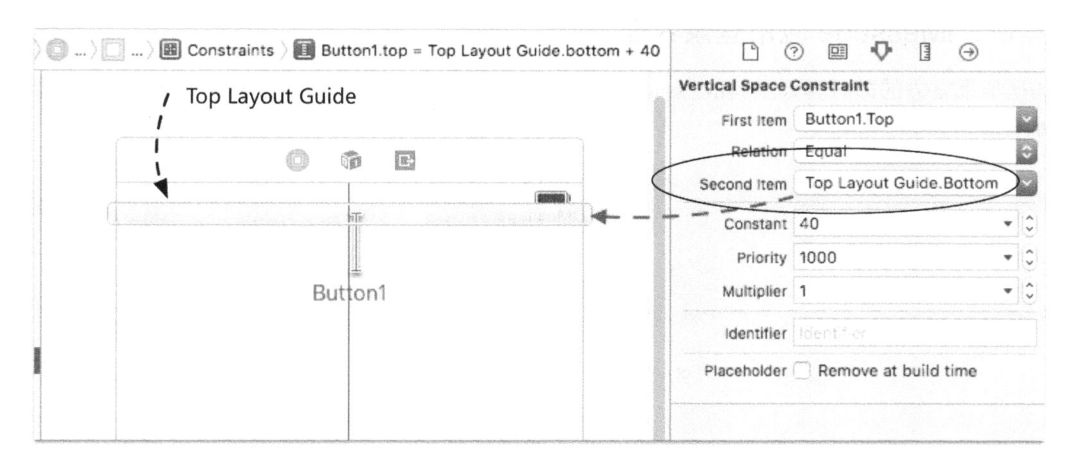

图7-18　选中Top Layout Guide.Bottom

7.5　静态表布局

我们在第6章中介绍了表视图。事实上，还有一种用于布局的表视图——静态表，而第6章所介绍的属于动态表。

7.5.1　什么是静态表

静态表的布局类似于在HTML网页中使用Table标签进行页面布局。静态表不同于动态表，基本上不需要编写代码，不需要实现表视图的委托协议或数据源协议，只需要在故事板里设计即可。要想切换动态表与静态表，需要在Interface Builder中选择表视图，如图7-19所示，选中Table View Controller Scene→Table View，然后打开其属性检查器，在Table View→Content中将原来的Dynamic Prototypes改为Static Cells即可。

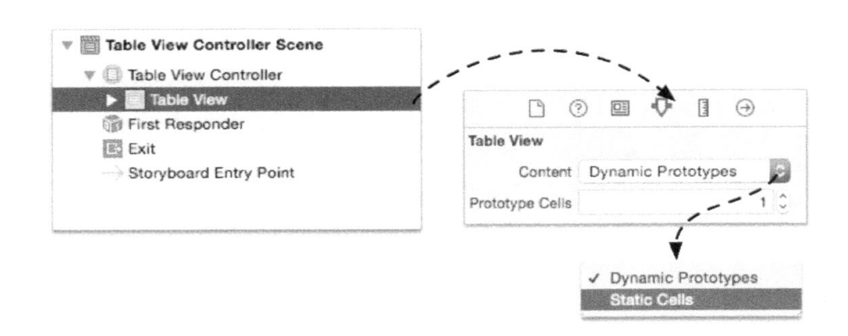

图7-19　动态表与静态表的切换

> **注意**　静态表使用界面布局有两个限制：一是，界面构建必须基于故事板技术，不能是XIB技术，更不能是代码实现；二是，视图控制器必须是表视图控制器。

7.5.2　案例：iMessage 应用登录界面

图7-20是苹果官方的即时聊天工具iMessage应用的登录界面，如果这个界面没有采用表视图来进行布局，界面会非常难看。

图7-20　iMessage应用登录界面

图7-20所示的界面很显然是表视图，分为3节，第1节有两个单元格，每一个单元格有一个文本框，文本框有输出口，上面的文本是节标题；第2节有一个单元格，其中放置一个“登录”按钮；第3节有一个单元格，其中包含标签控件和扩展指示器。

在iOS 5之前没有静态表，只能采用动态表。动态表本身不是为布局而设计的，如果通过动态表实现，则是一项非常繁重的工作。幸运的是，iOS 5之后的故事板技术可以使用静态表，通过静态表可以完全不用编写代码，只需要在Interface Builder中设计即可。

下面将图7-20的界面简化一下，采用静态表技术实现如图7-21所示的案例，具体实现步骤如下。

1. 创建工程

使用Single View Application模板创建一个名为StaticTable的工程。打开Interface Builder设计界面，在View Controller Scene中删除View Controller，然后从控件库中拖曳一个Table View Controller到设计界面。

2. 设置静态表

接着选择Table View Controller Scene→Table View，打开其属性检查器，如图7-22所示，从Content下拉列表中选择Static Cells，将Sections的值设为3（即3节），从Style下拉列表中选择Grouped。

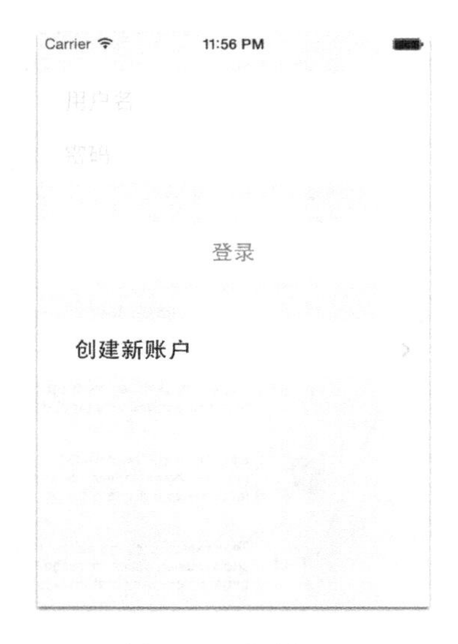

图7-21 登录界面

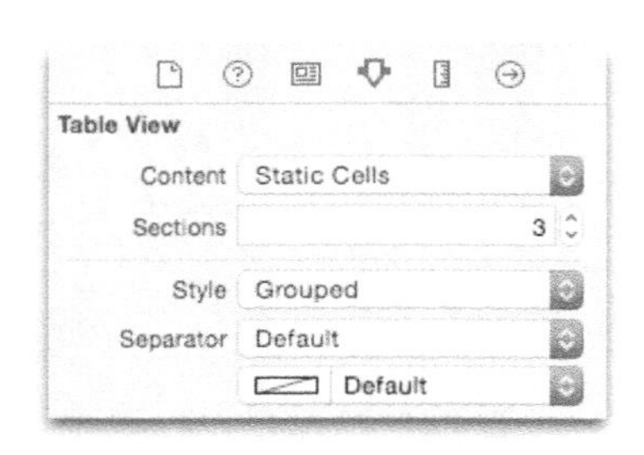

图7-22 静态表属性检查器

3. 设置第1节

选择Table View Controller Scene中的Section-1（选中第1节），打开它的属性检查器，如图7-23所示，将Rows的值设为2，即该节中包含两个单元格。还可以根据需要设定Header（节头）和Footer（节脚），本例中不设置它们。

然后，将两个TextField控件分别拖曳到该节中的单元格上，如图7-24所示。接着，设置TextField控件的属性。打开其属性检查器，如图7-25所示，设置Placeholder为“用户名”，Border Style为无边框样式。最后，不用忘记为TextField添加Auto Layout约束。

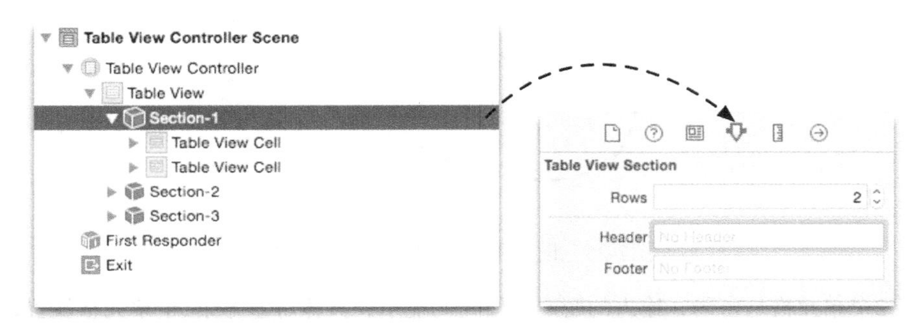

图7-23 静态表中的“节”属性检查器

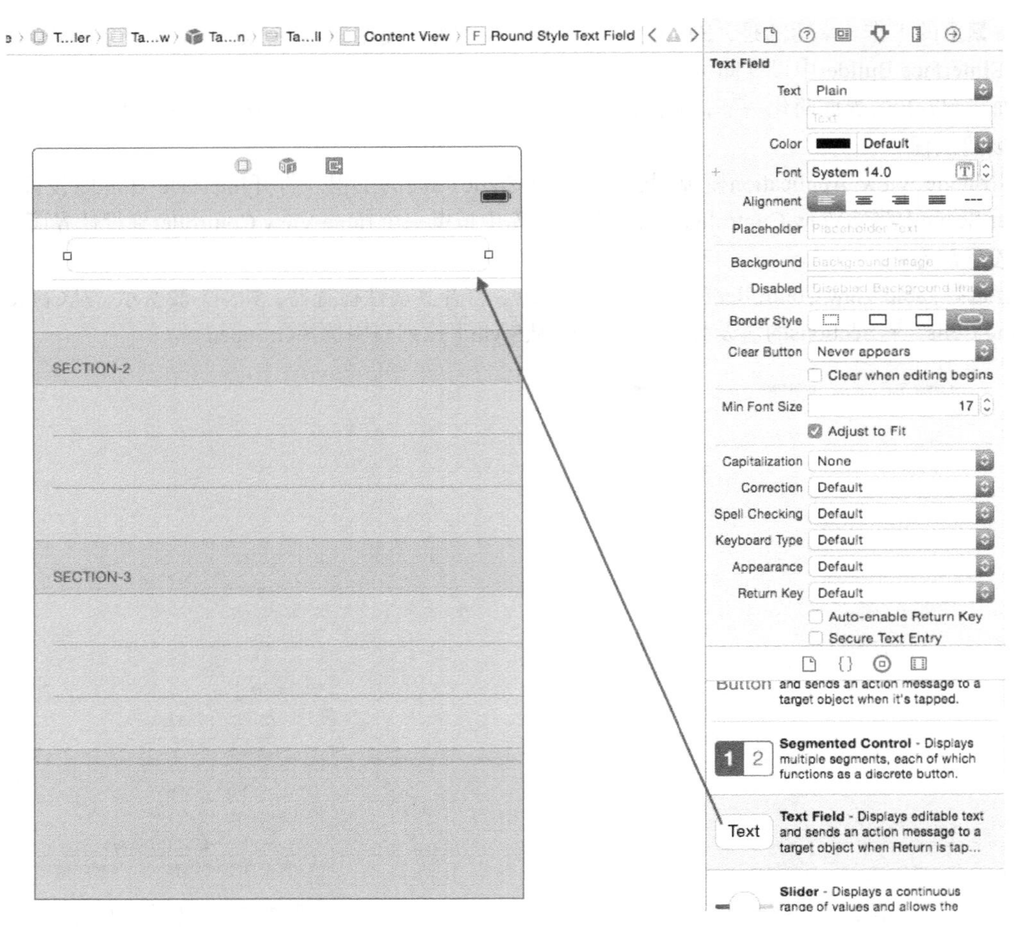

图7-24　拖曳Text Field到单元格

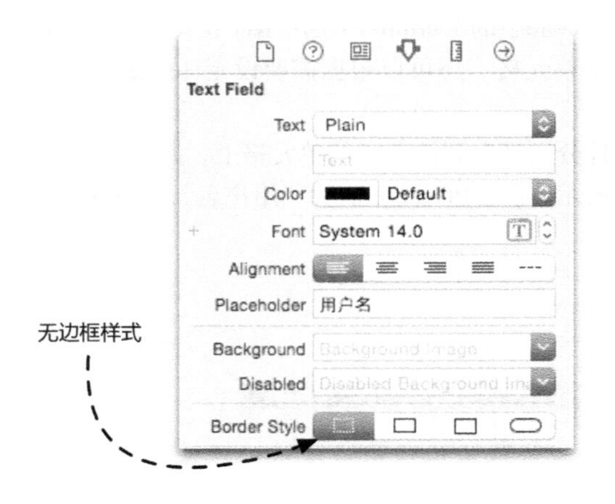

图7-25　设置用户名的Text Field属性

第1节中的第2个单元格是密码，输入的密码需要掩码显示，具体可参考第1个单元格设置，然后设置它的Secure Text Entry为选中，如图7-26所示。

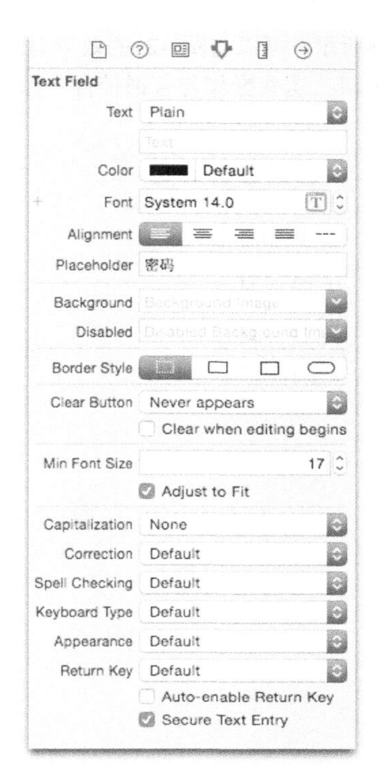

图7-26 设置“密码”文本框的属性

4. 设置第2节

静态表第2节中，有一个按钮，可以按照上面的方法设定。

5. 设置第3节

静态表第3节的单元格中有标签和扩展指示器，其中扩展指示器的设定如图7-27所示。选择Table View Controller Scene中的Table View Cell，打开其属性检查器，从Accessory下拉列表中选择Disclosure Indicator（扩展指示器）。最后，还要拖曳一个Label控件到单元格中，设置内容为“创建新账户”。

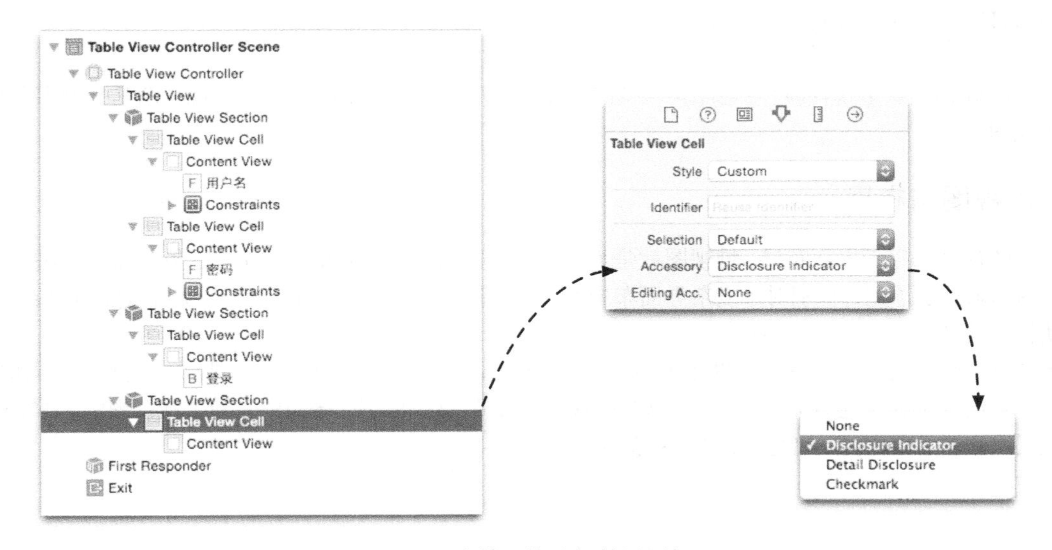

图7-27 为单元格选择扩展图标

这样整个界面就设计好了，如图7-28所示，可以与图7-21的效果对比一下。要完成该案例，还需要为“登录”按钮定义动作事件，为TextField定义输出口，这些操作与普通控件一致，不再赘述。

图7-28　设计完成的界面

7.6　使用堆视图 StackView

iOS 9和iOS 10提供了一种新容器视图——堆视图StackView，类是`UIStackView`。堆视图是一种容器视图，可以包含子视图。

7.6.1　堆视图与布局

容器视图都可以用于界面布局，堆视图就是为界面布局而设计的。Auto Layout技术虽然能够解决很多布局问题，但是用起来过于烦琐，特别对于复杂界面布局而言，更是不容易使用，而堆视图可以弥补Auto Layout的这些问题，使构建复杂界面布局变得简单。

堆视图可以管理两个不同方向的界面布局，其中图7-29a是垂直方向的堆视图，图7-29b是水平方向的堆视图。此外，两个方向的堆视图还可以任意嵌套。如图7-30所示，堆视图StackView2嵌套到了堆视图StackView1中。从界面构建层次来看，StackView1是StackView2的父视图。

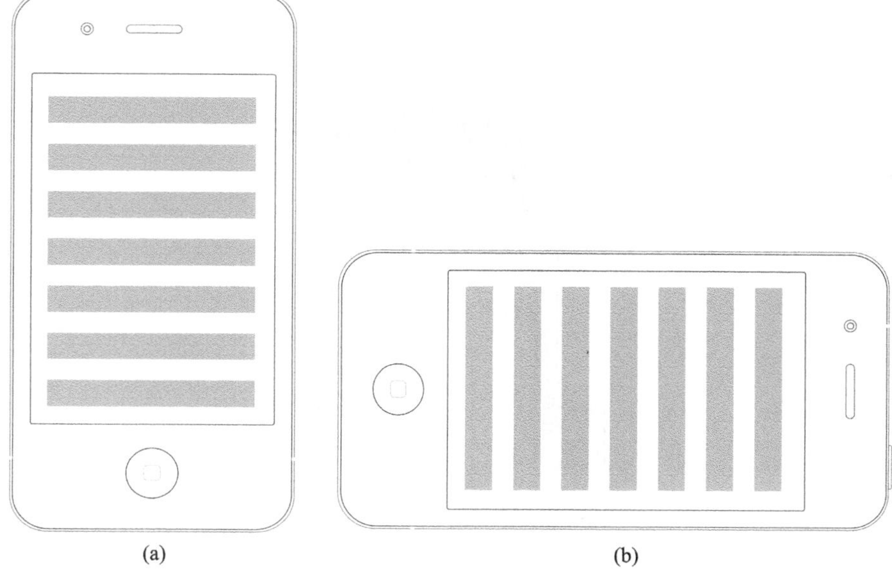

图7-29 堆视图界面布局

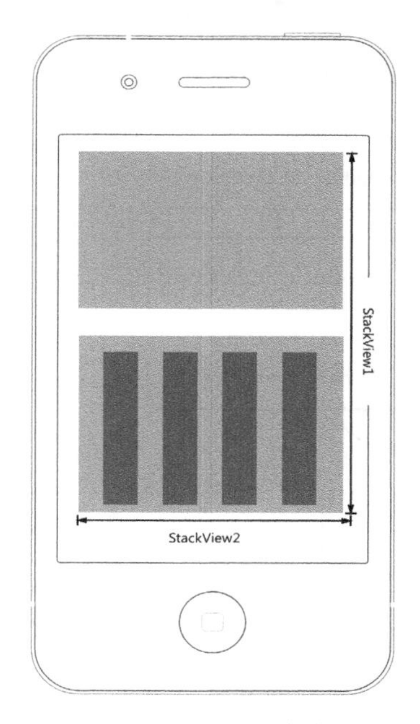

图7-30 堆视图嵌套

7.6.2 案例：堆视图布局

下面我们通过案例解释一下如图7-30所示的堆视图嵌套的情况。首先，使用Single View Application模板创建一个名为StackViewSample的工程。

使用堆视图构建如图7-31所示的界面，如果仅仅只采用一个堆视图，则实现起来比较困难，因为这3个按钮呈现三角形摆放。可以采用两个堆视图嵌套实现，界面层次结构如图7-31所示，其中整个界面是一个垂直

StackView，它的子视图是Button1和水平StackView，水平StackView的子视图是Button2和Button3。各视图对象间的关系如图7-32所示。

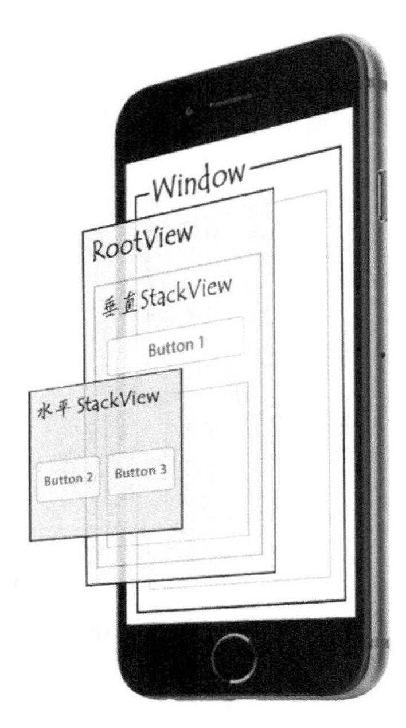

图7-31　应用界面的构建层次图

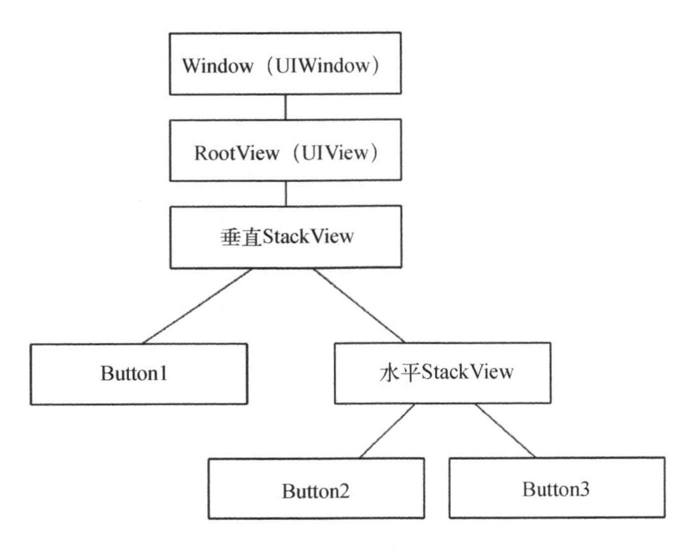

图7-32　各视图对象间的关系

下面我们介绍一下使用Interface Builder的具体实现过程。

1. 添加堆视图

首先，打开Main.storyboard文件，从对象库中拖曳一个Vertical Stack View到设计界面，如图7-33所示。然后从对象库中拖曳一个Horizontal Stack View到前一个Vertical Stack View中，如图7-34所示。

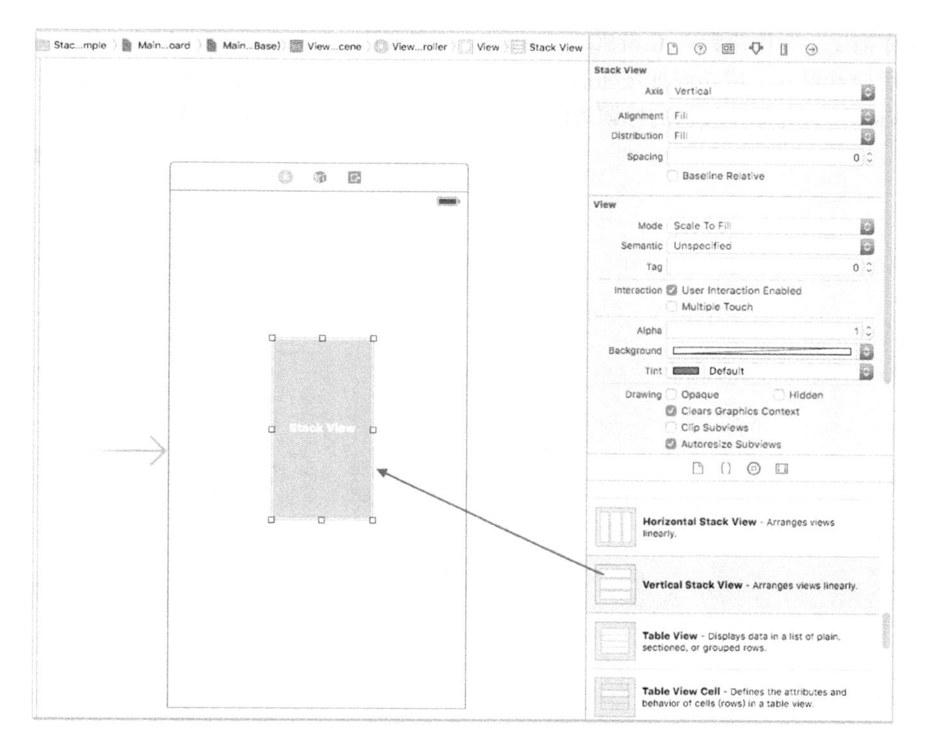

图7-33　添加Vertical Stack View

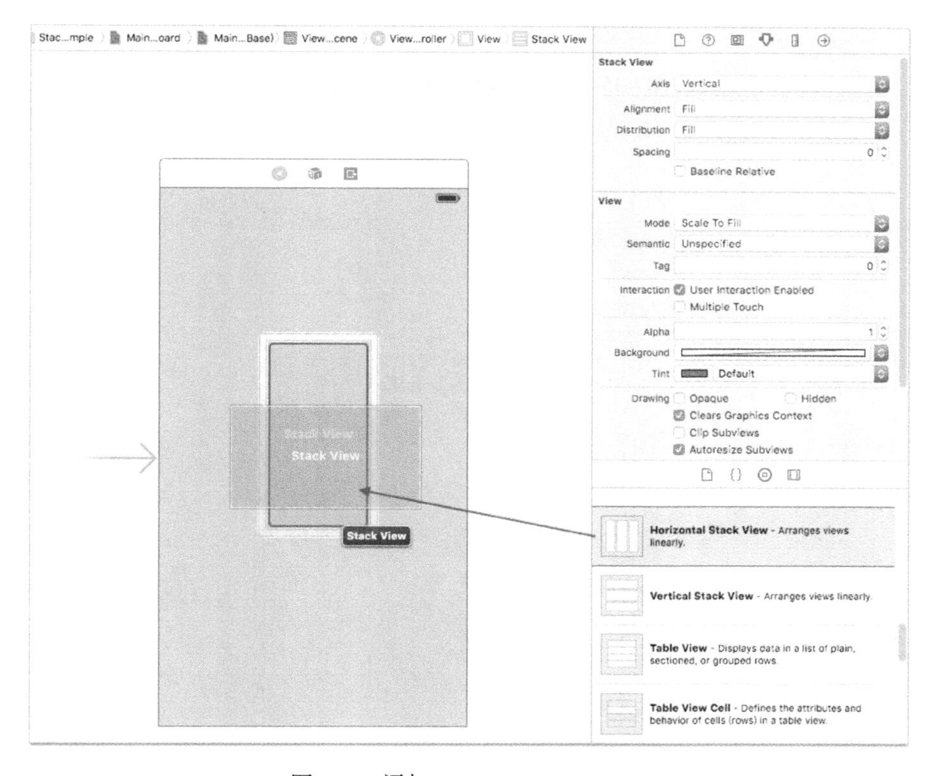

图7-34　添加Horizontal Stack View

接着从对象库拖曳一个按钮到Vertical Stack View中，注意要放在Horizontal Stack View的上方，如图7-35所示，当出现一个横线后松开鼠标。成功添加按钮后，请将按钮的标题改为Button1。通过类似的方法添加两个按钮到Horizontal Stack View中，并修改按钮的标题为Button2和Button3，最后的结果如图7-36所示。

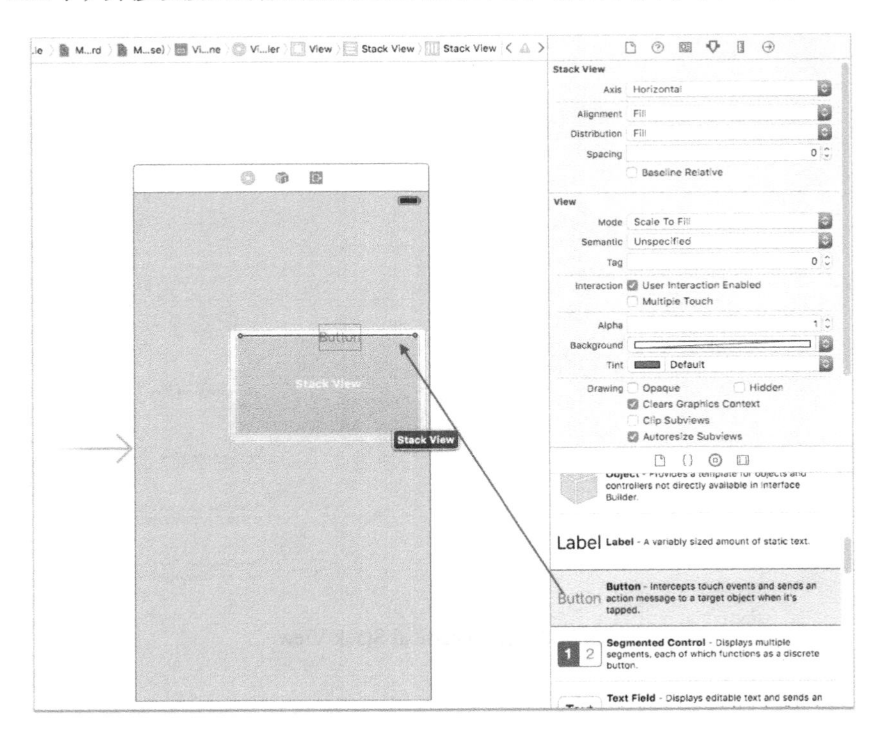

图7-35　添加按钮到Vertical Stack View

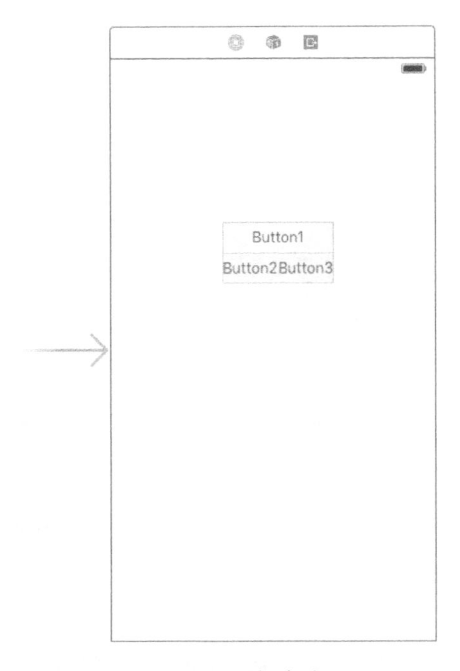

图7-36　添加完成

提示 堆视图中子视图的布局是由堆视图管理的，子视图不需要添加约束，但是堆视图本身是要添加约束的。

2. 设置堆视图属性

然后，需要设置堆视图属性。首先选中Vertical Stack View，如图7-37所示，打开其属性检查器，其中Axis属性用于设置堆视图是垂直还是水平方向，Spacing属性用于设置堆视图中子视图之间的距离，Alignment属性用于设置子视图之间的对齐方式，Distribution属性用于设置子视图的尺寸与位置，使这些子视图充分地利用堆视图空间。这里我们只是重新设置Spacing属性为400，其他属性为默认值就可以了。我们以同样的方式选中Horizontal Stack View，设置Spacing属性为150，如图7-38所示。

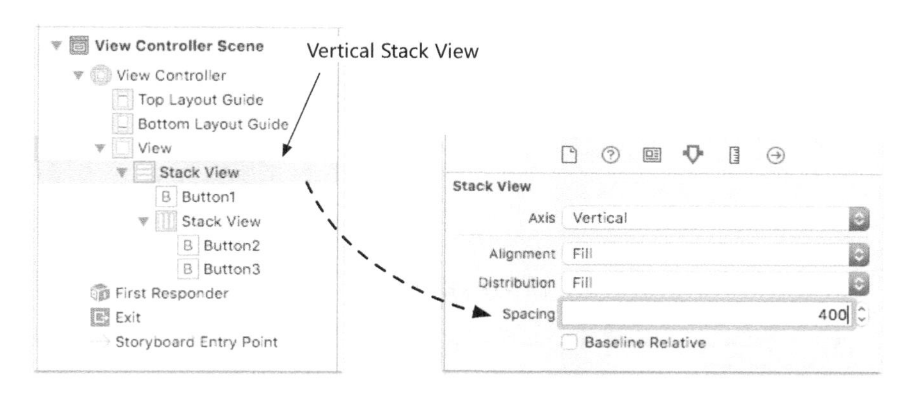

图7-37 设置垂直堆视图的属性

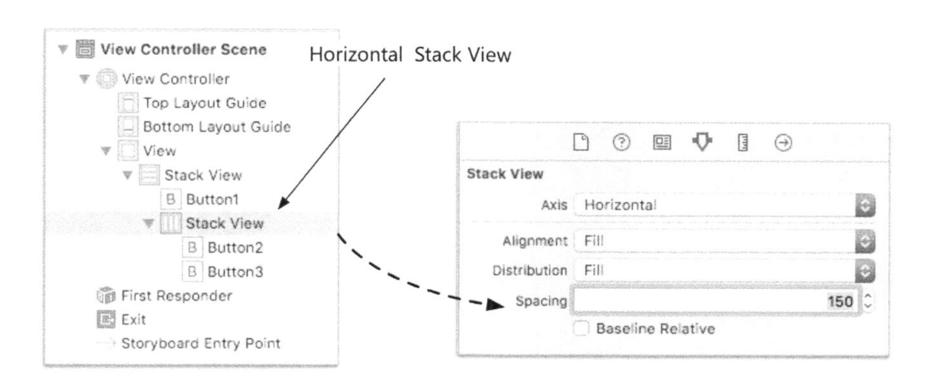

图7-38 设置水平堆视图的属性

3. 添加堆视图约束

设置完成堆视图属性后，可以为堆视图添加Auto Layout约束，拖曳垂直堆视图并将其摆放到屏幕中央，然后选中垂直堆视图并点击按钮，在弹出的菜单中选择Add Missing Constraints添加约束，结果如图7-39所示，其中有两个约束：Stack View.top和Stack View.centerX。

刚刚添加的两个约束还不够，还需要添加一个堆视图与下边距的绝对约束，这个约束可以保证当设备横屏时，Button2和Button3不会超出屏幕边界。选中垂直堆视图，点击Pin按钮，此时会弹出Pin菜单，如图7-40所示，我们将下边距设置为实线，设置完成后点击Add 1 Constraint按钮，添加约束并关闭菜单。

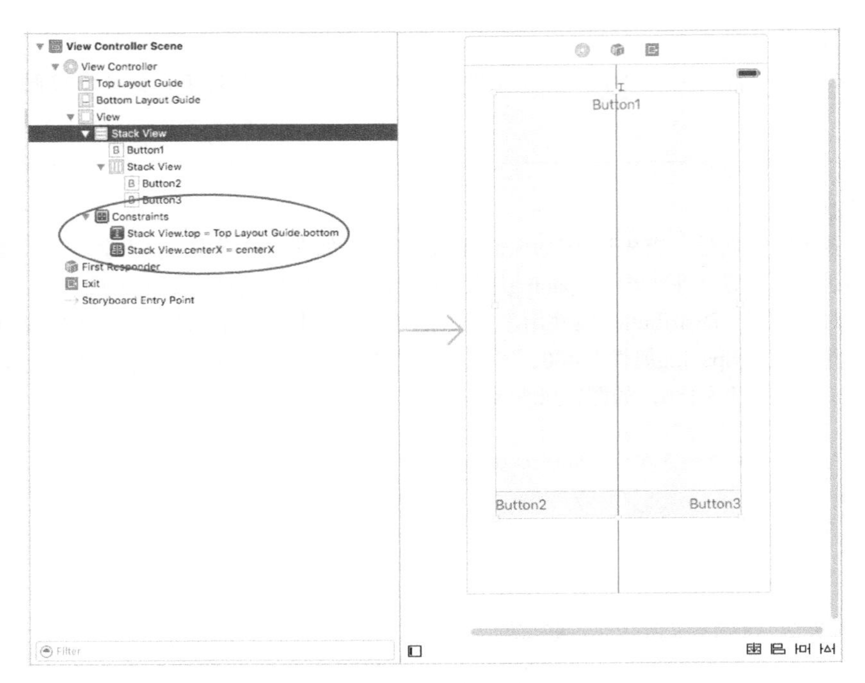

图7-39 添加约束

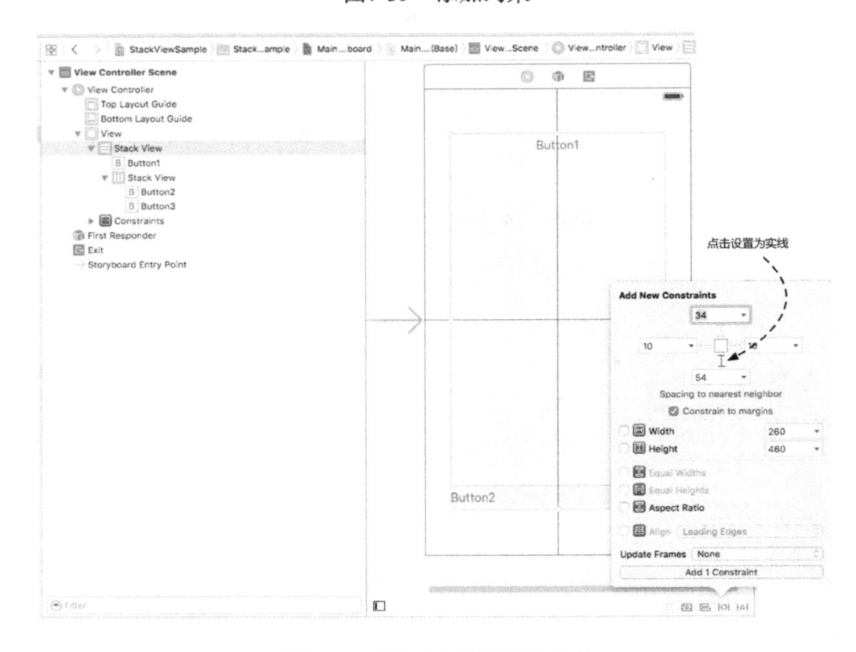

图7-40 添加下边距绝对约束

至此，我们的设计工作就完成了。

7.7 小结

本章首先介绍了3种主要的iOS界面布局UI设计模式，然后分别介绍了使用Auto Layout技术解决界面布局等问题，最后分别介绍了静态表视图和堆视图StackView。

第 8 章

屏幕适配

第7章介绍了使用Auto Layout技术可以解决大多数布局问题，但是随着iPhone 6/6 Plus/6s/6s Plus/7/7 Plus的发布，屏幕尺寸的差别越来越大，单纯使用Auto Layout技术就显得捉襟见肘了。为了适配多种不同的iOS设备屏幕，iOS 8之后推出了基于Auto Layout的Size Class技术。

8.1 iOS 屏幕的多样性

2014年9月9日，苹果公司发布了iPhone 6设备， iOS设备越来越多样化，其中屏幕的多样化是最为复杂的。因此，在介绍屏幕适配技术之前，有必要了解一下设备屏幕相关的信息。

8.1.1 iOS 屏幕介绍

到2016年10月10日为止，主流的iOS设备屏幕至少有6种，见图8-1和图8-2。

说明 由于本书的版本是基于iOS 10，iOS 10不再支持iPhone 4s设备，所以iOS主流设备中不再包括iPhone 4s。

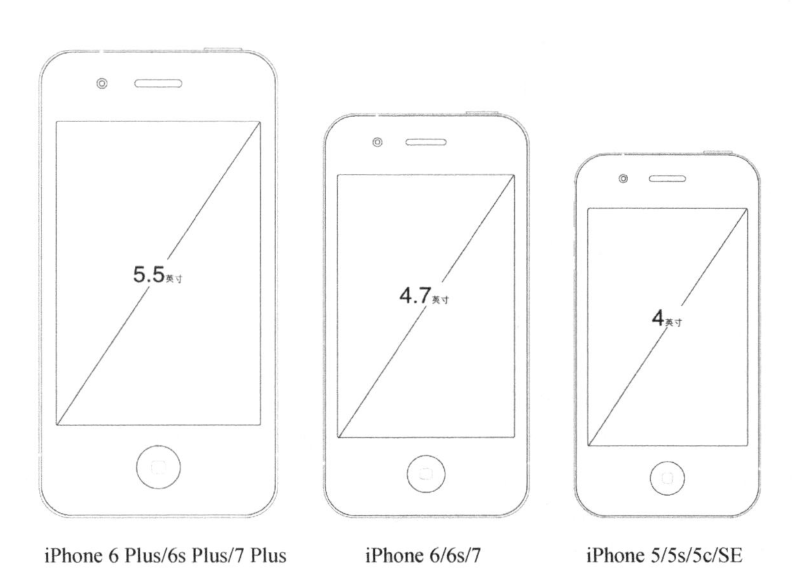

图8-1 iPhone设备的屏幕比较

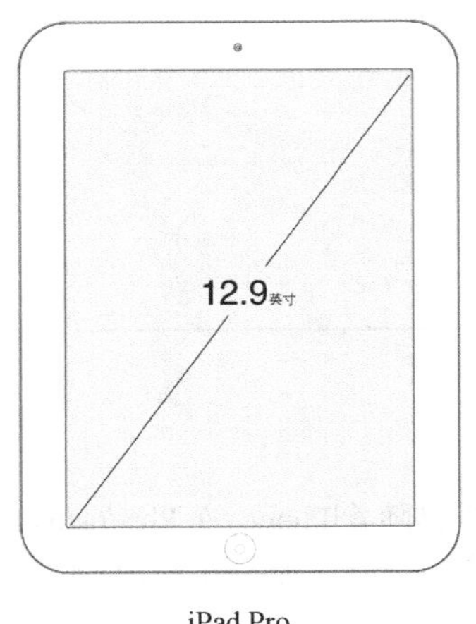

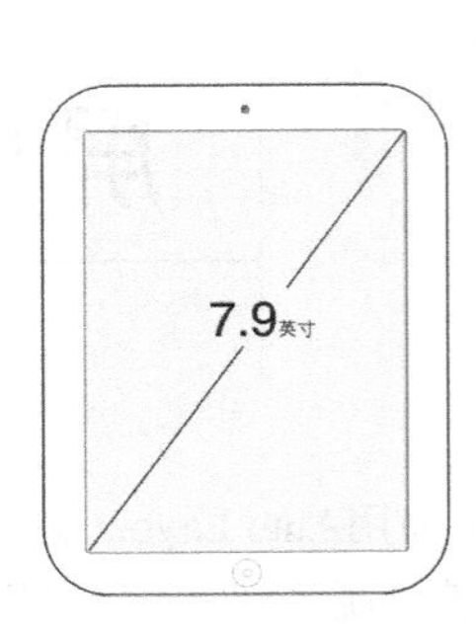

iPad Pro　　iPad Pro/Air 2/Air　　iPad mini 4/mini 2

图8-2　iPad设备的屏幕比较

更加详细的信息见表8-1。

表8-1　iOS设备的屏幕分辨率

设　备	屏幕尺寸（英寸）	屏幕分辨率（像素）	说　明
iPhone 6 Plus/6s Plus/7 Plus	5.5	1920 × 1080	Retina HD高清显示屏，401 ppi
iPhone 6/6s/7	4.7	1334 × 750	Retina HD高清显示屏，326 ppi
iPhone 5/5s/5c/SE	4	1136 × 640	Retina显示屏，326 ppi
iPad Pro	12.9	2732 × 2048	Retina显示屏，264 ppi
iPad Pro/Air 2/Air	9.7	2048 × 1536	Retina显示屏，264 ppi
iPad mini 4/mini 2	7.9	2048 × 1536	Retina显示屏，326 ppi

说明　表8-1中的ppi是像素密度单位，表示“像素/英寸”，如326 ppi表示每英寸上有326个像素。另外，iPad Pro有两个不同的尺寸，即12.9英寸和9.7英寸。

8.1.2　iOS 的 3 种分辨率

对于普通用户，了解表8-1所述的信息已经足够了，而对于设计人员和开发人员，大家还需要了解更深层的分辨率信息。

为了解决屏幕适配问题，一些游戏引擎中提出了3种分辨率：资源分辨率、设计分辨率和屏幕分辨率。

- **资源分辨率**。也就是资源图片的大小，单位是“像素”。
- **设计分辨率**。逻辑上的屏幕大小，单位是“点”。Interface Builder设计器和程序代码中的单位都是“点”。
- **屏幕分辨率**。是以像素为单位的屏幕大小，所有的应用都会渲染到这个屏幕上以展示给用户。

从表8-2可见，iPhone 6 Plus/6s Plus/7 Plus是最为特殊的设备，资源分辨率与屏幕分辨率的比例是1.15 : 1，而其他设备的比例是1 : 1。这3种分辨率对于不同的人群关注的方面也不同，UI设计人员主要关注资源分辨率，开发人员主要关注设计分辨率，而一般用户主要关注屏幕分辨率。

表8-2 iOS设备的3种分辨率

设 备	资源分辨率（像素）	设计分辨率（点）	屏幕分辨率（像素）	说 明
iPhone 6 Plus/6s Plus/7 Plus	2208 × 1242	736 × 414	1920 × 1080	1点 = 3倍像素，资源缩小1.15倍，渲染到屏幕上
iPhone 6/6s/7	1334 × 750	667 × 375	1334 × 750	1点 = 2倍像素
iPhone 5/5s/5c/SE	1136 × 640	568 × 320	1136 × 640	1点 = 2倍像素
iPad Pro	2732 × 2048	1366 × 1024	2732 × 2048	1点 = 2倍像素
iPad Pro/Air 2/Air	2048 × 1536	1024 × 768	2048 × 1536	1点 = 2倍像素
iPad mini 4/mini 2	2048 × 1536	1024 × 768	2048 × 1536	1点 = 2倍像素

8.1.3 获得 iOS 设备的屏幕信息

为了屏幕适配的需要，有时候我们需要获得iOS设备的屏幕信息，然后根据该信息判断是哪种iOS设备。视图控制器ViewController的主要代码如下：

```
//ViewController.swift文件
override func viewDidLoad() {
    super.viewDidLoad()

    let iOSDeviceScreenSize : CGSize = UIScreen.main.bounds.size              ①

    NSLog("%@ x %@", iOSDeviceScreenSize.width,
    ➥iOSDeviceScreenSize.height)
    var s : NSString = NSString(format: "%@ x %@",
    ➥iOSDeviceScreenSize.width, iOSDeviceScreenSize.height)
    self.label.text = s

    if (UIDevice.currentDevice().userInterfaceIdiom ==
    ➥UIUserInterfaceIdiom.phone) {                                            ②

        if (iOSDeviceScreenSize.height > iOSDeviceScreenSize.width)
        ➥{//竖屏情况                                                           ③

            if (iOSDeviceScreenSize.height == 568) {                           ④
                NSLog("iPhone 5/5s/5c/SE设备")
            } else if (iOSDeviceScreenSize.height == 667) {//iPhone 6/6s/7     ⑤
                NSLog("iPhone 6/6s/7设备")
            } else if (iOSDeviceScreenSize.height == 736) {
                //iPhone Plus                                                  ⑥
                NSLog("iPhone Plus设备")
            } else {//iPhone 4s等其他设备                                      ⑦
                NSLog("iPhone4s等其他设备")
            }
        }
        if (iOSDeviceScreenSize.width > iOSDeviceScreenSize.height)
            {//横屏情况
            if (iOSDeviceScreenSize.width == 568) {
                NSLog("iPhone 5/5s/5c/SE设备")
            } else if (iOSDeviceScreenSize.width == 667) {//iPhone 6/6s/7
                NSLog("iPhone 6/6s/7设备")
            } else if (iOSDeviceScreenSize.width == 736) {//iPhone Plus
                NSLog("iPhone Plus设备")
            } else {//其他设备
                NSLog("其他设备")
            }
        }
    }

}
```

```
//ViewController.m文件
- (void)viewDidLoad {
    [super viewDidLoad];

    CGSize iOSDeviceScreenSize = [UIScreen mainScreen].bounds.size;          ①

    NSLog(@"%f x %f", iOSDeviceScreenSize.width, iOSDeviceScreenSize.height);

    NSString *s = [NSString stringWithFormat:@"%f x %f",
    ➥iOSDeviceScreenSize.width, iOSDeviceScreenSize.height];
    self.label.text = s;

    if ([UIDevice currentDevice].userInterfaceIdiom ==
    ➥UIUserInterfaceIdiomPhone) {                                             ②

        if (iOSDeviceScreenSize.height > iOSDeviceScreenSize.width)
            {//竖屏情况                                                         ③

            if (iOSDeviceScreenSize.height == 568) {                           ④
                //iPhone 5/5s/5c/SE设备
                NSLog(@"iPhone 5/5s/5c/SE设备");
            } else if (iOSDeviceScreenSize.height == 667) {//iPhone 6/6s/7     ⑤
                NSLog(@"iPhone 6/6s/7设备");
            } else if (iOSDeviceScreenSize.height == 736) {                    ⑥
                //iPhone Plus
                NSLog(@"iPhone Plus设备");
            } else {//其他设备                                                 ⑦
                NSLog(@"其他设备");
            }
        }
        if (iOSDeviceScreenSize.width > iOSDeviceScreenSize.height)
        ➥{//横屏情况
            if (iOSDeviceScreenSize.width == 568) {
            ➥//iPhone 5/5s/5c/SE设备
                NSLog(@"iPhone 5/5s/5c/SE设备");
            } else if (iOSDeviceScreenSize.width == 667) {
                //iPhone 6/6s/7
                NSLog(@"iPhone 6/6s/7设备");
            } else if (iOSDeviceScreenSize.width == 736) {//iPhone Plus
                NSLog(@"iPhone Plus设备");
            } else {//其他设备
                NSLog(@"其他设备");
            }
        }
    }

}
```

第①行用于获得屏幕大小，返回值是CGSize类型的。第②行用于获得设备信息，然后判断是否为iPhone设备，其中UIDevice的静态方法+ currentDevice可以获得设备信息，UIUserInterfaceIdiom是枚举类型，枚举类型UIUserInterfaceIdiom中定义的成员如表8-3所示。

表8-3　UIUserInterfaceIdiom枚举成员

Swift枚举成员	Objective-C枚举成员	说　明
phone	UIUserInterfaceIdiomPhone	判断为iPhone设备
pad	UIUserInterfaceIdiomPad	判断为iPad设备
unspecified	UIUserInterfaceIdiomUnspecified	未知设备

获知是哪种设备后，还需要判断是横屏还是竖屏，第③行用于判断是否是竖屏的情况。第④行用于判断设备是否为iPhone 5/5s/5c/SE等设备，第⑤行用于判断设备是否为iPhone 6/6s/7，第⑥行用于判断设备是否为iPhone Plus，第⑦行用于判断是否为其他设备。

如果设备处于横屏情况，那么只需要判断屏幕的宽度。

读者测试这段代码时，可以打开本节的实例代码ScreenTest，然后在Xcode中选择不同的模拟器进行测试即可，如图8-3所示。

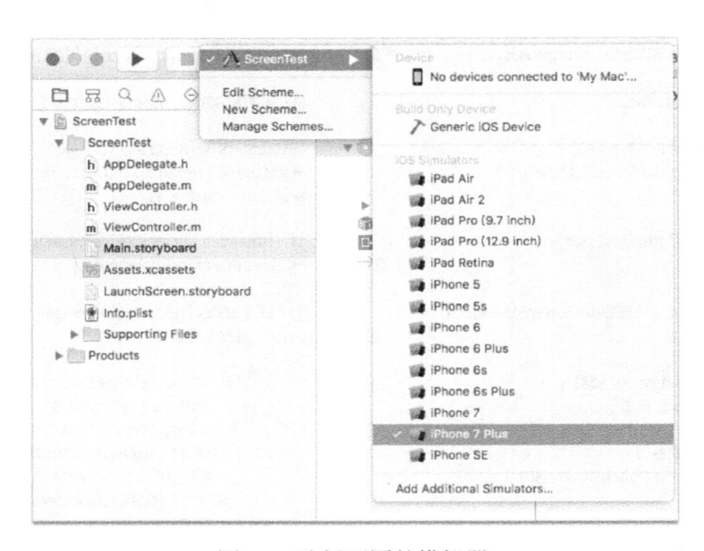

图8-3　选择不同的模拟器

8.2　Size Class 与 iOS 多屏幕适配

iOS 8之后，搭载设备越来越多，特别是iPhone 6和iPhone 6 Plus发布后，Auto Layout技术已经不能解决这么复杂的屏幕适配问题了。Auto Layout技术所能解决的只是界面差别小的情况，而界面差别很大时需要采用不同的用户界面文件。

8.2.1　在 Xcode 6 和 Xcode 7 中使用 Size Class 技术

为了应对新的变化，苹果公司在iOS 8中推出了新的屏幕适配技术——Size Class，它依赖并建立在Auto Layout技术之上。

在Xcode 6和Xcode 7的Interface Builder中，点击布局工具栏中的Size Class按钮“wAny hAny”将弹出Size Class面板，如图8-4所示。

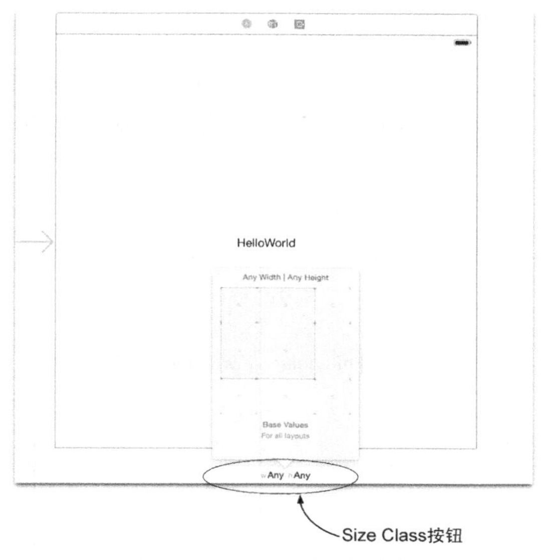

图8-4　Size Class按钮和面板

8.2.2　Size Class 的九宫格

图8-4所示的Size Class面板是一个九宫格，可以组合出9种情形，每一种情形对应9种不同Size Class取值中的一种。如图8-5所示，Size Class九宫格中有Width（宽）和Height（高）两个布局方向，坐标原点在左上角。Width和Height布局方向上还有3个类别：紧凑（Compact）、任意（Any）和标准（Regular）。所谓"紧凑"，就是屏幕空间相对比较小，例如iPhone竖屏时，宽度是"紧凑"的，而高度是"标准"的，取值为`wCompact | hRegular`；而在iPhone横屏时，宽度是"标准"的，而高度是"紧凑"的，取值为`wRegular | hCompact`。在"紧凑"和"标准"之间的值是"任意"。

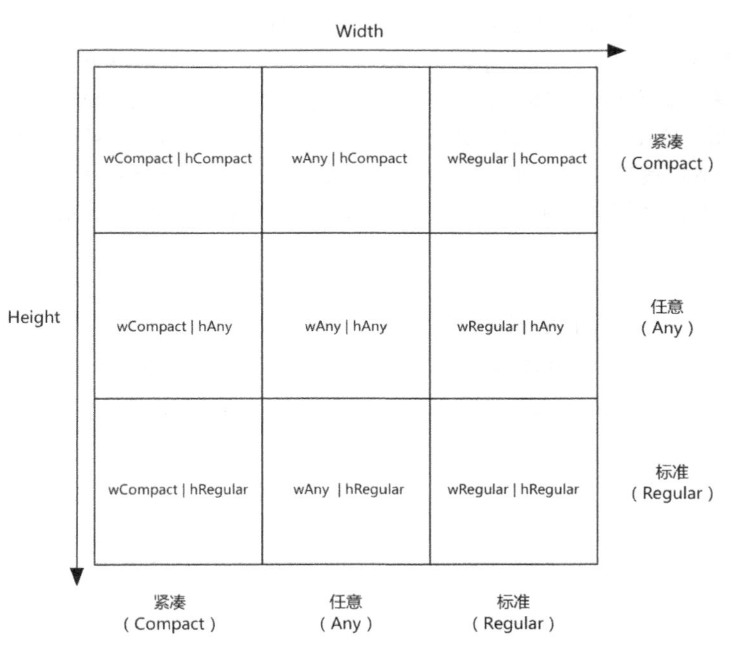

图8-5　Size Class九宫格

8.2.3 Size Class 的四个象限

这天书般的Size Class九宫格中组合出的9种情形，可以解决所有的iOS多屏幕适配。其中Any（任意）是默认情况，但通常情况下很少使用。因此，Size Class九宫格可以简化为Size Class四个象限，如图8-6所示。

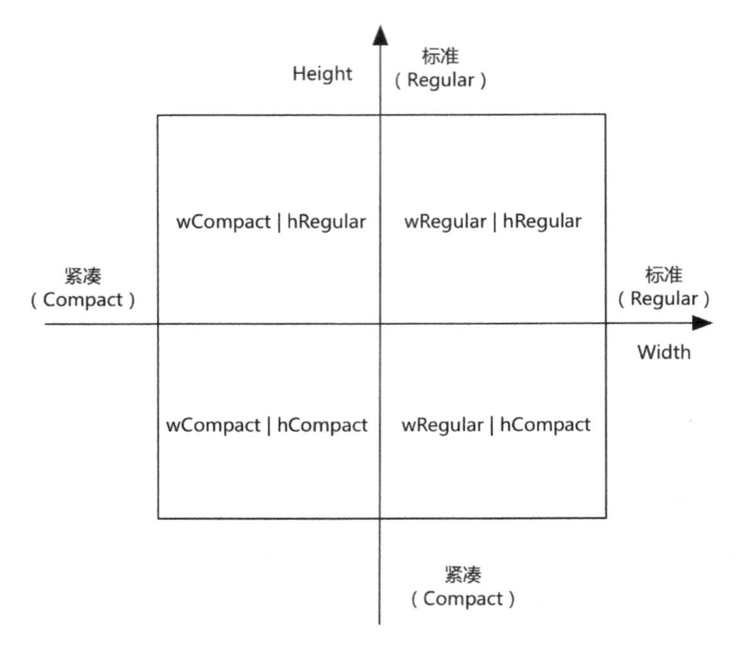

图8-6 Size Class四象限

如图8-6所示，Size Class需要考虑的值从9个简化到了4个，这在实际开发中意义重大。

各个设备的Size Class取值关系如图8-7所示，解释说明如下。

- **iPhone横屏**。宽度和高度都是“紧凑”的，所以Size Class取值为`wCompact | hCompact`。
- **iPhone竖屏**。宽度是“紧凑”的，高度是“标准”的，所以Size Class取值为`wCompact | hRegular`。
- **iPhone Plus横屏**。宽度是“标准”的，高度是“紧凑”的，所以Size Class取值为`wRegular | hCompact`。
- **iPhone Plus竖屏**。宽度“紧凑”的，高度是“标准”的，所以Size Class取值为`wCompact | hRegular`。
- **iPad横屏**。宽度和高度都是“标准”的，所以Size Class取值为`wRegular | hRegular`。
- **iPad竖屏**。宽度和高度都是“标准”的，所以Size Class取值为`wRegular | hRegular`。

提示 由于在iOS 9之后，在iPad Pro设备上，可以分屏运行两个不同的应用（屏幕分割为两不同部分，每一部分运行一个应用），这种情况下屏幕适配会更加复杂，我们将在10.6节中详细介绍。

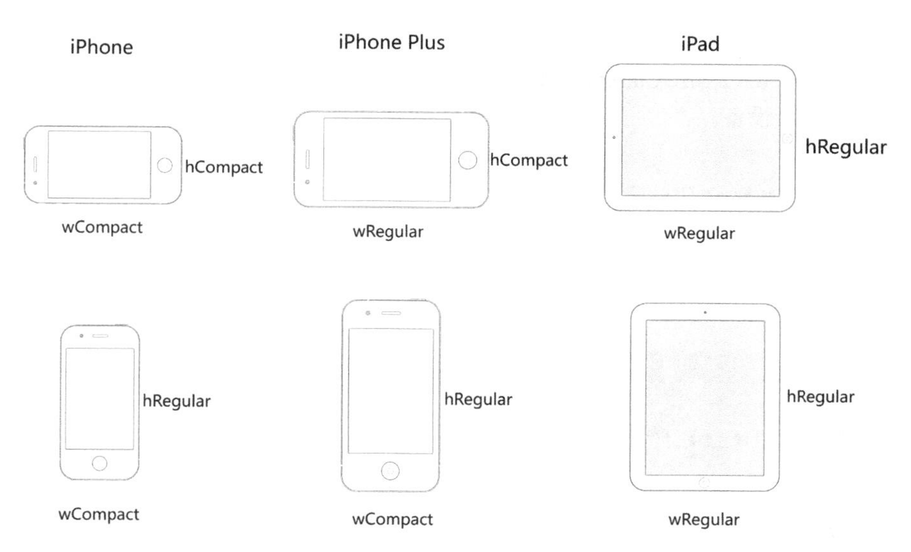

图8-7 各个设备的Size Class取值

8.2.4 在Xcode 8中使用Size Class

客观地说，Xcode 6和Xcode 7中采用九宫格方式设计界面，确实不方便。很多情况下，我们只需要考虑Size Class的4个值即可。为了应对这样的变化，Xcode 8提供了如图8-8所示的设备配置面板。要打开设备配置面板，可以点击Interface Builder设计界面下边的View as按钮。

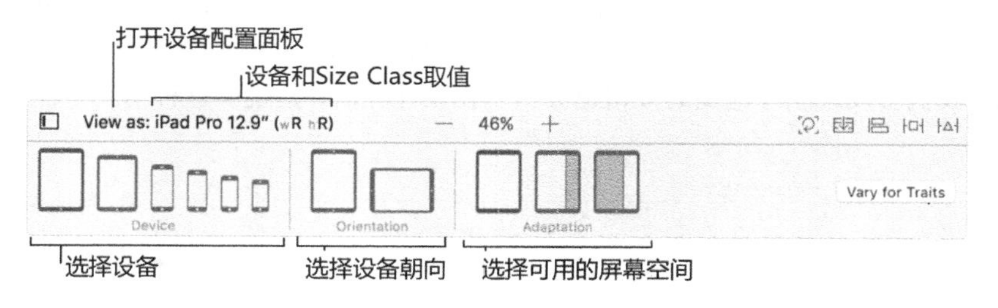

图8-8 设备配置面板

在设备配置面板中我们可以选择设备和设备朝向，如果选择了iPad设备，那么还可以选择可用的屏幕空间。点击Vary for Traits按钮，可以弹出“选择Size Class”对话框，如图8-9所示。

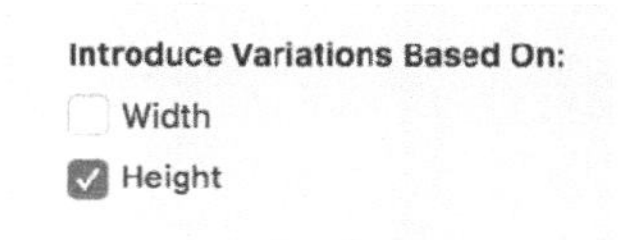

图8-9 “选择Size Class”对话框

那么，这个Vary for Traits按钮能够做什么呢？它能够改变UI特征，其中包含布局Size Class、约束和字体大小等内容。例如，如果我们在图8-8中选择设备iPhone 7，横屏朝向，如图8-10a所示，则当前设备的Size Class取值为`wCompact | hCompact`。然后点击Vary for Traits按钮后选中Height，则会出现如图8-10b所示的界面，其中的文字提示Varying 4 Compact Height Devices，说明这个变化会影响4个高度“紧凑”的设备，即`hCompact`。而8-10c所示的界面是选择了Width的情况，这个变化会影响14个`wCompact`设备。

综上所述，改变UI特征（点击Vary for Traits按钮）操作所影响的设备多少，与图8-10a的选择有关系。本例中选择iPhone 7，横屏朝向，那么Size Class选择Height，则只会影响相关的hCompact设备；而Size Class选择Width，则只会影响相关的wCompact设备。

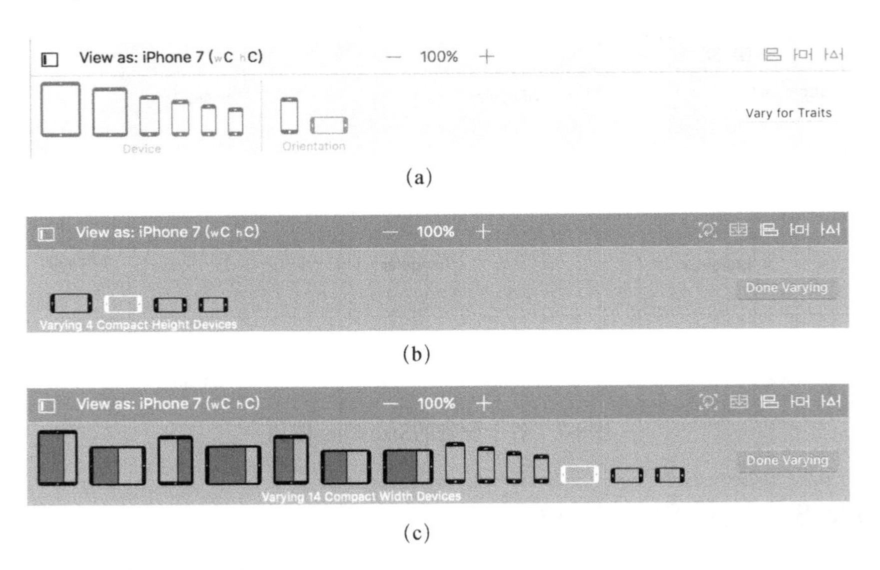

图8-10　改变UI特征

在设计过程中，我们可以选择改变UI特征，也可以不选择，那么两者有什么区别呢？没有选择改变UI特征时在Interface Builder设计界面中进行设计，设计结果会影响所有设备和朝向，例如选择设备iPhone 7，朝向横屏，Size Class的值是wCompact | hCompact，此时拖曳一个按钮到设计界面，并添加约束，如图8-11所示。然后选择竖屏，如图8-12所示，我们会发现刚才拖曳的按钮也会出现在设计界面，这说明这种情况下横屏和竖屏的变化会影响布局。

图8-11　没有改变UI特征，iPhone 7朝向横屏

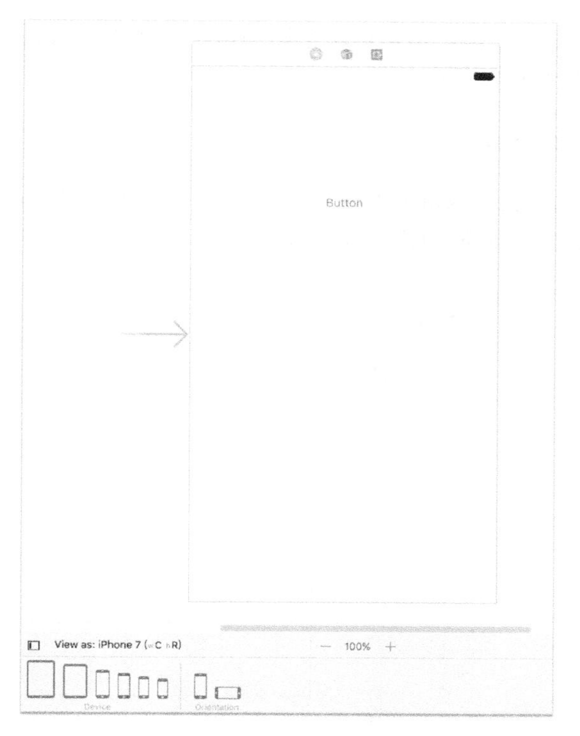

图8-12 没有改变UI特征，iPhone 7朝向竖屏

改变UI特征（点击Vary for Traits按钮），可以消除由于横屏和竖屏等特征变化产生的相互影响。同样是添加一个按钮，我们首先点击Vary for Traits按钮，在图8-9中选择Height，在设计界面中拖曳一个按钮，并添加约束，如图8-13所示。然后，点击Done Varying按钮完成，此时将设计界面切换到竖屏，会发现这里看不到刚刚在横屏时候添加的按钮。这说明使用改变UI特征，可以屏蔽特征变化产生的影响。

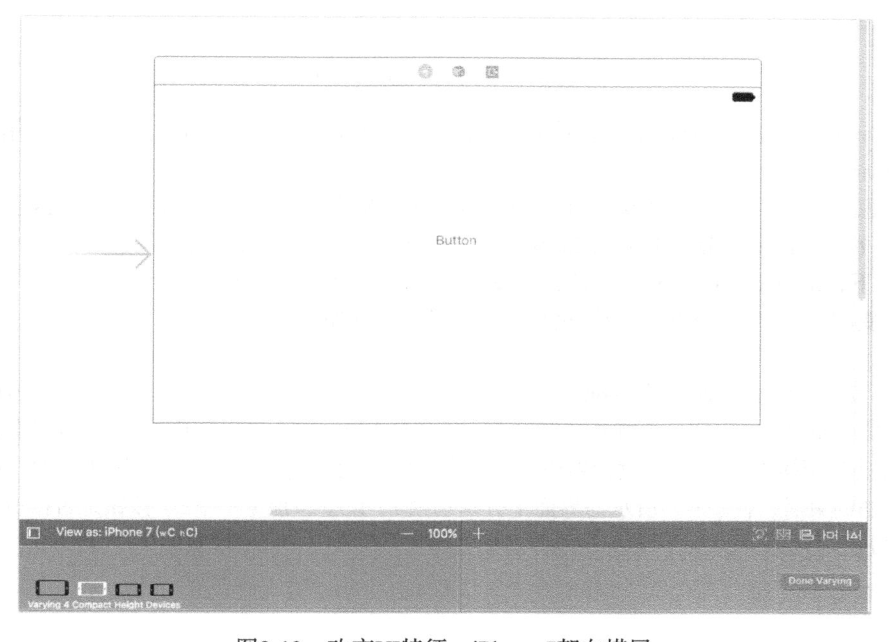

图8-13 改变UI特征，iPhone 7朝向横屏

另外，通过Xcode 8所提供的Size Class，我们可以直观地看到选择的Size Class对哪些设备会有影响，而Xcode 6和Xcode 7提供的“九宫格”则不那么直观。

8.2.5　案例：使用 Size Class

下面我们通过Xcode 8提供的Size Class重构7.4.2节所示的案例，界面如图8-14所示，界面上有3个按钮，要求在iPhone和iPhone Plus设备上都能实现横屏与竖屏的布局。

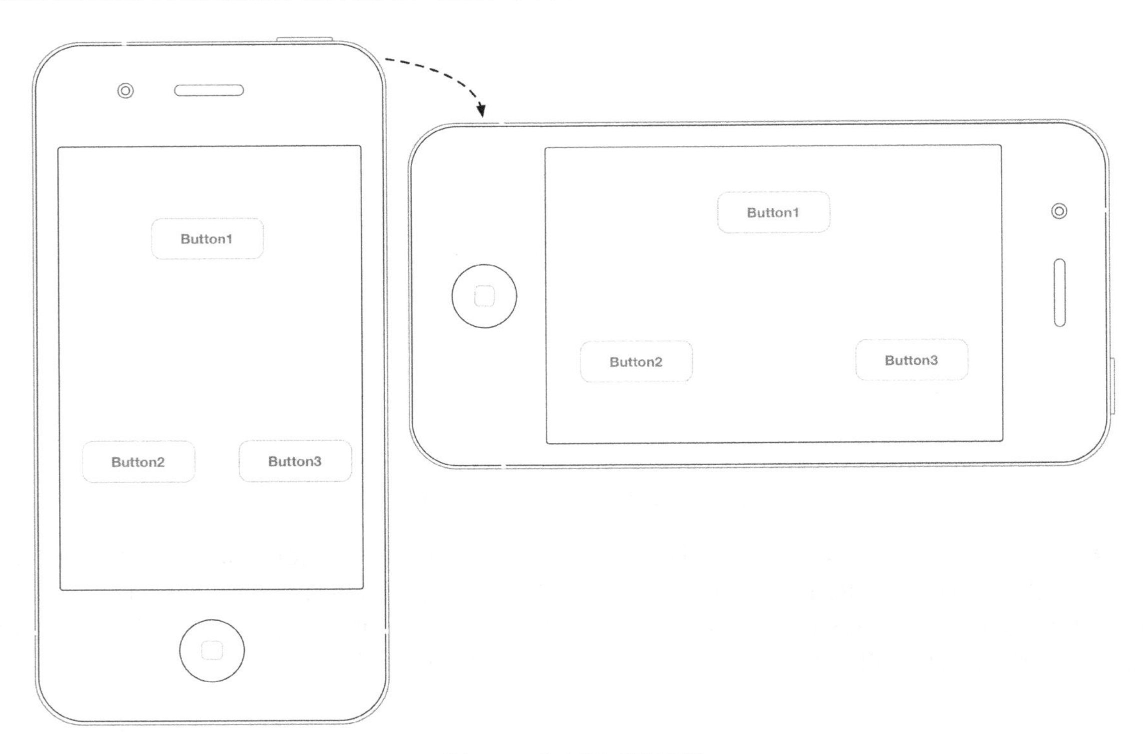

图8-14　案例原型设计图

首先，使用Single View Application模板创建一个名为SizeClassSample的工程。完成后，使用Interface Builder打开Main.storyboard。

然后需要考虑一下如何选择Size Class。由于iPhone Plus设备在布局方面与iPhone设备的区别在于：

- 横屏情况下，iPhone的Size Class是`wCompact | hCompact`，iPhone Plus的Size Class是`wRegular | hCompact`；
- 竖屏情况下，iPhone和iPhone Plus的Size Class都是`wCompact | hRegular`。

可以通过改变UI特征分别对横屏和竖屏进行设计，具体步骤如下。

1. 横屏设计

横屏情况下，如果考虑包含所有的iPhone和iPhone Plus设备，我们应该考虑`hCompact`。参考图8-10，选择iPhone 7横屏，然后点击Vary for Traits按钮，从打开的对话框中选择Height，这个选择会影响4个横屏设备。然后从对象库中拖曳3个按钮，按照图8-14所示摆放设计界面，并参考7.4.2节的案例分别为每一个按钮添加约束，具体添加过程不再赘述。添加约束后，在图8-15所示的界面中分别切换4个设备，预览一下3个按钮是否摆放整齐。最后，点击Done Varying按钮。

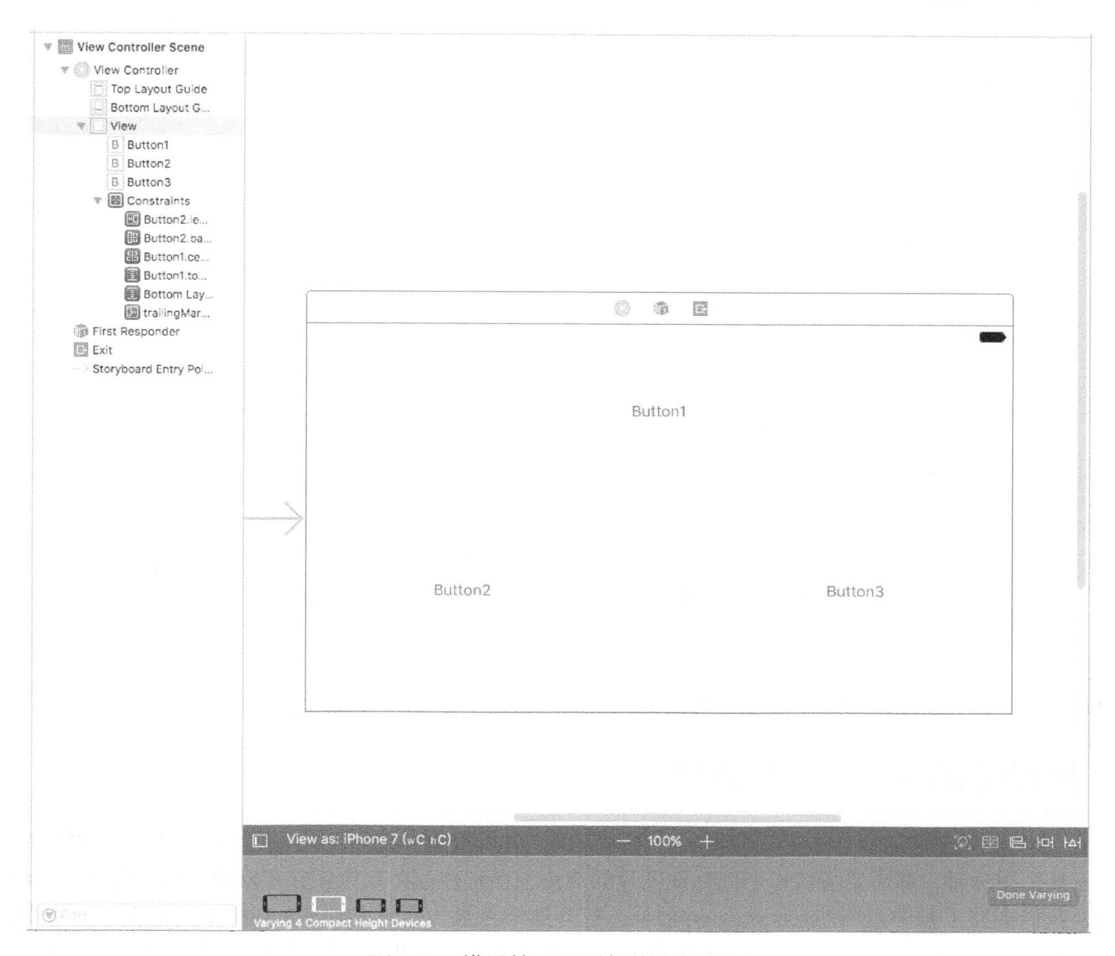

图8-15　横屏情况下添加按钮和约束

2. 竖屏设计

竖屏情况下，如果考虑包含所有的iPhone和iPhone Plus设备，我们应该考虑hRegular。选择iPhone 7竖屏，然后点击Vary for Traits按钮，从打开的对话框中选择Height，这个选择会影响18个横屏设备，如图8-16所示。

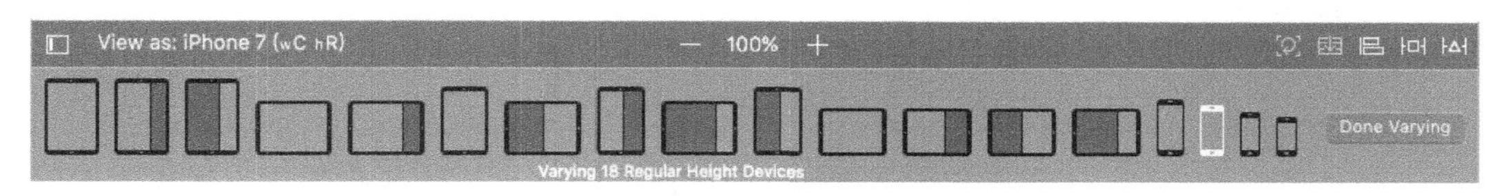

图8-16　影响18个横屏设备

然后从对象库中拖曳3个按钮，将其按照图8-14摆放设计界面，并参考7.4.2节的案例分别为每一个按钮添加约束，具体添加过程不再赘述。添加约束后，我们应该在图8-17所示的界面中分别切换18个设备，预览一下3个按钮是否摆放整齐。最后，点击Done Varying按钮。

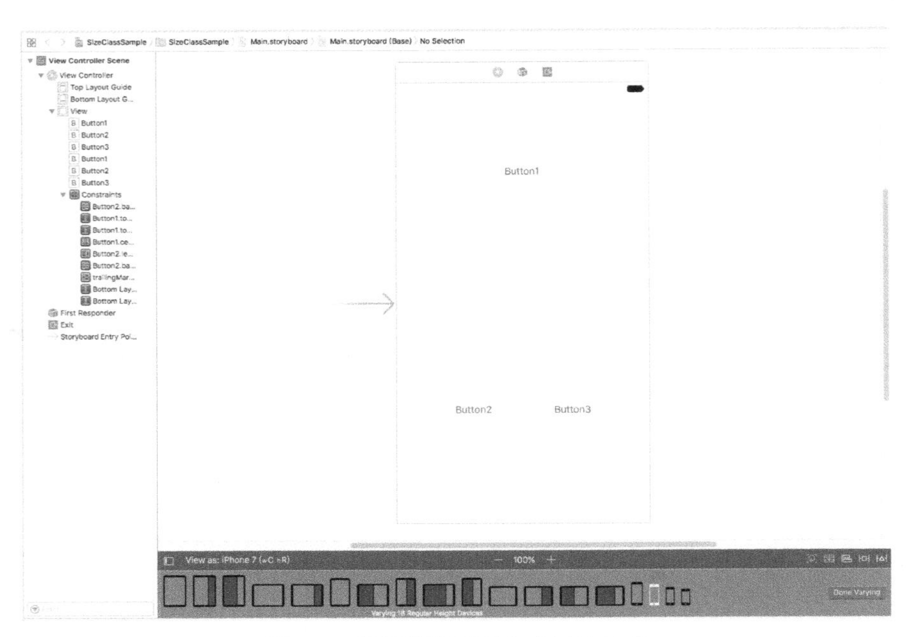

图8-17　竖屏情况下添加按钮和约束

到此为止，我们已经使用Size Class和Auto Layout技术实现了案例，可以运行看看效果了。

8.3　资源目录与图片资源适配

在Xcode 5及其之后的版本中开发应用时，图片资源文件管理有一些新变化。我们一起管理应用图片、启动界面、工具栏图标时，都需要清楚每个图标或图片的规格，而记住这些规格实在很麻烦。如果使用Xcode创建一个工程，我们会在工程中发现Assets.xcassets目录，打开这个目录，会看到如图8-18所示的界面，其中默认会有AppIcon项目。打开AppIcon项目，可以看到右边会有一些小虚框，这些小虚框下面有一些说明。需要说明的是，这里的AppIcon是应用图标。

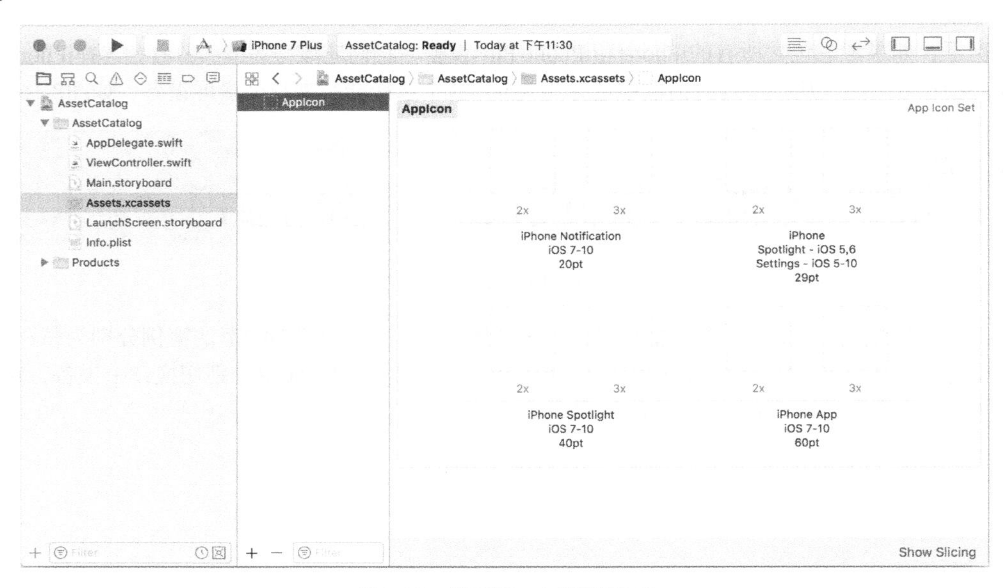

图8-18　使用Xcode资源目录

关于应用图标的内容，我们将在28.1.1节中介绍，本节介绍如何使用资源目录管理一般情况下使用的图片。资源目录最擅长管理在不同分辨率的设备上使用不同图片的情况。例如，一个应用的背景界面在普通显示屏和Retina显示屏下有不同的图片。如果我们有两张不同规格的图片——750 × 1334.png和1242 × 2208.png，其中750 × 1334.png用于所有iPhone设备，1242 × 2208.png用于所有iPhone Plus设备。打开Assets.xcassets，在右边的空白区域点击+按钮，此时将会弹出如图8-19所示的菜单，选择New Image Set菜单项，将会创建新的图片集。默认的图片集名为Image，请双击它，将其修改为background，如图8-20所示。

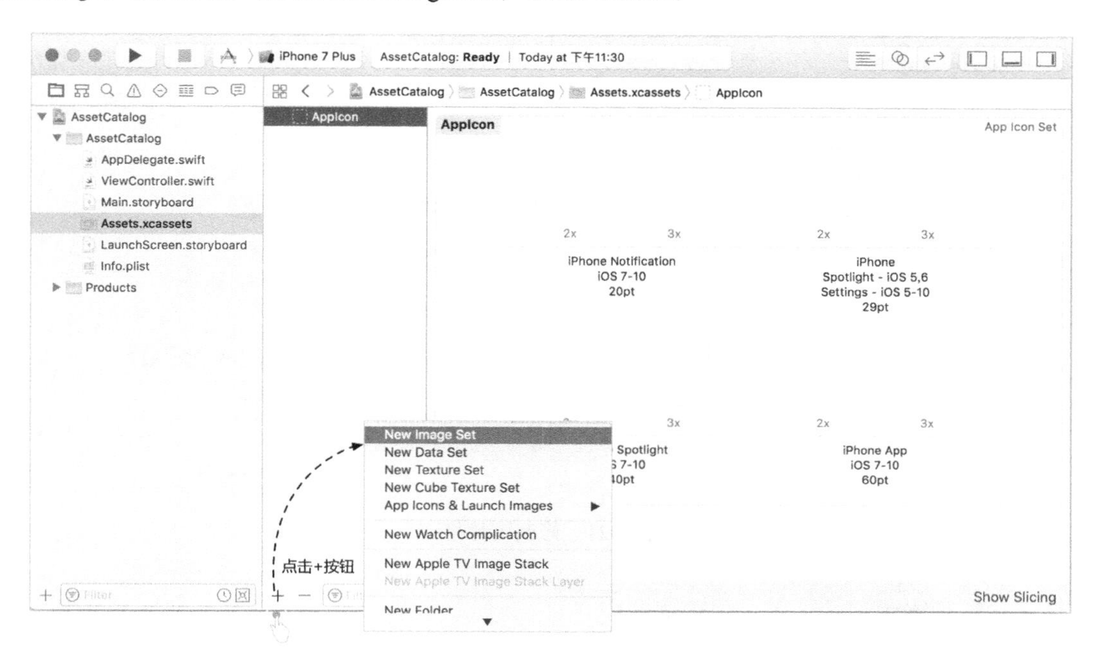

图8-19 添加图片集

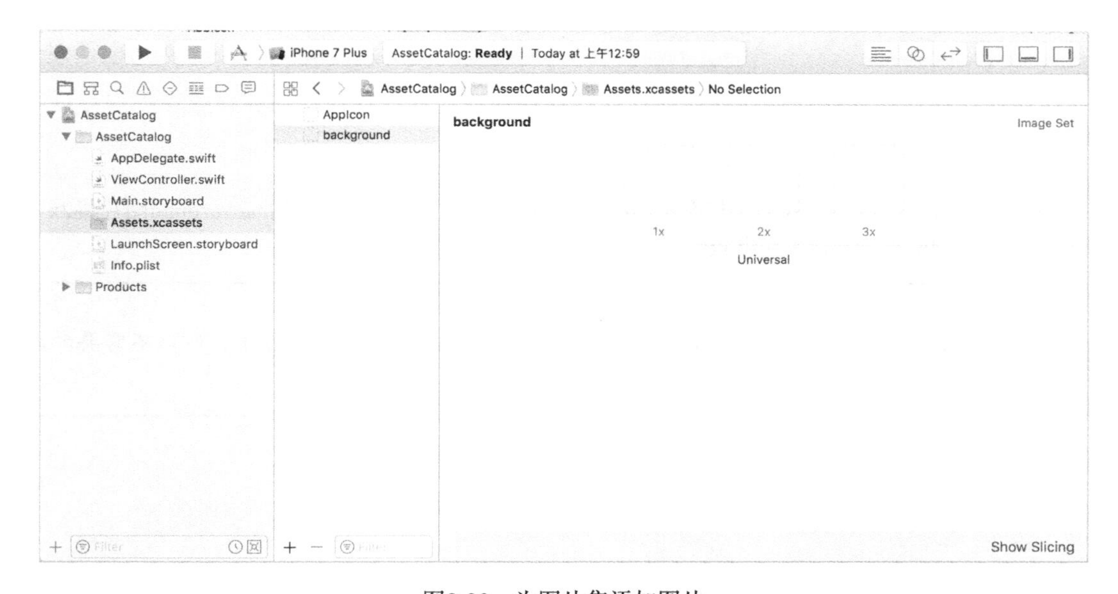

图8-20 为图片集添加图片

由于我们只考虑iPhone版本的图片集，而默认生成的是Universal版本（包括了iPhone和iPad）图片集，所以这里需要去掉Universal版本，添加iPhone版本的图片集。右击background图片集，在图8-21所示的弹出菜单中取消选中Devices→Universal。类似地，添加iPhone版本是在该菜单中选择iPhone项，如图8-22所示。

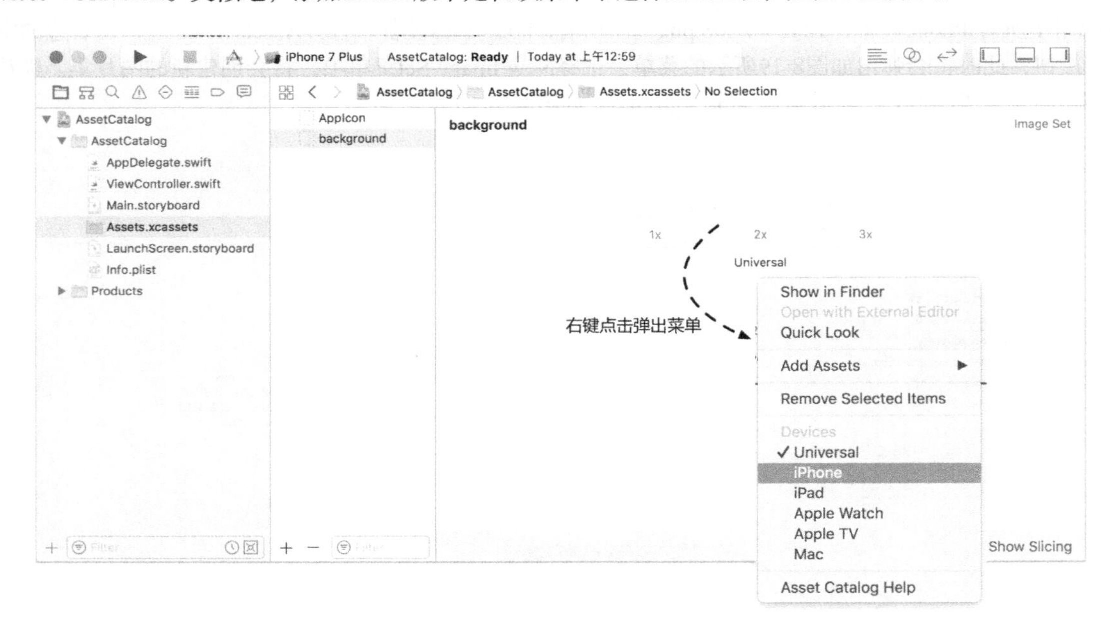

图8-21 修改图片集

图8-22 iPhone版的图片集

在Finder中打开图片所在的文件夹，如图8-23所示，拖曳750 × 1334.png图片到background图片集2x中，拖曳1242 × 2208.png图片到background图片集3x中。

提示 1x是iPhone普通显示屏设备所需的图片，iOS 10已经不再支持这种设备了，所以不需要考虑1x的情况。

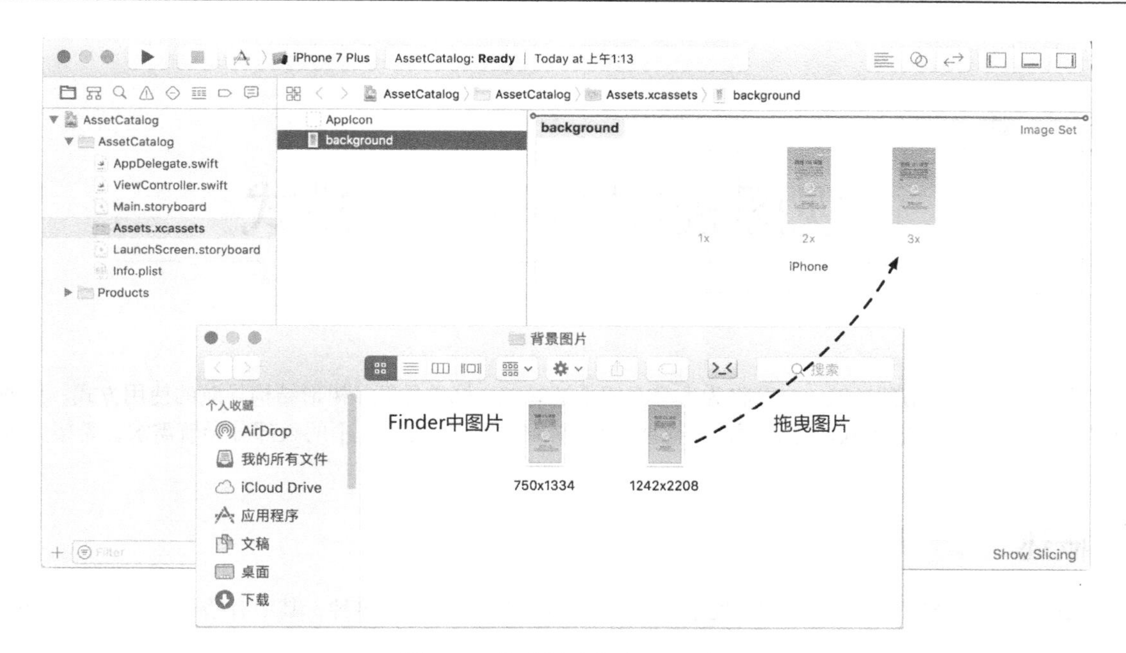

图8-23 修改图片集为iPhone版

添加完图片后，回到Interface Builder设计界面，为视图界面添加一个Image View控件。Image View控件也需要添加约束和屏幕适配，本例中我们只考虑iPhone和iPhone Plus设备竖屏的情况，具体过程可以参考8.2.5节。

屏幕适配完成后，打开Image View属性检查器，如图8-24所示，在Image View→Image中选择background，其中background为图片集名。

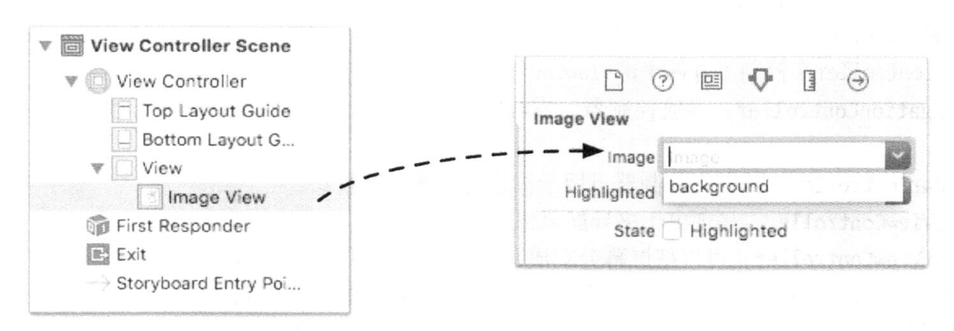

图8-24 选择图片

设置完成时，我们可以在不同的模拟器下运行，看看是否加载不同的图片。资源目录中的图片集不仅可以通过界面设计使用，还可以通过代码使用。下述代码用于从图片集background创建资源图片对象：

```
imageView.image = UIImage(named: "background")
```

```
imageView.image = [UIImage imageNamed:@"background"];
```

8.4 小结

本章首先向大家介绍iOS多分辨率屏幕适配方法，其中涉及的技术主要是Size Class，最后介绍了资源目录与图片资源适配。

第 9 章

视图控制器与导航模式

几乎每个应用都会用到导航，本章将为大家介绍平铺导航、标签导航、树形结构导航的使用方式。另外，本章还为大家讲解了3种导航模式的综合用法。这些知识基本包括了开发工作中的大部分导航需求，希望大家能有所收获。

9.1 概述

在Cocoa Touch的MVC设计模式中，处于重要地位的视图控制器有很多种，其中有些视图控制器与导航息息相关。

导航指引用户使用你的应用，没有有效的导航，用户就会迷失方向。

9.1.1 视图控制器的种类

在UIKit中，视图控制器有很多，其中有些负责显示视图，而有些起到导航的作用，有些还有其他用途，现在我们将与导航相关的视图控制器整理如下。

- **UIViewController**。用于自定义视图控制器的导航。例如，对于两个界面的跳转，我们可以用一个UIViewController来控制另外两个UIViewController。
- **UINavigationController**。导航控制器，它与UITableViewController结合使用，能够构建树形结构导航模式。
- **UITabBarController**。标签栏控制器，用于构建树标签导航模式。
- **UIPageViewController**。呈现电子书导航风格的控制器。
- **UISplitViewController**。可以把屏幕分割成几块的视图控制器，主要为iPad屏幕设计。
- **UIPopoverController**。呈现“气泡”风格视图的控制器，主要为iPad屏幕设计。

视图控制器随着iOS版本的变化而变化，例如UISplitViewController和UIPopoverController是随着iPad的出现而推出的，UIPageViewController则是iOS新推出的，主要用于构建电子书和电子杂志应用。

9.1.2 导航模式

如果火车站没有导航标牌，高速公路上没有路标，情况会怎样？毫无疑问，我们会无所适从，甚至手足无措。你的应用是否具备这些“标牌”和“路标”呢？完美的导航能够清晰地指引用户完成任务。导航是应用软件开发中极为重要的部分，想做好也是存在一定难度的。从内容组织形式上考虑，iPhone有3种导航模式，每一种导航模式都对应于不同的视图控制器。

- **平铺导航模式**。内容没有层次关系，展示的内容都放置在一个主屏幕上，采用分屏或分页控制器进行导航，可以左右或者上下滑动屏幕查看内容。图9-1展示了“iPhone天气预报”应用，它采用分屏进行导航。
- **标签导航模式**。内容被分成几个功能模块，每个功能模块之间没有什么关系。通过标签管理各个功能模块，点击标签可以切换功能模块。图9-2展示了“iPhone时钟”应用，它采用的就是标签导航模式。

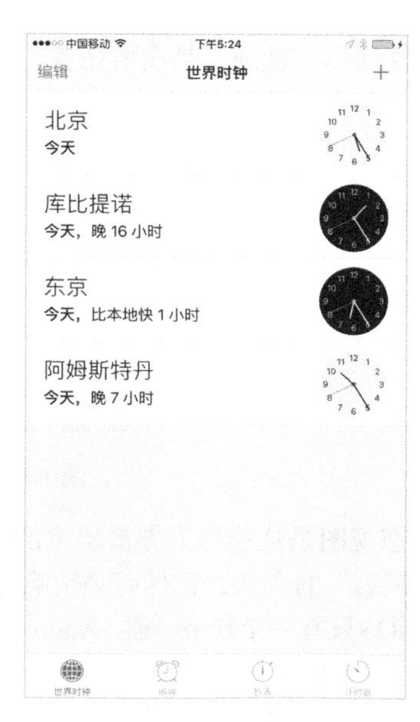

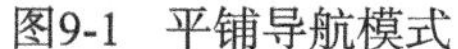
图9-1 平铺导航模式

图9-2 标签导航模式

- **树形结构导航模式。**内容是有层次的，从上到下细分或者具有分类包含等关系，例如黑龙江省包含了哈尔滨，哈尔滨又包含了道里区、道外区等。图9-3展示了“iPhone邮件”应用，它采用的就是树形结构导航模式。

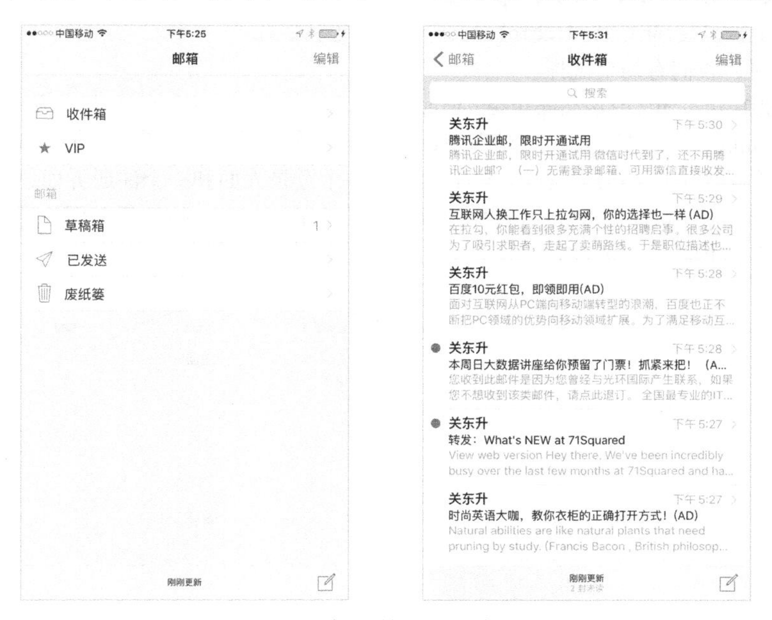

图9-3 树形结构导航模式

这3种导航模式基本可以满足大部分应用的导航需求，在实际应用中，我们有时会将几种导航模式组合在一起使用。

9.2 模态视图

在导航过程中，有时候需要放弃主要任务转而做其他次要任务，然后再返回到主要任务，这个“次要任务”就是在“模态视图”中完成的。图9-4为模态视图示意图，该图中的主要任务是登录后进入主界面，如果用户没有注册，就要先去“注册”。“注册”是次要任务，当用户注册完成后，他会关闭注册视图，回到登录界面继续进行主任务。

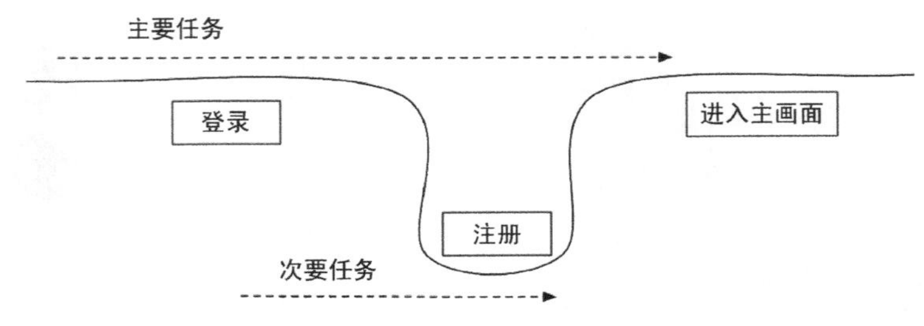

图9-4　模态视图示意图

默认情况下，模态视图是从屏幕下方滑出来的。完成的时候需要关闭这个模态视图，如果不关闭，就不能做别的事情，这就是“模态”的含义，它具有必须响应处理的意思。因此，模态视图中一定会有“关闭”或“完成”按钮，其根本原因是iOS只有一个Home键。Android和Window Phone就不会遇到这个问题，因为在这两个系统中遇到上述情况时，可以通过Back键返回。

负责控制模态视图的控制器称为模态视图控制器。模态视图控制器并非一个专门的类，它可以是上面提到的控制器的子类。负责主要任务视图的控制器称为主视图控制器，它与模态视图控制器之间是“父子”关系。由于`UIViewController`类中提供如下两个方法，所以任何视图控制器中都可以呈现和关闭模态视图。

- **`presentViewController:animated:completion`**。呈现模态视图。
- **`dismissViewControllerAnimated:completion`**。关闭模态视图。

在呈现模态视图时，我们有两个选择。

- 通过代码使用`UIViewController`的`presentViewController:animated:completion`方法实现。
- 通过Interface Builder在故事板的“过渡”（Segue）中实现，这个方式不需要编写代码。

下面我们通过一个案例来介绍模态视图。这个案例有一个登录界面和一个注册界面，在登录界面中点击“注册”按钮，屏幕下方会滑出注册模态视图。如图9-5所示，点击Cancel或Save按钮后关闭注册视图，而且在点击Save按钮时会把“用户ID”回传给登录界面。

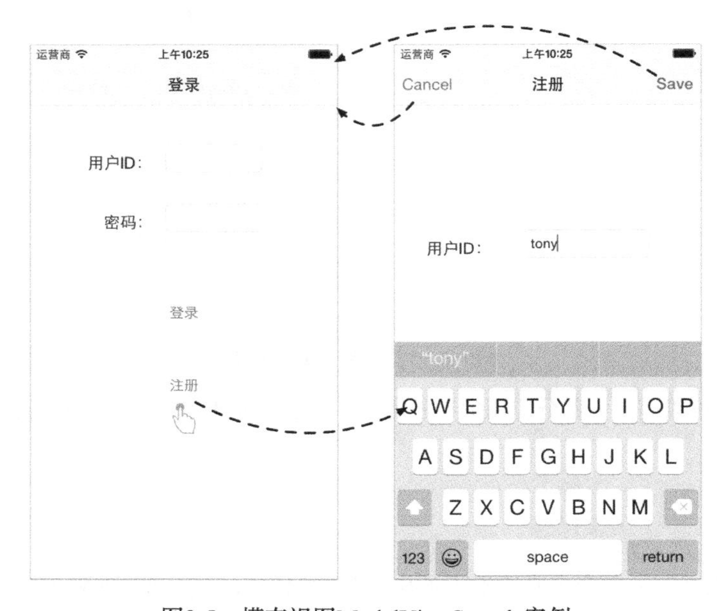

图9-5　模态视图ModalViewSample案例

9.2.1 Interface Builder 实现

使用Interface Builder实现ModalViewSample的具体步骤如下。

1. 创建工程

使用Xcode创建工程ModalViewSample，相关选项如下：模板采用Single View Application，Devices选择iPhone。

2. 屏幕适配

本案例中屏幕适配所有的iPhone竖屏，所以这里设置Size Class的值为`wCompact | hRegular`，如图9-6所示，具体操作大家可以参考8.2节。

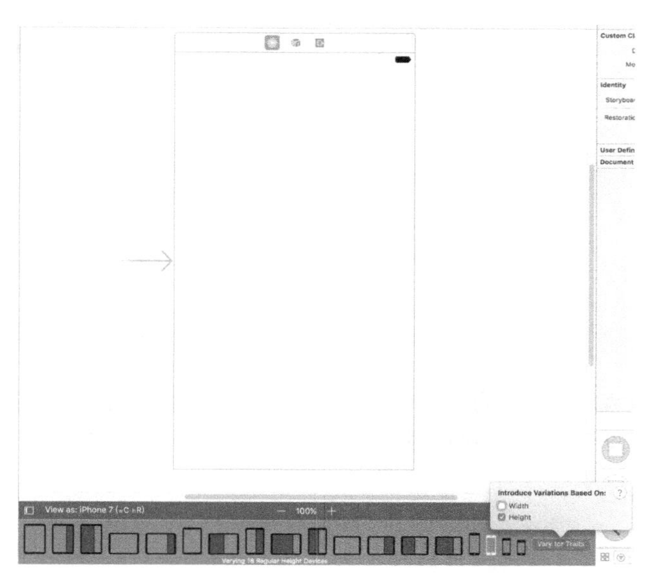

图9-6 当前视图控制器

3. 添加导航栏

本例的视图都是带有导航栏的（如图9-5所示），添加导航栏的方式有两种：一是从对象库中拖曳一个Navigation Bar到视图设计界面顶部，二是将当前视图控制器嵌入到一个导航控制器中。这里我们介绍第二种方式，具体步骤是：在故事板中选择View Controller，然后点击Editor→Embed→Navigation Controller菜单，添加完成后的设计界面如图9-7所示。此外，导航栏可以带有标题。双击导航栏中间部分，然后使导航栏标题处于编辑状态，并输入内容“登录”，如图9-8所示。

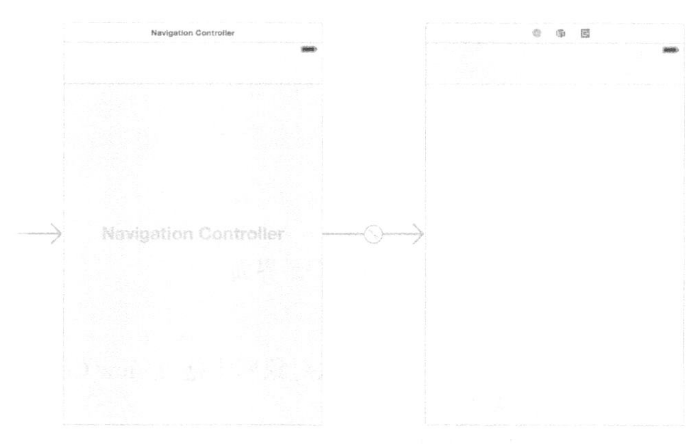

图9-7 将当前视图控制器嵌入到导航控制器

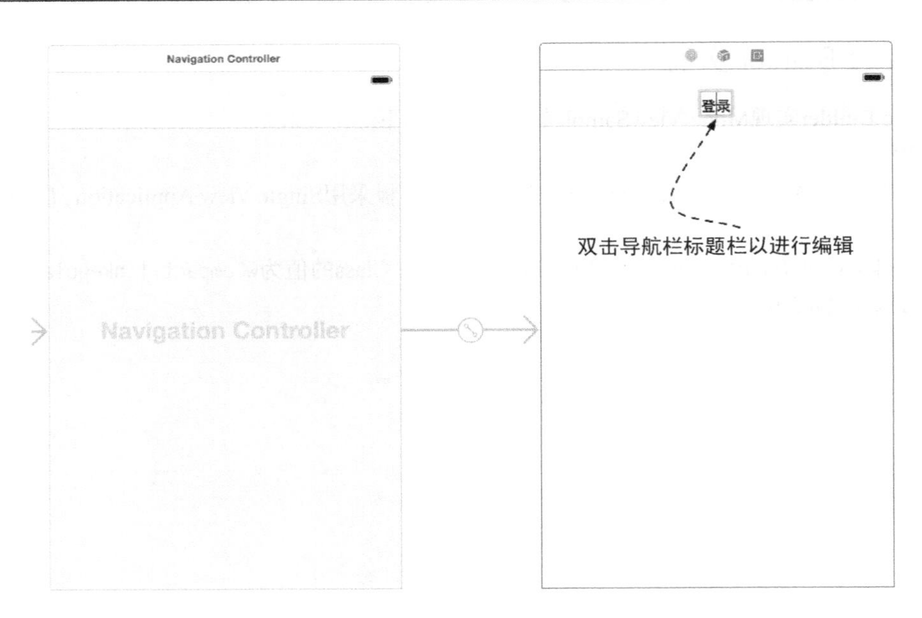

图9-8　编辑导航栏标题

4. 界面布局

参照如图9-5所示的登录视图界面布局，从对象库中拖曳视图到设计界面，并将其摆放到合适的位置，如图9-9所示，然后为每一个视图添加Auto Layout约束，添加约束的步骤可参考7.4节。

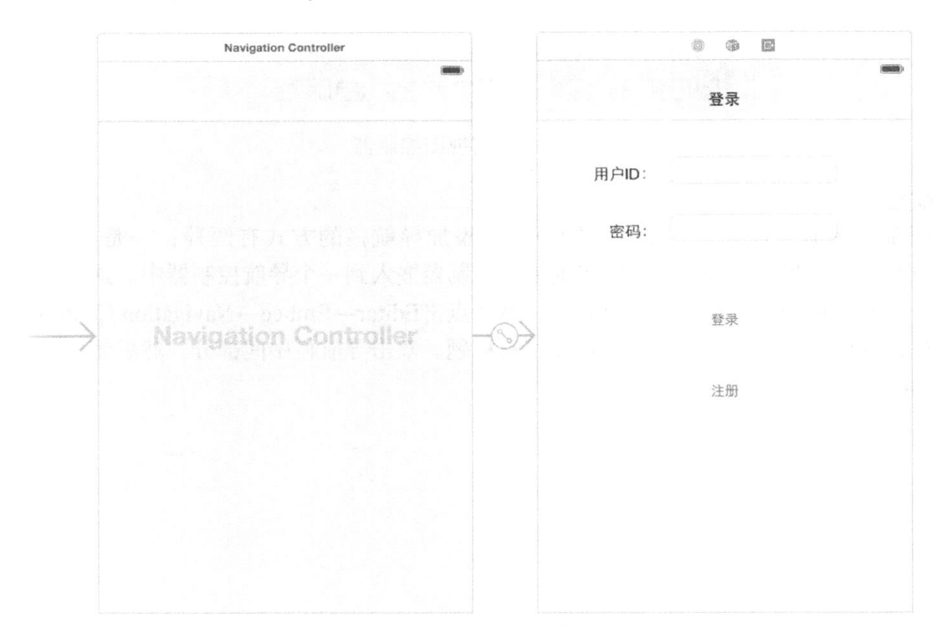

图9-9　设计登录界面

5. 添加注册视图

接着我们来设计第二个界面（注册视图）。我们需要从对象库中拖曳View Controller视图控制器到设计界面，参考上面的步骤将该视图控制器也嵌入到导航控制器中。

修改注册界面导航栏的标题为“注册”，然后从对象库中拖曳两个Bar Button Item到设计界面导航栏两边，如

图9-10所示。由于Cancel和Save按钮都是iOS系统按钮，所以我们可以设置左按钮的identifier属性为Cancel，右按钮的identifier属性为Save。

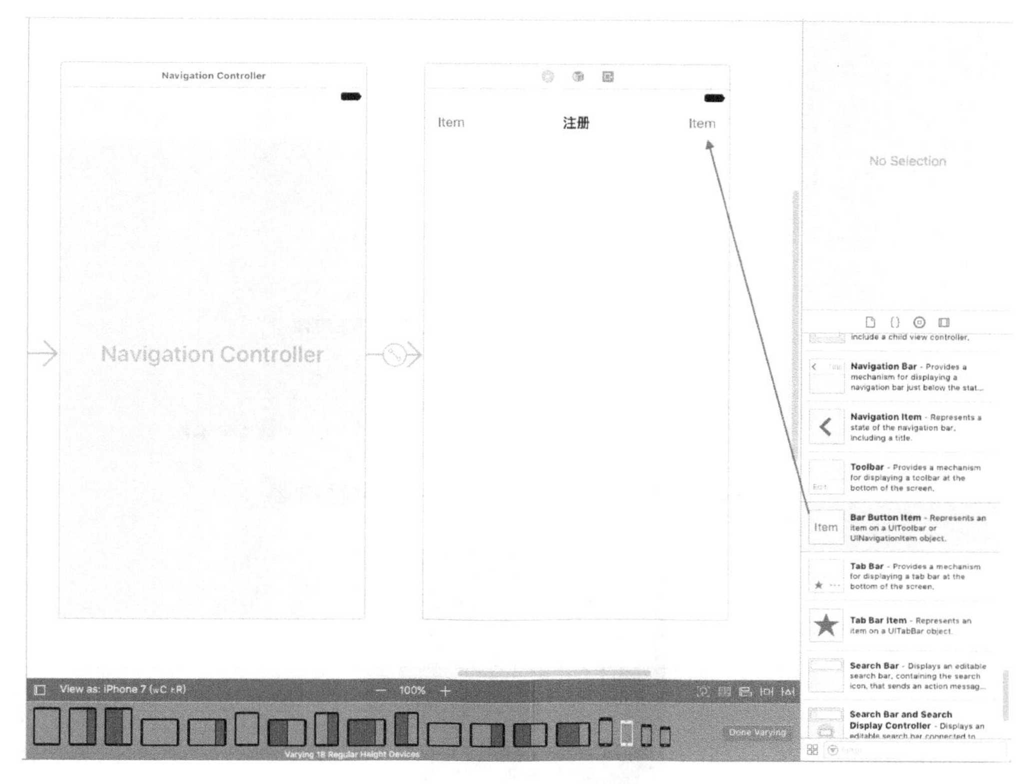

图9-10　添加左右导航栏按钮

6. 添加两个场景的过渡

到此为止，我们只是设计完成了两个独立的视图控制器，如图9-11所示。如果想点击登录界面中的“注册”按钮后，界面跳转到注册界面，则需要在登录场景和注册场景之间创建一个过渡，操作过程类似于连接输出口：按住control键，从“注册”按钮拖曳鼠标到注册导航控制器，如图9-12所示。然后松开鼠标，此时会弹出如图9-13所示的菜单，从中选择Present Modally菜单（这个菜单是模态类型的过渡）。

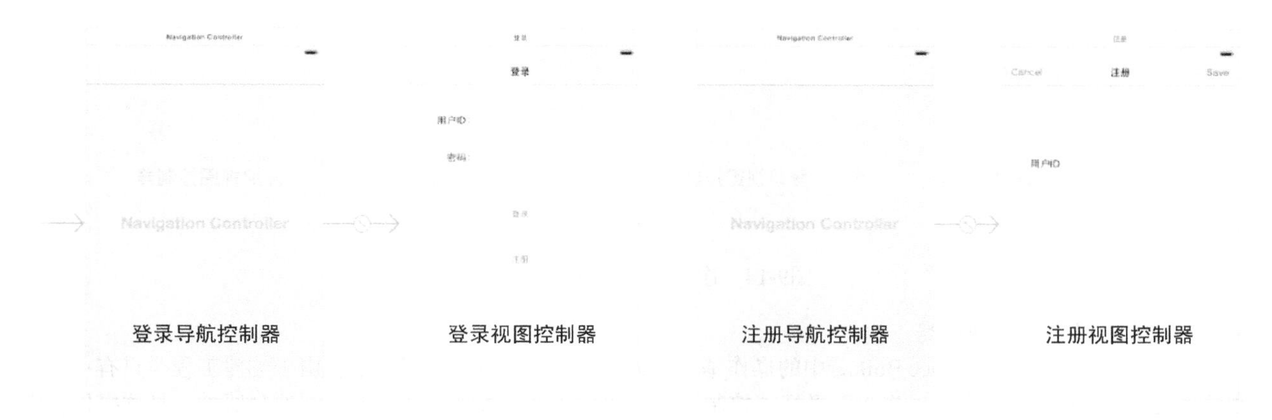

图9-11　两个独立的视图控制器

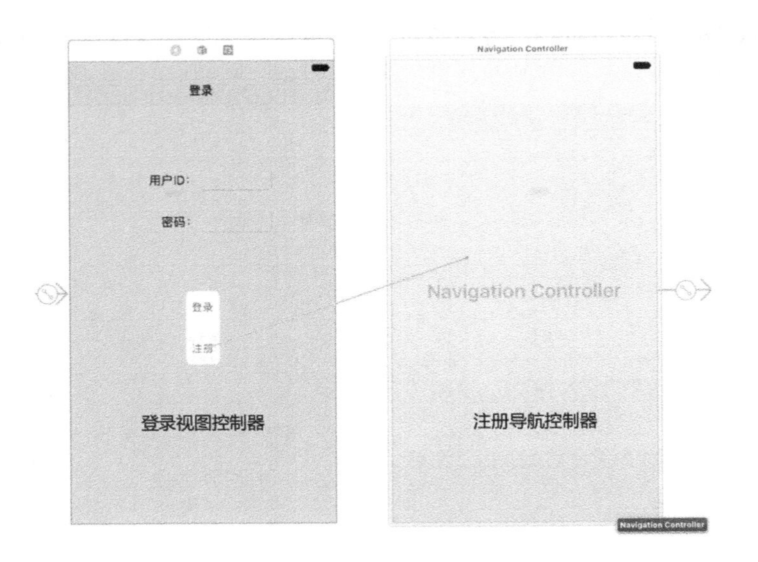

图9-12　从登录按钮拖曳鼠标到注册导航控制器

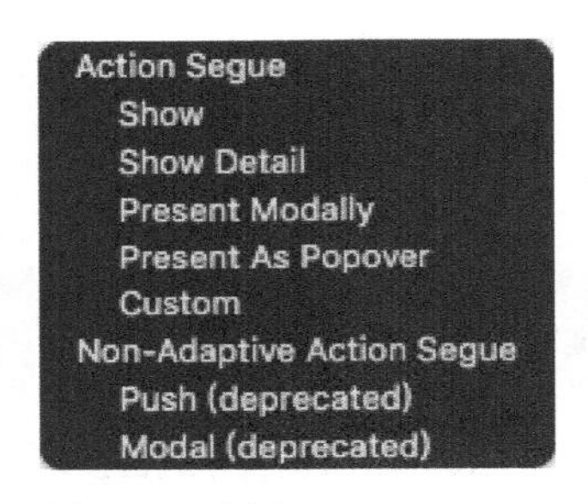

图9-13　选择Present Modally

两个场景之间成功创建过渡操作后，我们再看设计界面，其中4个控制器都连接在一起了，如图9-14所示。

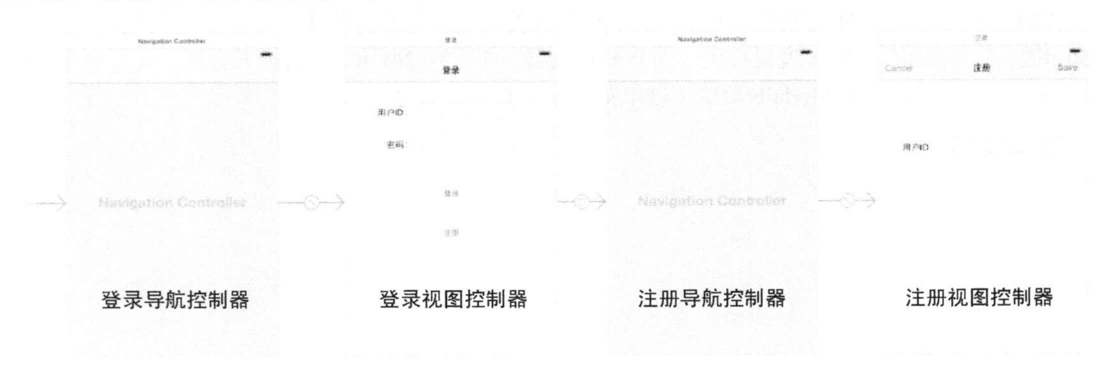

图9-14　连接在一起的视图控制器

7. 添加注册视图控制器类

到此为止，我们在Interface Builder中的操作基本完成，下面就可以编写代码了。由于创建工程时只有一个控制器类`ViewController`，它可以作为登录视图控制器类，我们还需要添加一个注册视图控制器类，具体操作步骤如下。

(1) 选择File→New→File...菜单项，在打开的Choose a template for your new file:对话框中选择Cocoa Touch

Class文件模板（如图9-15所示）。

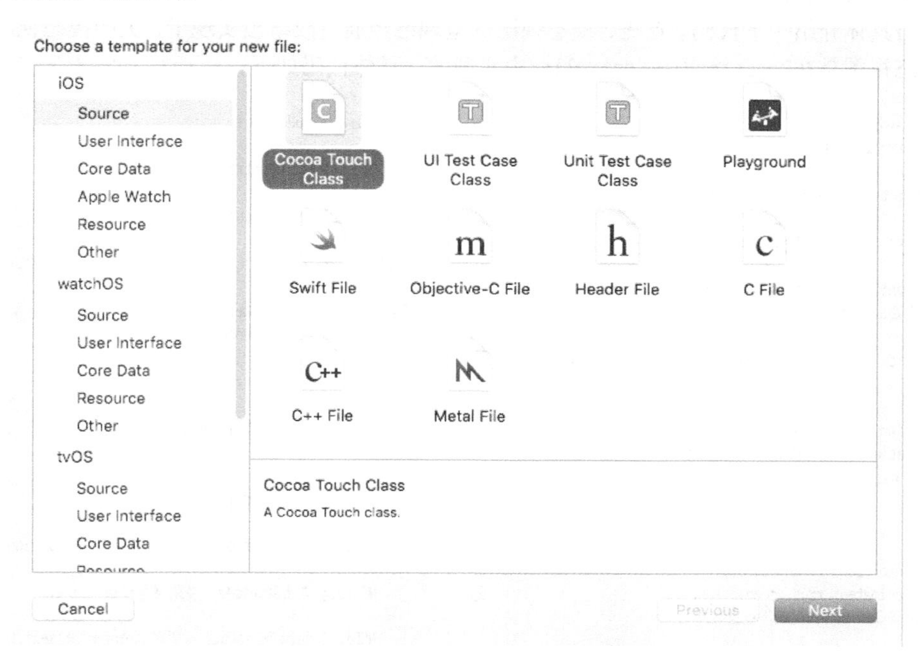

图9-15　选择文件模板

(2) 点击Next按钮，得到的界面如图9-16所示。在Class中输入RegisterViewController类名，从Subclass of下拉列表中选择UIViewController。注意，不要选中Also create XIB file复选框，这是因为本例中我们的界面是使用故事板而不是XIB文件设计的。

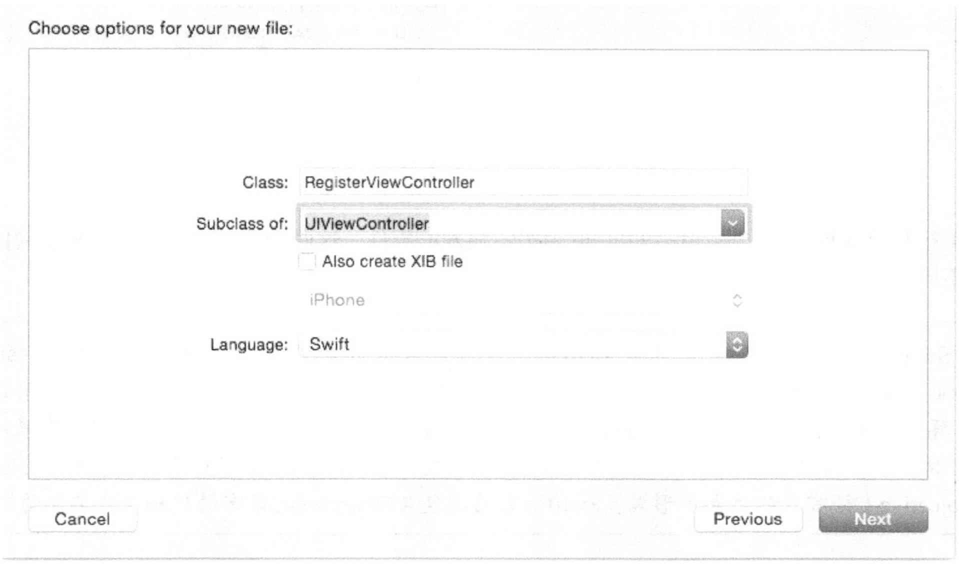

Choose options for your new file:

Class: RegisterViewController

Subclass of: UIViewController

Also create XIB file

iPhone

Language: Swift

Cancel　　Previous　　Next

图9-16　输入类名

(3) 点击Next按钮，此时`RegisterViewController`类就创建好了。然后回到Interface Builder中，选择注册视图控制器，打开其标识检查器，重新选择Class为RegisterViewController，这样故事板中的这个视图控制器就与代码中的RegisterViewController对应起来了。然后，我们再为注册界面导航栏中的左右按钮定义动作事件。

下面我们再具体介绍一下代码。首先，模态视图的呈现已经通过故事板实现了，关闭模态视图需要通过代码实现。关闭模态视图是在RegisterViewController中实现的，具体代码如下：

```
//RegisterViewController.swift文件
class RegisterViewController: UIViewController {

    @IBOutlet weak var txtUsername: UITextField!

    @IBAction func save(_ sender: AnyObject) {                                    ①

        self.dismiss(animated: true) { () -> Void in                              ②
            print("点击Save按钮，关闭模态视图")

            let dataDict = ["username" : self.txtUsername.text!]                  ③


            NotificationCenter.default.post(name:
            ➥Notification.Name(rawValue: "RegisterCompletionNotification"),
            ➥object: nil, userInfo: dataDict)                                    ④
        }
    }

    @IBAction func cancel(sender: AnyObject) {                                    ⑤
        self.dismiss(animated: true, completion: {                                ⑥
            print("点击Cancel按钮，关闭模态视图")
        })
    }

    ......
}
```

```
//RegisterViewController.m文件

#import "RegisterViewController.h"

@interface RegisterViewController ()

@property (weak, nonatomic) IBOutlet UITextField *txtUsername;

- (IBAction)save:(id)sender;
- (IBAction)cancel:(id)sender;

@end

@implementation RegisterViewController
......

- (IBAction)save:(id)sender {                                                     ①

    [self dismissViewControllerAnimated:TRUE completion:^{                        ②

        NSLog(@"点击Save按钮，关闭模态视图");

        NSDictionary *dataDict = @{@"username" : self.txtUsername.text};          ③
        [[NSNotificationCenter defaultCenter]
        ➥postNotificationName:@"RegisterCompletionNotification"
        ➥object:nil userInfo:dataDict];                                          ④

    }];

}

- (IBAction)cancel:(id)sender {                                                   ⑤
    [self dismissViewControllerAnimated:TRUE completion:^{                        ⑥
        NSLog(@"点击Cancel按钮，关闭模态视图");
    }];

}

@end
```

其中第①行和第⑤行是点击Save按钮和Cancel按钮触发的事件，这两个按钮都可以关闭模态视图，但是使用的方法稍有不同。

提示

在Swift版中，第②行采用的是Swift语言尾随闭包表示形式，尾随闭包是将dismiss(animated:completion:)方法的最后一个参数（闭包形式的）放到dismiss方法右小括号的后面。若读者想深入了解尾随闭包，可以参考我编写的《从零开始学Swift（第2版）》一书。第⑥行则没有采用尾随闭包表示形式，其中的completion参数值是一个闭包。而Objective-C版的第②行和第⑥行采用的是Objective-C语言代码块表示形式。Swift语言的闭包与Objective-C语言的代码块是类似的概念。

另外，第③行和第④行采用通知机制将参数回传给登录视图控制器。为了接收参数，我们需要在登录视图控制器ViewController中添加如下代码：

```
//ViewController.swift文件
override func viewDidLoad() {
    super.viewDidLoad()
```

```
//ViewController.m文件
- (void)viewDidLoad {
    [super viewDidLoad];
```

```
    NotificationCenter.default.addObserver(self,
    ➥selector: #selector(registerCompletion(_:)),
    ➥name: Notification.Name(rawValue: "RegisterCompletionNotification"),
    ➥object: nil)

}

override func didReceiveMemoryWarning() {
    super.didReceiveMemoryWarning()
    NotificationCenter.default.removeObserver(self)
}

func registerCompletion(notification: Notification) {

    let theData:NSDictionary = notification.userInfo!
    let username = theData["username"] as! String
    print("username = ", username)
}
```

```
    [[NSNotificationCenter defaultCenter] addObserver:self
    ➥selector:@selector(registerCompletion:)
    ➥name:@"RegisterCompletionNotification" object:nil];
}

- (void)didReceiveMemoryWarning {
    [super didReceiveMemoryWarning];
    [[NSNotificationCenter defaultCenter] removeObserver:self];
}


-(void)registerCompletion:(NSNotification*)notification {

    NSDictionary *theData = [notification userInfo];
    NSString *username = theData[@"username"];
    NSLog(@"username = %@",username);
}
```

首先需要在viewDidLoad中注册通知RegisterCompletionNotification，registerCompletion:是回调方法。关于通知机制，我们将在第11章中详细介绍。

9.2.2 代码实现

上一节中，我们在故事板中使用Interface Builder实现了ModalViewSample案例，这一节介绍如何通过代码来实现它。

从图9-5可见，ModalViewSample案例的根视图控制器是一个导航控制器，这需要修改AppDelegate代码，具体内容参考6.2.2节。我们重点解释视图控制器ViewController和RegisterViewController。

ViewController的代码如下：

```
//ViewController.swift文件
import UIKit

class ViewController: UIViewController {

    override func viewDidLoad() {
        super.viewDidLoad()

        let screen = UIScreen.main.bounds

        self.navigationItem.title = "登录"
        ......
        <添加控件到视图>

        NotificationCenter.default.addObserver(self,
        ➥selector: #selector(registerCompletion(_:)),
        ➥name: Notification.Name(rawValue: "RegisterCompletion-
        ➥Notification"), object: nil)
    }

    override func didReceiveMemoryWarning() {
        super.didReceiveMemoryWarning()
        NotificationCenter.default.removeObserver(self)
    }

    func onClick(sender: AnyObject) {

        let registerViewController = RegisterViewController()
        let navigationController = UINavigationController(rootViewController:
        ➥registerViewController)                                          ①
```

```
//ViewController.m文件
@implementation ViewController

- (void)viewDidLoad {
    [super viewDidLoad];

    CGRect screen = [[UIScreen mainScreen] bounds];

    self.navigationItem.title = @"登录";
    ......
    <添加控件到视图>

    [[NSNotificationCenter defaultCenter] addObserver:self
    ➥selector:@selector(registerCompletion:)
    ➥name:@"RegisterCompletionNotification" object:nil];
}

- (void)didReceiveMemoryWarning {
    [super didReceiveMemoryWarning];
    [[NSNotificationCenter defaultCenter] removeObserver:self];
}

- (void)onClick:(id)sender {

    RegisterViewController* registerViewController = [[RegisterViewController
    ➥alloc] init];
    UINavigationController* navigationController = [[UINavigationController
    ➥alloc] initWithRootViewController:registerViewController];          ①

    [self presentViewController:navigationController animated:TRUE
    ➥completion:nil];                                                    ②
```

```swift
        self.present(navigationController, animated: true, completion: nil) ②
    }

    func registerCompletion(notification: Notification) {

        let theData = notification.userInfo!
        let username = theData["username"] as! String
        print("username = ", username)
    }

}
```

```objectivec
}

-(void)registerCompletion:(NSNotification*)notification {

    NSDictionary *theData = [notification userInfo];
    NSString *username = theData[@"username"];

    NSLog(@"username = %@",username);
}

@end
```

上述代码中，第①行用于实例化导航控制器UINavigationController，构造函数initWithRootViewController:的参数是registerViewController，这会将RegisterViewController嵌入到导航控制器中。

第②行通过代码呈现模态视图，其中方法是presentViewController:animated:completion，第一个参数是呈现模态视图的控制器，注意这里是导航控制器navigationController。

9.3　平铺导航

平铺导航模式非常重要，一般用于简单的扁平化信息浏览。扁平化信息是指这些信息之间没有从属的层次关系，如北京、上海和哈尔滨之间就没有从属关系，而哈尔滨市与黑龙江省之间就是从属的层次关系。

9.3.1　应用场景

图9-17是iPhone的“天气”应用程序，每一个屏幕展示一个城市最近的天气信息，它是基于分屏导航实现的平铺导航模式。基于这种导航模式，可以构建iOS中的实用型应用程序。

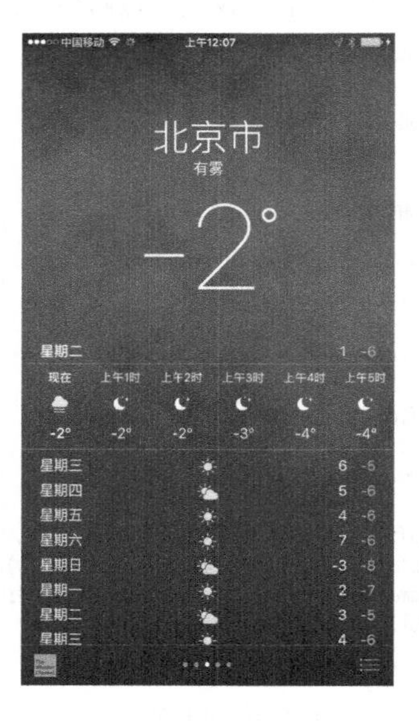

图9-17　iPhone“天气”应用

提示　实用型应用程序完成的简单任务对用户输入要求很低。用户打开实用型应用程序是为了快速查看信息摘要，或是在少数对象上执行简单任务。“天气”程序就是一个实用型应用程序的典型例子，它在一个易读的摘要中显示了重点明确的信息。——引自于苹果HIG（iOS Human Interface Guidelines，iOS人机界面设计指导手册）

图9-18所示是iPad的iBooks应用横屏显示界面，它是基于电子书导航实现的平铺导航模式，用户可以像翻书一样在页面之间导航，而且在翻动书页时还可以看到下一页或背面的内容，完全模拟真书的效果。

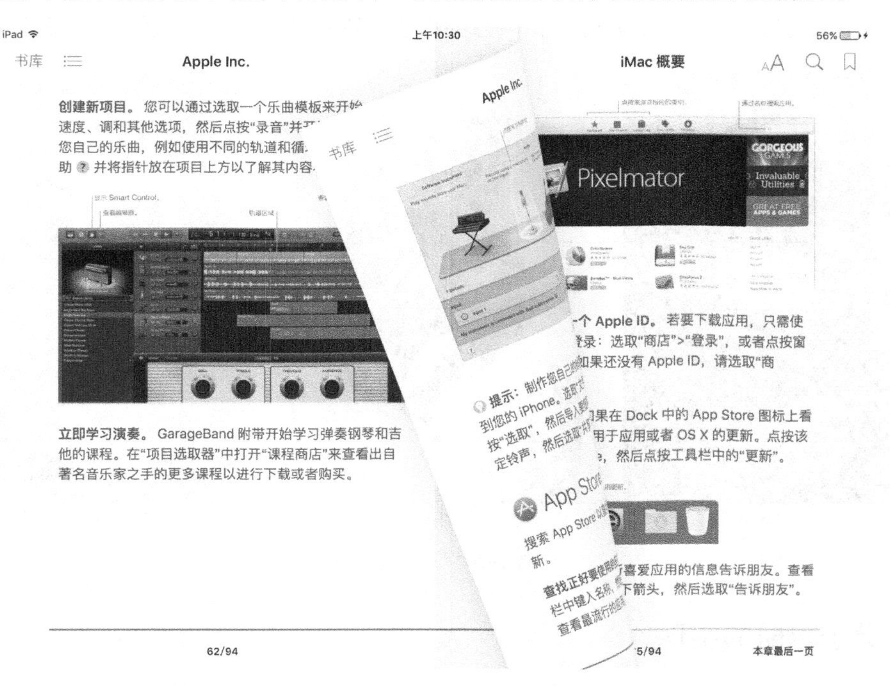

图9-18 iPad的iBooks应用（横屏双页显示）

图9-19所示是iPhone的iBooks应用竖屏时的单页显示情况。

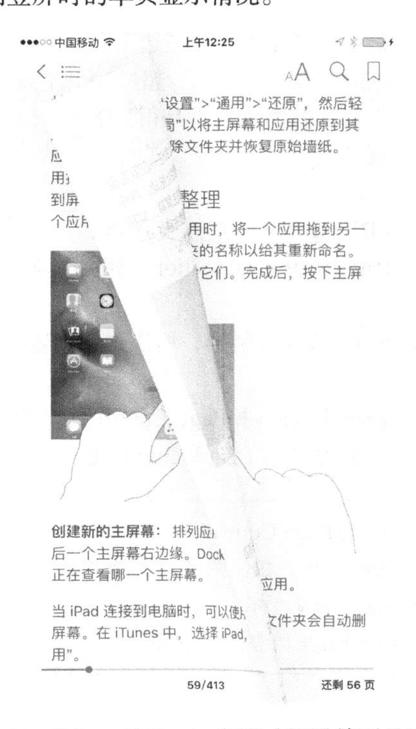

图9-19 iPhone的iBooks应用（竖屏单页显示）

为了进一步了解平铺导航，我们先从一个需求开始介绍。如果我想开发一个基于iPhone的“艺术品陈列室”应用，目前只有3件艺术品（如图9-20所示，图a是达芬奇的《蒙娜丽莎》名画，图b是罗丹的《思想者》雕塑，图c是保罗·克利的《肖像》名画）收录到应用中，这3件艺术品之间没有层次关系，而是扁平关系。

(a)

(b)

(c)

图9-20　3件艺术品

9.3.2　基于分屏导航的实现

基于分屏导航是平铺导航模式的主要实现方式，涉及的主要控件有分屏控件（UIPageControl）和屏幕滚动视图（UIScrollView）。其中分屏控件是iOS标准控件，分屏控件一般会在屏幕的下方，此处高亮的小点是当前屏幕的位置。

基于分屏导航的手势有两种，一个是点击高亮小点的左边（上边）或右边（下边）实现翻屏，另一个是用手在屏幕上滑动实现翻屏。屏幕的总数应该限制在20个以内，超过20个小点的分屏控件就会溢出。事实上，当一个应用超过10屏时，使用基于分屏导航的平铺导航模式就不是很方便了。

下面我们介绍基于分屏导航模式实现“艺术品陈列室”应用，首先介绍一下Interface Builder实现。

1. Interface Builder实现

使用Interface Builder实现PageControlNavigation的具体步骤如下。

(1) 创建工程。使用Xcode创建工程PageControlNavigation，相关选项如下：模板采用Single View Application，Devices选择iPhone。

(2) 屏幕适配。本案例中屏幕适配所有的iPhone竖屏，所以这里设置Size Class的值为`wCompact | hRegular`，具体操作大家可以参考8.2节。

(3) 界面布局。接着从对象库中拖曳Scroll View和Page Control到设计界面，并按照图9-21所示将其摆放到合适的位置，通过属性检查器将视图背景设置为黑色（Dark Text Color）。

提示　在视图层次关系中，Scroll View与Page Control是并列的，没有包含关系，它们都是View的同级别子视图。Page Control在Scroll View之上，在使用鼠标拖曳Page Control时一定要注意，因为这种操作很容易将Page Control拖曳到Scroll View内部，成为Scroll View的子视图。为了避免出现这种问题，可以使用键盘方向键移动Page Control并调整它的位置。

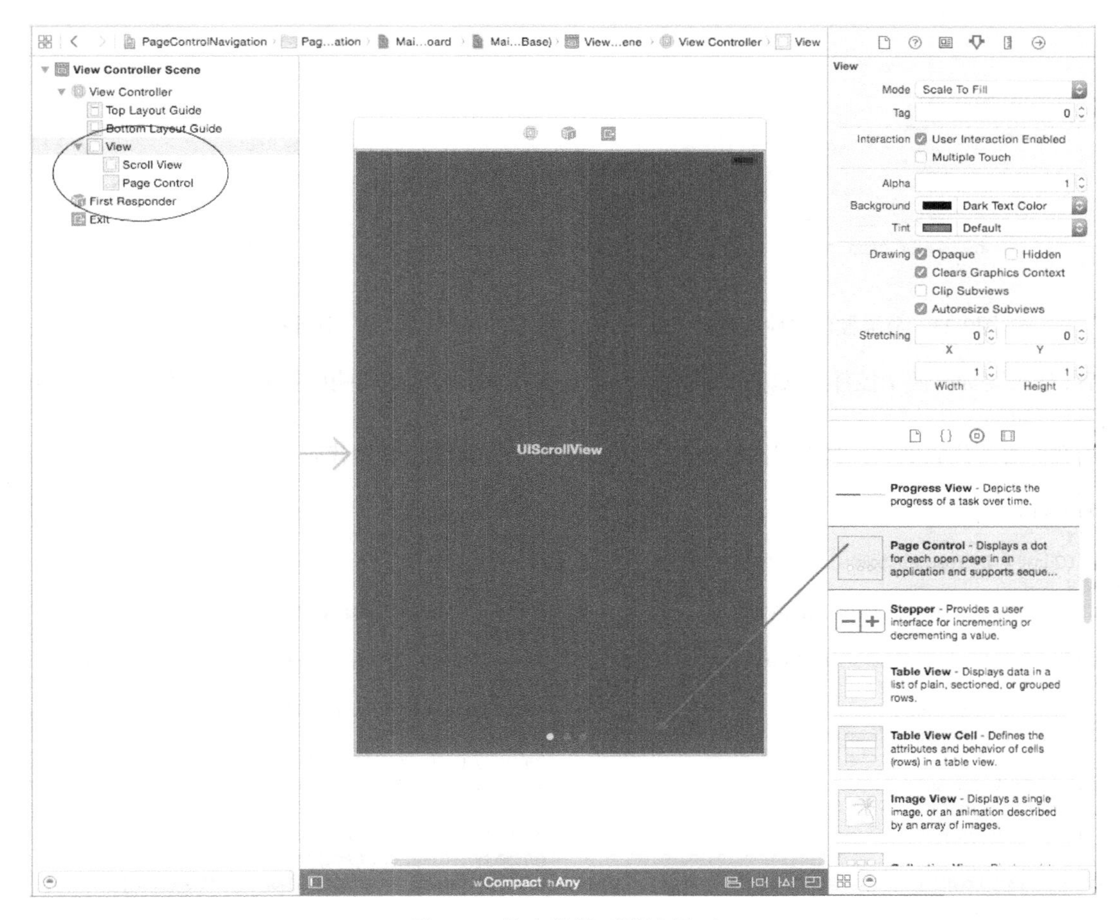

图9-21　拖曳控件到设计界面

(4) 设置Scroll View属性。选中Scroll View并打开其属性检查器，按照图9-22所示设置其中的属性。此时该Scroll View不显示水平滚动条（不选中Shows Horizontal Indicator）和垂直滚动条（不选中Shows Vertical Indicator），但可以滚动和分屏（选中Scrolling Enabled和Paging Enabled复选框）。分屏（Paging Enabled）属性可以使Scroll View每次滑动时翻一屏。

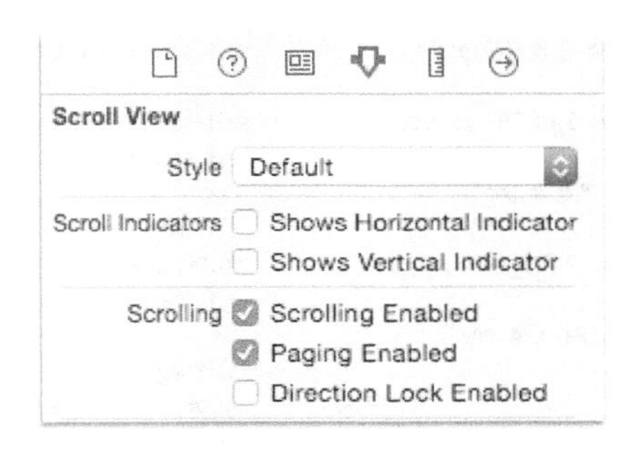

图9-22　设置Scroll View控件的属性

(5) 设置Page Control属性。然后选中Page Control控件，打开其属性检查器，设置Pages中的# of Pages（总屏数）属性为3，Current（当前屏）属性为0，如图9-23所示。再打开其尺寸检查器，如图9-24所示，修改Width（宽度）属性为300，而它的Height（高度）属性是不能修改的。将Width属性设置得宽一些是为了便于手指点击。

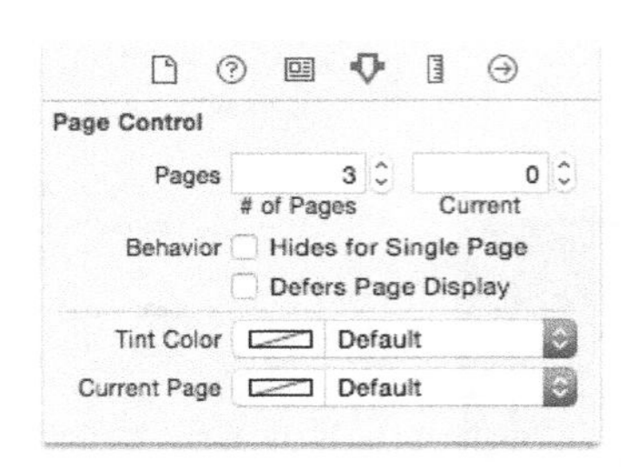

图9-23　分屏控件属性检查器

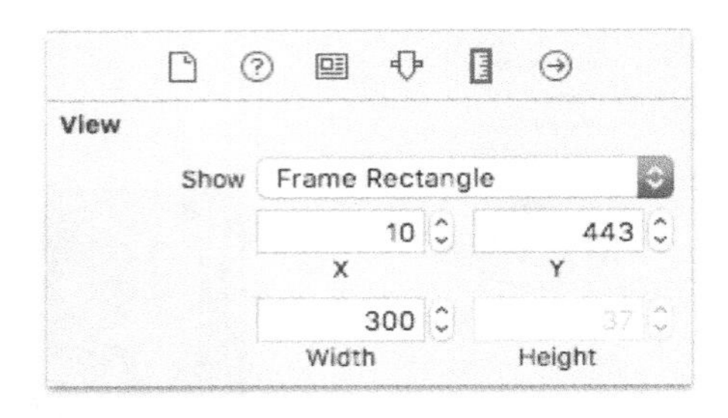

图9-24　分屏控件尺寸检查器

最后，还需要为这两个控件定义输出口并连线，而且要为分屏控件定义响应屏幕变化事件的方法changePage:并连线。

下面我们看看代码部分，其中视图控制器ViewController初始化的相关代码如下：

```
//ViewController.swift文件
//定义屏幕宽度
let S_WIDTH: CGFloat = UIScreen.main.bounds.size.width                         ①
//定义屏幕高度
let S_HEIGHT: CGFloat = UIScreen.main.bounds.size.height                       ②

class ViewController: UIViewController, UIScrollViewDelegate {                 ③

    @IBOutlet weak var scrollView: UIScrollView!
    @IBOutlet weak var pageControl: UIPageControl!

    var imageView1: UIImageView!
    var imageView2: UIImageView!
    var imageView3: UIImageView!

    override func viewDidLoad() {
        super.viewDidLoad()

        self.scrollView.delegate = self                                        ④

        self.scrollView.contentSize  = CGSize(width: S_WIDTH * 3,
        ➥height: S_HEIGHT)                                                    ⑤
        self.scrollView.frame = self.view.frame                                ⑥

        self.imageView1 = UIImageView(frame: CGRect(x: 0.0, y: 0.0,
        ➥width: S_WIDTH, height: S_HEIGHT))                                   ⑦
        self.imageView1.image = UIImage(named: "达芬奇-蒙娜丽莎.png")

        self.imageView2 = UIImageView(frame: CGRect(x: S_WIDTH, y: 0.0,
        ➥width: S_WIDTH, height: S_HEIGHT))
        self.imageView2.image = UIImage(named: "罗丹-思想者.png")

        self.imageView3 = UIImageView(frame: CGRect(x: 2 * S_WIDTH, y: 0.0,
        ➥width: S_WIDTH, height: S_HEIGHT))
        self.imageView3.image = UIImage(named: "保罗克利-肖像.png")

        self.scrollView.addSubview(self.imageView1)                            ⑧
        self.scrollView.addSubview(self.imageView2)
        self.scrollView.addSubview(self.imageView3)

    }
    ......
}
```

```
//ViewController.m文件
#import "ViewController.h"

//定义屏幕宽度宏
#define S_WIDTH [[UIScreen mainScreen] bounds].size.width                      ①
//定义屏幕高度宏
#define S_HEIGHT [[UIScreen mainScreen] bounds].size.height                    ②

@interface ViewController () <UIScrollViewDelegate>                            ③

@property (weak, nonatomic) IBOutlet UIScrollView *scrollView;
@property (weak, nonatomic) IBOutlet UIPageControl *pageControl;

@property (strong, nonatomic) UIImageView *imageView1;
@property (strong, nonatomic) UIImageView *imageView2;
@property (strong, nonatomic) UIImageView *imageView3;

- (IBAction)changePage:(id)sender;

@end

@implementation ViewController

- (void)viewDidLoad {
    [super viewDidLoad];

    self.scrollView.delegate = self;                                           ④

    self.scrollView.contentSize = CGSizeMake(S_WIDTH * 3, S_HEIGHT);           ⑤
    self.scrollView.frame = self.view.frame;                                   ⑥

    self.imageView1 = [[UIImageView alloc] initWithFrame:CGRectMake(0.0f,
    ➥0.0f, S_WIDTH, S_HEIGHT)];                                               ⑦
    self.imageView1.image = [UIImage imageNamed:@"达芬奇-蒙娜丽莎.png"];

    self.imageView2 = [[UIImageView alloc] initWithFrame:CGRectMake(S_WIDTH,
    ➥0.0f, S_WIDTH, S_HEIGHT)];
    self.imageView2.image = [UIImage imageNamed:@"罗丹-思想者.png"];

    self.imageView3 = [[UIImageView alloc] initWithFrame:CGRectMake(2 *
    ➥S_WIDTH, 0.0f, S_WIDTH, S_HEIGHT)];
    self.imageView3.image = [UIImage imageNamed:@"保罗克利-肖像.png"];

    [self.scrollView addSubview:self.imageView1];                              ⑧
    [self.scrollView addSubview:self.imageView2];
```

```
        [self.scrollView addSubview:self.imageView3];

    }
    ......

@end
```

上述代码中，第①行和第②行用于定义屏幕宽度和屏幕高度，Swift版中定义的是常量，Objective-C版定义的是宏。

第③行是在ViewController中声明实现UIScrollViewDelegate委托协议，这是为了响应屏幕滚动事件。

第④行用于设置ScrollView的委托对象为self（当前视图控制器），实现UIScrollViewDelegate委托协议的方法是scrollViewDidScroll:，当屏幕滚动时，会回调该方法。

第⑤行用于设置屏幕滚动视图的contentSize属性，该属性表示屏幕滚动视图中内容视图（Content View）的大小。第⑥行用于设置屏幕滚动视图的frame属性，该属性设置为当前屏幕大小（self.view.frame）。如图9-25所示，内容视图是图中灰色部分（320 × 544），而屏幕滚动视图的大小（frame指定的范围）只有320 × 460。正是因为内容视图超出了屏幕滚动视图的大小，才有滚动屏幕的必要。

提示 屏幕滚动视图的contentInset属性用于在内容视图周围添加边框，这往往是为了留出空间以放置工具栏、标签栏或导航栏等，如图9-26所示。contentInset属性有4个分量，分别是top、bottom、left和right，分别代表顶边距离、底边距离、左边距离和右边距离。

提示 屏幕滚动视图contentOffset属性用于设置内容视图坐标原点与屏幕滚动视图坐标原点的偏移量，返回CGPoint结构体类型。这个结构体类型包含x和y两个成员。如图9-27所示，内容视图沿Y轴负偏移（或者说屏幕滚动视图沿Y轴正偏移），X轴方向没有偏移。

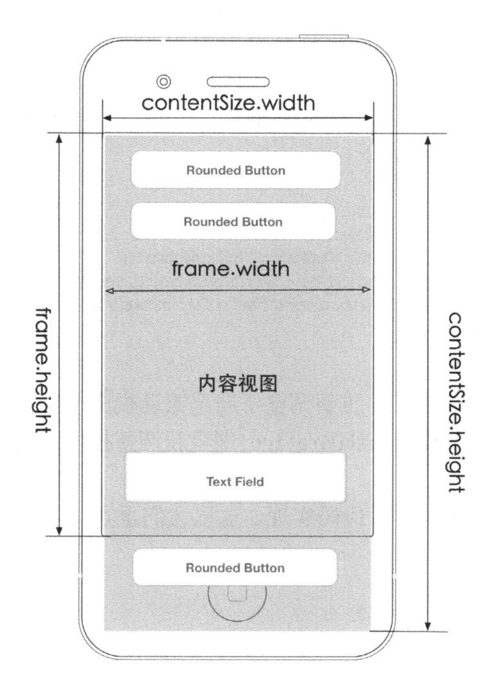

图9-25 contentSize属性

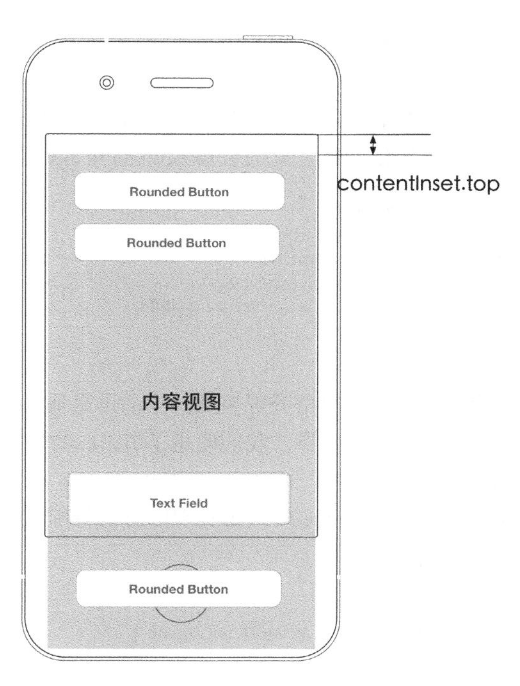

图9-26 contentInset属性

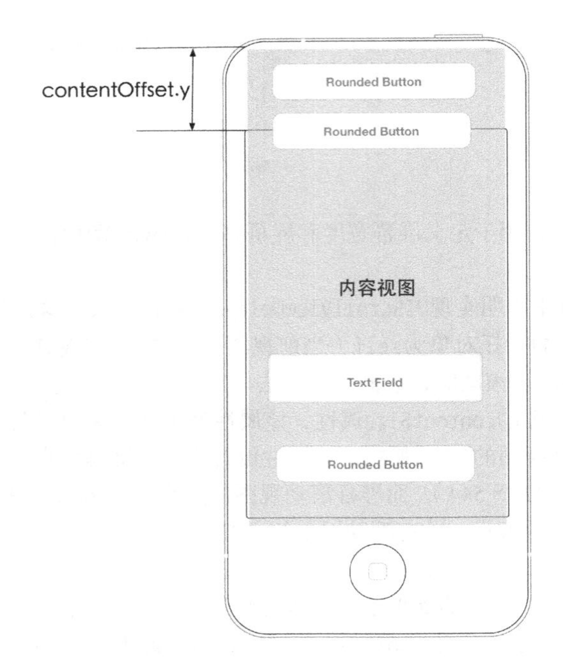

图9-27 内容视图沿Y轴负偏移

在ViewController中，图片视图UIImageView对象是通过代码创建的，见第⑦行，并且这些图片视图对象被添加到屏幕滚动视图中，见第⑧行。

屏幕滚动视图的事件处理方法scrollViewDidScroll:的代码如下：

```
//MARK: -- 实现UIScrollViewDelegate委托协议
func scrollViewDidScroll(scrollView: UIScrollView) {
    let offset = scrollView.contentOffset
    self.pageControl.currentPage = Int(offset.x / S_WIDTH)
}
```

```
#pragma mark -- 实现UIScrollViewDelegate委托协议
- (void) scrollViewDidScroll: (UIScrollView *) scrollView {
    CGPoint offset = scrollView.contentOffset;
    self.pageControl.currentPage = offset.x / S_WIDTH;
}
```

当左右滑动屏幕，屏幕滚动视图滚动完成时，我们需要计算和设定分屏控件的当前屏currentPage。当点击分屏控件时，屏幕发生变化，此时会触发changePage:方法，其代码如下：

```
//MARK: -- 实现UIPageControl事件处理
@IBAction func changePage(sender: AnyObject) {
    UIView.animateWithDuration(0.3, animations : {
        let whichPage = self.pageControl.currentPage
        self.pageControl.currentPage = Int(offset.x / S_WIDTH)
    })
}
```

```
#pragma mark -- 实现UIPageControl事件处理
- (IBAction)changePage:(id)sender {
    [UIView animateWithDuration:0.3f animations:^{
        NSInteger whichPage = self.pageControl.currentPage;
        self.scrollView.contentOffset = CGPointMake(S_WIDTH * whichPage, 0.0f);
    }];
}
```

在上述代码中，我们根据分屏控件的当前屏幕属性（currentPage）重新调整了屏幕滚动视图的偏移量。而且为了使屏幕变化产生动画效果，我们使用了UIView的静态方法animateWithDuration:来重新调整控件的偏移量。

2. 代码实现

前面介绍了使用Interface Builder实现的分屏导航案例，现在来看看代码实现。这里我们重点解释视图控制器ViewController的代码：

```
//ViewController.swift文件
class ViewController: UIViewController, UIScrollViewDelegate {

    ......

    override func viewDidLoad() {
```

```
//ViewController.m文件
......

@implementation ViewController

- (void)viewDidLoad {
```

```swift
        super.viewDidLoad()

        self.scrollView = UIScrollView()
        self.view.addSubview(self.scrollView)

        self.scrollView.delegate = self

        self.scrollView.contentSize  = CGSizeMake(self.view.frame.size.
        ➥width*3, self.scrollView.frame.size.height)
        self.scrollView.frame = self.view.frame
        self.scrollView.pagingEnabled = true                              ①
        self.scrollView.showsHorizontalScrollIndicator = false            ②
        self.scrollView.showsVerticalScrollIndicator = false              ③

        <添加图片视图>

        let pageControlWidth: CGFloat = 300.0
        let pageControlHeight: CGFloat = 37.0
        self.pageControl = UIPageControl(frame: CGRect(x: (S_WIDTH -
        ➥pageControlWidth) / 2,
        ➥y: S_HEIGHT - pageControlHeight,
        ➥width: pageControlWidth, height: pageControlHeight))             ④
        self.pageControl.numberOfPages = 3                                ⑤
        self.pageControl.addTarget(self,
        ➥action: #selector(changePage(_:)), for: .valueChanged)           ⑥
        self.view.addSubview(self.pageControl)
    }

    ......

}
```

```objectivec
    [super viewDidLoad];

    self.scrollView = [[UIScrollView alloc] init];
    [self.view addSubview:self.scrollView];

    self.scrollView.delegate = self;

    self.scrollView.contentSize  = CGSizeMake(self.view.frame.size.width*3,
    ➥self.scrollView.frame.size.height);
    self.scrollView.frame = self.view.frame;
    self.scrollView.pagingEnabled = TRUE;                                 ①
    self.scrollView.showsHorizontalScrollIndicator = FALSE;               ②
    self.scrollView.showsVerticalScrollIndicator = FALSE;                 ③

    <添加图片视图>
    CGFloat pageControlWidth  = 300.0;
    CGFloat pageControlHeight = 37.0;
    self.pageControl = [[UIPageControl alloc]
    ➥initWithFrame:CGRectMake((S_WIDTH - pageControlWidth) / 2,
    ➥S_HEIGHT - pageControlHeight, pageControlWidth, pageControlHeight)]; ④
    self.pageControl.numberOfPages = 3;                                   ⑤
    [self.pageControl addTarget:self
    ➥action:@selector(changePage:) forControlEvents:
    ➥UIControlEventValue Changed];
    [self.view addSubview:self.pageControl];                              ⑥
}

......

@end
```

上述代码中，第①行用于设置屏幕滚动视图每次滑动时翻一屏，第②行用于设置水平滚动条不显示，第③行用于设置垂直滚动条不显示。

第④行用于创建分屏控件对象，第⑤行用于设置总屏数属性numberOfPages为3，第⑥行用于为分屏控件添加事件ValueChanged和连接动作方法changePage:。

9.3.3 基于电子书导航的实现

在iOS 5之后，我们可以使用分页控制器（UIPageViewController）构建类似于电子书效果的应用。一个电子书效果的分页应用有很多相关的视图控制器，如图9-28所示。

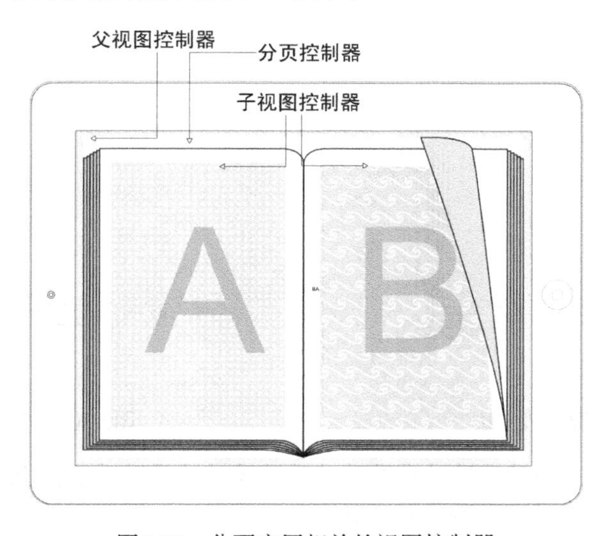

图9-28 分页应用相关的视图控制器

分页控制器需要放置在父视图控制器中，在分页控制器下面还要有子视图控制器，每个子视图控制器对应图中的一个页面。

在基于电子书导航实现的应用中，需要的类和协议有UIPageViewControllerDataSource协议、UIPageViewControllerDelegate协议和UIPageViewController类，其中UIPageViewController类没有对应的视图类。

UIPageViewControllerDataSource数据源协议中必须要实现的方法有以下两个。

- **pageViewController:viewControllerBeforeViewController:**。返回当前视图控制器之前的视图控制器，用于上一个页面的显示。
- **pageViewController:viewControllerAfterViewController:**。返回当前视图控制器之后的视图控制器，用于下一个页面的显示。

在UIPageViewControllerDelegate委托协议中，最重要的方法为pageViewController:spineLocationForInterfaceOrientation:，它根据屏幕旋转方向设置书脊位置（Spine Location）和初始化首页。

UIPageViewController中共有两个常用的属性：双面显示（doubleSided）和书脊位置（spineLocation）。

- **双面显示**。指在页面翻起时偶数页面会在背面显示。图9-18右侧为doubleSided设置为true的情况，图9-19所示为doubleSided设置为false（单面显示）的情况。单面显示在页面翻起的时候，可让用户看到页面的背面，背面的内容是当前页面透过去的，与当前内容是相反的镜像。
- **书脊位置**。书脊位置也是很重要的属性，但是它是只读的。要设置它，需要通过UIPageViewControllerDelegate委托协议中的pageViewController:spineLocationForInterfaceOrientation:方法来实现。书脊位置由枚举UIPageViewControllerSpineLocation定义，该枚举类型下的成员变量如下所示。
  - **min**。定义了书脊位置在书的最左边（或最上面），如图9-29所示，书将从右向左翻（或从下往上翻）。Objective-C版本为UIPageViewControllerSpineLocationMin。
  - **max**。定义了书脊位置在书的最右边（或最下面），如图9-30所示，书将从左向右翻（或从上往下翻）。Objective-C版本为UIPageViewControllerSpineLocationMax。

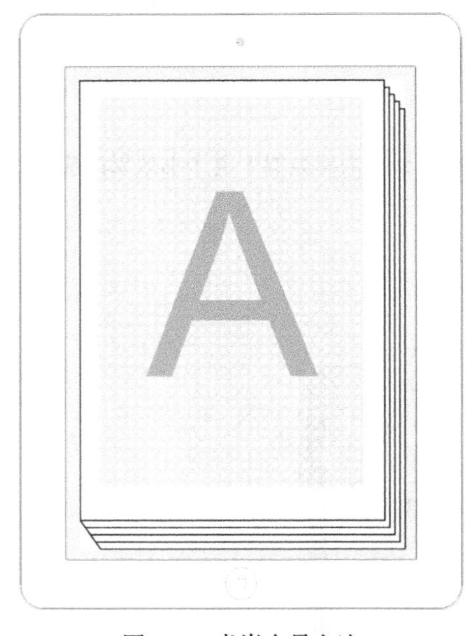

图9-29　书脊在最左边

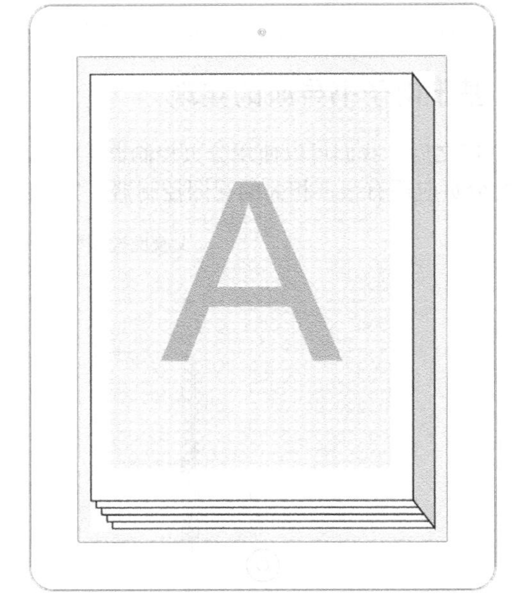

图9-30　书脊在最右边

  - **mid**。定义了书脊位置在书的中间，如图9-31所示，一般会在横屏下显示，屏幕分成两个页面。Objective-C版本为UIPageViewControllerSpineLocationMid。

图9-31 书脊在中间

下面我们使用电子书导航实现“艺术品陈列室”这个应用。因为UIPageViewController没有对应的视图，所以本节使用代码实现电子书导航的“艺术品陈列室”应用。

我们使用Single View Application模板创建一个名为 PageNavigation的工程。

提示 也可以采用Xcode工程模板Page-Based Application构建分页应用程序，但是会产生很多类，它们之间的关系比较复杂，这不利于初学者学习，因此不建议采用。

下面我们看看视图控制器ViewController的定义和属性：

```swift
//ViewController.swift文件
//翻页的方向
enum DirectionForward : Int {
    case before = 1 //向前
    case after = 2  //向后
}

class ViewController: UIViewController,
➥UIPageViewControllerDataSource, UIPageViewControllerDelegate {

    //当前Page的索引
    var pageIndex = 0
    //翻页的方向变量，其中Before表示向前，After表示向后
    var directionForward  =  DirectionForward.after

    var pageViewController: UIPageViewController!
    var viewControllers: [UIViewController]!

    ......
}
```

```objc
//ViewController.h文件
#import "ViewController.h"
//翻页的方向
enum DirectionForward
{
    ForwardBefore = 1 //向前
    ,ForwardAfter =2  //向后
};

@interface ViewController ()
➥<UIPageViewControllerDataSource,UIPageViewControllerDelegate> {
    //当前Page的索引
    int pageIndex;
    //翻页的方向变量，其中BeforeForward表示向前，AfterForward表示向后
    int  directionForward;
}

@property (strong, nonatomic) UIPageViewController *pageViewController;
@property (strong, nonatomic) NSArray *viewControllers;

@end
@implementation ViewController

- (void)viewDidLoad {
    [super viewDidLoad];

    //当前Page的索引
```

```
        pageIndex = 0;
        //翻页的方向变量
        directionForward = ForwardAfter;
        .....
    }
    .....
@end
```

在上述代码中，DirectionForward是我们定义的枚举类型，表示刚刚翻页的方向，其中ForwardBefore（Swift版是before）表示向前翻页，ForwardAfter（Swift版是after）表示向后翻页。ViewController类实现了UIPageViewControllerDataSource和UIPageViewControllerDelegate协议。变量pageIndex保存了当前页面的索引，变量directionForward保存了刚刚翻页的方向，pageViewController属性保存了UIPageViewController实例。viewControllers属性是数组类型，保存了3个界面对应的视图控制器。

下面我们看看ViewController中的viewDidLoad方法，其代码如下所示：

```
//ViewController.swift文件
override func viewDidLoad() {
    super.viewDidLoad()

    let page1ViewController = UIViewController()
    let page2ViewController = UIViewController()
    let page3ViewController = UIViewController()

    self.viewControllers = [
    ➥page1ViewController, page2ViewController, page3ViewController]        ①

    let imageView1 = UIImageView(frame: self.view.frame)
    imageView1.image = UIImage(named: "达芬奇-蒙娜丽莎.png")
    page1ViewController.view.addSubview(imageView1)

    let imageView2 = UIImageView(frame: self.view.frame)
    imageView2.image = UIImage(named: "罗丹-思想者.png")
    page2ViewController.view.addSubview(imageView2)

    let imageView3 = UIImageView(frame: self.view.frame)
    imageView3.image = UIImage(named: "保罗克利-肖像.png")
    page3ViewController.view.addSubview(imageView3)

    //设置UIPageViewController控制器
    self.pageViewController = UIPageViewController(transitionStyle:
    ➥.pageCurl, navigationOrientation: .horizontal, options: nil)          ②

    self.pageViewController.delegate = self                                 ③
    self.pageViewController.dataSource = self                               ④

    //设置首页
    self.pageViewController.setViewControllers([page1ViewController],
    ➥direction: .forward, animated: true, completion: nil)                 ⑤

    self.view.addSubview(self.pageViewController.view)

}
```

```
//ViewController.m文件
- (void)viewDidLoad {
    [super viewDidLoad];

    //当前Page的索引
    pageIndex = 0;
    //翻页的方向变量
    directionForward = ForwardAfter;

    UIViewController *page1ViewController = [[UIViewController alloc] init];
    UIViewController *page2ViewController = [[UIViewController alloc] init];
    UIViewController *page3ViewController = [[UIViewController alloc] init];

    self.viewControllers =
    ➥@[page1ViewController, page2ViewController, page3ViewController]; ①

    UIImageView *imageView1 = [[UIImageView alloc] initWithFrame:
    ➥self.view.frame];
    imageView1.image = [UIImage imageNamed:@"达芬奇-蒙娜丽莎.png"];
    [page1ViewController.view addSubview:imageView1];

    UIImageView *imageView2 = [[UIImageView alloc]
    ➥initWithFrame:self.view.frame];
    imageView2.image = [UIImage imageNamed:@"罗丹-思想者.png"];
    [page2ViewController.view addSubview:imageView2];

    UIImageView *imageView3 = [[UIImageView alloc]
    ➥initWithFrame:self.view.frame];
    imageView3.image = [UIImage imageNamed:@"保罗克利-肖像.png"];
    [page3ViewController.view addSubview:imageView3];

    //设置UIPageViewController控制器
    self.pageViewController = [[UIPageViewController alloc]
     ➥initWithTransitionStyle:UIPageViewControllerTransitionStylePageCurl
     ➥navigationOrientation:
     ➥UIPageViewControllerNavigationOrientation Horizontal options:nil]; ②

    self.pageViewController.delegate = self;                              ③
    self.pageViewController.dataSource = self;                            ④

    //设置首页
    [self.pageViewController setViewControllers:@[page1ViewController]
    ➥direction:UIPageViewControllerNavigationDirectionForward
    ➥animated:TRUE completion:nil];                                      ⑤

    [self.view addSubview:self.pageViewController.view];

}
```

在上述代码中，第①行用于将3个视图控制器UIViewController实例添加到viewControllers数组中。

第②行用于创建UIPageViewController实例，其中transitionStyle参数用于设定页面翻转的样式。其中取值是UIPageViewControllerTransitionStyle枚举类型，该枚举类型的成员如表9-1所示。

表9-1 UIPageViewControllerTransitionStyle枚举成员

Swift枚举成员	Objective-C枚举成员	说　明
pageCurl	UIPageViewControllerTransitionStylePageCurl	翻书效果样式
scroll	UIPageViewControllerTransitionStyleScroll	滑屏效果样式

参数navigationOrientation设定了翻页方向，它的取值是在UIPageViewControllerNavigationOrientation枚举类型中定义的，该枚举类型的成员如表9-2所示。

表9-2 UIPageViewControllerNavigationOrientation枚举成员

Swift枚举成员	Objective-C枚举成员	说　明
horizontal	UIPageViewControllerNavigationOrientationHorizontal	水平方向
vertical	UIPageViewControllerNavigationOrientationVertical	垂直方向

第③行用于设置pageViewController的委托对象为当前视图控制器，第④行用于设置pageViewController的数据源对象为当前视图控制器。

第⑤行的setViewControllers:direction:animated:completion:方法用于设定首页中显示的视图，其中第一个参数用于设置首页中显示的视图控制器集合。首页中显示几个视图与书脊类型有关：如果书脊类型是min或max，首页中将显示一个视图；如果是mid，首页中会显示两个视图。

setViewControllers:direction:animated:completion:方法中的参数direction是翻页的动画方向，它的取值为UIPageViewControllerNavigationDirection枚举类型，该枚举类型的成员如表9-3所示。

表9-3 UIPageViewControllerNavigationDirection枚举成员

Swift枚举成员	Objective-C枚举成员	说　明
forward	UIPageViewControllerNavigationDirectionForward	向前
reverse	UIPageViewControllerNavigationDirectionReverse	向后

我们再看看ViewController中数据源协议UIPageViewControllerDataSource的实现代码：

```
//ViewController.swift文件
//MARK: -- 实现UIPageViewControllerDataSource协议
func pageViewController(_ pageViewController: UIPageViewController,
➥viewControllerBeforeViewController viewController: UIViewController) ->
➥UIViewController? {
    pageIndex -= 1

    if (pageIndex < 0){
        pageIndex = 0
        return nil
    }

    directionForward = .before
    return self.viewControllers[pageIndex]
}

func pageViewController(_ pageViewController: UIPageViewController,
➥viewControllerAfterViewController viewController: UIViewController) ->
➥UIViewController? {

    pageIndex += 1

    if (pageIndex > 2){
```

```
//ViewController.m文件
#pragma mark -- 实现UIPageViewControllerDataSource协议
- (UIViewController *)pageViewController:(UIPageViewController *)
➥pageViewController
➥viewControllerBeforeViewController:(UIViewController *)viewController {

    pageIndex--;

    if (pageIndex < 0){
        pageIndex = 0;
        return nil;
    }

    directionForward = ForwardBefore;
    return self.viewControllers[pageIndex];
}

- (UIViewController *)pageViewController:(UIPageViewController *)
➥pageViewController
➥viewControllerAfterViewController:(UIViewController *)viewController {

    pageIndex++;
```

```
        pageIndex = 2
        return nil
    }

    directionForward = .after
    return self.viewControllers[pageIndex]
}
```

```
    if (pageIndex > 2){
        pageIndex = 2;
        return nil;
    }

    directionForward = ForwardAfter;
    return self.viewControllers[pageIndex];
}
```

在ViewController中，有关委托协议UIPageViewControllerDelegate实现方法的代码如下：

```
//ViewController.swift文件
//MARK: -- 实现UIPageViewControllerDelegate协议
func pageViewController(_ pageViewController: UIPageViewController,
➥spineLocationFor orientation: UIInterfaceOrientation) ->
➥UIPageViewControllerSpineLocation {                                          ①
    self.pageViewController.isDoubleSided = false
    return .min
}

func pageViewController(_ pageViewController: UIPageViewController,
➥didFinishAnimating finished: Bool, previousViewControllers: [UIViewController],
➥transitionCompleted completed: Bool) {                                       ②
    if (completed == false) {
        if (directionForward == .after) {
            pageIndex -= 1
        }
        if (directionForward == .before) {
            pageIndex += 1
        }
    }
}
```

```
//ViewController.m文件
#pragma mark -- 实现UIPageViewControllerDelegate协议
- (UIPageViewControllerSpineLocation)pageViewController:
➥(UIPageViewController *)pageViewController
➥spineLocationForInterfaceOrientation:(UIInterfaceOrientation)orientation
➥{                                                                             ①
    self.pageViewController.doubleSided = FALSE;
    return UIPageViewControllerSpineLocationMin;
}


- (void)pageViewController:(UIPageViewController *)pageViewController
➥didFinishAnimating:(BOOL)finished
➥previousViewControllers:(NSArray *)previousViewControllers
➥transitionCompleted:(BOOL)completed {                                         ②
    if (!completed) {
        if (directionForward == ForwardAfter) {
            pageIndex--;
        }
        if (directionForward == ForwardBefore) {
            pageIndex++;
        }
    }
}
```

第①行用来设置是否双面显示以及书脊的位置。因为spineLocation属性是只读的，所以只能在这个方法中设置书脊位置。该方法可以根据屏幕旋转方向的不同来动态设定书脊位置。

第②行中的方法是在翻页动作完成后触发的，可以利用这个方法判断用户是否成功翻到了下一页。用户很有可能只是翻了一点点，然后又放弃翻页，我们可以通过completed参数判断是否成功进行了翻页，如果该参数为true（Objective-C为YES或TRUE）则说明成功，否则为放弃翻页。若最终用户放弃了翻页，这时我们需要将记录页码的变量pageIndex恢复到之前的状态，然而是pageIndex减一还是pageIndex加一，还要看用户刚刚的翻页方向。directionForward变量用来记录翻页方向，这样我们就可以通过判断它来恢复pageIndex之前的状态了。

代码编写完毕后，我们可以运行一下，得到的效果如图9-32所示。

图9-32　运行效果

9.4 标签导航

标签导航模式是非常重要的导航模式。使用标签栏时有一定的指导原则：标签栏位于屏幕下方，占有49点的屏幕空间，有时可以隐藏起来；为了点击方便，标签栏中的标签不能超过5个，如果超过5个，则最后一个显示为“更多”，点击“更多”标签会出现更多的列表，如图9-33所示。

图9-33 “更多”标签

9.4.1 应用场景

对于中国东北三省的城市信息数据，如果把它们分成3组，你会怎么分呢？首先考虑按照行政区划分组。

- **第一组：** 哈尔滨、齐齐哈尔、鸡西、鹤岗、双鸭山、大庆、伊春、佳木斯、七台河、牡丹江、黑河、绥化，这12个城市归黑龙江省管辖。
- **第二组：** 长春、吉林、四平、辽源、通化、白山、松原、白城，这8个城市归吉林省管辖。
- **第三组：** 沈阳、大连、鞍山、抚顺、本溪、丹东、锦州、营口、阜新、辽阳、盘锦、铁岭、朝阳、葫芦岛，这14个城市归辽宁省管辖。

小组内部的数据有一定的关联关系，它们同属于一个行政管辖区域，小组之间互相独立，这就是标签导航模式适用的情况。

按照这样的分组方式在iPhone上摆放这些城市，仍然会分成3个屏幕。如图9-34所示，标签名就是省的名字。当我们选中某个省的标签时，屏幕会显示该省的城市信息，而且标签是高亮显示的。

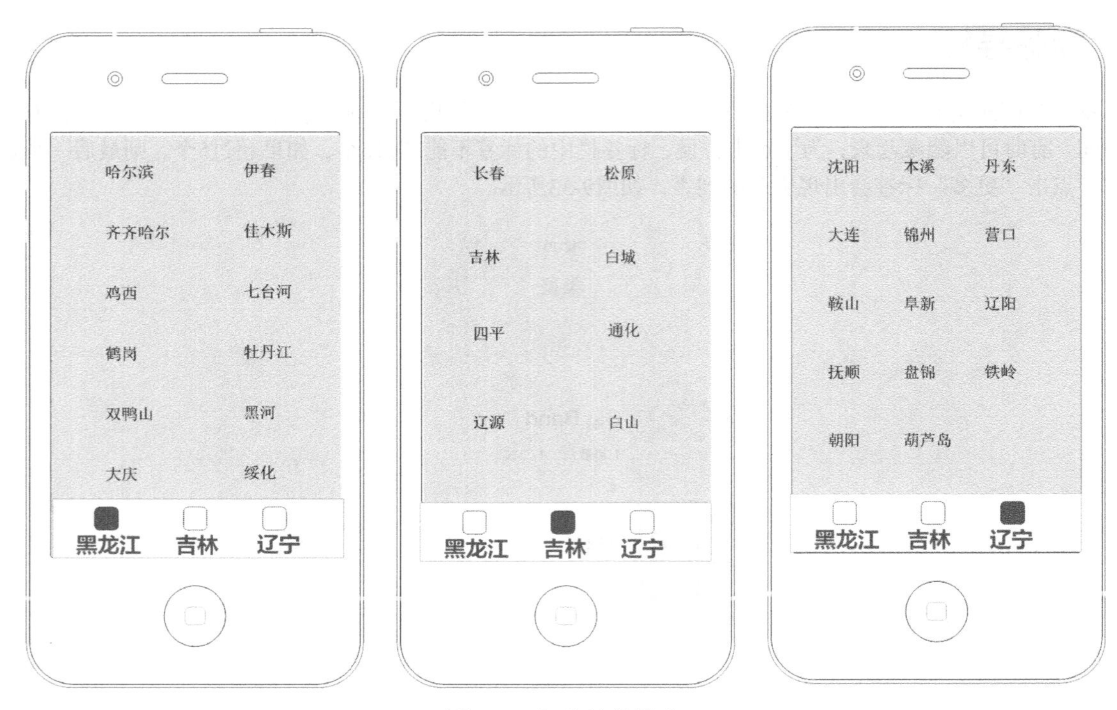

图9-34　标签导航模式

在开发具体应用的时候，标签导航模式的各个标签分别代表一个功能模块，各功能模块之间相对独立。

9.4.2 Interface Builder 实现

Xcode中提供Tabbed Application工程模板，可以创建标签导航模式的应用。默认情况下，Xcode中使用故事板技术来实现标签导航模式。用故事板技术实现标签导航很简单，我们不需要编写任何代码。

1. 创建Tabbed Application工程

请使用Tabbed Application模板创建一个名为TabNavigation的工程，创建完成之后，打开主故事板文件，如图9-35所示。

图9-35所示的3个场景（Scene）会由一些线连接起来，这些线就是过渡（Segue）。故事板开始的一端是Tab Bar Controller Scene，它是根视图控制器。图中有两个过渡，用来描述Tab Bar Controller Scene与First Scene和Second Scene之间的关系。

2. 添加场景

我们需要先修改两个现有的场景，然后再添加一个场景，才能满足业务需求。修改两个现有的场景很简单，直接修改视图控制器名就可以了，然后场景就会跟着变化。请添加一个场景到设计界面中，然后从对象库中拖曳一个View Controller到设计界面中，如图9-36所示。

此外，还需要添加场景和Tab Bar Controller Scene的连线，具体操作是：按住control键从Tab Bar Controller拖曳鼠标到View Controller，释放鼠标，此时会弹出如图9-37所示的菜单，从菜单中选择view controllers项，此时连线就做好了。

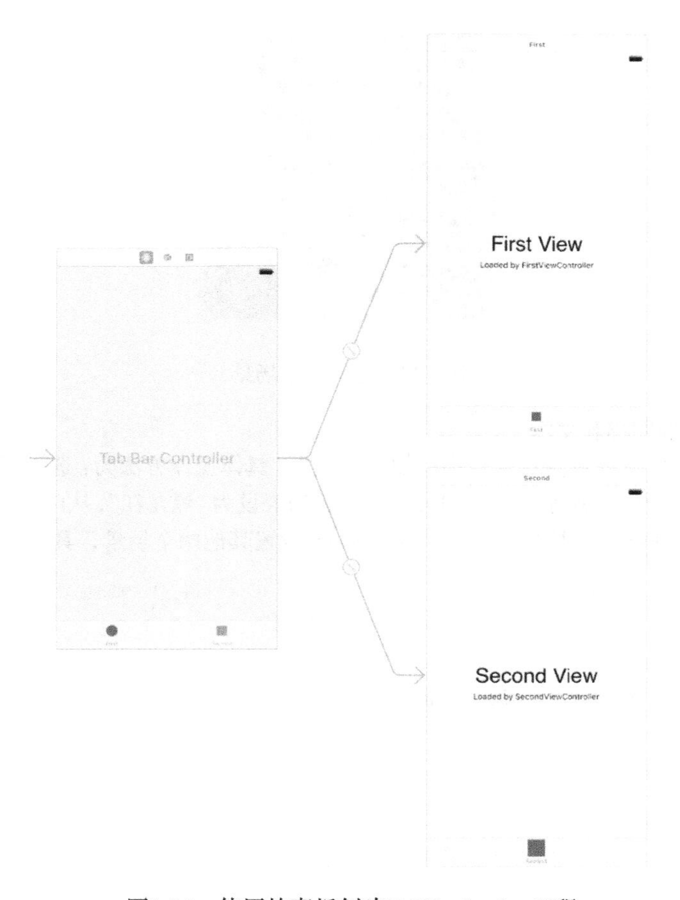

图9-35 使用故事板创建TabNavigation工程

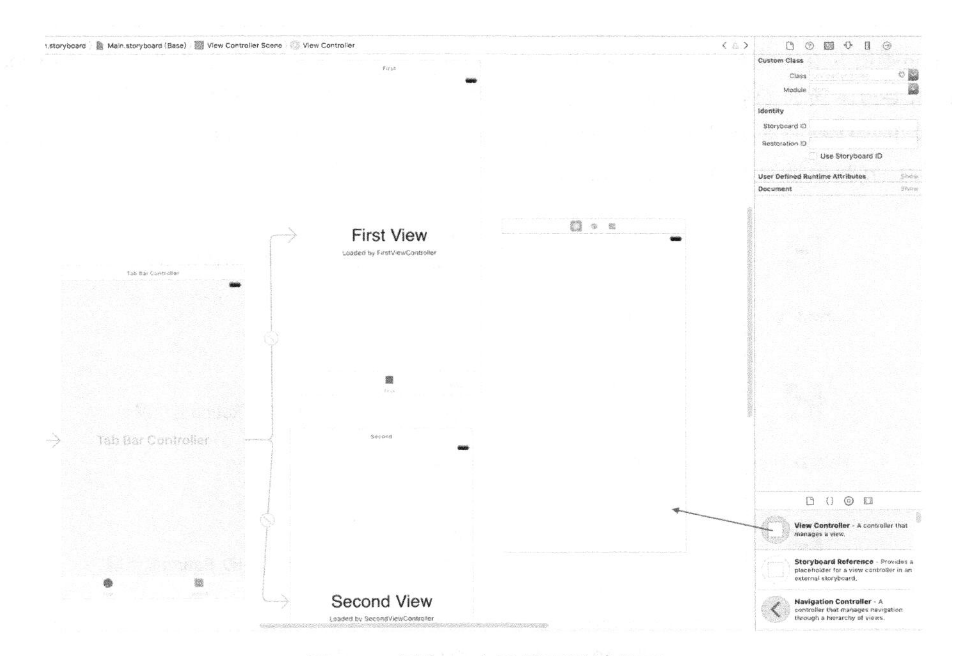

图9-36 添加一个场景到设计界面

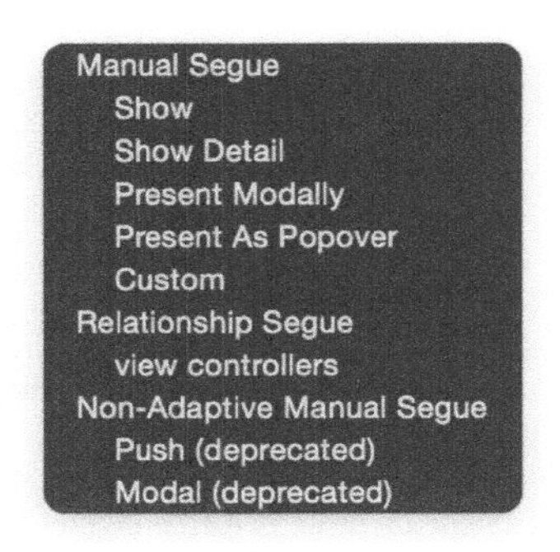

图9-37 连线两个场景

3. 添加标签栏项目中的图标和文本

然后添加图标到工程中，修改标签栏项目中的图标和文本，具体操作方法为：选择场景中的Hei Scene→Hei→First，打开其属性检查器。如图9-38所示，我们将Bar Item下的Title设为"黑龙江"，从Image下拉列表中选择Hei.png。请按照同样的办法修改其他两个视图控制器。参考Hei Scene设置其他两个场景，具体步骤不再赘述。

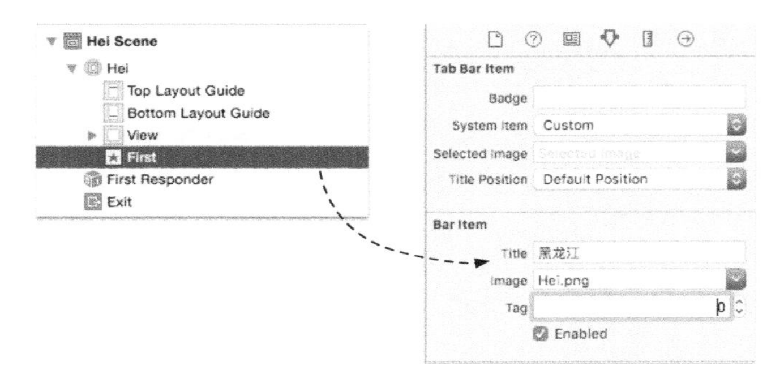

图9-38 修改标签栏项目中的图标和文本

提示 Hei.png等图片在本章代码的tabicons文件夹中，首先需要参考5.4节将这些图片添加到Xcode工程中。我们需要为不同的iOS设备准备不同规格的标签图标，它们的命名要遵守一定的规范。如图9-39所示，其中没有后缀的是普通显示屏需要的图标，@2x后缀的是为Retina显示屏（不包括iPhone Plus）准备的图标，@3x后缀的是为iPhone Plus Retina显示屏准备的图标。

图9-39 标签图标规格与命名规范

4. 界面布局

3个视图的内容可以参考图9-34实现，拖曳一些Label视图，摆放好位置，修改城市名字，然后再修改视图背景颜色即可，具体过程不再赘述。此时我们就实现了标签导航模式的一个实例，整个过程没有编写一行代码。

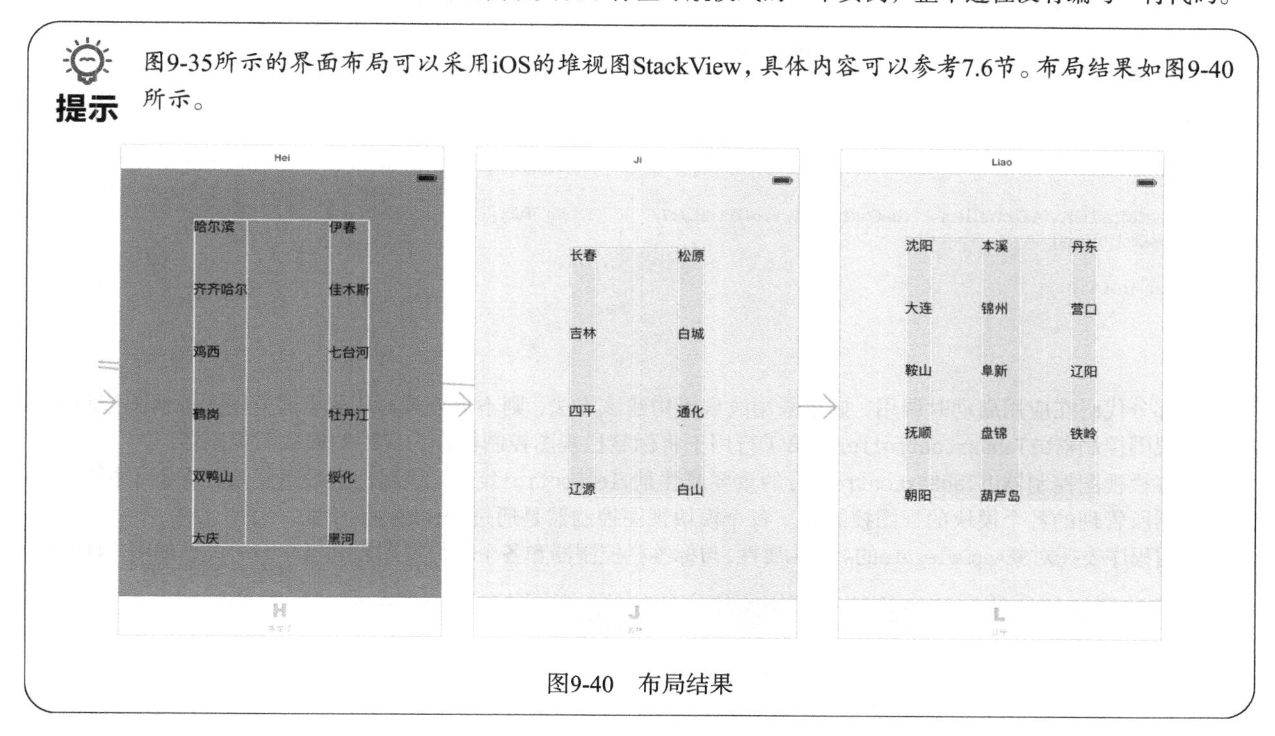

图9-40 布局结果

5. 添加视图控制器代码

在代码部分中，我们只需要3个视图控制器——HeiViewController、JiViewController和LiaoViewController，而目前只有两个视图控制器——FirstViewController和SecondViewController，现在可以把这两个视图控制器改为所需要的名字，然后再添加一个新的视图控制器。但是改名比较麻烦，我们还是推荐删除FirstViewController和SecondViewController，重新创建3个视图控制器。

创建过程是在菜单栏中选择File→New→File...菜单项，在文件模板中选择iOS→Cocoa Touch Class，此时将弹出新建文件对话框。我们在Class项目中输入HeiViewController，从Subclass of下拉列表中选择UIViewController，不选中Also create XIB file复选框。然后再回到Interface Builder中，选中View Controller Scene，打开其标识检查器，将Custom Class中的Class设为HeiViewController。参考HeiViewController设置其他两个视图控制器，具体步骤不再赘述。

9.4.3 代码实现

下面我们看看代码实现过程，其中的重点是应用程序委托对象AppDelegate类，其中设定了应用的根视图控制器。

AppDelegate的主要代码如下：

```
//AppDelegate.swift文件
class AppDelegate: UIResponder, UIApplicationDelegate {

    var window: UIWindow?

    func application(_ application: UIApplication,
    ➥didFinishLaunchingWithOptions launchOptions:
    ➥[UIApplicationLaunchOptionsKey: Any]?) -> Bool {
```

```
//AppDelegate.m文件
@implementation AppDelegate
- (BOOL)application:(UIApplication *)application
➥didFinishLaunchingWithOptions:(NSDictionary *)launchOptions {

    self.window = [[UIWindow alloc] initWithFrame:[[UIScreen mainScreen]
    ➥bounds]];
    [self.window makeKeyAndVisible];
```

```
        self.window = UIWindow(frame: UIScreen.main.bounds)
        self.window?.makeKeyAndVisible()

        let tabBarController = UITabBarController()
        self.window?.rootViewController = tabBarController                    ①

        let viewController1 = HeiViewController(nibName: "HeiViewController" ,
        ➥bundle: nil)                                                         ②
        let viewController2 = JiViewController(nibName: "JiViewController" ,
        ➥bundle: nil)
        let viewController3 = LiaoViewController(nibName: "LiaoViewController" ,
        ➥bundle: nil)

        tabBarController.viewControllers = [viewController1, viewController2,
        ➥viewController3]

        return true
    }
    ......
}
```

```
    UITabBarController* tabBarController = [[UITabBarController alloc] init];
    self.window.rootViewController = tabBarController;                        ①

    UIViewController *viewController1 = [[HeiViewController alloc]
    ➥initWithNibName:@"HeiViewController" bundle:nil];                        ②
    UIViewController *viewController2 = [[JiViewController alloc]
    ➥initWithNibName:@"JiViewController" bundle:nil];
    UIViewController *viewController3 = [[LiaoViewController alloc]
    ➥initWithNibName:@"LiaoViewController" bundle:nil];

    tabBarController.viewControllers =
    ➥@[viewController1, viewController2, viewController3];

    return TRUE;
}
......

@end
```

这部分代码在应用启动时调用，如果采用故事板构建该方法，则不需要编写代码。该应用的根视图控制器是标签栏视图控制器UITabBarController，第①行用于将标签栏视图控制器作为应用的根视图控制器。

标签栏视图控制器UITabBarController的重要属性是viewControllers，该属性是数组类型，保存了标签栏视图控制器所管理的各个模块的视图控制器。各个模块视图控制器是通过XIB文件创建的，见第②行。

应用程序委托对象AppDelegate的window属性，与标签栏控制器和各个模块视图控制器之间的关系如图9-41所示。

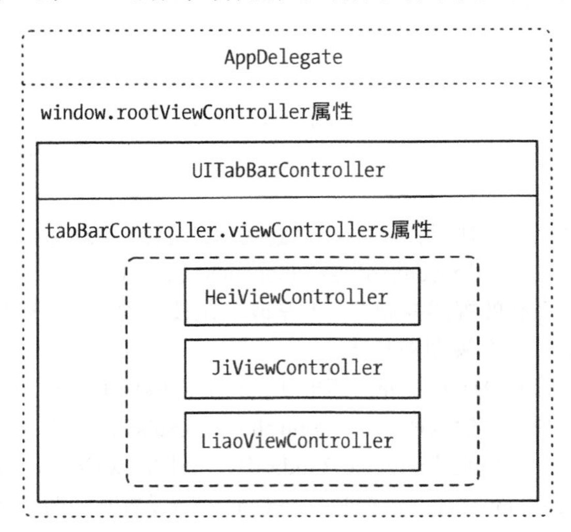

图9-41　window、标签栏控制器和模块视图控制器之间的关系

各个模块的视图控制器不是标签导航的重点，因此它们的界面构建采用XIB文件，为此需要重写这些视图控制器的构造函数initWithNibName:bundle:，代码如下：

```
class HeiViewController: UIViewController {

    override init(nibName nibNameOrNil: String?,
    ➥bundle nibBundleOrNil: Bundle?) {
        super.init(nibName: nibNameOrNil, bundle: nibBundleOrNil)
        self.title = "黑龙江"                                                 ①
        self.tabBarItem.image = UIImage(named: "Hei")                         ②
    }
    ......
}
```

```
- (id)initWithNibName:(NSString *)nibNameOrNil bundle:(NSBundle *)
➥nibBundleOrNil {
    self = [super initWithNibName:nibNameOrNil bundle:nibBundleOrNil];
    if (self) {
        self.title = @"黑龙江";                                               ①
        self.tabBarItem.image = [UIImage imageNamed:@"Hei"];                  ②
    }
    return self;
}
```

在上述方法中，第①行用于设定标签文字，第②行用于设定标签图标。注意，在Swift语言版中，还要重写如下构造函数：

```
required init?(coder aDecoder: NSCoder) {
    super.init(coder: aDecoder)
}
```

在Swift语言版中，该构造函数是父类要求重写的，本例不会调用该构造函数。

9.5　树形结构导航

树形结构导航模式也是非常重要的导航模式，它将导航视图控制器（`UINavigationController`）与表视图结合使用，主要用于构建有“从属关系”的导航。这种导航模式采用分层组织信息的方式，可以帮助我们构建iOS效率型应用程序。

提示　效率型应用程序具有组织和操作具体信息的功能，通常用于完成比较重要的任务，“相册”应用是其典型的例子。——引自于苹果HIG

9.5.1　应用场景

这里同样是按照行政区划来展示东北三省的城市信息，如下。

- 第一组：哈尔滨、齐齐哈尔、鸡西、鹤岗、双鸭山、大庆、伊春、佳木斯、七台河、牡丹江、黑河、绥化，这12个为黑龙江省管辖。
- 第二组：长春、吉林、四平、辽源、通化、白山、松原、白城，这8个为吉林省管辖。
- 第三组：沈阳、大连、鞍山、抚顺、本溪、丹东、锦州、营口、阜新、辽阳、盘锦、铁岭、朝阳、葫芦岛，这14个为辽宁省管辖。

对于每一个城市，如果还想看到更加详细的信息，比如想知道长春市在百度百科上的信息网址（http://baike.baidu.com/view/2172.htm），这种情况下“吉林省→长春→网址”就构成了一种从属关系，是一种层次模型，此时就可以使用树形导航模式。如果按照这样的分组在iPhone上展示这些城市信息，需要使用三级视图，如图9-42所示。

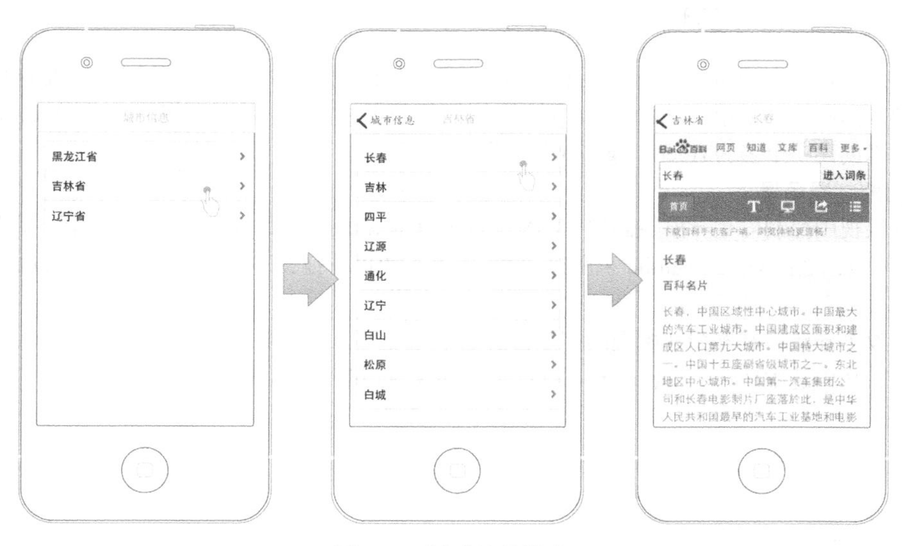

图9-42　树形导航模式

“iPhone邮件”应用如图9-43所示，它采用的就是树形结构的导航模式，所有界面的顶部都有一个导航栏。第一个界面是树形结构中的“树根”，我们称为“一级视图”或“根视图”；第二个界面是“二级视图”，它是“树干”；第三个界面是“三级视图”，是“树叶”。“树根”和“树干”采用表视图，因为表视图在分层组织信息方面的优势尤为突出。从理论上来讲，“树干”还可以有多级，但是注意不要太多，“树叶”一般是一个普通的视图，它能够完成具体展示的功能。

图9-43　“iPhone邮件”应用

我们可以为一级视图的导航栏添加左右按钮，但是在二级视图和三级视图中，导航栏的左按钮是由导航控制器自己添加的。它是汉泽尔与格莱特散在路上的“面包屑”[①]，你没有权利自己定义这个按钮，否则用户就会迷失在你的应用中。树形结构导航模式的缺点是你怎样导航进来的，就要怎样原路返回，这一点与标签导航模式不同，后者可以很快地在各个模块之间切换。

9.5.2　Interface Builder 实现

我们可以通过Xcode中的Master-Detail Application工程模板创建树形结构导航的应用，但是这种方式无法了解更多的细节问题，因此本书采用Single View Application工程模板来实现。首先，使用Single View Application模板并利用故事板技术创建一个名为TreeNavigation的工程。

1. 一级视图控制器场景

由于应用的根视图控制器是导航控制器（UINavigationController），我们需要打开故事板，删除用Single View Application工程模板创建的ViewController视图控制器，并从对象库中拖曳一个Navigation Controller到设计界面，如图9-44所示。

我们还需要设置导航控制器为初始视图控制器。如图9-45所示，选择场景中的Navigation Controller，然后打开其属性检查器，从中选中View Controller→is Initial View Controller复选框。

① 引自于格林兄弟所收录的德国童话《糖果屋》（德语：Hänsel und Gretel），又译《汉泽尔与格莱特》。

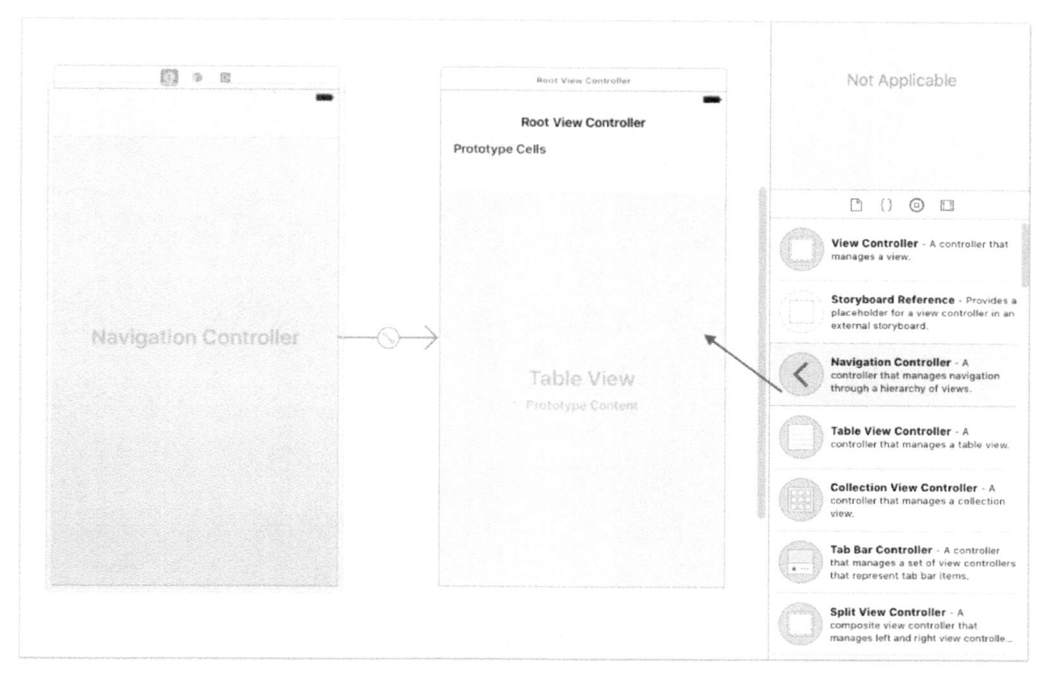

图9-44 添加导航控制器

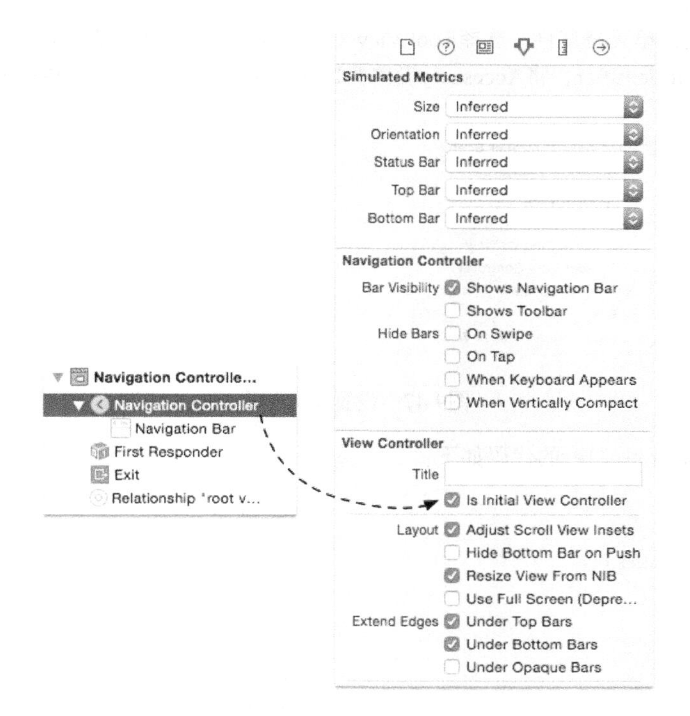

图9-45 设置集合视图控制为初始视图控制器

从对象库中直接拖曳导航控制器的一个好处是：同时为导航控制器提供一个一级视图控制器，如图9-46所示的Root View Controller就是一级视图控制器，它本身是一个表视图控制器。我们可以把代码中的`ViewController`作为一级视图控制器，此时需要修改`ViewController`继承的父类为`UITableViewController`。

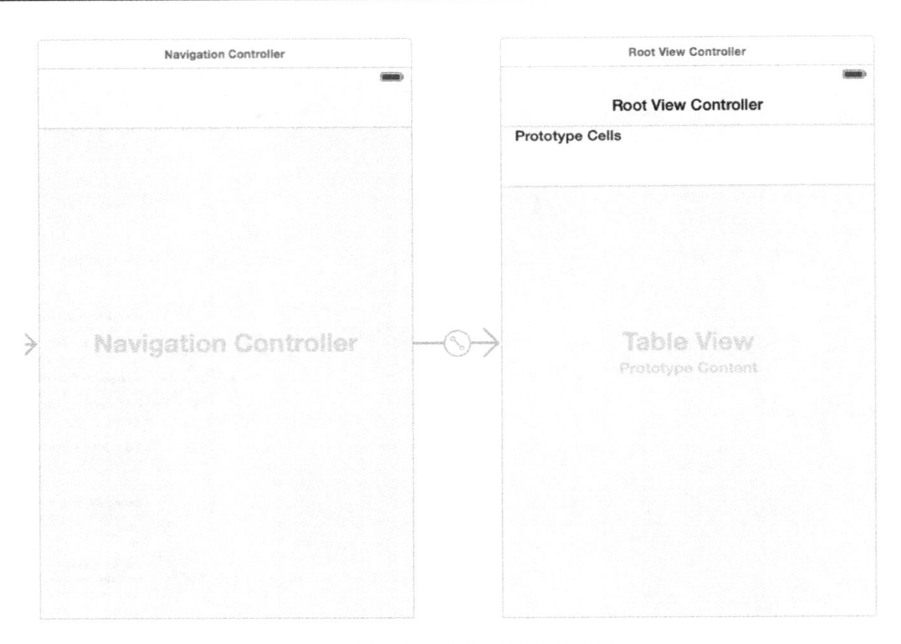

图9-46　导航控制器和根视图控制器

修改ViewController后，我们需要在故事板中选择Root View Controller，打开其标识检查器，从中选择Custom Class→Class下拉列表中的ViewController类。

此外，我们还需要设置单元格属性。选择Root View Controller Scene中的Table View Cell，打开其属性检查器，将Identifier属性设置为CellIdentifier，将Accessory设置为Disclosure Indicator，如图9-47所示。

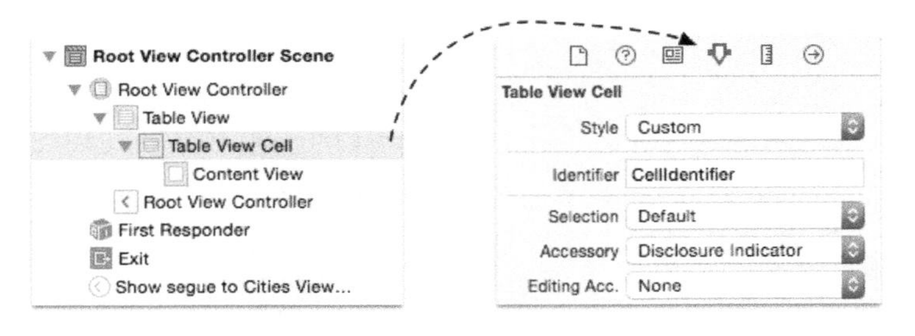

图9-47　设置单元格属性

一级视图控制器ViewController的代码如下：

```
import UIKit

class ViewController: UITableViewController {

    var dictData: NSDictionary!
    var listData: NSArray!

    override func viewDidLoad() {
        super.viewDidLoad()

        let plistPath = Bundle.main.path(forResource: "provinces_cities",
        ➥ofType: "plist")
        self.dictData = NSDictionary(contentsOfFile: plistPath!)
        self.listData = self.dictData.allKeys as NSArray
        self.title = "省份信息"
    }
```

```
//ViewController.h文件
#import <UIKit/UIKit.h>

@interface ViewController : UITableViewController

@end

//ViewController.m文件
#import "ViewController.h"
#import "CitiesViewController.h"

@interface ViewController ()

@property (strong, nonatomic) NSDictionary *dictData;
@property (strong, nonatomic) NSArray *listData;
```

```
    //MARK: -- 实现表视图数据源方法
    override func tableView(_ tableView: UITableView,
    ➥numberOfRowsInSection section: Int) -> Int {
        return self.listData.count
    }

    override func tableView(_ tableView: UITableView,
    ➥cellForRowAtindexPath: IndexPath) -> UITableViewCell  {

        let cellIdentifier = "CellIdentifier"

        let cell:UITableViewCell! = tableView
        ➥.dequeueReusableCellWithIdentifier(cellIdentifier,
        ➥forIndex Path:indexPath)

        let row = indexPath.row
        cell.textLabel?.text = self.listData[row] as? String

        return cell
    }

    //MARK: -- 场景过渡之前的预处理
    override func prepare(for segue: UIStoryboardSegue,
    ➥sender: Any?) {                                              ①

        if (segue.identifier == "ShowSelectedProvince") {

            let indexPath = self.tableView.indexPathForSelectedRow! as
            ➥NSIndexPath
            let selectedIndex = indexPath.row

            let citiesViewController = segue.destinationas!
            ➥CitiesViewController                                  ②
            let selectName = self.listData[selectedIndex] as! String
            citiesViewController.listData = self.dictData[selectName] as!
            ➥NSArray
            citiesViewController.title = selectName

        }
    }
}
```

```
@end

@implementation ViewController

- (void)viewDidLoad {
    [super viewDidLoad];

    self.tableView.delegate = self;
    self.tableView.dataSource = self;

    NSString *plistPath = [[NSBundle mainBundle]
    ➥pathForResource:@"provinces_cities" ofType:@"plist"];

    self.dictData = [[NSDictionary  alloc] initWithContentsOfFile:plistPath];
    self.listData = [self.dictData allKeys];
    self.title = @"省份信息";
}

#pragma mark -- 实现表视图数据源方法
- (NSInteger)tableView:(UITableView *)tableView
➥numberOfRowsInSection:(NSInteger)section {
    return [self.listData count];
}

- (UITableViewCell *)tableView:(UITableView *)tableView
➥cellForRowAtIndexPath:(NSIndexPath *)indexPath {

    static NSString *cellIdentifier = @"CellIdentifier";

    UITableViewCell *cell = [tableView
    ➥dequeueReusableCellWithIdentifier:cellIdentifier
    ➥forIndexPath:indexPath];

    NSInteger row = [indexPath row];
    cell.textLabel.text = self.listData[row];

    return cell;
}

#pragma mark -- 场景过渡之前的预处理
- (void)prepareForSegue:(UIStoryboardSegue *)segue sender:(id)sender {     ①

    if([segue.identifier isEqualToString:@"ShowSelectedProvince"]) {

        NSIndexPath *indexPath = [self.tableView indexPathForSelectedRow];
        NSInteger selectedIndex = indexPath.row;

        CitiesViewController *citiesViewController =
        ➥segue.destinationViewController;                                  ②
        NSString *selectName = self.listData[selectedIndex];
        citiesViewController.listData = self.dictData[selectName];
        citiesViewController.title = selectName;

    }
}

@end
```

上述代码中，第①行的方法是专门供故事板使用的，是UIViewController中的方法。当两个视图间进行跳转的时候，连接两个视图的过渡就会触发该方法，其中第②行中的segue.destinationViewController（Swift版是segue.destination）属性用于获得要跳转到的视图控制器对象。

2. 二级视图控制器场景

我们创建一个二级视图控制器CitiesViewController，具体操作方法是：选择菜单File→New→File...，在文件模板中选择iOS→Cocoa Touch Class，在弹出的“新建文件”对话框中将Class项目设置为CitiesViewController，从Subclass of下拉列表中选择UITableViewController，不选中Alse create XIB file复选框。

创建完二级视图控制器CitiesViewController后，回到设计界面，从对象库中拖曳一个Table View Controller对象到Interface Builder设计界面，作为二级视图控制器。然后按住control键，如图9-48所示，从上一个Root View Controller的单元格中拖动鼠标到当前添加的Table View Controller。释放鼠标，你将看到如图9-49所示的过渡（Segue）选择对话框，选择Selection Segue中的Show。

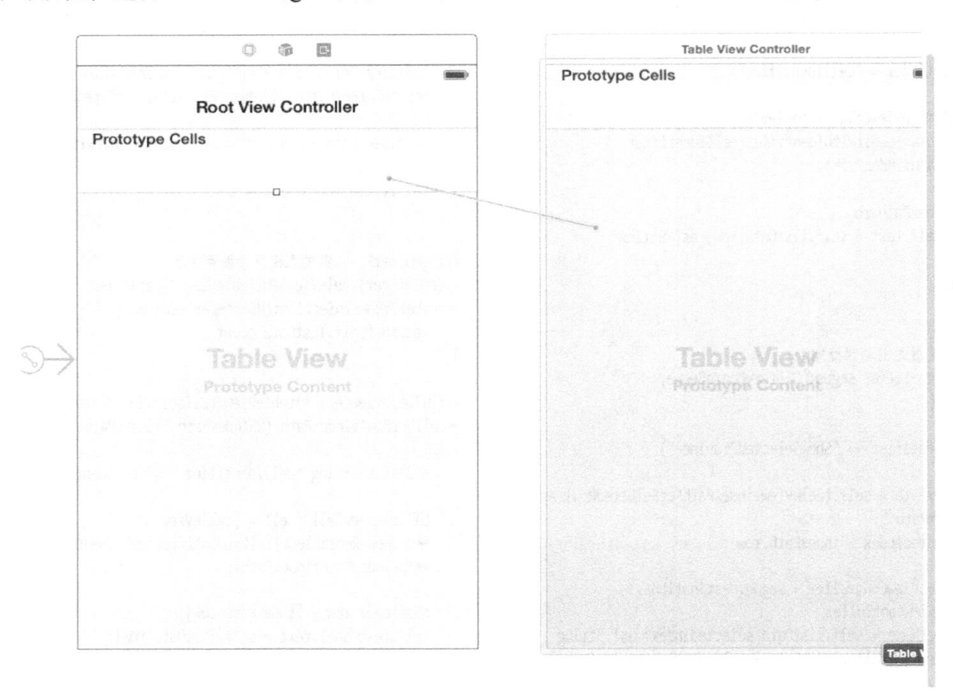

图9-48 两个视图控制器的连线

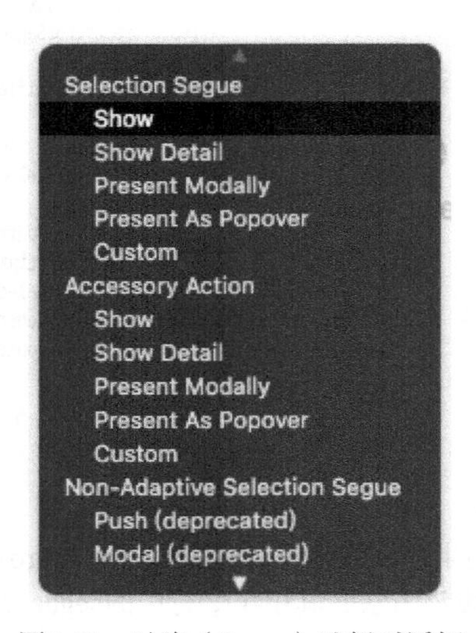

图9-49 过渡（Segue）选择对话框

另外，为了在代码中查询过渡（Segue）对象，需要为过渡设置Identifier属性，如图9-50所示，选中过渡，打开其属性检查器，然后在Identifier属性中输入ShowSelectedProvince。

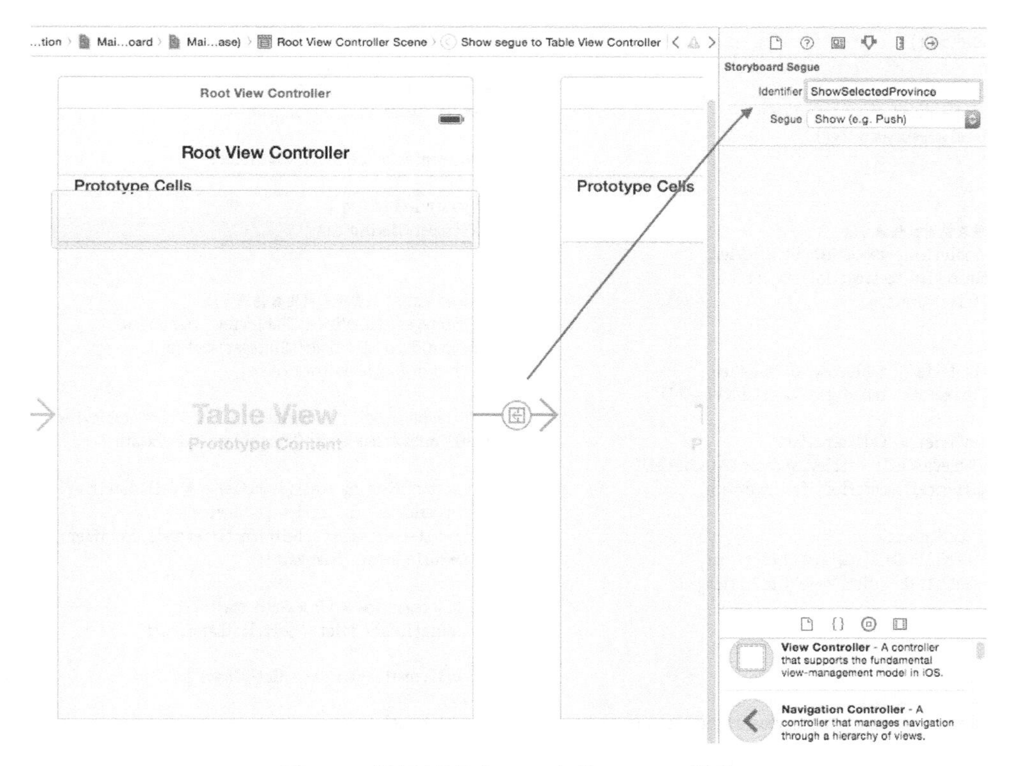

图9-50　设置过渡（Segue）的Identifier属性

选择Table View Controller，打开其标识检查器，在Custom Class的Class下拉列表中选择CitiesViewController。

此外，我们还需要设置单元格属性。选择Cities View Controller Scene中的Table View Cell，打开其属性检查器，将Identifier属性设置为CellIdentifier，将Accessory设置为Detail Disclosure，如图9-51所示。

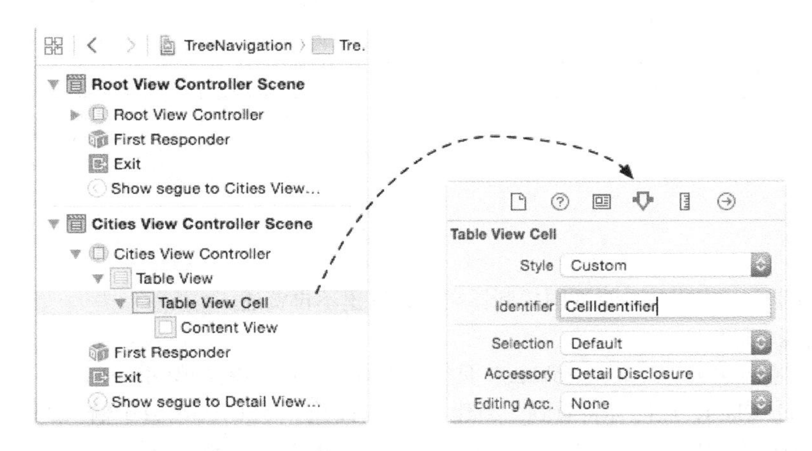

图9-51　设置单元格属性

二级视图控制器CitiesViewController的代码如下：

```
class CitiesViewController: UITableViewController {

    var listData: NSArray!

    override func viewDidLoad() {
```

```
//CitiesViewController.h文件
#import <UIKit/UIKit.h>
@interface CitiesViewController : UITableViewController
@property (weak, nonatomic) NSArray *listData;
@end
```

```
        super.viewDidLoad()
    }

    override func didReceiveMemoryWarning() {
        super.didReceiveMemoryWarning()
    }


    //MARK: -- 实现表视图数据源方法
    override func tableView(_ tableView: UITableView,
    ➥numberOfRowsInSection section: Int) -> Int {
        return self.listData.count
    }

    override func tableView(_ tableView: UITableView,
    ➥cellForRowAt indexPath: IndexPath) -> UITableViewCell {

        let cellIdentifier = "CellIdentifier"
        let cell:UITableViewCell! = tableView.dequeueReusableCell(
        ➥withIdentifier:cellIdentifier, for:indexPath)

        let row = indexPath.row
        let dict = self.listData[row] as! NSDictionary
        cell.textLabel?.text = dict["name"] as? String

        return cell
    }

    //MARK: -- 选择表视图行时触发
    override func prepare(for segue: UIStoryboardSegue, sender: Any?) {

        if (segue.identifier == "ShowSelectedCity") {

            let indexPath = self.tableView.indexPathForSelectedRow! as
            ➥IndexPath
            let selectedIndex = indexPath.row

            let dict = self.listData[selectedIndex] as! NSDictionary

            let detailViewController = segue.destination as! DetailView
            ➥Controller
            detailViewController.url = dict["url"] as! String
            detailViewController.title = dict["name"] as? String
        }

    }

}
```

```
//CitiesViewController.m文件
#import "CitiesViewController.h"
#import "DetailViewController.h"

@implementation CitiesViewController

- (void)viewDidLoad {
    [super viewDidLoad];
}

#pragma mark -- 实现表视图数据源方法
- (NSInteger)tableView:(UITableView *)tableView
➥numberOfRowsInSection:(NSInteger)section {
    return [self.listData count];
}
- (UITableViewCell *)tableView:(UITableView *)tableView
➥cellForRowAtIndexPath:(NSIndexPath *)indexPath {

    static NSString *cellIdentifier = @"CellIdentifier";
    UITableViewCell *cell = [tableView
    ➥dequeueReusableCellWithIdentifier:cellIdentifier
    ➥forIndexPath:indexPath];

    NSInteger row = [indexPath row];
    NSDictionary *dict = self.listData[row];

    cell.textLabel.text = dict[@"name"];

    return cell;
}

#pragma mark -- 选择表视图行时触发
- (void)prepareForSegue:(UIStoryboardSegue *)segue sender:(id)sender {

    if([segue.identifier isEqualToString:@"ShowSelectedCity"]) {

        NSIndexPath *indexPath = [self.tableView indexPathForSelectedRow];
        NSInteger selectedIndex = indexPath.row;

        NSDictionary *dict = self.listData[selectedIndex];

        DetailViewController *detailViewController =
        ➥segue.destinationViewController;
        detailViewController.url = dict[@"url"];
        detailViewController.title = dict[@"name"];
    }
}
@end
```

在二级视图控制器中，代码与一级视图控制器类似，这里不再赘述。

3. 三级视图控制器场景

新建三级视图控制器DetailViewController，具体操作方法是：选择菜单File→New→File…，在文件模板中选择iOS→Cocoa Touch Class，此时你将看到“新建文件”对话框，在Class项目中输入DetailViewController，从Subclass of下拉列表中选择UIViewController，不要选中Also create XIB file复选框。

然后回到设计界面，从对象库中拖曳一个View Controller对象到Interface Builder设计界面，作为三级视图控制器。然后按住control键，将鼠标从上一个CitiesViewController的单元格拖动到当前添加的View Controller，此时从弹出菜单中选择Selection Segue中的Show。

选中连线中间的过渡（Segue），打开其属性检查器，在Identifier属性中输入ShowSelectedCity。选择View Controller，打开其标识检查器，点击Custom Class→Class，将其设置为DetailViewController。

下面看一下详细视图控制器DetailViewController的代码：

```
import UIKit
import WebKit

class DetailViewController: UIViewController, WKNavigationDelegate {

    var webView: WKWebView!

    var url: String!

    override func viewDidLoad() {
        super.viewDidLoad()

        ///添加WKWebView
        self.webView = WKWebView(frame: self.view.frame)
        self.view.addSubview(self.webView)
        self.webView.navigationDelegate = self

        let url = URL(string: self.url)
        let request = URLRequest(url: url!)
        self.webView.load(request)

    }

    <实现WKNavigationDelegate委托协议>

}
```

```
//CitiesViewController.h文件
#import <UIKit/UIKit.h>

@interface DetailViewController : UIViewController

@property (weak, nonatomic) NSString *url;

@end

//CitiesViewController.m文件
#import "DetailViewController.h"
#import <WebKit/WebKit.h>

@interface DetailViewController () <WKNavigationDelegate>

@property(nonatomic, strong) WKWebView* webView;

@end

@implementation DetailViewController

- (void)viewDidLoad {
    [super viewDidLoad];

    ///添加WKWebView
    self.webView = [[WKWebView alloc] initWithFrame: self.view.frame];
    [self.view addSubview: self.webView];
    self.webView.navigationDelegate = self;

    NSURL * url = [NSURL URLWithString: self.url];
    NSURLRequest * request = [NSURLRequest requestWithURL:url];
    [self.webView loadRequest:request];

}

<实现WKNavigationDelegate委托协议>

@end
```

DetailViewController类中实现了WKNavigationDelegate协议，webView是WKWebView类型的属性，并定义为输出口。url属性是接收上一个视图控制器传递过来的参数，这里是选中城市的百度百科网址。

代码编写完成后，我们可以运行一下看看效果。

9.5.3 代码实现

下面我们看看代码实现过程，其中的重点是应用程序委托对象类AppDelegate，其主要代码如下：

```
class AppDelegate: UIResponder, UIApplicationDelegate {

    var window: UIWindow?
    func application(_ application: UIApplication,
    ➥didFinishLaunchingWithOptions launchOptions:
    ➥[UIApplicationLaunchOptionsKey: Any]?) -> Bool {

        self.window = UIWindow(frame: UIScreen.main.bounds)
        self.window?.makeKeyAndVisible()

        let navigationController = UINavigationController(rootViewController:
        ➥ViewController())
        self.window?.rootViewController = navigationController

        return true
```

```
@implementation AppDelegate

- (BOOL)application:(UIApplication *)application
➥didFinishLaunchingWithOptions:(NSDictionary *)launchOptions {

    self.window = [[UIWindow alloc] initWithFrame:[[UIScreen mainScreen]
    ➥bounds]];
    [self.window makeKeyAndVisible];

    ViewController* viewController = [[ViewController alloc] init];
    UINavigationController* navigationController =
    ➥[[UINavigationController alloc] initWithRootViewController:
    ➥viewController];

    self.window.rootViewController = navigationController;
```

```
    }
    ......
}
```

```
    return TRUE;
}
......

@end
```

这段代码读者应该不陌生，我们在6.6.2节的案例中采用了树形结构导航。树形结构导航的根视图控制器是导航控制器UINavigationController。UINavigationController的构造函数是-initWithRootViewController:，参数是导航控制器的根控制器，即导航控制器的一级视图控制器。

最后，导航控制器对象被赋值给self.window.rootViewController属性，self是应用程序委托对象AppDelegate，window属性与导航控制器和模块视图控制器之间的关系如图9-52所示。

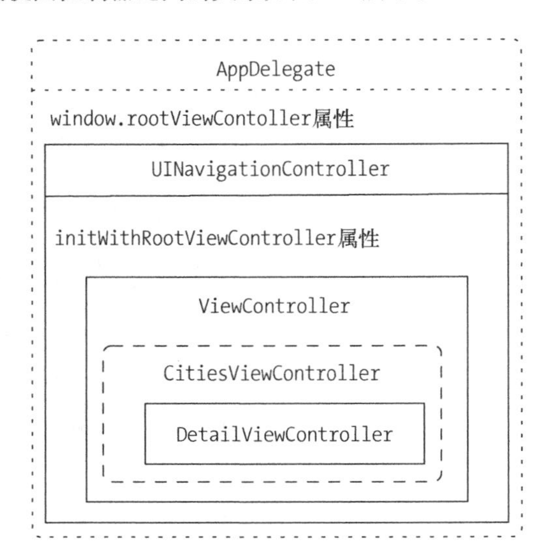

图9-52　window、导航控制器和模块视图控制器之间的关系

提示　树形结构导航中有两个根视图控制器：一个是应用程序委托对象中window的根视图控制器，它是UINavigationController实例，通过window的rootViewController属性指定；另一个是导航控制器的根视图控制器，通过构造函数-initWithRootViewController:指定给导航控制器。导航控制器的根视图控制器管理显示一级视图，即我们看到的第一个界面，因此也称为导航控制器的一级视图控制器。

下面我们看看一级视图控制器ViewController的代码。与Interface Builder实现相比，代码实现过程中不能使用prepareForSegue:sender:方法，因为该方法是故事板专用的，XIB和代码中不能使用它。我们可以实现表视图委托协议中的 tableView:didSelectRowAtIndexPath:方法，该方法在用户点击表视图单元格时触发。ViewController中的具体代码如下：

```
//MARK: -- 实现表视图委托方法
override func tableView(_ tableView: UITableView,
➥didSelectRowAt indexPath: IndexPath) {

    let selectedIndex = indexPath.row

    let citiesViewController = CitiesViewController()
    let selectName = self.listData[selectedIndex] as! String
    citiesViewController.listData = self.dictData[selectName] as! NSArray
```

```
#pragma mark -- 实现表视图委托协议方法
- (void)tableView:(UITableView *)tableView
➥didSelectRowAtIndexPath:(NSIndexPath *)indexPath {

    NSInteger selectedIndex = [indexPath row];
    NSDictionary *dict = self.listData[selectedIndex];
    DetailViewController *detailViewController = [[DetailViewController alloc]
    ➥init];
    detailViewController.url = dict[@"url"];
```

```swift
    citiesViewController.title = selectName

    self.navigationController?.pushViewController(citiesViewController,
    ➥animated: true)                                                      ①
}
```

```objc
    detailViewController.title = dict[@"name"];

    [self.navigationController pushViewController:detailViewController
    ➥animated:YES];                                                       ①

}
```

UINavigationController将要管理的视图控制器放入它的栈中，处于栈顶的视图控制器负责显示当前视图，如果要进入下一级视图，就将下一级视图控制器压栈，第①行的pushViewController:animated:方法实现这一目的。如果要回到上级视图，可以使用出栈方法。UINavigationController出栈方法有如下3个。

- popViewControllerAnimated:，回到上一级视图。
- popToRootViewControllerAnimated:，回到根视图。
- popToViewController:animated:，回到指定视图。

二级视图控制器CitiesViewController的代码，与一级视图控制器ViewController的代码比较类似，这里不再赘述。

三级视图控制器DetailViewController的代码，与Interface Builder实现完全一样，这里也不再赘述。

9.6　组合使用导航模式

有些情况下，我们会将3种导航模式综合到一起使用，其中还会用到模态视图。例如，Tweet是编写Twitter的应用，如图9-53所示。Tweet主要采用了标签导航模式和树形结构导航模式，有些地方（图9-53c的Bill Couch）还采用了平铺导航模式。点击导航栏右边的按钮，会打开一个模态视图，此时可以编辑Twitter。

(a)　(b)　(c)

图9-53　Tweet应用

9.6.1　应用场景

同样是划分东北三省的城市信息，我们可以采用组合方式实现，如图9-54所示。标签栏上是省名，标签导航可以进行省的切换。省信息中又采用树形结构导航，只不过树形结构中只有两级视图，二级视图（城市信息）导航栏右边的按钮可以实现添加城市信息的功能。

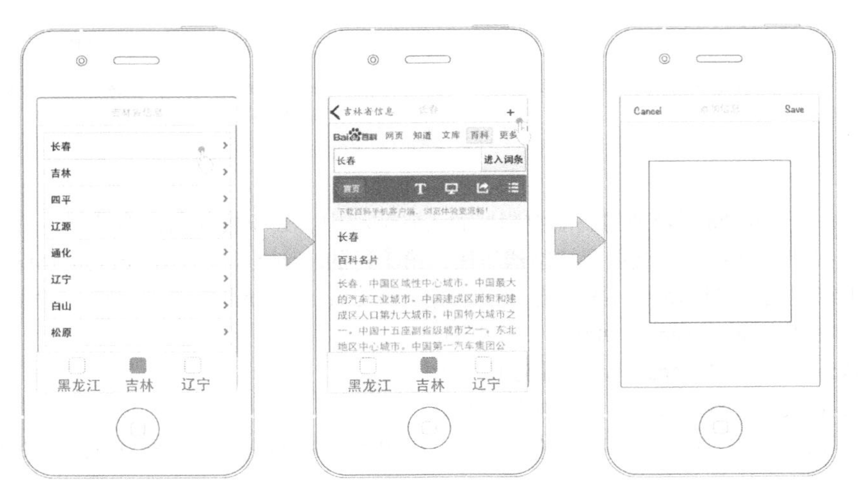

图9-54　组合导航模式

9.6.2　Interface Builder 实现

下面我们介绍如何用故事板实现组合导航模式。本例中虽然是组合导航模式，但本质上还是标签导航。我们在9.4节已经介绍了如何通过Tabbed Application模板创建标签导航，而本例我们介绍通过Single View Application模板创建标签导航，首先选择Single View Application模板创建一个名为NavigationComb的工程。

1. 标签控制器场景

工程创建完成后，打开故事板，删除ViewController视图控制器，并从对象库中拖曳一个Tab Bar Controller到设计界面，如图9-55所示。选择Item1和Item2，通过键盘上的Delete键删除它们。

此外，我们还需要设置Tab Bar Controller为初始视图控制器。选择Tab Bar Controller，打开其属性检查器，选中View Controller下的is Initial View Controller复选框。

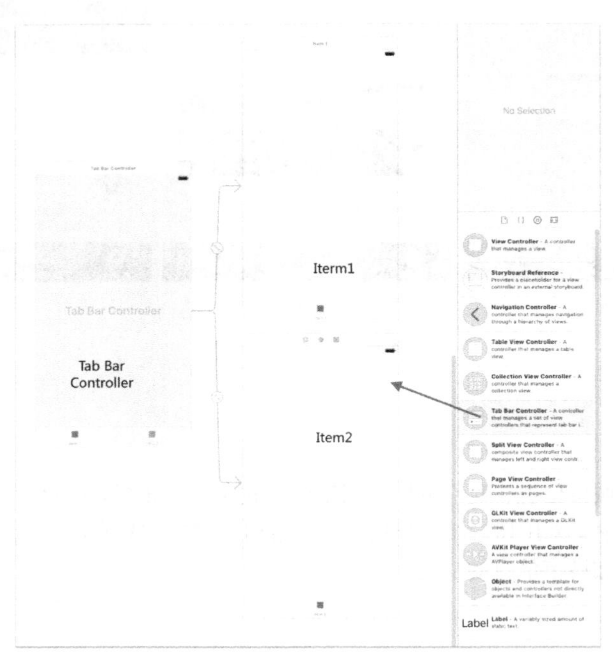

图9-55　添加Tab Bar Controller到设计界面

2. 一级视图控制器场景

一级视图控制器场景是指用户点击每一个标签进入的场景，它是树形结构导航的一级视图控制器。

具体设计过程是从对象库中拖曳Navigation Controller控制器到设计界面，如图9-56所示。然后再按住control键，从Tab Bar Controller中拖动鼠标到Navigation Controller，释放鼠标后，会弹出过渡（Segue）对话框。此时选择Relationship Segue中的view controllers，连接完成后的界面如图9-57所示。

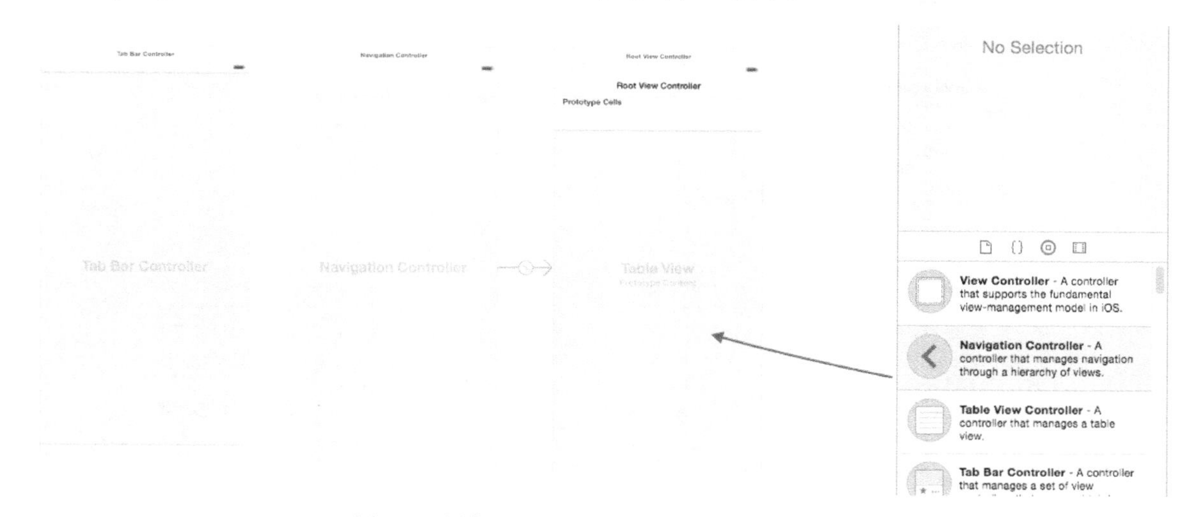

图9-56　添加Navigation Controller到设计界面

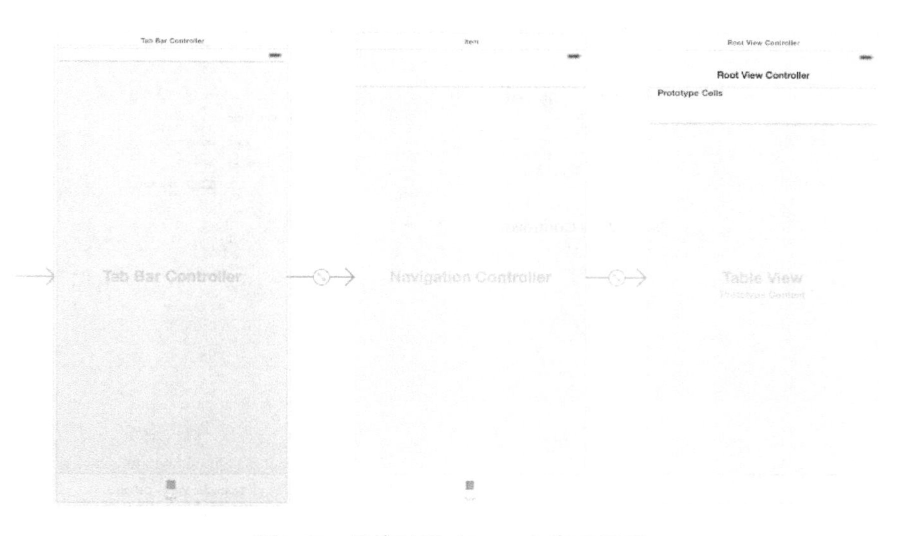

图9-57　连接过渡（Segue）完成界面

3. 二级视图控制器场景

二级视图控制器场景是指用户点击一级视图控制器单元格进入的场景，它是树形结构导航的二级视图控制器。

二级视图中有一个WebView，二级视图控制器与9.5节案例的三级视图控制器`DetailViewController`相同，我们可以直接拿来使用。

首先，我们需要从对象库中拖曳一个View Controller到故事板设计界面，然后按住control键，从上一个Root View Controller的单元格中拖动鼠标到当前添加的View Controller，选择Selection Segue中的Show。选中连线中间的过渡（Segue），打开其属性检查器，在Identifier属性中选ShowDetail（Identifier属性可以根据自己的情况命名，但要与程序代码中的Identifier保持一致）。

接着，我们需要在这个二级视图中拖曳一些控件。首先从对象库中拖曳一个Navigation Item到设计界面，如图9-58所示。然后我们在Navigation Item中添加一个右按钮，如图9-59所示，设置按钮的identifier为Add。为了方便管理和查看，我们可以把Navigation Item的Title属性设置为Detail View Controller。

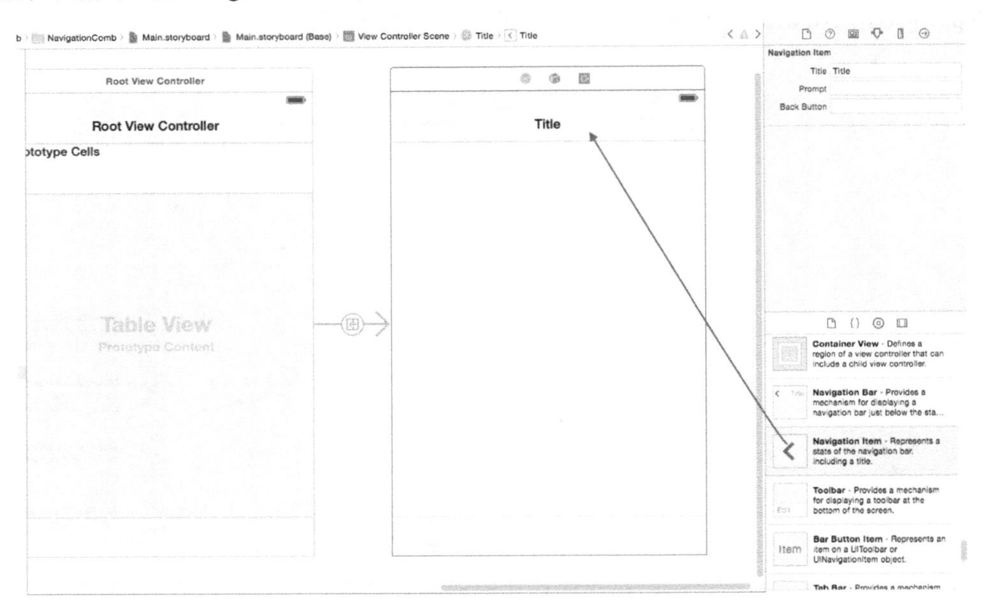

图9-58 添加Navigation Item到设计界面

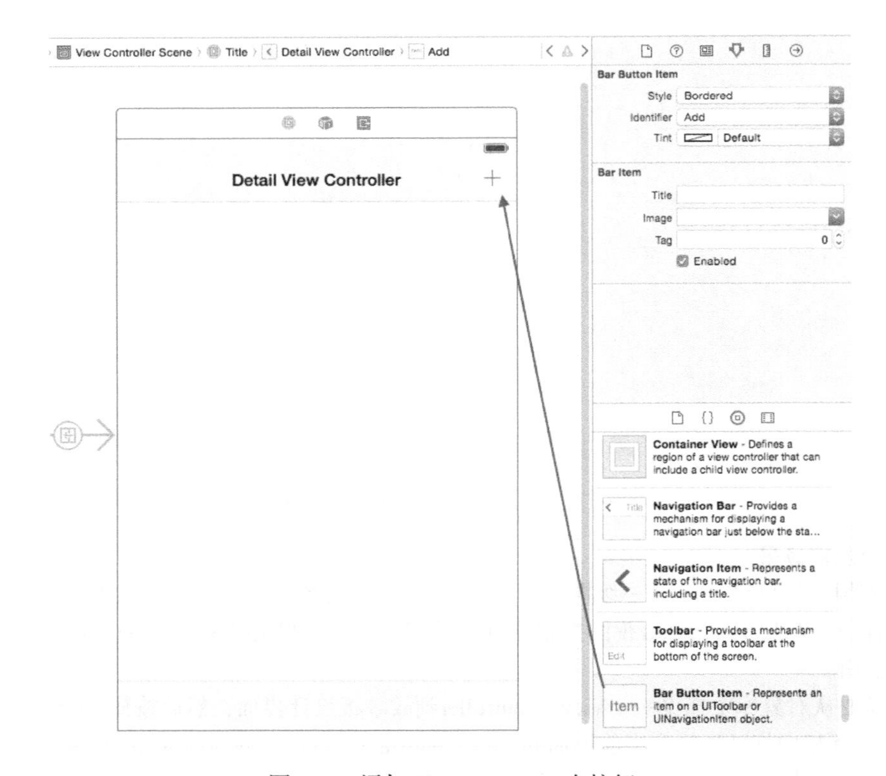

图9-59 添加Navigation Item右按钮

4. 模态视图控制器场景

在二级视图中点击导航栏右按钮，进入模态视图。下面我们来设计模态视图控制器。由于模态视图中也有导航栏和左右按钮，我们需要从对象库中拖曳一个View Controller到设计界面。默认情况下，View Controller不带导航栏，我们可以将View Controller嵌入到Navigation Controller中（参见图9-60），具体步骤是：选择菜单Editor→Embed In→Navigation Controller。

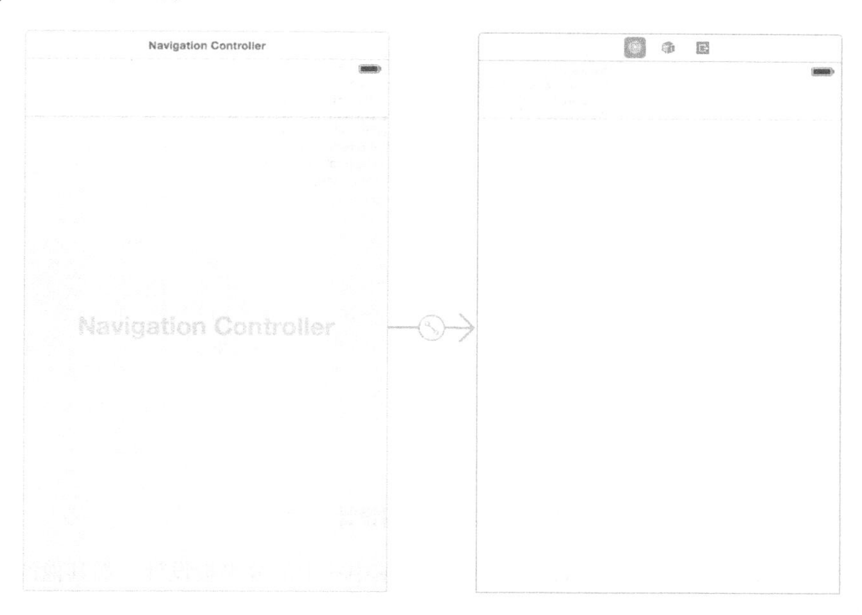

图9-60 将View Controller嵌入到Navigation Controller中

然后按住control键，将鼠标从Detail View Controller（二级视图控制器）导航栏右按钮拖动到刚刚新建的Navigation Controller上，在弹出的对话框中选择Present Modally，如图9-61所示。

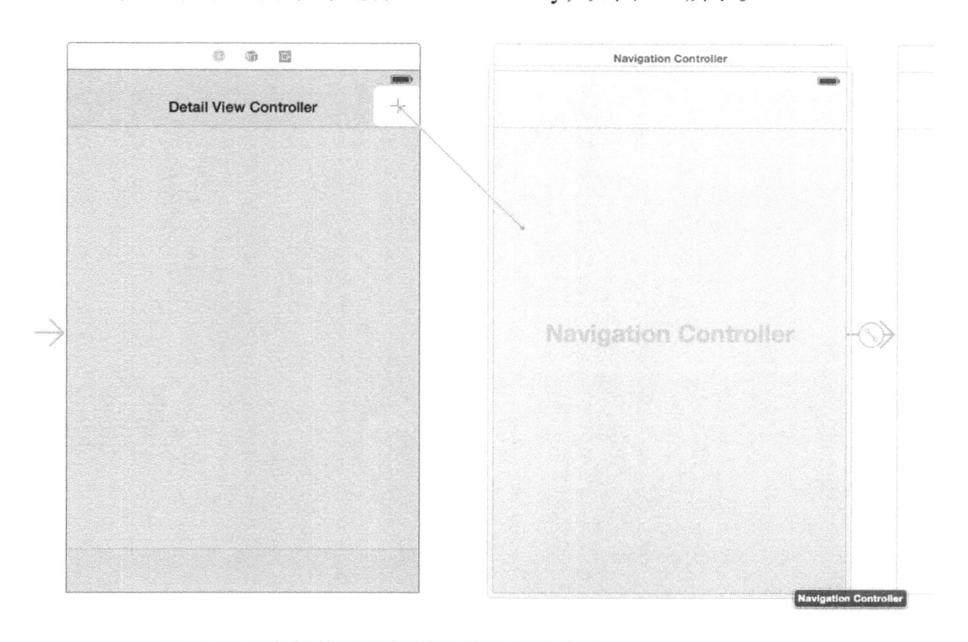

图9-61 用鼠标拖动导航栏右按钮到新建的Navigation Controller

我们还需要为模态视图添加一些视图，包括一个TextView和导航栏两个左右按钮，设计界面如图9-62所示。

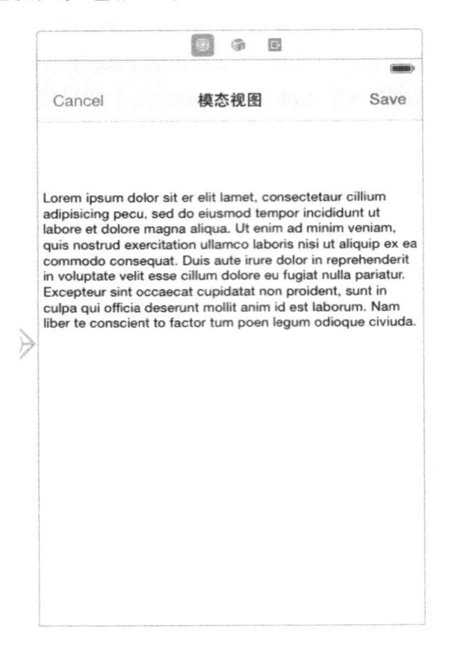

图9-62　模态视图设计界面

到此为止，故事板的设计如图9-63所示，这只是3个标签中一个的故事板设计，而其他两个标签与第一个类似，具体设计过程不再赘述。

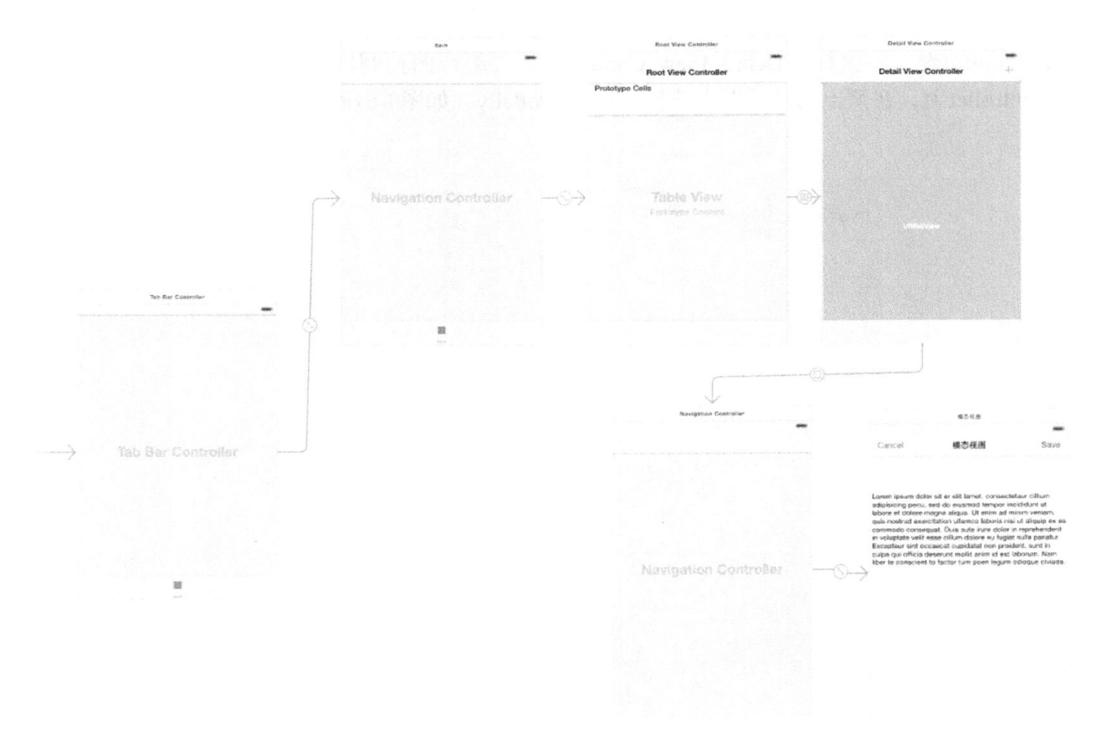

图9-63　故事板设计

5. 设置标签栏内容

标签栏中的标签包含标题和图标，我们可以参考9.4节中的案例设置一下。标签的标题分别是："黑龙江""吉林"和"辽宁"，图标分别是Hei.png、Ji.png和Liao.png。

到此为止，故事板的设计工作就完成了。完成之后的故事板如图9-64所示，其中有12个场景，很复杂吧?我们的业务还不是很复杂，就已经有这么多的场景（Sence）和过渡（Segue）。故事板的用意是想减少代码量，但是与此同时也增加了设置环节的工作量。虽然苹果主推使用故事板技术，但它并不是iOS解决编程问题的银弹[①]。

图9-64　最终故事板设计

6. 视图控制器代码

我们需要3个视图控制器文件：一级视图控制器ViewController、二级视图控制器DetailViewController和模态视图控制器ModalViewController。其中，ViewController的父类应该修改为UITableViewController。另外，DetailViewController和ModalViewController两个文件需要创建，我们设置它们的父类为UIViewController。这些视图控制器类创建好后，我们需要回到故事板设计界面，设置它们的Custom Class→Class属性。

下面看一下代码部分，一级视图控制器ViewController中的属性、视图加载等相关代码如下：

① 在西方古老的传说里，狼人是不死的，但是银弹可以杀死狼人，详情可参见http://baike.baidu.com/view/3413847.htm。

```
//ViewController.swift文件
class ViewController: UITableViewController {

    var dictData: NSDictionary!
    var listData: NSArray!

    override func viewDidLoad() {
        super.viewDidLoad()

        let plistPath = Bundle.main.path(forResource: "provinces_cities",
        ➥ofType: "plist")

        self.dictData = NSDictionary(contentsOfFile: plistPath!)
        let navigationController = self.parent as! UINavigationController       ①
        let selectProvinces = navigationController.tabBarItem.title!            ②

        print(selectProvinces)

        if (selectProvinces == "黑龙江") {
            self.listData = self.dictData["黑龙江省"] as! NSArray
            self.navigationItem.title = "黑龙江省信息"
        } else if (selectProvinces == "吉林") {
            self.listData = self.dictData["吉林省"] as! NSArray
            self.navigationItem.title = "吉林省信息"
        } else {
            self.listData = self.dictData["辽宁省"] as! NSArray
            self.navigationItem.title = "辽宁省信息"
        }

    }
    ......

}
```

```
//ViewController.m文件
#import "ViewController.h"

#import "DetailViewController.h"

@interface ViewController ()

@property (strong, nonatomic) NSDictionary *dictData;
@property (strong, nonatomic) NSArray *listData;

@end

@implementation ViewController


- (void)viewDidLoad {
    [super viewDidLoad];

    NSString *plistPath = [[NSBundle mainBundle]
    ➥pathForResource:@"provinces_cities" ofType:@"plist"];

    self.dictData = [[NSDictionary  alloc] initWithContentsOfFile:path];

    UINavigationController *navigationController=
    ➥(UINavigationController*)self.parentViewController;                        ①
    NSString *selectProvinces = navigationController.tabBarItem.title;          ②

    NSLog(@"%@", selectProvinces);

    if ([selectProvinces isEqualToString:@"黑龙江"]) {
        self.listData = self.dictData[@"黑龙江省"];
        self.navigationItem.title = @"黑龙江省信息";
    } else if ([selectProvinces isEqualToString:@"吉林"]) {
        self.listData = self.dictData[@"吉林省"];
        self.navigationItem.title = @"吉林省信息";
    } else {
        self.listData = self.dictData[@"辽宁省"];
        self.navigationItem.title = @"辽宁省信息";
    }
}
......

@end
```

ViewController是3个省共同使用的类。当然，我们可以为每一个省创建一个视图控制器，但是就本例而言，完全没有必要创建3个不同的类。如何区分是点击了哪个标签进入的呢？第①行用于获得当前视图控制器的父视图控制器，返回的是Navigation Controller。第②行用于获得选中的标签栏的标签名字，这个标签名字能够用于识别是点击哪个标签进入的。

在ViewController中，选择表视图行时，会触发过渡（Segue）方法：

```
//MARK: -- 选择表视图行时触发
override func prepare(for segue: UIStoryboardSegue, sender: Any?) {

    if (segue.identifier == "ShowDetail") {

        let indexPath = self.tableView.indexPathForSelectedRow! as IndexPath
        let selectedIndex = indexPath.row
        let dict = self.listData[selectedIndex] as! NSDictionary

        let detailViewController = segue.destination as! DetailViewController
        detailViewController.url = dict["url"] as! String
        detailViewController.title = dict["name"] as? String

    }

}
```

```
#pragma mark -- 选择表视图行时触发
- (void)prepareForSegue:(UIStoryboardSegue *)segue sender:(id)sender {

    if([segue.identifier isEqualToString:@"ShowDetail"]) {

        NSIndexPath *indexPath = [self.tableView indexPathForSelectedRow];
        NSInteger selectedIndex = indexPath.row;

        NSDictionary *dict = self.listData[selectedIndex];

        DetailViewController *detailViewController =
        ➥segue.destinationViewController;
        detailViewController.url = dict[@"url"];
        detailViewController.title = dict[@"name"];
    }
}
```

二级视图控制器DetailViewController与9.5节中的案例一样，没有什么变化，此处不再赘述。

因为本案例重点关注应用导航的实现，所以模态视图控制器ModalViewController中的代码非常少，在本例中只实现了Cancel按钮的功能，而没有实现Save按钮的功能。ModalViewController的主要代码如下：

```
class ModalViewController: UIViewController {
    ......
    @IBAction func cancel(_ sender: AnyObject) {
        self.dismiss(animated: true, completion: { () -> Void in
            NSLog("关闭模态视图")
        })
    }
}
```

```
#import "ModalViewController.h"

@implementation ModalViewController
......

- (IBAction)cancel:(id)sender {
    [self dismissViewControllerAnimated:TRUE completion:^{
        NSLog(@"关闭模态视图");
    }];
}

@end
```

代码编写完毕后，可以运行一下，效果如图9-65所示。

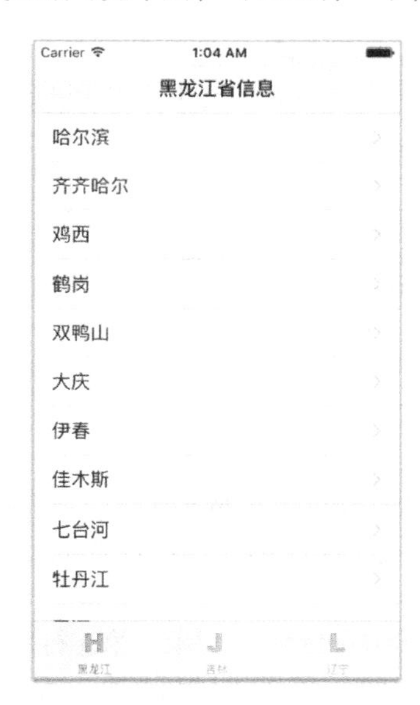

图9-65 运行效果

9.6.3 代码实现

下面我们看看代码实现过程，其中的重点是应用程序委托对象类AppDelegate，其中设定了应用的根视图控制器。

AppDelegate的主要代码如下：

```
//AppDelegate.swift文件
class AppDelegate: UIResponder, UIApplicationDelegate {

    var window: UIWindow?

    func application(_ application: UIApplication,
    ➥didFinishLaunchingWithOptions launchOptions:
    ➥[UIApplicationLaunchOptionsKey: Any]?) -> Bool {

        self.window = UIWindow(frame: UIScreen.main.bounds)
```

```
//AppDelegate.m文件
@implementation AppDelegate

- (BOOL)application:(UIApplication *)application
➥didFinishLaunchingWithOptions:(NSDictionary *)launchOptions {

    self.window = [[UIWindow alloc] initWithFrame:[[UIScreen mainScreen]
    ➥bounds]];
    [self.window makeKeyAndVisible];
```

```
        self.window?.makeKeyAndVisible()

        let tabBarController = UITabBarController()                              ①
        self.window?.rootViewController = tabBarController                       ②

        //设置黑龙江标签
        let viewController1 = ViewController()                                   ③
        let navigationController1 = UINavigationController(rootView
        ➥Controller: viewController1)                                            ④
        navigationController1.tabBarItem.title = "黑龙江"                         ⑤
        navigationController1.tabBarItem.image = UIImage(named: "Hei")           ⑥

        //设置吉林标签
        let viewController2 = ViewController()
        let navigationController2 = UINavigationController(rootView
        ➥Controller: viewController2)
        navigationController2.tabBarItem.title = "吉林"
        navigationController2.tabBarItem.image = UIImage(named: "Ji")

        //设置辽宁标签
        let viewController3 = ViewController()
        let navigationController3 = UINavigationController(rootView
        ➥Controller: viewController3)
        navigationController3.tabBarItem.title = "辽宁"
        navigationController3.tabBarItem.image = UIImage(named: "Liao")

        tabBarController.viewControllers = [navigationController1,
        ➥navigationController2, navigationController3]                          ⑦

        return true
    }
    ......
}
```

```
    UITabBarController* tabBarController = [[UITabBarController alloc]
    ➥init];                                                                     ①
    self.window.rootViewController = tabBarController;                           ②

    //设置黑龙江标签
    ViewController* viewController1 = [[ViewController alloc] init];             ③
    UINavigationController* navigationController1 =
    ➥[[UINavigationController alloc] initWithRootViewController:
    ➥viewController1];                                                           ④
    navigationController1.tabBarItem.title = @"黑龙江";                           ⑤
    navigationController1.tabBarItem.image = [UIImage imageNamed:@"Hei"];        ⑥

    //设置吉林标签
    ViewController* viewController2 = [[ViewController alloc] init];
    UINavigationController* navigationController2 =
    ➥[[UINavigationController alloc]
    ➥initWithRootViewController:viewController2];
    navigationController2.tabBarItem.title = @"吉林";
    navigationController2.tabBarItem.image = [UIImage imageNamed:@"Ji"];

    //设置辽宁标签
    ViewController* viewController3 = [[ViewController alloc] init];
    UINavigationController* navigationController3 =
    ➥[[UINavigationController alloc] initWithRootViewController:
    ➥viewController3];
    navigationController3.tabBarItem.title = @"吉林";
    navigationController3.tabBarItem.image = [UIImage imageNamed:@"Ji"];

    tabBarController.viewControllers =
    ➥@[navigationController1, navigationController2,
    ➥navigationController3];                                                     ⑦

    return TRUE;
}
......
@end
```

第①行用于实例化UITabBarController，第②行用于将UITabBarController实例赋值给self.window.rootViewController属性（Swift版本是self.window?.rootViewController），这使标签控制器UITabBarController成为应用的根控制器。

第③行~第⑥行是创建黑龙江标签的导航控制器，第③行用于实例化ViewController对象。第④行用于实例化UINavigationController对象，ViewController对象作为导航控制器的一级视图控制器。第⑤行的navigationController1.tabBarItem.title用于设定标签文字，第⑥行的navigationController1.tabBarItem.image用于设定标签图标。

一级视图控制器ViewController的实现代码如下：

```
//MARK: -- 实现表视图委托方法
override func tableView(_ tableView: UITableView,
➥didSelectRowAt indexPath: IndexPath) {

    let selectedIndex = indexPath.row
    let dict = self.listData[selectedIndex] as! NSDictionary

    let detailViewController = DetailViewController()
    detailViewController.url = dict["url"] as! String
    detailViewController.title = dict["name"] as? String

    self.navigationController?.pushViewController(detailViewController,
    ➥animated: true)
}
```

```
#pragma mark -- 实现表视图委托协议方法
- (void)tableView:(UITableView *)tableView
➥didSelectRowAtIndexPath:(NSIndexPath *)indexPath {

    NSInteger selectedIndex = [indexPath row];
    NSDictionary *dict = self.listData[selectedIndex];

    DetailViewController *detailViewController = [[DetailViewController alloc]
    ➥init];
    detailViewController.url = dict[@"url"];
    detailViewController.title = dict[@"name"];

    [self.navigationController pushViewController:detailViewController
    ➥animated:YES];

}
```

代码类似于9.5.3节，这里不再赘述。

二级视图控制器DetailViewController的代码如下：

```
//DetailViewController.swift文件
class DetailViewController: UIViewController, WKNavigationDelegate {

    var webView: WKWebView!

    var url: String!

    override func viewDidLoad() {
        super.viewDidLoad()

        let addButtonItem = UIBarButtonItem(barButtonSystemItem: .add,
        ➥target: self, action: #selector(add(_:)))                          ①
        self.navigationItem.rightBarButtonItem = addButtonItem               ②
        ......
    }


    func add(_ sender: AnyObject) {

        let modalViewController = ModalViewController()
        let navigationController = UINavigationController(rootViewController:
        ➥modalViewController)                                               ③

        self.present(navigationController, animated: true, completion: nil) ④
    }
    ......
}
```

```
//DetailViewController.m文件

@implementation DetailViewController

- (void)viewDidLoad {
    [super viewDidLoad];

    UIBarButtonItem* addButtonItem = [[UIBarButtonItem alloc]
    ➥initWithBarButtonSystemItem:UIBarButtonSystemItemAdd
    ➥target:self action:@selector(add:)];                                   ①
    self.navigationItem.rightBarButtonItem = addButtonItem;                  ②
    ......
}

- (void)add:(id)sender {

    ModalViewController* modalViewController = [[ModalViewController alloc]
    ➥init];
    UINavigationController* navigationController = [[UINavigationController
    ➥alloc] initWithRootViewController:modalViewController];                ③

    [self presentViewController:navigationController animated: TRUE
    ➥completion: nil];                                                      ④
}
......

@end
```

上述代码中，第①行用于创建UIBarButtonItem类型的Add按钮对象，第②行是导航栏的右按钮。

点击Add按钮时，将调用add:方法。在该方法中，第③行用于创建UINavigationController对象。UINavigationController对象的根视图控制器是modalViewController对象。

第④行通过presentViewController:animated:completion方法（Swift版是present(_:animated:completion:)）呈现UINavigationController对象所管理的视图，参数presentViewController是UINavigationController对象，而不是modalViewController对象。这是因为我们需要将modalViewController对象嵌入到导航控制器中。

模态视图控制器ModalViewController的代码如下：

```
//ModalViewController.swift文件

class ModalViewController: UIViewController {

    override func viewDidLoad() {
        super.viewDidLoad()

        self.view.backgroundColor = UIColor.white

        let saveButtonItem = UIBarButtonItem(barButtonSystemItem: .save,
        ➥target: self, action: #selector(save(_:)))
        self.navigationItem.rightBarButtonItem = saveButtonItem

        let cancelButtonItem = UIBarButtonItem(barButtonSystemItem: .cancel,
        ➥target: self, action: #selector(cancel(_:)))
        self.navigationItem.leftBarButtonItem = cancelButtonItem

        let screen = UIScreen.main.bounds

        let textViewWidth:CGFloat = 320
        let textViewHeight: CGFloat = 200
        let textViewTopView: CGFloat = 100
        let textView = UITextView(frame: CGRect(x: (screen.size.width -
```

```
//ModalViewController.m文件
@implementation ModalViewController

- (void)viewDidLoad {
    [super viewDidLoad];

    self.view.backgroundColor = [UIColor whiteColor];

    UIBarButtonItem* saveButtonItem = [[UIBarButtonItem alloc]
    ➥initWithBarButtonSystemItem:UIBarButtonSystemItemSave
    ➥target:self action:@selector(save:)];
    self.navigationItem.rightBarButtonItem = saveButtonItem;

    UIBarButtonItem* cancelButtonItem = [[UIBarButtonItem alloc]
    ➥initWithBarButtonSystemItem:UIBarButtonSystemItemCancel
    ➥target:self action:@selector(cancel:)];
    self.navigationItem.leftBarButtonItem = cancelButtonItem;

    CGRect screen = [[UIScreen mainScreen] bounds];

    CGFloat textViewWidth = 320;
    CGFloat textViewHeight = 200;
    CGFloat textViewTopView = 100;
```

```swift
        ➥textViewWidth)/2 ,
        ➥y: textViewTopView, width: textViewWidth, height: textViewHeight))
        ......
        self.view.addSubview(textView)
    }

    ......
    func cancel(_ sender: AnyObject) {
        self.dismiss(animated: true, completion: { () -> Void in
            print("关闭模态视图")
        })
    }
}
```

```objc
    UITextView* textView = [[UITextView alloc] initWithFrame:CGRectMake
    ➥((screen.size.width - textViewWidth)/2 , textViewTopView,
    ➥textViewWidth, textViewHeight)];
    ......

    [self.view addSubview:textView];
}

......

- (void)cancel:(id)sender {
    [self dismissViewControllerAnimated:TRUE completion:^{
        NSLog(@"关闭模态视图");
    }];
}

@end
```

ModalViewController的viewDidLoad方法创建两个UIBarButtonItem按钮对象，这两个按钮对象被添加到当前导航栏中，self.navigationItem可以获得当前导航栏对象。

此外，我们还创建了UITextView对象，并将其添加到当前视图中。

9.7　小结

通过本章的学习，我们已经可以判断应用是不是需要一个导航功能，并且知道在什么情况下选择平铺导航、标签导航、树形结构导航，或者同时综合使用这3种导航模式。针对标签导航和树形导航这两种相对复杂的导航模式，本章为大家提供了Interface Builder和代码实现两种实现方式。

第 10 章

iPad应用开发

10

iPhone应用完全可以在iPad设备上运行，这是苹果要求的。但是简单地将iPhone应用拿过来直接在iPad上运行，这种应用并不能称为iPad应用，也会被苹果禁止在App Store上发布。我们必须针对iPad设备的特点和应用场景开发iPad应用。在这一章中，我们来介绍iPad应用开发。

10.1 iPad 与 iPhone 应用开发的差异

iPad和iPhone应用在开发时需要注意：它们的屏幕尺寸不同、应用场景不同、导航模式不同和API不同。有关iPad和iPhone屏幕尺寸的不同，我们在8.1节已经介绍过，这里不再赘述。

10.1.1 应用场景不同

作为iOS开发者，我们应该熟悉iPhone和iPad应用的场景，然后才能开发出好的应用。iPhone是让用户一只手使用的设备，因此它适合在等车时拿出来看看天气、收发邮件、看看周围有哪些银行或者饭店，等等。而iPad是两只手使用的设备，它不太适合处理iPhone用户的场景。据调查，iPad多数用在家里，用来浏览网页、收发电子邮件、看照片、看视频、听音乐、玩电子游戏和看电子书等。作为平板电脑，它比笔记本电脑更轻便、更适合移动使用。

基于应用场景的不同，同样一款应用在iPhone和iPad上的功能选取和界面布局有着明显的不同。有些应用只能做成iPhone版本的，有些应用只能做成iPad版本的。与iPhone用户相比，iPad用户更期待具有高保真的、艺术品般的、高品质的应用，而绝非简单地放大iPhone应用的尺寸。

10.1.2 导航模式不同

在上一章中，我们介绍了iPhone的3种导航模式。在iPad中，平铺导航模式和标签导航模式与iPhone的基本一样，但树形结构导航模式与iPhone的差别比较大。

此外，两种设备上的模态视图导航也不同。iPhone呈现模态视图时，默认情况下会从屏幕下方滑出，占有整个屏幕，而iPad呈现模态视图时可以有多种样式供选择。

10.1.3 API 不同

iPhone和iPad都使用一个操作系统——iOS，因此，它们的API基本上是一样的，但有一些是iPad专用的。在iOS 8之前，`UISplitViewController`控制器是iPad专用的，`UISplitViewController`控制器用于将屏幕分栏。虽然在iOS 8及其之后的版本中，`UISplitViewController`也可以在iPhone中使用了，但是很多情况下还是应用于iPad场景，或是同时兼顾iPhone和iPad的应用中。`UIPopoverPresentationController`控制器用于呈现“浮动”类型的视图。此外，苹果还提供了一些iPad专用属性，以适应iPad屏幕的不同。随着学习的深入，我们会详细介绍。

10.2 iPad树形结构导航

iPad树形结构导航无论从风格上还是从具体实现上，都与iPhone有很大差别，本节就来介绍一下iPad树形结构导航的具体实现。

10.2.1 “邮件”应用中的树形结构导航

要了解iPad的树形结构导航模式，最好的示例就是iPad自带的“邮件”应用了，下面就来研究一下这个应用。

图10-1所示的是iPhone的“邮件”应用竖屏界面，导航采用了树形结构导航模式，新邮件编辑采用模态视图导航模式。

注意 iPhone Plus设备中“邮件”应用可以横屏显示，而且采用分栏显示方式，类似于iPad设备中的“邮件”应用。

图10-1 iPhone的“邮件”应用界面

iPad横屏时，“邮件”应用如图10-2所示。对比可以发现，iPhone版分成两个屏幕，而iPad版采用一个屏幕分成左右两栏，左栏是用于导航的菜单，iPad占用固定的320点，右栏是详细内容。iPad竖屏时，“邮件”应用如图10-3所示，默认只显示详细内容。左边的导航栏是隐藏的，需要时点击左上角的“收件箱”按钮，它会以Popover（浮动层）方式显示出来。

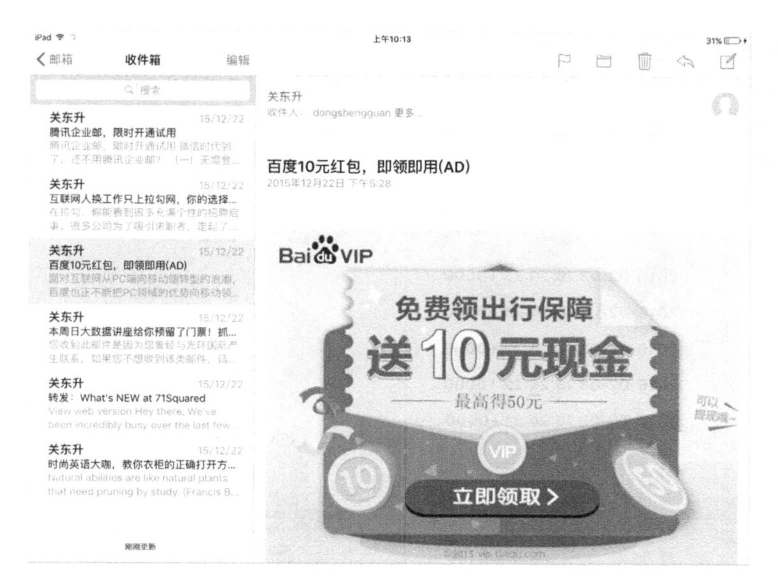

图10-2 横屏时iPad“邮件”应用界面

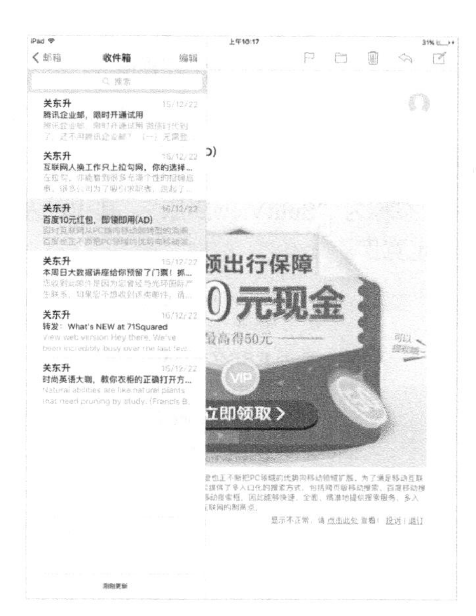

图10-3 竖屏时iPad的邮件应用界面

10.2.2 Master-Detail应用程序模板

在Xcode中，有一个Master-Detail Application模板，它是一个非常重要也比较复杂的模板，可以帮助我们构建树形结构导航模式的应用程序。

提示 这种树形结构导航模式与9.5节介绍的树形结构导航不同，它采用UI自适应设计，能够同时兼顾iPhone和iPad设备。它所采用的根视图控制器不是UINavigationController，而是UISplitViewController。

1. Master-Detail的UI自适应设计

图10-4所示的是iPhone下面Master-Detail应用程序的运行情况，其中图10-4a是主视图（MasterView），也就是`UISplitViewController`的Primary视图。点击主视图中的单元格，可以进入到如图10-4b所示的次视图（DetailView），也就是`UISplitViewController`的Secondary视图。

（a）主视图　　　　（b）次视图

图10-4　在iPhone上运行Master-Detail应用程序

在iPad横屏的情况下运行Master-Detail应用，由于屏幕比较宽（如图10-5所示），将采用分栏显示，分为左右两个视图，这种状态称为“SplitView膨胀”。其中，左侧是主视图，这是一个导航表视图，右侧是次视图。主视图占有320点的固定宽度。

图10-5　在iPad上运行Master-Detail应用程序（横屏）

图10-6所示的是iPad竖屏时的情况，此时只能看到次视图。如图10-6a所示，主视图会隐藏起来，当点击导航栏左边的Master按钮时，主视图从侧面滑出，如图10-6b所示。这种状态称为“SplitView收缩”。

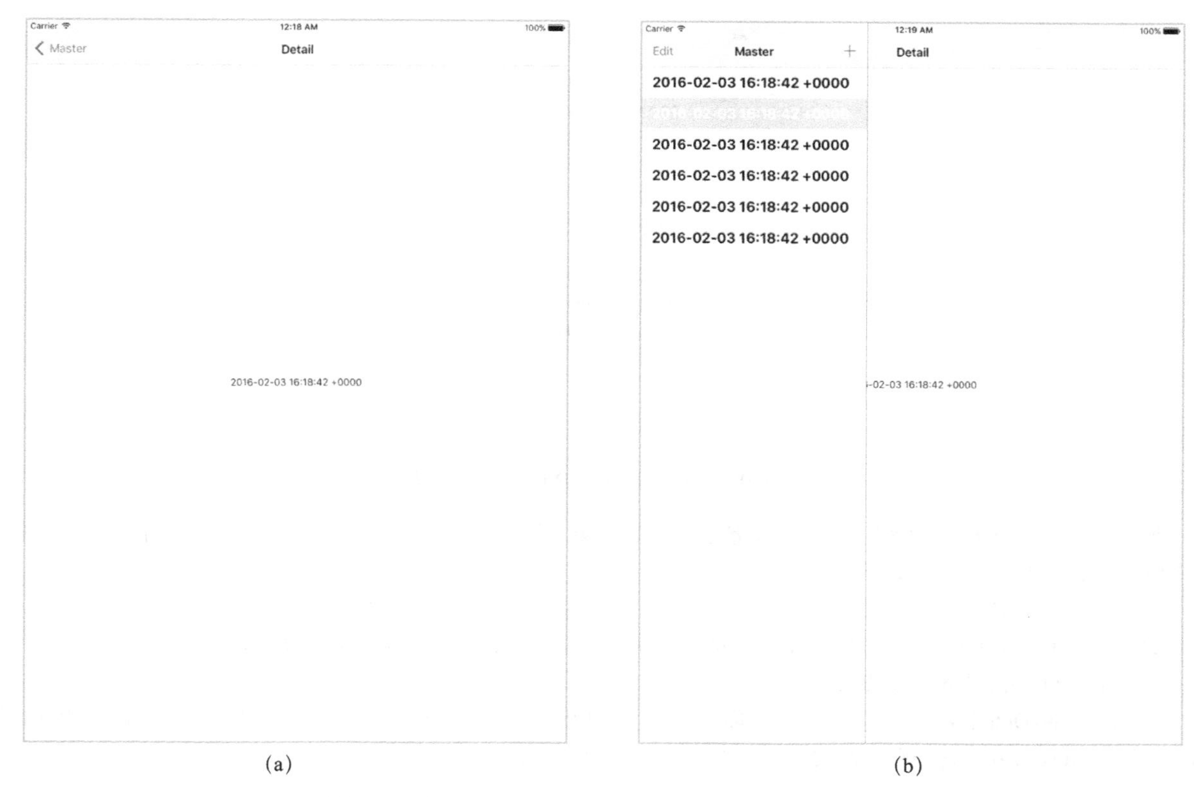

(a)　(b)

图10-6　在iPad上运行Master-Detail应用程序（竖屏）

如果Master-Detail应用程序在iPhone Plus设备上运行时，情况会比较特殊，竖屏时与在其他iPhone设备上运行没有差别，如图10-4所示。而横屏情况如图10-7所示，默认情况下与iPad横屏类似。如图10-7a所示，Master-Detail采用分栏显示，主视图占有295点的固定宽度，当点击次视图导航栏中的displayModeButtonItem按钮时，主视图会隐藏，如图10-7b所示，在此界面点击导航栏中的Master按钮将回到图10-7a所示的分栏界面。

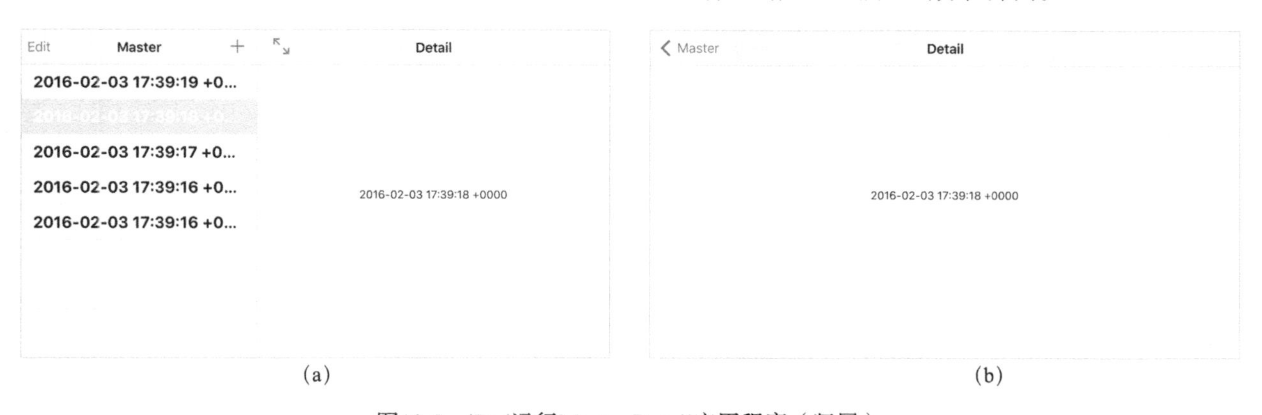

(a)　(b)

图10-7　iPad运行Master-Detail应用程序（竖屏）

displayModeButtonItem按钮根据它的作用有两种形式，如图10-8所示，其中⤡形式用来使次视图全屏，Master形式是“返回”按钮的作用。

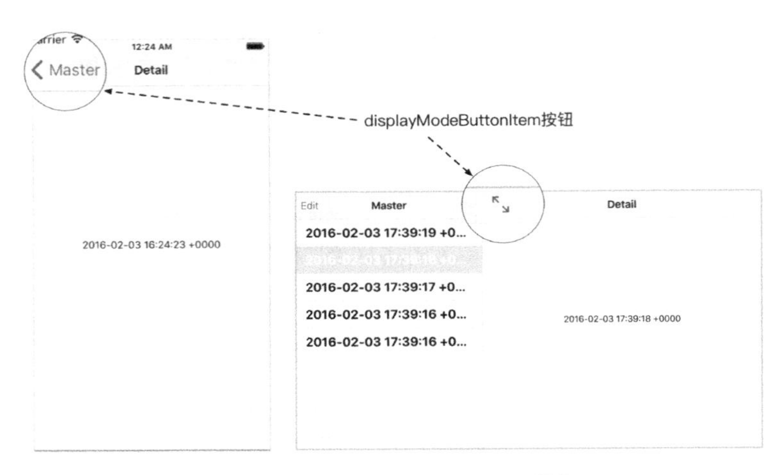

图10-8　displayModeButtonItem按钮

为什么Master-Detail应用程序在不同设备上运行的效果不同呢？这是因为采用了UI自适应设计，UI自适应要参考Size Class的取值，规则如下：

- 当Size Class为`wCompact`（宽度是"紧凑"的）时，SplitView收缩，见图10-4和图10-6；
- 当Size Class为`wRegular`（宽度是"标准"的）时，SplitView膨胀，见图10-5和图10-7。

2. Master-Detail的故事板设计

使用Master-Detail模板创建的应用界面是通过故事板实现的，如图10-9所示。故事板根视图控制器是`SplitViewController`，它分别控制两个导航控制器，这两个导航控制器分别管理主视图和次视图。

提示　从技术层面上看，`SplitViewController`控制的两个视图控制器并不一定是导航控制器，而可以是任何类型的视图控制器。但是使用导航控制器可以带来很多好处，例如导航控制器可以增加导航栏，有了导航栏，就可以在导航栏中添加左右按钮以及标题等。另外，导航控制器还可通过一个栈管理多个视图控制器，在主视图或从视图内进行导航。

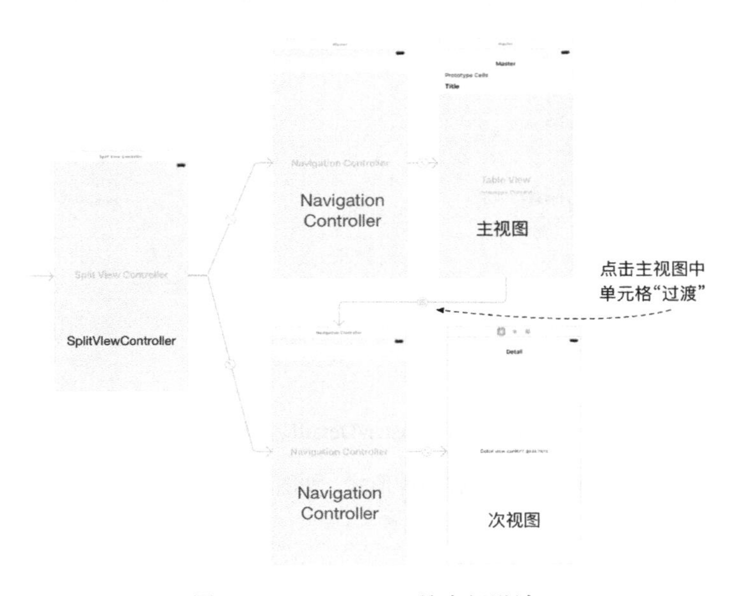

图10-9　Master-Detail故事板设计

3. Master-Detail的代码部分

Master-Detail模板所生成的代码核心部分是UISplitViewController和UISplitViewControllerDelegate。

在应用程序委托对象AppDelegate中设置UISplitViewController对象，并实现了UISplitViewController-Delegate委托协议，具体代码如下：

```
//AppDelegate.swift文件
import UIKit

@UIApplicationMain
class AppDelegate: UIResponder, UIApplicationDelegate,
➥UISplitViewControllerDelegate {                                          ①

    var window: UIWindow?

    func application(_ application: UIApplication,
    ➥didFinishLaunchingWithOptions launchOptions:
    ➥[UIApplicationLaunchOptionsKey: Any]?) -> Bool {

        let splitViewController = self.window!.rootViewController as!
        ➥UISplitViewController                                            ②
        let navigationController = splitViewController.
        ➥viewControllers[splitViewController.viewControllers.count-1] as!
        ➥UINavigationController                                           ③
        navigationController.topViewController!.navigationItem.
        ➥leftBarButton Item = splitViewController.displayModeButtonItem  ④
        splitViewController.delegate = self                               ⑤

        return true
    }
    ......
    //MARK: -- Split view
    func splitViewController(_ splitViewController: UISplitViewController,
    ➥collapseSecondary secondaryViewController:UIViewController, onto
    ➥primaryViewController:UIViewController) -> Bool {                   ⑥
        guard let secondaryAsNavController = secondaryViewController as?
        ➥UINavigationController else { return false }
        guard let topAsDetailController = secondaryAsNavController.topView
        ➥Controller as?
        ➥DetailViewController else { return false }
        if topAsDetailController.detailItem == nil {
            return true
        }
        return false
    }
}
```

```
//AppDelegate.m文件
#import "AppDelegate.h"
#import "DetailViewController.h"
@interface AppDelegate () <UISplitViewControllerDelegate>                 ①
@end

@implementation AppDelegate

- (BOOL)application:(UIApplication *)application
➥didFinishLaunchingWithOptions:(NSDictionary *)launchOptions {

    UISplitViewController *splitViewController =
    ➥(UISplitViewController *)self.window.rootViewController;            ②
    UINavigationController *navigationController =
    ➥[splitViewController.viewControllers lastObject];                   ③
    navigationController.topViewController.navigationItem.leftBarButtonItem
    ➥= splitViewController.displayModeButtonItem;                        ④
    splitViewController.delegate = self;                                  ⑤
    return TRUE;
}
......

#pragma mark -- Split view

- (BOOL)splitViewController:(UISplitViewController *)splitViewController
    ➥collapseSecondaryViewController:(UIViewController *)
    ➥secondaryViewController ontoPrimaryViewController:
    ➥(UIViewController *)primaryViewController {                         ⑥

    if ([secondaryViewController isKindOfClass:[UINavigationController class]]
    ➥&& [[(UINavigationController *)secondaryViewController
    ➥topViewController] isKindOfClass:[DetailViewController class]]
    ➥&& ([(DetailViewController *)[(UINavigationController *)
    ➥secondaryViewController topViewController] detailItem] == nil)) {
        return TRUE;
    } else {
        return FALSE;
    }
}

@end
```

上述代码中，第①行用于声明实现UISplitViewControllerDelegate委托协议。

第②行~第④行看起来很复杂，其中只是将UISplitViewController的displayModeButtonItem按钮放到次视图导航栏左侧。要分析清楚这些代码，就要熟悉UISplitViewController等控制器的层次关系，如图10-10所示。

从图10-10中可见，通过window的rootViewController属性可获得UISplitViewController对象，见第②行。UISplitViewController对象有一个viewControllers数组，其中保存了主视图控制器和次视图控制器，这两个控制器都是导航控制器（UINavigationController），通过第③行从viewControllers数组中取出最后一个元素，即次视图控制器。次视图控制器是导航控制器，它维护一个栈，当前视图是由处于栈顶的视图控制器所控制的，通过它的topViewController属性获得栈顶视图控制器，所以第④行中的navigationController.topViewController用于取得DetailViewController对象，它的navigationItem.leftBarButtonItem属性用于获得它所在导航栏的左按钮。

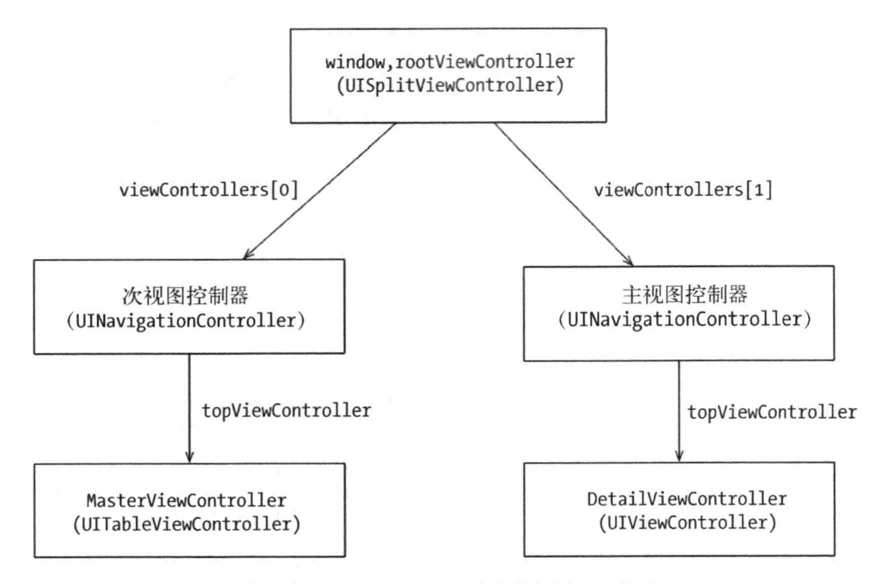

图10-10　Master-Detail控制器的层次关系

第⑤行用于将self赋值给splitViewController.delegate属性。

第⑥行是委托协议UISplitViewControllerDelegate要求实现的方法，该方法是在SplitView收缩合并次视图时调用，如果返回true（Objective-C中是YES或TRUE），则需要自己合并次视图，并移除次视图控制器；如果返回false（Objective-C中是NO或FALSE），则使用默认方式由系统合并。如果不实现该方法，系统会返回false（Objective-C是NO或FALSE）。

10.2.3　使用 Interface Builder 实现 SplitViewSample 案例

下面我们通过一个案例SplitViewSample来熟悉一下UISplitViewController控制器。图10-11是横屏情况下的SplitView视图，其中显示主视图和次视图。主视图中有Blue View和Yellow View选择项目，当选择其中的单元格时，右边的次视图就会显示相应的蓝色或黄色视图。点击Tap按钮，会弹出警告框，提示为蓝色还是黄色视图。

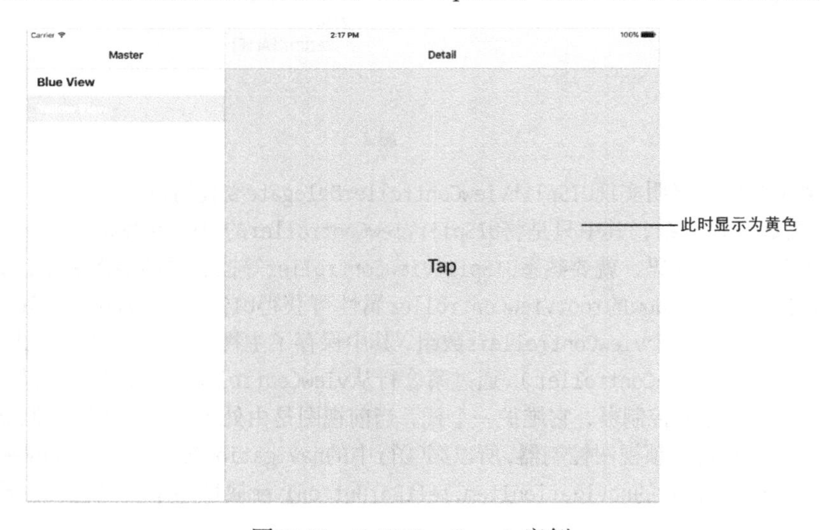

图10-11　SplitViewSample案例

本节中，我们首先介绍如何通过Interface Builder实现该案例，下一节再介绍如何通过代码实现该案例。

使用Xcode创建工程SplitViewSample，选择Master-Detail Application模板来创建工程，将Devices选择为iPad或Universal。

然后在工程中创建两个视图控制器UIViewController——BlueViewController和YellowViewController，它们的父类是UIViewController，创建过程中不要选中Also create XIB file复选框。

打开故事板文件，找到Detail视图控制器，删除Detail视图中的Detail view content goes here标签，如图10-12所示。

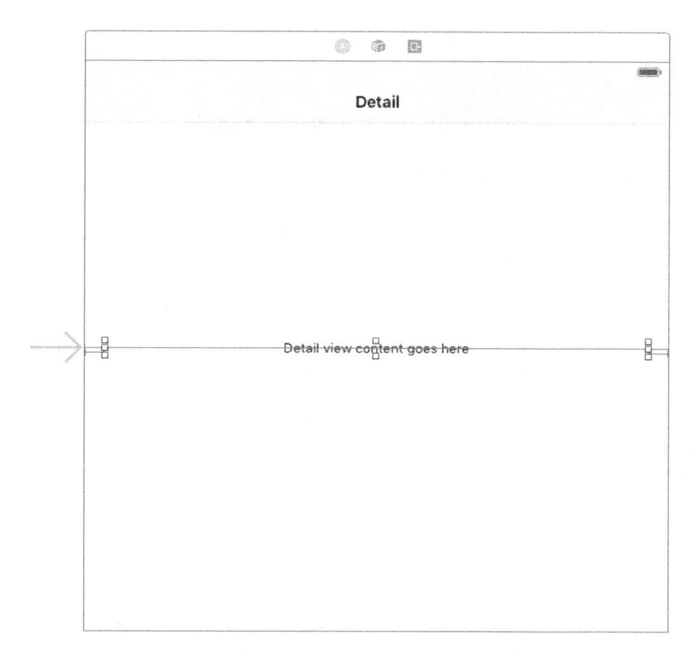

图10-12　Detail视图标签

然后从对象库中拖曳两个View Controller到设计界面，如图10-13所示，在每个视图中分别放置一个按钮。然后选择一个视图控制器中的View，将其背景改为蓝色；接着选择另外一个视图控制器中的View，将其背景改为黄色。

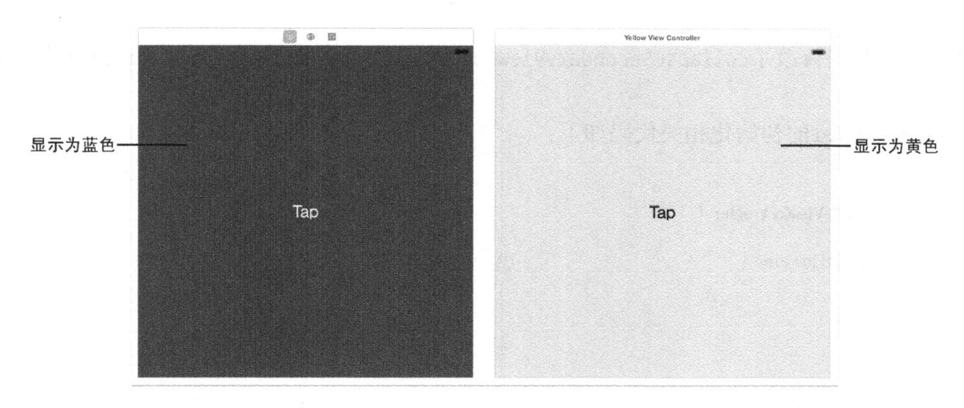

图10-13　视图控制器设计界面

首先，在故事板中选择蓝色视图控制器，打开其标识检查器（如图10-14所示），选择Custom Class→Class下拉列表中的BlueViewController类，修改Storyboard ID为blueViewController。然后我们为按钮添加动作事件。接下来，选择黄色视图控制器，打开其标识检查器，选择Custom Class→Class下拉列表中的YellowViewController类，修改

Storyboard ID为yellowViewController。然后，我们为按钮添加动作事件。

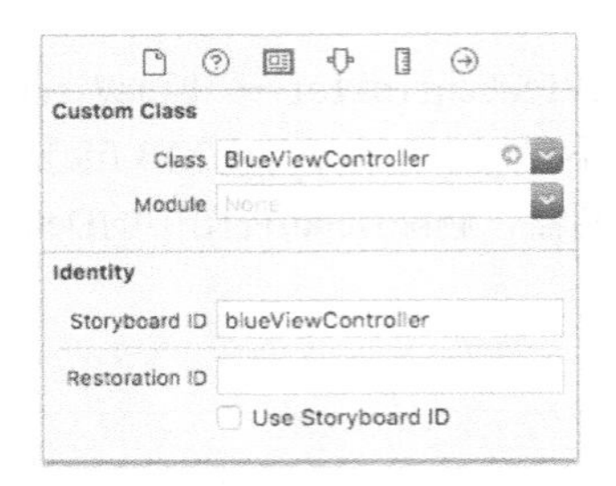

图10-14　设置视图控制器属性

下面我们看看代码部分，其中AppDelegate的代码如下：

```
//AppDelegate.swift文件
import UIKit

@UIApplicationMain
class AppDelegate: UIResponder, UIApplicationDelegate,
➥UISplitViewControllerDelegate {

    var window: UIWindow?

    func application(_ application: UIApplication,
    ➥didFinishLaunchingWithOptions launchOptions:
    ➥[UIApplicationLaunchOptionsKey: Any]?) -> Bool {
        ......
        return true
    }
    ......
    //MARK: -- Split view
    func splitViewController(_ splitViewController: UISplitViewController,
    ➥collapseSecondary secondaryViewController:UIViewController,
    ➥onto primaryViewController:UIViewController) -> Bool {
        return true
    }
}
```

```
//AppDelegate.m文件
#import "AppDelegate.h"
#import "DetailViewController.h"
@interface AppDelegate () <UISplitViewControllerDelegate>
@end

@implementation AppDelegate

- (BOOL)application:(UIApplication *)application
➥didFinishLaunchingWithOptions:(NSDictionary *)launchOptions {
    ......
    return TRUE;
}
......

#pragma mark -- Split view

- (BOOL)splitViewController:(UISplitViewController *)splitViewController
➥collapseSecondaryViewController:(UIViewController *)secondaryViewController
➥ontoPrimaryViewController:(UIViewController *)
➥primaryViewController{

    return TRUE;
}

@end
```

上述代码基本上只是修改了collapseSecondaryViewController:forSplitViewController:方法的返回值为true（Objective-C版为TRUE）。

MasterViewController的初始化相关代码如下：

```
//MasterViewController.swift文件
class MasterViewController: UITableViewController {

    var objects = ["Blue View", "Yellow View"]                ①

    override func viewDidLoad() {
        super.viewDidLoad()
    }

    override func viewWillAppear(_ animated: Bool) {
        self.clearsSelectionOnViewWillAppear =
        ➥self.splitViewController!.collapsed                   ②
        super.viewWillAppear(animated)
    }
    //MARK: -- Table View
    override func numberOfSections(in tableView: UITableView) -> Int {
```

```
//MasterViewController.m文件
#import "MasterViewController.h"
#import "DetailViewController.h"

@interface MasterViewController ()

@property NSArray *objects;                                    ①
@end

@implementation MasterViewController

- (void)viewDidLoad {
    [super viewDidLoad];
    self.objects = @[@"Blue View", @"Yellow View"];
}
```

```
        return 1
    }

    override func tableView(_ tableView: UITableView,
    ➥numberOfRowsInSection section: Int) -> Int {
        return objects.count
    }

    override func tableView(_ tableView: UITableView,
    ➥cellForRowAt indexPath: IndexPath) -> UITableViewCell {

        let object = objects[indexPath.row]
        cell.textLabel!.text = object
        return cell
    }
    ......

}
```

```
- (void)viewWillAppear:(BOOL)animated {
    self.clearsSelectionOnViewWillAppear =
    ➥self.splitViewController.isCollapsed;                    ②
    [super viewWillAppear:animated];
}

......

#pragma mark -- Table View

- (NSInteger)numberOfSectionsInTableView:(UITableView *)tableView {
    return 1;
}

- (NSInteger)tableView:(UITableView *)tableView
➥numberOfRowsInSection:(NSInteger)section {
    return self.objects.count;
}

- (UITableViewCell *)tableView:(UITableView *)tableView
➥cellForRowAtIndexPath:(NSIndexPath *)indexPath {
    UITableViewCell *cell = [tableView
    ➥dequeueReusableCellWithIdentifier:@"Cell" forIndexPath:indexPath];

    NSString *object = self.objects[indexPath.row];
    cell.textLabel.text = object;
    return cell;
}

@end
```

上述代码中，第①行用于声明objects集合属性。在Swift版中，objects属性的初始化是在声明时进行的，而Objective-C版中objects属性的初始化是在viewDidLoad方法中进行的。

第②行用于当SplitView处于收缩状态，而主视图显示时，清除表视图单元格的选中状态。其中表达式self.clearsSelectionOnViewWillAppear用于清除表视图单元格的选中状态，self.splitViewController.isCollapsed表达式用于设置获得SplitView收缩状态。

当点击主视图单元格时会触发过渡， MasterViewController中的相关代码如下：

```
override func prepare(for segue: UIStoryboardSegue, sender: Any?) {
    if segue.identifier == "showDetail" {
        if let indexPath = self.tableView.indexPathForSelectedRow {
            let controller = (segue.destination as!
            ➥UINavigationController)
            ➥.topViewController as! DetailViewController              ①

            controller.selectRow = indexPath.row                        ②
            controller.navigationItem.leftBarButtonItem =
            ➥self.splitViewController?.displayModeButtonItem

            controller.navigationItem.leftItemsSupplementBackButton =
            ➥true                                                      ③
        }
    }
}
```

```
- (void)prepareForSegue:(UIStoryboardSegue *)segue sender:(id)sender {
    if ([[segue identifier] isEqualToString:@"showDetail"]) {
        NSIndexPath *indexPath = [self.tableView indexPathForSelectedRow];
        DetailViewController *controller = (DetailViewController *)[[segue
        ➥destinationViewController] topViewController];                   ①
        controller.selectRow = indexPath.row;                               ②
        controller.navigationItem.leftBarButtonItem =
        ➥self.splitViewController.displayModeButtonItem;
        controller.navigationItem.leftItemsSupplementBackButton = TRUE;     ③
    }
}
```

上述代码中，第①行用于获得DetailViewController对象。注意，从segue的destinationViewController属性取出的视图控制器是导航控制器，因此我们还需要通过导航控制器的topViewController属性获取DetailViewController对象。

第②行用于将当前单元格行号（indexPath.row）赋值给DetailViewController对象的selectRow属性，在DetailViewController中需要使用selectRow属性判断显示哪个视图。

第③行用于设置导航栏左侧是否显示“返回”按钮。

DetailViewController的相关代码如下：

```
//DetailViewController.swift文件
import UIKit

class DetailViewController: UIViewController {
    //更新行号
    var selectRow = 0;                                                      ①

    override func viewDidLoad() {
        super.viewDidLoad()

        NSLog("选择的行号： %i", selectRow)


        if self.selectRow == 0 {
            //蓝色
            let blueViewController = self.storyboard!.
            ➥instantiateViewControllerWithIdentifier("blueViewController")
                                                                            ②
            self.addChildViewController(blueViewController)                 ③
            self.view.addSubview(blueViewController.view)                   ④
        } else {
            //黄色
            let yellowViewController = self.storyboard!.
            ➥instantiateViewControllerWithIdentifier(
            ➥"yellowViewController")
            self.addChildViewController(yellowViewController)
            self.view.addSubview(yellowViewController.view)
        }
    }
    ......
}
```

```
//DetailViewController.h文件
#import <UIKit/UIKit.h>

@interface DetailViewController : UIViewController

//更新行号
@property (nonatomic) NSInteger selectRow;                                  ①

@end

//DetailViewController.m文件
#import "DetailViewController.h"

@interface DetailViewController ()

@end

@implementation DetailViewController

- (void)viewDidLoad {
    [super viewDidLoad];

    NSLog(@"选择的行号： %li", (long)self.selectRow);

    if (self.selectRow == 0) {
        //蓝色
        UIViewController* blueViewController = [self.storyboard
        ➥instantiateViewControllerWithIdentifier:
        ➥@"blueViewController"];                                           ②
        [self addChildViewController:blueViewController];                   ③
        [self.view addSubview:blueViewController.view];                     ④
    } else {
        //黄色
        UIViewController* yellowViewController = [self.storyboard
        ➥instantiateViewControllerWithIdentifier:
        ➥@"yellowViewController"];
        [self addChildViewController:yellowViewController];
        [self.view addSubview:yellowViewController.view];
    }
}
......

@end
```

上述代码中，第①行声明了selectRow属性，这个属性用于保存从主视图控制器传递过来的行号。

第②行通过Storyboard ID分别创建蓝色视图控制器对象，第③行用于将蓝色视图控制器对象添加到当前视图控制器对象，第④行用于将黄色视图控制器中的视图添加到当前视图中。

提示 事实上，在iOS 5之前，第③行是可以省略的，只需要使用第④行的addSubview:方法来添加视图，但是iOS 5之后再这么做会有警告。

10.2.4　使用代码实现 SplitViewSample 案例

上一节中，我们在故事板中使用Interface Builder实现了SplitViewSample案例，这一节介绍如何通过代码实现SplitViewSample案例。

首先，请使用Single View Application模板创建一个名为SplitViewSample的工程，可以参考3.5.1节将SplitViewSample修改为纯代码工程。

下面看看代码部分，打开AppDelegate类，其代码如下：

```
//AppDelegate.swift文件
import UIKit

@UIApplicationMain
class AppDelegate: UIResponder, UIApplicationDelegate,
UISplitViewControllerDelegate {

    var window: UIWindow?

    func application(application: UIApplication,
    ➥didFinishLaunchingWithOptions launchOptions: [NSObject:
    ➥AnyObject]?) -> Bool {

        self.window = UIWindow(frame: UIScreen.main.bounds)
        self.window!.backgroundColor = UIColor.white
        self.window!.makeKeyAndVisible()

        let splitViewController = UISplitViewController()                ①
        splitViewController.delegate = self

        let masterViewController = MasterViewController()                ②
        let masterNavigationController = UINavigationController(
        ➥rootViewController: masterViewController)                      ③

        let detailViewController = DetailViewController()                ④
        let detailNavigationController = UINavigationController(
        ➥rootViewController: detailViewController)                      ⑤

        splitViewController.viewControllers =
        ➥[masterNavigationController, detailNavigationController]       ⑥
        detailViewController.navigationItem.leftBarButtonItem =
        ➥splitViewController.displayModeButtonItem()

        //创建的SplitView控制器，作为window的根视图控制器
        self.window!.rootViewController = splitViewController

        return true
    }

      ......

    func splitViewController(splitViewController: UISplitViewController,
    ➥collapseSecondaryViewController secondaryViewController:
    ➥UIViewController, ontoPrimaryViewController
    ➥primaryViewController:UIView Controller) -> Bool {

        return true
    }

}
```

```
//AppDelegate.m文件
......

#import "AppDelegate.h"

#import "MasterViewController.h"
#import "DetailViewController.h"

@interface AppDelegate () <UISplitViewControllerDelegate>

@end

@implementation AppDelegate


- (BOOL)application:(UIApplication *)application
➥didFinishLaunchingWithOptions:(NSDictionary *)launchOptions {

    self.window = [[UIWindow alloc] initWithFrame:[[UIScreen mainScreen]
    ➥bounds]];
    self.window.backgroundColor = [UIColor whiteColor];
    [self.window makeKeyAndVisible];

    UISplitViewController* splitViewController = [[UISplitViewController
    ➥alloc] init];                                                          ①
    splitViewController.delegate = self;

    MasterViewController *masterViewController = [[MasterViewController alloc]
    ➥init];                                                                 ②
    UINavigationController *masterNavigationController =
    ➥[[UINavigationController alloc]
    ➥initWithRootViewController:masterViewController];                      ③

    DetailViewController *detailViewController = [[DetailViewController alloc]
    ➥init];                                                                 ④
    UINavigationController *detailNavigationController = [[UINavigation
    ➥Controller alloc] initWithRootViewController:detailViewController];    ⑤

    splitViewController.viewControllers =
    ➥@[masterNavigationController, detailNavigationController];             ⑥
    detailViewController.navigationItem.leftBarButtonItem =
    ➥splitViewController.displayModeButtonItem;

    //创建的SplitView控制器，作为window的根视图控制器
    self.window.rootViewController = splitViewController;

    return TRUE;
}
......

- (BOOL)splitViewController:(UISplitViewController *)splitViewController
➥collapseSecondaryViewController:(UIViewController *)secondaryViewController
➥ontoPrimaryViewController:(UIViewController *)primaryViewController {
    return TRUE;
}

@end
```

上述代码中，第①行用于实例化UISplitViewController对象。第②行~第⑥行用于创建主视图控制器和次视图控制器，并将它们放到UISplitViewController对象中，这样会构建如图10-10所示的层次关系。

第③行用于创建主视图控制器masterNavigationController，它是导航控制器。导航控制器构造函数initWithRootViewController:中的参数是masterViewController，该参数是导航控制器管理的根视图控制器，是处于栈底的视图控制器。与之类似，第⑤行用于创建次视图控制器，它是导航控制器。

第⑥行用于将主视图控制器masterNavigationController和次视图控制器detailNavigationController放入到导航控制器的viewControllers属性中。注意它的前后顺序，第一个元素是主视图控制器，第二个元素是次视图控制器。

主视图控制器MasterViewController的主要代码如下：

```
//MasterViewController.swift文件
import UIKit

class MasterViewController: UITableViewController {

    <参考Interface Builder实现部分>
    //MARK: -- 表视图委托协议
    override func tableView(_ tableView: UITableView,
    ➥didSelectRowAt indexPath: IndexPath) {                                ①

        let controller = DetailViewController()
        controller.selectRow = indexPath.row
        controller.navigationItem.leftBarButtonItem =
        ➥self.splitViewController?.displayModeButtonItem
        controller.navigationItem.leftItemsSupplementBackButton = true
        let detailNavigationController = UINavigationController(
        ➥rootViewController: controller)

        self.showDetailViewController(detailNavigationController,
        ➥sender:self)                                                      ②
    }

}
```

```
//MasterViewController.m文件
@implementation MasterViewController

<参考Interface Builder实现部分>

#pragma mark -- 表视图委托协议

- (void)tableView:(UITableView *)tableView
➥didSelectRowAtIndexPath:(NSIndexPath *)indexPath {                          ①

    DetailViewController *controller = [[DetailViewController alloc] init];
    controller.selectRow = indexPath.row;
    controller.navigationItem.leftBarButtonItem =
    ➥self.splitViewController.displayModeButtonItem;
    controller.navigationItem.leftItemsSupplementBackButton = TRUE;
    UINavigationController *detailNavigationController =
    ➥[[UINavigationController alloc]
    ➥initWithRootViewController:controller];

    [self showDetailViewController:detailNavigationController
    ➥sender:self];                                                           ②

}

@end
```

上述代码中，第①行是实现表视图委托协议的方法tableView:didSelectRowAtIndexPath:，该方法在点击表视图单元格时触发，用来替换视图控制器的prepareForSegue:sender:方法（该方法只能在故事板中使用）。

第②行调用UISplitViewController的showDetailViewController:sender:方法呈现次视图。

次视图控制器DetailViewController的主要代码如下：

```
//DetailViewController.swift文件
class DetailViewController: UIViewController {
    //更新行号
    var selectRow = 0;

    override func viewDidLoad() {
        super.viewDidLoad()

        self.title = "Detail";                                              ①

        if self.selectRow == 0 {
            //蓝色
            let blueViewController = BlueViewController()                   ②
            self.addChildViewController(blueViewController)
            self.view.addSubview(blueViewController.view)
        } else {
            //黄色
            let yellowViewController = YellowViewController()               ③
            self.addChildViewController(yellowViewController)
            self.view.addSubview(yellowViewController.view)
        }
    }
    ......
}
```

```
//DetailViewController.m文件
@implementation DetailViewController

- (void)viewDidLoad {
    [super viewDidLoad];

    self.title = @"Detail";                                                   ①

    if (self.selectRow == 0) {
        //蓝色
        UIViewController* blueViewController = [[BlueViewController alloc]
        ➥init];                                                              ②
        [self addChildViewController:blueViewController];
        [self.view addSubview:blueViewController.view];
    } else {
        //黄色
        UIViewController* yellowViewController = [[YellowViewController alloc]
        ➥init];                                                              ③
        [self addChildViewController:yellowViewController];
        [self.view addSubview:yellowViewController.view];
    }
}
......

@end
```

上述代码中，第①行用于设置视图控制器的标题。第②行和第③行用于设置黄色和蓝色视图控制器，直接使用默认的构造函数。而在Interface Builder中，则使用了Storyboard ID创建视图控制器。

蓝色视图控制器BlueViewController和黄色视图控制器YellowViewController不再详细介绍。

10.3 iPad 模态视图

iPad模态视图与iPhone也有很大的差别，本节我们就来介绍一下iPad模态视图的具体实现。

10.3.1 “邮件”应用中的模态导航

图10-15所示的是iPhone中邮件的模态视图界面，默认情况下会从屏幕下方滑出，占满整个屏幕。

图10-15 iPhone模态视图

图10-16所示的是iPad和iPhone Plus的横屏模态视图界面，默认情况下会从屏幕下方滑出，显示在屏幕中间。图10-17所示的是iPad和iPhone Plus的竖屏模态视图界面，默认情况下会从屏幕下方滑出，占满整个屏幕。

图10-16 iPad和iPhone Plus的横屏模态视图

图10-17　iPad和iPhone Plus的竖屏模态视图

10.3.2　iPad 模态导航相关 API

UIViewController中有两个属性与模态视图相关：modalPresentationStyle和modalTransitionStyle。

- modalPresentationStyle属性用于设置呈现模态视图的样式，该属性是枚举类型UIModalPresentationStyle。UIModalPresentationStyle中的成员很多，与iPad模态视图有关的成员如表10-1所示。

表10-1　UIModalPresentationStyle枚举成员

Swift枚举成员	Objective-C枚举成员	说　明
fullScreen	UIModalPresentationFullScreen	全屏状态，是默认呈现样式
pageSheet	UIModalPresentationPageSheet	iPad横屏时覆盖屏幕一部分，未覆盖部分变暗，模态视图如图10-18所示。iPad竖屏时则全屏呈现，如图10-19所示
formSheet	UIModalPresentationFormSheet	无论是横屏（如图10-20所示）还是竖屏（如图10-21所示）情况下，模态视图的尺寸都是固定的，屏幕居中，呈现尺寸都不会变化
currentContext	UIModalPresentationCurrentContext	表示与父视图控制器有相同的呈现方式

- modalTransitionStyle属性用于设置呈现模态视图的动画效果，该属性是枚举类型UIModalTransitionStyle。UIModalTransitionStyle中的成员如表10-2所示。

表10-2　UIModalTransitionStyle枚举成员

Swift枚举成员	Objective-C枚举成员	说　明
coverVertical	UIModalTransitionStyleCoverVertical	默认动画效果，从屏幕底部滑出
flipHorizontal	UIModalTransitionStyleFlipHorizontal	水平翻转
crossDissolve	UIModalTransitionStyleCrossDissolve	交叉淡入淡出，当前视图淡出，模态视图淡入
partialCurl	UIModalTransitionStylePartialCurl	翻书效果，从右下角翻起

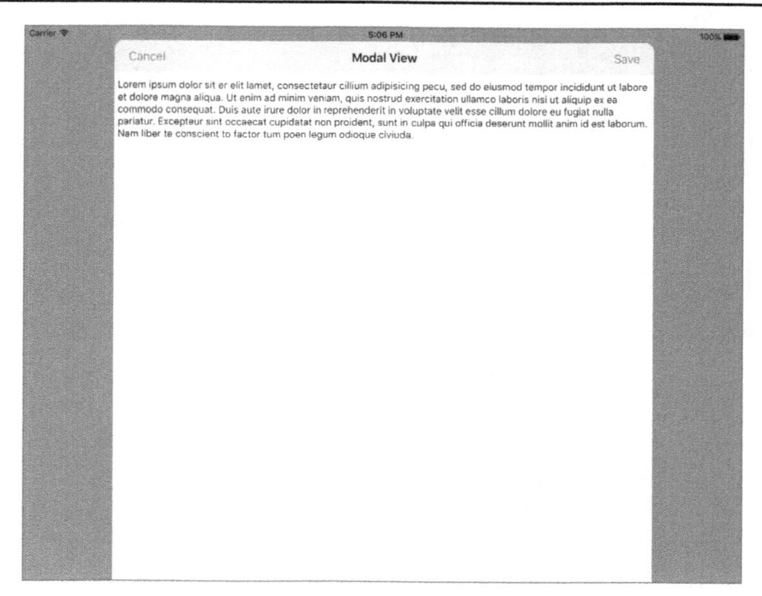

图10-18　PageSheet横屏呈现

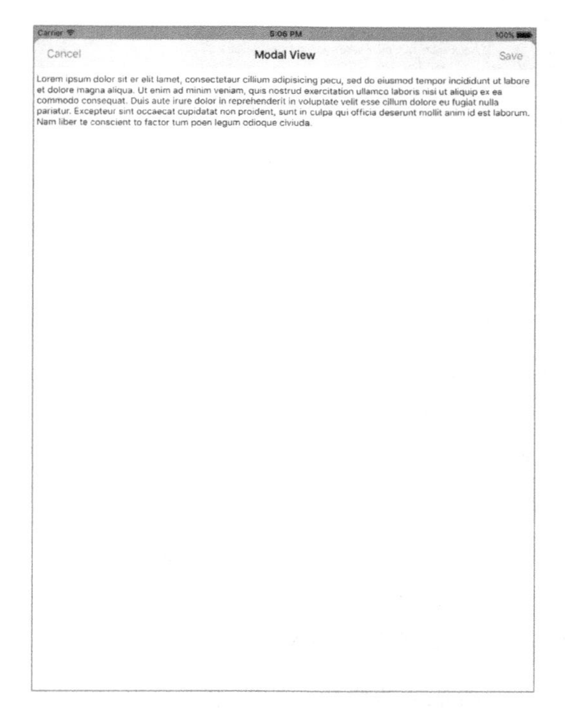

图10-19　PageSheet竖屏呈现

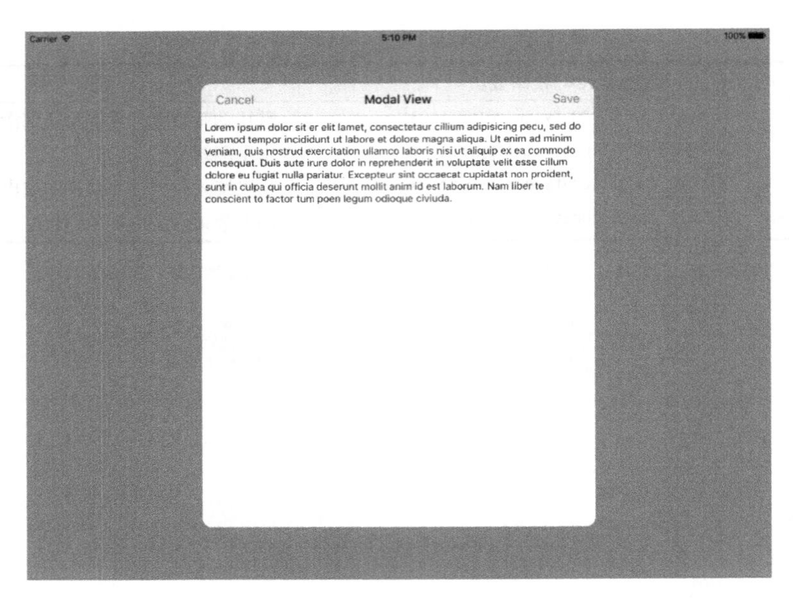

图10-20　FormSheet横屏呈现

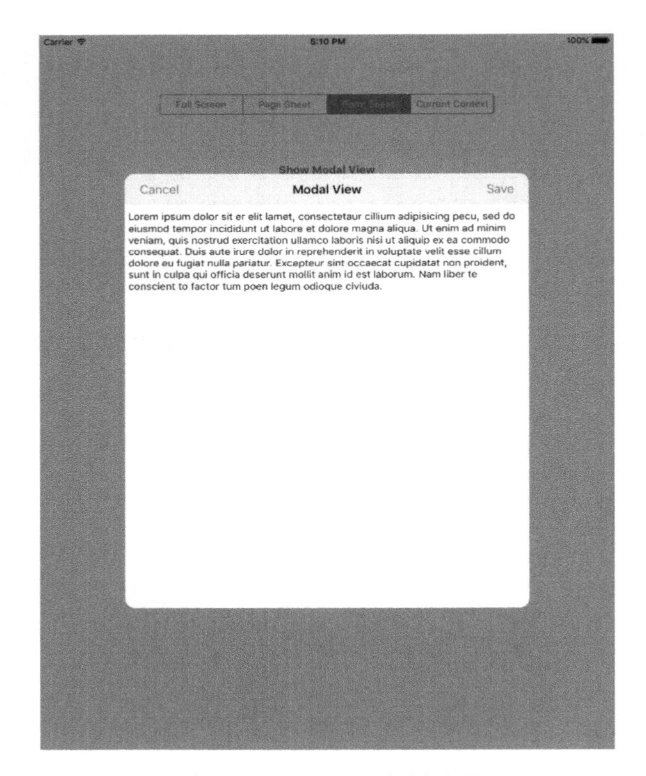

图10-21　FormSheet竖屏呈现

10.3.3　使用 Interface Builder 实现 ModalViewSample 案例

下面我们通过一个案例ModalViewSample来熟悉一下iPad模态导航。图10-22所示的是ModalViewSample案例，界面中有4个按钮，点击不同的按钮将呈现不同的模态视图。

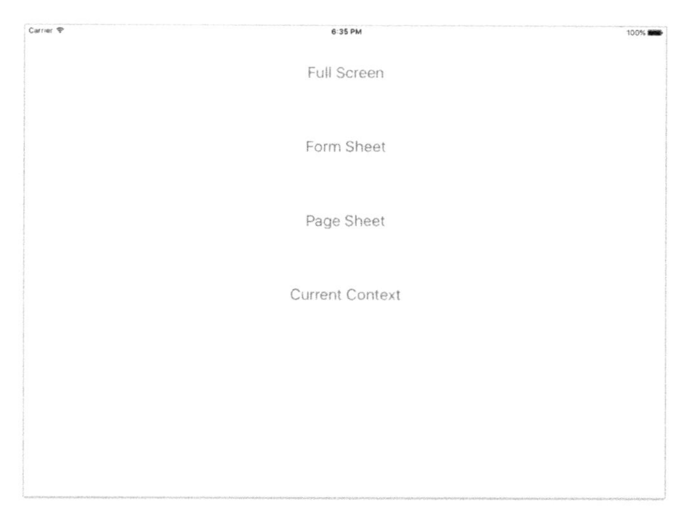

图10-22 ModalViewSample案例

本节中，我们首先介绍如何通过Interface Builder实现该案例，下一节再介绍如何通过代码实现该案例。

现在使用Xcode创建工程ModalViewSample，模板采用Single View Application，Devices选择iPad。然后打开故事板，从对象库中拖曳4个按钮控件到主视图设计界面，按图10-22所示摆放它们。然后，还要为它们添加约束，将按钮标题命名为Full Screen、Form Sheet、Page Sheet和Current Context。

提示 在图10-22所示的设计界面中，4个按钮从上到下垂直摆放，之间的间隔固定，我们可以使用垂直方向的堆视图UIStackView。关于堆视图的使用细节，读者可以参考7.6节。

我们还需要模态视图控制器，从对象库中拖曳View Controller到设计界面。由于模态视图需要一个导航栏，我们可以从对象库中拖曳一个Navigation Bar到模态视图设计界面顶部，也可以将当前模态视图控制器嵌入到一个导航控制器中，添加完成后的设计界面如图10-23所示。

然后在模态视图设计界面中，从对象库中拖曳TextView，并添加约束。接着为导航栏添加左右按钮，将左按钮设置为系统按钮Cancel，右按钮设置为系统按钮Save。最后的设计结果如图10-24所示。

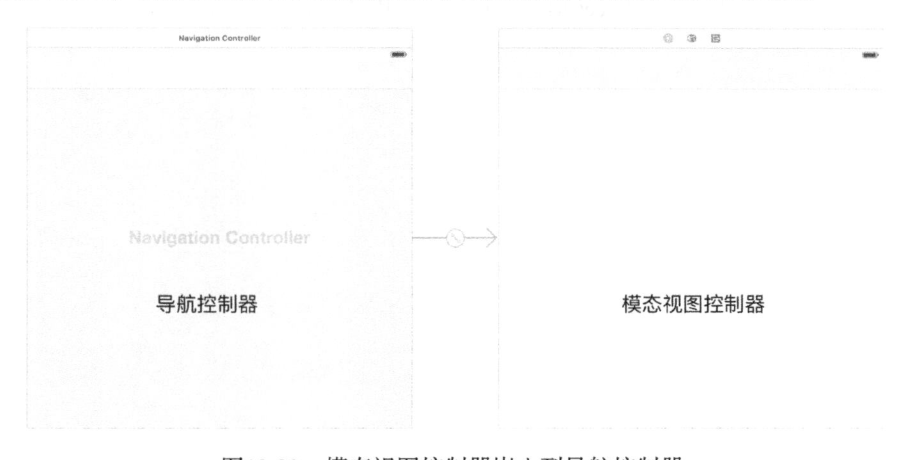

图10-23 模态视图控制器嵌入到导航控制器

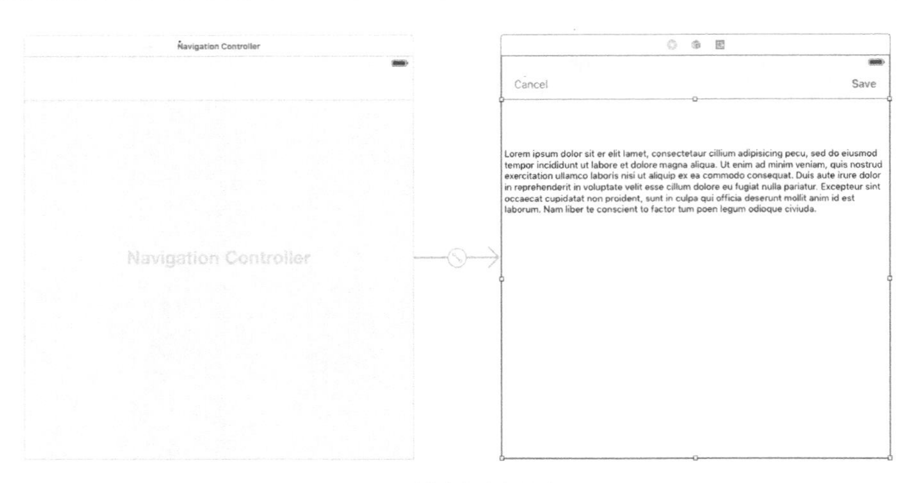

图10-24　模态视图设计界面

然后我们需要在工程中创建模态视图控制器类ModalViewController，选择它的父类为UIViewController。注意，创建过程中不要选择同时创建XIB文件。创建完成后回到故事板，选中模态视图控制器，然后再打开视图控制器的标识检查器（如图10-25所示），修改Class属性为ModalViewController。接着，为导航栏中的Cancel和Save按钮定义动作事件并连线。

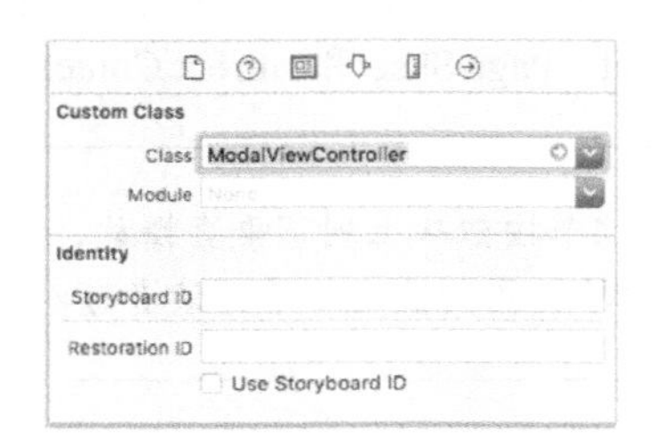

图10-25　ModalViewController标识检查器

下面为主视图中的按钮添加过渡呈现模态视图。如图10-26所示，选中Full Screen按钮，按住鼠标从Full Screen按钮拖曳到模态视图控制器，此时会弹出如图10-27所示的过渡菜单，请在菜单中选择Present Modally。然后选择过渡，打开其属性检查器（如图10-28所示），其中Presentation是呈现样式，Transition是呈现动画。要设置Full Screen按钮，你需要在Presentation中选择Full Screen项，在Transition中选择自己喜欢的动画。

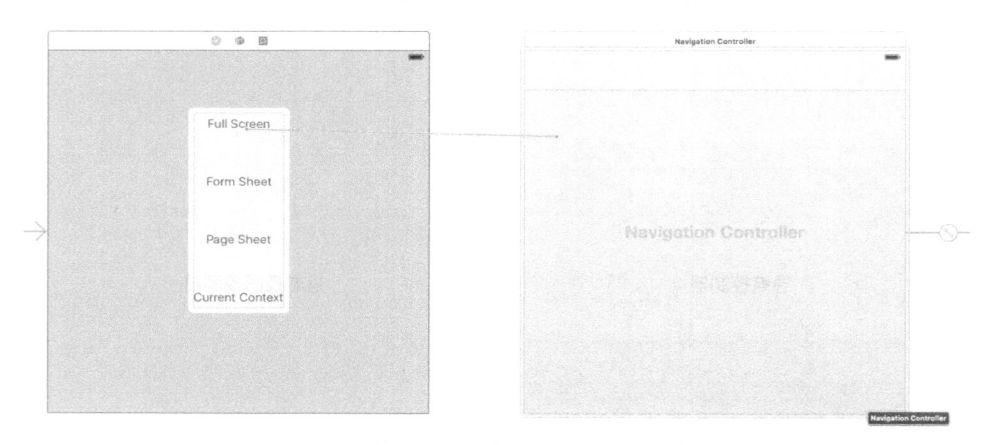

图10-26　通过故事板设计模态视图

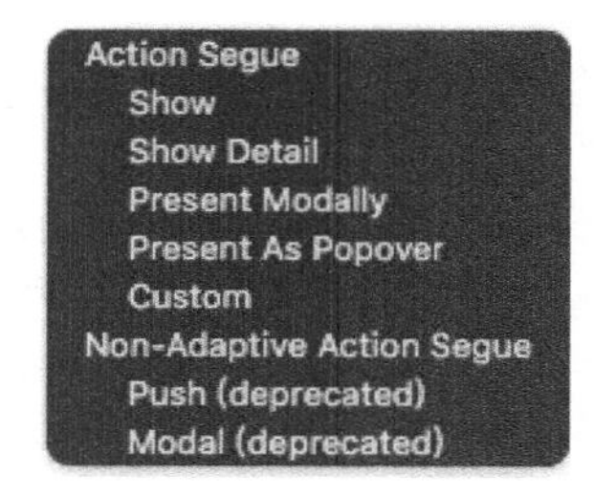

图10-27　过渡菜单

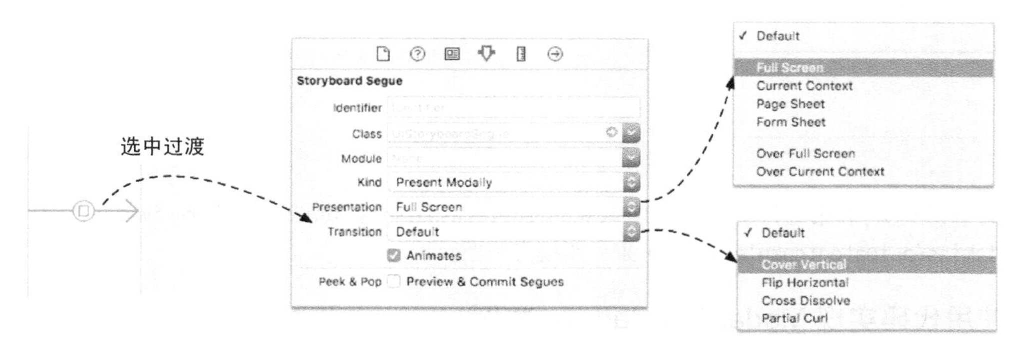

图10-28　过渡属性检查器

按照上述方法分别为其他3个按钮添加过渡，并选择合适的呈现样式和呈现动画。最后，设计完成的界面如图10-29所示。

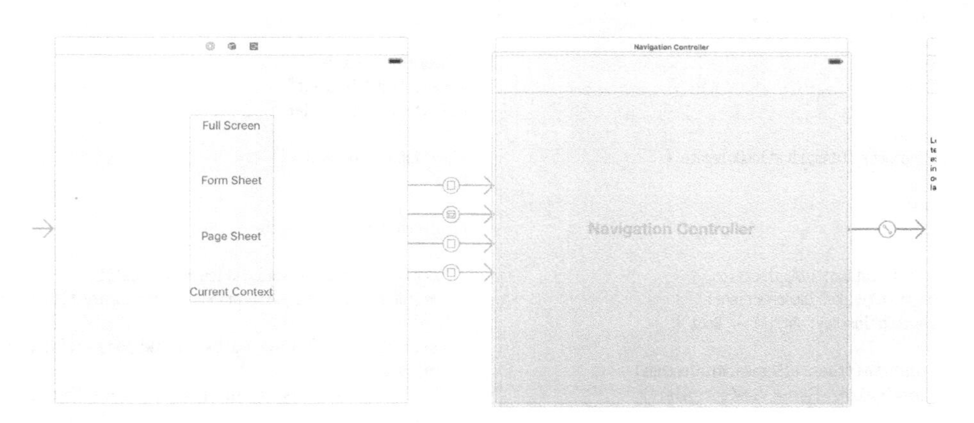

图10-29　模态视图设计完成

下面我们再看看代码部分。在主视图控制器ViewController中，我们并没有添加代码，只是在模态视图控制器ModalViewController中添加自己的代码。ModalViewController的代码如下：

```
//ModalViewController.swift文件
import UIKit

class ModalViewController: UIViewController {

    override func viewDidLoad() {
        super.viewDidLoad()
    }
```

```
//ModalViewController.m文件
#import "ModalViewController.h"

@interface ModalViewController ()

@end

@implementation ModalViewController
```

```swift
    override func didReceiveMemoryWarning() {
        super.didReceiveMemoryWarning()
    }

    @IBAction func save(_ sender: AnyObject) {
        self.dismiss(animated: true) { () -> Void in                    ①
            print("点击Save按钮, 关闭模态视图")
        }
    }

    @IBAction func cancel(_ sender: AnyObject) {
        self.dismiss(animated: true) { () -> Void in                    ②
            print("点击Cancel按钮, 关闭模态视图")
        }
    }
}
```

```objectivec
- (void)viewDidLoad {
    [super viewDidLoad];
}

- (void)didReceiveMemoryWarning {
    [super didReceiveMemoryWarning];
}
- (IBAction)save:(id)sender {
    [self dismissViewControllerAnimated:TRUE completion:^{              ①
        NSLog(@"点击Save按钮, 关闭模态视图");
    }];
}

- (IBAction)cancel:(id)sender {
    [self dismissViewControllerAnimated:TRUE completion:^{              ②
        NSLog(@"点击Cancel按钮, 关闭模态视图");
    }];
}

@end
```

上述代码中，第①行和第②行用于调用UIViewController的dismissViewControllerAnimated:completion:方法（Swift版是dismiss(animated:completion:)）来关闭模态视图。

10.3.4　使用代码实现 ModalViewSample 案例

上一节中，我们在故事板中实现了ModalViewSample案例，本节中，将介绍如何通过代码实现该案例。

首先，使用Single View Application模板创建一个名为ModalViewSample的工程，参考3.5.1节将ModalView-Sample修改为纯代码工程。

下面看看代码部分。打开AppDelegate类，其代码如下：

```swift
//AppDelegate.swift文件
import UIKit

@UIApplicationMain
class AppDelegate: UIResponder, UIApplicationDelegate {

    var window: UIWindow?


    func application(_ application: UIApplication,
    ➥didFinishLaunchingWithOptions launchOptions:
    ➥[UIApplicationLaunchOptionsKey: Any]?) -> Bool {

        self.window = UIWindow(frame: UIScreen.main.bounds)
        self.window?.rootViewController = ViewController()
        self.window?.backgroundColor = UIColor.white
        self.window?.makeKeyAndVisible()

        return true
    }
    ......
}
```

```objectivec
//AppDelegate.m文件
#import "AppDelegate.h"
#import "ViewController.h"

@interface AppDelegate ()
@end

@implementation AppDelegate

- (BOOL)application:(UIApplication *)application
    ➥didFinishLaunchingWithOptions:(NSDictionary *)launchOptions {

    self.window = [[UIWindow alloc] initWithFrame:[[UIScreen mainScreen]
    ➥bounds]];
    self.window.rootViewController = [[ViewController alloc] init];
    self.window.backgroundColor = [UIColor whiteColor];
    [self.window makeKeyAndVisible];

    return YES;
}
......
@end
```

ViewController的代码如下：

```swift
//ViewController.swift文件
import UIKit

class ViewController: UIViewController {

    override func viewDidLoad() {
```

```objectivec
//ViewController.m文件
#import "ViewController.h"
#import "ModalViewController.h"

@interface ViewController ()
```

```
    super.viewDidLoad()

    ///创建Full Screen按钮
    let buttonFullScreen = UIButton(type: .system)
    buttonFullScreen.setTitle("Full Screen", for: UIControlState())
    buttonFullScreen.titleLabel?.font = UIFont.systemFont(ofSize: 20)
    buttonFullScreen.addTarget(self, action: #selector(onclick(_:)),
    ➥for:.touchUpInside)
    //设置tag以区别其他按钮
    buttonFullScreen.tag = 100
    self.view.addSubview(buttonFullScreen)

    ///创建Form Sheet按钮
    let buttonFormSheet = UIButton(type: .system)
    buttonFormSheet.setTitle("Form Sheet", for: UIControlState())
    buttonFormSheet.titleLabel?.font = UIFont.systemFont(ofSize: 20)
    buttonFormSheet.addTarget(self, action: #selector(onclick(_:)),
    ➥for:.touchUpInside)
    //设置tag以区别其他按钮
    buttonFormSheet.tag = 200
    self.view.addSubview(buttonFormSheet)

    ///创建Page Sheet按钮
    let buttonPageSheet = UIButton(type: .system)
    buttonPageSheet.setTitle("Page Sheet", for: UIControlState())
    buttonPageSheet.titleLabel?.font = UIFont.systemFont(ofSize: 20)
    buttonPageSheet.addTarget(self, action: #selector(onclick(_:)),
    ➥for:.touchUpInside)
    //设置tag以区别其他按钮
    buttonPageSheet.tag = 300
    self.view.addSubview(buttonPageSheet)

    ///创建Current Context按钮
    let buttonCurrentContext = UIButton(type: .system)
    buttonCurrentContext.setTitle("Current Context", for:
    ➥UIControlState())
    buttonCurrentContext.titleLabel?.font = UIFont.systemFont(ofSize: 20)
    buttonCurrentContext.addTarget(self, action: #selector(onclick(_:)),
    ➥for:.touchUpInside)
    //设置tag以区别其他按钮
    buttonCurrentContext.tag = 400
    self.view.addSubview(buttonCurrentContext)

}

override func viewWillLayoutSubviews() {                         ①
    super.viewWillLayoutSubviews()

    NSLog("重新布局")

    let screen = UIScreen.main.bounds
    let buttonWidth: CGFloat = 200
    let buttonHeight: CGFloat = 20
    let buttonTopView: CGFloat = 115
    let buttonX: CGFloat = (screen.size.width - buttonWidth)/2

    ///创建Full Screen按钮
    let buttonFullScreen = self.view.viewWithTag(100)             ②
    buttonFullScreen!.frame = CGRect(x: buttonX, y: buttonTopView,
    ➥width: buttonWidth, height: buttonHeight)

    ///创建Form Sheet按钮
    let buttonFormSheet = self.view.viewWithTag(200)
    buttonFormSheet!.frame = CGRect(x: buttonX, y: buttonFullScreen!.
    ➥frame.origin.y + 100,
    ➥width: buttonWidth, height: buttonHeight)

    ///创建Page Sheet按钮
```

```
@end

@implementation ViewController

- (void)viewDidLoad {
    [super viewDidLoad];

    ///创建Full Screen按钮
    UIButton* buttonFullScreen = [UIButton buttonWithType:UIButtonTypeSystem];
    [buttonFullScreen setTitle:@"Full Screen" forState:UIControlStateNormal];
    buttonFullScreen.titleLabel.font = [UIFont systemFontOfSize:20];
    [buttonFullScreen addTarget:self action:@selector(onclick:)
    ➥forControlEvents:UIControlEventTouchUpInside];
    //设置tag以区别其他按钮
    buttonFullScreen.tag = 100;
    [self.view addSubview:buttonFullScreen];

    ///创建Form Sheet按钮
    UIButton* buttonFormSheet = [UIButton buttonWithType:UIButtonTypeSystem];
    [buttonFormSheet setTitle:@"Form Sheet" forState:UIControlStateNormal];
    buttonFormSheet.titleLabel.font = [UIFont systemFontOfSize:20];
    [buttonFormSheet addTarget:self action:@selector(onclick:)
    ➥forControlEvents:UIControlEventTouchUpInside];
    //设置tag以区别其他按钮
    buttonFormSheet.tag = 200;
    [self.view addSubview:buttonFormSheet];

    ///创建Page Sheet按钮
    UIButton* buttonPageSheet = [UIButton buttonWithType:UIButtonTypeSystem];
    [buttonPageSheet setTitle:@"Page Sheet" forState:UIControlStateNormal];
    buttonPageSheet.titleLabel.font = [UIFont systemFontOfSize:20];
    [buttonPageSheet addTarget:self action:@selector(onclick:)
    ➥forControlEvents:UIControlEventTouchUpInside];
    //设置tag以区别其他按钮
    buttonPageSheet.tag = 300;
    [self.view addSubview:buttonPageSheet];

    ///创建Current Context按钮
    UIButton* buttonCurrentContext = [UIButton buttonWithType:
    ➥UIButtonTypeSystem];
    [buttonCurrentContext setTitle:@"Current Context" forState:
    ➥UIControlStateNormal];
    buttonCurrentContext.titleLabel.font = [UIFont systemFontOfSize:20];
    [buttonCurrentContext addTarget:self action:@selector(onclick:)
    ➥forControlEvents:UIControlEventTouchUpInside];
    //设置tag以区别其他按钮
    buttonCurrentContext.tag = 400;
    [self.view addSubview:buttonCurrentContext];

}

-(void)viewWillLayoutSubviews {                                          ①
    [super viewWillLayoutSubviews];

    NSLog(@"重新布局");

    CGRect screen = [[UIScreen mainScreen] bounds];

    CGFloat buttonWidth = 200;
    CGFloat buttonHeight = 20;
    CGFloat buttonTopView = 115;
    CGFloat buttonX = (screen.size.width - buttonWidth)/2;

    ///创建Full Screen按钮
    UIButton* buttonFullScreen = [self.view viewWithTag:100];            ②
    buttonFullScreen.frame = CGRectMake(buttonX, buttonTopView, buttonWidth,
    ➥buttonHeight);
```

```swift
        let buttonPageSheet = self.view.viewWithTag(300)
        buttonPageSheet!.frame = CGRect(x: buttonX, y: buttonFormSheet!.frame.
        ➥origin.y + 100,
        ➥width: buttonWidth, height: buttonHeight)

        ///创建Current Context按钮
        let buttonCurrentContext = self.view.viewWithTag(400)
        buttonCurrentContext!.frame = CGRect(x: buttonX, y: buttonPageSheet!.
        ➥frame.origin.y + 100,
        ➥width: buttonWidth, height: buttonHeight)

    }

    override func didReceiveMemoryWarning() {
        super.didReceiveMemoryWarning()
    }

    func onclick(_ sender: AnyObject) {

        let modalViewController = ModalViewController()                    ③
        let navigationController = UINavigationController(rootViewController:
        ➥modalViewController)                                              ④
        navigationController.modalTransitionStyle = .coverVertical         ⑤

        let button = sender as! UIButton

        switch (button.tag) {
        case 100:
            navigationController.modalPresentationStyle = .fullScreen      ⑥
        case 200:
            navigationController.modalPresentationStyle = .formSheet
        case 300:
            navigationController.modalPresentationStyle = .pageSheet
        case 400:
            navigationController.modalPresentationStyle = .currentContext
        default :
            NSLog("默认分支")
        }

        self.present(navigationController, animated: true, completion: nil) ⑦
    }

}
```

```objectivec
    ///创建Form Sheet按钮
    UIButton* buttonFormSheet = [self.view viewWithTag:200];
    buttonFormSheet.frame = CGRectMake(buttonX,
    ➥buttonFullScreen.frame.origin.y + 100, buttonWidth, buttonHeight);

    ///创建Page Sheet按钮
    UIButton* buttonPageSheet = [self.view viewWithTag:300];
    buttonPageSheet.frame = CGRectMake(buttonX,
    ➥buttonFormSheet.frame.origin.y + 100, buttonWidth, buttonHeight);

    ///创建Current Context按钮
    UIButton* buttonCurrentContext = [self.view viewWithTag:400];
    buttonCurrentContext.frame = CGRectMake(buttonX,
    ➥buttonPageSheet.frame.origin.y + 100, buttonWidth, buttonHeight);
}

- (void)onclick:(id)sender {

    ModalViewController *modalViewController = [[ModalViewController alloc]
    ➥init];                                                                ③
    UINavigationController* navigationController = [[UINavigationController
    ➥alloc] initWithRootViewController:modalViewController];               ④
    navigationController.modalTransitionStyle =
    ➥UIModalTransitionStyleCoverVertical;                                  ⑤

    UIButton *button = (UIButton*)sender;

    switch (button.tag) {
        case 100:
            navigationController.modalPresentationStyle =
            ➥UIModalPresentationFullScreen;                                ⑥
            break;
        case 200:
            navigationController.modalPresentationStyle =
            ➥UIModalPresentationFormSheet;
            break;
        case 300:
            navigationController.modalPresentationStyle =
            ➥UIModalPresentationPageSheet;
            break;
        case 400:
            navigationController.modalPresentationStyle =
            ➥UIModalPresentationCurrentContext;
            break;
        default:
            NSLog(@"默认分支");
    }

    [self presentViewController:navigationController animated:TRUE
    ➥completion:nil];                                                      ⑦
}

@end
```

上述代码中，第①行重写viewWillLayoutSubviews方法，该方法在设备旋转时调用，我们在这个方法中重新设置界面中控件的位置。第②行中UIView的viewWithTag:方法通过tag获得视图对象。

第③行用于创建模态视图控制器ModalViewController，第④行用于将模态视图控制器对象放到导航控制器UINavigationController中，事实上模态视图控制器是UINavigationController。第⑤行设置模态视图呈现动画。第⑥行设置模态视图呈现样式。

第⑦行通过UIViewController的presentViewController:animated:completion:方法（Swift版是present:(_:animated:completion:)）弹出模态视图。

ModalViewController的代码如下：

```swift
import UIKit

class ModalViewController: UIViewController {

    override func viewDidLoad() {
        super.viewDidLoad()

        let saveButtonItem = UIBarButtonItem(barButtonSystemItem: .save,
        ➥target: self, action: #selector(save(_:)))                    ①
        let cancelButtonItem = UIBarButtonItem(barButtonSystemItem: .cancel,
        ➥target: self, action: #selector(cancel(_:)))                  ②

        self.navigationItem.rightBarButtonItem = saveButtonItem
        self.navigationItem.leftBarButtonItem = cancelButtonItem

        ///UITextView
        let textView = UITextView(frame: self.view.frame)
        textView.text = ...

        self.view.addSubview(textView)
    }

    override func didReceiveMemoryWarning() {
        super.didReceiveMemoryWarning()
    }

    func save(_ sender: AnyObject) {
        self.dismiss(animated: true) { () -> Void in
            print("点击Save按钮，关闭模态视图")
        }
    }

    func cancel(_ sender: AnyObject) {
        self.dismiss(animated: true) { () -> Void in
            print("点击Cancel按钮，关闭模态视图")
        }
    }
}
```

```objc
#import "ModalViewController.h"

@interface ModalViewController ()

@end

@implementation ModalViewController

- (void)viewDidLoad {
    [super viewDidLoad];

    UIBarButtonItem *saveButtonItem = [[UIBarButtonItem alloc]
    ➥initWithBarButtonSystemItem: UIBarButtonSystemItemSave
    ➥target:self action:@selector(save:)];                             ①

    UIBarButtonItem *cancelButtonItem = [[UIBarButtonItem alloc]
    ➥initWithBarButtonSystemItem: UIBarButtonSystemItemCancel
    ➥target:self action:@selector (cancel:)];                          ②

    self.navigationItem.rightBarButtonItem = saveButtonItem;
    self.navigationItem.leftBarButtonItem = cancelButtonItem;

    ///UITextView
    UITextView * textView = [[UITextView alloc] initWithFrame:self.view.frame];
    textView.text = ...;
    [self.view addSubview:textView];
}

- (IBAction)save:(id)sender {
    [self dismissViewControllerAnimated:TRUE completion:^{
        NSLog(@"点击Save按钮，关闭模态视图");
    }];
}

- (IBAction)cancel:(id)sender {
    [self dismissViewControllerAnimated:TRUE completion:^{
        NSLog(@"点击Cancel按钮，关闭模态视图");
    }];
}

@end
```

上述代码中，第①行和第②行通过代码创建导航栏左右按钮。

10.4 Popover视图

iPad应用中有一种Popover视图，这是一种临时视图，以“漂浮”的形式出现在视图表面，称为“浮动层”，如图10-30所示。触摸Popover视图的外边，将关闭视图。

由于Popover视图不会占满全屏，而且有一个气泡箭头指向其他视图或按钮，这些视图或按钮称为“锚点”。Popover视图中也常常包含一些控件，类似于表单，图10-31所示的是iPad中Safari浏览器的共享选项。

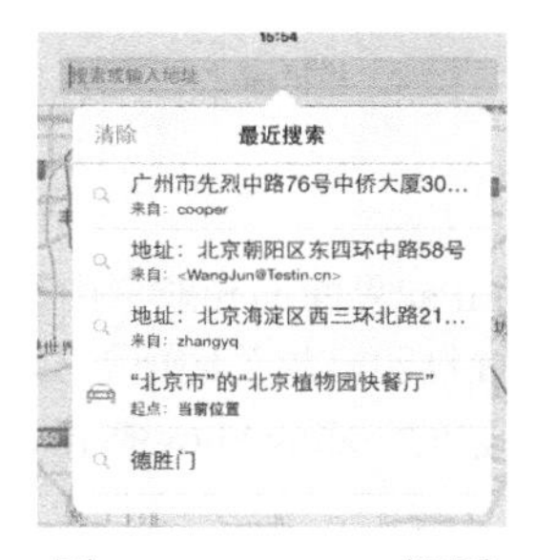

图10-30 Popover视图

图10-31 Safari浏览器的共享选项

10.4.1 Popover 相关 API

Popover相关的API主要包括：UIPopoverPresentationController类和UIPopoverPresentationControllerDelegate协议。另外，UIViewController中的modalPresentationStyle属性也与Popover视图有关。

UIPopoverPresentationController类相关的属性如下。

- **barButtonItem**。指定一个UIBarButtonItem类型的按钮作为锚点。
- **sourceView**。指定一个普通视图作为锚点。
- **sourceRect**。指定一个矩形区域作为锚点。
- **permittedArrowDirections**。指定锚点箭头的方向，属性类型是UIPopoverArrowDirection。UIPopoverArrowDirection是枚举类型，其中的成员如表10-3所示。

表10-3　UIPopoverArrowDirection枚举成员

Swift枚举成员	Objective-C枚举成员	说　明
up	UIPopoverArrowDirectionUp	向上
down	UIPopoverArrowDirectionDown	向下
left	UIPopoverArrowDirectionLeft	向左
right	UIPopoverArrowDirectionRight	向右
any	UIPopoverArrowDirectionAny	4个方向
unknown	UIPopoverArrowDirectionUnknown	未知

UIPopoverPresentationControllerDelegate类相关的方法如下。

- **popoverPresentationControllerShouldDismissPopover:**。返回布尔值，如果返回true（Objective-C中为TRUE或YES），则允许呈现Popover视图，否则不呈现Popover视图。
- **prepareForPopoverPresentation:**。Popover视图呈现时调用该方法。
- **popoverPresentationControllerDidDismissPopover:**。Popover视图关闭时，调用该方法。

UIViewController中的modalTransitionStyle属性用于设置呈现模态视图动画效果，该属性是枚举类型UIModalTransitionStyle的。UIModalTransitionStyle中的成员如表10-4所示。

表10-4　UIModalTransitionStyle枚举成员

Swift枚举成员	Objective-C枚举成员	说　明
coverVertical	UIModalTransitionStyleCoverVertical	默认动画效果，从屏幕底部滑出
flipHorizontal	UIModalTransitionStyleFlipHorizontal	水平翻转
crossDissolve	UIModalTransitionStyleCrossDissolve	交叉淡入淡出，当前视图淡出，模态视图淡入
partialCurl	UIModalTransitionStylePartialCurl	翻书效果，从右下角翻起

我们在10.3节介绍过modalPresentationStyle属性，其类型是UIModalPresentationStyle，它的UIModalPresentationPopover成员（Swift中为Popover）与Popover视图有关。

10.4.2 PopoverViewSample 案例

下面我们通过一个案例PopoverViewSample来熟悉一下Popover视图。在iPad界面的导航栏中有左右两个按钮（如图10-32所示），点击左边的Show按钮，会弹出Popover视图（其中可以设置打印机相关的项）。这是一个Popover表单视图，是通过故事板设定的，不用编写任何代码。如图10-33所示，点击右边的Coding Show按钮，会呈现Popover视图，这是一个选择列表，可以通过代码实现。

图10-32 Popover视图案例（点击Show按钮）

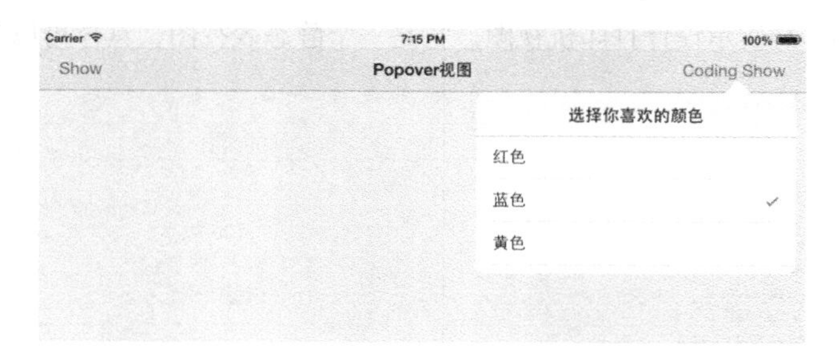

图10-33 Popover视图案例（点击Coding Show按钮）

采用Single View Application模板创建PopoverViewSample工程，其中Devices选择iPad。打开故事板文件，设计iPad界面。

1. 设计iPad主界面

打开故事板文件设计iPad界面，案例中iPad视图顶部有导航栏。我们可以将当前视图控制器嵌入到导航控制器中，也可以直接从对象库中拖曳一个Navigation Bar （导航栏）到设计界面顶部（与视图顶部距离为20点，这样不会遮挡状态栏），然后再从对象库拖曳两个Bar Button Item放到导航栏左右两边，设计样式如图10-34所示。然后，我们为右按钮Coding Show定义动作事件连线。

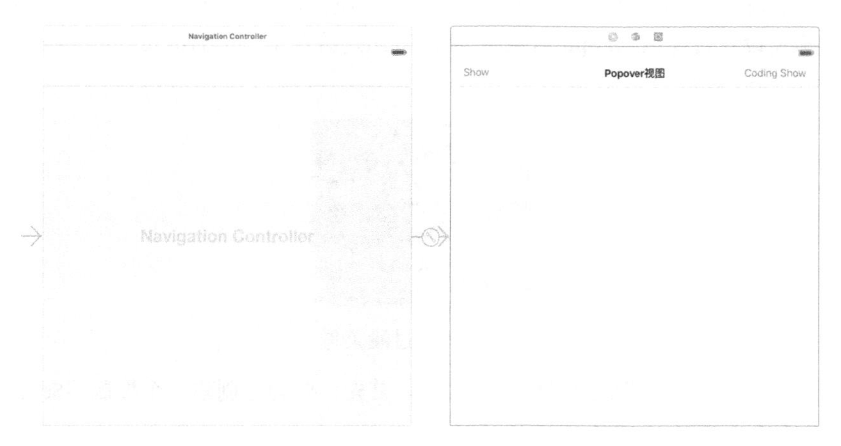

图10-34 iPad主界面

2. 设计打印机Popover视图

从对象库中拖曳一个新的Table View Controller，将其作为用于设置打印机的Popover视图控制器。由于从对象库拖曳进来的视图很大，我们需要重新设置它的大小。如图10-35所示，选中视图控制器，打开其尺寸检查器，在Simulated Size下拉列表中选择Freeform，选择这个选项后会出现Width和Height属性，我们可以根据自己的情况设置视图的高和宽。

图10-35　用于设置打印机的Popover视图

设置好大小之后，还需要设置Popover内容视图的大小。选择Table View Controller，打开其属性检查器（如图10-36所示），在Content Size中选中Use Preferred Explicit Size选项。

设置完成后，按照图10-37所示设计打印机视图。它是一个静态表视图，具体可以参考7.5节。

图10-36　设置Popover内容视图大小

图10-37　用于设置打印机的Popover视图

3. 添加过渡呈现Popover视图

点击Show按钮会呈现Popover视图，这是通过在故事板中添加过渡实现的。选中故事板主界面中的左按钮Show，按住control键拖曳设置打印机的Popover视图，此时弹出过渡菜单，如图10-38所示，请选择Present As Popover菜单。

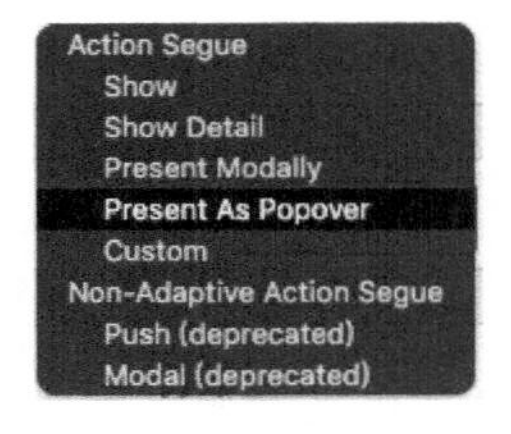

图10-38　过渡菜单

对于选择颜色的Popover视图，我们通过代码来实现。首先，要为它创建一个控制器SelectViewController，父类为UITableViewController。

下面我们看看代码部分，其中ViewController的代码如下：

```
//ViewController.swift文件
import UIKit

class ViewController: UIViewController,
➥UIPopoverPresentationControllerDelegate {                    ①

    override func viewDidLoad() {
        super.viewDidLoad()
```

```
//ViewController.m文件
#import "ViewController.h"
#import "SelectViewController.h"

@interface ViewController () <UIPopoverPresentationControllerDelegate>    ①

@end
```

```
    }

    @IBAction func show(_ sender: AnyObject) {

        let popoverViewController = SelectViewController()
        popoverViewController.modalPresentationStyle = .popover          ②
        self.present(popoverViewController, animated: true, completion: nil)
                                                                          ③

        //配置PopoverPresentationController
        let popController = popoverViewController.
        ➥popoverPresentationController                                   ④
        popController!.permittedArrowDirections = .any                    ⑤
        popController!.barButtonItem = sender as? UIBarButtonItem         ⑥
        popController!.delegate = self

    }

    //MARK: -- 实现UIPopoverPresentationControllerDelegate协议
    func prepareForPopoverPresentation(_ popoverPresentationController:
    ➥UIPopoverPresentationController) {                                  ⑦
        print("呈现Popover视图")
    }

    func popoverPresentationControllerDidDismissPopover(_ popoverPresentation
    ➥Controller: UIPopoverPresentationController) {                      ⑧
        print("关闭Popover视图")
    }
}
```

```
@implementation ViewController

- (void)viewDidLoad {
    [super viewDidLoad];
}

- (IBAction)show:(id)sender {

    SelectViewController *popoverViewController = [[SelectViewController
    ➥alloc] init];
    popoverViewController.modalPresentationStyle =
    ➥UIModalPresentationPopover;                                         ②
    [self presentViewController:popoverViewController animated:TRUE
    ➥completion:nil];                                                    ③

    //配置PopoverPresentationController
    UIPopoverPresentationController *popController =
    ➥[popoverViewController popoverPresentationController];              ④
    popController.permittedArrowDirections = UIPopoverArrowDirectionAny;  ⑤
    popController.barButtonItem = sender;                                 ⑥
    popController.delegate = self;
}

# pragma mark -- 实现UIPopoverPresentationControllerDelegate协议
- (void)prepareForPopoverPresentation:(
➥UIPopoverPresentationController *)popoverPresentation Controller {      ⑦
    NSLog(@"呈现Popover视图");
}

- (void)popoverPresentationControllerDidDismissPopover:(
➥UIPopoverPresentationController *)popoverPresentationController {       ⑧
    NSLog(@"关闭Popover视图");
}

@end
```

上述代码中，第①行用于声明实现委托协议UIPopoverPresentationControllerDelegate。第②行用于设置Popover视图控制器的modalPresentationStyle属性，这些属性需要设置为UIModalPresentationPopover（Swift版为Popover）。第③行通过UIViewController的presentViewController:animated:completion:方法（Swift版为present(_:animated:completion:)）呈现Popover视图。

第④行通过UIViewController的popoverPresentationController属性获得UIPopoverPresentationController对象。第⑤行设置Popover视图控制器的锚点箭头方向。第⑥行设置锚点对象，sender是当前按钮对象。

第⑦行实现UIPopoverPresentationControllerDelegate委托协议中的prepareForPopoverPresentation:方法。第⑧行实现UIPopoverPresentationControllerDelegate委托协议中的popoverPresentationControllerDidDismiss-Popover:方法。

SelectViewController的代码如下：

```
//SelectViewController.swift文件
import UIKit

class SelectViewController: UITableViewController {

    var listData: NSArray!
    var lastIndexPath: NSIndexPath!

    override func viewDidLoad() {
        super.viewDidLoad()

        self.listData = ["红色", "蓝色", "黄色"]
        self.preferredContentSize = CGSize(width: 200, height: 140)      ①
        self.tableView.scrollEnabled = false                             ②
```

```
//SelectViewController.m文件
#import "SelectViewController.h"
@interface SelectViewController ()
@property (nonatomic,strong) NSArray *listData;
@property (nonatomic, strong) NSIndexPath* lastIndexPath;
@end

@implementation SelectViewController

- (void)viewDidLoad {
    [super viewDidLoad];

    self.listData = @[@"红色", @"蓝色", @"黄色"];
    self.preferredContentSize = CGSizeMake(200, 140);                    ①
```

```
    }

    //MARK: -- UITableViewDataSource 协议方法
    override func tableView(_ tableView: UITableView,
    ➥numberOfRowsInSection section: Int) -> Int {
        return self.listData.count
    }

    override func tableView(_ tableView: UITableView,
    ➥cellForRowAt indexPath: IndexPath) -> UITableViewCell {

        let CellIdentifier = "Cell"
        var cell: UITableViewCell! = tableView.dequeueReusableCell(
        ➥withIdentifier: CellIdentifier)
        if cell == nil {
            cell = UITableViewCell(style: .default, reuseIdentifier:
            ➥CellIdentifier)
        }

        cell.textLabel?.text = self.listData[indexPath.row] as? String

        return cell
    }

    //MARK: -- UITableViewDelegate协议方法
    override func tableView(_ tableView: UITableView,
    ➥didSelectRowAt indexPath: IndexPath) {

        let newRow = indexPath.row

        let oldRow = (self.lastIndexPath != nil) ? self.lastIndexPath.row : -1

        if (newRow != oldRow) {
            let newCell = tableView.cellForRow(at: indexPath)
            newCell!.accessoryType = .checkmark                                ③

            if self.lastIndexPath != nil {
                let oldCell = tableView.cellForRow(at: self.lastIndexPath)
                oldCell!.accessoryType = .none                                 ④
            }

            self.lastIndexPath = indexPath
        }
    }
}
```

```
    self.tableView.scrollEnabled = FALSE;                                      ②
}

#pragma mark -- UITableViewDataSource 协议方法
- (NSInteger)tableView:(UITableView *)tableView numberOfRowsInSection:
➥(NSInteger)section {
    return [self.listData count];
}

- (UITableViewCell *)tableView:(UITableView *)tableView
➥cellForRowAtIndexPath:(NSIndexPath *)indexPath {

    static NSString *CellIdentifier = @"Cell";
    UITableViewCell *cell = [tableView dequeueReusableCellWithIdentifier:
    ➥CellIdentifier];
    if (cell == nil) {
        cell = [[UITableViewCell alloc]
        ➥initWithStyle:UITableViewCellStyleDefault reuse
        ➥Identifier:CellIdentifier];
    }

    cell.textLabel.text = self.listData[indexPath.row];
    return cell;
}

#pragma mark -- UITableViewDelegate协议方法
- (void)tableView:(UITableView *)tableView
➥didSelectRowAtIndexPath:(NSIndexPath *)indexPath {

    NSInteger newRow = indexPath.row;

    NSInteger oldRow = (self.lastIndexPath != nil) ?
    ➥[self.lastIndexPath row] :-1;

    if (newRow != oldRow) {
        UITableViewCell *newCell = [tableView cellForRowAtIndexPath:
        ➥indexPath];
        newCell.accessoryType = UITableViewCellAccessoryCheckmark;             ③

        if (self.lastIndexPath) {
            UITableViewCell *oldCell =
            ➥[tableView cellForRowAtIndexPath:self.lastIndexPath];
            oldCell.accessoryType = UITableViewCellAccessoryNone;              ④
        }
        self.lastIndexPath = indexPath;
    }
}
@end
```

上述代码中，第①行用于设置preferredContentSize属性，该属性用于设置Popover视图的大小。第②行用于设置Popover视图中的表视图不显示滚动条。

第③行设置当前单元格扩展图标为✓，第④行用于清除前一次选中的单元格扩展图标。

10.5 分屏多任务

从iOS 9开始，针对iPad最重要的变化莫过于分屏多任务了，这种多任务就是在屏幕上同时运行两个应用，甚至两个都是活动的，每个应用都有自己的屏幕区域。

iOS提供了3种多任务形式：Slide Over、分屏视图（Split View）和画中画（Picture in Picture）。

10.5.1 Slide Over 多任务

Slide Over功能能够使用户不必离开正在使用的应用，就可以打开另一个应用，例如回复微信或随手记个备忘事项。我们将手指放在iPad屏幕右边缘，然后向屏幕中间拖曳操作，则出现如图10-39所示的Slide Over多任务界面。此时原来的应用变暗，称为主应用，横屏情况下主应用占有70%屏幕，竖屏情况下占有60%屏幕；滑出来的应用高亮显示，称为次应用，横屏情况下次应用占有30%屏幕，竖屏情况占有40%屏幕。

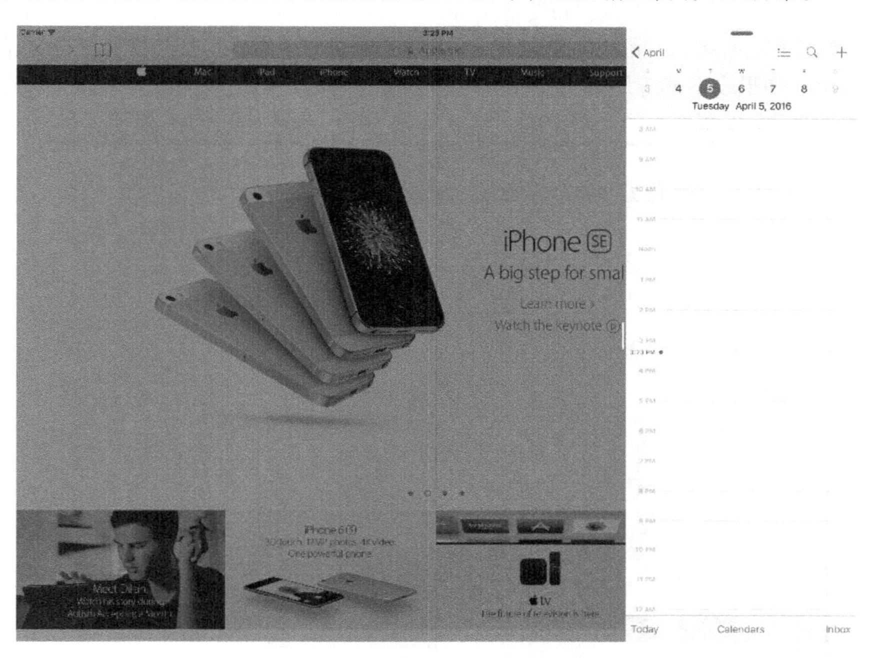

图10-39　Slide Over多任务

点击主应用屏幕可以关闭次应用。如果想切换其他的次应用，需要进入多任务列表，具体步骤是：在图10-39所示的界面中将手指放在次应用屏幕的上边缘，向下拖曳操作，此时出现如图10-40所示的界面。

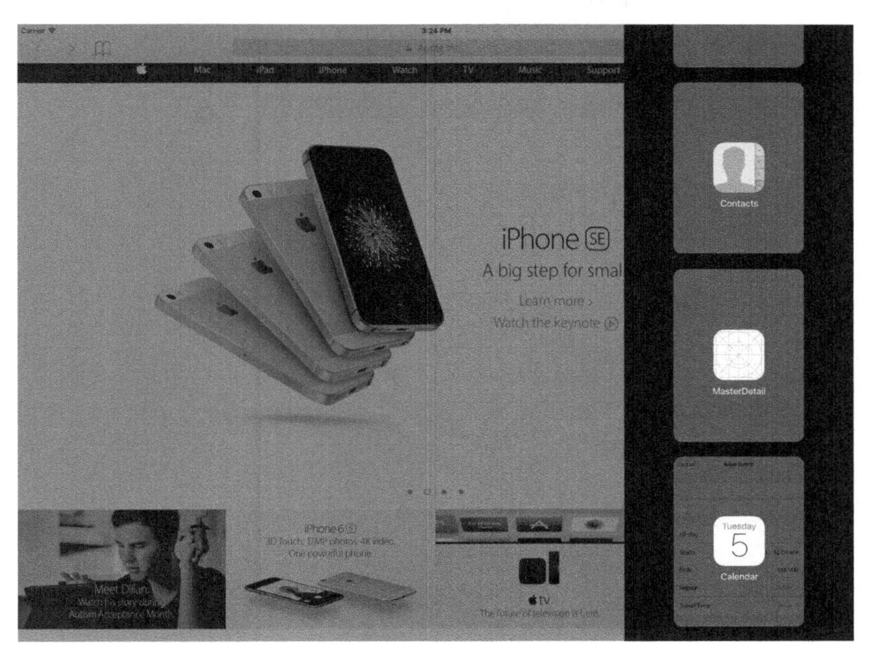

图10-40　多任务列表

10.5.2　分屏视图多任务

分屏视图多任务与Slide Over多任务类似，可以同时开启主应用和次应用。分屏视图多任务需要占用更多的CPU时间、内存空间，而且并不被所有的iPad设备支持，只有iPad mini 4、iPad Air 2和iPad Pro设备才支持。

要进入到分屏视图多任务状态，可以在Slide Over多任务界面中，用手指按住两个任务屏幕分界线中间的小竖线，如图10-41所示。如果设备和应用都支持分屏视图多任务，那么两个任务都处于高亮状态，我们可以左右拖曳小竖线，以此改变两个应用占用的屏幕空间，然后释放手指，界面如图10-42所示。

图10-41　两个任务屏幕分界线中间的小竖线

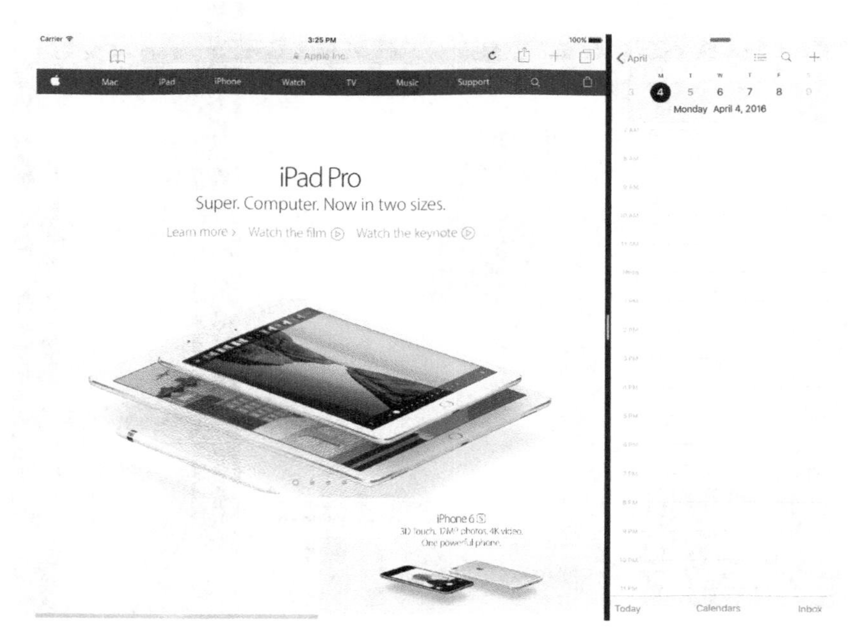

图10-42　分屏视图多任务

分屏视图多任务中两个应用占用的屏幕空间要比Slide Over多任务更加复杂。横屏情况下，如果主应用占有70%屏幕，次应用则占有30%屏幕，如图10-42所示；如果主应用占有50%屏幕，次应用则占有50%屏幕，如图10-43所示；在竖屏情况下只有一种比例，主应用占有60%屏幕，次应用则占有40%屏幕，如图10-44所示。

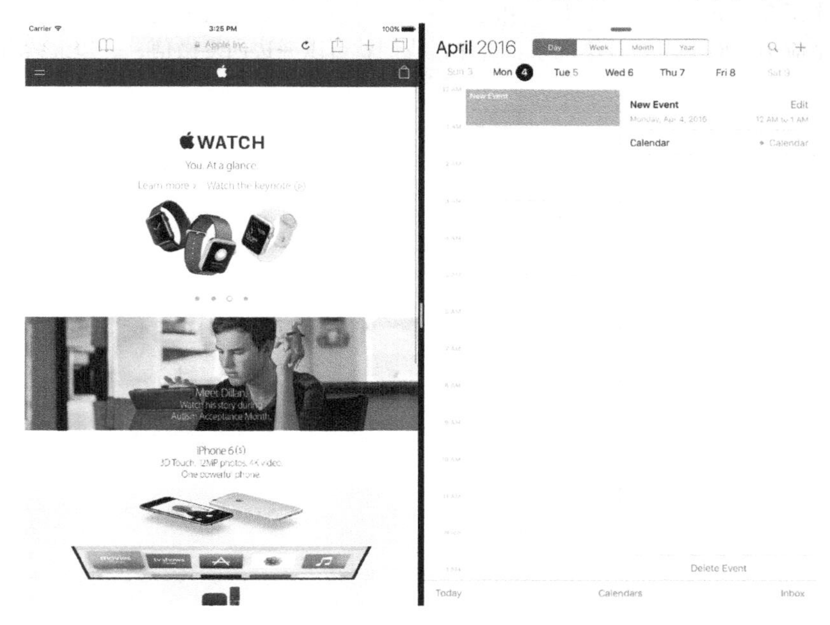

图10-43　分屏视图多任务横屏（50%:50%）

图10-44　分屏视图多任务竖屏（60%:40%）

10.5.3　画中画多任务

画中画功能可以让用户在主应用界面中悬浮播放视频窗口（如图10-45所示），并可移动视频窗口以及调整窗口大小。要进入画中画多任务，可以在观看视频时点按Home按钮，视频画面便会按比例缩小至显示屏的一角。这时可以打开另一个应用，而不需要暂停视频。这样，我们就可以一边回复微信，一边继续看视频了。

图10-45　画中画多任务

10.6　iPad 分屏多任务适配开发

我们在上一节中介绍了iPad多任务，这一节重点介绍开发分屏多任务应用时需要注意的问题。

10.6.1　分屏多任务前提条件

分屏多任务的前提条件有3个：

- iOS 9 以上；
- 设备支持所有方向；
- 启动屏幕界面设计，要使用故事板技术。

要让设备支持所有方向，具体的操作步骤是选择TARGETS→<你的目标>→General→Deployment Info，按照图10-46所示选择Device Orientation中的4个方向。

Xcode 7之后默认创建的工程都会带有一个启动屏幕故事板文件LaunchScreen.storyboard。如果没有这个文件，我们需要自己添加一个故事板文件，并且在如图10-46所示的Deployment Info→Main Interface中选择刚刚添加的故事板文件。

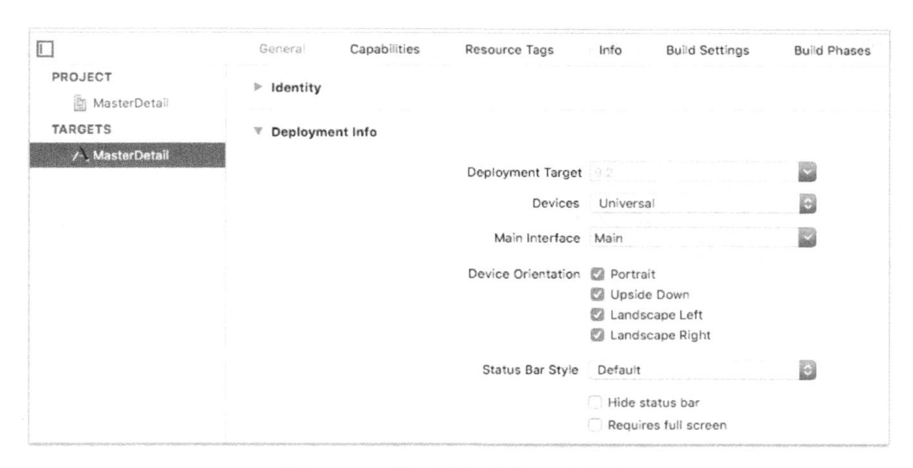

图10-46 设置设备支持所有方向

10.6.2 分屏多任务适配

要支持分屏多任务，我们需要考虑主应用和次应用的屏幕适配问题。

进行屏幕适配时，需要设置几种不同的Size Class值，这些Size Class的取值情况如图10-47所示。

- 图10-47a和图10-47b在没有分屏的情况下宽度和高度都是“正常”，Size Class的取值为`wRegular | hRegular`。
- 图10-47c是竖屏情况下，主应用和次应用都是宽度“紧凑”、高度“正常”，Size Class的取值为`wCompact | hRegular`。
- 图10-47d是横屏情况下，主应用和次应用的屏幕占比是70% : 30%，主应用Size Class的取值为`wRegular | hRegular`；次应用Size Class的取值为`wCompact | hRegular`。
- 图10-47e是横屏情况下，主应用和次应用的屏幕占比是50% : 50%，它们的Size Class都是`wCompact | hRegular`。

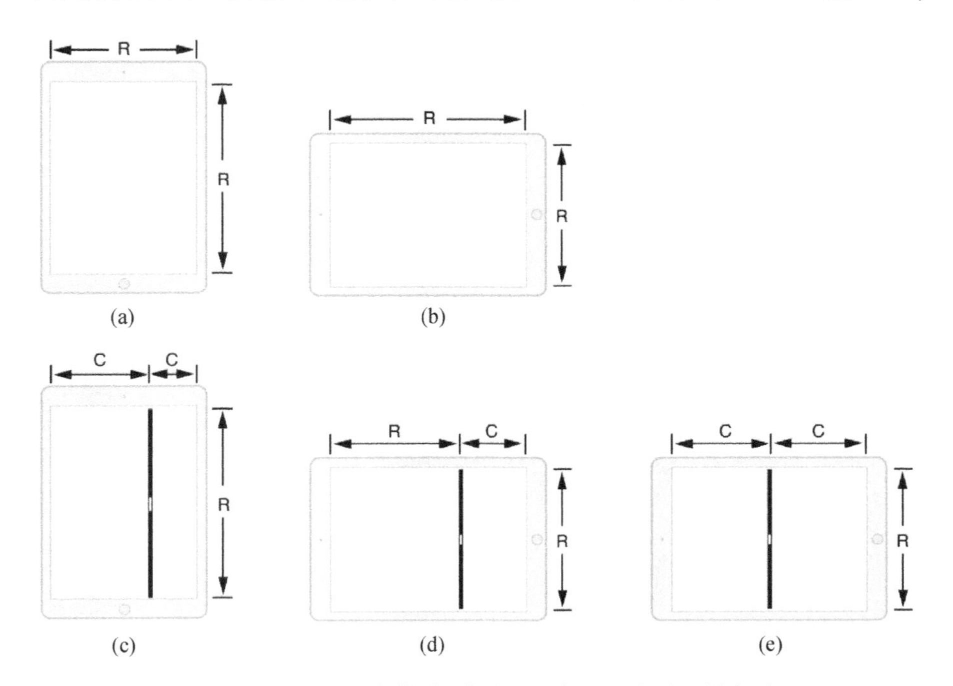

图10-47 Size Class取值（R表示“正常”，C表示“紧凑”）

我们可以归纳一下图10-47所示的情况，其实Size Class的值只需考虑wRegular | hRegular和wCompact | hRegular的情况。

10.7　小结

本章中，我们首先通过iPhone和iPad设备使用场景上的差异，介绍了iPad树形结构导航、iPad模态视图和Popover视图，最后介绍了iOS分屏多任务。

第 11 章

手势识别

11

电子触屏设备上的手势是用户与设备进行交流的特定语言。iOS能够识别这些手势，并且能够为开发人员提供开发接口。本章中，我们将介绍iOS手势识别。

11.1 手势种类

在移动设备上有极其丰富的手势，苹果为iOS设备提供了7种常用手势API，包括：Tap（单击）、Long Press（长按）、Pan（平移）、Swipe（滑动）、Rotation（旋转）、Pinch（手指的合拢和张开）和Screen Edge Pan（屏幕边缘平移）等。这些手势如表11-1所示。

表11-1　iOS设备手势

手　势　名	手　势　图	说　　明
Tap（单击）		选择、单击、碰触或连续碰触视图对象
Long Press（长按）		长时间按住屏幕上视图对象
Pan（平移）		拖曳屏幕上的一个视图对象平移到新的位置
Swipe（滑动）		快速拖曳屏幕上的视图对象，然后突然停在视图对象
Rotation（旋转）		用两个手指按住屏幕上的视图对象，然后旋转
Pinch（手指的合拢和张开）		多个手指按住屏幕上的视图对象，然后合并或张开
Screen Edge Pan（屏幕边缘平移）		在屏幕边缘平移、拖曳等操作

提示　除了表11-1所示的7种手势外，我们还可以自定义一些特殊的手势。我不赞同自定义那些鲜为人知的手势，因为鲜为人知的手势往往需要为用户提供一些操作文档。

11.2　手势识别器

在iOS设备上识别手势有两种实现方式：采用手势识别器（UIGestureRecognizer）和采用触摸事件（UITouch）。本节中，我们介绍采用手势识别器实现手势识别。

手势识别器类UIGestureRecognizer是一个抽象类，它有7个具体类：

- UITapGestureRecognizer
- UIPinchGestureRecognizer
- UIRotationGestureRecognizer
- UISwipeGestureRecognizer
- UIPanGestureRecognizer
- UILongPressGestureRecognizer
- UIScreenEdgePanGestureRecognizer

从上面这几个类的命名可以看出它们与表11-1介绍的7种手势的对应关系。如果这7种手势识别器不能满足要求，还可以直接继承UIGestureRecognizer实现自己的特殊手势识别。

11.2.1　视图对象与手势识别

手势识别一定发生在某一个视图对象上，它可能是常用标签、按钮、图片等视图或者控件。要对视图（UIView）对象进行手势识别，需要使用下面的语句添加手势识别器：

```
self.view.addGestureRecognizer(gestureRecognizer)
```

```
[self.view addGestureRecognizer: gestureRecognizer]
```

其中gestureRecognizer是具体的手势识别器对象。

此外，针对视图（UIView）对象，还需要设置一些属性，主要有以下两个属性。

- userInteractionEnabled。开启或关闭用户事件，为布尔值。Swift版本是isUserInteractionEnabled。
- multipleTouchEnabled。设置是否接收多点触摸事件，为布尔值。Swift版本是isMultipleTouchEnabled。

可以在程序代码中设置这两个属性，这通常是在视图控制器的viewDidLoad方法中完成的，示例代码如下：

```
override func viewDidLoad() {
    self.view.isMultipleTouchEnabled = false
    self.view.isUserInteractionEnabled = true
    ......
}
```

```
- (void)viewDidLoad {
    self.view.multipleTouchEnabled = NO;
    self.view.userInteractionEnabled = YES;
    ......
}
```

当然，也可以在Interface Builder中通过设计视图属性实现。在Interface Builder中选中要设置的视图对象，打开其属性检查器，在View→Interaction属性中设置这两个属性，如图11-1所示。

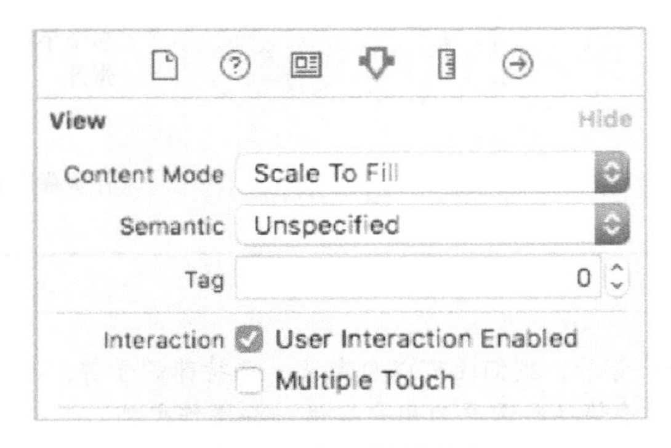

图11-1　设置视图属性

11.2.2 手势识别状态

UIGestureRecognizer类有一个state属性，用来表示手势识别过程中的状态。手势识别的状态分为7个，这些状态是在UIGestureRecognizerState枚举类型中定义的，如表11-2所示。

表11-2 UIGestureRecognizerState枚举成员

Swift枚举成员	Objective-C枚举成员	说　明
possible	UIGestureRecognizerStatePossible	默认样式，手势尚未识别
began	UIGestureRecognizerStateBegan	开始接收连续类型手势
changed	UGestureRecognizerStateChanged	接收连续类型手势状态变化
ended	UIGestureRecognizerStateEnded	结束接收连续类型手势
cancelled	UIGestureRecognizerStateCancelled	取消接收连续类型手势
failed	UIGestureRecognizerStateFailed	离散类型的手势识别失败

手势分为连续类型手势与离散类型手势。连续类型手势，如Pinch（手指的合拢和张开），整个过程中连续产生多个触摸点，其识别过程如图11-2所示，其中changed状态可能会多次变化，最后有ended（结束）和cancelled（取消）两种状态。离散类型手势，只发生一次，如Tap（单击）手势，如图11-3所示，识别过程只有两种状态：结束（ended）和失败（failed）。

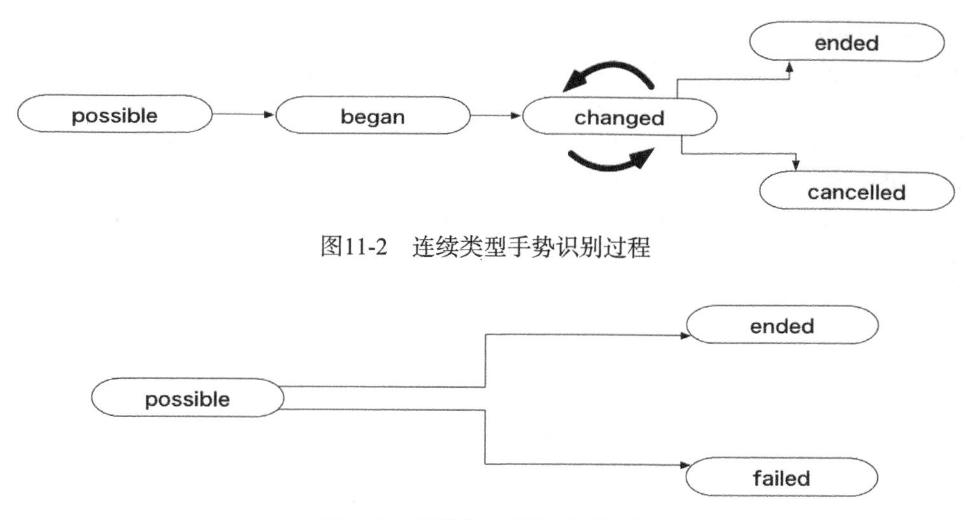

图11-2 连续类型手势识别过程

图11-3 离散类型手势识别过程

11.2.3 实例：识别 Tap 手势

使用UITapGestureRecognizer实现手势识别有两种方式：一种是在Interface Builder中进行设计实现，另一种是代码实现。

下面通过实例介绍一下，该实例如图11-4a所示，屏幕上有一个装满垃圾的垃圾桶，单击它倾倒垃圾，如图11-4b所示。再单击又装满，如此反复。

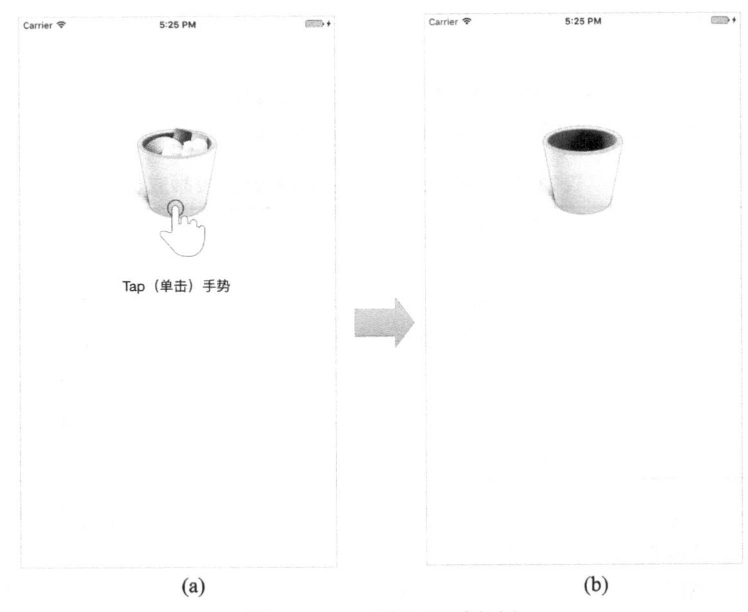

图11-4　Tap手势识别实例

1. Interface Builder实现

使用Single View Application模板创建一个工程，并将其命名为TapGestureRecognizer。

然后需要添加资源图片文件到工程，读者可以从本章的“资源”文件夹中找到资源文件（Blend Trash Empty.png和Blend Trash Full.png），并参考5.4节将它们添加到工程中。在设计界面上添加Image View控件，连接好输出口。

在Interface Builder的对象库中有7个手势识别器，如图11-5所示，这7个手势识别器对应表11-1中的7个手势。使用时，拖曳手势识别器对象到设计窗口中的视图对象上即可。如图11-6所示，拖曳Tap Gesture Recognizer对象到Image View上，注意不是View，这是因为要识别Image View上的Tap手势。添加完成后，视图设计界面的对象栏中会出现Tap手势识别器对象，如图11-7所示。

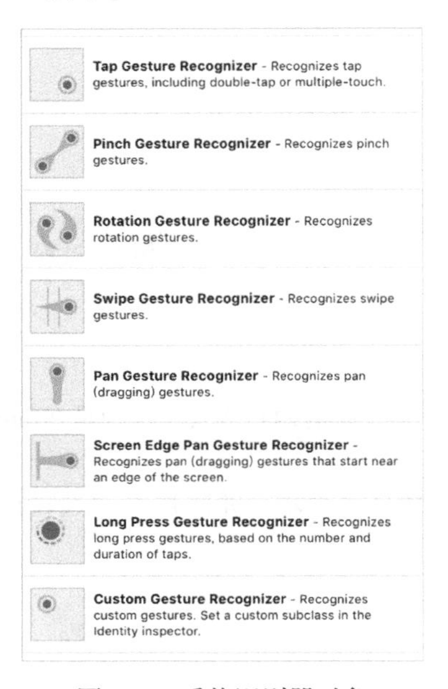

图11-5　手势识别器对象

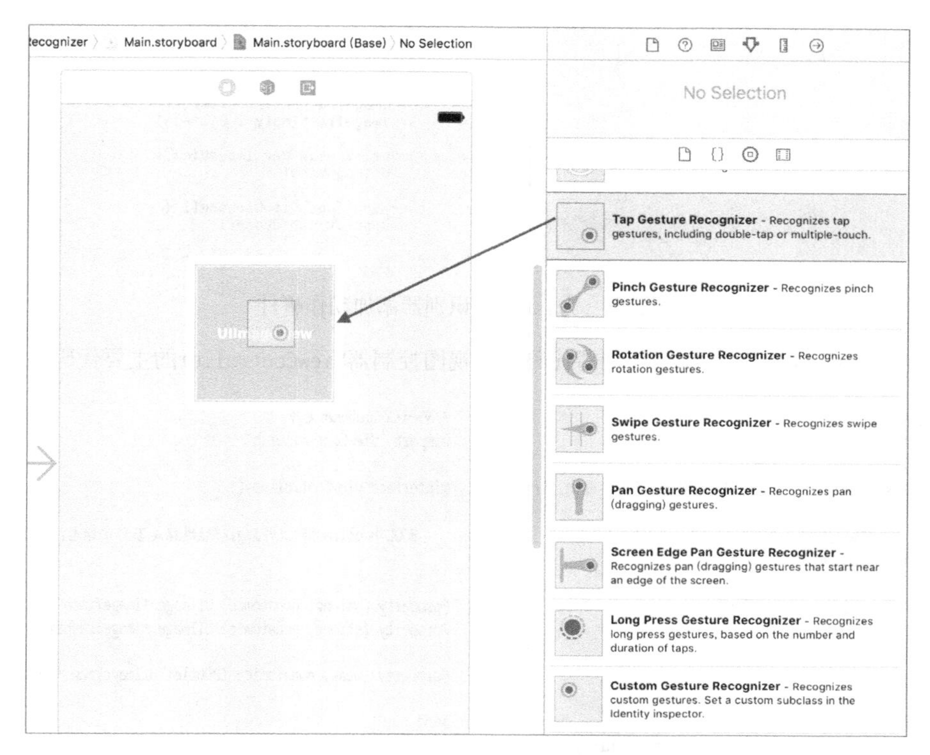

图11-6　添加Tap手势识别器对象

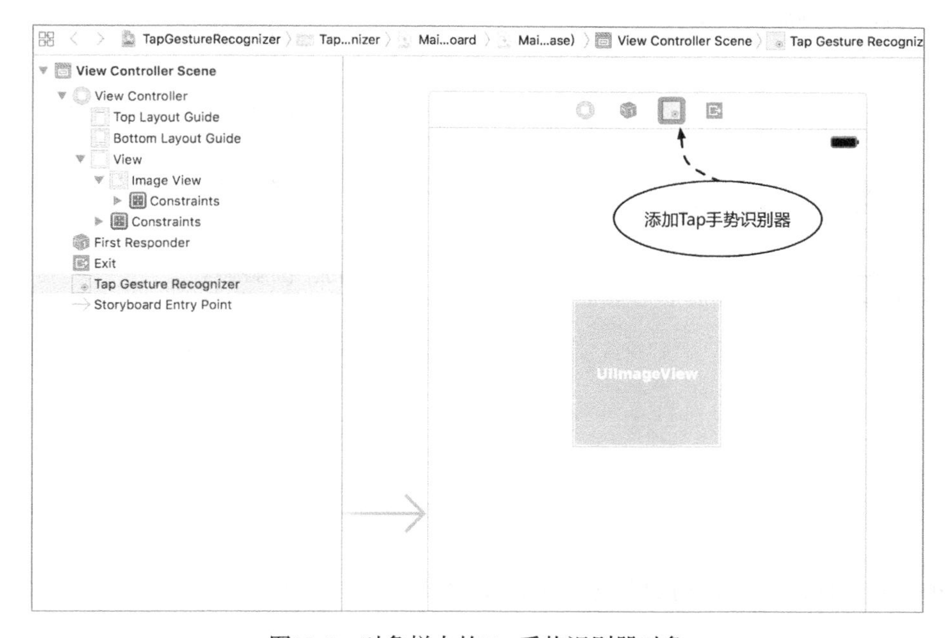

图11-7　对象栏中的Tap手势识别器对象

为Image View对象添加Tap手势识别器后，还需要添加动作事件，这个过程与为一般控件添加动作事件类似。打开辅助编辑器，选中对象栏中的Tap手势识别器对象，同时按住control键，将其拖曳到右边窗口，释放左键，会弹出一个对话框，如图11-8所示。在Connection列表框中选择Action，将Name命名为foundTap。

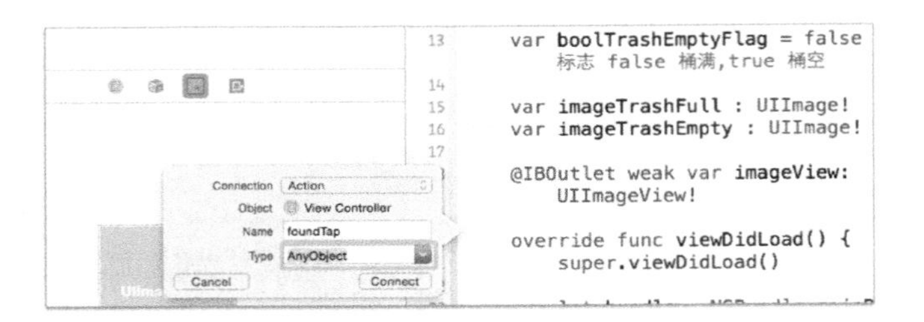

图11-8　为Tap手势识别器添加动作事件

设置完成后，就可以在视图控制器中编写代码了。视图控制器ViewController的主要代码如下：

```
//ViewController.swift文件
class ViewController: UIViewController {

    var boolTrashEmptyFlag = false //垃圾桶是否为空标志：false-桶满；true-桶空

    var imageTrashFull : UIImage!
    var imageTrashEmpty : UIImage!

    @IBOutlet weak var imageView: UIImageView!

    override func viewDidLoad() {
        super.viewDidLoad()

        //创建图片对象
        self.imageTrashFull = UIImage(named: "Blend Trash Full")          ①
        self.imageTrashEmpty = UIImage(named: "Blend Trash Empty")        ②

        self.imageView.image = self.imageTrashFull

    }

    //Tap手势处理方法
    @IBAction func foundTap(_ sender: AnyObject) {
        print("Tap")
        if boolTrashEmptyFlag {
            self.imageView.image = self.imageTrashFull
            boolTrashEmptyFlag = false
        } else {
            self.imageView.image = self.imageTrashEmpty
            boolTrashEmptyFlag = true
        }
    }

}
```

```
//ViewController.m文件
#import "ViewController.h"

@interface ViewController ()
{
    BOOL boolTrashEmptyFlag;//垃圾桶是否为空标志：NO-桶满；YES-桶空
}

@property (strong, nonatomic) UIImage *imageTrashFull;
@property (strong, nonatomic) UIImage *imageTrashEmpty;

@property (weak, nonatomic) IBOutlet UIImageView *imageView;

@end

@implementation ViewController

- (void)viewDidLoad {
    [super viewDidLoad];

    //创建图片对象
    self.imageTrashFull = [UIImage imageNamed:@"Blend Trash Full"];     ①
    self.imageTrashEmpty = [UIImage imageNamed:@"Blend Trash Empty"];   ②

    self.imageView.image = self.imageTrashFull;
}

//Tap手势处理方法
- (IBAction)foundTap:(id)sender {
    NSLog(@"Tap");
    if (boolTrashEmptyFlag) {
        self.imageView.image = self.imageTrashFull;
        boolTrashEmptyFlag = NO;
    } else {
        self.imageView.image = self.imageTrashEmpty;
        boolTrashEmptyFlag = YES;
    }
}

@end
```

上述代码中，第①行和第②行用于从资源文件中创建图片对象。注意，如果是PNG图片，可以省略文件后缀名。

2. 代码实现

首先，使用Single View Application模板创建一个名为TapGestureRecognizer的工程，然后参考3.5.1节将其修改为纯代码工程。

AppDelegate类与3.5.1节的案例基本一样，这里不再赘述，我们重点介绍一下视图控制器类ViewController，其代码如下：

```
//ViewController.swift文件
class ViewController: UIViewController {

    var boolTrashEmptyFlag = false //垃圾桶是否为空标志: false-桶满; true-桶空

    var imageTrashFull : UIImage!
    var imageTrashEmpty : UIImage!

    var imageView: UIImageView!

    override func viewDidLoad() {
        super.viewDidLoad()

        //界面初始化
        let screen = UIScreen.main.bounds
        let imageViewWidth: CGFloat = 128
        let imageViewHeight: CGFloat = 128
        let imageViewTopView: CGFloat = 148
        let frame = CGRect(x: (screen.size.width - imageViewWidth)/2,
        ➥y: imageViewTopView, width: imageViewWidth, height: imageViewHeight)
        self.imageView = UIImageView(frame: frame)
        self.view.addSubview(self.imageView)

        //创建图片对象
        self.imageTrashFull = UIImage(named: "Blend Trash Full")
        self.imageTrashEmpty = UIImage(named: "Blend Trash Empty")

        self.imageView.image = self.imageTrashFull

        //创建Tap手势识别器
        let tapRecognizer = UITapGestureRecognizer(target: self, action:
        ➥#selector(foundTap(_:)))                                          ①
        //设置Tap手势识别器属性
        tapRecognizer.numberOfTapsRequired = 1                              ②
        tapRecognizer.numberOfTouchesRequired = 1                           ③

        //将Tap手势识别器关联到imageView
        self.imageView.addGestureRecognizer(tapRecognizer)
        //设置imageView开启用户事件
        self.imageView.isUserInteractionEnabled = true                      ④
    }

    func foundTap(_ sender: AnyObject) {
        print("Tap")
        if boolTrashEmptyFlag {
            self.imageView.image = self.imageTrashFull
            boolTrashEmptyFlag = false
        } else {
            self.imageView.image = self.imageTrashEmpty
            boolTrashEmptyFlag = true
        }
    }

}
```

```
//ViewController.m文件
#import "ViewController.h"

@interface ViewController ()
{
    BOOL boolTrashEmptyFlag;//垃圾桶是否为空标志: NO-桶满; YES-桶空
}

@property (strong, nonatomic) UIImage *imageTrashFull;
@property (strong, nonatomic) UIImage *imageTrashEmpty;

@property (strong, nonatomic) UIImageView *imageView;

@end

@implementation ViewController

- (void)viewDidLoad {
    [super viewDidLoad];

    //界面初始化
    CGRect screen = [[UIScreen mainScreen] bounds];
    CGFloat imageViewWidth = 128;
    CGFloat imageViewHeight = 128;
    CGFloat imageViewTopView = 148;
    CGRect frame = CGRectMake((screen.size.width - imageViewWidth)/2 ,
    ➥imageViewTopView, imageViewWidth, imageViewHeight);
    self.imageView = [[UIImageView alloc] initWithFrame:frame];
    [self.view addSubview:self.imageView];

    //创建图片对象
    self.imageTrashFull = [UIImage imageNamed:@"Blend Trash Full"];
    self.imageTrashEmpty = [UIImage imageNamed:@"Blend Trash Empty"];

    self.imageView.image = self.imageTrashFull;

    //创建Tap手势识别器
    UITapGestureRecognizer *tapRecognizer =[[UITapGestureRecognizer alloc]
                                            initWithTarget:self
                                            action:@selector(foundTap:)]; ①
    //设置Tap手势识别器属性
    tapRecognizer.numberOfTapsRequired = 1;                               ②
    tapRecognizer.numberOfTouchesRequired = 1;                            ③

    //将Tap手势识别器关联到imageView
    [self.imageView addGestureRecognizer:tapRecognizer];
    //设置imageView开启用户事件
    self.imageView.userInteractionEnabled = YES;                          ④
}

- (void)foundTap:(id)sender {
    NSLog(@"Tap");
    if (boolTrashEmptyFlag) {
        self.imageView.image = self.imageTrashFull;
        boolTrashEmptyFlag = NO;
    } else {
        self.imageView.image = self.imageTrashEmpty;
        boolTrashEmptyFlag = YES;
    }
}

@end
```

这里在viewDidLoad方法中初始化视图和手势识别器。其中，第①行用于实例化手势识别器UITapGesture-Recognizer，这里使用了构造函数initWithTarget:action:。在该方法中，target参数是指定回调方法所在的目标对象，action参数用来设置手势识别后回调的方法。

第②行设置触发Tap的单击次数，1就是单击一下触发，如果是2就是双击。第③行设置触发Tap的触点个数，即有几个手指按在屏幕上。第④行用于设置imageView开启用户事件。

11.2.4　实例：识别 Long Press 手势

识别Long Press手势使用的手势识别器是UILongPressGestureRecognizer。也有两种实现方式：一种是在Interface Builder中进行设计实现，另一种是代码实现。由于Interface Builder实现与Tap手势实现类似，本节只介绍代码实现，其他手势亦是如此。

下面通过实例介绍一下，这个实例如图11-9a所示，在屏幕上有一个装满垃圾桶，长按后倾倒垃圾，如图11-9b所示。再长按后又装满，如此反复。

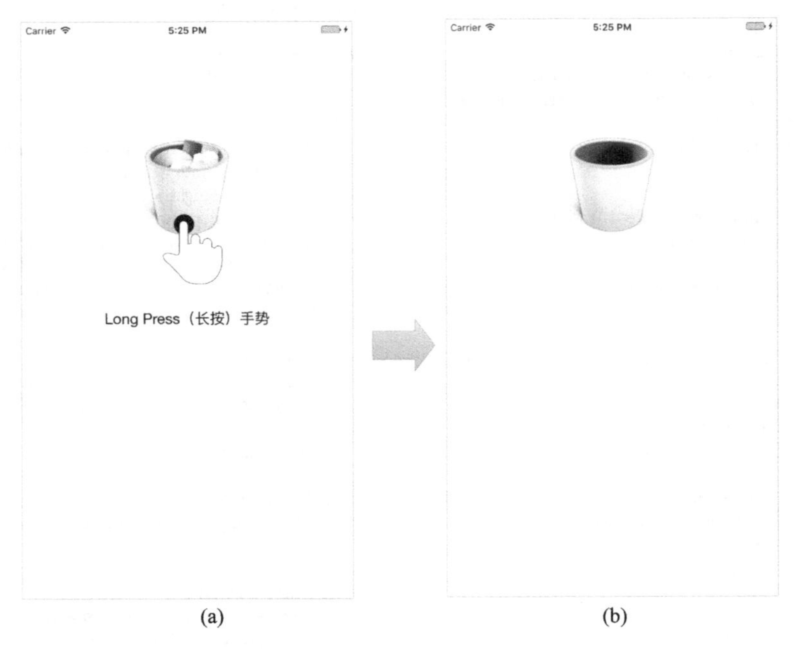

(a)　　　　(b)

图11-9　长按手势识别实例

首先，使用Single View Application模板创建一个名为LongPressGestureRecognizer的工程，然后参考3.5.1节将其修改为纯代码工程。

我们重点介绍一下视图控制器类ViewController，其代码如下：

```
//ViewController.swift文件
class ViewController: UIViewController {

    var boolTrashEmptyFlag = false //垃圾桶是否为空标志：false-桶满；true-桶空

    var imageTrashFull : UIImage!
    var imageTrashEmpty : UIImage!

    var imageView: UIImageView!

    override func viewDidLoad() {
        super.viewDidLoad()
        ......

        //创建Long Press手势识别器
        let recognizer = UILongPressGestureRecognizer(target: self,
```

```
//ViewController.m文件
#import "ViewController.h"

@interface ViewController ()
{
    BOOL boolTrashEmptyFlag;//垃圾桶是否为空标志：NO-桶满；YES-桶空
}

@property (strong, nonatomic) UIImage *imageTrashFull;
@property (strong, nonatomic) UIImage *imageTrashEmpty;

@property (strong, nonatomic) UIImageView *imageView;

@end

@implementation ViewController
```

```
        ➥action: #selector(foundLongPress(_:)))                        ①

        //设置Long Press手势识别器属性
        recognizer.allowableMovement = 100.0                           ②
        recognizer.minimumPressDuration = 1.0                          ③

        //将Long Press手势识别器关联到imageView
        self.imageView.addGestureRecognizer(recognizer)
        //设置imageView开启用户事件
        self.imageView.isUserInteractionEnabled = true
    }

    func foundLongPress(_ sender: UITapGestureRecognizer) {

        print("长按 state = ", sender.state.rawValue)     ④

        if (sender.state == .began) { //手势开始
            if boolTrashEmptyFlag {
                self.imageView.image = self.imageTrashFull
                boolTrashEmptyFlag = false
            } else {
                self.imageView.image = self.imageTrashEmpty
                boolTrashEmptyFlag = true
            }
        }
    }

}
```

```
- (void)viewDidLoad {
    [super viewDidLoad];
    ......

    //创建Long Press手势识别器
    UILongPressGestureRecognizer *recognizer =
    ➥[[UILongPressGestureRecognizer alloc]
                                    initWithTarget:self
                                    action:@selector(foundLongPress:)];     ①
    //设置Long Press手势识别器属性
    recognizer.allowableMovement = 100.0f;                                  ②
    recognizer.minimumPressDuration = 1.0;                                  ③
    //将Long Press手势识别器关联到imageView
    [self.imageView addGestureRecognizer:recognizer];

    //设置imageView开启用户事件
    self.imageView.userInteractionEnabled = YES;
}

- (void)foundLongPress:(UILongPressGestureRecognizer*)sender {

    NSLog(@"长按 state = %li",(long)sender.state);                         ④

    if (sender.state == UIGestureRecognizerStateBegan) { //手势开始
        if (boolTrashEmptyFlag) {
            self.imageView.image = self.imageTrashFull;
            boolTrashEmptyFlag = NO;
        } else {
            self.imageView.image = self.imageTrashEmpty;
            boolTrashEmptyFlag = YES;
        }
    }
}

@end
```

这里在viewDidLoad方法中初始化视图和手势识别器。其中，第①行实例化手势识别器UILongPressGestureRecognizer，这里使用了构造方法initWithTarget:action:，其中target参数指定回调方法所在的目标对象，action参数用来设置手势识别后回调的方法。

第②行用于设置手势识别之前最小移动的距离，单位是points（点）。

第③行设置手势识别的最短持续时间，单位是秒。

在foundLongPress:方法中，第④行中的sender.state取得手势识别器的状态。在Swift版中，sender.state.rawValue属性是状态的原始值（int类型整数）。在执行手势的过程中，状态变化会有日志输出。当手指长按超过1秒时，输出state = 1，即began状态；当手指在屏幕上移动时，state = 2，即changed状态；当手指抬起时state = 3，即ended状态。事实上，Tap等手势也有这些状态，只是上一节的实例我们不需要判断状态。

11.2.5 实例：识别Pan手势

识别Pan手势使用的手势识别器是UIPanGestureRecognizer，下面通过一个实例介绍一下。该实例如图11-10所示，在屏幕上有一个装满垃圾的垃圾桶，可以用手指按着它让其在屏幕上移动。

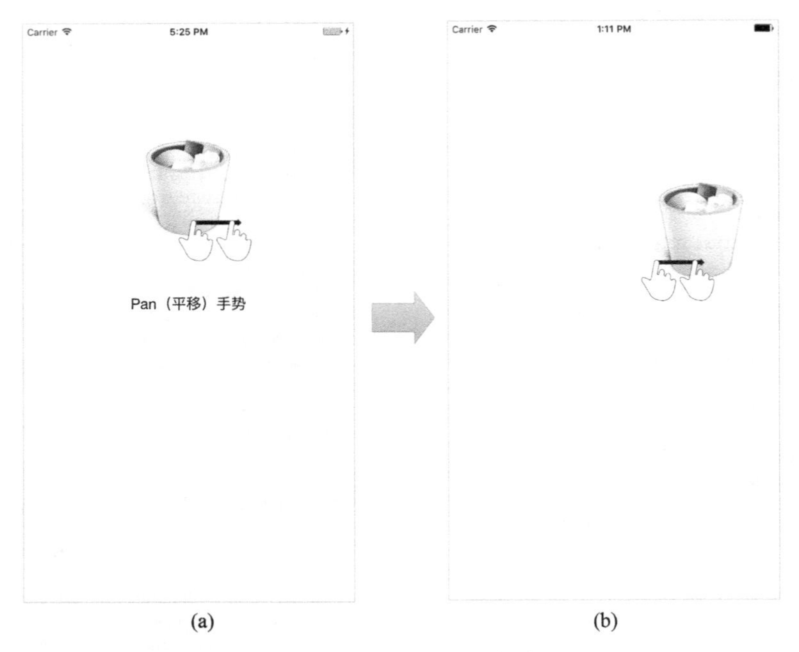

(a)　　(b)

图11-10　平移手势识别实例

首先，使用Single View Application模板创建一个名为PanGestureRecognizer的工程，然后参考3.5.1节将其修改为纯代码工程。

这里重点介绍一下视图控制器类ViewController，其代码如下：

```
//ViewController.swift文件
class ViewController: UIViewController {

    var boolTrashEmptyFlag = false
    //垃圾桶是否为空标志: false-桶满; true-桶空

    var imageTrashFull: UIImage!
    var imageTrashEmpty: UIImage!

    var imageView: UIImageView!

    override func viewDidLoad() {
        super.viewDidLoad()

        ......

        //创建Pan手势识别器
        let recognizer = UIPanGestureRecognizer(target: self, action:
        ➥#selector(foundPan(_:)))
        //设置Pan手势识别器属性
        recognizer.minimumNumberOfTouches = 1                                ①
        recognizer.maximumNumberOfTouches = 1                                ②

        //将Pan手势识别器关联到imageView
        self.imageView.addGestureRecognizer(recognizer)
        //设置imageView开启用户事件
        self.imageView.isUserInteractionEnabled = true
    }

    func foundPan(_ sender: UIPanGestureRecognizer) {

        print("平移 state = ", sender.state.rawValue)
```

```
//ViewController.m文件
#import "ViewController.h"

@interface ViewController ()
{
    BOOL boolTrashEmptyFlag;//垃圾桶是否为空标志: NO-桶满; YES-桶空
}

@property (strong, nonatomic) UIImage *imageTrashFull;
@property (strong, nonatomic) UIImage *imageTrashEmpty;

@property (strong, nonatomic) UIImageView *imageView;

@end

@implementation ViewController
- (void)viewDidLoad {
    [super viewDidLoad];

    ......

    //创建Pan手势识别器
    UIPanGestureRecognizer *recognizer = [[UIPanGestureRecognizer alloc]
                    initWithTarget:self
                    action:@selector(foundPan:)];
    //设置Pan手势识别器属性
    recognizer.minimumNumberOfTouches = 1;                                   ①
    recognizer.maximumNumberOfTouches = 1;                                   ②

    //将Pan手势识别器关联到imageView
    [self.imageView addGestureRecognizer:recognizer];
    //设置imageView开启用户事件
    self.imageView.userInteractionEnabled = YES;
```

```
        if sender.state != .ended && sender.state != .failed {          ③
            let location = sender.location(in: sender.view!.superview)  ④
            sender.view!.center = location                              ⑤
        }
    }
}
```

```
}

- (void)foundPan:(UIPanGestureRecognizer *)sender {

    NSLog(@"拖动 state = %li", (long) sender.state);

    if (sender.state != UIGestureRecognizerStateEnded
    ➥&& sender.state != UIGestureRecognizerStateFailed) {                ③
        CGPoint location = [sender locationInView:sender.view.superview]; ④
        sender.view.center = location;                                    ⑤
    }
}

@end
```

这里在viewDidLoad方法中初始化视图和手势识别器。第①行设置最小个数的触点，第②行设置最大个数的触点。

在foundPan:方法中，第③行判断手势识别器的状态在ended和failed的情况下，可以平移imageView。第④行返回触点在imageView父视图中的坐标。第⑤行用于将当前触点坐标作为imageView新的中心点位置。通过不断地改变imageView位置，就会看到平移的效果。

11.2.6　实例：Swipe 手势

识别Swipe手势使用的手势识别器是UISwipeGestureRecognizer，下面通过一个实例介绍一下。该实例如图11-11a所示，用户将手指放到屏幕上，分别进行上、下、左、右滑动。待检测到Swipe手势后，会在屏幕的标签中显示滑动方向信息，如图11-11b所示。

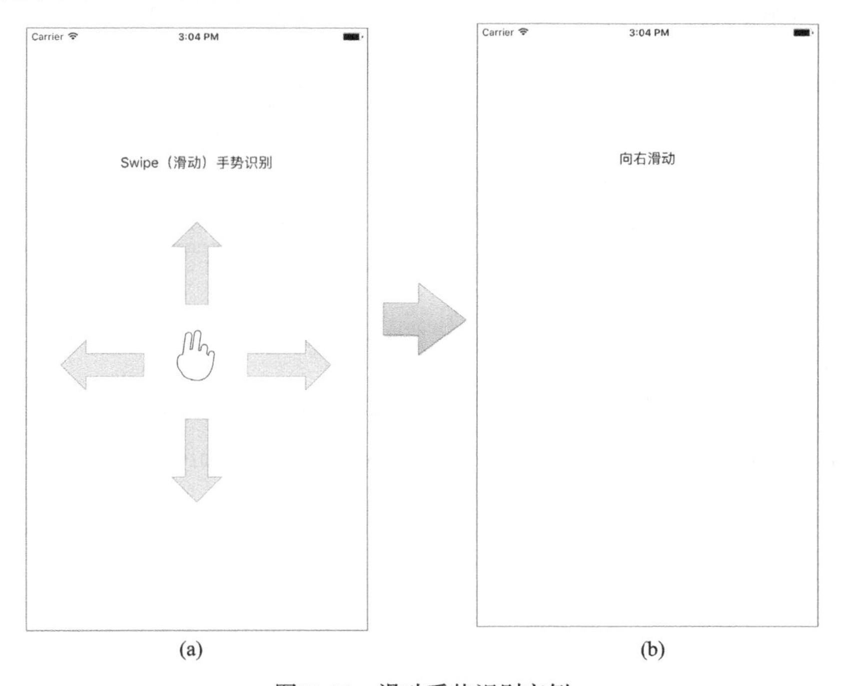

图11-11　滑动手势识别实例

首先，使用Single View Application模板创建一个名为SwipeGestureRecognizer的工程，然后参考3.5.1节将其修改为纯代码工程。

这里重点介绍一下视图控制器类ViewController，其代码如下：

```
//ViewController.swift文件
class ViewController: UIViewController {

    var label: UILabel!

    override func viewDidLoad() {
        super.viewDidLoad()

        //初始化界面
        let screen = UIScreen.main.bounds
        let labelWidth:CGFloat = 300
        let labelHeight:CGFloat = 30
        let labelTopView:CGFloat = 150
        let frame = CGRect(x: (screen.size.width - labelWidth)/2,
        ➥y: labelTopView, width: labelWidth, height: labelHeight)
        self.label = UILabel(frame: frame)
        self.label.text = "Swipe (滑动) 手势识别"
        //字体左右居中
        self.label.textAlignment = NSTextAlignment.center
        self.view.addSubview(self.label)

        //创建4个Swipe手势识别器
        let directions: [UISwipeGestureRecognizerDirection] =
            [.right, .left, .up, .down]                                  ①

        for direction in directions {
            let recognizer = UISwipeGestureRecognizer(target: self, action:
            ➥#selector(foundSwipe(_:)))                                   ②
            //设置识别滑动方向
            recognizer.direction = direction                              ③
            //将Swipe手势识别器关联到View
            self.view.addGestureRecognizer(recognizer)                    ④
        }

        //设置View开启用户事件
        self.view.isUserInteractionEnabled = true
    }

    func foundSwipe(_ sender: UISwipeGestureRecognizer) {

        print("direction = ", sender.direction.rawValue)

        switch sender.direction {
        case UISwipeGestureRecognizerDirection.down:
            self.label.text = "向下滑动"
        case UISwipeGestureRecognizerDirection.left:
            self.label.text = "向左滑动"
        case UISwipeGestureRecognizerDirection.right:
            self.label.text = "向右滑动"
        case UISwipeGestureRecognizerDirection.up:
            self.label.text = "向上滑动"
        default:
            self.label.text = "未知"
        }
    }
}
```

```
//ViewController.m文件
#import "ViewController.h"

@interface ViewController ()

@property (strong, nonatomic)UILabel* label;

@end

@implementation ViewController

- (void)viewDidLoad {
    [super viewDidLoad];

    //初始化界面
    CGRect screen = [[UIScreen mainScreen] bounds];
    CGFloat labelWidth = 300;
    CGFloat labelHeight = 30;
    CGFloat labelTopView = 150;
    CGRect frame = CGRectMake((screen.size.width - labelWidth)/2 ,
    ➥labelTopView, labelWidth, labelHeight);
    self.label = [[UILabel alloc] initWithFrame:frame];
    self.label.text = @"Swipe (滑动) 手势识别";
    //字体左右居中
    self.label.textAlignment = NSTextAlignmentCenter;
    [self.view addSubview:self.label];

    //创建4个Swipe手势识别器
    NSInteger directions[4] = {UISwipeGestureRecognizerDirectionRight,
        UISwipeGestureRecognizerDirectionLeft,
        UISwipeGestureRecognizerDirectionUp,
        UISwipeGestureRecognizerDirectionDown};                          ①

    for (int i = 0; i < 4; i++) {
        UISwipeGestureRecognizer *recognizer =
        ➥[[UISwipeGestureRecognizer alloc]
                initWithTarget:self
                        action:@selector(foundSwipe:)];                  ②
        //设置识别滑动方向
        recognizer.direction = directions[i];                            ③
        //将Swipe手势识别器关联到View
        [self.view addGestureRecognizer:recognizer];                     ④
    }

    //设置View开启用户事件
    self.view.userInteractionEnabled = YES;
}

- (void)foundSwipe:(UISwipeGestureRecognizer *)sender {

    NSLog(@"direction = %li", sender.direction);

    switch (sender.direction) {
        case UISwipeGestureRecognizerDirectionDown:
            self.label.text = @"向下滑动";
            break;
        case UISwipeGestureRecognizerDirectionLeft:
            self.label.text = @"向左滑动";
            break;
        case UISwipeGestureRecognizerDirectionRight:
            self.label.text = @"向右滑动";
            break;
        case UISwipeGestureRecognizerDirectionUp:
            self.label.text = @"向上滑动";
            break;
        default:
            self.label.text = @"未知";
            break;
```

```
        }
    }

@end
```

这里在viewDidLoad方法中初始化视图和手势识别器。注意为了识别4个不同的方向，我们需要创建创建4个Swipe手势识别器，它们的不同之处是direction属性。该属性的取值是枚举类型UISwipeGestureRecognizer-Direction，它有4个成员，如表11-3所示。

表11-3 UISwipeGestureRecognizerDirection枚举成员

Swift枚举成员	Objective-C枚举成员	说 明
right	UISwipeGestureRecognizerDirectionRight	向右滑动
left	UISwipeGestureRecognizerDirectionLeft	向左滑动
up	UISwipeGestureRecognizerDirectionUp	向上滑动
down	UISwipeGestureRecognizerDirectionDown	向下滑动

第①行定义集合directions，这里将UISwipeGestureRecognizerDirection枚举的4个成员放到directions集合中。

第②行至第④行是循环体，第②行用于创建Swipe手势识别器。第③行用于设置Swipe手势识别器的direction属性。第④行用于将Swipe手势识别器关联到View，这个View是当前屏幕上显示的默认视图（即根视图）。

11.2.7 实例：Rotation 手势

识别Rotation手势使用的手势识别器是UIRotationGestureRecognizer，下面通过一个实例介绍一下。该实例如图11-12a所示，在屏幕上有垃圾桶，垃圾桶会跟着两个手指一起旋转，如图11-12b所示。

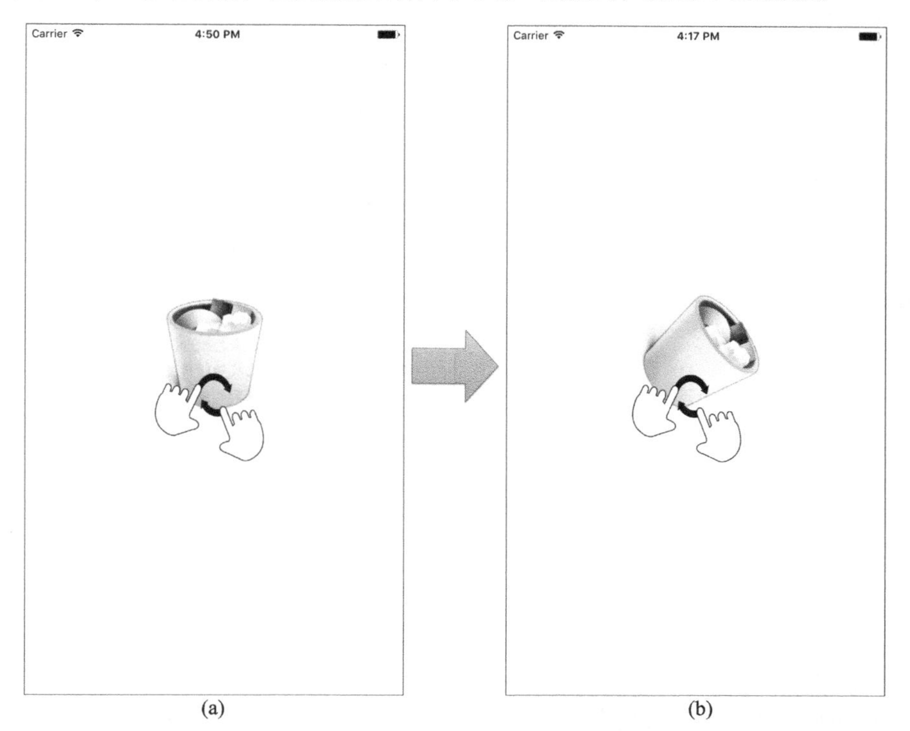

图11-12 Rotation手势识别实例

首先，使用Single View Application模板创建一个名为RotationGestureRecognizer的工程，然后参考3.5.1节将其修改为纯代码工程。

这里重点介绍一下视图控制器类ViewController，其代码如下：

```
//ViewController.swift文件
class ViewController: UIViewController {

    //垃圾桶旋转角度
    var rotationAngleInRadians: CGFloat = 0                                    ①

    var imageTrashFull: UIImage!
    var imageView: UIImageView!

    override func viewDidLoad() {
        super.viewDidLoad()

        //界面初始化
        let screen = UIScreen.main.bounds
        let imageViewWidth: CGFloat = 128
        let imageViewHeight: CGFloat = 128
        let imageViewTopView: CGFloat = 300
        let frame = CGRect(x: (screen.size.width - imageViewWidth) / 2,
        ➥y: imageViewTopView, width: imageViewWidth, height: imageViewHeight)
        self.imageView = UIImageView(frame: frame)
        self.view.addSubview(self.imageView)

        //创建图片对象
        self.imageTrashFull = UIImage(named: "Blend Trash Full")
        self.imageView.image = self.imageTrashFull

        //创建Rotation手势识别器
        let recognizer = UIRotationGestureRecognizer(target: self,
        ➥action: #selector(foundRotation(_:)))
        //将Rotation手势识别器关联到imageView
        self.imageView.addGestureRecognizer(recognizer)
        //设置imageView开启用户事件
        self.imageView.isUserInteractionEnabled = true
    }

    func foundRotation(_ sender: UIRotationGestureRecognizer) {
        //上一次角度加上本次旋转的角度
        self.imageView.transform = CGAffineTransform(rotationAngle:
        ➥self.rotationAngleInRadians + sender.rotation)                      ②

        //手势识别完成，保存旋转的角度
        if (sender.state == .ended) {
            self.rotationAngleInRadians += sender.rotation                    ③
        }
    }

}
```

```
//ViewController.m文件
#import "ViewController.h"

@interface ViewController () {
    //垃圾桶旋转角度
    CGFloat rotationAngleInRadians;                                            ①
}
@property(strong, nonatomic) UIImage *imageTrashFull;
@property(strong, nonatomic) UIImageView *imageView;

@end

@implementation ViewController

- (void)viewDidLoad {
    [super viewDidLoad];

    //界面初始化
    CGRect screen = [[UIScreen mainScreen] bounds];
    CGFloat imageViewWidth = 128;
    CGFloat imageViewHeight = 128;
    CGFloat imageViewTopView = 300;
    CGRect frame = CGRectMake((screen.size.width - imageViewWidth) / 2,
    ➥imageViewTopView, imageViewWidth, imageViewHeight);
    self.imageView = [[UIImageView alloc] initWithFrame:frame];
    [self.view addSubview:self.imageView];

    //创建图片对象
    self.imageTrashFull = [UIImage imageNamed:@"Blend Trash Full"];
    self.imageView.image = self.imageTrashFull;

    //创建Rotation手势识别器
    UIRotationGestureRecognizer *recognizer =
    ➥[[UIRotationGestureRecognizer alloc]
            initWithTarget:self
                    action:@selector(foundRotation:)];

    //将Rotation手势识别器关联到imageView
    [self.imageView addGestureRecognizer:recognizer];
    //设置imageView开启用户事件
    self.imageView.userInteractionEnabled = YES;
}

- (void)foundRotation:(UIRotationGestureRecognizer *)sender {
    //上一次角度加上本次旋转的角度
    self.imageView.transform =
    ➥CGAffineTransformMakeRotation(rotationAngleInRadians +
    ➥sender.rotation);                                                         ②

    //手势识别完成，保存旋转的角度
    if (sender.state == UIGestureRecognizerStateEnded) {
        rotationAngleInRadians += sender.rotation;                             ③
    }
}

@end
```

上述代码中，第①行的rotationAngleInRadians变量用来保存垃圾桶上一次旋转的角度。第②行将上一次角度加上本次旋转的角度作为本次旋转的角度，然后通过放射变换函数CGAffineTransformMakeRotation作用于ImageView对象使之旋转。

第③行是在手势识别完成后，保存旋转的角度到rotationAngleInRadians变量中。

提示 Rotation和Pinch等手势都会用到多点触摸（多个手指按住屏幕），真机设备很容易，而在模拟器上模拟两个手指按住屏幕却有些麻烦，这需要我们在按住鼠标的同时，按下Alt键，这时屏幕会出现两个小点，表示两个触点，如图11-13所示。

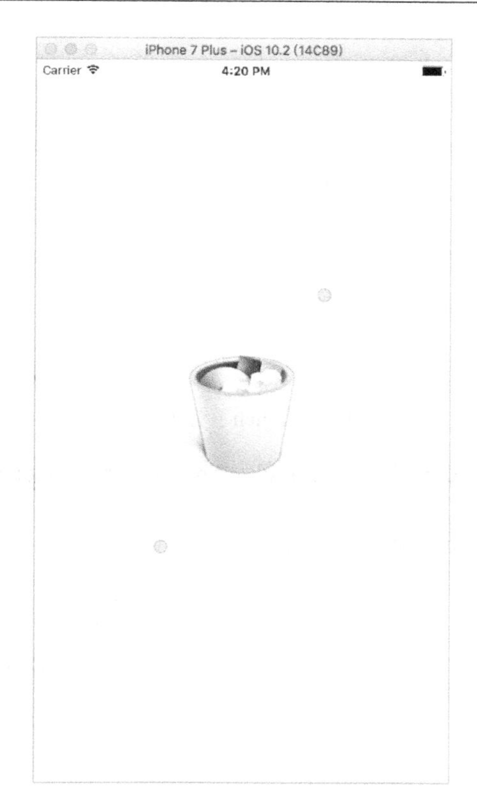

图11-13 模拟器中多点触摸

11.2.8 实例：Pinch 手势

识别Pinch手势使用的手势识别器是UIPinchGestureRecognizer，下面通过一个实例介绍一下。该实例如图11-14a所示，在屏幕上有一个垃圾桶，两个手指张开，垃圾桶会放大，如图11-14b所示。相反，如果两个手指合拢，则垃圾桶会缩小。

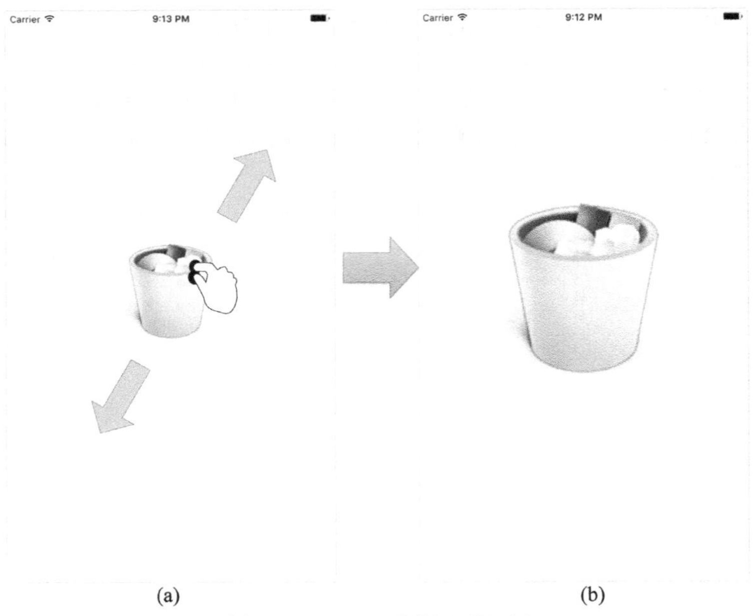

(a)　　(b)

图11-14　Pinch手势识别实例

首先，使用Single View Application模板创建一个名为PinchGestureRecognizer的工程，然后参考3.5.1节将其修改为纯代码工程。

这里重点介绍一下视图控制器类ViewController，其代码如下：

```
//ViewController.swift文件
class ViewController: UIViewController {

    //缩放因子
    var currentScale : CGFloat  = 1.0                                   ①
    var imageTrashFull : UIImage!
    var imageView: UIImageView!

    override func viewDidLoad() {
        super.viewDidLoad()

        //界面初始化
        let screen = UIScreen.main.bounds
        let imageViewWidth: CGFloat = 128
        let imageViewHeight: CGFloat = 128
        let imageViewTopView: CGFloat = 300
        let frame = CGRect(x: (screen.size.width - imageViewWidth)/2,
        ➥y: imageViewTopView, width: imageViewWidth, height: imageViewHeight)
        self.imageView = UIImageView(frame: frame)
        self.view.addSubview(self.imageView)

        //创建图片对象
        self.imageTrashFull = UIImage(named: "Blend Trash Full")
        self.imageView.image = self.imageTrashFull

        //创建Pinch手势识别器
        let recognizer = UIPinchGestureRecognizer(target: self, action:
        ➥#selector(foundPinch(_:)))
        //将Pinch手势识别器关联到imageView
        self.imageView.addGestureRecognizer(recognizer)
        //设置imageView开启用户事件
        self.imageView.isUserInteractionEnabled = true
    }
```

```
//ViewController.m文件
#import "ViewController.h"

@interface ViewController () {
    //缩放因子
    CGFloat currentScale;                                               ①
}

@property (strong, nonatomic) UIImage *imageTrashFull;
@property (strong, nonatomic) UIImageView *imageView;

@end

@implementation ViewController

- (void)viewDidLoad {
    [super viewDidLoad];

    //界面初始化
    CGRect screen = [[UIScreen mainScreen] bounds];
    CGFloat imageViewWidth = 128;
    CGFloat imageViewHeight = 128;
    CGFloat imageViewTopView = 300;
    CGRect frame = CGRectMake((screen.size.width - imageViewWidth)/2 ,
    ➥imageViewTopView, imageViewWidth, imageViewHeight);
    self.imageView = [[UIImageView alloc] initWithFrame:frame];
    [self.view addSubview:self.imageView];

    //创建图片对象
    self.imageTrashFull = [UIImage imageNamed:@"Blend Trash Full"];
    self.imageView.image = self.imageTrashFull;

    //创建Pinch手势识别器
    UIPinchGestureRecognizer *recognizer =[[UIPinchGestureRecognizer alloc]
```

```swift
    func foundPinch(_ sender: UIPinchGestureRecognizer) {

        print("缩放因子 = ", sender.scale)

        if sender.state == .ended {                                    ②
            currentScale = sender.scale                                ③
        } else if sender.state == .began && currentScale != 0.0 {      ④
            sender.scale = currentScale                                ⑤
        }

        self.imageView.transform = CGAffineTransform(scaleX: sender.scale, y:
        ➥sender.scale)                                                ⑥
    }

}
```

```objectivec
                                    initWithTarget:self
                                    action:@selector(foundPinch:)];
    //将Pinch手势识别器关联到imageView
    [self.imageView addGestureRecognizer:recognizer];
    //设置imageView开启用户事件
    self.imageView.userInteractionEnabled = YES;
}

- (void)foundPinch:(UIPinchGestureRecognizer*)sender {

    NSLog(@"缩放因子 = %f",sender.scale);

    if (sender.state == UIGestureRecognizerStateEnded) {                      ②
        currentScale = sender.scale;                                          ③
    } else if (sender.state == UIGestureRecognizerStateBegan && currentScale !=
        0.0f) {                                                               ④
        sender.scale = currentScale;                                          ⑤
    }
    self.imageView.transform = CGAffineTransformMakeScale(sender.scale,
    ➥sender.scale);                                                          ⑥
}

@end
```

上述代码中，第①行的currentScale变量用来记录上次缩放因子。

第②行用于判断是否是手势识别完成状态（ended），这种状态下通过第③行记录当前缩放因子，其中sender.scale属性可以获得缩放因子。

第④行表示如果在手势识别开始状态，可以通过第⑤行将上次保存的缩放因子作为当前缩放因子使用。这样可以保证视图缩放的连续变化，避免发生忽大忽小的情况。

第⑥行通过放射变换函数CGAffineTransformMakeScale进行缩放变换，该函数的第一个参数是x轴缩放因子，第二个参数是y轴缩放因子。

11.2.9 实例：Screen Edge Pan 手势

识别Screen Edge Pan手势使用的手势识别器是UIScreenEdgePanGestureRecognizer，下面通过一个实例介绍一下。该实例如图11-15a所示，用户用手指在屏幕从右边缘向左平移时，界面中的标签会显示“从左边缘向右平移”（见图11-15b）；用户手指在屏幕从左边缘向右平移时，界面中的标签会显示“从右边缘向左平移”（见图11-15c）。

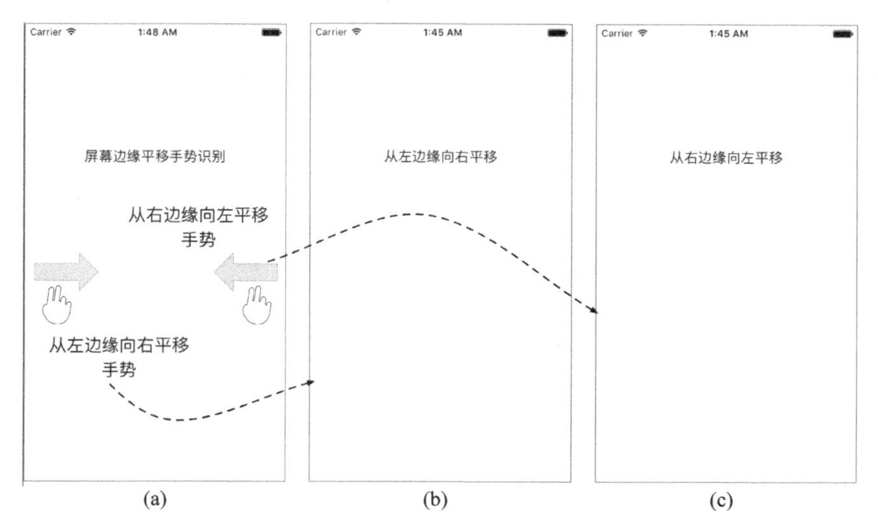

图11-15 屏幕边缘平移手势识别实例

首先，使用Single View Application模板创建一个名为ScreenEdgePanGestureRecognizer的工程，然后参考3.5.1节将其修改为纯代码工程。

这里重点介绍一下视图控制器类ViewController，其代码如下：

```
//ViewController.swift文件
class ViewController: UIViewController {

    var label: UILabel!

    override func viewDidLoad() {
        super.viewDidLoad()

        //初始化界面
        let screen = UIScreen.main.bounds
        let labelWidth: CGFloat = 400
        let labelHeight: CGFloat = 30
        let labelTopView: CGFloat = 150
        let frame = CGRect(x: (screen.size.width - labelWidth) / 2,
        ➥y: labelTopView, width: labelWidth, height: labelHeight)
        self.label = UILabel(frame: frame)
        self.label.text = "屏幕边缘平移手势识别"
        //字体左右居中
        self.label.textAlignment = NSTextAlignment.center
        self.view.addSubview(self.label)

        //创建两个ScreenEdgePan手势识别器
        let edges: [UIRectEdge] = [.right, .left]                        ①

        for edge in edges {
            let recognizer = UIScreenEdgePanGestureRecognizer(target: self,
            ➥action: #selector(foundScreenEdgePan(_:)))                 ②
            //设置屏幕边缘平移的方向
            recognizer.edges = edge                                      ③
            //将ScreenEdgePan手势识别器关联到View
            self.view.addGestureRecognizer(recognizer)                   ④
        }
        //设置View开启用户事件
        self.view.isUserInteractionEnabled = true
    }

    override func didReceiveMemoryWarning() {
        super.didReceiveMemoryWarning()
    }

    func foundScreenEdgePan(_ sender: UIScreenEdgePanGestureRecognizer) {

        print("edge = ", sender.edges.rawValue)

        switch sender.edges {
        case UIRectEdge.left:
            self.label.text = "从左边缘向右平移"
        case UIRectEdge.right:
            self.label.text = "从右边缘向左平移"
        default:
            self.label.text = ""
        }
    }

}
```

```
//ViewController.m文件
#import "ViewController.h"

@interface ViewController ()
@property(strong, nonatomic) UILabel *label;
@end

@implementation ViewController

- (void)viewDidLoad {
    [super viewDidLoad];

    //初始化界面
    CGRect screen = [[UIScreen mainScreen] bounds];
    CGFloat labelWidth = 300;
    CGFloat labelHeight = 30;
    CGFloat labelTopView = 150;
    CGRect frame = CGRectMake((screen.size.width - labelWidth) / 2,
    ➥labelTopView, labelWidth, labelHeight);
    self.label = [[UILabel alloc] initWithFrame:frame];
    self.label.text = @"屏幕边缘平移手势识别";
    //字体左右居中
    self.label.textAlignment = NSTextAlignmentCenter;
    [self.view addSubview:self.label];

    //创建两个ScreenEdgePan手势识别器
    NSInteger edges[2] = {UIRectEdgeRight, UIRectEdgeLeft};              ①

    for (int i = 0; i < 2; i++) {
        UIScreenEdgePanGestureRecognizer *recognizer =
        ➥[[UIScreenEdgePanGestureRecognizer alloc]
        ➥initWithTarget:self
        ➥action:@selector(foundScreenEdgePan:)];                         ②
        //设置屏幕边缘平移的方向
        recognizer.edges = edges[i];                                     ③
        //将ScreenEdgePan手势识别器关联到View
        [self.view addGestureRecognizer:recognizer];                     ④
    }

    //设置View开启用户事件
    self.view.userInteractionEnabled = YES;
}

- (void)foundScreenEdgePan:(UIScreenEdgePanGestureRecognizer*)sender {

    NSLog(@"edge = %li", sender.edges);

    switch (sender.edges) {
        case UIRectEdgeLeft:
            self.label.text = @"从左边缘向右平移";
            break;
        case UIRectEdgeRight:
            self.label.text = @"从右边缘向左平移";
            break;
        default:
            self.label.text = @"";
    }
}

@end
```

这里在viewDidLoad方法中初始化视图和手势识别器。注意，为了识别两个不同的方向，我们需要创建创建两个屏幕边缘平移手势识别器，它们之间的不同是edges属性。该属性的取值是枚举类型UIRectEdge，其主要成员有4个，如表11-4所示。

表11-4　UIRectEdge枚举成员

Swift枚举成员	Objective-C枚举成员	说　明
top	UIRectEdgeTop	从顶边缘向下平移
left	UIRectEdgeLeft	从左边缘向右平移
bottom	UIRectEdgeBottom	从底边缘向上平移
right	UIRectEdgeRight	从右边缘向左平移

第①行定义了集合edges，这里将UIRectEdge枚举的两个成员放到edges集合中。

第②行至第④行是循环体，其中第③行设置屏幕边缘平移的方向，第④行用于将识别器关联到View，这个View是当前屏幕上显示的默认视图（即根视图）。

11.3　小结

通过对本章的学习，读者可以了解手势的概念和种类，掌握手势识别器的编程过程等。

第 12 章

Quartz 2D绘图技术

在iOS中绘图技术主要有UIKit、Quartz 2D、Core Animation和OpenGL ES，下面简要介绍一下它们。

- UIKit。它是高级别的图形接口，其API都是基于Objective-C的。它能够访问绘图、动画、字体、图片等内容。
- Quartz 2D。它是iOS和macOS环境下的2D绘图引擎，涉及的内容包括：基于路径的绘图、透明度绘图、遮盖、阴影、透明层、颜色管理、抗锯齿渲染、生成PDF以及PDF元数据相关的处理。Quartz 2D也称为Core Graphics，缩写前缀为CG，它与Quartz Compositor统称为Quartz，Quartz原本是macOS的Darwin核心之上的绘图技术。它的API接口都是基于C的。
- Core Animation。它是动画技术，动画也属于绘图技术。
- OpenGL ES。它是OpenGL针对嵌入式设备的简化版本，可以绘制高性能的2D和3D图形。OpenGL ES超出了本书的介绍范围，这里不再讨论。

12.1 绘制技术基础

在iOS上，无论采用哪种绘图技术，都离不开UIView，绘制都发生在UIView对象的区域内。如果是默认视图（之前的所有实例都是默认视图），绘制工作是由iOS系统自动处理的；如果是自定义视图，则必须重写drawRect:方法，在此提供相应的绘制代码。

12.1.1 视图绘制周期

在iOS上绘制时比较麻烦，不会简单地调用一个方法就可以绘制出来。而是首先为需要绘制的视图或视图的部分区域设置一个需要绘制的标志，在事件循环的每一轮中，绘图引擎会检查是否有需要更新的内容，如果有，就会调用视图的drawRect:方法绘制，因此我们需要在绘制的视图中重写drawRect:方法。

一旦drawRect:方法被调用，就可以使用任何的UIKit、Quartz 2D 、OpenGL ES等技术对视图的内容进行绘制。

在绘图过程中除了使用drawRect:方法外，还会使用setNeedsDisplay和setNeedsDisplayInRect:方法。setNeedsDisplay和setNeedsDisplayInRect:方法用于设置视图或者视图部分区域是否需要重新绘制，其中前者用于重新绘制整个视图，后者用于重新绘制视图的部分区域。原则上，尽量不要绘制整个视图，以减少绘制带来的开销。触发视图重新绘制的动作有如下几种。

- 当遮挡你的视图的其他视图被移动或删除时。
- 将视图的hidden属性声明设置为false，使其从隐藏状态变为可见。
- 将视图滚出屏幕，然后再重新回到屏幕上。
- 显式调用视图的setNeedsDisplay或者setNeedsDisplayInRect:方法。

12.1.2 实例：填充屏幕

下面通过一个简单的实例了解一下视图绘制的过程。实例如图12-1所示，我们将整个屏幕所在的视图填充为灰色。视图轮廓是一个矩形，事实上我们只需要将视图所在的矩形填充即可。

图12-1 填充屏幕实例

将整个屏幕（整个视图）填充为灰色有很多办法，但是这里我们要求采用自定义视图，重写drawRect:方法来实现。具体的实现过程主要有两个步骤。

- ❑ 添加一个自定义的视图类。
- ❑ 自定义视图替换默认视图，默认视图是视图控制器自带的视图。

替换过程可以通过Interface Builder实现或者代码实现，下面我们分别介绍这两种实现方式。

1. Interface Builder实现

使用Single View Application模板创建一个工程，将工程命名为FillScreen。

- 添加一个自定义的视图类

创建完成工程后，我们需要添加一个自定义的视图类。选择菜单中的New File...菜单项，此时会弹出选择创建文件模板对话框，从中选择iOS→Source→Cocoa Touch class，出现如图12-2所示的对话框，在Class中输入MyView，在Subclass of 中选择UIView，在Language中选择Objective-C或Swift，然后点击Next按钮创建自定义视图。

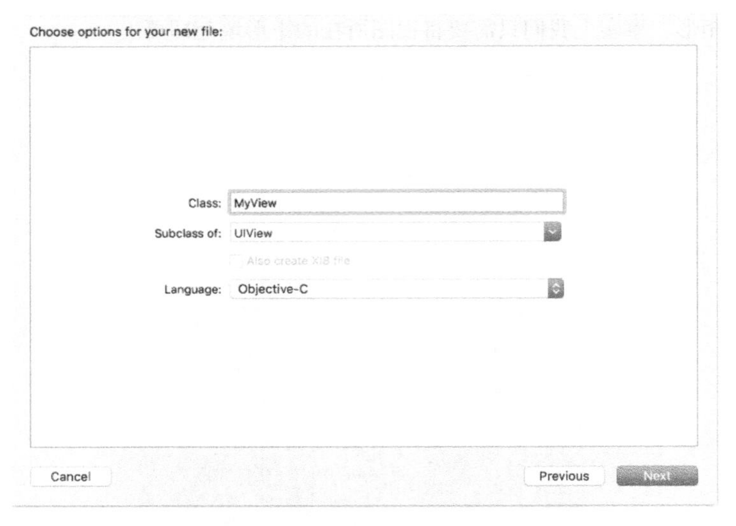

图12-2　创建自定义视图类

创建完成之后，修改MyView的代码，重写drawRect:方法（Swift版中是draw(_ rect:)方法）：

```
//MyView.swift文件
import UIKit
class MyView: UIView {
    override func draw(_ rect: CGRect) {
        UIColor.gray.setFill()                                   ①
        UIRectFill(rect)                                         ②
    }
}
```

```
//MyView.h文件
#import <UIKit/UIKit.h>
@interface MyView : UIView
@end

//MyView.m文件
#import "MyView.h"
@implementation MyView

- (void)drawRect:(CGRect)rect {
    [[UIColor grayColor] setFill];                               ①
    UIRectFill(rect);                                            ②
}

@end
```

在上述代码中，第①行用于为当前的图形上下文设置要填充的颜色，在Swift版本中通过UIColor静态属性gray获得灰颜色对象，而Objective-C版本中通过UIColor静态方法grayColor获得灰颜色对象。setFill方法用于设置颜色要进行填充处理，此时并没有真正绘制视图。当调用第②行的UIRectFill(rect)函数后，才按照设置颜色填充rect矩形。关于图形上下文的相关内容，我们将在下一节中详细介绍。

● 用自定义视图替换当前默认视图

代码部分就这么多，应用要想运行，还需要将默认视图替换为自定义视图（MyView）。打开主故事板文件Main.storyboard，选择View对象，打开其标识检查器，如图12-3所示，在Custom Class中选择Class为MyView。

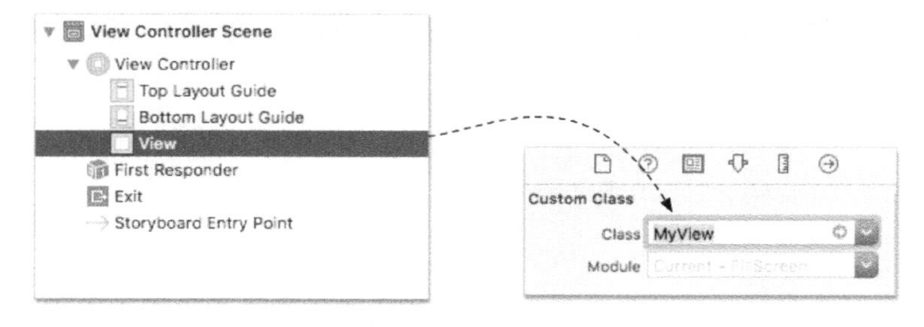

图12-3　选择自定义视图

2. 代码实现

通过代码实现时，添加自定义视图类与Interface Builder实现完全一样，MyView的代码也一样，这里不再赘述。这两种实现的主要区别在于自定义视图替换当前视图控制器中的默认视图。

首先，使用Single View Application模板创建一个名为FillScreen的工程，然后参考3.5.1节将该修改为纯代码工程，并添加MyView类。

接着修改AppDelegate的代码，具体如下：

```
//AppDelegate.swift文件
class AppDelegate: UIResponder, UIApplicationDelegate {

    var window: UIWindow?

    func application(_ application: UIApplication,
    ➥didFinishLaunchingWithOptions
    ➥launchOptions: [UIApplicationLaunchOptionsKey: Any]?) -> Bool {

        //创建根视图控制器
        let rootViewController = ViewController()
        //创建自定义视图
        let view = MyView(frame: UIScreen.main.bounds)                ①
        //用自定义视图替换默认视图
        rootViewController.view = view;                               ②

        self.window = UIWindow(frame: UIScreen.main.bounds)
        self.window?.rootViewController = rootViewController
        self.window?.makeKeyAndVisible()

        return true
    }
    ......
}
```

```
//AppDelegate.m文件
@implementation AppDelegate

- (BOOL)application:(UIApplication *)application
➥didFinishLaunchingWithOptions:(NSDictionary *)launchOptions {

    //创建根视图控制器
    ViewController* rootViewController = [[ViewController alloc] init];
    //创建自定义视图
    MyView* view = [[MyView alloc] initWithFrame:[[UIScreen mainScreen]
    ➥bounds]];                                                        ①
    //用自定义视图替换默认视图
    rootViewController.view = view;                                    ②

    self.window = [[UIWindow alloc] initWithFrame:[[UIScreen mainScreen]
    ➥bounds]];
    [self.window makeKeyAndVisible];
    self.window.rootViewController = rootViewController;

    return YES;
}
......

@end
```

上述代码中，第①行用于创建自定义视图MyView对象，该对象的大小与屏幕大小一样。第②行用自定义视图MyView对象替换系统默认视图。

12.1.3 填充与描边

UIKit提供了非常基本的绘图功能，主要的API如下。

- **UIRectFill(CGRect rect)**。填充矩形函数，这在前面的例子中介绍过。
- **UIRectFrame(CGRect rect)**。绘制矩形边框函数。
- **UIBezierPath**。绘制常见路径类，包括线段、弧线、矩形、圆角矩形和椭圆。

UIKit虽然提供了UIBezierPath等类，但是对线段、渐变、阴影、反锯齿等高级特性的支持还不及Quartz 2D。

下面我们看看UIRectFrame(CGRect rect)函数的用法。修改MyView视图的drawRect:方法，具体如下所示：

```
override func draw(_ rect: CGRect) {

    UIColor.black.setFill()
    UIRectFill(rect)

    UIColor.white.setStroke()                                         ①
    let frame = CGRect(x: 20, y: 30, width: 100, height: 300)
    UIRectFrame(frame)                                                ②
}
```

```
- (void)drawRect:(CGRect)rect {

    [[UIColor blackColor] setFill];
    UIRectFill(rect);

    [[UIColor whiteColor] setStroke];                                 ①
    CGRect frame = CGRectMake(20, 30, 100, 300);
    UIRectFrame(frame);                                               ②
}
```

第①行至第②行是在视图上绘制一个白色的矩形边框，绘制边框的过程称为描边（stroke）。第①行用于设置当前的图形上下文要描边的颜色为白色，如图12-4所示。第②行调用UIRectFrame(frame)函数后进行描边。

图12-4　矩形描边

12.1.4　绘制图像和文本

除了可以绘制几何图形外，还可以绘制文本和图像，绘制的效果就像是使用标准控件（UILabel和UIImageView）一样。这些绘制可以使用UIImage和NSString实现，对应的绘制方法如下。

UIImage类中绘制图像的主要方法如下。

- **drawAtPoint:**。在指定的绘制点绘制图片，如图12-5a所示，如果图片宽度超过屏幕宽度，则图片的部分内容无法显示。
- **drawInRect:**。在指定的矩形里绘制图片，如图12-5b所示。
- **drawAsPatternInRect:**。在指定的矩形里平铺绘制图片，如果图片大小超出了指定的矩形，形式上与drawAtPoint:方法类似，如果图片大小小于指定的矩形，效果如图12-5c所示，就会有平铺的效果。

在UIKit框架中提供了一个NSString类的扩展UIStringDrawing，它提供了绘制文本的主要方法。

- **drawAtPoint:withAttributes:**。文本在指定点绘制。
- **drawInRect:withAttributes:**。文本在指定的矩形里绘制。

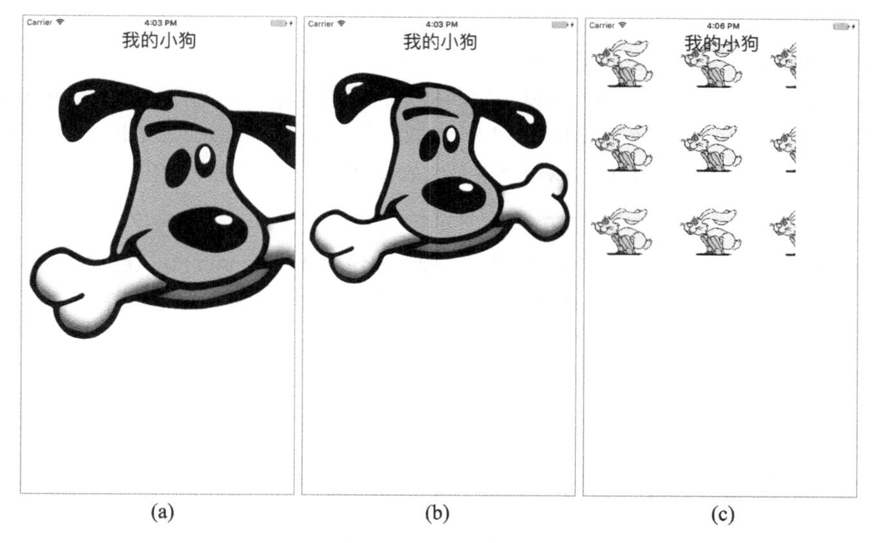

(a)　(b)　(c)

图12-5　绘制方法

下面通过一个实例介绍一下图像和文本的绘制，实例界面如图12-5b所示。请参考12.1.2节构建工程**ImageStringSample**，自定义视图类MyView的drawRect:方法（Swift版中是draw(_ rect:)方法）如下所示：

```swift
override func draw(_ rect: CGRect) {

    //填充白色背景
    UIColor.white.setFill()
    UIRectFill(rect)

    let image = UIImage(named: "dog")

    //设置一个rect矩形区域
    let imageRect = CGRect(x: 0, y: 40, width: UIScreen.main.bounds.size.width,
    ➥height: UIScreen.main.bounds.size.width)                                   ①
    //绘制图片
    image!.draw(in: imageRect)                                                   ②
    //image!.draw(at: CGPoint(x: 0, y: 40))                                      ③
    //image!.drawAsPattern(in: CGRect(x: 0, y: 0, width: 320, height: 400))
                                                                                 ④

    let title: NSString = "我的小狗"
    let font = UIFont.systemFont(ofSize: 28)                                     ⑤
    let attr = [NSFontAttributeName: font]                                       ⑥
    //获得字符串大小
    let size = title.size(attributes: attr)                                      ⑦
    //水平居中时x轴坐标
    let xpos = UIScreen.main.bounds.midX - size.width / 2;                       ⑧
    //绘制字符串
    title.draw(at: CGPoint(x: xpos, y: 20), withAttributes:attr)                 ⑨
    //let stringRect = CGRect(x: xpos, y: 60, width: 100, height: 40)
    //title.draw(in: stringRect, withAttributes: attr)

}
```

```objc
- (void)drawRect:(CGRect)rect {

    //填充白色背景
    [[UIColor whiteColor] setFill];
    UIRectFill(rect);

    UIImage* image = [UIImage imageNamed:@"dog"];

    //设置一个rect矩形区域
    CGRect imageRect = CGRectMake(0, 40, [[UIScreen mainScreen]
    ➥bounds].size.width,
    ➥[[UIScreen mainScreen] bounds].size.width);                                ①
    //绘制图片
    [image drawInRect:imageRect];                                                ②
    //[image drawAtPoint:CGPointMake(0, 40)];                                    ③
    //[image drawAsPatternInRect:CGRectMake(0, 0, 320, 400)];                    ④

    NSString *title = @"我的小狗";
    UIFont *font = [UIFont systemFontOfSize:28];                                 ⑤
    NSDictionary *attr = @{NSFontAttributeName:font};                            ⑥
    //获得字符串大小
    CGSize size = [title sizeWithAttributes:attr];                               ⑦
    //水平居中时x轴坐标
    CGFloat xpos = [[UIScreen mainScreen] bounds].size.width / 2 -
    ➥size.width / 2;                                                            ⑧
    //绘制字符串
    [title drawAtPoint:CGPointMake(xpos, 20) withAttributes:attr];               ⑨
    //CGRect stringRect = CGRectMake(xpos, 60, 100, 40);
    //[title drawInRect:stringRect withAttributes:attr];
}
```

上述代码中，第①行至第④行用于绘制图片，第⑤行至第⑨行用于绘制字符串。其中第①行用于设置一个矩形区域，这个区域用来绘制图片，该区域是一个与屏幕同宽的正方形；第②行通过drawInRect:方法（Swift版中是draw(in:)方法）绘制图片；第③行通过drawAtPoint:方法（Swift版中是draw(at:)方法）绘制图片；第④行通过drawAsPatternInRect:方法（Swift版中是drawAsPattern(in:)方法）绘制图片。

第⑤行用于创建字号28的系统字体UIFont对象。第⑥行按照NSFontAttributeName键将UIFont对象放到attr字典集合中。attr字典集合用来设置绘制字符串时的一些属性，例如字体、颜色和样式等。第⑦行用于获得字符串大小。由于绘制字符串时想让其水平居中，所以必须获得字符串的宽度。第⑧行计算水平居中时X轴坐标，其中[[UIScreen mainScreen] bounds].size.width / 2（Swift版本是UIScreen.main.bounds.midX）表达式表示屏幕宽度的一半。第⑨行通过drawAtPoint:方法（Swift版中是draw(at:)方法）绘制字符串。

> **注意** UIKit的坐标系原点在左上角，drawAtPoint:和drawAtPoint: withAttributes:等基于坐标点的绘制方法也是从左上角开始绘制的。上述代码中，第⑨行绘制字符串时，为了让字符串水平居中，我们必须要找到它的左上角坐标，注意它的X轴坐标不是"屏幕宽度一半"，而是"屏幕宽度一半"减"字符串所占空间宽度的一半"。

12.2　Quartz 图形上下文

我们在前面的实例中多次提到图形上下文，这是一个什么概念呢？

图形上下文包含绘制系统执行后绘制命令所需要的信息，它定义了各种基本的绘制参数，如绘制使用的颜色、裁剪区域、线段的宽度及风格信息、字体信息等。这些信息被封装到CGContext对象中，CGContext就是图形上下文。在前面的实例中，我们采用的都是默认的图形上下文。在复杂情况下，我们可以自定义图形上下文，这可以通过UIGraphicsGetCurrentContext函数实现。

下面通过一个简单的实例了解如何自定义图形上下文。如图12-6所示，我们在整个视图中绘制一个红色三角型。

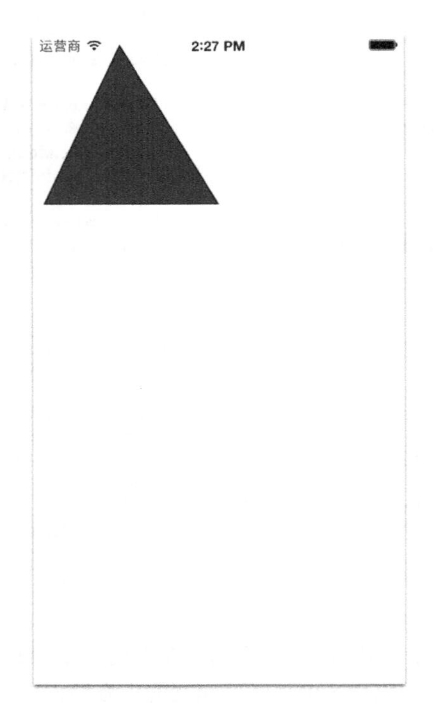

图12-6　绘制三角形

这个实例看似简单，实际上则比较复杂，它需要进行如下几个步骤。

(1) 通过路径描述三角形轮廓。

(2) 三角形轮廓描边。

(3) 三角形填充红色。

(4) 绘制路径。

关于路径的内容，我们将在下一节中介绍。这样复杂的步骤已经不能使用默认的图形上下文，而是需要使用自定义图形上下文。因为自定义图形上下文可以设置复杂的绘制参数和步骤，然后绘制执行时，一次性绘制出来。

首先，请参考12.1.2节创建StrokedFilledTriangle工程，并创建自定义视图类MyView。自定义视图类MyView的drawRect:方法（Swift版中是draw(_ rect:)方法）如下所示：

```
override func draw(_ rect: CGRect) {

    //填充白色背景
    UIColor.white.setFill()
    UIRectFill(rect)

    //自定义图形上下文
    let context = UIGraphicsGetCurrentContext()                        ①
    context!.move(to: CGPoint(x: 75, y: 10))                           ②
    context!.addLine(to: CGPoint(x: 10, y: 150))
    context!.addLine(to: CGPoint(x: 160, y: 150))
    context!.closePath()                                               ③

    //设置黑色描边参数
    UIColor.black.setStroke()
    //设置红色条填充参数
    UIColor.red.setFill()
    //绘制路径
    context!.drawPath(using: CGPathDrawingMode.fillStroke)            ④

}
```

```
- (void)drawRect:(CGRect)rect {

    //填充白色背景
    [[UIColor whiteColor] setFill];
    UIRectFill(rect);

    //自定义图形上下文
    CGContextRef context = UIGraphicsGetCurrentContext();              ①
    CGContextMoveToPoint (context, 75, 10);                            ②
    CGContextAddLineToPoint (context, 10, 150);
    CGContextAddLineToPoint (context, 160, 150);
    CGContextClosePath(context);                                       ③
    //设置黑色描边参数
    [[UIColor blackColor] setStroke];
    //设置红色条填充参数
    [[UIColor redColor] setFill];
    //绘制路径
    CGContextDrawPath(context, kCGPathFillStroke);                     ④
}
```

在上述代码中。第①行至第④行用于获得图形上下文和设置参数。第①行的UIGraphicsGetCurrentContext()函数用于创建图形上下文CGContext对象，第②行至第③行与路径有关，相关内容留在下一节介绍。

注意　图形上下文类型在Objective-C版本中使用的是CGContextRef，它是CGContext的引用类型。

12.3　Quartz 路径

在绘制复杂图形时会用到路径，上一节的实例已经用到了路径相关的代码，本节中我们来介绍Quartz路径。

12.3.1　Quartz 路径概述

Quartz路径可以用来描述矩形、圆及其他想要画的2D几何图形。通过路径，我们可以对这些几何图形进行描边、填充和描边填充处理。Core Graphics（Quartz 2D）中有4个基本图元用于描述路径：点、线段、弧和贝塞尔（Bézier）曲线。

1. 点

点是二维空间中的一个位置，不要把它想成像素，一个点完全不占空间，所以画一个点不会在屏幕上显示任何东西。我们可以在路径里加入很多的点，想加多少加多少。

2. 线段

线段由两个点定义：起点和终点。线段可以通过描边绘制出来，我们可以通过设置图形上下文，如画笔宽度或者颜色等参数，绘制出两点之间的线段。线段没有面积，所以它们不能被填充。我们可以用一组线段或曲线组成一个具有闭合路径的几何图形，然后填充它。

3. 弧

弧可以由一个圆心点、半径、起始角和结束角描述。圆是弧的特例，只需要设置起始角为0度，结束角为360度就可以了。因为弧是占有一定面积的路径，所以可以被填充、描边和描边填充出来。

4. 贝塞尔曲线

贝塞尔曲线是法国数学家贝塞尔在工作中发现的，任何一条曲线都可以通过与它相切的控制线两端的点的位置来定义。因此，贝塞尔曲线可以用4个点描述，其中两个点描述两个端点（图12-7中的P1和P2，图12-8中的P2和P3），另外两个点描述每一端的切线（图12-7中的P0，图12-8中的P1和P0）。贝塞尔曲线可以分为：二次方贝塞尔曲线（如图12-7所示）和高阶贝塞尔曲线（图12-8是三次方贝塞尔曲线）。

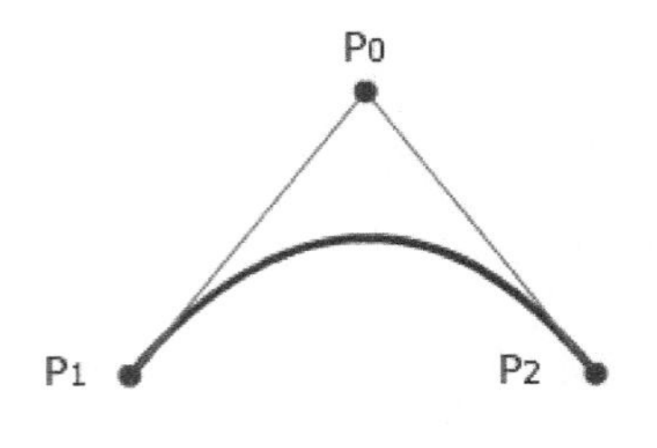

图12-7　二次方贝塞尔曲线

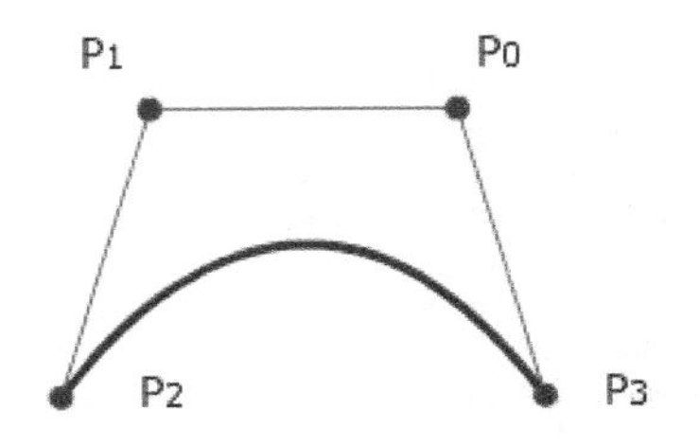

图12-8　三次方贝塞尔曲线

下面我们解释一下上节实例有关路径的代码，其中下面4条语句是绘制路径：

```
context!.move(to: CGPoint(x: 75, y: 10))          ①
context!.addLine(to: CGPoint(x: 10, y: 150))      ②
context!.addLine(to: CGPoint(x: 160, y: 150))     ③
context!.closePath()                              ④
```

```
CGContextMoveToPoint (context, 75, 10);           ①
CGContextAddLineToPoint (context, 10, 150);       ②
CGContextAddLineToPoint (context, 160, 150);      ③
CGContextClosePath(context);                      ④
```

上述代码中，第①行的`CGContextMoveToPoint`函数（Swift版`move(to:)`方法）表示以(75, 10)作为绘制起始点。第②行的`CGContextAddLineToPoint`函数（Swift版`addLine(to:)`方法）表示绘制从(75, 10)到(10, 150)的线段。类似地，第③行用于绘制从(10, 150)到(160, 150)的线段。这样，通过上述3条语句就定义了一个三角形路径，如图12-9所示。路径是按照(75, 10) →(10, 150)→ (160, 150)顺序定义的，因此它不是闭合的。如果需要闭合，可以调用第④行的`CGContextClosePath(context)`函数（Swift版本是`context!.closePath()`）实现，如图12-10所示。

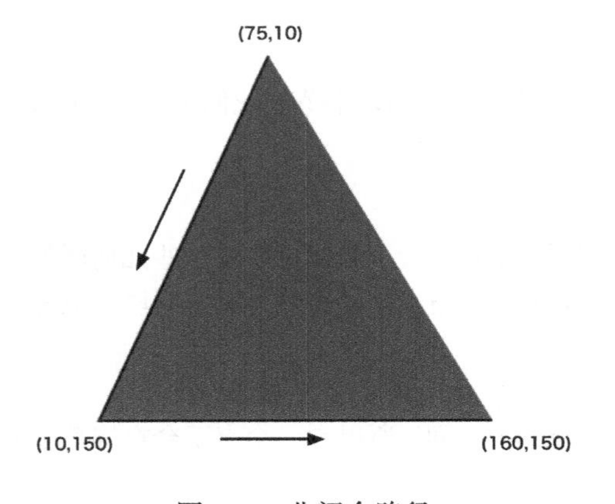

图12-9　非闭合路径

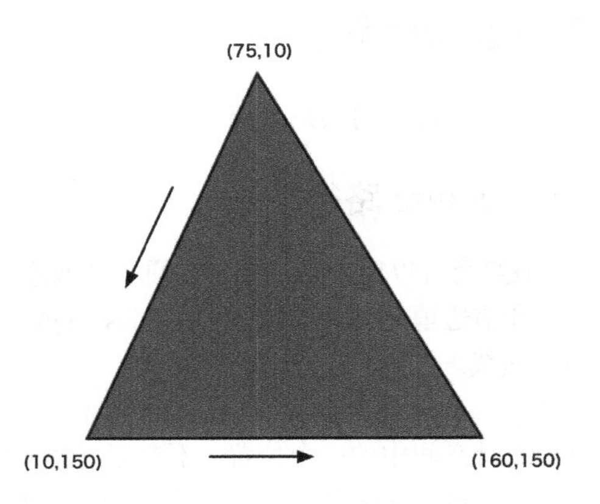

图12-10　闭合路径

定义和绘制路径是两个不同的操作，一般先定义路径，再绘制它。在Objecitve-C版本中，CGContextDrawPath(context, kCGPathFillStroke)函数实现绘制路径，其中kCGPathFillStroke参数是填充描边处理，它是CGPathDrawingMode枚举类型中定义的成员，其中常用的成员有kCGPathFill和kCGPathStroke，分别代表填充和描边处理。在Swift版本中，context!.drawPath(using: CGPathDrawingMode.fillStroke)语句实现了绘制路径，其中CGPathDrawingMode.fillStroke参数是填充描边处理，fillStroke是CGPathDrawingMode枚举类型中定义的成员，其中常用的成员为fill和stroke，分别代表填充和描边处理。

12.3.2　实例：使用贝塞尔曲线

下面我们介绍一个使用贝塞尔曲线定义路径的实例。图12-11是半个花瓶的轮廓，它的轮廓很复杂，很多都不是直线，所以需要分成很多贝塞尔曲线，然后找到贝塞尔曲线的控制点，就可以绘制它了。最好在坐标纸上找到各个曲线控制点，当然这需要一系列复杂的计算。

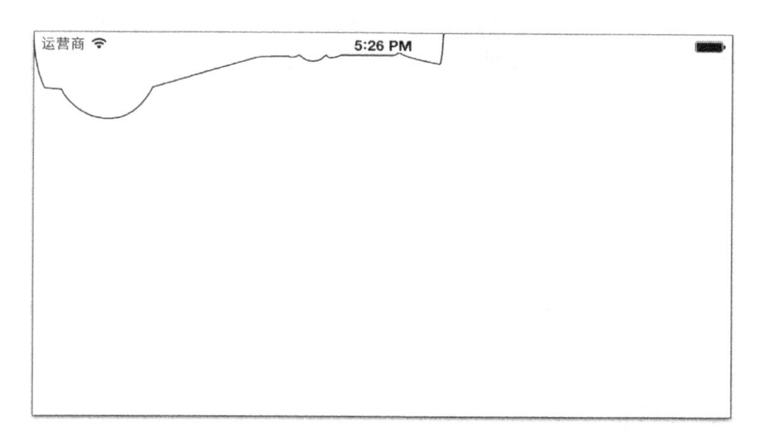

图12-11　曲线定义路径实例

首先，请参考12.1.2节创建BezierCurve工程，并创建自定义视图类MyView，该类的drawRect:方法（Swift版是draw(_ rect:)方法）如下所示：

```
override func draw(_ rect: CGRect) {

    //填充白色背景
    UIColor.white.setFill()
    UIRectFill(rect)

    let context = UIGraphicsGetCurrentContext()

    context!.move(to: CGPoint(x: 333, y: 0))                                    ①
    context!.addCurve(to:CGPoint(x: 330, y: 26) , control1: CGPoint(x: 333, y: 0),
    ➥control2: CGPoint(x: 332, y: 26))                                          ②
    context!.addCurve(to: CGPoint(x: 299, y: 17), control1: CGPoint(x: 330, y:
    ➥26),
    ➥control2: CGPoint(x: 299, y: 20))                                          ③
    context!.addLine(to: CGPoint(x: 296, y: 17))
    context!.addCurve(to: CGPoint(x: 291, y: 19), control1: CGPoint(x: 296, y: 17),
    ➥control2: CGPoint(x: 296, y: 19))
    context!.addLine(to: CGPoint(x: 250, y: 19))
    context!.addCurve(to: CGPoint(x: 238, y: 19), control1: CGPoint(x: 250, y: 19),
    ➥control2: CGPoint(x: 241, y: 24))
    context!.addCurve(to: CGPoint(x: 227, y: 24), control1: CGPoint(x: 236, y: 20),
    ➥control2: CGPoint(x: 234, y: 24))
    context!.addCurve(to: CGPoint(x: 216, y: 19), control1: CGPoint(x: 220, y: 24),
    ➥control2: CGPoint(x: 217, y: 19))
```

```
- (void)drawRect:(CGRect)rect {

    //填充白色背景
    [[UIColor whiteColor] setFill];
    UIRectFill(rect);

    CGContextRef context = UIGraphicsGetCurrentContext();

    CGContextMoveToPoint(context, 333, 0);                                       ①
    CGContextAddCurveToPoint(context, 333, 0, 332, 26, 330, 26);                 ②
    CGContextAddCurveToPoint(context, 330, 26, 299, 20, 299, 17);                ③
    CGContextAddLineToPoint(context, 296, 17);
    CGContextAddCurveToPoint(context, 296, 17, 296, 19, 291, 19);
    CGContextAddLineToPoint(context, 250, 19);
    CGContextAddCurveToPoint(context, 250, 19, 241, 24, 238, 19);
    CGContextAddCurveToPoint(context, 236, 20, 234, 24, 227, 24);
    CGContextAddCurveToPoint(context, 220, 24, 217, 19, 216, 19);
    CGContextAddCurveToPoint(context, 214, 20, 211, 22, 207, 20);
    CGContextAddCurveToPoint(context, 207, 20, 187, 20, 182, 21);
    CGContextAddLineToPoint(context, 100, 45);
    CGContextAddLineToPoint(context, 97, 46);
    CGContextAddCurveToPoint(context, 97, 46, 86, 71, 64, 72);
    CGContextAddCurveToPoint(context, 42, 74, 26, 56, 23, 48);
    CGContextAddLineToPoint(context, 9, 47);
```

```
    context!.addCurve(to: CGPoint(x: 207, y: 20), control1: CGPoint(x: 214, y: 20),
    ➥control2: CGPoint(x: 211, y: 22))
    context!.addCurve(to: CGPoint(x: 182, y: 21), control1: CGPoint(x: 207, y: 20),
    ➥control2: CGPoint(x: 187, y: 20))
    context!.addLine(to: CGPoint(x: 100, y: 45))
    context!.addLine(to: CGPoint(x: 97, y: 46))
    context!.addCurve(to: CGPoint(x: 64, y: 72), control1: CGPoint(x: 97, y: 46),
    ➥control2: CGPoint(x: 86, y: 71))
    context!.addCurve(to: CGPoint(x: 23, y: 48), control1: CGPoint(x: 42, y: 74),
    ➥control2: CGPoint(x: 26, y: 56))
    context!.addLine(to: CGPoint(x: 9, y: 47))
    context!.addCurve(to: CGPoint(x: 0, y: 0), control1: CGPoint(x: 9, y: 47),
    ➥control2: CGPoint(x: 0, y: 31))                                          ④

    context!.strokePath()

}
```

```
    CGContextAddCurveToPoint(context, 9, 47, 0, 31, 0, 0);                     ④
    CGContextStrokePath(context);
}
```

上述代码中，第①行至第④行都是在定义这个瓶子的轮廓，其中Objective-C版本中使用CGContextAddCurve-ToPoint函数定义贝塞尔曲线，在Swift版本中使用CGContext类的addCurve方法定义贝塞尔曲线，比较说明如下：

```
func addCurve(to end: CGPoint,   //端点
    control1: CGPoint,           //第一控制点
    control2: CGPoint)           //第二控制点
```

```
void CGContextAddCurveToPoint(CGContextRef c,
    CGFloat cp1x, CGFloat cp1y,      //第一控制点
    CGFloat cp2x, CGFloat cp2y,      //第二控制点
    CGFloat x, CGFloat y)            //端点
```

注意　addCurve方法的第一个参数是端点，而CGContextAddCurveToPoint函数的最后两个参数才是端点。

12.4　Quartz 坐标变换

图形的另外一种操作就是变换，主要包括平移、缩放和旋转等形式的变换。变换离不开坐标，不同的绘图系统对于坐标系的定义也有所区别。

12.4.1　坐标系

苹果的2D图形技术是Quartz 2D和UIKit。Quartz 2D是macOS和iOS环境下的2D绘图引擎，涉及的内容包括：基于路径的绘图、透明度绘图、遮盖、阴影、透明层、颜色管理、抗锯齿渲染、生成PDF以及PDF元数据相关的处理。在iOS中，我们还可以通过UIKit进行图形绘制，但是Quartz 2D和UIKit的坐标系不同。

Quartz 2D中，坐标系的原点在左下角，X轴方向向右为正方向，Y轴方向向上为正方向，如图12-12所示。

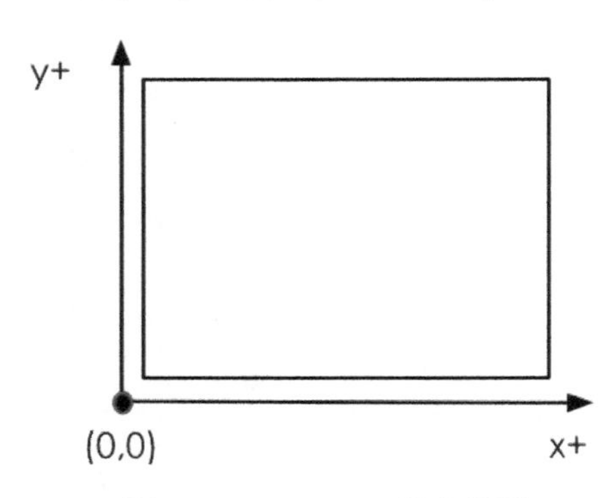

图12-12　Quartz 2D的坐标系

UIKit中，坐标系的原点在左上角，X轴方向向右为正方向，Y轴方向向下为正方向，如图12-13所示。

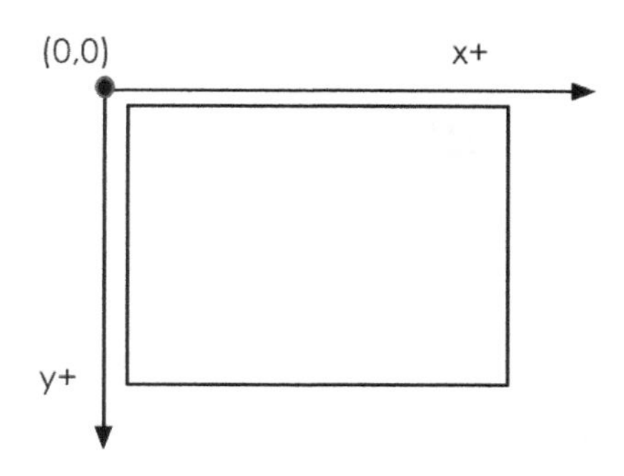

图12-13　UIKit的坐标系

下面我们通过实例介绍它们的不同，图12-14是一张可爱的招财猫图片。

图12-14　原始图片

绘制这样的图片，在不同坐标系下会有什么不同呢？绘制图片可以通过上一节介绍的UIImage类中几个绘制图像的方法实现，这里就不再介绍了。我们看看下面的实现方式：

```
override func draw(_ rect: CGRect) {

    //填充白色背景
    UIColor.white.setFill()
    UIRectFill(rect)

    //创建UIImage图片对象
    let uiImage  = UIImage(named: "cat")

    //将UIImage图片对象转换为CGImage图片对象
    let cgImage = uiImage!.cgImage

    let context = UIGraphicsGetCurrentContext()

    let imageRect = CGRect(x: 0, y: 0, width: uiImage!.size.width, height:
    ➥uiImage!.size.height)
    context!.draw(cgImage!, in: imageRect)
}
```

```
- (void)drawRect:(CGRect)rect {

    //填充白色背景
    [[UIColor whiteColor] setFill];
    UIRectFill(rect);

    //创建UIImage图片对象
    UIImage *uiImage = [UIImage imageNamed:@"cat"];
    //将UIImage图片对象转换为CGImage图片对象
    CGImageRef cgImage = uiImage.CGImage;

    CGContextRef context = UIGraphicsGetCurrentContext();

    CGRect imageRect = CGRectMake(0, 0, uiImage.size.width,
    ➥uiImage.size.height);
    CGContextDrawImage(context, imageRect, cgImage);
}
```

上面的实现方式全部通过Quartz 2D函数来实现绘制，其中CGContextDrawImage函数（Swift版本中CGContext类的draw方法）相当于UIImage类中的绘制图像方法（drawAtPoint:和drawInRect:等）。但是绘制出来的结果令人沮丧，图像倒过来了，如图12-15所示。这是由于Quartz 2D坐标系和UIKit坐标系不同所导致的，我们使用的绘制函数CGContextDrawImage都是基于Quartz 2D坐标系的。

图12-15　绘制图片

为了正确显示，我们需要对其进行坐标变换。这里需要修改的代码如下：

```swift
override func draw(_ rect: CGRect) {

    //填充白色背景
    UIColor.white.setFill()
    UIRectFill(rect)

    //创建UIImage图片对象
    let uiImage  = UIImage(named: "cat")
    //将UIImage图片对象转换为CGImage图片对象
    let cgImage = uiImage!.cgImage

    let context = UIGraphicsGetCurrentContext()

    context!.scaleBy(x: 1, y: -1)                                    ①
    context!.translateBy(x: 0, y: -uiImage!.size.height)             ②

    let imageRect = CGRect(x: 0, y: 0, width: uiImage!.size.width, height:
    ➥uiImage!.size.height)

    context!.draw(cgImage!, in: imageRect)
}
```

```objc
- (void)drawRect:(CGRect)rect {

    //填充白色背景
    [[UIColor whiteColor] setFill];
    UIRectFill(rect);

    //创建UIImage图片对象
    UIImage *uiImage = [UIImage imageNamed:@"cat"];
    //将UIImage图片对象转换为CGImage图片对象
    CGImageRef cgImage = uiImage.CGImage;

    CGContextRef context = UIGraphicsGetCurrentContext();

    CGContextScaleCTM(context, 1, -1);                                  ①
    CGContextTranslateCTM(context, 0, -uiImage.size.height);            ②

    CGRect imageRect = CGRectMake(0, 0, uiImage.size.width,
    ➥uiImage.size.height);
    CGContextDrawImage(context, imageRect, cgImage);

}
```

这里我们添加了第①行和第②行代码，它们的含义先不用理解。这是关于坐标变换的问题，下一节会介绍。添加之后的运行结果如图12-16所示。

提示

有的读者可能已经发现，如果使用UIImage类提供的绘制图像方法（drawAtPoint:和drawInRect:等），不会出现这个问题。这是因为UIImage采用的坐标是UIKit坐标，已经进行了转换，不需要再进行转换了。

图12-16 坐标变换之后的运行效果

12.4.2 2D 图形的基本变换

在图形变换的过程中，需要大量使用矢量、矩阵及其运算。图形变换本质上是矩阵计算，每一种变换都有一个矩阵，原始图形中的像素点坐标与矩阵进行计算得到新的像素点坐标，然后重新绘制这些像素到视图。这里涉及线性代数和解析几何的相关知识，这些内容超出了本书的范围，这里不再介绍。2D图形的基本变换包括平移、缩放和旋转这3种变换。

- **平移变换**。平移是物体从一个位置到另一位置所做的直线移动。如果要把一个位于 $P(x, y)$ 的点移到新位置 $P'(x', y')$，只要在原坐标上加上平移距离 T_x 及 T_y 即可（如图12-17所示）。

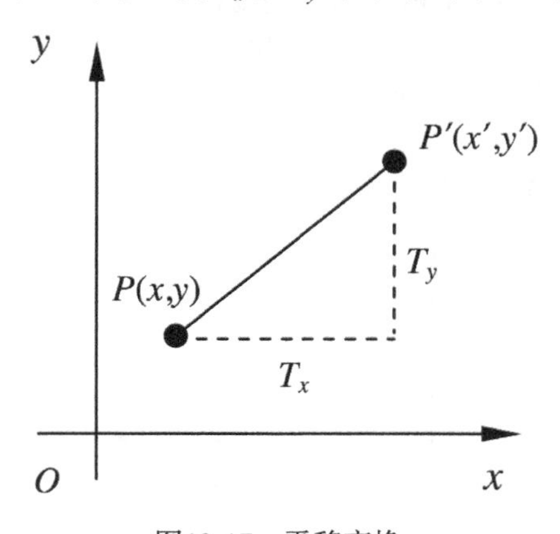

图12-17 平移变换

- **缩放变换**。用来改变一物体大小的变换称为缩放变换。如果要对一个多边形进行等比例变换，那么可以把各顶点的坐标(x, y)均乘以缩放因子S_x和S_y，以得到变换后的坐标(x', y')。其中，S_x 及 S_y 可以是任意正数，S_x、S_y可以相等或不等。如果缩放因子的数值小于1，则物体会缩小；大于1，则物体会放大；S_x及S_y都等于 1，则物体的大小和形状保持不变。需要注意的是，图12-18表示的缩放变换是针对坐标原点的。

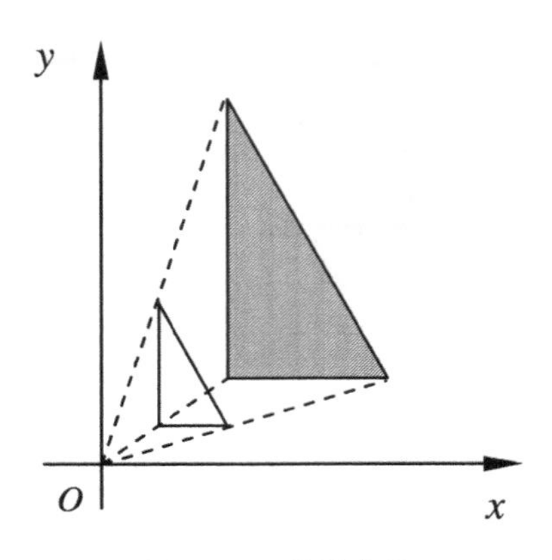

图12-18　缩放变换

- **旋转变换**。物体上的各点绕一固定点沿圆周路径做转动称为旋转变换。我们可用旋转角表示旋转量的大小。一个点由位置(x, y)旋转到(x', y')的角度为自水平轴算起的角度，如图12-19所示，θ为旋转角。

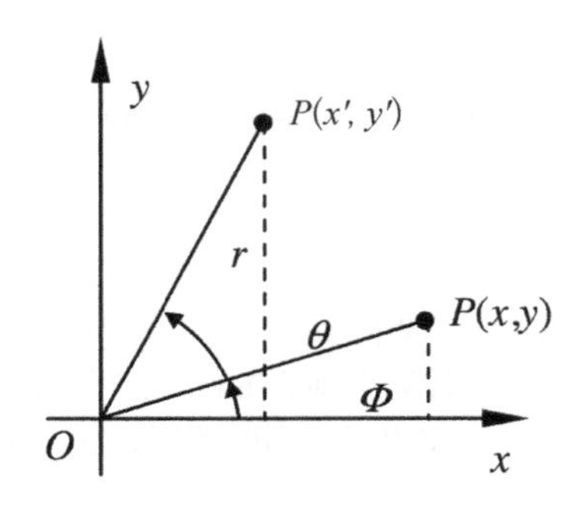

图12-19　旋转变换

有的图形系统还提供另外几种很有用的变换，如反射变换及错切变换等。这里重点介绍一下反射变换。反射是用来产生物体的镜像的一种变换。物体的镜像一般是相对于一个对称轴生成的，因此反射变换可以分为：X轴对称变换、Y轴对称变换和坐标原点的对称变换。

1. X轴对称变换

X轴对称变换是一种特殊形式的缩放变换，其中缩放因子 S_x和S_y分别为S_x=1和S_y= –1，如图12-20所示。

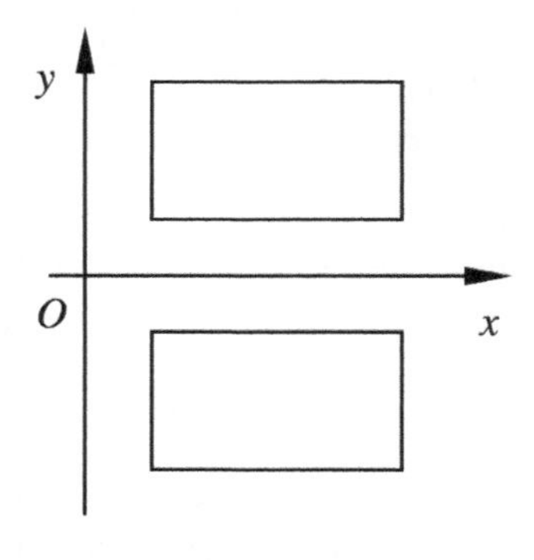

图12-20　X轴对称变换

2. Y轴对称变换

Y轴对称变换是一种特殊形式的缩放变换，其中缩放因子 S_x和S_y分别为S_x=–1和S_y= 1，如图12-21所示。

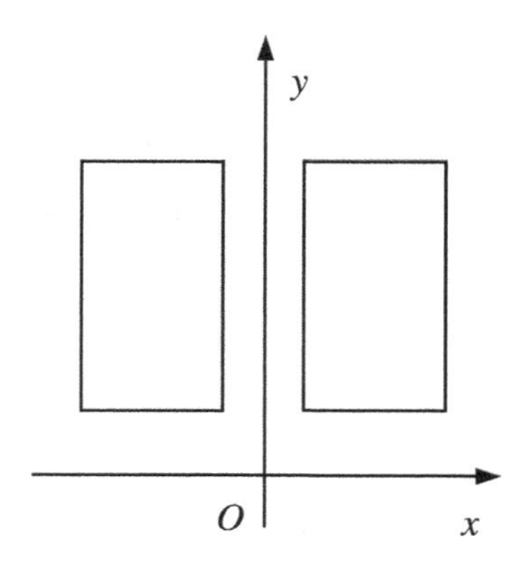

图12-21 Y轴对称变换

3. 坐标原点的对称变换

坐标原点的对称变换是一种特殊形式的缩放变换，其中缩放因子 S_x和S_y分别为S_x=–1和S_y=–1，如图12-22所示。

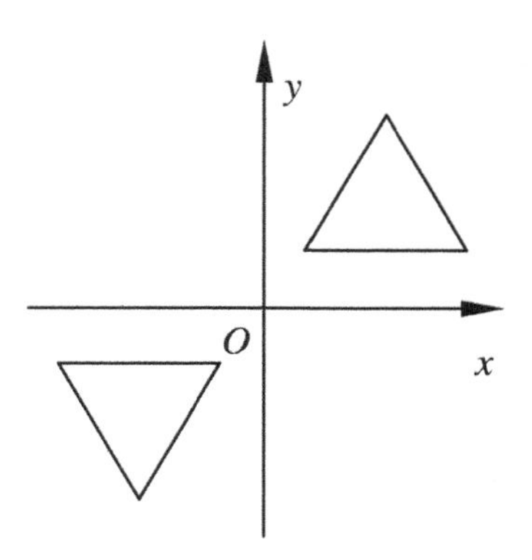

图12-22 坐标原点的对称变换

12.4.3 CTM 变换

Quartz 2D提供了多种形式的变换，其中主要有当前变换矩阵（Current Transformation Matrix，CTM）变换和仿射（affine）变换。CTM变换比较简单，主要的函数有如下几个。

- **CGContextRotateCTM**：旋转变换。
- **CGContextScaleCTM**：缩放变换。
- **CGContextTranslateCTM**：平移变换。

1. 平移变换

平移变换根据指定的T_x、T_y值移动坐标系统的原点。我们通过Objective-C中CGContextTranslateCTM函数实现平移变换，在Swift中使用CGContext类的translateBy方法实现平移变换，比较说明如下：

```
func translateBy(x tx: CGFloat, y ty: CGFloat)
```

```
void CGContextTranslateCTM(CGContextRef c, CGFloat tx, CGFloat ty)
```

图12-23右图显示了一幅图片沿X轴移动了100个单位，沿Y轴移动了50个单位，具体代码如下：

```
override func draw(_ rect: CGRect) {

    //填充白色背景
```

```
- (void)drawRect:(CGRect)rect {

    //填充白色背景
```

```
    UIColor.white.setFill()
    UIRectFill(rect)

    //创建UIImage图片对象
    let uiImage  = UIImage(named: "cat")
    //将UIImage图片对象转换为CGImage图片对象
    let cgImage = uiImage!.cgImage

    let context = UIGraphicsGetCurrentContext()

    //平移变换
    context!.translateBy(x: 100, y: 50)                          ①

    let imageRect = CGRect(x: 0, y: 0, width: uiImage!.size.width, height:
    ➥uiImage!.size.height)
    context!.draw(cgImage!, in: imageRect)

}
```

```
    [[UIColor whiteColor] setFill];
    UIRectFill(rect);

    //创建UIImage图片对象
    UIImage *uiImage = [UIImage imageNamed:@"cat"];
    //将UIImage图片对象转换为CGImage图片对象
    CGImageRef cgImage = uiImage.CGImage;

    CGContextRef context = UIGraphicsGetCurrentContext();

    //平移变换
    CGContextTranslateCTM (context, 100, 50);      ①

    CGRect imageRect = CGRectMake(0, 0, uiImage.size.width,
    ➥uiImage.size.height);
    CGContextDrawImage(context, imageRect, cgImage);

}
```

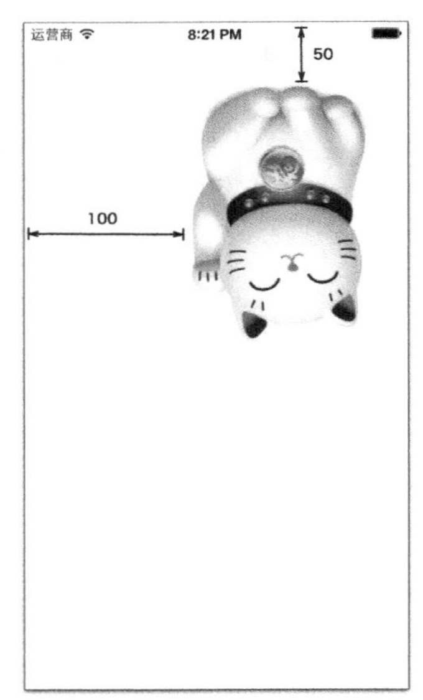

图12-23　平移变换

2. 缩放变换

缩放操作根据指定的S_x、S_y缩放因子来改变图像的大小，从而放大或缩小图像。S_x、S_y缩放因子的大小决定了新的坐标系是否比原始坐标系大或者小。另外，通过指定S_x因子为负数来指定变换是X轴对称变换，同样可以指定S_y因子为负数来指定变换是Y轴对称变换。

在Objective-C中，我们通过CGContextScaleCTM函数实现缩放变换。在Swift中，使用CGContext类的scaleBy方法实现缩放变换。比较说明如下：

```
func scaleBy(x sx: CGFloat, y sy: CGFloat)
```

```
void CGContextScaleCTM(CGContextRef c, CGFloat sx, CGFloat sy)
```

图12-24显示了指定S_x因子为0.5，S_y因子为0.75后的缩放效果，具体代码如下：

```
override func draw(_ rect: CGRect) {

    //填充白色背景
    UIColor.white.setFill()
    UIRectFill(rect)
```

```
- (void)drawRect:(CGRect)rect {

    //填充白色背景
    [[UIColor whiteColor] setFill];
    UIRectFill(rect);
```

```
    //创建UIImage图片对象
    let uiImage  = UIImage(named: "cat")
    //将UIImage图片对象转换为CGImage图片对象
    let cgImage = uiImage!.cgImage

    let context = UIGraphicsGetCurrentContext()

    //缩放变换
    context!.scaleBy(x: 0.5, y: 0.75)

    let imageRect = CGRect(x: 0, y: 0, width: uiImage!.size.width, height:
    ➥uiImage!.size.height)
    context!.draw(cgImage!, in: imageRect)

}
```

```
    //创建UIImage图片对象
    UIImage *uiImage = [UIImage imageNamed:@"cat"];
    //将UIImage图片对象转换为CGImage图片对象
    CGImageRef cgImage = uiImage.CGImage;

    CGContextRef context = UIGraphicsGetCurrentContext();

    //缩放变换
    CGContextScaleCTM (context, 0.5, 0.75);

    CGRect imageRect = CGRectMake(0, 0, uiImage.size.width,
    ➥uiImage.size.height);
    CGContextDrawImage(context, imageRect, cgImage);

}
```

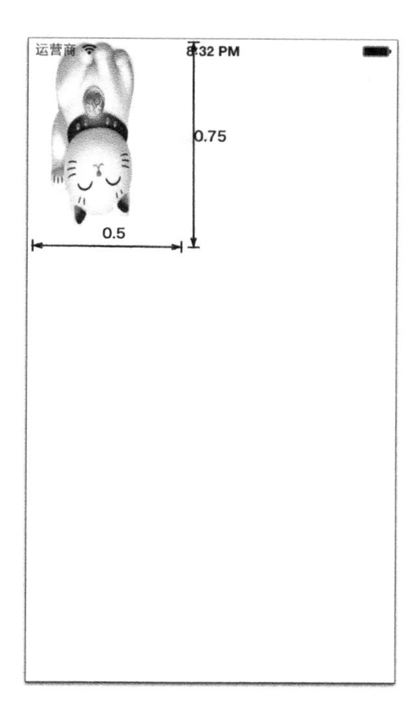

图12-24　缩放变换

3. 旋转变换

S_x、S_y旋转变换根据指定的角度来旋转坐标。在Objective-C中，我们通过CGContextRotateCTM函数实现旋转变换。在Swift中，使用CGContext类的rotate方法实现旋转变换。比较说明如下：

```
func rotate(by angle: CGFloat)
```

```
void CGContextRotateCTM(CGContextRef c, CGFloat angle)
```

其中angle是旋转角度，以弧度为单位。

图12-25显示了图片以原点为中心顺时针旋转45度，相关代码如下所示：

```
override func draw(_ rect: CGRect) {

    //填充白色背景
    UIColor.white.setFill()
    UIRectFill(rect)

    //创建UIImage图片对象
    let uiImage  = UIImage(named: "cat")
```

```
- (void)drawRect:(CGRect)rect {

    //填充白色背景
    [[UIColor whiteColor] setFill];
    UIRectFill(rect);

    //创建UIImage图片对象
    UIImage *uiImage = [UIImage imageNamed:@"cat"];
```

```
    //将UIImage图片对象转换为CGImage图片对象
    let cgImage = uiImage!.cgImage

    let context = UIGraphicsGetCurrentContext()

    //旋转变换
    context!.rotate(by: CGFloat(45.0 * M_PI / 180.0))                ①

    let imageRect = CGRect(x: 0, y: 0, width: uiImage!.size.width, height:
    ➥uiImage!.size.height)
    context!.draw(cgImage!, in: imageRect)

}
```

```
    //将UIImage图片对象转换为CGImage图片对象
    CGImageRef cgImage = uiImage.CGImage;

    CGContextRef context = UIGraphicsGetCurrentContext();

    //旋转变换
    CGContextRotateCTM (context, (45.0 * M_PI / 180.0));             ①

    CGRect imageRect = CGRectMake(0, 0, uiImage.size.width,
    ➥uiImage.size.height);
    CGContextDrawImage(context, imageRect, cgImage);

}
```

上述代码中，第①行的(45.0 * M_PI / 180.0)表达式用于将弧度转化为度。

图12-25　旋转变换

提示　由于旋转操作使图片的部分区域置于屏幕之外，所以区域外的部分被裁减。如果旋转的弧度为负数，则图形是逆时针旋转。

4. 组合变换

有些情况下，我们需要组合变换，从而得到累加效果。默认情况下，采用Quartz绘制的图片都是倒置的（如图12-26a所示）。要想绘制出正常效果的图片（如图12-26b所示），则需要进行一系列的组合变化。其中主要代码如下，其他代码可参考12.4.1节：

```
//组合变换
context!.scaleBy(x: 1, y: -1)                                   ①
context!.translateBy(x: 0, y: -uiImage!.size.height)            ②
```

```
//组合变换
CGContextScaleCTM(context, 1, -1);                              ①
CGContextTranslateCTM(context, 0, uiImage.size.height);         ②
```

本例中是先进行了Y轴对称变换（缩放因子为Sx=-1和Sy= 1的缩放变换），见第①行。再进行平移变换，见第②行。

(a)

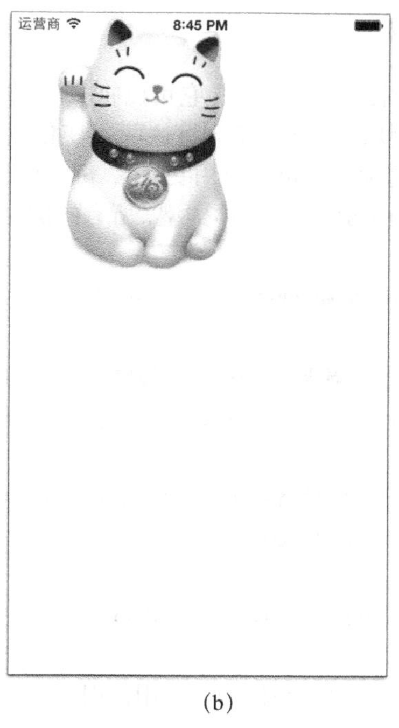

(b)

图12-26 组合变换

提示

当相同的绘制程序在一个UIView对象和Quartz图形上下文上进行绘制时，需要做一个变换，使Quartz图形上下文与UIView具有相同的坐标系。要达到这一目的，需要将Quartz图形上下文绘制的图形进行Y轴对称变换，即Y轴乘以–1，如图12-27b所示。然后再进行平移变化，向下移动该图形，如图12-27c所示。

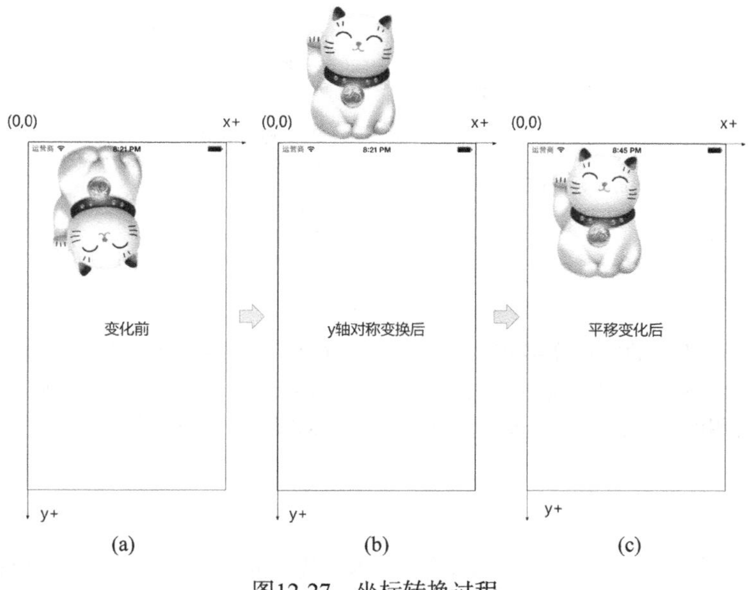

图12-27 坐标转换过程

12.4.4 仿射变换

仿射变换也是一种2D坐标变换，它可以将多次变换的效果累加起来，但当前变换矩阵却不能。例如，先平移(100,200)再旋转30度，仿射变换会将这两次变化效果累加起来，当前变换矩阵只旋转30度，不会进行平移(100,200)。因此说，仿射变换可以重用变换，，每一种变换都可以用矩阵表示，通过多次矩阵相乘得到最后结果。仿射变换是CGAffineTransform结构体类型，所有的仿射变换函数的返回值都是CGAffineTransform实例。仿射变换函数如下所示。

- **CGAffineTransformMakeRotation**：创建并初始化旋转矩阵，对应的Swift版本是init(rotationAngle: CGFloat)。
- **CGAffineTransformMakeScale**：创建新的缩放矩阵，对应的Swift版本是init(scaleX: CGFloat, y: CGFloat)。
- **CGAffineTransformMakeTranslation**：创建新的平移矩阵，对应的Swift版本是init(translationX: CGFloat, y: CGFloat)。
- **CGAffineTransformRotate**：旋转矩阵，对应的Swift版本是CGAffineTransform的rotated(by: CGFloat)方法。
- **CGAffineTransformScale**：缩放矩阵，对应的Swift版本是CGAffineTransform的scaledBy(x: CGFloat, y: CGFloat)方法。
- **CGAffineTransformTranslate**：平移矩阵，对应的Swift版本是CGAffineTransform的translatedBy(x: CGFloat, y: CGFloat)方法。

如图12-26b所示的效果，我们可以使用仿射变换，具体代码如下：

```
override func draw(_ rect: CGRect) {

    //填充白色背景
    UIColor.white.setFill()
    UIRectFill(rect)

    //创建UIImage图片对象
    let uiImage  = UIImage(named: "cat")
    //将UIImage图片对象转换为CGImage图片对象
    let cgImage = uiImage!.cgImage

    let context = UIGraphicsGetCurrentContext()

    //缩放变换
    var myAffine = CGAffineTransform(scaleX: 1, y: -1)                    ①
    //平移变换
    myAffine = myAffine.translatedBy(x: 0, y: -uiImage!.size.height)      ②
    //连接到CTM矩阵
    context!.concatenate(myAffine)                                        ③

    let imageRect = CGRect(x: 0, y: 0, width: uiImage!.size.width, height:
    ➥uiImage!.size.height)
    context!.draw(cgImage!, in: imageRect)

}
```

```
- (void)drawRect:(CGRect)rect {

    //填充白色背景
    [[UIColor whiteColor] setFill];
    UIRectFill(rect);

    //创建UIImage图片对象
    UIImage *uiImage = [UIImage imageNamed:@"cat"];
    //将UIImage图片对象转换为CGImage图片对象
    CGImageRef cgImage = uiImage.CGImage;

    CGContextRef context = UIGraphicsGetCurrentContext();

    //缩放变换
    CGAffineTransform myAffine = CGAffineTransformMakeScale(1, -1);       ①
    //平移变换
    myAffine = CGAffineTransformTranslate(myAffine, 0, -uiImage.size.height);
                                                                          ②
    //连接到CTM矩阵
    CGContextConcatCTM(context, myAffine);                                ③

    CGRect imageRect = CGRectMake(0, 0, uiImage.size.width,
    ➥uiImage.size.height);
    CGContextDrawImage(context, imageRect, cgImage);

}
```

首先，第①行用于创建新的缩放变换矩阵。第②行是在平移变换矩阵上乘以缩放变换矩阵，第③行通过CGContextConcatCTM函数（Swift版本中是CGContext类的concatenate(_ transform: CGAffineTransform)方法）连接到CTM矩阵。需要说明的是，仿射变换最后需要连接到CTM矩阵才能输出结果。

12.5 小结

通过本章的学习，我们可以了解到Quartz 2D绘图技术，其中包括UIKit绘图技术、绘制视图的路径、绘制图像和文本、坐标、Quartz坐标和坐标变换。

第 13 章

动画技术

13

作为移动应用，绚丽的动画是必不可少的。iOS应用实现动画的核心技术是Core Animation框架，简称CA。有时一些简单的动画不会直接使用Core Animation框架，而是使用UIView动画。UIView动画本质上也是Core Animation框架实现的，只不过进行了封装和优化。本章中，我们重点介绍视图动画和Core Animation框架。

13.1 视图动画

如果需要在视图（UIView）上进行一些简单的动画，可以使用视图动画。视图动画底层还是使用Core Animation，但是动画的实现细节都封装起来了。而且UIKit类通常都有animated布尔类型参数，它可以开启动画效果。下面的代码是开关控件（UISwitch）设置状态的方法，其中animated参数用于设定是否开启动画效果：

```
func setOn(_ on: Bool, animated: Bool)
```

```
- (void)setOn:(BOOL)on animated:(BOOL)animated
```

每个视图都关联到一个图层（CALayer）对象，视图主要用来处理事件，图层用来处理动画，视图上所有的动画、绘制和可视效果都直接或间接地由图层处理。

视图有一系列支持动画的属性，包括frame、bounds、center、alpha和transform等。此外，还有一些属性，比如动画延迟事件、动画曲线（淡入/淡出、淡入、淡出和线性等）、动画过渡、重复次数和自动反转等属性。

13.1.1 动画块

视图（UIView）动画采用代码块（Swift为闭包）形式的方法，其主要方法如下。

- + animateWithDuration:delay:options:animations:completion:
- + animateWithDuration:animations:completion:
- + animateWithDuration:animations:

这些方法之间是重载关系，相同名字的参数含义相同。下面看看+ animateWithDuration:delay:options:animations:completion:方法的声明：

```
class func animate(withDuration duration: TimeInterval,
        delay: TimeInterval,
        options: UIViewAnimationOptions = [],
        animations: @escaping () -> Void,
        completion: ((Bool) -> Void)? = nil)
```

```
+ (void)animateWithDuration:(NSTimeInterval)duration
    delay:(NSTimeInterval)delay
    options:(UIViewAnimationOptions)options
    animations:(void (^)(void))animations
    completion:(void (^)(BOOL finished))completion
```

这个方法是静态方法，其中参数duration是动画持续的时间；参数delay是动画延迟执行的时间，如果为0，则动画马上执行；options是执行动画选项的类型（UIViewAnimationOptions类型）。动画选项可以包括：动画曲线、动画过渡、重复次数和自动反转等内容设置。animations参数是一个代码块（Swift为闭包），用来设置动画的属性；completion参数也是一个代码块（Swift为闭包），是在动画结束时调用的。

下面通过一个实例介绍动画块的用法以及动画处理的过程。这个实例如图13-1所示。在屏幕上有一个球体图片和一个Tap Me按钮，当用户点击Tap Me按钮时，球体向下移动100个点，持续时间是1.5秒。当再次点击Tap Me

按钮时，球体向上移动100个点，就这样反复运动。

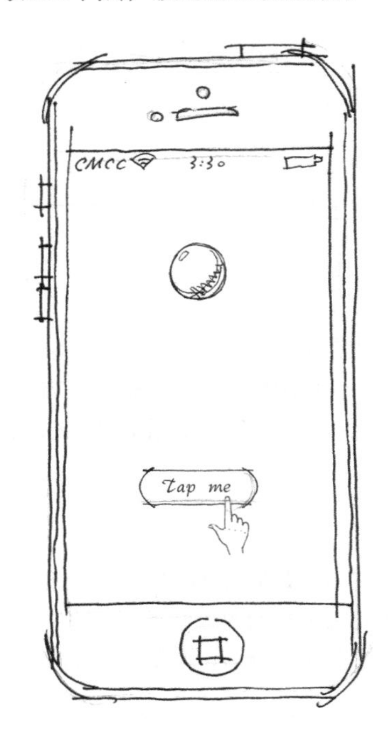

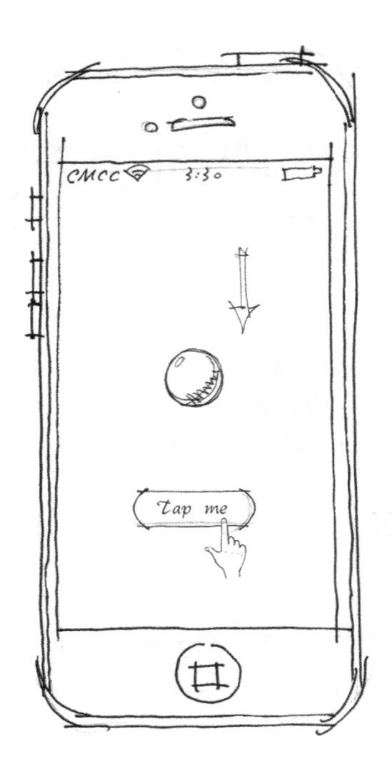

图13-1　动画块实例

首先，使用Single View Application模板创建一个名为AnimationBlock的工程，然后参考3.5.1节将其修改为纯代码工程。然后，添加资源图片文件到工程。读者可以从本章的“资源”文件夹中添加所有图片，详情可参考5.4节。

这里重点介绍一下视图控制器类ViewController，其代码如下：

```
//ViewController.swift文件
class ViewController: UIViewController {

    //小球运动方向标志，1表示向下运行，-1表示向上运行
    var flag = 1;

    var ball: UIImageView!

    override func viewDidLoad() {
        super.viewDidLoad()

        //界面初始化
        let screen = UIScreen.main.bounds

        let imageWidth: CGFloat = 86
        let imageHeight: CGFloat = 86
        let imageTopView: CGFloat = 150
        let imageFrame = CGRect(x: (screen.size.width - imageWidth) / 2,
        ➥y: imageTopView, width: imageWidth, height: imageHeight)
        //创建Image View对象
        self.ball = UIImageView(frame: imageFrame)
        //设置Image View的图片属性
        self.ball.image = UIImage(named: "Ball.png")
        //添加Image View到当前视图
        self.view.addSubview(self.ball)

        //创建按钮对象
```

```
//ViewController.m文件
#import "ViewController.h"

@interface ViewController () {
    //小球运动方向标志，1表示向下运行，-1表示向上运行
    int flag;
}

@property(strong, nonatomic) UIImageView *ball;

@end

@implementation ViewController

- (void)viewDidLoad {
    [super viewDidLoad];

    //初始化小球运动方向标志
    flag = 1;

    //界面初始化
    CGRect screen = [[UIScreen mainScreen] bounds];
    CGFloat imageWidth = 86;
    CGFloat imageHeight = 86;
    CGFloat imageTopView = 150;
    CGRect imageFrame = CGRectMake((screen.size.width - imageWidth) / 2,
    ➥imageTopView, imageWidth, imageHeight);
```

```
        let button = UIButton(type: .custom)
        //设置按钮正常状态时显示的图片
        button.setImage(UIImage(named: "ButtonOutline.png"), for: .normal)
        //设置按钮高亮状态时显示的图片
        button.setImage(UIImage(named: "ButtonOutlineHighlighted.png"),
        ➥for: .highlighted)
        //设置按钮触摸动作输出口
        button.addTarget(self, action: #selector(onClick(_:)),
        ➥for: .touchUpInside)

        let buttonWidth: CGFloat = 130
        let buttonHeight: CGFloat = 50
        let buttonTopView: CGFloat = 500
        button.frame = CGRect(x: (screen.size.width - buttonWidth) / 2,
        ➥y: buttonTopView, width: buttonWidth, height: buttonHeight)
        //添加按钮到当前视图
        self.view.addSubview(button)

    }

    func onClick(_ sender: AnyObject) {

        UIView.animate(withDuration: 1.5, animations: { () -> Void in      ①
            var frame = self.ball.frame                                     ②
            frame.origin.y += CGFloat(200 * self.flag)                      ③
            self.flag *= -1 //取反                                          ④
            self.ball.frame = frame                                         ⑤
        })

    }
}
```

```
    //创建Image View对象
    self.ball = [[UIImageView alloc] initWithFrame:imageFrame];
    //设置Image View的图片属性
    self.ball.image = [UIImage imageNamed:@"Ball.png"];
    //添加Image View到当前视图
    [self.view addSubview:self.ball];

    //创建按钮对象
    UIButton *button = [UIButton buttonWithType:UIButtonTypeCustom];
    //设置按钮正常状态时显示的图片
    [button setImage:[UIImage imageNamed:@"ButtonOutline.png"]
    ➥forState:UIControlStateNormal];
    //设置按钮高亮状态时显示的图片
    [button setImage:[UIImage imageNamed:@"ButtonOutlineHighlighted.png"]
    ➥forState:UIControlStateHighlighted];
    //设置按钮触摸动作输出口
    [button addTarget:self action:@selector(onClick:)
    ➥forControlEvents:UIControlEventTouchUpInside];

    CGFloat buttonWidth = 130;
    CGFloat buttonHeight = 50;
    CGFloat buttonTopView = 500;
    button.frame = CGRectMake((screen.size.width - buttonWidth) / 2,
    ➥buttonTopView, buttonWidth, buttonHeight);
    //添加按钮到当前视图
    [self.view addSubview:button];
}

- (void)onClick:(id)sender {

    [UIView animateWithDuration:1.5 animations:^{                    ①
        CGRect frame = self.ball.frame;                               ②
        frame.origin.y += 200 * flag;                                 ③
        flag *= -1; //取反                                            ④
        self.ball.frame = frame;                                      ⑤
    }];
}
@end
```

在onClick:方法中实现动画触发，使用了UIView的最简单形式的动画方法animateWithDuration:animations:。第①行中的1.5秒用于设置动画持续的时间，第②行用于获得小球的frame属性，该属性决定了视图的大小和位置。本例中的动画就是在1.5秒的时间内移动它的位置，由于只是上下移动，所以只需要改变frame中的y轴坐标即可。第③行中的表达式frame.origin.y就是y轴坐标，每次动画移动200点。第④行中的flag确定了小球向上还是向下移动。第⑤行将重新计算后的frame属性设置给小球。

13.1.2 动画结束的处理

有的时候，我们需要捕获动画开始与结束事件。如果采用下面的UIView方法实现动画功能，可以在completion代码块中结束动画：

- + animateWithDuration:animations:completion:
- + animateWithDuration:delay:options:animations:completion:

下面通过实例介绍一下它们的用法。如图13-2所示，这里的实例基于上一节的实例进行了修改。当用户点击Tap Me按钮时，动画开始，同时按钮消失。当动画结束时，Tap Me按钮又显示出来了。

视图的alpha属性称为“透明度”，用于设置视图显示还是消失。该属性的取值范围是0.0～1.0，如果alpha = 0.0，则视图完全透明，即消失；如果alpha = 1.0，则视图完全不透明，即完全显示。该属性可以在animations动画代码块中改变，从而实现视图渐变的效果。

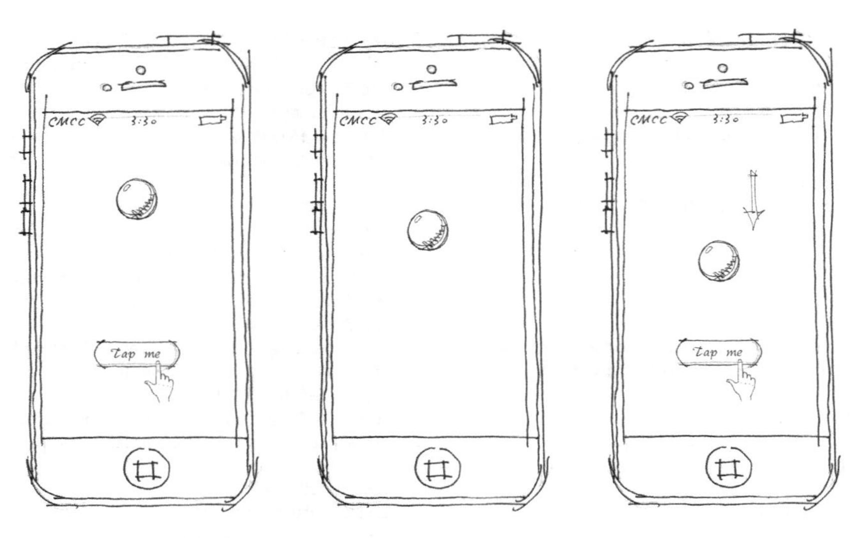

图13-2　动画实例

下面看看核心代码，其中AnimationCallBack工程中ViewController的主要代码如下：

```
func onClick(_ sender: AnyObject) {
    //动画开始时，将按钮设置为不可见
    self.button.alpha = 0.0

    UIView.animate(withDuration: 1.5, animations: { () -> Void in
        var frame = self.ball.frame
        frame.origin.y += CGFloat(200 * self.flag)
        self.flag *= -1 //取反
        self.ball.frame = frame
    }, completion: { (finished) -> Void in                          ①
        print("动画结束")
        self.viewAnimationDone()
    })
}

//动画结束之后的处理
func viewAnimationDone() {                                          ②
    //为按钮显示过程添加动画
    UIView.animate(withDuration: 1.0, animations: { () -> Void in  ③
        //动画结束后，将按钮设置为可见
        self.button.alpha = 1.0
    })
}
```

```
- (void)onClick:(id)sender {
    //动画开始时，将按钮设置为不可见
    self.button.alpha = 0.0;

    [UIView animateWithDuration:1.5 animations:^{
        CGRect frame = self.ball.frame;
        frame.origin.y += 200 * flag;
        flag *= -1; //取反
        self.ball.frame = frame;
    } completion:^(BOOL finished) {                                 ①
        NSLog(@"动画结束了。");
        [self viewAnimationDone];
    }];

}

//动画结束之后的处理
- (void)viewAnimationDone {                                         ②
    //为按钮显示过程添加动画
    [UIView animateWithDuration:1. animations:^{                    ③
        //动画结束后，将按钮设置为可见
        self.button.alpha = 1.0;
    }];
}
```

从上面的代码可以看到，在onClick:方法中我们使用了+ animateWithDuration:animations:completion:方法。其中第①行是completion代码块（Swift是闭包），该代码块在动画结束时调用，在该代码块中调用了自定义的viewAnimationDone方法。注意，在viewAnimationDone方法中又定义了一个动画（见第③行），这个动画是让按钮显示出来（alpha=1.0）。

13.1.3　过渡动画

前面提到过与动画相关的属性还有很多，例如动画曲线、过渡（界面跳转）动画、重复次数和自动反转等重要的属性，其中有一些是关于过渡动画[①]（Transition Animation）的，如动画曲线等。

---

① 过渡动画就是界面或视图之间跳转时的动画，也有人称为转场动画。

UIView类有两个动画过渡方法：

- **+ transitionWithView:duration:options:animations:completion:**。在指定的视图容器内创建动画过渡。该方法的声明如下：

```
class func transition(with view: UIView,
    duration: TimeInterval,
    options: UIViewAnimationOptions = [],
    animations: (() -> Void)?,
    completion: ((Bool) -> Void)? = nil)
```

```
+ (void)transitionWithView:(UIView *)view
    duration:(NSTimeInterval)duration
    options:(UIViewAnimationOptions)options
    animations:(void (^)(void))animations
    completion:(void (^)(BOOL finished))completion
```

其中参数view是视图；参数duration是动画持续的时间；options是执行动画的选项；animations参数是一个代码块；completion是在动画结束时调用的代码块。

- **+ transitionFromView:toView:duration:options:completion:**。在指定的两个视图之间创建动画过渡。该方法的声明如下：

```
+ (void)transitionFromView:(UIView *)fromView
    toView:(UIView *)toView
    duration:(NSTimeInterval)duration
    options:(UIViewAnimationOptions)options
    completion:(void (^)(BOOL finished))completion
```

```
class func transition(from fromView: UIView,
    to toView: UIView,
    duration: TimeInterval,
    options: UIViewAnimationOptions = [],
    completion: ((Bool) -> Void)? = nil)
```

该方法的参数与+ transitionWithView:duration:options:animations:completion:方法意义相同，这里不再赘述。

下面我们通过一个实例介绍过渡动画属性的用法，如图13-3所示。界面中的4个按钮会触发4个不同的视图过渡动画，图13-3b和图13-3c所示的是向上翻页的过渡动画效果。

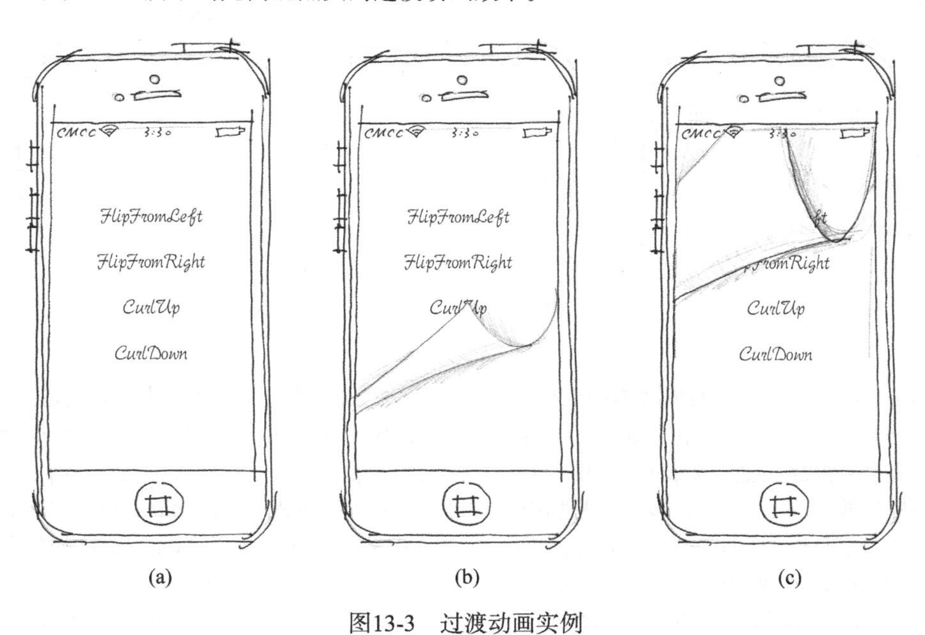

图13-3 过渡动画实例

下面我们看看核心代码部分。打开本书配套的代码库中的AnimationTransition工程，其中视图控制器ViewController的主要代码如下：

```
//ViewController.swift文件
class ViewController: UIViewController {

    ......
```

```
//ViewController.m文件
#import "ViewController.h"

@implementation ViewController
```

```
@IBAction func doUIViewAnimation(_ sender: AnyObject) {

    let button = sender as! UIButton
    NSLog("tag = %i", button.tag)

    switch button.tag {                                                    ①
    case 1:
        UIView.transition(with: self.view, duration: 1.5,
        ➥options: [.curveEaseOut,                                          ②
        ➥.transitionFlipFromLeft],                                         ③
        ➥animations: { () -> Void in
            NSLog("动画开始...")
        }, completion: { (finished) -> Void in
            NSLog("动画完成.")
        })
    case 2:
        UIView.transition(with: self.view, duration: 1.5,
        ➥options: [.curveEaseOut, .transitionFlipFromRight], animations:
        ➥{ () -> Void in
            NSLog("动画开始...")
        }, completion: { (finished) -> Void in
            NSLog("动画完成.")
        })

    case 3:
        UIView.transition(with: self.view, duration: 1.5,
        ➥options: [.curveEaseOut, .transitionCurlUp], animations: { () ->
        ➥Void in
            NSLog("动画开始...")
        }, completion: { (finished) -> Void in
            NSLog("动画完成.")
        })
    default:
        UIView.transition(with: self.view, duration: 1.5,
        ➥options: [.curveEaseOut, .transitionCurlDown], animations: { ()
        ➥-> Void in
            NSLog("动画开始...")
        }, completion: { (finished) -> Void in
            NSLog("动画完成.")
        })
    }
  }

}
```

```
......

- (IBAction)doUIViewAnimation:(id)sender {

    UIButton *button = (UIButton *)sender;
    NSLog(@"tag = %li", (long)button.tag);

    switch (button.tag) {                                                  ①
        case 1:
            [UIView transitionWithView:self.view duration:1.5
                        options:UIViewAnimationOptionCurveEaseOut |       ②
                        UIViewAnimationOptionTransitionFlipFromLeft       ③
                        animations:NULL completion:NULL];
            break;
        case 2:
            [UIView transitionWithView:self.view duration:1.5
            ➥options:UIViewAnimationOptionCurveEaseOut | UIViewAnimation
            ➥OptionTransitionFlipFromRight
            ➥animations:NULL completion:NULL];
            break;
        case 3:
            [UIView transitionWithView:self.view duration:1.5
            ➥options:UIViewAnimationOptionCurveEaseOut | UIViewAnimation
            ➥OptionTransitionCurlUp
            ➥animations:NULL completion:NULL];
            break;
        case 4:
            [UIView transitionWithView:self.view duration:1.5
            ➥options:UIViewAnimationOptionCurveEaseOut | UIViewAnimation
            ➥OptionTransitionCurlDown
            ➥animations:NULL completion:NULL];
            break;
    }

}

@end
```

doUIViewAnimation:方法用于实现动画的处理，界面中的4个按钮都会触发该方法。我们通过设置按钮的tag属性（在Interface Builder中设计按钮的tag属性）来区分这4个不同的按钮，其中第①行的button.tag表达式可以获得按钮的tag属性，然后通过switch语句判断。

switch语句中的每一个分支都是通过UIView的+ transitionWithView:duration:options:animations:completion:方法实现过渡动画的。过渡动画方法中的options参数（UIViewAnimationOptions类型）可以设置不同的动画过渡类型和动画曲线类型等，其中第②行中的UIViewAnimationOptionCurveEaseOut（Swift版是.curveEaseOut）用于设置动画曲线。动画曲线用于设置动画过渡时速度的变化情况，它是在枚举UIViewAnimationOptions中定义的成员。UIViewAnimationOptions中定义了22个成员，其中与动画曲线相关的成员如下。

- **UIViewAnimationOptionCurveEaseInOut**。缓入缓出，即开始和结束时减速。Swift版本是curveEaseInOut。
- **UIViewAnimationOptionCurveEaseIn**。缓入，开始时减速。Swift版本是curveEaseIn。
- **UIViewAnimationOptionCurveEaseOut**。缓出，结束时减速。Swift版本是curveEaseOut。
- **UIViewAnimationOptionCurveLinear**。线性，即匀速运动。Swift版本是curveLinear。

另外，第③行的UIViewAnimationOptionTransitionFlipFromLeft（Swift版是.transitionFlipFromLeft）用于设置过渡动画，它也是在枚举UIViewAnimationOptions中定义的成员，其中过渡动画相关的成员如下。

- **UIViewAnimationOptionTransitionFlipFromLeft**。设置从左往右翻转。Swift版本是transitionFlipFromLeft。
- **UIViewAnimationOptionTransitionFlipFromRight**。设置从右往左翻转。Swift版本是transitionFlipFromRight。
- **UIViewAnimationOptionTransitionCurlUp**。设置向上翻页。Swift版本是transitionCurlUp。
- **UIViewAnimationOptionTransitionCurlDown**。设置向下翻页。Swift版本是transitionCurlDown。
- **UIViewAnimationOptionTransitionCrossDissolve**。设置交叉溶解效果。Swift版本是transitionCrossDissolve。
- **UIViewAnimationOptionTransitionFlipFromTop**。设置从上往下翻转。Swift版本是transitionFlipFromTop。
- **UIViewAnimationOptionTransitionFlipFromBottom**。设置从下往上翻转。Swift版本是transitionFlipFromBottom。

提示

options参数是选项类型（UIViewAnimationOptions类型），选项类型中的成员值是位掩码[①]，这些成员可以进行位或运算。例如，Objective-C版本中的UIViewAnimationOptionCurveEaseOut | UIViewAnimationOptionTransitionFlipFromLeft（Swift版本中是[.curveEaseOut, .transitionFlipFromLeft]），其中UIViewAnimationOptionCurveEaseOut（Swift版中是curveEaseOut）用于设置缓出速度，UIViewAnimationOptionTransitionFlipFromLeft（Swift版中是transitionFlipFromLeft）设置从左往右翻转，那么这个位或运算表达式的计算结果是两种成员所代表的效果叠加，即过渡动画以缓出速度从左往右翻转。

13.2 Core Animation 框架

我们前面学习的UIView动画其底层也是由Core Animation实现的，因此这里有必要介绍一下Core Animation框架。

13.2.1 图层

图层（CALayer）是动画发生的场所，它的很多属性与视图十分相似，有位置、大小、变换和内容等属性。它也有层次关系，有自己的子图层。我们也可以继承CALayer自定义图层，重写绘制方法，从而实现自己绘制图层的目的。

但图层与视图还有本质的不同，视图是重量级的对象，它负责界面的绘制和事件响应，图层负责动画处理。一个视图中包含一个图层，视图依赖于图层实现动画。

在视图中有很多图层，它们是通过zPosition属性来区别它们在立体空间中的深度，即z轴上的顺序。

下面我们通过实例介绍一下图层创建和图层的一些属性。如图13-4所示，在图层上绘制一张图片。

首先，使用Single View Application模板创建一个名为LayerSample的工程，然后参考3.5.1节将其修改为纯代码工程。然后，添加资源图片文件到工程。

① 掩码是一串二进制代码对目标字段进行位与运算，屏蔽当前的输入位。——引自百度百科http://baike.baidu.com/view/68.htm

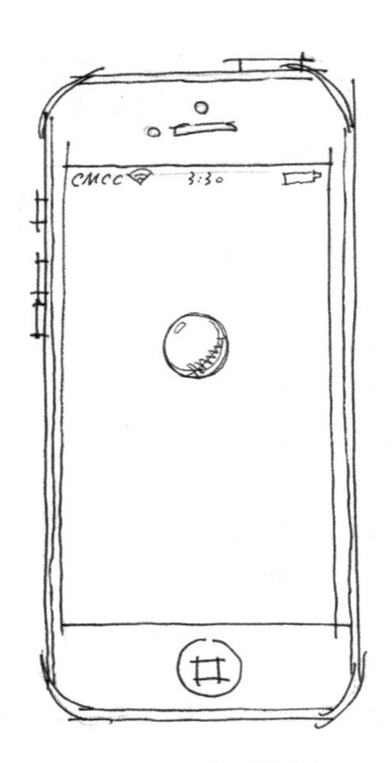

图13-4　图层实例

这里重点介绍一下视图控制器类ViewController，其代码如下：

```
//ViewController.swift文件
class ViewController: UIViewController {

    var ballLayer: CALayer!

    override func viewDidLoad() {
        super.viewDidLoad()

        //创建层对象ballLayer
        self.ballLayer = CALayer()                                          ①

        //读取图片创建UIImage对象
        let image = UIImage(named: "Ball2.png")

        //设置层对象ballLayer
        //设置层内容
        self.ballLayer.contents = image?.cgImage                            ②
        //设置层内容的布局方式
        self.ballLayer.contentsGravity = kCAGravityResizeAspect             ③
        //设置层的边界
        self.ballLayer.bounds = CGRect(x: 0.0, y: 0.0, width: 125.0, height:
        ➥125.0)                                                            ④
        //设置层的位置
        self.ballLayer.position = CGPoint(x: self.view.bounds.midX,
                                          y: self.view.bounds.midY)         ⑤
        //添加ballLayer层到当前层
        self.view.layer.addSublayer(self.ballLayer)                         ⑥

    }

}
```

```
//ViewController.m文件
#import "ViewController.h"

@interface ViewController ()

@property(nonatomic) CALayer *ballLayer;

@end

@implementation ViewController

- (void)viewDidLoad {
    [super viewDidLoad];

    //创建层对象ballLayer
    self.ballLayer = [CALayer layer];                                       ①

    //读取图片创建UIImage对象
    UIImage *image = [UIImage imageNamed:@"Ball2.png"];

    //设置层对象ballLayer
    //设置层内容
    self.ballLayer.contents = (__bridge id)(image.CGImage);                 ②
    //设置层内容的布局方式
    self.ballLayer.contentsGravity = kCAGravityResizeAspect;                ③
    //设置层的边界
    self.ballLayer.bounds = CGRectMake(0.0, 0.0, 125.0, 125.0);             ④
    //设置层的位置
    self.ballLayer.position = CGPointMake(CGRectGetMidX(self.view.bounds),
                                          CGRectGetMidY(self.view.bounds));
                                                                            ⑤
    //添加ballLayer层到当前层
    [self.view.layer addSublayer:self.ballLayer];

}

@end
```

在viewDidLoad方法中，第①行用于创建图层对象。在Objective-C版本中，CALayer调用静态+layer方法获得图层对象；而在Swift版本中，则调用CALayer的构造函数创建并初始化图层对象。

第②行至第⑤行用于设置层对象ballLayer。

第②行的self.ballLayer.contents用于设置层内容。层的contents属性可以接收CGImage类型的图片对象或者是其他图层的内容，CGImage类型在Objective-C版中是CGImageRef，即CGImage的引用类型。

提示 在Objective-C版本中，第②行中的(__bridge id)类型转换表达式用于将Core Foundation对象转换为Objective-C对象，CGImageRef是Core Foundation对象。进行ARC内存管理时，我们使用(__bridge id)将Core Foundation对象转换为Objective-C对象，该转换不会改变对象的所有权。

第③行用于设置图层内容的布局方式，其中kCAGravityResizeAspect常量可以使内容保持高宽比原样重新调整。

第④行用于设定图层的边界，Swift版本通过CGRect构造函数创建CGRect对象，Objective-C版本通过CGRectMake函数创建CGRect对象。

第⑤行设置图层的position（位置）属性，Swift版本通过CGPoint构造函数创建CGPoint对象，self.view.bounds.midX是视图中心点的x坐标，self.view.bounds.midY是视图中心点的y坐标。Objective-C版本通过CGPointMake函数创建CGPoint对象，使用CGRectGetMidX(self.view.bounds)函数计算出视图的中心点的x坐标，使用CGRectGetMidY(self.view.bounds)函数计算出视图的中心点的y坐标。

13.2.2 隐式动画

了解了图层的概念后，我们再来介绍Core Animation动画技术。在Core Animation框架中，有两种形式的动画：隐式动画和显式动画。

- **隐式动画**：这是一种最简单的动画，不用设置定时器，不用考虑线程或者重画，它的很多属性都是默认的。前面介绍的视图动画属于隐式动画。
- **显式动画**：这是一种使用CABasicAnimation创建的动画，通过CABasicAnimation，可以更明确地定义属性如何改变动画。显式动画还有更复杂的显式动画类型——关键帧动画（CAKeyframeAnimation），它可以定义动画的起点和终点，还可以定义某些帧之间的动画。

本节通过一个简单实例介绍一下隐式动画。实例如图13-5所示，点击Tap Me按钮，飞机从左上角飞到右下角，透明度也由模糊变得清楚起来。

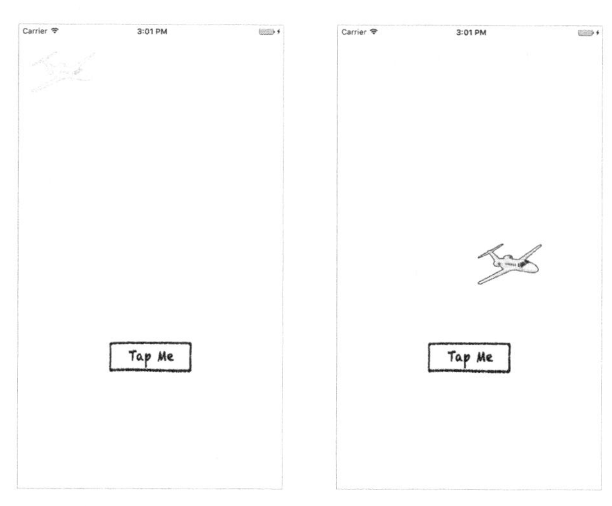

图13-5 隐式动画实例

首先，使用Single View Application模板创建一个名为ImplicitAnimation的工程，然后参考3.5.1节将其修改为纯代码工程。然后，添加资源图片文件到工程。

这里重点介绍一下视图控制器类ViewController，其代码如下：

```
//ViewController.swift文件
class ViewController: UIViewController {

    var plane: UIImageView!

    override func viewDidLoad() {
        super.viewDidLoad()

        //界面初始化
        let screen = UIScreen.main.bounds

        let imageWidth: CGFloat = 100
        let imageHeight: CGFloat = 100
        let imageTopView: CGFloat = 25
        let imageLeftView: CGFloat = 20
        let imageFrame = CGRect(x: imageLeftView,
        ➥y: imageTopView, width: imageWidth, height: imageHeight)
        //创建Image View对象plane
        self.plane = UIImageView(frame: imageFrame)
        //设置plane的图片属性
        self.plane.image = UIImage(named: "clipartPlane.png")
        //设置plane上图层的opacity属性
        self.plane.layer.opacity = 0.25                                    ①
        //添加plane到当前视图
        self.view.addSubview(self.plane)

        //创建按钮对象
        let button = UIButton(type: .custom)
        //设置按钮在正常状态时显示的图片
        button.setImage(UIImage(named: "ButtonOutline.png"), for: .normal)
        //设置按钮在高亮状态时显示的图片
        button.setImage(UIImage(named: "ButtonOutlineHighlighted.png"),
        ➥for: .highlighted)
        //设置按钮触摸动作输出口
        button.addTarget(self, action: #selector(movePlane(_:)),
        ➥for: .touchUpInside)

        let buttonWidth: CGFloat = 130
        let buttonHeight: CGFloat = 50
        let buttonTopView: CGFloat = 500
        button.frame = CGRect(x: (screen.size.width - buttonWidth) / 2,
        ➥y: buttonTopView, width: buttonWidth, height: buttonHeight)
        //添加按钮到当前视图
        self.view.addSubview(button)

    }

    func movePlane(_ sender: AnyObject) {

        //创建平移仿射变换
        let moveTransform = CGAffineTransform(translationX: 200, y: 300)   ②
        //将仿射变换作用于plane视图上的层
        self.plane.layer.setAffineTransform(moveTransform)                 ③
        //设置图层的opacity属性
        self.plane.layer.opacity = 1                                       ④

    }

}
```

```
//ViewController.m文件
#import "ViewController.h"

@interface ViewController ()

@property(strong, nonatomic) UIImageView *plane;

@end

@implementation ViewController

- (void)viewDidLoad {
    [super viewDidLoad];

    //界面初始化
    CGRect screen = [[UIScreen mainScreen] bounds];

    CGFloat imageWidth = 100;
    CGFloat imageHeight = 100;
    CGFloat imageTopView = 25;
    CGFloat imageLeftView = 20;
    CGRect imageFrame = CGRectMake(imageLeftView,
    ➥imageTopView, imageWidth, imageHeight);
    //创建Image View对象plane
    self.plane = [[UIImageView alloc] initWithFrame:imageFrame];
    //设置plane的图片属性
    self.plane.image = [UIImage imageNamed:@"clipartPlane.png"];
    //设置plane上图层的opacity属性
    self.plane.layer.opacity = 0.25;                                       ①
    //添加plane到当前视图
    [self.view addSubview:self.plane];

    //创建按钮对象
    UIButton *button = [UIButton buttonWithType:UIButtonTypeCustom];
    //设置按钮在正常状态时显示的图片
    [button setImage:[UIImage imageNamed:@"ButtonOutline.png"]
    ➥forState:UIControlStateNormal];
    //设置按钮在高亮状态时显示的图片
    [button setImage:[UIImage imageNamed:@"ButtonOutlineHighlighted.png"]
    ➥forState:UIControlStateHighlighted];
    //设置按钮触摸动作输出口
    [button addTarget:self action:@selector(movePlane:)
    ➥forControlEvents:UIControlEventTouchUpInside];

    CGFloat buttonWidth = 130;
    CGFloat buttonHeight = 50;
    CGFloat buttonTopView = 500;
    button.frame = CGRectMake((screen.size.width - buttonWidth) / 2,
    ➥buttonTopView, buttonWidth, buttonHeight);
    //添加按钮到当前视图
    [self.view addSubview:button];

}

- (void)movePlane:(id)sender {

    //创建平移仿射变换
    CGAffineTransform moveTransform = CGAffineTransformMakeTranslation(200,
    ➥300);                                                                ②
    //将仿射变换作用于plane视图上的层
    self.plane.layer.affineTransform = moveTransform;                      ③
    //设置层的opacity属性
```

```
    self.plane.layer.opacity = 1;                               ④
}
@end
```

在viewDidLoad方法中，第①行用于设置plane视图的图层不透明度（opacity）为0.25。opacity的取值范围是0.0~1.0，其中0.0表示完全透明，1.0表示完全不透明。

注意 13.1.2节介绍视图动画时，视图的可见和消失是通过alpha属性实现的，本例中是通过opacity属性实现的。opacity属性是图层（CALayer）的属性，alpha是视图（UIView）的属性。

movePlane:方法是用户点击按钮时触发的方法，在这个方法中我们开始隐式动画，plane上图层的opacity属性初始为0.25，然后在动画过程中opacity从0.25变为1.0。第②行用于创建平移仿射变换变量moveTransform。第③行用于设置plane上层的仿射变换，即从当前位置平移到(200, 300)，飞机就可以动起来了。第④行用于设置plane上层的opacity为1。

13.2.3 显式动画

为了实现对动画的精准控制，可以使用显式动画。在使用显式动画时，不必定义图层属性的变化，也不必执行它们，而是通过CABasicAnimation逐个定义动画。其中每个动画都含有各自的持续时间、重复次数等属性。然后使用addAnimation:forKey:方法分别将每个动画应用到图层的特定属性中。

要创建显式动画对象，可以通过如下代码实现：

```
let opAnim = CABasicAnimation(keyPath: "opacity")
```

```
CABasicAnimation *opAnim = [CABasicAnimation animationWithKeyPath:@"opacity"];
```

其中参数keyPath指定图层的属性名，opacity是图层的不透明度属性，因此上述代码创建的动画是针对图层opacity变化的动画。这些图层属性有：

- transform.scale
- transform.scale.x
- transform.scale.y
- transform.rotation.z
- opacity
- margin
- zPosition
- backgroundColor
- cornerRadius
- borderWidth
- bounds
- contents
- contentsRect
- cornerRadius
- frame
- hidden
- mask
- masksToBounds
- position

- ❑ shadowColor
- ❑ shadowOffset
- ❑ shadowOpacity
- ❑ shadowRadius
- ❑ transform

这些属性就不一一解释了，但是需要注意的是keyPath参数接收的是字符串。

下面使用显式动画实现上一节中的飞机实例，这里我们可以对它进行更多的控制。ViewController的主要代码如下：

```
class ViewController: UIViewController {

    var plane: UIImageView!

    ......

    func movePlane(_ sender: AnyObject) {

        //创建opacity动画
        let opAnim = CABasicAnimation(keyPath: "opacity")                        ①
        //设置动画持续时间
        opAnim.duration = 3.0
        //设置opacity开始值
        opAnim.fromValue = 0.25
        //设置opacity结束值
        opAnim.toValue = 1.0
        //设置累计上次值
        opAnim.isCumulative = true
        //设置动画重复2次
        opAnim.repeatCount = 2
        //设置动画结束时的处理方式
        opAnim.fillMode = kCAFillModeForwards
        //设置动画结束时是否停止
        opAnim.isRemovedOnCompletion = false
        //添加动画到层
        self.plane.layer.add(opAnim, forKey : "animateOpacity")                   ②

        //创建平移仿射变换
        let moveTransform = CGAffineTransform(translationX: 200, y: 300)          ③
        //创建平移动画
        let moveAnim = CABasicAnimation(keyPath: "transform")
        moveAnim.duration = 6.0
        //设置结束位置
        moveAnim.toValue = NSValue(caTransform3D:
        ➥CATransform3DMakeAffineTransform(moveTransform))                        ④
        moveAnim.fillMode = kCAFillModeForwards
        moveAnim.isRemovedOnCompletion = false
        self.plane.layer.add(moveAnim, forKey : "animateTransform")               ⑤

    }
}
```

```
#import "ViewController.h"
......

@implementation ViewController

......

- (void)movePlane:(id)sender {

    //创建opacity动画
    CABasicAnimation *opAnim = [CABasicAnimation animationWithKeyPath:
    ➥@"opacity"];                                                               ①
    //设置动画持续时间
    opAnim.duration = 3.0;
    //设置opacity开始值
    opAnim.fromValue = @0.25;    //数值为0.25的NSNumber对象
    //设置opacity结束值
    opAnim.toValue= @1.0;        //数值为1.0的NSNumber对象
    //设置累计上次值
    opAnim.cumulative = YES;
    //设置动画重复2次
    opAnim.repeatCount = 2;
    //设置动画结束时的处理方式
    opAnim.fillMode = kCAFillModeForwards;
    //设置动画结束时是否停止
    opAnim.removedOnCompletion = NO;
    //添加动画到层
    [self.plane.layer addAnimation:opAnim forKey:@"animateOpacity"];            ②

    //创建平移仿射变换
    CGAffineTransform moveTransform = CGAffineTransformMakeTranslation(200,
    ➥300);                                                                      ③
    //创建平移动画
    CABasicAnimation *moveAnim = [CABasicAnimation
    ➥animationWithKeyPath:@"transform"];
    moveAnim.duration = 6.0;
    //设置结束位置
    moveAnim.toValue= [NSValue valueWithCATransform3D:
    ➥CATransform3DMakeAffineTransform(moveTransform)];                          ④
    moveAnim.fillMode = kCAFillModeForwards;
    moveAnim.removedOnCompletion = NO;
    [self.plane.layer addAnimation:moveAnim forKey:@"animateTransform"];        ⑤

}

@end
```

在movePlane:方法中，第①行至第②行定义了不透明度（opacity）变化动画，opAnim.duration = 3.0定义动画持续时间是3秒钟，它将opacity从0.25变化到1.0。属性opAnim.fromValue用于定义开始值，属性opAnim.toValue用于定义结束值。opAnim.repeatCount = 2定义了动画重复2次。opAnim.cumulative属性用于指定累计上次值。

注意　下面的代码可以防止闪回问题（动画完成后，飞机的不透明度应该是1.0，闪一下就变回到初始的0.25）：

```
opAnim.fillMode = kCAFillModeForwards
opAnim.removedOnCompletion = false
```

```
opAnim.fillMode = kCAFillModeForwards;
opAnim.removedOnCompletion = NO;
```

其中opAnim的fillMode属性用于设置动画结束时的处理方式，kCAFillModeForwards代表保持动画结束值，opAnim.removedOnCompletion用于设置动画结束时是否停止，这样设置前面的fillMode属性才能起作用。

第③行至第⑤行定义了位置变化的显式动画，其中第③行用于创建变换类型的显式动画对象。

第④行用于设定结束值，它的内容是仿射变换矩阵，其中CATransform3DMakeAffineTransform函数创建的是仿射变换矩阵数据结构。由于仿射变换矩阵数据结构不是一个对象类型，不能赋值给toValue属性，所以需要将其封装为NSValue对象。NSValue对象是用来存储一个C或者Objective-C数据的简单容器，所有的基本数据类型都可以封装成NSValue对象。

13.2.4　关键帧动画

关键帧是一种常用的动画技术，其基本原理是将动画序列中比较关键的帧提取出来，而其他帧根据时间用这些关键帧插值计算得到。关键帧动画也是显式动画，它能够比普通显式动画更精准地控制动画效果。同样是前面介绍的飞机实例，我们采用关键帧动画来实现。如图13-6所示，有3个关键帧0秒、0.5秒、1.0秒，不同时间点上的不透明度是0.25、0.75、1.0，也就是说在动画运行的0秒～0.5秒时间内，不透明度从0.25到0.75，在0.5秒～1.0秒时间内不透明度从0.75到1.0。

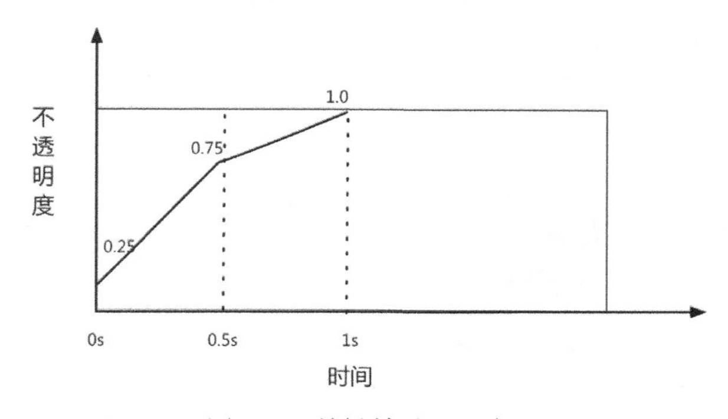

图13-6　关键帧动画示意图

修改ViewController代码中的movePlane方法，具体如下：

```
func movePlane(_ sender: AnyObject) {

    //创建opacity的关键帧动画
    let opAnim = CAKeyframeAnimation(keyPath: "opacity")          ①
    //设置动画持续时间
    opAnim.duration = 6.0
    //设置关键帧时间点
    opAnim.keyTimes = [0.0, 0.5, 1.0]                             ②
    //设置每个关键帧上的opacity值
    opAnim.values = [0.25, 0.75, 1.0]                             ③
    //设置动画结束时的处理方式
    opAnim.fillMode = kCAFillModeForwards
    //设置动画结束时是否停止
```

```
- (void)movePlane:(id)sender {

    //创建opacity的关键帧动画
    CAKeyframeAnimation *opAnim = [CAKeyframeAnimation
    ➥animationWithKeyPath:@"opacity"];                            ①
    //设置动画持续时间
    opAnim.duration = 6.0;
    //设置关键帧时间点
    opAnim.keyTimes = @[@0.0, @0.5, @1.0];                        ②
    //设置每个关键帧上的opacity值
    opAnim.values = @[@0.25, @0.75, @1.0];                        ③
    //设置动画结束时的处理方式
    opAnim.fillMode = kCAFillModeForwards;
```

```swift
    opAnim.isRemovedOnCompletion = false
    //添加动画到层
    self.plane.layer.add(opAnim, forKey : "animateOpacity")                ④

    //创建平移仿射变换
    let moveTransform = CGAffineTransform(translationX: 200, y: 300)
    //创建平移动画
    let moveAnim = CABasicAnimation(keyPath: "transform")
    moveAnim.duration = 6.0
    //设置结束位置
    moveAnim.toValue = NSValue(caTransform3D:
    ➥CATransform3DMakeAffineTransform(moveTransform))
    moveAnim.fillMode = kCAFillModeForwards
    moveAnim.isRemovedOnCompletion = false
    self.plane.layer.add(moveAnim, forKey : "animateTransform")

}
```

```objectivec
    //设置动画结束时是否停止
    opAnim.removedOnCompletion = NO;
    //添加动画到层
    [self.plane.layer addAnimation:opAnim forKey:@"animateOpacity"];        ④

    //创建平移仿射变换
    CGAffineTransform moveTransform = CGAffineTransformMakeTranslation(200,
    ➥300);
    //创建平移动画
    CABasicAnimation *moveAnim = [CABasicAnimation animationWithKeyPath:
    ➥@"transform"];
    moveAnim.duration = 6.0;
    //设置结束位置
    moveAnim.toValue= [NSValue valueWithCATransform3D:
    ➥CATransform3DMakeAffineTransform(moveTransform)];
    moveAnim.fillMode = kCAFillModeForwards;
    moveAnim.removedOnCompletion = NO;
    [self.plane.layer addAnimation:moveAnim forKey:@"animateTransform"];
}
```

在movePlane:方法中，第①行至第④行实现了关键帧动画。第②行定义了每个关键帧时间点序列，第③行设置每个关键帧上的不透明度值序列，这两个序列是相互对应的。

提示

上述Objective-C代码中，第②行和第③行中的表达式采用“字面量”形式初始化数组，@[@0.0, @0.5, @1.0]表达式相当于如下语句：

```objectivec
[NSArray arrayWithObjects:
    [NSNumber numberWithFloat:0.25],
    [NSNumber numberWithFloat:0.75],
    [NSNumber numberWithFloat:1.0], nil]
```

在字面量中，@[...]表示数组（NSArray）对象，@{...}表示字典（NSDictionary）对象，字面量@0.25表示数值为0.25的NSNumber对象。

13.2.5　使用路径

对于位置变化的属性，也可以设置路径，然后动画可以按照这个路径运动。修改13.1.1节的实例，让屏幕中的小球沿着一个五角形轨迹运动，如图13-7所示。

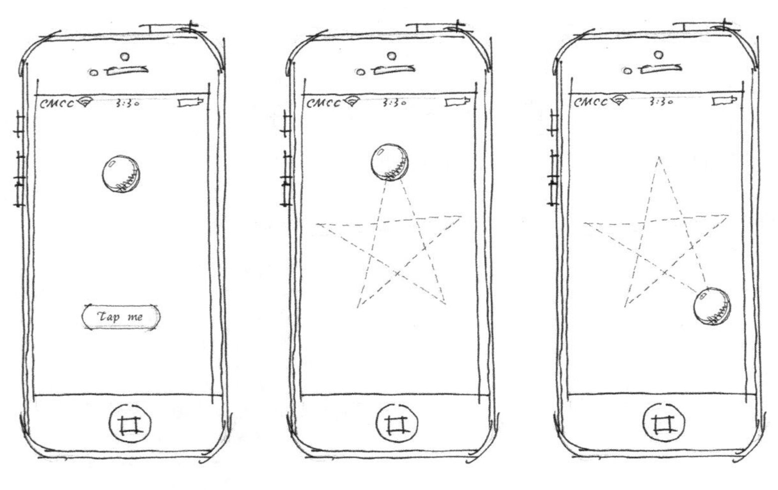

图13-7　关键帧路径

ViewController的代码如下：

```
//ViewController.swift文件
import UIKit

class ViewController: UIViewController, CAAnimationDelegate {                ①

    var ball: UIImageView!
    var button: UIButton!

    ......

    func onClick(_ sender: AnyObject) {
        //设置按钮不可见
        self.button.alpha = 0.0

        //创建路径
        let starPath = CGMutablePath()                                        ②
        starPath.move(to: CGPoint(x: 160.0, y: 100.0))                        ③
        starPath.addLine(to: CGPoint(x: 100.0, y: 280.0))                     ④
        starPath.addLine(to: CGPoint(x: 260.0, y: 170.0))
        starPath.addLine(to: CGPoint(x: 60.0, y: 170.0))
        starPath.addLine(to: CGPoint(x: 220.0, y: 280.0))
        //闭合路径
        starPath.closeSubpath()                                               ⑤

        //创建位置变化的帧动画
        let animation = CAKeyframeAnimation(keyPath: "position")
        //设置动画持续时间
        animation.duration = 10.0
        //设置self为动画委托对象
        animation.delegate = self                                             ⑥
        //设置动画路径
        animation.path = starPath                                             ⑦

        self.ball.layer.add(animation, forKey : "position")

    }

    // MARK: -- 实现委托协议CAAnimationDelegate
    //动画开始方法
    func animationDidStart(_ anim: CAAnimation) {                             ⑨
        print("动画开始...")
    }

    // 动画结束方法
    func animationDidStop(_ anim: CAAnimation, finished flag: Bool) {        ⑩
        print("动画开始...")
    }

}
```

```
//ViewController.m文件
#import "ViewController.h"

@interface ViewController () <CAAnimationDelegate>                           ①

@property(strong, nonatomic) UIImageView *ball;
@property (strong, nonatomic) UIButton *button;

@end

@implementation ViewController
......

- (void)onClick:(id)sender {
    //设置按钮不可见
    self.button.alpha = 0.0;

    //创建路径
    CGMutablePathRef starPath = CGPathCreateMutable();                        ②
    CGPathMoveToPoint(starPath,NULL,160.0, 100.0);                            ③
    CGPathAddLineToPoint(starPath, NULL, 100.0, 280.0);                       ④
    CGPathAddLineToPoint(starPath, NULL, 260.0, 170.0);
    CGPathAddLineToPoint(starPath, NULL, 60.0, 170.0);
    CGPathAddLineToPoint(starPath, NULL, 220.0, 280.0);
    //闭合路径
    CGPathCloseSubpath(starPath);                                             ⑤

    //创建位置变化的帧动画
    CAKeyframeAnimation *animation = [CAKeyframeAnimation
    ➥animationWithKeyPath:@"position"];
    //设置动画持续时间
    animation.duration = 10.0;
    //设置self为动画委托对象
    animation.delegate = self;                                                ⑥
    //设置动画路径
    animation.path = starPath;                                                ⑦
    //释放动画路径对象
    CFRelease(starPath);                                                      ⑧

    [self.ball.layer addAnimation:animation forKey:@"position"];

}

#pragma mark -- 实现委托协议CAAnimationDelegate
// 动画开始方法
- (void)animationDidStart:(CAAnimation *)anim {                              ⑨
    NSLog(@"动画开始...");
}

// 动画结束方法
- (void)animationDidStop:(CAAnimation *)anim finished:(BOOL)flag {          ⑩
    NSLog(@"动画结束...");
    [UIView animateWithDuration:1.0 animations:^{
        //设置按钮完全可见
        self.button.alpha = 1.0;
    }];
}

@end
```

第①行声明实现CAAnimationDelegate协议，第⑥行声明当前视图控制器为动画委托对象。CAAnimation-Delegate协议是iOS 10之后提供的API，它有两个可选的实现方法。

- **animationDidStart:**。动画开始时调用的方法，见第⑨行。

- ❑ **animationDidStop:finished:**。动画结束时调用的方法，finished参数是布尔类型，true表示动画结束，false表示动画没有结束，见第⑩行。

第②行至第⑤行创建并设定路径，CGPathCreateMutable()用于创建可变路径对象CGMutablePath。

提示　在Objective-C版中，我们使用可变路径对象的引用类型CGMutablePathRef。CGMutablePathRef属于Core Foundation框架，该框架是基于C语言级别的API，需要开发人员自己管理内存。第⑧行就是释放动画路径对象的内存，其中CFRelease是Core Foundation框架中释放对象内存的函数。

第③行将(160.0,100.0)坐标作为路径的起点。第④行中的CGPathAddLineToPoint函数（Swift版是CGMutablePath的addLine方法）是将线段添加到路径对象中。第⑤行中的CGPathCloseSubpath函数（Swift版是CGMutablePath的closeSubpath方法）用于关闭路径。关闭路径会将路径起点和终点连接起来形成一个闭合路径。我们知道，绘制一个五角形至少需要5个线段，但上述代码只是绘制了4个线段，最后一个线段可以不用绘制，直接关闭路径即可（这部分工作底层会帮我们完成）。关闭路径后，说明路径添加完成，此时需要通过第⑦行设置动画路径。需要说明的是，只有关键帧动画（CAKeyframeAnimation）才有path属性。

13.3　小结

通过对本章的学习，读者可以了解如何使用iOS中的动画技术：视图动画和Core Animation框架。

Part 2

第二部分

数据与网络通信篇

本部分内容

- 第 14 章　数据持久化
- 第 15 章　数据交换格式
- 第 16 章　REST Web Service

第 14 章

数据持久化

信息和数据在现代社会中扮演着至关重要的角色，已成为生活中不可或缺的一部分。我们经常接触的信息有电话号码本、QQ通信录、消费记录等，而智能手机就是这些信息和数据的载体和传播工具。在本章中，我们将向大家介绍iOS系统中数据的多种持久化方式。

14.1 概述

iOS有一套完整的数据安全体系，iOS应用程序只能访问自己的目录，这个目录称为沙箱目录，而应用程序间禁止数据的共享和访问。访问一些特殊的应用时，如联系人应用，必须通过特定的API访问。现在，iOS支持主流的数据持久化方式，本章将探讨这些持久化方式。

14.1.1 沙箱目录

沙箱目录是一种数据安全策略，很多系统都采用沙箱设计，实现HTML5规范的一些浏览器也采用沙箱设计。沙箱目录设计的原理就是只允许自己的应用访问目录，而不允许其他应用访问。在Android平台中，我们通过Content Provider技术将数据共享给其他应用；而在iOS系统中，特有的应用（联系人等）需要特定的API才可以共享数据，而其他应用之间都不能共享数据。

下面的目录是iOS平台的沙箱目录，我们可以在模拟器中看到，在真实设备上也是这样存储的：

```
/Users/<用户>/Library/Developer/CoreSimulator/Devices/
25516FE8-A71E-447C-879A-60EA83DEA1C7/data/Containers/Data/Application/5EC9A38C-EFD1-4E95-8574-17F0C27448D3
```

每个应用安装之后都有类似的目录。在4EC70796-3C2C-4FF4-B869-1E84E9C2A22B目录下面有Documents、Library和tmp子目录，它们都是沙箱目录的子目录，其目录结构如下所示：

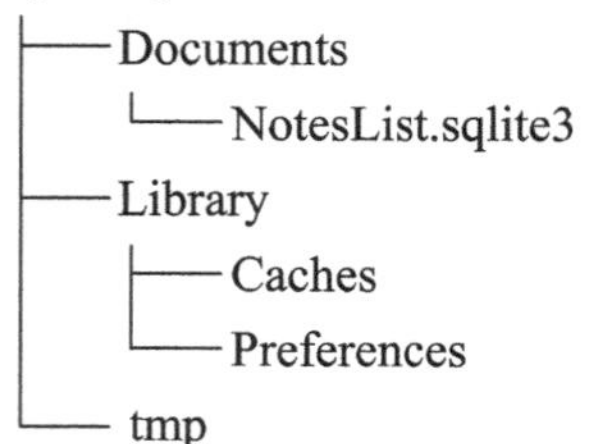

下面我们分别介绍这3个子目录，它们有不同的用途、场景和访问方式。

1. Documents目录

该目录用于存储非常大的文件或需要经常频繁更新的数据，能够进行iTunes或iCloud的备份。获取目录位置的代码如下所示：

```swift
let documentDirectory: NSArray =
➥NSSearchPathForDirectoriesInDomains(.DocumentDirectory, .UserDomainMask,true)
```

```objc
NSArray * documentDirectory =
➥NSSearchPathForDirectoriesInDomains(NSDocumentDirectory,
➥NSUserDomainMask, YES);
```

其中documentDirectory是只有一个元素的数组，因此还需要使用下面的代码取出一个路径来：

```
let myDocPath = documentDirectory[0] as NSString
```

```
NSString * myDocPath = [documentDirectory objectAtIndex:0];
```

或：

```
let myDocPath = documentDirectory.lastObject as NSString
```

```
NSString * myDocPath = [documentDirectory lastObject];
```

因为documentDirectory数组只有一个元素，所以取第一个元素和最后一个元素都是一样的，都可以取出Documents目录。

2. Library目录

在Library目录下面有Preferences和Caches目录，其中前者用于存放应用程序的设置数据，后者与Documents很相似，可以存放应用程序的数据，用来存储缓存文件。

3. tmp目录

这是临时文件目录，用户可以访问它。它不能够进行iTunes或iCloud的备份。要获取该目录的位置，可以使用如下代码：

```
let tmpDirectory = NSTemporaryDirectory()
```

```
NSString *tmpDirectory = NSTemporaryDirectory();
```

14.1.2 持久化方式

持久化方式就是数据存取方式。iOS支持本地存储和云端存储，本章主要介绍本地存储，主要涉及如下3种机制。

- **属性列表**。集合对象可以读写到属性列表文件中。
- **SQLite数据库**。SQLite是一个开源嵌入式关系型数据库。
- **Core Data**。它是一种对象关系映射技术（ORM），本质上也是通过SQLite存储的。

采用什么技术要视具体情况而定。属性列表文件一般用于存储少量数据，Foundation框架中的集合对象都有对应的方法读写属性列表文件。SQLite数据库和Core Data一般用于有几个简单表关系的大量数据的情况。如果是复杂表关系而且数据量很大，我们应该考虑把数据放在远程云服务器中。

14.2 实例：MyNotes 应用

在具体介绍各种持久化技术之前，我们先介绍一下本章所使用的实例——MyNotes应用。这个实例基于iOS（iPhone和iPad两个平台），具有增加、删除和查询备忘录的基本功能。图14-1所示的是MyNotes应用的用例图。

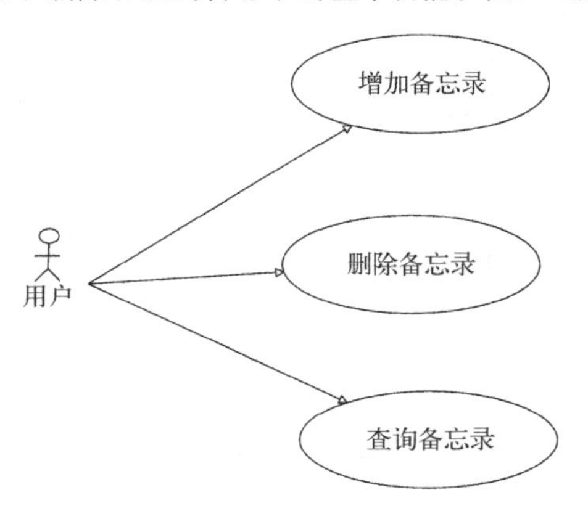

图14-1 MyNotes应用的用例图

考虑到iOS有iPhone和iPad两个平台，我们针对不同的平台绘制了相应的设计原型草图，如图14-2、图14-3和图14-4所示。

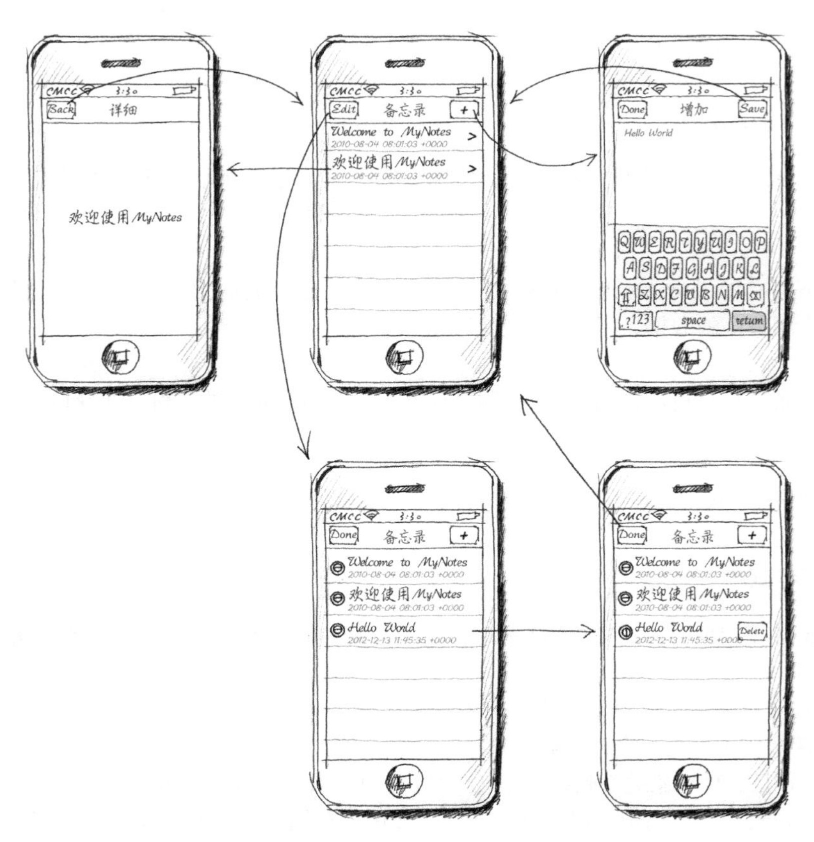

图14-2　iPhone版本的MyNotes设计原型草图

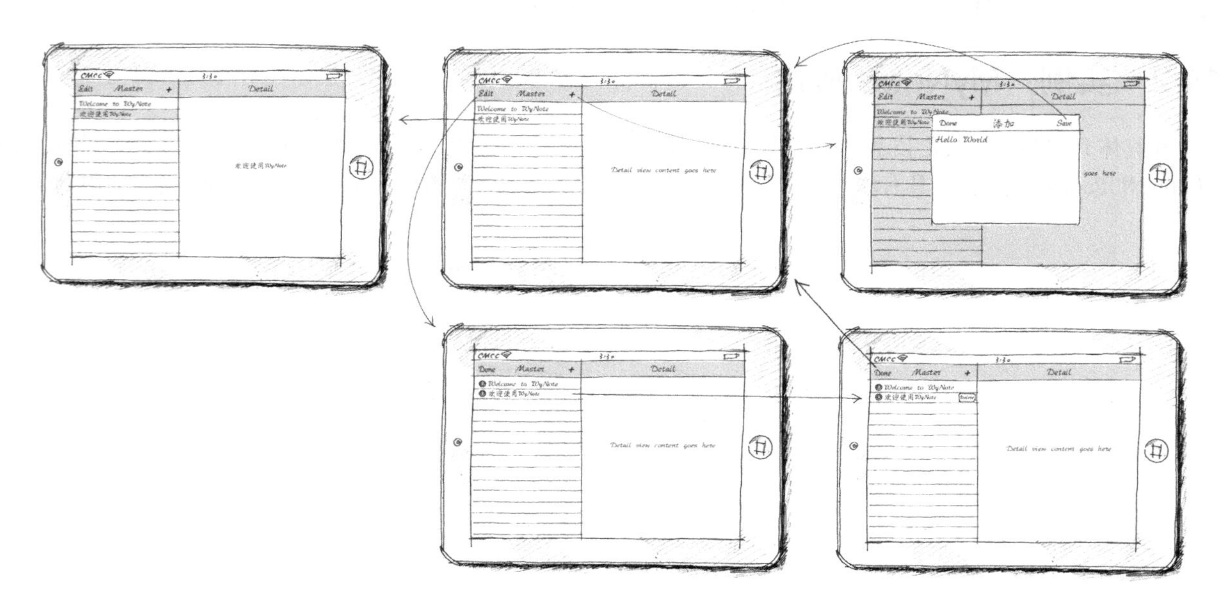

图14-3　iPad版本的MyNotes横屏设计原型草图

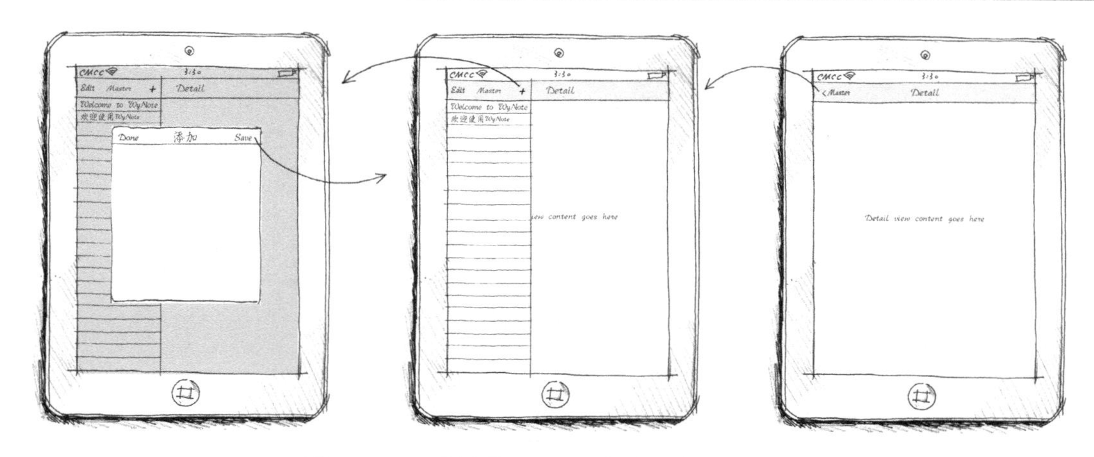

图14-4 iPad版本的MyNotes竖屏设计原型草图

在设计MyNotes应用时，为了构建松耦合的应用，应该遵守软件设计的基本原则：表示组件①应该是轻薄的，不应该包含业务逻辑②和数据持久化逻辑③。如果业务逻辑比较复杂，应用中应该设计业务逻辑组件，同样业务逻辑组件中也不应该包含数据持久化逻辑。

MyNotes应用的业务逻辑比较简单，可以不需要业务逻辑组件，直接在表示逻辑（主要是视图控制器）中直接访问数据持久化逻辑组件。数据持久化逻辑组件又可以细分为DAO和domain两个组件。DAO是数据访问对象，数据库中每一个数据表(table)对应一个DAO对象，每一个DAO对象中有访问数据表的CRUD④ 这4类方法。domain组件是业务领域中的实体类，实体是应用中的“人”“事”“物”等，MyNotes中的“备忘录信息”Note就是实体类。另外，在数据库设计时，实体会被设计成数据库中的数据表。

MyNotes应用采用树形结构导航，读者可以参考9.5节采用Master-Detail Application工程模板创建该应用，具体表示逻辑不再赘述。至于数据持久化逻辑，主要是修改DAO组件。采用何种数据化技术，只是影响了DAO，不会影响表示逻辑。下面我们展开介绍不同的数据化技术如何实现数据持久化逻辑。

14.3 属性列表

属性列表文件是一种XML文件，Foundation框架中的数组和字典等都可以与属性列表文件互相转换，如图14-5所示。

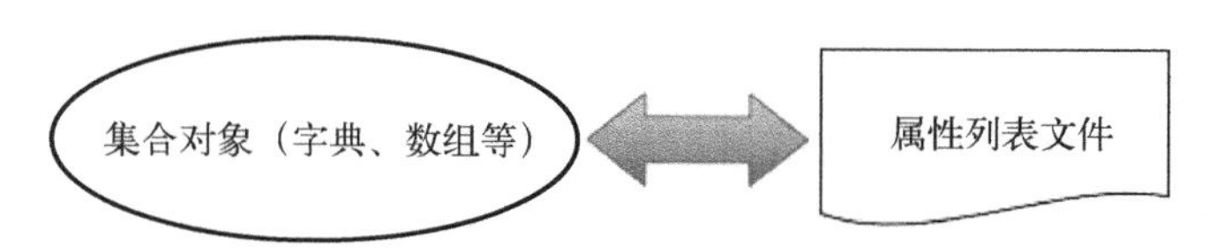

图14-5 集合对象与属性列表文件的对应关系

图14-6所示的是属性列表文件NotesList.plist，它是一个数组，其中有两个元素，其元素结构是字典类型。图14-7所示的是对应的NSArray，它是与NotesList.plist属性列表文件对应的集合对象。

① 表示组件。用户与应用交互的组件，在iOS应用中它主要由UIKit Framework构成，包括前面学习的视图、控制器、控件和事件处理等内容。

② 应用的核心业务规则、商业规则和算法等，例如银行利息的计算、个人所得税的计算等。

③ 实现数据持久化的代码，根据数据持久化方式的不同，它需要采用不同的技术实现，可能是访问SQLite数据的API函数，也可能是Core Data技术，或是访问文件的NSFileManager，或是网络通信技术。

④ CRUD方法是访问数据的4个方法，即增加、读取、更新和删除：C为Create，表示增加数据；R是Read，表示读取数据；U是Update，表示更新数据；D是Delete，表示删除数据。

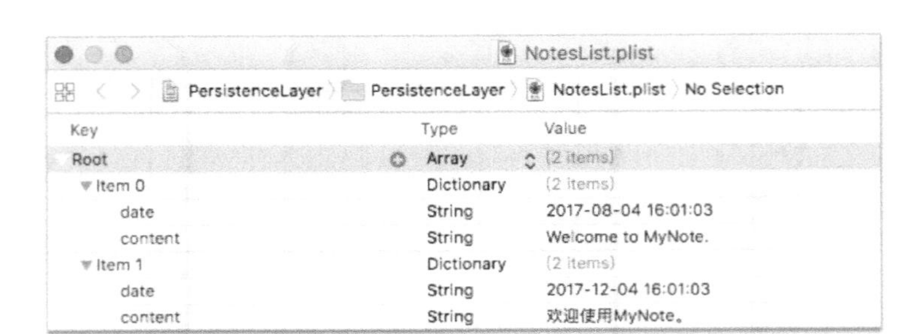

图14-6　属性列表文件NotesList.plist

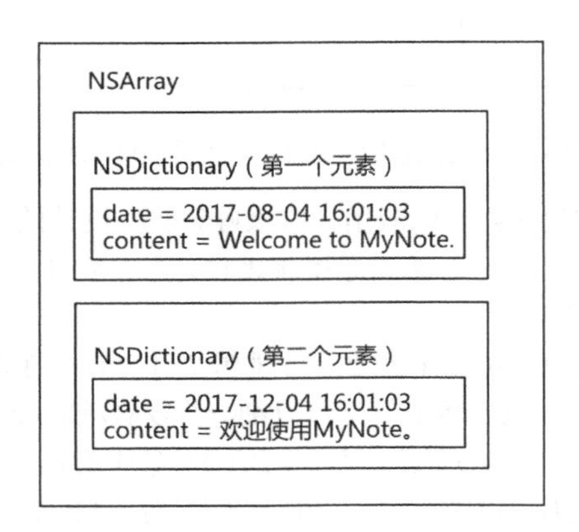

图14-7　与NotesList.plist对应的NSArray集合对象

数组类NSArray和字典类NSDictionary提供了读写属性列表文件的方法，其中NSArray类的方法如下所示。

- **+ arrayWithContentsOfFile:**。静态创建工厂方法，用于从属性列表文件中读取数据，创建NSArray对象。Swift语言没有对应的构造函数。
- **- initWithContentsOfFile:**。构造函数，用于从属性列表文件中读取数据，创建NSArray对象。Swift语言表示为convenience init?(contentsOfFile aPath:String)。
- **- writeToFile:atomically:**。该方法把NSArray对象写入属性列表文件，它的第一个参数是文件名，第二个参数表明是否使用辅助文件：如果为true，则先写入辅助文件，然后将辅助文件重新命名为目标文件；如果为false，则直接写入目标文件。

NSDictionary类的方法如下所示。

- **+ dictionaryWithContentsOfFile:**。静态工厂方法，用于从属性列表文件中读取数据并创建NSDictionary对象。Swift语言没有对应的构造函数。
- **- initWithContentsOfFile:**。构造函数，用于从属性列表文件中读取数据并初始化NSDictionary对象。Swift语言表示成convenience init?(contentsOfFile Path:String)。
- **- writeToFile:atomically:**。将NSDictionary对象写入属性列表文件，它的第一个参数是文件名。第二个参数表明是否使用辅助文件：如果为true，则先写入辅助文件，然后将辅助文件重新命名为目标文件；如果为false，则直接写入目标文件。

Note的代码如下，它只有两个属性，其中date是创建备忘录的日期，content是备忘录的内容：

```
//Note.swift文件
import Foundation
class Note {

    var date: Date
    var content: String
```

```
//
//Note.h文件
//
#import <Foundation/Foundation.h>

@interface Note : NSObject
```

```
    init(date: Date, content: String) {
        self.date = date
        self.content = content
    }

    init() {
        self.date = Date()
        self.content = ""
    }
}
```

```
@property(nonatomic, strong) NSDate* date;
@property(nonatomic, strong) NSString* content;

-(instancetype)initWithDate:(NSDate*)date content:(NSString*)content;

-(instancetype)init;

@end

//
//Note.m文件
//
#import "Note.h"

@implementation Note

-(instancetype)initWithDate:(NSDate*)date content:(NSString*)content {

    self = [super init];

    if (self) {
        self.date = date;
        self.content = content;
    }

    return self;
}

-(instancetype)init {

    self = [super init];

    if (self) {
        self.date = [[NSDate alloc] init];
        self.content = @"";
    }

    return self;
}

@end
```

访问属性文件的DAO是NoteDAO，其主要方法和属性如下：

```
public class NoteDAO {

    //私有的DateFormatter属性
    private var dateFormatter = DateFormatter()
    //私有的沙箱目录中的属性列表文件路径
    private var plistFilePath: String!

    public static let sharedInstance: NoteDAO = {
        let instance = NoteDAO()

        //初始化沙箱目录中的属性列表文件路径
        instance.plistFilePath = instance.applicationDocumentsDirectoryFile()
        //初始化DateFormatter
        instance.dateFormatter.dateFormat = "yyyy-MM-dd HH:mm:ss"
        //初始化属性列表文件
        instance.createEditableCopyOfDatabaseIfNeeded()

        return instance
    }()

    //初始化属性列表文件
```

```
//NoteDAO.h文件
#import "Note.h"

@interface NoteDAO : NSObject

......

@end

//NoteDAO.m文件
#import "NoteDAO.h"

@interface NoteDAO ()    //声明NoteDAO扩展

//NoteDAO扩展中DateFormatter属性是私有的
@property (nonatomic,strong) NSDateFormatter *dateFormatter;

//NoteDAO扩展中沙箱目录中的属性列表文件路径是私有的
@property (nonatomic,strong) NSString *plistFilePath;

@end
```

```
    private func createEditableCopyOfDatabaseIfNeeded() {                ①

        let fileManager = FileManager.default
        let dbexits = fileManager.fileExists(atPath: self.plistFilePath)
        if (!dbexits) {

            let frameworkBundle = Bundle(for: NoteDAO.self)                ②
            let frameworkBundlePath = frameworkBundle.resourcePath as NSString?
            let defaultDBPath = frameworkBundlePath!.appendingPathComponent
            ➥("NotesList.plist")

            do {
                try fileManager.copyItem(atPath: defaultDBPath, toPath:
                ➥self.plistFilePath)                                       ③
            } catch {
                let nserror = error as NSError
                NSLog("数据保存错误：%@", nserror.localizedDescription)
                assert(false, "错误写入文件")                                 ④
            }
        }
    }

    private func applicationDocumentsDirectoryFile() -> String {            ⑤
        let documentDirectory: NSArray =
        ➥NSSearchPathForDirectoriesInDomains(.documentDirectory,
        ➥.userDomainMask, true) as NSArray
        let path = (documentDirectory[0] as AnyObject)
        ➥.appendingPathComponent("NotesList.plist") as String
        return path
    }
    ......
}
```

```
@implementation NoteDAO

static NoteDAO *sharedSingleton = nil;

+ (NoteDAO *)sharedInstance {
    static dispatch_once_t once;
    dispatch_once(&once, ^{

        sharedSingleton = [[self alloc] init];

        //初始化沙箱目录中的属性列表文件路径
        sharedSingleton.plistFilePath = [sharedSingleton
        ➥applicationDocumentsDirectoryFile];
        //初始化DateFormatter
        sharedSingleton.dateFormatter = [[NSDateFormatter alloc] init];
        [sharedSingleton.dateFormatter setDateFormat:@"yyyy-MM-dd HH:mm:ss"];
        //初始化属性列表文件
        [sharedSingleton createEditableCopyOfDatabaseIfNeeded];

    });
    return sharedSingleton;
}

//初始化属性列表文件
- (void)createEditableCopyOfDatabaseIfNeeded {                              ①

    NSFileManager *fileManager = [NSFileManager defaultManager];
    BOOL dbexits = [fileManager fileExistsAtPath:self.plistFilePath];
    if (!dbexits) {

        NSBundle *frameworkBundle = [NSBundle bundleForClass:
        ➥[NoteDAO class]];                                                 ②
        NSString *frameworkBundlePath = [frameworkBundle resourcePath];
        NSString *defaultDBPath = [frameworkBundlePath
        ➥stringByAppendingPathComponent:@"NotesList.plist"];

        NSError *error;
        BOOL success = [fileManager copyItemAtPath:defaultDBPath
        ➥toPath:self.plistFilePath error:&error];                          ③
        NSAssert(success, @"错误写入文件");                                    ④

    }
}

- (NSString *)applicationDocumentsDirectoryFile {                           ⑤
    NSString *documentDirectory =
    ➥[NSSearchPathForDirectoriesInDomains(NSDocumentDirectory,
    ➥NSUserDomainMask, TRUE) lastObject];
    NSString *path = [documentDirectory
    ➥stringByAppendingPathComponent:@"NotesList.plist"];
    return path;
}

......

@end
```

上述代码中，第①行声明的`createEditableCopyOfDatabaseIfNeeded`方法用于判断沙箱Documents目录中是否存在NotesList.plist文件。如果不存在，则从资源目录中复制一个，资源目录中的NotesList.plist文件如图14-8所示。资源目录NotesList.plist文件中预先有两条数据，如图14-7所示。

提示

获得NSBundle（Swift版为Bundle）对象方法。通过NSBundle对象，可以访问当前程序的资源目录。要获得该对象，主要有两个方法——+ mainBundle:和+ bundleForClass:，前者获得主NSBundle对象，即当前运行的应用程序NSBundle对象；后者获得当前类所在的NSBundle对象。在Swift中，要获得框架资源Bundle，可以通过init(for aClass: AnyClass)构造函数实现，见第②行。在业务逻辑和数据持久化逻辑组件中，我们推荐使用第二种方法，因为有时候这些组件可能在另外的工程或框架中。

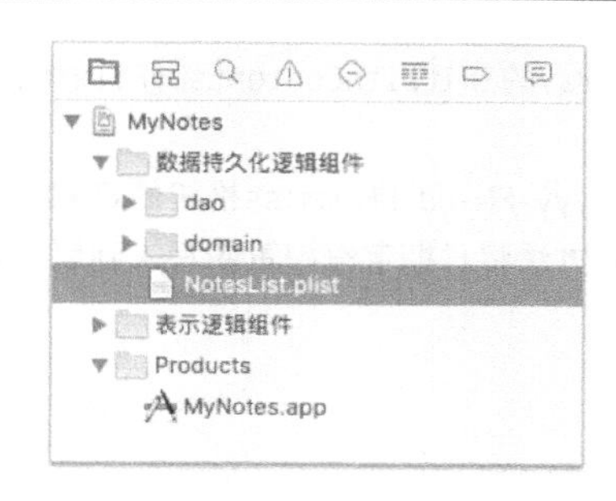

图14-8　资源目录中的NotesList.plist文件

第③行实现文件复制，这主要通过文件管理器类NSFileManager（Swift是FileManager）的复制方法来实现，具体如下：

```
func copyItem(atPath srcPath: String, toPath dstPath: String) throws
```

```
- (BOOL)copyItemAtPath:(NSString *)srcPath toPath:(NSString *)dstPath
➥error:(NSError * _Nullable *)error;
```

提示

复制方法在Swift版本采用了do-try-catch错误处理模式，其中do { try ...}语句是尝试复制操作，如果失败，则执行catch {...}语句。要详细了解Swift的错误处理模式，可以参考我编写的《从零开始学Swift（第2版）》一书。

第④行的assert是断言函数（Objective-C语言中使用NSAssert宏），它在第一个参数为false时抛出异常。

第⑤行声明了applicationDocumentsDirectoryFile方法，用于获得沙箱Documents目录中NotesList.plist文件的完整路径。

提示

在Swift语言中，我们可以将一些没有参数但有返回值的方法设计成“计算属性”。在上述代码Swift版中，第⑤行的applicationDocumentsDirectoryFile方法可以设计为applicationDocumentsDirectoryFile计算属性，参考代码如下：

```
let applicationDocumentsDirectoryFile: String = {
    let documentDirectory: NSArray =
    ➥NSSearchPathForDirectoriesInDomains(.DocumentDirectory,
    ➥.UserDomainMask, true)
    let path = documentDirectory[0]
    ➥stringByAppendingPathComponent("NotesList.plist") as String
    return path
}()
```

在NoteDAO中插入数据的代码如下所示：

```
public func create(_ model: Note) -> Int {
```

```
- (int)create:(Note *)model {
```

```
    let array = NSMutableArray(contentsOfFile: self.plistFilePath)!        ①
    let strDate = self.dateFormatter.string(from: model.date as Date)      ②
    let dict = NSDictionary(objects: [strDate, model.content],
    ➥forKeys: ["date" as NSCopying, "content" as NSCopying])

    array.add(dict)

    array.write(toFile: self.plistFilePath, atomically: true)              ③

    return 0
}
```

```
    NSMutableArray *array = [[NSMutableArray alloc] initWithContentsOf
    ➥File:self.plistFilePath];                                              ①
    NSString *strDate = [self.dateFormatter stringFromDate:model.date];    ②
    NSDictionary *dict = @{@"date" : strDate, @"content" : model.content};

    [array addObject:dict];

    [array writeToFile:self.plistFilePath atomically:TRUE];                ③

    return 0;
}
```

上述代码中，第①行通过NSMutableArray的initWithContentsOfFile:构造函数读取属性列表文件内容，并初始化NSMutableArray对象。

第②行是将NSDate日期对象转换成为yyyy-MM-dd HH:mm:ss格式的日期字符串。

第③行中的writeToFile:atomically:方法将日期字符串重新写入到属性列表文件中。

在NoteDAO中删除数据的代码如下：

```
public func remove(_ model: Note) -> Int {

    let array = NSMutableArray(contentsOfFile: self.plistFilePath)!

    for item in array {
        let dict = item as! NSDictionary
        let strDate = dict["date"] as! String
        let date = dateFormatter.date(from: strDate)

        //比较日期主键是否相等
        if date! == model.date as Date {                                   ①
            array.remove(dict)
            array.write(toFile: self.plistFilePath, atomically: true)
            break
        }
    }
    return 0
}
```

```
- (int)remove:(Note *)model {

    NSMutableArray *array = [[NSMutableArray alloc] initWithContentsOfFile:
    ➥self.plistFilePath];

    for (NSDictionary *dict in array) {
        NSString *strDate = dict[@"date"];
        NSDate *date = [self.dateFormatter dateFromString:strDate];

        //比较日期主键是否相等
        if ([date isEqualToDate:model.date]) {                             ①
            [array removeObject:dict];
            [array writeToFile:self.plistFilePath atomically:TRUE];
            break;
        }
    }
    return 0;
}
```

在上述代码中，我们需要注意第①行代码，它用于判断两个日期是否相等。因为备忘录的日期字段是主键[①]，所以只有在日期相等的情况下，才能从array集合中移除这个对象，最后再重新写回属性列表文件。

在NoteDAO中修改数据的代码如下所示：

```
public func modify(_ model: Note) -> Int {

    let array = NSMutableArray(contentsOfFile: self.plistFilePath)!

    for item in array {
        let dict = item as! NSDictionary
        let strDate = dict["date"] as! String
        let date = dateFormatter.date(from: strDate)

        //比较日期主键是否相等
        if date! == model.date as Date {
            dict.setValue(model.content, forKey: "content")
            array.write(toFile: self.plistFilePath, atomically: true)
            break
        }
    }
    return 0
}
```

```
- (int)modify:(Note *)model {

    NSArray *array = [[NSArray alloc] initWithContentsOfFile:
    ➥self.plistFilePath];

    for (NSDictionary *dict in array) {

        NSDate *date = [self.dateFormatter dateFromString:dict[@"date"]];

        //比较日期主键是否相等
        if ([date isEqualToDate:model.date]) {
            [dict setValue:model.content forKey:@"content"];
            [array writeToFile:self.plistFilePath atomically:TRUE];
            break;
        }
    }
    return 0;
}
```

① 主键（primary key），也称为“主关键字”或“主码”，是表中的一个或多个字段，它用于唯一地标识表中的某一条记录。主键不能为空值。

在NoteDAO中查询所有数据的代码如下所示：

```
public func findAll() -> NSMutableArray {

    let array = NSMutableArray(contentsOfFile: self.plistFilePath)!

    let listData = NSMutableArray()

    for item in array {
        let dict = item as! NSDictionary
        let strDate = dict["date"] as! String
        let date = dateFormatter.date(from: strDate)!
        let content = dict["content"] as! String

        let note = Note(date: date, content: content)

        listData.add(note)
    }

    return listData
}
```

```
- (NSMutableArray *)findAll {

    NSArray *array = [[NSArray alloc] initWithContentsOfFile:
    ➥self.plistFilePath];

    NSMutableArray *listData = [[NSMutableArray alloc] init];

    for (NSDictionary *dict in array) {

        NSString *strDate = dict[@"date"];
        NSDate *date = [self.dateFormatter dateFromString:strDate];
        NSString *content = dict[@"content"];

        Note *note = [[Note alloc] initWithDate:date content:content];

        [listData addObject:note];

    }
    return listData;
}
```

在NoteDAO中按照主键查询数据的代码如下所示：

```
public func findById(_ model: Note) -> Note? {

    let array = NSMutableArray(contentsOfFile: self.plistFilePath)!

    for item in array {

        let dict = item as! NSDictionary
        let strDate = dict["date"] as! String
        let date = dateFormatter.date(from: strDate)!
        let content = dict["content"] as! String

        //比较日期主键是否相等
        if date == model.date as Date {
            let note = Note(date: date, content: content)
            return note
        }
    }
    return nil
}
```

```
- (Note *)findById:(Note *)model {

    NSArray *array = [[NSArray alloc] initWithContentsOfFile:
    ➥self.plistFilePath];

    for (NSDictionary *dict in array) {

        NSString *strDate = dict[@"date"];
        NSDate *date = [self.dateFormatter dateFromString:strDate];
        NSString *content = dict[@"content"];

        //比较日期主键是否相等
        if ([date isEqualToDate:model.date]) {
            Note *note = [[Note alloc] initWithDate:date content:content];
            return note;
        }
    }
    return nil;
}
```

修改完成后，试运行一下看看效果。

14.4 使用 SQLite 数据库

2000年，D. 理查德·希普开发并发布了嵌入式系统使用的关系数据库SQLite，目前的主流版本是SQLite 3。SQLite是开源的，它采用C语言编写，具有可移植性强、可靠性高、小而易用的特点。SQLite运行时与使用它的应用程序之间共用相同的进程空间，而不是单独的两个进程。

SQLite提供了对SQL-92标准的支持，支持多表、索引、事务、视图和触发。SQLite是无数据类型的数据库，就是字段不用指定类型。下面的代码在SQLite中是合法的：

```
CREATE TABLE mytable
(  a VARCHAR(10),
   b NVARCHAR(15),
   c TEXT,
   d INTEGER,
   e FLOAT,
```

```
    f BOOLEAN,
    g CLOB,
    h BLOB,
    i TIMESTAMP,
    j NUMERIC(10,5)
    k VARYING CHARACTER (24),
    l NATIONAL VARYING CHARACTER(16)
);
```

14.4.1　SQLite 数据类型

虽然SQLite可以忽略数据类型，但从编程规范上讲，我们还是应该在Create Table语句中指定数据类型。因为数据类型可以表明这个字段的含义，便于阅读和理解代码。SQLite支持的常见数据类型如下所示。

- **INTEGER**。有符号的整数类型。
- **REAL**。浮点类型。
- **TEXT**。字符串类型，采用UTF-8和UTF-16字符编码。
- **BLOB**。二进制大对象类型，能够存放任何二进制数据。

SQLite中没有Boolean类型，可以采用整数0和1替代。SQLite中也没有日期和时间类型，它们存储在TEXT、REAL和INTEGER类型中。

为了兼容SQL-92中的其他数据类型，我们可以将它们转换为上述几种数据类型：

- 将VARCHAR、CHAR和CLOB转换为TEXT类型；
- 将FLOAT、DOUBLE转换为REAL类型；
- 将NUMERIC转换为INTEGER或者REAL类型。

14.4.2　添加 SQLite3 库

首先添加SQLite3库到工程环境中：选择工程中的TARGETS→Build Phases→Link Binary With Libraries(0 items)，点击左下角的+按钮，从弹出界面中选择libsqlite3.0.tbd或libsqlite3.tbd，然后点击Add按钮添加，如图14-9所示。

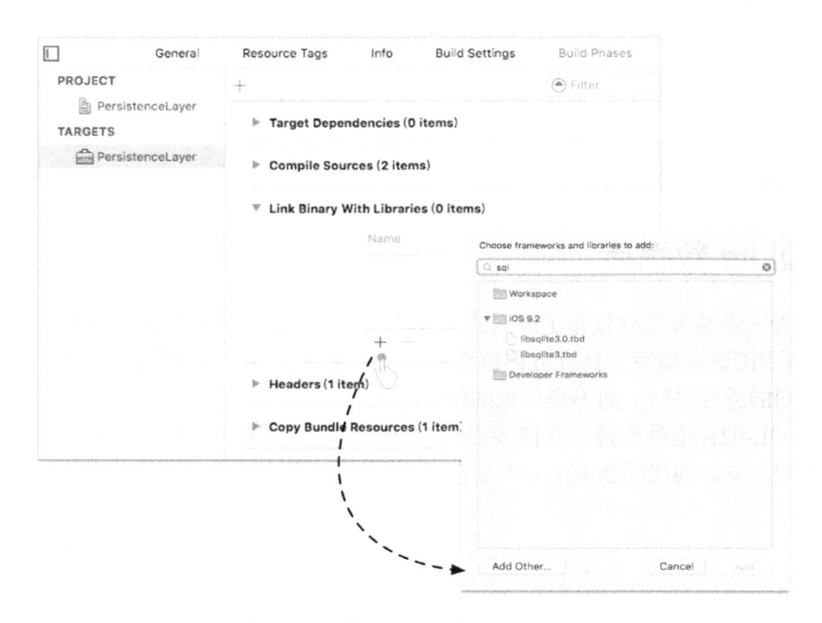

图14-9　添加SQLite3库到工程环境

14.4.3 配置 Swift 环境

如果我们采用Swift语言开发，则会更加复杂。这是因为SQLite API是基于C语言的，Swift语言要想调用C语言API，则需要桥接头文件。有关桥接头文件的相关内容，可以参考我编写的《从零开始学Swift（第2版）》。在本例中，桥接头文件MyNotes-Bridging-Header.h位于PersistenceLayer目录下，如图14-10所示，其中当前路径是MyNotes.xcodeproj工程文件所在的目录。

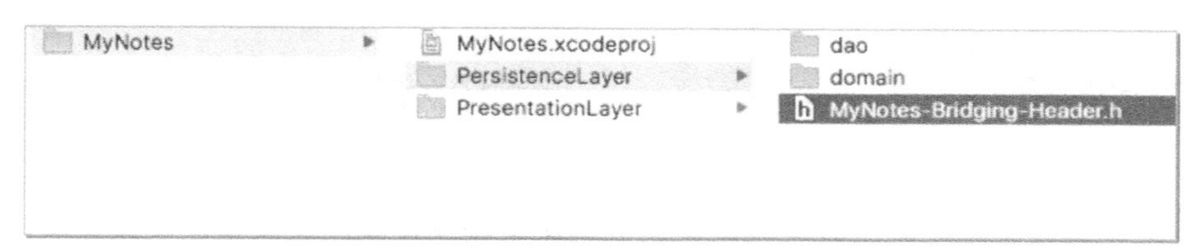

图14-10 桥接头文件MyNotes-Bridging-Header.h

如果按照图14-10所示，桥接头文件MyNotes-Bridging-Header.h放置在PersistenceLayer目录下，那么手工配置桥接头文件的方法是：选择TARGETS→Build Settings→Swift Compiler - General，打开如图14-11所示的对话框，从中输入PersistenceLayer/MyNotes-Bridging-Header.h。

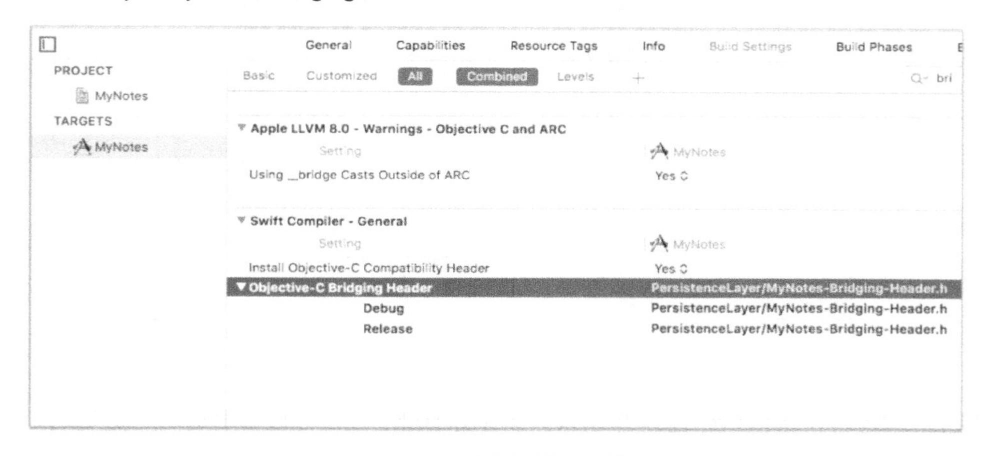

图14-11 配置桥接头文件

配置完成之后，需要在MyNotes-Bridging-Header.h中引入SQLite头文件，代码如下：

```
#import "sqlite3.h"
```

14.4.4 创建数据库

配置完环境后，我们就可以创建数据库了。需要经过如下3个步骤：

(1) 使用`sqlite3_open`函数打开数据库；

(2) 使用`sqlite3_exec`函数执行`Create Table`语句，创建数据库表；

(3) 使用`sqlite3_close`函数释放资源。

在这个过程中，我们使用了3个SQLite3函数，它们都是C语言函数。

下面来看看具体的代码，NoteDAO文件中的属性、常量和初始化相关的代码如下所示：

```
//桥接头文件MyNotes.Bridging.Header.h
#import "sqlite3.h"                                  ①
```

```
//NoteDAO.m文件
#import "NoteDAO.h"
#import "sqlite3.h"                                  ①
```

```
//NoteDAO.swift文件
import Foundation

let DBFILE_NAME = "NotesList.sqlite3"

public class NoteDAO {

    private var db: COpaquePointer = nil                                    ②

    //私有DateFormatter属性
    private var dateFormatter = DateFormatter()
    //私有沙箱目录中的属性列表文件路径
    private var plistFilePath: String!

    //初始化文件
    private func createEditableCopyOfDatabaseIfNeeded() {

        let cpath = self.plistFilePath.cString(using: String.Encoding.utf8) ③

        if sqlite3_open(cpath!, &db) != SQLITE_OK {                         ④
            NSLog("数据库打开失败。")
        } else {
            let sql = "CREATE TABLE IF NOT EXISTS
            ➥Note (cdate TEXT PRIMARY KEY, content TEXT)"                    ⑤
            let cSql = sql.cStringUsingEncoding(NSUTF8StringEncoding)        ⑥

            if (sqlite3_exec(db,cSql!, nil, nil, nil) != SQLITE_OK) {        ⑦
                NSLog("建表失败。")
            }
        }
        sqlite3_close(db)                                                    ⑧
    }
    ......
}
```

```
#define DBFILE_NAME @"NotesList.sqlite3"

//声明NoteDAO扩展
@interface NoteDAO () {
    sqlite3 *db;                                                             ②
}

//NoteDAO扩展中的DateFormatter属性是私有的
@property(nonatomic, strong) NSDateFormatter *dateFormatter;

//NoteDAO扩展中沙箱目录的属性列表文件路径是私有的
@property(nonatomic, strong) NSString *plistFilePath;

@end

@implementation NoteDAO
......

//初始化文件
- (void)createEditableCopyOfDatabaseIfNeeded {

    const char *cpath = [self.plistFilePath UTF8String];                     ③

    if (sqlite3_open(cpath, &db) != SQLITE_OK) {                             ④
        NSLog(@"数据库打开失败。");
    } else {
        NSString *sql = [NSString stringWithFormat:@"CREATE TABLE IF NOT EXISTS
        ➥Note (cdate TEXT PRIMARY KEY, content TEXT);"];                     ⑤
        const char *cSql = [sql UTF8String];                                 ⑥

        if (sqlite3_exec(db, cSql, NULL, NULL, NULL) != SQLITE_OK) {         ⑦
            NSLog(@"建表失败。");
        }
    }
    sqlite3_close(db);                                                       ⑧
}
......

@end
```

上述代码中，第①行在Objective-C版本中引入头文件sqlite3.h，在Swift版本中是在桥接头文件中引入头文件sqlite3.h。

第②行在Objective-C版本中声明C指针类型变量db，在Swift版本中是COpaquePointer类型。COpaquePointer类型映射到C指针类型。

第③行用于将NSString类型转换为C接受的char*类型数据。

第④行用于打开数据库，其中sqlite3_open函数的第一个参数是数据库文件的完整路径，需要注意的是，在SQLite3函数中，接受的是char*类型的数据。第二个参数为sqlite3指针变量db的地址；返回值是整数类型。在SQLite3中，我们定义了很多常量，如果返回值等于常量SQLITE_OK，则说明创建成功。

如果成功打开数据库，则需要创建数据库中的表，其中第⑤行是建表的SQL语句。SQL语句CREATE TABLE IF NOT EXISTS Note (cdate TEXT PRIMARY KEY, content TEXT)在表Note不存在时创建，否则不创建。第⑥行用于将Swift的String类型转换为C接受的char*类型。

第⑦行的sqlite3_exec函数执行SQL语句，该函数的第一个参数是sqlite3指针变量db的地址，第二个参数是要执行的SQL语句，第三个参数是要回调的函数，第四个参数是要回调函数的参数，第五个参数是表示执行出错的字符串。

最后，数据操作执行完成。第⑧行的sqlite3_close函数用于释放资源，代码中多次使用了该函数。在数据库打开失败、SQL语句执行失败，以及SQL语句执行成功等情况下，都会调用该函数。原则上，无论正常结束还是异常结束，必须使用sqlite3_close函数释放资源。

14.4.5 查询数据

数据查询一般会带有查询条件，这可以使用SQL语句的where子句实现，但是在程序中需要动态绑定参数给where子句。查询数据的具体操作步骤如下所示：

(1) 使用sqlite3_open函数打开数据库；

(2) 使用sqlite3_prepare_v2函数预处理SQL语句；

(3) 使用sqlite3_bind_text函数绑定参数；

(4) 使用sqlite3_step函数执行SQL语句，遍历结果集；

(5) 使用sqlite3_column_text等函数提取字段数据；

(6) 使用sqlite3_finalize和sqlite3_close函数释放资源。

在NoteDAO中按照主键查询数据的方法如下：

```
//按照主键查询数据的方法
public func findById(_ model: Note) -> Note? {

    let cpath = self.plistFilePath.cString(using: String.Encoding.utf8)

    if sqlite3_open(cpath!, &db) != SQLITE_OK {                        ①
        NSLog("数据库打开失败。")
    } else {
        let sql = "SELECT cdate,content FROM Note where cdate =?"
        let cSql = sql.cString(using: String.Encoding.utf8)

        var statement:OpaquePointer? = nil                             ②
        //预处理过程
        if sqlite3_prepare_v2(db, cSql!, -1, &statement, nil) == SQLITE_OK { ③
            //准备参数
            let strDate = self.dateFormatter.string(from: model.date as Date)
            let cDate = strDate.cString(using: String.Encoding.utf8)

            //开始绑定参数
            sqlite3_bind_text(statement, 1, cDate!, -1, nil)           ④

            //执行查询
            if sqlite3_step(statement) == SQLITE_ROW {                 ⑤

                let note = Note()
                if let strDate = getColumnValue(index:0, stmt:statement!) { ⑥
                    let date : Date = self.dateFormatter.date(from: strDate)!
                    note.date = date
                }
                if let strContent = getColumnValue(index:1, stmt:statement!) {
                    note.content = strContent
                }
                sqlite3_finalize(statement)                            ⑦
                sqlite3_close(db)                                      ⑧

                return note
            }

        }
        sqlite3_finalize(statement)
    }
    sqlite3_close(db)
    return nil
}

//获得字段数据
private func getColumnValue(index:CInt, stmt:OpaquePointer)->String? {
```

```
- (Note *)findById:(Note *)model {

    const char *cpath = [self.plistFilePath UTF8String];

    if (sqlite3_open(cpath, &db) != SQLITE_OK) {                       ①
        NSLog(@"数据库打开失败。");
    } else {
        NSString *sql = @"SELECT cdate,content FROM Note where cdate =?";
        const char *cSql = [sql UTF8String];

        sqlite3_stmt *statement;                                       ②
        //预处理过程
        if (sqlite3_prepare_v2(db, cSql, -1, &statement, NULL) ==
        ➥SQLITE _OK) {                                                ③
            //准备参数
            NSString *strDate = [self.dateFormatter
            ➥stringFromDate:model.date];
            const char *cDate = [strDate UTF8String];

            //开始绑定参数
            sqlite3_bind_text(statement, 1, cDate, -1, NULL);          ④

            //执行查询
            if (sqlite3_step(statement) == SQLITE_ROW) {               ⑤

                char *bufDate = (char *)
                ➥sqlite3_column_text(statement, 0);                   ⑥
                NSString*strDate = [[NSString alloc]
                ➥initWithUTF8String:bufDate];
                NSDate *date = [self.dateFormatter dateFromString:strDate];

                char *bufContent = (char *) sqlite3_column_text(statement, 1);
                NSString *strContent = [[NSString alloc]
                ➥initWithUTF8String:bufContent];

                Note *note = [[Note alloc] initWithDate:date
                ➥content:strContent];

                sqlite3_finalize(statement);                           ⑦
                sqlite3_close(db);                                     ⑧

                return note;
            }
        }
        sqlite3_finalize(statement);
    }
    sqlite3_close(db);
```

```
    if let ptr = UnsafeRawPointer.init(sqlite3_column_text(stmt, index)) {  ⑨
        let uptr = ptr.bindMemory(to:CChar.self, capacity:0)               ⑩
        let txt = String(validatingUTF8:uptr)                              ⑪
        return txt
    }
    return nil
}
```

```
        return nil;
    }
```

该方法执行了6个步骤，其中第(1)步如第①行所示，它与创建数据库的第一个步骤一样，这里就不再介绍了。

第(2)步如第③行所示。sqlite3_prepare_v2函数是SQL预处理语句，预处理的目的是将SQL编译成二进制代码，提高SQL语句的执行速度。sqlite3_prepare_v2函数的第三个参数代表全部SQL字符串的长度；第四个参数是sqlite3_stmt指针的地址，它是语句对象，我们可通过该语句对象执行SQL语句；第五个参数是SQL语句没有执行的部分语句。

第(3)步如第④行所示，sqlite3_bind_text函数用于绑定SQL语句的参数，其中第一个参数是statement指针，第二个参数为序号（从1开始），第三个参数为字符串值，第四个参数为字符串长度，第五个参数为一个函数指针。如果SQL语句中带有问号，则这个问号（它是占位符）就是要绑定的参数，示例代码如下：

```
let sql = "SELECT cdate,content FROM Note where cdate =?"
```

```
NSString *sql = @"SELECT cdate,content FROM Note where cdate =?";
```

第(4)步为使用sqlite3_step(statement)执行SQL语句，如第⑤行所示。如果sqlite3_step函数的返回值等于SQLITE_ROW，则说明还有其他行没有遍历。

第(5)步为提取字段数据，如Objectvie-C代码第⑥行（Swift代码为第⑨行）所示，它使用sqlite3_column_text函数读取字符串类型的字段。需要说明的是，sqlite3_column_text函数的第二个参数用于指定select字段的索引（从0开始）。

在Swift版本中，第⑥行调用自定义的getColumnValue函数实现从字段中提取数据。在getColumnValue函数中，第⑨行中的UnsafeRawPointer类表示任意类型的C指针，第⑩行的bindMemory(to:CChar.self, capacity:0)方法用于将任意类型的C指针绑定到Char类型指针，第一个参数是要绑定的指针类型，第二个参数是容量，若不能确定容量，可以将其设置为0。第⑪行的String(validatingUTF8:uptr)是从指针所指内容创建String，validatingUTF8参数表示验证指针内容是否包含以null结尾的UTF8字符串。

读取字段函数的采用与字段类型有关，SQLite3中类似的常用函数还有：

- sqlite3_column_blob()
- sqlite3_column_double()
- sqlite3_column_int()
- sqlite3_column_int64()
- sqlite3_column_text()
- sqlite3_column_text16()

关于其他API，读者可以参考http://www.sqlite.org/cintro.html。

第(6)步是释放资源，与14.4.4节创建数据库的过程不同，这里不仅使用sqlite3_close函数关闭数据库（如第⑧行所示），而且要使用sqlite3_finalize函数释放语句对象statement（如第⑦行所示）。

另外，第②行中的statement是SQL语句对象，用来执行SQL语句。

在NoteDAO中查询所有数据的方法如下：

```
//查询所有数据的方法
public func findAll() -> NSMutableArray {

    let cpath = self.plistFilePath.cString(using: String.Encoding.utf8)
    let listData = NSMutableArray()
```

```
- (NSMutableArray *)findAll {

    const char *cpath = [self.plistFilePath UTF8String];
    NSMutableArray *listData = [[NSMutableArray alloc] init];

    if (sqlite3_open(cpath, &db) != SQLITE_OK) {
```

```
    if sqlite3_open(cpath!, &db) != SQLITE_OK {
        NSLog("数据库打开失败。")
    } else {
        let sql = "SELECT cdate,content FROM Note"
        let cSql = sql.cString(using: String.Encoding.utf8)

        //语句对象
        var statement:OpaquePointer? = nil
        //预处理过程
        if sqlite3_prepare_v2(db, cSql!, -1, &statement, nil) == SQLITE_OK {
            //执行查询
            while sqlite3_step(statement) == SQLITE_ROW {                ①

                let note = Note()
                if let strDate = getColumnValue(index:0, stmt:statement!) {
                    let date : Date = self.dateFormatter.date(from:
                    ➥strDate)!
                    note.date = date
                }
                if let strContent = getColumnValue(index:1,
                ➥stmt:statement!) {
                    note.content = strContent
                }
                listData.add(note)
            }

            sqlite3_finalize(statement)
            sqlite3_close(db)

            return listData
        }
        sqlite3_finalize(statement)
    }
    sqlite3_close(db)
    return listData
}
```

```
        NSLog(@"数据库打开失败。");
    } else {

        NSString *sql = @"SELECT cdate,content FROM Note";
        const char *cSql = [sql UTF8String];

        //语句对象
        sqlite3_stmt *statement;
        //预处理过程
        if (sqlite3_prepare_v2(db, cSql, -1, &statement, NULL) == SQLITE_OK) {
            //执行查询
            while (sqlite3_step(statement) == SQLITE_ROW) {              ①
                char *bufDate = (char *) sqlite3_column_text(statement, 0);
                NSString *strDate = [[NSString alloc]
                ➥initWithUTF8String:bufDate];
                NSDate *date = [self.dateFormatter dateFromString:strDate];

                char *bufContent = (char *) sqlite3_column_text(statement, 1);
                NSString *strContent = [[NSString alloc]
                ➥initWithUTF8String:bufContent];

                Note *note = [[Note alloc] initWithDate:date
                ➥content:strContent];

                [listData addObject:note];
            }

            sqlite3_finalize(statement);
            sqlite3_close(db);

            return listData;
        }
        sqlite3_finalize(statement);
    }
    sqlite3_close(db);
    return listData;
}
```

查询所有数据的方法与按照主键查询数据的方法类似，区别在于本方法没有查询条件，不需要绑定参数。第①行用于遍历查询结果集，使用while循环语句，不是if判断语句。sqlite3_step函数执行查询，常量SQLITE_ROW表示结果集中还有数据。

14.4.6 修改数据

修改数据时涉及的SQL语句有insert、update和delete，这3个SQL语句都可以带参数。修改数据的具体步骤如下所示：

(1) 使用sqlite3_open函数打开数据库；

(2) 使用sqlite3_prepare_v2函数预处理SQL语句；

(3) 使用sqlite3_bind_text函数绑定参数；

(4) 使用sqlite3_step函数执行SQL语句；

(5) 使用sqlite3_finalize和sqlite3_close函数释放资源。

这与查询数据相比，少了提取字段数据这个步骤，其他步骤一样。下面看看代码部分。

在NoteDAO中插入数据的代码如下：

```
public func create(_ model: Note) -> Int {

    let cpath = self.plistFilePath.cString(using: String.Encoding.utf8)
```

```
- (int)create:(Note *)model {

    const char *cpath = [self.plistFilePath UTF8String];
```

```
    if sqlite3_open(cpath!, &db) != SQLITE_OK {
        NSLog("数据库打开失败。")
    } else {
        let sql = "INSERT OR REPLACE INTO note (cdate, content) VALUES (?,?)"
        let cSql = sql.cString(using: String.Encoding.utf8)

        //语句对象
        var statement:OpaquePointer? = nil
        //预处理过程
        if sqlite3_prepare_v2(db, cSql!, -1, &statement, nil) == SQLITE_OK {

            let strDate = self.dateFormatter.string(from: model.date as Date)
            let cDate = strDate.cString(using: String.Encoding.utf8)

            let cContent = model.content.cString(using: String.Encoding.utf8)

            //开始绑定参数
            sqlite3_bind_text(statement, 1, cDate!, -1, nil)
            sqlite3_bind_text(statement, 2, cContent!, -1, nil)

            //执行插入
            if sqlite3_step(statement) != SQLITE_DONE {                    ①
                NSLog("插入数据失败。")
            }
        }
        sqlite3_finalize(statement)
    }
    sqlite3_close(db)
    return 0
}
```

```
    if (sqlite3_open(cpath, &db) != SQLITE_OK) {
        NSLog(@"数据库打开失败。");
    } else {
        NSString *sql = @"INSERT OR REPLACE INTO note (cdate, content)
        ➥VALUES (?,?)";
        const char *cSql = [sql UTF8String];

        //语句对象
        sqlite3_stmt *statement;
        //预处理过程
        if (sqlite3_prepare_v2(db, cSql, -1, &statement, NULL) == SQLITE_OK) {

            NSString *strDate = [self.dateFormatter
            ➥stringFromDate:model.date];
            const char *cDate = [strDate UTF8String];

            const char *cContent = [model.content UTF8String];

            //绑定参数开始
            sqlite3_bind_text(statement, 1, cDate, -1, NULL);
            sqlite3_bind_text(statement, 2, cContent, -1, NULL);

            //执行插入数据
            if (sqlite3_step(statement) != SQLITE_DONE) {                    ①
                NSLog(@"插入数据失败。");
            }
        }
        sqlite3_finalize(statement);
    }
    sqlite3_close(db);
    return 0;
}
```

第①行中的sqlite3_step(statement)语句用于执行插入操作，其中常量SQLITE_DONE表示执行完成。

在NoteDAO中删除数据的代码如下：

```
public func remove(_ model: Note) -> Int {

    let cpath = self.plistFilePath.cString(using: String.Encoding.utf8)

    if sqlite3_open(cpath!, &db) != SQLITE_OK {
        NSLog("数据库打开失败。")
    } else {
        let sql = "DELETE from note where cdate =?"
        let cSql = sql.cString(using: String.Encoding.utf8)

        //语句对象
        var statement:OpaquePointer? = nil
        //预处理过程
        if sqlite3_prepare_v2(db, cSql!, -1, &statement, nil) == SQLITE_OK {

            let strDate = self.dateFormatter.string(from: model.date as
            ➥Date)
            let cDate = strDate.cString(using: String.Encoding.utf8)

            //开始绑定参数
            sqlite3_bind_text(statement, 1, cDate!, -1, nil)
            //删除数据
            if sqlite3_step(statement) != SQLITE_DONE {
                NSLog("删除数据失败。")
            }
        }
        sqlite3_finalize(statement)
    }
```

```
- (int)remove:(Note *)model {

    const char *cpath = [self.plistFilePath UTF8String];

    if (sqlite3_open(cpath, &db) != SQLITE_OK) {
        NSLog(@"数据库打开失败。");
    } else {
        NSString *sql = @"DELETE  from note where cdate =?";
        const char *cSql = [sql UTF8String];

        //语句对象
        sqlite3_stmt *statement;
        //预处理过程
        if (sqlite3_prepare_v2(db, cSql, -1, &statement, NULL) == SQLITE_OK) {

            NSString *strDate = [self.dateFormatter
            ➥stringFromDate:model.date];
            const char *cDate = [strDate UTF8String];

            //开始绑定参数
            sqlite3_bind_text(statement, 1, cDate, -1, NULL);
            //删除数据
            if (sqlite3_step(statement) != SQLITE_DONE) {
                NSLog(@"删除数据失败。");
            }
        }
        sqlite3_finalize(statement);
    }
```

```
    sqlite3_close(db)
    return 0
}
```

```
    sqlite3_close(db);
    return 0;
}
```

在NoteDAO中修改数据的代码如下：

```
public func modify(_ model: Note) -> Int {

    let cpath = self.plistFilePath.cString(using: String.Encoding.utf8)

    if sqlite3_open(cpath!, &db) != SQLITE_OK {
        NSLog("数据库打开失败。")
    } else {
        let sql = "UPDATE note set content=? where cdate =?"
        let cSql = sql.cString(using: String.Encoding.utf8)

        //语句对象
        var statement:OpaquePointer? = nil
        //预处理过程
        if sqlite3_prepare_v2(db, cSql!, -1, &statement, nil) == SQLITE_OK {

            let strDate = self.dateFormatter.string(from: model.date as
            ➥Date)
            let cDate = strDate.cString(using: String.Encoding.utf8)

            let cContent = model.content.cString(using: String.
            ➥Encoding.utf8)

            //开始绑定参数
            sqlite3_bind_text(statement, 1, cContent!, -1, nil)
            sqlite3_bind_text(statement, 2, cDate!, -1, nil)

            //执行修改操作
            if sqlite3_step(statement) != SQLITE_DONE {
                NSLog("修改数据失败。")
            }
        }
        sqlite3_finalize(statement)
    }
    sqlite3_close(db)
    return 0
}
```

```
- (int)modify:(Note *)model {

    const char *cpath = [self.plistFilePath UTF8String];

    if (sqlite3_open(cpath, &db) != SQLITE_OK) {
        NSLog(@"数据库打开失败。");
    } else {

        NSString *sql = @"UPDATE note set content=? where cdate =?";
        const char *cSql = [sql UTF8String];

        //语句对象
        sqlite3_stmt *statement;
        //预处理过程
        if (sqlite3_prepare_v2(db, cSql, -1, &statement, NULL) == SQLITE_OK) {

            NSString *strDate = [self.dateFormatter
            ➥stringFromDate:model.date];
            const char *cDate = [strDate UTF8String];

            const char *cContent = [model.content UTF8String];

            //开始绑定参数
            sqlite3_bind_text(statement, 1, cContent, -1, NULL);
            sqlite3_bind_text(statement, 2, cDate, -1, NULL);
            //执行修改操作
            if (sqlite3_step(statement) != SQLITE_DONE) {
                NSLog(@"修改数据失败。");
            }
        }
        sqlite3_finalize(statement);
    }
    sqlite3_close(db);
    return 0;
}
```

14.5　iOS 10 中的 Core Data 技术

Core Data是苹果为macOS和iOS系统应用开发提供的对象关系映射技术。它基于高级数据持久化API，其底层最终是SQLite数据库、二进制文件和内存数据保存，使开发人员不用再关心数据的存储细节问题，不用再使用SQL语句，不用面对SQLite的C语言函数。

14.5.1　对象关系映射技术

听说过Hibernate[①]的人对对象关系映射技术（ORM）应该不会感到陌生，ORM是关系数据模型和对象模型类之间的一个纽带。

无论哪一种模型，都是为了描述和构建应用系统。在应用系统中，一个非常基础的概念是“实体”。“实体”是应用系统中的“人”“事”和“物”，它们能够在关系模型和对象模型中以不同的形态存在。如图14-12所示，实体在关系模型中代表表的一条数据，该表描述了实体的结构有哪些属性和关系。实体在对象模型中代表类的一

① Hibernate是一个开放源代码的对象关系映射Java EE框架。

个对象，类描述了实体的结构，实体是类的对象。因此，表是与类对应的概念，记录是与对象对应的概念。

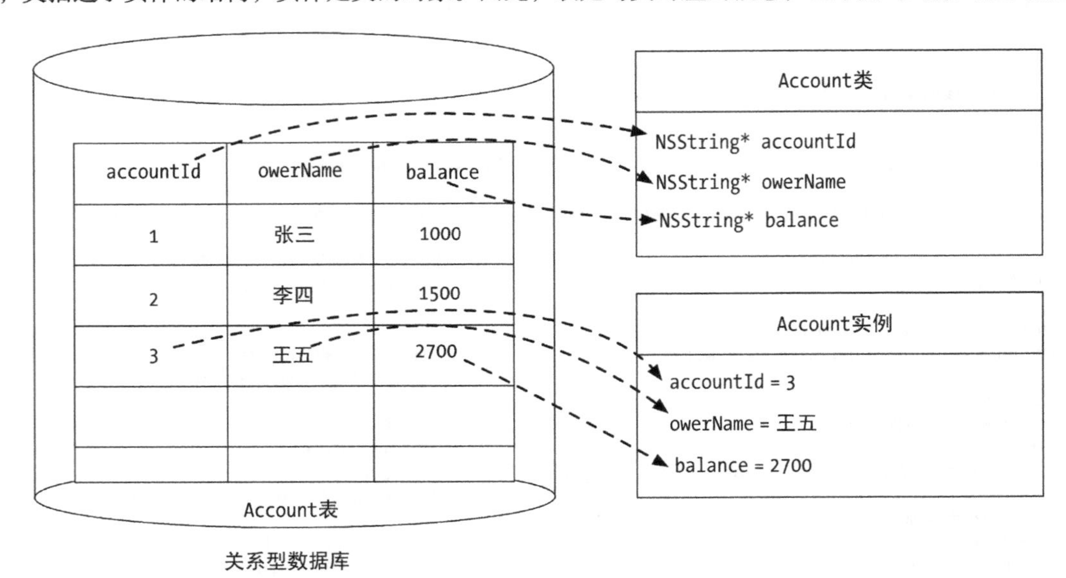

图14-12 ORM

关系模型和对象模型是有区别的，对象模型更加先进，能够描述继承、实现、关联、聚合和组成等复杂的关系，而关系模型只能描述一对一、一对多和多对多的关系。这两种模型之间的不和谐问题称为“阻抗不匹配”问题，而ORM可以解决“阻抗不匹配”问题。

14.5.2 添加 Core Data 支持

使用Xcode工具时，我们可以很方便地为工程添加Core Data支持。在Xcode的工程模板中，有两个模板（即Master-Detail Application和Single View Application模板）可以直接为工程添加Core Data支持，具体方法是创建工程时选中Use Core Data复选框，如图14-13所示。

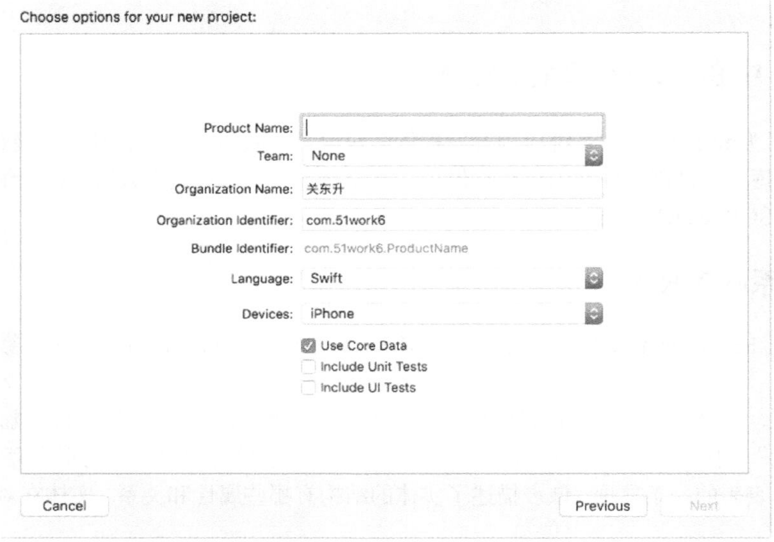

Choose options for your new project:

Product Name:

Team: None

Organization Name: 关东升

Organization Identifier: com.51work6

Bundle Identifier: com.51work6.ProductName

Language: Swift

Devices: iPhone

Use Core Data

Include Unit Tests

Include UI Tests

Cancel

Previous

Next

图14-13 添加Core Data支持

其他的模板需要我们自己添加Core Data支持。下面看看通过模板生成的代码，这些代码主要生成在AppDelegate中。AppDelegate中的主要代码如下：

```
import UIKit
import CoreData
@UIApplicationMain
class AppDelegate: UIResponder, UIApplicationDelegate {
    var window: UIWindow?
    ......
    lazy var persistentContainer: NSPersistentContainer = {                  ①
        let container = NSPersistentContainer(name: "HelloWorld")             ②
        container.loadPersistentStores(completionHandler:
        ➥{ (storeDescription, error) in
            if let error = error as NSError? {
                fatalError("Unresolved error \(error), \(error.userInfo)")
            }
        })
        return container
    }()

    //MARK: -- Core Data数据保存方法
    func saveContext () {                                                    ③
        let context = persistentContainer.viewContext
        if context.hasChanges {
            do {
                try context.save()
            } catch {
                let nserror = error as NSError
                fatalError("Unresolved error \(nserror), \(nserror.
                ➥userInfo)")
            }
        }
    }

}
```

```
//AppDelegate.h中的代码
#import <UIKit/UIKit.h>
#import <CoreData/CoreData.h>
@interface AppDelegate : UIResponder <UIApplicationDelegate>
@property (strong, nonatomic) UIWindow *window;
@property (readonly, strong) NSPersistentContainer *persistentContainer;
- (void)saveContext;
@end

//AppDelegate.m中的代码
#import "AppDelegate.h"
@implementation AppDelegate
......
#pragma mark - Core Data stack

@synthesize persistentContainer = _persistentContainer;

- (NSPersistentContainer *)persistentContainer {                             ①
    @synchronized (self) {
        if (_persistentContainer == nil) {
            _persistentContainer = [[NSPersistentContainer alloc]
            ➥initWithName:@"HelloWorld"];                                    ②
            [_persistentContainer loadPersistentStoresWithCompletionHandler:
            ➥^(NSPersistentStoreDescription *storeDescription, NSError *error) {
                if (error != nil) {
                    NSLog(@"Unresolved error %@, %@", error, error.userInfo);
                    abort();
                }
            }];
        }
    }
    return _persistentContainer;
}

#pragma mark -- Core Data数据保存方法

- (void)saveContext {                                                        ③
    NSManagedObjectContext *context = self.persistentContainer.viewContext;
    NSError *error = nil;
    if ([context hasChanges] && ![context save:&error]) {
        NSLog(@"Unresolved error %@, %@", error, error.userInfo);
        abort();
    }
}

@end
```

上述代码中，第①行定义了NSPersistentContainer类型的属性persistentContainer。NSPersistentContainer是iOS 10新增加的类，称为“持久化容器”，用来简化之前的Core Data栈创建过程，使用了一个新类NSPersistentStoreDescription来描述配置信息。第②行用于创建NSPersistentContainer对象，构造函数的参数name是容器的名字，它也是放到MainBundle资源目录中模型文件的名字。第③行声明保存数据的方法。

14.5.3 Core Data 栈

Core Data实现数据持久化是通过Core Data栈。Core Data栈如图14-14所示，它有一个或多个被管理的对象上下文，它连接到一个持久化存储协调器，一个持久化存储协调器连接到一个或多个持久化对象存储，持久化对象存储与底层存储文件关联。一个持久化存储协调器也可以管理多个被管理对象模型。一个持久化存储协调器就意味着一个Core Data栈。通过Core Data栈，我们可以实现数据查询、插入、删除和修改等操作。

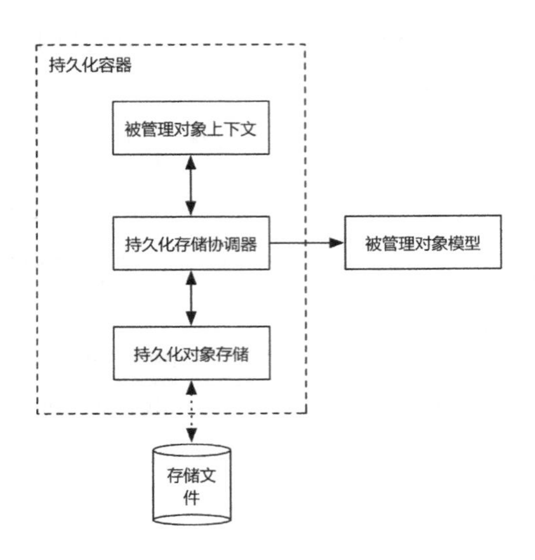

图14-14　Core Data栈

这些概念说明如下。

- 被管理对象模型（Managed Object Model，MOM），是系统中的“实体”，与数据库中的表等对象对应。对应类是`NSManagedObjectModel`。
- 持久化存储协调器（Persistent Store Coordinator，PSC），在持久化对象存储之上提供了一个接口，我们可以把它考虑成数据库的连接。对应类是`NSPersistentStoreCoordinator`。
- 被管理对象上下文（Managed Object Context，MOC），在被管理对象上下文中可以查找、删除和插入对象，然后通过栈同步到持久化对象存储。对应类是`NSManagedObjectContext`。
- 持久化对象存储（Persistent Object Store，POS），执行所有底层的从对象到数据的转换，并负责打开和关闭数据文件。它有3种持久化实现方式：SQLite、二进制文件和内存形式。

在iOS 10之前创建Core Data栈的过程非常复杂，为了简化创建过程，我们在iOS 10推出了持久化容器`NSPersistentContainer`类，见图14-14，它帮助维持Core Data栈中的被管理对象模型、持久化存储协调器和被管理对象上下文，使得开发人员不用关心创建Core Data栈的细节问题。

14.6　案例：采用 Core Data 重构 MyNotes 应用

Core Data栈的创建默认是在应用程序委托对象`AppDelegate`中实现的。Core Data栈属于数据持久逻辑，按照14.2节所介绍的设计原则，Core Data栈相关的对象只出现在数据持久化逻辑组件中（这些Core Data栈相关的对象包括`NSManagedObjectContext`、`NSPersistentStoreCoordinator`、`NSManagedObjectModel`和`NSManagedObject`等）。

14.6.1　建模和生成实体

14.5节的代码中有xcdatamodeld文件，它是模型文件[①]。Core Data可以利用它可视化地设计数据库，生成实体类代码和SQLite数据库文件。下面我们介绍建模和生成实体这两个过程。

1. 建模

默认情况下，如果采用模板生成方式，会创建一个与工程名相同的数据模型文件——<工程名>.xcdatamodeld。但是如果不采用模板，可以选择菜单File→New→File，从打开的选择文件模板对话框中选择iOS→Core Data→Data Model，如图14-15所示。

① 模型文件一般是在数据库设计阶段用可视化建模工具创建的数据库模型描述文件。

图14-15　创建数据模型文件

然后点击Next按钮，输入相应的文件名，这里输入CoreDataNotes，这样创建的数据模型文件就是CoreDataNotes.xcdatamodeld。

打开CoreDataNotes.xcdatamodeld文件，会看到如图14-16所示的模型对话框，在这个对话框中可以创建实体（entity）、实体属性和实体关系等。

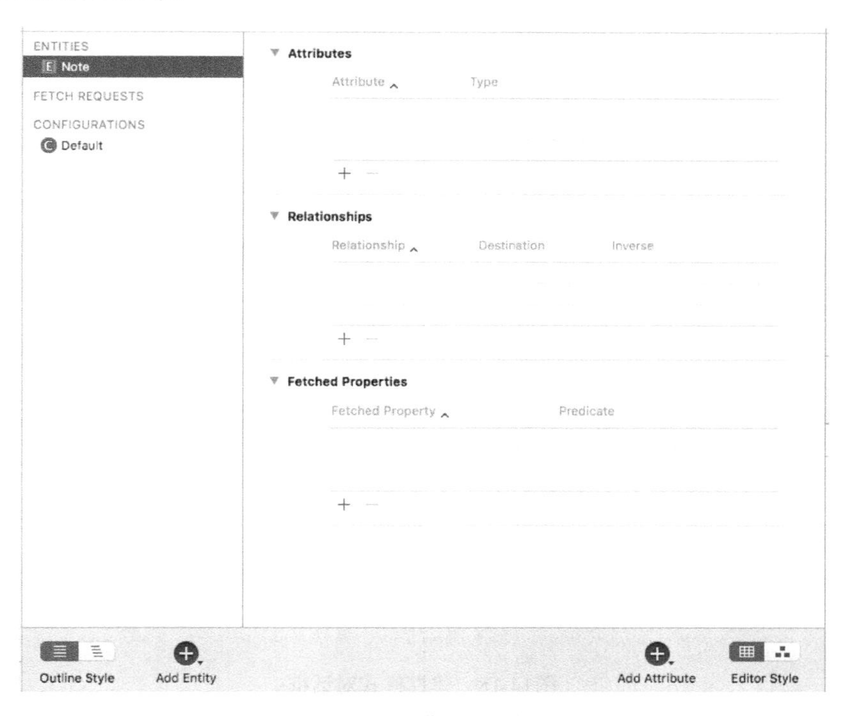

图14-16　模型对话框

创建实体的过程如图14-17所示：点击Add Entity按钮添加实体，将其名称修改为Note，在右边的Attributes列表框中添加属性date和content，并将它们的Type分别设为Date和String。

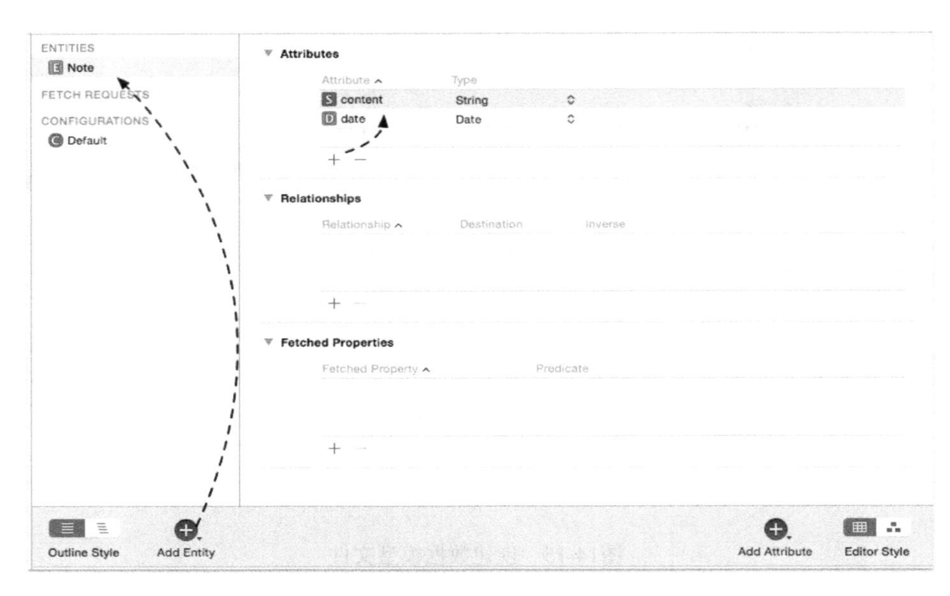

图14-17　创建实体

接着选择右下角的Editor Style按钮，得到的建模样式对话框如图14-18所示。

图14-18　建模样式对话框

2. 生成实体

首先选择实体Note，打开其数据模型检查器（如图14-19所示），在Entity的Class输入框中输入NoteManagedObject（它是被Core Data栈管理的对象）。

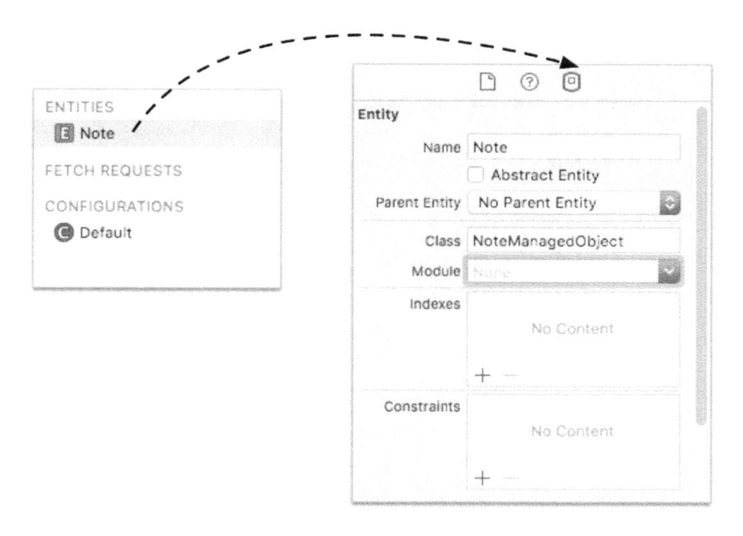

图14-19 输入实体类名

然后选择Note实体，再选择Editor→Create NSManagedObject Subclass...菜单项，打开如图14-20所示的界面，从中选择CoreDataNotes数据模型文件。接着点击Next按钮，进入如图14-21所示的界面，从中选择Note实体。接着点击Next按钮，进入选择生成文件位置界面，从中选择合适的文件夹，然后点击Create按钮创建文件。

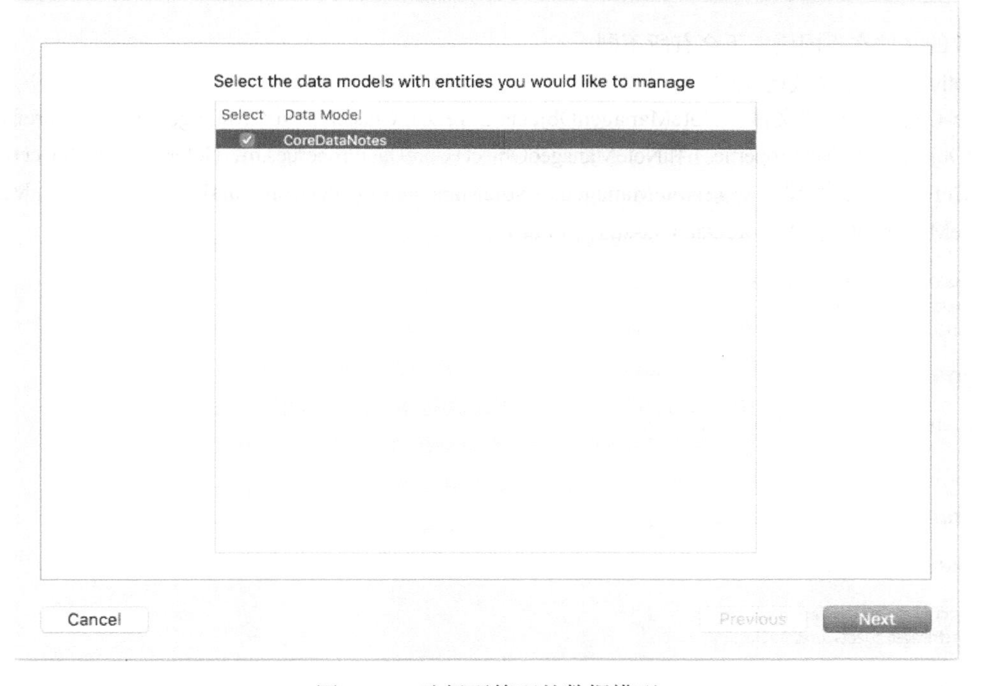

图14-20 选择要管理的数据模型

图14-21　选择要管理的实体

之后生成的文件在不同语言下会有所不同。

1. Objective-C版本生成的文件

Objective-C版本生成的文件：NoteManagedObject+CoreDataClass.h、NoteManagedObject+CoreDataClass.m、NoteManagedObject+CoreDataProperties.h和NoteManagedObject+CoreDataProperties.m，NoteManagedObject+CoreDataClass中声明了被Core Data栈管理的对象NoteManaged。NoteManagedObject+CoreDataProperties是NoteManagedObject的类别。NoteManagedObject+CoreDataClass的代码如下：

```
//NoteManagedObject+CoreDataClass.h文件
#import <Foundation/Foundation.h>
#import <CoreData/CoreData.h>

NS_ASSUME_NONNULL_BEGIN                         ①

@interface NoteManagedObject : NSManagedObject  ②

@end

NS_ASSUME_NONNULL_END                           ③

#import "NoteManagedObject+CoreDataProperties.h"

//NoteManagedObject+CoreDataClass.m文件
#import "NoteManagedObject+CoreDataClass.h"

@implementation NoteManagedObject

@end
```

第①行和第③行是两个宏，它们之间的成员变量、参数、属性和返回值等类型都被标志为nonnull。第②行声明继承NSManagedObject类，这使得NoteManagedObject成为被Core Data栈管理的对象。

类别NoteManagedObject+CoreDataProperties的代码如下：

```
//NoteManagedObject+CoreDataProperties.h文件

#import "NoteManagedObject+CoreDataClass.h"


NS_ASSUME_NONNULL_BEGIN

@interface NoteManagedObject (CoreDataProperties)

+ (NSFetchRequest<NoteManagedObject *> *)fetchRequest;

@property (nullable, nonatomic, copy) NSString *content;
@property (nullable, nonatomic, copy) NSDate *date;

@end

NS_ASSUME_NONNULL_END


//NoteManagedObject+CoreDataProperties.m文件

#import "NoteManagedObject+CoreDataProperties.h"

@implementation NoteManagedObject (CoreDataProperties)

+ (NSFetchRequest<NoteManagedObject *> *)fetchRequest {
    return [[NSFetchRequest alloc] initWithEntityName:@"Note"];
}

@dynamic content;                                    ①
@dynamic date;                                       ②

@end
```

类别NoteManagedObject+CoreDataProperties中主要声明了一些属性，我们不需要在该分类中修改代码。第①行和第②行都使用@dynamic关键字限定属性，这是因为在Objective-C分类中声明的属性是动态的，也说明访问属性存取方法（getter和setter）会在运行时由Core Data动态创建。

2. Swift版本生成的文件

Swift版本生成的文件：NoteManagedObject+CoreDataClass和NoteManagedObject+CoreDataProperties。NoteManagedObject+CoreDataClass中定义NoteManagedObject类，它是被Core Data栈管理的对象，NoteManagedObject+CoreDataProperties是NoteManagedObject的扩展。NoteManagedObject的代码如下：

```
//NoteManagedObject+CoreDataClass.swift
import Foundation
import CoreData

@objc(NoteManagedObject)                        ①
class NoteManagedObject: NSManagedObject {

//添加自己的代码

}
```

上述代码中，第①行的@objc(NoteManagedObject)属性是为了防止出现如下的类加载错误：

```
Unable to load class named 'NoteManagedObject' for entity 'Note'. Class not found, using default NSManagedObject instead.
```

扩展NoteManagedObject+CoreDataProperties的代码如下：

```
import Foundation
import CoreData

extension NoteManagedObject {
```

```
    @nonobjc public class func fetchRequest() -> NSFetchRequest<NoteManagedObject> {
        return NSFetchRequest<NoteManagedObject>(entityName: "Note");
    }

    @NSManaged public var date: NSDate?                    ①
    @NSManaged public var content: String?                 ②

}
```

上述代码中，第①行和第②行都使用@NSManaged关键字限定属性，这说明该属性是被Core Data管理的。

注意

默认情况下，模型文件生成的代码是Swift语言，如果想生成Objective-C语言，可以选中CoreDataNotes.xcdatamodeld文件，打开右边的文件检查器，如图14-22所示，在Code Generation→Language中选择所需要的语言。

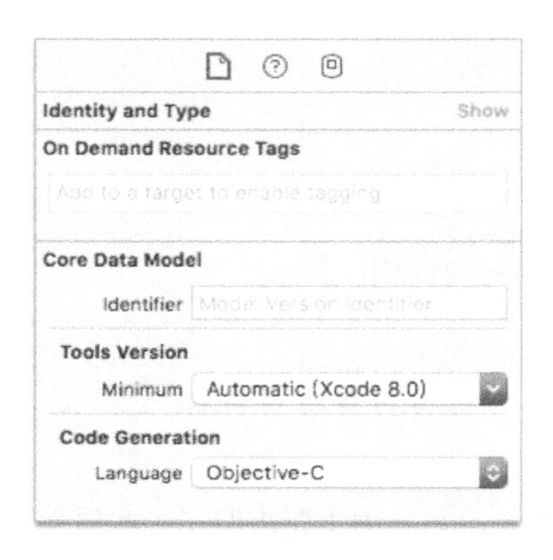

图14-22　选择生成语言

14.6.2　Core Data 栈 DAO

Core Data栈需要在数据持久逻辑组件中创建，我们可以考虑设计一个能够创建Core Data栈的DAO父类——CoreDataDAO。CoreDataDAO类的代码如下：

```
//CoreDataDAO.swift
import Foundation
import CoreData

open class CoreDataDAO: NSObject {

    //返回持久化存储容器
    lazy var persistentContainer: NSPersistentContainer = {
        let container = NSPersistentContainer(name: "CoreDataNotes")
        container.loadPersistentStores(completionHandler:
        ➥{ (storeDescription, error) in
            if let error = error as NSError? {
                print("持久化存储容器错误：", error.localizedDescription)
            }
        })
        return container
    }()

    ///MARK: -- 保存数据
    //保存数据
    func saveContext () {
        let context = persistentContainer.viewContext
```

```
//CoreDataDAO.h
#import <Foundation/Foundation.h>
#import <CoreData/CoreData.h>

@interface CoreDataDAO : NSObject

//返回持久化存储容器
@property (readonly, strong) NSPersistentContainer *persistentContainer;

//保存数据
- (void)saveContext;

@end

//CoreDataDAO.m
#import "CoreDataDAO.h"

@implementation CoreDataDAO

@synthesize persistentContainer = _persistentContainer;

#pragma mark -- Core Data堆栈
```

```
        if context.hasChanges {
            do {
                try context.save()
            } catch {
                let nserror = error as NSError
                print("数据保存错误: ", nserror.localizedDescription)

            }
        }
    }
}
```

```
//返回持久化存储容器
- (NSPersistentContainer *)persistentContainer {
    @synchronized (self) {
        if (_persistentContainer == nil) {
            _persistentContainer = [[NSPersistentContainer alloc]
            ➥initWithName:@"CoreDataNotes"];
            [_persistentContainer loadPersistentStoresWithCompletionHandler:
            ➥^(NSPersistentStoreDescription *storeDescription, NSError *error) {
                if (error != nil) {
                    NSLog(@"持久化存储容器错误: %@",
                    ➥error.localizedDescription);
                    abort();
                }
            }];
        }
    }
    return _persistentContainer;
}


#pragma mark -- 保存数据
//保存数据
- (void)saveContext {
    NSManagedObjectContext *context = self.persistentContainer.viewContext;
    NSError *error = nil;
    if ([context hasChanges] && ![context save:&error]) {
        NSLog(@"数据保存错误: %@", error.localizedDescription);
        abort();
    }
}


@end
```

CoreDataDAO代码类似于14.5.2节通过Xcode模板生成的Core Data栈代码。

然后我们让NoteDAO继承CoreDataDAO，并且增加了NoteManagedObject被管理实体类，这样数据持久层工程中的类如表14-1所示。

表14-1　数据持久层工程中的类

类　　名	说　　明
CoreDataDAO	DAO基类
NoteDAO	NoteDAO类
Note	未被管理的实体类
NoteManagedObject	被管理的实体类

提示　Note和NoteManagedObject看起来有点儿重复，但是它们有不同的角色，这是一个非常重要的问题。由于这里采用了分层设计，NoteManagedObject对象必须被严格限定在持久层中使用，不能出现在表示层和业务逻辑层，而Note对象则可以出现在表示层、业务逻辑层和持久层中。因此，在持久层中将Core Data栈查询出的NoteManagedObject数据转换为Note对象，返回给表示层和业务逻辑层。这个工作看起来比较麻烦，但基于Core Data实现的分层架构必须遵守这个规范。

14.6.3　查询数据

下面我们具体看看在分层架构下如何采用Core Data技术查询和修改（insert、update和delete）数据。先来看看查询，它分为无条件查询和有条件查询。

1. 无条件查询

在NoteDAO中查询所有数据的方法如下：

```
public func findAll() -> NSMutableArray {

    let context = persistentContainer.viewContext

    let entity = NSEntityDescription.entity(forEntityName: "Note", in:
    ➥context)                                                              ①

    let fetchRequest = NSFetchRequest<NSFetchRequestResult>()              ②
    fetchRequest.entity = entity                                           ③

    let sortDescriptor = NSSortDescriptor(key:"date", ascending:true)      ④
    let sortDescriptors = [sortDescriptor]
    fetchRequest.sortDescriptors = sortDescriptors                         ⑤
    let resListData = NSMutableArray()

    do {
        let listData = try context.fetch(fetchRequest)                     ⑥

        if listData.count > 0 {

            for item in listData {                                         ⑦
                let mo = item as! NoteManagedObject
                let note = Note(date: mo.date!, content: mo.content!)      ⑧
                resListData.add(note)                                      ⑨
            }
        }
        return resListData
    } catch {
        NSLog("查询数据失败")
        return resListData
    }
}
```

```
- (NSMutableArray *)findAll {

    NSManagedObjectContext *context = self.persistentContainer.viewContext;

    NSEntityDescription *entity = [NSEntityDescription
    ➥entityForName:@"Note" inManagedObjectContext:context];               ①

    NSFetchRequest *fetchRequest = [[NSFetchRequest alloc] init];          ②
    fetchRequest.entity = entity;                                          ③

    NSSortDescriptor *sortDescriptor = [[NSSortDescriptor alloc]
    ➥initWithKey:@"date" ascending: TRUE];                                ④
    NSArray *sortDescriptors = @[sortDescriptor];
    fetchRequest.sortDescriptors = sortDescriptors;                        ⑤

    NSError *error = nil;
    NSArray *listData = [context executeFetchRequest:fetchRequest
    ➥error:&error];                                                       ⑥

    if (error != nil) {
        return nil;
    }

    NSMutableArray *resListData = [[NSMutableArray alloc] init];
    for (NoteManagedObject *mo in listData) {                              ⑦
        Note *note = [[Note alloc] initWithDate:mo.date
        ➥content:mo.content];                                             ⑧
        [resListData addObject:note];                                      ⑨
    }

    return resListData;
}
```

第①行中的NSEntityDescription是实体关联的描述类，通过指定实体的名字获得NSEntityDescription实例对象，实体的名字是在数据模型文件中定义的。第②行中的NSFetchRequest是数据提取请求类，用于查询。第③行把实体描述设定到请求对象中。第④行中的NSSortDescriptor是排序描述类，它可以指定排序字段以及排序方式。第⑤行把排序描述对象的数组赋值给请求对象中。

第⑥行根据前面设置的请求对象执行查询，返回数组集合。数组集合中放置的是被管理的NoteManagedObject实体对象，我们需要把它们转换到Note实体对象中，并放置在数组集合里，然后返回给业务逻辑层。第⑦行~第⑨行实现了这个转换。

2. 有条件查询

在NoteDAO中按照主键查询数据的代码如下：

```
public func findById(_ model: Note) -> Note? {

    let context = persistentContainer.viewContext

    let entity = NSEntityDescription.entity(forEntityName: "Note", in:
    ➥context)

    let fetchRequest = NSFetchRequest<NSFetchRequestResult>()
    fetchRequest.entity = entity
    fetchRequest.predicate = NSPredicate(format: "date = %@", model.
    ➥date)                                                                ①

    do {
        let listData = try context.fetch(fetchRequest)
```

```
- (Note *)findById:(Note *)model {

    NSManagedObjectContext *context = self.persistentContainer.viewContext;

    NSEntityDescription *entity = [NSEntityDescription entityForName:@"Note"
    ➥inManagedObjectContext:context];

    NSFetchRequest *fetchRequest = [[NSFetchRequest alloc] init];
    fetchRequest.entity = entity;
    fetchRequest.predicate = [NSPredicate
    ➥predicateWithFormat: @"date = % @",model.date];                      ①

    NSError *error = nil;
    NSArray *listData = [context executeFetchRequest:fetchRequest
    ➥error:&error];
```

```
        if listData.count > 0 {
            let mo = listData[0] as! NoteManagedObject
            let note = Note(date: mo.date!, content: mo.content!)
            return note
        }
    } catch {
        NSLog("查询数据失败")
    }
    return nil
}
```

```
    if (error == nil && [listData count] > 0) {
        NoteManagedObject *mo = [listData lastObject];
        Note *note = [[Note alloc] initWithDate:mo.date content:mo.content];
        return note;
    }
    return nil;
}
```

与无条件查询不同的是增加了第①行，NSPredicate用来定义一个逻辑查询条件的谓词对象，可以过滤集合对象。我们在6.4节中介绍过它的用法，上面定义的查询条件是查询日期字段。

14.6.4 修改数据

这里的修改数据也是指insert、update和delete操作。在NoteDAO中插入数据的方法如下：

```
//插入数据的方法
public func create(_ model: Note) -> Int {

    let context = persistentContainer.viewContext

    let note = NSEntityDescription.insertNewObject(forEntityName: "Note",
    ➥into:context) as! NoteManagedObject                                    ①

    note.date = model.date                                                   ②
    note.content = model.content                                             ③

    //保存数据
    self.saveContext()                                                       ④

    return 0
}
```

```
//插入数据的方法
- (int)create:(Note *)model {

    NSManagedObjectContext *context = self.persistentContainer.viewContext;

    NoteManagedObject *note = [NSEntityDescription
    ➥insertNewObjectForEntityForName:@"Note"
    ➥inManagedObjectContext:context];                                       ①
    note.date = model.date;                                                  ②
    note.content = model.content;                                            ③

    //保存数据
    [self saveContext];                                                      ④

    return 0;
}
```

第①行用于创建一个被管理的Note实体对象，其中insertNewObjectForEntityForName方法的返回类型是NSManagedObject。本例事实上是NoteManagedObject，因为NoteManagedObject是NSManagedObject的子类。

第②行用于设定date属性的值，第③行用于设定content属性的值。NSManagedObject对象属性的赋值也可以使用NSManagedObject的setValue: forKey:方法实现，替代代码如下：

```
note.setValue(model.date, forKey: "date")       //等价于note.date = model.date

note.setValue(model.content, forKey: "content")  //等价于note.content =
                                                 //model.content
```

```
[note setValue: model.content forKey:@"content"];  //等价于note.date =
                                                   //model.date

[note setValue: model.date forKey:@"date"];        //等价于note.content =
                                                   //model.content
```

第④行中的saveContext方法用来保存数据变化，并将其同步到持久化数据文件中。

在NoteDAO中删除数据的方法如下：

```
//删除数据的方法
public func remove(_ model: Note) -> Int {

    let context = persistentContainer.viewContext

    let entity = NSEntityDescription.entity(forEntityName: "Note", in:
    ➥context)

    let fetchRequest = NSFetchRequest<NSFetchRequestResult>()
    fetchRequest.entity = entity
    fetchRequest.predicate = NSPredicate(format: "date = %@", model.date)
```

```
//删除数据的方法
- (int)remove:(Note *)model {

    NSManagedObjectContext *context = self.persistentContainer.viewContext;

    NSEntityDescription *entityDescription = [NSEntityDescription
    ➥entityForName:@"Note" inManagedObjectContext:context];

    NSFetchRequest *request = [[NSFetchRequest alloc] init];
    [request setEntity:entityDescription];
```

```
    do {
        let listData = try context.fetch(fetchRequest)
        if listData.count > 0 {
            let note = listData[0] as! NSManagedObject
            context.delete(note)                                        ①
            //保存数据
            self.saveContext()
        }
    } catch {
        NSLog("删除数据失败")
    }
    return 0
}
```

```
    NSPredicate *predicate = [NSPredicate predicateWithFormat: @"date = %@",
    ➥model.date];
    [request setPredicate:predicate];

    NSError *error = nil;
    NSArray *listData = [context executeFetchRequest:request error:&error];

    if (error == nil && [listData count] > 0) {
        NoteManagedObject *note = [listData lastObject];
        [context deleteObject:note];                                    ①
        //保存数据
        [self saveContext];
    }

    return 0;
}
```

进行删除操作时，首先要查询出要删除的实体，然后使用第①行删除实体。

在NoteDAO中修改数据的方法如下：

```
//修改数据的方法
public func modify(_ model: Note) -> Int {

    let context = persistentContainer.viewContext

    let entity = NSEntityDescription.entity(forEntityName: "Note", in:
    ➥context)

    let fetchRequest = NSFetchRequest<NSFetchRequestResult>()
    fetchRequest.entity = entity
    fetchRequest.predicate = NSPredicate(format: "date = %@", model.date)

    do {
        let listData = try context.fetch(fetchRequest)
        if listData.count > 0 {
            let note = listData[0] as! NoteManagedObject
            //note.setValue(model.content, forKey: "content")
            note.content = model.content

            //保存数据
            self.saveContext()
        }
    } catch {
        NSLog("修改数据失败")
    }
    return 0
}
```

```
//修改数据的方法
- (int)modify:(Note *)model {

    NSManagedObjectContext *context = self.persistentContainer.viewContext;

    NSEntityDescription *entityDescription = [NSEntityDescription
    ➥entityForName:@"Note" inManagedObjectContext:context];

    NSFetchRequest *request = [[NSFetchRequest alloc] init];
    [request setEntity:entityDescription];

    NSPredicate *predicate = [NSPredicate predicateWithFormat: @"date = %@",
    ➥model.date];
    [request setPredicate:predicate];

    NSError *error = nil;
    NSArray *listData = [context executeFetchRequest:request error:&error];

    if (error == nil && [listData count] > 0) {
        NoteManagedObject *note = [listData lastObject];
        note.content = model.content;
        //保存数据
        [self saveContext];
    }
    return 0;
}
```

要进行修改操作，首先需要查询出要修改的实体，然后修改实体中的属性，最后调用saveContext方法保存修改，并将其同步到持久化数据文件中。

14.7　小结

根据数据的规模和使用特点，我们可以选择将其放在本地或者云服务器中，而本章主要讨论了本地数据持久化。我们分析了数据存取的几种方式，以及每种数据存取方式适合什么样的场景，并分别举例介绍了每种存取方式的实现。

第 15 章

数据交换格式

数据交换格式就像两个人在聊天一样，采用彼此都能“听”得懂的语言，你来我往，其中的语言就相当于通信中的数据交换格式。有时候，为了防止聊天被人偷听，可以采用暗语。同理，计算机程序之间也可以通过数据加密技术防止“偷听”。

数据交换格式主要分为纯文本格式、XML格式和JSON格式，其中纯文本格式是一种简单的、无格式的数据交换方式。

例如，为了告诉别人一些事情，我会写下如图15-1所示的留言条。留言条有一定的格式，共有4部分——称谓、内容、落款和时间。

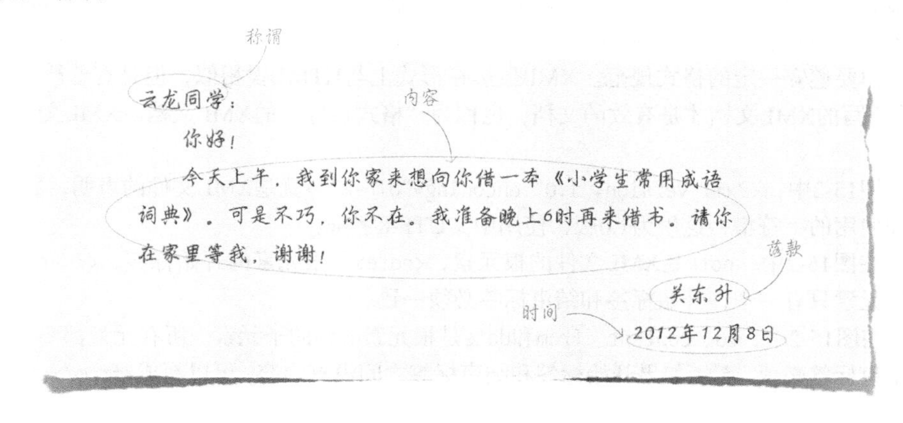

图15-1　留言条格式

如果我们用纯文本格式描述留言条，可以采用如下形式：

```
"云龙同学","你好！\n今天上午，我到你家来想向你借一本《小学生常用成语词典》。可是不巧，你不在。我准备晚上6时再来借书。请你在家里等我，谢谢！","关东升","2012年12月08日"
```

留言条中的4部分数据按照顺序存放，各部分之间用逗号分割。数据量小的时候可以采用这种格式，但是随着数据量增加，问题也会暴露出来，我们可能搞乱它们的顺序，如果各个数据部分能有描述信息就好了。而XML格式和JSON格式可以带有描述信息，它们叫作“自描述的”结构化文档。

将上面的留言条写成XML格式，具体如下：

```
<?xml version="1.0" encoding="UTF-8"?>
<note>
    <to>云龙同学</to>
    <conent>你好！\n今天上午，我到你家来想向你借一本《小学生常用成语词典》。
        可是不巧，你不在。我准备晚上6时再来借书。请你在家里等我，谢谢！</conent>
    <from>关东升</from>
    <date>2012年12月08日</date>
</note>
```

我们看到位于尖括号中的内容（<to>...</to>等）就是描述数据的标识，在XML中称为“标签”。

将上面的留言条写成JSON格式，具体如下：

```
{to:"云龙同学",conent:"你好！\n今天上午，我到你家来想向你借一本《小学生常用成语词典》。可是不巧，你不在。我准备晚上6时再来借书。请你在家里等我，谢谢！",from:"关东升",date:"2012年12月08日"}
```

数据放置在大括号（{}）之中，每个数据项目之前都有一个描述名（如to等），描述名和数据项目之间用冒号（:）分开。

可以发现，JSON所用的字节数一般要比XML少，这也是很多人喜欢采用JSON格式的主要原因，因此JSON也称为“轻量级”的数据交换格式。接下来，我们将重点介绍XML和JSON数据交换格式。

15.1 XML 数据交换格式

XML是一种自描述的数据交换格式。虽然XML数据交换格式不如JSON“轻便”，但也非常重要，多年来一直被用于各种计算机语言中，是老牌的、经典的、灵活的数据交换方式。

15.1.1 XML 文档结构

在读写XML文档之前，我们需要了解XML文档结构。前面提到的留言条XML文档由开始标签<note>和结束标签</note>组成，它们就像括号一样，把数据项括起来。这样不难看出，标签<to></to>之间是“称谓”，标签<content></content>之间是“内容”，标签<from></from>之间是“落款”，标签<date></date>之间是“日期”。

XML文档结构要遵守一定的格式规范。XML虽然在形式上与HTML很相似，但是有着严格的语法规则。只有严格按照规范编写的XML文档才是有效的文档，也称为“格式良好”的XML文档。XML文档的基本架构可以分为下面几部分。

- **声明**。在图15-2中，<?xml version="1.0" encoding="UTF-8"?>就是XML文件的声明，它定义了XML文件的版本和使用的字符集，这里为1.0版，使用中文UTF-8字符。
- **根元素**。在图15-2中，note是XML文件的根元素，<note>是根元素的开始标签，</note>是根元素的结束标签。根元素只有一个，开始标签和结束标签必须一致。
- **子元素**。在图15-2中，to、content、from和date是根元素note的子元素。所有元素都要有结束标签，开始标签和结束标签必须一致。如果开始标签和结束标签之间没有内容，可以写成<from/>，这称为“空标签”。

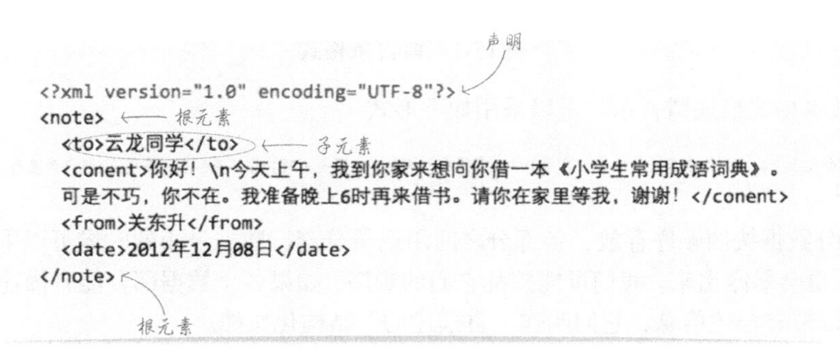


```
<?xml version="1.0" encoding="UTF-8"?>
<note>
  <to>云龙同学</to>
  <conent>你好！\n今天上午，我到你家来想向你借一本《小学生常用成语词典》。
  可是不巧，你不在。我准备晚上6时再来借书。请你在家里等我，谢谢！</conent>
  <from>关东升</from>
  <date>2012年12月08日</date>
</note>
```

图15-2　XML文档结构

- **属性**。图15-3所示是具有属性的XML文档，而留言条的XML文档中没有属性。属性定义在开始标签中。在开始标签<Note id="1">中，id="1"是Note元素的一个属性，id是属性名，1是属性值，其中属性值必须放置在双引号或单引号之间。一个元素不能有多个相同名字的属性。
- **命名空间**。用于为XML文档提供名字唯一的元素和属性。例如，在一个学籍信息的XML文档中需要引用到教师和学生，它们都有一个子元素id，这时直接引用id元素会造成名称冲突，但是将两个id元素放到不同的命名空间中就会解决这个问题。图15-4中以xmlns:开头的内容都属于命名空间。

❑ **限定名**。它是由命名空间引出的概念，定义了元素和属性的合法标识符。限定名通常在XML文档中用作特定元素或属性引用。图15-4中的标签<soap:Body>就是合法的限定名，前缀soap是由命名空间定义的。

```
<?xml version="1.0" encoding="UTF-8"?>
<Notes>
  <Note id="1">  ← 属性
    <CDate>2012-12-21</CDate>
    <Content>早上8点钟到公司</Content>
    <UserID>tony</UserID>
  </Note>
  <Note id="2">
    <CDate>2012-12-22</CDate>
    <Content>发布iOSBook1</Content>
    <UserID>tony</UserID>
  </Note>
</Notes>
```

图15-3 有属性的XML文档

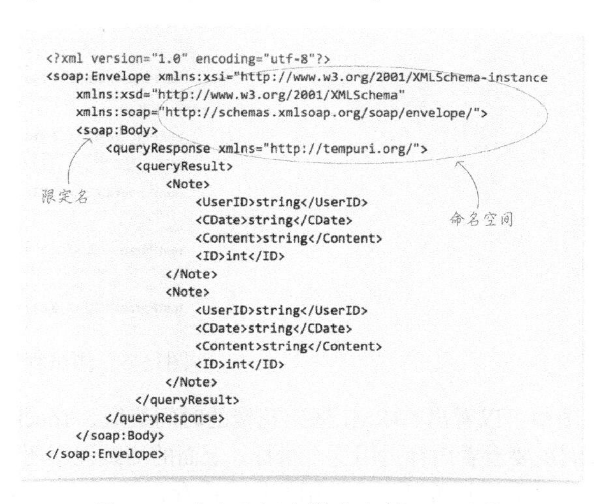


```
<?xml version="1.0" encoding="utf-8"?>
<soap:Envelope xmlns:xsi="http://www.w3.org/2001/XMLSchema-instance
    xmlns:xsd="http://www.w3.org/2001/XMLSchema"
    xmlns:soap="http://schemas.xmlsoap.org/soap/envelope/">
  <soap:Body>
    <queryResponse xmlns="http://tempuri.org/">
      <queryResult>
        <Note>
          <UserID>string</UserID>
          <CDate>string</CDate>
          <Content>string</Content>
          <ID>int</ID>
        </Note>
        <Note>
          <UserID>string</UserID>
          <CDate>string</CDate>
          <Content>string</Content>
          <ID>int</ID>
        </Note>
      </queryResult>
    </queryResponse>
  </soap:Body>
</soap:Envelope>
```

图15-4 命名空间和限定名的XML文档

15.1.2 解析 XML 文档

XML文档操作有“读”与“写”，读入XML文档并分析的过程称为“解析”。事实上，在使用XML进行开发的过程中，“解析”XML文档占很大的比重。

解析XML文档在目前有两种流行的模式：SAX和DOM。SAX是一种基于事件驱动的解析模式。解析XML文档时，程序从上到下读取XML文档，如果遇到开始标签、结束标签和属性等，就会触发相应的事件。但是这种解析XML文件的方式有一个弊端，那就是只能读取XML文档，不能写入XML文档，它的优点是解析速度快。iOS重点推荐使用SAX模式解析。

DOM模式将XML文档作为一棵树状结构进行分析，获取节点的内容以及相关属性，或是新增、删除和修改节点的内容。XML解析器在加载XML文件以后，DOM模式将XML文件的元素视为一个树状结构的节点，一次性读入内存。如果文档比较大，解析速度就会变慢。但是DOM模式有一点是SAX无法取代的，那就是DOM能够修改XML文档。

iOS SDK提供了两个XML框架，具体如下所示。

❑ **NSXML**。它是基于Objective-C语言的SAX解析框架，是iOS SDK默认的XML解析框架，不支持DOM模式。

❑ **libxml2**。它（http://xmlsoft.org/）是基于C语言的XML解析器，被苹果整合在iOS SDK中，支持SAX和DOM模式。

此外，在iOS中解析XML时，还有很多第三方框架可以采用，具体如下所示。

❑ **TBXML**。它是轻量级的DOM模式解析库，不支持XML文档验证和XPath，只能读取XML文档，不能写XML文档，但是解析XML是最快的。

❑ **TouchXML**。它是基于DOM模式的解析库。与TBXML类似，只能读取XML文档，不能写XML文档。

❑ **KissXML**。它是基于DOM模式的解析库，基于TouchXML，主要的不同是可以写入XML文档。

❑ **TinyXML**。它是基于C++语言的DOM模式解析库，可以读写XML文档，不支持XPath。

❑ **GDataXML**。它是基于DOM模式的解析库，由谷歌开发，可以读写XML文档，支持XPath查询。

这么多框架我们选择哪一个呢？解析性能是我们进行选择的主要指标，我编写了一个测试程序来测试这些框架。测试文件是MyNotes应用中的XML文档，我准备了10 000条Note数据的XML文档，保存后文件大小达到了1.2 MB。测试程序的运行结果如图15-5所示。

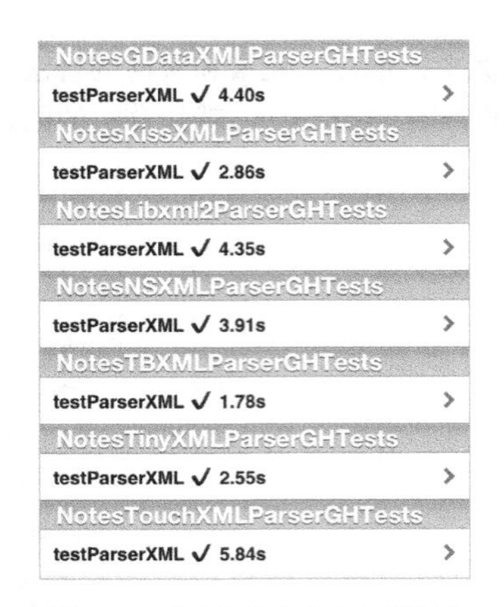

图15-5　测试程序的运行结果

从图中可以看出TBXML框架花费的时间最短，TouchXML框架花费的时间最长。当然，速度并不能说明一切，我们还要看看内存占用这个指标。上面的测试程序还可以进行内存占用峰值和执行后驻留内存的比较，内存占用峰值是衡量解析过程中占用的最大内存，它会影响应用程序的当前运行状况，影响是暂时的。而执行后驻留内存是衡量解析完成之后内存的驻留情况，它会影响应用程序运行后的状况，影响是长期的。

图15-6所示的是使用Xcode的Instruments工具进行检测的结果。有关Instruments工具的具体用法，我们将在第25章中介绍。

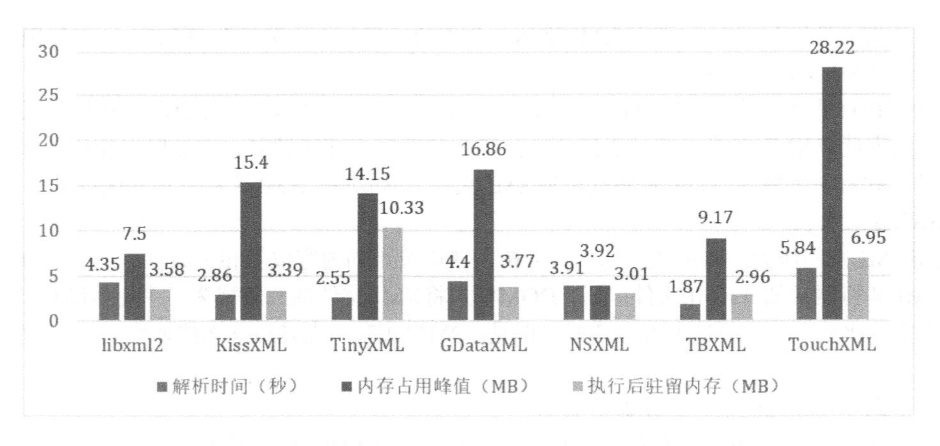

图15-6　XML解析框架性能图表

图15-6展示了内存占用峰值、驻留内存和解析速度这3个指标的比较。TouchXML应该是最差的了，TBXML虽然是DOM解析模式，但解析速度最快，可是内存占用峰值比较高，而驻留内存较低。而KissXML和TinyXML也是一个不错的选择，还有iOS SDK中的NSXML在速度和内存占用上都比较优秀，如果这几个指标都想兼顾，NSXML是不错的选择。

15.2　案例：MyNotes 应用读取 XML 数据

下面我们通过一个案例介绍使用NSXML和TBXML框架解析XML的过程。现在有一个记录MyNotes（备忘录）信息的Notes.xml文件，它的内容如下：

```
<?xml version="1.0" encoding="UTF-8"?>
<Notes>
    <Note id="1">
```

```
        <CDate>2012-12-21</CDate>
        <Content>早上8点钟到公司</Content>
        <UserID>tony</UserID>
    </Note>
    <Note id="2">
        <CDate>2012-12-22</CDate>
        <Content>发布iOSbook1</Content>
        <UserID>tony</UserID>
    </Note>
    <Note id="3">
        <CDate>2012-12-23</CDate>
        <Content>发布iOSbook2</Content>
        <UserID>tony</UserID>
    </Note>
    <Note id="4">
        <CDate>2012-12-24</CDate>
        <Content>发布iOSbook3</Content>
        <UserID>tony</UserID>
    </Note>
    <Note id="5">
        <CDate>2012-12-25</CDate>
        <Content>发布2016奥运会应用</Content>
        <UserID>tony</UserID>
    </Note>
</Notes>
```

文档中的根元素是Notes，其中有很多子元素Note，而每个Note元素都有一个id属性（表示“备忘录”的序号），以及CDate（表示“备忘录”的日期）、Content（表示“备忘录”的内容）和UserID（表示“备忘录”的创建人ID）这3个子元素。

下面我们以MyNotes（备忘录）应用作为案例，案例运行界面如图15-7所示，其中数据来源于本地资源文件中的Notes.xml文件。我们需要使用NSXMLParser框架解析XML文档，并将数据放置于界面表视图中。

图15-7 MyNotes应用

15.2.1　使用 NSXML 解析

NSXML是iOS SDK自带的，也是苹果默认的解析框架，它采用SAX模式解析，是SAX解析模式的代表。NSXML框架的核心是NSXMLParser及其委托协议NSXMLParserDelegate，其中主要的解析工作是在NSXMLParserDelegate实现类中完成的。委托中定义了很多回调方法，在SAX解析器从上到下遍历XML文档的过程中，遇到开始标签、结束标签、文档开始、文档结束和字符串时，就会触发这些方法。这些方法有很多，下面我们列出5个常用的。

- **parserDidStartDocument:**。在文档开始的时候触发。
- **parser:didStartElement:namespaceURI:qualifiedName:attributes:**。遇到一个开始标签时触发，其中namespaceURI部分是命名空间，qualifiedName是限定名，attributes是字典类型的属性集合。
- **parser:foundCharacters:**。遇到字符串时触发。
- **parser:didEndElement:namespaceURI:qualifiedName:**。遇到结束标签时触发。
- **parserDidEndDocument:**。在文档结束时触发。

上面这5个方法都是按照解析文档的顺序触发的，理解它们的先后顺序很重要。下面我们再通过图15-8所示的UML时序图来了解它们的触发顺序。

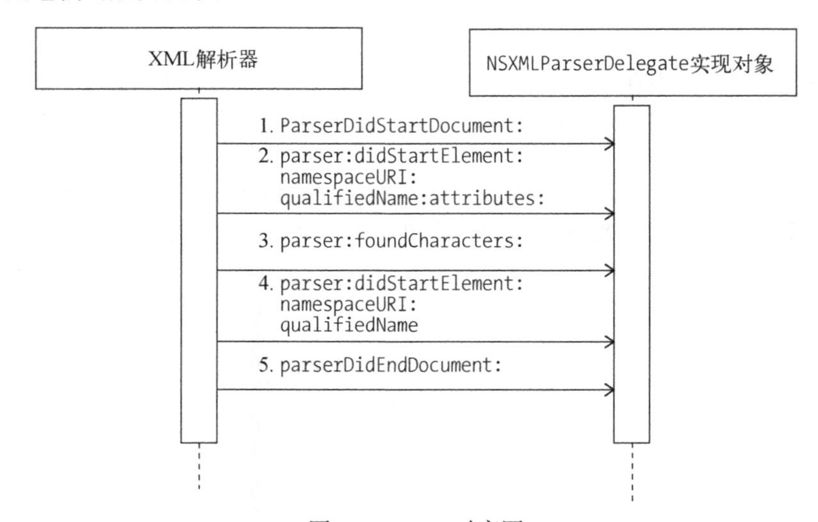

图15-8　UML时序图

就同一个元素而言，触发顺序是按照图15-8所示进行的。在整个解析过程中，它们的触发次数是：方法1和方法5为一对，都只触发一次；方法2和方法4为一对，触发多次；方法3在方法2和方法4之间触发，也会多次触发。触发的字符包括换行符和回车符等特殊字符，编程时需要注意。

表示逻辑组件的代码我不想过多修改，所以编写了一个专门的解析类NotesXMLParser。NotesXMLParser文件中类定义和属性相关的代码如下：

```
import Foundation

class NotesXMLParser: NSObject, NSXMLParserDelegate {

    //解析出的数据内部是字典类型
    private var listData: NSMutableArray!

    //当前标签的名字
    private var currentTagName: String!
    ……
}
```

```
//NotesXMLParser.h文件的代码
@interface NotesXMLParser : NSObject <NSXMLParserDelegate>

//解析出的数据内部是字典类型
@property (strong,nonatomic) NSMutableArray * listData;
//当前标签的名字
@property (strong,nonatomic) NSString *currentTagName;

//开始解析
-(void)start;

@end
```

NotesXMLParser类实现了NSXMLParserDelegate协议，还定义了currentTagName属性，其目的是在图15-8所示的方法2到方法4执行期间临时存储正在解析的元素名。在方法3（parser:foundCharacters:）触发时，我们能够知道目前解析器处于哪个元素中。

在NotesXMLParser中，start方法的代码如下：

```swift
//开始解析
func start() {

    let path = Bundle.main.path(forResource: "Notes", ofType: "xml")!
    let url = URL(fileURLWithPath: path)

    //开始解析XML
    let parser = XMLParser(contentsOf: url)!
    parser.delegate = self
    parser.parse()

}
```

```objc
- (void)start {
    NSString *path = [[NSBundle mainBundle] pathForResource:@"Notes"
    ➥ofType:@"xml"];

    NSURL *url = [NSURL fileURLWithPath:path];
    //开始解析XML
    NSXMLParser *parser = [[NSXMLParser alloc] initWithContentsOfURL:url];
    parser.delegate = self;
    [parser parse];
}
```

NSXMLParser是解析类，有3个构造函数，具体如下所示。

- **initWithContentsOfURL:**。可以使用URL对象创建解析对象。本例中采用该方法先从资源文件中加载对象，获得URL对象，再使用URL对象构建解析对象。Swift语言中表示为init?(contentsOf: URL)。
- **initWithData:**。可以使用NSData（Swift版是Data）创建解析对象。Swift语言中表示为init(data: Data)。
- **initWithStream:**。可以使用IO流对象创建解析对象。Swift语言中表示为init(stream:)。

解析对象创建好后，我们需要指定委托属性delegate为self，然后发送parse消息，开始解析文档。在NotesXMLParser中，parserDidStartDocument:方法的代码如下：

```swift
func parserDidStartDocument(parser: NSXMLParser!) {
    self.istData = NSMutableArray()
}
```

```objc
- (void)parserDidStartDocument:(NSXMLParser *)parser {
    self.listData = [[NSMutableArray alloc] init];
}
```

由于parserDidStartDocument:方法只在解析开始时触发一次，因此可以在这个方法中初始化解析过程中用到的一些成员变量。

在NotesXMLParser中，parser:parseErrorOccurred:方法的代码如下：

```swift
func parser(parser: NSXMLParser, parseErrorOccurred parseError: NSError) {
    print(parseError)
}
```

```objc
- (void)parser:(NSXMLParser *)parser parseErrorOccurred:(NSError *)parseError
{
    NSLog(@"%@",parseError);
}
```

出错方法前面没有介绍，这主要是因为该方法一般在调试阶段使用，实际发布时意义不大。更不要对用户使用UIAlertView提示，用户会被这些专业的错误信息吓坏的。

在NotesXMLParser中，parser:didStartElement:namespaceURI:qualifiedName:attributes:方法的代码如下：

```swift
//遇到一个开始标签时触发
func parser(_ parser: XMLParser, didStartElement elementName: String,
➥namespaceURI: String?, qualifiedName qName: String?,
➥attributes attributeDict: [String:String]) {

    self.currentTagName = elementName
    if self.currentTagName == "Note" {
        let id = attributeDict["id"]                 ①
        let dict = NSMutableDictionary()             ②
        dict["id"] = id                              ③
        self.listData.add(dict)                      ④
    }
}
```

```objc
//遇到一个开始标签时触发
- (void)parser:(NSXMLParser *)parser didStartElement:(NSString *)elementName
➥namespaceURI:(NSString *)namespaceURI
➥qualifiedName:(NSString *)qualifiedName
➥attributes:(NSDictionary *)attributeDict {

    self.currentTagName = elementName;
    if ([self.currentTagName isEqualToString:@"Note"]) {
        NSString *identifier = [attributeDict objectForKey:@"id"];          ①
        NSMutableDictionary *dict = [[NSMutableDictionary alloc] init];     ②
        [dict setObject:identifier forKey:@"id"];                           ③
        [self.listData addObject:dict];                                     ④
    }

}
```

在该方法中，我们把elementName参数赋值给成员变量currentTagName，其中elementName参数是正在解析的元素名字。如果元素名字为Note，取出它的属性id。属性是从attributeDict参数中传递过来的，它是一个字典类型，其中键的名字就是属性名字，值是属性的值。第①行代码负责从字典中取出id属性。在第②行代码中，我们实例化一个可变字典对象，用来存放解析出来的Note元素数据。成功解析之后，字典中应该有4对数据，即id、CDate、Content和UserID。第③行代码把id放入可变字典中。第④行代码把可变字典放入可变数组变量notes中。

在NotesXMLParser中，parser:foundCharacters:方法的代码如下：

```
//遇到字符串时触发
func parser(_ parser: XMLParser, foundCharacters string: String) {

    //替换回车符和空格
    let s1 = string.trimmingCharacters(in: CharacterSet.whitespacesAnd
    ➥Newlines)                                                          ①

    if s1 == "" {
        return
    }

    let dict = self.listData.lastObject as! NSMutableDictionary
    if (self.currentTagName == "CDate") {
        dict["CDate"] = string
    }

    if (self.currentTagName == "Content") {
        dict["Content"] = string
    }

    if (self.currentTagName == "UserID") {
        dict["UserID"] = string
    }
}
```

```
//遇到字符串时触发
- (void)parser:(NSXMLParser *)parser foundCharacters:(NSString *)string {

    //替换回车符和空格
    string = [string stringByTrimmingCharactersInSet:[NSCharacterSet
    ➥whitespaceAndNewlineCharacterSet]];                                ①
    if ([string isEqualToString:@""]) {
        return;
    }
    NSMutableDictionary *dict = [self.listData lastObject];

    if ([self.currentTagName isEqualToString:@"CDate"] && dict) {
        [dict setObject:string forKey:@"CDate"];
    }

    if ([self.currentTagName isEqualToString:@"Content"] && dict) {
        [dict setObject:string forKey:@"Content"];
    }

    if ([self.currentTagName isEqualToString:@"UserID"] && dict) {
        [dict setObject:string forKey:@"UserID"];
    }
}
```

该方法主要用于解析元素的文本内容。由于换行符和回车符等特殊字符也会触发该方法，所以第①行用来过滤掉换行符和回车符。其中stringByTrimmingCharactersInSet:方法是剔除这些字符的方法，NSCharacterSet类的静态方法whitespaceAndNewlineCharacterSet指定字符集为换行符和回车符。

在NotesXMLParser中，parser:didEndElement:namespaceURI:qualifiedName:方法的代码如下：

```
//遇到结束标签时触发
func parser(_ parser: XMLParser, didEndElement elementName: String,
➥namespaceURI: String?, qualifiedName qName: String?) {
    self.currentTagName = nil
}
```

```
//遇到结束标签时触发
- (void)parser:(NSXMLParser *)parser didEndElement:(NSString *)elementName
➥namespaceURI:(NSString *)namespaceURI
➥qualifiedName:(NSString *)qName {
    self.currentTagName = nil;
}
```

在NotesXMLParser中，parserDidEndDocument:方法的代码如下：

```
//遇到文档结束时触发
func parserDidEndDocument(_ parser: XMLParser) {
    print("NSXML解析完成...")
    NotificationCenter.default.post(name:
    ➥Notification.Name(rawValue: "reloadViewNotification"), object:
    ➥self.listData)
}
```

```
//遇到文档结束时触发
- (void)parserDidEndDocument:(NSXMLParser *)parser {
    NSLog(@"NSXML解析完成...");
    [[NSNotificationCenter defaultCenter]
    ➥postNotificationName:@"reloadViewNotification" object:self.listData
    ➥userInfo:nil];
}
```

进入该方法就意味着解析完成，需要清理一些成员变量，同时要将数据返回给视图控制器。这里我们使用通知机制将数据通过广播通知投送回视图控制器。

在表示逻辑组件中，需要修改的视图控制器主要是MasterViewController类，其代码如下：

```swift
override func viewDidLoad() {
    ......

    NotificationCenter.default.addObserver(self,
            selector: #selector(reloadView(_:)),
            name: Notification.Name(rawValue:
            ➥"reloadViewNotification"),
            object: nil)                                         ①

    let parser = NotesTBXMLParser() //NotesXMLParser()           ②
    parser.start()                                               ③

}
//MARK: -- 处理通知
func reloadView(_ notification: Notification) {                  ④
    let resList = notification.object as! [[String:String]]
    self.listData = resList
    self.tableView.reloadData()
}
```

```objectivec
- (void)viewDidLoad {
    ......

    [[NSNotificationCenter defaultCenter] addObserver:self
    ➥selector:@selector(reloadView:)
    ➥name:@"reloadViewNotification" object:nil];                ①

    NotesTBXMLParser *parser = [[NotesTBXMLParser alloc] init];   ②
    //开始解析
    [parser start];                                              ③

}

#pragma mark -- 处理通知
-(void)reloadView:(NSNotification*)notification {                ④
    NSMutableArray *resList = [notification object];
    self.listData  = resList;
    [self.tableView reloadData];
}
```

有标号的代码是刚添加的，其中第①行代码用于注册一个通知，这样MasterViewController才能在解析完成后接收到投送回来的通知。一旦投送成功，就会触发第④行的reloadView:方法，在该方法中取出数据并重新加载表视图。有关表视图其他方法的实现，我们就不再介绍了。

15.2.2 使用 TBXML 解析

使用TBXML解析XML文档时，采用的是DOM模式。通过上面的比较可以发现，它是非常好的解析框架。下面我们介绍一下TBXML框架的用法。

提示

TBXML是第三方开源库，使用时需要下载源代码，然后配置环境。这些源代码也会依赖很多其他的第三方库或框架，所以配置环境比较复杂，这是使用第三库令人头疼的地方。在开源社区有人提供了一个管理Xcode项目依赖库的工具——CocoaPods。为了对比使用CocoaPods前后的区别，本章我们还是采用手动配置环境。有关CocoaPods工具的内容，我们将在第27章中介绍。

首先，我们从https://github.com/71squared/TBXML处下载TBXML。另外，TBXML的技术支持网站是http://www.tbxml.co.uk/TBXML/TBXML_Free.html。

下载完成并解压后，请将TBXML-Headers和TBXML-Code文件夹添加到工程中。我们需要在工程中添加TBXML所依赖的Framework和库，具体如下：

- CoreGraphics.framework
- libz.tbd

然后需要做一些配置，这个过程在Objective-C与Swift中是有区别的。

1. Objective-C版本配置

在Objective-C版本中，需要在工程预编译头文件MyNotes-Prefix.pch中添加如下代码：

```objectivec
#import <Foundation/Foundation.h>
#define ARC_ENABLED
```

由于默认情况下TBXML支持MRC[①]内存管理，在TBXML中定义ARC_ENABLED宏可以打开ARC[②]的开关，那么TBXML能够支持ARC工程。

---

① MRC（Manual Reference Counting，手动引用计数），就是由程序员自己负责管理对象生命周期，负责对象的创建和销毁。

② ARC（Automatic Reference Counting）是自动引用计数内存管理。

> **提示**　预编译头文件可以预先编译，其内容可以添加到工程中所有的Objective-C、C和C++代码模块，所以预先编译中所引入的头文件和宏，会作用于工程中所有的Objective-C、C和C++代码模块。

默认情况下，从Xcode 6开始创建的工程并没有预编译头文件，此时可以在Xcode菜单中点击File→New→File，打开创建文件模板界面（如图15-9所示），选择PCH File文件模板，点击Next按钮，进入如图15-10所示的界面。然后在Save As中输入要创建的预编译头文件名MyNotes-Prefix.pch，然后将文件保存在与.xcodeproj工程文件相同的目录下。

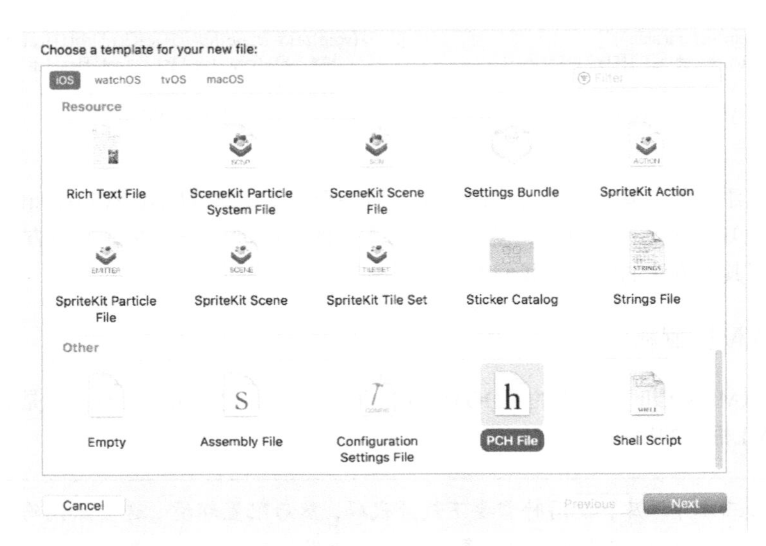

图15-9　创建文件模板界面

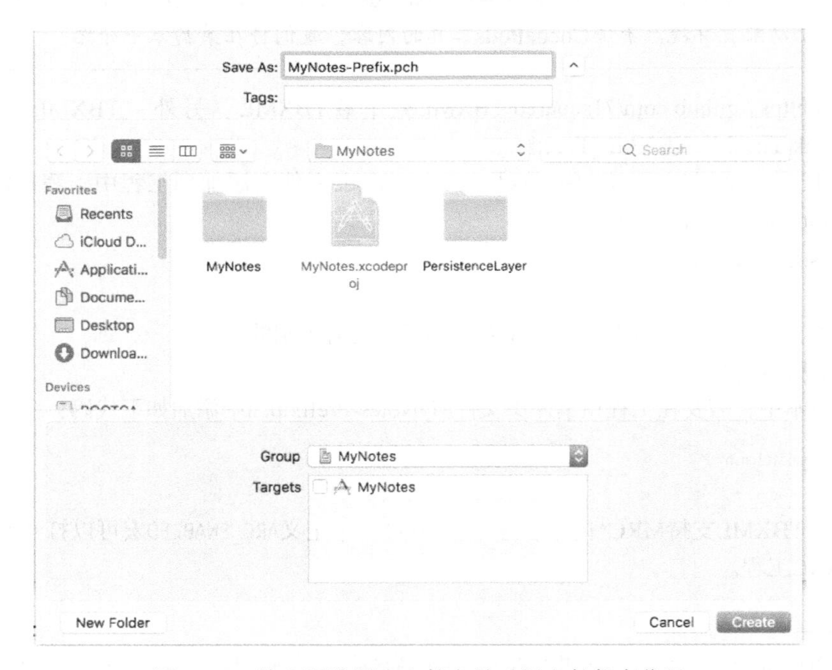

图15-10　输入预编译头文件名并选择文件保存位置

当然，这只是创建了一个预编译头文件，我们还需要把这个文件配置到工程中。打开工程，选择TARGETS→MyNotes→Build Settings→Apple LLVM 8.0 - Language→Prefix Header，输入刚才创建的MyNotes-Prefix.pch文件名，如图15-11所示。

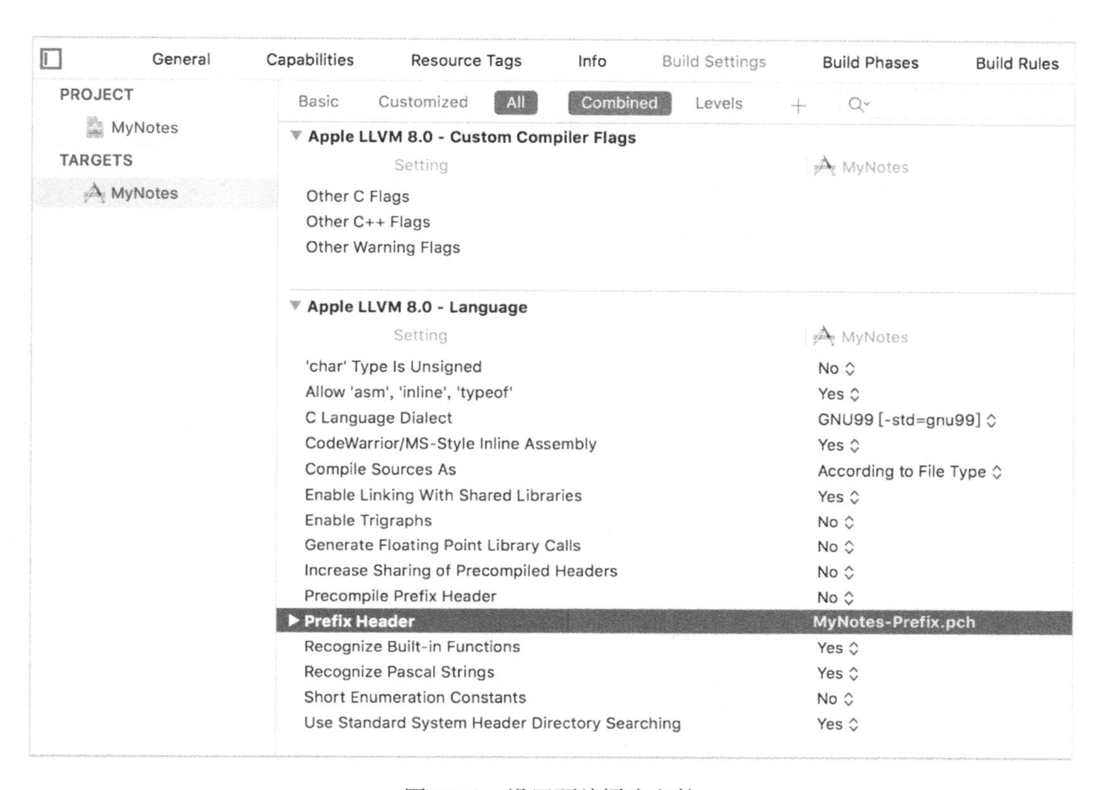

图15-11　设置预编译头文件

2. Swift版本配置

在Swift版本中，工程预编译头文件对Swift代码模块不起作用，我们需要在每个头文件（即TBXML.h和TBXML+Compression.h）中添加如下内容：

```
#import <Foundation/Foundation.h>
#define ARC_ENABLED
```

由于需要在Swift中调用Objective-C，我们需要在桥接头文件MyNotes-Bridging-Header.h中引入头文件TBXML.h。完成之后再进行编译就可以了。

如果工程中没有生成过桥接头文件，我们需要自己创建和配置。大家可以通过Xcode菜单点击File→New→File，以此打开创建文件模板界面（如图15-12所示），从中选择Header File文件模板，点击Next按钮，进入如图15-13所示的界面。然后我们在Save As中输入要创建的桥接文件名MyNotes-Bridging-Header.h，然后将文件保存在与.xcodeproj工程文件相同的目录下。

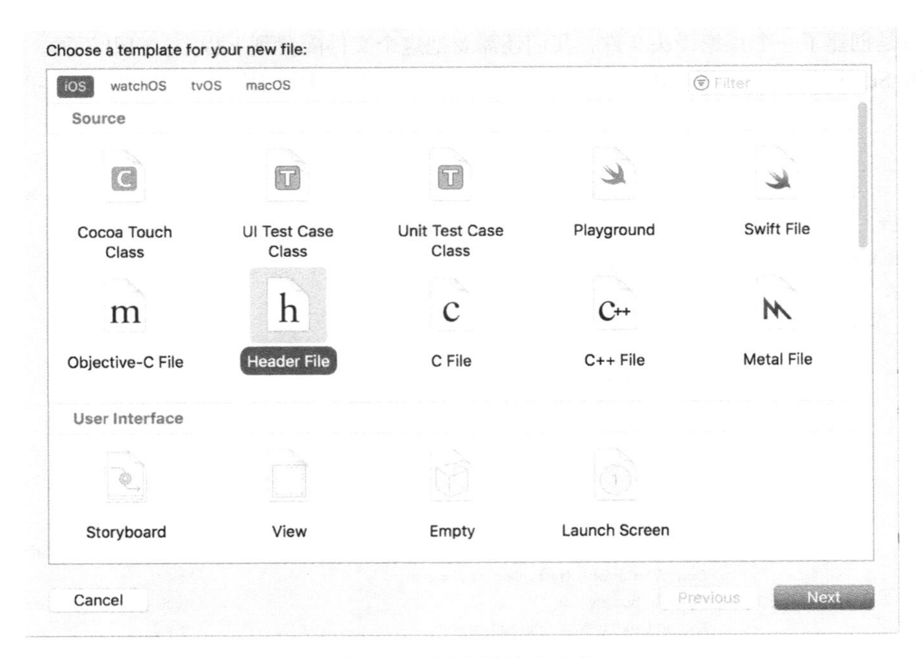

图15-12　创建桥接头文件

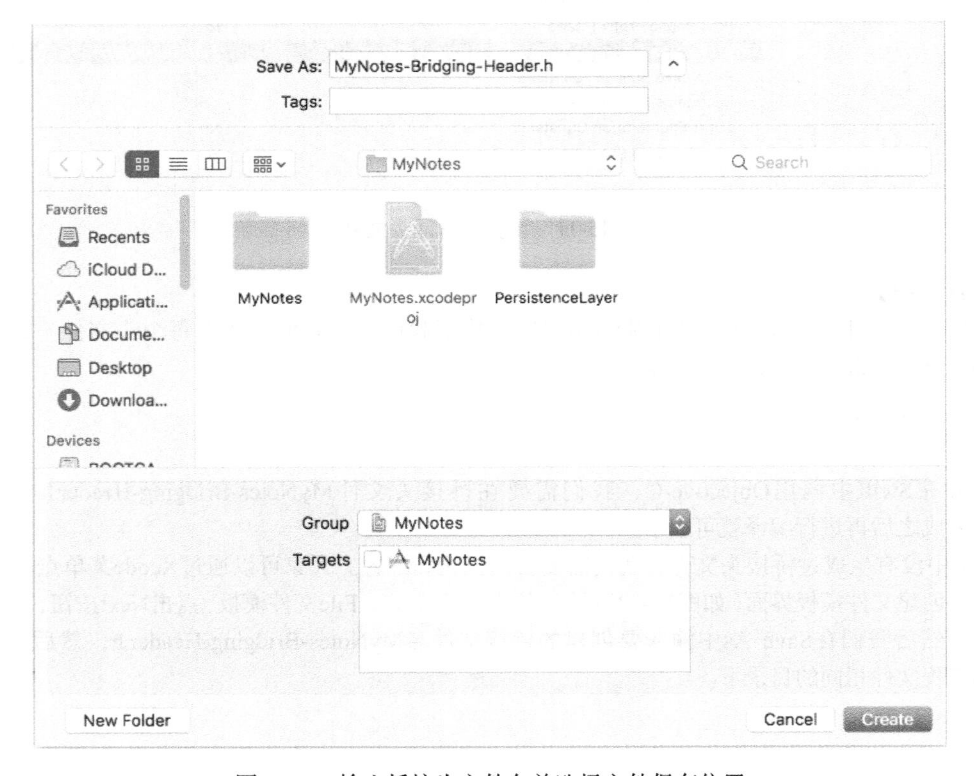

图15-13　输入桥接头文件名并选择文件保存位置

我们还需要把这个文件配置到工程中。打开工程，选择TARGETS→MyNotes→Build Settings→ Swift Compiler - Code General→Objective-C Bridging Header，输入刚才创建的MyNotes-Bridging-Header.h文件名，如图15-14所示。

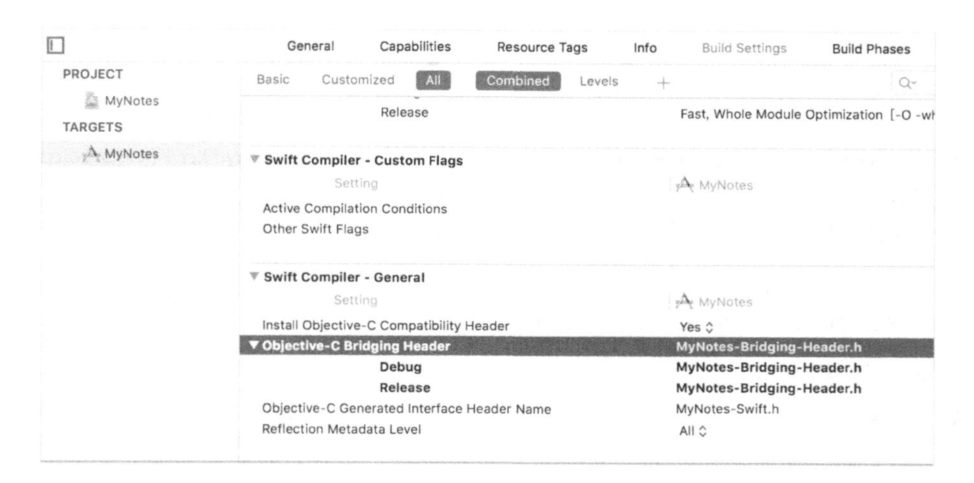

图15-14　设置桥接头文件

提示

桥接头文件的作用是为Swift语言调用Objective-C对象搭建一个"桥"，在桥接头文件中引入所需要的Objective-C Public头文件。这些Public头文件会暴露给同一应用目标的Swift文件，这样在Swift文件中就可以访问这些Public头文件所声明的Objective-C类等内容。关于Swift与Objective-C、C++和C混合调用的问题，读者可参考我编写的另外一本书《从零开始学Swift（第2版）》。

我们再看一下代码实现部分，先创建一个NotesTBXMLParser类来解析XML文档。NotesTBXMLParser的代码如下：

```
class NotesTBXMLParser: NSObject {

    //解析出的数据内部是字典类型
    private var listData: NSMutableArray!

    //开始解析
    func start() {

        self.listData = NSMutableArray()

        let tbxml = TBXML(xmlFile: "Notes.xml")                          ①

        let root = tbxml?.rootXMLElement                                 ②

        if root != nil {

            var noteElement = TBXML.childElementNamed("Note", parentElement:
            ➥root)                                                      ③

            while noteElement != nil {

                let dict = NSMutableDictionary()

                let dateElement = TBXML.childElementNamed("CDate",
                ➥parentElement: noteElement)
                if dateElement != nil {
                    let date = TBXML.text(for: dateElement)              ④
                    dict["CDate"] = date
                }

                let contentElement = TBXML.childElementNamed("Content",
                ➥parentElement: noteElement)
```

```
//NotesTBXMLParser.h文件
#import "TBXML.h"

@interface NotesXMLParser : NSObject <NSXMLParserDelegate>

//解析出的数据内部是字典类型
@property(strong, nonatomic) NSMutableArray *listData;
//当前标签的名字
@property(strong, nonatomic) NSString *currentTagName;

//开始解析
- (void)start;

@end

//NotesTBXMLParser.m文件
#import "NotesTBXMLParser.h"

@implementation NotesTBXMLParser

//开始解析
-(void)start {

    self.listData = [[NSMutableArray alloc] init];

    TBXML* tbxml = [[TBXML alloc] initWithXMLFile:@"Notes.xml" error:nil]; ①

    TBXMLElement * root = tbxml.rootXMLElement;                            ②

    //如果root元素有效
    if (root) {
```

```
            if contentElement != nil {
                let content = TBXML.text(for: contentElement)
                dict["Content"] = content
            }

            let userIDElement = TBXML.childElementNamed("UserID",
            ➥parentElement: noteElement)
            if userIDElement != nil {
                let userID = TBXML.text(for: userIDElement)
                dict["UserID"] = userID
            }

            let identifier = TBXML.value(ofAttributeNamed: "id", for:
            ➥noteElement)                                                  ⑤
            dict["id"] = identifier

            self.listData.add(dict)

            noteElement = TBXML.nextSiblingNamed("Note",
            ➥searchFrom: noteElement)                                      ⑥

        }
    }

    print("TBXML解析完成...")
    NotificationCenter.default.post(name:
    ➥Notification.Name(rawValue: "reloadViewNotification"), object:
    ➥self.listData)

    self.listData = nil
  }

}
```

```
    TBXMLElement * noteElement = [TBXML childElementNamed:@"Note"
    ➥parentElement:root];                                                  ③

      while ( noteElement != nil) {

          NSMutableDictionary *dict = [NSMutableDictionary new];

          TBXMLElement *dateElement = [TBXML childElementNamed:@"CDate"
          ➥parentElement:noteElement];
          if ( dateElement != nil) {
              NSString *date = [TBXML textForElement:dateElement];      ④
              dict[@"CDate"] = date;
          }

          TBXMLElement *contentElement = [TBXML childElementNamed:@"Content"
          ➥parentElement:noteElement];
          if ( contentElement != nil) {
              NSString *content = [TBXML textForElement:contentElement];
              dict[@"Content"] = content;
          }

          TBXMLElement *userIDElement = [TBXML childElementNamed:@"UserID"
          ➥parentElement:noteElement];
          if ( userIDElement != nil) {
              NSString *userID = [TBXML textForElement:userIDElement];
              dict[@"UserID"] = userID;
          }

          //获得ID属性
          NSString *identifier = [TBXML valueOfAttributeNamed:@"id"
          ➥forElement:noteElement error:nil];                              ⑤
          dict[@"id"] = identifier;

          [self.listData addObject:dict];

          noteElement = [TBXML nextSiblingNamed:@"Note"
          ➥searchFromElement:noteElement];                                 ⑥

      }
    }

    NSLog(@"TBXML解析完成...");

    [[NSNotificationCenter defaultCenter]
    ➥postNotificationName:@"reloadViewNotification" object:self.
    ➥listData userInfo:nil];
    self.listData = nil;

}

@end
```

与NSXML不同，TBXML解析采用DOM模式，不需要事件驱动，使用起来比较简单。第①行代码使用构造函数`initWithXMLFile:error:`创建一个TBXML对象，这个构造函数是从文件中构造TBXML对象的。TBXML提供了丰富的构造函数，下面是它的几个构造函数。

- **`initWithXMLString:error:`**。通过XML字符串构造TBXML对象。
- **`initWithXMLString:`**。通过XML字符串构造TBXML对象。
- **`initWithXMLData:error:`**。通过`NSData`数据构造TBXML对象，这个方法非常适用于在网络通信下解析。

第②行代码用于获得文档的根元素对象。由于没有提供XPath支持，解析文档需要从根元素开始，这个过程有点像“剥洋葱皮”。第③行代码是查找`root`元素下面的`Note`元素。在Notes.xml文档中，`Note`元素应该有很多，但是`childElementNamed:parentElement:`方法只是返回第一个`Note`元素，那么如何循环得到其他`Note`元素呢？第⑥行代码就是获得同层下一个`Note`元素的方法，`Sibling`意为“兄弟”元素，即非父子关系的同层元素。

第④行代码中TBXML的textForElement:方法用于取得元素的文本内容，这就相当于“剥洋葱皮”已经剥到了“洋葱心”，这个方法就是取出这个“洋葱心”。第⑤行代码中TBXML的valueOfAttributeNamed:forElement:error:方法用于获得属性值。

在视图控制器MasterViewController中修改viewDidLoad方法就可以运行了：

```
override func viewDidLoad() {
    super.viewDidLoad()
    ......
    var parser = NotesTBXMLParser()
    parser.start()

}
```

```
- (void)viewDidLoad {
    ......
    NotesTBXMLParser *parser = [[NotesTBXMLParser alloc] init];
    [parser start];
}
```

15.3 JSON 数据交换格式

JSON（JavaScript Object Notation）是一种轻量级的数据交换格式。所谓“轻量级”，是相对XML文档结构而言的，描述项目字符少，所以描述相同的数据所需的字符个数要少，那么传输的速度就会提高，而流量也会减少。

如果留言条采用JSON描述，可以设计成下面这样：

```
{"to":"云龙同学",
    "conent": "你好！\n今天上午，我到你家来想向你借一本《小学生常用成语词典》。可是
        不巧，你不在。我准备晚上6时再来借书。请你在家里等我，谢谢！",
    "from": "关东升",
    "date": "2012年12月08日"}
```

由于Web和移动平台开发对流量的要求是要尽可能少，对速度的要求是要尽可能快，而轻量级的数据交换格式JSON就成为了理想的数据交换格式。

15.3.1 JSON 文档结构

构成JSON文档的两种结构为对象和数组。对象是“名称–值”对集合，它类似于Objective-C中的字典类型，而数组是一连串元素的集合。

对象是一个无序的“名称–值”对集合，一个对象以左括号（{）开始，以右括号（}）结束。每个“名称”后跟一个冒号（:），“名称–值”对之间使用逗号（,）分隔。JSON对象的语法表如图15-15所示。

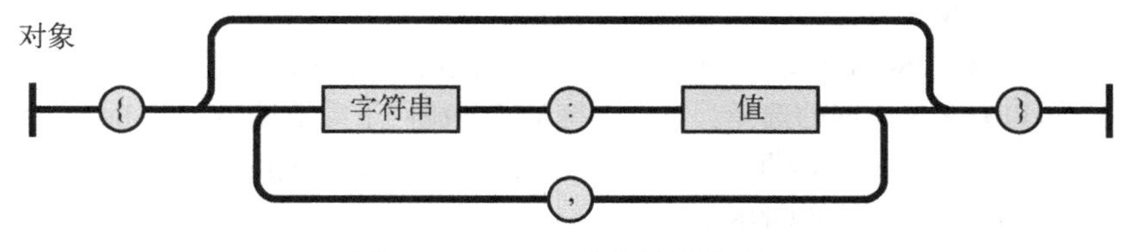

图15-15　JSON对象的语法表

下面是一个JSON对象的例子：

```
{
    "name":"a.htm",
    "size":345,
    "saved":true
}
```

数组是值的有序集合，以左中括号（[）开始，以右中括号（]）结束，值之间使用逗号（,）分隔。JSON数组的语法表如图15-16所示。

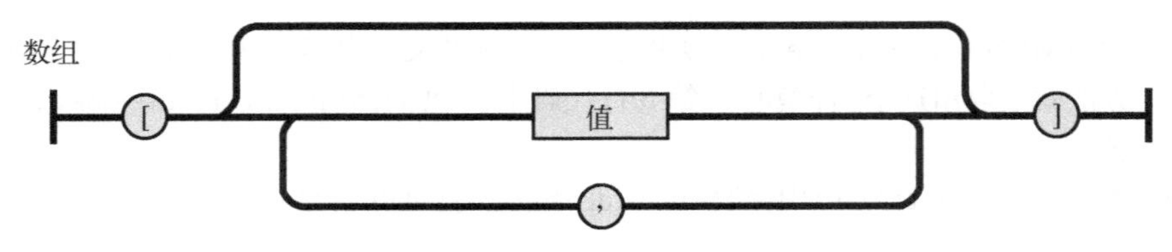

图15-16　JSON数组的语法表

下面是一个JSON数组的例子：

```
["text","html","css"]
```

在数组中，值可以是双引号括起来的字符串、数值、true、false、null、对象或者数组，而且这些结构可以嵌套。数组中值的JSON语法结构如图15-17所示。

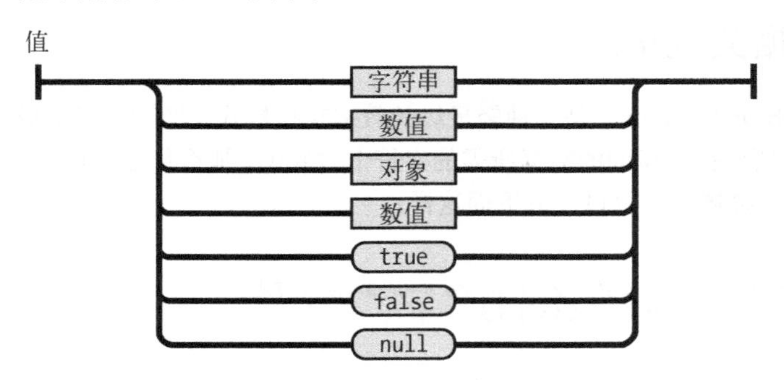

图15-17　JSON值的语法结构图

15.3.2　JSON 数据编码/解码

把数据从JSON文档中读取处理的过程称为“解码”，即解析和读取过程。由于JSON技术比较成熟，在iOS平台上也会有很多框架可以进行JSON的编码/解码，具体如下所示。

- SBJson。它是比较老的JSON编码/解码框架，原名是json-framework。这个框架现在更新仍然很频繁，支持ARC，源码下载地址为https://github.com/stig/json-framework。
- TouchJSON。它也是比较老的JSON编码/解码框架，支持ARC和MRC，源码下载地址为https://github.com/TouchCode/TouchJSON。
- YAJL。它是比较优秀的JSON框架，基于SBJson进行了优化，底层API使用C编写，上层API使用Objective-C编写，使用者可以有多种不同的选择。它不支持ARC，源码下载地址为http://lloyd.github.com/yajl/。
- JSONKit。它是更为优秀的JSON框架，它的代码很小，但是解码速度很快，不支持ARC，源码下载地址为https://github.com/johnezang/JSONKit。
- NextiveJson。它也是非常优秀的JSON框架，与JSONKit的性能差不多，但是在开源社区中没有JSONKit知名度高，不支持ARC，源码下载地址为https://github.com/nextive/NextiveJson。
- NSJSONSerialization。它是iOS 5之后苹果提供的API，是目前非常优秀的JSON编码/解码框架，支持ARC。iOS 5之后的SDK就已经包含这个框架了，不需要额外安装和配置。如果你的应用要兼容iOS 5之前的版本，这个框架不能使用。

为了解析这些框架的性能，我们可以参照XML实现MyNotes应用。我准备了10 000条Note数据的JSON文档，保存后文件大小仅为700 KB，而同样信息的XML文档是1.2 MB，这也印证了JSON是轻量级的数据交互格式。测试程序的运行结果如图15-18所示。

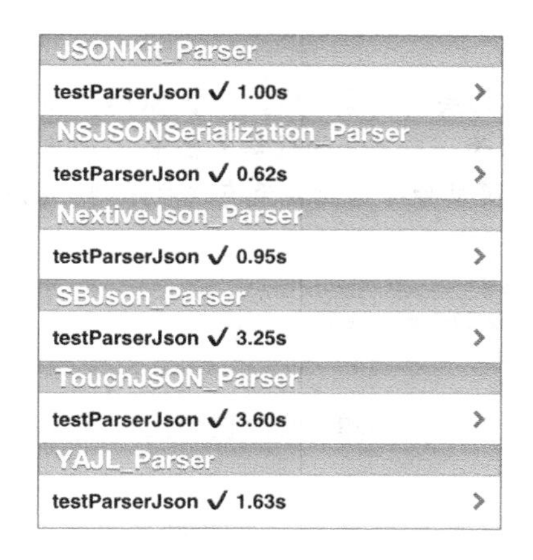

图15-18　测试程序的运行结果

从图15-18中可以看出，苹果提供的NSJSONSerialization框架花费的时间最短，TouchJSON和SBJson框架花费的时间比较长。我们再来看看内存占用指标，使用Instruments检查工具来检查内存占用情况，最后将内存占用峰值、驻留内存占用和上面的解码花费时间一起绘制成图表，如图15-19所示。

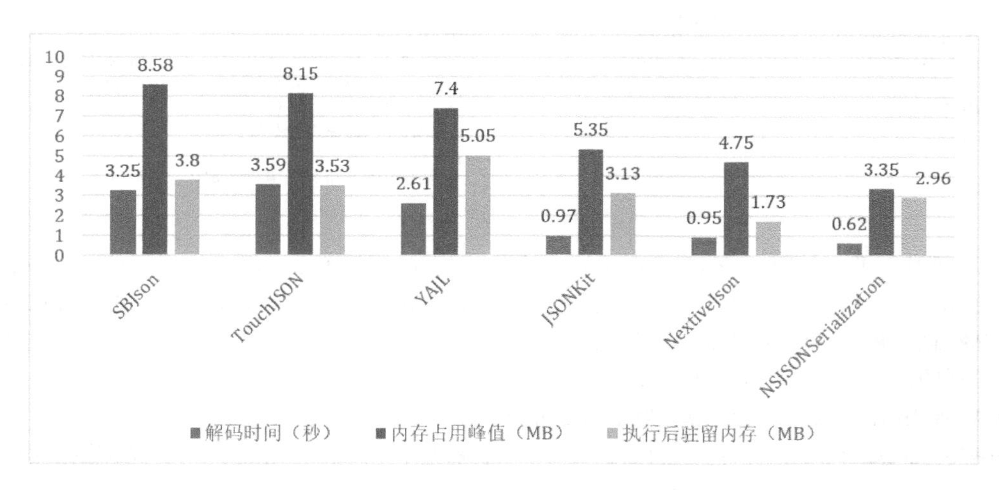

图15-19　JSON解码框架性能图表

图15-19展示了内存占用峰值、驻留内存和解析速度这3个指标的比较情况。TouchJSON和SBJson应该是最差的，NSJSONSerialization是解码速度最快的，内存占用峰值是最低的。可见，NSJSONSerialization是一个非常优秀的JSON解码框架。

15.4　案例：MyNotes 应用 JSON 解码

下面我们通过一个案例MyNotes学习一下NSJSONSerialization的用法。这里重新设计数据结构为JSON格式，其中备忘录信息Notes.json文件的内容如下：

```
{"ResultCode":0,"Record":[
    {"ID":"1","CDate":"2016-12-23","Content":"发布iOSBook0","UserID":"tony"},
    {"ID":"2","CDate":"2016-12-24","Content":"发布iOSBook1","UserID":"tony"},
    {"ID":"3","CDate":"2016-12-25","Content":"发布iOSBook2","UserID":"tony"},
    {"ID":"4","CDate":"2016-12-26","Content":"发布iOSBook3","UserID":"tony"},
    {"ID":"5","CDate":"2016-12-27","Content":"发布iOSBook4","UserID":"tony"}]}
```

提示

iOS平台中对于JSON文档的结构要求比较严格，每个JSON数据项目的“名称”必须使用双引号括起来，不能使用单引号或没有引号。在下面的代码文档中，“名称”省略了双引号，该文档在iOS平台解析时会出现异常，而在Java等其他平台上就没有这些限制，也不会出现异常：

```
{ResultCode:0,Record:[
    {ID:'1',CDate:'2016-12-23',Content:'发布iOSBook0',UserID:'tony'},
    {ID:'2',CDate:'2016-12-24',Content:'发布iOSBook1',UserID:'tony'}]}
```

事实上，NSJSONSerialization使用起来更为简单。修改视图控制器MasterViewController的viewDidLoad方法，具体代码如下：

```
override func viewDidLoad() {
    super.viewDidLoad()

    ......
    let path = Bundle.main.path(forResource: "Notes", ofType: "json")!
    let jsonData = try! Data(contentsOf: URL(fileURLWithPath: path))

    do {
        let jsonObj = try JSONSerialization.jsonObject(with: jsonData,
        ➥options: .mutableContainers) as! NSDictionary                    ①
        self.listData = jsonObj["Record"] as! [[String: String]]
    } catch {
        print("JSON解码失败")
    }

}
```

```
- (void)viewDidLoad {
    [super viewDidLoad];

    ......

    NSString *path = [[NSBundle mainBundle] pathForResource:@"Notes"
    ➥ofType:@"json"];
    NSData *jsonData = [[NSData alloc] initWithContentsOfFile:path];

    NSError *error;
    id jsonObj = [NSJSONSerialization JSONObjectWithData:jsonData
    ➥options:NSJSONReadingMutableContainers error:&error];                ①
    if (!jsonObj || error) {
        NSLog(@"JSON解码失败");
    }
    self.listData = jsonObj[@"Record"];
}
```

在第①行代码中，我们使用NSJSONSerialization的类方法JSONObjectWithData:options:error:进行解码，其中options参数指定了解析JSON的模式。该参数是枚举类型NSJSONReadingOptions定义的，共有如下3个常量。

- **NSJSONReadingMutableContainers**。指定解析返回的是可变的数组或字典。如果以后需要修改结果，这个常量是合适的选择。Swift版是mutableContainers。
- **NSJSONReadingMutableLeaves**。指定叶节点是可变字符串。Swift版是MutableLeaves。
- **NSJSONReadingAllowFragments**。允许解析器可以解析那些不是JSON数组或字典的JSON数据片段，如123、"ABC"等。Swift版是AllowFragments。

此外，NSJSONSerialization还提供了JSON编码的方法：dataWithJSONObject:options:error:和writeJSONObject:toStream:options:error:。JSON编码方法的用法与解码方法非常类似，这里我们就不再介绍了，留待后面章节中遇到时再介绍。

15.5　小结

通过对本章的学习，我们了解了数据交换格式，其中XML和JSON是主要的方式。

第 16 章

REST Web Service

Web Service技术通过Web协议提供服务，保证不同平台上的应用服务可以相互操作，为客户端程序提供不同的服务。类似Web Service的服务不断问世，如Java的RMI（Remote Method Invocation，远程方法调用）、Java EE的EJB（Enterprise JavaBean，企业级JavaBean）、CORBA（Common Object Request Broker Architecture，公共对象请求代理体系结构）和微软的DCOM（Distributed Component Object Model，分布式组件对象模型）等。

16.1 概述

目前，3种主流的Web Service实现方案有REST①、SOAP②和XML-RPC③。XML-RPC和SOAP都是比较复杂的技术，XML-RPC是SOAP的前身。与复杂的SOAP和XML-RPC相比，REST风格的Web Service更加简洁，越来越多的Web Service开始采用REST风格设计和实现。例如，亚马逊已经提供了REST风格的Web Service进行图书查找，雅虎提供的Web Service也是REST风格的。因此，本书会重点介绍REST风格的Web Service。

16.1.1 REST Web Service 概念

REST被翻译为"表征状态转移"，听起来很抽象，"表征"指客户端看到的页面，页面的跳转就是状态的转移，客户端通过请求URI获得要显示的页面。通常，REST使用HTTP、URI、XML以及HTML这些现有的协议和标准。

REST Web Service是一个使用HTTP并遵循REST原则的Web Service，使用URI来定位资源。Web Service的数据交互格式主要采用JSON和XML等。Web Service所支持的HTTP请求方法包括POST、GET、PUT或DELETE等。

提示

REST只是一种设计风格，不是设计规范，更不是行业标准，因此它的设计很灵活。REST Web Service的概念也很宽泛，可以泛指采用HTTP和HTTPS等传输协议并通过URI定位资源的Web Service。数据交互格式通常是JSON或XML格式。

REST Web Service应用层采用的是HTTP和HTTPS等传输协议，因此我们有必要介绍一下它们。

16.1.2 HTTP 协议

HTTP是"Hypertext Transfer Protocol"的缩写，即"超文本传输协议"。网络中使用的基本协议是TCP/IP协

① REST（Representational State Transfer，表征状态转移）是Roy Fielding博士在2000年的博士论文中提出的一种软件架构风格。——引自维基百科：http://zh.wikipedia.org/zh-cn/REST

② SOAP（Simple Object Access Protocol，简单对象访问协议）是交换数据的一种协议规范，用于在计算机网络Web服务中交换带结构的信息。——引自维基百科：http://zh.wikipedia.org/wiki/SOAP

③ XML-RPC是一个远程过程调用（Remote Procedure Call，RPC）的分布式计算协议，通过XML封装调用函数，并使用HTTP协议作为传送机制。——引自维基百科：http://zh.wikipedia.org/wiki/XML-RPC

议，目前被广泛采用的HTTP、HTTPS、FTP、Archie和Gopher等是建立在TCP/IP协议之上的应用层协议，不同的协议对应着不同的应用。

Web Service主要使用HTTP协议，这是一个属于应用层的面向对象的协议，其简洁、快速的方式适用于分布式超文本信息的传输。它于1990年被提出，经过多年的使用与发展，得到了不断完善和扩展。HTTP协议支持C/S网络结构，是无连接协议，即每一次请求时建立连接，服务器处理完客户端的请求后，应答给客户端，然后断开连接，不会一直占用网络资源。

HTTP/1.1协议共定义了8种请求方法：OPTIONS、HEAD、GET、POST、PUT、DELETE、TRACE和CONNECT。GET和POST方法在具体应用中用得比较多，下面重点介绍这两个方法。

GET方法是向指定的资源发出请求，发送的信息"显式"地跟在URL后面。GET方法应该只用于读取数据，例如静态图片等。GET方法有点儿像使用明信片给别人写信，"信内容"写在外面，接触到的人都可以看到，因此是不安全的。

POST方法是向指定资源提交数据，请求服务器进行处理，例如提交表单或者上传文件等，数据被包含在请求体中。POST方法像是把"信内容"装入信封中，接触到的人看不到，因此是安全的。

16.1.3　HTTPS 协议

HTTPS是"Hypertext Transfer Protocol Secure"的缩写，即"超文本传输安全协议"，是超文本传输协议和SSL的组合，用以提供加密通信及对网络服务器身份的鉴定。

简单地说，HTTPS是HTTP的升级版，区别在于：HTTPS使用https://代替http://，且使用端口443，而HTTP使用端口80来与TCP/IP进行通信。SSL使用40位关键字作为RC4流加密算法，这对于商业信息的加密是合适的。HTTPS和SSL支持使用X.509数字认证，如果需要的话，用户可以确认发送者是谁。

16.1.4　苹果 ATS 限制

自iOS 9开始，苹果为了提高网络安全，对应用发出的所有网络请求通信，必须采用HTTPS协议。否则，应用在运行时就会抛出错误：The resource could not be loaded because the ATS App Transport Security policy requires the use of a secure connection。这个限制就是苹果ATS（App Transport Security）限制。ATS限制我们在4.6节中介绍Web视图的时候就遇到了。

注意

在iOS 9中，默认非HTTPS的网络是被禁止的，我们可以在info.plist（工程配置文件）文件中设置NSAllowsArbitraryLoads为YES而跳过ATS限制。但在iOS 10中，从2017年1月1日起苹果不允许通过这个方法跳过ATS限制，也就是强制使用HTTPS，如果不这样的话，提交到App Store的应用可能会被拒绝。

一方面出于ATS的限制，另一方面确实应该增强网络安全来使用HTTPS协议，但是毕竟很多情况下我们只有别人提供的Web Service，特别是国内的很多Web Service都还是基于HTTP。考虑到这个问题，苹果允许我们设置ATS白名单（例外名单）。参考4.6节，在Xcode工程属性文件Info.plist中添加App Transport Security Settings键，并在该键下面添加Exception Domains键，如图16-1所示，设置例外的域名，在具体域名下可以添加更加详细的设置键，这些主要的键说明如下。

- NSIncludesSubdomains。是否包括子域名。
- NSExceptionRequiresForwardSecrecy。是否支持正向保密[①]。

① 正向保密是信息安全中提出的观点。要求一个密钥只能访问由它所保护的数据；用来产生密钥的元素一次一换，不能再产生其他的密钥；一个密钥被破解，并不影响其他密钥的安全性。设计旨在避免长期使用一个密钥所引起的不安全问题。

- NSExceptionAllowsInsecureHTTPLoads。是否允许不安全HTTP加载。
- NSExceptionMinimumTLSVersion。设置传输层安全协议最小版本，默认值是TLSV1.2。

Key	Type	Value
▼ Information Property List	Dictionary	(17 items)
▼ App Transport Security Settings	Dictionary	(1 item)
▼ Exception Domains	Dictionary	(2 items)
▼ 51work6.com	Dictionary	(3 items)
NSIncludesSubdomains	Boolean	YES
NSExceptionRequiresForwardSecrecy	Boolean	NO
NSExceptionAllowsInsecureHTTPLoads	Boolean	YES
▼ sina.com	Dictionary	(3 items)
NSIncludesSubdomains	Boolean	YES
NSExceptionRequiresForwardSecrecy	Boolean	NO
NSExceptionAllowsInsecureHTTPLoads	Boolean	YES
Localization native development region	String	en

图16-1　工程属性设置

提示　Xcode工程属性文件Info.plist是文本文件，我们可以通过文本编辑工具打开和编辑info.plist文件。它的结构都是键值对，即字典结构。图16-1添加的键，使用文本编辑工具打开，其内容如下，如果你对这些键值很熟悉，可以直接在这里修改：

```
<?xml version="1.0" encoding="UTF-8"?>
<!DOCTYPE plist PUBLIC "-//Apple//DTD PLIST 1.0//EN" "http://www.apple.com/DTDs/PropertyList-1.0.dtd">
<plist version="1.0">
<dict>
    <key>NSAppTransportSecurity</key>
    <dict>
        <key>NSExceptionDomains</key>
        <dict>
            <key>51work6.com</key>
            <dict>
                <key>NSIncludesSubdomains</key>
                <true/>
                <key>NSExceptionRequiresForwardSecrecy</key>
                <false/>
                <key>NSExceptionAllowsInsecureHTTPLoads</key>
                <true/>
            </dict>
            <key>sina.com</key>
            <dict>
                <key>NSIncludesSubdomains</key>
                <true/>
                <key>NSExceptionRequiresForwardSecrecy</key>
                <false/>
                <key>NSExceptionAllowsInsecureHTTPLoads</key>
                <true/>
            </dict>
        </dict>
    </dict>
 ......
</dict>
</plist>
```

16.2 使用 NSURLSession

NSURLSession是苹果于iOS 7之后提供的网络通信API，用来替代NSURLConnection网络通信库。NSURL-

Session可以为每个用户会话（session）配置缓存、协议和cookie[①]，支持安全证书策略；支持FTP、HTTP和HTTPS等网络通信协议；支持实现后台下载和上传，以及断点续传功能；还可以取消、恢复、挂起网络请求操作。

16.2.1 NSURLSession API

NSURLSession API所指的内容不仅仅是同名类NSURLSession，还包括一系列相互关联的类和协议。主要类有：

- NSURLSession
- NSURLSessionConfiguration
- NSURLSessionTask

主要协议有：

- NSURLSessionDelegate
- NSURLSessionTaskDelegate
- NSURLSessionDataDelegate
- NSURLSessionDownloadDelegate

注意　在Swift 3版本中，NSURLSession API中的类和协议命名去掉了NS前缀，例如：NSURLSession在Swift 3中是URLSession，本章中不再做特殊说明。

1. 会话

会话是指应用程序与服务器建立的通信对象，一个应用程序可以建立多个会话，每一个会话协调一组相关的数据传输任务。在NSURLSession API中，会话类是NSURLSession。具体使用时，我们可以创建4种形式的会话。

- **简单会话**。不可配置会话，只能执行基本的网络情况，通过NSURLSession的静态方法+sharedSession获得NSURLSession对象。Swift版通过shared静态属性获得该对象。
- **默认会话（default session）**。与简单会话类似，但是可以配置会话，通过NSURLSession的静态方法+ sessionWithConfiguration:或+ sessionWithConfiguration:delegate:delegateQueue:获得NSURLSession对象，其中的configuration参数是NSURLSessionConfiguration类型。NSURLSessionConfiguration是会话配置类，通过静态方法+ defaultSessionConfiguration获得默认会话配置对象。
- **短暂会话（ephemeral session）**。不存储任何数据在磁盘中，所有的数据都是缓存的。证书存储等都保存在内存中。当应用结束会话时，它们被自动释放。获得对话对象的方式与默认会话类似，不同的是会话配置对象是通过NSURLSessionConfiguration静态方法+ ephemeralSessionConfiguration获得的。
- **后台会话（background session）**：当应用程序在后台运行时，可以在后台执行上传或下载数据任务。获得对话对象的方式与默认会话类似，不同的是会话配置对象是通过NSURLSessionConfiguration静态方法+ backgroundSessionConfigurationWithIdentifier:获得的。

2. 会话任务

会话总是基于任务的，会话任务类是NSURLSessionTask，会话通过了3种形式的任务，如下。

- **数据任务（data task）**。请求网络资源，从服务器返回一个或多个NSData对象。数据任务支持简单会话、默认会话和短暂会话，但是不支持后台会话。数据任务是NSURLSessionDataTask类。
- **上传任务（upload task）**。通常以文件的形式发送数据，支持后台上传。上传任务是NSURLSessionUploadTask类。
- **下载任务（download task）**。以文件形式接收数据，支持后台下载。下载任务是NSURLSessionDownloadTask类。

---

① cookie是网站服务器端为了辨别用户身份，在服务器端生成并存储在用户本地终端设备（电脑、智能手机等）上的数据。

事实上，NSURLSessionTask是一个抽象类，它有NSURLSessionDataTask和NSURLSessionDownloadTask这两个子类，而NSURLSessionUploadTask是NSURLSessionDataTask的子类。类图如图16-2所示。

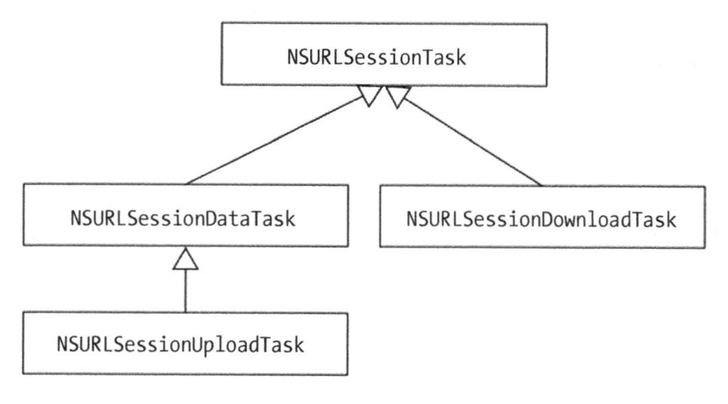

图16-2　任务类图

所有任务都可以取消、暂停或恢复。默认情况下，任务是暂停的，所以创建任务后，需要恢复任务才能执行操作。

16.2.2　简单会话实现 GET 请求

NSURLSession提供了HTTP和HTTPS的异步请求，支持GET、POST、PUT和DELETE等方法请求，这里我们先了解其中最为简单的GET方法请求。

为了学习这些API的用法，我们重构MyNotes应用案例。这次数据来源不是本地Notes.xml或Notes.json文件，而是网络服务器。

我们首先实现查询业务，查询业务请求可以在主视图控制器MasterViewController类中实现。MasterView-Controller中的主要相关代码如下：

```
class MasterViewController: UITableViewController {

    //保存数据列表
    var listData = NSArray()

    override func viewDidLoad() {
        ......
        self.startRequest()                                           ①
    }
    ......

    //开始请求Web Service
    func startRequest() {

        var strURL = NSString(format: "http://www.51work6.com/service/
        ➥mynotes/WebService.php?email=%@&type=%@&action=%@",
        ➥"<你的51work6.com用户邮箱>", "JSON", "query")                ②
        strURL = strURL.addingPercentEncoding(withAllowedCharacters:
        ➥CharacterSet.urlQueryAllowed)!                               ③

        let url = URL(string: strURL)!
        let request = URLRequest(url: url)
        let session = URLSession.shared                               ④


        let task: URLSessionDataTask = session.dataTask(with: request,
        ➥completionHandler:
        ➥{ (data, response, error) in                                 ⑤
```

```
//MasterViewController.m代码

- (void)viewDidLoad {
    ......

    [self startRequest];                                              ①
}

#pragma mark -- 开始请求Web Service
-(void)startRequest {

    NSString *strURL = [[NSString alloc]
    ➥initWithFormat:@"http://www.51work6.com/service/mynotes/
    ➥WebService.php?email=%@&type=%@&action=%@",
    ➥@"<你的51work6.com用户邮箱>",@"JSON",@"query"];                  ②

    strURL = [strURL stringByAddingPercentEncodingWithAllowedCharacters:
    ➥[NSCharacterSet URLQueryAllowedCharacterSet]];                   ③

    NSURL *url = [NSURL URLWithString:strURL];
    NSURLRequest *request = [[NSURLRequest alloc] initWithURL:url];

    NSURLSession*session = [NSURLSession sharedSession];              ④

    NSURLSessionDataTask *task = [session dataTaskWithRequest:request
    ➥completionHandler:^(NSData * data, NSURLResponse * response,
```

```
            print("请求完成...")
            if error == nil {
                do {
                    let resDict = try JSONSerialization.jsonObject(with:
                    ➥data!, options: JSONSerialization.ReadingOptions.
                    ➥allowFragments) as! NSDictionary

                    DispatchQueue.main.async(execute: {
                        self.reloadView(resDict)                          ⑥
                    })

                } catch {
                    print("返回数据解析失败")
                }
            } else {
                print("error : ", error!.localizedDescription)
            }
        })
        task.resume()                                                     ⑦
    }

    //MARK: -- 重新加载表视图
    func reloadView(_ res : NSDictionary) {

        let resultCode: NSNumber = res["ResultCode"] as! NSNumber         ⑧

        if (resultCode.intValue >= 0) {
            self.listData = res["Record"] as! NSArray                     ⑨
            self.tableView.reloadData()
        } else {
            let errorStr = resultCode.errorMessage                        ⑩
            let alertController: UIAlertController = UIAlertController(title:
            ➥"错误信息",
            ➥message: errorStr, preferredStyle: .alert)
            let okAction = UIAlertAction(title: "OK", style: .default, handler:
            ➥nil)
            alertController.addAction(okAction)
            //显示
            self.present(alertController, animated: true, completion: nil)

        }
    }
}
```

```
    ➥NSError * error) {                                                  ⑤
        NSLog(@"请求完成...");

        if (!error) {
            NSDictionary *resDict = [NSJSONSerialization JSONObjectWithData:
            ➥data options:NSJSONReadingAllowFragments error:nil];

            dispatch_async(dispatch_get_main_queue(), ^{
                [self reloadView:resDict];                               ⑥
            });

        } else {
            NSLog(@"error : %@", error.localizedDescription);
        }
    }];

    [task resume];                                                       ⑦

}

#pragma mark -- 重新加载表视图
-(void)reloadView:(NSDictionary*)res {

    NSNumber *resultCode = res[@"ResultCode"];                           ⑧

    if ([resultCode integerValue] >=0) {
        self.listData = res[@"Record"];                                  ⑨
        [self.tableView reloadData];
    } else {
        NSString *errorStr = [resultCode errorMessage];                  ⑩
        UIAlertController* alertController  = [UIAlertController
        ➥alertControllerWithTitle:@"错误信息"
        ➥message: errorStr
        ➥preferredStyle:UIAlertControllerStyleAlert];
        UIAlertAction* okAction = [UIAlertAction actionWithTitle:@"OK"
        ➥style:UIAlertActionStyleDefault handler: nil];
        [alertController addAction:okAction];
        //显示
        [self presentViewController:alertController animated:true
        ➥completion:nil];
    }
}
```

上述代码中，第①行代码调用自己的startRequest方法实现请求Web Service。在startRequest方法的代码中，第②行用于指定请求的URL，这时URL所指向的Web Service由本书服务器51work6.com提供，请求的参数全部暴露在URL后面，这是GET请求方法的典型特征。为了能够正确请求数据，需要开发人员提供合适的参数，具体如下所示。

- **email**。它是www.51work6.com网站的注册用户邮箱。如果用户使用这些Web Service，首先需要到这个网站注册成为会员，然后提供自己的注册邮箱。
- **type**。它是数据交互类型。Web Service提供了3种方式的数据：JSON、XML和SOAP。
- **action**。它指定调用Web Service的一些方法，这些方法有add、remove、modify和query，分别代表插入、删除、修改和查询处理。

第③行使用NSString的stringByAddingPercentEncodingWithAllowedCharacters:（Swift版是addingPercentEncoding(withAllowedCharacters:)）方法将字符串转换为URL字符串。在网上传输的时候，URL中不能有中文等特殊字符，比如特殊字符“<”必须进行URL编码才能传输，“<”字符的URL编码是“%3C”。Objective-C版中方法的参数是字符集类型NSCharacterSet，通过静态方法URLQueryAllowedCharacterSet创建一个带有请求参数的URL字符串允许的字符集；在Swift版中方法的参数是字符集类型CharacterSet，通过静态属性urlQueryAllowed创建一个带有请求参数的URL字符串允许的字符集。

第④行用于创建简单会话对象。第⑤行是通过NSURLSession的dataTaskWithRequest:completionHandler:方法获得数据任务（NSURLSessionDataTask）对象，第一个参数是NSURLRequest请求对象，第二个参数completionHandler是请求完成回调的代码块（Swift中是闭包）。代码块completionHandler中的data参数是从服务器返回的数据，response是从服务器返回的应答对象，error是错误对象，如果error为nil，则说明请求过程没有错误发生。

第⑥行是在请求完成时调用reloadView:方法，该方法用于重新加载表视图中的数据。

提示

调用reloadView:方法是在GCD主队列中执行的，创建主队列Objective-C是通过dispatch_async(dispatch_get_main_queue(), ^{...})语句实现的，而Swift 3变化很大，是通过DispatchQueue.main.async(execute: {...})语句实现的。由于默认情况下，简单会话是在非主队列中执行的，当遇到表视图刷新这种更新UI界面的操作时，要切换回主队列执行。有关GCD技术的内容，我们会在第25章中详细介绍。

第⑦行是在会话任务对象上调用resume方法开始执行任务，新创建的任务默认情况下是暂停的。

在reloadView:方法中，第⑧行用于返回ResultCode。当ResultCode大于等于0时，说明服务器端操作成功。

第⑨行用于取得从服务端返回的数据。从服务器返回的JSON格式数据有两种情况，一种是成功返回，相关代码如下：

```
{"ResultCode":0,"Record":[{"ID":1,"CDate":"2012-12-23","Content":
    "这只是一条测试数据，http://www.51work6.com注册。"}]}
```

另一种是失败返回，相关代码如下：

```
{"ResultCode":-1}
```

其中ResultCode数据项用于说明调用Web Service的结果。为了减少网络传输，我们只传递消息代码，不传递消息内容。上述代码中的第⑩行是根据结果编码获得结果消息，编码是NSNumber的扩展类（Objective-C中称为类别），它是在NSNumber+Message中定义的，具体代码如下：

```
import Foundation

extension NSNumber {
    var errorMessage : String {
        var errorStr = ""
        switch (self) {
        case -7:
            errorStr = "没有数据。"
        case -6:
            errorStr = "日期没有输入。"
        case -5:
            errorStr = "内容没有输入。"
        case -4:
            errorStr = "ID没有输入。"
        case -3:
            errorStr = "数据访问失败。"
        case -2:
            errorStr = "你的账号最多能插入10条数据。"
        case -1:
            errorStr = "用户不存在，请到51work6.com注册。"
        default:
            errorStr = ""
        }
        return errorStr
    }
}
```

```
//NSNumber+Message.h文件代码
#import <Foundation/Foundation.h>

@interface NSNumber (Message)

-(NSString *)errorMessage;

@end

//NSNumber+Message.m文件代码
@implementation NSNumber (Message)

-(NSString *)errorMessage {
    NSString *errorStr = @"";
    switch ([self integerValue]) {
        case -7:
            errorStr = @"没有数据。";
            break;
        case -6:
            errorStr = @"日期没有输入。";
            break;
        case -5:
            errorStr = @"内容没有输入。";
            break;
        case -4:
            errorStr = @"ID没有输入。";
            break;
        case -3:
            errorStr = @"数据访问失败。";
```

```
            break;
        case -2:
            errorStr = @"你的账号最多能插入10条数据。";
            break;
        case -1:
            errorStr = @"用户不存在，请到51work6.com注册。";
        default:
            break;
    }

    return errorStr;
}

@end
```

NSNumber扩展中的代码很简单，我们就不再赘述了。注意，如果返回的结果代码小于0，则说明操作失败了。

提示　如果是在www.51work6.com网站刚刚注册的新用户，运行本节程序，会发现没有数据，读者可以先运行16.3节的具有插入和删除功能的MyNotes案例，插入测试数据，然后再来运行本节案例。

16.2.3　默认会话实现 GET 请求

使用简单会话实现的网络请求不能进行任何配置，不是很灵活。例如，我们想在主队列（主线程所在队列）中发出网络请求，由于不能进行配置，所以只能把更新UI的代码放在dispatch_async中执行。很多情况下我们可以使用默认会话，然后进行配置 。

修改16.2.2节的MyNotes案例来使用默认会话，其中主视图控制器MasterViewController类中主要的相关代码如下：

```
//MARK: -- 开始请求Web Service
func startRequest() {

    var strURL = NSString(format: "http://www.51work6.com/service/
    ➥mynotes/WebService.php?email=%@&type=%@&action=%@",
    ➥"<你的51work6.com用户邮箱>", "JSON", "query")
    strURL = strURL.addingPercentEncoding(withAllowedCharacters:
    ➥CharacterSet.urlQueryAllowed)!

    let url = URL(string: strURL)!
    let request = URLRequest(url: url)

    let defaultConfig = URLSessionConfiguration.default                    ①
    let session = URLSession(configuration: defaultConfig,
    ➥delegate: nil, delegateQueue: OperationQueue.main)                   ②

    let task: URLSessionDataTask = session.dataTask(with: request,
    ➥completionHandler: { (data, response, error) in
        print("请求完成...")
        if error == nil {
            do {
                let resDict = try JSONSerialization.jsonObject(with:
                ➥data!, options: JSONSerialization.ReadingOptions.
                ➥allowFragments) as! NSDictionary

                //DispatchQueue.main.async(execute: {
                    self.reloadView(resDict)                               ③
                //})

            } catch {
                print("返回数据解析失败")
```

```
#pragma mark -- 开始请求Web Service

- (void)startRequest {

    NSString *strURL = [[NSString alloc]
    ➥initWithFormat:@"http://www.51work6.com/service/mynotes/
    ➥WebService.php?email=%@&type=%@&action=%@",
    ➥@"<你的51work6.com用户邮箱>",@"JSON",@"query"];

    strURL = [strURL stringByAddingPercentEncodingWithAllowedCharacters:
    ➥[NSCharacterSet URLQueryAllowedCharacterSet]];

    NSURL *url = [NSURL URLWithString:strURL];
    NSURLRequest *request = [[NSURLRequest alloc] initWithURL:url];

    NSURLSessionConfiguration *defaultConfig =
    ➥[NSURLSessionConfiguration defaultSessionConfiguration];             ①
    NSURLSession *session = [NSURLSession
    ➥sessionWithConfiguration: defaultConfig
    ➥delegate: nil delegateQueue: [NSOperationQueue mainQueue]];          ②

    NSURLSessionDataTask *task = [session dataTaskWithRequest:request
    ➥completionHandler: ^(NSData *data, NSURLResponse *response, NSError
    ➥*error) {
        NSLog(@"请求完成...");
        if (!error) {
            NSDictionary *resDict = [NSJSONSerialization
            ➥JSONObjectWithData:data options:NSJSONReadingAllowFragments
            ➥error:nil];
            //dispatch_async(dispatch_get_main_queue(), ^{
                [self reloadView:resDict];                                 ③
```

```
        }
    } else {
        print("error : ", error!.localizedDescription)
    }
})
task.resume()
}
```

```
        //});
    } else {
        NSLog(@"error : %@", error.localizedDescription);
    }
}];
[task resume];
}
```

上述代码与16.2.2节中MyNotes的实现非常类似，代码编号部分是不同之处。第①行通过NSURLSession-Configuration的静态方法+ defaultSessionConfiguration创建默认配置对象。第②行用于创建默认会话对象，其中delegate参数用于指定会话委托对象。由于本例中网络请求任务之后回调的是代码块，而非委托对象的方法，因此本例中delegate参数被赋值为nil，会话委托方式一般在下载会话中用得很多。delegateQueue参数是设置会话任务执行所在的操作队列，通过[NSOperationQueue mainQueue]（Swift版通过OperationQueue.main）语句获得主操作队列（即主线程），NSOperationQueue（Swift版是OperationQueue）是操作队列，内部封装了线程。

由于会话任务是在主线程中执行的，所以第③行不需要再放到并发队列方法dispatch_async（Swift版是DispatchQueue.main.async）中执行了。

16.2.4 实现 POST 请求

前面两节采用的都是GET方法请求，下面我们介绍POST方法请求。使用POST方法请求的关键是用NSMutableURLRequest类替代NSURLRequest类。

这里我们把MyNotes应用变成POST方法，此时MasterViewController中startRequest方法的代码如下：

```
//MARK: -- 开始请求Web Service
func startRequest() {

    let strURL = "http://www.51work6.com/service/mynotes/WebService.php"    ①

    let url = URL(string: strURL)!

    let post = NSString(format: "email=%@&type=%@&action=%@",
    ➥"<你的51work6.com用户邮箱>", "JSON", "query"))    ②

    let postData: Data = post.data(using: String.Encoding.utf8.rawValue)!    ③

    let mutableURLRequest = NSMutableURLRequest(url: url)    ④
    mutableURLRequest.httpMethod = "POST"    ⑤
    mutableURLRequest.httpBody = postData    ⑥
    let request = mutableURLRequest as URLRequest    ⑦

    ......
}
```

```
- (void)startRequest {

    NSString *strURL =
    ➥@"http://www.51work6.com/service/mynotes/WebService.php";    ①

    NSURL *url = [NSURL URLWithString:strURL];

    NSString *post = [NSString
    ➥stringWithFormat:@"email=%@&type=%@&action=%@",
    ➥@"<你的51work6.com用户邮箱>", @"JSON", @"query"];    ②
    NSData *postData = [post dataUsingEncoding:NSUTF8StringEncoding];    ③

    NSMutableURLRequest *request = [NSMutableURLRequest
    ➥requestWithURL:url];    ④
    [request setHTTPMethod:@"POST"];    ⑤
    [request setHTTPBody:postData];    ⑥

    NSURLSessionConfiguration *defaultConfig = [NSURLSessionConfiguration
    ➥defaultSessionConfiguration];
    NSURLSession *session = [NSURLSession sessionWithConfiguration:
    ➥defaultConfig delegate: nil delegateQueue:
    ➥[NSOperationQueue mainQueue]];

    ......
}
```

第①行用于创建一个URL字符串，在这个URL字符串后面没有参数（即没有?号后面的内容）。请求参数放到请求体中，如第⑥行所示，其中postData就是请求参数，是NSData类型（Swift中是Data类型）。参数字符串是在第②行创建的，最后的参数字符串如下所示：

```
email=<你的51work6.com用户邮箱>&type=JSON&action=query
```

第③行将参数字符串转换成NSData类型（Swift中是Data类型），编码一定要采用UTF-8。

第④行用于创建可变的请求对象NSMutableURLRequest。因为它是可变对象，所以可以通过属性设置其内容。第⑤行中的HTTPMethod属性用于设置HTTP请求方法为POST，这里不能使用GET方法请求。第⑥行中的HTTPBody属性用于设置请求数据。

注意

由于session.dataTask方法只能接收URLRequest。在Swift版中，第⑦行用于将NSMutableURLRequest转换为URLRequest，URLRequest是结构体，NSMutableURLRequest是一个类，它不能自动转换，需要强制转换。而在Objective-C版本不需要这样的转换，这是因为在Objective-C中NSMutableURLRequest是NSURLRequest的子类。

16.2.5　下载数据

在NSURLSession API中，下载数据需要使用下载任务类NSURLSessionDownloadTask实现。前面介绍的3种会话形式都可以使用，具体采用哪一种要视用户的需求而定。例如：在下载过程中想知道下载的进度，或能够支持断点续传，我们需要实现NSURLSessionDownloadDelegate委托协议来接收服务器回调事件，注意这时候不能使用代码块接收服务器回调事件。这个过程需要使用默认会话。

下面通过一个案例介绍如何采用默认会话实现下载数据。我们设计了如图16-3所示的应用，当用户点击GO按钮时，它将从服务器下载一张图片并将其显示在界面中。

图16-3　设计原型图

UI部分的设计步骤就不再介绍了，下面我们直接看看主视图控制器。先来看看ViewController文件中类声明和属性等的相关代码：

```
class ViewController: UIViewController, URLSessionDownloadDelegate {

    @IBOutlet weak var progressView: UIProgressView!
    @IBOutlet weak var imageView1: UIImageView!
    ......
}
```

```
//ViewController.m文件

#import "ViewController.h"

@interface ViewController () <NSURLSessionDownloadDelegate>

@property(weak, nonatomic) IBOutlet UIProgressView *progressView;
@property(weak, nonatomic) IBOutlet UIImageView *imageView1;

@end

@implementation ViewController
......
@end
```

其中，progressView是与界面中进度条对应的属性，imageView1是与界面对应的图片视图控件。

下面我们直接看看主视图控制器ViewController中GO按钮的调用方法，具体如下：

```
@IBAction func onClick(sender: AnyObject) {

    let strURL = String(format: "http://www.51work6.com/
➥service/download.php?email=%@&FileName=test1.jpg",
➥"<你的51work6.com用户邮箱>")                                    ①
    let url = URL(string: strURL)!

    let defaultConfig = URLSessionConfiguration.default
    let session = URLSession(configuration: defaultConfig,
➥delegate: self, delegateQueue: OperationQueue.main)            ②

    let downloadTask = session.downloadTask(with: url)           ③

    downloadTask.resume()
}
```

```
- (IBAction)onClick:(id)sender {

    NSString *strURL = [[NSString alloc]
➥initWithFormat:@"http://www.51work6.com/
➥service/download.php?email=%@&FileName=test1.jpg",
➥@"<你的51work6.com用户邮箱>"];                                  ①

    NSURL *url = [NSURL URLWithString:strURL];

    NSURLSessionConfiguration *defaultConfig =
➥[NSURLSessionConfiguration defaultSessionConfiguration];
    NSURLSession *session = [NSURLSession
➥sessionWithConfiguration:defaultConfig
➥delegate:self delegateQueue:[NSOperationQueue mainQueue]];    ②

    NSURLSessionDownloadTask *downloadTask = [session
➥downloadTaskWithURL:url];                                      ③

    [downloadTask resume];
}
```

上述代码中，第①行用于设置图片下载地址。第②行用于创建默认会话对象，其中delegate为self，即将当前视图控制器作为下载任务会话的委托对象。第③行用于创建下载会话任务，然后调用resume方法开始执行。

一旦开始执行下载会话任务，就会回调实现NSURLSessionDownloadTask协议的委托对象。实现NSURLSession-DownloadTask协议的主要代码如下：

```
//MARK: -- 实现NSURLSessionDownloadDelegate委托协议
func urlSession(_ session: URLSession, downloadTask: URLSessionDownloadTask,
➥didWriteData bytesWritten: Int64, totalBytesWritten: Int64,
➥totalBytesExpectedToWrite: Int64) {                              ①

    let progress = Float(totalBytesWritten) / Float(totalBytesExpectedToWrite)
                                                                     ②
    self.progressView .setProgress(progress, animated: true)        ③
    NSLog("进度= %f", progress)
    NSLog("接收: %lld 字节 (已下载: %lld 字节) 期待: %lld 字节.",
    ➥bytesWritten, totalBytesWritten, totalBytesExpectedToWrite)
}

func urlSession(_ session: URLSession, downloadTask:
➥URLSessionDownloadTask, didFinishDownloadingTo location: URL) {   ④

    NSLog("临时文件: %@", location.description)

    let downloadsDir = NSSearchPathForDirectoriesInDomains
    ➥(.documentDirectory,
    ➥.userDomainMask, true)[0]

    let downloadStrPath = downloadsDir + "/test1.jpg"
    let downloadURLPath = URL(fileURLWithPath: downloadStrPath)

    let fileManager = FileManager.default
    if fileManager.fileExists(atPath: downloadStrPath) == true {   ⑤
        do {
            try fileManager.removeItem(atPath: downloadStrPath)    ⑥
        } catch let error as NSError {
            NSLog("删除文件失败: %@", error.localizedDescription)
        }
    }

    do {
```

```
#pragma mark -- 实现NSURLSessionDownloadDelegate委托协议

- (void)URLSession:(NSURLSession *)session
➥downloadTask:(NSURLSessionDownloadTask *)downloadTask
➥didWriteData:(int64_t)bytesWritten
➥totalBytesWritten:(int64_t)totalBytesWritten
➥totalBytesExpectedToWrite:(int64_t)totalBytesExpectedToWrite {    ①

    float progress = totalBytesWritten * 1.0 / totalBytesExpectedToWrite;  ②
    [self.progressView setProgress:progress animated:TRUE];                ③
    NSLog(@"进度= %f", progress);
    NSLog(@"接收: %lld 字节 (已下载: %lld 字节) 期待: %lld 字节.",
    ➥bytesWritten, totalBytesWritten, totalBytesExpectedToWrite);
}

- (void)URLSession:(NSURLSession *)session
➥downloadTask:(NSURLSessionDownloadTask *)downloadTask
➥didFinishDownloadingToURL:(NSURL *)location {                     ④

    NSLog(@"临时文件: %@\n", location);

    NSString *downloadsDir = [NSSearchPathForDirectoriesInDomains(
    ➥NSDocumentDirectory, NSUserDomainMask, TRUE) objectAtIndex:0];

    NSString *downloadStrPath = [downloadsDir
    ➥stringByAppendingPathComponent:@"/test1.jpg"];
    NSURL *downloadURLPath = [NSURL fileURLWithPath:downloadStrPath];

    NSFileManager *fileManager = [NSFileManager defaultManager];

    NSError *error = nil;
    if ([fileManager fileExistsAtPath:downloadStrPath]) {               ⑤
        [fileManager removeItemAtPath:downloadStrPath error:&error];   ⑥
        if (error) {
            NSLog(@"删除文件失败: %@", error.localizedDescription);
```

```
        try fileManager.moveItem(at: location, to: downloadURLPath)        ⑦
        NSLog("文件保存: %@", downloadStrPath)
        let img = UIImage(contentsOfFile: downloadStrPath)                 ⑧
        self.imageView1.image = img                                        ⑨

    } catch let error as NSError {
        NSLog("保存文件失败: %@", error.localizedDescription)
    }
}
```

```
        }
    }

    error = nil;

    if ([fileManager moveItemAtURL:location toURL:downloadURLPath error:&
    error]) {                                                              ⑦
        NSLog(@"文件保存: %@", downloadStrPath);
        UIImage *img = [UIImage imageWithContentsOfFile:downloadStrPath];  ⑧
        self.imageView1.image = img;                                       ⑨

    } else {
        NSLog(@"保存文件失败: %@", error.localizedDescription);
    }
}
```

上述代码中，第①行的委托方法是获知下载进度的关键，其中参数bytesWritten是当前从服务器接收的字节数，totalBytesWritten是累计已经接收的字节数，totalBytesExpectedToWrite是期待接收的字节数。所以代码第②行中的totalBytesWritten * 1.0 / totalBytesExpectedToWrite表达式可以计算下载进度。第③行用于更新进度条的进度，属于更新UI操作，需要在主队列（主线程所在队列）中执行。由于配置会话时设置的是主队列（主线程所在队列），所以更新UI的操作不必放在dispatch_async中执行。

第④行的委托方法是下载完成时回调的，其中location参数是下载过程中保存数据的本地临时文件。第⑤行用于判断在沙箱Documents目录下是否存在test1.jpg文件，如果存在，则通过第⑥行删除test1.jpg文件，这可以防止最新下载的文件不能覆盖之前遗留的test1.jpg文件。

第⑦行用于将下载时保存数据的本地临时文件移动到沙箱Documents目录下，并命名为test1.jpg文件，这个过程中沙箱Documents目录下不能有test1.jpg名字的文件，否则无法完成移动操作，所以我们要在第⑥行中删除之前的test1.jpg文件。

文件下载并移动成功后，第⑧行用于构建UIImage对象，然后再把UIImage对象赋值给图片视图imageView1，见第⑨行。这样，在界面中我们就可以看到下载的图片了。

上传数据可以采用NSURLSession API中的NSURLSessionUploadTask任务实现。由于上传数据时需要模拟HTML表单上传数据，数据采用multipart/form-data格式，即将数据分割成小段进行上传，具体实现非常复杂。NSURLSessionUploadTask任务没有屏蔽这些复杂性。在上传数据的实现中，我不推荐使用NSURLSessionUploadTask任务，而推荐采用后面要介绍的网络请求框架。

16.3　实例：使用 NSURLSession 重构 MyNotes 案例

在前面几节中，我们介绍的都是调用REST Web Service的查询方法。本节中，我们将介绍MyNotes应用其他的功能实现，包括：插入、修改和删除备忘录信息。

这3个操作调用的Web Service与查询一样，都是http://www.51work6.com/service/mynotes/WebService.php。因为采用的是HTTP请求方法，但我们建议使用POST方法。这是因为GET请求的是静态资源，数据传输过程也不安全，而POST主要请求动态资源。与查询类似，每个方法调用都要传递很多参数，它们之间的关系见表16-1。

表16-1　方法调用与参数关系

调用方法	type参数	action参数	id参数	date参数	content参数	email参数
add	需要	需要	不需要	需要	需要	需要
modify	需要	需要	需要	需要	需要	需要
remove	需要	需要	需要	不需要	不需要	需要

表16-1中各个参数的说明如下。

- **type**。同“查询”调用，是数据交互类型。
- **action**。同“查询”调用，指定调用Web Service的哪些方法。
- **id**。备忘录信息中的主键，隐藏在界面之后。当删除和修改时，需要把它传给Web Service。
- **date**。备忘录信息中的日期字段数据。
- **content**。备忘录信息中的内容字段数据。
- **email**。备忘录信息中的用户邮箱字段，通过它可以查询与当前邮箱关联的用户数据。

16.3.1 插入方法

插入方法主要是在插入视图控制器AddViewController中实现的，具体实现过程与查询业务非常类似。AddViewController中开始请求Web Service以及回调方法的代码如下：

```
//MARK: -- 开始请求Web Service
func startRequest() {

    //准备参数
    let date = Date()
    let dateFormatter: DateFormatter = DateFormatter()
    dateFormatter.dateFormat = "yyyy-MM-dd"
    let dateStr = dateFormatter.string(from: date)

    //设置参数
    let post = NSString(format: "email=%@&type=%@&action=%@&date=
    ➥%@&content=%@",
    ➥"<你的51work6.com用户邮箱>", "JSON", "add", dateStr, self.txtView.text)
    let postData: Data = post.data(using: String.Encoding.utf8.rawValue)!

    let strURL = "http://www.51work6.com/service/mynotes/WebService.php"
    let url = URL(string: strURL)!

    let mutableURLrequest = NSMutableURLRequest(url: url)
    mutableURLrequest.httpMethod = "POST"
    mutableURLrequest.httpBody = postData
    let request =  mutableURLrequest as URLRequest

    let defaultConfig = URLSessionConfiguration.default
    let session = URLSession(configuration: defaultConfig, delegate: nil,
    ➥delegateQueue: OperationQueue.main)

    let task: URLSessionDataTask = session.dataTask(with: request,
    ➥completionHandler: {
        (data, response, error) -> Void in
        NSLog("请求完成...")
        if error == nil {
            do {
                let resDict = try JSONSerialization.jsonObject(with: data!,
                ➥options: JSONSerialization.ReadingOptions.allowFragments)
                ➥as! NSDictionary

                let resultCode: NSNumber = resDict["ResultCode"] as! NSNumber
                var message = "操作成功。"
                if (resultCode.intValue < 0) {
                    message = resultCode.errorMessage
                }

                let alertController: UIAlertController = UIAlertController
                ➥(title: "提示信息",
                ➥message: message, preferredStyle: .alert)
                let okAction = UIAlertAction(title: "OK", style: .default,
                ➥handler: nil)
                alertController.addAction(okAction)
```

```
- (void)startRequest {

    //准备参数
    NSDate *date = [[NSDate alloc] init];
    NSDateFormatter *dateFormatter = [[NSDateFormatter alloc] init];
    [dateFormatter setDateFormat:@"yyyy-MM-dd"];
    NSString *dateStr = [dateFormatter stringFromDate:date];
    //设置参数
    NSString *post = [NSString stringWithFormat:
    ➥@"email=%@&type=%@&action=%@&date=%@&content=%@",
    ➥@"<你的51work6.com用户邮箱>", @"JSON", @"add", dateStr,
    ➥self.txtView.text];

    NSData *postData = [post dataUsingEncoding:NSUTF8StringEncoding];

    NSString *strURL = @"http://www.51work6.com/service/mynotes/
    ➥WebService.php";
    NSURL *url = [NSURL URLWithString:strURL];

    NSMutableURLRequest *request = [NSMutableURLRequest requestWithURL:url];
    [request setHTTPMethod:@"POST"];
    [request setHTTPBody:postData];

    NSURLSessionConfiguration *defaultConfig = [NSURLSessionConfiguration
    ➥defaultSessionConfiguration];
    NSURLSession *session = [NSURLSession sessionWithConfiguration:default
    ➥Config delegate:nil delegateQueue:[NSOperationQueue mainQueue]];

    NSURLSessionDataTask *task = [session dataTaskWithRequest:request
    ➥completionHandler:^(NSData *data, NSURLResponse *response, NSError
    ➥*error) {
        NSLog(@"请求完成...");
        if (!error) {
            NSDictionary *resDict = [NSJSONSerialization
            ➥JSONObjectWithData:data options:
            ➥NSJSONReadingAllowFragments error:nil];

            NSNumber *resultCode = resDict[@"ResultCode"];
            NSString *message = @"操作成功。";
            if ([resultCode integerValue] < 0) {
                message = [resultCode errorMessage];
            }

            UIAlertController *alertController = [UIAlertController
            ➥alertControllerWithTitle:@"提示信息"
            ➥message:message
            ➥preferredStyle:UIAlertControllerStyleAlert];
            UIAlertAction *okAction = [UIAlertAction actionWithTitle:@"OK"
            ➥style:UIAlertActionStyleDefault handler:nil];
```

```
            //显示
            self.present(alertController, animated: true, completion: nil)

        } catch {
            NSLog("返回数据解析失败")
        }

    } else {
        NSLog("error : %@", error!.localizedDescription)
    }
})
task.resume()
}
```

```
        [alertController addAction:okAction];
        //显示
        [self presentViewController:alertController animated:true
➥completion:nil];

    } else {
        NSLog(@"error : %@", error.localizedDescription);
    }
}];
[task resume];
}
```

插入方法的调用关键是请求服务的URL，其中一定要将action设定为add方法的参数，其他的参考表16-1。插入成功后，请返回到主视图界面。

16.3.2 修改方法

修改方法与插入方法非常相似，请求过程是一样的，差别在于调用Web Service的参数，因此我们重点看看它的参数部分。修改方法是在视图控制器DetailViewController中实现的，其中startRequest方法的代码如下：

```
//MARK: -- 开始请求Web Service
func startRequest() {
    //准备参数
    let date = Date()
    let dateFormatter: DateFormatter = DateFormatter()
    dateFormatter.dateFormat = "yyyy-MM-dd"
    let dateStr = dateFormatter.string(from: date)

    let dict = self.detailItem as! NSDictionary
    let id = dict["ID"] as! NSNumber

    //设置参数
    let post = NSString(format:
➥"email=%@&type=%@&action=%@&date=%@&content=%@&id=%@",
➥"<你的51work6.com用户邮箱>", "JSON", "modify", dateStr,
➥self.txtView.text, id)

    let strURL = "http://www.51work6.com/service/mynotes/WebService.php"

    let url = URL(string: strURL)!

    let mutableURLrequest = NSMutableURLRequest(url: url)
    mutableURLrequest.httpMethod = "POST"
    mutableURLrequest.httpBody = postData
    let request =  mutableURLrequest as URLRequest

    <参考插入部分代码>
}
```

```
- (void)startRequest {
    //准备参数
    NSDate *date = [[NSDate alloc] init];
    NSDateFormatter *dateFormatter = [[NSDateFormatter alloc] init];
    [dateFormatter setDateFormat:@"yyyy-MM-dd"];
    NSString *dateStr = [dateFormatter stringFromDate:date];

    NSDictionary *dict = (NSDictionary*)self.detailItem;
    NSNumber* _id = (NSNumber*)dict[@"ID"];

    //设置参数
    NSString *post = [NSString stringWithFormat:
➥@"email=%@&type=%@&action=%@&date=%@&content=%@&id=%@",
➥@"<你的51work6.com用户邮箱>", @"JSON", @"modify",
➥dateStr, self.txtView.text, _id];
    NSData *postData = [post dataUsingEncoding:NSUTF8StringEncoding];

    NSString *strURL =
➥@"http://www.51work6.com/service/mynotes/WebService.php";

    NSURL *url = [NSURL URLWithString:strURL];

    NSMutableURLRequest *request = [NSMutableURLRequest requestWithURL:url];
    [request setHTTPMethod:@"POST"];
    [request setHTTPBody:postData];

    <参考插入部分代码>
}
```

修改方法与插入方法类似，只是请求服务的URL不同，修改方法需要提供表16-1所示的全部参数。其中id参数是从前一个视图控制器传递来的，日期是取的当前系统时间，只有内容是从界面的TextView控件中取出来的。

16.3.3 删除方法

删除方法的调用过程与前面两个方法非常相似，差别在于调用Web Service的参数。但麻烦的是删除方法调用与查询方法调用是在同一个视图控制器MasterViewController中完成的，我们需要做一些判断。下面先看看MasterViewController文件中类声明、属性、事件处理等的相关代码：

```
import UIKit

enum ActionTypes {                                         ①
    case query      //查询操作
    case remove     //删除操作
    case add        //添加操作
    case mod        //修改操作
}

class MasterViewController: UITableViewController {

    //请求动作标识
    var action = ActionTypes.query
    //删除行号
    var deleteRowId = -1

    var detailViewController: DetailViewController? = nil

    //保存数据列表
    var listData = NSArray()

    override func viewWillAppear(animated: Bool) {         ②
        super.viewWillAppear(true)
        action = ActionTypes.query
        self.startRequest()
    }
    ......
}
```

```
//MasterViewController.m文件的代码
#import "MasterViewController.h"
#import "DetailViewController.h"
#import "NSNumber+Message.h"

enum ActionTypes {                                         ①
    QUERY,      //查询操作
    REMOVE,     //删除操作
    ADD,        //添加操作
    MOD         //修改操作
};

@interface MasterViewController () {
    enum ActionTypes action;    //请求动作标识
    NSInteger deleteRowId;      //删除行号
}

@property(strong, nonatomic) DetailViewController *detailViewController;
//保存数据列表
@property(nonatomic, strong) NSMutableArray *listData;

//重新加载表视图
- (void)reloadView:(NSDictionary *)res;

//开始请求Web Service
- (void)startRequest;

@end

@implementation MasterViewController

- (void)viewWillAppear:(BOOL)animated {                   ②
    [super viewWillAppear:YES];
    action = QUERY;
    [self startRequest];
}
......

end
```

上述代码中，第①行用于声明枚举类型ActionTypes，其中定义了4个成员，这是为了提高程序可读性。第②行是viewWillAppear:方法，之所以将请求Web Service的startRequest方法放到该方法中，是为了保证每次该视图出现时，都会请求服务器返回数据。请求方法之前添加了action = ActionTypes.query（Objective-C版本是action = QUERY）语句，表示当前调用操作是查询。

删除方法是在表视图数据源的tableView:commitEditingStyle:forRowAtIndexPath:方法中实现的，其代码如下：

```
override func tableView(_ tableView: UITableView,
➥commit editingStyle: UITableViewCellEditingStyle,
➥forRowAt indexPath: IndexPath) {

    if editingStyle == .delete {
        //删除数据
        action = ActionTypes.remove                        ①
        deleteRowId = indexPath.row
        self.startRequest()
    }
}
```

```
- (void)tableView:(UITableView *)tableView
➥commitEditingStyle:(UITableViewCellEditingStyle)editingStyle
➥forRowAtIndexPath:(NSIndexPath *)indexPath {

    if (editingStyle == UITableViewCellEditingStyleDelete) {
        //删除数据
        action = REMOVE;                                   ①
        deleteRowId = indexPath.row;
        [self startRequest];
    }
}
```

第①行表示当前调用的是删除操作。请求Web Service的startRequest方法也需要做一些修改，该方法主要用于判断请求动作标识，其代码如下：

```
func startRequest() {

    let strURL = "http://www.51work6.com/service/mynotes/WebService.php"
    var post = ""
    if action == ActionTypes.query {                                    ①
        //查询处理
        post = String(format: "email=%@&type=%@&action=%@",
        ➥"<你的51work6.com用户邮箱>", "JSON", "query")
    } else if action == ActionTypes.remove{                             ②
        //删除处理
        let dict = self.listData[deleteRowId] as! NSDictionary
        let id = dict.objectForKey("ID") as! NSNumber
        post = String(format: "email=%@&type=%@&action=%@&id=%@",
        ➥"<你的51work6.com用户邮箱>", "JSON", "remove", id)
    }
    let postData: Data = post.data(using: String.Encoding.utf8)!

    let url = URL(string: strURL)!

    let mutableURLrequest = NSMutableURLRequest(url: url)
    mutableURLrequest.httpMethod = "POST"
    mutableURLrequest.httpBody = postData
    let request =  mutableURLrequest as URLRequest


    let defaultConfig = URLSessionConfiguration.default
    let session = URLSession(configuration: defaultConfig, delegate: nil,
    ➥delegateQueue: OperationQueue.main)

    let task: URLSessionDataTask = session.dataTask(with: request,
    ➥completionHandler: {  (data, response, error) -> Void in
        NSLog("请求完成...")
        if error == nil {
            do {
                let resDict = try JSONSerialization.jsonObject(with: data!,
                ➥options: JSONSerialization.ReadingOptions.allowFragments)
                ➥as! NSDictionary

                if self.action == ActionTypes.query {                   ③
                    //查询处理
                    self.reloadView(resDict)
                } else if self.action == ActionTypes.remove {           ④
                    //删除处理
                    <参考插入部分代码>
                }

            } catch {
                NSLog("返回数据解析失败")
            }
        } else {
            NSLog("error : %@", error!.localizedDescription)
        }
    })
    task.resume()
}
```

```
- (void)startRequest {

    NSString *strURL =
    ➥@"http://www.51work6.com/service/mynotes/WebService.php";

    NSURL *url = [NSURL URLWithString:strURL];

    NSString *post;
    if (action == QUERY) {//查询处理                                      ①
        post = [NSString stringWithFormat:@"email=%@&type=%@&action=%@",
        ➥@"<你的51work6.com用户邮箱>", @"JSON", @"query"];
    } else if (action == REMOVE) {//删除处理                              ②
        NSMutableDictionary *dict = self.listData[deleteRowId];
        post = [NSString stringWithFormat:@"email=%@&type=%@&action=
        ➥%@&id=%@",@"<你的51work6.com用户邮箱>", @"JSON", @"remove",
        ➥dict[@"ID"]];
    }

    NSData *postData = [post dataUsingEncoding:NSUTF8StringEncoding];

    NSMutableURLRequest *request = [NSMutableURLRequest requestWithURL:url];
    [request setHTTPMethod:@"POST"];
    [request setHTTPBody:postData];

    NSURLSessionConfiguration *defaultConfig =
    ➥[NSURLSessionConfiguration defaultSessionConfiguration];
    NSURLSession *session = [NSURLSession sessionWithConfiguration:
    ➥defaultConfig delegate:nil delegateQueue:[NSOperationQueue mainQueue]];

    NSURLSessionDataTask *task = [session dataTaskWithRequest:request
    ➥completionHandler: ^(NSData *data,
    ➥NSURLResponse *response, NSError *error) {
        NSLog(@"请求完成...");
        if (!error) {
            NSDictionary *resDict = [NSJSONSerialization
            ➥JSONObjectWithData:data
            ➥options:NSJSONReadingAllowFragments error:nil];

            if (action == QUERY) {//查询处理                               ③
                [self reloadView:resDict];
            } else if (action == REMOVE) {//删除处理                       ④
                <参考插入部分代码>
            }
        } else {
            NSLog(@"error : %@", error.localizedDescription);
        }
    }];

    [task resume];
}
```

上述代码实现了查询和删除两个功能，但是具体执行哪个动作要根据action判断。其中第①行用于判断查询请求处理，第②行用于判断删除请求处理。

但是请求处理返回之后，我们还需要通过第③行判断是否为查询请求返回，用第④行判断是否为删除请求返回。

16.4 使用 AFNetworking 框架

苹果提供的NSURLSession网络请求API比之前的NSURLConnectionn网络请求API有很大进步。但是，在易用性方面，NSURLSession仍然不是很理想。基于易用性和安全性等方面的考虑，我们可以选用第三方网络请求框架，目前比较流行的主要有3种：ASIHTTPRequest、AFNetworking和MKNetworkKit。

16.4.1 比较 ASIHTTPRequest、AFNetworking 和 MKNetworkKit

在这3个框架中，ASIHTTPRequest是最早设计的框架，它功能强大但不支持ARC，现在原作者已经停止更新了。AFNetworking在ASIHTTPRequest之后出现，相较而言更加简单。MKNetworkKit的设计思想来源于ASIHTTPRequest和AFNetworking，结合了这两个框架的共同特点，并且增加了一些新特性。相比前两个框架，MKNetworkKit是最为轻量级的框架，但是对于HTTPS等协议的支持不好。

我们通过表16-2比较一下这3个框架的优缺点。

表16-2 比较ASIHTTPRequest、AFNetworking和MKNetworkKit

	ASIHTTPRequest	AFNetworking	MKNetworkKit
支持iOS和macOS	是	是	是
支持ARC	否	是	是
断点续传	是	是	是
同步异步请求	支持同步/异步	只支持异步	只支持异步
图片缓存到内存	否	是	是
后台下载	是	是	是
下载进度	是	是	是
上传进度	否	是	否
缓存离线请求	否	是	是
cookie	是	否	否
HTTPS	是	是	是（需要插件）

从表16-2可见AFNetworking框架的优势。AFNetworking框架是Alamofire基金会（Alamofire Software Foundation，http://alamofire.org/）支持的项目，因此能够获得比较稳定的技术支持，以及后续的升级和维护，这也是很多项目选择它的一个原因。另外，AFNetworking 3在底层采用了NSURLSession，能够发挥NSURLSession的优势，能够很好地与NSURLSession结合使用，且弥补NSURLSession的不足之处。

16.4.2 安装和配置 AFNetworking 框架

AFNetworking框架（http://afnetworking.com）是CocoaPods的主要追随者，其安装手册只提供了CocoaPods安装方式，但是为帮助大家了解手工配置的原理，本章还是会介绍如何手工配置AFNetworking框架。而安装CocoaPods和配置AFNetworking框架的过程我们会在第27章中详细介绍。

首先，从https://github.com/AFNetworking/AFNetworking下载AFNetworking框架代码，然后打开AFNetworking目录。主要的目录结构如下：

```
├──AFNetworking
├──AFNetworking.xcodeproj
├──AFNetworking.xcworkspace
├──Example
├──Framework
├──Tests
└──UIKit+AFNetworking
```

其中AFNetworking和UIKit+AFNetworking目录是框架源代码，AFNetworking.xcodeproj是框架的工程文件，而工作空间AFNetworking.xcworkspace可以包括：框架的工程和示例工程内容。

我们可以通过两种方式将AFNetworking添加到自己工程中：添加源代码方式和添加AFNetworking.xcodeproj方式。

1. 添加源代码方式

我们可以将AFNetworking压缩包中的AFNetworking和UIKit+AFNetworking目录添加到新工程中（作为Xcode的“组”），添加之后如图16-4所示。

那么，在程序中使用时需要采用如下方式引入头文件：

```
#import "AFNetworking.h"
```

2.添加AFNetworking.xcodeproj方式

添加AFNetworking.xcodeproj方式是将AFNetworking压缩包中的AFNetworking.xcodeproj工程文件添加到新工程中。AFNetworking.xcodeproj工程事实上编译的目标是AFNetworking.framework框架文件。

添加AFNetworking.xcodeproj的过程是选择新工程，如图16-5所示，在右键快捷菜单中选择Add Files to “MyNotes”...菜单项，然后在弹出的对话框中选择AFNetworking.xcodeproj文件，如图16-6所示。

图16-4　添加AFNetworking和UIKit+AFNetworking目录到工程

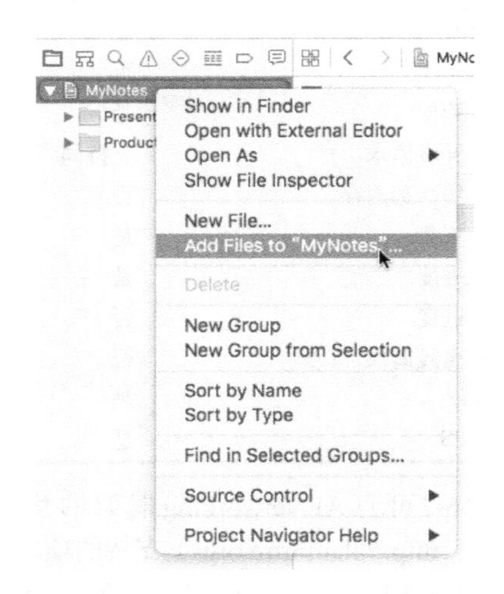

图16-5　添加AFNetworking.xcodeproj到新工程快捷菜单

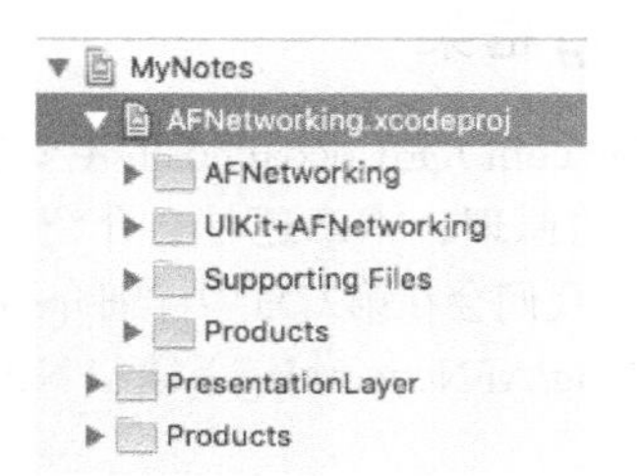

图16-6　添加AFNetworking.xcodeproj到新工程

由于AFNetworking.xcodeproj是一个框架工程，编译的结果是AFNetworking.framework框架文件，我们还需要在工程中添加AFNetworking.framework。打开MyNotes工程，选择TARGETS→MyNotes→Build Phases→Link Binary With Libraries（3 items），如图16-7所示。然后点击左下角的+按钮，从弹出界面中选择AFNetworking.framework from 'AFNetworking iOS'，再点击Add按钮，这样依赖关系就添加好了。

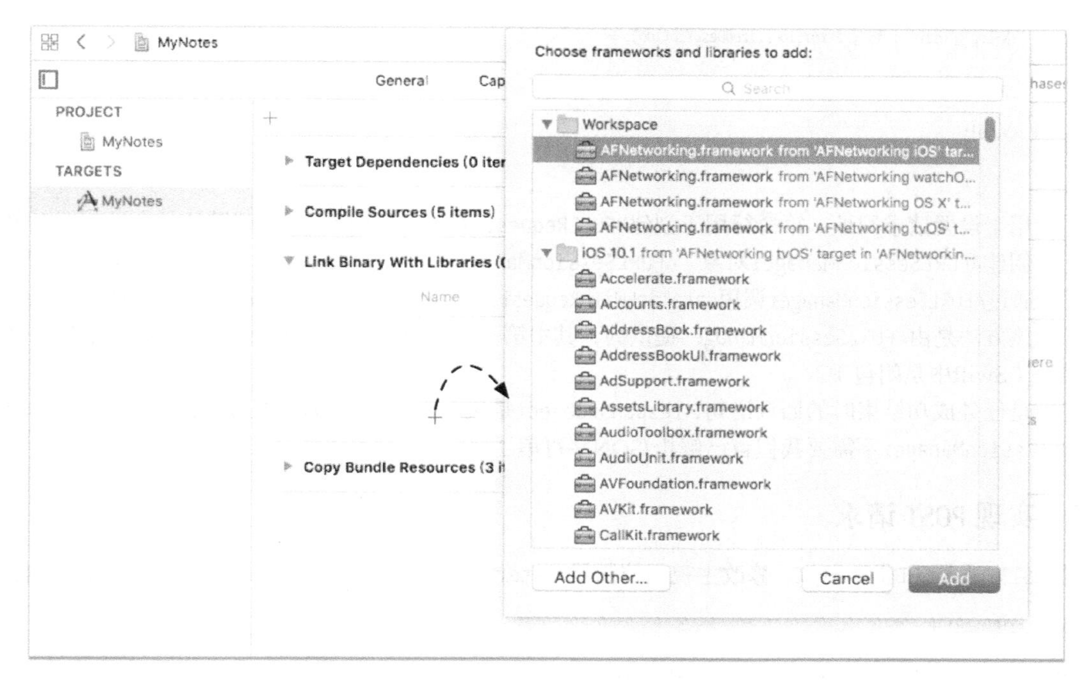

图16-7 添加AFNetworking.framework框架到新工程

那么，在程序中使用AFNetworking.framework时，需要采用如下方式引入头文件：

```
#import <AFNetworking/AFNetworking.h>
```

16.4.3 实现 GET 请求

配置好环境之后，我们就可以使用AFNetworking开发网络应用了，这里先介绍下GET请求的实现。

下面还是以MyNotes应用为例来介绍，不过只考虑查询功能的实现。修改主视图控制器MasterViewController中的startRequest方法，具体如下：

```
- (void)startRequest {

    NSString *strURL = [[NSString alloc]
    ➥initWithFormat:@"http://www.51work6.com/service/mynotes/
    ➥WebService.php?email=%@&type=%@&action=%@",
    ➥@"<你的51work6.com用户邮箱>", @"JSON", @"query"];                    ①

    strURL = [strURL stringByAddingPercentEncodingWithAllowedCharacters:
    ➥[NSCharacterSet URLQueryAllowedCharacterSet]];

    NSURL *url = [NSURL URLWithString:strURL];
    NSURLRequest *request = [[NSURLRequest alloc] initWithURL:url];          ②

    NSURLSessionConfiguration *defaultConfig =
    ➥[NSURLSessionConfiguration defaultSessionConfiguration];                ③
    AFURLSessionManager *manager = [[AFURLSessionManager alloc]
    ➥initWithSessionConfiguration:defaultConfig];                            ④

    NSURLSessionDataTask *task = [manager dataTaskWithRequest:request
    ➥completionHandler:^(NSURLResponse *response, id responseObject, NSError *error) {   ⑤
        NSLog(@"请求完成...");
        if (!error) {
            [self reloadView:responseObject];                                 ⑥
        } else {
```

```
            NSLog(@"error : %@", error.localizedDescription);
        }
    }];

    [task resume];

}
```

第①行用于设置请求网址，第②行用于创建NSURLRequest对象，第③行用于创建默认的会话配置对象。

第④行创建AFURLSessionManager对象。AFURLSessionManager是由AFNetworking框架提供的，负责管理会话。

第⑤行通过AFURLSessionManager调用dataTaskWithRequest:completionHandler:方法，创建NSURLSessionDataTask数据任务。该方法是由AFURLSessionManager提供的，其中第一个参数是NSURLRequest对象，第二个参数是任务回调的代码块（Swift中是闭包）。

第⑥行是任务成功结束时的回调语句，responseObject是从服务器返回的JSON对象，它可以是字典或数组类型，AFURLSessionManager不需要我们自己解析JSON字符串了。

16.4.4　实现 POST 请求

发送POST方法与GET非常相似。修改主视图控制器MasterViewController中的startRequest方法，具体如下：

```
- (void)startRequest {

    NSString *strURL = @"http://www.51work6.com/service/mynotes/WebService.php";
    NSURL *url = [NSURL URLWithString:strURL];

    //设置参数
    NSString *post = [NSString stringWithFormat:@"email=%@&type=%@&action=%@",
    ➥@"<你的51work6.com用户邮箱>", @"JSON", @"query"];
    NSData *postData = [post dataUsingEncoding:NSUTF8StringEncoding];

    NSMutableURLRequest *request = [NSMutableURLRequest requestWithURL:url];        ①
    [request setHTTPMethod:@"POST"];
    [request setHTTPBody:postData];

    NSURLSessionConfiguration *defaultConfig =
    ➥[NSURLSessionConfiguration defaultSessionConfiguration];
    AFURLSessionManager *manager = [[AFURLSessionManager alloc]
    ➥initWithSessionConfiguration:defaultConfig];

    NSURLSessionDataTask *task = [manager dataTaskWithRequest:request
    ➥completionHandler:^(NSURLResponse *response, id responseObject, NSError *error) {
        NSLog(@"请求完成...");
        if (!error) {
            [self reloadView:responseObject];
        } else {
            NSLog(@"error : %@", error.localizedDescription);
        }
    }];

    [task resume];

}
```

POST请求是将请求对象换成NSMutableURLRequest，它是可变的请求对象，具体细节读者可以参考16.2.4节，这里不再赘述。

16.4.5　下载数据

AFNetworking 3基于NSURLSession API，我们在16.2.5节介绍了如何使用NSURLSession下载数据。使用NSURLSession API下载数据的功能已经非常健壮了，再加上AFNetworking对其进行了封装，使用AFNetworking编写下载应用程序非常简单。

下面我们重构16.2.5节的DownloadSample下载案例，其中UI部分不再介绍了，直接看看主视图控制器。先是ViewController中的主要代码：

```
#import "AFNetworking.h"

@interface ViewController ()

@property(weak, nonatomic) IBOutlet UIProgressView *progressView;
@property(weak, nonatomic) IBOutlet UIImageView *imageView1;

@end

@implementation ViewController

- (IBAction)onClick:(id)sender {

    NSString *strURL = [[NSString alloc] initWithFormat:@"http://www.51work6.com/
    ➥service/download.php?email=%@&FileName=test1.jpg",
    ➥@"<你的51work6.com用户邮箱>"];

    NSURL *url = [NSURL URLWithString:strURL];
    NSURLRequest *request = [NSURLRequest requestWithURL:url];

    NSURLSessionConfiguration *defaultConfig = [NSURLSessionConfiguration
    ➥defaultSessionConfiguration];
    AFURLSessionManager *manager = [[AFURLSessionManager alloc]
    ➥initWithSessionConfiguration:defaultConfig];

    NSURLSessionDownloadTask *downloadTask = [manager
    ➥downloadTaskWithRequest:request
    ➥progress:^(NSProgress *downloadProgress) {                              ①

        NSLog(@"%@", [downloadProgress localizedDescription]);               ②
        dispatch_async(dispatch_get_main_queue(), ^{
            [self.progressView
            ➥setProgress:downloadProgress.fractionCompleted animated:TRUE];  ③
        });

    } destination:^NSURL *(NSURL *targetPath, NSURLResponse *response) {      ④

        NSString *downloadsDir = [NSSearchPathForDirectoriesInDomains(
        ➥NSDocumentDirectory, NSUserDomainMask, TRUE) objectAtIndex:0];
        NSString *downloadStrPath = [downloadsDir
        ➥stringByAppendingPathComponent:[response suggestedFilename]];       ⑤
        NSURL *downloadURLPath = [NSURL fileURLWithPath:downloadStrPath];

        return downloadURLPath;

    } completionHandler:^(NSURLResponse *response, NSURL *filePath, NSError *error) {   ⑥

        NSLog(@"File downloaded to: %@", filePath);
        NSData *imgData = [[NSData alloc] initWithContentsOfURL:filePath];
        UIImage *img = [UIImage imageWithData:imgData];
        self.imageView1.image = img;

    }];
    [downloadTask resume];
}
......

@end
```

上述代码中，第①行是通过NSURLSessionManager对象的如下方法获得下载会话任务对象：

```
- downloadTaskWithRequest:(NSURLRequest *)request
    progress:(void (^)(NSProgress *downloadProgress)) downloadProgressBlock
    destination:(NSURL * (^)(NSURL *targetPath, NSURLResponse *response))destination
    completionHandler:(void (^)(NSURLResponse *response,
    ➥NSURL *filePath, NSError *error))completionHandler
```

在该方法中，参数request是请求对象；progress是参数代码块，用来获得当前运行进度，代码块的参数downloadProgress是NSProgress类型（NSProgress是Foundation框架提供的，用来描述某个任务的进展情况）；destination参数是下载文件的保存路径；response是应答对象；completionHandler参数也是代码库，是从服务器返回应答数据时回调的。

在下载进度代码块中，第②行中的[downloadProgress localizedDescription]语句可以获得下载进度对象downloadProgress的本地化信息。第③行用于更新进度条的进度，通过NSProgress的fractionCompleted属性获得下载进度的分数表示方式。

第④行用于保存下载文件的代码块。第⑤行用于获得应用沙箱目录下推荐的文件名，其中[response suggestedFilename]是获得推荐的文件名，这个文件名就是服务器端存储的文件名。

第⑥行是从服务器返回应答数据时回调的代码块，这里设置图片视图，用来显示下载的图片。

16.4.6　上传数据

AFNetworking框架也可以实现上传功能，下面通过一个实例介绍一下如何实现上传。我们设计了如图16-8所示的应用，当用户点击GO按钮时，应用从本地上传图片，然后再下载它并将其显示在界面中。

图16-8　上传应用的设计原型图

本案例的UI部分不再介绍了，下面直接看看主视图控制器。先来看看ViewController中的主要代码：

```
#import "ViewController.h"
#import "AFNetworking.h"

@interface ViewController ()

@property(weak, nonatomic) IBOutlet UIProgressView *progressView;
@property(weak, nonatomic) IBOutlet UIImageView *imageView1;
@property(weak, nonatomic) IBOutlet UILabel *label;

@end

@implementation ViewController
```

```
......

- (IBAction)onClick:(id)sender {

    NSString *uploadStrURL = @"http://www.51work6.com/service/upload.php";
    NSDictionary *params = @{@"email" : @"<你的51work6.com用户邮箱>"};

    NSString *filePath = [[NSBundle mainBundle] pathForResource:@"test2" ofType:@"jpg"];

    NSMutableURLRequest *request = [[AFHTTPRequestSerializer serializer]            ①
    ➥multipartFormRequestWithMethod:@"POST"URLString:uploadStrURL parameters:params
    ➥constructingBodyWithBlock:^(id <AFMultipartFormData> formData) {
        [formData appendPartWithFileURL:[NSURL
        ➥fileURLWithPath:filePath] name:@"file"
        ➥fileName:@"1.jpg" mimeType:@"image/jpeg" error:nil];                        ②
    } error:nil];                                                                      ③

    AFURLSessionManager *manager = [[AFURLSessionManager alloc]
    ➥initWithSessionConfiguration:
    ➥[NSURLSessionConfiguration defaultSessionConfiguration]];
    NSURLSessionUploadTask *uploadTask;
    uploadTask = [manager uploadTaskWithStreamedRequest:request                      ④
        progress:^(NSProgress *uploadProgress) {                                      ⑤
            NSLog(@"上传: %@", [uploadProgress localizedDescription]);
            dispatch_async(dispatch_get_main_queue(), ^{
                [self.progressView
                ➥setProgress:uploadProgress.fractionCompleted];
            });

        }
        completionHandler:^(NSURLResponse *response,
        ➥id esponseObject, NSError *error) {                                          ⑥
            if (!error) {
                NSLog(@"上传成功");
                [self download];                                                       ⑦
            } else {
                NSLog(@"上传失败: %@", error.localizedDescription);
            }
        }];

    [uploadTask resume];

    self.label.text = @"上传进度";
    self.progressView.progress = 0.0;
}

......

@end
```

上述代码中，第①行~第③行用于创建NSMutableURLRequest对象，这个对象是通过AFHTTPRequestSerializer对象的如下方法实现的，其中AFHTTPRequestSerializer是HTTP请求序列化对象，封装了HTTP请求参数（放在URL问号之后的部分）和表单数据：

```
- multipartFormRequestWithMethod:(NSString *)method
    URLString:(NSString *)URLString
    parameters:(NSDictionary *)parameters
    constructingBodyWithBlock:(void (^)(id <AFMultipartFormData> formData))block
    error:(NSError *__autoreleasing *)error
```

该方法可以发送multipart/form-data格式的表单数据；method参数的请求方法一般是POST和PUT；URLString参数是上传时的服务器地址；parameters是请求参数，它是字典结构；constructingBodyWithBlock是请求体代码块；error参数是错误对象。

第②行是代码块constructingBodyWithBlock中的代码，其中只有一条语句：

```
[formData appendPartWithFileURL:[NSURL fileURLWithPath:filePath]
➥name:@"file" fileName:@"1.jpg" mimeType:@"image/jpeg" error:nil];
```

该语句调用AFMultipartFormData的appendPartWithFileURL:name:error:方法，添加multipart/form-data格式的表单数据。这种格式的数据被分割成小段进行上传，该方法的第一个参数是要上传的文件路径；第二个参数name是与数据相关的名称，一般是file，相当于HTML中表单内的选择文件控件<input type="file" >类型；第三个参数fileName是文件名，是放在请求头中的文件名，不能为空，可以与本地文件名不同；第四个参数mimeType是数据相关的MIME类型；最后一个参数error是错误对象。

上述代码中，第④行~第⑥行用于创建NSURLSessionUploadTask上传会话任务，其中使用了AFURLSessionManager的如下方法实现：

```
- uploadTaskWithStreamedRequest:(NSURLRequest *)request
    progress:(void (^)(NSProgress *uploadProgress)) uploadProgressBlock
    completionHandler:(void (^)(NSURLResponse *response,
    ➥id responseObject, NSError *error))completionHandler
```

该方法中的request参数是请求对象；progress参数代码块用来获得当前运行的进度，代码块的参数uploadProgress也是NSProgress类型；completionHandler参数也是代码库，是从服务器返回应答数据时回调的。

第⑤行用于执行progress代码块。第⑥行用于执行completionHandler代码块，如果error为nil，则上传成功。

上传成功后，接着执行第⑦行的[self download]语句，download是我们自定义的方法，用来下载刚刚上传的图片，这主要是出于测试目的，看看是否成功上传了图片。有关download方法的内容，大家可以参考16.4.5节，这里不再赘述。

16.5　使用为 Swift 设计的网络框架：Alamofire

AFNetworking框架可以为Objective-C语言开发网络应用提供服务，也可以为Swift语言开发网络应用提供服务。然而，因为AFNetworking并非专为Swift语言而设计，所以使用时需要引入桥接头文件，并进行一些配置工作。Alamofire基金会推出了专为Swift语言设计的网络请求框架——Alamofire，该框架能够充分发挥Swift语法的特点。本节中，我们就来介绍Alamofire框架。

16.5.1　安装和配置 Alamofire 框架

Alamofire框架也可以使用CocoaPods工具进行安装，但是本章采用手工配置。

首先，从https://github.com/Alamofire/Alamofire下载Alamofire框架代码，这里我们选用Alamofire 4版本。下载完成后，打开Alamofire目录，可以看到主要的目录结构如下：

```
.
├──Alamofire.xcodeproj
├──iOSExample.xcodeproj
├──Alamofire.xcworkspace
├──Documentation
├──Example
├──Source
└──Tests
```

其中Source目录是框架源代码，Alamofire.xcodeproj是框架的工程文件，iOSExample.xcodeproj 是示例工程，而工作空间Alamofire.xcworkspace 可以包括：框架的工程和示例工程内容。

我们可以将Alamofire.xcodeproj添加到自己的工程中，具体步骤请参考16.4.2节。

Alamofire.xcodeproj的编译结果也是Alamofire.framework框架文件，需要添加到工程中，具体过程参考16.4.2节，添加结果如图16-9所示。

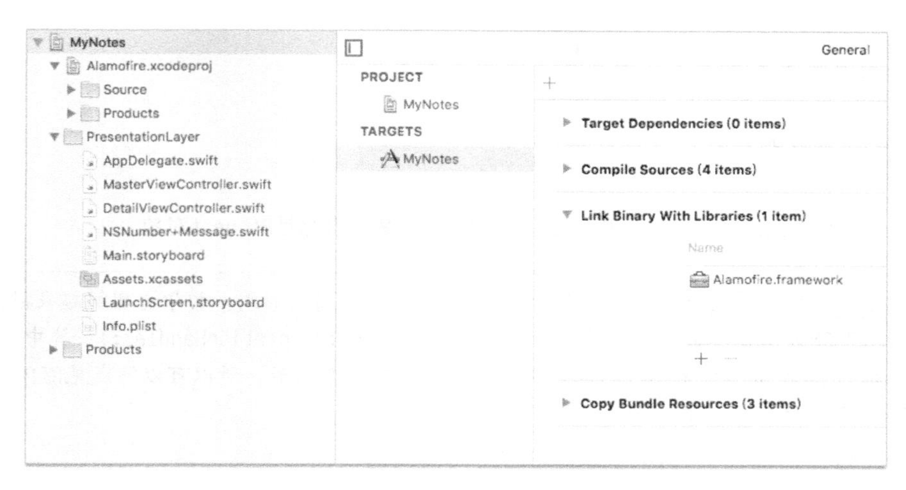

图16-9 添加Alamofire.framework框架

使用Alamofire.framework时，我们在Swift文件中添加如下语句来引入Alamofire模块：

```
import Alamofire
```

16.5.2 实现 GET 请求

配置好环境之后，就可以使用Alamofire开发网络应用了，我们先介绍GET请求。

下面还是以MyNotes应用为例来进行介绍，这里只考虑查询功能的实现。修改主视图控制器MasterView-Controller中的startRequest方法，具体如下：

```
//MARK: -- 开始请求Web Service
func startRequest() {

    let strURL = "http://www.51work6.com/service/mynotes/WebService.php"
    let params = ["email": "<你的51work6.com用户邮箱>",
    ➥"type": "JSON", "action": "query"]                              ①

    Alamofire.request(strURL, method: .get, parameters: params)       ②
        .responseJSON { response in                                   ③
            self.reloadView(response.result.value as! NSDictionary)   ④
        }

}
```

将使用Alamofire进行GET网络请求的代码与16.2.2节和16.4.3节实现的GET方法进行比较，应该不难发现Alamofire框架实现的代码非常简洁。第①行用于设置请求参数，类型是字典。

提示 第②行的Alamofire.request()语句并不是调用Alamofire类的静态方法request，Alamofire不是一个类，而是一个命名空间。在Swift语言中，虽然没有类似于C++声明命名空间的关键字，但是命名空间的概念还是有的，模块名就是命名空间，所以Alamofire.framework的命名空间是Alamofire。request不是静态方法，而是函数。函数与方法的区别是：函数不属于任何类型，而方法一定是在某个具体的类型中声明的。Alamofire.request()语句的解释是：调用Alamofire命名空间（Alamofire.framework）中的request()函数，其返回值是Request对象。

第②行~第④行执行GET请求，第②行返回Request对象，第③行调用Request对象的responseJSON方法，而responseJSON方法返回的还是Request对象，这样就会形成一个调用链。示例代码如下：

```
Alamofire.request(strURL, method: .get, parameters: params)
    .responseString { response in

    }
    .responseJSON { response in

    }
```

其中Alamofire.request、responseString和responseJSON等的返回值都是Request对象。

提示

代码中responseString和responseJSON方法都采用了尾随闭包形式，初学者理解起来很困难。例如，responseJSON方法带参数的形式是responseJSON(options:completionHandler:)，其中options可以省略，completionHandler是参数闭包。由于闭包为最后一个参数，所以可以写成尾随闭包形式。代码中都采用了尾随闭包形式，如下：

```
responseJSON(options:) { //没有省略options参数
}
```

第④行中的response为应答对象，response.result.value用于获得从服务器返回的数据。由于是通过responseJSON方法调用的，这个数据被封装到字典或数组中。那么，如果是responseString方法的调用，则返回String类型的数据。

16.5.3　实现 POST 请求

发送POST方法与GET非常相似。修改主视图控制器MasterViewController中的startRequest方法，具体如下：

```
//MARK: -- 开始请求Web Service
func startRequest() {

    let strURL = "http://www.51work6.com/service/mynotes/WebService.php"
    let params = ["email": "",
    ➥"type": "JSON", "action": "query"]

    Alamofire.request(strURL, method: .post, parameters: params)         ①
        .responseJSON {
            response in
            self.reloadView(response.result.value as! NSDictionary)
        }

}
```

POST请求是将request函数中的.get换为.post。

16.5.4　下载数据

Alamofire 4也基于NSURLSession API，因此使用Alamofire框架编写实现下载功能的应用更加健壮和简单。

下面我们重构16.2.5节的DownloadSample下载案例，其中UI部分不再介绍，直接看看主视图控制器。先来看看ViewController中的主要代码：

```
@IBAction func onClick(_ sender: AnyObject) {

    let strURL = String(format: "http://www.51work6.com/service/download.php?
    ➥email=%@&FileName=test1.jpg", "<你的51work6.com用户邮箱>")

    let destination = DownloadRequest.suggestedDownloadDestination()         ①

    Alamofire.download(strURL, method: .post, to: destination)
        .downloadProgress { progress in
            print("下载进度: \(progress.fractionCompleted)")                    ②
```

```
            self.progressView.setProgress(Float(progress.fractionCompleted),
            ➥animated: true)                                              ③
        }
        .responseJSON { response in
            print("下载成功，保存文件：\(response.destinationURL!)")          ④
            let data = NSData(contentsOf: response.destinationURL!)        ⑤
            let img = UIImage(data: data as! Data)                         ⑥
            self.imageView1.image = img
    }
}
```

上述代码中，第①行中的DownloadRequest是Alamofire 提供的下载请求类。suggestedDownloadDestination()方法返回Alamofire推荐的下载目标文件的闭包。

第②行中的progress方法能够获取下载进度，progress.fractionCompleted用于获得下载进度的小数表示方式。第③行用于更新进度视图progressView。

第④行中的response.destinationURL表达式用于返回下载目标的文件路径，第⑤行通过文件路径创建NSData缓存对象。第⑥行通过缓存对象创建UIImage对象。

16.5.5 上传数据

Alamofire 4框架也可以实现上传功能。下面我们采用Alamofire 4框架重构16.4.6节的UploadSample上传案例，其中UI部分不再介绍了，直接看看主视图控制器。先来看看ViewController中的主要代码：

```
@IBAction func onClick(_ sender: AnyObject) {

    self.label.text = "上传进度"
    self.progressView.progress = 0.0

    let uploadStrURL = String(format: "http://www.51work6.com/service/upload.php?
    ➥email=%@", "<你的51work6.com用户邮箱>")
    let fileURL =  Bundle.main.url(forResource: "test2", withExtension: "jpg")!

    Alamofire.upload(                                                    ①
        multipartFormData: { multipartFormData in                        ②
            multipartFormData.append(fileURL, withName: "file",
            ➥fileName: "1.jpg", mimeType: "image/jpeg")                  ③
        },
        usingThreshold: UInt64(1024),                                    ④
        to: uploadStrURL,
        encodingCompletion: { result in                                  ⑤
            switch result {
            case .success(let upload, _, _):                             ⑥
                upload.uploadProgress { progress in
                    print("上传进度 = \(progress.fractionCompleted)")      ⑦
                    self.progressView.setProgress(Float(progress.fractionCompleted),
                    ➥animated: true)
                }.response { resp in                                     ⑧
                    print("上传成功")
                    self.download()
                }
            case .failure(let encodingError):                            ⑨
                print("上传失败：\(encodingError)")
            }
        }
    )

}
```

上述代码中，第①行使用Alamofire.upload函数上传数据。第②行是提供给封装multipart/form-data格式的表单数据。第③行通过append方法向表单数据中添加数据。第④行设置内存占用阈值。第⑤行是表单数据编码完成之后调用的闭包，第⑥行表示成功返回数据的情况，第⑦行是打印上传进度，第⑧行是成功结束回调的闭包，此时self.download()方法将刚刚上传的图片下载到本地，来验证下载是否成功。第⑨行用于设置失败结束回调的闭包。

提示　方法中提到的"编码"，是将表单中的multipart/form-data格式数据序列化（编码）为可以进行网络传输的数据。

16.6　反馈网络信息改善用户体验

在使用有网络服务的应用时，用户希望看到应用运行的进度、网络状态等反馈信息。作为网络应用的开发者，我们必须注意这些问题，以便给用户提供良好的用户体验。

16.6.1　使用下拉刷新控件改善用户体验

首先，看看MyNotes应用中的一个缺陷。在显示备忘录数据的表视图界面中，数据请求放在viewWillAppear:方法中，相关代码可以查看16.3节。viewWillAppear:方法的代码如下：

```
override func viewWillAppear(animated: Bool) {
    super.viewWillAppear(true)
    action = ActionTypes.query
    self.startRequest()                                    ①
}
```

```
-(void)viewWillAppear:(BOOL)animated {
    [super viewWillAppear: TRUE];
    action = ACTION_QUERY;
    [self startRequest];                                   ①
}
```

在上述代码中，第①行用于网络请求。这条语句放在viewWillAppear:方法中的最大问题是：每次显示这个界面时，都会发起网络请求。这样设计的目的是修改、删除和插入完成后，我们想重新刷新一下界面看看更新的数据。但是这种设计有副作用，即进入详细信息界面后，从图16-10②号界面的"备忘录"按钮再回到①号界面时，也会发起网络请求，如果数据量很大，界面就会比较"卡"。

有没有一种方法让用户在需要的时候自己刷新呢？在主界面视图（图16-10所示的①号界面）的导航栏中，我们已经设置两个按钮（Edit和+按钮），不能再放置按钮了。现在有一种交互方式，即向下拉动表视图可以触发刷新动作，而iOS 6之后的版本提供了这种刷新控件。

图16-10　查看MyNotes应用的详细信息

图16-11所示为iOS 6中的下拉刷新，有点儿像是在拉"胶皮糖"，当"胶皮糖"拉断后，就会出现活动指示器，

而图16-12所示为iOS 7及其之后版本的下拉刷新。比较这两张图可以发现，iOS 7及其之后版本的下拉刷新很简单，动画效果也是扁平化设计了。

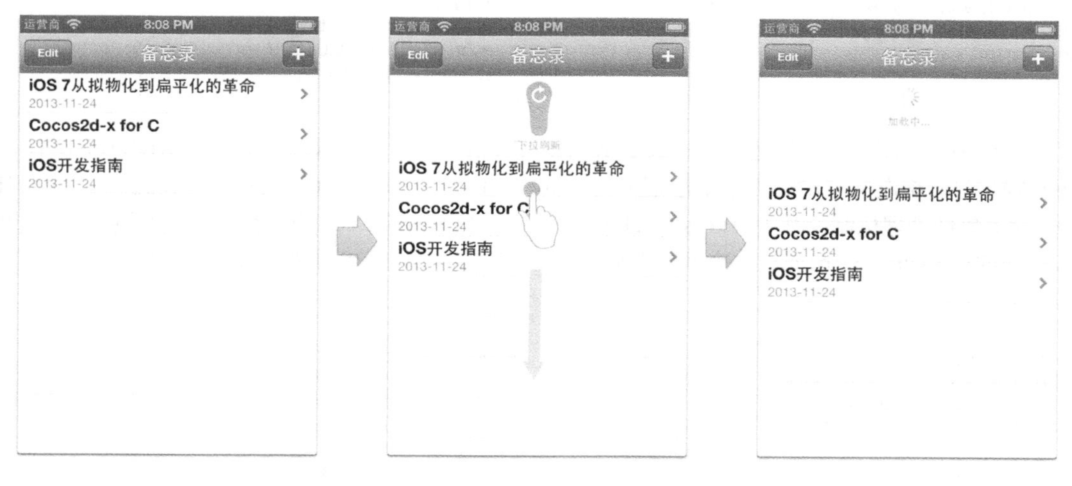

图16-11 iOS 6中的下拉刷新

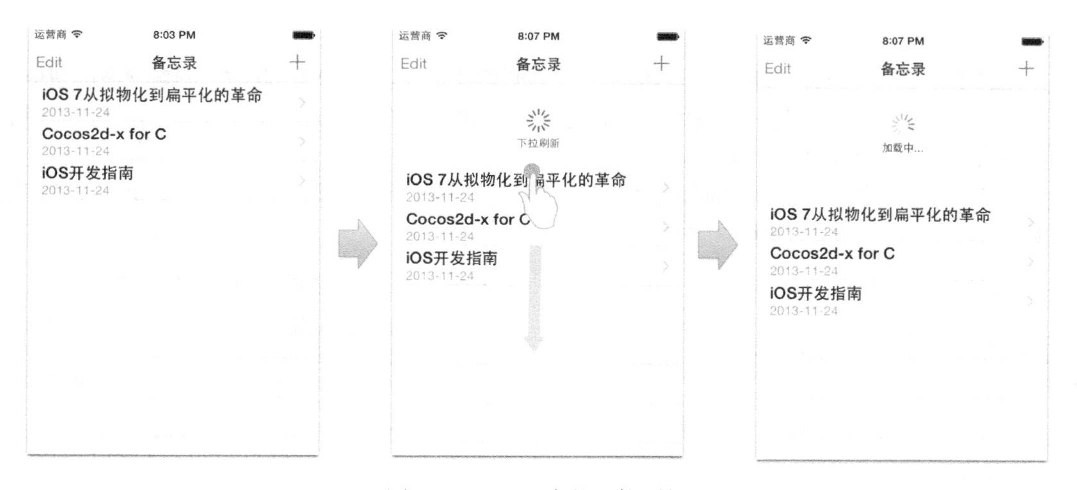

图16-12 iOS 7中的下拉刷新

在iOS 6之后的版本中，表视图控制器中添加了refreshControl属性，这个属性是UIRefreshControl类型。UIRefreshControl就是为表视图提供的下拉刷新控件。目前，UIRefreshControl类只能用于表视图中，不能用于其他视图。使用下拉刷新控件（UIRefreshControl）时，我们不需要考虑控件布局等问题，表视图控制器会将其自动放置于表视图中。

在主视图控制器MasterViewController中删除viewWillAppear:方法的代码，修改主视图控制器MasterViewController中的viewDidLoad:方法：

```swift
override func viewDidLoad() {
    super.viewDidLoad()
    ......

    //查询请求数据
    action = ActionTypes.query
    self.startRequest()
```

```objc
- (void)viewDidLoad {
    [super viewDidLoad];
    ......

    //查询请求数据
    action = ACTION_QUERY;
    [self startRequest];
```

```
    //初始化UIRefreshControl
    let rc = UIRefreshControl()                                                    ①
    rc.attributedTitle = NSAttributedString(string: "下拉刷新")                    ②
    rc.addTarget(self, action: #selector(refreshTableView),
    ➥for: UIControlEvents.valueChanged)                                           ③
    self.refreshControl = rc                                                       ④
}
```

```
    //初始化UIRefreshControl
    UIRefreshControl *rc = [[UIRefreshControl alloc] init];                        ①
    rc.attributedTitle = [[NSAttributedString alloc]initWithString:@"
    ➥下拉刷新"];                                                                  ②
    [rc addTarget:self action:@selector(refreshTableView)
    ➥forControlEvents:UIControlEventValueChanged];                                ③
    self.refreshControl = rc;                                                      ④
}
```

在上述代码中，第①行用于构造`UIRefreshControl`对象，第②行用于设置它的`attributedTitle`属性，该属性用于为下拉控件显示标题文本。第③行为刷新控件添加`UIControlEventValueChanged`（Swift版是`UIControlEvents.valueChanged`）事件处理机制，其中`refreshTableView`是`UIControlEventValueChanged`事件的处理方法。第④行用于设置表视图的`refreshControl`属性，这里把刚刚创建的`UIRefreshControl`对象赋值给该属性。`refreshTableView`方法的代码如下：

```
func refreshTableView() {
    if (self.refreshControl?.refreshing == true) {                                 ①
        self.refreshControl?.attributedTitle =
        ➥NSAttributedString(string: "加载中...")                                  ②
        //查询请求数据
        action = ActionTypes.query
        self.startRequest()                                                        ③
    }
}
```

```
-(void) refreshTableView {
    if (self.refreshControl.refreshing) {                                          ①
        self.refreshControl.attributedTitle = [[NSAttributedString
        ➥alloc]initWithString:@"加载中..."];                                      ②
        //查询请求数据
        action = ACTION_QUERY;
        [self startRequest];                                                       ③
    }
}
```

在上述代码中，第①行通过控件的`refreshing`属性判断控件是否还处于刷新状态。刷新状态的图标是我们常见的活动指示器，而显示的文字是“加载中...”，这是通过第②行设置控件的`attributedTitle`属性实现的。接下来应该是网络请求操作，如第③行所示。

由于异步请求成功返回之后需要回调`reloadView:`方法，在这个方法中我们需要停止刷新控件。这里修改主视图控制器`MasterViewController`中的`reloadView:`方法：

```
func reloadView(_ res: NSDictionary) {

    self.refreshControl?.endRefreshing()                                           ①
    self.refreshControl?.attributedTitle = NSAttributedString(string:
    ➥"下拉刷新")                                                                  ②

    let resultCode: NSNumber = res["ResultCode"] as! NSNumber

    if (resultCode.intValue >= 0) {
        self.listData = res["Record"] as! NSArray
        self.tableView.reloadData()
    } else {
        let errorStr = resultCode.errorMessage
        let alertController: UIAlertController = UIAlertController(title:
        ➥"错误信息",
        ➥message: errorStr, preferredStyle: .alert)
        let okAction = UIAlertAction(title: "OK", style: .default, handler: nil)
        alertController.addAction(okAction)
        //显示
        self.present(alertController, animated: true, completion: nil)

    }
}
```

```
- (void)reloadView:(NSDictionary *)res {

    [self.refreshControl endRefreshing];                                           ①
    self.refreshControl.attributedTitle =
    ➥[[NSAttributedString alloc]initWithString:@"下拉刷新"];                      ②

    NSNumber *resultCode = res[@"ResultCode"];

    if ([resultCode integerValue] >= 0) {
        self.listData = res[@"Record"];
        [self.tableView reloadData];
    } else {
        NSString *errorStr = [resultCode errorMessage];
        UIAlertController *alertController =
        ➥[UIAlertController alertControllerWithTitle:@"错误信息"
        ➥message:errorStr preferredStyle:UIAlertControllerStyleAlert];
        UIAlertAction *okAction = [UIAlertAction actionWithTitle:@"OK"
        ➥style:UIAlertActionStyleDefault handler:nil];
        [alertController addAction:okAction];
        //显示
        [self presentViewController:alertController animated:true
        ➥completion:nil];
    }
}
```

其中第①行调用刷新控件的`endRefreshing`方法来停止刷新。第②行设置显示文本为“下拉刷新”。此时我们就将刷新控件添加到表视图的主视图控制器中了，大家可以测试一下看看这种用户体验是不是很好呢！

16.6.2 使用活动指示器控件

当应用请求网络资源时，请求的数据要过一会儿才能返回，这段时间我们可以给用户看点儿东西，消除他们的心理等待时间，从而给用户提供更好的体验，这种控件叫作活动指示器。iOS提供了两种活动指示器：活动指示器控件（UIActivityIndicatorView）和网络活动指示器。本节先介绍活动指示器控件。

活动指示器控件用起来比较灵活。从技术的角度来说，活动指示器控件可以放置在视图中。当然，从设计规范上讲，活动指示器控件应该放置在工具栏、导航栏以及弹出的对话框中，请求结束时应该消失。

那么，在MyNotes案例的主视图界面中，活动指示器控件应该放在哪里呢？MyNotes案例有一个导航栏，我们只能将活动指示器控件放在图16-13所示的3个位置。

图16-13 活动指示器控件的放置位置

因为这个应用左右两个位置已经被Edit和+按钮占据了，所以活动指示器控件只能放置在中间（②号）位置。下面我们将活动指示器控件添加到MyNotes案例中。修改主视图控制器MasterViewController中的viewDidLoad:方法：

```
override func viewDidLoad() {
    super.viewDidLoad()
    ......
    //查询请求数据
    action = ActionTypes.query
    self.startRequest()

    //在导航栏中显示活动指示器
    self.showActivityIndicatorViewInNavigationItem()                ①
}
```

```
- (void)viewDidLoad {
    [super viewDidLoad];
    ......
    //查询请求数据
    action = ACTION_QUERY;
    [self startRequest];

    //在导航栏中显示活动指示器
    [self showActivityIndicatorViewInNavigationItem];               ①
}
```

在上述代码中，第①行调用showActivityIndicatorViewInNavigationItem方法，它是我们自己定义的方法，其代码如下：

```
func showActivityIndicatorViewInNavigationItem() {
    var aiview = UIActivityIndicatorView(activityIndicatorStyle:
    ➥UIActivityIndicatorViewStyle.gray)                              ①
    self.navigationItem.titleView = aiview                           ②
    aiview.startAnimating()                                          ③
    self.navigationItem.prompt = "数据加载中..."                      ④
}
```

```
-(void) showActivityIndicatorViewInNavigationItem {
    UIActivityIndicatorView *aiview = [[UIActivityIndicatorView alloc]
    ➥initWithActivityIndicatorStyle: UIActivityIndicatorViewStyleGray];
                                                                     ①
    self.navigationItem.titleView = aiview;                          ②
    [aiview startAnimating];                                         ③
    self.navigationItem.prompt = @"数据加载中...";                    ④
}
```

在上述代码中，第①行用于创建活动指示器控件。我们也可以在Interface Builder中使用故事板或XIB文件创建活动指示器控件，并设置它的属性。在第①行中，我们设置活动指示器的样式是UIActivityIndicator-ViewStyle.gray（Objective-C中是UIActivityIndicatorViewStyleGray）。第②行设置导航栏项目的titleView属性，该属性的位置如图16-14所示。注意，设置titleView属性之后，就不能显示title属性了。title属性如图16-15所示，它与titleView属性的位置是重合的、互斥的。导航栏项目的prompt属性常用于提示用户，第④行使得数据加载时会提示“数据加载中...”。

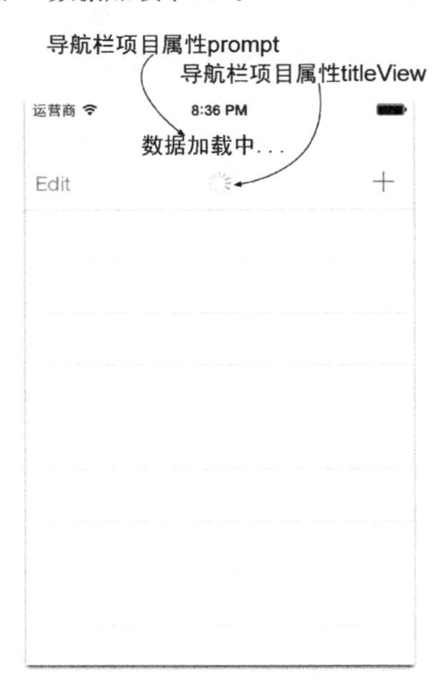

图16-14　导航栏项目的prompt和titleView属性

图16-15　导航栏项目的title属性

活动指示器控件有4个重要的方法和属性，具体如下所示。

- **startAnimating方法**。用于开始动画，即旋转起来，如第③行所示。
- **stopAnimating方法**。用于停止动画，即停止旋转。
- **isAnimating方法**。判断是否在旋转。
- **hidesWhenStopped属性**。它是布尔值，用于设置控件停止时是否隐藏。

当接收请求时，应该停止活动指示器的“旋转”，这需要在reloadView:方法中停止，其代码如下：

```
func reloadView(_ res: NSDictionary) {

    //停止活动指示器，恢复导航栏
    self.navigationItem.titleView = nil;                    ①
    self.navigationItem.prompt = nil;                       ②
    ......
}
```

```
-(void)reloadView:(NSDictionary*)res {

    //停止等待指示器，恢复导航栏
    self.navigationItem.titleView = nil;                    ①
    self.navigationItem.prompt = nil;                       ②
    ......
}
```

停止指示器控件本应该调用stopAnimating方法，但放在导航栏项目中的活动指示器控件有所不同。更重要的是，要移除这个控件，让原来的title内容显示出来，可以使用上述代码的第①行。另外，我们还需要通过第②行设置prompt属性为nil。

16.6.3　使用网络活动指示器

网络活动指示器能够在状态栏中以经典旋转小图标的形式出现。

它使用UIApplication类的networkActivityIndicatorVisible属性（布尔值）设置。因为UIApplication采用单例设计模式，所以我们可以在程序的任何地方使用[UIApplication sharedApplication]方法调用获得的UIApplication对象，Swift版是用UIApplication.shared静态属性获得UIApplication对象。

下面我们为MyNotes应用添加网络活动指示器。修改主视图控制器MasterViewController中的startRequest方法：

```
func startRequest() {
    ......
    UIApplication.shared.networkActivityIndicatorVisible = true        ①
}
```

```
-(void)startRequest {
    ......
    [UIApplication sharedApplication].networkActivityIndicatorVisible =
    ➥TRUE;                                                            ①
}
```

在上述代码中，第①行设置networkActivityIndicatorVisible属性为true（Objective-C版为TRUE或YES），即会在状态栏中显示网络活动指示器图标。另外，我们需要在请求结束时停止，这需要在reloadView:方法中添加代码，具体如下：

```
func reloadView(res : NSDictionary) {
    ......
    UIApplication.shared.networkActivityIndicatorVisible =
    ➥false                                                            ①
}
```

```
-(void)reloadView:(NSDictionary*)res {
    ......
    [UIApplication sharedApplication].networkActivityIndicatorVisible =
    ➥FALSE;                                                           ①
}
```

其中第①行用于设置networkActivityIndicatorVisible属性为false（Objective-C版为FALSE或NO），并且其图标会在状态栏中消失。

16.7　小结

本章重点介绍了REST风格Web Service的访问，还在访问Web Service的过程中介绍了NSURLSession API、AFNetworking框架和Alamofire框架。

Part 3 第三部分

进 阶 篇

本部分内容

- 第 17 章　定位服务
- 第 18 章　苹果地图应用
- 第 19 章　访问通讯录
- 第 20 章　iOS 10 应用扩展
- 第 21 章　重装上阵的 iOS 10 用户通知

第 17 章

定位服务

在这个时代，我们已经越来越离不开手机了。手机上越来越多的应用是基于地图的，而大部分地图又使用了定位服务。手机的优势就是能够到处移动，我们往往需要知道自己所在的位置，然后再查询自己周围的饭店、影院以及交通路线等。查找自己的位置时，可以使用GPS等方式提供的定位服务。找到饭店、影院和交通路线等信息时，再通过地图标注出来。

17.1 定位服务概述

iOS提供了4种不同的途径进行定位，具体如下所示。

- **Wi-Fi**。通过Wi-Fi路由器的地理位置信息查询，比较省电。iPhone、iPod touch和iPad都可以采用这种方式定位。
- **蜂窝式移动电话基站**。通过移动运营商基站定位。只有iPhone、3G版本的iPod touch和iPad可以采用这种方式定位。
- **GPS卫星**。通过GPS卫星位置定位，这种方式最为准确，但是耗电量大，不能遮挡。iPhone、iPod touch和iPad都可以采用这种方式定位。
- **iBeacon微定位**。苹果公司在iOS 7后支持iBeacon技术。iBeacon技术是苹果研发的，它使用低功耗蓝牙技术，通过多个iBeacon基站创建一个信号区域（地理围栏），当设备进入该区域时，相应的应用程序便会提示用户进入了这个地理围栏。

在对定位服务编程时，iOS不像Android系统那样可以指定采用哪种途径进行定位。iOS的API把底层这些细节屏蔽掉了，开发人员和用户并不知道现在设备是采用哪种方式进行定位的（微定位除外），iOS系统会根据设备的情况和周围的环境采用一套最佳的解决方案。也就是说，如果能够接收GPS信息，那么设备优先采用GPS定位，否则采用Wi-Fi或蜂窝基站定位。在Wi-Fi和蜂窝基站之间，优先使用Wi-Fi，如果无法连接到Wi-Fi，才使用蜂窝基站定位。

提示 GPS（全球定位系统）是20世纪70年代由美国陆、海、空三军联合研制的新一代空间卫星导航定位系统，其主要目的是为陆、海、空三大领域提供实时、全天候和全球性的导航服务，并用于情报收集、核爆监测和应急通信等一些军事目的。经过20余年的研究实验，耗资300亿美元，到1994年3月，全球覆盖率高达98%的24颗GPS卫星已布设完成。到2016年3月30日，中国也成功发射22颗北斗导航卫星。这些导航卫星分为军用频道和民用频道，前者是加密的，定位精度极高，而后者的定位精度要低一些。

总体来说，GPS定位的优点是准确、覆盖面广，缺点是不能被遮挡（例如在建筑物里面收不到GPS卫星信号）、GPS开启后比较费电。蜂窝基站不仅误差比较大，而且会耗费用户流量。而Wi-Fi定位是最经济实惠的。

17.1.1 定位服务编程

在iOS中，定位服务API主要使用Core Location框架，定位时主要使用CLLocationManager、CLLocation-ManagerDelegate和CLLocation这3个类，下面简要介绍一下它们。

- **CLLocationManager**。用于定位服务管理类，它能够给我们提供位置信息和高度信息，也可以监控设备进入或离开某个区域，还可以获得设备的运行方向等。
- **CLLocationManagerDelegate**。它是CLLocationManager类的委托协议。
- **CLLocation**。该类封装了位置和高度信息。

用户所在的位置是比较私密的信息，应用获取这些信息时，用户是有知情权和否定权的。如果应用在用户不知情的情况下获得其位置信息，这在某些国家是违法的。在iOS 8及其之后的版本中，可以在定位服务的应用中设置应用定位服务为“永不”“使用应用期间”和“始终”等选项，如图17-1所示。设置界面可以通过iOS中“设置”→“隐私”→“定位服务”→“*XX*应用”打开。

当第一次访问用户位置信息时，可以要求用户授权（授权对话框如图17-2所示）。一旦授权，该应用以后再使用的时候不会弹出授权对话框。图17-2中的“智捷课堂iOS开发指南示例”信息是可以自定义的。

下面我们通过一个案例介绍一下iOS定位服务API。如图17-3所示，在应用启动时，会获得位置信息，并显示在对应的文本框中。如果设备位置发生变化，也会重新获取位置信息，并更新对应的文本框。

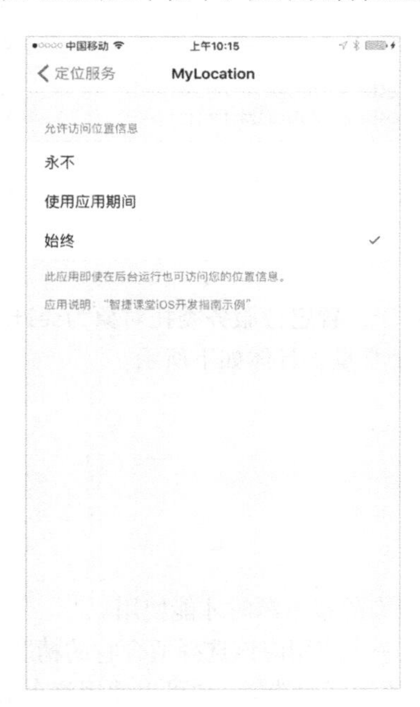

图17-1 设置定位服务

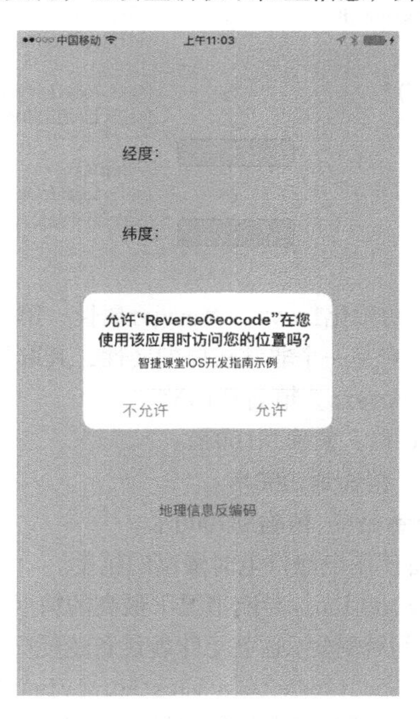

图17-2 访问位置信息授权对话框

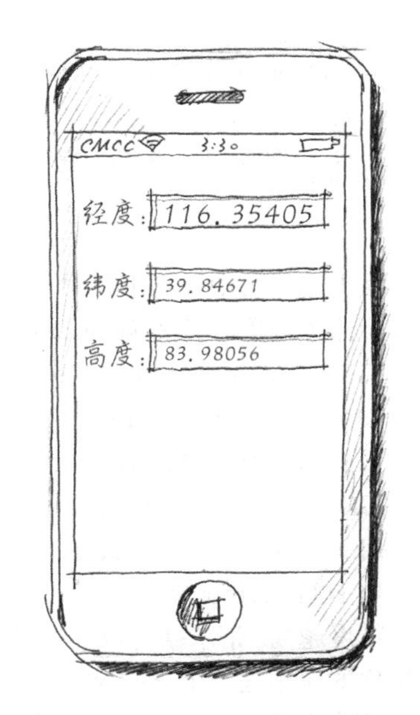

图17-3 定位服务案例原型设计

通过Xcode创建一个Single View Application工程MyLocation。然后参考图17-3的原型设计应用界面，这里不再赘述。下面我们直接看看实现代码，其中主要代码是在视图控制器ViewController中编写的。ViewController中类的定义、属性声明等的代码如下：

```
//ViewController.swift文件
import UIKit
import CoreLocation                                                    ①

class ViewController: UIViewController, CLLocationManagerDelegate {   ②

    //经度
```

```
//ViewController.m文件
#import "ViewController.h"
#import <CoreLocation/CoreLocation.h>                                  ①

@interface ViewController ()<CLLocationManagerDelegate>               ②

//经度
```

```swift
    @IBOutlet weak var txtLng: UITextField!
    //纬度
    @IBOutlet weak var txtLat: UITextField!
    //高度
    @IBOutlet weak var txtAlt: UITextField!

    var locationManager: CLLocationManager!                        ③
    ......
}
```

```objc
@property (weak, nonatomic) IBOutlet UITextField *txtLng;
//纬度
@property (weak, nonatomic) IBOutlet UITextField *txtLat;
//高度
@property (weak, nonatomic) IBOutlet UITextField *txtAlt;

@property(nonatomic, strong) CLLocationManager *locationManager;     ③


@end
```

在上述代码中，第①行引入了CoreLocation模块，第②行定义ViewController类时声明遵守CLLocation-ManagerDelegate协议。此外，我们还在第③行中定义了CLLocationManager类型的locationManager属性。

在ViewController中，viewDidLoad方法的代码如下：

```swift
//ViewController.swift文件
override func viewDidLoad() {
    super.viewDidLoad()

    //定位服务管理对象初始化
    self.locationManager = CLLocationManager()                                   ①
    self.locationManager.delegate = self                                         ②
    self.locationManager.desiredAccuracy = kCLLocationAccuracyBest               ③
    self.locationManager.distanceFilter = 1000.0                                 ④

    self.locationManager.requestWhenInUseAuthorization()                         ⑤
    self.locationManager.requestAlwaysAuthorization()                            ⑥

    //开始定位
    self.locationManager.startUpdatingLocation()                                 ⑦
    print("开始定位")
}
```

```objc
//ViewController.m文件
- (void)viewDidLoad {
    [super viewDidLoad];

    //定位服务管理对象初始化
    self.locationManager = [[CLLocationManager alloc] init];                     ①
    self.locationManager.delegate = self;                                        ②
    self.locationManager.desiredAccuracy = kCLLocationAccuracyBest;              ③
    self.locationManager.distanceFilter = 1000.0f;                               ④

    [self.locationManager requestWhenInUseAuthorization];                        ⑤
    [self.locationManager requestAlwaysAuthorization];                           ⑥

    //开始定位
    [self.locationManager startUpdatingLocation];                                ⑦
    NSLog(@"开始定位");
}
```

在上述代码中，第①行用于创建并初始化locationManager属性，第②行用于设置定位服务委托对象为self。第③行用于设置desiredAccuracy属性，它是一个非常重要的属性，其取值有6个常量，具体如下所示。

- kCLLocationAccuracyNearestTenMeters。精确到10米。
- kCLLocationAccuracyHundredMeters。精确到100米。
- kCLLocationAccuracyKilometer。精确到1000米。
- kCLLocationAccuracyThreeKilometers。精确到3000米。
- kCLLocationAccuracyBest。设备使用电池供电时最高的精度。
- kCLLocationAccuracyBestForNavigation。导航情况下最高的精度，一般有外接电源时才能使用。

精度越高，请求获得位置信息的时间就越短，这就意味着设备越耗电，因此一个应用应该选择适合它的精度。如果你的应用是一个车载导航应用，kCLLocationAccuracyBestForNavigation是比较好的选择，你可以使用汽车上的电瓶为设备供电。如果你的应用是为徒步旅行者提供的导航应用，kCLLocationAccuracyHundredMeters是一个不错的选择。

第④行用于设置distanceFilter属性，它是距离过滤器，定义了设备移动后获得位置信息的最小距离，单位是米，本例中设置为1000米。

第⑤行和第⑥行是iOS 8新添加的，它们会弹出用户授权对话框，其中第⑤行的requestWhenInUse-Authorization方法是要求用户“使用应用期间”授权，requestAlwaysAuthorization方法则要求用户“始终”授权。为了能够弹出授权对话框，我们需要修改工程配置。打开工程，找到Info.plist属性列表文件，添加两个键NSLocationAlwaysUsageDescription和NSLocationWhenInUseUsageDescription，对应的文字内容可以自定义，结果如图17-4所示。

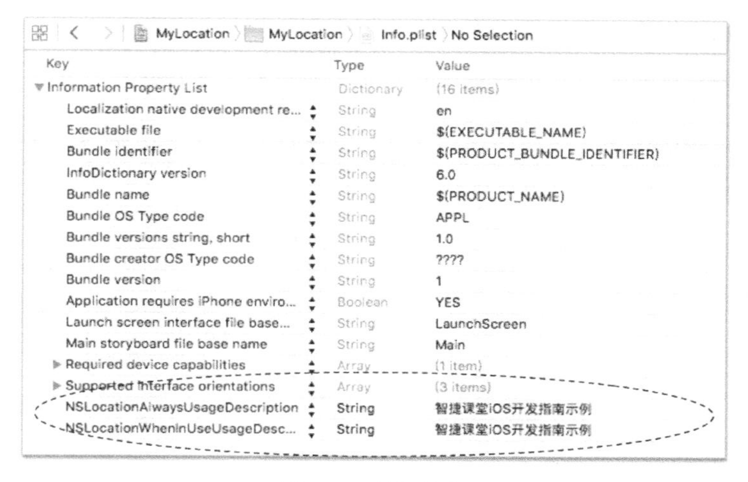

图17-4 添加键值

第⑦行通过CLLocationManager对象的startUpdatingLocation方法开始定位服务。根据设定的条件，它不断请求回调新的位置信息。因此，开启这个方法一定要慎重，在视图控制器的声明周期方法viewWillAppear:中使用这个方法是最合适的。与停止定位服务对应的方法是stopUpdatingLocation方法，它是在视图控制器的viewWillDisappear:方法中调用的。viewWillDisappear:方法的代码如下：

```
override func viewWillDisappear(_ animated: Bool) {
    super.viewWillDisappear(animated)
    //停止定位
    self.locationManager.stopUpdatingLocation()
}
```

```
- (void)viewWillDisappear:(BOOL)animated {
    [super viewWillDisappear:animated];
    //停止定位
    [self.locationManager stopUpdatingLocation];
}
```

这个方法在视图消失（应用退到后台）时调用，能够保证最及时地关闭定位服务。在iOS 6中，请求有所变化，定位服务应用退到后台后，可以延迟更新位置信息，这可以通过allowDeferredLocationUpdatesUntil-Traveled:timeout:方法实现。要关闭延迟更新，可以使用disallowDeferredLocationUpdates方法实现。此外，在iOS 6之后，新增了pausesLocationUpdatesAutomatically属性，它能设定自动暂停位置更新，而把定位服务的开启和暂停管理权交给系统，这样更加合理和简单。

一旦定位服务开启，并设置好CLLocationManager委托属性delegate后，当用户设备移动的距离超过一定距离后，就会回调委托方法。与定位服务有关的方法有如下3个。

- ❑ locationManager:didUpdateLocations:。定位成功。这是iOS 6之后新增的方法。
- ❑ locationManager:didFailWithError:。定位失败。
- ❑ locationManager:didChangeAuthorizationStatus:。授权状态发生变化时调用。

实现CLLocationManager委托的代码如下：

```
func locationManager(_ manager: CLLocationManager,
➥didUpdateLocations locations: [CLLocation]) {

    var currLocation = locations[locations.count - 1] as CLLocation          ①

    self.txtLat.text = String(format:"%3.5f",
    ➥currLocation.coordinate.latitude)                                       ②
    self.txtLng.text = String(format:"%3.5f",
    ➥currLocation.coordinate.longitude)                                      ③
    self.txtAlt.text = String(format:"%3.5f", currLocation.altitude)          ④

}
```

```
- (void)locationManager:(CLLocationManager *)manager
➥didUpdateLocations:(NSArray *)locations {
    CLLocation *currLocation = [locations lastObject];                        ①

    self.txtLat.text = [NSString stringWithFormat:@"%3.5f",
    ➥currLocation.coordinate.latitude];                                      ②
    self.txtLng.text = [NSString stringWithFormat:@"%3.5f",
    ➥currLocation.coordinate.longitude];                                     ③
    self.txtAlt.text = [NSString stringWithFormat:@"%3.5f",
    ➥currLocation.altitude];                                                 ④
}
```

```
func locationManager(_ manager: CLLocationManager!, didFailWithError error:
➥NSError!) {
    print("error: \(error.localizedDescription)")
}

func locationManager(_ manager: CLLocationManager,
➥didChangeAuthorizationStatus status: CLAuthorizationStatus) {          ⑤

    switch status {
    case .authorizedAlways :
        print("已经授权")
    case .authorizedWhenInUse :
        print("使用时授权")
    case .denied :
        print("拒绝")
    case .restricted :
        print("受限")
    case .notDetermined:
        print("用户还没有确定")
    }

}
```

```
- (void)locationManager:(CLLocationManager *)manager
➥didFailWithError:(NSError *)error {
    NSLog(@"error: %@",error);
}

- (void)locationManager:(CLLocationManager *)manager
➥didChangeAuthorizationStatus:(CLAuthorizationStatus)status{          ⑤

    if (status == kCLAuthorizationStatusAuthorizedAlways) {
        NSLog(@"已经授权");
    } else if (status == kCLAuthorizationStatusAuthorizedWhenInUse) {
        NSLog(@"使用时授权");
    } else if (status == kCLAuthorizationStatusDenied) {
        NSLog(@"拒绝");
    } else if (status == kCLAuthorizationStatusRestricted) {
        NSLog(@"受限");
    } else if (status == kCLAuthorizationStatusNotDetermined) {
        NSLog(@"用户还没有确定");
    }
}
```

在locationManager:didUpdateLocations:方法中，参数locations是位置变化的集合，它按照时间变化的顺序存放。如果想获得当前设备的位置，可以使用第①行获得集合中的最后一个元素，它就是设备的当前位置。从集合中返回的对象类型是CLLocation，CLLocation封装了位置、高度等信息。在上面的代码中，我们使用了它的两个属性altitude和coordinate，其中前者是高度值，后者是封装经度和纬度的结构体CLLocationCoordinate2D。该结构体的定义如下：

```
struct CLLocationCoordinate2D {
    var latitude: CLLocationDegrees          //纬度
    var longitude: CLLocationDegrees         //经度
}
```

```
typedef struct {
    CLLocationDegrees latitude;          //纬度
    CLLocationDegrees longitude;         //经度
} CLLocationCoordinate2D;
```

其中latitude为纬度信息，longitude为经度信息，它们都是CLLocationDegrees类型。CLLocationDegrees是使用typedef定义的double类型。

在locationManager:didUpdateLocations:方法中，第②行中的currLocation.coordinate.latitude表达式用于获得设备当前的纬度，第③行中的currLocation.coordinate.longitude表达式用于获得设备当前的经度，而第④行中的currLocation.altitude表达式用于获得高度。

在第⑤行中，locationManager:didChangeAuthorizationStatus:方法的参数status是获得的授权信息。另外，也可以使用CLLocationManager的静态方法authorizationStatus获得授权信息。

17.1.2　测试定位服务

一般情况下，定位服务应用的测试和运行有两个选择：模拟器和设备。原则上，我们先通过模拟器，然后再使用设备测试。由于定位服务的特点，使用设备测试时我们需要到现场进行测试，所以有的时候有一定的局限性。因此，使用模拟器测试有时是不可替代的。

在Xcode早期版本中，模拟器是不能模拟位置信息变化的，默认获取的位置信息是固定的苹果公司总部地址。而现在的Xcode版本预先设置了几个地址，我们可以模拟改变位置。如果想让模拟器一开始运行的时候就能够获得模拟数据，可以在启动参数中设置。首先，在Xcode工具的左上角编辑应用的Scheme（方案），如图17-5所示。

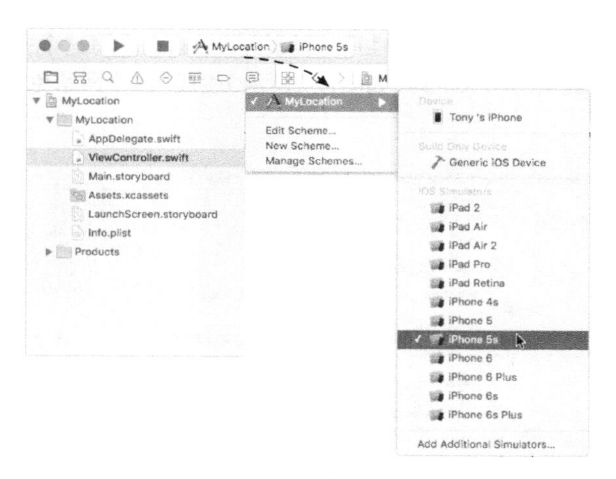

图17-5 编辑方案

选择Edit Scheme菜单项后，会弹出如图17-6所示的对话框，从中选择Run MyLocation.app→Options，在Core Location项目中选中Allow Location Simulation复选框，然后在下面的Default Location下拉列表中选择你感兴趣的城市。

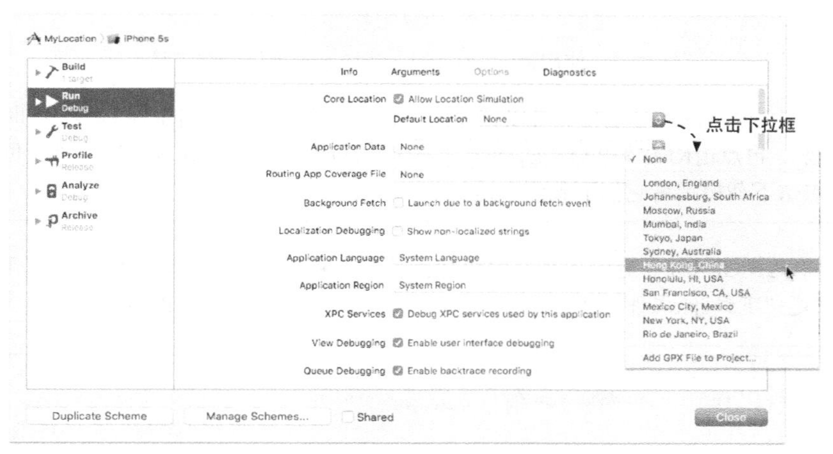

图17-6 设置启动参数

这样应用启动时，就会模拟定位到你选择的城市了。如果列表中没有我们需要的地点，可以使用最下面的Add GPX File to Project...菜单项为工程添加一个GPX[①]文件。下面是GPX文件的内容：

```
<?xml version="1.0" encoding="UTF-8" standalone="no" ?>
<gpx xmlns="http://www.topografix.com/GPX/1/1"
    creator="MyGeoPosition.com" version="1.1"
    xmlns:xsi="http://www.w3.org/2001/XMLSchema-instance"
    xsi:schemaLocation="http://www.topografix.com/GPX/1/1
        http://www.topografix.com/GPX/1/1/gpx.xsd">
    <wpt lat="40.002240" lon="117.323328">
        <name>中国北京 东城区北京站东街北京 邮政编码: 100005</name>
        <src>MyGeoPosition.com</src>
        <link>http://mygeoposition.com</link>
    </wpt>
</gpx>
```

① GPX（GPS eXchange Format，GPS交换格式）是一个XML格式，是为应用软件设计的通用GPS数据格式。

——引自维基百科http://zh.wikipedia.org/wiki/Gpx

<wpt>标签中的lat属性用于设置纬度，lon属性用于设置经度。自己手写这个文件还是比较麻烦的，我一般使用www.mygeoposition.com网站提供的GPX工具（如图17-7所示）生成。这个网站免费提供地理信息编码和反编码、生成KML和GPX文件等服务。

图17-7　使用www.mygeoposition.com提供GPX工具

如果想找到“清华大学”的地理坐标，可以在地图上找到地点，然后点击鼠标右键，此时会出现一个描述该地点经纬度的“气泡”，再点击KML/GPX标签，就会生成如图17-8所示的KML和GPX文件，可以直接把这些内容复制出来，也可以点击下载按钮将它们下载到本地。

提示　在图17-8中选择GPX标签后，下面的按钮标签是“下载kml文件”，这应该是本站开发者的笔误，我猜测应该是“下载gpx文件”才对，但是这不影响我们使用它下载GPX文件。

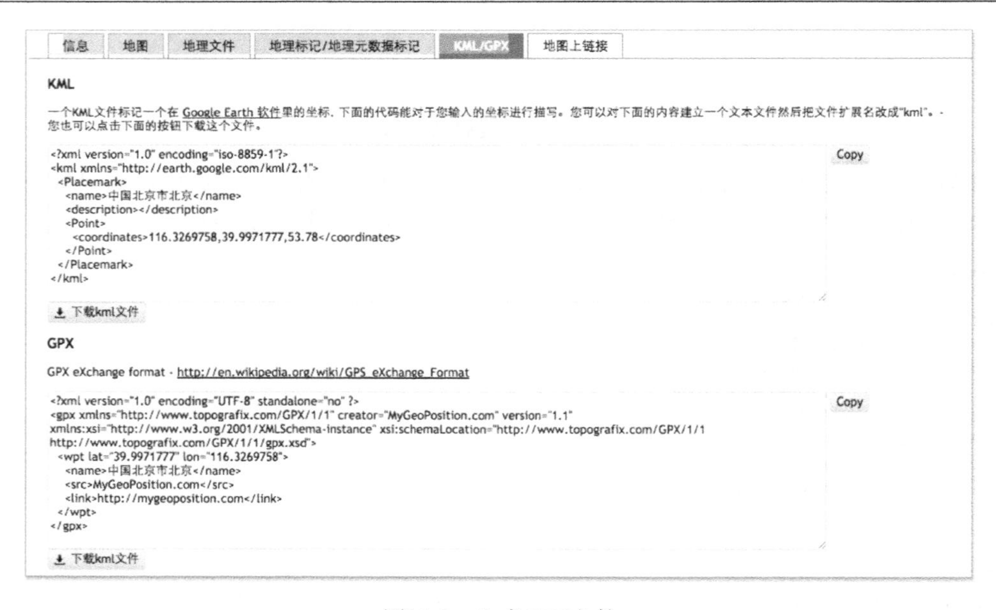

图17-8　生成GPX文件

得到GPX文件后，可以通过图17-9所示的Add GPX File to Project...菜单项将它添加到Xcode工程中，此时菜单中就会出现GPX文件了。

如果在应用启动参数中没有设置初始的模拟位置数据，我们还可以在运行之后设置。在调试工具栏中选择“模拟定位”按钮，即可选择模拟位置，如图17-10所示。

Xcode中的模拟器还提供了连续位置变化测试能力。如果想开发导航应用，这个功能对我们有很大的帮助。此外，模拟器有几个固定的模式，可以发出连续变化的位置数据。打开模拟菜单的Debug→Location菜单项，可以发现共有6个菜单项，如图17-11所示。其中Apple菜单项是苹果公司总部位置，Custom Location菜单项是自己定义的位置，选择该菜单项会弹出如图17-12所示的对话框，在此可以输入经度和纬度。

图17-11中后面3个菜单项都能发出连续的位置变化数据，它们的起始点从苹果公司总部开始，按照一个固定的线路运动，这三者的区别是City Bicycle Ride是最慢的，City Run要快一些，Freeway Drive最快。

图17-9 添加GPX文件

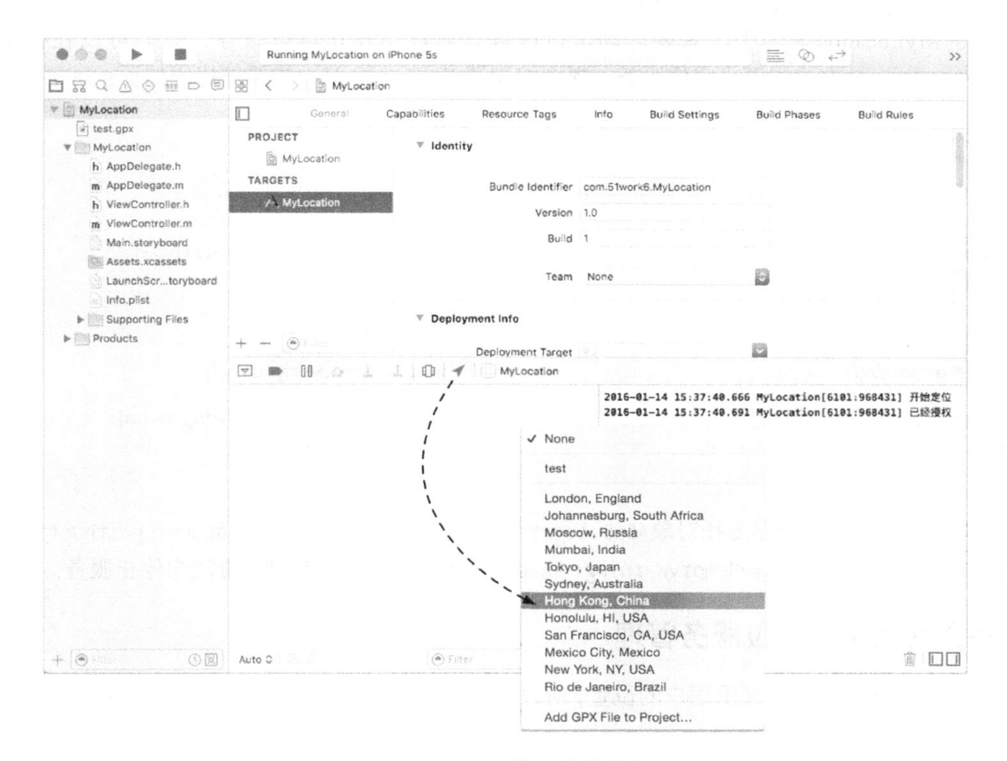

图17-10 设置模拟定位数据

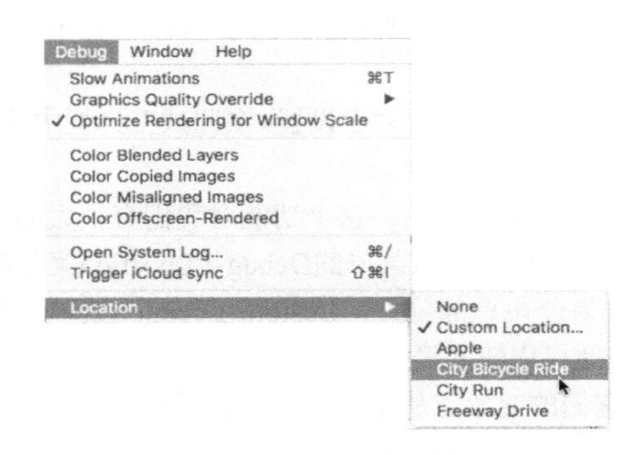

图17-11　模拟器位置菜单

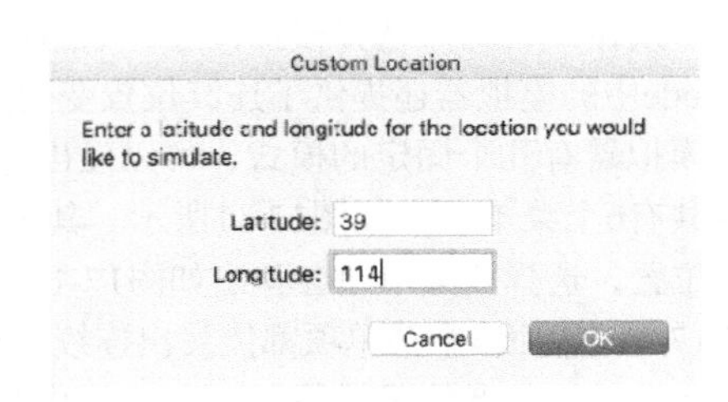

图17-12　模拟器中自定义位置

17.2　管理定位服务

定位服务比较耗电，我们不仅在精度和距离过滤器上要加以考虑，也要管理好开启或停止的合适时间，这是非常重要的。

什么时候开启定位服务，什么时候停止定位服务，这要根据用户的具体需求而定。由于具体业务的复杂性，这里只介绍几种管理定位服务开启和停止模式代码。

17.2.1　应用启动与停止下的定位服务管理

默认情况下，定位服务是可以在后台运行的，例如车载导航应用、健身类型应用，都需要在后台花比较长的时间启动定位服务。当然，这些应用比较耗电。

在17.1节的MyLocation案例中，只要启动定位服务，而没用停止定位服务，我们可以在视图控制器的didReceiveMemoryWarning方法中停止定位服务。视图控制器ViewController主要的代码如下：

```
override func viewDidLoad() {
    super.viewDidLoad()
    ......

    //开始定位
    self.locationManager.startUpdatingLocation()
    print("开始定位")
}

override func didReceiveMemoryWarning() {
    super.didReceiveMemoryWarning()
    //停止定位
    self.locationManager.stopUpdatingLocation()
    print("停止定位")
}
```

```
- (void)viewDidLoad {
    [su per viewDidLoad];
    ......

    //开始定位
    [self.locationManager startUpdatingLocation];
    NSLog(@"开始定位");
}

- (void)didReceiveMemoryWarning {
    [super didReceiveMemoryWarning];
    //停止定位
    [self.locationManager stopUpdatingLocation];
    NSLog(@"停止定位");
}
```

当然，我们也可以在应用程序委托对象AppDelegate的application:didFinishLaunchingWithOptions:方法中开启服务，在applicationDidReceiveMemoryWarning: 或applicationWillTerminate:方法中停止服务。

17.2.2　视图切换下的定位服务管理

有些应用需要在地图视图上显示用户的位置，所以当视图可见时，应用应该开启定位服务；当视图不可见时，则停止定位服务。

在图17-13所示的案例中，图17-13a中界面可见，会启动定位服务，而进入到图17-13b所示的界面时，则会停止定位服务。

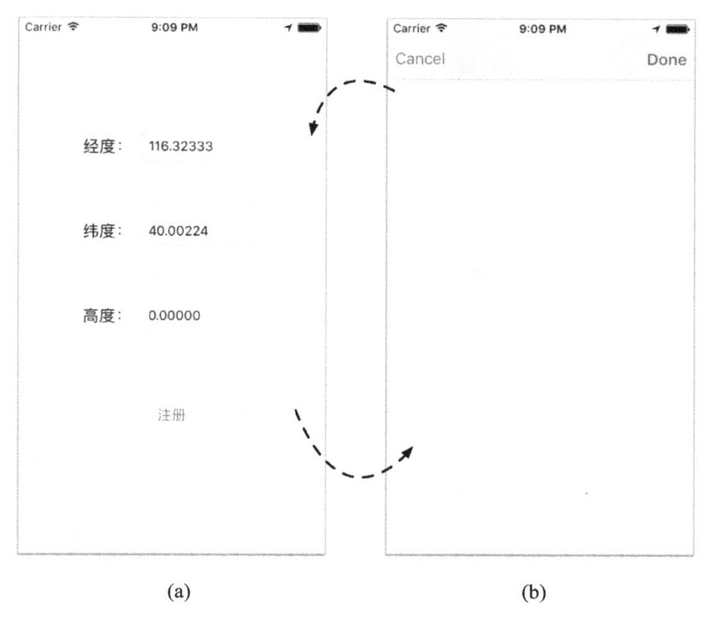

(a)　　　(b)

图17-13　视图切换

实现图17-13所示的案例时，ViewController中主要的代码如下：

```
override func viewWillAppear(_ animated: Bool) {
    super.viewWillAppear(animated)
    //开始定位
    self.locationManager.startUpdatingLocation()
    print("开始定位")
}

override func viewWillDisappear(_ animated: Bool) {
    super.viewWillDisappear(animated)
    //停止定位
    self.locationManager.stopUpdatingLocation()
    print("停止定位")
}
```

```
- (void)viewWillAppear:(BOOL)animated {
    [super viewWillAppear:animated];
    //开始定位
    [self.locationManager startUpdatingLocation];
    NSLog(@"开始定位");
}

- (void)viewWillDisappear:(BOOL)animated {
    [super viewWillDisappear:animated];
    //停止定位
    [self.locationManager stopUpdatingLocation];
    NSLog(@"停止定位");
}
```

17.2.3　应用前后台切换下的定位服务管理

很多基于位置服务的应用，在进入前台时启动定位服务，在退到后台时停止定位服务。这种情况下管理启动和停止定位服务有很多种方法，如果一个应用中有多个界面需要显示定位信息，则定位服务的管理可以在应用程序委托对象AppDelegate中进行，然后位置发生变化后，通过通知机制通知所有显示定位信息的视图控制器。当然，这需要在相应的视图控制器中注册观察位置的通知。

这里还用17.1节的MyLocation案例，首先在AppDelegate中添加代码：

```
//AppDelegate.swift文件

let UpdateLocationNotification = "kUpdateLocationNotification"

import UIKit
import CoreLocation
```

```
//ViewController.m文件
#import "AppDelegate.h"
#import <CoreLocation/CoreLocation.h>

#define UpdateLocationNotification @"kUpdateLocationNotification"

@interface AppDelegate () <CLLocationManagerDelegate>
```

```
@UIApplicationMain
class AppDelegate: UIResponder, UIApplicationDelegate,
CLLocationManagerDelegate {

    var window: UIWindow?
    var locationManager: CLLocationManager!
    func application(_ application: UIApplication,
    ➥didFinishLaunchingWithOptions launchOptions:
    ➥[UIApplicationLaunchOptionsKey: Any]?) -> Bool { {

        //定位服务管理对象初始化
        self.locationManager = CLLocationManager()
        self.locationManager.delegate = self
        self.locationManager.desiredAccuracy = kCLLocationAccuracyBest
        self.locationManager.distanceFilter = 1000.0

        self.locationManager.requestWhenInUseAuthorization()
        self.locationManager.requestAlwaysAuthorization()

        return true
    }

    func applicationDidBecomeActive(_ application: UIApplication) {          ①
        //开始定位
        self.locationManager.startUpdatingLocation()
        print("定位开始")
    }

    func applicationDidEnterBackground(_ application: UIApplication) {       ②
        //停止定位
        self.locationManager.stopUpdatingLocation()
        print("定位停止")
    }

    //MARK: -- Core Location委托方法用于实现位置的更新

    func locationManager(_ manager: CLLocationManager,
    ➥didUpdateLocations locations: [CLLocation]) {                           ③

        let currLocation = locations.last! as CLLocation
        NotificationCenter.default.post(name: Notification.Name(rawValue:
        ➥UpdateLocationNotification), object: currLocation)                  ④
    }

    func locationManager(_ manager: CLLocationManager,
    ➥didFailWithError error: NSError) {
        print("error: \(error.localizedDescription)")
    }

    func locationManager(_ manager: CLLocationManager,
    ➥didChangeAuthorization status: CLAuthorizationStatus) {

        switch status {
        case .authorizedAlways:
            print("已经授权")
        case .authorizedWhenInUse:
            NSLog("使用时授权")
        case .denied:
            print("拒绝")
        case .restricted:
            print("受限")
        case .notDetermined:
            print("用户还没有确定")
        }

    }
}
```

```
@property(nonatomic, strong) CLLocationManager *locationManager;

@end

@implementation AppDelegate

- (BOOL)application:(UIApplication *)application
➥didFinishLaunchingWithOptions:(NSDictionary *)launchOptions {

    //定位服务管理对象初始化
    self.locationManager = [[CLLocationManager alloc] init];
    self.locationManager.delegate = self;
    self.locationManager.desiredAccuracy = kCLLocationAccuracyBest;
    self.locationManager.distanceFilter = 1000.0f;

    [self.locationManager requestWhenInUseAuthorization];
    [self.locationManager requestAlwaysAuthorization];

    return TRUE;
}

- (void)applicationDidBecomeActive:(UIApplication *)application {            ①
    //开始定位
    [self.locationManager startUpdatingLocation];
    NSLog(@"开始定位");
}

- (void)applicationDidEnterBackground:(UIApplication *)application {         ②
    //停止定位
    [self.locationManager stopUpdatingLocation];
    NSLog(@"停止定位");
}

#pragma mark -- Core Location委托方法用于实现位置的更新

- (void)locationManager:(CLLocationManager *)manager
➥didUpdateLocations:(NSArray *)locations {                                  ③

    CLLocation *currLocation = [locations lastObject];
    [[NSNotificationCenter defaultCenter] postNotificationName:
    ➥UpdateLocationNotification object:currLocation];                       ④
}

- (void)locationManager:(CLLocationManager *)manager didFailWithError:
➥(NSError *)error {
    NSLog(@"error: %@", error);
}

- (void)locationManager:(CLLocationManager *)manager
➥didChangeAuthorizationStatus:(CLAuthorizationStatus)status {

    if (status == kCLAuthorizationStatusAuthorizedAlways) {
        NSLog(@"已经授权");
    } else if (status == kCLAuthorizationStatusAuthorizedWhenInUse) {
        NSLog(@"使用时授权");
    } else if (status == kCLAuthorizationStatusDenied) {
        NSLog(@"拒绝");
    } else if (status == kCLAuthorizationStatusRestricted) {
        NSLog(@"受限");
    } else if (status == kCLAuthorizationStatusNotDetermined) {
        NSLog(@"用户还没有确定");
    }
}

@end
```

应用启动的过程中（也包括从后台恢复时）都会调用applicationDidBecomeActive:方法，我们在该方法中开启定位服务，见第①行。

当应用退到后台时，调用applicationDidEnterBackground:方法，我们在该方法中通知定位服务，见第②行。

应用程序委托对象AppDelegate也实现了CLLocationManagerDelegate协议，当位置变化之后，系统会回调第③行的locationManager:didUpdateLocations:方法。我们需要在这个方法中获得位置信息，第④行用于将位置信息通过通知投送给显示位置信息的视图控制器。

显示位置信息的视图控制器的主要代码如下：

```
//ViewController.swift文件
import UIKit
import CoreLocation

class ViewController: UIViewController {

    //经度
    @IBOutlet weak var txtLng: UITextField!
    //纬度
    @IBOutlet weak var txtLat: UITextField!
    //高度
    @IBOutlet weak var txtAlt: UITextField!

    override func viewDidLoad() {
        super.viewDidLoad()
        NotificationCenter.default.addObserver(self,
        ➥selector: #selector(updateLocation(_:)),
        ➥name: Notification.Name(rawValue: UpdateLocationNotification,)
        ➥object: nil)
    }

    override func didReceiveMemoryWarning() {
        super.didReceiveMemoryWarning()
        NotificationCenter.default.removeObserver(self)
    }

    //接收位置变化通知
    func updateLocation(_ notification: NSNotification) {
        let currLocation = notification.object as! CLLocation
        self.txtLat.text = String(format: "%3.5f",
        ➥currLocation.coordinate.latitude)
        self.txtLng.text = String(format: "%3.5f",
        ➥currLocation.coordinate.longitude)
        self.txtAlt.text = String(format: "%3.5f", currLocation.altitude)
    }
}
```

```
//ViewController.m文件

#import "ViewController.h"
#import <CoreLocation/CoreLocation.h>

#define UpdateLocationNotification @"kUpdateLocationNotification"

@interface ViewController ()

//经度
@property(weak, nonatomic) IBOutlet UITextField *txtLng;
//纬度
@property(weak, nonatomic) IBOutlet UITextField *txtLat;
//高度
@property(weak, nonatomic) IBOutlet UITextField *txtAlt;

@end

@implementation ViewController

- (void)viewDidLoad {
    [super viewDidLoad];
    [[NSNotificationCenter defaultCenter] addObserver:self
    ➥selector:@selector(updateLocation:)
    ➥name:UpdateLocationNotification object:nil];
}

- (void)didReceiveMemoryWarning {
    [super didReceiveMemoryWarning];
    [[NSNotificationCenter defaultCenter] removeObserver:self];
}


- (void)updateLocation:(NSNotification *)notification {
    CLLocation *currLocation = [notification object];
    self.txtLat.text = [NSString stringWithFormat:@"%3.5f",
    ➥currLocation.coordinate.latitude];
    self.txtLng.text = [NSString stringWithFormat:@"%3.5f",
    ➥currLocation.coordinate.longitude];
    self.txtAlt.text = [NSString stringWithFormat:@"%3.5f",
    ➥currLocation.altitude];
}

@end
```

上述代码中的updateLocation:方法是接收到位置变化通知时调用的方法，我们在这个方法中获取位置CLLocation信息的对象，并将其显示到界面中。

17.2.4　设置自动暂停位置服务

前面介绍的方法都是手动管理位置服务的开启和停止，iOS 6之后CLLocationManager类新增pausesLocationUpdatesAutomatically属性，它能设置自动暂停位置服务，定位服务的开启和暂停管理权交给系统，这样会更加合理和简单。

设置自动暂停位置服务的功能，既可以通过设置pausesLocationUpdatesAutomatically属性实现，也可以通过设置activityType属性实现。activityType是应用活动类型，activityType属性是枚举类型CLActivityType。CLActivityType定义的成员如表17-1所述。

表17-1　CLActivityType枚举中的成员

Swift枚举成员	Objective-C枚举成员	说　明
other	CLActivityTypeOther	默认值，未知的类型
automotiveNavigation	CLActivityTypeAutomotiveNavigation	车载导航类型，位置变化快
fitness	CLActivityTypeFitness	步行导航类型，位置变化慢
otherNavigation	CLActivityTypeOtherNavigation	其他导航类型，位置变化适中

在设置CLLocationManager对象的代码中添加如下语句：

```
//设置自动暂停位置服务
self.locationManager.pausesLocationUpdatesAutomatically = true
self.locationManager.activityType = .automotiveNavigation
```

```
//设置自动暂停位置服务
self.locationManager.pausesLocationUpdatesAutomatically = TRUE;
self.locationManager.activityType = CLActivityTypeAutomotiveNavigation;
```

17.2.5　后台位置服务管理

出于安全性和灵活性的考虑，iOS 9和iOS 10又进一步增强了后台的应用位置服务管理，能够在后台灵活管理多个CLLocationManager对象，并且设置参数。

首先，需要开启支持应用后台运行的模式。在Xcode工程中选择TARGETS→MyLocation→Capabilities，如图17-14所示，开启Background Modes后面的开关控件，并在Modes中选中Location updates复选框。

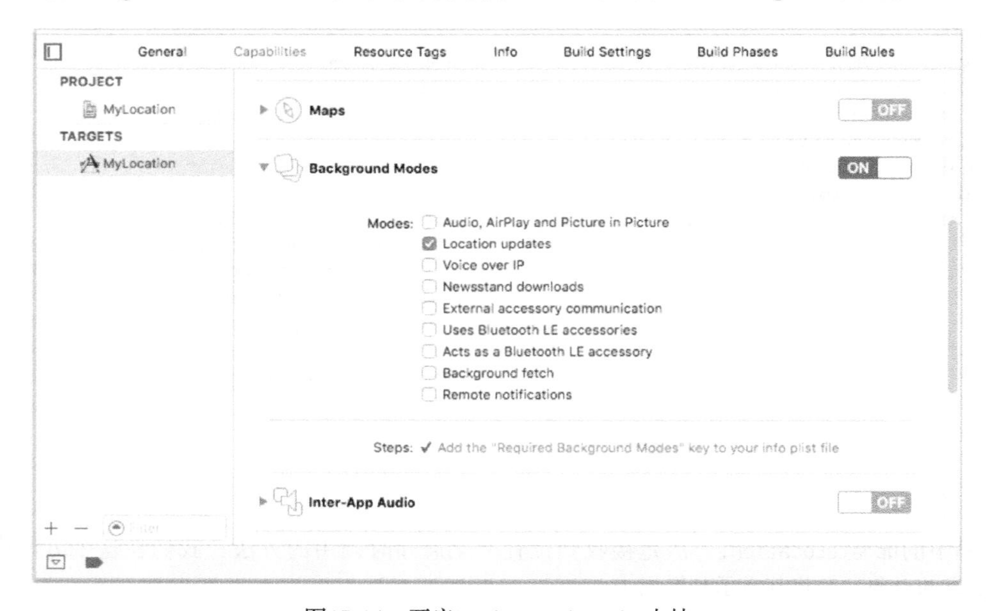

图17-14　开启Background Modes支持

在设置CLLocationManager对象的代码中添加如下语句：

```
self.locationManager.allowsBackgroundLocationUpdates = true
```

```
self.locationManager.allowsBackgroundLocationUpdates = TRUE;
```

将该属性设置为true，则表示可以在后台运行，设置为false表示不能，其默认值是false。

17.3 地理信息编码与反编码

有时候我们需要把经纬度转化为地点信息，而有时候需要通过地点的描述信息查询出它的经纬度。这些处理就是本节要介绍的地理信息编码与反编码。

17.3.1 地理信息反编码

在上一节的实例中，我们虽然知道了经度和纬度，那又能怎么样呢？一般人是很难知道这些数字（114.15816，22.28468）代表的是什么地方。地理信息反编码就是根据给定地点的地理坐标，返回这个地点的相关文字描述信息，这些文字描述信息被封装在CLPlacemark类中，我们把这个类叫作“地标”类。地标类有很多属性，下面简要介绍一下。

- **name**。有关地址的名字属性。
- **addressDictionary**。地址信息的字典，包含一些键值对，其中的键是在AddressBook.framework（地址簿框架）中定义好的。
- **ISOcountryCode**。ISO国家代号。
- **country**。国家信息。
- **postalCode**。邮政编码。
- **administrativeArea**。行政区域信息。
- **subAdministrativeArea**。行政区域附加信息。
- **locality**。指定城市信息。
- **subLocality**。指定城市信息附加信息。
- **thoroughfare**。指定街道级别信息。
- **subThoroughfare**。指定街道级别的附加信息。

地理信息反编码使用CLGeocoder类实现，这个类能够实现在地理坐标与地理文字描述信息之间的转换。CLGeocoder类中进行地理信息反编码的方法是：

```
func reverseGeocodeLocation(_ location: CLLocation,
➥completionHandler completionHandler: CLGeocodeCompletionHandler)
```

```
- (void)reverseGeocodeLocation:(CLLocation *)location
➥completionHandler:(CLGeocodeCompletionHandler)completionHandler
```

其中参数location是要定位的地理位置对象，completionHandler参数指定了一个代码块CLGeocodeCompletion-Handler对象，用于地理信息反编码之后的回调。

17.3.2 实例：地理信息反编码

图17-15是在上一节实例的基础上添加了地理信息反编码功能。在界面添加了一个“地理信息反编码”按钮，反编码的结果就会显示在下面的TextView控件中。

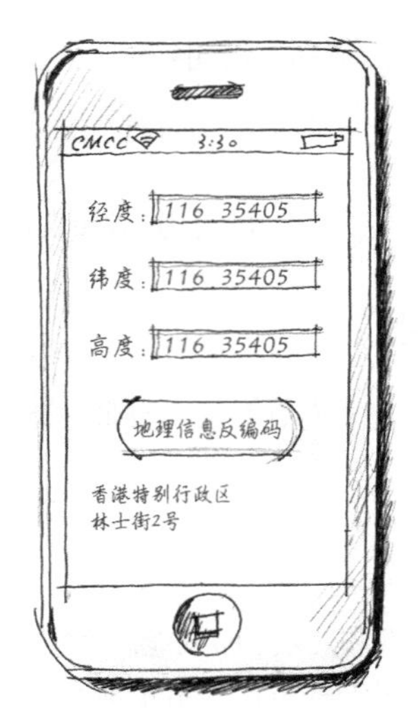

图17-15　地理信息反编码实例

UI设计部分我们不再介绍，直接看实现代码，主要是在视图控制器ViewController中添加按钮事件处理方法reverseGeocode:

```
//ViewController.swift文件
@IBAction func reverseGeocode(_ sender: AnyObject) {

    let geocoder = CLGeocoder()

    geocoder.reverseGeocodeLocation(self.currLocation,
    ➥completionHandler: { (placemarks, error) -> Void in                ①

        if error != nil {
            print(error!.localizedDescription)
        } else if placemarks != nil && placemarks!.count  > 0 {         ②

            let placemark = placemarks![0]                               ③
            let name = placemark.name                                    ④
            self.txtView.text = name
        }
    })

}
```

```
//ViewController.m文件
- (IBAction)reverseGeocode:(id)sender {

    CLGeocoder *geocoder = [[CLGeocoder alloc] init];

    [geocoder reverseGeocodeLocation:self.currLocation
    ➥completionHandler:^(NSArray<CLPlacemark *> *placemarks,
    ➥NSError *error) {                                                  ①

        if (error) {
            NSLog(@"Error is %@",error.localizedDescription);
        } else if ([placemarks count] > 0) {                             ②

            CLPlacemark *placemark = placemarks[0];                      ③
            NSString *name = placemark.name;                             ④
            self.txtView.text = name;

        }
    }];
}
```

第①行调用reverseGeocodeLocation:completionHandler:方法进行地理信息编码，其中self.currLocation是CLLocation类型的属性，它是在委托方法locationManager:didUpdateLocations:中初始化的。在回调代码块（Swift版本为闭包）completionHandler:中，参数placemarks是返回的地标集合。事实上，一个地理坐标点泛指一个范围，因此在这个范围中可能有多种不同的描述信息，这些信息被放在地标集合中。另一个参数error描述了出错信息，如果error非空，则说明反编码失败。

第②行用于判断地标集合长度大于0（即反编码成功并且有返回的描述信息）的情况。第③行用于从地标集合中取出第一个地标CLPlacemark对象，如果用户需要查看所有的地标信息，循环遍历出来就可以了。第④行取出地标对象的name属性。

17.3.3　地理信息编码查询

地理信息编码查询与反编码刚好相反，它给定地理信息的文字描述，查询出相关的地理坐标，这种查询结果也是一个集合。例如，查询“城南”，结果会找出很多地点，因为这个关键词太普遍了，这种情况下可以指定区域范围查询。地理信息编码查询也是采用CLGeocoder类，其中有关地理信息编码的方法如下。

- **geocodeAddressDictionary:completionHandler**。通过指定一个地址信息字典对象参数进行查询。
- **geocodeAddressString:completionHandler**。通过指定一个地址字符串参数进行查询。
- **geocodeAddressString:inRegion:completionHandler**。通过指定地址字符串和查询范围作为参数进行查询，其中inRegion部分的参数是指定的查询范围，它是CLRegion类型。

17.3.4 实例：地理信息编码查询

下面我们通过一个实例（如图17-16所示）介绍一下地理信息编码查询。在“输入查询地点关键字”文本框中输入关键字，多个关键字之间可以用“,”分隔。使用“地理信息编码查询”按钮查询出结果后，取出第一个地址显示在界面中。

图17-16 地理信息编码查询案例

视图控制器ViewController中按钮事件处理方法geocodeQuery:的代码如下：

```
//ViewController.swift文件
@IBAction func geocodeQuery(_ sender: AnyObject) {

    if self.txtQueryKey.text == nil {
        return
    }

    let geocoder = CLGeocoder()

    geocoder.geocodeAddressString(self.txtQueryKey.text!,
    ➥completionHandler: { (placemarks, error) -> Void in          ①

        if error != nil {
            print("\(error?.localizedDescription)")
        } else if placemarks!.count > 0 {

            let placemark = placemarks![0] as CLPlacemark

            let name = placemark.name!

            let location = placemark.location!                      ②

            let lng = location.coordinate.longitude
            let lat = location.coordinate.latitude
```

```
//ViewController.m文件
- (IBAction)geocodeQuery:(id)sender {

    if (self.txtQueryKey.text == nil) {
        return;
    }

    CLGeocoder *geocoder = [[CLGeocoder alloc] init];

    [geocoder geocodeAddressString:self.txtQueryKey.text
    ➥completionHandler:^(NSArray<CLPlacemark *> * placemarks,
    ➥NSError * error) {                                            ①

        if (error) {
            NSLog(@"Error is %@",error.localizedDescription);
        } else if ([placemarks count] > 0) {

            CLPlacemark* placemark = placemarks[0];

            NSString* name = placemark.name;

            CLLocation *location = placemark.location;              ②

            double lng = location.coordinate.longitude;
```

```
            self.txtView.text = String(format: "经度: %3.5f \n纬度: %3.5f
            ➥\n%@", lng, lat, name)

        }
    })

    //关闭键盘
    self.txtQueryKey.resignFirstResponder()

}
```

```
            double lat = location.coordinate.latitude;

            self.txtView.text = [NSString stringWithFormat:
            ➥@"经度: %3.5f \n纬度: %3.5f \n%@", lng, lat, name];

        }

    }];
    //关闭键盘
    [self.txtQueryKey resignFirstResponder];
}
```

第①行调用geocodeAddressString:completionHandler:方法进行地理信息编码查询，这里没有指定查询范围。查询返回结果保存到placemarks参数中，它是数值类型，取出集合中的第一个地标对象。第②行使用placemark.location属性获得地标的经纬度信息。

这样就可以查询了。如果想指定查询范围，可以使用geocodeAddressString:inRegion:completionHandler:方法，代码如下：

```
let location = CLLocation(latitude: 40.002240, longitude: 117.323328)     ①
let region = CLCircularRegion(center:location.coordinate,
➥radius: 5000, identifier:"GeocodeRegion")                                 ②
let geocoder = CLGeocoder()
geocoder.geocodeAddressString(self.txtQueryKey.text!, in:region,
➥completionHandler: { (placemarks, error) -> Void in                       ③
    //TODO
})
```

```
CLLocation *location = [[CLLocation alloc]
➥initWithLatitude:40.002240 longitude:117.323328];                        ①

CLCircularRegion* region = [[CLCircularRegion alloc]
➥initWithCenter:location.coordinate
➥radius:5000 identifier:@"GeocodeRegion"];                                  ②

[geocoder geocodeAddressString:self.txtQueryKey.text inRegion:region
➥completionHandler:^(NSArray<CLPlacemark *> * placemarks,
➥NSError * error) {                                                         ③
    //TODO
}];
```

在上述代码中，第①行创建了一个CLLocation对象，该对象指定查询范围中心点。当然，这个对象可以是从其他模块传递过来的。

第②行用于构造一个圆形区域对象CLCircularRegion，其构造函数为initWithCenter:radius: identifier:，其中第一个参数是指定区域的中心点，类型是CLLocationCoordinate2D，第二个参数radius指定区域半径的单位为米，第三个参数identifier用于为区域指定一个标识，保证在你的应用中是唯一的，这个参数不能为空。第③行用于执行查询。

17.4　小结

通过对本章的学习，我们了解了iOS中的定位服务技术，包括地理信息编码和反编码查询。

第 18 章

苹果地图应用

18

当使用GPS定位服务获得位置信息后，很多应用的进一步需求是在地图上添加标注，这样才能使定位的位置或线路看起来一目了然。在移动开发中，定位服务与地图应用是完全不同的两套API，但是它们结合得很紧密，上一章介绍了定位，本章就来介绍地图应用开发。

18.1 使用 iOS 苹果地图

在iOS 6之后，苹果用自己的地图代替了谷歌地图，但是API编程接口没有太大的变化，地图应用开发主要使用Map Kit API，其核心是MKMapView类。

18.1.1 显示地图

在Map Kit API中，显示地图的视图是MKMapView，其委托协议是MKMapViewDelegate。使用Map Kit API时，需要导入MapKit框架。

MKMapView的mapType属性可以设置地图类型。地图类型共有5种，是在枚举类型MKMapType中定义的成员，如表18-1所示。

表18-1 MKMapType枚举成员

Swift枚举成员	Objective-C枚举成员	说　明
standard	MKMapTypeStandard	标准地图类型，如图18-1a所示
satellite	MKMapTypeSatellite	卫星地图类型。如图18-1b所示，卫星地图中没有街道名称等信息
hybrid	MKMapTypeHybrid	混合地图类型，如图18-1c所示，卫星地图上标注出街道等信息
satelliteFlyover	MKMapTypeSatelliteFlyover	卫星Flyover地图类型。Flyover地图能够提供从空中俯瞰的3D效果。如图18-2所示，但是目前中国区还不支持
hybridFlyover	MKMapTypeHybridFlyover	在卫星Flyover地图上标注出街道等信息，但是目前中国区还不支持

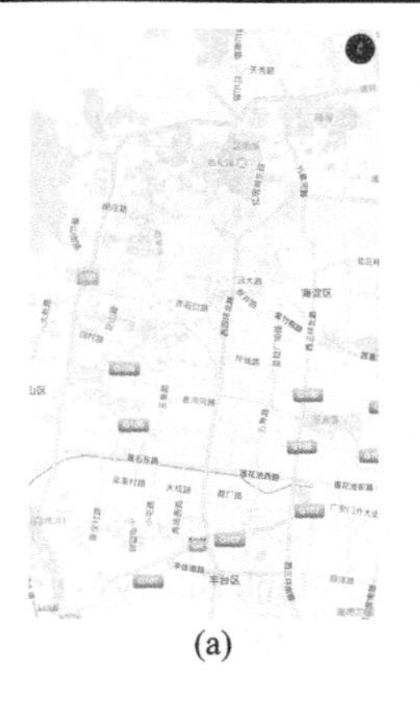
(a)

(b)

(c)

图18-1 地图类型

图18-2　iOS 9 Flyover地图类型

下面我们通过一个案例介绍一下显示地图的控制，图18-3所示的界面有一个地图视图，我们可以通过分段控件中的按钮切换3种地图类型。

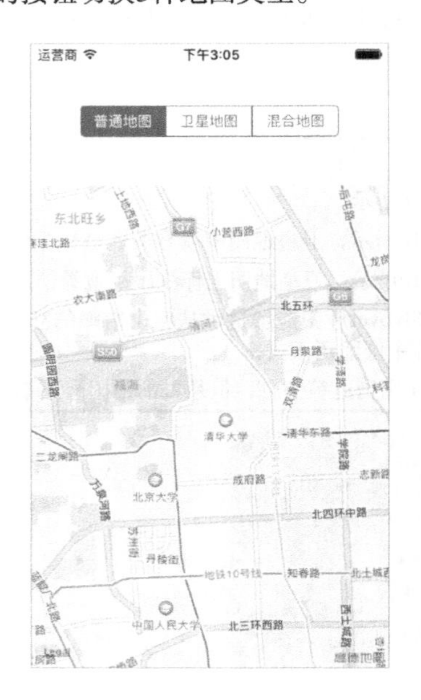

图18-3　显示iOS地图应用案例

首先，使用Single View Application模板创建一个名为MapSample的工程。然后为工程添加MapKit.framework框架，具体步骤是选择工程中的TARGETS→MapSample→Build Phases→Link Binary With Libraries（0 items），选择右下角的+按钮，打开Choose frameworks and libraries to add（选择要添加的框架和库）对话框，如图18-4所示。我们在该对话框中选择MapKit.framework，点击Add按钮完成添加。

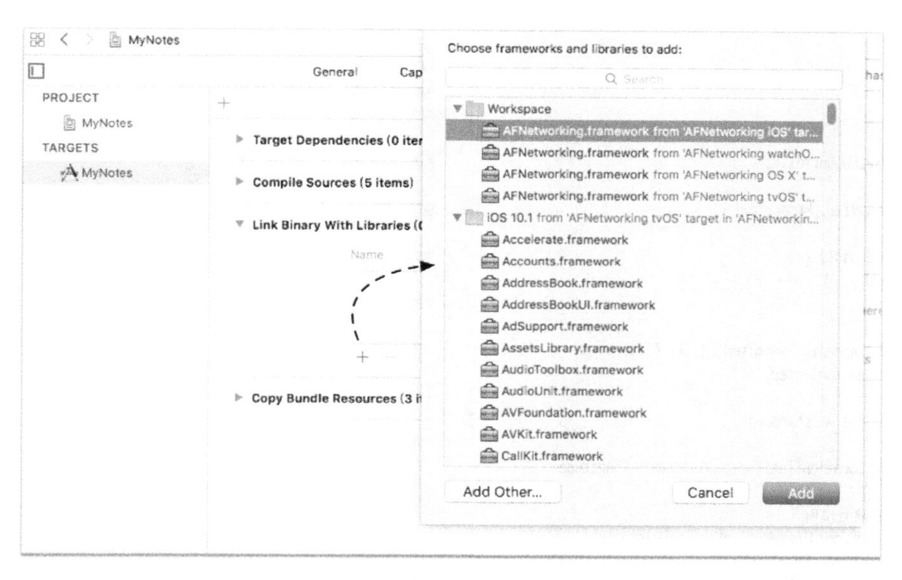

图18-4　添加框架到工程

添加完MapKit.framework框架后，就可以设计界面了。打开Main.storyboard文件，如图18-5所示，从对象库中拖曳MapKit View到设计界面，然后为地图视图添加约束和输出口。有关添加分段控件到设计界面的过程这里不再赘述。

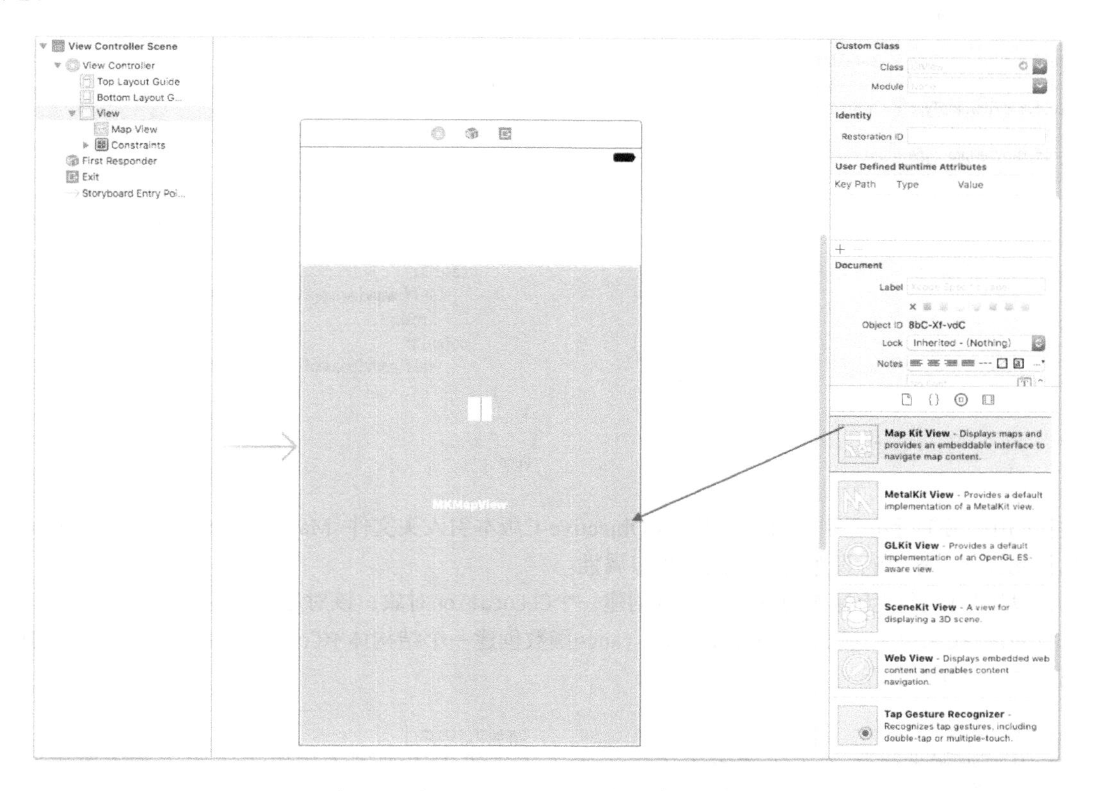

图18-5　在Interface Builder中设计Map View

下面我们看看代码部分，主视图控制器ViewController中与类定义和属性声明相关的代码如下：

```
//ViewController.swift文件
import UIKit
import MapKit                                                          ①

class ViewController: UIViewController {

    @IBOutlet weak var mapView: MKMapView!                             ②

    override func viewDidLoad() {
        super.viewDidLoad()
    }

    override func viewWillAppear(_ animated: Bool) {
        super.viewWillAppear(animated)

        self.mapView.mapType = .standard                               ③

        let location = CLLocation(latitude: 40.002240, longitude:
➥116.323328)                                                           ④
        //调整地图位置和缩放比例
        let viewRegion = MKCoordinateRegionMakeWithDistance
➥(location.coordinate, 10000, 10000)                                   ⑤
        self.mapView.region = viewRegion                               ⑥
    }

    @IBAction func selectMapViewType(_ sender: AnyObject) {

        let sc = sender as! UISegmentedControl

        switch (sc.selectedSegmentIndex) {
        case 1:
            self.mapView.mapType = .satellite
        case 2:
            self.mapView.mapType = .hybrid
        default:
            self.mapView.mapType = .standard
        }

    }
}
```

```
//ViewController.m文件
#import "ViewController.h"
#import <MapKit/MapKit.h>                                              ①

@interface ViewController ()

@property (weak, nonatomic) IBOutlet MKMapView *mapView;               ②

@end

@implementation ViewController

- (void)viewDidLoad {
    [super viewDidLoad];
}

- (void)viewWillAppear:(BOOL)animated {

    [super viewWillAppear: animated];

    self.mapView.mapType = MKMapTypeStandard;                          ③

    CLLocation *location = [[CLLocation alloc] initWithLatitude:40.002240
➥longitude:116.323328];                                                ④
    //调整地图位置和缩放比例
    MKCoordinateRegion viewRegion =
➥MKCoordinateRegionMakeWithDistance(location.coordinate, 10000,
➥10000);                                                               ⑤
    self.mapView.region = viewRegion;                                  ⑥

}

- (IBAction)selectMapViewType:(id)sender {

    UISegmentedControl* sc = (UISegmentedControl*)sender;

    switch (sc.selectedSegmentIndex) {
        case 1:
            self.mapView.mapType = MKMapTypeSatellite;
            break;
        case 2:
            self.mapView.mapType = MKMapTypeHybrid;
            break;
        default:
            self.mapView.mapType = MKMapTypeStandard;
    }
}

@end
```

上述代码中，第①行用于引入MapKit模块（Objective-C版本引入头文件<MapKit/MapKit.h>），使用Map Kit API时需要执行此操作。第②行设置地图视图输出口属性。

第③行设置地图类型为标准地图。第④行创建一个CLLocation对象，该对象指定地图中心点。

第⑤行使用MKCoordinateRegionMakeWithDistance函数创建一个结构体MKCoordinateRegion，该结构体封装了一个地图区域，其定义如下：

```
struct MKCoordinateRegion {
    var center: CLLocationCoordinate2D              //中心点
    var span: MKCoordinateSpan                      //跨度
}
```

```
typedef struct {
    CLLocationCoordinate2D center;                  //中心点
    MKCoordinateSpan span;                          //跨度
} MKCoordinateRegion;
```

MKCoordinateRegion结构体的成员center定义了区域中心点，是CLLocationCoordinate2D结构体类型。span成

员定义了区域的跨度，是MKCoordinateSpan结构体类型。MKCoordinateSpan结构体封装了地图上的跨度信息，它的定义如下：

```
struct MKCoordinateSpan {
    var latitudeDelta: CLLocationDegrees        //区域的南北跨度
    var longitudeDelta: CLLocationDegrees       //区域的东西跨度
}
```

```
typedef struct {
    CLLocationDegrees latitudeDelta;        //区域的南北跨度
    CLLocationDegrees longitudeDelta;       //区域的东西跨度
} MKCoordinateSpan;
```

在上述代码中，latitudeDelta为南北跨度，longitudeDelta为东西跨度，它们的单位是米。

第⑥行用于重新设置地图视图的显示区域，这会影响地图缩放。

18.1.2 显示 3D 地图

在iOS 7之后，苹果提供了3D地图显示功能，我们可以在应用中添加3D视图，其中核心是摄像机类MKMapCamera。为了支持3D地图，MKMapView增加了camera属性，该属性是MKMapCamera类型。

摄像机是3D空间中非常重要的内容，摄像机的属性决定了我们看到的3D地图样式。摄像机有如下4个重要的属性。

- **centerCoordinate**。设置地图视图的中心坐标。
- **pitch**。摄像机俯视角，0° 垂直指向地图，90° 平行于地图。
- **altitude**。摄像机海拔高度。
- **heading**。摄像头前进方向，相对于地理北方的角度，0° 则说明地图视图顶边为地理北方，90° 则说明地图视图的顶部为地理东方。

下面我们为上一节案例添加3D地图显示功能。修改主视图控制器ViewController的代码如下：

```
//ViewController.swift文件
class ViewController: UIViewController {
    ......

    override func viewWillAppear(_ animated: Bool) {
        super.viewWillAppear(animated)
        self.mapView.mapType = .Standard
        self.placeCamera()                                              ①
    }

    @IBAction func selectMapViewType(_ sender: AnyObject) {

        let sc = sender as! UISegmentedControl
        switch (sc.selectedSegmentIndex) {
        case 1:
            self.mapView.mapType = .satellite
        case 2:
            self.mapView.mapType = .hybrid
        default:
            self.mapView.mapType = .standard
        }
        self.placeCamera()                                              ②
    }

    func placeCamera() {                                                ③

        //设置3D摄像机
        let mapCamera = MKMapCamera()                                   ④
        mapCamera.centerCoordinate =
        ➥CLLocationCoordinate2DMake(40.002240, 116.323328)              ⑤
        mapCamera.pitch = 45                                            ⑥
        mapCamera.altitude = 500                                        ⑦
        mapCamera.heading = 45                                          ⑧
```

```
//ViewController.m文件
@implementation ViewController
......

- (void)viewWillAppear:(BOOL)animated {
    [super viewWillAppear: animated];
    self.mapView.mapType = MKMapTypeStandard;
    [self placeCamera];                                                 ①
}

- (IBAction)selectMapViewType:(id)sender {
    UISegmentedControl* sc = (UISegmentedControl*)sender;
    switch (sc.selectedSegmentIndex) {
        case 1:
            self.mapView.mapType = MKMapTypeSatellite;
            break;
        case 2:
            self.mapView.mapType = MKMapTypeHybrid;
            break;
        default:
            self.mapView.mapType = MKMapTypeStandard;
    }
    [self placeCamera];                                                 ②
}

-(void) placeCamera {                                                   ③
    //设置3D摄像机
    MKMapCamera* mapCamera = [MKMapCamera camera];                      ④
    mapCamera.centerCoordinate =
    ➥CLLocationCoordinate2DMake(40.002240, 116.323328);                 ⑤
    mapCamera.pitch = 45;                                               ⑥
    mapCamera.altitude = 500;                                           ⑦
    mapCamera.heading = 45;                                             ⑧
```

```
        //设置地图视图的camera属性
        self.mapView.camera = mapCamera
    }
}
```

```
    //设置地图视图的camera属性
    self.mapView.camera = mapCamera;
}

@end
```

我们需要在视图显示时和切换地图类型时设置摄像机，如第①行和第②行调用placeCamera方法。第③行定义了placeCamera方法，该方法实现了摄像机的设置。第④行用于创建摄像机对象MKMapCamera。Objective-C版中是采用静态方法camera创建摄像机对象。第⑤行设置centerCoordinate属性。第⑥行设置pitch属性，这里设置为45°，一般情况下设置为45是比较理想的。第⑦行用于设置altitude属性，这里设置为500。第⑧行设置heading属性为45。

18.2　添加标注

出于多种目的，我们需要在地图上添加一些标注点。在苹果地图中默认样式是一个大头针，如图18-6所示。大头针还可以带有主标题和副标题。

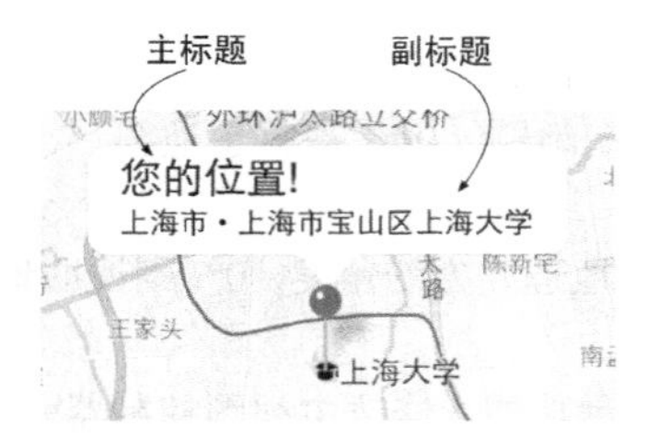

图18-6　默认样式的地图标注点

下面我们通过案例介绍一下如何在地图上添加标注点。如图18-7所示，在“输入查询地点关键字”文本框中输入关键字，点击“查询”按钮，通过地理信息编码查询，获得地标信息，然后在地图上标注出来。

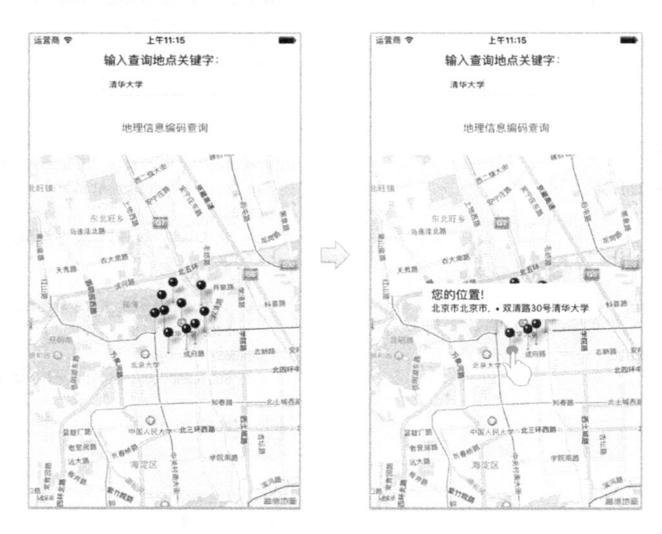

图18-7　在地图上添加标注点的案例

首先，使用Single View Application模板创建一个名为MapAnnotationSample的工程。参考18.1.1节的案例添加MapKit.framework到工程，并在Interface Builder中设计如图18-7所示的界面，具体过程不再赘述。

我们需要两个步骤：第一步是实现查询，第二步是在地图上添加标注。

18.2.1 实现查询

我们通过“查询”按钮触发查询动作，实现查询可以通过17.3.3节介绍的地理信息编码查询，也可以通过MKLocalSearch类实现。本例使用地理信息编码查询实现方式。

通过地理信息编码查询实现的相关代码如下：

```
//ViewController.swift文件
@IBAction func geocodeQuery(_ sender: AnyObject) {

    if (self.txtQueryKey.text == nil) {
        return
    }

    let geocoder = CLGeocoder()
    geocoder.geocodeAddressString(self.txtQueryKey.text!,
    ➥completionHandler: { (placemarks, error) -> Void in

        print("查询记录数：", placemarks!.count)
        self.mapView.removeAnnotations(self.mapView.annotations)          ①

        for placemark in placemarks! {

            let annotation = MyAnnotation(coordinate: placemark.location!.
            ➥coordinate)                                                   ②
            annotation.city = placemark.locality
            annotation.state = placemark.administrativeArea
            annotation.streetAddress = placemark.thoroughfare
            annotation.zip = placemark.postalCode                          ③

            self.mapView.addAnnotation(annotation)                         ④

        }

        if placemarks!.count  > 0 {
            //取出最后一个地标点
            let lastPlacemark = placemarks!.last                           ⑤
            //调整地图位置和缩放比例
            let viewRegion = MKCoordinateRegionMakeWithDistance(
            ➥lastPlacemark!.location!.coordinate, 10000, 10000)
            self.mapView.setRegion(viewRegion, animated: true)             ⑥
        }

        //关闭键盘
        self.txtQueryKey.resignFirstResponder()
    })
}
```

```
//ViewController.m文件
- (IBAction)geocodeQuery:(id)sender {

    if (self.txtQueryKey.text == nil || [self.txtQueryKey.text length] == 0) {
        return;
    }

    CLGeocoder *geocoder = [[CLGeocoder alloc] init];
    [geocoder geocodeAddressString:_txtQueryKey.text
    ➥completionHandler:^(NSArray *placemarks, NSError *error) {

        NSLog(@"查询记录数：%lu", [placemarks count]);
        [self.mapView removeAnnotations:self.mapView.annotations];          ①

        for (CLPlacemark *placemark in placemarks) {

            MyAnnotation *annotation = [[MyAnnotation alloc]
            ➥initWithCoordinate:placemark.location.coordinate];            ②
            annotation.streetAddress = placemark.thoroughfare;
            annotation.city = placemark.locality;
            annotation.state = placemark.administrativeArea;
            annotation.zip = placemark.postalCode;                          ③

            [self.mapView addAnnotation:annotation];                        ④
        }

        if ([placemarks count] > 0) {
            //取出最后一个地标点
            CLPlacemark *lastPlacemark = placemarks.lastObject;             ⑤
            //调整地图位置和缩放比例
            MKCoordinateRegion viewRegion = MKCoordinateRegionMakeWith
            ➥Distance(lastPlacemark.location.coordinate, 10000, 10000);
            [self.mapView setRegion:viewRegion animated:TRUE];              ⑥
        }

        //关闭键盘
        [_txtQueryKey resignFirstResponder];
    }];
}
```

上述代码中，第①行用于移除目前地图上所有的标注点，否则反复查询后，你会发现地图上的标注点越来越多。

第②行用于实例化MyAnnotation对象。MyAnnotation类是我们自定义的实现MKAnnotation协议的地图标注点类。因为地图上的标注点是MKPinAnnotationView（大头针标注视图）类型，这个视图要求标注点信息由实现MKAnnotation协议的类提供。如果标注点上显示的信息是固定的，可以使用Map Kit API实现MKPointAnnotation标注类，而不需要自己编写MyAnnotation类。第②行~第③行将地标CLPlacemark对象信息取出，放入MyAnnotation对象中。为什么要这样导来导去呢？这是因为在MKPinAnnotationView视图中，只能接收实现MKAnnotation协议的类，而地标类CLPlacemark没有实现MKAnnotation协议。

第④行把标注点对象MyAnnotation添加到地图视图上。一旦该方法被调用，地图视图委托方法mapView:viewForAnnotation:就会被回调。

第⑤行取出最后一个地标点。然后，第⑥行将这个地标点设置为地图中心点，类似的设置也可以通过语句self.mapView.region = viewRegion实现，但是使用该语句设置中心点的过程中没有动画效果。使用setRegion:

animated:方法不仅可以设置中心点，而且可以在将animated设置为true（或YES）时，会使地图有“飞”过去的动画效果。

自定义标注类MyAnnotation的代码如下：

```swift
//MyAnnotation.swift文件
import MapKit

class MyAnnotation: NSObject, MKAnnotation {
    //街道信息属性
    var streetAddress: String!
    //城市信息属性
    var city: String!
    //州、省、市信息
    var state: String!
    //邮编
    var zip: String!
    //地理坐标
    var coordinate: CLLocationCoordinate2D

    init(coordinate: CLLocationCoordinate2D) {
        self.coordinate = coordinate
    }

    var title: String? {
        return "你的位置!"
    }
    var subtitle: String? {

        let res = NSMutableString()
        if  (self.state != nil) {
            res.appendFormat("%@", self.state)
        }

        if  (self.city != nil) {
            res.appendFormat(" • %@", self.state)
        }

        if  (self.zip != nil) {
            res.appendFormat(" • %@", self.zip)
        }

        if  (self.streetAddress != nil) {
            res.appendFormat(" • %@", self.streetAddress)
        }
        return res as String
    }
}
```

```objc
//MyAnnotation.h文件
#import <MapKit/MapKit.h>

@interface MyAnnotation : NSObject <MKAnnotation>

-(instancetype)initWithCoordinate:(CLLocationCoordinate2D) coordinate;

//街道信息属性
@property(nonatomic, copy) NSString *streetAddress;
//城市信息属性
@property(nonatomic, copy) NSString *city;
//州、省、市信息
@property(nonatomic, copy) NSString *state;
//邮编
@property(nonatomic, copy) NSString *zip;
//地理坐标
@property(nonatomic, readwrite) CLLocationCoordinate2D coordinate;

@end

//MyAnnotation.m文件
#import "MyAnnotation.h"

@implementation MyAnnotation

-(instancetype)initWithCoordinate:(CLLocationCoordinate2D) coordinate {
    self = [super init];
    if (self) {
        self.coordinate = coordinate;
    }
    return self;
}

- (NSString *)title {
    return @"你的位置!";
}

- (NSString *)subtitle {

    NSMutableString *res = [[NSMutableString alloc] init];
    if  (self.state != nil) {
        [res appendFormat:@"%@", self.state];
    }

    if  (self.city != nil) {
        [res appendFormat:@" • %@", self.state];
    }

    if  (self.zip != nil) {
        [res appendFormat:@" • %@", self.zip];
    }

    if  (self.streetAddress != nil) {
        [res appendFormat:@" • %@", self.streetAddress];
    }

    return res;
}

@end
```

地图上的标注点类必须实现MKAnnotation协议。MKAnnotation协议需要重写如下两个属性。

- ❑ **title**。标注点上的主标题。
- ❑ **subtitle**。标注点上的副标题。

在重写subtitle属性时，我们将它的相关信息拼接成字符串赋值给它。这里，我们可以根据自己的需要和习惯拼接在这个字符串的前后。

18.2.2 在地图上添加标注

查询出结果后，我们通过如下代码将标注点添加到地图视图中：

```
self.mapView.addAnnotation(annotation)
```

```
[self.mapView addAnnotation:annotation];
```

这会回调地图委托协议MKMapViewDelegate的mapView:viewForAnnotation:方法，我们需要实现这个方法才能实现在地图上添加标注点的操作。

实现MKMapViewDelegate协议的相关代码如下：

```
//ViewController.swift文件
override func viewDidLoad() {
    super.viewDidLoad()

    self.mapView.mapType = .standard
    self.mapView.delegate = self                                        ①
}

//MARK: -- 实现MKMapViewDelegate协议
func mapView(_ mapView: MKMapView, viewFor annotation: MKAnnotation) ->
➥MKAnnotationView? {

    var annotationView = self.mapView
    ➥.dequeueReusableAnnotationViewWithIdentifier("PIN_ANNOTATION")
    ➥as? MKPinAnnotationView                                            ②

    if annotationView == nil {
        annotationView = MKPinAnnotationView(annotation: annotation,
        ➥reuseIdentifier: "PIN_ANNOTATION")                             ③
    }
    annotationView!.pinTintColor = UIColor.red                          ④
    annotationView!.animatesDrop = true                                 ⑤
    annotationView!.canShowCallout = true                               ⑥

    return annotationView!
}

func mapViewDidFailLoadingMap(_ mapView: MKMapView,
➥withError error:NSError) {
    print("error : \(error.localizedDescription)")
}
```

```
//ViewController.m文件
- (void)viewDidLoad {
    [super viewDidLoad];

    self.mapView.mapType = MKMapTypeStandard;
    self.mapView.delegate = self;                                       ①
}

#pragma mark -- 实现MKMapViewDelegate协议
- (MKAnnotationView *)mapView:(MKMapView *)theMapView
➥viewForAnnotation:(id <MKAnnotation>)annotation {

    MKPinAnnotationView *annotationView = (MKPinAnnotationView *)
    [self.mapView
    ➥dequeueReusableAnnotationViewWithIdentifier:@"PIN_ANNOTATION"];   ②
    if (annotationView == nil) {
        annotationView = [[MKPinAnnotationView alloc]
        ➥initWithAnnotation:annotation reuseIdentifier:
        ➥@"PIN_ANNOTATION"];                                            ③
    }

    annotationView.pinTintColor = [UIColor redColor];                   ④
    annotationView.animatesDrop = TRUE;                                 ⑤
    annotationView.canShowCallout = TRUE;                               ⑥

    return annotationView;
}

- (void)mapViewDidFailLoadingMap:(MKMapView *)theMapView
➥withError:(NSError *)error {
    NSLog(@"error : %@", [error localizedDescription]);
}
```

在上述代码中，第①行将当前视图控制器self作为地图视图委托对象。

mapView:viewForAnnotation:在为地图视图添加标注时回调。给地图视图添加标注的方法是self.mapView.addAnnotation(annotation)，其中annotation是地图标注对象。

第②行~第③行用于获得地图标注对象MKPinAnnotationView，其中采用了可重用对象MKPinAnnotationView设计。这里使用可重用对象是为了节约内存。一般情况下，我们建议尽可能使用已有对象，减少实例化对象。首先，在第②行中，我们使用dequeueReusableAnnotationViewWithIdentifier:方法通过一个可重用标识符PIN_ANNOTATION获得MKPinAnnotationView对象；如果这个对象不存在，则需要使用第③行的initWithAnnotation:reuseIdentifier:构造函数初始化，其中reuseIdentifier参数是可重用标识符。

第④行设置大头针标注视图的颜色为紫色。第⑤行说明设置标注视图时，是否以动画效果的形式显示在地图上。第⑥行用于在标注点上显示一些附加信息。如果canShowCallout为true（或YES），则点击“大头针”头时，会出现一个气泡（如图18-6所示）。而气泡中的文字信息封装在MyAnnotation对象中，其中第一行文字（大一点的文字）保存在title属性中，而第二行文字（小一点儿的文字）保存在subtitle属性中。

18.3　跟踪用户位置变化

一些导航应用需要在地图上显示用户的位置，苹果地图为此提供了一个蓝色小圆点图标，来表示用户当前所处的位置，如图18-8a所示。另外，如果带有导航功能，则蓝色小圆点图标会有探照灯效果，如图18-8b所示，同时还会在屏幕的右上角显示一个指南针图标。

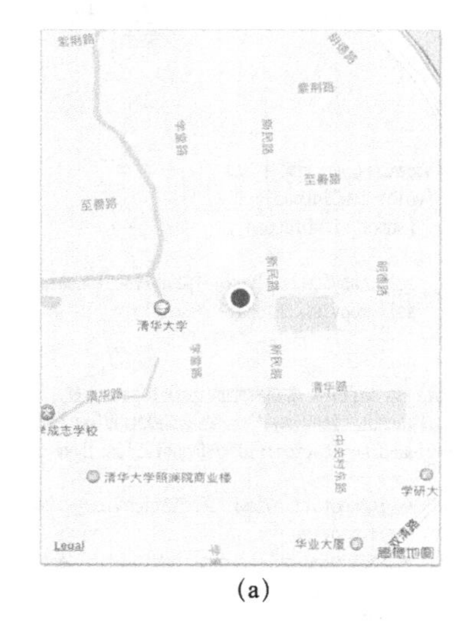

(a)

(b)

图18-8　显示用户位置

MapKit提供了跟踪用户位置和方向变化的API，这样我们就不用自己编写定位服务代码了。MKMapView的属性showsUserLocation可以开启在地图视图上显示用户位置功能（即出现蓝色小圆点）；MKMapView的setUserTrackingMode:animated:方法可以设置用户跟踪模式，其中第二个参数是MKUserTrackingMode枚举类型。枚举类型MKUserTrackingMode中定义的成员，如表18-2所示。

表18-2　MKUserTrackingMode枚举成员

Swift枚举成员	Objective-C枚举成员	说　明
none	MKUserTrackingModeNone	没有用户跟踪模式
follow	MKUserTrackingModeFollow	跟踪用户的位置变化
followWithHeading	MKUserTrackingModeFollowWithHeading	跟踪用户的位置和前进方向变化

下面我们将18.1.1节的案例修改一下，添加跟踪用户位置功能。相关UI部分与18.1.1节的案例一样，不需要修改。下面我们重点看看代码，主视图控制器ViewController的代码如下：

```
//ViewController.swift文件
import UIKit
import MapKit
import CoreLocation

class ViewController: UIViewController, MKMapViewDelegate {
```

```
//ViewController.m文件
#import "ViewController.h"
#import <MapKit/MapKit.h>
#import <CoreLocation/CoreLocation.h>

@interface ViewController () <MKMapViewDelegate>
```

```swift
    @IBOutlet weak var mapView: MKMapView!

    var locationManager: CLLocationManager!

    override func viewWillAppear(_ animated: Bool) {
        ......

        //授权
        self.locationManager = CLLocationManager()                              ①
        self.locationManager.requestWhenInUseAuthorization()
        self.locationManager.requestAlwaysAuthorization()                       ②

        self.mapView.showsUserLocation = true                                   ③
        self.mapView.userLocation.title = "我在这里！"                            ④
        self.mapView.delegate = self                                            ⑤
    }

    ......

    //MARK: -- 实现MKMapViewDelegate协议
    func mapViewDidFinishLoadingMap(_ mapView: MKMapView) {                     ⑥
        mapView.userTrackingMode = .followWithHeading                           ⑦
    }
}
```

```objc
@property (weak, nonatomic) IBOutlet MKMapView *mapView;
@property(nonatomic, strong) CLLocationManager *locationManager;

@end

@implementation ViewController

- (void)viewWillAppear:(BOOL)animated {
    ......

    //授权
    self.locationManager = [[CLLocationManager alloc] init];                    ①
    [self.locationManager requestWhenInUseAuthorization];
    [self.locationManager requestAlwaysAuthorization];                          ②

    self.mapView.showsUserLocation = TRUE;                                      ③
    self.mapView.userLocation.title = @"我在这里！";                              ④
    self.mapView.delegate = self;                                               ⑤
}
......

#pragma mark -- 实现MKMapViewDelegate协议
- (void)mapViewDidFinishLoadingMap:(MKMapView *)mapView {                       ⑥
    mapView.userTrackingMode = MKUserTrackingModeFollowWithHeading;             ⑦
}

@end
```

上述代码中，第①行~第②行是用户授权服务位置信息，这是因为跟踪用户位置需要定位服务，所以需要定位授权。另外，还需要修改工程配置文件Info.plist，添加定位服务请求授权键——NSLocationAlwaysUsageDescription和NSLocationWhenInUseUsageDescription，具体参考17.1.1节。

第③行的showsUserLocation属性用于设置在地图上显示用户位置。第④行的userLocation是从地图视图中获得用户位置对象MKUserLocation。MKUserLocation的title属性可以设置主标题，subtitle属性可以设置副标题。

第⑤行将当前视图控制器self作为地图视图委托对象。

第⑥行是实现委托协议MKMapViewDelegate的mapViewDidFinishLoadingMap:方法，该方法会在地图视图加载成功时回调。

第⑦行设置用户跟踪模式，用户跟踪模式是通过设置MapView实例方法setUserTrackingMode: animated:实现的。

跟踪用户位置的变化测试起来比较麻烦，我们可以用模拟器测试一下。应用运行之后，选择模拟器的Location菜单项，通过如图17-11所示的后面3个菜单项（City Bicycle Ride、City Run和Freeway Drive）可以改变用户的位置。

18.4 使用程序外地图

有时候我们只使用地图的一些基本功能，如查看位置、导航、参考路线等，此时不必自己设计地图界面，也不需要面对复杂的API，只需在应用中调用程序外地图。

iOS设备中都带有一个“地图”应用，如图18-9所示，我们可以使用它完成地图相关的大部分工作。事实上，我们可以在自己的应用程序中调用它，并且给它传递一些参数进行初始化显示。

在图18-10a所示的界面中输入查询地点关键字，然后进行地理信息编码查询，结果会调入设备自带的苹果地图，如图18-10b所示，界面中的标注点是我们传递给它的。我们可以像使用自带地图应用那样进行查询线路、设置地图类型等操作。

图18-9　iPhone自带的地图应用

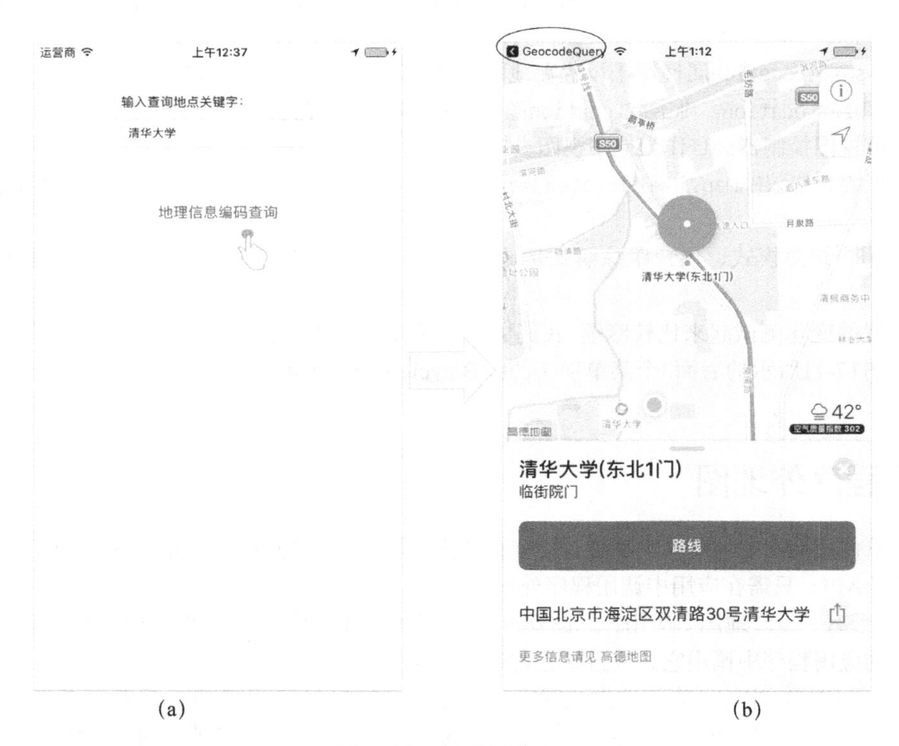

(a)　　(b)

图18-10　调用自带地图

提示 在iOS 9和iOS 10中，从一个应用进入另外一个应用时，屏幕的左上角会有一个返回按钮，可以返回到前一个应用。例如，点击如图18-10b所示的 GeocodeQuery 按钮可以返回GeocodeQuery。

要实现如图18-10所示的案例，我们可以在17.3.4节GeocodeQuery案例的基础上进行修改。查询过程可以使用17.3.4节介绍的地理信息编码实现，也可以使用MKLocalSearch类实现。这一节我们用MKLocalSearch类实现GeocodeQuery案例。

首先，修改17.3.4节中的GeocodeQuery案例，在工程中添加MapKit.framework。修改ViewController的geocodeQuery方法，代码如下：

```
//ViewController.swift文件
@IBAction func geocodeQuery(sender: AnyObject) {

    if self.txtQueryKey.text == nil {
        return
    }

    let request = MKLocalSearchRequest()                                    ①
    request.naturalLanguageQuery = self.txtQueryKey.text                    ②
    let search = MKLocalSearch(request: request)                            ③

    search.start(completionHandler: { (response, error) in                  ④

        if response?.mapItems.count > 0 {
            //取出最后一个地标点
            let lastMapItem = response?.mapItems.last                       ⑤
            lastMapItem!.openInMapsWithLaunchOptions(nil)                   ⑥
        }
    }
    //关闭键盘
    self.txtQueryKey.resignFirstResponder()
}
```

```
//ViewController.m文件
- (IBAction)geocodeQuery:(id)sender {

    if (self.txtQueryKey.text == nil) {
        return;
    }

    MKLocalSearchRequest *request = [[MKLocalSearchRequest alloc] init];    ①
    request.naturalLanguageQuery = self.txtQueryKey.text;                   ②
    MKLocalSearch *search = [[MKLocalSearch alloc] initWithRequest:
    ➥request];                                                             ③

    [search startWithCompletionHandler:^(
    ➥MKLocalSearchResponse *response, NSError *error) {                    ④

        if ([response.mapItems count] > 0) {
            //取出最后一个地标点
            MKMapItem *lastMapItem = response.mapItems.lastObject;          ⑤
            [lastMapItem openInMapsWithLaunchOptions:nil];                  ⑥
        }
    }];

    //关闭键盘
    [self.txtQueryKey resignFirstResponder];
}
```

在上述代码中，第①行用于实例化查询请求对象MKLocalSearchRequest，第②行设置请求对象的natural-LanguageQuery属性（该属性是查询的关键字），第③行用于实例化查询对象MKLocalSearch。

第④行调用查询对象的startWithCompletionHandler:方法执行查询，该方法的参数是一个代码块（Swift版是闭包），用来执行查询之后的回调。代码块有两个参数，即response和error，response是查询应答对象，error是查询过程中的错误对象。

第⑤行中的response.mapItems属性可以返回一个查询结果MKMapItem的集合，那么response.mapItems.lastObject属性就是获得最后一个MKMapItem元素。

第⑥行执行MKMapItem的openInMapsWithLaunchOptions:方法，会调用iOS自带的苹果地图应用，参数是字典类型。这个参数可以控制显示地图的初始化信息，它包含一些键，具体如下。

- **MKLaunchOptionsDirectionsModeKey**。设定路线模式，它有两个值MKLaunchOptionsDirectionsModeDriving（驾车路线）和MKLaunchOptionsDirectionsModeWalking（步行路线）。
- **MKLaunchOptionsMapTypeKey**。设定地图类型。
- **MKLaunchOptionsMapCenterKey**。设定地图中心点。
- **MKLaunchOptionsMapSpanKey**。设置地图跨度。
- **MKLaunchOptionsShowsTrafficKey**。设置显示交通状况。

例如，我们可以使用下面的代码在地图上设置行车路线：

```swift
let lastMapItem = response?.mapItems.last
let options = [MKLaunchOptionsDirectionsModeKey :
➥MKLaunchOptionsDirectionsModeDriving]
lastMapItem!.openInMaps(launchOptions: options)
```

```objectivec
MKMapItem *lastMapItem = response.mapItems.lastObject;
NSDictionary* options = @{MKLaunchOptionsDirectionsModeKey :
➥MKLaunchOptionsDirectionsModeDriving};
[lastMapItem openInMapsWithLaunchOptions:options];
```

设置行车路线后，应用会在地图上标注出路线，如图18-11所示。默认情况下，起点是当前位置，这个位置通过定位服务获得，终点则是我们查询的地点。

图18-11　标注行车路线

18.5　小结

本章中，我们介绍了iOS苹果地图的使用，包括显示地图、添加标注以及跟踪用户位置变化等。最后，我们介绍了程序外地图的使用，主要是调用iOS苹果地图。

第 19 章 访问通讯录

移动设备上都有一个很重要的内置数据库——通讯录，苹果把它扩展到了iCloud上，使苹果设备间可以共享通讯录信息。在iOS上，通讯录放在SQLite3数据库中，但是应用之间不能直接访问，也就是说我们自己编写的应用不能采用数据持久化技术直接访问通讯录数据库。为了支持访问通讯录数据库，苹果开放了一些专门的API 。

19.1 通讯录的安全访问设置

在iOS 9和iOS 10中，有些访问通讯录的API（例如：Contacts框架）可以在后台访问通讯录，出于安全方面的考虑，这些API的应用在第一次访问通讯录时需要获得用户的授权，如图19-1所示。注意：一个应用只授权一次，即便是这个应用删除后重新安装，也不必再次授权。

图19-1 用户授权对话框

iOS 10中，开发人员需要在工程配置文件Info.plist中添加NSContactsUsageDescription键，自定义对话框提示信息，打开工程找到Info.plist属性列表文件，添加NSContactsUsageDescription键，并修改后面的描述文字，结果如图19-2所示。

Key	Type	Value
▼ Information Property List	Dictionary	(15 items)
CFBundleDevelopmentRegion	String	en
CFBundleExecutable	String	$(EXECUTABLE_NAME)
CFBundleIdentifier	String	$(PRODUCT_BUNDLE_IDENTIFIER)
CFBundleInfoDictionaryVersion	String	6.0
CFBundleName	String	$(PRODUCT_NAME)
CFBundlePackageType	String	APPL
CFBundleShortVersionString	String	1.0
CFBundleSignature	String	????
CFBundleVersion	String	1
LSRequiresIPhoneOS	Boolean	YES
UILaunchStoryboardName	String	LaunchScreen
UIMainStoryboardFile	String	Main
▶ UIRequiredDeviceCapabilities	Array	(1 item)
NSContactsUsageDescription	String	此应用需要访问通讯录
▶ UISupportedInterfaceOrientations	Array	(3 items)

图19-2 添加键值

此外，用户还可以在设置应用中禁止或允许某个应用访问通讯录。设置应用中有一个隐私设置项目，可以借由设置通讯录的安全访问，如图19-3所示。

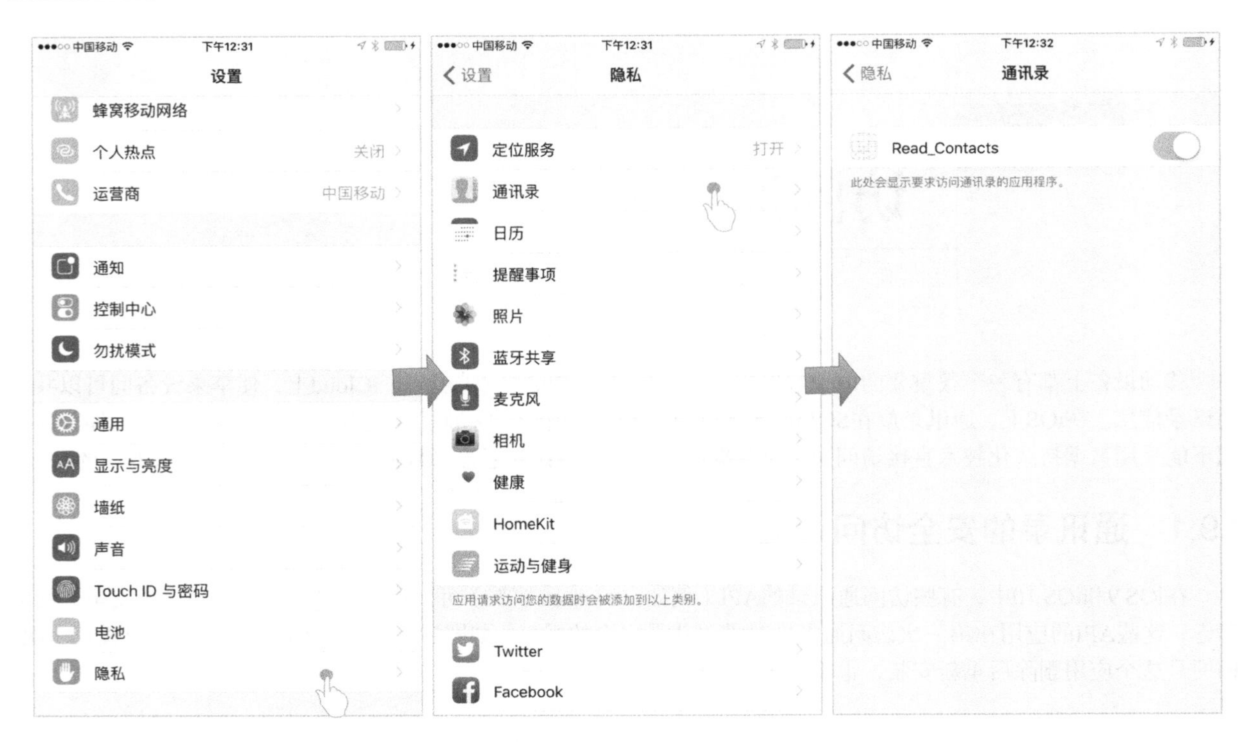

图19-3　设置通讯录的安全访问

19.2　使用 Contacts 框架读取联系人信息

在iOS 9之前，开发访问通讯录的应用使用两个框架——AddressBook和AddressBookUI，而之后则要求使用Contacts和ContactsUI框架了，这主要是因为AddressBook和AddressBookUI框架提供的是C风格API，开发起来非常麻烦，而且AddressBook和AddressBookUI框架中很多API已经不能适用于新的iOS系统了。

Contacts框架提供了直接访问通讯录数据中的联系人以及联系人属性等的API。Contacts框架中常用的类见表19-1。

表19-1　Contacts框架中常用的类

类　名	说　明
CNContactStore	封装访问通讯录的接口，可以查询、保存通讯录信息
CNContact	封装通讯录中联系人信息数据，是数据库的一条记录
CNGroup	封装通讯录组信息数据，一个组包含了多联系人的信息，一个联系人也可以隶属于多个组
CNContainer	封装通讯录容器信息数据，一个容器包含多联系人的信息，但一个联系人只能隶属于一个容器

这一节首先介绍如何使用Contacts框架读取联系人信息，下一节介绍如何使用Contacts框架写入联系人信息。

读取通讯录中联系人的一般过程是先查找联系人记录，然后再访问记录的属性。属性又可以分为单值属性和多值属性。本节通过图19-4所示的例子介绍联系人的查询、单值属性和多值属性的访问，以及如何读取联系人中的图片数据。

Read_Contacts案例是从iOS设备上读取通讯录中的联系人并将其显示在一个表视图中，可以在通讯录中查询联系人，点击联系人可以进入详细信息界面。

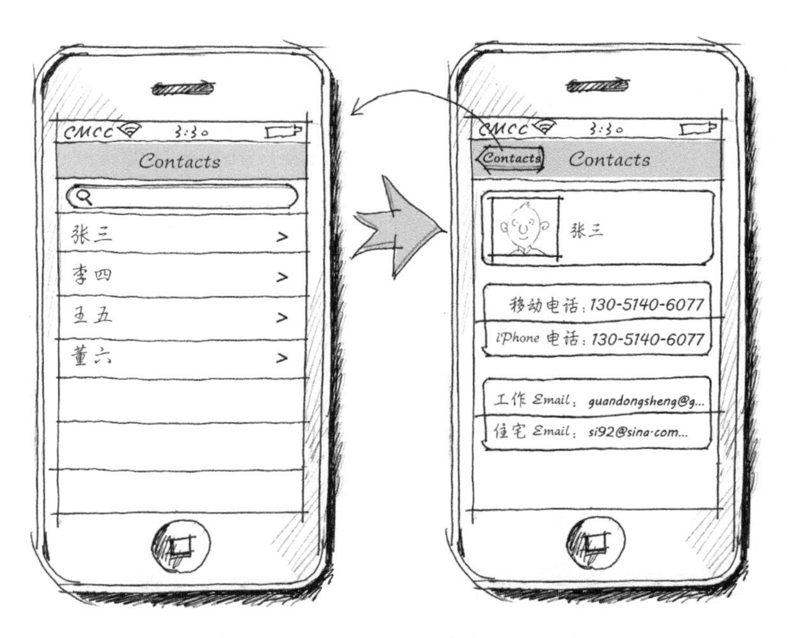

图19-4 Read_Contacts案例原型设计

19.2.1 查询联系人

从通讯录中查询联系人数据，主要是通过CNContactStore类的3个查询方法实现的，这3个方法如下。

- **unifiedContactWithIdentifier:keysToFetch:error:**。通过联系人标识（CNContact的identifier属性）查询单个联系人（CNContact）。第二个参数是要查询联系人的属性集合，第三个参数error是返回的错误。Swift版没有error参数，如果出错，则抛出错误。
- **unifiedContactsMatchingPredicate:keysToFetch:error:**。通过一个谓词（NSPredicate）定义逻辑查询条件，查询出单个联系人。
- **enumerateContactsWithFetchRequest:error:usingBlock:**。通过一个联系人读取对象（CNContactFetchRequest）查询联系人集合。第三个参数是代码块（Swift中称为闭包），查询成功后会回调这个代码块。

下面先看一下ViewController类中属性和初始化等的相关代码：

```swift
import UIKit
import Contacts

class ViewController: UITableViewController, UISearchBarDelegate,
➥UISearchResultsUpdating {

    var searchController: UISearchController!
    var listContacts: [CNContact]!                                    ①

    override func viewDidLoad() {
        super.viewDidLoad()

        //实例化UISearchController
        self.searchController = UISearchController(searchResultsController: nil)
        //设置self为更新搜索结果对象
        self.searchController.searchResultsUpdater = self
        //在搜索时将背景设置为灰色
        self.searchController.dimsBackgroundDuringPresentation = false
        //将搜索栏放到表视图的表头中
        self.tableView.tableHeaderView = self.searchController.searchBar
```

```objc
//ViewController.h文件
#import <UIKit/UIKit.h>
@interface ViewController : UITableViewController

@end

//ViewController.m文件
#import "ViewController.h"
#import "DetailViewController.h"

#import <Contacts/Contacts.h>

@interface ViewController () <UISearchBarDelegate, UISearchResultsUpdating>

@property(strong, nonatomic) UISearchController *searchController;

@property(strong, nonatomic) NSArray *listContacts;                   ①

@end

@implementation ViewController
```

```
        DispatchQueue.global(qos: .utility).async {                    ②
            //查询通讯录中的所有联系人
            self.listContacts = self.findAllContacts()                 ③
            DispatchQueue.main.async {                                 ④
                self.tableView.reloadData()                            ⑤
            }
        }

    }
    ......
}
```

```
- (void)viewDidLoad {
    [super viewDidLoad];

    //实例化UISearchController
    self.searchController = [[UISearchController alloc]
    ➥initWithSearchResultsController:nil];
    //设置self为更新搜索结果对象
    self.searchController.searchResultsUpdater = self;
    //在搜索时将背景设置为灰色
    self.searchController.dimsBackgroundDuringPresentation = FALSE;
    //将搜索栏放到表视图的表头中
    self.tableView.tableHeaderView = self.searchController.searchBar;

    dispatch_async(dispatch_get_global_queue(
    ➥DISPATCH_QUEUE_PRIORITY_DEFAULT, 0), ^{                         ②
        //查询通讯录中的所有联系人
        self.listContacts = [self findAllContacts];                    ③
        dispatch_async(dispatch_get_main_queue(), ^{                   ④
            [self.tableView reloadData];                               ⑤
        });
    });
}

......

@end
```

上述代码中，第①行用于声明属性listContacts，这是装载联系人的数组集合。

第②行~第⑤行是查询通讯录中的所有联系人，并刷新表视图。事实上，执行查询的核心语句是第③行中调用self的findAllContacts方法。但是这个查询方法被放在dispatch_async中，这是GCD（Grand Central Dispatch）技术中的并发访问队列，这会使findAllContacts查询在后台的并发队列中执行。第④行是dispatch_async，用于执行一个主队列。第⑤行是表视图刷新操作，更新或刷新UI界面的功能必须放到主队列中。有关GCD技术的内容，我们会第25章中详细介绍。

提示　在后台并发队列中执行通讯录查询：一方面可以提高查询的执行效率，另一方面可以防止通讯录被首次访问时出现“白屏”现象。所谓“白屏”现象，是指首次访问通讯录时没有出现如图19-1所示的授权对话框，而是显示如图19-5a所示的白屏界面，这种情况下如果点击设备的Home 键使应用回到后台，则会出现授权对话框，如图19-5b所示。如果在如图19-5b所示的界面中点击OK按钮，授权用户访问通讯录，那么以后再次运行该应用，就不会再看到授权对话框，这说明授权完成了。这种首次访问出现“白屏”现象的原因是：默认情况下通讯录查询和通讯录授权都是串行执行的（即同步执行），出现一种死锁状态，如果把通讯录查询放到后台并发队列中执行，则不会出现死锁状态，这样就可以解决“白屏”现象了。

第③行调用findAllContacts方法查询通讯录中所有的联系人数据。另外，当前表视图界面中还添加了搜索栏，这部分代码可以参考6.4节。

(a)

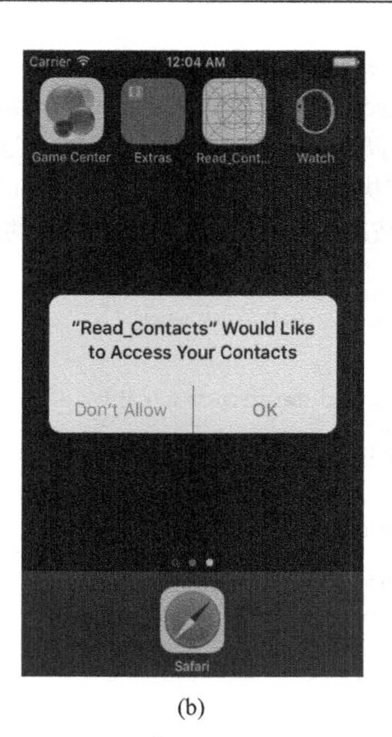

(b)

图19-5　首次访问通讯录时出现“白屏”现象

ViewController中查询通讯录中所有联系人的方法findAllContacts:的代码如下：

```swift
//MARK: -- 查询通讯录中所有联系人
func findAllContacts() -> [CNContact] {
    //返回的联系人集合
    var contacts = [CNContact]()

    let keysToFetch = [CNContactFamilyNameKey, CNContactGivenNameKey]        ①

    let fetchRequest = CNContactFetchRequest(keysToFetch: keysToFetch)       ②

    let contactStore = CNContactStore()                                      ③
    do {
        try contactStore.enumerateContacts(with: fetchRequest, usingBlock: {
                                                                             ④
            (contact, stop) -> Void in
            contacts.append(contact)                                         ⑤
        })
    } catch let error as NSError {
        print(error.localizedDescription)
    }

    return contacts
}
```

```objc
#pragma mark -- 查询通讯录中所有联系人

- (NSArray*)findAllContacts {

    //返回的联系人集合
    id contacts = [[NSMutableArray alloc] init];

    NSArray *keysToFetch = @[CNContactFamilyNameKey,
    ➥CNContactGivenNameKey];                                                 ①
    CNContactFetchRequest *fetchRequest =
    ➥[[CNContactFetchRequest alloc] initWithKeysToFetch:keysToFetch];       ②

    CNContactStore *contactStore = [[CNContactStore alloc] init];            ③
    NSError *error = nil;
    [contactStore enumerateContactsWithFetchRequest:fetchRequest
    ➥error:&error usingBlock:^(CNContact *_Nonnull contact, BOOL *_Nonnull
    ➥stop) {                                                                 ④
        if (!error) {
            [contacts addObject:contact];                                    ⑤
        } else {
            NSLog(@"error : %@", error.localizedDescription);
        }
    }];

    return contacts;
}
```

上述代码中，第①行声明要查询的联系人的属性集合，CNContactFamilyNameKey和CNContactGivenNameKey 都是联系人属性。按照中国人的习惯，CNContactFamilyNameKey一般是“姓氏”，CNContactGivenNameKey一般是“名字”。

第②行用于实例化联系人读取对象CNContactFetchRequest，构造函数的参数是联系人的属性集合。

第③行用于实例化CNContactStore对象。通过CNContactStore对象，可以执行查询、插入和更新联系人信息的操作。CNContactStore是通讯录访问的核心类。

第④行通过CNContactStore对象的enumerateContactsWithFetchRequest:error:usingBlock:方法查询联系人。在Swift版本中，该语句要放到do-try-catch语句中，这是Swift 2.0之后引入的错误处理机制。

第⑤行是查询语句执行成功的情况，将查询的contact对象添加到contacts集合中。

ViewController中按照姓名查询通讯录中联系人的findContactsByName:方法的代码如下:

```
//MARK: -- 按照姓名查询通讯录中的联系人
func findContactsByName(_ searchName: String?) -> [CNContact] {

    //没有输入任何字符
    if (searchName == nil || searchName!.characters.count == 0) {          ①
        //返回通讯录中的所有联系人
        return self.findAllContacts()
    }

    let contactStore = CNContactStore()

    let keysToFetch = [CNContactFamilyNameKey, CNContactGivenNameKey]
    let predicate = CNContact.predicateForContactsMatchingName(searchName!)
                                                                            ②

    do {
        //没有错误的情况下返回查询结果
        let contacts = try contactStore.unifiedContacts(matching: predicate,
        ➥keysToFetch: keysToFetch as [CNKeyDescriptor])                     ③
        return contacts
    } catch let error as NSError {
        print(error.localizedDescription)
        //如果有错误发生，返回通讯录中的所有联系人
        return self.findAllContacts()
    }
}
```

```
#pragma mark -- 按照姓名查询通讯录中的联系人

- (NSArray*)findContactsByName:(NSString *)searchName {

    //没有输入任何字符
    if ([searchName length] == 0) {                                          ①
        //返回通讯录中的所有联系人
        return [self findAllContacts];
    }

    CNContactStore *contactStore = [[CNContactStore alloc] init];

    NSArray *keysToFetch = @[CNContactFamilyNameKey, CNContactGivenNameKey];
    NSPredicate *predicate = [CNContact predicateForContactsMatchingName:
    ➥searchName];                                                           ②
    NSError *error = nil;
    id contacts = [contactStore unifiedContactsMatchingPredicate:predicate
    ➥keysToFetch:keysToFetch error:&error];                                 ③

    if (!error) {
        //没有错误的情况下返回查询结果
        return contacts;
    } else {
        //如果有错误发生，返回通讯录中的所有联系人
        return [self findAllContacts];
    }
}
```

上述代码中，第①行是判断搜索栏中没有任何输入情况下的查询处理，这种情况下我们认为是无条件查询，可以调用findAllContacts方法来实现。

第②行通过CNContact对象调用predicateForContactsMatchingName:方法返回一个谓词对象NSPredicate。谓词对象封装了一个逻辑查询条件，这个条件是匹配所有的姓名属性。

第③行通过CNContactStore对象调用unifiedContactsMatchingPredicate:keysToFetch:error:方法执行查询，查询的结果是单个联系人信息。

19.2.2　读取单值属性

一个联系人（CNContact）信息中有很多属性，这些属性有单值属性和多值属性之分。单值属性是只有一个值的属性，如姓氏和名字等。常用的几个属性如下。

- **givenName**。名字。
- **familyName**。姓氏。
- **middleName**。中间名。
- **namePrefix**。前缀。
- **nameSuffix**。后缀。
- **nickname**。昵称。
- **phoneticGivenName**。名字汉语拼音或音标。
- **phoneticFamilyName**。姓氏汉语拼音或音标。

- **phoneticMiddleName**。中间名汉语拼音或音标。
- **organizationName**。组织名。
- **jobTitle**。头衔。
- **departmentName**。部门。
- **note**。备注。

这些属性全部是只读的字符串类型（Swift版是String类型，Objective-C版是NSString类型）。

ViewController中的tableView:cellForRowAtIndexPath:方法主要实现了访问单值属性，相关代码如下：

```
override func tableView(_ tableView: UITableView,
➥cellForRowAt indexPath: IndexPath) -> UITableViewCell {
    let cell = tableView.dequeueReusableCell(withIdentifier: "Cell", for:
    ➥indexPath)

    let contact = self.listContacts[indexPath.row]                     ①
    let firstName = contact.givenName                                  ②
    let lastName = contact.familyName                                  ③

    let name = "\(firstName) \(lastName)"
    cell.textLabel!.text = name

    return cell
}
```

```
- (UITableViewCell *)tableView:(UITableView *)tableView
➥cellForRowAtIndexPath:(NSIndexPath *)indexPath {

    UITableViewCell *cell = [tableView
    ➥dequeueReusableCellWithIdentifier:@"Cell" forIndexPath:indexPath];

    CNContact *contact = self.listContacts[indexPath.row];             ①
    NSString *firstName = contact.givenName;                           ②
    NSString *lastName = contact.familyName;                           ③

    NSString *name = [NSString stringWithFormat:@"%@ %@", firstName, lastName];
    cell.textLabel.text = name;

    return cell;
}
```

在为表视图单元格提供数据源的方法中，第①行是从集合属性listContacts取出CNContact对象，第②行是取出联系人的givenName属性。第③行是取出联系人的familyName属性，为能够取出该属性内容，我们需要将该属性名包含到keysToFetch（unified-Contacts-Matching-Predicat:keysToFetch:error:方法中第二个参数）集合中，如下面的代码所示：

```
let keysToFetch = [CNContactFamilyNameKey, CNContactGivenNameKey]
```

```
NSArray *keysToFetch = @[CNContactFamilyNameKey, CNContactGivenNameKey];
```

此外，要把选中的联系人传递给详细界面，这个过程是在ViewController中的prepareForSegue:sender:方法中完成的，相关代码如下所示：

```
override func prepare(for segue: UIStoryboardSegue, sender: Any?) {
    if segue.identifier == "showDetail" {
        let indexPath = self.tableView.indexPathForSelectedRow
        let detailViewController = segue.destination as!
        ➥DetailViewController
        detailViewController.selectContact =
        ➥self.listContacts[indexPath!.row]
    }
}
```

```
- (void)prepareForSegue:(UIStoryboardSegue *)segue sender:(id)sender {
    if ([[segue identifier] isEqualToString:@"showDetail"]) {
        NSIndexPath *indexPath = [self.tableView indexPathForSelectedRow];
        DetailViewController *detailViewController = [segue
        ➥destination ViewController];
        detailViewController.selectContact =
        ➥self.listContacts[indexPath.row];
    }
}
```

上述代码是从Segue中取出DetailViewController视图控制器，然后将选中的联系人对象赋值给它的selectContact属性，通过这种方式将参数传递给视图控制器。

19.2.3　读取多值属性

多值属性是包含多个值的集合类型，如电话号码、电子邮箱和URL等，它们主要有下面的常量定义。

- **phoneNumbers**。电话号码属性。
- **emailAddresses**。电子邮箱属性。
- **urlAddresses**。URL属性。
- **postalAddresses**。地址属性。

- **instantMessageAddresses**。即时聊天属性。
- **socialProfiles**。社交账号属性。

这些属性全部是只读的CNLabeledValue集合类型（Swift版是String类型，Objective-C版是NSString类型）。CNLabeledValue类型中包含label（标签）、value（值）和identifier（ID）等部分，其中标签和值都是可以重复的，而ID是不能重复的，如图19-6所示。

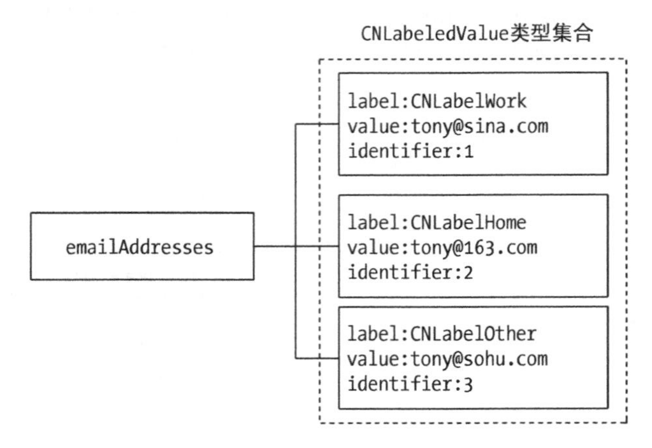

图19-6　多值属性中的标签、值和ID

在DetailViewController的viewDidLoad方法中取得多值属性，示例代码如下：

```
override func viewDidLoad() {
    super.viewDidLoad()

    let contactStore = CNContactStore()

    let keysToFetch = [CNContactFamilyNameKey, CNContactGivenNameKey,
    ➥CNContactEmailAddressesKey, CNContactPhoneNumbersKey,
    ➥CNContactImageDataKey]                                                    ①

    do {
        let contact = try contactStore.unifiedContact(withIdentifier:
        ➥self.selectContact.identifier,
        ➥keysToFetch: keysToFetch as [CNKeyDescriptor])                         ②

        //保存查询出的联系人
        self.selectContact = contact                                            ③

        //取得姓名属性
        let firstName = contact.givenName
        let lastName = contact.familyName

        let name = "\(firstName) \(lastName)"
        self.lblName.text = name

        //取得电子邮件属性
        let emailAddresses = contact.emailAddresses                             ④

        for emailProperty in emailAddresses {

            if emailProperty.label == CNLabelWork {                             ⑤
                self.lblWorkEmail.text = emailProperty.value as String
            } else if emailProperty.label == CNLabelHome {                      ⑥
                self.lblHomeEmail.text = emailProperty.value as String
            } else {
                print("其他Email : \(emailProperty.value)")
            }
```

```
- (void)viewDidLoad {
    [super viewDidLoad];

    CNContactStore *contactStore = [[CNContactStore alloc] init];

    NSArray *keysToFetch = @[CNContactFamilyNameKey, CNContactGivenNameKey,
    ➥CNContactEmailAddressesKey, CNContactPhoneNumbersKey,
    ➥CNContactImageDataKey];                                                   ①
    NSError *error = nil;
    CNContact *contact = [contactStore unifiedContactWithIdentifier:
    ➥self.selectContact.identifier keysToFetch:keysToFetch
    ➥error:&error];                                                            ②

    //保存查询出的联系人
    self.selectContact = contact;                                               ③

    if (!error) {

        //取得姓名属性
        NSString *firstName = contact.givenName;
        NSString *lastName = contact.familyName;

        NSString *name = [NSString stringWithFormat:@"%@ %@", firstName,
        ➥lastName];
        [self.lblName setText:name];

        //取得电子邮件属性
        NSArray<CNLabeledValue<NSString*>*> *emailAddresses =
        ➥contact.emailAddresses;                                               ④

        for (CNLabeledValue<NSString*>* emailProperty in emailAddresses) {

            if ([emailProperty.label isEqualToString:CNLabelWork]) {            ⑤
                [self.lblWorkEmail setText:emailProperty.value];
            } else if ([emailProperty.label isEqualToString:CNLabelHome]) {     ⑥
                [self.lblHomeEmail setText:emailProperty.value];
```

```
        }
        ......

    } catch let error as NSError {
        print(error.localizedDescription)
    }
}
```

```
            } else {
                NSLog(@"%@: %@", @"其他Email", emailProperty.value);
            }
        }

        ......
    }

}
```

上述代码中，第①行声明要查询的联系人的属性集合。第②行通过CNContactStore对象调用unifiedContactWithIdentifier:keysToFetch:error:方法执行查询，其中第一个参数是联系人标识，self.selectContact.identifier表达式用于获得联系人标识。第③行是将查询出来的联系人重新保存到self.selectContact属性中，这是为了后面的修改而做的准备。self.selectContact属性是从前面ViewController视图控制器传递过来的，其中所包含的联系人中包含的属性只有familyName和givenName，而这次查询后联系人中包含的属性有familyName、givenName、emailAddresses、phoneNumbers和imageData。

第④行从联系人对象中取得电子邮件属性（emailAddresses），在Objective-C版中可见的返回值类型是NSArray<CNLabeledValue<NSString*>*>，这说明返回值是CNLabeledValue类型的数组集合，泛型类型CNLabeledValue<NSString*>中的NSString*类型限定了value属性的类型。

第⑤行通过emailProperty标签判断是否为工作电子邮件，第⑥行通过emailProperty标签判断是否为家庭电子邮件。

在DetailViewController的viewDidLoad方法中，取得电话号码多值属性的代码如下：

```
//取得电话号码属性
let phoneNumbers = contact.phoneNumbers                                          ①
for phoneNumberProperty in phoneNumbers {

    let phoneNumber = phoneNumberProperty.value as! CNPhoneNumber                ②

    if phoneNumberProperty.label == CNLabelPhoneNumberMobile {                   ③
        self.lblMobile.text = phoneNumber.stringValue
    } else if phoneNumberProperty.label == CNLabelPhoneNumberiPhone {           ④
        self.lblIPhone.text = phoneNumber.stringValue
    } else {
        print("其他电话 : \(phoneNumber.stringValue)")
    }
}
```

```
//取得电话号码属性
NSArray<CNLabeledValue<CNPhoneNumber*>*> *phoneNumbers =
➥contact.phoneNumbers;                                                          ①

for (CNLabeledValue<CNPhoneNumber*>* phoneNumberProperty in phoneNumbers) {

    CNPhoneNumber *phoneNumber = phoneNumberProperty.value;                      ②

    if ([phoneNumberProperty.label isEqualToString:
    ➥CNLabelPhoneNumberMobile]) {                                               ③
        [self.lblMobile setText:phoneNumber.stringValue];
    } else if ([phoneNumberProperty.label
    ➥isEqualToString:CNLabelPhoneNumberiPhone]) {                               ④
        [self.lblIPhone setText:phoneNumber.stringValue];
    } else {
        NSLog(@"%@: %@", @"其他电话", phoneNumber.stringValue);
    }
}
```

在上述代码中，第①行从联系人对象中取得电话号码属性（phoneNumbers），在Objective-C版中可见返回值类型是NSArray<CNLabeledValue<CNPhoneNumber*>*>，这说明返回值是CNLabeledValue类型的数组集合，泛型类型CNLabeledValue<CNPhoneNumber*>中的CNPhoneNumber*类型限定了value属性类型。第②行用于取出value属性，类型是CNPhoneNumber*，CNPhoneNumber有一个stringValue属性用来获取电话号码字符串。

第③行通过phoneNumberProperty标签判断是否为移动电话号码，第④行通过phoneNumberProperty标签判断是否为iPhone电话号码。

此外，还有下面几个主要的电话标签。

- **CNLabelPhoneNumberMain**。主要电话号码标签。
- **CNLabelPhoneNumberHomeFax**。家庭传真电话号码标签。
- **CNLabelPhoneNumberWorkFax**。工作传真电话号码标签。

19.2.4　读取图片属性

通讯录中的联系人可以有一张照片，我们可以通过联系人的imageData属性读取联系人照片。imageData属性是NSData类型。

在DetailViewController的viewDidLoad方法中，取得联系人照片的代码如下：

```
//取得个人图片
if let photoData = contact.imageData {
    self.imageView.image = UIImage(data: photoData)
}
```

```
//取得个人图片
NSData *photoData = contact.imageData;
if(photoData){
    [self.imageView setImage:[UIImage imageWithData:photoData]];
}
```

19.3　使用 Contacts 框架写入联系人信息

写入联系人信息到通讯录时，涉及创建联系人、修改联系人和删除联系人等操作。写入联系人相关的API有CNMutableContact和CNSaveRequest。其中，CNMutableContact是可变联系人对象，所有的写入联系人操作都必须放到CNMutableContact中；CNSaveRequest是保存请求对象，通过该对象指定对联系人进行创建、修改和删除操作。

下面我们通过一个Write_Contacts实例介绍联系人的创建、修改和删除操作。

图19-7所示的是创建联系人信息的原型图。在首页上点击导航栏右边的+按钮，可以呈现添加联系人模态视图，添加完成之后，点击Save按钮保存并返回首页；如果点击Cancel按钮，则不保存。

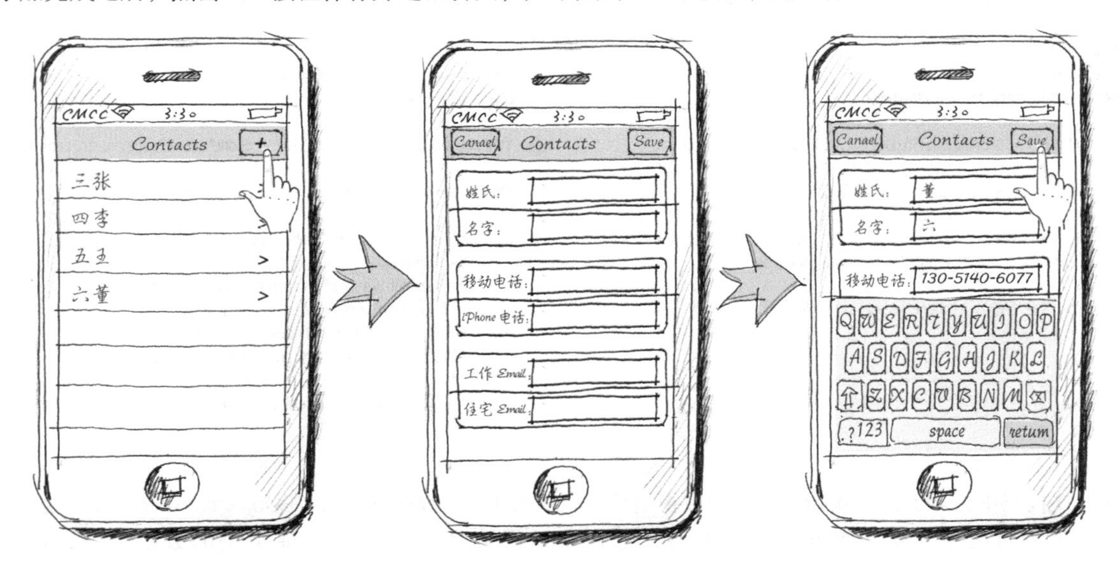

图19-7　Write_Contacts案例原型设计

图19-8所示的是修改联系人信息的原型图。在首页点击某个联系人单元格，会导航到该联系人详细信息界面，在这个界面中可以修改电话号码和电子邮件地址，但不能修改姓名。修改完成之后，点击Save按钮保存并导航回首页。

图19-9所示的是删除联系人信息的原型图。在首页点击某个联系人单元格，导航到该联系人详细信息界面，在这个界面中点击“删除联系人”按钮，就可以删除该联系人并导航回首页。

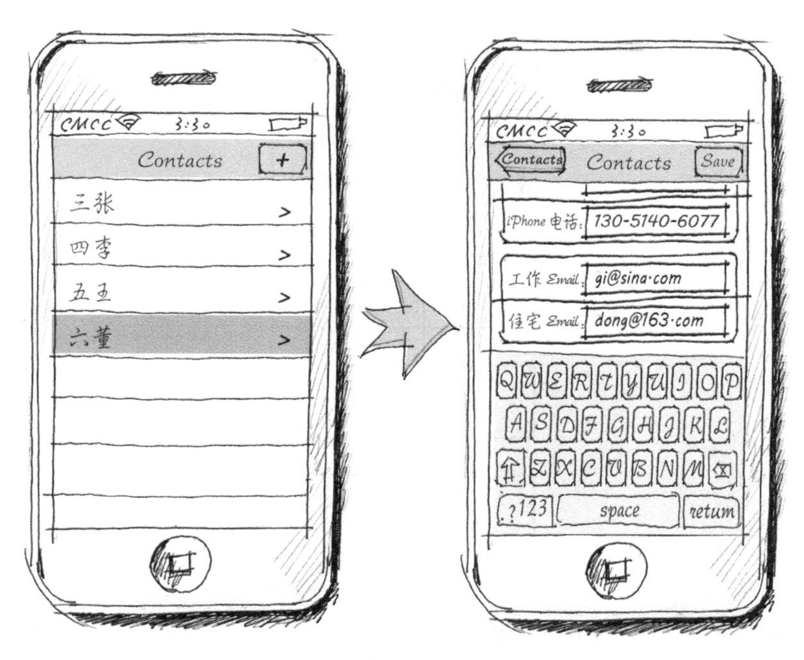

图19-8 修改联系人信息

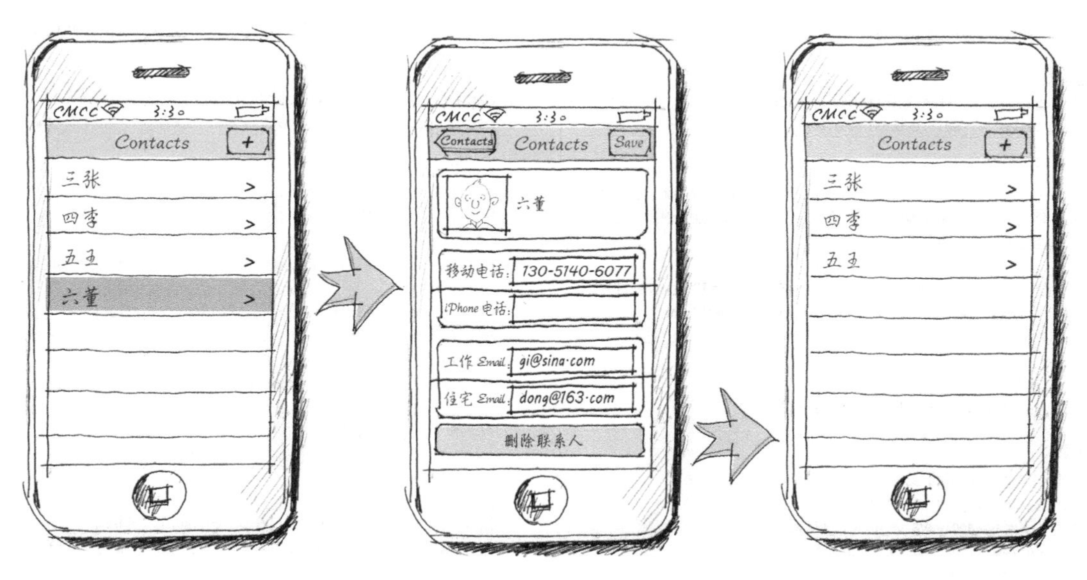

图19-9 删除联系人信息

19.3.1 创建联系人

如图19-10所示，创建联系人基本上都会经历5个步骤。

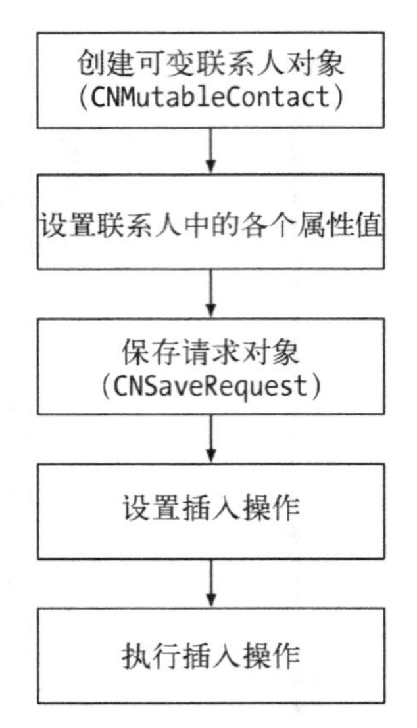

图19-10　创建联系人流程图

下面我们看看添加界面中添加按钮的触发方法，即AddViewController的saveClick:方法：

```
@IBAction func saveClick(_ sender: AnyObject) {

    let contact = CNMutableContact()                                          ①

    //设置姓名属性
    contact.familyName =  self.txtFirstName.text!                             ②
    contact.givenName =  self.txtLastName.text!

    //设置电话号码
    let mobilePhoneValue = CNPhoneNumber(stringValue: self.txtMobile.text!)  ③
    let mobilePhone = CNLabeledValue(label: CNLabelPhoneNumberMobile,
    ➥value: mobilePhoneValue)                                                ④

    let iPhoneValue = CNPhoneNumber(stringValue: self.txtIPhone.text!)
    let iPhone = CNLabeledValue(label: CNLabelPhoneNumberiPhone,
    ➥value: iPhoneValue)

    //添加电话号码到数据库
    contact.phoneNumbers = [mobilePhone, iPhone]                              ⑤

    //设置电子邮件属性
    let homeEmail = CNLabeledValue(label: CNLabelHome, value:
    ➥self.txtHomeEmail.text! as NSString)
    let workEmail = CNLabeledValue(label: CNLabelWork, value:
    ➥self.txtWorkEmail.text! as NSString)
    //添加电子邮件到数据库
    contact.emailAddresses = [homeEmail, workEmail]

    //最后保存
    let request = CNSaveRequest()                                             ⑥
    request.addContact(contact, toContainerWithIdentifier: nil)               ⑦

    let contactStore = CNContactStore()

    do {
        try contactStore.execute(request)                                     ⑧
        //关闭模态视图
        self.dismiss(animated: true, completion:nil)
    } catch let error as NSError {
        print(error.localizedDescription)
```

```
- (IBAction)saveClick:(id)sender {

    CNMutableContact* contact = [[CNMutableContact alloc] init];              ①

    //设置姓名属性
    contact.familyName =  self.txtFirstName.text;                             ②
    contact.givenName =  self.txtLastName.text;

    //设置电话号码
    CNPhoneNumber* mobilePhoneValue = [[CNPhoneNumber alloc]
    ➥initWithStringValue:self.txtMobile.text];                               ③
    CNLabeledValue* mobilePhone = [[CNLabeledValue alloc]
    ➥initWithLabel:CNLabelPhoneNumberMobile value:mobilePhoneValue];         ④

    CNPhoneNumber* iPhoneValue = [[CNPhoneNumber alloc]
    ➥initWithStringValue:self.txtIPhone.text];
    CNLabeledValue* iPhone = [[CNLabeledValue alloc]
    ➥initWithLabel:CNLabelPhoneNumberiPhone value:iPhoneValue];

    //添加电话号码到数据库
    contact.phoneNumbers = @[mobilePhone, iPhone];                            ⑤

    //设置电子邮件属性
    CNLabeledValue* homeEmail = [[CNLabeledValue alloc]
    ➥initWithLabel:CNLabelHome value:self.txtHomeEmail.text];
    CNLabeledValue* workEmail = [[CNLabeledValue alloc]
    ➥initWithLabel:CNLabelWork value:self.txtWorkEmail.text];
    //添加电子邮件到数据库
    contact.emailAddresses = @[homeEmail, workEmail];

    //最后保存
    CNSaveRequest* request = [[CNSaveRequest alloc] init];                    ⑥
    [request addContact:contact toContainerWithIdentifier:nil];               ⑦

    CNContactStore *contactStore = [[CNContactStore alloc] init];
    NSError* error;
    [contactStore executeSaveRequest:request error:&error];                   ⑧

    if (!error) {
        //关闭模态视图
```

```swift
    }
}
```

```objc
        [self dismissViewControllerAnimated:TRUE completion:nil];
    } else {
        NSLog(@"error : %@", error.localizedDescription);
    }

}
```

在上述代码中，第①行创建可变联系人对象CNMutableContact。第②行设置姓名中的familyName属性，它是一个单值属性，设置起来比较简单。第③行~第⑤行设置多值属性phoneNumbers。多值属性phoneNumbers是一个集合，每一个元素都是CNLabeledValue<CNPhoneNumber*>形式的泛型类型，CNLabeledValue中的值是CNPhoneNumber类型，而非普通的字符串类型。第③行用于创建CNPhoneNumber对象。第④行用于创建移动电话CNLabeledValue对象，其中CNLabelPhoneNumberMobile是标签。最后，第⑤行将CNLabeledValue对象集合赋值给phoneNumbers属性。

第⑥行创建保存请求对象CNSaveRequest。

第⑦行设置插入操作的类型，方法addContact:toContainerWithIdentifier:表示插入联系人的操作，其中第一个参数是联系人对象，第二个参数是所在容器的ID，其中nil表示默认容器。

第⑧行通过CNContactStore对象调用executeSaveRequest:error:方法，执行第⑦行设置的插入操作。

19.3.2 修改联系人

如图19-11所示，修改联系人一般也会经历5个步骤。修改联系人与创建联系人非常类似，差别在于后者中“可变联系人对象”是创建的，而前者中“可变联系人对象”是修改之前获得的。

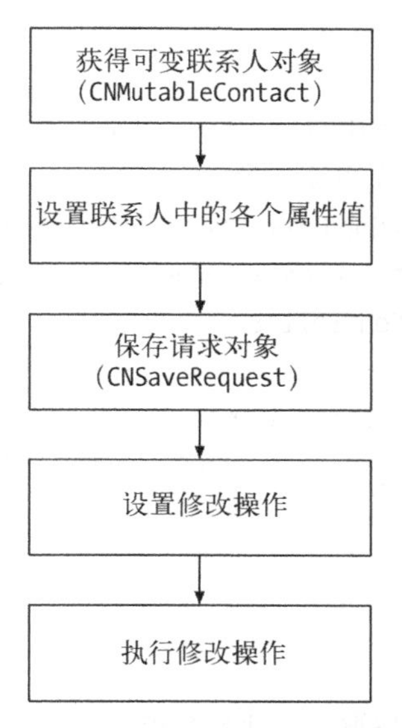

图19-11 修改联系人的流程图

下面我们看看修改界面中“保存”按钮的代码，即DetailViewController中的saveClick:方法：

```swift
@IBAction func saveClick(_ sender: AnyObject) {

    let contact = self.selectContact.mutableCopy() as! CNMutableContact      ①

    //设置电话号码
    let mobilePhoneValue = CNPhoneNumber(stringValue: self.txtMobile.text!)
    let mobilePhone = CNLabeledValue(label: CNLabelPhoneNumberMobile,
    ➥value: mobilePhoneValue)

    let iPhoneValue = CNPhoneNumber(stringValue: self.txtIPhone.text!)
    let iPhone = CNLabeledValue(label: CNLabelPhoneNumberiPhone, value:|
```

```objc
- (IBAction)saveClick:(id)sender {

    CNMutableContact* contact = [self.selectContact mutableCopy];      ①

    //设置电话号码
    CNPhoneNumber* mobilePhoneValue = [[CNPhoneNumber alloc]
    ➥initWithStringValue:self.txtMobile.text];
    CNLabeledValue* mobilePhone = [[CNLabeledValue alloc]
    ➥initWithLabel:CNLabelPhoneNumberMobile value:mobilePhoneValue];

    CNPhoneNumber* iPhoneValue = [[CNPhoneNumber alloc]
```

```
➥iPhoneValue)

//添加电话号码到数据库
contact.phoneNumbers = [mobilePhone, iPhone]

//设置电子邮件属性
let homeEmail = CNLabeledValue(label: CNLabelHome, value:
➥self.txtHomeEmail.text! as NSString)
let workEmail = CNLabeledValue(label: CNLabelWork,.
➥value: selftxtWorkEmail.text! as NSString)

//添加电子邮件到数据库
contact.emailAddresses = [homeEmail, workEmail]

//最后保存
let request = CNSaveRequest()
request.update(contact)                                          ②

let contactStore = CNContactStore()

do {
    try contactStore.execute(request)                            ③
    //导航回根视图控制器ViewController
    _ = self.navigationController?.popToRootViewController (true)   ④
} catch let error as NSError {
    print(error.localizedDescription)
}
}
```

```
➥initWithStringValue:self.txtIPhone.text];
CNLabeledValue* iPhone = [[CNLabeledValue alloc]
➥initWithLabel:CNLabelPhoneNumberiPhone value:iPhoneValue];

//添加电话号码到数据库
contact.phoneNumbers = @[mobilePhone, iPhone];

//设置电子邮件属性
CNLabeledValue* homeEmail = [[CNLabeledValue alloc]
➥initWithLabel:CNLabelHome value:self.txtHomeEmail.text];
CNLabeledValue* workEmail = [[CNLabeledValue alloc]
➥initWithLabel:CNLabelWork value:self.txtWorkEmail.text];
//添加电子邮件到数据库
contact.emailAddresses = @[homeEmail, workEmail];

//最后保存
CNSaveRequest* request = [[CNSaveRequest alloc] init];
[request updateContact:contact];                                 ②

CNContactStore *contactStore = [[CNContactStore alloc] init];
NSError* error;
[contactStore executeSaveRequest:request error:&error];          ③

if (!error) {
    //导航回根视图控制器ViewController
    [self.navigationController popToRootViewControllerAnimated:TRUE]; ④
} else {
    NSLog(@"error : %@", error.localizedDescription);
}
}
```

第①行用于获得可变联系人对象CNMutableContact，它不是实例化获得的，而是通过self.selectContact属性复制出来的，mutableCopy方法会复制出一个可变类型。

第②行用于设置修改操作的类型，方法updateContact:是修改联系人的操作，其参数是联系人对象。

第③行通过CNContactStore对象调用executeSaveRequest:error:方法，来执行第②行所设置的修改操作。

第④行直接导航回根视图控制器ViewController。

注意　Swift版本中，第④行的self.navigationController?.popToRootViewController(true)表达式是有返回值的，返回值赋值给下划线“_”，下划线在Swift语言中表示可以省略。在Swift 3中，如果没有变量或常量接受返回值，则会有警告。

19.3.3　删除联系人

与添加联系人和修改联系人相比，删除联系人比较简单，不需要“设置联系人中的各个属性值”，它的流程如图19-12所示，一般会经历4个步骤。

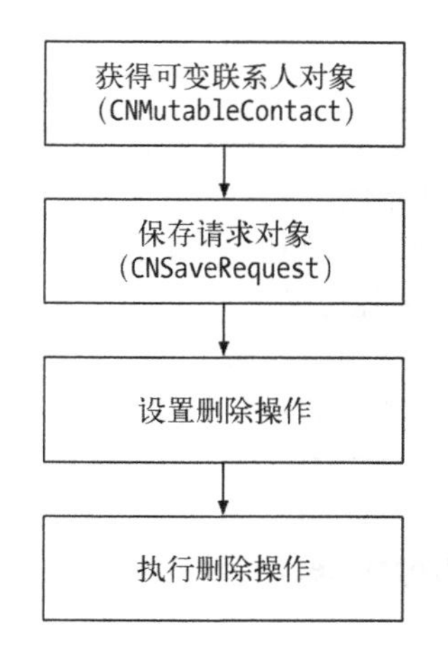

图19-12 删除联系人流程图

下面看看删除界面中删除按钮触发的方法，即DetailViewController中的deleteClick:方法：

```
@IBAction func deleteClick(_ sender: AnyObject) {

    let contact = self.selectContact.mutableCopy() as! CNMutableContact

    let request = CNSaveRequest()
    request.delete(contact)                                              ①

    let contactStore = CNContactStore()

    do {
        try contactStore.execute(request)
        //导航回根视图控制器ViewController
        _ = self.navigationController?.popToRootViewController (true)   ②

    } catch let error as NSError {
        print(error.localizedDescription)
    }
}
```

```
- (IBAction)deleteClick:(id)sender {

    CNMutableContact* contact = [self.selectContact mutableCopy];

    CNSaveRequest* request = [[CNSaveRequest alloc] init];
    [request deleteContact:contact];                                     ①

    CNContactStore *contactStore = [[CNContactStore alloc] init];
    NSError* error;
    [contactStore executeSaveRequest:request error:&error];

    if (!error) {
        //导航回根视图控制器ViewController
        [self.navigationController popToRootViewControllerAnimated:TRUE]; ②
    } else {
        NSLog(@"error : %@", error.localizedDescription);
    }

}
```

第①行用于设置删除操作的类型，方法deleteContact:表示删除联系人操作，其参数是联系人对象。

第②行可以导航到首页（即当前界面的根视图），其效果与点击导航栏左侧的返回按钮一样。

19.4 使用系统提供的界面

很多应用在访问联系人时未必都需要自己制作界面，我们可以使用iOS系统提供的界面，就像使用通讯录应用一样。ContactsUI框架提供这些界面视图及视图控制器，其中包括两个视图控制器和两个对应的委托协议，如表19-2所示。

表19-2 ContactsUI框架中的视图控制器

视图控制器	说 明
CNContactPickerViewController	它是从通讯录中选取联系人的控制器，对应的委托协议为CNContactPickerDelegate
CNContactViewController	查看、创建、编辑单个联系人信息的控制器，对应的委托协议为CNContactViewControllerDelegate

19.4.1　选择联系人

选择联系人时，需要使用CNContactPickerViewController控制器，它是选取联系人的控制器。图19-13所示的是模态视图形式的联系人选取界面视图。使用CNContactPickerViewController控制器，不仅可以选择联系人，而且可以选择联系人中的属性。如图19-14所示，当选中一个联系人之后，可进入到属性选择界面。

CNContactPickerViewController对应的委托协议是CNContactPickerDelegate，它定义了5个主要的方法。

- **contactPickerDidCancel:**。点击联系人选取界面中的Cancel按钮时调用。
- **contactPicker:didSelectContact:**。选中单个联系人时调用。
- **contactPicker:didSelectContactProperty:**。选中单个联系人的属性时调用。
- **contactPicker:didSelectContacts:**。选中多个联系人时调用。
- **contactPicker:didSelectContactProperties:**。选中多个联系人的属性时调用。

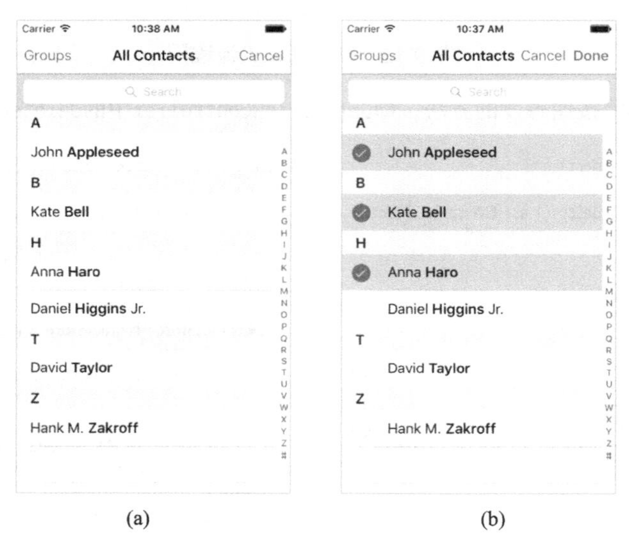

图19-13　联系人选取界面（图a为单选联系人，图b为多选联系人）

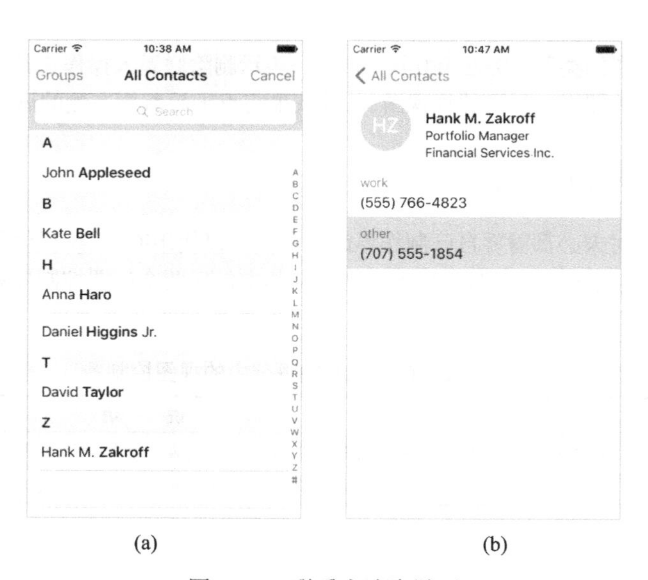

图19-14　联系人选取界面

下面我们通过MyFriend案例介绍如何使用CNContactPickerViewController从通讯录中选择联系人。如图19-15所示，当用户点击首页中导航栏右侧的+按钮时，会看到模态视图形式的联系人选取界面，添加完成后可回到首页并把选择的用户添加到表视图中。

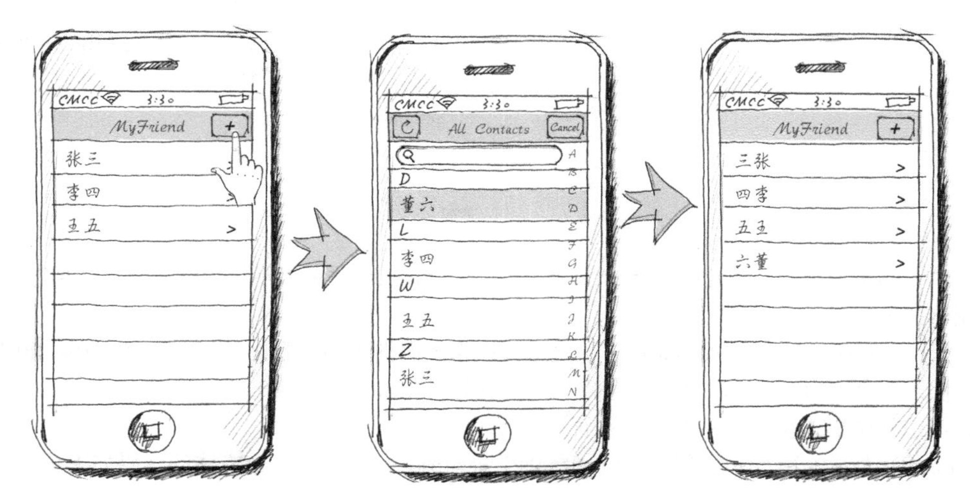

图19-15　MyFriend案例原型设计

在MyFriend工程中，ViewController中的属性和viewDidLoad相关的代码如下所示：

```
//ViewController.swift文件
import UIKit
import ContactsUI

class ViewController: UITableViewController, CNContactPickerDelegate {

    var listContacts: [CNContact]!

    override func viewDidLoad() {
        super.viewDidLoad()

        self.listContacts = [CNContact]()
    }

    @IBAction func selectContacts(_ sender: AnyObject) {

        let contactPicker = CNContactPickerViewController()                    ①
        contactPicker.delegate = self                                          ②
        contactPicker.displayedPropertyKeys = [CNContactPhoneNumbersKey]       ③

        self.present(contactPicker, animated: true, completion: nil)           ④
    }

    //MARK: -- 表视图数据源
    override func tableView(_ tableView: UITableView,
    ➥numberOfRowsInSection section: Int) -> Int {
        return self.listContacts.count
    }

    override func tableView(_ tableView: UITableView,
    ➥cellForRowAt indexPath: IndexPath) -> UITableViewCell {

        let cell = tableView.dequeueReusableCell(withIdentifier: "Cell", for:
        ➥indexPath)
        let contact = self.listContacts[indexPath.row]
        let firstName = contact.givenName
        let lastName = contact.familyName
```

```
//ViewController.m文件
#import "ViewController.h"
#import <ContactsUI/ContactsUI.h>

@interface ViewController () <CNContactPickerDelegate>
@property(strong, nonatomic) NSMutableArray *listContacts;
@end

@implementation ViewController

- (void)viewDidLoad {
    [super viewDidLoad];
    self.listContacts = [[NSMutableArray alloc] init];
}

- (IBAction)selectContacts:(id)sender {

    CNContactPickerViewController *contactPicker =
    ➥[[CNContactPickerViewController alloc] init];                            ①
    contactPicker.delegate = self;                                             ②
    contactPicker.displayedPropertyKeys = @[CNContactPhoneNumbersKey];         ③

    [self presentViewController:contactPicker animated:TRUE
    ➥completion:nil];                                                         ④

}
#pragma mark -- 实现表视图数据源协议

- (NSInteger)tableView:(UITableView *)tableView
➥numberOfRowsInSection:(NSInteger)section {
    return [self.listContacts count];
}

- (UITableViewCell *)tableView:(UITableView *)tableView
➥cellForRowAtIndexPath:(NSIndexPath *)indexPath {

    UITableViewCell *cell = [tableView
```

```swift
        let name = "\(firstName) \(lastName)"
        cell.textLabel!.text = name

        return cell
    }

    ......
}
```

```objc
    ➥dequeueReusableCellWithIdentifier:@"Cell" forIndexPath:indexPath];

    CNContact *contact = self.listContacts[indexPath.row];
    NSString *firstName = contact.givenName;
    NSString *lastName = contact.familyName;

    NSString *name = [NSString stringWithFormat:@"%@ %@", firstName, lastName];
    cell.textLabel.text = name;

    return cell;
}
......

@end
```

上述代码中，第①行用于实例化CNContactPickerViewController联系人视图控制器对象，第②行用于设置联系人视图控制器的委托对象为self。

第③行设置displayedPropertyKeys属性，它是选择联系人属性时能够看到的属性。在图19-14b中，会看到联系人的电话属性，这是因为displayedPropertyKeys被设置为[CNContactPhoneNumbersKey]。

第④行以模态视图形式呈现联系人选择界面。

实现CNContactPickerDelegate委托协议中contactPicker:didSelectContact:方法的代码如下：

```swift
//ViewController.swift文件
func contactPicker(_ picker: CNContactPickerViewController,
➥didSelect contact: CNContact) {

    if !self.listContacts.contains(contact) {
        self.listContacts.append(contact)
        self.tableView.reloadData()
    }
}
```

```objc
//ViewController.m文件
- (void)contactPicker:(CNContactPickerViewController *)picker
➥didSelectContact:(CNContact *)contact {

    if (![self.listContacts containsObject:contact]) {
        [self.listContacts addObject:contact];
        [self.tableView reloadData];
    }
}
```

实现CNContactPickerDelegate委托协议中contactPicker:didSelectContacts:方法的代码如下：

```swift
//ViewController.swift文件
func contactPicker(_ picker: CNContactPickerViewController,
➥didSelect contacts: [CNContact]) {

    for contact in contacts where !self.listContacts.contains(contact) {   ①
        self.listContacts.append(contact)
    }
    self.tableView.reloadData()
}
```

```objc
//ViewController.m文件
- (void)contactPicker:(CNContactPickerViewController *)picker
➥didSelectContacts:(NSArray<CNContact *> *)contacts {

    for (CNContact *contact in contacts) {                                  ①
        if (![self.listContacts containsObject:contact]) {
            [self.listContacts addObject:contact];
        }
    }
    [self.tableView reloadData];
}
```

上述代码Swift版中，第①行使用了Swift的where语句过滤for循环。

19.4.2　显示和修改联系人

显示和修改联系人时，我们会使用CNContactViewController控制器，其中可以显示和修改联系人，如图19-16所示。

CNContactViewController控制器必须在一个导航控制器（UINavigationController）内管理。我们使用下面的代码管理CNContactViewController控制器：

```swift
self.navigationController?.pushViewController(controller, animated: true)
```

```objc
[self.navigationController pushViewController:controller animated:TRUE];
```

CNContactViewController对应的委托协议是CNContactViewControllerDelegate，该委托协议定义了一个必须

实现的方法contactViewController:shouldPerformDefaultActionForContactProperty:，该方法在选择联系人属性时调用，返回true时则调用该属性的默认动作，例如选择的是电子邮件属性，则调用iOS内置的电子邮件程序发送电子邮件。如果该方法返回false，则不做任何动作。

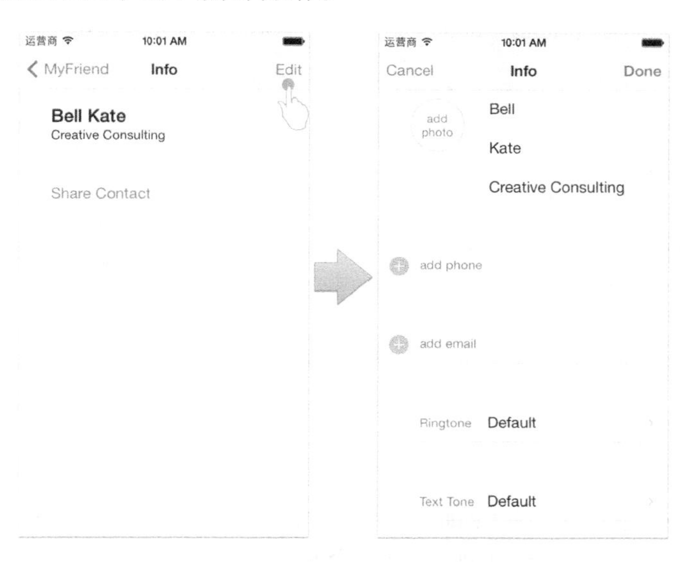

图19-16　显示和修改联系人

现在我们进一步完善上一节的案例。点击首页中的某个联系人，进入该联系人的详细信息界面，其中可以显示和修改联系人，如图19-17所示。

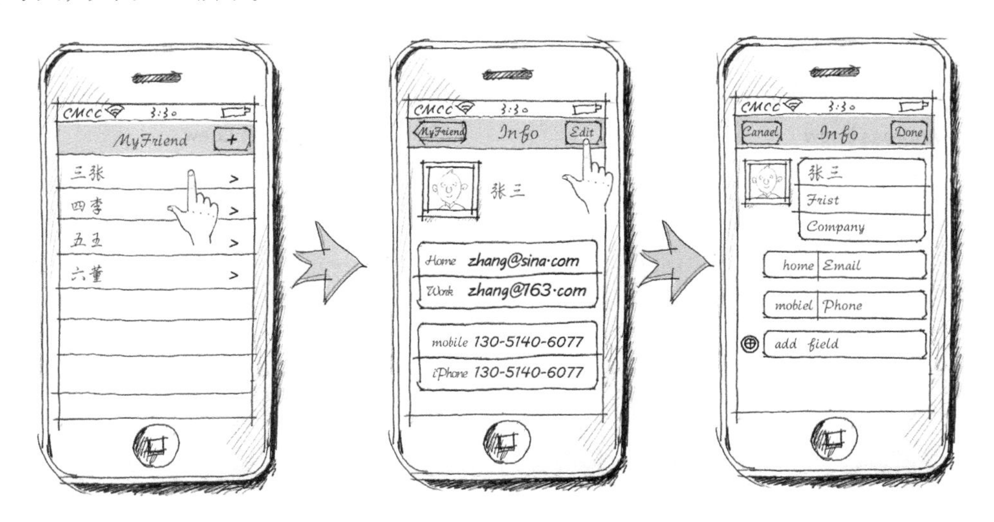

图19-17　显示和修改联系人

在MyFriend工程的ViewController中显示联系人的代码如下：

```
override func tableView(_ tableView: UITableView,
➥didSelectRowAt indexPath: IndexPath) {

let contactStore = CNContactStore()
let selectedContact = self.listContacts[indexPath.row]          ①

let keysToFetch= [CNContactViewController.descriptorForRequiredKeys()] ②
```

```
- (void)tableView:(UITableView *)tableView
➥didSelectRowAtIndexPath:(NSIndexPath *)indexPath {

    CNContactStore* contactStore = [[CNContactStore alloc] init];

    CNContact *selectedContact = self.listContacts[indexPath.row];          ①
```

```
    do {
        let contact = try contactStore.unifiedContact(withIdentifier:
        ➥selectedContact.identifier, keysToFetch: keysToFetch)          ③

        let controller = CNContactViewController(forContact: contact)   ④
        controller.delegate = self                                       ⑤
        controller.contactStore = contactStore
        controller.allowsEditing = true                                  ⑥
        controller.allowsActions = true                                  ⑦

        controller.displayedPropertyKeys =
        ➥[CNContactPhoneNumbersKey, CNContactEmailAddressesKey]

        self.navigationController?.pushViewController(controller,
        ➥animated: true)

    } catch let error as NSError {
        print(error.localizedDescription)
    }
}
```

```
    id keysToFetch = @[[CNContactViewController
    ➥descriptorForRequiredKeys]];                                        ②
    NSError *error;
    CNContact *contact = [contactStore
    ➥unifiedContactWithIdentifier:selectedContact.identifier
    ➥keysToFetch:keysToFetch error:&error];                              ③

    if (!error) {
        CNContactViewController* controller = [CNContactViewController
        ➥viewControllerForContact:contact];                              ④
        controller.delegate = self;                                       ⑤
        controller.contactStore = contactStore;
        controller.allowsEditing = TRUE;                                  ⑥
        controller.allowsActions = TRUE;                                  ⑦

        controller.displayedPropertyKeys = @[CNContactPhoneNumbersKey,
        ➥CNContactEmailAddressesKey];

        [self.navigationController pushViewController:controller animated:
          TRUE];
    } else {
        NSLog(@"error : %@", error.localizedDescription);
    }
}
```

第①行从listContacts属性中取出选中的联系人，第②行用于获得CNContactViewController视图控制器所需要的联系人属性集合，descriptorForRequiredKeys是静态方法。

第④行用于创建CNContactViewController对象，第⑤行用于设置CNContactViewController的委托对象为self。

第⑥行设置allowsEditing属性，该属性设定联系人视图是否可以编辑。如果该属性为true，则视图导航栏右边会出现Edit按钮（如图19-18所示）；如果该属性为false，则不会出现该按钮，如图19-19所示。

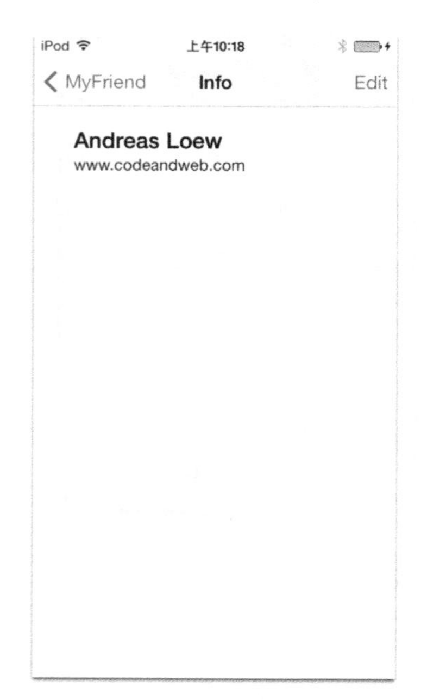

图19-18　allowsEditing为true

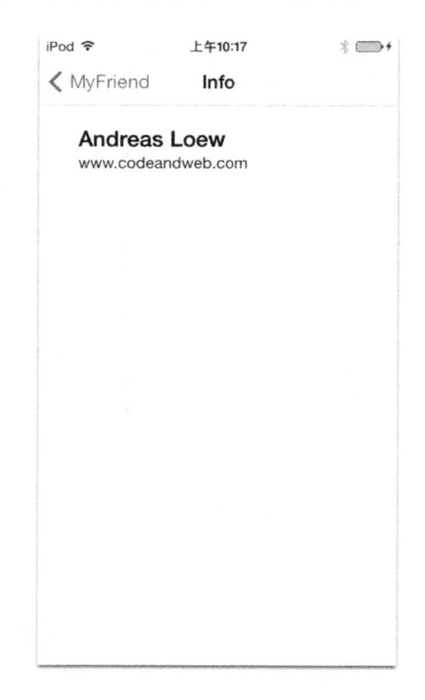

图19-19　allowsEditing为false

第⑦行设置allowsActions属性，该属性设定是否可以显示动作按钮，如果该属性为true，则联系人信息界面下方会有Share Contact按钮用于共享联系人信息（如图19-20所示）。在编辑完联系人并返回后，在这个界面中还可以看到如E-mail、FaceTime、Share Contact和Add to Favorites按钮等（如图19-21所示）。

在MyFriend工程的ViewController中，实现CNContactViewControllerDelegate委托方法的代码如下：

```
func contactViewController(_ viewController: CNContactViewController,
➥shouldPerformDefaultActionFor property: CNContactProperty) -> Bool {
    return true
}
```

```
- (BOOL)contactViewController:(CNContactViewController *)viewController
➥shouldPerformDefaultActionForContactProperty:(CNContactProperty *)
➥property {
    return TRUE;
}
```

上述委托方法在选中联系人属性时调用，如果返回true，则执行该属性的默认操作，如果是false，则不执行任何操作。例如：选中的电话号码属性，它的默认属性是拨打电话，但需要注意拨打电话动作只有在iPhone设备上才能执行。

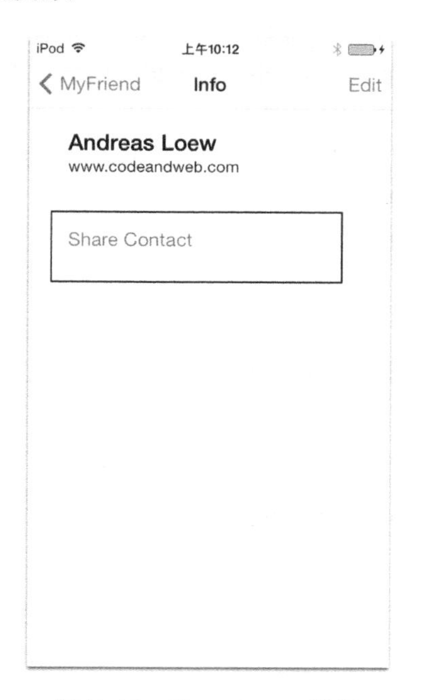

图19-20 Share Contact按钮

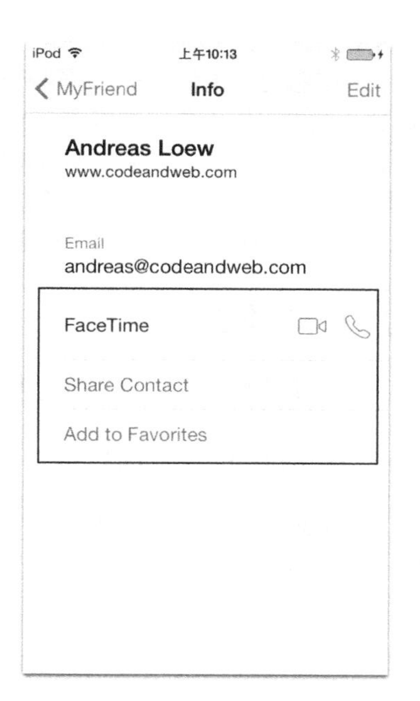

图19-21 FaceTime、Share Contact和Add to Favorites按钮

19.5 小结

在iOS开发中，很多需求都涉及通讯录的访问。本章首先介绍了访问通讯录所需要的框架，然后介绍了如何使用Contacts框架访问联系人信息，具体包括联系人的单值属性、多值属性和图片属性的访问。接下来，我们介绍了如何使用Contacts框架将联系人信息写入通讯录数据库。最后，本章介绍了如何使用ContactsUI框架提供系统界面实现选择联系人、显示和修改联系人。

第 20 章 iOS 10应用扩展

苹果在iOS 8之后推出了一种新的技术——应用扩展（App Extension），它能够为用户提供系统特定的扩展功能。有了扩展，我们编写应用程序时，不用所有功能都自己实现。例如，编写一个图像处理应用时，需要编辑图片功能，此时这个功能可以由应用扩展来完成。

iOS 8平台只推出了6个应用扩展，而到iOS 10已经累计有19个应用扩展了，涉及方方面面。作为iOS开发人员，有必要掌握一些应用扩展。本章不打算介绍全部的应用扩展，而是选择了几个有代表性的应用扩展——Today、表情包（Sticker Pack）和Messages，其中表情包和Messages应用扩展是iOS 10新增的。

20.1 应用扩展概述

这一节中，我们先介绍一下应用扩展的种类，再剖析一下它的工作原理和生命周期。

20.1.1 iOS 10 应用扩展种类

第一次使用Xcode 8创建应用扩展目标时，我惊奇地发现iOS平台可以创建如此众多的应用扩展，共有19个应用扩展，如图20-1所示。

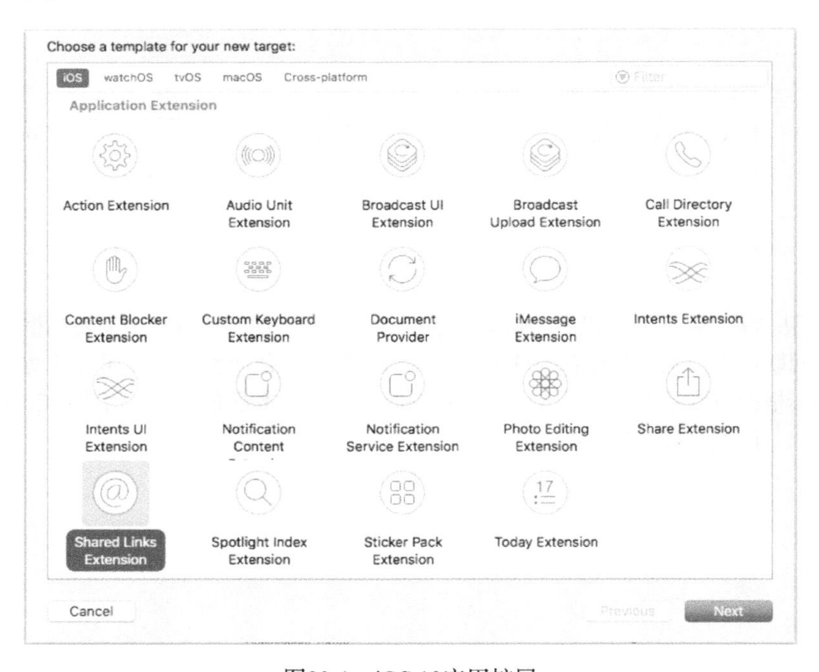

图20-1　iOS 10应用扩展

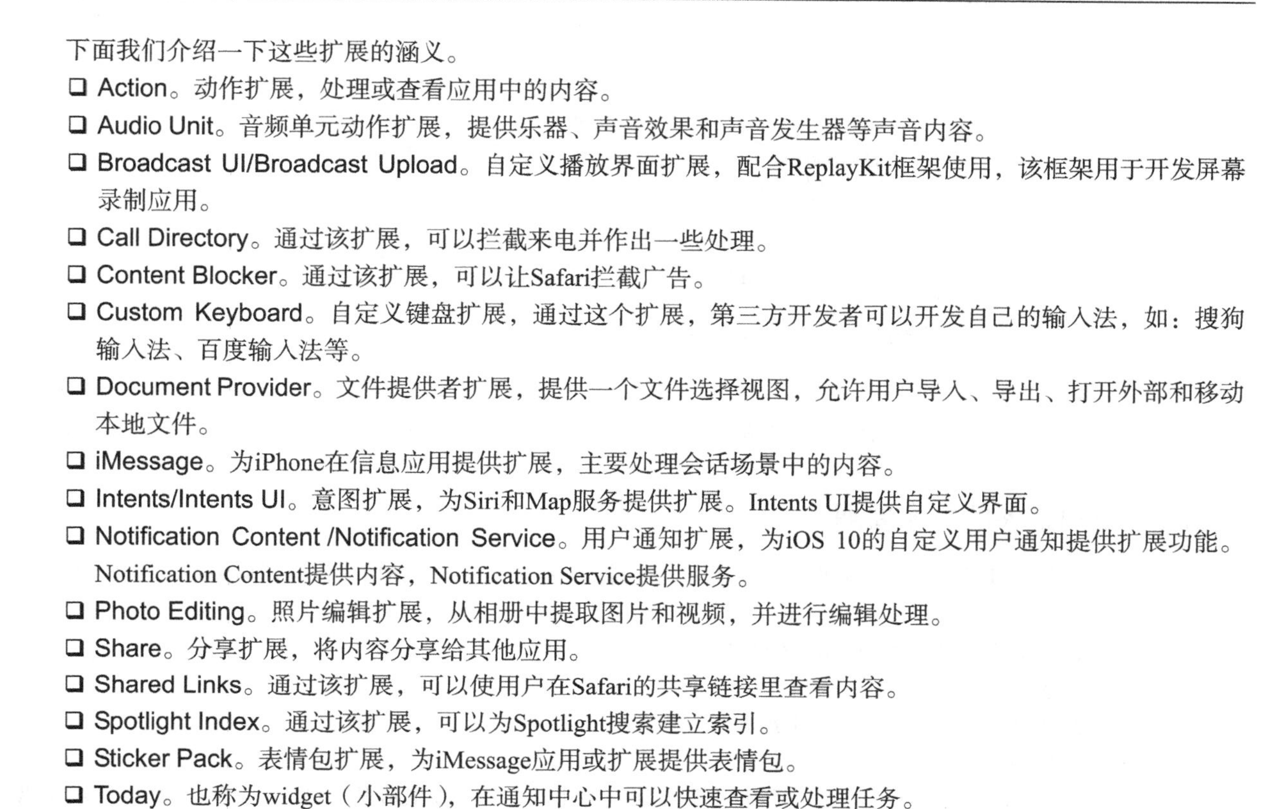

下面我们介绍一下这些扩展的涵义。

- Action。动作扩展，处理或查看应用中的内容。
- Audio Unit。音频单元动作扩展，提供乐器、声音效果和声音发生器等声音内容。
- Broadcast UI/Broadcast Upload。自定义播放界面扩展，配合ReplayKit框架使用，该框架用于开发屏幕录制应用。
- Call Directory。通过该扩展，可以拦截来电并作出一些处理。
- Content Blocker。通过该扩展，可以让Safari拦截广告。
- Custom Keyboard。自定义键盘扩展，通过这个扩展，第三方开发者可以开发自己的输入法，如：搜狗输入法、百度输入法等。
- Document Provider。文件提供者扩展，提供一个文件选择视图，允许用户导入、导出、打开外部和移动本地文件。
- iMessage。为iPhone在信息应用提供扩展，主要处理会话场景中的内容。
- Intents/Intents UI。意图扩展，为Siri和Map服务提供扩展。Intents UI提供自定义界面。
- Notification Content /Notification Service。用户通知扩展，为iOS 10的自定义用户通知提供扩展功能。Notification Content提供内容，Notification Service提供服务。
- Photo Editing。照片编辑扩展，从相册中提取图片和视频，并进行编辑处理。
- Share。分享扩展，将内容分享给其他应用。
- Shared Links。通过该扩展，可以使用户在Safari的共享链接里查看内容。
- Spotlight Index。通过该扩展，可以为Spotlight搜索建立索引。
- Sticker Pack。表情包扩展，为iMessage应用或扩展提供表情包。
- Today。也称为widget（小部件），在通知中心中可以快速查看或处理任务。

20.1.2 应用扩展工作原理

除iMessage 和Sticker Pack这两个扩展外，其他扩展不能独立存在，必须依附于一个应用才可以使用。这个被依附的扩展称为容器应用（Containing App）。通过容器应用，扩展可以在App Store上发布并销售。需要说明的是，苹果强调容器应用不能过于简单。

提示 为了能够在App Store上发布iMessage 和Sticker Pack扩展，苹果允许创建iMessage 和Sticker Pack类型的应用程序，这两个应用可以在App Store上发布。iMessage提供一个App Store按钮，可以直接点击该按钮，将应用下载并安装到本地。

使用扩展的往往不是容器应用，而是另外一个应用，它被称为宿主应用（Host App）。扩展、容器应用和宿主应用之间的关系如图20-2所示。

1. 扩展与容器应用的关系

扩展与容器应用不能直接通信，它们是两个独立的可执行文件，各自运行在自己的独立进程中，扩展可以通过`UIApplication`的`openURL`方法指定URL打开，`openURL`方法的打开方式是进程调用，不能传递参数和数据。如果扩展与容器应用之间需要数据传递，可以通过数据共享的方式。

2. 扩展与宿主应用的关系

扩展与宿主应用之间可以进行通信，相互之间可以传递数据，数据封装称为`NSExtensionItem`对象。

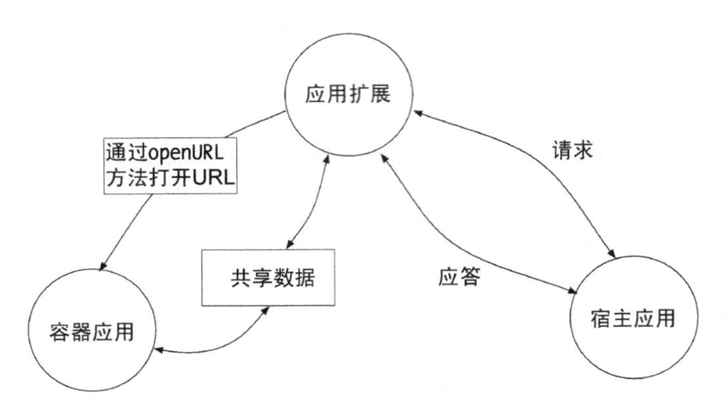

图20-2　扩展、容器应用和宿主应用之间的关系

20.1.3　应用扩展的生命周期

应用扩展的生命周期与应用程序的生命周期完全不同，前者与所在的容器应用没有任何关系，只是与宿主应用有关系。

应用扩展的生命周期如图20-3所示，分为4个阶段。

1. 用户选择应用扩展

用户在宿主应用的UI界面中选择应用扩展。

2. 系统启动应用扩展

系统实例化应用扩展，宿主应用与应用扩展建立通信通道。宿主应用定义了一个上下文环境对象`ExtensionContext`，通过上下文宿主应用与对象应用扩展可以进行数据交换。

3. 应用扩展运行

在应用扩展中运行代码，完成一些功能。

4. 系统结束应用扩展

扩展执行完成，系统结束应用扩展，回到宿主应用。

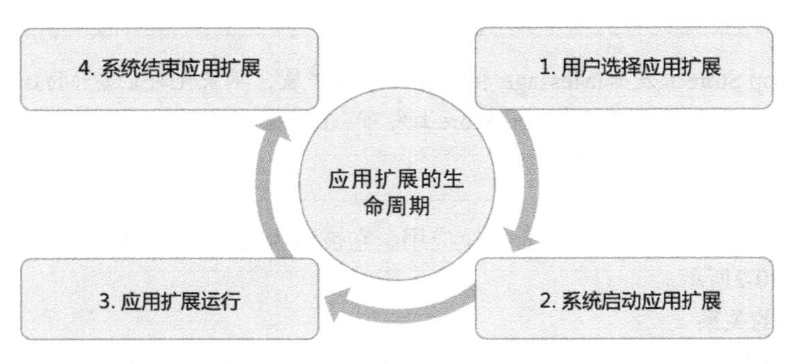

图20-3　应用扩展的生命周期

20.2　Today 应用扩展

苹果提供了多样的应用扩展，每一个扩展都有自己独立的一套API，全部掌握非常耗时并且也没有必要。下面我们介绍一个有代表性的应用扩展——Today。

20.2.1　使用 Today 应用扩展

在iOS 10中进入Today扩展的方式和界面与iOS 9有很大的不同。如图20-4所示，在通过桌面首屏时，再向右滑动屏幕，此时会进入通知中心的Today扩展界面。

图20-4　进入Today扩展界面

默认情况下，Today扩展界面会有天气、股市和日程等扩展。如果想添加和删除扩展，可以将Today扩展屏幕拖曳到底部，此时会看到“编辑”按钮，如图20-5a所示，点击该按钮，会进入如图20-5b所示的界面，在这里可以删除现有扩展，添加下面列出的扩展。

(a)　　　　　　　　　　(b)

图20-5　添加和删除扩展

20.2.2　实例：奥运会倒计时牌

下面我们通过实例介绍一下Today应用扩展的开发过程。首先根据自己的需要，创建一个工程，然后再创建一个目标，具体过程可以参见2.3.1节。选择Xcode菜单File→New→Target，此时会弹出一个选择模板对话框，从中选择iOS→Application Extension，如图20-6所示，选中Today Extension，接着点击Next按钮，然后在产品名中输入CountDown，选择合适的设置内容后，创建Today应用扩展，成功后如图20-7所示。

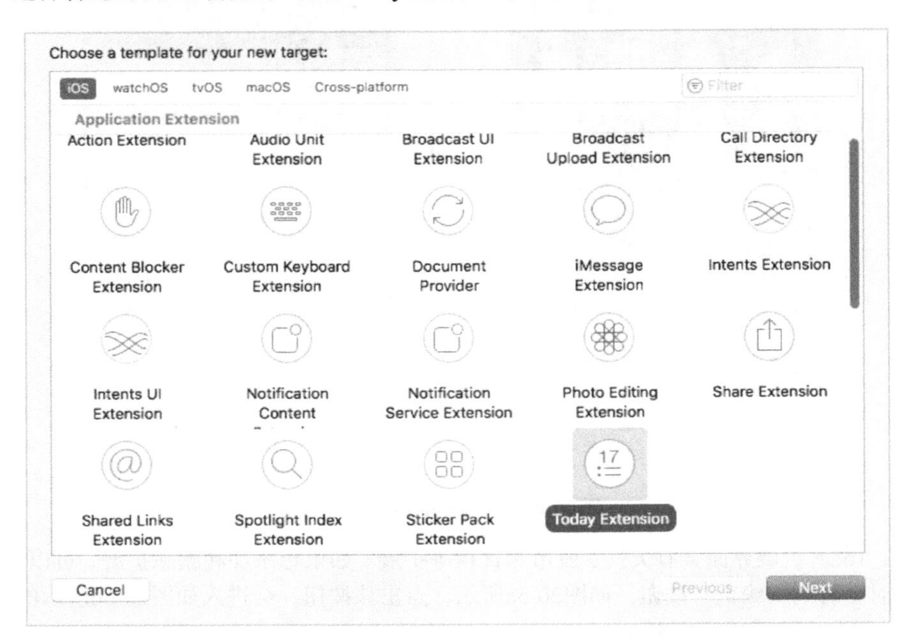

图20-6　创建Today扩展目标

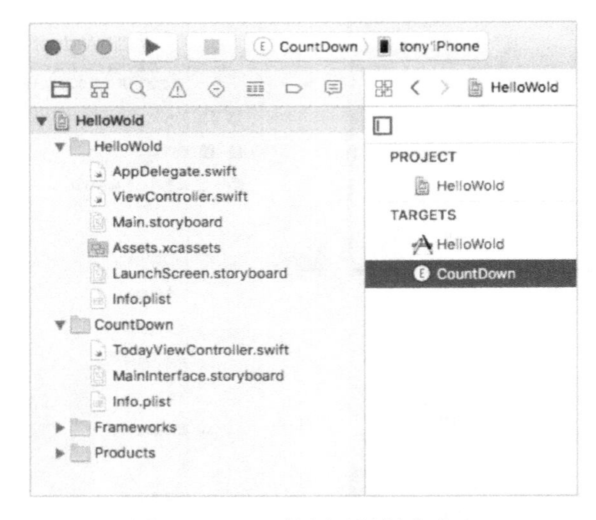

图20-7　Today扩展目标创建成功

打开CountDown扩展中的MainInterface.storyboard故事板文件，按照图20-8所示设计界面，这里我们可以设计Today应用扩展的界面。最终的设计结果如图20-9所示，界面背景为蓝色，文字为白色，具体设计过程与前面介绍的应用程序设计没有区别，这里不再赘述。另外，不要忘记屏幕适配问题。

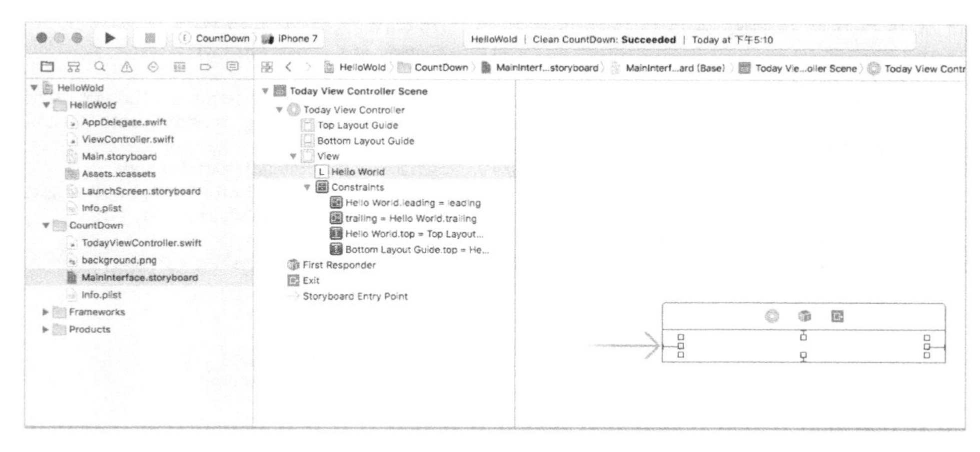

图20-8 MainInterface.storyboard故事板文件

图20-9 最终设计界面

界面设计完成后，为HelloWorld标签定义输出口。因为我们需要在程序中改变这个标签的内容，让它显示2020奥运会开幕倒计时的时间。

再来看代码部分，这部分比较简单，只有一个视图控制器TodayViewController，其代码如下：

```
//TodayViewController.swift文件
import UIKit
import NotificationCenter

class TodayViewController: UIViewController, NCWidgetProviding {            ①

    @IBOutlet weak var lblCountDown: UILabel!

    override func viewDidLoad() {
        super.viewDidLoad()

        //创建DateComponents对象
        let comps = DateComponents()                                          ②
        //设置DateComponents对象的日期属性
        comps.day = 24
        //设置DateComponents对象的月属性
        comps.month = 7
        //设置DateComponents对象的年属性
        comps.year = 2020                                                     ③

        //创建日历对象
        let calender = NSCalendar(calendarIdentifier:NSCalendar.Identifier.
        ➥gregorian)                                                          ④

        //获得2020-7-24的Date日期对象
        let destinationDate = calender!.date(from: comps as DateComponents)
                                                                              ⑤

        let date = Date()

        //获得当前日期到2020-7-24的DateComponents对象
        let components = calender!.components(.day, from: date,
```

```
//TodayViewController.m文件
#import "TodayViewController.h"
#import <NotificationCenter/NotificationCenter.h>

@interface TodayViewController () <NCWidgetProviding>                        ①
//显示倒计时
@property (weak, nonatomic) IBOutlet UILabel *lblCountDown;
@end

@implementation TodayViewController

- (void)viewDidLoad {
    [super viewDidLoad];

    //创建NSDateComponents对象
    NSDateComponents *comps = [[NSDateComponents alloc] init];               ②
    //设置NSDateComponents对象的日期属性
    [comps setDay:24];
    //设置NSDateComponents对象的月属性
    [comps setMonth:7];
    //设置NSDateComponents对象的年属性
    [comps setYear:2020];                                                    ③
    //创建日历对象
    NSCalendar *calender = [[NSCalendar alloc]
    ➥initWithCalendarIdentifier:NSCalendarIdentifierGregorian];             ④
    //获得2020-7-24的NSDate日期对象
    NSDate *destinationDate = [calender dateFromComponents:comps];           ⑤
    NSDate *date = [NSDate date];
    //获得当前日期到2020-7-24的NSDateComponents对象
    NSDateComponents *components = [calender components:NSCalendarUnitDay
    ➥fromDate:date toDate:destinationDate
    ➥options:NSCalendarWrapComponents];                                     ⑥
```

```
        ➥to:destinationDate!, options: .wrapComponents)          ⑥

        //获得当前日期到2020-7-24相差的天数
        let days = components.day                                ⑦

        let strLabel = NSMutableAttributedString(string: String(format:
        ➥"%i天", days!))                                          ⑧

        strLabel.addAttribute(NSFontAttributeName,
        ➥value: UIFont.preferredFont (forTextStyle: .footnote),
        ➥range: NSMakeRange(strLabel.length - 1, 1))              ⑨

        self.lblCountDown.attributedText = strLabel              ⑩

    }

    override func didReceiveMemoryWarning() {
        super.didReceiveMemoryWarning()
    }

    func widgetPerformUpdate(completionHandler: (@escaping (NCUpdateResult) ->
    ➥Void)) {
        completionHandler(NCUpdateResult.newData)
    }

}
```

```
    //获得当前日期到2020-7-24相差的天数
    NSInteger days = components.day;                              ⑦

    NSMutableAttributedString *strLabel = [[NSMutableAttributedString alloc]
    ➥initWithString:[NSString stringWithFormat:@"%li天",(long)days]];   ⑧

    [strLabel addAttribute:NSFontAttributeName
    ➥value:[UIFont preferredFontForTextStyle: UIFontTextStyleFootnote]
    ➥range:NSMakeRange(strLabel.length - 1, 1)];                  ⑨

    self.lblCountDown.attributedText = strLabel;                  ⑩

}

- (void)didReceiveMemoryWarning {
    [super didReceiveMemoryWarning];
}

- (void)widgetPerformUpdateWithCompletionHandler:(void (^)(NCUpdateResult))
➥completionHandler {
    completionHandler(NCUpdateResultNewData);
}

@end
```

上述代码中，第①行用于声明Today扩展的视图控制器，其父类UIViewController与普通的应用没有区别。另外，还实现了NCWidgetProviding协议，该协议能够让用户自定义Today扩展的外观和行为。

第②行用于创建NSDateComponents（Swift版是DateComponents）对象，该对象类是一个封装了日期的可扩展组件，通过这个类可以计算出两个日期的年、月、日、时、分、秒的差别。第②行~第③行用于设置日期可扩展组件类的日、月、年。

第④行用于创建NSCalendar日历对象。NSCalendar封装了系统时间类，用于与时间有关的计算，其中initWithCalendarIdentifier:构造函数通过一个日历标识创建日历对象。在本应用中，使用的日历标识是NSCalendarIdentifierGregorian（Swift版本是NSCalendar.Identifier.gregorian），它代表格里历（我们现行的公历）。此外，还有很多日历，如NSCalendarIdentifierChinese、NSCalendarIdentifierJapanese和NSCalendarIdentifierHebrew等。

第⑤行借助日历对象的dateFromComponents:方法从可扩展组件获得日期的NSDate对象，这个NSDate对象代表的时间是2020年7月24日，即2020东京奥运会的开幕时间。

第⑥行又创建了一个NSDateComponents（Swift版是DateComponents）对象。与第②行不同，它通过指定开始时间和结束时间来创建。在components:fromDate:toDate:方法中，components参数指定时间返回标志，这些标志是在NSCalendarUnit枚举类型中定义的，Swift版本是在NSCalendar.Unit结构体中定义的，其中常用的成员如下所示。

- **NSCalendarUnitYear**。Swift版本中是year。
- **NSCalendarUnitMonth**。Swift版本中是month。
- **NSCalendarUnitDay**。Swift版本中是day。
- **NSCalendarUnitHour**。Swift版本中是hour。
- **NSCalendarUnitMinute**。Swift版本中是minute。
- **NSCalendarUnitSecond**。Swift版本中是second。

本例是NSCalendarUnitDay（Swift版本中是day），只计算相差的天数。如果对两个日期的年、月、日、时、分、秒的差别进行位运算，可以使用如下代码：

```
NSInteger units = NSCalendarUnitYear | NSCalendarUnitMonth | NSCalendarUnitDay |
➥NSCalendarUnitHour | NSCalendarUnitMinute | NSCalendarUnitSecond;
```

```
let units: Set<Calendar.Component> =
➥[.year, .month, .day, .hour, .minute, .second]
```

在上述代码中，第⑥行的components:fromDate:toDate:options:方法中，options参数是NSCalendarWrap-Components枚举类型，Swift版本是NSCalendar.Options结构体类型，它表示按照算术加法计算日历。

第⑦行用于计算相差的天数。类似地，components.year用于获得相差的年数，components.month用于获得相差的月数。

第⑧行~第⑩行使用iOS 7提供的TextKit技术，实现了同一个字符串的不同显示字体。由于倒计时牌中最后的"天"与数字的字体不同，我们使用UILabel控件的attributedText属性接收NSAttributedString字符串，这种字符串有别于我们以前使用的NSString字符串，NSAttributedString字符串可以让字符具有不同的风格和样式。

20.3　开发表情包

为了增强社交功能，苹果在iOS 10中增加了一种新应用类型——iMessage应用，而开发表情包能够为iMessage应用提供丰富的、有趣的贴纸功能。

20.3.1　iMessage 应用

iMessage是苹果的即时通信应用，使用iMessage需要注册和激活你的Apple ID。在iOS 10之前，苹果没有提供iMessage应用程序开发接口，我们之只能在iOS中的"消息"应用中使用iMessage功能。

在iOS 10 中，苹果提供了一个Message框架以用来开发iMessage应用，而且在iOS 10之后App Store上可以下载iMessage类型的应用程序了。进入到iMessage的聊天界面（可以是iOS自带的消息应用），点击图标，打开如图20-10a所示的界面，这是一个处于活动状态的iMessage应用，点击左下角的图标，打开如图20-10b所示的iMessage应用抽屉，这是所有已经安装iMessage的应用。如果点击Store图标 +，则会进入App Store的iMessage应用下载页面，如图20-10c所示。

注意　图20-10b所示的iMessage应用抽屉中的Store图标 +，只有设备上才有，模拟器中没有。

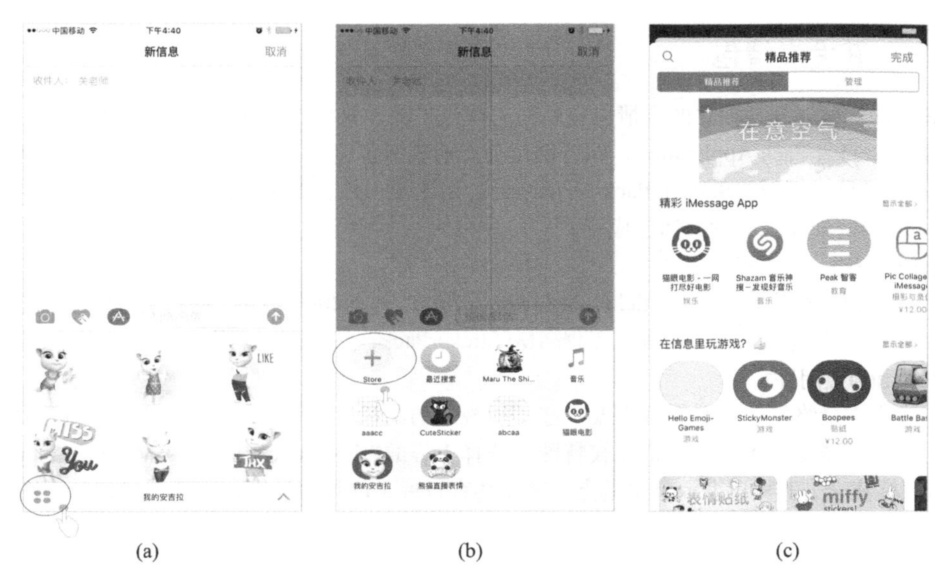

(a)　(b)　(c)

图20-10　iMessage应用

20.3.2　表情包

表情包为iMessage应用提供贴纸功能。如图20-11所示，在iMessage应用中，可以给好友发送贴纸。贴纸本质上就是图片，并且支持GIF动画图片。这些贴纸被打成包，就是表情包了。

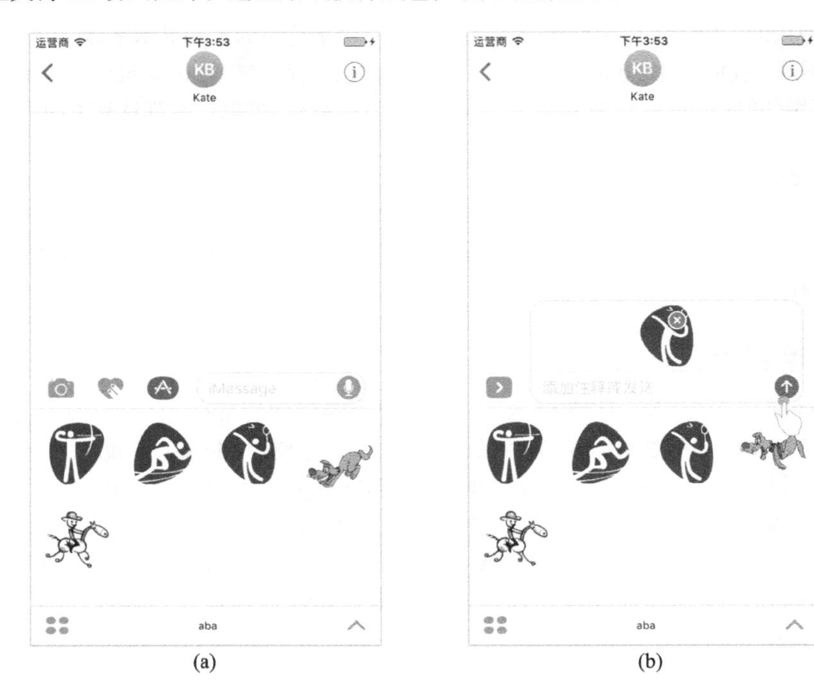

图20-11　表情包

在iMessage应用中，不仅可以给好友发送表情包中的贴纸，还可以发送文本、声音和视频等内容。也可以认为表情包是一种特殊的iMessage应用。

20.3.3　实例：开发表情包

使用Xcode工具，可以开发独立的表情包应用和表情包扩展。要创建独立的表情包应用，可以参考2.1.4节。选择工程模板类型Sticker Pack Application，即可创建独立的表情包应用。要创建表情包扩展，可以参考20.2.1节，创建一个目标，选择目标模板为Sticker Pack Extension。无论哪一种表情包，都不需要编写代码。下面我们通过实例具体介绍一下如何开发独立的表情包应用。

1. 创建表情包应用

参考2.1.4节，选择工程模板类型为Sticker Pack Application。如图20-12所示，创建工程模板，选择iOS下Sticker Pack Application模板，点击Next按钮后输入工程名MyStickerPack，然后创建工程。

2. 添加iMessage应用图标

每一个iOS应用在最终发布的时候都需要一些不同规格的图标，iMessage应用也需要，而且iMessage应用与普通的iOS应用不同，需要的图标的规格比较特殊。选择工程的Stickers.xcstickers→iMessage App Icon，可见如图20-13所示的图标集合，其中pt表示单位是点，这里共有11种不同规格的图标，请美工设计好图标后，参考8.3节添加图标。

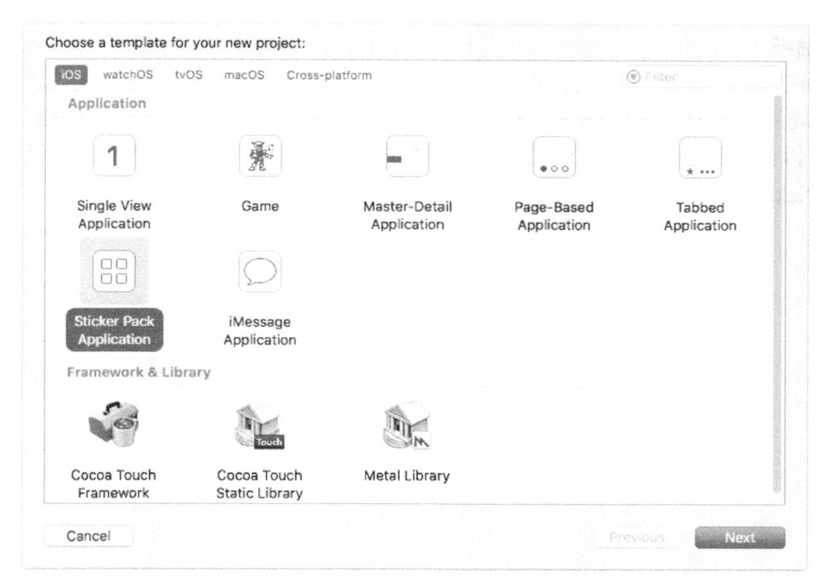

图20-12　表情包应用工程模板

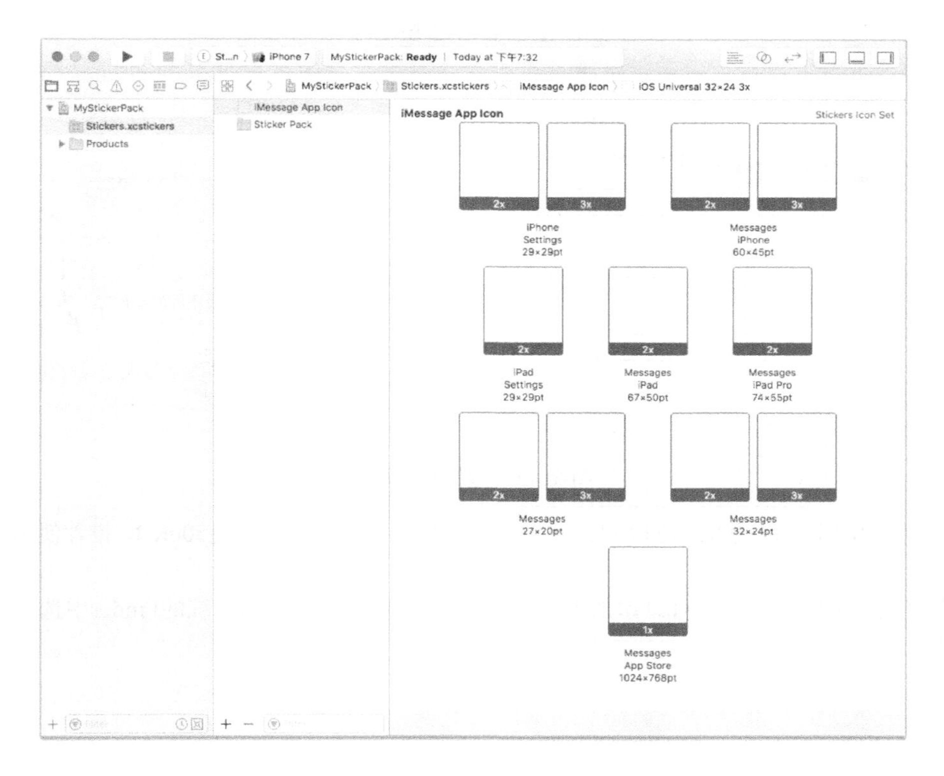

图20-13　iMessage应用图标集合

注意　设计iMessage应用时需要注意，这11种图标中有些高宽比例是不一致的，所以在设计时就不能通过一张图片的缩放而得到其他图标，而是需要根据高宽比重新设计。

3. 添加表情包贴纸

表情包贴纸布局是一种网格形式，有3种不同尺寸的布局，如图20-14所示，小尺寸布局的最大图片是300×300像素，中尺寸布局的最大图片是408×408像素，大尺寸布局的最大图片是618×618像素。根据自己的实际情况选择相应的布局，以便达到最好的显示效果。要选择不同的布局尺寸，可以按照图20-15所示，选择Sticker Pack的属性检查器，在Sticker Size下拉列表中选择尺寸。

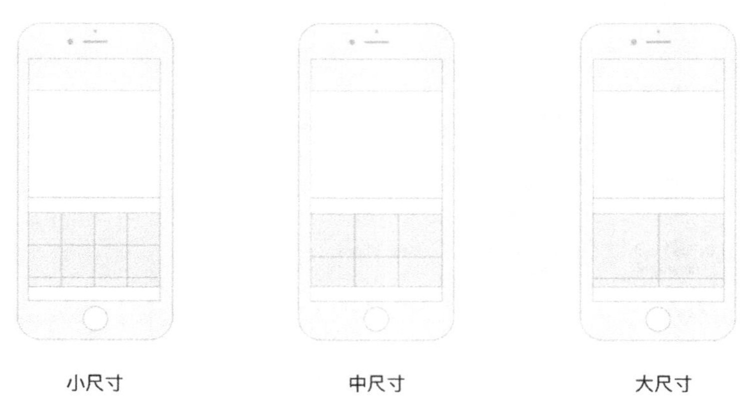

图20-14　表情包贴纸布局

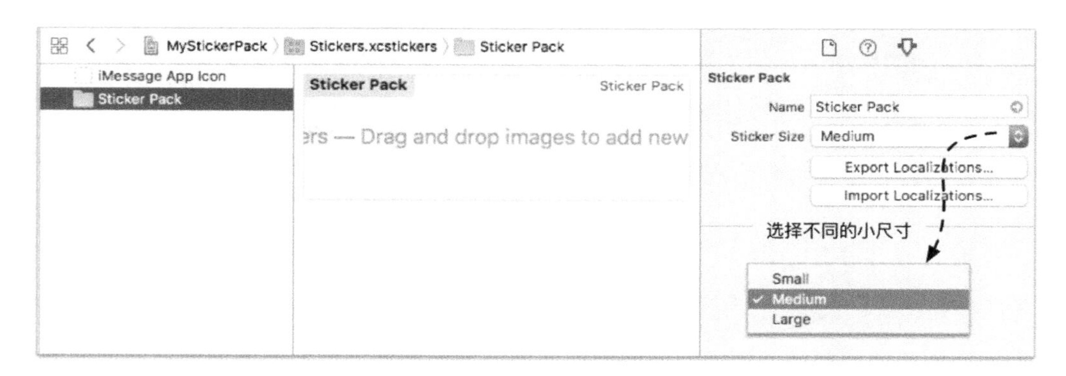

图20-15　选择布局尺寸

表情包贴纸图片的文件格式可以是PNG、APNG[①]、GIF或JPG，大小不超过500KB。推荐使用PNG和APNG，大小不超过500KB。

设计好表情包贴纸图片后，就可以添加了。如图20-16所示，在macOS系统的Finder中选中图片，拖曳到Sticker Pack中。

① APNG（Animated Portable Network Graphics）格式是PNG的位图动画扩展。

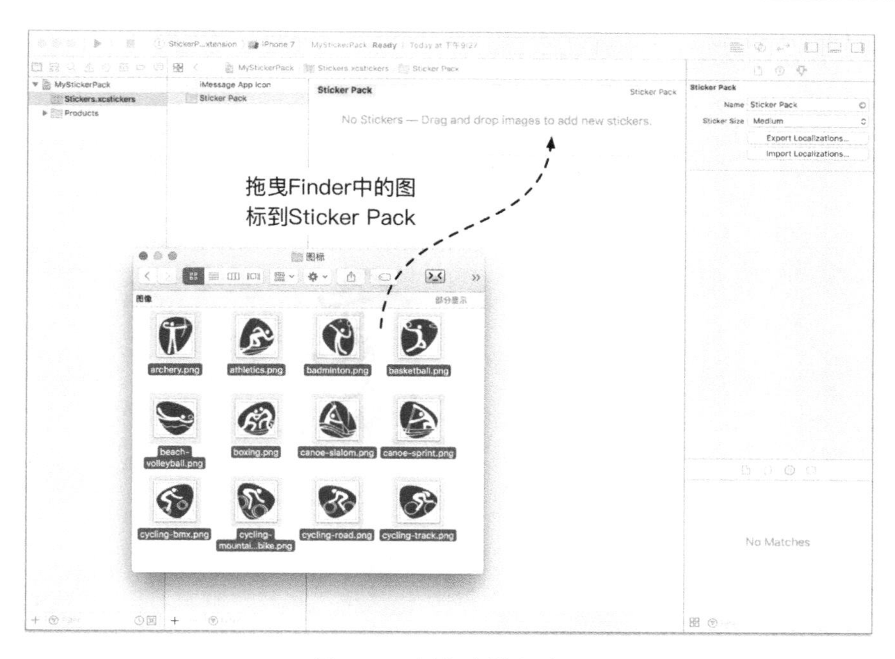

图20-16　添加表情包贴纸

表情包的贴纸动画除了可以使用APNG和GIF文件外，还可以使用序列帧动画。序列帧动画由多个帧（图片）构成，这些帧连续播放就会展现动画效果了。在表情包中添加序列帧动画的具体步骤是右击Sticker Pack区域，在弹出的右键菜单中选择Add Assets → New Sticker Sequence，然后添加一个Sticker Sequence，这就是序列帧动画表情包。我们需要将序列帧文件拖曳到Sticker Sequence，如图20-17所示。添加成功后的Sticker Pack区域如图20-18所示。

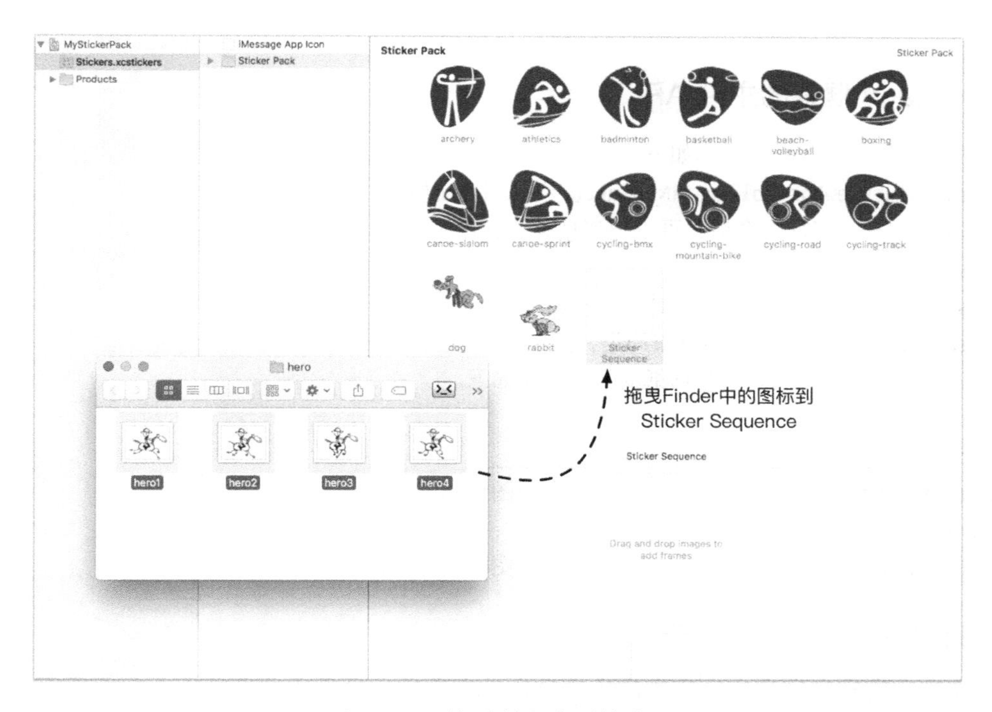

图20-17　添加表情包序列帧动画

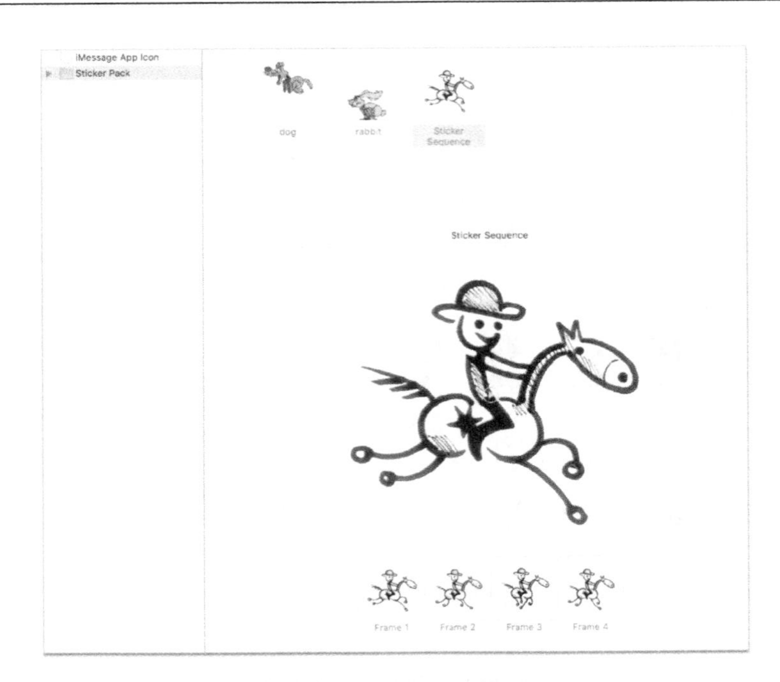

图20-18　成功添加表情包序列帧动画

添加完成后，就可以运行一下看看效果了。

20.4　Message 框架

如果想自己的iMessage应用不仅可以发表情贴纸，还想有一些其他功能，此时就需要通过Message框架开发iMessage应用了。

20.4.1　Message 框架的主要 API

Message框架提供了一些类，具体如下。

- **`MSMessagesAppViewController`**。iMessage应用视图控制器。
- **`MSConversation`**。当前的会话，可以把消息添加到输入框中。
- **`MSMessage`**。消息对象。
- **`MSMessageLayout`**。消息布局类，是一个抽象类。
- **`MSMessageTemplateLayout`**。消息布局类，是一个具体类。

此外，Message框架中有关表情包贴纸的类和协议如下。

- **`MSSticker`**。贴纸对象。
- **`MSStickerView`**。贴纸视图。
- **`MSStickerBrowserViewController`**。贴纸容器视图控制器。
- **`MSStickerBrowserView`**。贴纸容器视图。
- **`MSStickerBrowserViewDataSource`**。贴纸容器视图数据源协议。

20.4.2　消息布局

消息`MSMessage`创建成功后，就会涉及布局的问题。在`MSMessage`中，有一个`MSMessageLayout`类型的`layout`属性用于设置布局。目前，iOS只提供一个具体的默认布局类MSMessageTemplateLayout。默认布局如图20-19所示。

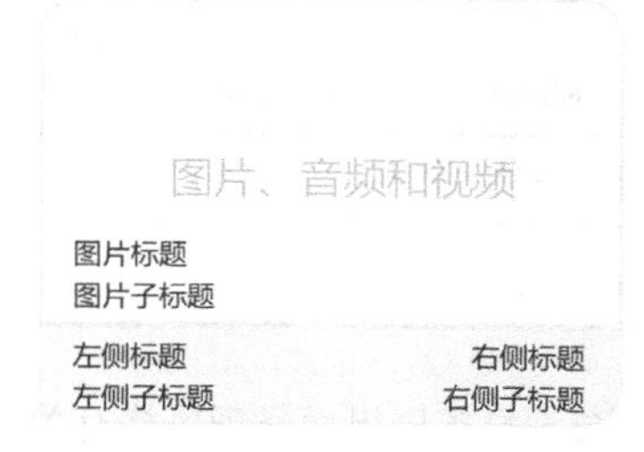

图20-19　消息默认布局

从图20-19中可见，消息默认布局中有很多内容，它们与MSMessageTemplateLayout类的属性对应，这些属性说明如下。

- **image**。消息图片。
- **imageTitle**。消息图片标题。
- **imageSubtitle**。消息图片子标题。
- **caption**。消息左侧标题。
- **subcaption**。消息左侧子标题。
- **trailingCaption**。消息右侧标题。
- **trailingSubcaption**。消息右侧子标题。
- **mediaFileURL**。音频或视频的URL地址，如果image属性被设置，该属性将被忽略。

20.4.3　消息扩展界面的收缩和展开

消息扩展界面有两种呈现风格——紧缩（compact）和展开（expanded）。默认的消息扩展界面呈现风格是紧缩，如图20-20a所示，点击右下角的^按钮展开，如图20-20b所示，再点击右上角的∨按钮收缩。

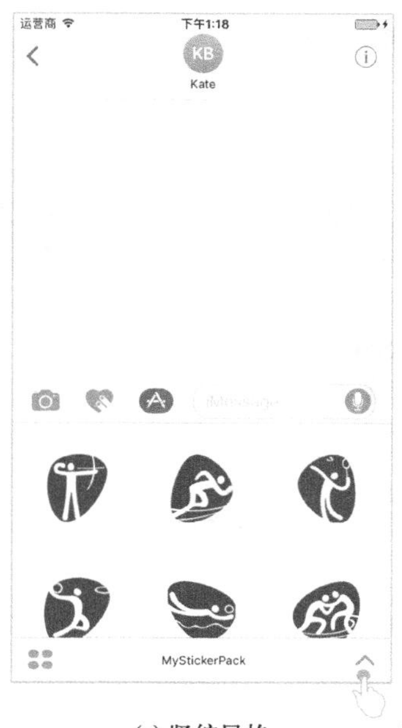

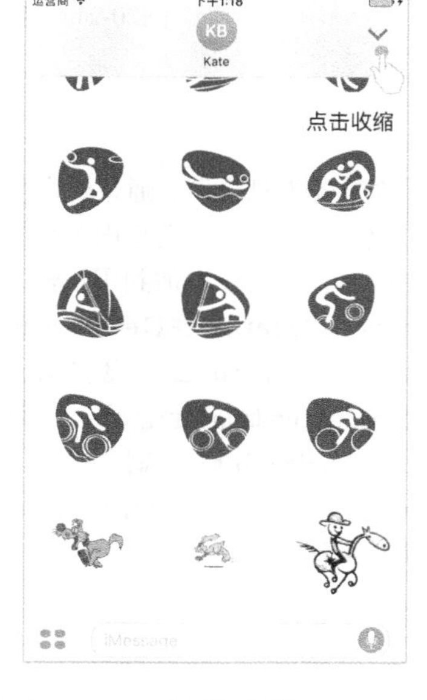

(a) 紧缩风格　　点击展开　　(b) 展开风格

图20-20　消息扩展界面的收缩和展开

MSMessagesAppViewController视图控制器的presentationStyle属性保存了消息扩展界面的呈现状态，该属性是MSMessagesAppPresentationStyle枚举类型。枚举MSMessagesAppPresentationStyle的成员如表20-1所示。

表20-1　MSMessagesAppPresentationStyle枚举成员

Swift枚举成员	Objective-C枚举成员	说　明
compact	MSMessagesAppPresentationStyleCompact	收缩状态，是默认的呈现样式
expanded	MSMessagesAppPresentationStyleExpanded	展开状态

在MSMessagesAppViewController中，与消息扩展界面呈现风格有关的方法还有如下几个。

- **requestPresentationStyle:**。切换界面呈现风格。
- **willTransitionToPresentationStyle:**。切换界面之前调用。
- **didTransitionToPresentationStyle:**。切换界面之后调用。

20.4.4　消息应用的生命周期

消息应用的生命周期分为：启动过程和销毁过程。在生命周期的不同阶段，我们会回调MSMessagesAppViewController中的相关方法，如图20-21所示。

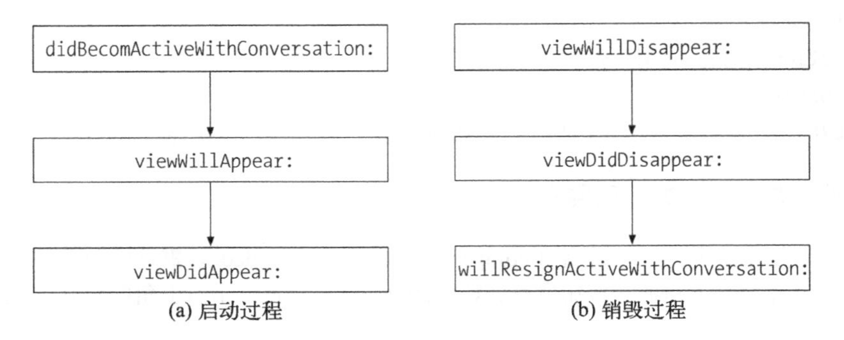

图20-21　消息应用的生命周期

20.4.5　消息会话

MSConversation类用来描述消息会话，通过这个类，可以将消息内容添加到用户输入框中。如图20-22a所示，选中要发送的消息内容（本例是图片），消息内容会出现在用户输入框中，点击发送按钮即可发送。

MSConversation类中将消息等内容添加到用户输入框的方法有如下4个。

- **insertAttachment:withAlternateFilename:completionHandler:**。添加一个附件。
- **insertMessage:completionHandler:**。添加Message对象消息。
- **insertSticker:completionHandler:**。添加表情包。
- **insertText:completionHandler:**。添加文本。

如果是在MSMessagesAppViewController中添加消息，可以通过成员变量activeConversation获取当前会话对象，然后调用上述方法添加消息。

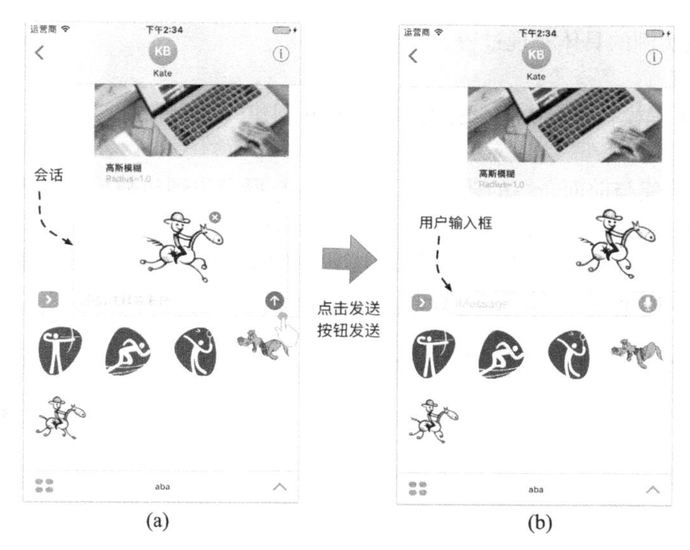

图20-22　消息会话

20.4.6　实例：高斯模糊滤镜

使用Xcode工具，可以开发独立的iMessage应用和iMessage扩展。创建独立的iMessage应用时，可以参考2.1.4节，选择工程模板类型iMessage Application。创建iMessage扩展时，可以参考20.2.1节，创建一个目标，然后选择目标模板为iMessage Extension。

下面我们通过一个实例具体介绍一下如何开发独立的iMessage应用。该实例如图20-23所示，Message扩展界面有Stepper控件，单击增加或减少按钮，可以改变Stepper数值，然后用Stepper数值作为高斯模糊①半径②参数对图片进行处理，然后将处理之后的图片作为iMessage消息发送。

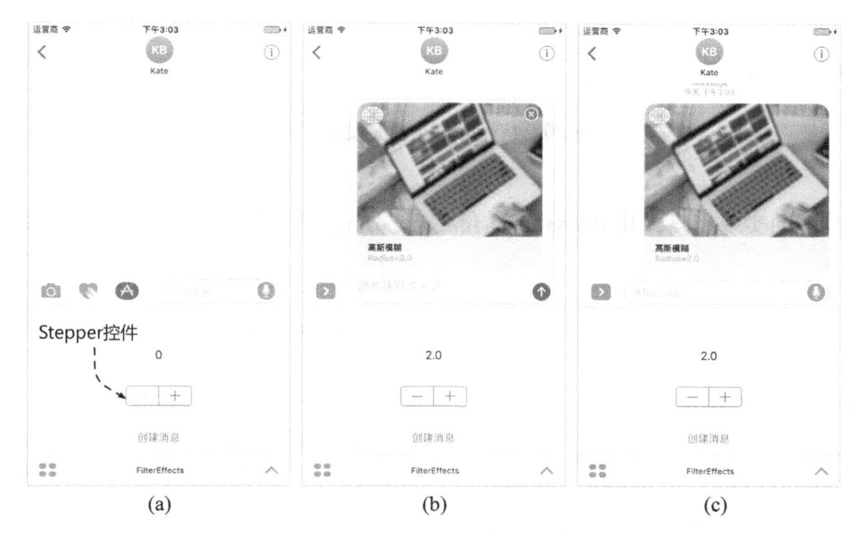

图20-23　iMessage应用实例

① 高斯模糊是Photoshop中的一种滤镜，通过滤镜处理的图片可以获得意想不到的效果。iOS和macOS中也提供了几十种滤镜。
② 高斯模糊最重要的参数就是模糊半径Radius，其取值是0.0 ~ 10.0。

下面我们介绍一下该实例的具体实现过程。

1. 创建iMessage应用

参考20.3.3节，选择工程模板类型iMessage Application并创建FilterEffects工程。

2. 构建界面

iMessage应用界面的构建与前面介绍的普通应用没有区别，既可以通过故事板实现，也可以通过纯代码实现。本例中我们采用故事板来构建界面。故事板文件是MainInterface.storyboard，打开该故事板文件，然后拖曳一个Label（显示Stepper数值）、一个Stepper和一个按钮，如图20-24所示，接着在设计界面中摆放好它们的位置，然后添加约束和适配，最后为Label定义输出口、Stepper和按钮定义动作事件。

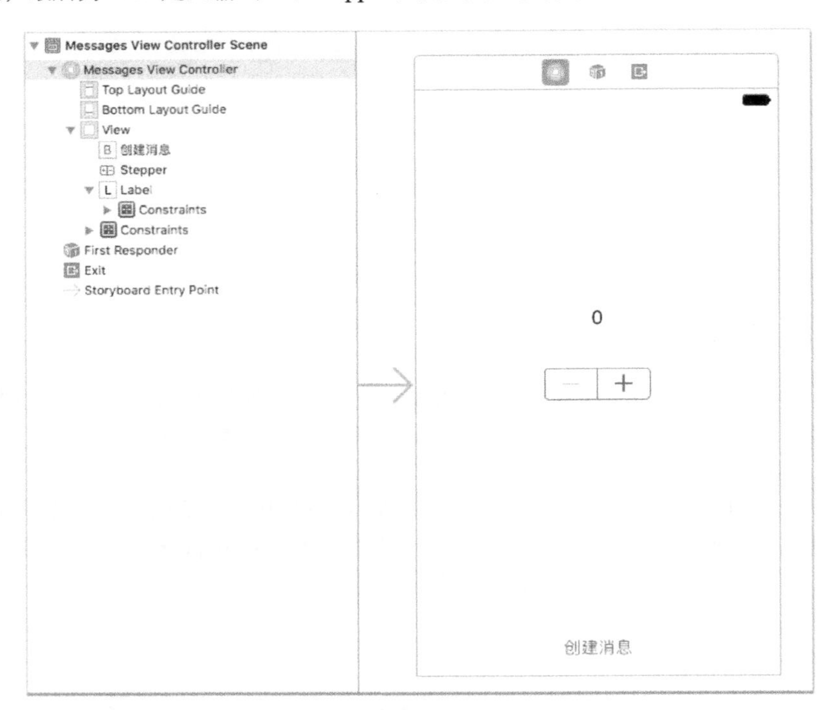

图20-24　通过故事板设计界面

3. 实现代码

下面我们看看具体的实现代码，其中MessagesViewController的代码如下：

```
//MessagesViewController.swift文件

import UIKit
import Messages

class MessagesViewController: MSMessagesAppViewController {

    @IBOutlet weak var label: UILabel!
    @IBOutlet weak var stepper: UIStepper!

    @IBAction func onValueChanged(_ sender: Any) {                    ①
        let strValue = String(format: "%.1f", self.stepper.value)
        self.label.text = strValue
    }

    @IBAction func onclick(_ sender: Any) {

        if let image = filterGaussianBlur(self.stepper.value), let conversation
```

```
//MessagesViewController.m文件

#import "MessagesViewController.h"

@interface MessagesViewController ()

@property (weak, nonatomic) IBOutlet UILabel *label;
@property (weak, nonatomic) IBOutlet UIStepper *stepper;

@end

@implementation MessagesViewController

- (IBAction)onValueChanged:(id)sender {                                ①
    NSString* strValue = [NSString stringWithFormat:@"%.1f",
    ➥self.stepper.value];
    self.label.text = strValue;
```

```
    ➥= activeConversation {                                              ②
        let layout = MSMessageTemplateLayout()                           ③
        layout.image = image                                             ④
        layout.caption = "高斯模糊"
        layout.subcaption = "Radius=\(self.stepper.value)"               ⑤

        let message = MSMessage()                                        ⑥
        message.layout = layout

        conversation.insert(message, completionHandler:
        ➥{ (error: Error?) in                                           ⑦
            if error != nil {
                print(error!)
            }

        })
    }

}

func filterGaussianBlur(_ value : Double) -> UIImage? {

    if let image = UIImage(named: "SkyDrive340.png"),
    ➥let gaussianBlur = CIFilter(name: "CIGaussianBlur") {               ⑧
        let cImage = CIImage(cgImage: image.cgImage!)                    ⑨

        gaussianBlur.setValue(cImage, forKey: "inputImage")              ⑩
        gaussianBlur.setValue(value, forKey: "inputRadius")              ⑪
        let result = gaussianBlur.value(forKey: "outputImage") as! CIImage
                                                                         ⑫

        let context = CIContext(options: nil)                            ⑬
        let imageRef = context.createCGImage(result, from: CGRect(x: 0, y: 0,
        ➥width: image.size.width, height: image.size.height))           ⑭

        let outputImage =  UIImage(cgImage: imageRef!)                   ⑮
        return outputImage
    }

    return nil
}
......

}
```

```
}

- (IBAction)onclick:(id)sender {

    UIImage *image = [self filterGaussianBlur:self.stepper.value];             ②

    MSMessageTemplateLayout* layout = [[MSMessageTemplateLayout alloc] init];
                                                                                ③
    layout.image = image;                                                       ④
    layout.caption = @"高斯模糊";
    layout.subcaption = [NSString stringWithFormat:@"Radius=%.1f",
    ➥self.stepper.value];                                                     ⑤

    MSMessage* message = [[MSMessage alloc] init];                              ⑥
    message.layout = layout;

    MSConversation* conversation = self.activeConversation;
    [conversation insertMessage:message completionHandler:^(NSError*
    ➥_Nullable error) {                                                       ⑦
        if (error != nil) {
            NSLog(@"%@", error);
        }
    }];

}

- (UIImage*)filterGaussianBlur: (double)value {

    UIImage *image = [UIImage imageNamed:@"SkyDrive340.png"];

    CIFilter *gaussianBlur = [CIFilter filterWithName: @"CIGaussianBlur"];⑧
    CIImage *cImage = [CIImage imageWithCGImage:[image CGImage]];              ⑨

    [gaussianBlur setValue: cImage forKey: @"inputImage"];                     ⑩
    [gaussianBlur setValue: [NSNumber numberWithDouble:value] forKey:
    ➥@"inputRadius"];                                                         ⑪
    CIImage *result = [gaussianBlur valueForKey:@"outputImage"];               ⑫

    CIContext *context = [CIContext contextWithOptions:nil];                   ⑬
    CGImageRef imageRef = [context createCGImage:result
    ➥fromRect:CGRectMake(0, 0, image.size.width, image.size.height)];         ⑭

    UIImage *outputImage = [[UIImage alloc] initWithCGImage:imageRef];         ⑮
    return outputImage;
}

......

@end
```

上述代码中，第①行用于声明UIStepper的事件处理方法onValueChanged:。第②行中的filterGaussianBlur:方法用于处理图片，返回的结果是UIImage对象，它作为消息中的图片发送。第③行用于实例化MSMessage-TemplateLayout默认布局对象，第④行～第⑤行用于设置默认布局对象。第⑥行用于创建MSMessage对象，并通过message.layout = layout语句设置布局属性。第⑦行通过会话对象conversation调用insertMessage:completion-Handler:方法添加消息到用户输入框，其中message参数是消息对象，completionHandler参数是完成之后回调的代码块（Swift中的闭包）。

第⑧行～第⑮行通过滤镜处理图片。一般情况下，滤镜使用流程可以分成以下3个步骤。

(1) 创建滤镜CIFilter对象。

(2) 设置滤镜参数。

(3) 输出结果。

第⑧行通过指定滤镜名创建滤镜对象，CIGaussianBlur是高斯模糊滤镜名。第⑨行通过CGImage创建CIImage

图像对象。CIImage和CIFilter都属于Core Image框架，滤镜能够处理CIImage图像对象。第⑩行用于设置滤镜的输入参数（inputImage）。第⑪行用于设置高斯模糊半径参数的取值，其中inputRadius是输入参数名。第⑫行是取得滤镜之后的图像对象，类型为CIImage。

第⑬行创建CIContext对象。CIContext构造方法是一个字典类型参数，它规定了各种选项，包括颜色格式以及内容是否应该运行在CPU或是GPU上。对于本例，默认值就可以了，所以只需要传入nil。

第⑭行使用CIContext的createCGImage:fromRect:方法创建CGImage对象，Objective-C版是CGImageRef类型，CGImageRef是CGImage引用类型。参数fromRect用于设置图像大小。

第⑮行通过CGImage图像对象创建UIImage图像对象，UIImage图像对象是Message消息所需要的类型。

20.5 小结

本章中，我们主要介绍了iOS 10应用扩展技术，然后介绍了Today应用扩展技术。此外，还讨论了iOS 10新推出的表情包和iMessage应用开发。

第 21 章

重装上阵的iOS 10用户通知

随着移动互联网和智能手机的发展，iOS和Android等系统都提供了用户通知功能。用户通知是非常重要的功能，它能够提醒用户接收信息、查看内容以及完成一些操作。iOS早期版本就提供了用户通知，但在iOS 10中，用户通知有很大的变化，可以在通知中显示图片和视频等内容。本章中，我们将介绍如何开发用户通知。

21.1 用户通知概述

用户通知就是能够给用户一种提示，不仅仅是iOS系统，苹果的macOS也有用户通知。只要安装了应用，并且应用没有处于前台活动状态，那么该应用就可以接收通知。当然，如果你在设置中关闭了通知，就没法接收通知了。

21.1.1 通知种类

用户通知根据通知信息的来源，可以分为本地通知和远程通知。本地通知是由iOS操作系统根据条件在本机上触发的，例如闹钟就是基于时间触发提醒通知的。远程通知是第三方（我们安装了它的客户端）远程推送给用户的iOS设备的，这种通知常用于商家推销自己的产品。通过这个渠道推送自己的产品信息是不错的选择，但是作为商家，一定要遵守起码的道德规范，不要在用户睡觉的时候推送通知，否则用户就会毫不留情地屏蔽你这个应用的通知，甚至卸载应用。

21.1.2 通知界面

本地通知和远程通知的用户界面形式都一样。

用户通知从视觉效果上看有两种形式：横幅通知和提醒通知。图21-1所示的是微信用户通知，通知会显示在屏幕顶部的通知栏中。横幅通知和提醒通知的区别在于：横幅通知在通知栏中显示一会儿就会消失；而提醒通知不会消失，会一直出现在通知栏中，除非点击它进行默认的操作。

提示 如果Mac电脑和iPhone上都安装了微信，当需要通过Mac登录微信客户端时，iPhone上有一个确认的过程，这个过程会给用户一个通知，如图21-1所示。

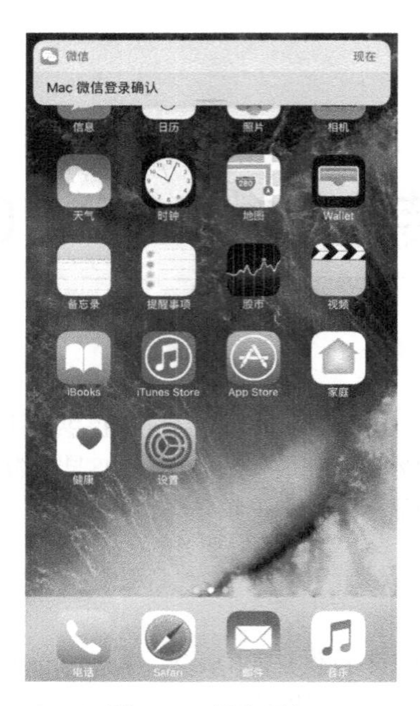

图21-1　用户通知

当通知在通知栏中消失后，它会保留在通知中心里。要想查看通知中心，可以用手向下拉动状态栏，如图21-2所示，这个设计是在iOS 5之后的新功能。单击通知后面的关闭按钮可以清除通知。单击具体通知项，可以进入应用。

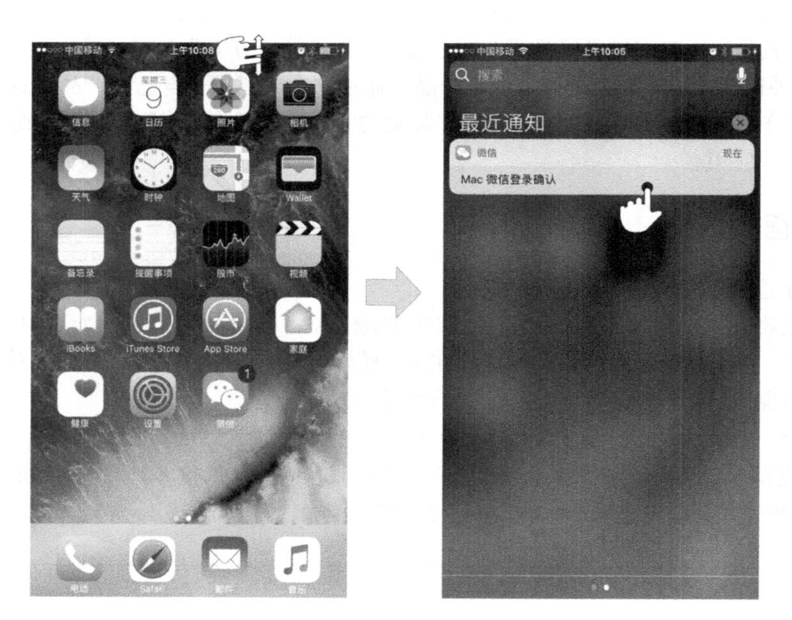

图21-2　查看通知

应用接收到通知时，会在其图标上添加标记，即图标右上角显示的红色数字，如图21-3所示。标记中的数字可以通过程序设置，标记的数据量一般代表有多少条未读通知，当用户看完这些通知后，图标上的标记会一起变化。

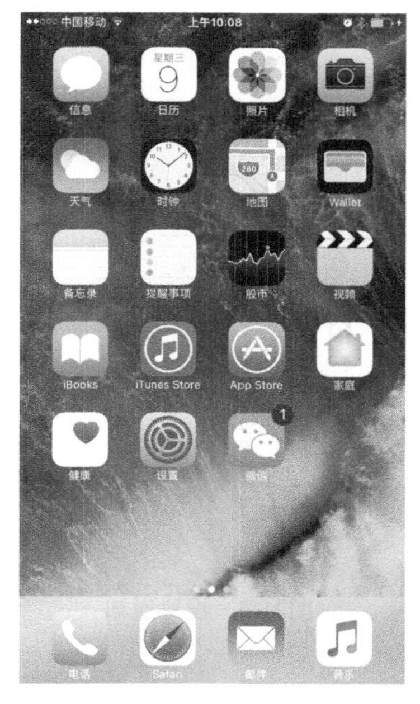

图21-3　应用图标标记

即使在锁屏情况下，也可以接收通知。如图21-4a所示，从右向左滑动通知项目，在通知项目的后面会出现“查看”和“清除”按钮，如图21-4b所示，此时可以查看或清除通知。

(a)　　(b)

图21-4　应用图标标记

21.1.3　设置通知

用户通知可以在系统设置中进行。启动“设置”应用，进入通知，选择“通知”中心下面的应用，如图21-5所示，从中可以设置通知的样式、是否关闭通知、是否关闭提示声音和是否关闭标记显示等。

图21-5　设置通知

21.2　开发本地通知

开发iOS本地通知，主要分为如下3步。

(1) 请求授权。

(2) 通知创建与发送。

(3) 通知接收后的处理。

21.2.1　开发本地通知案例

下面我们通过一个案例介绍一下开发本地通知的过程。该案例如图21-6a所示，启动应用界面中有两种触发器，点击“时间中断触发器”→“开启通知”按钮，则开启计划5秒后到达的通知。点击“日历触发器”→“开启通知”按钮，则开启计划每天上午7:30分到达的通知。通知到达后的界面如图21-6b所示，此时会在通知栏中显示通知。下拉通知如图21-6c所示，其中会看见通知的附件，还有一个“点赞”操作按钮。

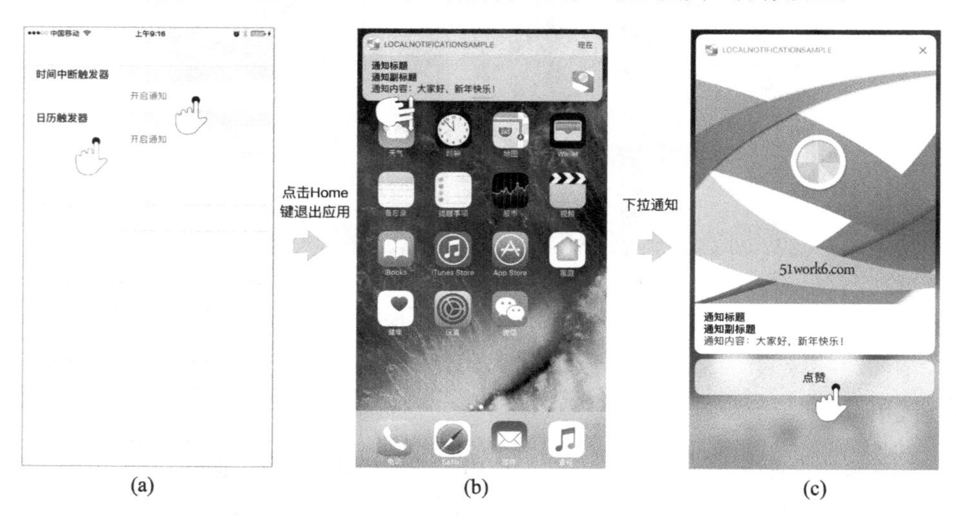

(a)　(b)　(c)

图21-6　开发本地通知案例

注意 通知能够在通知栏中显示的前提是，接收通知的应用处于非活动状态，也就是应用退到后台。所以在测试上面的案例时，当点击开启按钮后，需要点击设备的Home键退出应用。

从图21-6a可见，界面布局采用的是表视图，参考7.5节设计该界面。注意两个“开启通知”按钮都共用一个事件处理方法。如图21-7所示，两个按钮的触摸事件（Touch Up Inside）都连接到事件处理方法`scheduleNotification:`，具体连接过程是选中按钮，按住ctrl键，拖曳鼠标到方法`scheduleNotification:`。

图21-7 两个按钮共用一个事件处理方法

两个按钮共用一个事件处理方法时，如何区分是点击了哪个按钮呢？这可以通过设置按钮的Tag属性来搞定。如图21-8所示，选中第一个按钮，打开其属性检查器，找到View→Tag属性，然后将其设置为200。接着，将第二个按钮的Tag属性设置为300。

参考代码如下：

```
@IBAction func scheduleNotification(_ sender: UIButton) {

    if (sender.tag == 300) {//300是点击日历触发器按钮，200是点击时间中断触
                        //发器按钮
        trigger = self.createCalendarTrigger()
    }
}
```

```
- (IBAction)scheduleNotification:(id)sender {

    UIButton* button = (UIButton*)sender;
    if (button.tag == 300) {//300是点击日历触发器按钮，200是点击时间中断触
                            //发器按钮
        trigger = [self createCalendarTrigger];
    }
}
```

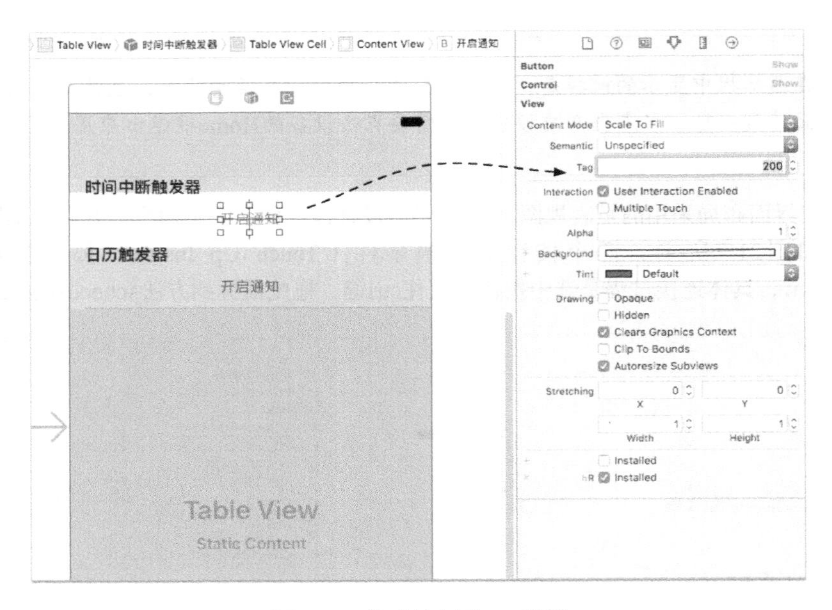

图21-8　修改按钮的Tag属性

21.2.2　请求授权

应用要能够接收系统的通知，则要求用户对该应用接收通知的行为进行授权。具体来讲，授权包括：是否出现提示框、是否有声音和是否在应用图标上显示标记（红色的数字）。UNUserNotification框架中这些授权类型被定义在一个枚举类型UNAuthorizationOptions中，其中与用户通知授权相关的成员如表21-1所示。

表21-1　用户通知授权相关的成员

Swift枚举成员	Objective-C枚举成员	说　明
alert	UNAuthorizationOptionAlert	是否出现提示框
sound	UNAuthorizationOptionSound	是否有声音
badge	UNAuthorizationOptionBadge	是否显示应用图标标记

具体授权过程是通过UNUserNotificationCenter的requestAuthorizationWithOptions:completionHandler:方法实现的。一般情况下，授权代码是在AppDelegate的application:didFinishLaunchingWithOptions:方法中，具体如下：

```
func application(_ application: UIApplication, didFinishLaunchingWithOptions
➥launchOptions: [UIApplicationLaunchOptionsKey: Any]?) -> Bool {

    UNUserNotificationCenter.current().requestAuthorization(options:
    ➥[.alert, .sound]) {(granted, error) in                              ①
        if granted {
            print("授权成功")
        } else {
            print("授权失败")
        }
    }

    return true
}
```

```
- (BOOL)application:(UIApplication *)application
➥didFinishLaunchingWithOptions:(NSDictionary *)launchOptions {

    [[UNUserNotificationCenter currentNotificationCenter]
     requestAuthorizationWithOptions:(UNAuthorizationOptionAlert |
     ➥UNAuthorizationOptionSound)                                        ①
     completionHandler:^(BOOL granted, NSError * _Nullable error) {
        if (!error) {
            NSLog(@"授权成功");
        } else {
            NSLog(@"授权失败");
        }
    }];


    return YES;
}
```

第①行用于设置授权类型，这些成员都是整数常量，可以进行逻辑或运算，其中表达式UNAuthorization-OptionAlert | UNAuthorizationOptionSound（Swift版本是[.alert，.sound]）表示两种授权类型都支持。

21.2.3　通知的创建与发送

通知的创建与发送又可以细化为如下4个步骤。

(1) 创建通知内容。

(2) 创建触发器。

(3) 创建通知请求对象。

(4) 添加通知请求到通知中心。

1. 创建通知内容

用户通知内容由UNNotificationContent和UNMutableNotificationContent两个类描述，其中前者是不可变的用户通知内容类，后者是可变的用户通知内容类，主要的属性如下。

- **title**。通知主标题。
- **subtitle**。通知副标题。
- **body**。通知内容。
- **badge**。应用图标标记数字，NSNumber类型。
- **sound**。通知到达的音效，UNNotificationSound类型。
- **attachments**。通知内容附件集合，可以保存多个UNNotificationAttachment对象。

视图控制器ViewController中scheduleNotification:方法的代码如下：

```
let content = UNMutableNotificationContent()
content.body = "通知内容：大家好，新年快乐！"
//默认的通知提示音
content.sound = UNNotificationSound.default()
content.subtitle = "通知副标题"
content.title = "通知标题"
```

```
UNMutableNotificationContent * content =
➥[[UNMutableNotificationContent alloc] init];
content.body = @"通知内容：大家好，新年快乐！";
//默认的通知提示音
content.sound = [UNNotificationSound defaultSound];
content.subtitle = @"通知副标题";
content.title = @"通知标题";
```

2. 创建触发器

本地通知触发器有3种类型：时间中断触发器、日历触发器和位置触发器。

- 时间中断触发器

通过UNTimeIntervalNotificationTrigger类的静态工厂方法+ triggerWithTimeInterval:repeats:创建并初始化UNTimeIntervalNotificationTrigger对象；Swift版本是通过构造函数init(timeInterval:repeats:)初始化该对象，其中timeInterval参数是从现在开始算起的时间间隔，单位是秒，repeats参数为布尔类型，用于设置是否重复触发。

视图控制器ViewController中createTimeTrigger方法的代码如下：

```
let timeTrigger = UNTimeIntervalNotificationTrigger(timeInterval: 5,
➥repeats:false)
```

```
UNTimeIntervalNotificationTrigger* timeTrigger =
➥[UNTimeIntervalNotificationTrigger
➥triggerWithTimeInterval:(5) repeats: NO];
```

时间中断触发器的repeats参数都应该设置为NO（Swift为false），这是因为这种触发器不能重复执行。

- 日历触发器

通过UNCalendarNotificationTrigger类的静态工厂方法+ triggerWithDateMatchingComponents:repeats:创建UNCalendarNotificationTrigger对象；Swift版本是通过构造函数init(dateMatching:repeats:)初始化该对象，其中dateMatching参数是DateComponents日期的可扩展组件，可以设置触发时间点，repeats参数为布尔类型，用于设置是否重复触发。

视图控制器ViewController中createTimeTrigger方法的代码如下：

```swift
var components = DateComponents()
components.hour = 7
components.minute = 30
let calendarTrigger = UNCalendarNotificationTrigger(dateMatching: components,
➥repeats: true)
```

```objc
NSDateComponents *components= [[NSDateComponents alloc]init];
components.hour = 7;
components.minute= 30;

UNCalendarNotificationTrigger *calendarTrigger=
➥[UNCalendarNotification Trigger
➥triggerWithDateMatchingComponents: components
➥repeats:YES];
```

其中NSDateComponents类是一个封装了日期的可扩展组件，只是设置了小时和分钟，没有设置具体日期，而且repeats设置为YES（Swift为true），那么该触发器就是每天上午7:30触发。如果省略了小时设置，那么就是每小时的30分触发。

● 位置触发器

通过UNLocationNotificationTrigger类的静态工厂方法+ triggerWithRegion:repeats:创建UNLocation-NotificationTrigger对象；Swift版本是通过构造函数init(region: CLRegion, repeats: Bool)初始化该对象，其中region参数是一个地理区域，repeats参数为布尔类型，用于设置是否重复触发。示例代码如下：

```swift
let center = CLLocationCoordinate2D(latitude: 39.80783, longitude: 116.35445)
let region = CLCircularRegion(center: center, radius: 5000, identifier:
➥"beijing")                                                                  ①
region.notifyOnEntry = true                                                   ②
region.notifyOnExit = false                                                   ③
let locationTrigger = UNLocationNotificationTrigger(region: region, repeats:
➥true)
```

```objc
CLLocationCoordinate2D center = CLLocationCoordinate2DMake(39.80783,
➥116.35445);
CLCircularRegion *region= [[CLCircularRegion alloc]initWithCenter:center
➥radius:5000
➥identifier:@"beijing"];                                                     ①
region.notifyOnEntry= YES;                                                    ②
region.notifyOnExit= NO;                                                      ③
//region表示位置信息；repeats表示是否重复（CLRegion可以是地理位置信息）
UNLocationNotificationTrigger *locationTrigger= [UNLocationNotification
➥Trigger triggerWithRegion:region repeats:YES];
```

上述代码中，第①行用于设置原型的地理区域，其中center是圆点，radius是半径，单位是米。第②行用于设置进入区域时触发通知，第③行表示退出区域时触发。

3. 创建通知请求对象

我们需要将前面创建的内容和触发器封装到通知请求对象UNNotificationRequest中，其中通知请求对象用于管理本地通知内容和触发计划。

视图控制器ViewController中scheduleNotification:方法的代码如下：

```swift
let request = UNNotificationRequest(identifier:
➥"com.51work6.local.Notification" ,
➥content: content, trigger: trigger)
```

```objc
UNNotificationRequest *request = [UNNotificationRequest
➥requestWithIdentifier:@"com.51work6.local.IntervalNotification"
➥content:content
➥trigger:[self createLocationTrigger]];
```

4. 添加通知请求到通知中心

事实上，现在我们还没有开启通知计划，这需要将通知添加到通知中心。

视图控制器ViewController中scheduleNotification:方法的代码如下：

```swift
UNUserNotificationCenter.current().add(request) {(error) in
    if let error = error {
        print("添加通知: \(error)")
    } else {
        print("添加通知请求到通知中心")
    }
}
```

```objc
[[UNUserNotificationCenter currentNotificationCenter]
➥addNotificationRequest:request
➥withCompletionHandler:^(NSError * _Nullable error) {
    if (!error) {
        NSLog(@"添加通知请求到通知中心");
    }
}];
```

上述代码将通知请求对象添加到通知中心，这样就开启了通知计划，等到触发器条件满足时则发送通知。

21.2.4　通知接收后的处理

用户接收到通知后，有可能对通知进行一些操作，其中会涉及的类有：UNNotificationAction和UNNotificationCategory，以及委托协议UNUserNotificationCenterDelegate。下面简要说明一下。

- UNNotificationAction是动作，封装了用户要进行的操作。它还有一个子类UNTextInputNotification-Action，该子类能够接收用户输入文本操作。
- UNNotificationCategory是类别，它定义了一组操作。
- UNUserNotificationCenterDelegate是响应动作的委托对象方法，当用户接收到通知并且执行了操作后，会回调该委托对象方法。

示例代码如下：

```
//设置通知内容的类别标识
content.categoryIdentifier = "myCategory"
......

let action = UNNotificationAction(identifier: "myAction", title: "OK",
➥options: [])                                                              ①
let category = UNNotificationCategory(identifier: "myCategory",
➥actions: [action], intentIdentifiers: [],options: [])                     ②
let categories: Set = [category]                                            ③
UNUserNotificationCenter.current().setNotificationCategories(categories)    ④
UNUserNotificationCenter.current().delegate = self                          ⑤
......
//实现委托协议UNUserNotificationCenterDelegate
func userNotificationCenter(_ center: UNUserNotificationCenter, didReceive
➥response: UNNotificationResponse, withCompletionHandler completionHandler:
➥@escaping () -> Void) {                                                   ⑥
    if response.actionIdentifier == "myAction" {                            ⑦
        print("OK! ")
    }
    completionHandler()                                                     ⑧
}
```

```
//设置通知内容的类别标识
content.categoryIdentifier = "myCategory"
......

UNNotificationAction *action = [UNNotificationAction
actionWithIdentifier:@"myAction"
➥title:@"点赞" options: UNNotificationActionOptionNone];                    ①

UNNotificationCategory* category = [UNNotificationCategory
➥categoryWithIdentifier:@"myCategory"
➥actions:@[action] intentIdentifiers:@[@""]
➥options: UNNotificationCategoryOptionNone];                               ②

NSSet* categories = [NSSet setWithObject:category];                         ③
[[UNUserNotificationCenter currentNotificationCenter]
➥setNotificationCategories:categories];                                    ④

[UNUserNotificationCenter currentNotificationCenter].delegate = self;       ⑤
......

//实现委托协议UNUserNotificationCenterDelegate
-(void)userNotificationCenter:(UNUserNotificationCenter *)center
➥didReceiveNotificationResponse:(UNNotificationResponse *)response
➥withCompletionHandler:(void (^)())completionHandler {                     ⑥

    if ([response.actionIdentifier isEqual: @"myAction"]) {                 ⑦
        NSLog(@"点赞了！ ");

    }
    completionHandler();                                                    ⑧
}
```

上述代码中，第①行用于创建动作对象，其中identifier参数是为动作分配的标识，title参数是动作按钮上显示的文本。第②行用于创建类别对象，其中identifier参数是为类别分配的标识，注意类别标识必须与通知内容的类别标识（content.categoryIdentifier）一致；actions参数是动作集合，intentIdentifiers是意图标识集合，意图可以通过Siri处理动作，一般情况下设置为空内容。

第③行用于创建一个类别对象的Set集合，这是因为第④行的setNotificationCategories:只能结合Set集合。

第⑤行用于设置self为委托对象，这需要当前类实现委托协议的相关方法。第⑥行实现的方法是用户执行操作时回调的方法。第⑦行用于判断动作标识是否为myAction，从而进行相应的处理。第⑧行调用代码块completionHandler。

21.3　开发推送通知

远程通知也叫推送通知，虽然从用户角度看到的通知形式和管理都是一样的，但是推送通知的运行机理与本地通知完全不同。

21.3.1　推送通知机理

下面分析一下推送通知的运行机理，如图21-9所示。

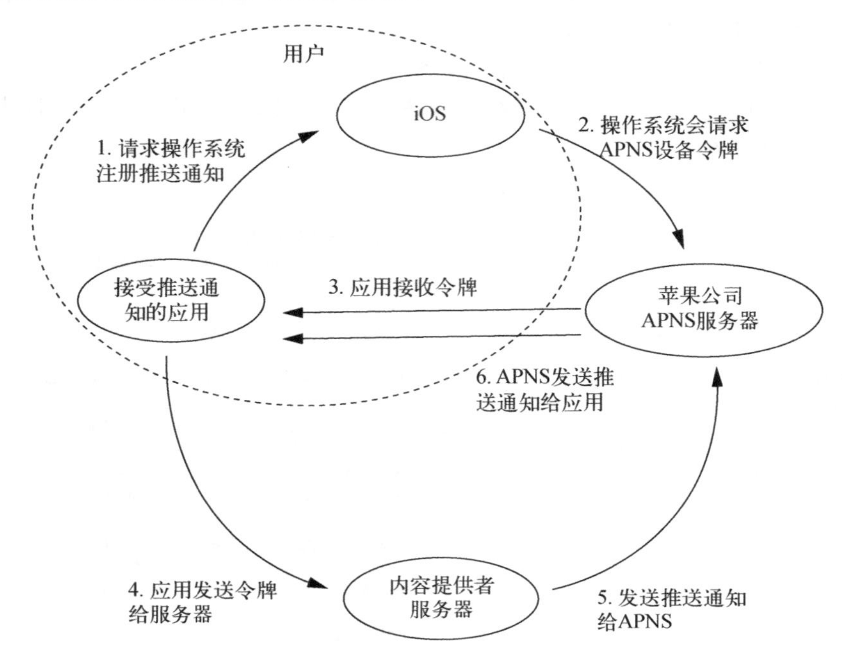

图21-9　推送通知的运行机理

参与使用推送通知的角色有3个，每一方缺一不可，他们是：

- 使用应用的用户
- 提供推送内容的提供者
- 苹果公司

用户使用iOS设备安装了具有推送功能的应用，如查看股票信息的应用，内容提供者是给用户提供推送信息服务商，如股票公司、证券公司等。苹果公司为推送通知提供了一个Apple Push Notification Service（缩写，APNS）服务，即苹果推送服务。APNS会与所有苹果设备建立安全持久连接通道，内容提供者并不是直接把通知发送给用户的设备，而是通过APNS发送的。

首先，用户安装了应用，应用运行的时候请求操作系统，操作系统会请求APNS。如果成功，APNS会返回令牌给应用，这个过程由操作系统完成，不需要开发人员实现。但是要为应用做一些设置，并在开发者网站生成App ID和描述文件（Provisioning Profile）。应用获得令牌后，把它发送给服务器内容提供者，内容提供者接收到这个令牌后，再与APNS通信，APNS认证通过后，内容提供者发送通知给APNS，然后再由APNS发送通知给用户设备。

21.3.2　生成 SSL 证书

由于涉及内容提供者与苹果公司APNS服务器之间的远程通信，所以APNS服务器要求认证内容提供者的身份，才能进行推送通知，这个认证采用SSL（Secure Sockets Layer，安全套接层）认证方式。SSL认证需要内容提供者提供公开密钥文件，这个公开密钥文件是苹果公司针对内容提供者的特定应用进行数字签名后的文件，我们称为SSL证书。

下面我们来介绍一下如何生成SSL证书。这个过程比较麻烦，需要iOS开发者使用账号登录开发者网站，还要使用macOS系统中的钥匙串访问应用请求证书签名。图21-10所示的是生成SSL证书的流程图。

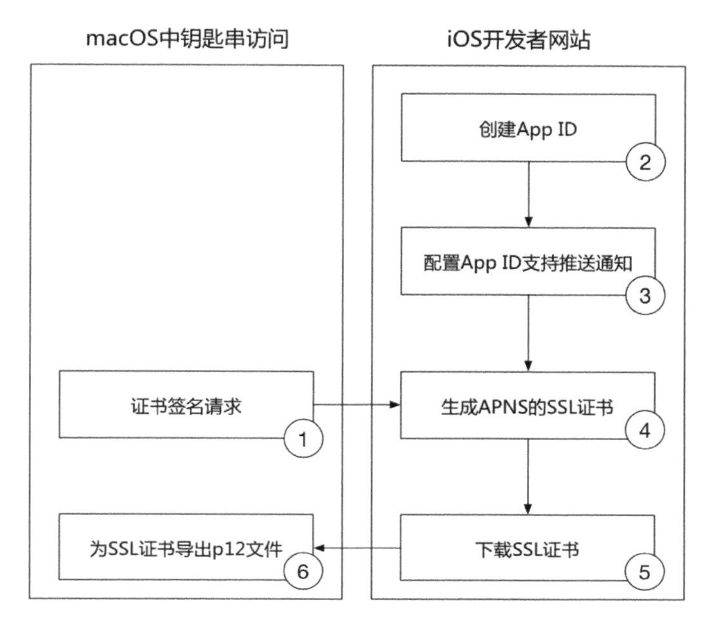

图21-10　生成SSL证书的流程图

1. 证书签名请求

SSL证书的生成过程是，首先要在macOS系统的钥匙串访问应用中进行证书签名请求，证书签名请求就相当于给苹果公司写一个“申请书”。

通过在macOS系统中选择“实用工具”→“钥匙串访问”，然后在打开的窗口中选择“钥匙串访问”→“证书助理”→“从证书颁发机构请求证书...”菜单项，如图21-11所示。

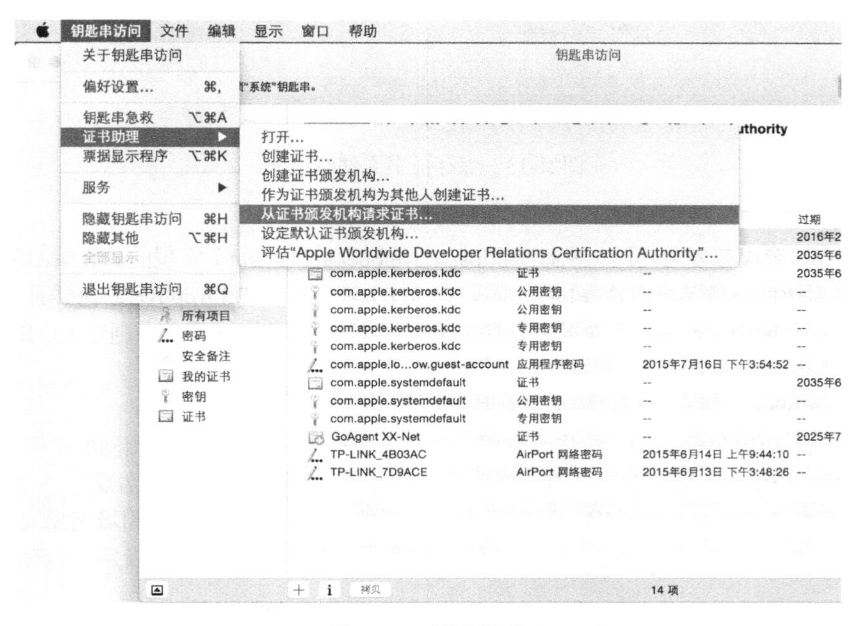

图21-11　请求证书

此时会弹出“证书助理”对话框，如图21-11所示。在“用户电子邮件地址”中输入你的邮件地址，在“常用名称”中输入你的名字，然后在“请求是”选项中勾选“存储到磁盘”单选按钮。

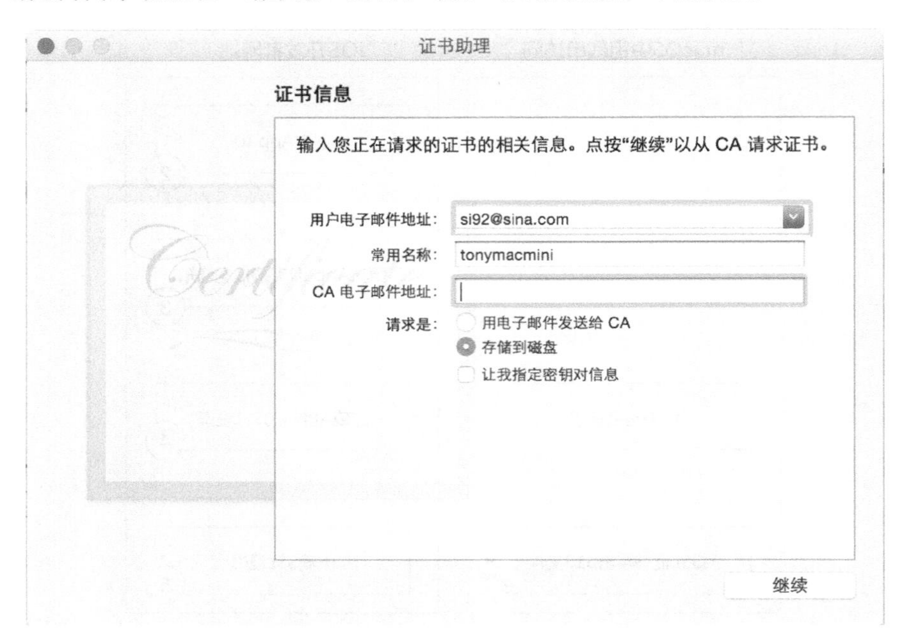

图21-12　“证书助理”对话框

接着单击“继续”按钮，然后会弹出保存证书文件对话框，如图21-13所示。可以选择保存文件的位置和文件名，最后单击“存储”按钮即可。

图21-13　保存证书文件对话框

2. 创建App ID

创建App ID是通过苹果开发者网站（网址为https://developer.apple.com）实现的。登录iOS开发者网站，如图21-14所示，接着在左边的导航菜单中选择Identifiers→App IDs项，打开iOS App IDs管理页面，如图21-15所示。

在图21-15所示的页面中，点击右上角的添加按钮 + ，会打开如图21-16所示的创建App ID详细页面，下面简要介绍一下该页面中各项的含义。

- App ID Description。描述，可以输入一些描述应用的信息。
- App ID Prefix。应用包种子ID，它作为应用的前缀，所描述的应用共享了相同的公钥。
- App ID Suffix→Explicit App ID。适用于单个应用的后缀，苹果推荐使用域名反写。
- App ID Suffix→Wildcard App ID。适用于多个应用的后缀，苹果推荐使用域名反写。本例中输入的是com.51work6.PushNotificationSample，与图21-17所示的应用程序目标中设定的包标识符保持一致就可以了。
- App Services。可以选择应用中包含的服务。

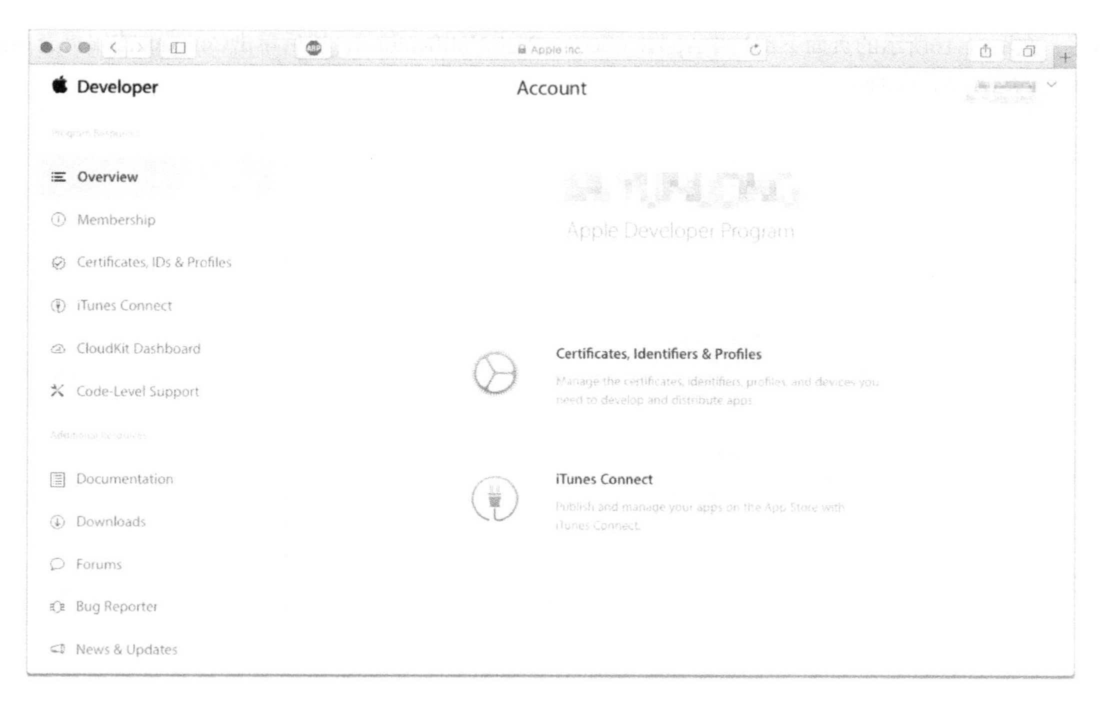

图21-14　成功登录苹果开发者网站

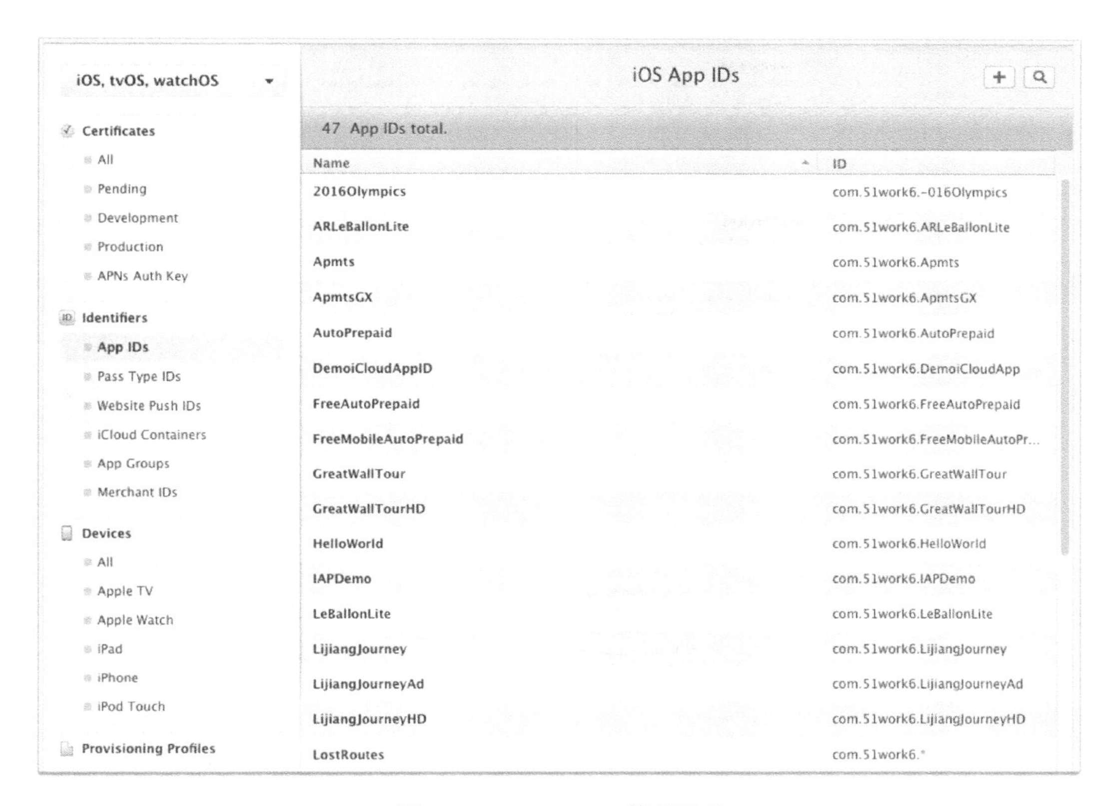

图21-15　iOS App IDs管理页面

最后，在图21-16所示的页面下部选中App Services→Push Notifications，然后单击Continue按钮提交信息，此时会跳转到创建App ID页面。

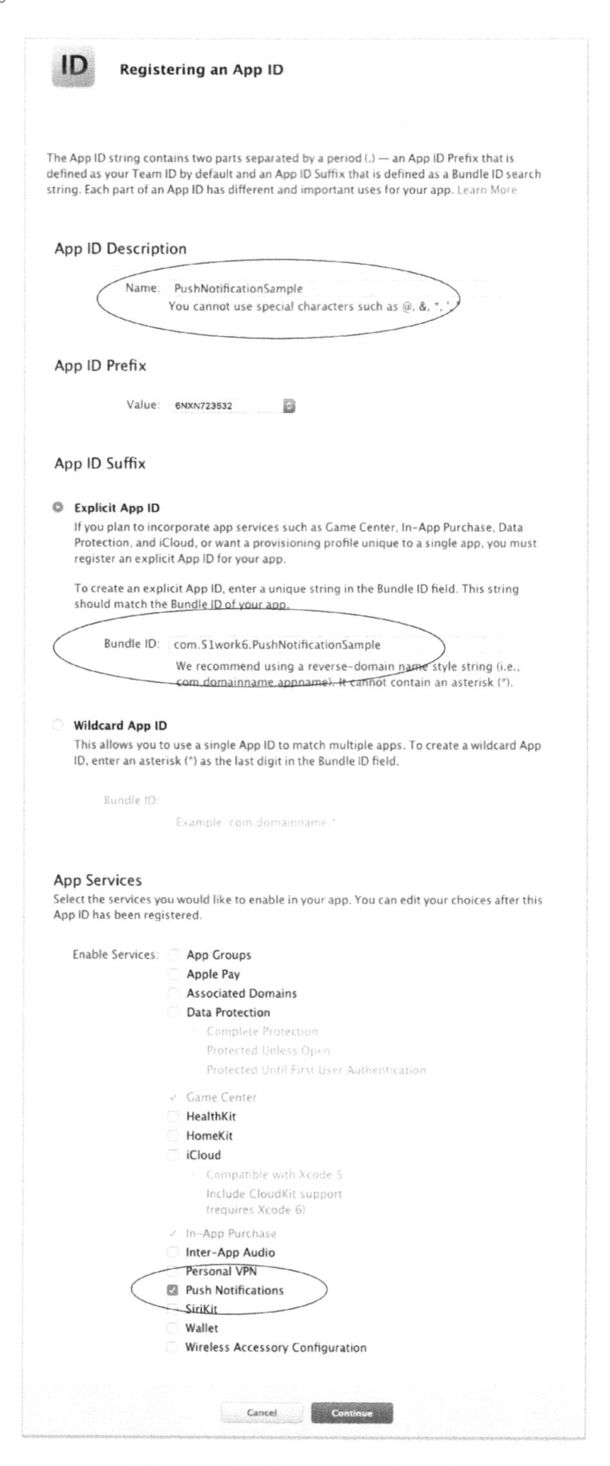

图21-16　创建App ID的详细页面

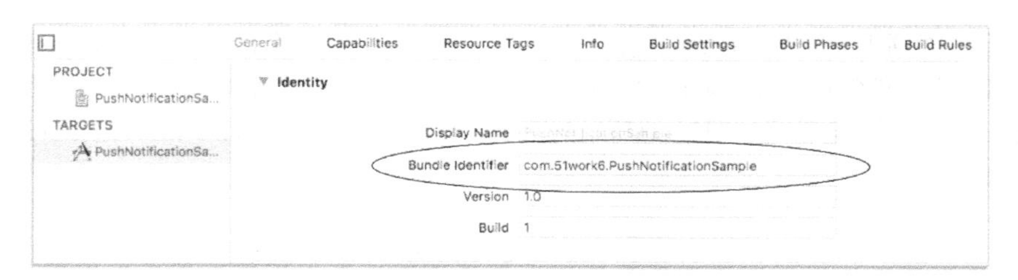

图21-17　应用程序目标中设定的包标识符

3. 配置App ID支持推送通知

对于一般的应用而言，到此为止App ID就创建成功了，而对于具有推送通知功能的应用，还要进行配置。在App ID列表中找到PushNotificationSample，展开相关信息，可见Push Notifications（推送通知）后面都是Configurable（可配置的），单击Edit按钮，进入如图21-18所示的创建证书页面。

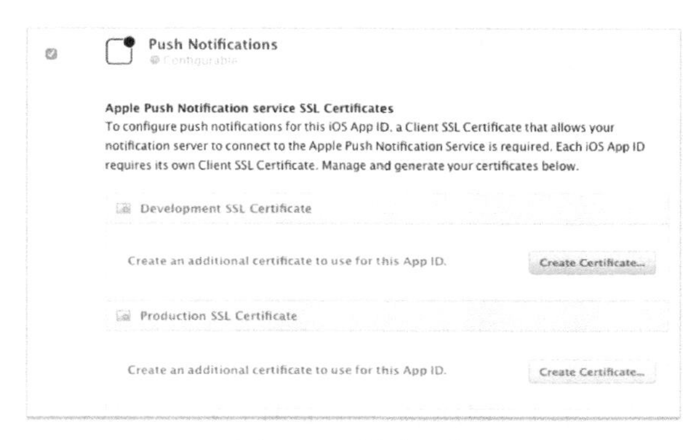

图21-18　创建证书

4. 生成APNS的SSL证书

在图21-18所示的页面中，Push Notification中有两个可以配置，它们是：Development SSL Certificate（为开发配置证书）和Production SSL Certificate（为发布产品配置证书）。选择Push Notifications→Development SSL Certificate，单击Create Certificate按钮，就会弹出如图21-19所示的页面。

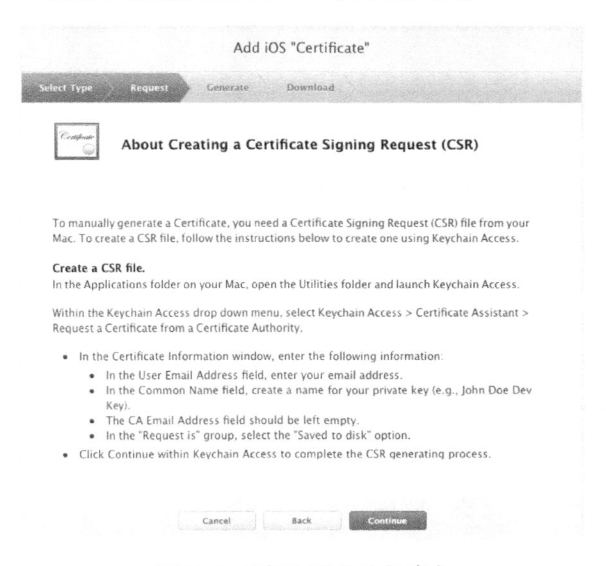

图21-19　产生证书签名请求

在图21-19所示的对话框中选择Continue按钮，会进入如图21-20所示的对话框，通过Choose File按钮选择签名生成的证书签名请求文件CertificateSigningRequest.certSigningRequest。然后单击Generate按钮，开始上传文件，上传成功并生成SSL证书后，会看到如图21-21所示的页面，在这个页面中单击Download按钮可以下载生成的SSL证书，以便于后面使用。

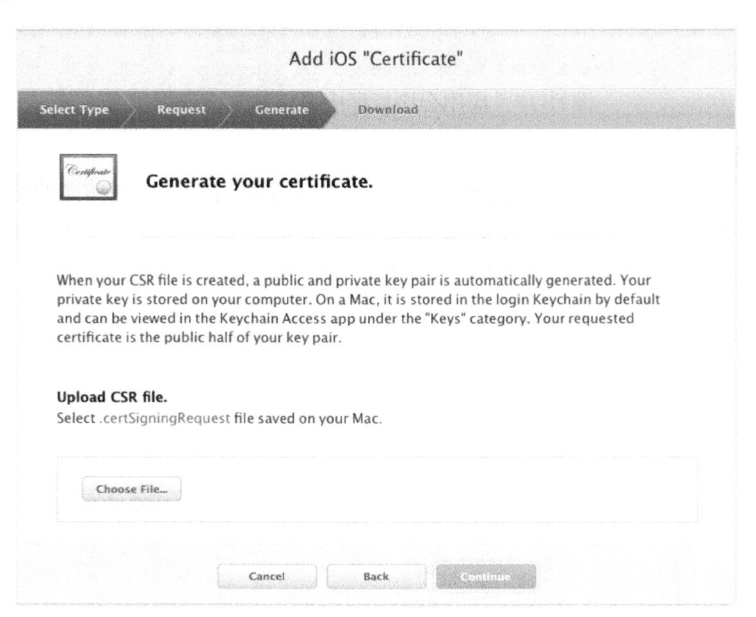

图21-20　选择证书文件

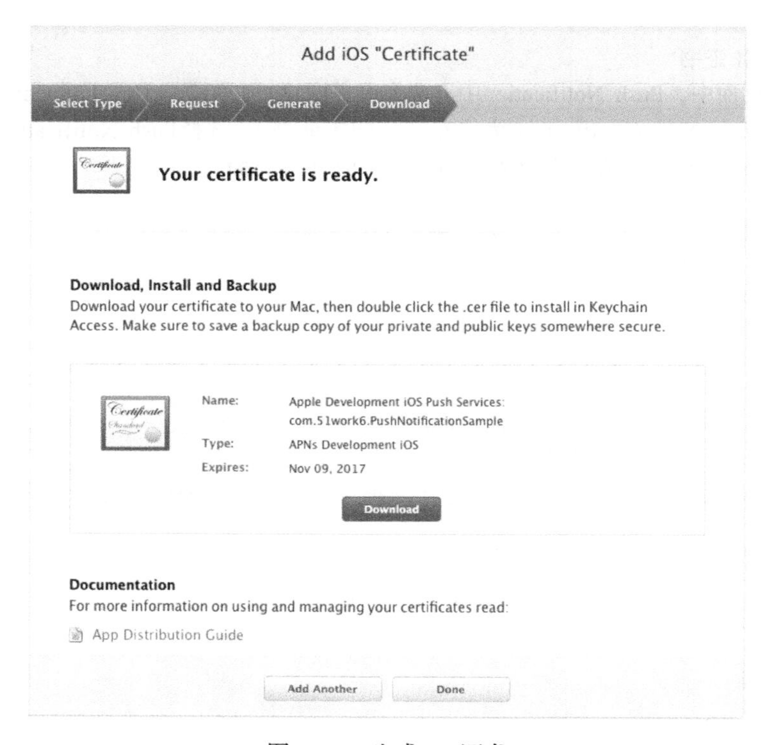

图21-21　生成SSL证书

单击Done按钮，关闭对话框回到配置页面。在配置页面中单击Done按钮，然后回到App ID的列表页面。这时再回来看一下我们刚刚配置的应用，它的Push Notifications→Development下的状态变为了绿色，表示可以使用了，如图21-22所示。

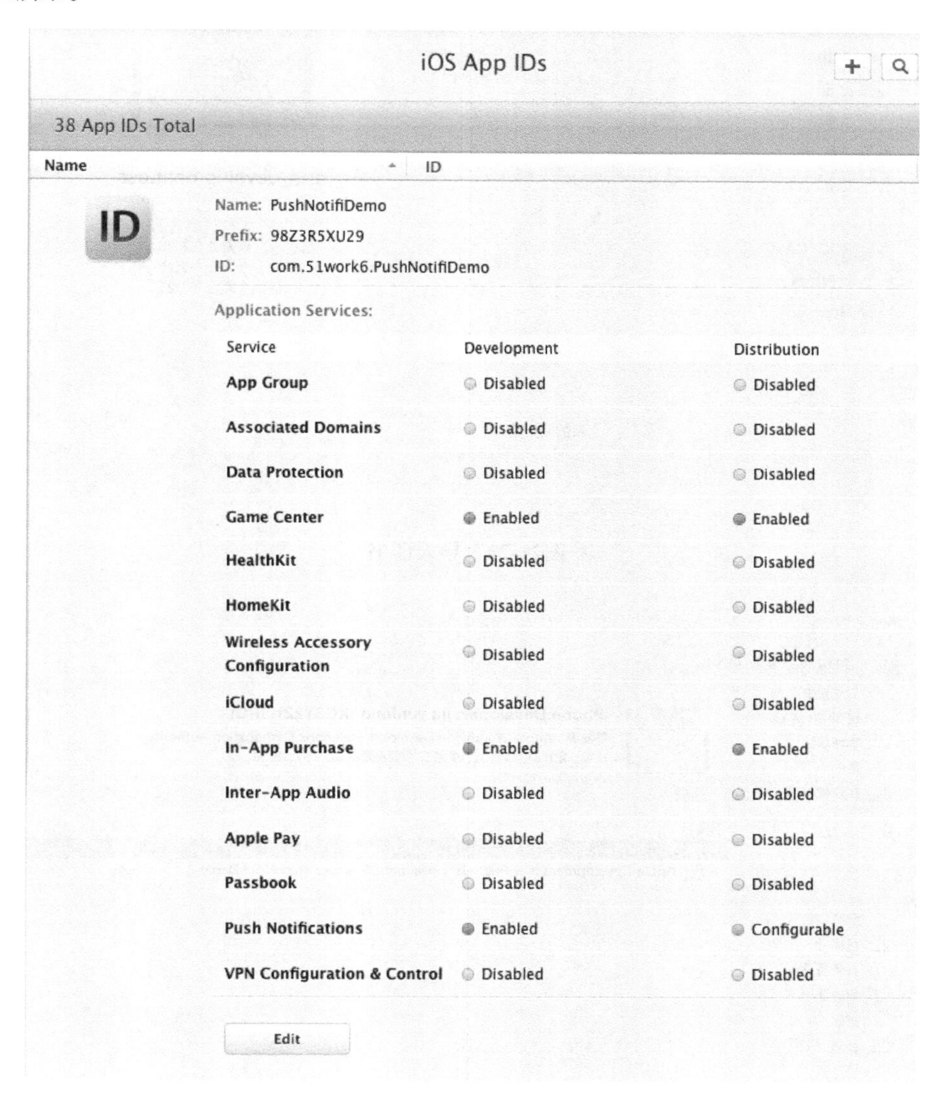

图21-22　推送状态可用

5. 下载SSL证书

在图21-21所示的页面中下载证书文件aps_development.cer到本地，然后把文件导入到macOS的钥匙串访问应用中。在“钥匙串访问”界面中打开“文件”→“导入项目”菜单，此时会打开如图21-23所示的对话框，从中选择“目的钥匙串”为“登录”，然后单击“打开”按钮导入证书文件。证书导入成功后，可以在“登录”→“我的证书”列表中看到，如图21-24所示。

图21-23　导入证书

图21-24　成功导入证书

6. 为SSL证书导出p12文件

在编程时，需要使用p12格式的文件。p12用于存放个人证书和私钥，通常包含保护密码，是二进制格式。

打开钥匙串访问工具，右击刚才导入的证书，如图21-25所示，然后选择导出。此时会弹出如图21-26所示的对话框，从中将“文件格式”选为“个人信息交换(.p12)”，再单击“存储”按钮。

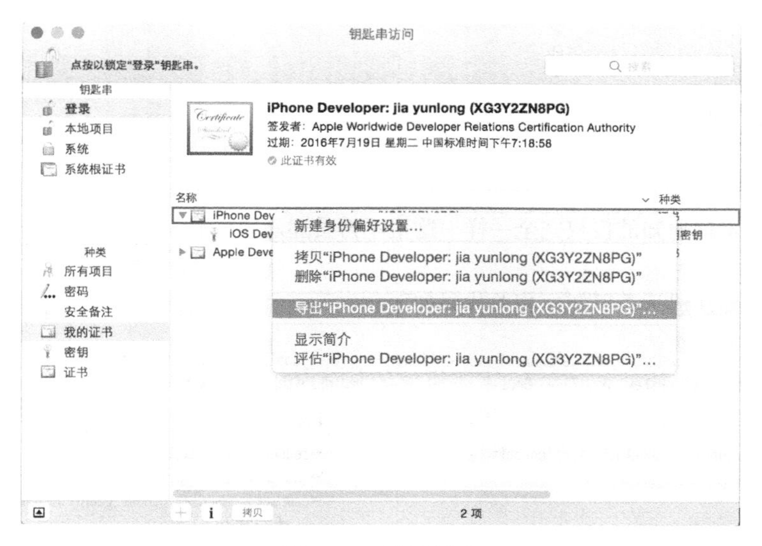

图21-25 选择证书

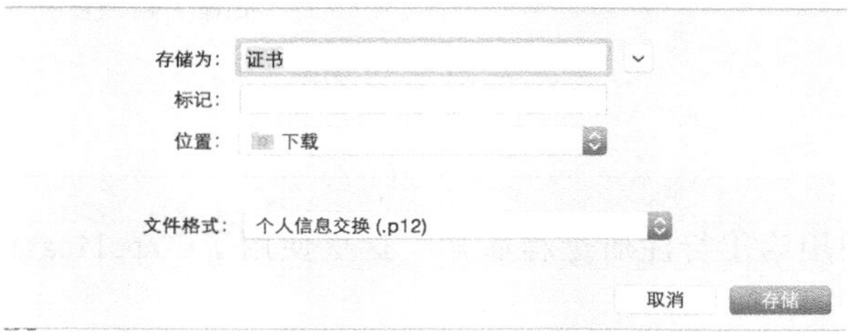

图21-26 导出证书

在存储的过程中需要选择文件的保存位置和密码保护，需要提供密码，如图21-27所示。输入密码后，单击“好”按钮，成功导出文件“证书.p12”，把这个文件保管好以备后面编程时使用。

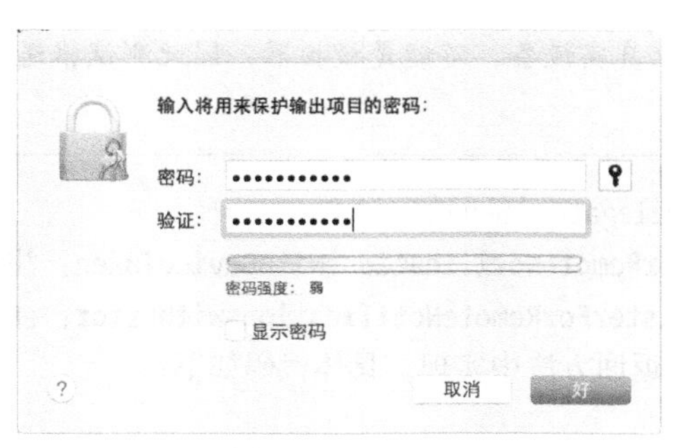

图21-27 设置保护密码

21.3.3 iOS 客户端编程

但是对于iOS客户端而言，推送通知的编程要比本地通知简单，这主要是因为推送通知客户端不需要编写通知内容、发送通知和编写触发器。推送通知触发器由iOS系统触发，程序员不需要编写代码。

推送通知的iOS客户端开发有如下4个步骤。

(1) 请求授权。

(2) 注册通知。

(3) 获得设备令牌。

(4) 通知接收后的处理。

下面简要介绍这4个步骤。

1. 请求授权

推送通知的授权与本地通知的授权完全一样，这里不再赘述。

2. 注册通知

授权成功之后，需要注册远程通知，相关代码如下：

```
func application(_ application: UIApplication, didFinishLaunchingWithOptions
➥launchOptions: [UIApplicationLaunchOptionsKey: Any]?) -> Bool {

    //授权
    UNUserNotificationCenter.current().requestAuthorization(options:
    ➥[.badge, .alert, .sound]) {(granted, error) in
        if granted {
            print("授权成功")
            application.registerForRemoteNotifications()                ①
        } else {
            print("授权失败")
        }
    }

    return true
}
```

```
- (BOOL)application:(UIApplication *)application
➥didFinishLaunchingWithOptions:(NSDictionary *)launchOptions {

    [[UNUserNotificationCenter currentNotificationCenter]
     ➥requestAuthorizationWithOptions:(UNAuthorizationOptionBadge |
     ➥UNAuthorizationOptionAlert | UNAuthorizationOptionSound)
     ➥completionHandler:^(BOOL granted, NSError * _Nullable error) {
        if (!error) {
            NSLog(@"授权成功");
            [application registerForRemoteNotifications];         ①
        } else {
            NSLog(@"授权失败");
        }
    }];

    return YES;
}
```

在授权成功的情况下调用第①行注册远程通知，这里使用了UIApplication对象的registerForRemoteNotifications方法注册。

3. 获得设备令牌

调用registerForRemoteNotifications方法注册远程通知的过程，就是应用请求操作系统，由操作系统请求APNS服务器返回针对当前设备和应用的令牌（token）。

提示

由于设备令牌的申请要求真实设备，不能是模拟器，因此测试推送通知不能是在模拟器上进行。

注册成功返回后，会回调AppDelegate对象的如下方法。

- **application:didRegisterForRemoteNotificationsWithDeviceToken**：注册成功后回调的方法。
- **application:didFailToRegisterForRemoteNotificationsWithError**：注册失败后回调的方法。

要获得设备令牌，可以在成功返回方法中实现，具体代码如下：

```
func application(_ application: UIApplication,
➥didRegisterForRemoteNotificationsWithDeviceToken deviceToken: Data) {

    //获得设备令牌
    let deviceTokenString = deviceToken.reduce("", {$0 + String(format: "%02X",
    ➥$1)})                                                              ①
    print(deviceTokenString)
}
```

```
-(void)application:(UIApplication *)application didRegisterForRemote
➥NotificationsWithDeviceToken:(NSData *)deviceToken {

    NSString *tokeStr = deviceToken.description;                          ①
    NSLog(@"tokeStr: %@", deviceToken);

    tokeStr = [tokeStr stringByReplacingOccurrencesOfString:@"<"
    ➥withString:@""];                                                    ②
    tokeStr = [tokeStr stringByReplacingOccurrencesOfString:@">"
    ➥withString:@""];                                                    ③
    tokeStr = [tokeStr stringByReplacingOccurrencesOfString:@" "
    ➥withString:@""];                                                    ④
    NSLog(@"设备令牌: %@", tokeStr);
}
```

从上述代码中可见，deviceToken参数是返回的令牌，但是在Swift语言中它是Data类型，Objective-C是NSData类型。由于API的差别，不同语言从deviceToken获得令牌的方式也不同。Swift版本通过第①行的deviceToken.reduce("", {$0 + String(format: "%02X", $1)})表达式获得，其中reduce是一个比较抽象的函数，该函数将一个序列（或数组）中的元素连接成为一个值，第一个参数是初始值，第二个参数是闭包，指定连接的规则。本例中的{$0 + String(format: "%02X", $1)}闭包表达式中，$0和$1是序列（或数组）的两个元素，String(format: "%02X", $1)是将二进制数字格式化为两位的十六进制字符串，不足两位用0补位。

在Objective-C版本中，第①行的deviceToken.description表达式可以将NSData转化为NSString字符串类型，此时我们输出的deviceToken的结果如下：

```
<297cb240 6d86216f 99299ded 1fe1b17b 4d07cb86 4b56b026 e896b38f bdfda705>
```

可见，令牌是由十六进制字符串表示的，但是前后还有<和>符号，中间还有空格。我们需要删除这些字符，第②行～第④行是用空字符串替换这些字符，最后所得的字符串就是我们需要的设备令牌了。

提示　令牌字符串是为内容提供者准备的，内容提供者需要通过客户端程序获得用户设备的令牌字符串和应用程序App ID，并保存到自己的数据库中，以备需要推送通知时使用。

4. 通知接收后的处理

推送通知接收后的处理与本地通知完全一样，这里不再赘述。

21.3.4　在iOS设备上运行客户端

推送通知iOS客户端的程序必须在iOS设备上运行，而且作为推送通知的Xcode工程也需要进行一些设置。

1. 设置Xcode工程

打开Xcode工程，选中TARGETS→PushNotificationSample→Capabilities，如图21-28所示，开启Push Notifications，这表明当前应用支持推送通知。另外，最后也开启Background Modes，并在Background Modes中选中Remote notifications，这表示开启后台模式，允许接收远程通知在后台运行。

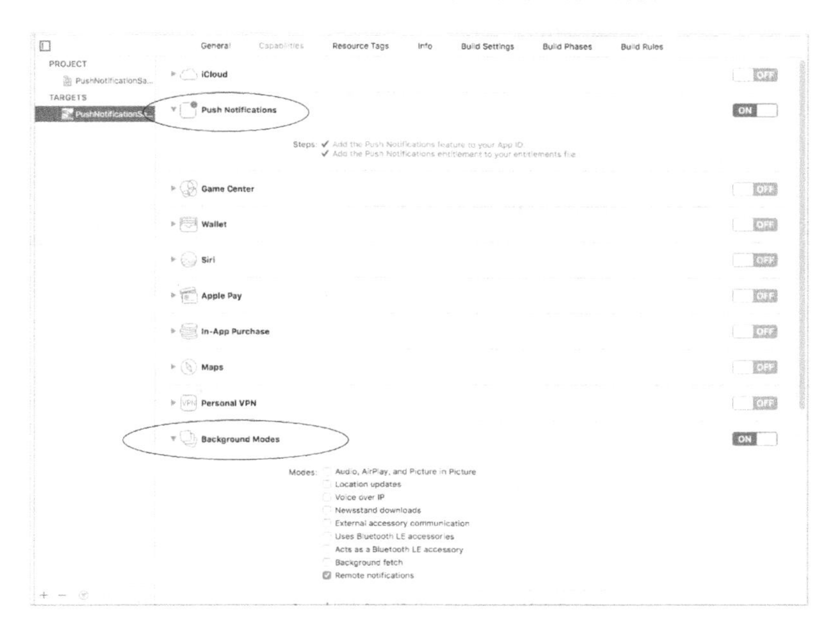

图21-28　设置Xcode工程

2. 在设备上编译和运行

首先，需要在Xcode中设置账号，推送通知必须是付费的开发者账号。具体步骤是打开Xcode菜单Xcode→Preferences，打开Xcode使用偏好，选择Accounts标签，如图21-29所示。点击左下角的+按钮，选择Add Apple ID菜单，此时会弹出如图21-30所示的对话框，在其中输入Apple ID（付费的开发者账号）和密码，如果账号和密码匹配，则添加成功，界面如图21-31所示。

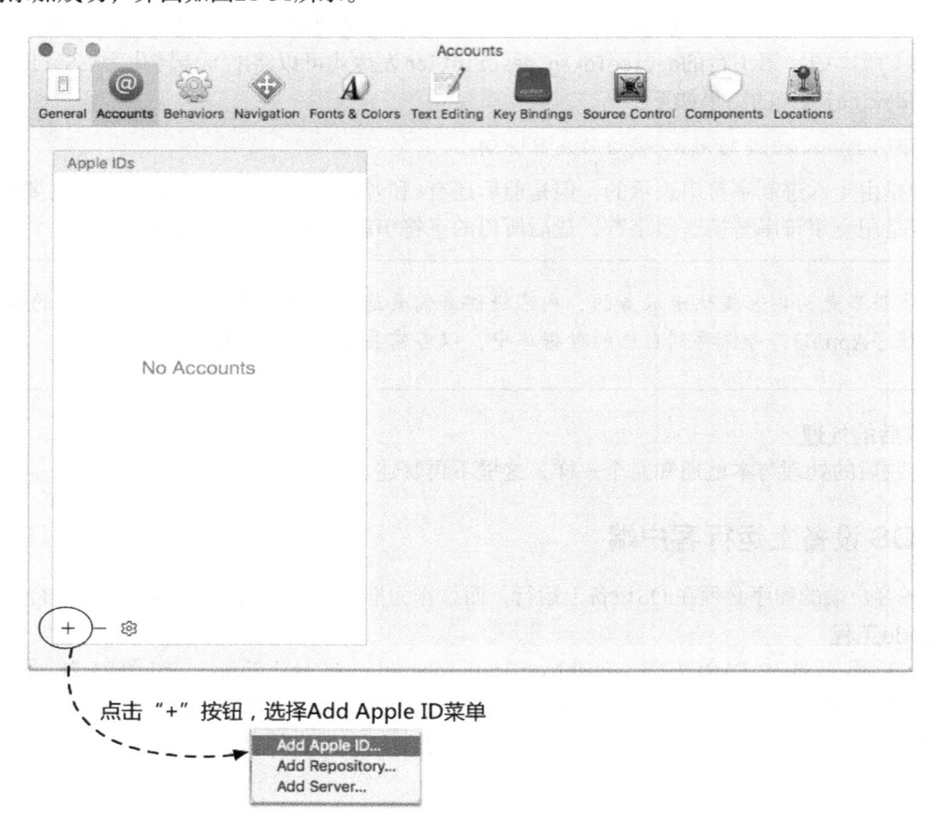

图21-29　添加Apple ID

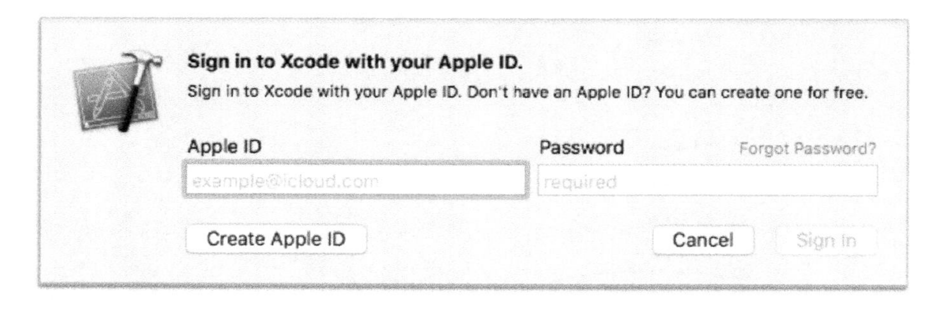

图21-30　账号和密码输入对话框

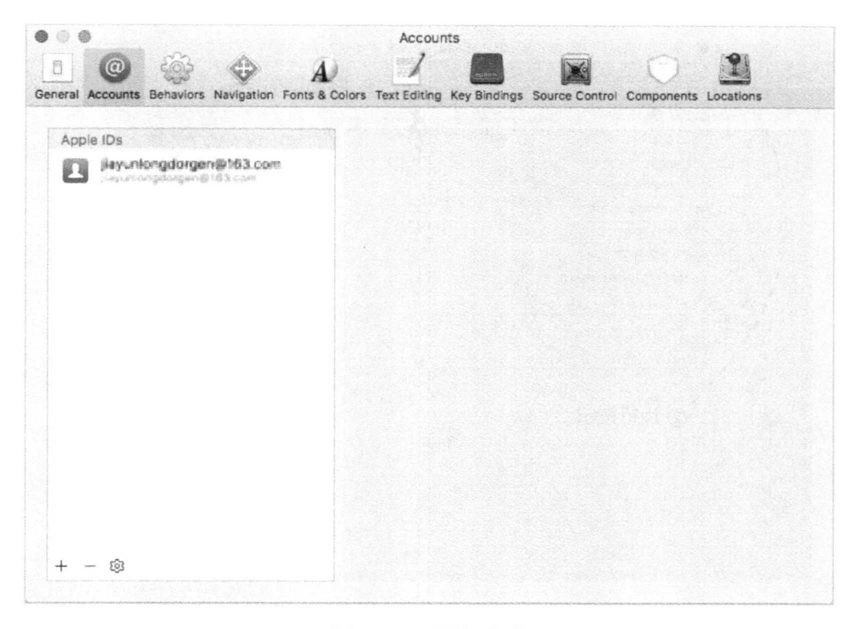

图21-31 添加完成

接下来，就可以对应用进行签名了。这个过程就是通过设置的Apple ID生成一个特殊的数字证书，这个证书对当前这个应用进行签名，这样应用就可以编译了。如图21-32所示，选择TARGETS→PushNotificationSample→General→Signing，在Team的下拉列表中选择你设置的账号。

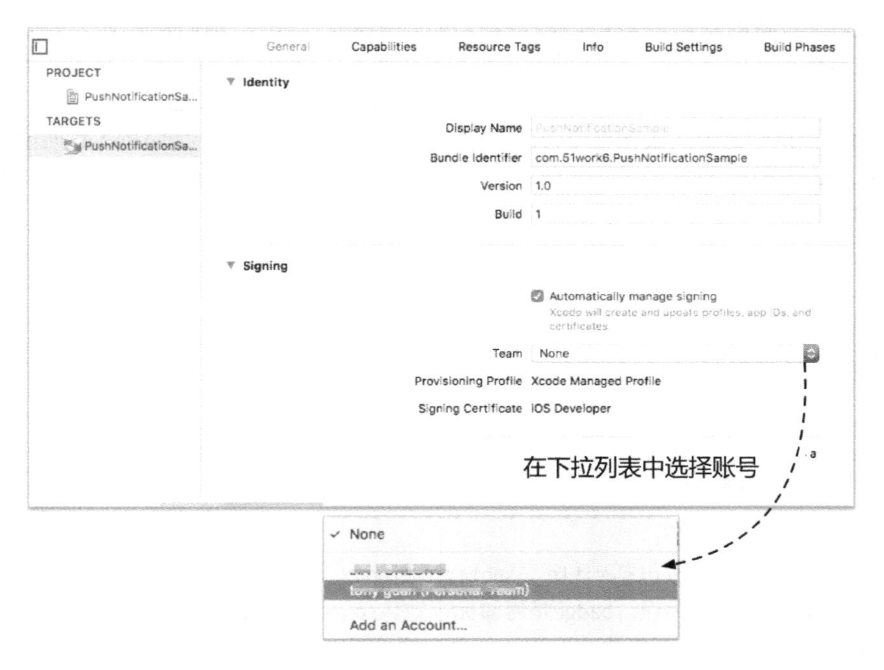

图21-32 应用签名

应用签名成功后，就可以运行了。如图21-33所示，选择运行的设备，然后点击左上角的“运行”按钮▶，即可看到运行结果。

图21-33　运行应用

21.3.5　内容提供者推送通知

内容提供者接收到设备的令牌并保存起来，在有新的内容需要推送时，他们将启动一个服务程序逐个设备推送他们的内容。在具体推送的过程中，并非直接由内容提供者发送给用户设备，而是由服务程序与APNS通信建立信任连接，然后把数据推送给APNS，再由APNS利用安全通道推送给用户设备。

如果要编写内容提供者的推送服务程序，需要进行SSL认证编程，并且构建APNS数据包。数据包分为3个主要部分：Command（命令）、deviceToken（令牌）和Payload（载荷）。载荷不能超过256字节，是JSON格式，例如：

```
{
    "aps": {
        "alert": {
            "title":"通知标题",
            "subtitle":"通知副标题"
            "body":"通知内容：大家好，新年快乐！"
        },
        "badge":1,
        "sound":"default",
        "category":'custom_category_id',
    }
}
```

alert是推送的信息，其中还包括title、subtitle和body。category是通知类别的标识，就是在创建UNNotificationCategory时指定的标识。badge是通知标记数，sound是通知声音。

推送服务程序可以使用很多计算机语言实现，如果从便于管理的角度看，可以使用PHP、Java和.NET，甚至Node.js。编写这些推送通知程序超出了本书范围，我推荐使用第三方已经编好的推送通知工具。我习惯使用Pusher（https://github.com/noodlewerk/NWPusher/releases/tag/0.7.0），它是基于macOS的小应用，并且提供了Xcode源代码。Pusher的运行界面如图21-34所示。

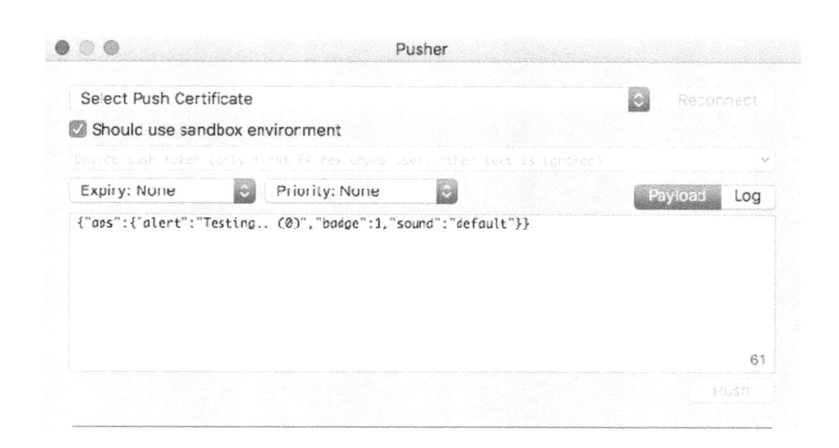

图21-34 Pusher工具

在图21-34中，Select Push Certificate可以选择21.3.2节生成的SSL证书（证书.p12文件）。选中Should use sandbox environment，表示推送给苹果APNS测试服务器。苹果APNS服务器有两个：gateway.push.apple.com是产品发布后使用的APNS服务器，gateway.sandbox.push.apple.com是应用测试时使用的APNS服务器。在Should use sandbox environment复选框的下面是设备令牌字符串。Payload中输入载荷JSON字符串，这里的JSON字符串是经过压缩的，即去掉空格、换行和回车字符。输入完这些内容之后，点击Push按钮就可以推送了。

21.4 小结

通过对本章的学习，我们了解了苹果的本地通知和推送通知，熟悉了本地通知的开发流程，掌握了推送通知的机理和开发流程，讨论了App ID创建和SSL证书的生成等内容。

Part 4 第四部分

测试、调试和优化篇

本部分内容

- 第 22 章 找出程序中的 bug——调试
- 第 23 章 iOS 测试驱动与单元测试
- 第 24 章 iOS 应用 UI 测试
- 第 25 章 让你的程序“飞”起来—— 性能优化

第 22 章

找出程序中的bug——调试

编码过程中出现bug[①]在所难免，有时在找出这些bug上耗费的精力和时间不比重新创建一个工程少多少。调试可以帮我们找出程序中的bug，熟练掌握各种调试工具，能够快速找出程序中的bug，提高开发效率。

22.1 Xcode 调试工具

Xcode提供了强大的代码编辑、性能分析和调试功能，我们应该熟练掌握。为了便于学习，我们把12.4节的案例拿到本章中使用。

22.1.1 定位编译错误

使用Xcode工具进行开发时，很容易出现定位编译错误，而大部分编译错误都会在编写代码时提示给程序员，一小部分编译错误等到编译时才能够显示出来。如图22-1所示，出现错误的位置会显示红色圆形感叹号❶。

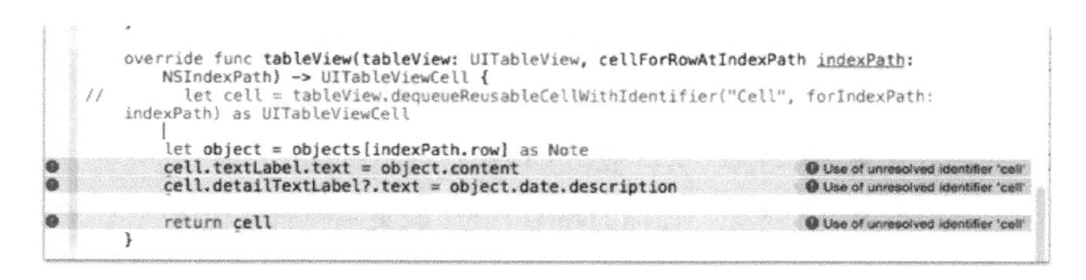

图22-1 Xcode中显示编译错误

提示

程序错误分为编译错误和逻辑错误：前者是在程序编译时暴露出来的错误，可以通过Xcode定位，编译器还会给出错误原因提示；后者是指程序运行的结果与我们期望的不一致，这些错误可以通过调试和测试找出。

编译Xcode工程时，可以使用control+B快捷键完成，也可以通过Product→Build菜单项来完成。除了显示编译错误，Xcode还可以显示警告。每一个警告我们都不应该忽略，它可能引起应用运行时的崩溃。Xcode中显示警告时会显示黄色三角形感叹号⚠，如图22-2所示。

点击导航面板中的⚠按钮，可以打开问题导航面板，如图22-3所示。

在导航面板中，可以查看所有工程中出现的所有错误和警告。此外，导航面板还给出了相关的提示和解决方案。

① bug是指程序中的缺陷和漏洞。在本书中，bug也包括错误和异常。

```
// MARK: - Segues
override func prepareForSegue(segue: UIStoryboardSegue, sender: AnyObject?) {
    if segue.identifier == "showDetail" {
        if let indexPath = self.tableView.indexPathForSelectedRow() {

            let object = objects[indexPath.row]
                                        ⚠ Constant 'object' inferred to have type 'AnyObject', which may be unexpected
            let controller = (segue.destinationViewController as UINavigationController).
                topViewController as DetailViewController
            //传递数据给详细视图控制器
            controller.detailItem = object
            controller.navigationItem.leftBarButtonItem = self.splitViewController?.
                displayModeButtonItem()
            controller.navigationItem.leftItemsSupplementBackButton = true
        }
    }
}
```

图22-2　Xcode中显示警告

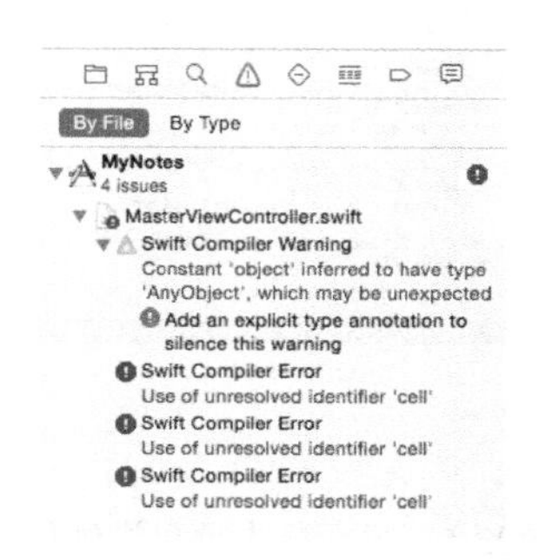

图22-3　问题导航面板

22.1.2 查看和显示日志

在Xcode中编译、运行、测试和分析工程时都会有日志输出，查看和分析这些日志对我们的编程工作非常重要。在导航面板中点击 按钮即可显示日志导航面板，如图22-4所示。

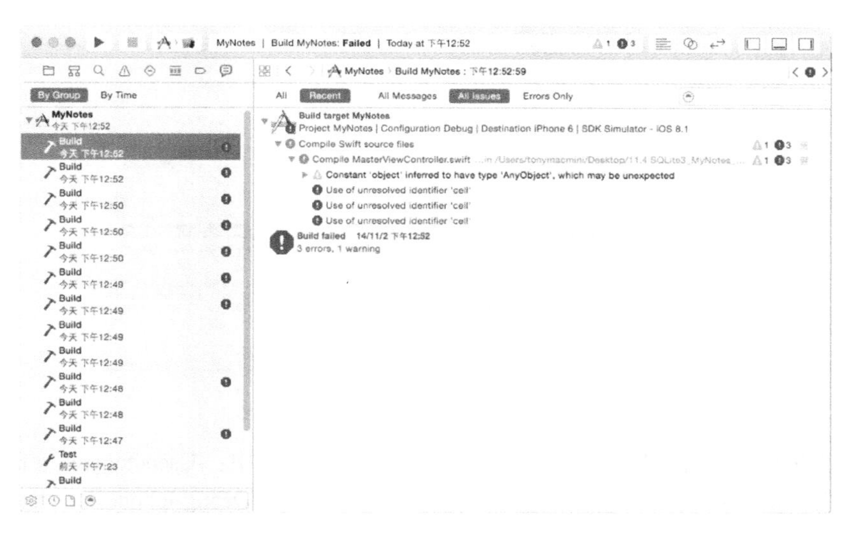

图22-4　显示日志

在日志导航面板中，左侧显示每次的操作，右边是对应操作的日志。图22-4所示的是进行的编译操作，日志中会显示正常消息和问题消息，正常消息是绿色圆形对号 ，问题消息包括了警告和错误，它们的图标与上一节介绍的一样。点击每一行结尾的显示列表图标 ，可列出该项目的详细信息，如图22-5所示。

图22-5显示了编译MasterViewController日志的过程，告诉用户编写时采用了哪些命令、成功还是失败、失败的原因等。

图22-5　显示日志项目详细信息

日志分析很有用。例如，在上传应用到App Store时，需要找到编译之后的.app文件，此时可以打开Touch命令日志项目，如图22-6所示。

图22-6　Touch命令日志项目

从图22-6中可以看到，生成的MyNotes.app文件被复制到下面的目录中：/Users/tonymacmini/Library/Developer/Xcode/DerivedData/MyNotes-eklosuulemjghbbopynqivljctlg/Build/Products/Debug-iphonesimulator/。

22.1.3　设置和查看断点

第一次运行编写成功的程序时，往往会出现始料未及的结果。为了找出原因，我们需要在程序中设置断点进行调试。断点指在条件满足的情况下程序会挂起在哪里，我们可以在这里查看变量、单步执行代码等操作内容。断点可以分为以下6种类型。

- **文件行断点**。执行到特定文件某一行时触发。

- **符号断点**。调用某一个函数或方法时触发，程序挂起在函数或方法的第一行。
- **异常断点**。产生异常时触发，可以设置捕获Objective-C异常、C++异常和所有异常。Objective-C异常断点会在遇到Objective-C异常时触发，C++异常断点在遇到C++异常时触发。
- **Swift错误断点**。产生Swift错误时触发。
- **OpenGL ES断点**。产生OpenGL ES异常时触发。
- **单元测试失败断点**。当进行单元测试时，测试失败的情况下断点会停留在测试失败的代码行。

鉴于后面3种类型用得比较少，这里就不介绍了，下面我们简要介绍一下前3种类型。

1. 文件行断点设置

设置文件行断点很简单，直接点击该文件中的行号即可。如图22-7所示，我们在MasterViewController文件中设置第43行为断点。

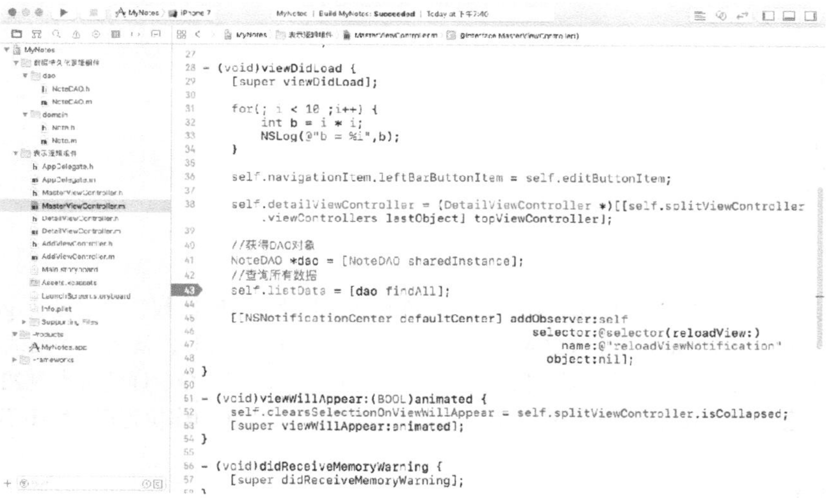

图22-7 设置文件行断点

断点可以删除、禁止使用和编辑。在断点上点击鼠标右键，弹出的快捷菜单如图22-8所示。当选择Disable Breakpoint菜单项时，会禁止使用断点，这时断点处于灰色状态；选择Delete Breakpoint菜单项时，可以删除该断点。此外，我们也可以拖曳断点离开行号列，当鼠标指针变为时释放鼠标，这样也可以删除断点。

选择Edit Breakpoint...菜单项，会弹出断点编辑对话框，如图22-9所示。在这个对话框中，我们可以为断点设定触发条件和忽略次数，并添加动作。

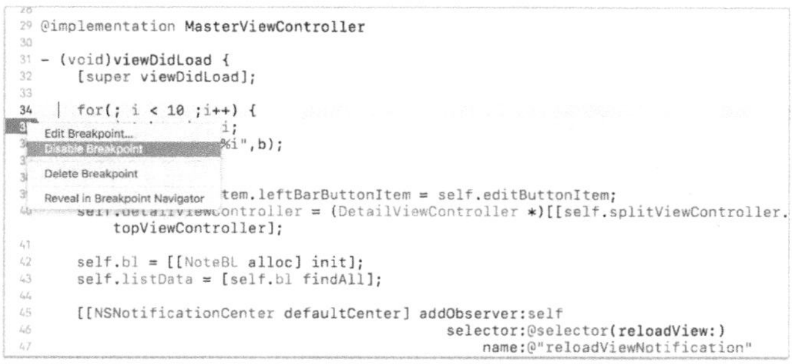

图22-8 断点管理

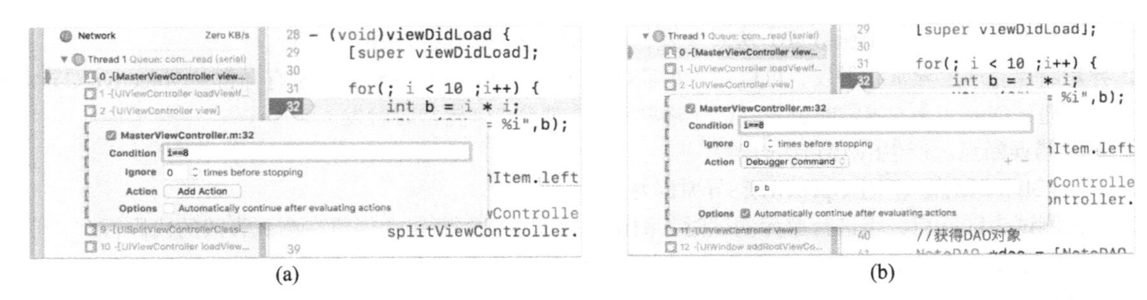

图22-9　编辑断点

如果有下面这段程序代码，我们只是想看看i==8是什么情况，可以在第35行中设置断点。默认情况下，循环体10次都触发断点，我们需要在图22-9a所示的Condition中设置i==8。

```
static int i = 0;
......
- (void)viewDidLoad {
    [super viewDidLoad];

    for(; i < 10 ;i++) {
        int b = i * i;
        NSLog(@"b = %i",b);
    }
    ......
}
```

这样，当程序运行到i==8时就会挂起。我们还可以在Ignore中设置忽略次数，比如设置Ignore为8，从而达到同样的效果。我们可以在图22-9a中点击Add Action来添加动作。所谓"动作"，就是一些调试命令。如图22-9b所示，在Action输入框中输入p b，该命令是在调试窗口输出b变量，p是调试命令，详情可参见22.2节。默认情况下，动作执行一次，程序会挂起。Action下面还有一个Options，选中它，可以在动作执行后，程序不再挂起，按图22-9b自动继续执行。

2. 符号断点设置

符号断点就是把方法或函数名作为断点。与设置文件行断点不同，设置符号断点时，需要点击导航面板中的 ▷ 按钮，打开断点导航面板，如图22-10所示。在断点导航面板中，我们可以看到所有的断点。

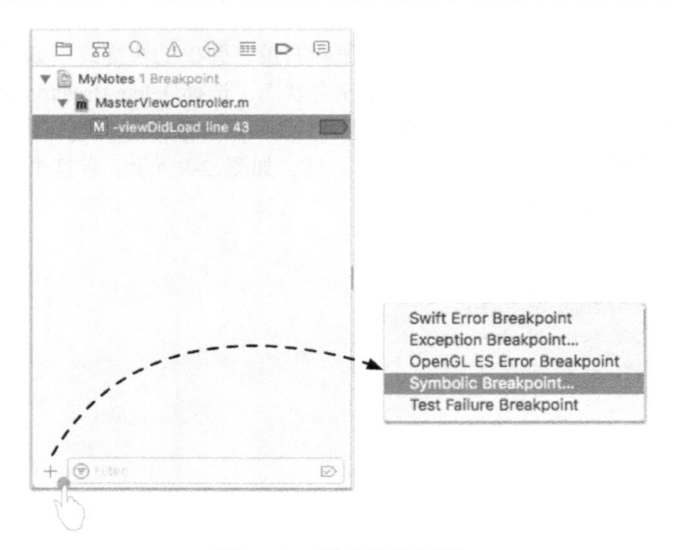

图22-10　断点导航面板

点击图22-10中左下角的+按钮，会弹出如图22-10右图所示的菜单，其中有5个菜单项。这里我们选择Symbolic Breakpoint...菜单项，此时可以弹出创建符号断点对话框，如图22-11所示。

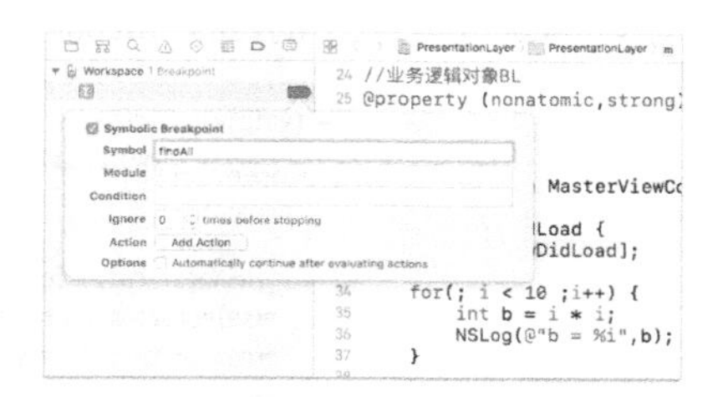

图22-11 创建符号断点对话框

在Symbol中输入findAll方法。NoteDAO类中有这个方法，当调用findAll方法时，都会触发断点。图22-12所示的断点挂起在findAll方法的第一行。

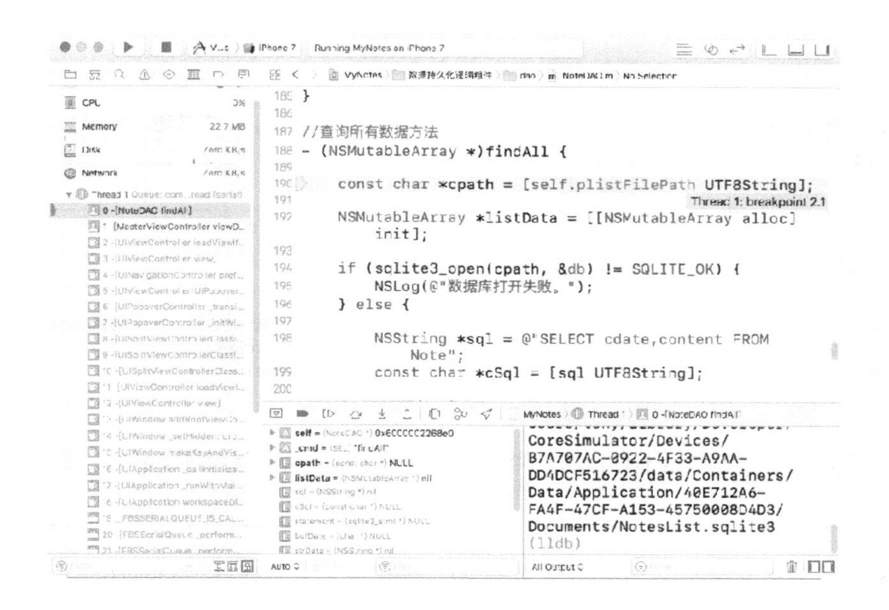

图22-12 断点挂起在NoteDAO中findAll方法的第一行

3. 异常断点设置

在图22-10所示的界面中点击左下角的+按钮，从弹出菜单中选择Exception Breakpoint菜单项，此时会弹出创建异常断点对话框，如图22-13所示。

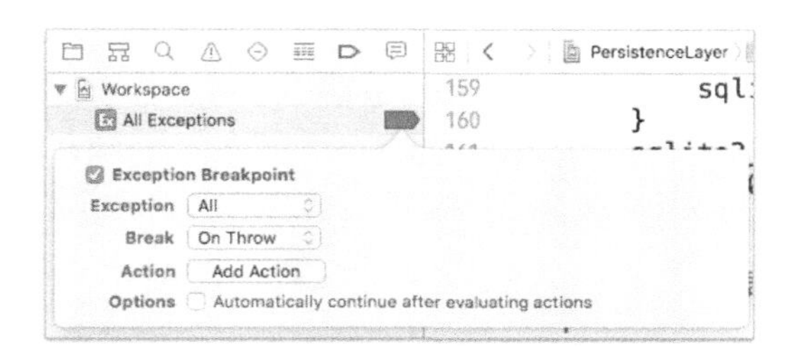

图22-13 设置异常断点

在Exception项中可以选择All、Objective-C或C++异常断点，Break项可以设定On Throw还是On Catch，即断点是在抛出时触发还是在捕获时触发。

为了测试，我们可以修改一下MasterViewController中的viewDidLoad方法，人为制造一个运行期异常：

```
override func viewDidLoad() {
    super.viewDidLoad()

    ......

    //查询所有的数据
    self.listData = self.bl.findAll()

    NotificationCenter.default.addObserver(self,
    ➥selector: Selector("reloadView:"),
    ➥name: Notification.Name(rawValue:
    ➥"reloadViewNotification"),
    ➥object: nil)                                    ①

}
```

```
- (void)viewDidLoad {
    [super viewDidLoad];
    ......

    [[NSNotificationCenter defaultCenter] addObserver:self
    ➥selector:@selector(reloadView:)
    ➥name:@"reloadViewNotification"
    ➥object:nil];                          ①
}
```

注意　在Swift版中，通过构造函数Selector("reloadView:")创建选择器，也可以通过#selector(reloadView(_:))创建而Swift 3推荐使用后面的写法。

我们将第①行中的selector修改为reloadView1:，相关代码如下：

```
NotificationCenter.default.addObserver(self,
➥selector: Selector("reloadView1:"),
➥name: Notification.Name(rawValue: "reloadViewNotification"),
➥object: nil)
```

```
[[NSNotificationCenter defaultCenter] addObserver:self
➥selector:@selector(reloadView1:)
➥name:@"reloadViewNotification"
➥object:nil];
```

也就是说，reloadView1:方法是不存在的，当点击保存操作时，这里就会抛出异常。默认情况下，没有捕获和设置断点。

提示　对于Swift版，抛出异常后，程序会直接跳到AppDelegate.swift；对于Objective-C版，抛出异常后，程序会直接跳到main.m中的main函数里面，如图22-14所示。

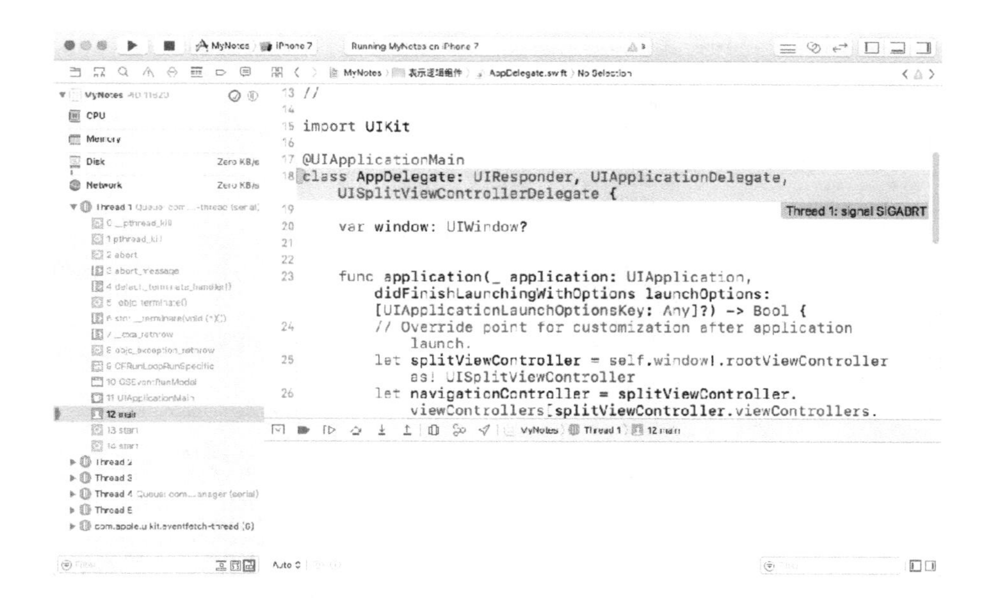

(a)

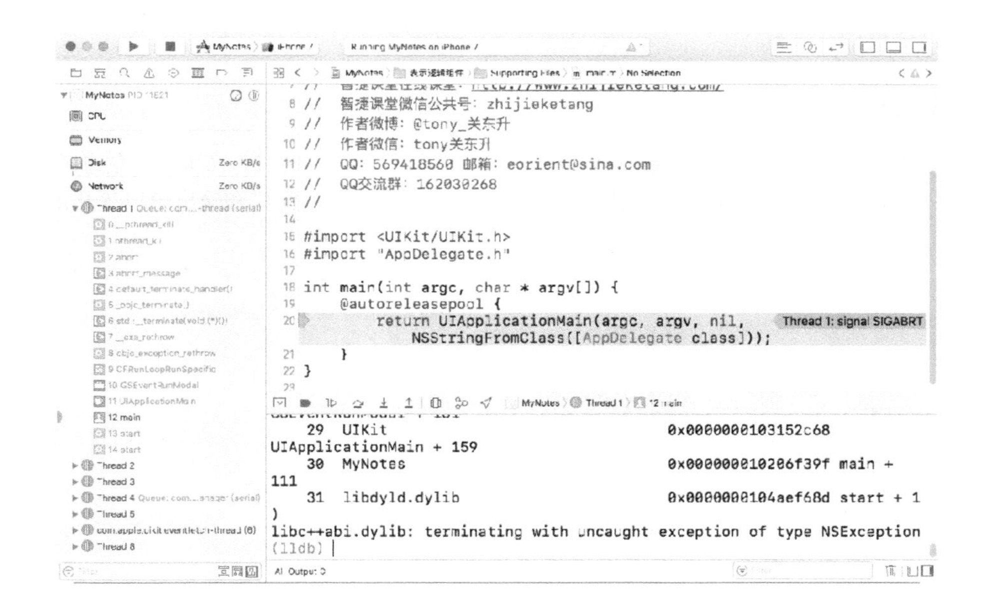

(b)

图22-14　异常的默认处理（图a为Swift版，图b为Objective-C版）

如果我们设置了异常断点，程序会挂起在AddViewController中出现异常的代码行上，如图22-15所示。默认情况下，异常会在输出窗口中输出。本例的输出结果如下：

```
2017-03-08 19:51:59.082 MyNotes[11453:912678] *** Terminating app due to uncaught exception 'NSInvalidArgumentException', reason: '-[MasterViewController reloadView1:]: unrecognized selector sent to instance 0x7fc93dc04100'2016-11-05 03:23:41.528 PresentationLayer[19065:3252760] *** Terminating app due to uncaught exception 'NSInvalidArgumentException', reason:  '-[MasterViewController reloadView1:]: unrecognized selector sent to instance 0x7fc7c9504020'
```

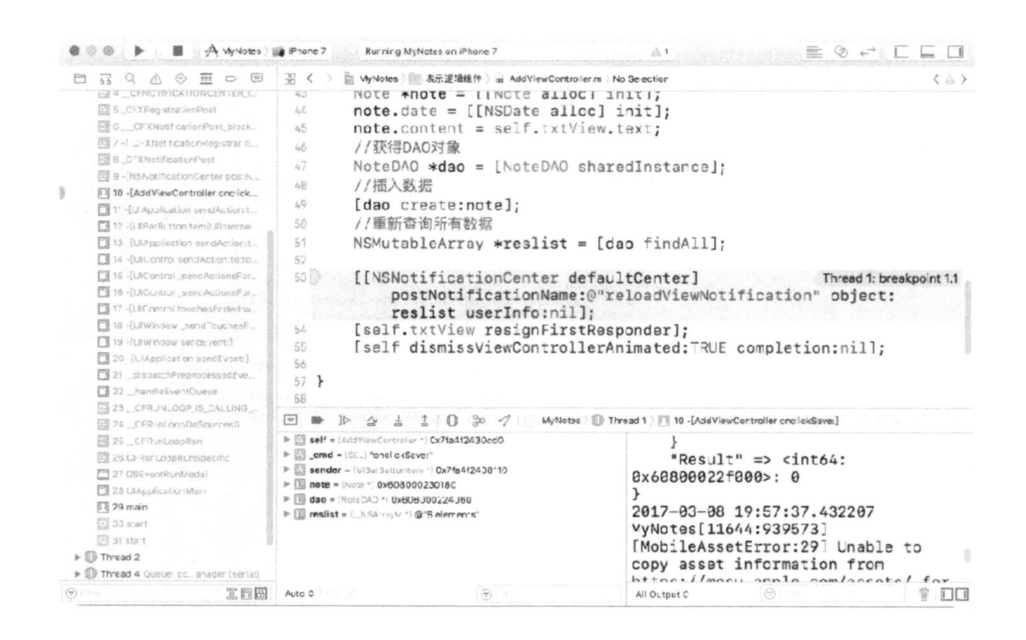

图22-15　设置了异常断点处理

22.1.4　调试工具

Xcode提供了强大的调试功能，如果断点挂起，就会进入调试界面，如图22-16所示。左边区域是调试导航面板，右下方有一个调试窗口，分为调试工具栏、窗口显示按钮、变量查看窗口和输出窗口等几部分。

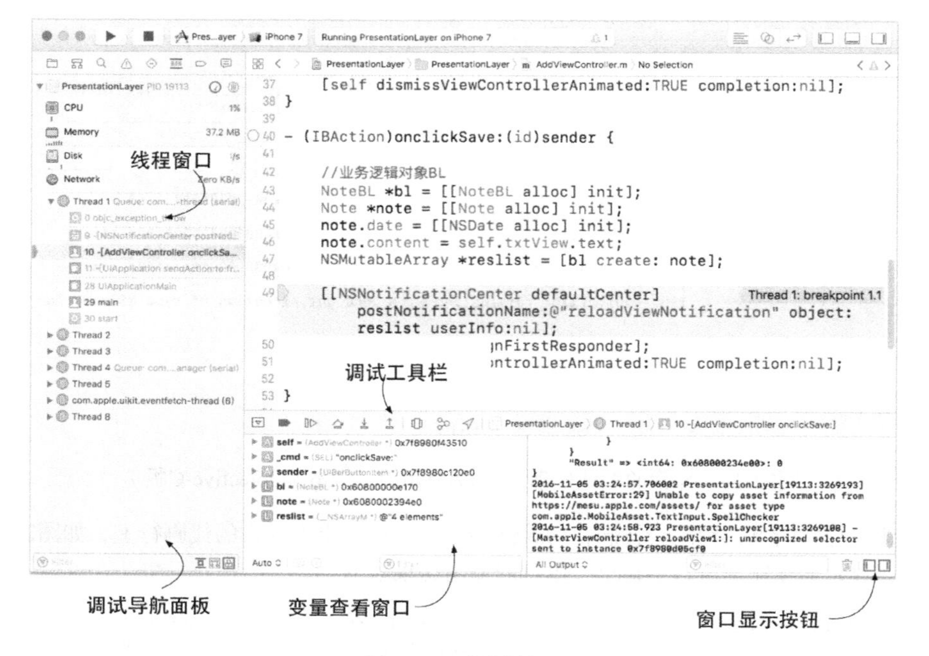

图22-16　调试界面

图22-17是调试工具栏，其中隐藏按钮可以隐藏或者显示调试窗口；视图调试按钮可以调试视图，显示视图层次结构；模拟位置按钮可以向模拟器设备发送虚拟的位置坐标，它用于位置服务应用中的测试；使用跳转栏，可以跳转到具体工程下某个类的方法中，能够跟踪程序的运行过程。

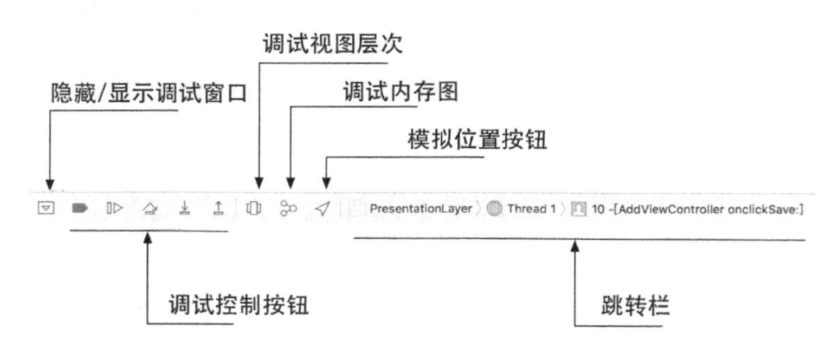

图22-17 调试工具栏

调试控制按钮中有5个是调试过程中最为常用的，如图22-18所示。在断点挂起之后，点击继续执行按钮可以继续执行。单步跳过按钮是单步执行，遇到方法和函数时不进入。使用单步进入按钮，则进入方法或者函数里。单步跳出按钮在从方法或函数里面跳回到原来调用它的地方时使用。

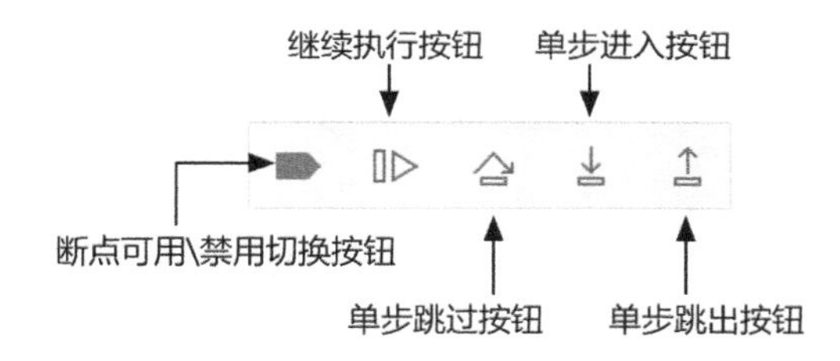

图22-18 执行控制按钮

例如，当程序挂起在viewDidLoad方法的self.listData = [dao findAll]语句时，如图22-19所示，点击继续执行按钮，程序就会接着运行，直到运行到下一个断点，或者结束。如果点击单步跳过按钮，则程序执行会跳过第43行。如果点击单步进入按钮，则进入NoteDAO.m的findAll方法中。

```
31 - (void)viewDidLoad {
32     [super viewDidLoad];
33 
34     for(; i < 10 ;i++) {
35         int b = i * i;
36         NSLog(@"b = %i",b);
37     }
38 
39     self.navigationItem.leftBarButtonItem = self.editButtonItem;
40     self.detailViewController = (DetailViewController *)[[self.
           splitViewController.viewControllers lastObject]
           topViewController];
41     |
42     self.bl = [[NoteBL alloc] init];
43     self.listData = [self.bl findAll];              Thread 1: breakpoint 2.1
44 
45     [[NSNotificationCenter defaultCenter] addObserver:self
46                                              selector:@selector
                                               (reloadView:)
47 
                                               name:@"reloadViewNoti
                                               fication"
48                                                 object:nil];
49 }
50 
```

图22-19 程序挂起

22.1.5　输出窗口

使用窗口显示按钮，可以控制同时显示左右两个窗口（变量查看窗口和输出窗口），或者只显示其中一个窗口。输出窗口有3个选择——All Output、Debugger Output和Target Output，如图22-20所示。

调试程序时，可以在Debugger Output窗口中执行编译器的调试命令。如图22-21所示。

Target Output窗口中可以显示程序出错和异常等信息，以及通过一些函数（如NSLog和assert函数）输出的信息，如图22-22所示。

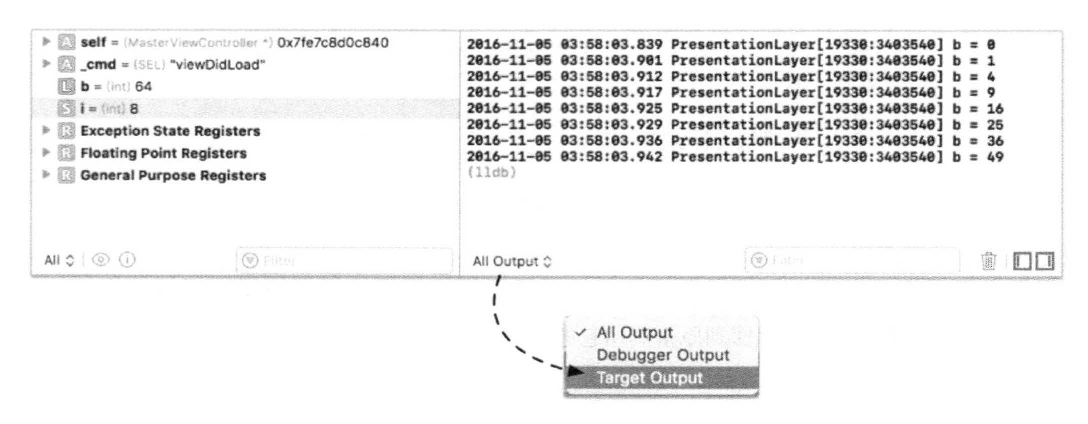

图22-20　输出窗口

图22-21　Debugger Output窗口

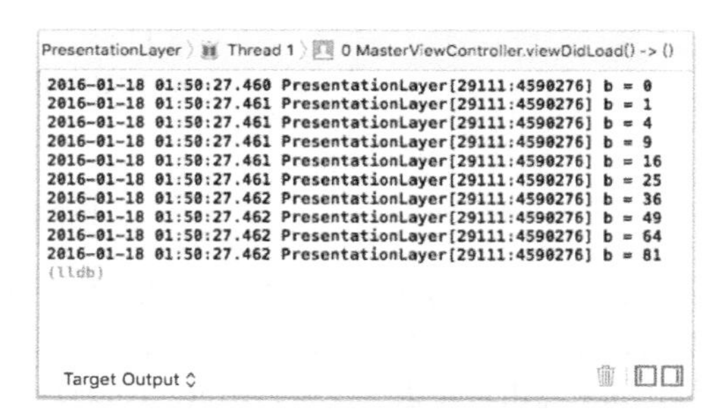

图22-22　Target Output窗口

22.1.6　变量查看窗口

变量查看窗口位于调试窗口的左侧，用于查看变量和寄存器的内容。通过点击该窗口左下角的小三角，可以选择查看变量的范围——Auto、Local Variables和All Variables, Registers, Globals and Statics，如图22-23所示，其中各项的含义如下所示。

- Auto。查看经常使用的变量。
- Local Variables。查看本地变量。
- All Variables, Registers, Globals and Statics。查看全部变量，包括寄存器和全局变量等，如图22-24所示，其中图标A是自动变量、S是静态变量、R是寄存器、L是本地变量。

在变量查看窗口中选择变量后，点击鼠标右键，会弹出一个快捷菜单，从中可以进行一些操作。下面先演示一下如何编辑变量。如图22-25所示，选择变量`i= (int) 10`，点击鼠标右键，从弹出的快捷菜单中选择Edit Value...菜单项，此时该行代码进入修改状态，我们可以在其中输入相关值。

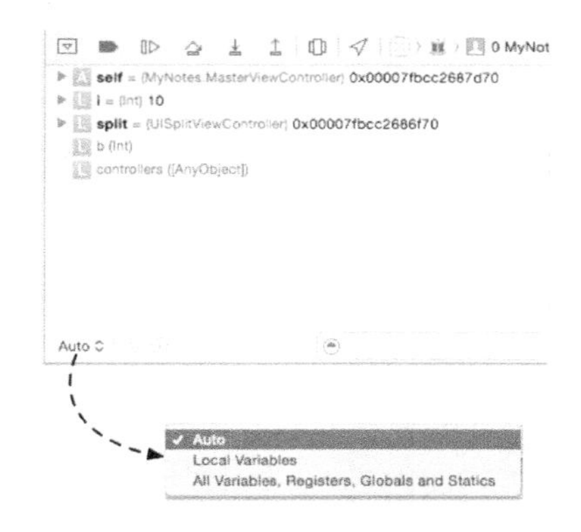

图22-23 变量查看窗口

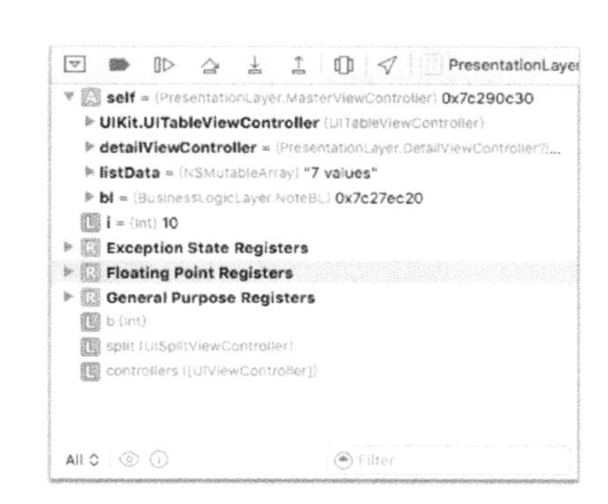

图22-24 变量查看窗口的全部变量选项

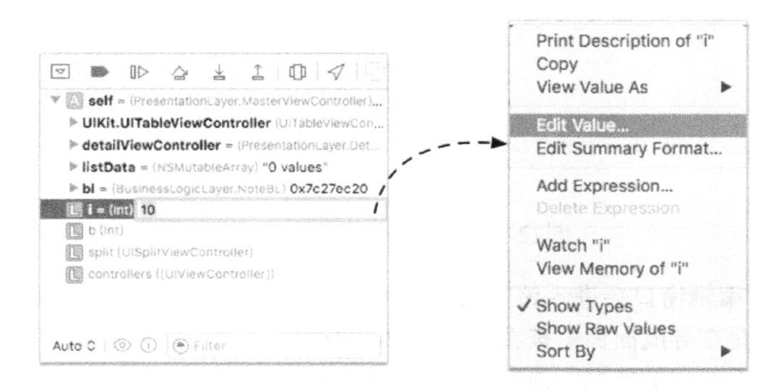

图22-25 修改变量

此外，我们也可以通过快捷菜单打印变量。如图22-26所示，右击listData变量（它是NSMutableArray类型的对象），从弹出的快捷菜单中选择Print Description of“listData”菜单项，此时会在变量输出窗口中打印变量。

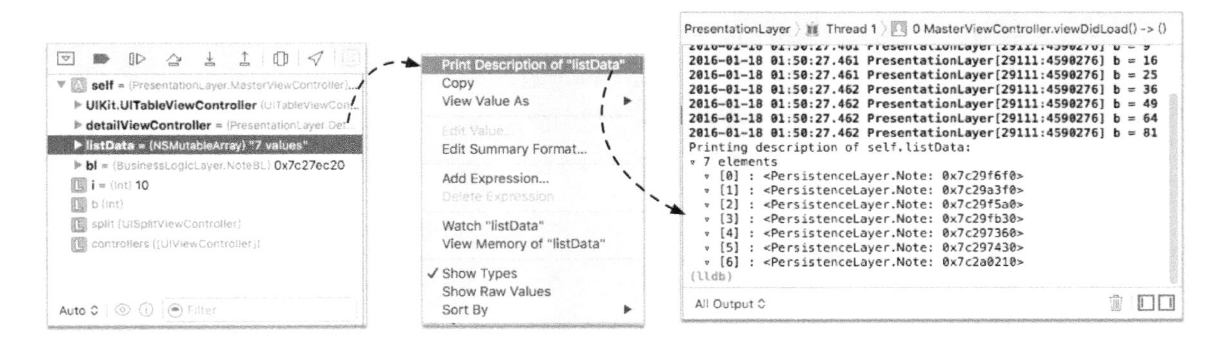

图22-26 打印变量

22.1.7 查看线程

iOS和macOS是支持多线程的。作为开发工具的Xcode，当然要支持查看线程的情况。在Xcode中，有两种方式可以查看线程，一种是在跳转栏中选择线程下拉列表，如图22-27所示。选择某个线程后，Xcode会显示一个代码运行的栈。

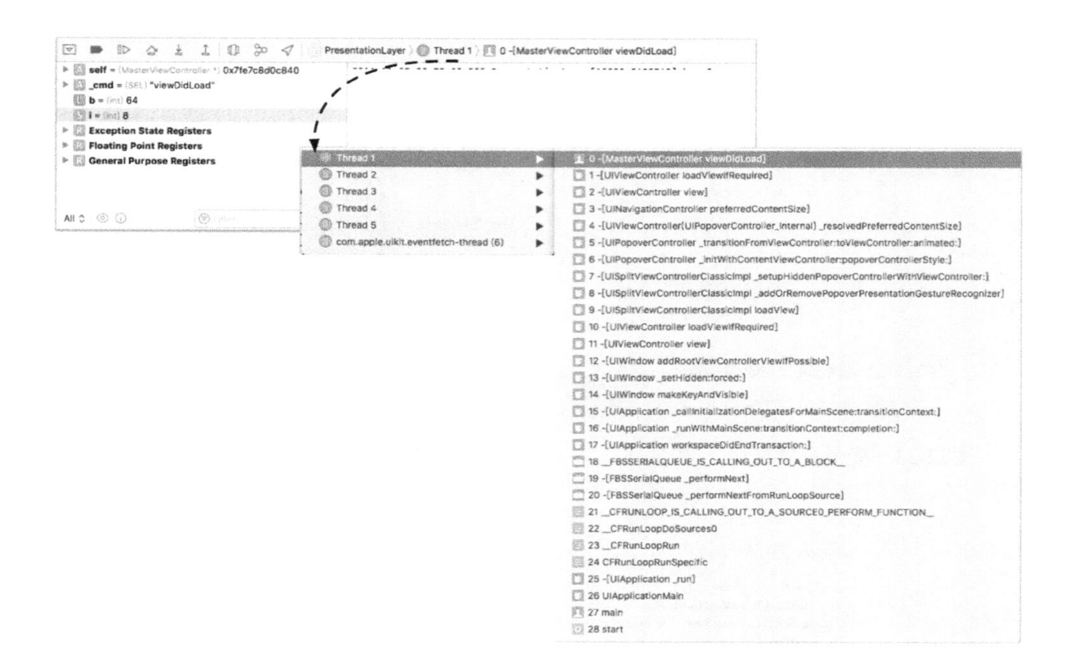

图22-27　在跳转栏中查看线程

选择栈中的方法，此时编辑窗口会进入该方法，如果该方法没有源代码，将显示汇编语言。

另一种查看线程的方法是在导航面板中查看。如图22-28所示，点击“显示调试导航面板”按钮，打开显示调试导航面板，从中可以查看线程的情况。该面板只显示大概的调用栈，没有上一种查看方式反应的情况详细。

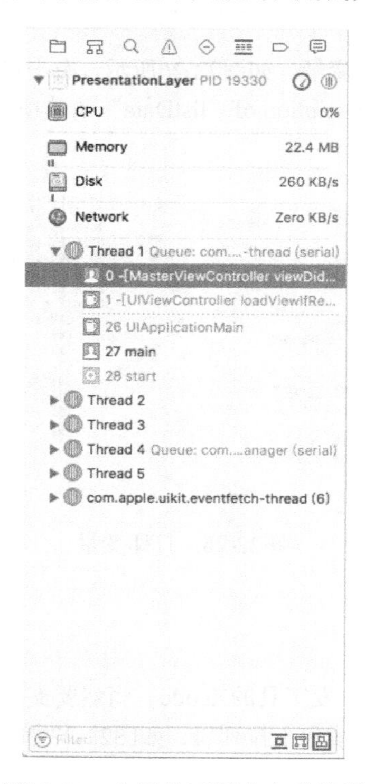

图22-28　在导航面板中查看线程

22.2 LLDB调试工具

我们在22.1.5节中使用过p和po命令在调试输出窗口中计算并输出表达式的内容，事实上p和po就是一种LLDB调试工具的命令，LLDB调试工具的编译器相对独立于Xcode。Objective-C有过3种编译器——GCC、LLVM[①]GCC和Apple LLVM，其中GCC是比较古老的编译器，现在主要使用LLVM GCC和Apple LLVM。GCC的调试工具是GDB，它是GCC Debug工具的缩写。LLVM GCC和Apple LLVM的调试工具是LLDB（或lldb）。GDB和LLDB命令有一些差别，本书只介绍LLDB命令。LLDB命令有很多，本节只选取了几个常用的命令。

如何进入LLDB进行调试呢？进入LLDB调试工具的一种方式是从终端进入，还有一种是从Xcode进入。Xcode工具我们比较熟悉，因此这里主要介绍这种方式。具体做法很简单，在程序中设置断点，运行时挂起，在输出窗口中选择Debugger Output，此时输出窗口有(lldb)命令提示符，这就进入LLDB调试工具了。

22.2.1 断点命令

关于断点命令，我们主要介绍设置断点、查看断点、删除断点、单步进入、单步跳过和继续运行等内容。

1. 设置断点

设置断点使用命令breakpoint set，该命令可以设置文件行（file line）断点和符号（symbolic）断点，并且都有简略写法。

为了在MasterViewController.m中第41行设置断点（这是一种文件行断点），我们可以使用如下命令：

```
(lldb) breakpoint set --file MasterViewController.m --line 41
(lldb) br s -f MasterViewController.m -l 41
(lldb) b MasterViewController.m:41
```

这3条命令可以实现同样的效果。我们测试一下命令，可以发现，如图22-29所示，程序执行到第41行后挂起了。

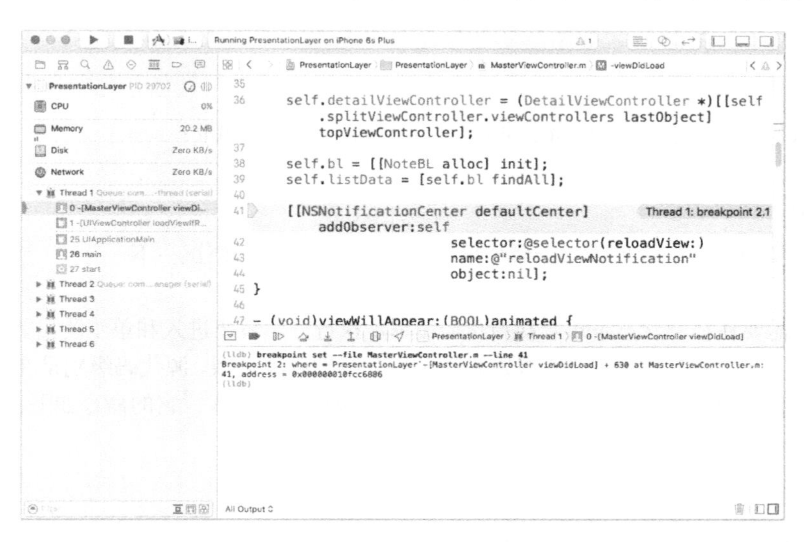

图22-29 测试breakpoint set命令

上面介绍的是文件行断点设置。如果要在所有的findAll方法调用时挂起，则属于符号断点设置，我们可以使用如下命令：

```
(lldb) breakpoint set --selector findAll
(lldb) br s -S findAll
```

① LLVM是Low Level Virtual Machine（低级虚拟机）的简称。

这两条命令可以实现同样的效果。这个命令很简单，大家可以自己测试一下。

2. 查看断点

断点设置太多了之后，我们需要查看一下断点设置情况，使用的命令如下：

```
(lldb) breakpoint list
(lldb) br l
```

这两条命令可以实现同样的效果。测试这个命令的结果如图22-30所示。其中有3个断点，每个断点前面都有标号，1.1:号断点通过Xcode常规方式设置，2.1:号和2.2:号断点是使用`breakpoint set --selector findAll`命令设置的。每个断点后面都有一些关于本断点的描述。

```
(lldb) breakpoint set --selector findAll
Breakpoint 2: 2 locations.
(lldb) breakpoint list
Current breakpoints:
1: file = '/Users/[illegible]/code/ch20/20.2/ObjC/MyNotesWorkspace/PresentationLayer/
PresentationLayer/MasterViewController.m', line = 26, locations = 1, resolved = 1, hit count = 1

  1.1: where = PresentationLayer`-[MasterViewController viewDidLoad] + 23 at MasterViewController.m:26,
address = 0x000000010abb8627, resolved, hit count = 1

2: name = 'findAll', locations = 2, resolved = 2, hit count = 1
  2.1: where = BusinessLogicLayer`-[NoteBL findAll] + 16 at NoteBL.m:32, address = 0x000000010ac4ae50,
resolved, hit count = 1
  2.2: where = PersistenceLayer`-[NoteDAO findAll] + 19 at NoteDAO.m:183, address = 0x000000010cf1aaa3,
resolved, hit count = 0

(lldb)

All Output
```

图22-30　查看断点命令

提示　使用breakpoint set设置的断点，既不能显示在Xcode工具的断点导航面板中，也不能通过Xcode来管理。只有通过Xcode常规方式设置的断点才可以。

3. 删除断点

`breakpoint set`设置的断点需要使用命令来删除。删除断点的命令如下：

```
(lldb) breakpoint delete 断点编号
(lldb) br del 断点编号
```

这两条命令可以实现同样的效果。这个命令很简单，大家可以自己测试一下。

4. 单步进入

有的时候，我们需要在挂起状态下单步执行代码，单步执行分为单步进入和单步跳出。单步进入就是能够进入到方法或者函数中，该命令有基于源代码级别和基于指令级别两种类型。源代码级别是在源代码中进行单步执行，指令级别是在汇编指令中单步执行。我们只介绍源代码级别单步进入，它的命令如下：

```
(lldb) thread step-in
(lldb) step
(lldb) s
```

这3条命令可以实现同样的效果。第一条命令中有`thread`，这说明单步进入命令是在同一个线程中的单步运行，不涉及跨线程问题。

5. 单步跳过

单步跳过与单步进入类似，它遇到方法或函数时不会进入，直接跳过。该命令也有基于源代码级别和基于指令级别的两种类型。源代码级别单步跳过命令如下：

```
(lldb) thread step-over
(lldb) next
(lldb) n
```

这3条命令可以实现同样的效果，后两个是简略写法。

6. 继续运行

继续运行与单步跳过和单步进入类似，执行该命令时，程序会运行到下一个断点或结束。该命令也有基于源代码级别和基于指令级别两种类型。源代码级别的继续运行命令如下：

```
(lldb) thread continue
(lldb) continue
(lldb) c
```

这3条命令可以实现同样的效果，后两个是简略写法。

7. 当前函数或方法返回

当调试进入到某个函数或方法时，由于一些原因不想往下执行，而是直接返回函数或方法的结果，此时可以使用thread return。源代码级别的继续运行命令如下：

```
(lldb) thread return @"abc"
```

这条命令在一个返回值为字符串类型函数或方法中执行，执行结果是跳出该函数或方法，并返回@"abc"字符串。

22.2.2 观察点命令

什么是观察点呢？就是在程序中对于某个要观察的变量设置一个观察点，当这个变量变化的时候，程序就会挂起。它与断点很相似，只是触发条件不同。这里我们分别介绍设置观察点、查看观察点和删除观察点这3个命令。

1. 设置观察点

首先，看看MasterViewController.m的viewDidLoad方法的代码：

```
- (void)viewDidLoad {
    [super viewDidLoad];
    for(; i < 10 ;i++) {
        int b = i * i;
        NSLog(@"b = %i",b);
    }
    ......
}
```

如果为循环体变量b设置观察点，则命令如下：

```
(lldb) watchpoint set variable b
(lldb) wa s v b
```

这两条命令可以实现同样的效果，后一个是简略写法。为变量设置观察点时，变量不能超过它的作用域，变量b的作用域是for循环体，否则命令会出现下面的错误：

```
error: no variable or instance variable named 'b' found in this frame
```

2. 查看观察点

观察点设置的数量较多时，我们需要查看一下观察点设置情况，使用的命令如下：

```
(lldb) watchpoint list
(lldb) watch l
```

这两条命令可以实现同样的效果，后一个是简略写法。测试这个命令的结果如图22-31所示。

```
(lldb) watchpoint set variable b
Watchpoint created: Watchpoint 1: addr = 0x7fff5d85047c size = 4 state = enabled type = w
    declare @ '/Users/[illegible]/code/ch20/20.2/ObjC/MyNotesWorkspace/
PresentationLayer/PresentationLayer/MasterViewController.m:30'
    watchpoint spec = 'b'
    new value: 0

Watchpoint 1 hit:
old value: 0
new value: 0
(lldb)
```

图22-31　查看观察点命令

3. 删除观察点

使用watchpoint set设置的观察点需要使用命令删除，删除观察点的命令如下：

```
(lldb) watchpoint delete 观察点编号
(lldb) watch del 观察点编号
```

这两条命令可以实现同样的效果，后一个是简略写法。这个命令很简单，大家可以自己测试一下，如图22-32所示。

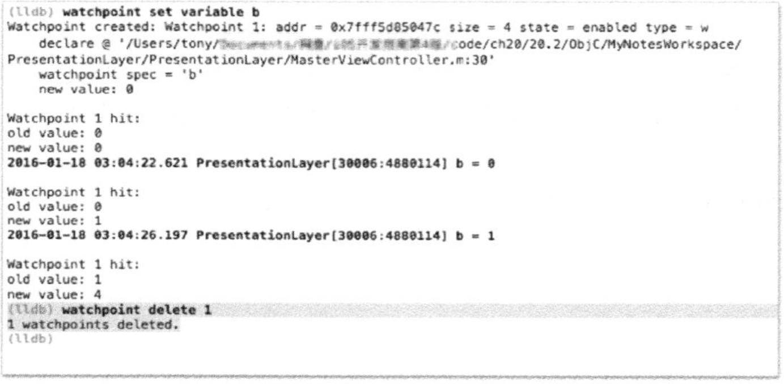

```
(lldb) watchpoint set variable b
Watchpoint created: Watchpoint 1: addr = 0x7fff5d85047c size = 4 state = enabled type = w
    declare @ '/Users/tony/[illegible]code/ch20/20.2/ObjC/MyNotesWorkspace/
PresentationLayer/PresentationLayer/MasterViewController.m:30'
    watchpoint spec = 'b'
    new value: 0

Watchpoint 1 hit:
old value: 0
new value: 0
2016-01-18 03:04:22.621 PresentationLayer[30006:4880114] b = 0

Watchpoint 1 hit:
old value: 0
new value: 1
2016-01-18 03:04:26.197 PresentationLayer[30006:4880114] b = 1

Watchpoint 1 hit:
old value: 1
new value: 4
(lldb) watchpoint delete 1
1 watchpoints deleted.
(lldb)
```

图22-32　删除观察点命令

22.2.3　查看变量和计算表达式命令

使用调试工具少不了查看变量和计算表达式，下面我们分别介绍几个有关的命令，其中包括frame、target、expr和print等。

1. 查看本地变量

使用如下命令可以查看当前栈帧[①]的所有本地变量：

```
(lldb) frame variable
(lldb) fr v
```

这两条命令可以实现同样的效果，后一个是简略写法。我们测试一下该命令，程序挂起后，在输出窗口中输入frame variable命令，输出结果如图22-33所示。

有时候，我们只想看看某个具体变量，此时可以使用下面的命令：

```
(lldb) frame variable bar
(lldb) fr v bar
(lldb) p bar
```

① 栈帧表示对当前线程的调用栈的一个函数调用。

其中bar是变量名。这3条命令可以实现同样的效果，第二条语句是第一条语句的简略写法，第三条语句中的p bar是print bar的缩写，print命令是打印和计算表达式。

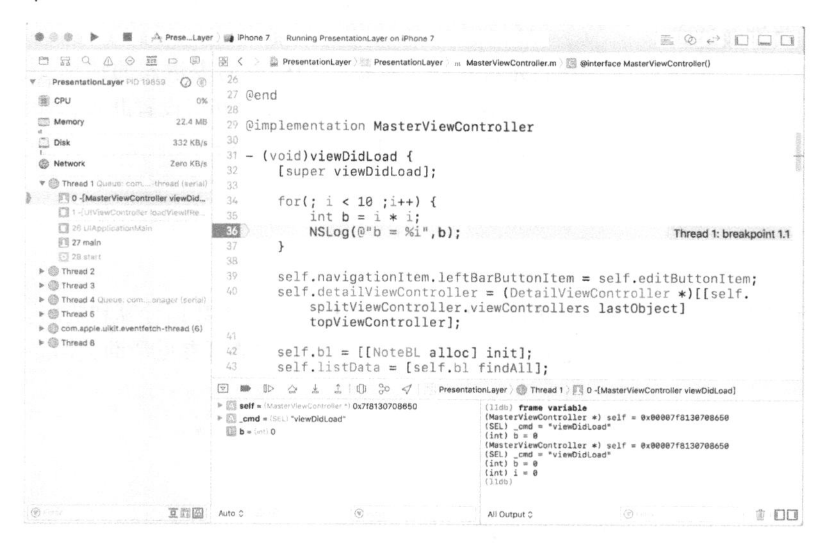

图22-33　测试frame命令

2. 查看全局变量

C语言家族中有全局变量的概念，它包括global/static变量，可以使用下面的命令查看这些变量的内容：

```
(lldb) target variable
(lldb) ta v
```

这两条命令可以实现同样的效果，后一个是简略写法。我们测试一下该命令，程序挂起后，在输出窗口中输入target variable命令，输出结果如图22-34所示。

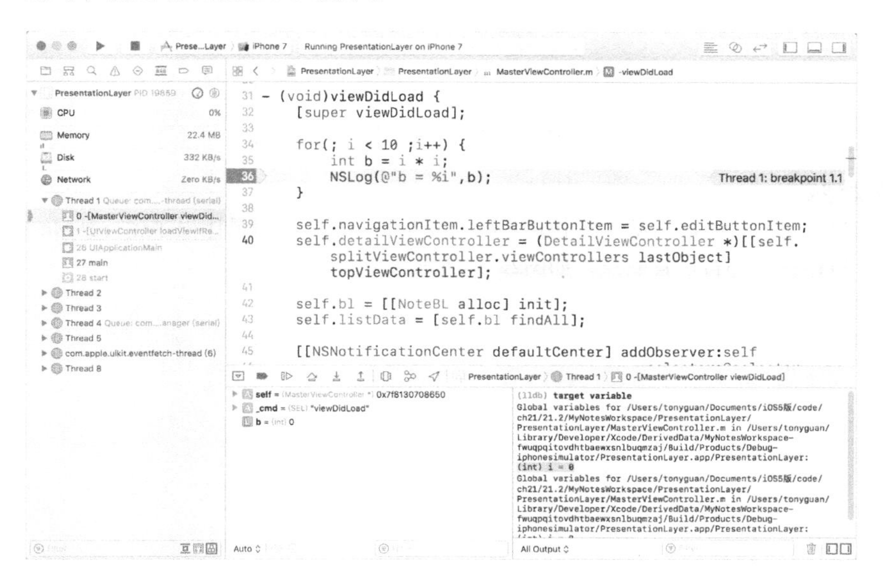

图22-34　测试target命令

有的时候我们只是想看看某个具体变量，此时可以使用下面的命令：

```
(lldb) target variable baz
```

```
(lldb) ta v baz
```

其中baz是变量名。这两条命令可以实现同样的效果，第二条语句是第一条语句的简略写法。

3. 计算基本数据类型表达式

计算表达式在调试时是很重要的，我们先看看基本数据类型构成的表达式的计算。为了计算程序运行到MasterViewController.m中的第36行时表达式`i * i`的结果，我们可以在第36行设置断点，如图22-35所示。

如图22-35所示，为了计算`i * i`表达式，我们可以使用下面几个命令：

```
(lldb) expr (int) i * i
(lldb) expr  i * i
(lldb) print  i * i
(lldb) p  i * i
```

其中expr和print都可以计算表达式。print还可以省略为p，一般用于基本数据计算和输出。expr是expression的简写，最简单的时候可以省略为e，其功能强大，可以有很多参数，详情可参见帮助文档。

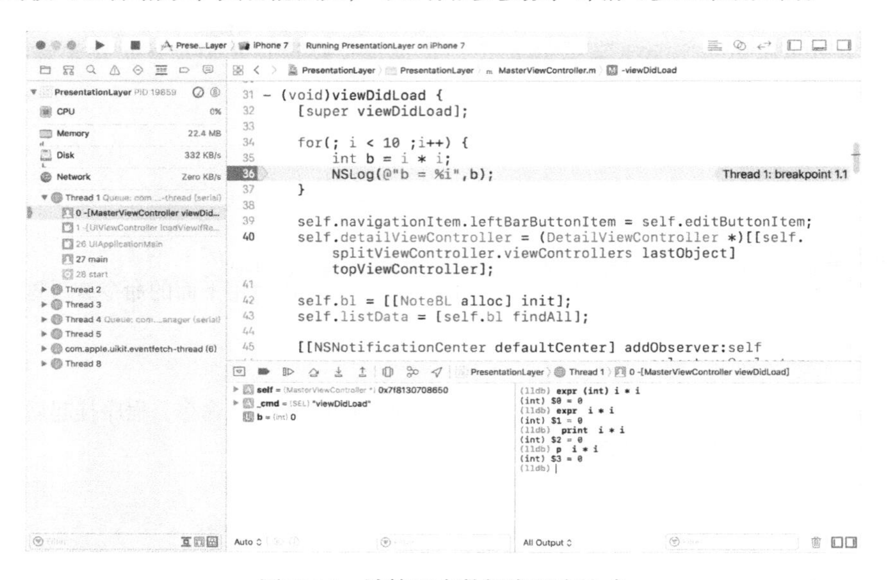

图22-35　计算基本数据类型表达式

4. 计算对象类型表达式

Objective-C作为面向对象语言，表达式中会用到对象类型。如果我们想让程序运行到MasterViewController.m中第45行时挂起，可以在第45行设置断点，如图22-36所示。

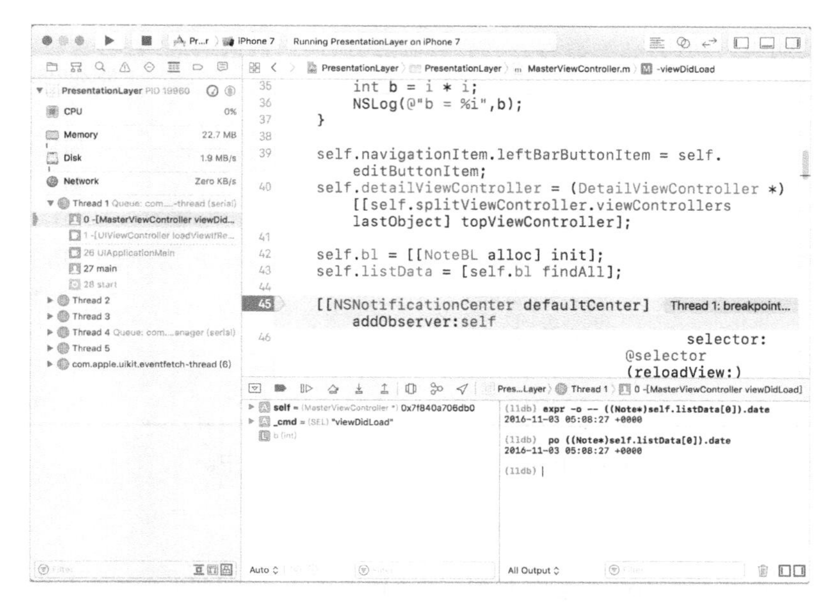

图22-36　计算对象数据类型表达式

如图22-36所示，为了计算listData属性中第一个元素Note对象的date属性表达式，我们可以使用下面几个命令：

```
(lldb) expr -o -- ((Note*)self.listData[0]).date
2016-11-03 05:08:27 +0000

(lldb) po ((Note*)self.listData[0]).date
2016-11-03 05:08:27 +0000
```

表达式中的self.listData[0]是取得NSArray数组的第一个元素，为了使其是一个合法的表达式，需要使用强制类型转换((Note*)self.listData[0])，然后再调用date属性。上述命令中expr -o --和po都可以计算并输出Note对象的date属性，其中-o表示打印对象参数，po命令是print object的缩写。

22.3　日志与断言输出

在程序运行的过程中，日志可以输出一些信息到输出窗口。断言是程序调试执行过程中设置的一些条件，当条件满足时，正常执行；当条件不满足时，终止程序，抛出错误栈信息。根据这些信息，我们可以分析程序出了什么问题以及程序的运行情况。

22.3.1　使用 NSLog 函数

NSLog是Foundation框架提供的Objective-C日志输出函数，与C中的printf函数类似，可以格式化输出。对于不同的数据类型，NSLog函数中的格式化字符串是不同的，如表22-1所示。

表22-1　NSLog函数中的格式化字符串

类　型	实　例	NSLog中的格式化字符串
char	'a'、'\n'	%c
short int	-10	%hi、%hx、%ho
unsigned short int	9	%hu、%hx、%ho
int	17、-99、0xFFAE、0878	%i、%x、%o

（续）

类　型	实　例	NSLog中的格式化字符串
unsigned int	17u、101U、0XFFu	%u、%x、%o
long int	17L、-2998、0xffffL	%li、%lx、%lo
unsigned long int	17UL、-100ul、0xffeeUL	%li、%lx、%lo
long long int	0xe5e5e5e5LL、500ll	%lli、%llx、%llo
unsigned long long int	17ull、oxffefULL	%llu、%llx、%llo
float	12.3f、3.1e-5f	%f、%e、%g
double	12.34、3.1e-5	%f、%e、%g
long double	12.34l、3.1e-5l	%Lf、%Le、%Lg
对象指针	"<Note: 0x7697220>"、"<Note: 0x769ad70>"	%p、%@

与NSLog函数类似的还有NSLogv函数，它可以将输出重新定向到文件中。

22.3.2　使用断言

NSLog函数是无条件输出，即程序运行到该语句，就会输出结果。如果想有条件输出结果，在Swift中可以使用断言函数assert和assertionFailure，它们的定义如下所示。

- **assert (condition, message)**。condition为false时抛出异常，在日志中输出message信息。
- **assertionFailure (message)**。抛出异常，在日志中输出message信息。

在Objective-C中，断言函数被定义为NSAssert宏，它的定义如下：

```
#define NSAssert(condition, desc, ...)
```

其中第一个参数condition是布尔表达式，第二个参数desc是描述信息，参数后面的...是格式化desc描述信息的。如果condition为NO（或FALSE），则输出desc描述信息，并抛出异常NSInternalInconsistencyException；如果condition为YES（或TRUE），则不输出信息。

在MasterViewController的viewDidLoad方法中，使用断言的代码如下：

```
override func viewDidLoad() {
    super.viewDidLoad()

    for i in 0 ..< 10 {
        assert(i >= 0 && i < 9, "i变量超出了范围。")                    ①
        let b = i * i
        NSLog("b = %i", b)
    }
    ......

    //查询所有的数据
    self.listData = self.bl.findAll()

    NotificationCenter.default.addObserver(self,
    ➥selector: #selector(reloadView(_:)),
    ➥name: Notification.Name(rawValue: "reloadViewNotification"),
    ➥object: nil)
}
```

```
- (void)viewDidLoad {
    [super viewDidLoad];

    for(; i < 10 ;i++) {
        NSAssert(i >= 0 && i < 9, @"i = %i变量超出了范围。", i);        ①
        int b = i * i;
        NSLog(@"b = %i",b);
    }

    ......

    self.bl = [[NoteBL alloc] init];
    self.listData = [self.bl findAll];

    [[NSNotificationCenter defaultCenter] addObserver:self
    ➥selector:@selector(reloadView:)
}
```

在第①行中，当变量i等于9时，条件（i >= 0 && i < 9）为“假”，程序会抛出异常，输出窗口输出的内容如图22-37所示。

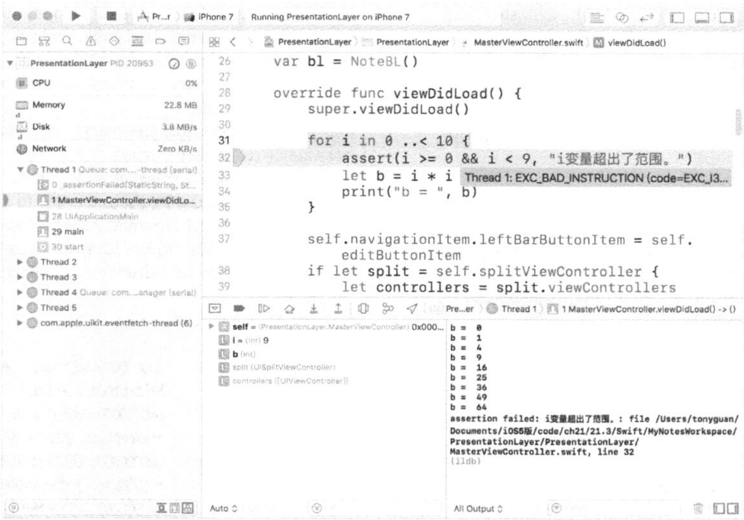

图22-37 断言输出信息

提示

在Objective-C中，与NSAssert类似的宏还有NSAssert1、NSAssert2、NSAssert3、NSAssert4和NSAssert5，其中宏后面的数字代表格式化描述信息参数的个数，它们的定义如下：

```
#define NSAssert1(condition, desc, arg1)
#define NSAssert2(condition, desc, arg1, arg2)
#define NSAssert3(condition, desc, arg1, arg2, arg3)
#define NSAssert4(condition, desc, arg1, arg2, arg3, arg4)
#define NSAssert5(condition, desc, arg1, arg2, arg3, arg4, arg5)
```

22.4 异常栈报告分析

在使用Xcode工具的开发过程中，面对运行异常，很多初学者往往毫无头绪，不知道如何跟踪异常栈，以及如何分析异常栈报告，本节就来解决这两个问题。

提示

异常栈是程序抛出异常之前，对象之间方法（或函数）调用的“路径”，它是程序运行的“黑匣子”。异常栈的输出内容包括很多信息，如调用顺序、方法所属框架（或库）、方法所属类、方法（或函数）地址等。

22.4.1 跟踪异常栈

默认情况下，使用Xcode工具进行开发时，若产生异常，会有信息输出，也会有异常栈输出，我们可以分析这些栈信息。

为了学习这些知识点，我们使用14.4节的MyNotes案例，修改一下MasterViewController中的`viewDidLoad`方法，参考22.1.3节制造一个运行期异常。

当点击添加功能中的保存操作时，程序运行到这里就会抛出异常。默认情况下，运行产生异常时输出到输出窗口的内容如下：

```
2016-11-05 11:03:17.825083 PresentationLayer[1439:28079] [MC] System group
➥container for systemgroup.com.apple.configurationprofiles path is
➥/Users/tonyguan/Library/Developer/CoreSimulator/Devices/6CC84B2B-7EC2-44EC
➥-8C53-A3CDCFE542C0/data/Containers/Shared/SystemGroup/systemgroup.com.appl
➥e.configurationprofiles
2016-11-05 11:03:17.825909 PresentationLayer[1439:28079] [MC] Reading from
➥private effective user settings.
2016-11-05 11:03:20.706 PresentationLayer[1439:28079]
➥-[PresentationLayer.MasterViewController reloadView1:]: unrecognized
➥selector sent to instance 0x7faabc4067a0
2016-11-05 11:03:20.739 PresentationLayer[1439:28079] *** Terminating app due
➥to uncaught exception 'NSInvalidArgumentException', reason:
➥'-[PresentationLayer.MasterViewController reloadView1:]: unrecognized
➥selector sent to instance 0x7faabc4067a0'
*** First throw call stack:
(
    0   CoreFoundation      0x0000000103ebe34b __exception
                            ➥Preprocess + 171
    1   libobjc.A.dylib     0x000000010392621e objc_exception
                            ➥_throw + 48
    2   CoreFoundation      0x0000000103f2df34 -[NSObject
                            ➥(NSObject) doesNotRecognizeSelector:] + 132
    3   CoreFoundation      0x0000000103e43c15 ___forwar
                            ➥ding___ + 1013
    4   CoreFoundation      0x0000000103e43798 _CF_forwar
                            ➥ding_prep_0 + 120
    5   CoreFoundation      0x0000000103e5c19c __CFNOTIFICAT
                            ➥IONCENTER_IS_CALLING_OUT_TO_AN_OBSERVER__ + 12
    6   CoreFoundation      0x0000000103e5c09b _CFXRegistra
                            ➥tionPost + 427
    7   CoreFoundation      0x0000000103e5be02 ___CFXNotifi
                            ➥cationPost_block_invoke + 50
    8   CoreFoundation      0x0000000103e1eea2 -[_CFXNotifica
                            ➥tionRegistrar find:object:observer:enumerator:] + 2018
    9   CoreFoundation      0x0000000103e1df3b _CFXNotifica
                            ➥tionPost + 667
    10  Foundation          0x00000001033ee0ab -[NSNotifica
                            ➥tionCenter postNotificationName:object:userInfo:] + 66
    11  PresentationLayer   0x00000001030fb901 _TFC17Presen
                            ➥tationLayer17AddViewController11onclickSavefPs9Any
                            ➥Object_T_ + 1281
    12  PresentationLayer   0x00000001030fbd66 _TToFC17Presen
                            ➥tationLayer17AddViewController11onclickSavefPs9Any
                            ➥Object_T_ + 54
    13  UIKit               0x00000001042e35b8 -[UIApplica
                            ➥tion sendAction:to:from:forEvent:] + 83
    14  UIKit               0x0000000104724405 -[UIBar
                            ➥ButtonItem(UIInternal) _sendAction:withEvent:] + 149
    15  UIKit               0x00000001042e35b8 -[UIApplica
                            ➥tion sendAction:to:from:forEvent:] + 83
    16  UIKit               0x0000000104468edd -[UIControl
                            ➥sendAction:to:forEvent:] + 67
    17  UIKit               0x00000001044691f6 -[UIControl
                            ➥_sendActionsForEvents:withEvent:] + 444
    18  UIKit               0x0000000104469380 -[UIControl
                            ➥_sendActionsForEvents:withEvent:] + 838
    19  UIKit               0x00000001044680f2 -[UIControl
                            ➥touchesEnded:withEvent:] + 668
    20  UIKit               0x0000000104350ce1 -[UIWindow
                            ➥_sendTouchesForEvent:] + 2747
    21  UIKit               0x00000001043523cf -[UIWindow
                            ➥sendEvent:] + 4011
    22  UIKit               0x00000001042ff63f -[UIAppli
                            ➥cation sendEvent:] + 371
    23  UIKit               0x0000000104af171d __dispatch
                            ➥PreprocessedEventFromEventQueue + 3248
    24  UIKit               0x0000000104aea3c7 __handle
                            ➥EventQueue + 4879
```

```
2016-11-05 11:02:20.674727 PresentationLayer[1380:27052] [MC] System group
➥container for systemgroup.com.apple.configurationprofiles path is
➥/Users/tonyguan/Library/Developer/CoreSimulator/Devices/6CC84B2B-7EC2-44EC
➥-8C53-A3CDCFE542C0/data/Containers/Shared/SystemGroup/systemgroup.com.appl
➥e.configurationprofiles
2016-11-05 11:02:20.675648 PresentationLayer[1380:27052] [MC] Reading from
➥private effective user settings.
2016-11-05 11:02:34.252 PresentationLayer[1380:27052] -[MasterViewController
➥reloadView1:]: unrecognized selector sent to instance 0x7fd594404ae0
2016-11-05 11:02:34.257 PresentationLayer[1380:27052] *** Terminating app due
➥to uncaught exception 'NSInvalidArgumentException', reason:
➥'-[MasterViewController reloadView1:]: unrecognized selector sent to instance
➥0x7fd594404ae0'
*** First throw call stack:
(
    0   CoreFoundation      0x000000010d5f834b __exception
                            ➥Preprocess + 171
    1   libobjc.A.dylib     0x000000010d05921e objc_
                            ➥exception_throw + 48
    2   CoreFoundation      0x000000010d667f34 -[NSObject
                            ➥(NSObject) doesNotRecognizeSelector:] + 132
    3   CoreFoundation      0x000000010d57dc15 ___forwar
                            ➥ding___ + 1013
    4   CoreFoundation      0x000000010d57d798 _CF_forwarding
                            ➥_prep_0 + 120
    5   CoreFoundation      0x000000010d59619c __CFNOTIFICAT
                            ➥IONCENTER_IS_CALLING_OUT_TO_AN_OBSERVER__ + 12
    6   CoreFoundation      0x000000010d59609b _CFXRegistra
                            ➥tionPost + 427
    7   CoreFoundation      0x000000010d595e02 ___CFXNotifi
                            ➥cationPost_block_invoke + 50
    8   CoreFoundation      0x000000010d558ea2 -[_CFXNotifi
                            ➥cationRegistrar find:object:observer:enumerator:] +
                            ➥2018
    9   CoreFoundation      0x000000010d557f3b _CFXNotifica
                            ➥tionPost + 667
    10  Foundation          0x000000010cb210ab -[NSNotifi
                            ➥cationCenter postNotificationName:object:userInfo:] +
                            ➥66
    11  PresentationLayer   0x000000010ca7ad3c -[AddView
                            ➥Controller onclickSave:] + 444
    12  UIKit               0x000000010da1d5b8
                            ➥-[UIApplication sendAction:to:from:forEvent:] + 83
    13  UIKit               0x000000010de5e405 -[UIBarButton
                            ➥Item(UIInternal) _sendAction:withEvent:] + 149
    14  UIKit               0x000000010da1d5b8
                            ➥-[UIApplication sendAction:to:from:forEvent:] + 83
    15  UIKit               0x000000010dba2edd
                            ➥-[UIControl sendAction:to:forEvent:] + 67
    16  UIKit               0x000000010dba31f6
                            ➥-[UIControl _sendActionsForEvents:withEvent:] + 444
    17  UIKit               0x000000010dba3380
                            ➥-[UIControl _sendActionsForEvents:withEvent:] + 838
    18  UIKit               0x000000010dba20f2
                            ➥-[UIControl touchesEnded:withEvent:] + 668
    19  UIKit               0x000000010da8ace1 -[UIWindow
                            ➥_sendTouchesForEvent:] + 2747
    20  UIKit               0x000000010da8c3cf
                            ➥-[UIWindow sendEvent:] + 4011
    21  UIKit               0x000000010da3963f
                            ➥-[UIApplication sendEvent:] + 371
    22  UIKit               0x000000010e22b71d __dispatch
                            ➥PreprocessedEventFromEventQueue + 3248
    23  UIKit               0x000000010e2243c7 __handle
                            ➥EventQueue + 4879
    24  CoreFoundation      0x000000010d59d311 __CFRUNLOOP
                            ➥_IS_CALLING_OUT_TO_A_SOURCE0_PERFORM_FUNCTION__ + 17
    25  CoreFoundation      0x000000010d58259c __CFRunLoop
```

```
25 CoreFoundation      0x0000000103e63311 __CFRUNLOOP_
                       ➥IS_CALLING_OUT_TO_A_SOURCE0_PERFORM_FUNCTION__ + 17
26 CoreFoundation      0x0000000103e4859c __CFRunLoop
                       ➥DoSources0 + 556
27 CoreFoundation      0x0000000103e47a86 __CFRun
                       ➥LoopRun + 918
28 CoreFoundation      0x0000000103e47494 CFRunLoopRun
                       ➥Specific + 420
29 GraphicsServices    0x000000010b94fa6f GSEvent
                       ➥RunModal + 161
30 UIKit               0x00000001042e1964 UIApplication
                       ➥Main + 159
31 PresentationLayer   0x00000001030fdbbf main + 111
32 libdyld.dylib       0x000000010335468d start + 1
33 ???                 0x0000000000000001 0x0 + 1
)
libc++abi.dylib: terminating with uncaught exception of type NSException
```

```
                       ➥DoSources0 + 556
26 CoreFoundation      0x000000010d581a86 __CFRun
                       ➥LoopRun + 918
27 CoreFoundation      0x000000010d581494 CFRunLoopRun
                       ➥Specific + 420
28 GraphicsServices    0x0000000111369a6f GSEventRun
                       ➥Modal + 161
29 UIKit               0x000000010da1b964 UIApplication
                       ➥Main + 159
30 PresentationLayer   0x000000010ca7c50f main + 111
31 libdyld.dylib       0x000000010f50668d start + 1
32 ???                 0x0000000000000001 0x0 + 1
)
libc++abi.dylib: terminating with uncaught exception of type NSException
```

这些日志由系统输出，其中`*** First throw call stack:(...)`之间的内容就是异常栈信息。

22.4.2 分析栈报告

下面我们介绍一下如何分析栈报告，以及它里面的内容是什么含义。

一条栈信息由5部分构成，如图22-38所示。

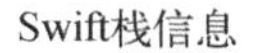

Objective-C栈信息

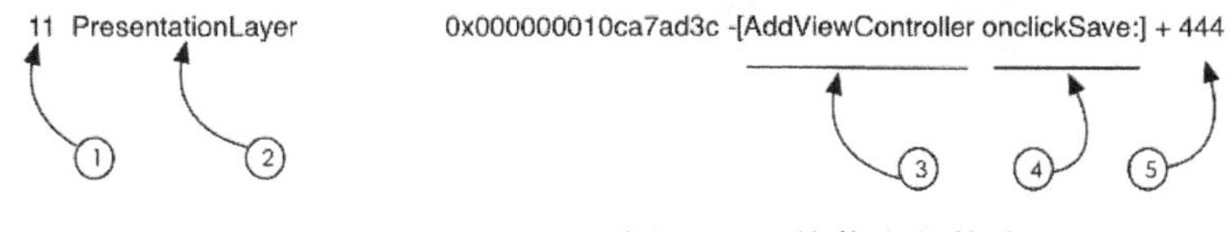

图22-38 栈信息的构成

- **第①部分**：栈输出序号，序号越大，表示越早被调用。
- **第②部分**：调用方法（或函数）所属的框架（或库），图中所示的PresentationLayer是我们自己编写的表示层工程。
- **第③部分**：调用方法（或函数）所属的类名。
- **第④部分**：调用方法（或函数）名，这个信息对我们很重要。
- **第⑤部分**：调用方法（或函数）编译之后的代码偏移量，这个信息很多人误认为是行号，对我们基本没有帮助。

此外，栈信息是要从下往上看的，程序运行的过程是从下面的方法（或函数）调用项目的方法（或函数）。下面是调用`AddViewController`类的`onclickSave`方法：

```
11 PresentationLayer     0x00000001030fb901 _TFC17Presenta
                         ➥tionLayer17AddViewController11onclick
                         ➥SavefPs9AnyObject_T_ + 1281
```

```
11 PresentationLayer 0x000000010ca7ad3c
                     ➥-[AddViewController onclickSave:] + 444
```

在程序运行的过程中，栈信息可能很长，我们不需要每一行都去看，只需关注自己的工程代码。这是因为首先要假定别人提供给我们的框架（或库）是正确的，先看自己工程中的方法（或函数），找到那条调用语句看看是不是有问题。

这就是栈报告分析过程了。当然，是否能够找出问题还会因人而异，有调试经验的人可能很快就能找出问题。

22.5　在 iOS 设备上调试

所有应用在发布之前一定要在iOS设备上调试，本节我们就来简要介绍一下。在Xcode 7之前，一个应用要想在iOS设备上运行，需要付费的苹果开发者账号（也是一个Apple ID）对应用进行数字签名才行。在Xcode 7及其之后的版本中，我们不需要付费的苹果开发者账号，只需要一个Apple ID连接iOS设备进行运行和调试了。

事实上，无论是付费开发者账号，还是免费的Apple ID，都需要进行数字签名。数字签名需要使用描述文件。生成描述文件的过程：注册设备→创建开发证书→生成描述文件。这个过程可以一次性地由Xcode自动帮我们完成，也可以自己手动来完成。

下面介绍一下设备调试流程，这个流程分为两步：Xcode设置和设备设置。

22.5.1　Xcode 设置

首先，需要在Xcode中设置账号，这个账号既可以是付费的开发者Apple ID，也可以是免费的Apple ID。具体步骤参考21.3.4节，这里不再赘述。

22.5.2　设备设置

22.5.1节的设置完成后，可以编译应用，但运行时会出现如图22-39所示的错误提示。我们会发现该应用已经安装到iOS设备上了，但是点击运行，会出现如图22-40所示的不信任提示。接下来，我们需要在iOS设备上做一些设置，来“信任”这个应用。

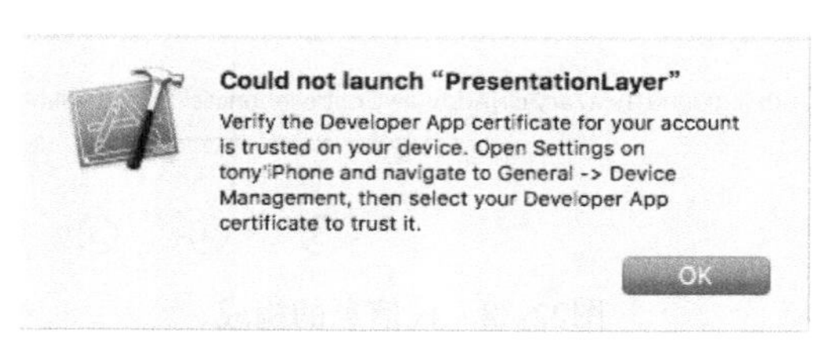

图22-39　Xcod错误提示

打开iOS的“设置”应用，选择“通用”→“设备管理”，然后在“开发商应用”中选择你的账号，接着点击“信任*XXX*”按钮，如图22-41所示。然后，在弹出的对话框（如图22-42所示）中点击“信任”按钮即可。最后，再回到桌面运行刚才的应用，会发现此时就可以运行了。

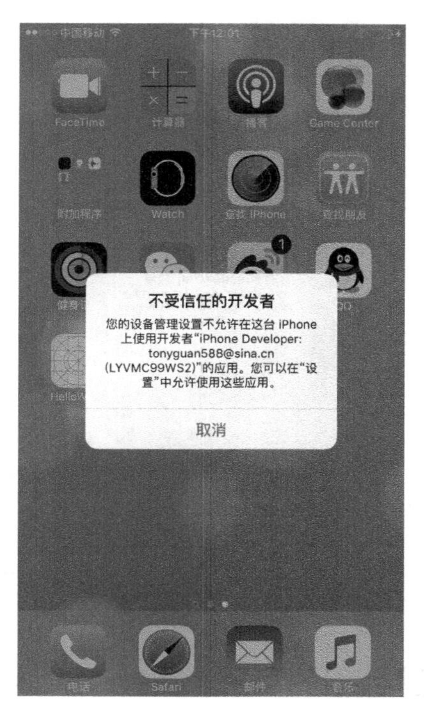

图22-40　iOS不信任提示

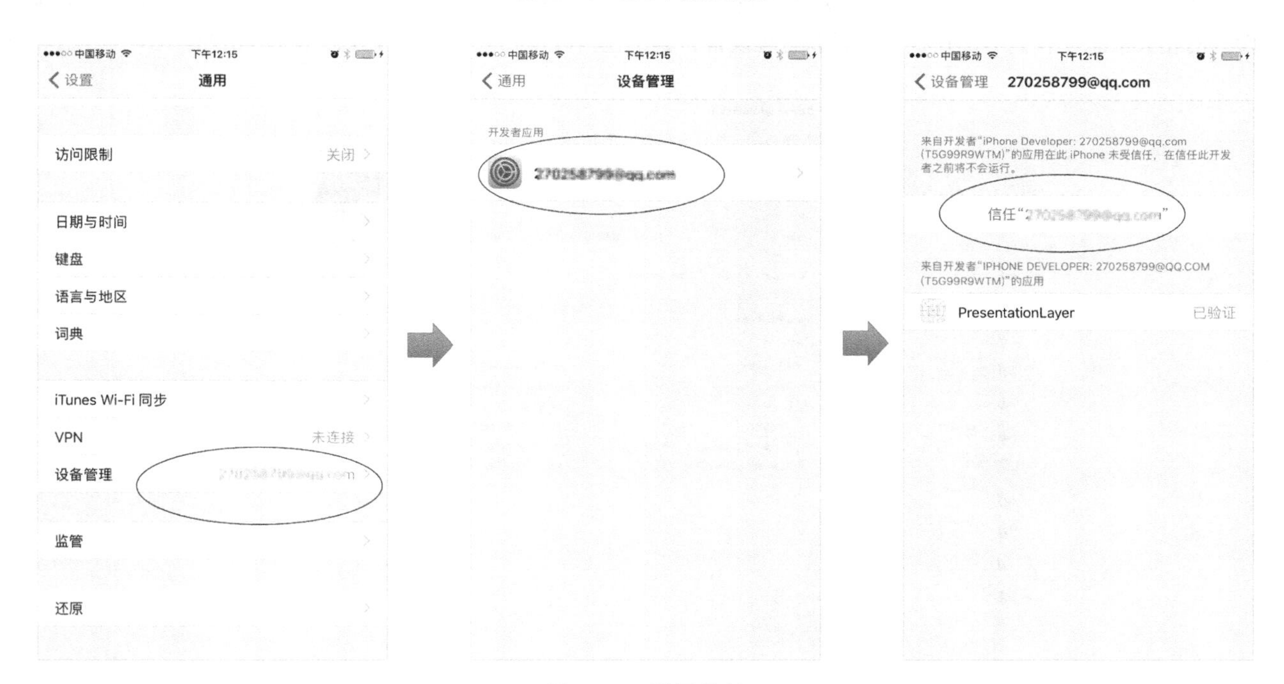

图22-41　设置信任

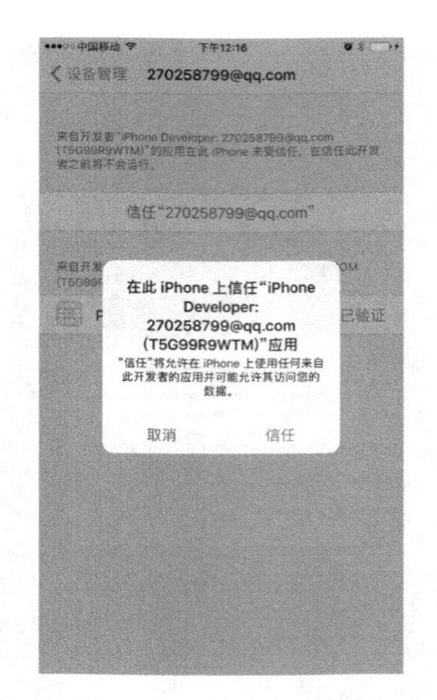

图22-42　设置信任对话框

22.6　Xcode 设备管理工具

选择Window→Devices菜单项，进入如图22-43所示的界面，从中选择Tony's iPhone设备，右边界面会显示该设备的相关信息，我们可以在这里查看和管理设备，其中包括设备的应用、设备日志和屏幕快照。

图22-43　设备信息

22.6.1　查看设备上的应用程序

在图22-43中，Installed Apps中是已经安装到设备上的应用，我们可以在这个界面中安装和删除应用，具体可

以通过下面的 + 和 − 按钮实现。在该界面的左下角，还有一个⚙按钮，点击该按钮，将弹出如图22-44所示的菜单，其中Show Container菜单项用于查看应用的沙箱目录。单击该菜单项，将得到如图22-45所示的界面。

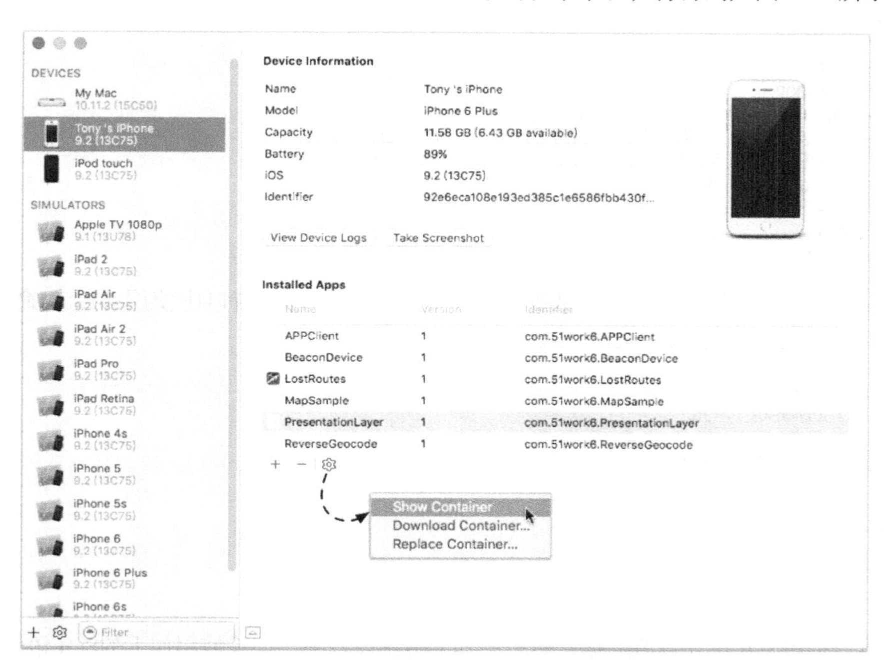

图22-44 查看设备上的应用

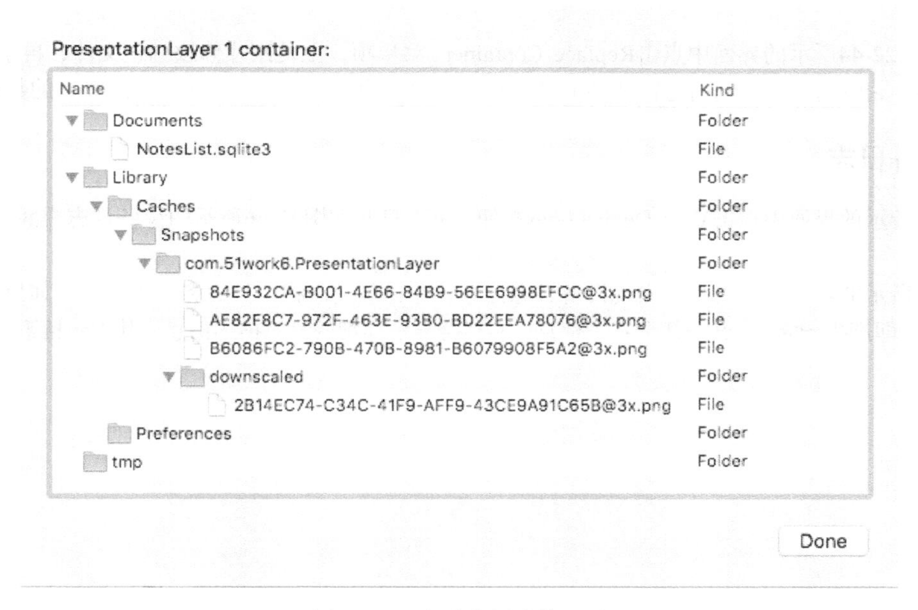

图22-45 查看应用沙箱目录

Download Container...菜单项（如图22-44所示）可以将设备中的沙箱目录复制到当前的开发电脑中。下载的数据文件命名类似于com.51work6.PresentationLayer 2017-01-18 10/46.59.504.xcappdata，xcappdata是包文件。右击该文件，将显示包内容，具体如下：

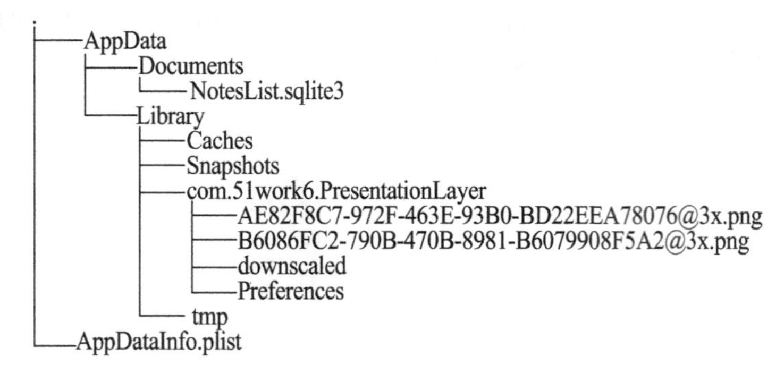

其中NotesList.sqlite3是应用中的SQLite数据文件。我们可以在终端窗口中使用sqlite3命令查看NotesList.sqlite3中的数据。下面是查询Note表中数据的SQL语句：

```
$ sqlite3 NotesList.sqlite3
SQLite version 3.7.12 2017-01-03 19:43:07
Enter ".help" for instructions
Enter SQL statements terminated with a ";"
sqlite> .tables
Note
sqlite> select * from Note;
2016-10-18 17:05:39|ab  c
2016-10-18 17:13:07|x  y  z
sqlite>
```

当然，我们可以通过SQL语句修改数据，然后将其上传到设备中。如果使用如下的SQL语句插入两个数据：

```
sqlite> insert into Note (cdate,content) values ('2016-10-28 10:13:07','我的iOS');
sqlite> insert into Note (cdate,content) values ('2016-10-20 1:13:07','我的Java');
sqlite>
```

然后在如图22-44所示的界面中点击Replace Container...菜单项，上传刚才修改的包文件，再重新运行设备上的PresentationLayer应用。从得到的运行结果中可以发现，设备的界面中显示了我们刚刚插入的数据。

22.6.2　设备日志

在图22-43所示的界面中点击View Device Logs按钮，可以打开如图22-46所示的设备日志界面，从中可以查看设备的信息。

在图22-46所示的界面中，右栏会出现设备日志信息，左栏包括进程、类型和日志时间。如果类型是Crash，代表应用崩溃（即产生异常）。点击某个日志信息，右栏会显示详细的日志描述信息，其中包括跟踪异常栈的信息输出。

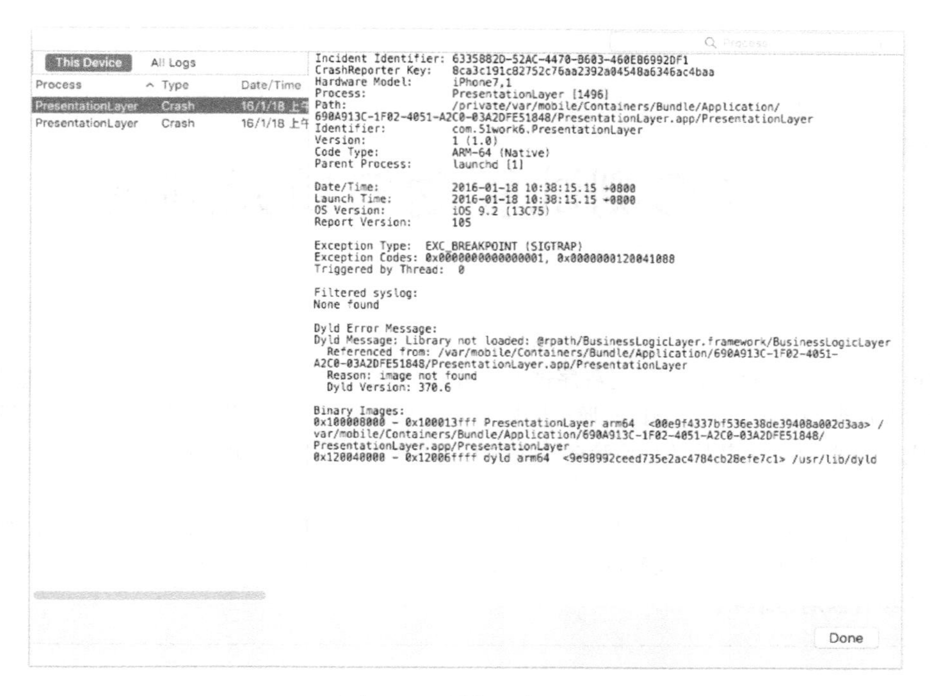

图22-46　设备日志界面

22.7　小结

在本章中，我们介绍了iOS中都有哪些调试工具并重点介绍了几个常用的工具，具体包括日志与断言的输出、异常栈报告分析等内容。接下来，我们讲解了如何在设备上调试应用，最后分析了Xcode设备管理工具的使用。

本章内容十分重要，希望大家能够好好消化这部分知识。

第 23 章

iOS测试驱动与单元测试

为了找出程序中的错误与缺陷，需要对程序进行测试。测试也是保证产品质量、安全性和完整性的重要手段，更是软件开发生命周期的重要阶段。测试按照阶段划分为：单元测试、集成测试、系统测试和回归测试。单元测试是对软件组成单元进行测试，其目的是检验软件基本组成单元的正确性，其测试对象是软件设计的最小单位——模块。

单元测试是一种白盒测试。白盒测试是一种细粒度的测试，具体到方法、函数和异常测试，因此是由能够看懂编程语言、了解程序结构的程序员发起的。为了验证程序的正确性，程序员需要编写测试程序，按照测试用例测试程序是否能够有预期的结果。

提示 测试用例是一组条件和场景，根据它来确定程序的正确性。

软件工程方法强调，没有进行单元测试并给出单元测试报告的程序是不能提交给测试团队进行其他测试的，更不能发布应用程序，否则将是一场灾难！它使我们投入更多的时间，浪费更多的资源。

单元测试也是软件工程方法学的一个重要手段。近年来，有一种很流行的软件工程方法学——极限编程（eXtreme Programming，XP），它是一种敏捷软件开发方法，其实践核心是测试驱动开发（Test-Driven Development，TDD），或称为测试先行（Test First）。本章中，我们将围绕测试驱动这个主题介绍基于测试驱动的iOS软件开发。

23.1 测试驱动的软件开发概述

单元测试是测试驱动的核心，可以通过手动测试或者自动测试来完成，但自动测试需要使用测试控件。本节中，我们就来介绍单元测试、测试驱动的概念以及iOS单元测试框架。

23.1.1 测试驱动的软件开发流程

传统的开发流程如图23-1所示，先是程序编码，然后设计单元测试用例，编写单元测试程序，进行单元测试，最后出具单元测试报告。如果测试没有通过，要根据测试修改程序代码，然后再重新走单元测试流程。

而极限编程的测试驱动的软件开发流程如图23-2所示，先是设计单元测试用例程序，编写单元测试程序，然后编写程序代码和进行单元测试，最后出具单元测试报告。如果测试没有通过，则需要根据测试结果修改程序代码，然后重新走单元测试流程。如果通过测试，再设计其他的单元测试用例。

在测试驱动开发流程中，各个阶段都是一个可逆的反复迭代过程。用例的设计可以先是功能说明书中的一个功能，然后针对该功能进行测试驱动的开发流程，再编写其他的功能。

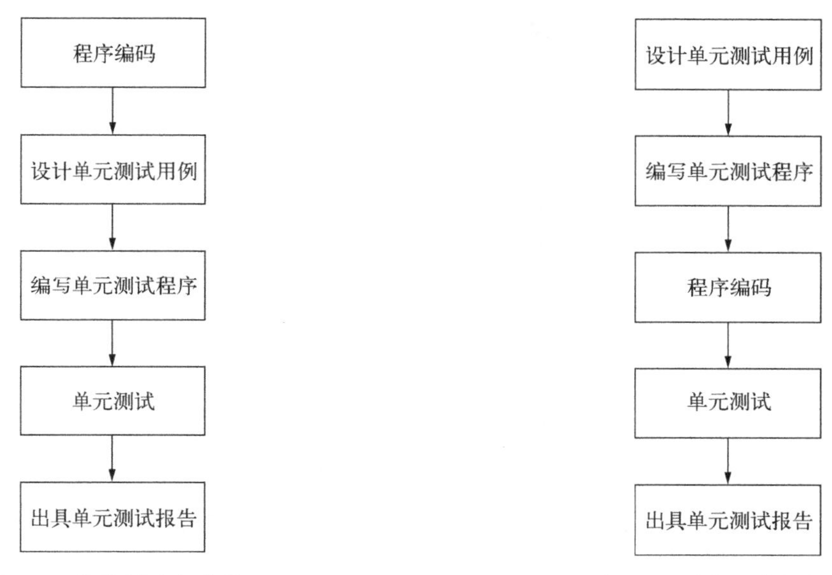

图23-1　传统的开发流程　　图23-2　测试驱动开发流程

比较这两种方式，我们可以发现测试驱动开发的优势很明显，它能够及时发现程序中的问题，从而少犯错误，减少资源浪费。

23.1.2　测试驱动的软件开发案例

下面我们通过案例介绍测试驱动的软件开发流程。该案例是一个iPhone版的计算个人月工资和奖金所得税的简易工具。根据最新的个人所得税法，月工资和奖金所得税分成7个级别，如表23-1所示，起征点为3500元，计算公式为：

月应纳个人所得税税额=月应纳税所得额 × 适用税率 – 速算扣除数

其中，月应纳税所得额=工资和奖金 – 三险一金 – 起征点。在本案例中，“三险一金”我们设为0，“起征点”是3500。

表23-1　个人所得税的7个级别

月应纳税所得额	适用税率	速算扣除数（元）
月应纳税额不超过1500元	3%	0
月应纳税额在1500元至4500元之间	10%	105
月应纳税额在4500元至9000元之间	20%	555
月应纳税额在9000元至35 000元之间	25%	1005
月应纳税额在35 000元至55 000元之间	30%	2755
月应纳税额在55 000元至80 000元之间	35%	5505
月应纳税额超过80 000元	45%	13505

该应用的设计原型草图如图23-3所示，用户可以在文本框中输入月收入总额，点击“计算”按钮，即可计算出“月应纳个人所得税税额”，并将其显示在下面的标签中。

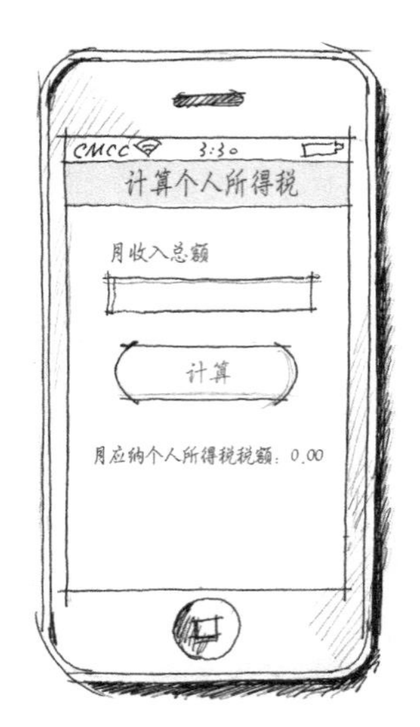

图23-3　计算个人所得税应用的设计原型草图

根据需求，我们设计了7个测试用例（如表23-2所示）。在输入条件中，我们采用常见值和边界值作为测试数据进行测试，从而来增加测试用例的测试覆盖率。

表23-2　个人所得税应用的单元测试用例

测试用例	输入条件：月收入总额（元）	输出结果：月应纳个人所得税税额（元）	说　明
1	5000	45	测试月应纳税额不超过1500元
2	8000	345	测试月应纳税额在1500元至4500元之间
3	12 500	1245	测试月应纳税额在4500元至9000元之间
4	38 500	7745	测试月应纳税额在9000元至35 000元之间
5	58 500	13 745	测试月应纳税额在35 000元至55 000元之间
6	83 500	22 495	测试月应纳税额在55 000元至80 000元之间
7	103 500	31 495	测试月应纳税额超过80 000元

个人所得税的计算是在业务逻辑层实现的，我们设计了个人所得税的业务逻辑类TaxRevenueBL，其代码如下：

```
//TaxRevenueBL.swift文件
import UIKit

class TaxRevenueBL: NSObject {

    //计算个人所得税
    func calculate(_ revenue: Double) -> Double {

        //月应纳个人所得税税额
        var tax = 0.0

        //月应纳税所得额
        let dbTaxRevenue = revenue - 3500

        //月应纳税所得额不超过1500元
        if dbTaxRevenue <= 1500 {//失败 if (dbTaxRevenue >= 1500)
            tax = dbTaxRevenue * 0.03
            //月应纳税所得额在1500元至4500元之间
        } else if dbTaxRevenue > 1500 && dbTaxRevenue <= 4500 {
            tax = dbTaxRevenue * 0.1 - 105
            //月应纳税所得额在4500元至9000元之间
        } else if dbTaxRevenue > 4500 && dbTaxRevenue <= 9000 {
            tax = dbTaxRevenue * 0.2 - 555
```

```
//TaxRevenueBL.h文件
#import <Foundation/Foundation.h>

@interface TaxRevenueBL : NSObject

-(double) calculate:(double)revenue;

@end
//TaxRevenueBL.m文件
#import "TaxRevenueBL.h"
@implementation TaxRevenueBL

//计算个人所得税
-(double) calculate:(double)revenue {

    //月应纳个人所得税税额
    double tax = 0.0;

    //月应纳税所得额
    double dbTaxRevenue = revenue - 3500;

    //月应纳税所得额不超过1500元
    if (dbTaxRevenue <= 1500) {//失败 if (dbTaxRevenue >= 1500)
```

```
            //月应纳税所得额在9000元至35000元之间
        } else if dbTaxRevenue > 9000 && dbTaxRevenue <= 35000 {
            tax = dbTaxRevenue * 0.25 - 1005
            //月应纳税所得额在35000元至55000元之间
        } else if dbTaxRevenue > 35000 && dbTaxRevenue <= 55000 {
            tax = dbTaxRevenue * 0.3 - 2755
            //月应纳税所得额在55000元至80000元之间
        } else if dbTaxRevenue > 55000 && dbTaxRevenue <= 80000 {
            tax = dbTaxRevenue * 0.35 - 5505
            //月应纳税所得额超过80000元
        } else if dbTaxRevenue > 80000 {
            tax = dbTaxRevenue * 0.45 - 13505
        }
        return tax
    }
}
```

```
        tax = dbTaxRevenue * 0.03;
        //月应纳税所得额在1500元至4500元之间
    } else if (dbTaxRevenue > 1500 && dbTaxRevenue <=4500) {
        tax = dbTaxRevenue * 0.1 - 105;
        //月应纳税所得额在4500元至9000元之间
    } else if (dbTaxRevenue > 4500 && dbTaxRevenue <=9000) {
        tax = dbTaxRevenue * 0.2 - 555;
        //月应纳税所得额在9000元至35000元之间
    } else if (dbTaxRevenue > 9000 && dbTaxRevenue <=35000) {
        tax = dbTaxRevenue * 0.25 - 1005;
        //月应纳税所得额在35000元至55000元之间
    } else if (dbTaxRevenue > 35000 && dbTaxRevenue <=55000) {
        tax = dbTaxRevenue * 0.3 - 2755;
        //月应纳税所得额在55000元至80000元之间
    } else if (dbTaxRevenue > 55000 && dbTaxRevenue <=80000) {
        tax = dbTaxRevenue * 0.35 - 5505;
        //月应纳税所得额超过80000元
    } else if (dbTaxRevenue > 80000) {
        tax = dbTaxRevenue * 0.45 - 13505;
    }
    return tax;
}
@end
```

TaxRevenueBL类中只有一个方法，见第①行的calculate方法，该方法用来计算个人所得税，其参数revenue是月应纳税所得额，返回值是月应纳税金额。

23.1.3 iOS 单元测试框架

原则上，是否使用测试框架都不会影响单元测试的结果，但是“工欲善其事，必先利其器”，使用单元测试框架更便于我们测试和分析结果。

主要的Cocoa单元测试框架有以下几种。

- OCUnit。它是开源测试框架，与Xcode 4工具集成在一起使用非常方便。测试报告以文本形式输出到输出窗口中。
- GHUnit。它是开源测试框架，可以将测试报告以应用的形式可视化输出到设备或模拟器上，也可以以文本的形式输出到输出窗口中。使用GHUnit，可以测试用OCUnit编写的测试用例。
- XCTest。它是基于OCUnit的苹果下一代测试框架。Xcode 5创建的工程默认使用XCTest作为单元测试框架。

在上述3种框架中，GHUnit属于第三方开发的，它支持图形界面测试和真机测试，而OCUnit不能进行真机测试。但是GHUnit与XCTest相比，就没有这些优势了。XCTest支持图形界面测试和真机测试，它继承了OCUnit的优点。在版本更新方面，GHUnit已经很久没有更新了，XCTest则是由苹果负责更新和维护。因此，本书会重点介绍XCTest测试框架的用法。

23.2 使用 XCTest 测试框架

我们在上一节中实现了一个基于测试驱动的软件开发案例，但是其中的单元测试程序没有采用任何框架，使用起来比较烦琐。本节中，我们将介绍Cocoa单元测试框架XCTest的用法。

23.2.1 添加 XCTest 到工程

如果我们使用的是Xcode开发工具，添加XCTest到工程中有两种方法：一种是在创建工程时添加；另一种是在现有工程中添加iOS Unit Testing Bundle目标（Target）来实现。下面我们详细介绍这两种添加过程。

1. 创建工程时添加XCTest

如果使用的是Xcode 创建工程，则在创建工程对话框中（如图23-4所示）选中Include Unit Tests，便会在创建工程目标的同时创建一个测试用例目标，如图23-5所示。之后，在生成的Xcode导航面板中，有一个PITaxTests组，这个组中的类就是生成的测试类，并且还在目标列表中生成了测试目标PITaxTests。

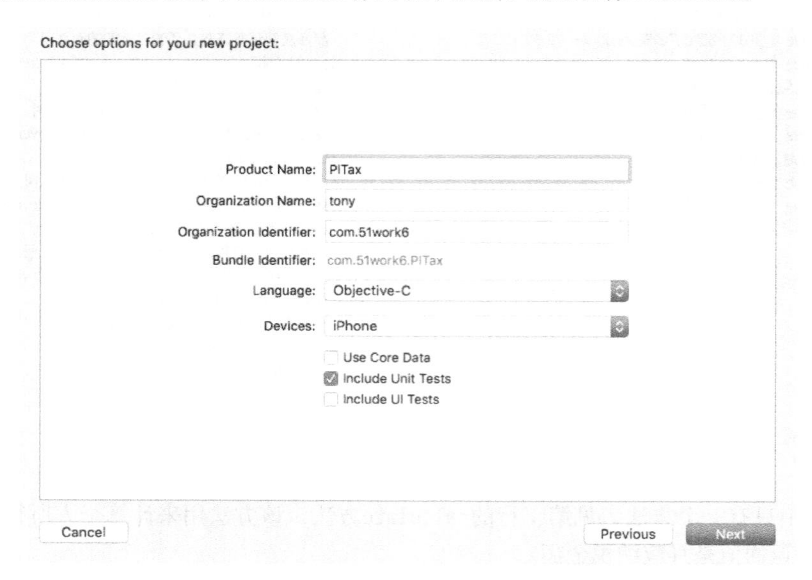

图23-4　创建工程对话框

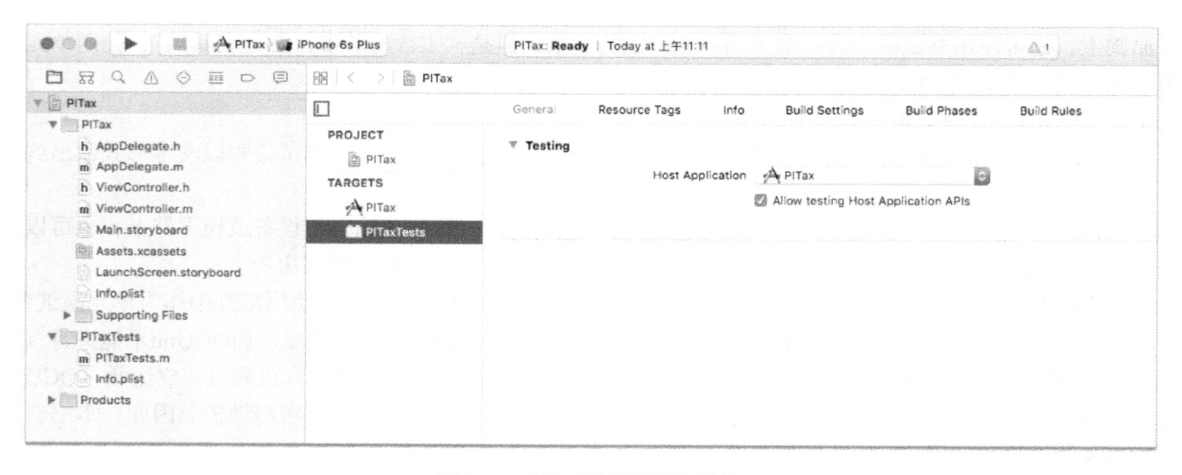

图23-5　添加的测试用例目标

2. 在现有工程中添加测试用例目标

在一个现有的工程中，选择File→New→Target...菜单项，此时打开的界面如图23-6所示，从中选择iOS→Test中的iOS Unit Testing Bundle模板。点击Next按钮，进入下一个目标相关项界面，如图23-7所示。在Product Name中输入PITaxTests；Language中选择语言，原则上要测试的类使用何种语言，测试用例类也应该采用这种语言；在Project中选择我们当前的测试工程，Target to be Tested中选择要测试的目标。

Choose a template for your new target:
iOS
watchOS
tvOS
macOS
Cross-platform
Application Extension
Intents UI Extension
Notification Content
Notification Service Extension
Photo Editing Extension
Share Extension
Shared Links Extension
Spotlight Index Extension
Sticker Pack Extension
Today Extension
Test
iOS UI Testing Bundle
iOS Unit Testing Bundle
Cancel
Previous
Next

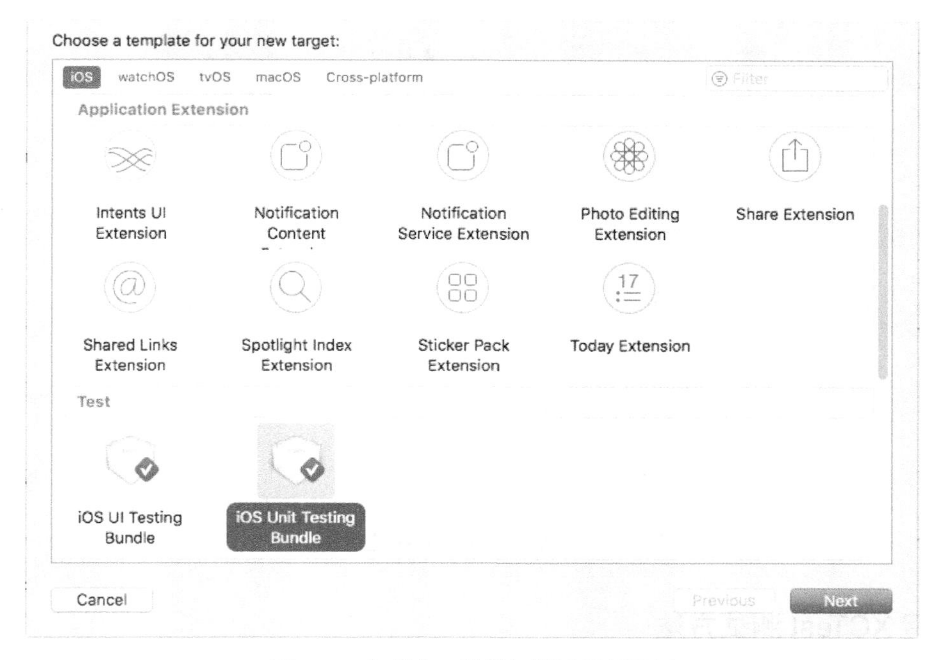

图23-6 为现有工程添加测试用例目标

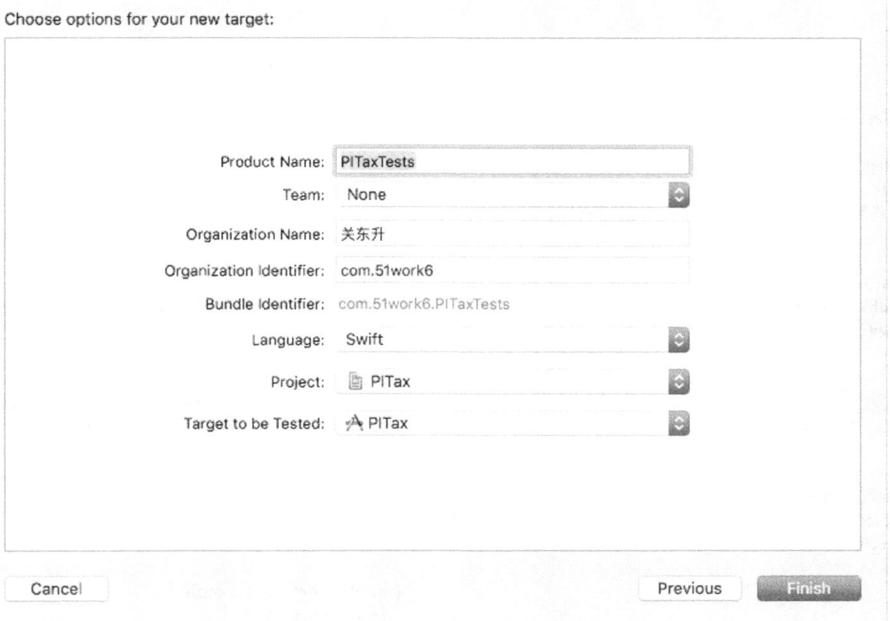

Choose options for your new target:
Product Name: PITaxTests
Team: None
Organization Name: 关东升
Organization Identifier: com.51work6
Bundle Identifier: com.51work6.PITaxTests
Language: Swift
Project: PITax
Target to be Tested: PITax
Cancel
Previous
Finish

图23-7 为现有工程添加测试用例目标

设置完相关项后，点击Finish按钮创建测试用例目标。添加完成后的工程如图23-8所示，此时在导航面板中多出了一个PITaxTests目标。

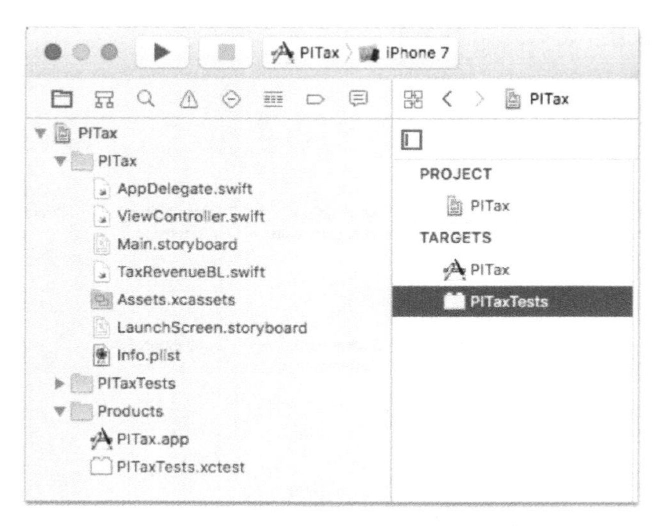

图23-8　添加测试用例目标之后的工程

23.2.2　编写 XCTest 测试方法

无论哪种方式，默认生成的测试用例类基本都是一样的。下面的代码是PITaxTests目标默认生成的PITaxTests测试用例类：

```
import UIKit
import XCTest

class PITaxTests: XCTestCase {

    override func setUp() {                                        ①
        super.setUp()
        //在此添加测试用例初始化代码，此方法在每个测试用例执行前执行
    }

    override func tearDown() {                                     ②
        //在此添加释放测试用例资源代码，此方法在每个测试用例执行后执行
        super.tearDown()
    }

    func testExample() {                                           ③
        XCTAssert(true, "Pass")                                    ④
    }

    func testPerformanceExample() {                                ⑤
        self.measureBlock() {                                      ⑥
            //性能测试代码
        }
    }

}
```

```
#import <UIKit/UIKit.h>
#import <XCTest/XCTest.h>

@interface PITaxTests: XCTestCase

@end

@implementation PITaxTests

- (void)setUp {                                                    ①
    [super setUp];
    //在此添加测试用例初始化代码，此方法在每个测试用例执行前执行
}

- (void)tearDown {                                                 ②
    //在此添加释放测试用例资源代码，此方法在每个测试用例执行后执行
    [super tearDown];
}

- (void)testExample {                                              ③
    XCTAssert(TRUE, @"Pass");                                      ④
}

- (void)testPerformanceExample {                                   ⑤
    [self measureBlock:^{                                          ⑥
        //性能测试代码
    }];
}

@end
```

作为XCTest框架的测试用例类，PITaxTests类需要继承XCTestCase父类。第①行中，setUp方法是初始化方法，在每个测试用例执行前执行。第②行中，tearDown方法是释放资源的方法，在每个测试用例执行后执行。因此，在测试类运行的生命周期中，这两个方法可能多次运行，它们的时序图如图23-9所示。

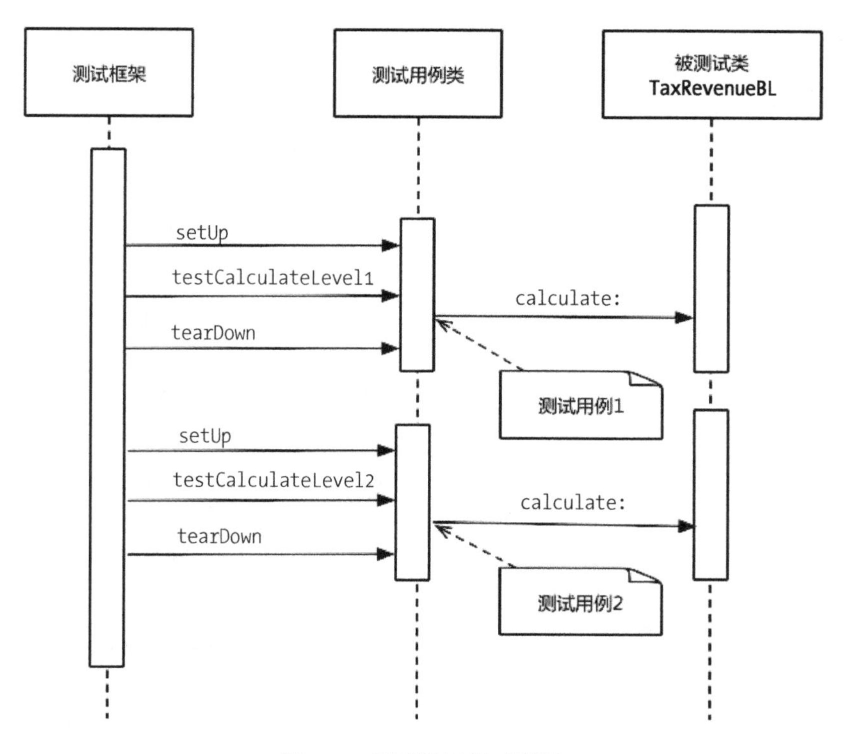

图23-9 测试类运行时序图

第③行的testExample方法是一般的测试用例方法，其中第④行的XCTAssert是XCTestCase框架定义的一个函数，它的第一个参数为布尔表达式，如果为true（Objective-C为YES或TRUE），表示断言通过，第二个参数是断言的描述。注意测试用例方法必须以test开头。

第⑤行的testPerformanceExample方法是性能测试用例方法。第⑥行中，Swift版本中的self.measureBlock(){...}语句是闭包，而Objective-C版本中的[self measureBlock:^{...}]语句是代码块，它们的作用是类似的，需要测试性能的代码编写在{...}内部。

每一个单元测试用例对应于测试类中的一个方法，下面我们还是通过计算个人所得税应用介绍它们的编写过程。我们使用PITaxTests测试用例类测试ViewController类中的calculate:方法，其测试用例的规划和设计与23.1.2节一样，具体可以参考表23-2。

测试用例类PITaxTests的代码如下：

```
//PITaxTests.swift文件
import XCTest
import PITax

class PITaxTests: XCTestCase {

    var bl: TaxRevenueBL!                                    ①

    override func setUp() {
        super.setUp()
        self.bl = TaxRevenueBL()                             ②
    }

    override func tearDown() {
        self.bl = nil                                        ③
        super.tearDown()
    }
```

```
//PITaxTests.m文件

#import <XCTest/XCTest.h>
#import "TaxRevenueBL.h"

@interface PITaxTests : XCTestCase

@property (nonatomic,strong) TaxRevenueBL *bl;                ①

@end

@implementation PITaxTests

- (void)setUp {
    [super setUp];
    self.bl = [[TaxRevenueBL alloc] init];                   ②
}
```

```swift
    //用例1：测试月应纳税额不超过1500元
    func testCalculateLevel1() {
        let dbRevenue = 5000.0
        let tax = self.bl.calculate(dbRevenue)
        XCTAssertEqual(tax, 45.0, "用例1测试失败")
    }

    //用例2：测试月应纳税额在1500元至4500元之间
    func testCalculateLevel2() {
        let dbRevenue = 8000.0
        let tax = self.bl.calculate(dbRevenue)
        XCTAssertEqual(tax, 345.0, "用例2测试失败")
    }

    //用例3：测试月应纳税额在4500元至9000元之间
    func testCalculateLevel3() {
        let dbRevenue = 12500.0
        let tax = self.bl.calculate(dbRevenue)
        XCTAssertEqual(tax, 1245.0, "用例3测试失败")
    }

    //用例4：测试月应纳税额在9000元至35000元之间
    func testCalculateLevel4() {
        let dbRevenue = 38500.0
        let tax = self.bl.calculate(dbRevenue)
        XCTAssertEqual(tax, 7745.0, "用例4测试失败")
    }

    //用例5：测试月应纳税额在35000元至55000元之间
    func testCalculateLevel5() {
        let dbRevenue = 58500.0
        let tax = self.bl.calculate(dbRevenue)
        XCTAssertEqual(tax, 13745.0, "用例5测试失败")
    }

    //用例6：测试月应纳税额在55000元至80000元之间
    func testCalculateLevel6() {
        let dbRevenue = 83500.0
        let tax = self.bl.calculate(dbRevenue)
        XCTAssertEqual(tax, 22495.0, "用例6测试失败")
    }

    //用例7：测试月应纳税额超过80000元
    func testCalculateLevel7() {
        let dbRevenue = 103500.0
        let tax = self.bl.calculate(dbRevenue)
        XCTAssertEqual(tax, 31495.0, "用例7测试失败")
    }

}
```

```objc
- (void)tearDown {
    self.bl = nil;                                              ③
    [super tearDown];
}

//用例1：测试月应纳税额不超过1500元
- (void)testCalculateLevel1 {
    double dbRevenue = 5000;
    double tax =[self.bl calculate:dbRevenue];
    XCTAssertEqual(tax, 45.0, @"用例1测试失败");
}

//用例2：测试月应纳税额在1500元至4500元之间
- (void)testCalculateLevel2 {
    double dbRevenue = 8000;
    double tax =[self.bl calculate:dbRevenue];
    XCTAssertEqual(tax, 345.0, @"用例2测试失败");
}

//用例3：测试月应纳税额在4500元至9000元之间
- (void)testCalculateLevel3 {
    double dbRevenue = 12500;
    double tax =[self.bl calculate:dbRevenue];
    XCTAssertEqual(tax, 1245.0, @"用例3测试失败");
}

//用例4：测试月应纳税额在9000元至35000元之间
- (void)testCalculateLevel4 {
    double dbRevenue = 38500;
    double tax =[self.bl calculate:dbRevenue];
    XCTAssertEqual(tax, 7745.0, @"用例4测试失败");
}

//用例5：测试月应纳税额在35000元至55000元之间
- (void)testCalculateLevel5 {
    double dbRevenue = 58500;
    double tax =[self.bl calculate:dbRevenue];
    XCTAssertEqual(tax, 13745.0, @"用例5测试失败");
}

//用例6：测试月应纳税额在55000元至80000元之间
- (void)testCalculateLevel6 {
    double dbRevenue = 83500;
    double tax =[self.bl calculate:dbRevenue];
    XCTAssertEqual(tax, 22495.0, @"用例6测试失败");
}

//用例7：测试月应纳税额超过80000元
- (void)testCalculateLevel7 {
    double dbRevenue = 103500;
    double tax =[self.bl calculate:dbRevenue];
    XCTAssertEqual(tax, 31495.0, @"用例7测试失败");
}

@end
```

在上述代码中，第①行定义了TaxRevenueBL类型的bl属性，在setUp方法中初始化bl（见第②行），在tearDown方法中释放bl（见第③行）。测试方法testCalculateLevel1至 testCalculateLevel7对应测试用例1至测试用例7，测试方法中的XCTAssertEqual是XCTest框架中定义的断言函数。在XCTest框架中，与断言有关的常用函数如表23-3所示。

表23-3 XCTest框架定义的常用断言函数

函 数	说 明
XCTAssertEqualObjects	当两个对象内容不相等或者是其中一个对象为nil时，断言失败
XCTAssertEqual	当参数1不等于参数2时断言失败，用于C中的基本数据测试
XCTAssertNil	当参数不是nil时，断言失败
XCTAssertNotNil	当参数是nil时，断言失败
XCTAssertTrue	当表达式为false时，断言失败
XCTAssertFalse	当表达式为true时，断言失败
XCTAssertThrows	如果表达式没有抛出异常，则断言失败
XCTAssertNoThrow	如果表达式抛出异常，则断言失败

23.2.3 运行测试用例目标

在Xcode 中，我们有多种方法可以运行测试用例目标。多个测试用例目标可以一起运行，也可以分别运行。此外，既可以运行测试用例目标中的单个测试类，也可以只运行测试类中的单个测试方法。总之，非常灵活。

运行这个测试用例目标的方法与运行应用不同，选择方案（Scheme）中运行的平台（见图23-10）后，可以通过菜单项Product→Test运行，也可以通过点击工具栏中的Test按钮（位于Run下拉菜单中）运行，如图23-11所示，还可以使用快捷键command+u运行。

提示

编译测试用例目标时，可以通过Xcode菜单Product→Build for→Testing，或通过快捷键shift+command+u实现编译。

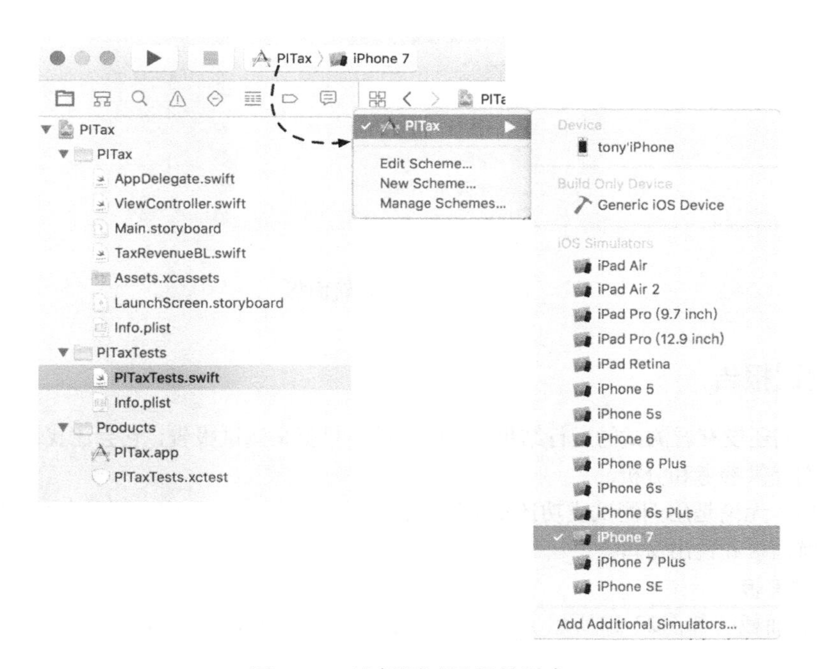

图23-10 选择测试运行的平台

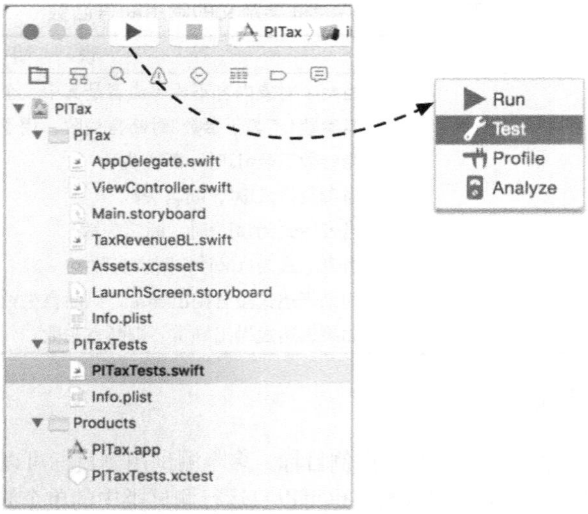

图23-11　运行测试用例目标

当然，我们也可以单独运行测试类。打开导航面板中的测试导航面板，如图23-12所示，在这些测试导航面板中，将鼠标放在测试用例类后面或测试方法后面，这时候会出现一个小箭头，点击这个小箭头就可以运行这个测试类了。此外，我们也可以将鼠标放在具体的测试方法后面，单独运行其中的一个测试方法。

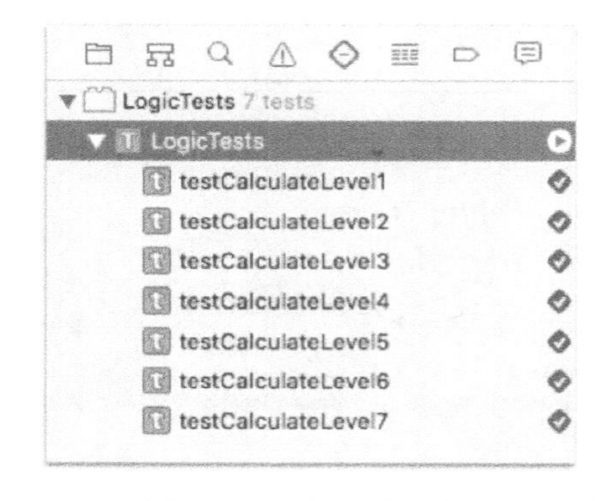

图23-12　测试导航面板

23.2.4　分析测试报告

到目前为止，我们还没有看测试的运行结果。这些测试结果就是测试报告，它会比我们自己编写测试类输出更多的信息，为我们提供参考和分析。

测试程序运行后，无论是用例测试成功还是失败，都会输出信息，这些信息可以在3个地方看到：错误警告导航面板、测试导航面板和输出窗口。

1. 错误警告导航面板

在错误警告导航面板（如图23-13所示）中看到的是用例测试失败信息，用例测试成功的信息不会在这里输出。

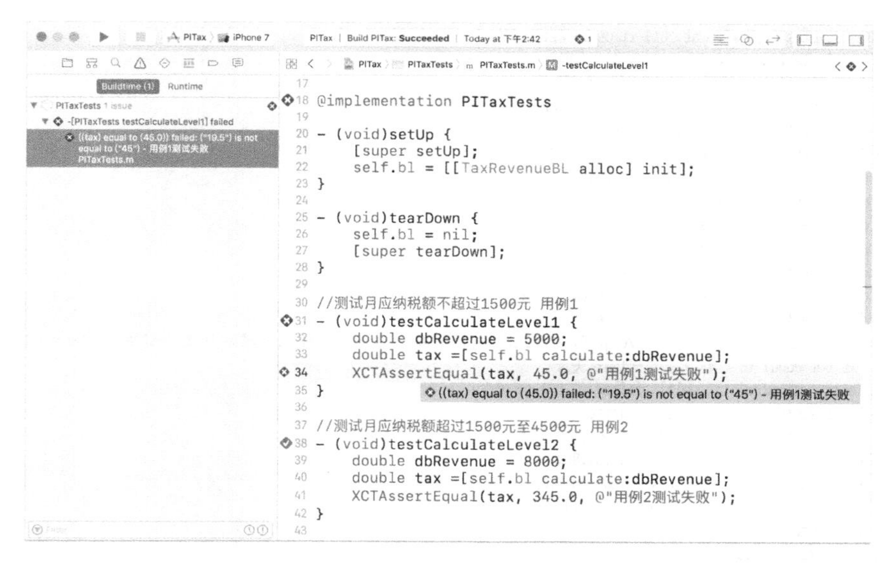

图23-13　错误警告导航面板

在图23-13中，显示的警告信息如下：

```
PITaxTests.m: test failure: -[PITaxTests testCalculateLevel1] failed: ((tax) equal to (45.0)) failed: ("19.5") is not equal to ("45") - 用例1测试失败
```

这些信息告诉我们testCalculateLevel1测试方法中的期望值和实际值不一致，断言失败。

2. 测试导航面板

在测试导航面板中，我们可以看到更多信息，包括输出成功和失败的信息，如图23-14所示。其中绿色的◆图标表示测试成功，红色的◆图标代表测试失败，点击这些图标可以导航到测试方法。

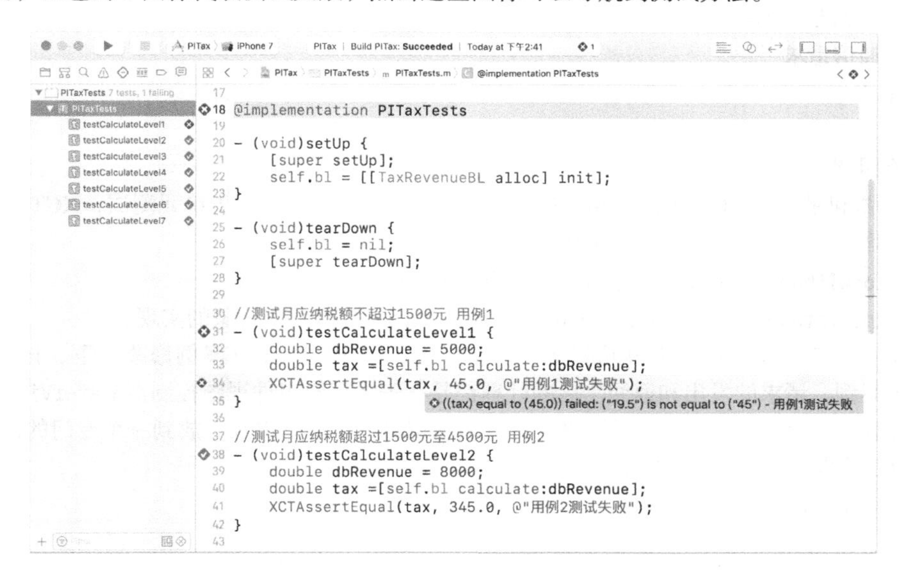

图23-14　测试导航面板

3. 输出窗口

我们也可以在输出窗口中看到测试信息，下面的内容为运行测试用例PITaxTests所输出的内容：

```
Test Suite 'Selected tests' started at 2016-11-05 14:42:09.824
```

```
Test Suite 'PITaxTests.xctest' started at 2016-11-05 14:42:09.826
Test Suite 'PITaxTests' started at 2016-11-05 14:42:09.827
Test Case '-[PITaxTests testCalculateLevel2]' started.
Test Case '-[PITaxTests testCalculateLevel2]' passed (0.002 seconds).
Test Case '-[PITaxTests testCalculateLevel6]' started.
Test Case '-[PITaxTests testCalculateLevel6]' passed (0.001 seconds).
Test Case '-[PITaxTests testCalculateLevel3]' started.
Test Case '-[PITaxTests testCalculateLevel3]' passed (0.000 seconds).
Test Case '-[PITaxTests testCalculateLevel7]' started.
Test Case '-[PITaxTests testCalculateLevel7]' passed (0.005 seconds).
Test Case '-[PITaxTests testCalculateLevel4]' started.
Test Case '-[PITaxTests testCalculateLevel4]' passed (0.001 seconds).
Test Case '-[PITaxTests testCalculateLevel1]' started.
/Users/tonyguan/Documents/iOS5版/code/ch23/23.2/ObjC/PITax/PITaxTests/PITaxTests.m:34: error: -[PITaxTests testCalculateLevel1] : ((tax) equal to (45.0))
➥failed: ("19.5") is not equal to ("45") - 用例1测试失败
Test Case '-[PITaxTests testCalculateLevel1]' failed (0.002 seconds).
Test Case '-[PITaxTests testCalculateLevel5]' started.
Test Case '-[PITaxTests testCalculateLevel5]' passed (0.022 seconds).
Test Suite 'PITaxTests' failed at 2016-11-05 14:42:09.864.
     Executed 7 tests, with 1 failure (0 unexpected) in 0.032 (0.038) seconds
Test Suite 'PITaxTests.xctest' failed at 2016-11-05 14:42:09.865.
     Executed 7 tests, with 1 failure (0 unexpected) in 0.032 (0.039) seconds
Test Suite 'Selected tests' failed at 2016-11-05 14:42:09.866.
     Executed 7 tests, with 1 failure (0 unexpected) in 0.032 (0.042) seconds
```

我们会看到这些信息以Test Suite或Test Case开头，其中以Test Suite开头的是测试用例集合，就是一个测试类。以Test Case开头的是测试用例，每个测试用例都包括一个开始日志和一个结束日志。此外，信息中还包括运行每一测试用例的时间以及测试用例集合的时间。在Test Case…中，passed说明测试通过。

23.3　异步单元测试

在Xcode 5及其之前的版本中，XCTest测试框架不能进行异步单元测试，异步单元测试是单元测试的盲区。需要进行异步单元测试的应用包括：

- 异步网络通信请求
- 后台处理
- 闭包调用
- 委托对象回调

Xcode 6之后提供的XCTest框架已经能够进行异步单元测试了。针对异步单元测试，XCTest框架增加的API添加了期望对象类XCTestExpectation和XCTestCase的实例方法，具体如下。

- **expectationWithDescription:**。返回一个期望对象XCTestExpectation。
- **waitForExpectationsWithTimeout: handler:**。XCTestCase等待期望对象的实现。

为了方便介绍XCTest框架的异步单元测试，我们将16.2.4节的MyNotes案例修改一下，该案例是一个异步HTTP POST请求案例，请求的发出和返回之后的处理都是在表示逻辑组件视图控制器MasterViewController中实现的，这种做法不符合设计规范，我们需要将它们从表示逻辑组件剥离出来，放到一个专门的HTTP请求类中进行处理，这样的设计也便于单元测试的实施。

剥离出来的HTTP请求类的代码如下：

```
//NotesURLConnection.swift代码
import Foundation

protocol NotesURLConnectionDelegate {                                ①

    //成功查询所有数据的方法
    func findAllFinished(_ res: NSDictionary)

    //查询所有数据失败的方法
```

```
//NotesURLConnection.h代码
#import <Foundation/Foundation.h>

@protocol NotesURLConnectionDelegate                                 ①

//成功查询所有数据的方法
- (void)findAllFinished:(NSDictionary *)res;

//查询所有数据失败的方法
```

```
    func findAllFailed(_ error: NSError)

}

class NotesURLConnection: NSObject, NSURLConnectionDataDelegate {          ②

    //保存数据列表
    var listData: NSMutableArray!

    //接收从服务器返回数据
    var datas: NSMutableData!

    var delegate: NotesURLConnectionDelegate!                              ③

    //查询所有数据的方法
    func findAll() {
        let strURL = "http://www.51work6.com/service/mynotes/WebService.php"

        let url = URL(string: strURL)!

        let post = NSString(format: "email=%@&type=%@&action=%@",
        ➥"<你的51work6.com用户邮箱>", "JSON", "query")

        let postData: Data = post.data(using: String.Encoding.utf8.rawValue)!

        let mutableURLrequest = NSMutableURLRequest(url: url)
        mutableURLrequest.httpMethod = "POST"
        mutableURLrequest.httpBody = postData
        let request =  mutableURLrequest as URLRequest

        let defaultConfig = URLSessionConfiguration.default
        let session = URLSession(configuration: defaultConfig,
        ➥delegate: nil, delegateQueue: OperationQueue.main)

        let task: URLSessionDataTask = session.dataTask(with: request,
        ➥completionHandler: {
            (data, response, error) -> Void in
            NSLog("请求完成...")
            if error == nil {
                do {
                    let resDict = try JSONSerialization.jsonObject(with:
                    ➥data!,options: JSONSerialization.ReadingOptions.
                    ➥allowFragments) as! NSDictionary

                    self.delegate.findAllFinished(resDict)                 ④

                } catch {
                    NSLog("返回数据解析失败")
                }
            } else {
                NSLog("error : %@", error!.localizedDescription)
                self.delegate.findAllFailed(error as! NSError)             ⑤
            }
        })

        task.resume()
    }

}
```

```
- (void)findAllFailed:(NSError *)error;

@end

@interface NotesURLConnection : NSObject                                   ②

//保存数据列表
@property(nonatomic, strong) NSMutableArray *listData;

@property(nonatomic, weak) id <NotesURLConnectionDelegate> delegate;       ③

//查询所有数据的方法
- (void)findAll;

@end

//NotesURLConnection.m代码

#import "NotesURLConnection.h"

@implementation NotesURLConnection

//查询所有数据的方法
- (void)findAll {

    NSString *strURL =
    ➥@"http://www.51work6.com/service/mynotes/WebService.php";

    NSURL *url = [NSURL URLWithString:strURL];

    NSString *post = [NSString
stringWithFormat:@"email=%@&type=%@&action=%@",
    ➥@"<你的51work6.com用户邮箱>", @"JSON", @"query"];
    NSData *postData = [post dataUsingEncoding:NSUTF8StringEncoding];

    NSMutableURLRequest *request = [NSMutableURLRequest requestWithURL:url];
    [request setHTTPMethod:@"POST"];
    [request setHTTPBody:postData];

    NSURLSessionConfiguration *defaultConfig =
    ➥[NSURLSessionConfiguration defaultSessionConfiguration];
    NSURLSession *session = [NSURLSession sessionWithConfiguration:
    ➥default Config delegate:nil delegateQueue:[NSOperationQueue mainQueue]];

    NSURLSessionDataTask *task = [session dataTaskWithRequest:request
    ➥completionHandler: ^(NSData *data, NSURLResponse *response, NSError
    ➥*error) {
        NSLog(@"请求完成...");
        if (!error) {
            NSDictionary *dict = [NSJSONSerialization JSONObjectWith
            ➥Data:data options:NSJSONReadingAllowFragments
            ➥error:nil];
            [self.delegate findAllFinished:dict];                          ④
            } else {
        NSLog(@"error : %@", error.localizedDescription);
            [self.delegate findAllFailed:error];                           ⑤
        }
    }];

    [task resume];
}

@end
```

上述代码基于委托设计模式的HTTP请求对象NotesURLConnection，其中第①行定义了委托协议NotesURL-ConnectionDelegate，第②行定义了HTTP请求对象类NotesURLConnection，第③行声明遵守NSURLConnection

DataDelegate委托协议类型的delegate属性。

如果异步请求成功，则会回调委托对象的findAllFinished方法，详见第④行。如果异步请求失败，会回调委托对象的findAllFailed方法，详见第⑤行。

测试用例代码：

```
//MyNotesTests.swift文件
import XCTest

class MyNotesTests: XCTestCase, NotesURLConnectionDelegate {

    var expectation: XCTestExpectation!                                    ①

    override func setUp() {
        super.setUp()
    }

    override func tearDown() {
        super.tearDown()
    }

    func testFindAll() {                                                   ②

        self.expectation = self.expectation(description: "Request NotesURL")
                                                                           ③
        let conn = NotesURLConnection()                                    ④
        conn.delegate = self                                               ⑤
        //开始查询
        conn.findAll()
        self.waitForExpectations(timeout: 5.0, handler: nil)               ⑥
    }

    //MARK: -- 成功查询所有数据的方法
    func findAllFinished(_ res: NSDictionary) {                            ⑦

        self.expectation.fulfill()                                         ⑧
        XCTAssert(true, "成功查询所有数据的方法")

        let resultCode: NSNumber = res.object(forKey: "ResultCode") as!
        ➥NSNumber
        XCTAssertEqual(resultCode.intValue, 0)

        let objects = res.object(forKey: "Record") as! NSArray
        XCTAssertEqual(objects.count, 4)

    }

    //MARK: -- 查询所有数据失败的方法
    func findAllFailed(_ error: NSError) {                                 ⑨
        self.expectation.fulfill()
        XCTAssertNotNil(error)                                             ⑩
        XCTFail("查询所有数据失败的方法")
    }
}
```

```
//MyNotesTests.m文件
#import <XCTest/XCTest.h>
#import "NotesURLConnection.h"

@interface MyNotesTests : XCTestCase <NotesURLConnectionDelegate>

@property(strong, nonatomic) XCTestExpectation *expectation;               ①

@end

@implementation MyNotesTests

- (void)setUp {
    [super setUp];
}

- (void)tearDown {
    [super tearDown];
}

- (void)testFindAll {                                                      ②

    self.expectation = [self expectationWithDescription:@"Request NotesURL"]; ③
    NotesURLConnection *conn = [[NotesURLConnection alloc] init];          ④
    conn.delegate = self;                                                  ⑤

    //开始查询
    [conn findAll];
    [self waitForExpectationsWithTimeout:5.0 handler:nil];                 ⑥
}


#pragma mark -- NotesURLConnection委托协议

//成功查询所有数据的方法
- (void)findAllFinished:(NSDictionary *)res {                              ⑦

    [self.expectation fulfill];                                            ⑧
    XCTAssert(true, @"成功查询所有数据的方法");

    NSNumber *resultCode = (NSNumber *) [res objectForKey:@"ResultCode"];
    XCTAssertEqual([resultCode integerValue], 0);

    NSMutableArray *objects = [res objectForKey:@"Record"];
    XCTAssertEqual(objects.count, 4);
}

//查询所有数据失败的方法
- (void)findAllFailed:(NSError *)error {                                   ⑨
    [self.expectation fulfill];                                            ⑩
    XCTAssertNotNil(error);
    XCTFail(@"查询所有数据失败的方法");
}

@end
```

上述代码用于定义XCTestCase类MyNotesTests。为了进行异步测试，MyNotesTests类需要声明遵守NotesURLConnectionDelegate协议。第①行用于定义期望对象XCTestExpectation的属性expectation。

第②行的testFindAll方法是测试用例方法，它必须以test开头。第③行创建并初始化期望对象属性

expectation。第④行用于创建NotesURLConnection对象，第⑤行用于将当前控制器作为NotesURLConnection的委托对象。第⑥行用于设置期望对象延迟5秒，该行代码会使测试暂停，运行Run Loop[①]循环，直到超时或所有的期望对象完成实现。这个超时的设计是XCTest能够实现异步单元测试的关键，没有超时等待，程序就会马上结束。

最后运行测试，在输出窗口中看到的测试日志信息如下：

```
2016-11-05 15:09:27.045 MyNotes[4701:208195] 请求完成...
Test Suite 'All tests' started at 2016-11-05 15:09:27.197
Test Suite 'MyNotesTests.xctest' started at 2016-11-05 15:09:27.198
Test Suite 'MyNotesTests' started at 2016-11-05 15:09:27.199
Test Case '-[MyNotesTests testFindAll]' started.
2016-11-05 15:09:27.387 MyNotes[4701:208195] 请求完成...
Test Case '-[MyNotesTests testFindAll]' passed (0.191 seconds).
Test Suite 'MyNotesTests' passed at 2016-11-05 15:09:27.391.
     Executed 1 test, with 0 failures (0 unexpected) in 0.191 (0.192) seconds
Test Suite 'MyNotesTests.xctest' passed at 2016-11-05 15:09:27.392.
     Executed 1 test, with 0 failures (0 unexpected) in 0.191 (0.194) seconds
Test Suite 'All tests' passed at 2016-11-05 15:09:27.397.
     Executed 1 test, with 0 failures (0 unexpected) in 0.191 (0.199) seconds
```

23.4 性能测试

在Xcode 6及其之后的版本中，XCTest测试框架中还增加了性能测试的功能，我们可以将要测试的代码放入到一个测试代码块中执行10次，然后给出测试报告。

下面我们通过一个实例介绍一下如何使用XCTest进行性能测试。在15.2节的实例中，我们是从XML文件中解析数据显示到表视图中，现在就来测试一下15.2节中使用NSXML和TBXML这两种技术解析XML在性能上有多大的差别。

由于TBXML是第三方框架，我们需要设置测试用例，具体过程请参考15.2.2节。需要注意的是，基于Swift的测试用例目标与应用目标类似，也需要桥接头文件，如图23-15所示，可以与应用目标共用桥接头文件MyNotes-Bridging-Header.h。

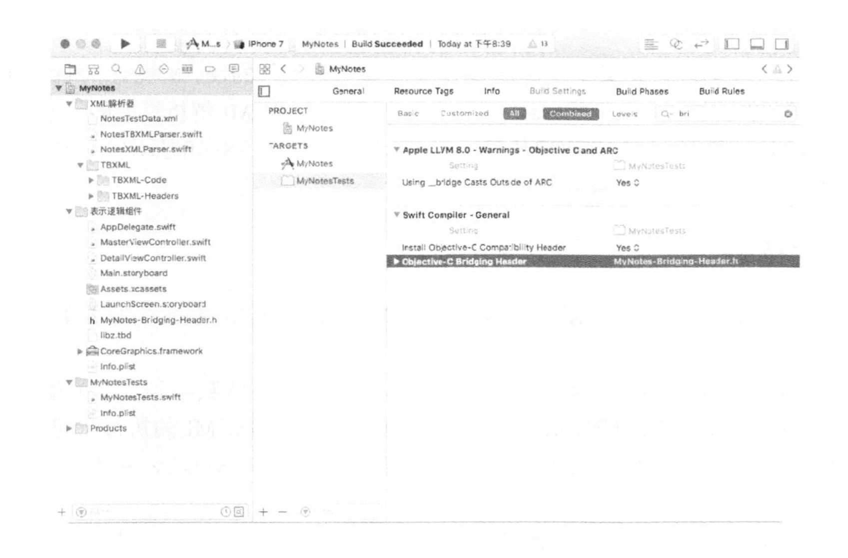

图23-15 测试用例目标桥接头文件

① Run Loop是iOS事件处理循环机制，用来不停地调配工作以及处理输入事件。

23.4.1　测试用例代码

下面我们看看MyNotesTests中的MyNotesTests文件，相关代码如下：

```
//MyNotesTests.swift文件
import XCTest

class MyNotesTests: XCTestCase {

    override func setUp() {
        super.setUp()
    }

    override func tearDown() {
        super.tearDown()
    }

    func testNotesTBXMLParser() {                                    ①
        self.measureBlock() {                                         ②
            let parser = NotesTBXMLParser()
            parser.start()
        }
    }

    func testNotesXMLParser() {                                       ③
        self.measureBlock() {
            let parser = NotesXMLParser()
            parser.start()

        }
    }
}
```

```
//MyNotesTests.h文件
#import <XCTest/XCTest.h>

#import "NotesTBXMLParser.h"

@interface MyNotesTests : XCTestCase

@end

@implementation MyNotesTests

- (void)setUp {
    [super setUp];
}

- (void)tearDown {
    [super tearDown];
}

- (void)testNotesTBXMLParser {                                         ①
    [self measureBlock:^{                                              ②
        NotesTBXMLParser* parser =[[NotesTBXMLParser alloc] init];
        [parser start];
    }];
}

- (void)testNotesXMLParser {                                           ③
    [self measureBlock:^{
        NotesXMLParser* parser = [[NotesXMLParser alloc] init];
        [parser start];
    }];
}

@end
```

上述代码中，第①行和第③行定义了性能测试方法，其中第①行的testNotesTBXMLParser方法是为了测试TBXML解析性能的，第③行的testNotesXMLParser方法是为了测试NSXML解析性能的。

在性能测试的过程中，需要测试的代码要放到measureBlock代码块（Swift是闭包）中，见第②行代码所示。measureBlock代码块（Swift是闭包）会被反复执行10次。

23.4.2　分析测试结果

我们为测试准备了一万条数据（见工程中的NotesTestData.xml文件），性能测试时如果数据量太少，测试结果差别不明显。

运行测试用例后的界面如图23-16所示，每个测试方法后面都有一些数字和文字描述信息，其中Time是执行需要的时间，越短说明速度越快，TBXML的执行时间为0.052sec，NSXML的执行时间为0.128sec，显而易见TBXML解析XML的速度要快。STDEV基于样本估算标准偏差[①]，0.1% ≤STDEV ≤ 10%说明测试可以通过，否则不会通过。

在图23-16所示的界面中，我们只能看到测试的概括结果。如果想查看每次测试的详细结果，可以在输出日志中查看。性能测试日志如下：

---

① 标准偏差（Std Dev，Standard Deviation），统计学术语，用以衡量数据值偏离算术平均值的程度。标准偏差越小，这些值偏离平均值就越少，反之亦然。

```
2017-03-08 20:38:05.514 MyNotes[12994:1097698] NSXML解析完成...
Test Suite 'All tests' started at 2017-03-08 20:38:05.840
Test Suite 'MyNotesTests.xctest' started at 2017-03-08 20:38:05.841
Test Suite 'MyNotesTests' started at 2017-03-08 20:38:05.843
Test Case '-[MyNotesTests testNotesTBXMLParser]' started.
2017-03-08 20:38:06.224 MyNotes[12994:1097698] TBXML解析完成...
2017-03-08 20:38:06.275 MyNotes[12994:1097698] TBXML解析完成...
2017-03-08 20:38:06.331 MyNotes[12994:1097698] TBXML解析完成...
2017-03-08 20:38:06.390 MyNotes[12994:1097698] TBXML解析完成...
2017-03-08 20:38:06.433 MyNotes[12994:1097698] TBXML解析完成...
2017-03-08 20:38:06.503 MyNotes[12994:1097698] TBXML解析完成...
2017-03-08 20:38:06.569 MyNotes[12994:1097698] TBXML解析完成...
2017-03-08 20:38:06.643 MyNotes[12994:1097698] TBXML解析完成...
2017-03-08 20:38:06.714 MyNotes[12994:1097698] TBXML解析完成...
2017-03-08 20:38:06.755 MyNotes[12994:1097698] TBXML解析完成...
/Users/tony/Documents/iOS5/修订/code/ch23/13.2/ObjC/MyNotes/MyNotesTests/MyNotesTests.m:34: Test Case '-[MyNotesTests testNotesTBXMLParser]' measured
[Time, seconds] average: 0.060, relative standard deviation: 20.365%, values: [0.066093, 0.050351, 0.061817, 0.055497, 0.039731, 0.071552, 0.069297, 0.079359,
0.059930, 0.042251], performanceMetricID:com.apple.XCTPerformanceMetric_WallClockTime, baselineName: "", baselineAverage: , maxPercentRegression: 10.000%,
maxPercentRelativeStandardDeviation: 10.000%, maxRegression: 0.100, maxStandardDeviation: 0.100
Test Case '-[MyNotesTests testNotesTBXMLParser]' passed (0.931 seconds).
Test Case '-[MyNotesTests testNotesXMLParser]' started.
2017-03-08 20:38:07.221 MyNotes[12994:1097698] NSXML解析完成...
2017-03-08 20:38:07.372 MyNotes[12994:1097698] NSXML解析完成...
2017-03-08 20:38:07.561 MyNotes[12994:1097698] NSXML解析完成...
2017-03-08 20:38:07.693 MyNotes[12994:1097698] NSXML解析完成...
2017-03-08 20:38:07.873 MyNotes[12994:1097698] NSXML解析完成...
2017-03-08 20:38:08.014 MyNotes[12994:1097698] NSXML解析完成...
2017-03-08 20:38:08.124 MyNotes[12994:1097698] NSXML解析完成...
2017-03-08 20:38:08.246 MyNotes[12994:1097698] NSXML解析完成...
2017-03-08 20:38:08.348 MyNotes[12994:1097698] NSXML解析完成...
2017-03-08 20:38:08.451 MyNotes[12994:1097698] NSXML解析完成...
/.../MyNotesTests/MyNotesTests.m:41: Test Case '-[MyNotesTests testNotesXMLParser]' measured [Time, seconds] average: 0.136, relative standard deviation:
21.027%, values: [0.129907, 0.150545, 0.188758, 0.133044, 0.180979, 0.138490, 0.110170, 0.122989, 0.100988, 0.103738],
performanceMetricID:com.apple.XCTPerformanceMetric_WallClockTime, baselineName: "", baselineAverage: , maxPercentRegression: 10.000%,
maxPercentRelativeStandardDeviation: 10.000%, maxRegression: 0.100, maxStandardDeviation: 0.100
Test Case '-[MyNotesTests testNotesXMLParser]' passed (1.683 seconds).
Test Suite 'MyNotesTests' passed at 2017-03-08 20:38:08.459.
    Executed 2 tests, with 0 failures (0 unexpected) in 2.614 (2.617) seconds
Test Suite 'MyNotesTests.xctest' passed at 2017-03-08 20:38:08.460.
    Executed 2 tests, with 0 failures (0 unexpected) in 2.614 (2.619) seconds
Test Suite 'All tests' passed at 2017-03-08 20:38:08.461.
    Executed 2 tests, with 0 failures (0 unexpected) in 2.614 (2.621) seconds


Test session log:
    /Users/tony/Library/Developer/Xcode/DerivedData/MyNotes-hbakcbarntksmmaiaryoilgpsbky/Logs/Test/456C53BA-40C5-4121-928D-EBA29B3AA0C9/Session-
MyNotesTests-2017-03-08_203757-qefYAZ.log
```

其中`values`是每次测试的时间，从中可见很多详细信息。文字表述的测试结果不是很友好，Xcode还提供测试结果的可视化图形界面。在图23-16所示的界面中，点击测试结果的那条灰线，会弹出如图23-17所示的详细结果对话框，其中黑色横线就是时间平均值，10个柱状图是10次测试的结果，点击下面的数字可以查看某一次的具体运行时间。

由于性能测试与其他单元测试不同，没有绝对的对与错，那么如何来衡量测试成功还是失败呢？这时我们需要为测试结果设立一个标准，这个标准就是Baseline（基线），每次运行的平均值（Average）与基线进行比较，计算公式Result（结果）= (Average - Baseline) / Baseline，如果 ± 0.1% ≤ Result ≤ ± 10%说明测试通过，否则不会通过。图23-18所示的Result为4.241%是比较好的运行结果。

```
27 }
28
29 - (void)tearDown {
30     [super tearDown];
31 }
32
33 - (void)testNotesTBXMLParser {
34     [self measureBlock:^{                              Time: 0.060 sec (21% STDEV)
35         NotesTBXMLParser* parser = [[NotesTBXMLParser alloc] init];
36         [parser start];
37     }];
38 }
39
40 - (void)testNotesXMLParser {
41     [self measureBlock:^{                              Time: 0.136 sec (22% STDEV)
42         NotesXMLParser* parser = [[NotesXMLParser alloc] init];
43         [parser start];
44     }];
45 }
46
47 @end
48
```

```
baselineName: "", baselineAverage: , maxPercentRegression: 10.000%,
maxPercentRelativeStandardDeviation: 10.000%, maxRegression: 0.100,
maxStandardDeviation: 0.100
Test Case '-[MyNotesTests testNotesXMLParser]' passed (1.683 seconds).
Test Suite 'MyNotesTests' passed at 2017-03-08 20:38:08.459.
     Executed 2 tests, with 0 failures (0 unexpected) in 2.614 (2.617) seconds
Test Suite 'MyNotesTests.xctest' passed at 2017-03-08 20:38:08.460.
     Executed 2 tests, with 0 failures (0 unexpected) in 2.614 (2.619) seconds
Test Suite 'All tests' passed at 2017-03-08 20:38:08.461.
     Executed 2 tests, with 0 failures (0 unexpected) in 2.614 (2.621) seconds

Test session log:
```

图23-16　运行测试后的界面

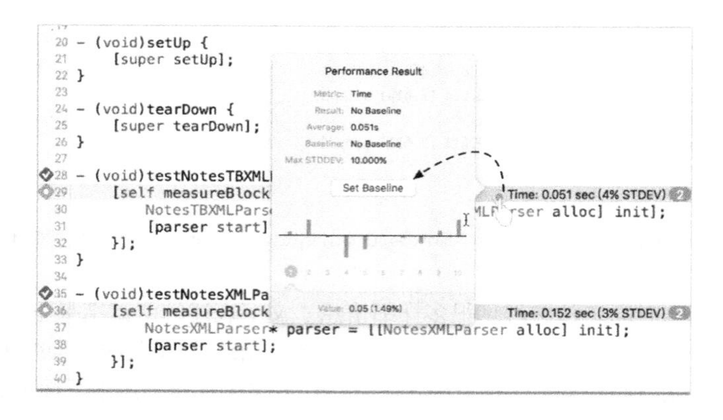

图23-17　查询性能测试结果

那么，如何设置基线呢？如果此前没有设置过基线，如图23-17所示，Result为No Baseline，点击Set Baseline按钮，则可编辑基线数值。如果已经设置过基线，可以按图23-18所示，点击Edit按钮编辑基线数值。基线一旦处于编辑状态，如图23-19所示，会出现Accept、Cancel和Save按钮，点击Accept按钮是接受当前平均值作为基线数值，Cancel是取消当前修改，Save是保存此次修改。

图23-18 设置基线运行性能测试结果

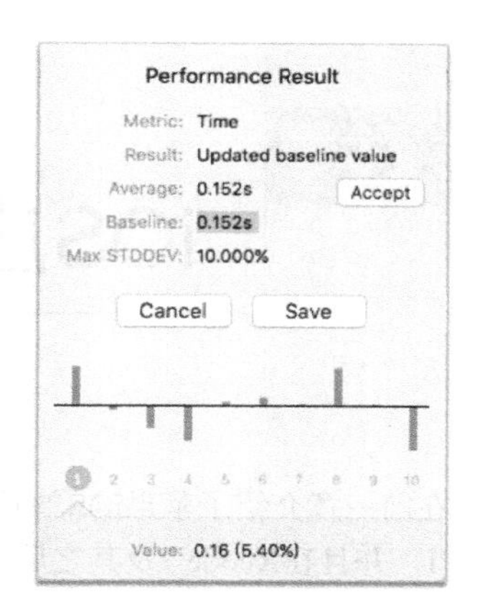

图23-19 编辑基线

23.5 小结

通过本章的学习，我们了解了测试驱动的iOS开发，掌握了测试驱动开发流程，以及单元测试框架XCTest，以及异步测试和性能测试等。

第 24 章 iOS应用UI测试

我们已经在前一章介绍了苹果iOS单元测试框架XCTest，在iOS 9和iOS 10中，苹果在XCTest框架的基础上增加了UI测试API，并且在Xcode 7及其之后的版本中提供了UI测试工具。

24.1 UI 测试概述

UI测试从来都是开发人员和测试人员的梦魇，它是最麻烦的。UI测试的过程涉及很多方面，包括UI事件处理、表示逻辑、控件输入验证和获取UI环境对象，其中最困难的是获取UI环境对象。例如，在使用故事板的iOS应用中，视图控制器和视图等UI对象不是在测试程序中创建的，而是由运行环境通过故事板创建的，如果直接实例化它们，会丢失很多状态。

要解决这种尴尬的局面，只能改善测试方法，提高工具的易用性。由于UI测试用例都是围绕界面操作而设计的，一些测试工具可以将这些操作录制下来，生成测试代码，测试人员可以适当修改测试代码，使之能够适用于更加普通的情况，然后就可以运行这些测试程序，出具测试报告。这就是“自动化测试”，测试代码被称为“测试脚本”。

事实上，苹果提供了两种自动化测试框架：UIAutomation框架和XCTest测试框架。UIAutomation框架是在iOS 4之后发布的，但是至今仍然鲜为人知，很少人使用。这主要是因为它非常难用，要结合Xcode的Instruments工具一起使用，它所录制生成的脚本是JavaScript语言的，测试人员还要学习JavaScript语法形式的UIAutomation API。Xcode提供的UI测试功能，可以生成Objective-C或Swift语言的测试脚本，新增的API是基于XCTest的，提供了Objective-C和Swift两种语言支持。

本章重点介绍如何使用XCTest测试框架实现UI测试。为了方便，本书将XCTest测试框架中的UI测试相关部分称为UI测试框架。

24.2 添加 UI 测试到工程

使用Xcode 7工具添加UI测试框架到工程中，有两种方法：一种是在创建工程时添加；另一种是在现有工程中添加iOS UI Testing Bundle目标（Target）。下面我们详细介绍一下这两种方法。

24.2.1 创建工程时添加 UI 测试框架

在Xcode 7及其之后的版本中创建工程时，在创建工程信息对话框（如图24-1所示）中，选中Include UI Tests复选框，这会使在创建工程目标的同时，创建一个UI测试用例目标。如图24-2所示，在生成的Xcode导航面板中，有一个PITaxUITests组，这个组中的类就是生成的测试类，并且还在目标列表中生成了测试目标PITaxUITests。

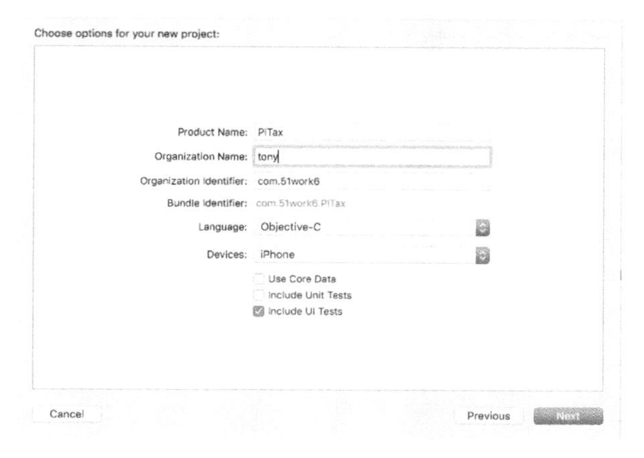

图24-1 创建工程信息对话框

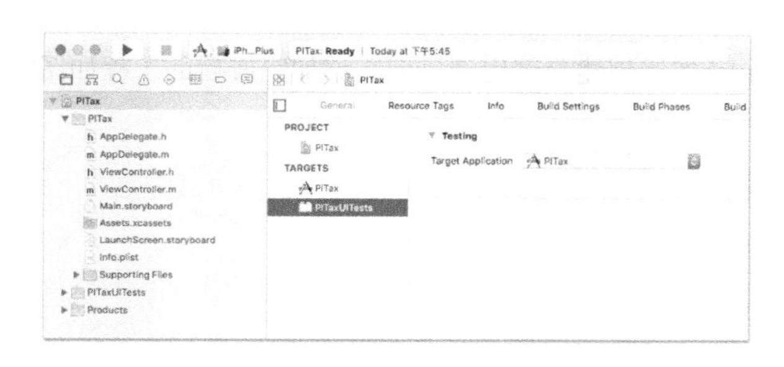

图24-2 添加完成的UI测试用例目标

24.2.2 在现有工程中添加UI测试用例目标

在一个现有的工程中，选择File→New→Target...菜单项，此时打开的界面如图24-3所示，从中选择iOS→Test中的iOS UI Testing Bundle模板。点击Next按钮，进入下一个目标相关项界面，如图24-4所示，在Product Name中输入PITaxUITests；Language中选择语言，原则上要测试的类使用何种语言，测试用例类也应该采用这种语言；Project中选择我们当前的测试工程，Target to be Tested中选择要测试的目标。

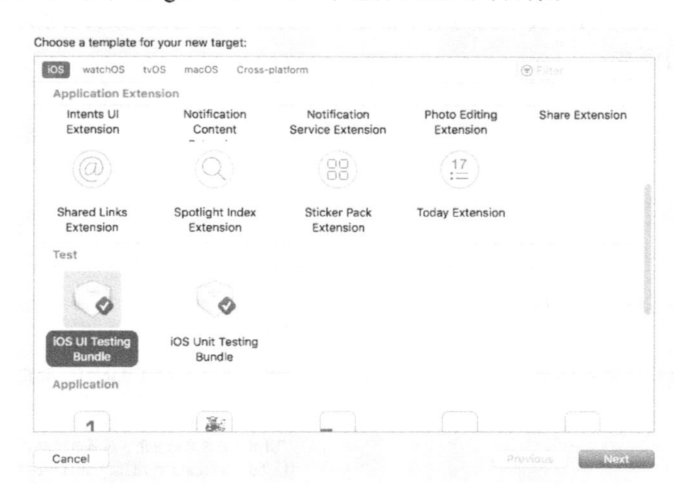

图24-3 选择模板

图24-4　为现有工程添加测试用例目标

设置完相关项后，点击Finish按钮创建测试用例目标。添加完成后的工程如图24-5所示，此时在导航面板中多出了一个PITaxUITests目标。

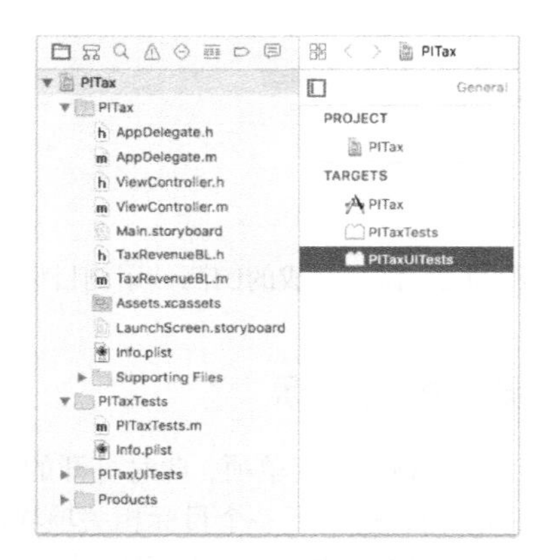

图24-5　添加测试用例目标之后的工程

无论哪种方式，默认生成的UI测试用例类基本都是一样的。下面的代码是UI测试用例PITaxUITests目标默认生成的PITaxUITests测试用例类：

```
//PITaxUITests.swift文件
import XCTest

class PITaxUITests: XCTestCase {

    override func setUp() {
        super.setUp()

        //测试用例出错后是否继续执行，设置为true表示继续执行，设置为false
        //表示终止
        continueAfterFailure = false
        //UI测试必须启动应用，该语句可以启动应用程序
        XCUIApplication().launch()
    }

    override func tearDown() {

        super.tearDown()
    }
```

```
//PITaxUITests.m文件
#import <XCTest/XCTest.h>

@interface PITaxUITests : XCTestCase
@end

@implementation PITaxUITests

- (void)setUp {
    [super setUp];

    //测试用例出错后是否继续执行，设置为true表示继续执行，设置为false
    //表示终止
    self.continueAfterFailure = NO;
    //UI测试必须启动应用，该语句可以启动应用程序
    [[[XCUIApplication alloc] init] launch];

}
```

```
    func testExample() {
    }

}
```

```
- (void)tearDown {

    [super tearDown];
}

- (void)testExample {

}

@end
```

从上述代码可见，UI测试用例类与单元测试用例类非常类似。

24.3 录制脚本

在Xcode 7及其之后的版本中，它提供测试脚本录制工具UI Recording，通过该工具可以生成Objective-C或Swift语言的测试脚本。

24.3.1 录制之前的准备

下面我们为PITax案例的UI测试用例目标PITaxUITests录制脚本。在开始录制脚本之前，我们需要为界面中的控件添加“辅助功能①”，苹果为“辅助功能”开发提供一套API，我们现在使用的UI测试框架基于辅助功能和XCTest两种技术。辅助功能提供了访问界面中UI元素的能力。在Interface Builder中打开故事板或XIB文件，如图24-6所示，选中控件，然后打开右边的标识检查器，在Accessibility→ Label中设置控件的标签。此外，也可以设置Identifier（标识）属性。

本例中，我们只需要设置Txt Revenue文本框的标签属性为TextField Revenue，计算按钮的标签属性为Calculate Button即可。

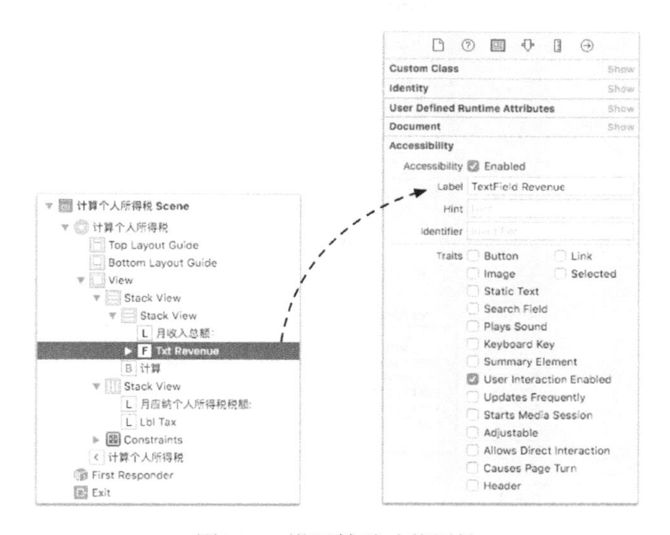

图24-6 设置辅助功能属性

24.3.2 录制过程

下面我们可以开始录制了。打开PITaxUITests测试用例类，如图24-7所示，将光标置于测试方法中，UI

① 辅助功能是指软件为残障人士提供了一些特殊功能，如屏幕放大和语音提示等功能。

Recording会在这里生成代码。然后，点击调试栏中的“录制”按钮●开始录制。

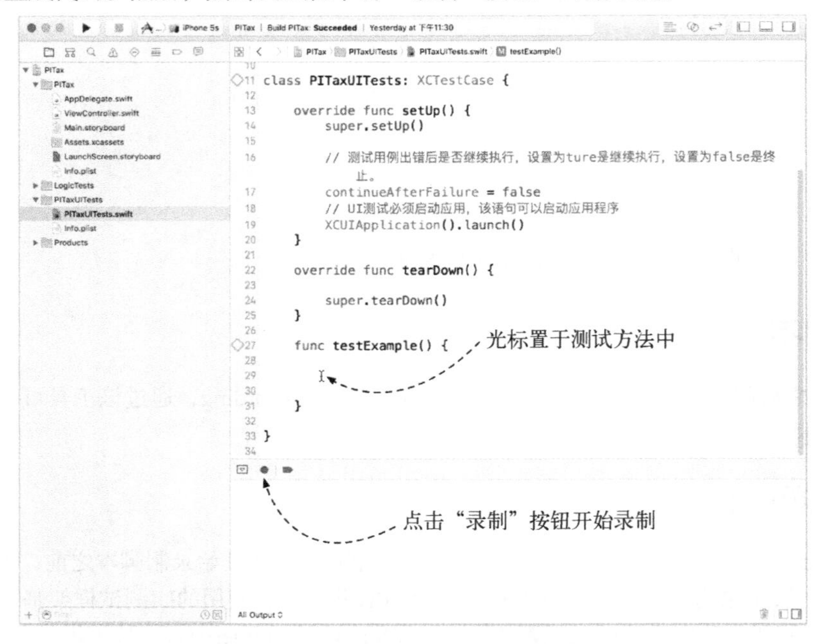

图24-7　开始录制脚本

录制一旦开始，我们会发现应用启动起来了，然后就可以在应用上进行常规的操作了。例如，我们可以在文本框中输入5000，然后点击“计算”按钮，计算获得的结果45.00会显示到标签上。这些操作都被UI Recording记录下来，并且在测试用例类中生成了代码，如图24-8所示。当然，我们也可以点击“停止”按钮⦿停止录制。

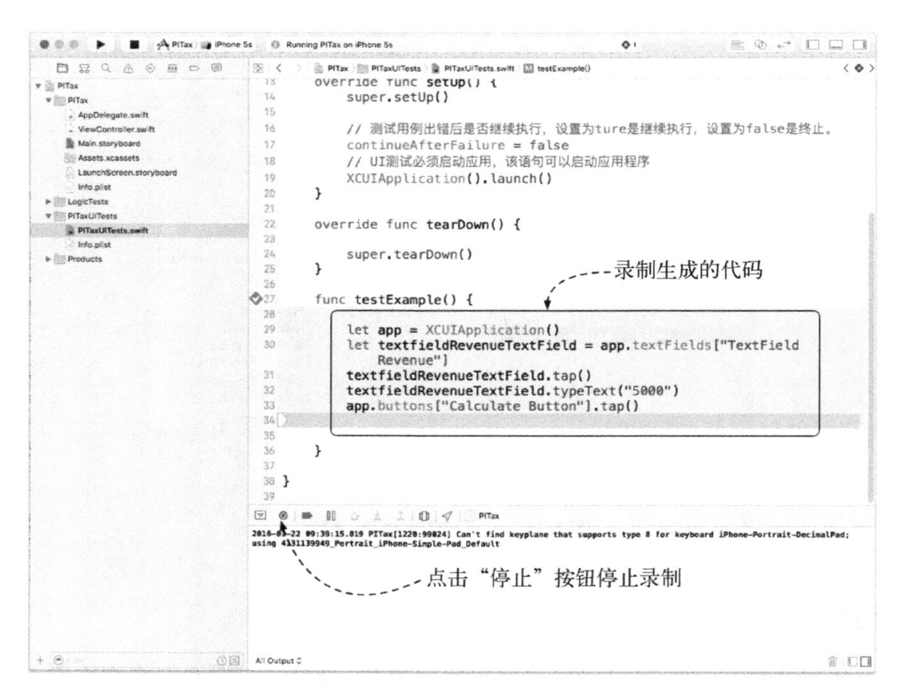

图24-8　停止录制脚本

24.3.3 修改录制脚本

UI Recording工具将我们的操作录制下来，录制的脚本不会有if和for等逻辑分支和循环语句，录制脚本只是针对特定控件的。例如，我们要想删除表视图中所有的单元格，但是录制过程中只是删除了其中一个单元格，此时需要修改脚本，为其添加for和if等语句，使之适用于所有的单元格。

另外，录制的脚本也不会生成断言。需要什么断言，一般是在设计测试用例时确定下来的。在本例中，我们需要断言计算结果标签是45.00。修改代码如下：

```
func testExample() {

    /////////////录制生成的脚本开始//////////////////
    let app = XCUIApplication()
    let textfieldRevenueTextField = app.textFields["TextField Revenue"]
    textfieldRevenueTextField.tap()
    textfieldRevenueTextField.typeText("5000")
    app.buttons["Calculate Button"].tap()
    /////////////录制生成的脚本结束/////////////////

    //自己添加的代码
    let lbl = app.staticTexts["45.00"]                              ①
    XCTAssert(lbl.exists)                                           ②

}
```

```
- (void)testExample {

    /////////////录制生成的脚本开始//////////////////
    XCUIApplication *app = [[XCUIApplication alloc] init];
    XCUIElement *textfieldRevenueTextField = app.textFields[@"TextField
    ➥Revenue"];
    [textfieldRevenueTextField tap];
    [textfieldRevenueTextField typeText:@"5000"];
    [app.buttons[@"Calculate Button"] tap];
    /////////////录制生成的脚本结束/////////////////

    //自己添加的代码
    XCUIElement* lbl = app.staticTexts[@"45.00"];                   ①
    XCTAssert(lbl.exists);                                          ②

}
```

上述代码中，第①行~第②行是自己添加的代码，第①行的app对象是XCUIApplication实例。关于XCUIApplication类，我们将在下一节中详细介绍。app.staticTexts["45.00"]（Objective-C版为app.staticTests[@"45.00"]）是获得45.00静态文本元素对象，如果计算的结果不是45.00，那么lbl对象会不存在。第②行用于断言lbl对象是否存在，其中exists属性是布尔值。

提示 XCTAssert断言函数是在XCTest中提供的。在UI测试时，可以使用XCTest框架中的所有断言函数。至于运行测试用例，可参考第23章，这里不再赘述。

24.4 访问 UI 元素

在测试过程中需要访问UI元素，而这些UI元素的API是从UIAutomation API传承下来的，本节就简要介绍一下它们。

24.4.1 UI 元素的层次结构树

在介绍相关的API之前，先来了解一下UI元素层次树。前面介绍iOS界面构建时，提到过界面是由视图按照层次结构构建起来的，UI测试框架也采用类似的层次结构管理UI元素。

图24-9是第6章介绍的简单表视图的SimpleTable案例。

界面中的UI元素构成如下所示的层次结构树：

图24-9 Simple Table案例

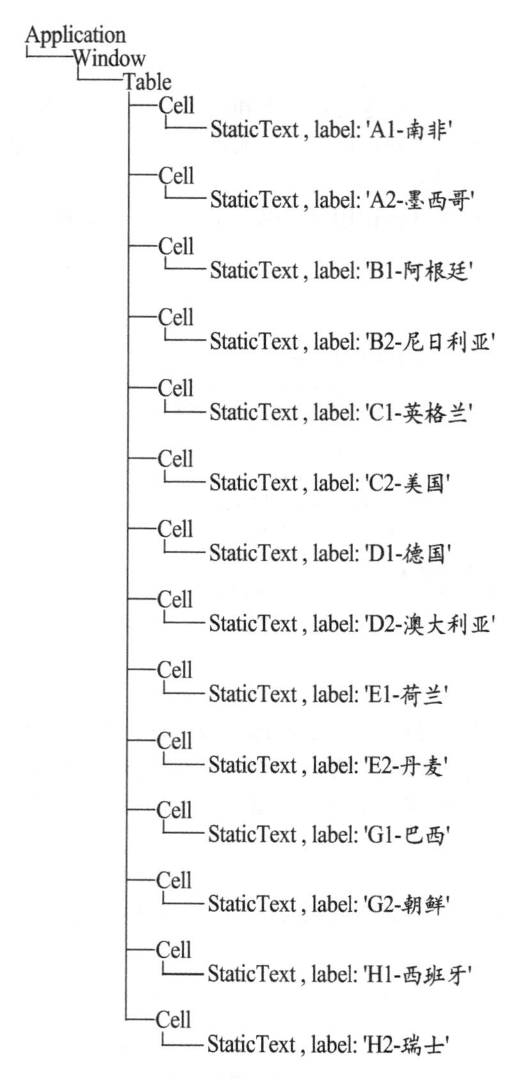

Application是根元素，其他元素都是它的后代（descendant）元素。Window是Application的直接子（child）元素，Table是Window的直接子元素，Table当然也是Application的后代元素。以此类推，Cell是Table的直接子元素，StaticText是Cell的直接子元素。

Application、Window、Table、Cell和StaticText是构成界面的UI元素类型。在UI测试框架中，定义了81种元素类型，有一些是iOS特有的，有些是macOS特有的，这里我们重点介绍几个iOS中常用的元素类型，如表24-1所示。

表24-1　iOS常用元素类型

元素类型	说　明
Application	应用程序类型
Window	Window类型
Table	表视图类型
Cell	表视图中的单元格类型
StaticText	静态文本类型，如标签等视图
TextField	文本框类型

（续）

元素类型	说　明
SecureTextField	密码框类型
Button	按钮类型
Slider	滑块类型
Switch	开关类型

24.4.2 UI 测试中相关 API

在iOS 9和iOS 10中，XCTest框架中增加了与UI测试相关的API，其中主要的类有XCUIElement、XCUIApplication和XCUIElementQuery。XCUIElement表示被测试应用中的元素，其主要方法和属性如下。

- **exists属性**。判断元素是否存在。
- **descendantsMatchingType:方法**。指定类型查找层次结构中的所有后代元素，返回值是XCUIElementQuery类型。
- **childrenMatchingType:方法**。指定类型查找层次结构中的所有直接子元素，返回值是XCUIElementQuery类型。
- **typeText:方法**。获得焦点后，模拟键盘输入字符到元素。
- **tap方法**。点击元素动作。
- **doubleTap方法**。双击元素动作。

XCUIApplication是XCUIElement的子类，表示被测试的应用，主要方法只有下面这个。

- **launch方法**。启动被测试的应用程序。

XCUIElementQuery类用来查询XCUIElement元素，它事实上是集合类型。查询单个元素相关的方法如下。

- **elementBoundByIndex:**。通过索引访问元素。
- **elementMatchingPredicate:**。通过谓词NSPredicate指定查询条件进行查询。
- **elementMatchingType:identifier:**。指定类型，通过id进行查询。
- **objectForKeyedSubscript:**。通过下标访问元素，它的简化形式为[]，例如app.textFields["TextField Revenue"]（Objective-C版本为app.textFields[@"TextField Revenue"]）。

XCUIElementQuery中查询多个元素相关的方法如下。

- **descendantsMatchingType:**。指定类型查找层次结构中的所有后代元素。
- **childrenMatchingType:**。指定类型查找层次结构中的所有直接子元素。
- **matchingPredicate:**。通过谓词NSPredicate指定查询条件进行查询。
- **matchingType:identifier:**。指定类型，通过id进行查询。
- **matchingIdentifier:**。通过id进行查询。
- **containingPredicate:**。在后代元素中，通过谓词NSPredicate指定查询条件进行查询。
- **containingType:identifier:**。在后代元素中，通过指定类型，通过id进行查询。

XCUIElementQuery中的属性如下。

- **element**。获取单个XCUIElement元素。
- **count**。获取元素的个数。
- **debugDescription**。调试信息。

24.5 表示逻辑组件测试最佳实践

本节我们将MyNotes案例的表示逻辑组件进行测试。首先，参考24.2.2节为14.3节的MyNotes案例添加UI测试用例目标。

表示逻辑组件中的UI测试用例设计是与界面操作有关的，我们总结了如下界面操作。

- 备忘录查询操作。
- 增加备忘录操作。
- 删除备忘录操作。
- 显示备忘录详细信息操作。

24.5.1　备忘录查询操作

备忘录查询操作就是首页显示时，从业务逻辑层获得数据显示到表视图单元格中。我们需要断言表视图单元格的行数，以及每一个单元格内容与数据库中对应的字段是否一致。

测试用例PresentationLayerUITests中相关的测试代码如下：

```
//PresentationLayerUITests.swift文件
//MARK: -- 查询操作
func testMasterViewControllerTableViewCellFindAll() {

    let app = XCUIApplication()
    //获得当前界面中的表视图
    let tableView = app.tables.element(boundBy: 0)                    ①
    //断言表视图存在
    XCTAssert(tableView.exists)
    //断言表视图单元格数为0
    XCTAssertEqual(tableView.cells.count, 2)

    let cell1 = tableView.cells.element(boundBy: 0)                   ②
    XCTAssertTrue(cell1.staticTexts["Welcome to MyNote."].exists)     ③

    let cell2 = tableView.cells.element(boundBy: 1)
    XCTAssertTrue (cell2.staticTexts["欢迎使用MyNote。"].exists)
}
```

```
//PresentationLayerUITests.m文件
#pragma mark -- 查询操作
- (void)testMasterViewControllerTableViewCellFindAll {

    XCUIApplication* app = [[XCUIApplication alloc] init];
    //获得当前界面中的表视图
    XCUIElement* tableView = [app.tables elementBoundByIndex:0];                ①
    //断言表视图存在
    XCTAssertTrue(tableView.exists);
    //断言表视图单元格数为0
    XCTAssertEqual(tableView.cells.count, 2);

    XCUIElement* cell1 = [tableView.cells elementBoundByIndex:0];               ②
    XCTAssertTrue(cell1.staticTexts[@"Welcome to MyNote."].exists);             ③

    XCUIElement* cell2 = [tableView.cells elementBoundByIndex:1];
    XCTAssertTrue(cell2.staticTexts[@"欢迎使用MyNote。"].exists);

}
```

上述代码中，第①行用于获得当前界面中的第一个表视图，其中app.tables用于获得当前界面中所有的表视图。app.tables这个写法事实上是app.descendantsMatchingType(.Table)（Objective-C是[app descendantsMatchingType: XCUIElementTypeTable]）简化写法，即查找app中所有的后代元素。

第②行用于获得表视图中的第一个单元格元素，其中表达式tableView.cells是tableView.descendants-MatchingType(.Cell)（Objective-C是[tableView descendantsMatchingType:XCUIElementTypeCell]）的简化写法。

第③行用于断言第一个单元格中文本内容是否为Welcome，其中表达式cell1.staticTexts是cell1.descendants-MatchingType(.StaticText)（Objective-C是[cell1 descendantsMatchingType:XCUIElementTypeStatic Text]）的简化写法。表达式staticTexts["Welcome"]（Objective-C是staticTexts[@"Welcome"]）是通过下标访问元素的，是objectForKeyedSubscript:方法的简化形式。

提示　上述测试用例方法运行之前，应该先启动应用修改数据，要保证只有两条数据，内容是Welcome和“欢迎”，并且保证第一个单元格的内容是Welcome，第二个单元格的内容是“欢迎”。

24.5.2　增加备忘录操作

增加备忘录操作的过程是，用户点击首页中的Add按钮，界面跳转到增加界面，在TextView中输入内容，点击Save按钮回到首页。我们需要断言表视图中是否增加了一个单元格。如果在增加界面中点击Cancel按钮，可以取消操作，返回首页。

测试用例PresentationLayerUITests中相关的测试代码如下：

```
//MARK: -- 增加备忘录操作
func testAddViewControllerSave() {

    let app = XCUIApplication()
    //获得当前界面中的表视图
    let tableView = app.tables.element(boundBy: 0)
    var cellsCount = tableView.cells.count

    //点击Add按钮，跳转到增加界面
    app.navigationBars["备忘录"].buttons["Add"].tap()                    ①

    //获得增加界面中的TextView对象
    let textView = app.textViews["Text View"]
    textView.tap()
    textView.typeText("HelloWorld")                                      ②

    //在增加界面中添加Save按钮，跳转到备忘录界面
    app.navigationBars["增加"].buttons["Save"].tap()
    //断言备忘录界面中表视图单元格数为+1
    cellsCount += 1
    XCTAssertEqual(tableView.cells.count, cellsCount)

}
//MARK: -- 增加备忘录时取消操作
func testAddViewControllerCancel() {

    let app = XCUIApplication()
    //获得当前界面中的表视图
    let tableView = app.tables.element(boundBy: 0)
    let cellsCount = tableView.cells.count

    //点击Add按钮，跳转到增加界面
    app.navigationBars["备忘录"].buttons["Add"].tap()

    //在增加界面中添加Save按钮，跳转到备忘录界面
    app.navigationBars["增加"].buttons["Cancel"].tap()
    //断言备忘录界面中表视图单元格数没有+1
    XCTAssertEqual(tableView.cells.count, cellsCount)

}
```

```
#pragma mark -- 增加备忘录操作
- (void)testAddViewControllerSave {

    XCUIApplication* app = [[XCUIApplication alloc] init];
    //获得当前界面中的表视图
    XCUIElement* tableView = [app.tables elementBoundByIndex:0];
    NSUInteger cellsCount = tableView.cells.count;

    //点击Add按钮，跳转到增加界面
    [app.navigationBars[@"备忘录"].buttons[@"Add"] tap];                 ①

    //获得增加界面中的TextView对象
    XCUIElement* textView = app.textViews[@"Text View"];
    [textView tap];
    [textView typeText:@"HelloWorld"];                                   ②

    //在增加界面中添加Save按钮，跳转到备忘录界面
    [app.navigationBars[@"增加"].buttons[@"Save"] tap];
    //断言备忘录界面中表视图单元格数为+1
    XCTAssertEqual(tableView.cells.count, cellsCount + 1);

}

#pragma mark -- 增加备忘录时取消操作
- (void)testAddViewControllerCancel {

    XCUIApplication* app = [[XCUIApplication alloc] init];
    //获得当前界面中的表视图
    XCUIElement* tableView =  [app.tables elementBoundByIndex:0];
    NSUInteger cellsCount = tableView.cells.count;

    //点击Add按钮，跳转到增加界面
    [app.navigationBars[@"备忘录"].buttons[@"Add"] tap];

    //在增加界面中添加Save按钮，跳转到备忘录界面
    [app.navigationBars[@"增加"].buttons[@"Cancel"] tap];
    //断言备忘录界面中表视图单元格数没有+1
    XCTAssertEqual(tableView.cells.count, cellsCount);

}
```

上述代码中，第①行表示点击Add按钮，结果会跳转到增加界面。第②行表示通过键盘在textView中输入字符串。

24.5.3 删除备忘录操作

删除备忘录操作的过程是，用户点击首页中的Edit按钮，使表视图处于编辑状态，然后点击单元格删除按钮，再点击单元格后面的删除确定按钮Delete，删除之后再点击Done按钮完成操作。删除成功后我们会断言，单元格是否减少了一行，还要断言删除的单元格数据已经不存在了。

测试用例PresentationLayerUITests中相关的测试代码如下：

```
//MARK: -- 删除最后一个单元格的操作
func testMasterViewControllerTableViewCellRemove() {

    let app = XCUIApplication()
    let navigationBar = app.navigationBars["备忘录"]
    navigationBar.buttons["Edit"].tap()

    let tableView = app.tables.element(boundBy: 0)
    var cellsCount = tableView.cells.count

    tableView.buttons.element(boundBy: 0).tap()
    tableView.buttons["Delete"].tap()
```

```
#pragma mark -- 删除最后一个单元格的操作
- (void)testMasterViewControllerTableViewCellRemove {

    XCUIApplication* app = [[XCUIApplication alloc] init];
    XCUIElement* navigationBar = app.navigationBars[@"备忘录"];
    [navigationBar.buttons[@"Edit"] tap];

    XCUIElement* tableView =  [app.tables elementBoundByIndex:0];
    NSUInteger cellsCount = tableView.cells.count;

    [[tableView.buttons elementBoundByIndex:0] tap];
    [tableView.buttons[@"Delete"] tap];
```

```
    navigationBar.buttons["Done"].tap()

    cellsCount -= 1
    XCTAssertEqual(tableView.cells.count, cellsCount - 1)

    let staticTexts = tableView.cells.staticTexts["Welcome to MyNote."]   ①
    //tableView.staticTexts["Welcome to MyNote."]                         ②
    XCTAssert(!staticTexts.exists)                                        ③

}
```

```
    [navigationBar.buttons[@"Done"] tap];

    XCTAssertEqual(tableView.cells.count, cellsCount - 1);

    XCUIElement* staticTexts = tableView.cells.staticTexts[@"Welcome to
➥MyNote."];                                                              ①
    //tableView.staticTexts[@"Welcome to MyNote."];                       ②
    XCTAssertTrue(!staticTexts.exists);                                   ③
}
```

第①行用于试图获得删除单元格中的静态文本，第③行代码用于断言这个对象不存在。需要注意的是，第①行也可以使用第②行替换，这是因为它们都是descendantsMatchingType:的简化写法，即从后代元素中查找。单元格中的静态文本元素是单元格的子元素，更是表视图的子元素。

提示　上述测试用例方法运行之前，我们应该先启动应用修改数据，要保证第一个单元格的内容是Welcome to MyNote.。

24.5.4　显示备忘录详细信息操作

显示备忘录详细信息的操作是，用户点击首页中的某个单元格，界面跳转到详细界面，我们需要断言详细界面TextView中的内容是否等于选中单元格中的文本。然后，点击左上角的返回按钮回到首页。

测试用例PresentationLayerUITests中相关的测试代码如下：

```
//MARK: -- 显示详细信息操作
func testDetailViewControllerShowDetail() {

    let app = XCUIApplication()
    //获得当前界面中的表视图
    let tableView = app.tables.element(boundBy: 0)
    let cell1 = tableView.cells.element(boundBy: 0)
    cell1.tap()

    let welcomeStaticText = app.staticTexts["Welcome to MyNote."]
    //断言
    XCTAssertTrue(welcomeStaticText.exists)

    app.navigationBars["详细"].buttons["备忘录"].tap()

}
```

```
#pragma mark -- 显示详细信息操作
- (void)testDetailViewControllerShowDetail {

    XCUIApplication* app = [[XCUIApplication alloc] init];
    //获得当前界面中的表视图
    XCUIElement* tableView =  [app.tables elementBoundByIndex:0];
    XCUIElement* cell1 = [tableView.cells elementBoundByIndex:0];
    [cell1 tap];

    XCUIElement* welcomeStaticText = app.staticTexts[@"Welcome to MyNote."];
    //断言
    XCTAssertTrue(welcomeStaticText.exists);

    [app.navigationBars[@"详细"].buttons[@"备忘录"] tap];
}
```

提示　上述测试用例方法运行之前，我们应该先启动应用修改数据，保证点击的单元格内容是Welcome。

24.6　小结

通过本章的学习，我们了解了UI测试框架，它基于XCTest和UIAutomation框架。最后，我们还介绍了如何在分层架构中实施UI测试。

第 25 章 让你的程序“飞”起来——性能优化

相对电脑而言，移动设备具有内存少、CPU速度慢等特点，因此iOS开发人员需要尽可能优化应用的性能。性能优化需要考虑的问题很多，本章就来介绍几个重要的优化方法。本章是非常难掌握的一章，当然也是非常重要的一章。

25.1 内存优化

在Swift语言中，内存管理采用ARC（Automatic Reference Counting，自动引用计数）。ARC是与MRC（Manual Reference Counting，手动引用计数）相对而言的，这些概念来源于Objective-C的内存管理方式。

25.1.1 内存管理

这里我们有必要先介绍一下Objective-C的内存管理方法，共有3种，具体如下。

- MRC。就是由程序员自己负责管理对象生命周期，负责对象的创建和销毁。
- ARC。采用与MRC一样的内存引用计数管理方法，但不同的是，它在编译时会在合适的位置插入对象内存释放（如release、autorelease和retain等），程序员不用关心对象释放的问题。苹果推荐在新项目中使用ARC，但在iOS 5之前的系统中不能采用ARC。
- GC。在Objective-C 2.0之后，内存管理出现了类似于Java和C#的内存垃圾收集技术，但是垃圾收集与ARC完全不同，垃圾收集是后台有一个线程负责检查已经不再使用的对象，然后释放之。由于后台有一个线程一直运行，所以会严重影响性能，这也是Java和C#程序的运行速度无法超越C++的主要原因。GC技术不能应用于iOS开发，只能应用于macOS开发。

从上面的介绍可知，iOS采用MRC和ARC这两种方式，ARC是苹果推荐的方式，MRC方式相对比较原始，对于程序员的能力要求很高，但是它很灵活、方便，很不容易驾驭好。Swift采用ARC方式管理内存，因此使用起来比较简单。

25.1.2 使用 Analyze 工具检查内存泄漏

内存泄漏指一个对象或变量在使用完成后没有释放掉，这个对象一直占用这部分内存，直到应用停止。如果这种对象过多，内存就会耗尽，其他应用就无法运行。这个问题在C++、C和Objective-C的MRC中是比较普遍的问题。

在Objective-C中，释放对象的内存时，可以发送release和autorelease消息，release消息马上将引用计数减1，autorelease消息会将对象放入内存释放池，会延迟到内存释放池周期到后，内存释放池会将池中所有对象的引用计数减1。当引用计数为0时，对象所占用的内存才被释放。

下面我们看看本节配套的Objective-C工程中ViewController的代码片段：

```
- (void)viewDidLoad {
    [super viewDidLoad];

    NSBundle *bundle = [NSBundle mainBundle];
    NSString *plistPath = [bundle pathForResource:@"team" ofType:@"plist"];
    //获取属性列表文件中的全部数据
    self.listTeams = [[NSArray alloc] initWithContentsOfFile:plistPath];

}

- (UITableViewCell *)tableView:(UITableView *)tableView
➥cellForRowAtIndexPath: (NSIndexPath *)indexPath {
    static NSString *CellIdentifier = @"CellIdentifier";
    UITableViewCell *cell = [tableView dequeueReusableCellWithIdentifier:CellIdentifier];
    if (cell == nil) {
        cell = [[UITableViewCell alloc] initWithStyle:UITableViewCellStyleDefault
        ➥reuseIdentifier:CellIdentifier];
    }

    NSUInteger row = [indexPath row];
    NSDictionary *rowDict = [self.listTeams objectAtIndex:row];
    cell.textLabel.text =  [rowDict objectForKey:@"name"];

    NSString *imagePath = [rowDict objectForKey:@"image"];
    imagePath = [imagePath stringByAppendingString:@".png"];
    cell.imageView.image = [UIImage imageNamed:imagePath];

    cell.accessoryType = UITableViewCellAccessoryDisclosureIndicator;

    return cell;
}

- (void)tableView:(UITableView *)tableView
➥didSelectRowAtIndexPath:(NSIndexPath *)indexPath {

    NSUInteger row = [indexPath row];
    NSDictionary *rowDict = [self.listTeams objectAtIndex:row];
    NSString *rowValue  =  [rowDict objectForKey:@"name"];


    NSString *message = [[NSString alloc] initWithFormat:@"您选择了%@队。", rowValue];
    UIAlertView *alert = [[UIAlertView alloc]initWithTitle:@"请选择球队"
                                    message:message
                                    delegate:self
                                    cancelButtonTitle:@"Ok"
                                    otherButtonTitles:nil];
    [alert show];

    [tableView deselectRowAtIndexPath:indexPath animated:TRUE];
}
```

大家看看，上面的这3个方法会有什么问题呢？如果代码基于ARC，这是没有问题的，但遗憾的是这是基于MRC的，都存在内存泄漏的可能性。从理论上讲，内存泄漏是由对象或变量没有释放引起的，但实践证明并非所有的未释放对象或变量都会导致内存泄漏，这与硬件环境和操作系统环境有关，因此我们需要检测工具帮助我们找到这些“泄漏点”。

在Xcode中，共提供了两种工具帮助查找泄漏点：Analyze和Instruments。Analyze是静态分析工具，Instruments是动态分析工具。这一节我们重点介绍Analyze工具。

Analyze用来检查MRC代码的内存泄漏问题，通过对代码进行分析，查找release、autorelease、retain、alloc、new、copy和mutableCopy等与管理引用计数相关的方法，检查那些引用计数不为0的对象，从而找到可疑的泄漏点。由于ARC代码没用这些管理引用计数的消息，所以它无法分析出ARC代码的内存泄漏问题。

我们通过Product→Analyze菜单项启动Analyze，图25-1所示为使用Analyze工具进行静态分析之后的代码界面。

```
- (void)viewDidLoad {
    [super viewDidLoad];

    NSBundle *bundle = [NSBundle mainBundle];
    NSString *plistPath = [bundle pathForResource:@"team"
                                           ofType:@"plist"];
    //获取属性列表文件中的全部数据
    self.listTeams = [[NSArray alloc] initWithContentsOfFile:plistPath];

}

- (void)didReceiveMemoryWarning {
    [super didReceiveMemoryWarning];
    // Dispose of any resources that can be recreated.
}

#pragma mark --UITableViewDataSource 协议方法
- (NSInteger)tableView:(UITableView *)tableView numberOfRowsInSection:(NSInteger)section {
    return [self.listTeams count];
}

- (UITableViewCell *)tableView:(UITableView *)tableView cellForRowAtIndexPath:(NSIndexPath *)indexPath {

    static NSString *CellIdentifier = @"CellIdentifier";
    UITableViewCell *cell = [tableView dequeueReusableCellWithIdentifier:CellIdentifier];
    if (cell == nil) {
        cell = [[UITableViewCell alloc] initWithStyle:UITableViewCellStyleDefault reuseIdentifier:
            CellIdentifier] ;
    }

    NSUInteger row = [indexPath row];
    NSDictionary *rowDict = [self.listTeams objectAtIndex:row];
    cell.textLabel.text =  [rowDict objectForKey:@"name"];

    NSString *imagePath = [rowDict objectForKey:@"image"];
    imagePath = [imagePath stringByAppendingString:@".png"];
    cell.imageView.image = [UIImage imageNamed:imagePath];

    cell.accessoryType = UITableViewCellAccessoryDisclosureIndicator;

    return cell;
}



#pragma mark --UITableViewDelegate 协议方法
- (void)tableView:(UITableView *)tableView didSelectRowAtIndexPath:(NSIndexPath *)indexPath {

    NSUInteger row = [indexPath row];
    NSDictionary *rowDict = [self.listTeams objectAtIndex:row];
    NSString *rowValue  =  [rowDict objectForKey:@"name"];

    NSString *message = [[NSString alloc] initWithFormat:@"您选择了%@队。", rowValue];
                                                  1. Method returns an Objective-C object with a +1 retain count
    UIAlertView *alert = [[UIAlertView            2. Object leaked: object allocated and stored into 'message' i...
        alloc]initWithTitle:@"请选择球队"           Potential leak of an object stored into 'message'
                            message:message
                            delegate:self
                            cancelButtonTitle:@"Ok"
                            otherButtonTitles:nil];
    [alert show];
//    [alert release];
//    [message release];
    [tableView deselectRowAtIndexPath:indexPath animated:YES];
}

@end
```

图25-1 使用Analyze进行静态分析之后的代码界面

在图25-1中，凡是有图标的行都是工具发现的疑似泄漏点。点击viewDidLoad方法中疑似泄漏点行首的图标，会展开分析结果，具体如图25-2所示。

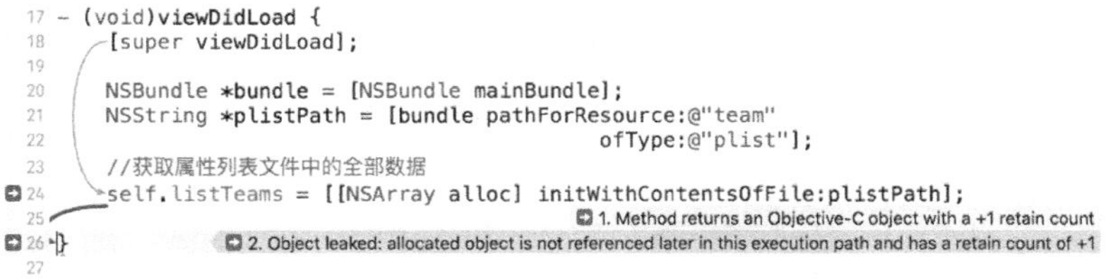

```
- (void)viewDidLoad {
    [super viewDidLoad];

    NSBundle *bundle = [NSBundle mainBundle];
    NSString *plistPath = [bundle pathForResource:@"team"
                                           ofType:@"plist"];
    //获取属性列表文件中的全部数据
    self.listTeams = [[NSArray alloc] initWithContentsOfFile:plistPath];
                       1. Method returns an Objective-C object with a +1 retain count
}   2. Object leaked: allocated object is not referenced later in this execution path and has a retain count of +1
```

图25-2 viewDidLoad方法疑似泄漏点的展开结果

图25-2中的线表明了程序执行的路径。在这个路径中，第1处说明在第24行中，Objective-C对象的引用计数是1，说明在这里创建了一个Objective-C对象。第2处说明在第26行中引用计数为1，该对象没有释放，怀疑有泄漏。这样的说明已经很明显地告诉我们问题所在了，[[NSArray alloc] initWithContentsOfFile:plistPath]创建了一个对象，并赋值给listTeams属性所代表的成员变量，然而完成了赋值工作之后，创建的对象并没有显式地发送release和autorelease消息。这里可以将代码修改如下：

```
NSArray *array = [[NSArray alloc] initWithContentsOfFile:plistPath];
self.listTeams = array;
[array release];
```

点击tableView:cellForRowAtIndexPath:方法中疑似泄漏点行首的图标，展开分析结果，如图25-3所示。

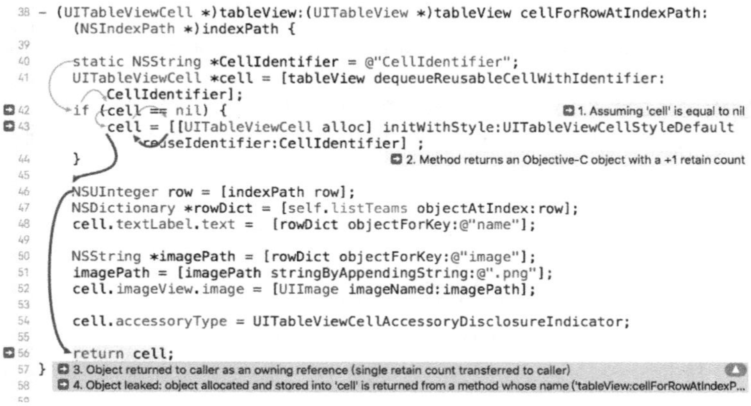

```
38 - (UITableViewCell *)tableView:(UITableView *)tableView cellForRowAtIndexPath:
     (NSIndexPath *)indexPath {
39
40     static NSString *CellIdentifier = @"CellIdentifier";
41     UITableViewCell *cell = [tableView dequeueReusableCellWithIdentifier:
           CellIdentifier];
42     if (cell == nil) {                          1. Assuming 'cell' is equal to nil
43         cell = [[UITableViewCell alloc] initWithStyle:UITableViewCellStyleDefault
               reuseIdentifier:CellIdentifier] ;
44     }                            2. Method returns an Objective-C object with a +1 retain count
45
46     NSUInteger row = [indexPath row];
47     NSDictionary *rowDict = [self.listTeams objectAtIndex:row];
48     cell.textLabel.text =  [rowDict objectForKey:@"name"];
49
50     NSString *imagePath = [rowDict objectForKey:@"image"];
51     imagePath = [imagePath stringByAppendingString:@".png"];
52     cell.imageView.image = [UIImage imageNamed:imagePath];
53
54     cell.accessoryType = UITableViewCellAccessoryDisclosureIndicator;
55
56     return cell;
57 }  3. Object returned to caller as an owning reference (single retain count transferred to caller)
58     4. Object leaked: object allocated and stored into 'cell' is returned from a method whose name ('tableView:cellForRowAtIndexP...
```

图25-3　tableView:cellForRowAtIndexPath:方法的疑似泄漏点展开结果

这主要说明UITableViewCell *类型的cell对象在第56行有可能存在泄漏。在表视图中，tableView:cellForRowAtIndexPath:方法用于实例化表视图单元格并设置数据，因此cell对象实例化后不能马上释放，而应该使用autorelease延迟释放。可以在创建cell对象时发送autorelease消息，将代码修改如下：

```
if (cell == nil) {
    cell = [[[UITableViewCell alloc] initWithStyle:UITableViewCellStyleDefault
    ➥reuseIdentifier:CellIdentifier] autorelease];
}
```

我们再看一下tableView:didSelectRowAtIndexPath:方法中的疑似泄漏点，共有两个。点击行首的图标，展开分析结果，具体如图25-4和图25-5所示。

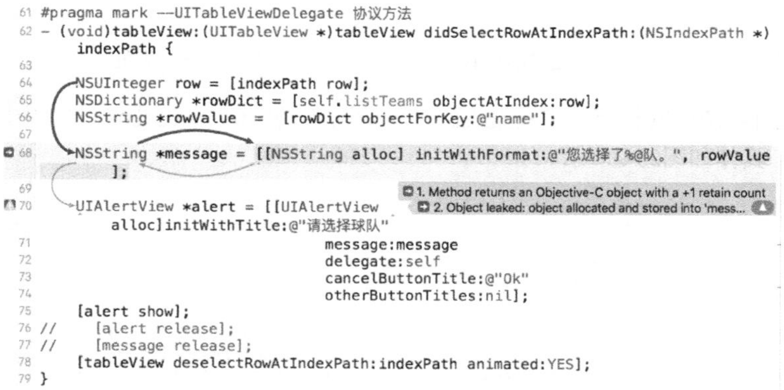

```
61 #pragma mark --UITableViewDelegate 协议方法
62 - (void)tableView:(UITableView *)tableView didSelectRowAtIndexPath:(NSIndexPath *)
     indexPath {
63
64     NSUInteger row = [indexPath row];
65     NSDictionary *rowDict = [self.listTeams objectAtIndex:row];
66     NSString *rowValue  =  [rowDict objectForKey:@"name"];
67
68     NSString *message = [[NSString alloc] initWithFormat:@"您选择了%@队。", rowValue
           ];
69                                 1. Method returns an Objective-C object with a +1 retain count
70     UIAlertView *alert = [[UIAlertView      2. Object leaked: object allocated and stored into 'mess...
           alloc]initWithTitle:@"请选择球队"
71                                message:message
72                                delegate:self
73                                cancelButtonTitle:@"Ok"
74                                otherButtonTitles:nil];
75     [alert show];
76 //     [alert release];
77 //     [message release];
78     [tableView deselectRowAtIndexPath:indexPath animated:YES];
79 }
```

图25-4　tableView:didSelectRowAtIndexPath:方法疑似泄漏点1的展开结果

```
61 #pragma mark --UITableViewDelegate 协议方法
62 - (void)tableView:(UITableView *)tableView didSelectRowAtIndexPath:(NSIndexPath *)
       indexPath {
63
64     NSUInteger row = [indexPath row];
65     NSDictionary *rowDict = [self.listTeams objectAtIndex:row];
66     NSString *rowValue  =  [rowDict objectForKey:@"name"];
67
68     NSString *message = [[NSString alloc] initWithFormat:@"您选择了%@队。", rowValue
           ];
69                                            1. Method returns an Objective-C object with a +1 retain count
70     UIAlertView *alert = [[UIAlertView     2. Object leaked: object allocated and stored into 'mess...
           alloc]initWithTitle:@"请选择球队"
71                                 message:message
72                                 delegate:self
73                                 cancelButtonTitle:@"Ok"
74                                 otherButtonTitles:nil];
75     [alert show];
76 //    [alert release];
77 //    [message release];
78     [tableView deselectRowAtIndexPath:indexPath animated:YES];
79 }
```

图25-5　tableView:didSelectRowAtIndexPath:方法疑似泄漏点2的展开结果

图25-4所示的是message对象创建之后没有释放，我们只需要在[alert show]之后添加[message release]语句就可以了。

在Objective-C中，实例化对象有如下两种方式：

```
NSString *message = [[NSString alloc] initWithFormat:@"您选择了%@队。", rowValue];    ①
NSString *message = [NSString stringWithFormat:@"您选择了%@队。", rowValue];          ②
```

第①行所示的以init开头的构造方法在alloc之后调用，我们将其称为"实例构造方法"。对于使用该方法创建的对象，其所有权是调用者，调用者需要对它的生命周期负责，具体说就是负责创建和释放。第②行所示的以string开头的方法，它是静态工厂方法，通过类直接调用。对于使用该方法创建的对象，其所有权非调用者所有，调用者无权释放它，否则就会因过度释放而"僵尸化"，这个问题我们会在下一节中介绍。

> **提示** 采用alloc、new、copy和mutableCopy所创建的对象，所有权属于调用者，它的生命周期由调用者管理，调用者负责通过release或autorelease方法释放对象。

图25-5所示的是UIAlertView *类型的alert对象创建后没有释放，我们只需要在[alert show]之后添加[alert release]语句就可以了。修改之后的代码如下：

```
- (void)tableView:(UITableView *)tableView
➥didSelectRowAtIndexPath:(NSIndexPath *)indexPath {
    NSUInteger row = [indexPath row];
    NSDictionary *rowDict = [self.listTeams objectAtIndex:row];
    NSString *rowValue  =  [rowDict objectForKey:@"name"];


    NSString *message = [[NSString alloc] initWithFormat:@"您选择了%@队。", rowValue];
    UIAlertView *alert = [[UIAlertView alloc]initWithTitle:@"请选择球队"
                                                    message:message
                                                   delegate:self
                                          cancelButtonTitle:@"Ok"
                                          otherButtonTitles:nil];
    [alert show];
    [alert release];
    [message release];
    [tableView deselectRowAtIndexPath:indexPath animated:TRUE];
}
```

25.1.3 使用 Instruments 工具检查内存泄漏

上面介绍的是使用Analyze静态分析工具查找MRC代码的可疑泄漏点，而ARC代码则可以使用Instruments动态分析工具中的Leaks和Allocations分析模板。

提示　Instruments中Leaks分析模板在Xcode 7之前可以检查出MRC和ARC代码中的内存泄漏问题，但Xcode 7及其之后的版本不能检查出MRC代码中的内存泄漏问题。关于MRC代码的内存泄漏问题，推荐使用Analyze查找。

在Xcode中通过Product→Profile菜单项启动Instruments动态分析工具，如图25-6所示，我们可以看到Instruments中还有很多分析模板可以动态分析跟踪内存、CPU、文件、电池和动画等。

选择Leaks模板，打开的界面如图25-7所示。

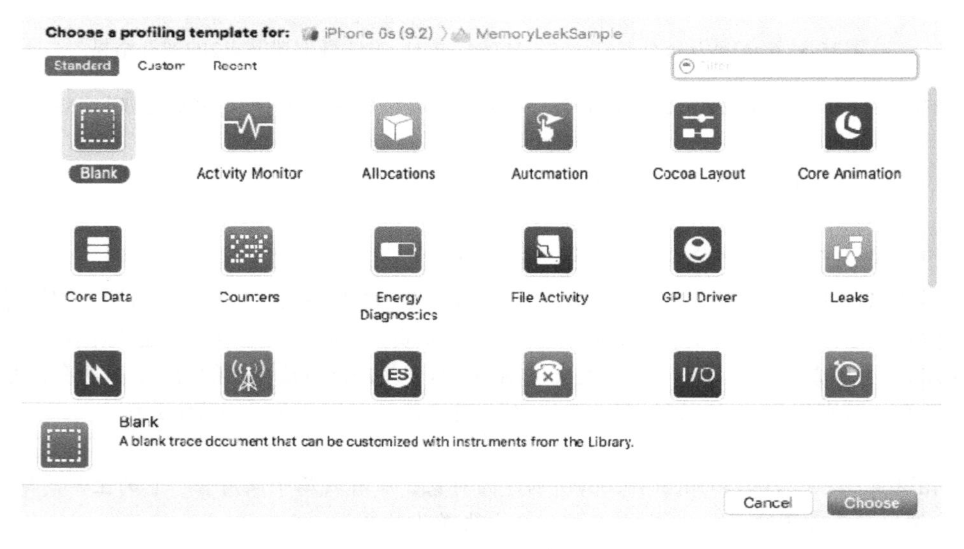

图25-6　Instruments分析工具

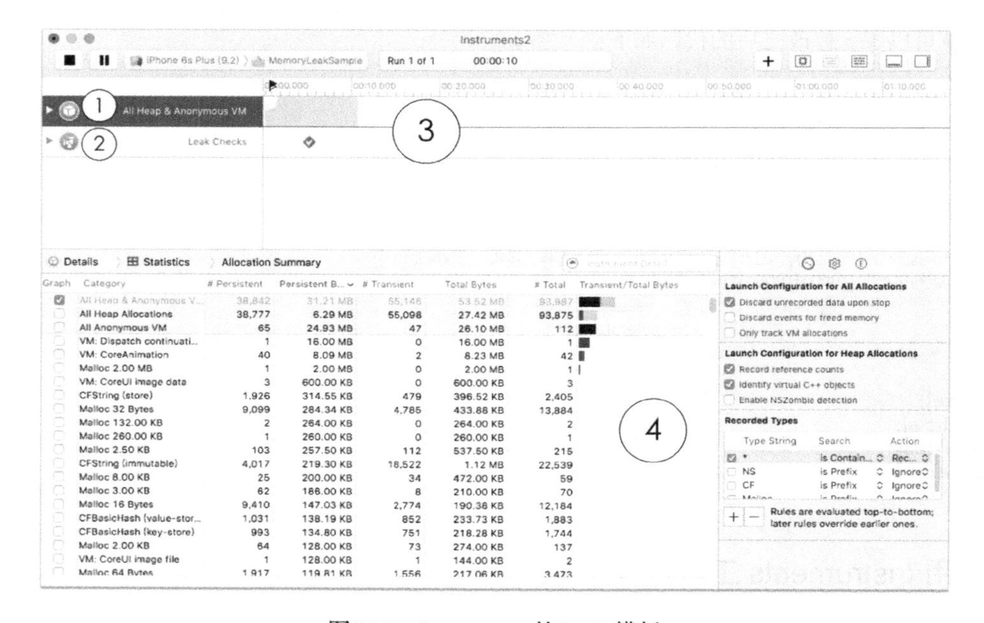

图25-7　Instruments的Leaks模板

在Instruments中，虽然选择了Leaks模板，但默认情况下也会添加Allocations模板。基本上只要分析内存，都会使用Allocations模板，它可以监控内存分布情况。选中Allocations模板（图中①区域），右边的③区域会显示随着时间的变化内存使用的折线图，同时在④区域会显示内存使用的详细信息以及对象分配情况。点击Leaks模板（图中②区域），可以查看内存泄漏情况。如图25-8所示，如果在③区域有红色菱形图标出现，则有内存泄漏，此时④区域则会显示泄漏的对象。可以发现，里面有两个对象，还列出了它们的内存地址、占用字节、所属框架和响应方法等信息。如果④区域是绿色的菱形图标出现，则没有检查出内存泄漏。

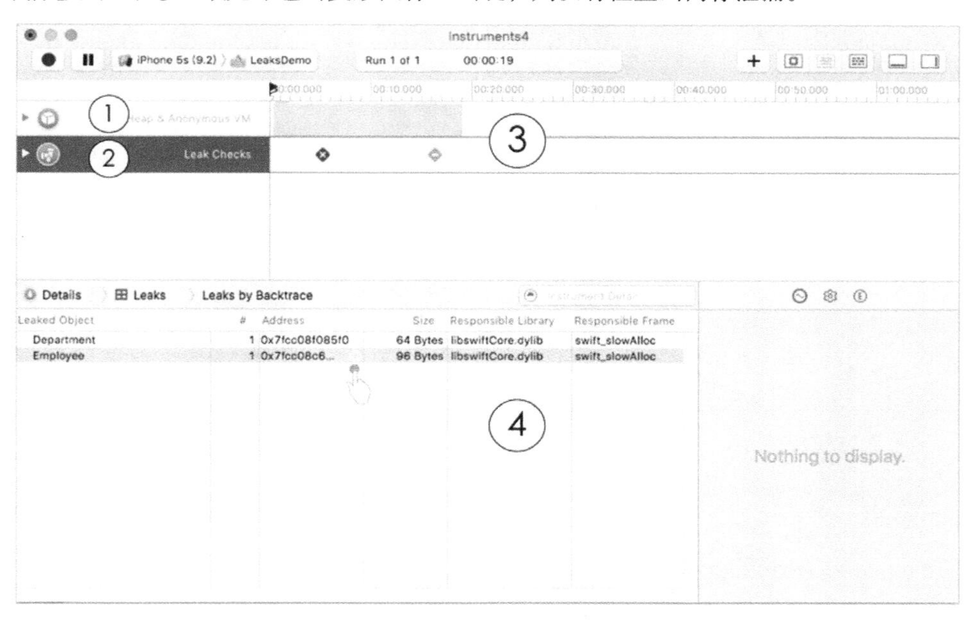

图25-8 Instruments检测到的内存泄漏

在图25-8所示的界面中，点击泄漏对象Address列后面的按钮，会进入如图25-9所示的详细界面。其中，RetCt是引用计数列。可以发现，最后的引用计数不为零，这说明该对象内存没有释放。

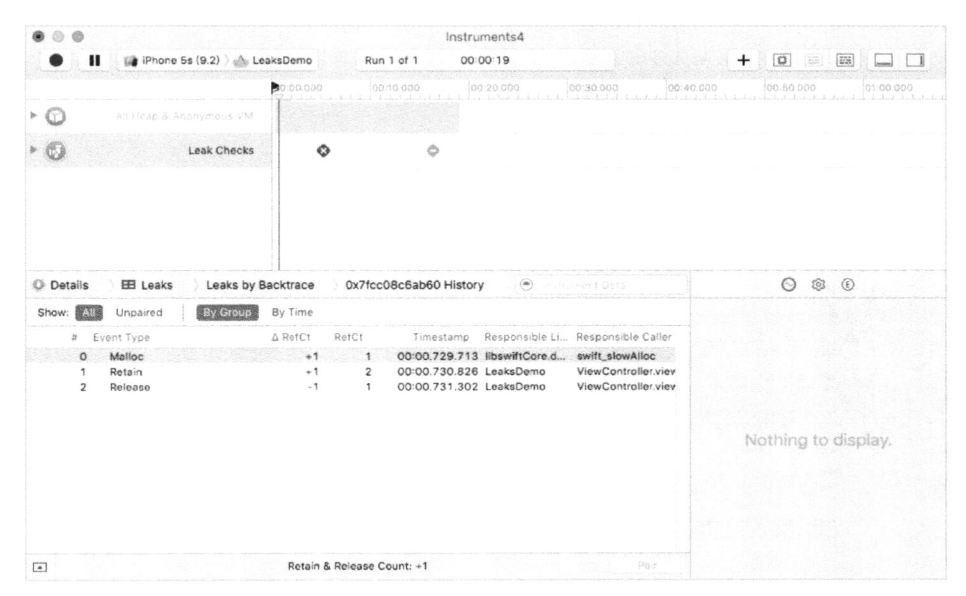

图25-9 查看泄漏的详细信息

在图25-9中，点击右边的跟踪栈信息按钮Ⓔ，如图25-10所示。在打开的界面中，Stack Trace中图标所示的条目是我们自己应用的代码，点击它即可进入程序代码，如图25-11所示。

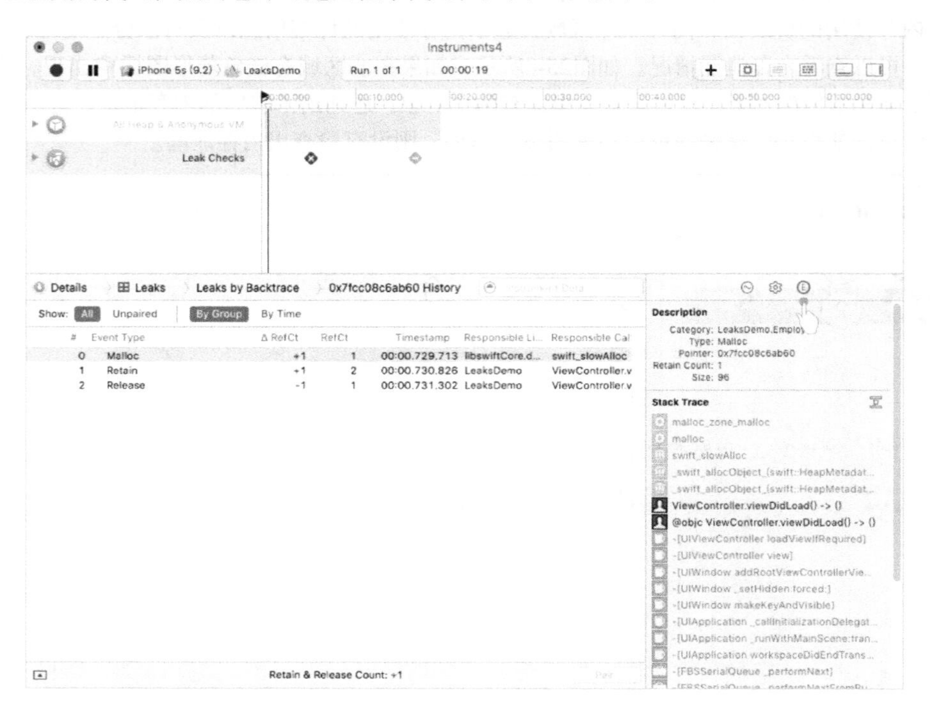

图25-10　查看栈信息

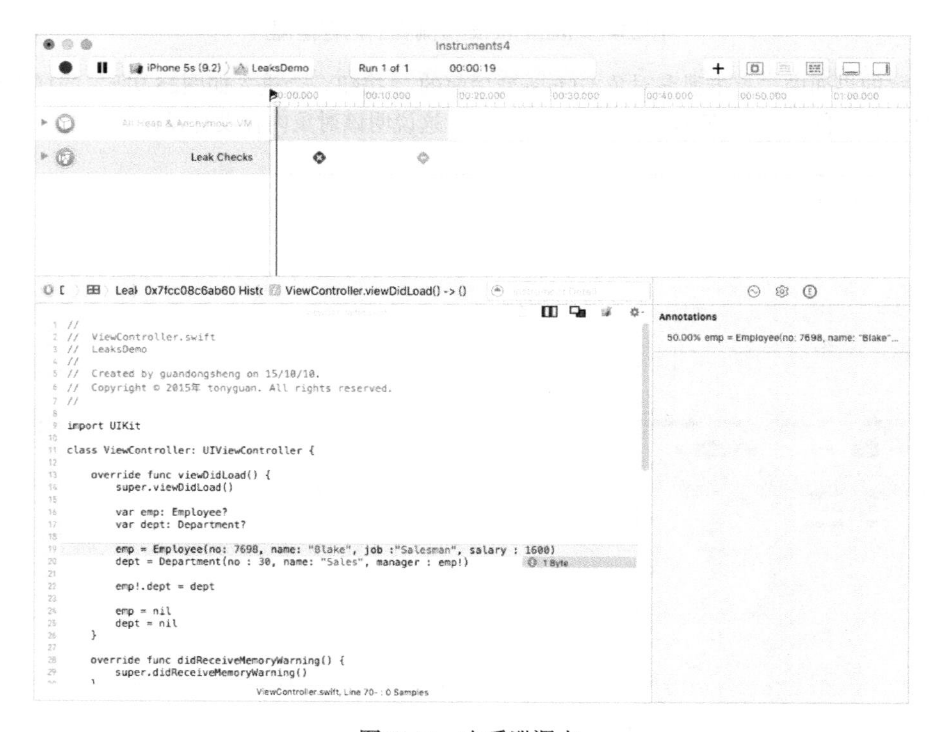

图25-11　查看泄漏点

图25-11所示的第19行代码是可能的泄漏点。ARC代码中的内存泄漏多半是由于强引用循环引起的，Leaks模板提供了查看引用循环视图。如图25-12所示，点击Leaks，选择Cycles & Roots菜单项，会打开如图25-13所示的界面。

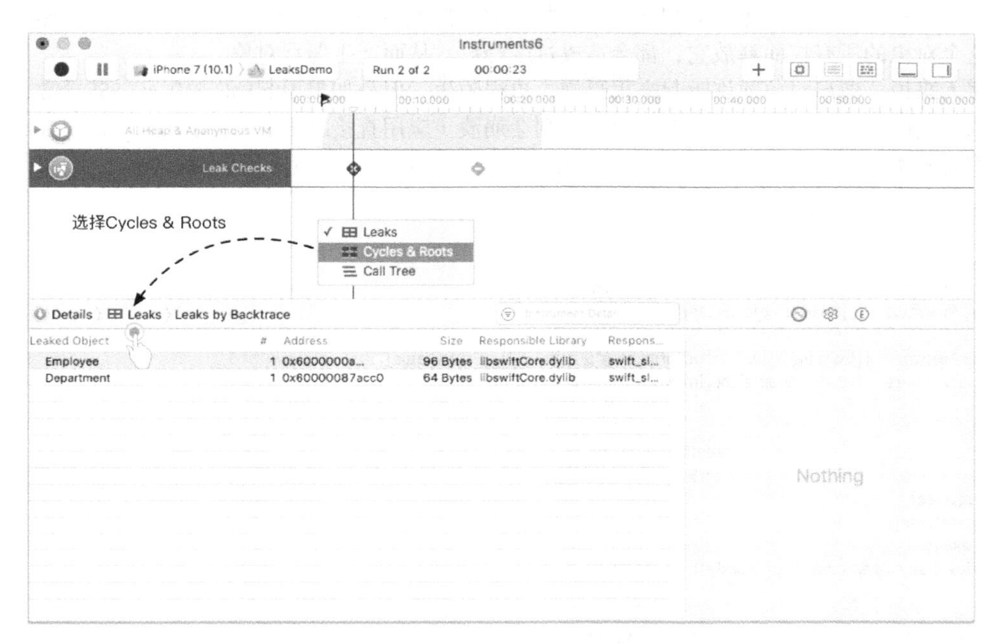

图25-12 选择Cycles & Roots菜单项

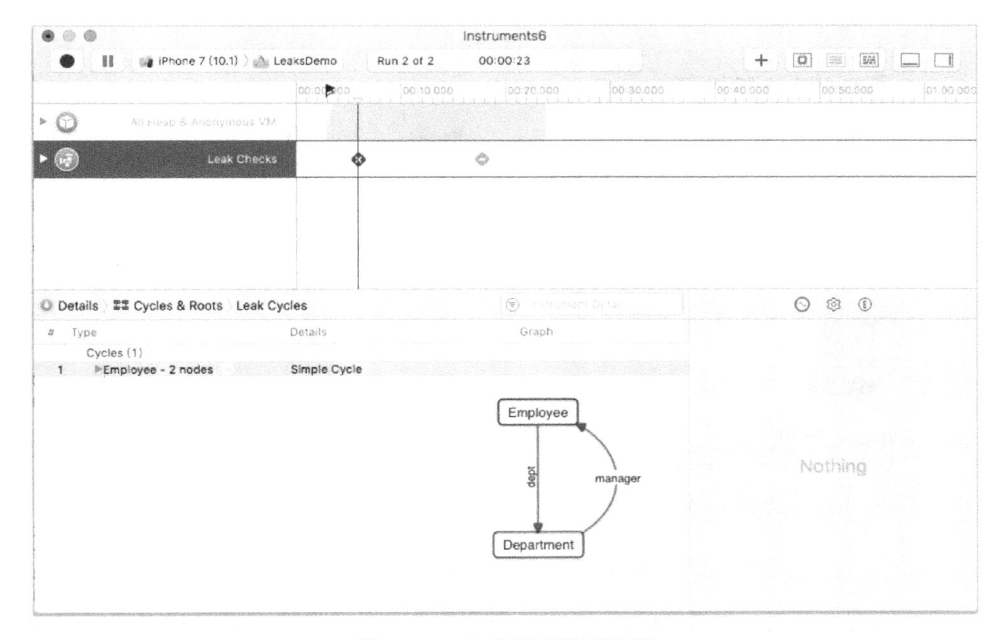

图25-13 查看引用循环视图

从图25-13中可见，Employee与Department之间的引用关系。

事实上，内存泄漏是极其复杂的问题，工具使用是一方面，经验是另一方面。提高经验，然后借助于工具才是解决内存泄漏的根本。

25.1.4　使用 Instruments 工具检查僵尸对象

内存泄漏指一个对象或变量在使用完成后没有释放掉。如果我们走了另外一个极端情况，会是什么样呢？这就导致过度释放问题，从而使对象“僵尸化”，该对象则被称为僵尸对象。如果一个对象已经被释放过了，或者调用者没有这个对象的所有权而释放它，都会造成过度释放，从而产生僵尸对象。

对于很多人来说，僵尸对象或许听起来很恐怖，也很陌生，但是如果说起EXC_BAD_ACCESS异常，可能大家并不陌生。如果应用的某个方法试图调用僵尸对象，则会崩溃（应用直接跳出），并抛出EXC_BAD_ACCESS异常。

下面我们看看本节配套的Objective-C工程中ViewController的代码片段：

```
- (void)tableView:(UITableView *)tableView
➥didSelectRowAtIndexPath:(NSIndexPath *)indexPath {
    NSUInteger row = [indexPath row];
    NSDictionary *rowDict = [self.listTeams objectAtIndex:row];
    NSString *rowValue  =  [rowDict objectForKey:@"name"];

    NSString *message = [[NSString alloc] initWithFormat:@"您选择了%@队。", rowValue];
    UIAlertView *alert = [[UIAlertView alloc]initWithTitle:@"请选择球队"
                                                   message:message
                                                  delegate:self
                                         cancelButtonTitle:@"Ok"
                                         otherButtonTitles:nil];
    [alert release];                                                    ①
    [message release];
    [alert show];                                                       ②
    [tableView deselectRowAtIndexPath:indexPath animated:TRUE];
}
```

注意看上述代码中的第①行和第②行，你会发现什么问题吗？程序运行时，抛出EXC_BAD_ACCESS异常。假设我们现在无法找到问题，可以使用Instruments工具的Zombies分析模板。按照图25-14所示选择Zombies模板，接着点击Choose按钮就可以进入了。

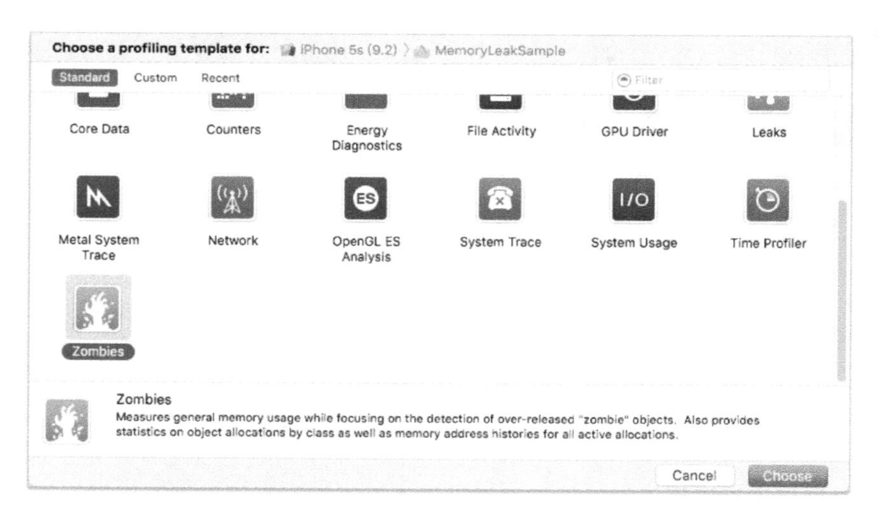

图25-14　Instruments的Zombies模板

当点击录制按钮开始运行时，如果发现僵尸对象，就会弹出一个对话框，如图25-15所示，点击其中的按钮，便会在屏幕下方显示僵尸对象的详细信息（如图25-16所示）。

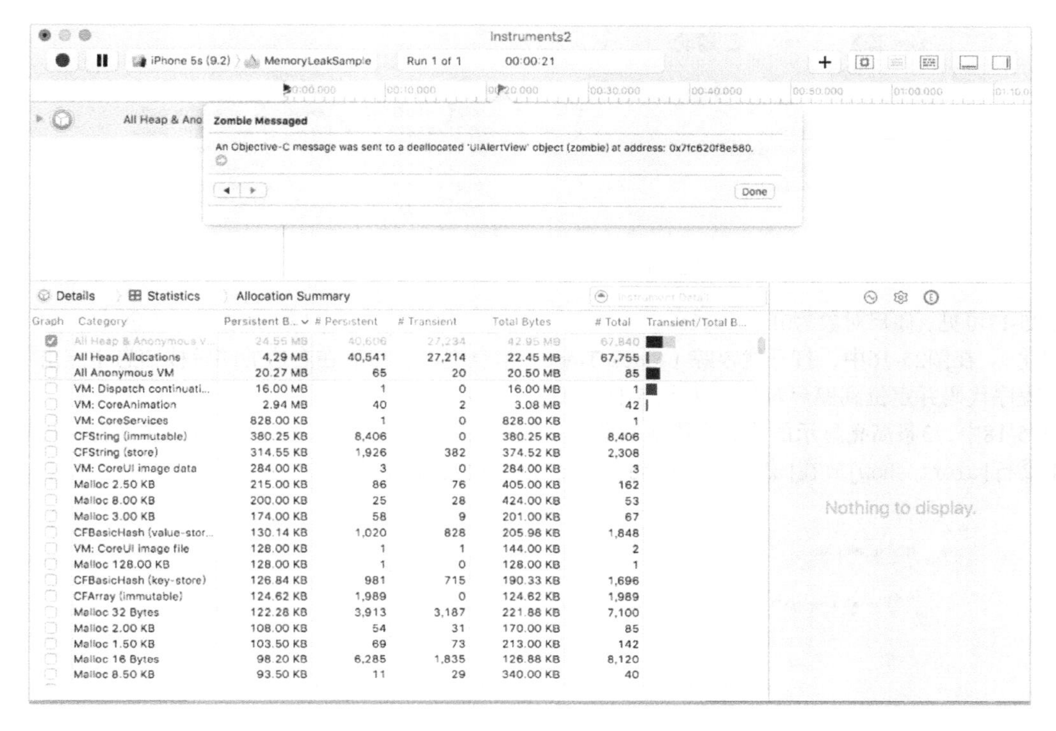

图25-15 僵尸对象信息

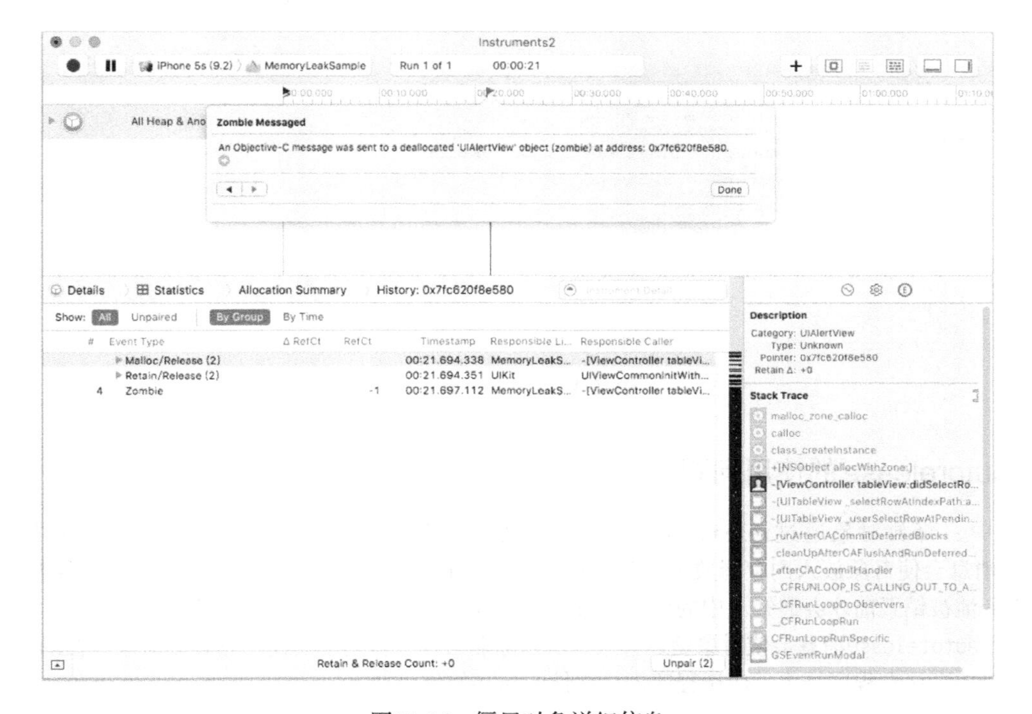

图25-16 僵尸对象详细信息

我们还可以在图25-16左下角，点击Event Type列的▶按钮，展开Malloc/Release和Retain/Release，查看僵尸对象引用计数的变化。如图25-17所示，其中△RefCt列表示引用计数的变化，RefCt列表示当前引用计数值。

Show: All　Unpaired　By Group　By Time

#	Event Type	Δ RefCt	RefCt	Timestamp	Responsible Li...	Responsible Caller
	▼Malloc/Release (2)			00:21.694.338	MemoryLeakS...	-[ViewController tableV
0	Malloc	+1	1	00:21.694.338	MemoryLeakS...	-[ViewController tableV
3	Release	-1	0	00:21.695.819	MemoryLeakS...	-[ViewController tableV
	▼Retain/Release (2)			00:21.694.351	UIKit	UIViewCommonInitWith
1	Retain	+1	2	00:21.694.351	UIKit	UIViewCommonInitWith
2	Release	-1	1	00:21.694.411	UIKit	UIViewCommonInitWith
4	Zombie		-1	00:21.697.112	MemoryLeakS...	-[ViewController tableV

图25-17　僵尸对象引用计数的变化

从图25-17可见，僵尸对象为UIAlertView类型，从上到下僵尸对象的引用计数变化是：1（创建）→0（释放）→-1（僵尸化）。在图25-16中，打开栈跟踪（Stack Trace）信息视图，然后在右边的跟踪栈信息中点击▣条目，进入我们的程序代码并定位到僵尸对象，如图25-18所示。

在图25-18中，3条高亮显示的代码会影响对象的引用计数，从中不难发现问题。就本例而言，我们需要将本节开头第②行[alert show]放在[alert release]语句之前调用就可以了。

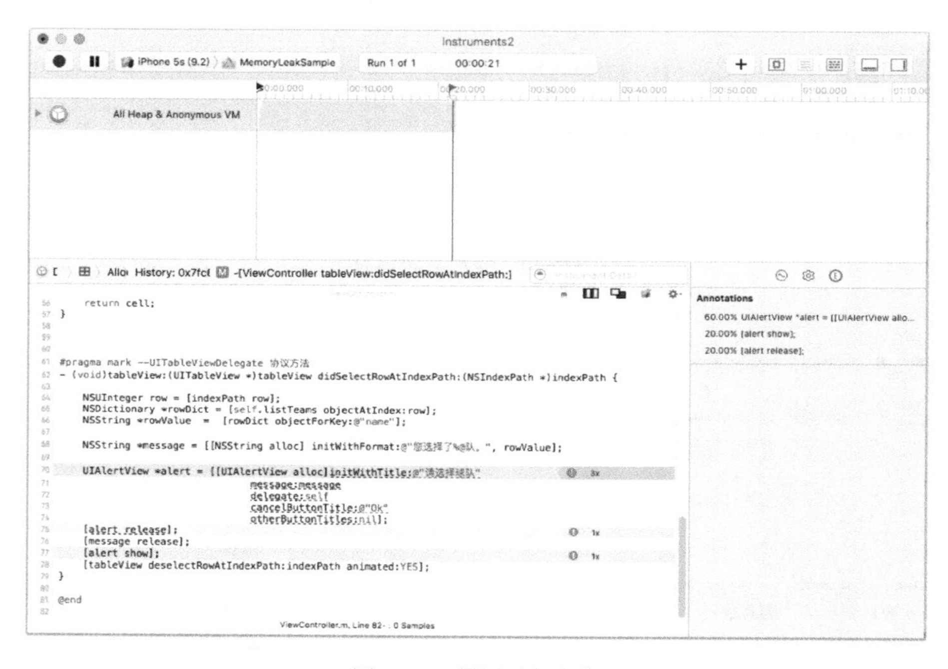

图25-18　僵尸对象定位

25.1.5　autorelease的使用问题

在MRC中，释放对象通过release或autorelease消息实现，其中release消息会立刻使引用计数减1，autorelease消息会使对象放入内存释放池中延迟释放，对象的引用计数并不变化，而是向内存释放池中添加一条记录，直到池被销毁前才会通知池中的所有对象全部发送release消息，才真正将引用计数减少。

由于使用autorelease消息会使对象延迟释放，所以除非必须，否则不要使用它释放对象。在iOS程序中，内存释放池的释放默认在程序结束。应用程序入口文件main.m的代码如下：

```
int main(int argc, char *argv[]) {
    @autoreleasepool {
        return UIApplicationMain(argc, argv, nil, NSStringFromClass([AppDelegate class]));
    }
}
```

代码被包裹在@autoreleasepool {...}之间，这是池的作用范围，默认是整个应用。如果产生大量对象，采用autorelease释放也会导致内存泄漏。那么什么时候才必须使用autorelease呢？我们看看下面的代码：

```
- (UITableViewCell *)tableView:(UITableView *)tableView
➥cellForRowAtIndexPath: (NSIndexPath *)indexPath {

    static NSString *CellIdentifier = @"CellIdentifier";
    UITableViewCell *cell = [tableView dequeueReusableCellWithIdentifier:CellIdentifier];
    if (cell == nil) {
        cell = [[[UITableViewCell alloc] initWithStyle:UITableViewCellStyleDefault
        ➥reuseIdentifier:CellIdentifier] autorelease];
    }

    NSUInteger row = [indexPath row];
    NSDictionary *rowDict = [self.listTeams objectAtIndex:row];
    cell.textLabel.text =  [rowDict objectForKey:@"name"];

    NSString *imagePath = [rowDict objectForKey:@"image"];
    imagePath = [imagePath stringByAppendingString:@".png"];
    cell.imageView.image = [UIImage imageNamed:imagePath];

    cell.accessoryType = UITableViewCellAccessoryDisclosureIndicator;

    return cell;
}
```

在上述代码中，cell对象不能马上释放，我们需要使用它设置表视图界面。autorelease一般用在为其他调用者提供对象的方法中，对象在该方法中不能马上释放，而需要延迟释放。

此外，还有一种情况需要使用autorelease，就是使用静态工厂方法获得对象时，这是因为静态工厂方法内部使用了autorelease。使用静态工厂方法的代码如下：

```
NSString *message = [NSString stringWithFormat:@"您选择了%@队。", rowValue];
```

该对象的所有权虽然不是当前调用者，但它是由iOS系统通过发送autorelease消息放入到池中的。当然，这一切对于开发者都是不可见的，我们也要注意减少使用这样的语句。

25.1.6 响应内存警告

好的应用应该在系统内存警告的情况下释放一些可以重新创建的资源。在iOS中，我们可以在应用程序委托对象、视图控制器以及其他类中获得系统内存警告消息。

1. 应用程序委托对象

在应用程序委托对象中接收内存警告消息，需要重写applicationDidReceiveMemoryWarning:方法，具体可参考本节实例代码中AppDelegate的代码片段：

```
- (void)applicationDidReceiveMemoryWarning:(UIApplication *)application {
    NSLog(@"AppDelegate中调用applicationDidReceiveMemoryWarning:");
}
```

2. 视图控制器

在视图控制器中接收内存警告消息，需要重写didReceiveMemoryWarning方法，具体可参考本节实例代码中ViewController的代码片段：

```
- (void)didReceiveMemoryWarning {
    NSLog(@"ViewController中didReceiveMemoryWarning调用");
    [super didReceiveMemoryWarning];
    //释放成员变量
    [_listTeams release];
}
```

注意，释放资源代码应该放在[super didReceiveMemoryWarning]语句后面。

3. 其他类

在其他类中可以使用通知。在发生内存警告时，iOS系统会发出UIApplicationDidReceiveMemoryWarning-Notification通知，凡是在通知中心注册了该通知的类都会接收到内存警告通知，具体可参考本节实例代码中ViewController的代码片段：

```
- (void)viewDidLoad {
    [super viewDidLoad];

    NSBundle *bundle = [NSBundle mainBundle];
    NSString *plistPath = [bundle pathForResource:@"team"
                                           ofType:@"plist"];
    //获取属性列表文件中的全部数据
    NSArray *array = [[NSArray alloc] initWithContentsOfFile:plistPath];
    self.listTeams = array;
    [array release];

    //接收内存警告通知，调用handleMemoryWarning方法处理
    NSNotificationCenter *center = [NSNotificationCenter defaultCenter];
    [center addObserver:self
               selector:@selector(handleMemoryWarning)
                   name:UIApplicationDidReceiveMemoryWarningNotification
                 object:nil];

}
//处理内存警告
-(void) handleMemoryWarning {
    NSLog(@"ViewController中调用handleMemoryWarning");
}
```

在上述代码中，我们在viewDidLoad方法中注册UIApplicationDidReceiveMemoryWarningNotification消息，接收到报警信息后调用handleMemoryWarning方法。这些代码完全可以写在其他类中，直接在ViewController中重写didReceiveMemoryWarning方法就可以了。本例只是示意性地介绍一下UIApplicationDidReceiveMemoryWarning-Notification报警消息。

内存警告在设备上并不经常出现，一般我们没有办法模拟，但模拟器上有一个功能可以模拟内存警告。启动模拟器，选择Hardware→Simulate Memory Warning模拟器菜单，这时我们会在输出窗口中看到内存警告发生了，具体如下所示：

```
2017-1-19 15:58:51.032 MemoryLeakSample[1396:41574] Received memory warning.
2017-1-19 15:58:51.033 MemoryLeakSample[1396:41574] AppDelegate中调用applicationDidReceiveMemoryWarning:
2017-1-19 15:58:51.034 MemoryLeakSample[1396:41574] ViewController中handleMemoryWarning调用
2017-1-19 15:58:51.034 MemoryLeakSample[1396:41574] ViewController中didReceiveMemoryWarning调用
```

25.2 优化资源文件

从狭义上讲，资源文件是放置在应用程序本地与应用程序一起编译、打包和发布的非程序代码文件，如应用中用到的声音、视频、图片和文本文件等。从广义上讲，资源文件可以放置于任何地方，既可以放置于本地，也可以放在云服务器中。

在iOS中，本地资源文件编译后，会放置于应用程序包文件中（即<应用名>.app文件）。下列代码用于访问如图25-19所示的team.plist本地资源文件：

```
let plistPath = Bundle.main.path(forResource: "team", ofType: "plist")
```

```
NSString *plistPath = [[NSBundle mainBundle] pathForResource:@"team"
➥ofType:@"plist"];
```

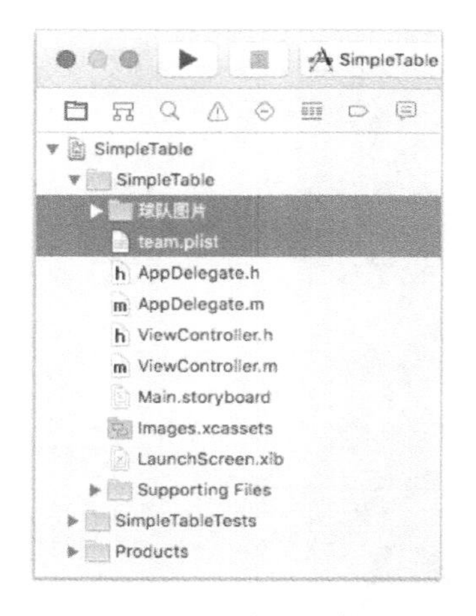

图25-19　资源文件

图25-19所示的“球队图片”组也放置了一些资源文件。添加资源文件的方法是通过右键添加文件到工程中。资源文件在使用的过程中需要优化，包括文件格式、文件类型、文件大小和文件结构等方面，使得它更适合某个应用。“适合”这两个字很重要。当然，优化方向有很多，下面我们从图片文件优化和音频文件优化这两个方面介绍一下。

25.2.1　图片文件优化

图片文件优化包括文件格式和文件大小的优化。在移动设备中，支持的图片格式主要是PNG、GIF和JPEG格式，苹果推荐使用PNG格式。在Xcode中，集成了第三方PNG优化工具pngcrush[①]，它可以在编译的时候对PNG格式文件进行优化和压缩，而我们只需要设定如图25-20所示的编译参数Compress PNG Files为Yes就可以了。

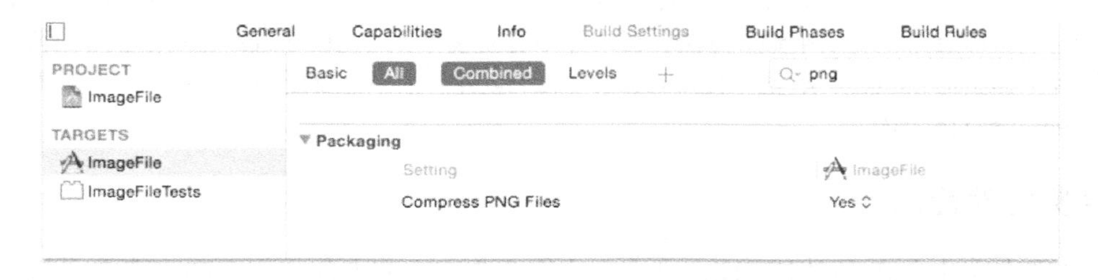

图25-20　设定编译参数Compress PNG Files

打开本节案例中“测试图片”目录中的background.png文件，在Finder中查看该文件的属性，它是一个320 × 480px、大小为317 KB的PNG图片，如图25-21a所示。

使用Xcode编译工程，在编译之后的目录中找到ImageFile.app包文件。打开包文件，查看目录中background.png文件的属性，可以发现该文件是205 KB的PNG图片了，如图25-21b所示。

Xcode工具可以在编译时优化PNG图片，但是即便经过优化和压缩的PNG图片文件，也比JPEG图片文件大得多。打开“测试图片”目录中的background-8(优化压缩).png文件和background-h.jpg文件，比较可以发现，前者是经过优化的质量最低的PNG-8（8位PNG格式）文件，其大小是61KB；后者是经过优化的质量最高的JPEG格式文件，其大小是22 KB。在本例中，PNG文件几乎是JPEG文件的3倍。

① 它是PNG图形文件优化工具（http://pmt.sourceforge.net/pngcrush/），提供了基于微软Windows、Unix和Linux的命令行工具。

(a)

(b)

图25-21　PNG文件编译前后对比

如果是本地资源文件，这样的差别不是很大，但如果是分布在网络云服务器中的资源文件，应用在加载这些图片时，会从网络上下载到本地，这时候JPEG就很有优势了。

综上所述，如果在本地资源的情况下，我们应该优先使用PNG格式文件，如果资源来源于网络，最好采用JPEG格式文件。

另外，图片是一种很特殊的资源文件。创建`UIImage`对象时，可以使用静态工厂方法`+imageNamed:`和构造函数`-initWithContentsOfFile:`。`+imageNamed:`方法会在内存中建立缓存，这些缓存直到应用停止才清除。如果是贯穿整个应用的图片（如图标、logo等），推荐使用`+imageNamed:`创建。如果是仅使用一次的图片，推荐使用构造函数`-initWithContentsOfFile:`创建。

25.2.2　音频文件优化

在讨论音频文件优化之前，我们先讨论一下音频文件格式。在iOS平台中，主要的音频文件格式有以下4种。

- **WAV文件**。WAV是一种由微软和IBM联合开发的用于音频数字存储的文件格式。WAV文件的格式灵活，可以存储多种类型的音频数据。由于文件较大，不太适合于移动设备这些存储容量小的设备。
- **MP3（MPEG Audio Layer 3）文件**。MP3利用MPEG Audio Layer 3技术，将数据以1∶10甚至1∶12的压缩率压缩成容量较小的文件。MP3是一种有损压缩格式，它尽可能地去掉人耳无法感觉的部分和不敏感的部分。这么高的压缩比率非常适合于移动设备这些存储容量小的设备，现在非常流行。
- **CAFF（Core Audio File Format）文件**。CAFF是苹果开发的专门用于macOS和iOS系统的无压缩音频格式，它被设计用来替换老的WAV格式。
- **AIFF（Audio Interchange File Format）文件**。AIFF是苹果开发的专门用于macOS系统的专业的音频文件格式。AIFF的压缩格式是AIFF-C（或AIFC），将数据以4∶1压缩率进行压缩，应用于macOS和iOS系统。

音频文件优化包括文件格式和文件大小的优化，但也要考虑到文件使用场景、采用的技术（OpenAL、AVAudioPlayer）等因素。在iOS应用中，使用本地音频资源文件的主要应用场景是背景音乐和音乐特效，下面我们从这两个方面介绍相关的优化技术。

1. 背景音乐优化

背景音乐会在应用中反复播放，它会一直驻留在内存中并耗费CPU，所以更合适比较小的文件，而压缩文件是不错的选择。压缩文件主要有AIFC和MP3这两种格式，一般我们首选AIFC，因为这是苹果推荐的格式。但是我们获得的原始文件格式不一定是AIFC，这种情况下我们需要使用afconvert工具[①]将其转换为AIFC格式。在终端中执行如下命令：

```
$ afconvert -f AIFC -d ima4 Fx08822_cast.wav
```

其中-f AIFC参数用于转换为AIFC格式，-d ima4参数指定解码方式，Fx08822_cast.wav是要转换的源文件。转换成功后，会在相同目录下生成Fx08822_cast.aifc文件。本例中的源文件Fx08822_cast.wav的大小是295 KB，转换之后的Fx08822_cast.aifc文件的大小是82 KB。当然，afconvert工具也可以转换MP3等其他压缩格式文件。如果我们同时有WAV文件，就应该优先采用WAV文件。MP3本身是有损压缩，如果再经过afconvert转换，音频的质量会受到影响。

2. 音乐特效优化

音乐特效用于很多游戏中，如发射子弹、敌人被打死或按钮点击等发出的声音，这些声音都是比较短的。如果追求震撼的3D效果，可以采用苹果专用的无压缩CAFF格式文件，其他格式的文件尽量不要考虑。一般不要使用压缩音频文件，这主要是因为音乐特效通常采用OpenAL技术，它只接受无压缩的音频文件。另外，压缩音频文件都会造成音质的丢失。如果我们没有CAFF格式的文件，也可以使用afconvert工具将其转换为CAFF格式。在终端中执行如下命令：

```
$ afconvert -f caff -d LEI16 Fx08822_cast.wav
```

其中-f caff参数用于转换为CAFF格式，-d LEI16参数指定解码方式，Fx08822_cast.wav是要转换的源文件。默认音频的采样频率为22 050Hz，如果想提高音频采样频率，可以通过如下命令：

```
$ afconvert -f caff -d LEI16@44100  Fx08822_cast.wav
```

其中-d LEI16@44100参数中的44100表示音频采样频率为44100 Hz。

如果我们采用的资源文件不在本地，而是在分布在网络云服务器中，那么情况就另当别论了。应用在加载这些音频文件时，带宽往往是要考虑的问题，减小文件大小胜过对音质的要求，这种情况下MP3格式是非常适合的。

综上所述，音频文件在使用本地资源的情况下，应用于背景音乐时AIFC格式是首选，应用于音乐特效时CAFF格式是首选。如果资源来源于网络，最好采用MP3格式文件。

25.3 延迟加载

延迟加载（lazy load）指一些对象不是在应用和视图等初始化时创建，而是在用到它的时候创建。当应用中有一些对象并不经常使用时，延迟加载可以提高程序性能。

25.3.1 资源文件的延迟加载

首先，我们要考虑的就是对资源文件的延迟加载。由于资源文件的访问涉及IO操作，这本身就会耗费一定的CPU时间，如果文件比较大而且加载时机又不合适，就会造成内存浪费。前面我们了解到的资源文件包括图片、音频和文本文件等，无论是什么类型的文件，有些情况下采用延迟加载是很有必要的。

例如，我们有如图25-22所示的需求，可以使用分屏控件（UIPageControl）左右滑动屏幕来浏览这3张图片。

① 苹果在macOS中提供的音频转换命令行工具，位于/usr/bin目录下。

图25-22　图片延迟加载实例

PageControlNavigation实例是没有采用延迟加载的实现代码，其中ViewController的代码如下：

```
//ViewController.swift文件
import UIKit

//定义屏幕宽度
let S_WIDTH: CGFloat = UIScreen.main.bounds.size.width
//定义屏幕高度
let S_HEIGHT: CGFloat = UIScreen.main.bounds.size.height

class ViewController: UIViewController, UIScrollViewDelegate {

    var page1: UIImageView!
    var page2: UIImageView!
    var page3: UIImageView!

    @IBOutlet weak var scrollView: UIScrollView!
    @IBOutlet weak var pageControl: UIPageControl!

    override func viewDidLoad() {
        super.viewDidLoad()

        self.scrollView.delegate = self
        self.scrollView.contentSize = CGSizeMake(S_WIDTH * 3, S_HEIGHT)

        self.page1 = UIImageView(frame: CGRectMake(0.0, 0.0, S_WIDTH,
        ➥S_HEIGHT))
        self.page1.image = UIImage(named: "达芬奇-蒙娜丽莎.png")

        self.page2 = UIImageView(frame:
        ➥CGRectMake(S_WIDTH, 0.0, S_WIDTH, S_HEIGHT))
        self.page2.image = UIImage(named: "罗丹-思想者.png")

        self.page3 = UIImageView(frame:
        ➥CGRectMake(2 * S_WIDTH, 0.0, S_WIDTH, S_HEIGHT))
        self.page3.image = UIImage(named: "保罗克利-肖像.png")

        self.scrollView.addSubview(page1)
        self.scrollView.addSubview(page2)
        self.scrollView.addSubview(page3)

    }
```

```
//ViewController.m文件
#import "ViewController.h"

//定义屏幕宽度的宏
#define S_WIDTH [[UIScreen mainScreen] bounds].size.width
//定义屏幕高度的宏
#define S_HEIGHT [[UIScreen mainScreen] bounds].size.height

@interface ViewController () <UIScrollViewDelegate>

@property (strong, nonatomic) UIImageView *page1;
@property (strong, nonatomic) UIImageView *page2;
@property (strong, nonatomic) UIImageView *page3;

@property (weak, nonatomic) IBOutlet UIScrollView *scrollView;
@property (weak, nonatomic) IBOutlet UIPageControl *pageControl;

- (IBAction)changePage:(id)sender;

@end

@implementation ViewController

- (void)viewDidLoad {
    [super viewDidLoad];

    self.scrollView.delegate = self;
    self.scrollView.contentSize = CGSizeMake(S_WIDTH * 3, S_HEIGHT);

    self.page1 = [[UIImageView alloc] initWithFrame:
    ➥CGRectMake(0.0f, 0.0f, S_WIDTH, S_HEIGHT)];
    self.page1.image = [UIImage imageNamed:@"达芬奇-蒙娜丽莎.png"];

    self.page2 = [[UIImageView alloc] initWithFrame:
    ➥CGRectMake(S_WIDTH, 0.0f, S_WIDTH, S_HEIGHT)];
    self.page2.image = [UIImage imageNamed:@"罗丹-思想者.png"];

    self.page3 = [[UIImageView alloc] initWithFrame:
    ➥CGRectMake(2 * S_WIDTH, 0.0f, S_WIDTH, S_HEIGHT)];
    self.page3.image = [UIImage imageNamed:@"保罗克利-肖像.png"];
```

```
    //实现UIScrollViewDelegate协议中的方法
    func scrollViewDidScroll(scrollView: UIScrollView) {
        let offset = scrollView.contentOffset
        self.pageControl.currentPage = Int(offset.x) / Int(S_WIDTH)
    }

    @IBAction func changePage(sender: AnyObject) {
        UIView.animateWithDuration(0.3, animations: {
            let whichPage = self.pageControl.currentPage
            self.scrollView.contentOffset = CGPointMake(S_WIDTH *
            ➥CGFloat(whichPage), 0)
        })
    }
}
```

```
    [self.scrollView addSubview:self.page1];
    [self.scrollView addSubview:self.page2];
    [self.scrollView addSubview:self.page3];
}
//实现UIScrollViewDelegate协议中的方法
- (void) scrollViewDidScroll: (UIScrollView *) aScrollView {
    CGPoint offset = aScrollView.contentOffset;
    self.pageControl.currentPage = offset.x / S_WIDTH;
}

#pragma mark --
#pragma mark PageControl stuff
- (IBAction)changePage:(id)sender {
    [UIView animateWithDuration:0.3f animations:^{
        NSInteger whichPage = self.pageControl.currentPage;
        self.scrollView.contentOffset = CGPointMake(S_WIDTH * whichPage,
        ➥0.0f);
    }];
}

@end
```

我们是在viewDidLoad方法中一次加载全部3张图片，但是有的时候用户不一定会浏览后面的图片，他可能只看到第一张或第二张，而第三张并没有看，此时后面的两张图片仍然加载内存的话，会造成内存浪费。

采用延迟加载实现时（见实例LazyLoadPageControlNavigation），ViewController的代码如下：

```
//ViewController.swift文件
import UIKit

//定义屏幕宽度
let S_WIDTH: CGFloat = UIScreen.main.bounds.size.width
//定义屏幕高度
let S_HEIGHT: CGFloat = UIScreen.main.bounds.size.height

class ViewController: UIViewController, UIScrollViewDelegate {

    var page1: UIImageView!
    var page2: UIImageView!
    var page3: UIImageView!

    @IBOutlet weak var scrollView: UIScrollView!
    @IBOutlet weak var pageControl: UIPageControl!

    override func viewDidLoad() {
        super.viewDidLoad()

        self.scrollView.delegate = self
        self.scrollView.contentSize = CGSizeMake(S_WIDTH * 3, S_HEIGHT)

        self.page1 = UIImageView(frame: CGRectMake(0.0, 0.0, S_WIDTH,
        ➥S_HEIGHT))
        self.page1.image = UIImage(named: "达芬奇-蒙娜丽莎.png")          ①

        self.scrollView.addSubview(page1)

    }

    //实现UIScrollViewDelegate协议中的方法
    func scrollViewDidScroll(scrollView: UIScrollView) {
        let offset = scrollView.contentOffset
        self.pageControl.currentPage = Int(offset.x) / Int(S_WIDTH)

        self.loadImage(self.pageControl.currentPage + 1)

    }
```

```
//ViewController.m文件
#import "ViewController.h"

//定义屏幕宽度的宏
#define S_WIDTH [[UIScreen mainScreen] bounds].size.width
//定义屏幕高度的宏
#define S_HEIGHT [[UIScreen mainScreen] bounds].size.height
@interface ViewController () <UIScrollViewDelegate>

@property(strong, nonatomic) UIImageView *page1;
@property(strong, nonatomic) UIImageView *page2;
@property(strong, nonatomic) UIImageView *page3;

@property(weak, nonatomic) IBOutlet UIScrollView *scrollView;
@property(weak, nonatomic) IBOutlet UIPageControl *pageControl;

- (IBAction)changePage:(id)sender;

- (void)loadImage:(NSInteger)nextPage;

@end

@implementation ViewController

- (void)viewDidLoad {
    [super viewDidLoad];

    self.scrollView.delegate = self;
    self.scrollView.contentSize = CGSizeMake(S_WIDTH * 3, S_HEIGHT);

    self.page1 = [[UIImageView alloc] initWithFrame:
    ➥CGRectMake(0.0f, 0.0f, S_WIDTH, S_HEIGHT)];
    self.page1.image = [UIImage imageNamed:@"达芬奇-蒙娜丽莎.png"];          ①

    [self.scrollView addSubview:self.page1];
}
//实现UIScrollViewDelegate协议中的方法
- (void)scrollViewDidScroll:(UIScrollView *)aScrollView {
    CGPoint offset = aScrollView.contentOffset;
```

```
    @IBAction func changePage(sender: AnyObject) {
        UIView.animateWithDuration(0.3, animations: {
            let whichPage = self.pageControl.currentPage
            self.scrollView.contentOffset = CGPointMake(S_WIDTH * CGFloat
            ➥(whichPage), 0)
            self.loadImage(self.pageControl.currentPage + 1)
        })
    }

    func loadImage(nextPage: Int) {

        if nextPage == 1 && self.page2 == nil {
            self.page2 = UIImageView(frame:
            ➥CGRectMake(S_WIDTH, 0.0, S_WIDTH, S_HEIGHT))
            self.page2.image = UIImage(named: "罗丹-思想者.png")
            self.scrollView.addSubview(page2)
        }


        if nextPage == 2 && self.page3 == nil {
            self.page3 = UIImageView(frame:
            ➥CGRectMake(2 * S_WIDTH, 0.0, S_WIDTH, S_HEIGHT))
            self.page3.image = UIImage(named: "保罗克利-肖像.png")
            self.scrollView.addSubview(page3)
        }
    }
}
```

```
    self.pageControl.currentPage = offset.x / S_WIDTH;
    [self loadImage:self.pageControl.currentPage + 1];
}

#pragma mark --
#pragma mark PageControl stuff

- (IBAction)changePage:(id)sender {
    [UIView animateWithDuration:0.3f animations:^{
        NSInteger whichPage = self.pageControl.currentPage;
        self.scrollView.contentOffset = CGPointMake(S_WIDTH * whichPage,
        ➥0.0f);
        [self loadImage:self.pageControl.currentPage + 1];
    }];
}

- (void)loadImage:(NSInteger)nextPage {
    if (nextPage == 1 && self.page2 == nil) {
        self.page2 = [[UIImageView alloc] initWithFrame:
        ➥CGRectMake(S_WIDTH, 0.0f, S_WIDTH, S_HEIGHT)];
        self.page2.image = [UIImage imageNamed:@"罗丹-思想者.png"];
        [self.scrollView addSubview:self.page2];
    }

    if (nextPage == 2 && self.page3 == nil) {
        self.page3 = [[UIImageView alloc] initWithFrame:
        ➥CGRectMake(2 * S_WIDTH, 0.0f, S_WIDTH, S_HEIGHT)];
        self.page3.image = [UIImage imageNamed:@"保罗克利-肖像.png"];
        [self.scrollView addSubview:self.page3];
    }
}

@end
```

我们重新修改了这个实例，在viewDidLoad方法中只加载第一张图片，见第①行。如果用户滑动屏幕或点击分屏控件进入第二个屏幕，则调用loadImage:方法加载第二张图片。类似地，如果要进入第三个屏幕，则调用loadImage:方法加载第三张图片。

在这两种实现方式中，LazyLoadPageControlNavigation实现了延迟加载。很显然，LazyLoadPageControl-Navigation的延迟加载友好很多。那么，两者究竟有多大的差别，这是可以量化的。通过Instruments工具的Allocations模板，可以分析ViewController视图控制器加载时内存占用方面的差别。图25-23是无延迟加载实现案例的Allocations模板跟踪，图25-24是采用延迟加载实现案例的Allocations模板跟踪。

如图25-23所示，界面启动时，内存占用马上达到9.20 MB。如图25-24所示，界面启动时，内存占用为4.50 MB，当我们滑动到第二和第三屏幕时，内存占用达到9.18 MB，内存变化会有明显的两个阶梯。

在上面的案例中可以发现，延迟加载的优势很明显。如果一定会访问到资源文件，则延迟加载这些资源文件时，在内存占用方面就没有优势了，但是在界面加载速度方面还是有优势的。

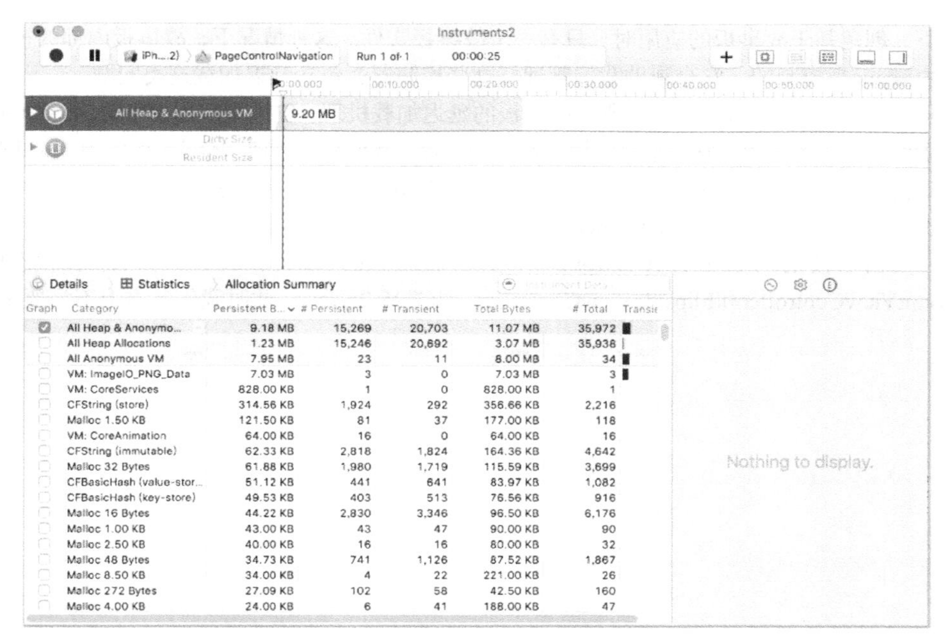

图25-23 采用无延迟加载实现的案例的Allocations模板跟踪

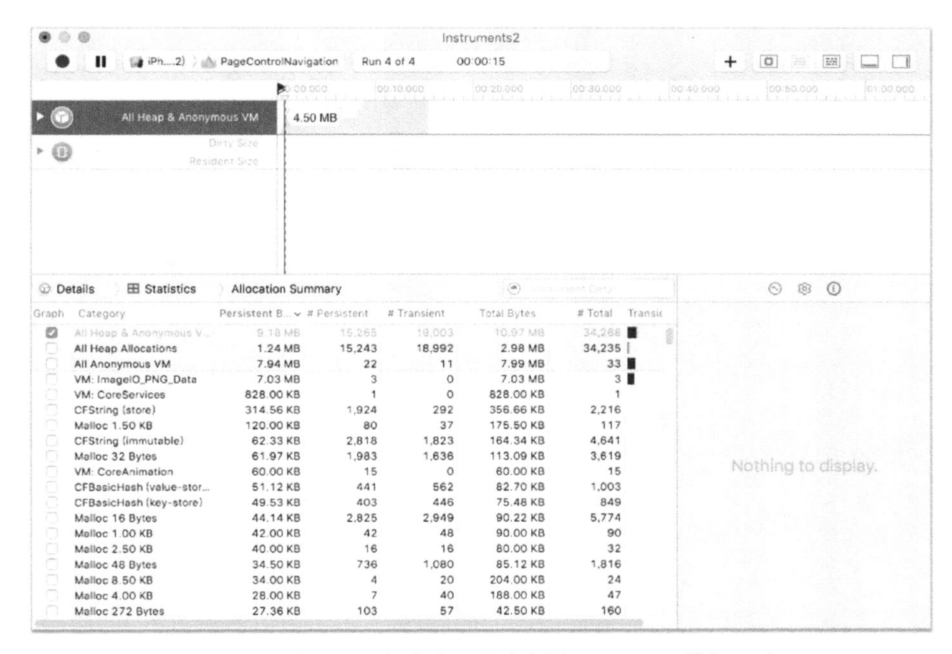

图25-24 采用延迟加载实现的案例的Allocations模板跟踪

25.3.2 故事板文件的延迟加载

XIB和故事板也都属于资源文件，它是非常特殊的资源文件，应用不仅需要读取它，而且要根据里面描述的信息创建视图和子视图，以及它们的视图控制器等对象。创建这么多对象会耗费很多时间，占用很多内存，因此，它们的延迟加载问题非常重要。

默认情况下，创建基于故事板的应用时，只有一个故事板文件。这种情况下，故事板内部的视图控制器的创建和加载都是由Segue来控制的，它会帮助我们管理好这些控制器，包括延迟加载这类问题。

这里我们创建一个实用型应用程序[①]，研究故事板的延迟加载机理。实用型应用一般会有两个视图：主视图，它显示应用的主要功能；子视图，它用来对应用进行一些设置。我们自己创建一个实用型应用，如图25-25所示。

在Xcode 5之前，可以使用Utility Application模板创建，Xcode 5之后就没有这个模板了，此时可以通过Single View Application模板创建StoryboardLazyLoadDemo工程。

在主视图中点击ⓘ按钮，MainViewController会延迟加载FlipsideViewController，然后弹出模态模式。使用模态Segue连接MainViewController和FlipsideViewController，如图25-26所示，此时我们基本上不需要编写什么代码。

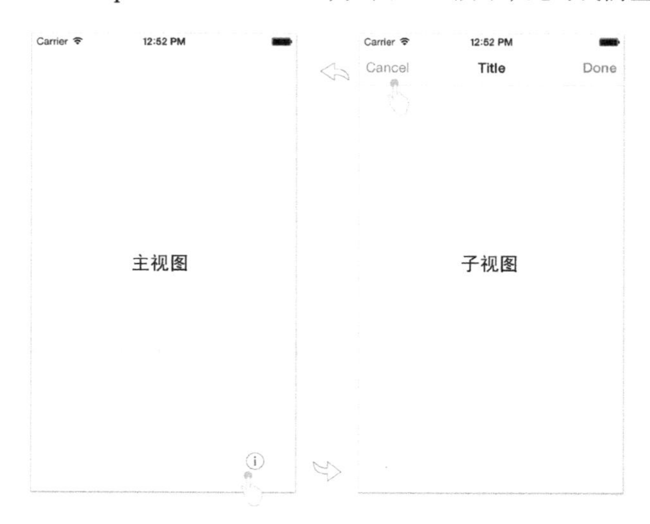

图25-25　实用型应用

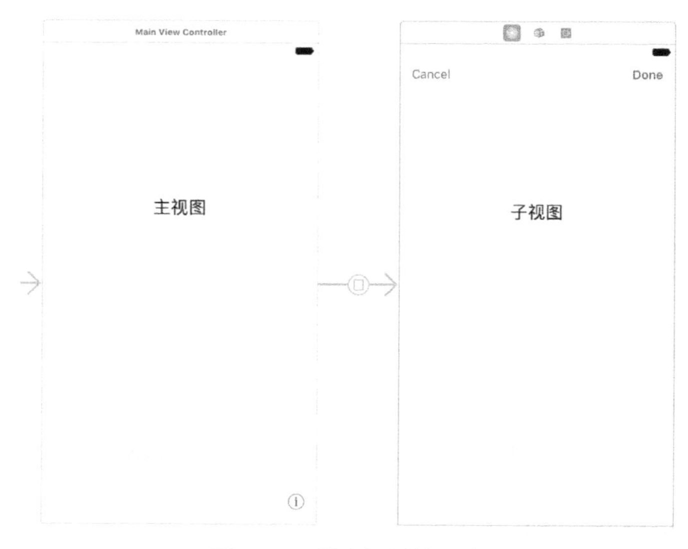

图25-26　模态视图的过渡

① 实用型应用程序完成的简单任务对用户输入要求很低。用户打开实用型应用程序，是为了快速查看信息摘要或是在少数对象上执行简单任务。天气程序就是一个实用型应用程序的典型例子，它在一个易读的摘要中显示了重点明确的信息。

Segue定义了两个视图控制器的导航关系，也用来维护和管理下一个视图控制器的延迟加载时机，这种情况下我们无法“插手”视图控制器的延迟加载。但是一种情况除外，那就是使用了故事板，而控制器之间没有定义导航关系，也没有定义过渡，如图25-27所示。

图25-27 没有定义过渡的故事板

这种情况下，添加showInfo:方法响应主视图的ⓘ按钮点击事件，具体可参考StoryboardLazyLoadNoSegue-Demo工程的MainViewController，相关代码如下：

```swift
@IBAction func showInfo(sender: AnyObject) {

    //获得当前故事板对象
    let mainStoryboard = self.storyboard                                    ①

    //从故事板文件中创建故事板对象
    //let mainStoryboard = UIStoryboard(name: "Main", bundle: nil)          ②

    //通过名为flipsideViewController的Storyboard ID创建视图控制器对象
    let flipsideViewController = mainStoryboard?.
    ➥instantiateViewController(withIdentifier: "flipsideViewController")
    ➥as! FlipsideViewController


    flipsideViewController.modalTransitionStyle =
    ➥UIModalTransitionStyle.flipHorizontal

    self.present(flipsideViewController, animated: true, completion: nil)
}
```

```objc
- (IBAction)showInfo:(id)sender {

    //获得当前故事板对象
    UIStoryboard *mainStoryboard = [self storyboard];                       ①
    //从故事板文件创建Main故事板对象
    //UIStoryboard *mainStoryboard = [UIStoryboard storyboardWithName:
    ➥@"Main" bundle:nil];                                                  ②

    //通过名为flipsideViewController的Storyboard ID创建视图控制器对象
    FlipsideViewController* controller = [mainStoryboard
    ➥instantiateViewControllerWithIdentifier:@"flipsideViewController"];

    controller.modalTransitionStyle = UIModalTransitionStyleFlipHorizontal;
    [self presentViewController:controller animated:TRUE completion:nil];
}
```

在单一故事板文件中，第①行可以获得当前的故事板对象。如果想在多故事板的情况下获得非当前故事板对象，可以通过第②行的UIStoryboard构造函数创建。本例中不用使用该语句，使用它会多创建一个故事板对象，就会占用更多的内存。

25.3.3 XIB 文件的延迟加载

相对于故事板而言，XIB文件要灵活很多。XIB文件有两种：一种是描述视图控制器的，另一种是描述视图的，它们的加载方式有所区别。无论是哪一种，分散管理的XIB文件使我们通过编程方式访问它更加方便。

在Xcode 5之后，不能创建基于XIB文件的工程了，需要通过Single View Application模板创建NibLazyLoadDemo工程，然后删除主故事板文件。接着创建视图控制器，如图25-28所示，一定要选择Also create XIB file复选框，这会帮助我们创建与视图控制器对应的XIB文件。

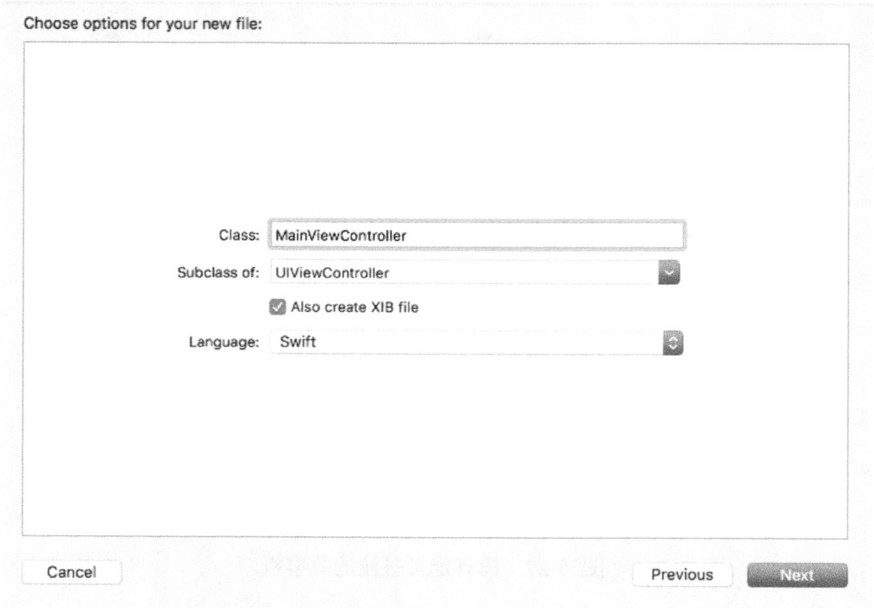

图25-28　创建视图控制器

创建好视图控制器后，我们需要修改AppDelegate使应用启动时能够加载MainViewController。AppDelegate的主要代码如下：

```
@UIApplicationMain
class AppDelegate: UIResponder, UIApplicationDelegate {

    var window: UIWindow?

    var mainViewController: MainViewController?

    func application(_ application: UIApplication,
    ➥didFinishLaunchingWithOptions launchOptions:
    ➥[UIApplicationLaunchOptionsKey: Any]?) -> Bool {

        self.window = UIWindow()                                    ①
        self.window!.frame = UIScreen.main.bounds

        self.mainViewController = MainViewController(nibName:
        ➥"MainViewController", bundle: nil)                        ②
        self.window!.rootViewController = self.mainViewController
        self.window!.makeKeyAndVisible()                            ③

        return true
    }
    ......
}
```

```
#import "AppDelegate.h"
#import "MainViewController.h"

@interface AppDelegate ()

@property (strong, nonatomic) MainViewController *mainViewController;

@end

@implementation AppDelegate

- (BOOL)application:(UIApplication *)application
➥didFinishLaunchingWithOptions:(NSDictionary *)launchOptions {

    self.window = [[UIWindow alloc] init];                           ①
    self.window.frame = [UIScreen mainScreen].bounds;

    self.mainViewController = [[MainViewController alloc]
    ➥initWithNibName:@"MainViewController"bundle:nil];              ②
    self.window.rootViewController = self.mainViewController;
    [self.window makeKeyAndVisible];                                 ③

    return TRUE;
}

......

@end
```

上述代码中，第①行~第③行是我们添加的代码。第①行用于创建Window对象，该对象表示屏幕，所有视图都放到Window中。第②行通过XIB文件创建视图控制器对象，然后再把视图控制器添加到Window对象中。

主视图控制器MainViewController中showInfo:方法的代码如下：

```swift
@IBAction func showInfo(_ sender: AnyObject) {

    //通过XIB文件创建视图控制器对象
    let flipsideViewController =
    ➥FlipsideViewController(nibName: "FlipsideViewController",
    ➥bundle: nil)

    flipsideViewController.modalTransitionStyle =
    ➥UIModalTransitionStyle.FlipHorizontal
    self.present(flipsideViewController, animated: true, completion:nil)
}
```

```objc
- (IBAction)showInfo:(id)sender {
    FlipsideViewController* controller = [[FlipsideViewController alloc]
    ➥initWithNibName:@"FlipsideViewController" bundle:nil];

    controller.modalTransitionStyle = UIModalTransitionStyleFlipHorizontal;
    [self presentViewController:controller animated:TRUE completion:nil];
}
```

本例中的XIB文件是视图控制器XIB文件，我们可以使用视图控制器的initWithNibName:bundle:构造函数从XIB文件中创建视图控制器对象。

有些情况下，故事板和XIB文件会混合使用。在有故事板的工程中，有时候需要使用别人已经编写好的XIB文件和对应类（视图或视图控制器）。当然，通过上面的两种方式也是可以的。

25.4 数据持久化的优化

在iOS中，数据持久化的载体主要有文件、SQLite数据库和Core Data。本节中，我们就从这几个方面入手讨论数据持久化的优化问题。

25.4.1 使用文件

文件是数据持久化的重要载体。文件优化可以包括很多方面，下面我们就从文件访问、文件结构和文件大小这3个方面来介绍。

1. 文件访问优化

避免多次写入很少的数据，最好是当数据积攒到一定数量时一次写入。因为文件访问涉及IO操作，我们知道频繁的IO操作会影响性能，所以最好将文件读写访问从主线程中剥离出来，由一个子线程负责。另外，过于频繁地写入数据会影响设备中闪存的寿命。

文件的写入应该采用增量方式，每次只写入变化的部分，不要为改变几个字节写入整个文件。这样就要求不能采用简单的属性列表对象写入方式。这是一个很复杂的问题，文件内容的变化可以是追加、删除和修改。文件追加很容易实现，删除就比较麻烦了，需要找到要删除的数据，这样访问文件就采用随机访问方式了。修改与删除的问题是一样的。与其这么麻烦，不如采用别的持久化技术了。

2. 文件结构优化

文件要保存数据，它就应该是结构化的。苹果中的.plist文件就是很好的结构化文件，其结构是层次模型的树形结构，层次的深浅会影响读取/写入的速度。在能够满足用户需求的情况下，要减少层次深度。下面是一个世界杯足球赛部分小组信息的属性列表文件team(5层次).plist：

```xml
<?xml version="1.0" encoding="UTF-8"?>
<!DOCTYPE plist PUBLIC "-//Apple//DTD PLIST 1.0//EN" "http://www.apple.com/DTDs/
    PropertyList-1.0.dtd">
<plist version="1.0">
<array>
    <dict>
        <key>teamname</key>
        <string>A</string>
        <key>teamlist</key>
```

```
        <array>
            <dict>
                <key>name</key>
                <string>南非</string>
                <key>image</key>
                <string>SouthAfrica</string>
            </dict>
            <dict>
                <key>name</key>
                <string>墨西哥</string>
                <key>image</key>
                <string>Mexico</string>
            </dict>
        </array>
    </dict>
    <dict>
        <key>tearmname</key>
        <string>B</string>
        <key>tearmlist</key>
        <array>
            <dict>
                <key>name</key>
                <string>阿根廷</string>
                <key>image</key>
                <string>Argentina</string>
            </dict>
            <dict>
                <key>name</key>
                <string>尼日利亚</string>
                <key>image</key>
                <string>Nigeria</string>
            </dict>
        </array>
    </dict>
</array>
</plist>
```

该文件有5个层次，具体如图25-29所示，其中第一层是数组类型集合；第二层是字典集合，其中描述了小组名和小组中的球队列表；第三层是数组类型集合，描述了小组中的球队列表；第四层是字典集合；第五层是字符串，描述了球队名和球队图标信息。

Key	Type	Value	
▼Root	Array	(2 items)	①
▼Item 0	Dictionary	(2 items)	②
tearmname	String	A	
▼tearmlist	Array	(2 items)	③
▼Item 0	Dictionary	(2 items)	④
name	String	南非	⑤
image	String	SouthAfrica	
▼Item 1	Dictionary	(2 items)	
name	String	墨西哥	
image	String	Mexico	
▼Item 1	Dictionary	(2 items)	
tearmname	String	B	
▼tearmlist	Array	(2 items)	
▼Item 0	Dictionary	(2 items)	
name	String	阿根廷	
image	String	Argentina	
▼Item 1	Dictionary	(2 items)	
name	String	尼日利亚	
image	String	Nigeria	

图25-29　5个层次的team.plist文件

这个文件访问起来很不方便，遍历起来也很不方便，也很影响性能。我们重新设计了这个属性列表文件，其内容如下：

```
<?xml version="1.0" encoding="UTF-8"?>
<!DOCTYPE plist PUBLIC "-//Apple//DTD PLIST 1.0//EN" "http://www.apple.com/DTDs/
➥PropertyList-1.0.dtd">
<plist version="1.0">
<array>
    <dict>
        <key>name</key>
        <string>A1-南非</string>
        <key>image</key>
        <string>SouthAfrica</string>
    </dict>
    <dict>
        <key>name</key>
        <string>A2-墨西哥</string>
        <key>image</key>
        <string>Mexico</string>
    </dict>
    <dict>
        <key>name</key>
        <string>B1-阿根廷</string>
        <key>image</key>
        <string>Argentina</string>
    </dict>
    <dict>
        <key>name</key>
        <string>B2-尼日利亚</string>
        <key>image</key>
        <string>Nigeria</string>
    </dict>
</array>
</plist>
```

此时这个文件有3个层次，其中第一层是数组类型集合，第二层是字典集合，第三层是字符串，描述了球队名和球队图标信息，如图25-30所示。

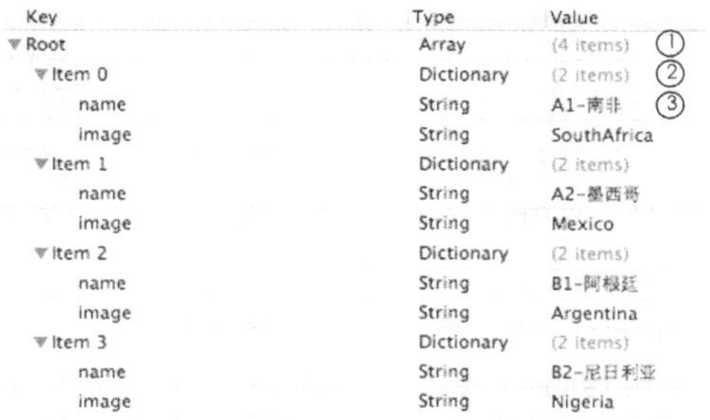

Key	Type	Value
▼Root	Array	(4 items) ①
▼Item 0	Dictionary	(2 items) ②
name	String	A1-南非 ③
image	String	SouthAfrica
▼Item 1	Dictionary	(2 items)
name	String	A2-墨西哥
image	String	Mexico
▼Item 2	Dictionary	(2 items)
name	String	B1-阿根廷
image	String	Argentina
▼Item 3	Dictionary	(2 items)
name	String	B2-尼日利亚
image	String	Nigeria

图25-30 3层次的team.plist文件

与上面的5层次文件相比，3层次访问起来比较方便，性能会比较好。此外，在文件大小方面，3层次文件是647 KB，5层次文件是893 KB。

3. 文件大小优化

文件大小也是优化的一个重要指标。从上面的比较可以看到，调整文件结构可以减少文件大小。此外，我们也可以通过序列化.plist文件来减少文件大小。Foundation框架提供了NSPropertyListSerialization类，它就是为此而设计的。NSPropertyListSerialization类中有两个常用方法，具体如下所示。

- **+ dataWithPropertyList:format:options:error:**。按照指定的格式和操作参数，序列化属性列表对象到NSData对象。
- **+ propertyListWithData:options:format:error:**。按照指定的格式和操作参数，从NSData对象反序列化到属性列表对象中。

为了介绍NSPropertyListSerialization类，我们修改一下14.3节的MyNotes案例，数据原来保存在NotesList.plist属性列表文件中，现在我们换成序列化二进制文件NotesList.binary。下面我们修改NoteDAO类。首先，添加如下两个方法：

```
//从文件中读取数据到NSMutableArray
private func readFromArray(_ path: String) -> NSMutableArray? {
    //从文件读取数据到data对象
    let data = NSMutableData(contentsOfFile: path)!
    var array: NSMutableArray!
    do {
        //反序列化到属性列表对象 (NSMutableArray)
        array = try PropertyListSerialization.propertyList(from: data as Data,
        ➥options: .mutableContainersAndLeaves, format: nil)
        ➥as! NSMutableArray                                          ①
    } catch let err as NSError {
        NSLog("读取文件失败, %@", err.description)
    }
    return array
}

//写入NSMutableArray数据到文件中
private func write(_ array: NSMutableArray, toFilePath path: String) {
    do {
        //把属性列表对象 (NSMutableArray) 序列化为NSData
        let data = try PropertyListSerialization.data(fromPropertyList: array,
        ➥format: PropertyListSerialization.
        ➥PropertyListFormat.binary, options: 0)                      ②
        //写入到沙箱目录的序列化文件
        try data.write(to: URL(fileURLWithPath: path), options: .atomic)
    } catch let err as NSError {
        NSLog("写入文件失败, %@", err.description)
    }
}
```

```
//从文件中读取数据到NSMutableArray
-(NSMutableArray*) readFromArray: (NSString*) path {
    //从文件读取数据到data对象
    NSMutableData *data = [[NSMutableData alloc] initWithContentsOfFile:path];
    //反序列化到属性列表对象 (NSMutableArray)
    NSMutableArray* array = [NSPropertyListSerialization
    ➥propertyListWithData:data
    ➥options:NSPropertyListMutableContainersAndLeaves format: NULL
    ➥error:NULL];                                                    ①

    return array;
}

//写入NSMutableArray数据到文件中
-(void) write:(NSMutableArray*)array toFilePath: (NSString*) path {
    //把属性列表对象 (NSMutableArray) 序列化为NSData
    NSData * data = [NSPropertyListSerialization dataWithPropertyList:array
    ➥format:NSPropertyListBinaryFormat_v1_0
    ➥options:NSPropertyListMutableContainersAndLeaves error:NULL];    ②

    //写入到沙箱目录的序列化文件
    BOOL success = [data writeToFile:path atomically:TRUE];

    if (!success) {
        NSAssert(0, @"错误写入文件");
    }
}
```

在上述代码中，readFromArray:方法从文件中读取数据到NSMutableArray，其流程是读取文件到NSMutableData对象，然后对NSMutableData对象反序列化，获得NSMutableArray属性列表对象。本例中的属性列表对象是NSMutableArray类型，其中第①行用于处理这一过程，data参数是反序列化的数据来源，它是NSData类型；options参数是NSPropertyListReadOptions类型。NSPropertyListReadOptions是枚举类型，其成员值如下。

- **NSPropertyListImmutable**。属性列表包含不可变对象。
- **NSPropertyListMutableContainers**。属性列表父节点是可变类型，子节点是不可变类型。
- **NSPropertyListMutableContainersAndLeaves**。属性列表父节点和子节点都是可变类型。

另外，在第①行中，format参数为nil（或NULL），说明格式是自动识别的。

> **提示**　属性列表对象是与属性列表文件结构对应的，它可以是NSData、NSString、NSArray和NSDictionary类型以及它们的可变类型。此外，还可以是NSDate和NSNumber类型。

write:toFilePath:方法把NSMutableArray数据序列化后写入到文件中，其流程是先序列化NSMutableArray数据到NSData对象中，然后把NSData对象写入到文件中。第②行就是完成序列化处理的，+dataWithPropertyList:format:options:error:方法中array参数是要序列化的属性列表对象，format参数是NSPropertyListFormat枚举类型。NSPropertyListFormat枚举类型包含的常量有如下几个。

- **XMLFormat_v1_0**。指定属性列表文件格式是XML格式，仍然是纯文本类型，不会压缩文件。Objective-C版本为NSPropertyListXMLFormat_v1_0。
- **BinaryFormat_v1_0**。指定属性列表文件格式为二进制格式，文件是二进制类型，会压缩文件。Objective-C

版本为NSPropertyListBinaryFormat_v1_0。

- OpenStepFormat。指定属性列表文件格式为ASCII码格式，对于旧格式的属性列表文件，不支持写入操作。Objective-C版本为NSPropertyListOpenStepFormat。

本例中，我们设置的是BinaryFormat_v1_0，大小减少了，加载速度提高了，这样就达到了优化的效果。关于本例中其他方法的改动，这里就不再介绍了。

25.4.2 使用 SQLite 数据库

当需要处理较大的数据集合时，就不能采用文件了。因为文件不支持事务处理，这时候我们可以选择SQLite数据库或Core Data。本节中，我们先从表结构、查询和插入（或删除）这几个方面介绍一下SQLite数据库方面的优化。

1. 表结构优化

SQLite是嵌入式关系型数据，它可以建立多表之间复杂的关系，但是如果放在iOS、Android等这些移动设备上时，我们需要考虑设备上本地表能建多少，表中字段有多少，表之间关系的复杂程度等问题。

在CPU处理能力低、内存少、存储空间少的情况下，我们不能在本地建立复杂表关系，表的个数不要超过5个，表中的字段数也不宜太多。移动设备中的数据不可能是企业级系统数据的全部，它只是企业级系统的补充和扩展。例如，在你的iPhone手机中，不可能有全部的新浪微博用户信息，一方面是不安全，另一方面是数据量很大，最高配置的iPhone也不可能存放下这么多数据。这是我们在开发移动应用时始终要牢记的：移动设备在整个应用系统中的角色是什么？

2. 查询优化

查询是衡量数据库性能的重要指标之一。在查询方面可优化的有很多，例如建立索引、限制返回记录数和where条件子句等。

使用索引，能够提高查询的性能。具体哪些字段需要创建索引，这很关键。只有在表连接或where条件子句中使用字段时，才能提高查询性能。在INTEGER PRIMARY KEY字段上，一般不用建索引。如果表中的数据很少，则建索引的效果不大。

由于移动设备屏幕相对来说比较小，屏幕上能显示的数据不多，如果一次查询出的记录数超过屏幕能显示的行数，这就没有必要了。因为这样反而会占用更多的内存，耗费宝贵的CPU时间。因此，我们需要为查询添加返回记录数的限制。下面的语句是SQLite支持的写法：

```
SELECT * FROM Note Limit 10 Offset 5;
```

以上语句表示从Note表查询数据，其中10表示查询的最大记录数不超过10个，5表示偏移量，即跳过5行取10个。

在where条件子句的优化方面，就有更多优化方式了。比如，尽量不要使用LIKE模糊匹配查询，如果可能，则使用=查询；尽量不要使用IN语句，可以使用=和or替代。此外，在多个条件中，要把非文本的条件放在前面，文本条件放在后面，示例代码如下：

```
(salary > 5000000) AND (lastName LIKE 'Guan') 优于 (lastName LIKE 'Guan') AND (salary > 5000000)
```

这是因为非文本的条件判断比较快，如果不满足，就不用再计算后面的条件表达式了。

3. 插入（或删除）优化

索引可以提供查询性能，但是对于插入和删除是有负面影响的。索引就像是书中的目录，插入和删除数据必然造成索引重排，所以创建索引要慎重。

在SQLite中，有一些PRAGMA指令可以改变数据库的行为。PRAGMA synchronous指令用于设置数据同步操作。同步是指在插入数据时，将数据同时保存到存储介质中。如果PRAGMA synchronous = OFF，则表示关闭了数据同步，不等待数据保存到存储介质就可继续执行插入操作，这在大量数据插入时可以大大提高速度。在Objective-C中，可以调用sqlite3_exec函数设置数据是否同步，相关语句如下：

```
sqlite3_exec(db, "PRAGMA synchronous = OFF", nil, nil, nil)
```

插入完成后，也可以重新设置PRAGMA synchronous = NORMAL或PRAGMA synchronous = FULL。

25.4.3　使用 Core Data

Core Data是面向对象的ORM技术，苹果公司推荐使用。它提供了缓冲、延迟加载等技术，其性能比较好，但有时候我们会发现它的性能要比SQLite差，这主要与存储类型的设置有关。Core Data的存储类型有NSSQLiteStoreType、NSBinaryStoreType和NSInMemoryStoreType，我们主要采用NSSQLiteStoreType类型，这样底层存储就采用了SQLite数据库，SQLite数据库的优点也能发挥出来。

使用Core Data时，还要考虑查询优化问题。它的查询是通过NSFetchRequest执行Predicate定义的逻辑查询条件实现的，在优化规则上与SQLite的where条件子句是一样的。此外，如果要查询返回记录数的限制，可以使用如下语句：

```
let fetchRequest = NSFetchRequest()
//限制一次提取数据的记录数
fetchRequest.fetchLimit = 50
//限制提取记录的偏移量
fetchRequest.fetchOffset = 10
```

```
NSFetchRequest *request = [[NSFetchRequest alloc] init];
//限制一次提取数据的记录数
[request setFetchLimit:10];
//限制提取记录的偏移量
[request setFetchOffset:5];
```

这两条语句相当于SELECT * FROM Note Limit 10 Offset 5;。

此外，还可以设置pragma指令，相关语句如下：

```
//设置持久化存储描述对象
let description = NSPersistentStoreDescription()                          ①
description.setValue("OFF" as NSObject?, forPragmaNamed: "synchronous")
description.setValue("OFF" as NSObject?, forPragmaNamed: "count_changes")
description.setValue("MEMORY" as NSObject?, forPragmaNamed: "journal_mode")
description.setValue("MEMORY" as NSObject?, forPragmaNamed: "temp_store")

let container = NSPersistentContainer(name: "CoreDataNotes")
container.persistentStoreDescriptions = [description]                     ②
```

```
//设置持久化存储描述对象
NSPersistentStoreDescription *description = [[NSPersistentStoreDescription
➥alloc] init];                                                            ①
[description setValue:@"OFF" forPragmaNamed:@"synchronous"];
[description setValue:@"OFF" forPragmaNamed:@"count_changes"];
[description setValue:@"MEMORY" forPragmaNamed:@"journal_mode"];
[description setValue:@"MEMORY" forPragmaNamed:@"temp_store"];

_persistentContainer.persistentStoreDescriptions = @[description];        ②
```

在上述代码中，我们首先把这些pragma指令放置于可变字典pragmaOptions中。上述代码中，第①行用于设置持久化存储描述对象NSPersistentStoreDescription，其中NSPersistentStoreDescription是iOS 10新推出的API。通过NSPersistentStoreDescription类，设置底层SQLite数据库的pragma指令。第②行中，Swift版的container和Objective-C的_persistentContainer都是NSPersistentContainer对象。

为了方便分析Core Data的执行情况，我们可以使用Instruments工具中的Core Data分析模板，如图25-31所示。

进入Core Data分析模板后，如图25-32所示，可以看到其内部有3个子模板：Fetches、Cache和Saves。

当我们执行查询、插入和删除操作时，在Fetches和Saves模板右边会产生很多线。其中，①部分的线段为Fetch duration（执行查询的持续时间）；②部分的线段为Fetch count（查询的记录数据），将虚线拉到上面可以看到这些内容的具体数值；如果有数据要保存，③部分产生的线段为Save duration（保存所持续的时间）；④部分是更具体的信息，Fetch entity列是查询的实体类（Note），Fetch count列是查询的记录数，Fetch duration列是查询的执行时间。

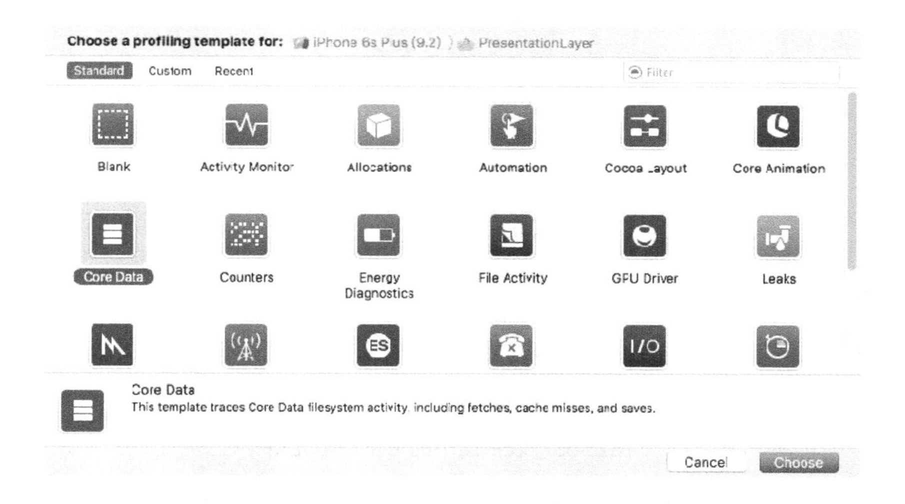

图25-31　选择Core Data分析模板

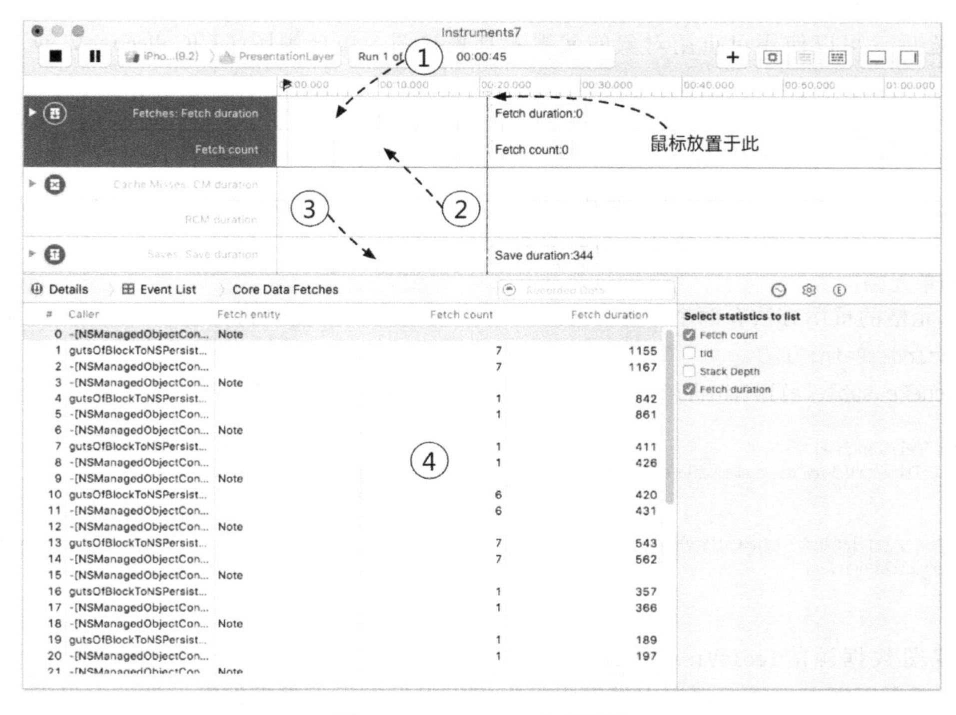

图25-32　Core Data分析模板

25.5　可重用对象的使用

在iOS的一些视图中，它们的内部包含了子视图，当父视图显示区域发生变化时（如用手滑动屏幕），原来在屏幕中的子视图就会滑出到屏幕之外，而原来在屏幕之外的子视图就有机会进入屏幕中。如图25-33所示，当屏幕向上滑动时，Cupertino单元格滑出屏幕之外，Sherman Oaks单元格滑入到屏幕中。

图25-33 表视图中的可重用单元格

这些操作会有什么问题吗？这在第6章中讨论过。如果每次新单元格进入到屏幕中都去实例化一个新的单元格，必然增加内存开销。采用可重用对象设计可以不去实例化新的单元格，而是先使用可重用单元格标识到视图中去找，如果找到则使用，没有则创建。

在iOS 6之后，可以使用可重用对象的父视图有表视图、集合视图（UICollectionView）和地图视图（MKMapView）。

25.5.1 表视图中的可重用对象

在iOS 6之后，表视图中有两种子视图采用可重用对象设计，它们是表视图单元格（UITableViewCell）和表视图节头节脚视图（UITableViewHeaderFooterView）。

1. 表视图单元格

表视图单元格的重用方法有两个：dequeueReusableCellWithIdentifier:方法和dequeueReusableCellWith-Identifier:forIndexPath:方法。

通过dequeueReusableCellWithIdentifier:方法，可以用标识符从表视图中获得可重用单元格，模式代码如下：

```
let cellIdentifier = "CellIdentifier"
var cell:UITableViewCell! = tableView.dequeueReusableCellWithIdentifier
➥(cellIdentifier)
if (cell == nil) {
    cell = UITableViewCell(style: UITableViewCellStyle.Default,
    ➥reuseIdentifier:cellIdentifier)
}
```

```
static NSString *CellIdentifier = @"CellIdentifier";
UITableViewCell *cell = [tableView
➥dequeueReusableCellWithIdentifier:CellIdentifier];
if (cell == nil) {
    cell = [[UITableViewCell alloc] initWithStyle:UITableViewCellStyleDefault
    ➥reuseIdentifier:CellIdentifier];
}
```

要在表视图数据源的tableView:cellForRowAtIndexPath:方法中使用可重用单元格设计，首先通过dequeueReusableCellWithIdentifier:方法从表视图中找，如果cell为空，则需要使用initWithStyle:reuseIdentifier:构造函数创建。

dequeueReusableCellWithIdentifier:forIndexPath:方法是iOS 6之后提供的方法。与上一个方法相比，该方法的签名多了forIndexPath:部分。它可以通过指定单元格位置获得可重用单元格，不需要判断，模式代码如下：

```
let CellIdentifier = "CellIdentifier"
let cell = tableView.dequeueReusableCellWithIdentifier(CellIdentifier,
➥forIndexPath: indexPath)let cellIdentifier = "CellIdentifier"
```

```
static NSString *CellIdentifier = @"CellIdentifier";
UITableViewCell *cell = [tableView dequeueReusableCellWithIdentifier:
➥CellIdentifier forIndexPath:indexPath];
```

这个方法的使用有一些限制，一般不应用于XIB技术中，主要应用于故事板和纯代码中。图25-34在故事板中设置表视图单元格的Identifier属性为CellIdentifier。

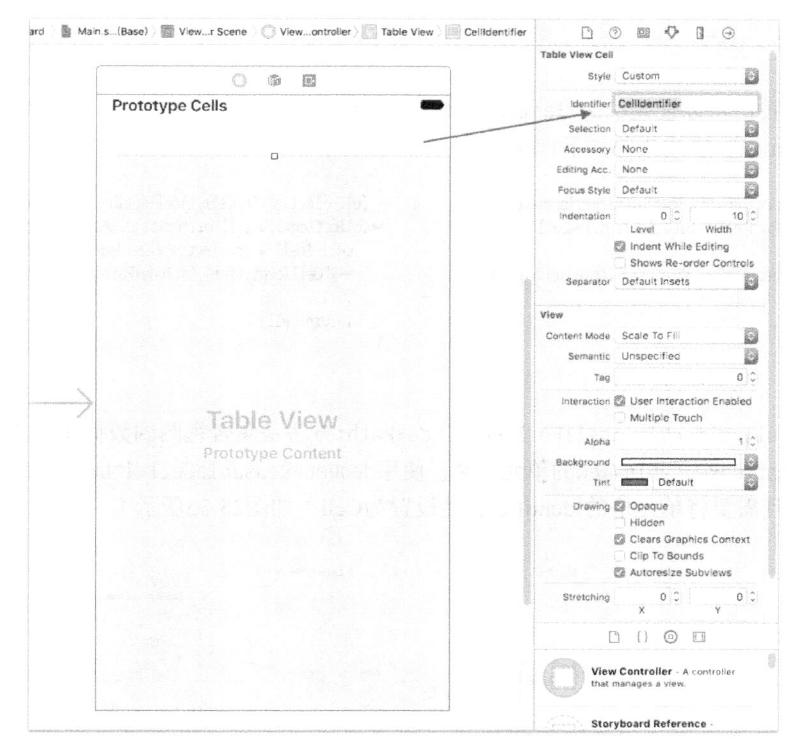

图25-34　设置表视图单元格的属性

2. 表视图节头节脚视图

UITableViewHeaderFooterView也是iOS 6之后新加的内容，节头和节脚也会反复出现，它也需要可重用设计。使用表视图的dequeueReusableHeaderFooterViewWithIdentifier:方法获得UITableViewHeaderFooterView对象后，如果没有可重用的UITableViewHeaderFooterView对象，则使用initWithReuseIdentifier:构造函数创建。其模式代码如下：

```
override func tableView(tableView: UITableView,
➥viewForHeaderInSection section: Int) -> UIView? {

    let headerReuseIdentifier  = "TableViewSectionHeaderViewIdentifier"

    var sectionHeaderView :UITableViewHeaderFooterView! =
    ➥tableView.dequeueReusableHeaderFooterVie(wWithIdentifie:
    ➥(headerReuseIdentifier)

    if sectionHeaderView == nil {
       sectionHeaderView =
       ➥UITableViewHeaderFooterView(reuseIdentifier:headerReuseIdentifier)
    }
    ......
    return sectionHeaderView
}
```

```
- (UIView *)tableView:(UITableView *)tableView
➥viewForHeaderInSection:(NSInteger)section {

    static NSString *headerReuseIdentifier =
    ➥@"TableViewSectionHeaderViewIdentifier";

    UITableViewHeaderFooterView *sectionHeaderView = [tableView
    ➥dequeueReusableHeaderFooterViewWithIdentifier:headerReuseIdentifier];

    if (!sectionHeaderView) {
        sectionHeaderView  = [[UITableViewHeaderFooterView alloc]
        ➥initWithReuseIdentifier:headerReuseIdentifier];
    }
    ......
    return sectionHeaderView;
}
```

需要在表视图委托协议UITableViewDelegate中的tableView:viewForHeaderInSection:方法中使用可重用对象设计。

25.5.2　集合视图中的可重用对象

集合视图在iOS 6之后才可以使用。它也有两种子视图采用可重用对象设计，它们是单元格视图和补充视图，这两个视图都继承自UICollectionReusableView，使用时需要自己编写相关代码。

1. 单元格视图

在集合视图中，我们可以使用UICollectionView的dequeueReusableCellWithReuseIdentifier:forIndexPath:方法获得可重用的单元格，模式代码如下：

```
override func collectionView(_ collectionView: UICollectionView,
➥cellForItemAt indexPath: IndexPath) -> UICollectionViewCell {

    let cell = collectionView.dequeueReusableCell(withReuseIdentifier:
    ➥"Cell",for: indexPath) as! Cell
    ......
    return cell
}
```

```
- (UICollectionViewCell *)collectionView:(UICollectionView *)
➥collectionView cellForItemAtIndexPath:(NSIndexPath *)indexPath {
    Cell *cell = [collectionView dequeueReusableCellWithReuseIdentifier:
    ➥@"CellIdentifier" forIndexPath:indexPath];
    ......
    return cell;
}
```

在上述代码中，collectionView:cellForItemAtIndexPath:方法是集合视图的数据源方法，其中Cell是我们自定义的继承自UICollectionReusableView的单元格类。使用dequeueReusableCellWithReuseIdentifier:时，需要使用故事板设计UI，并且需要将单元格的Identifier属性设置为Cell（如图25-35所示）。

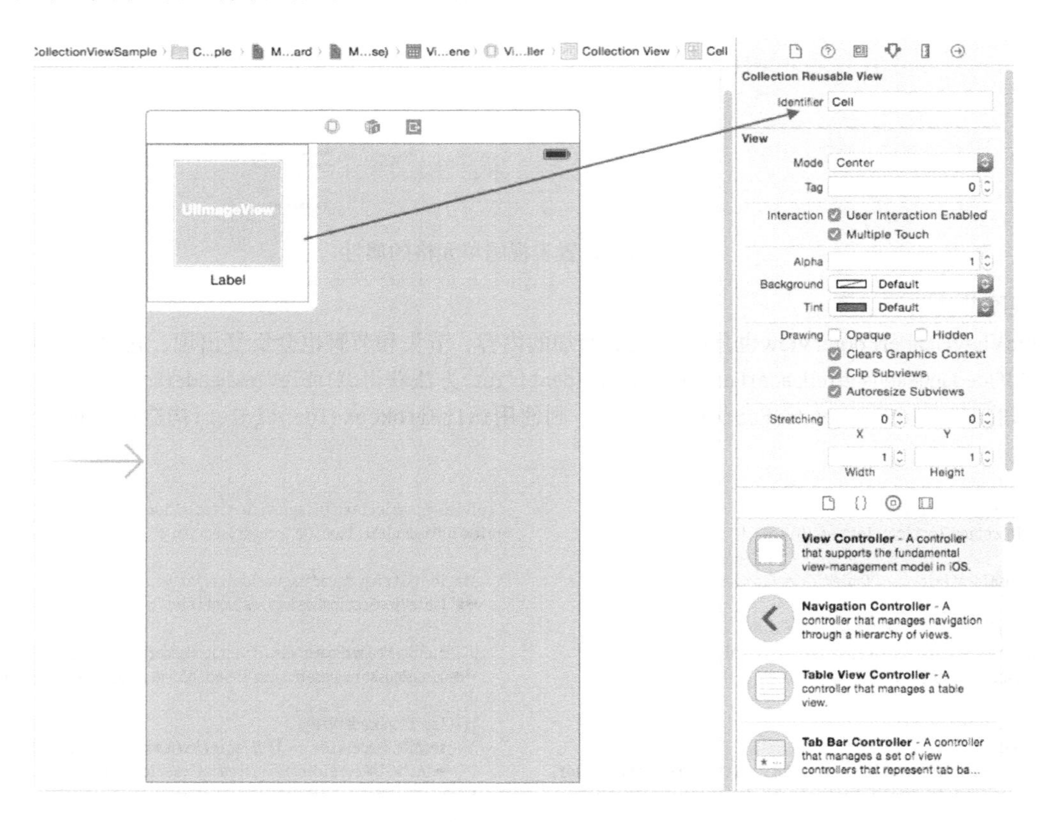

图25-35　设置集合视图单元格的Identifier属性

2. 补充视图

集合视图单元格可以使用UICollectionView的dequeueReusableSupplementaryViewOfKind:withReuseIdentifier:forIndexPath:方法获得可重用的补充视图，模式代码如下：

```
override func collectionView(_ collectionView: UICollectionView,
➥viewForSupplementaryElementOfKind kind: String,
➥at indexPath: IndexPath) -> UICollectionReusableView {
```

```
- (UICollectionReusableView *)collectionView:(UICollectionView *)
➥collectionView viewForSupplementaryElementOfKind:(NSString *)kind
➥atIndexPath: (NSIndexPath *)indexPath {
```

```
    let headerView: UICollectionReusableView  = collectionView
    ➥.dequeueReusableSupplementaryView(
    ➥ofKind: UICollectionElementKindSectionHeader,
    ➥withReuseIdentifier : "HeaderIdentifier", for:indexPath)
    ......
    return headerView
}
```

```
    HeaderView *headerView = [collectionView
    ➥dequeueReusableSupplementaryViewOfKind:
    ➥UICollectionElementKindSectionHeader
    ➥withReuseIdentifier:@"HeaderIdentifier" forIndexPath:indexPath];

    ......

    return headerView;
}
```

collectionView:viewForSupplementaryElementOfKind:atIndexPath:方法是集合视图的数据源方法，其中HeaderView是我们自定义的继承自UICollectionReusableView的补充视图类。使用dequeueReusableSupplementary-ViewOfKind:withReuseIdentifier:forIndexPath:方法时，需要使用故事板设计UI，并将补充视图的Identifier属性设置为HeaderIdentifier，如图25-36所示。

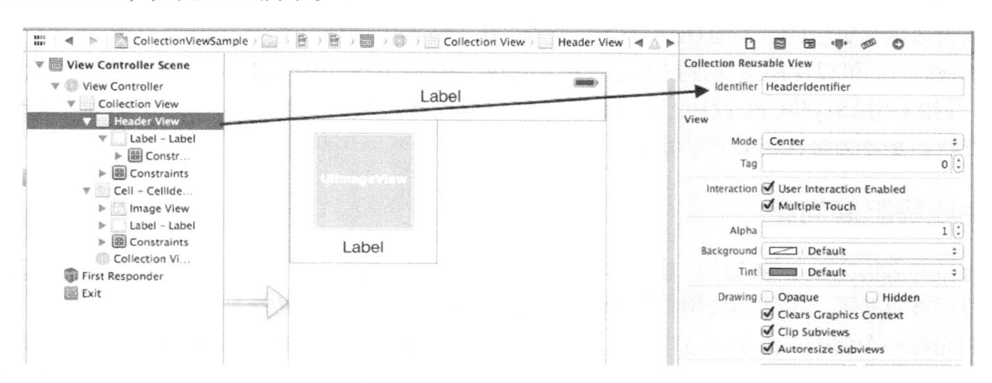

图25-36 设置补充视图的属性

25.5.3 地图视图中的可重用对象

在开发地图应用时，也有一个可重用对象MKPinAnnotationView，它是地图上的一个标注。使用地图视图的dequeueReusableAnnotationViewWithIdentifier:方法时，可以获得MKPinAnnotationView对象。如果没有可重用的MKPinAnnotationView对象，则使用initWithAnnotation:reuseIdentifier:构造函数创建。其模式代码如下：

```
func mapView(_ mapView: MKMapView,
➥viewForAnnotation annotation: MKAnnotation) -> MKAnnotationView? {

    var annotationView = self.mapView
    ➥.dequeueReusableAnnotationView(withIdentifier: "PIN_ANNOTATION")
    ➥as? MKPinAnnotationView

    ➥as? MKPinAnnotationView
    if annotationView == nil {
        annotationView = MKPinAnnotationView(annotation: annotation,
        ➥reuseIdentifier: "PIN_ANNOTATION")
    }
    annotationView!.pinTintColor = UIColor.red
    annotationView!.animatesDrop = true
    annotationView!.canShowCallout = true

    return annotationView!
}
```

```
- (MKAnnotationView *)mapView:(MKMapView *)theMapView
➥viewForAnnotation:(id <MKAnnotation>)annotation {

    MKPinAnnotationView *annotationView = (MKPinAnnotationView *)
    ➥[self.mapView dequeueReusableAnnotationViewWithIdentifier:
    ➥@"PIN_ANNOTATION"];
    if (annotationView == nil) {
        annotationView = [[MKPinAnnotationView alloc]
        ➥initWithAnnotation:annotation reuseIdentifier:@"PIN_ANNOTATION"];
    }

    annotationView.pinTintColor = [UIColor redColor];
    annotationView.animatesDrop = TRUE;
    annotationView.canShowCallout = TRUE;

    return annotationView;
}
```

这段处理代码是地图视图中常用的处理方式，请大家牢记。

25.6　并发处理

并发处理能够同时处理多个任务，本节就来介绍主线程阻塞问题以及GCD技术实现并发处理等相关内容。

25.6.1　一些概念

并发处理涉及很多概念，其中主要有线程和进程。

- 进程就是相互隔离的、独立运行的程序。一个进程就是一个执行中的程序，而每一个进程都有自己独立的一块内存空间和一组系统资源。
- 线程是轻量级的进程，就像进程一样，线程在程序中是独立的、并发执行的，每个线程有它自己的局部变量。同一个进程中的多个线程之间共享相同的内存地址空间，这意味着它们可以访问相同的变量和常量。

每个程序都至少有一个线程，这个线程就是主线程。当一个程序启动时，主线程被创建，主线程控制程序的主要流程，负责显示和更新UI，所有UI元素的更新必须在主线程中进行。

在CPU单核时代，可以使用多线程技术进行并发处理。现在，iOS设备的CPU已经进入多核时代，从iPhone 4s和iPad 2之后开始采用A5双核CPU设计。异步设计方法可以充分发挥多核优势。GCD（Grand Central Dispatch）是一种异步方法，是专为多核CPU而设计的并发处理技术。

25.6.2　主线程阻塞问题

主线程所做的事情应该是响应用户输入、事件处理、更新UI，而耗时的任务不要在主线程中处理。由于耗时任务使得主线程被阻塞了，不能响应用户的请求，这样应用的用户体验会很差。

下面我们先看一个例子。如图25-37所示，点击Load Image按钮，会从http://www.51work6.com/book/test2.jpg中加载图片。当我们点击Load Image按钮时，按钮会一直处于按下状态而不弹起，直到图片显示“完成”。这是因为主线程要进行耗时的处理（如进行网络通信、数据传输等任务），导致主线程不能响应用户的输入和请求，这就是我们要讨论的线程阻塞问题。

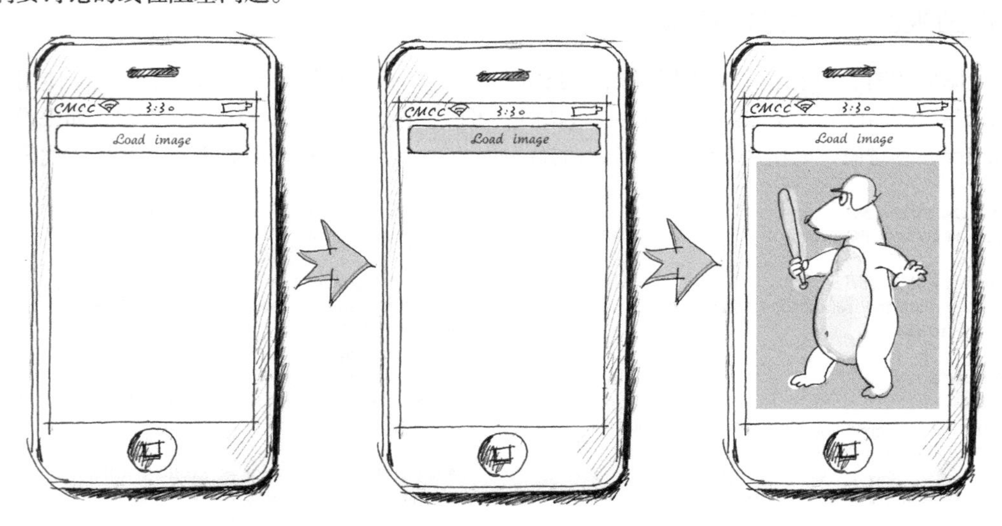

图25-37　主线程阻塞案例

下面我们看看代码部分。在BlockDemo工程中，ViewController中click:方法的代码如下：

```
@IBAction func click(_ sender: AnyObject) {

    let url = URL(string : "http://www.51work6.com/book/test2.jpg")
```

```
- (IBAction)click:(id)sender {

    NSURL *url = [NSURL
```

```
    let imgData = try? Data(contentsOf: url!)
    let img = UIImage(data : imgData!)
    self.imageView.image = img
}
```

```
➥URLWithString:@"http://www.51work6.com/book/test2.jpg"];
NSData *imgData = [NSData dataWithContentsOfURL: url ];
UIImage *img = [UIImage imageWithData:imgData];
self.imageView.image = img;

}
```

由于不能直接通过URL创建UIImage对象，所以先构建NSData对象，它是从URL请求回来的二进制数据对象，然后再用它来构建UIImage对象。最后，把UIImage对象赋值给UIImageView的image属性。

那么，如何解决主线程阻塞问题呢？那就是把这些执行比较耗时的阻塞线程的任务从主线程中移出到其他线程中处理。

25.6.3 选择 NSThread、NSOperation 还是 GCD

解决主线程阻塞问题时，我们可以使用NSThread、NSOperation和GCD等技术。

(1) NSThread是传统的线程类，需要自己管理线程的生命周期、线程同步、加锁、睡眠以及唤醒等。

(2) NSOperation是面向对象高级别线程类，不自己需要管理线程。

(3) GCD是基于C语句级别的API，它提供了C函数。不需要自己管理线程，支持多核CPU处理。GCD是苹果重点推荐的并发技术，唯一的缺陷是它基于C语言的API。

25.6.4 GCD 技术

在GCD中，有一个重要的概念，那就是派发队列（dispatch queue）。派发队列是一个对象，它可以接受任务，并将任务以先到先执行的顺序来执行。派发队列可以是并发的或串行的。并发队列可以执行多任务，串行队列同一时间只执行单一任务。

1. 3种类型的派发队列

在GCD中，有3种类型的派发队列。

- **串行队列**。串行队列通常用于同步访问一个特定的资源，每次只能执行一个任务。使用函数dispatch_queue_create，可以创建串行队列。
- **并发队列**。也称为全局派发队列，可以并发地执行一个或多个任务。当并发地执行多个任务时，必然涉及哪些任务先执行，哪些任务后执行的问题。在macOS 10.10和iOS 8之前，是通过并发队列优先级决定的，而在macOS 10.10和iOS 8之后，则通过QoS（Quality of Service）技术管理。使用dispatch_get_global_queue函数，可以创建并发队列。
- **主队列**。它在应用程序的主线程中，用于更新UI。其他两个队列不能更新UI。使用dispatch_get_ main_queue函数，可以获得主队列。

2. QoS技术

QoS是在macOS 10.10和iOS 8之前提供的新技术，我们通过QoS告诉操作系统并发队列如何工作，然后操作系统会通过合理的资源控制，从而以最高效的方式执行并发队列。这其中主要涉及CPU调度、IO操作优先级、任务执行在哪个线程以及执行的顺序等内容。我们通过一个抽象的Quality of Service参数来表明任务的意图以及类别。QoS提供NSQualityOfService（Swift版是QualityOfService）枚举类型，它有如下5个成员。

- **NSQualityOfServiceUserInteractive**。与用户交互的任务，这些任务通常跟UI刷新相关，例如动画，它会在一瞬间完成。
- **NSQualityOfServiceUserInitiated**。由用户发起的并且可以立即得到结果的任务。例如，翻动表视图时加载数据，然后显示单元格，这些任务通常跟后续的用户交互相关，会在几秒或者更短的时间内完成。
- **NSQualityOfServiceUtility**。一些耗时的任务，这些任务不会马上返回结果，例如下载任务，它可能花费几秒或者几分钟的时间。

- **NSQualityOfServiceBackground**。这些任务对用户不可见，可以长时间在后台运行。
- **NSQualityOfServiceDefault**。优先级介于NSQualityOfServiceUserInteractive和NSQualityOfServiceUtility之间。这个值是系统默认值，我们不应该使用它设置自己的任务。

把BlockDemo工程修改为GCD实现，具体代码如下：

```
@IBAction func click(_ sender: AnyObject) {

    self.imageView.image = nil

    let url = URL(string : "http://www.51work6.com/book/test2.jpg")

    DispatchQueue.global(qos: .utility).async {                        ①
        let imgData = try? Data(contentsOf: url!)
        let img = UIImage(data : imgData!)

        DispatchQueue.main.async {                                     ②
            self.imageView.image = img
        }
    }
}
```

```
- (IBAction)click:(id)sender {

    self.imageView.image = nil;

    NSURL *url = [NSURL
    ➥URLWithString:@"http://www.51work6.com/book/test2.jpg"];

    dispatch_async(dispatch_get_global_queue(NSQualityOfServiceUtility, 0), ^{
                                                                       ①

        NSData *imgData = [NSData dataWithContentsOfURL:url];
        UIImage *img = [UIImage imageWithData:imgData];

        dispatch_async(dispatch_get_main_queue(), ^{                   ②
            self.imageView.image = img;
        });

    });

}
```

从上述代码中可见，GCD代码中，Objective-C和Swift这两种语言差别很大，这主要是因为Swift 3的语法去Objectvie-C化的结果。在Swift 3之前，它们还是很相似的。其中第①行用于创建全局派发队列，Swift通过`DispatchQueue.global(qos: .utility). async {...}`语句实现，其中`.utility`是`QualityOfService`枚举类型成员；Objective-C通过`dispatch_async(dispatch_get_global_queue (NSQualityOfServiceUtility, 0), ^{...}`语句实现，其中`NSQualityOfServiceUtility`是枚举成员。由于我们需要的图片要从网络上下载，但是图片相对比较小，因此需要将并发队列设置为`NSQualityOfServiceUtility`。

更新UI的语句，必须在主队列中执行。第②行用于执行主队列，其中Objective-C是通过`dispatch_async(dispatch_get_main_queue(), ^{...}`语句切换到主队列，Swift是通过`DispatchQueue.main.async {...}`语句切换到主队列。可见，在Swift 3中，GCD变得非常简洁。

25.7 小结

通过对本章的学习，我们了解了性能优化方法，其中包括内存优化、资源文件优化、延迟加载、持久化优化、使用可重用对象和并发访问等，这些内容都是非常重要的，希望广大读者认真掌握。

Part 5 第五部分

实战篇

本部分内容

- 第 26 章　管理好你的程序代码——代码版本控制
- 第 27 章　项目依赖管理
- 第 28 章　把应用放到 App Store 上
- 第 29 章　iOS 开发项目实战——2020 东京奥运会应用开发及 App Store 发布

第 26 章

管理好你的程序代码——代码版本控制

我有个朋友曾经做了一个时长3个月的项目，就在项目即将结束的时候他的硬盘坏了，数据全部丢失，而他之前从不做备份，结果可想而知。这件事对我的触动很大，我后来经常备份程序代码并将其备份在不同的电脑上。于是，有一段时间我每天下班的时候，都把程序代码备份到公司的服务器。随着时间的推移，备份到服务器上的数据越来越多，我很难快速查到想要的资料。在我与同事之间进行代码整合时，我们用U盘互相复制，由于版本不能及时更新造成了很多问题，其中必须要解决的问题有以下两点。

- **程序代码的备份**。为了便于查到历史版本，能够进行比较，知道修改过什么地方，是谁修改了这些代码等。
- **代码共享与整合**。能很容易得到团队其他人的代码，也能够很容易地把代码共享给其他成员。

解决这几个问题时，我们可以使用版本控制工具。版本控制工具是一种软件，开发人员要习惯使用版本控制工具，每日提交程序代码，提交的代码应该有清晰的注释，成员之间应该及时沟通。

版本控制工具有很多，本章中我们为大家介绍Git。

26.1 概述

版本控制的重要性毋庸置疑，我们必须要使用代码版本控制工具，每一个程序员和项目管理人员都必须深刻认识到这一点。

26.1.1 版本控制历史

版本控制的最早方式是将文件复制到文件服务器上，命名为“××年××月××日×××备份”目录，这在现在看来太原始了。作为软件工具，版本控制经历过两个阶段：集中管理模式和分布式管理模式。

- **集中管理模式**。这以一个服务器作为代码库，团队人员本地没有代码库，只能与服务器进行交互。这种类型的版本控制工具有VSS（Visual Source Safe，微软开发的Microsoft Visual Studio套件中的软件之一）、CVS（Concurrent Versions System，并发版本系统）、SVN（Subversion）等，其中SVN是目前这种模式的佼佼者。
- **分布式管理模式**。这是更为先进的模式，不仅有一个中心代码库，而且每位团队人员本地也都有代码库，在不能上网的情况下也可以提交代码。该类型的版本控制工具有Git、Mercurial、Bazzar和Darcs。

Git是为了帮助管理Linux内核开发项目而开发的一个开源的版本控制工具。之前，BitKeeper工具是Linux内核开发人员使用的主要版本控制工具，它采用许可证管理版本，但Linus Torvalds①觉得BitKeeper不适合Linux开源社区的工作，在2005年开始着手开发Git来替代BitKeeper。虽然Git最初是为了辅助Linux内核开发，但我们发现在很多其他软件项目中也都可以使用Git。

① 著名的电脑程序员、黑客，Linux内核的发明人及该计划的合作者。

26.1.2 基本概念

版本控制工具涉及很多术语和概念，这里将常用的概念整理如下。

- **代码库**（repository）。存放项目代码以及历史备份的地方。
- **分支**（branch）。为了验证和实验一些想法、版本发布、缺陷修改等需要，建立一个开发主干之外的分支，这个分支被隔离在各自的开发线上。当改变一个分支中的文件时，这些更改不会出现在开发主干和其他分支中。
- **合并分支**（merging branch）。完成某分支工作后，将该分支上的工作成果合并到主分支上。
- **签出**（check out）。从代码库获得文件或目录，将其作为副本保存在工作目录下，此副本包含了指定代码库的最新版本。
- **提交**（commit）。将工作目录中修改的文件或目录作为新版本复制回代码库。
- **冲突**（conflict）。有时候提交文件或目录时可能会遇到冲突，当两个或多个开发人员更改文件中的一些相同行时，将发生冲突。
- **解决**（resolution）。遇到冲突时，需要人为干预解决，这必须通过手动编辑该文件进行处理，必须有人逐行检查该文件，以接受一组更改并删除另一组更改。除非冲突解决，否则存在冲突的文件无法成功提交到代码库中。
- **索引**（index）。Git工具特有的概念。在修改的文件提交到代码库之前做出一个快照，这个快照称为“索引”，它一般会暂时存储在一个临时存储区中。

26.2 Git 代码版本控制

基于分布式管理模式的代码版本控制工具是现在的主流，而且Git还有成熟的代码托管服务GitHub网站。

26.2.1 服务器搭建

如果项目不需要与其他开发人员协同开发，我们就不需要Git服务器。与SVN和CVS不同，Git管理模式是分布的。如图26-1所示，每个开发者的本地电脑都有代码库，为了实现与他人协同开发，需要有个共享的代码库。每个开发者都可以为其他开发者提供服务。他们之间需要安装一些通信协议并提供安全认证，使授权客户端能够访问他的代码库。因此，任何能够提供通信协议和安全认证的代码库，都可以认为是服务器。

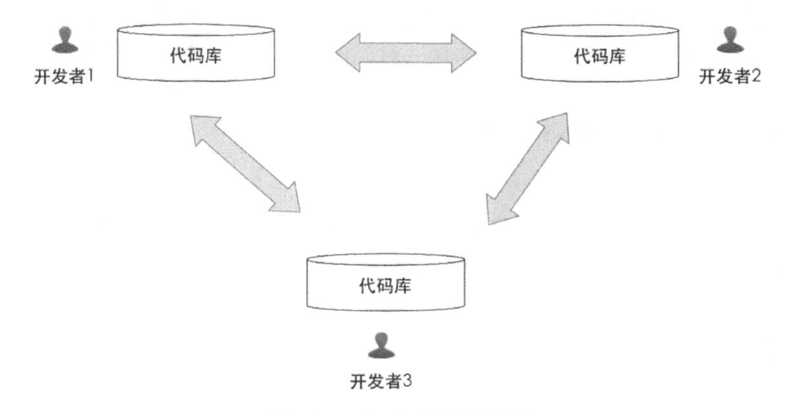

图26-1 Git分布管理模式

为了能够与彼此进行通信，需要安装合适的通信协议，Git可以选择的协议有很多，比如本地协议、Git协议、HTTP(S)协议和SSH协议，其中本地协议主要用于本地文件系统的访问，Git协议只支持读远程库且无需身份认证，在读取远程库时它的效率最高，HTTP(S)协议和SSH协议都可以支持远程库读写、身份认证等操作。

SSH协议由于被广泛安装，因此Git通信中普遍采用SSH。我们知道SSH协议下每个用户都有一个账号，随着项目规模的扩大，开发者增加到上百人之后，服务器维护这些用户账号是非常麻烦的。Gitosis和Gitolite是基于SSH协议的Git服务器软件，它们能够使所有用户都使用同一个专用的SSH账号访问代码库，各个用户通过公钥认证的方式使用此专用SSH账号访问代码库，而用户在连接时使用的不同的公钥可以区分用户的身份。Gitosis和Gitolite除了具有SSH传统的优势外，还支持企业级授权和远程建库。

与Gitolite相比，Gitosis有很多信息需要配置，且难度很大。Gitolite是Gitosis的下一代版本，功能更加强大，配置比较简单，安装方法有多种形式，有一些比较简单，非常适合于初学者。本章重点介绍Gitolite安装、配置和管理等操作。

Linux对于SSH协议支持得很好，因此我们配置的Git服务器基于Linux的Ubuntu版本。在Ubuntu下搭建Git服务器的流程如图26-2所示。

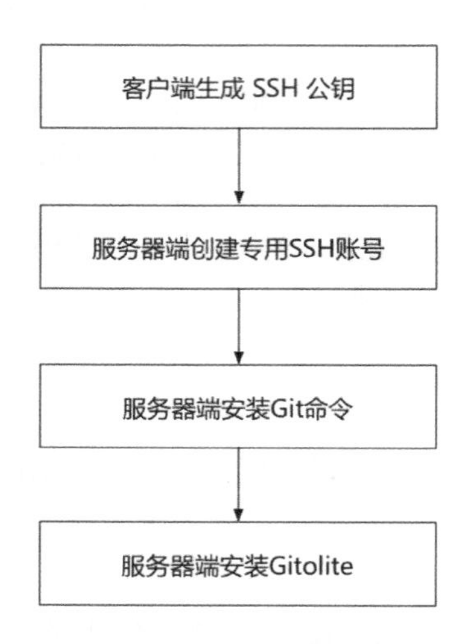

图26-2　搭建Git服务器的流程

1. 客户端生成SSH公钥

SSH采用公钥认证，通过ssh-keygen命令生成。首先，在当前用户主目录的.ssh目录下面生成两个文件。

- id_rsa。私钥文件，基于RSA算法创建，该私钥文件要妥善保管，它被保存到客户端中。
- id_rsa.pub。公钥文件，与上面的私钥文件是一对，该文件可以公开，放置于服务器端。

采用公钥认证SSH登录，可以实现无口令登录远程服务器，即用公钥认证取代口令认证。这是我们生成SSH公钥的主要原因。我们在终端窗口中输入如下指令：

```
$ ssh-keygen
```

创建了自己的公钥/私钥对后，需要将.ssh目录中的id_rsa.pub发送到服务器。如果我们想让当前客户端用户作为Git服务器管理者的话，可以在终端窗口中执行如下指令：

```
$ cd ~/.ssh
$ scp id_rsa.pub tonyguan@192.168.1.115:/tmp/admin.pub
tonyguan@192.168.1.115's password:
id_rsa.pub                                100%  420    0.4KB/s    00:00
```

其中tonyguan是服务器的用户，将本机上的公钥文件id_rsa.pub发送到服务器目录tmp，并重新命名为admin.pub。

提示

如果命令在执行过程中出现ssh: connect to host 192.168.1.115 port 22: Connection refused错误，这很有可能是服务器端没有安装SSH协议。在服务器终端中输入sudo apt-get install openssh- server命令，可以安装SSH协议。

2. 服务器端创建专用SSH账号

我们需要为Git服务器访问代码库创建一个专用SSH账号，用户名为git，此时可以在服务器终端中输入如下指令：

```
$ sudo adduser git
```

输入密码和相关信息后，即可成功创建git应用。如果这个用户已经创建，确认它没有其他的用途后，可以删除git用户再创建，删除命令如下：

```
$ sudo userdel git
```

3. 服务器端安装Git命令

要在服务器端安装Git，可以使用Ubuntu提供的apt-get安装包管理工具进行安装，即在服务器终端中输入如下指令：

```
$ sudo apt-get install git
```

4. 服务器端安装Gitolite

上面的准备工作完成后，就可以在服务器端安装Gitolite软件了。首先，使用su切换到git用户，在服务器终端中输入如下指令：

```
$ su - git
$ mkdir -p ~/bin

$ git clone git://github.com/sitaramc/gitolite
$ gitolite/install -ln ~/bin
$ gitolite setup -pk /tmp/admin.pub

Initialized empty Git repository in /home/git/repositories/gitolite-admin.git/
Initialized empty Git repository in /home/git/repositories/testing.git/
```

安装成功后，会在git主目录下创建repositories目录，在repositories目录下有gitolite-admin.git和testing.git目录。事实上，这两个子目录是两个Git代码库，其中gitolite-admin.git保存了Gitolite配置信息，它可以在客户端远程管理，testing.git代码库是为了测试而使用的。

26.2.2 Gitolite服务器管理

Gitolite服务器可在本地管理，然后将结果推送给远程服务器，服务器会执行这些配置，从而实现管理的目的。首先，要在管理员客户端电脑上克隆服务器端gitolite-admin.git库，具体指令如下：

```
$ git clone git@192.168.1.115:gitolite-admin
```

提示

在macOS下，可以到https://code.google.com/p/git-osx-installer/中下载.dmg文件并安装。

进入gitolite-admin目录后，可以发现它的结构如下：

```
├──conf
│   └──gitolite.conf
└──keydir
    └──admin.pub
```

该目录下有conf和keydir这两个目录，conf目录下面有gitolite.conf，keydir目录下面的admin.pub文件是之前管理员客户端传递给服务器的公钥文件。keydir目录可以存放多个用户的公钥文件，从而实现对其他用户的公钥认证。

进入gitolite-admin目录，使用vi等文本工具打开gitolite.conf文件，内容如下：

```
repo    gitolite-admin
        RW+     =   admin

repo    testing
        RW+     =   @all
```

在这个授权文件中，我们设置了gitolite-admin和testing这两个代码库。gitolite-admin只允许admin用户有读写和强制更新的权限，RW为读写权限，+号为强制更新权限。testing用于设置测试代码库，这里设置为所有人都可以读写以及强制更新。

下面我们通过一个实际应用场景来介绍Gitolite的添加用户和授权修改的实现过程。

我想为我们的iOS开发团队创建一个组ios_team，用户有zhang、tony和zhao。服务器端的代码库是ios_repo，ios_team的组成员对代码库ios_repo有读写和强制更新的权限。

修改后的授权文件gitolite.conf的内容如下：

```
@ios_team=zhang tony zhao
repo    gitolite-admin
        RW+     =   admin

repo    testing
        RW+     =   @all

repo    ios_repo
        RW+     =   @ios_team
```

保存后并退出。然后我们需要从zhang、tony和zhao那里获得他们的公钥文件id_rsa.pub，分别将其命名为zhang.pub、tony.pub和zhao.pub，然后将其复制到gitolite-admin目录下的keydir目录中。keydir的目录结构如下：

```
keydir
├──admin.pub
├──tony.pub
├──zhang.pub
└──zhao.pub
```

然后在客户端终端进入gitolite-admin目录，输入如下命令：

```
$ git add .
$ git commit -m 'add user'
[master 6e4f3ad] add user
 3 files changed, 3 insertions(+)
 create mode 100644 keydir/tony.pub
 create mode 100644 keydir/zhang.pub
 create mode 100644 keydir/zhao.pub
```

这两个命令都是Git中的基本命令，`git add .`用于添加文件到本地代码库缓存，`git commit`用于提交缓存中的修改，并将其保存到本地代码库，-m参数是注释说明信息。现在的修改还只是在本地代码库，并没有同步到远程服务器上。要同步到远程服务器上，我们需要在客户端终端进入gitolite-admin目录，输入如下命令：

```
$ git push
Counting objects: 7, done.
Delta compression using up to 4 threads.
Compressing objects: 100% (5/5), done.
Writing objects: 100% (5/5), 1.04 KiB, done.
```

```
Total 5 (delta 0), reused 0 (delta 0)
To git@192.168.1.115:gitolite-admin
   76eabfb..6e4f3ad  master -> master
```

git push命令用于推送本地代码库到远程服务器代码库。如果出现上面的运行结果，就修改完成了。如果打开服务器的repositories目录，此时会多一个iso_repo.git代码库，这就是我们新添加的代码库了。我们可以在zhao用户电脑上输入下面的指令测试一下：

```
$ git clone git@192.168.1.115:ios_repo
```

正确执行完成后，则会克隆一个ios_repo代码库到zhao的本地电脑了。

提示 如果命令在执行过程中出现Agent admitted failure to sign using the key错误，需要在本机上执行$ ssh-add ~/.ssh/id_rsa命令。

26.2.3 Git 常用命令

无论我们的项目是否需要与他人协同开发，都会用到Git的一些常用命令。我们在上一节中就用到了git add、git commit和git push命令。本节中，我们将介绍一些常用的Git命令，包括git help、git log、git init、git add、git rm、git commit和git status等命令。最后，还介绍了Git图形界面辅助工具gitk。

1. git help

第一个要掌握的是git help，通过它可以自己查找命令的帮助信息。在终端中执行如下命令：

```
$ git help <命令>
```

其中help后面是要查询的命令。

2. git log

该命令可以查看Git的日志信息。在终端中执行git log命令：

```
$ git log
```

得到的执行结果为：

```
commit 2f027fbad790fa3e61ec9965a18415203f5e9683
Author: tonyguan <si92@sina.com>
Date:   Wed Oct 31 12:57:13 2015 +0800

    a

commit ac29dd648c7e78bfed5682742855683b5242c27f
Author: git on zhao-VirtualBox <git@zhao-VirtualBox>
Date:   Wed Oct 31 12:53:27 2015 +0800

    start
```

3. git init

该命令可以创建一个新的代码库，或者是初始化一个已存在的代码库。例如，我想在本地创建一个ios_repo代码库，可以先使用mkdir创建这个目录，然后再执行git init命令。在终端中执行如下命令：

```
$ mkdir ios_repo
$ cd ios_repo
$ git init
Initialized empty Git repository in /Users/tonyguan/ios_repo/.git/
```

使用mkdir创建ios_repo目录（它只是一个普通的目录，并不是代码库）后，git init会在ios_repo目录下生成一个隐藏的.git目录。

4. git add

该命令用来更新索引，记录下哪些文件有修改，或者添加了哪些文件。该命令并没有更新代码库，只有在提交的时候才将这些变化更新到代码库中。在终端中执行如下命令：

```
$ git add .
```

可以将当前工作目录和子目录下所有新添加和修改的文件添加到索引中。如果只想将某个文件添加到索引中，可以使用如下命令：

```
$ git add filename
```

或

```
$ git add *.txt
```

这里可以指定文件名，也可以使用通配符。

5. git rm

该命令用于删除索引或代码库中的文件，然后通过提交命令将变化更新到代码库中。在终端中执行如下命令：

```
$ git rm filename 或 $ git rm *.txt
```

6. git commit

该命令用于更新缓存中的索引，但未将变化更新到代码库中。在终端中执行如下命令：

```
$ git commit -m 'tony commit'
```

其中-m设定提交注释信息。

7. git status

该命令可以显示当前git的状态，包括哪些文件修改、删除和添加了，但是没有提交的信息。在终端中执行如下命令：

```
$ git status
```

会显示类似如下的内容：

```
# On branch master
# Changes not staged for commit:
#   (use "git add <file>..." to update what will be committed)
#   (use "git checkout -- <file>..." to discard changes in working directory)
#   (commit or discard the untracked or modified content in submodules)
#
#   modified:   .DS_Store
#   modified:   HelloWorld (modified content, untracked content)
#
no changes added to commit (use "git add" and/or "git commit -a")
```

26.2.4　Git 分支

上一节介绍了Git中的常用命令，下面我们介绍Git分支。分支在版本控制中占有非常重要的地位。我们建立分支可能出于多种原因，但是无外乎基于验证一些想法、版本发布、bug修改等目的。

我们现在拿Objective-C版本的HelloWorld案例了解一下分支的用法，其中HelloWorld工程中视图控制器ViewController.m的代码如下：

```
//
//ViewController.m
//HelloWorld
```

```
//
//Created by tonyguan on 2017-1-22.
//Copyright (c) 2015年 tonyguan. All rights reserved.
//

#import "ViewController.h"

@interface ViewController ()

@end

@implementation ViewController

- (void)viewDidLoad
{
    [super viewDidLoad];
}

- (void)didReceiveMemoryWarning
{
    [super didReceiveMemoryWarning];
}

@end
```

提示　我在本地创建了一个名为ios_repo的代码库，把HelloWorld工程复制到ios_repo代码库中，进入ios_repo下的HelloWorld目录添加并提交该工程。

1. 创建分支

创建分支时，使用命令git branch <分支名>。如果我们要创建一个testing分支，可以在终端中执行如下命令：

```
$ git branch testing
```

创建完成后，我们需要查看一下分支情况，此时可以在终端中执行如下命令：

```
$ git branch
* master
  testing
```

其中*号的分支是当前分支，master是git默认创建的分支，testing是我们刚刚创建的分支。

2. 切换分支

在我们编辑工程文件时，需要在不同的分支间切换，此时可以在终端中执行如下命令：

```
$ git checkout testing
Switched to branch 'testing'
```

首先，我们要在master分支中修改ViewController.m文件，具体如下：

```
//
//ViewController.m
//HelloWorld
//
//Created by tonyguan on 2017-1-22.
//Copyright (c) 2015年 tonyguan. All rights reserved.
//

#import "ViewController.h"

@interface ViewController ()

@end
```

```
@implementation ViewController

- (void)viewDidLoad
{
    [super viewDidLoad];
    NSLog(@"Hello World.");
}

- (void)viewWillAppear
{
    NSLog(@"viewWillAppear call");
}

- (void)didReceiveMemoryWarning
{
    [super didReceiveMemoryWarning];
}

@end
```

在上述代码中，我们在viewDidLoad方法中添加输出Hello World.，并添加viewWillAppear方法。然后在终端中执行命令提交代码：

```
$ git add .
$ git commit -m 'master branch commit'
[master b62ea54] master branch commit
 2 files changed, 7 insertions(+), 2 deletions(-)
 rewrite HelloWorld.xcodeproj/project.xcworkspace/xcuserdata/
    tonyguan.xcuserdatad/UserInterfaceState.xcuserstate (83%)
```

切换到testing分支中，修改ViewController.m文件，具体如下：

```
//
//ViewController.m
//HelloWorld
//
//Created by tonyguan on 2017-1-22.
//Copyright (c) 2015年 tonyguan. All rights reserved.
//

#import "ViewController.h"

@interface ViewController ()

@end

@implementation ViewController


- (void)viewDidLoad
{
    [super viewDidLoad];
    NSLog(@"世界你好.");
}

- (void)didReceiveMemoryWarning
{
    [super didReceiveMemoryWarning];
}

@end
```

与master分支不同的是，在testing分支中，我们在viewDidLoad方法中添加输出“世界你好.”，而且也没有添加viewWillAppear方法。可见，master分支和testing分支在输出“世界你好.”还是输出Hello World.是存在冲突的。然后在终端中执行命令提交代码：

```
$ git add .
$ git commit -m 'testing branch commit'
[testing c27a331] testing branch commit
 2 files changed, 2 insertions(+), 2 deletions(-)
 rewrite HelloWorld.xcodeproj/project.xcworkspace/xcuserdata/
    tonyguan.xcuserdatad/UserInterfaceState.xcuserstate (83%)
```

3. 合并分支

如果我们把testing分支合并到master分支，需要先切换回master分支，具体命令如下：

```
$ git checkout master
Switched to branch 'master'
```

合并分支可以使用git merge命令。在终端中执行如下合并命令：

```
$ git merge testing
error: Your local changes to the following files would be overwritten by merge:HelloWorld.
    xcodeproj/project.xcworkspace/xcuserdata/tonyguan.xcuserdatad/UserInterface
        State.xcuserstate
Please, commit your changes or stash them before you can merge.
Aborting
```

如果在提交和合并之前有一些修改和变化，会报出错误，此时我们需要重新提交或放弃修改这个文件。在终端中执行如下命令可以放弃修改该文件：

```
$ git checkout HelloWorld.xcodeproj
```

这里使用了git checkout命令放弃修改，在前面切换分支时我们也使用了该命令。

然后再使用git merge命令合并，会出现如下问题：

```
$ git merge testing
warning: Cannot merge binary files: HelloWorld.xcodeproj/project.xcworkspace/xcuserdata/
    tonyguan.xcuserdatad/UserInterfaceState.xcuserstate (HEAD vs. testing)

Auto-merging HelloWorld/ViewController.m
CONFLICT (content): Merge conflict in HelloWorld/ViewController.m
Auto-merging HelloWorld.xcodeproj/project.xcworkspace/xcuserdata/tonyguan.xcuserdatad/
    UserInterfaceState.xcuserstate
CONFLICT (content): Merge conflict in HelloWorld.xcodeproj/project.xcworkspace/xcuserdata/
    tonyguan.xcuserdatad/UserInterfaceState.xcuserstate
Automatic merge failed; fix conflicts and then commit the result.
```

问题的本质是master分支和testing分支在输出时有冲突，这个时候我们再看看程序代码ViewController.m：

```
//
//ViewController.m
//HelloWorld
//
//Created by tonyguan on 2017-1-22.
//Copyright (c) 2015年 tonyguan. All rights reserved.
//

#import "ViewController.h"

@interface ViewController ()

@end

@implementation ViewController
```

```
- (void)viewDidLoad
{
    [super viewDidLoad];
<<<<<<< HEAD
    NSLog(@"Hello World.");
=======
    NSLog(@"世界你好.");
>>>>>>> testing
}

- (void)viewWillAppear
{
    NSLog(@"viewWillAppear call");
}


- (void)didReceiveMemoryWarning
{
    [super didReceiveMemoryWarning];
}

@end
```

其中viewWillAppear方法成功合并，而viewDidLoad方法中<<<<<<< HEAD ... >>>>>>>表示其中的内容有冲突。由于Git无法判断谁对谁错，所以需要人工解决。解决完后再次添加并提交，在终端中执行如下命令：

```
$ git commit -a -m 'testing branch merge commit'
[master 60eac40] testing branch merge commit
```

其中git commit -a -m相当于git add .和git commit -m的执行效果。

4. 删除分支

在分支合并完成且不再使用的情况下，可以使用git branch -d命令删除分支。要删除testing分支，可以在终端中执行如下命令：

```
$ git branch -d testing
Deleted branch testing (was c27a331).
```

也可以使用-D参数代替-d，其中-D表示强制删除分支。

26.2.5　Git 协同开发

如果我们的项目需要一个团队协同开发，就需要搭建Git服务器。协同开发涉及的常用命令有git clone、git fetch、git pull、git push和git remote add等。

下面我们通过解决实际工作中的一些问题来掌握这些命令的实现。如图26-3所示，开发者1和开发者2都共享一个服务器代码库ios_repo，服务器环境都已经搭建完成。

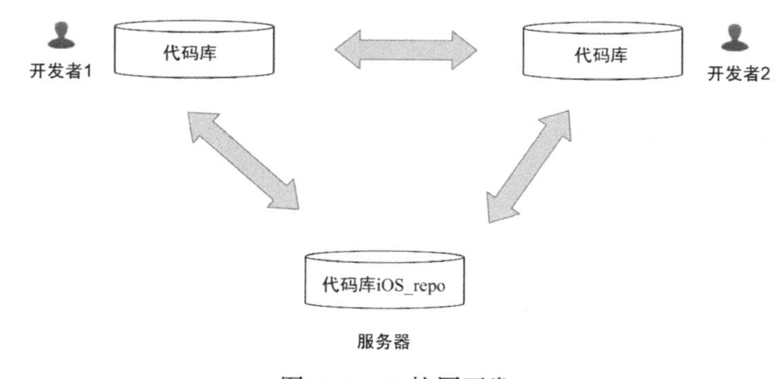

图26-3　Git协同开发

他们在工作中遇到的几个问题如下所示。

- 问题1。开发者1创建ios_repo代码库和HelloWorld工程，然后提交到远程的服务器代码库。
- 问题2。开发者2克隆服务器代码库ios_repo到本地代码库，对HelloWorld工程进行修改，并重新提交给服务器代码库。
- 问题3。当开发者2重新提交给服务器代码库时，开发者1如何获得本地代码库？

1. 问题1

开发者1的电脑上没有ios_repo代码库，需要自己创建，而不是从服务器上克隆过来。首先，创建ios_repo目录作为代码库，再将HelloWorld工程复制到ios_repo代码库下面，它们的目录结构如下：

```
ios_repo
└──────HelloWorld
        ├─────HelloWorld
        │      ├──────AppDelegate.h
        │      ├──────AppDelegate.m
        │      ├──────HelloWorld-Info.plist
        │      ├──────HelloWorld-Prefix.pch
        │      ├──────ViewController.h
        │      ├──────ViewController.m
        │      ├──────en.lproj
        │      │       ├──────InfoPlist.strings
        │      │       └──────ViewController.xib
        │      └──────main.m
        └─────HelloWorld.xcodeproj
```

进入ios_repo目录，会在终端中执行如下命令来初始化ios_repo代码库：

```
$ git init
```

如果修改了代码，需要提交内容到本地代码库，此时可以在终端执行如下命令：

```
$ git add .
$ git commit -m 'tony commit'
 [master 3609824] tony commit
 13 files changed, 658 insertions(+)
 create mode 100644 HelloWorld/.DS_Store
 create mode 100644 HelloWorld/HelloWorld.xcodeproj/project.pbxproj
 create mode 100644 HelloWorld/HelloWorld.xcodeproj/project.xcworkspace/
     contents.xcworkspacedata
 create mode 100644 HelloWorld/HelloWorld.xcodeproj/project.xcworkspace/
     xcuserdata/tonyguan.xcuserdatad/UserInterfaceState.xcuserstate
 create mode 100644 HelloWorld/HelloWorld.xcodeproj/xcuserdata/
     tonyguan.xcuserdatad/xcdebugger/Breakpoints.xcbkptlist
 create mode 100644 HelloWorld/HelloWorld.xcodeproj/xcuserdata/
     tonyguan.xcuserdatad/xcschemes/HelloWorld.xcscheme
 create mode 100644 HelloWorld/HelloWorld.xcodeproj/xcuserdata/
     tonyguan.xcuserdatad/xcschemes/xcschememanagement.plist
 create mode 100644 HelloWorld/HelloWorld/AppDelegate.h
 create mode 100644 HelloWorld/HelloWorld/HelloWorld-Info.plist
 create mode 100644 HelloWorld/HelloWorld/HelloWorld-Prefix.pch
 create mode 100644 HelloWorld/HelloWorld/ViewController.h
 create mode 100644 HelloWorld/HelloWorld/en.lproj/InfoPlist.strings
 create mode 100644 HelloWorld/HelloWorld/en.lproj/ViewController.xib
```

然后将本地代码库推送到远程服务器，此时可以使用命令git remote add和git push，具体如下所示：

```
$ git remote add HW git@192.168.1.115:ios_repo
$ git push HW master
Counting objects: 29, done.
Delta compression using up to 4 threads.
Compressing objects: 100% (22/22), done.
Writing objects: 100% (26/26), 21.36 KiB, done.
Total 26 (delta 0), reused 0 (delta 0)
To git@192.168.1.115:ios_repo
   43184b3..3609824  master -> master
```

其中git remote add HW git@192.168.1.115:ios_repo是给远程代码库一个名字HW，以便与远程代码库交互。git push HW master就是向刚才定义的远程代码库HW的master分支推送数据。

2. 问题2

开发者2的电脑上没有ios_repo代码库，他不需要自己创建，而是从服务器上克隆过来，此时可以在终端中执行如下命令：

```
$ git clone git@192.168.1.115:ios_repo
```

然后他也对HelloWorld做了一些修改，那么如何推送他的数据到服务器代码库呢？事实上，这个过程与开发者1刚刚的推送方式是一样的，这里就不再介绍了。

3. 问题3

开发者1再次被告知他们程序有新的版本，他需要从服务器代码库中获取新的程序。使用git fetch命令，可以从服务器代码库获取数据，相关命令如下：

```
$ git fetch HW
remote: Counting objects: 9, done.
remote: Compressing objects: 100% (4/4), done.
remote: Total 5 (delta 3), reused 0 (delta 0)
Unpacking objects: 100% (5/5), done.
From 192.168.1.115:ios_repo
    3609824..7680cce  master     -> HW/master
```

这时打开修改的文件，发现没有变化，这是因为还需要使用git merge命令合并HW/master到本地master分支，相关代码如下：

```
$ git merge HW/master
Updating 3609824..7680cce
Fast-forward
    HelloWorld/HelloWorld/ViewController.m |    1 -
    1 file changed, 1 deletion(-)
```

再看看修改的文件是否发生了变化。合并过程中也可能发生冲突，需要人为解决这些冲突再合并，这可以通过Git提供的更加简便的命令git pull来实现。git pull命令是git fetch和git merge命令的一个组合，相关代码如下：

```
$ git pull HW master
From 192.168.1.115:ios_repo
* branch            master     -> FETCH_HEAD
Updating 7680cce..26b89ea Fast-forward  .DS_Store         | Bin 0 -> 6148 bytes
 .../UserInterfaceState.xcuserstate                       | Bin 25632 -> 26270 bytes
 HelloWorld/HelloWorld/ViewController.m                   |   2 +-
3 files changed, 1 insertion(+), 1 deletion(-)
create mode 100644 .DS_Store
```

其中HW是远程代码库名，master是要合并的本地分支。

26.2.6　Xcode 中 Git 的配置与使用

我们在前面介绍过很多Git命令都是在命令行下运行的。在命令行下管理Git有很多优点，这就不用多说了，但最大的缺点是要求用户记住这些命令，此时Git图形界面的优势就体现出来了。作为集成开发环境工具Xcode，它也提供了一定的Git图形界面功能。但是要想在Xcode中使用Git管理工程代码，还要进行一些配置。

如果我们是使用Xcode创建一个iOS工程，在Xcode工程中能够列入代码版本控制的文件是有规定的，不能是编写的二进制文件、临时文件和用户特有的文件等。下面是在Xcode 中创建的HelloWorld工程的目录结构：

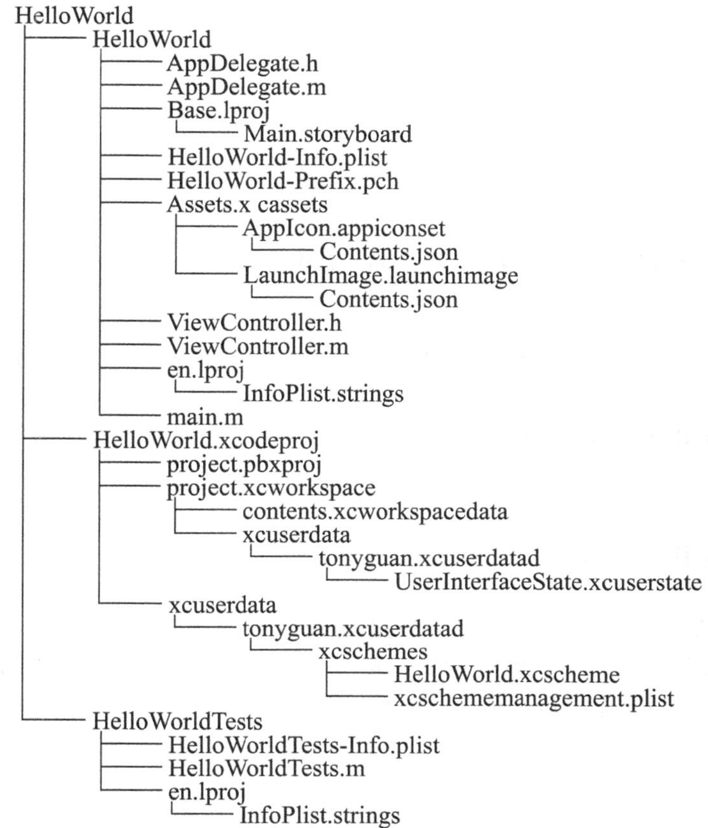

其中HelloWorld.xcodeproj属于包文件，它内部的很多东西是不能提交的，包括project.xcworkspace和xcuserdata，它们是与用户有关的。在Git中，有一个.gitignore配置文件，其中可以设置被忽略的文件。下面是一个.gitignore配置文件的内容：

```
# Exclude the build directory
build/*

# Exclude temp nibs and swap files
*~.nib
*.swp

# Exclude macOS folder attributes
.DS_Store

# Exclude user-specific XCode 3 and 5 files
*.mode1
*.mode1v3
*.mode2v3
*.perspective
*.perspectivev3
*.pbxuser
*.xcworkspace
xcuserdata
```

在这个文件中，#号表示注释，可以使用正则表达式。这个文件创建后，应该放在什么地方呢？如果只考虑忽略一个特定的工程，.gitignore文件应该放在代码库目录下面，目录结构如下所示：

如果考虑适用于所有的Xcode工程，则需要使用`git config`命令配置Git，具体代码如下：

```
$ git config -global core.excludesfile  ~/.gitignore
```

该命令会将配置信息写入到~/.gitconfig文件中，其中`-global`参数用于配置全局信息，~/.gitignore说明文件放置于当前用户目录下。

为了使本机上的所有代码库都默认使用git用户，还需要执行如下命令：

```
$ git config --global user.name git
$ git config --global user.email eorient@sina.com
```

此外，我们还需要在Xcode中进行一些配置，具体如下。

1. 配置远程账户信息

为了将本地代码库推送给远程服务器代码库，我们需要在Xcode中配置远程代码账户信息。通过Xcode→Preferences... →Accounts，进入如图26-4所示的账号管理界面。首先，我们要确认左边账户列表中是否有我们刚刚认证过的账户，它以git开头，如图26-4中的git@192.168.1.115:HelloWorld。

图26-4　管理账号界面

如果没有找到账户信息，就需要添加一个账户了。点击管理界面左下角的“+”按钮，弹出的菜单如图26-5所示，从中选择Add Repository菜单，此时会弹出如图26-6所示的添加账号对话框。我们需要在Address中输入地址，本例中是git@192.168.1.115:HelloWorld，Authentication中选择SSH Keys，Passphrase是你自己创建SSH时设置的私钥密码。

2. 配置远程代码库名

我们也需要在Xcode中配置远程代码库名，相当于git remote add命令。打开Xcode菜单，选择Source Control→<账户描述名>→Configure <账户描述名>项，此时弹出的对话框如图26-7所示。选择Remotes标签，然后点击左下角的+按钮，在弹出菜单中选择Add Remote…菜单，这时会弹出如图26-8所示的对话框。在Name中输入HW123，在Address中输入git@192.168.1.115:HelloWorld，完成后点击Add Remote按钮确定。

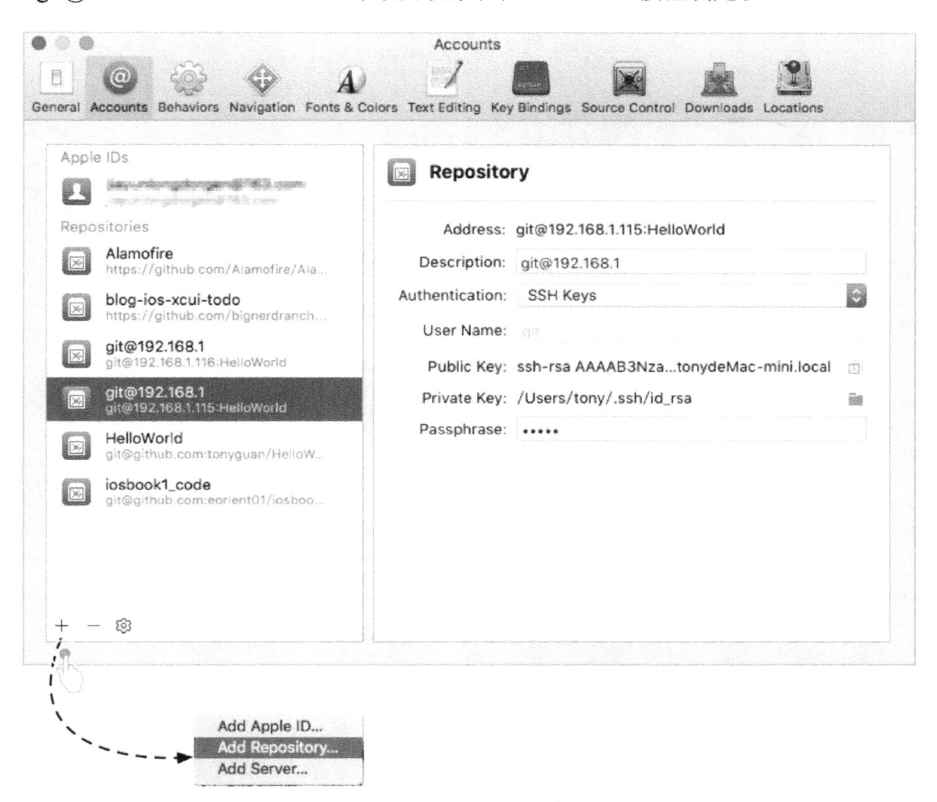

图26-5 添加账号

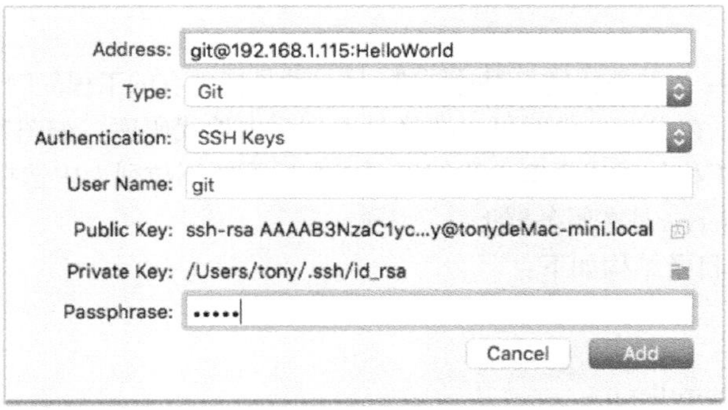

图26-6 添加账号对话框

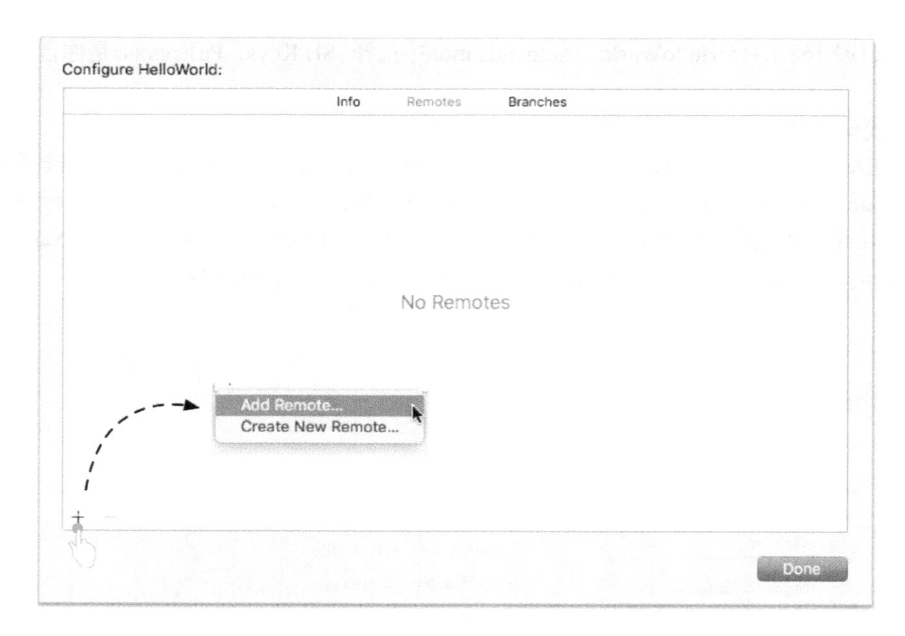

图26-7　配置远程代码库名

图26-8　添加账号对话框

配置工作完成后，再介绍一下如何在Xcode中使用Git。这里还是通过解决实际工作中的一些问题来学习，我们工作中遇到的几个问题如下所示。

- **问题1**。如何创建代码库，并将代码添加和提交到代码库？
- **问题2**。如何将本地代码库推送给远程服务器代码库？
- **问题3**。如何克隆远程服务器代码库到本地？
- **问题4**。如何获取远程代码库数据，并解决冲突问题？

- 问题1

创建代码库的方式有两种：一种是新建工程时创建；另一种是把现有的工程复制到代码库下，再初始化代码库。

如果是新建工程时创建，在保存文件时可以选择是否创建本地代码库。如图26-9所示，如果选中Create Git repository on复选框，就会为工程创建代码库。我们还可以在后面的下拉列表中选择创建代码库的位置，其中My Mac是指本机，Add To New Server是在服务器上。

注意，在Xcode中生成的目录结构如下：

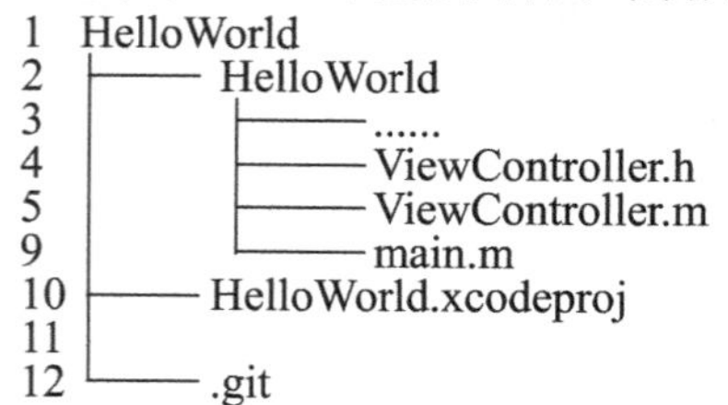

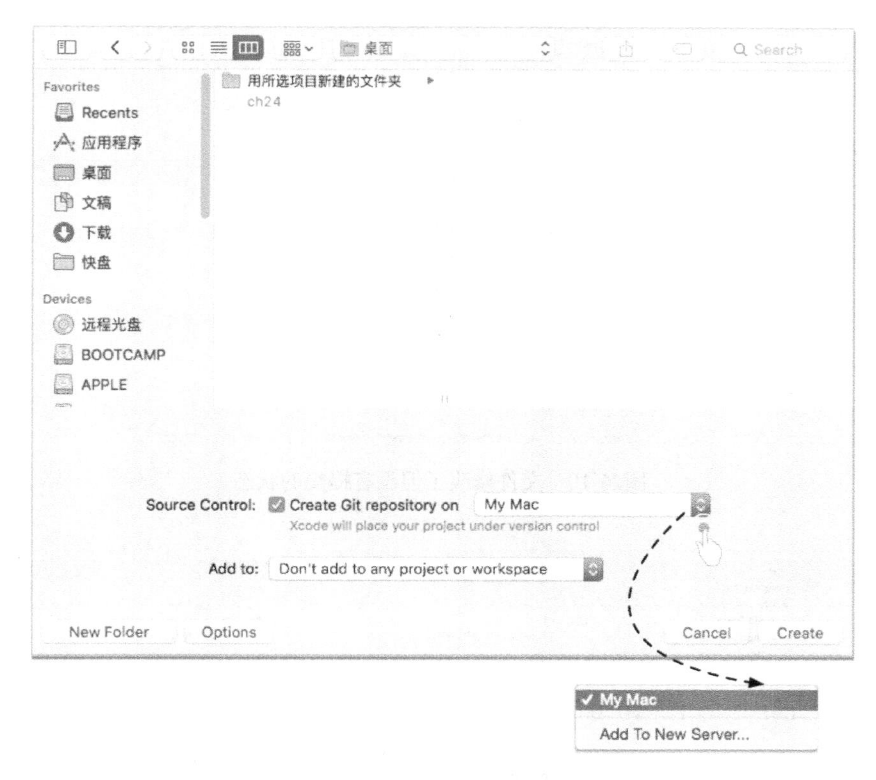

图26-9 创建代码库

第一行的HelloWorld是工程目录，也是代码库的根目录，第二行的HelloWorld目录是存放源程序的目录。而我们以前的目录结构与此不同，具体如下所示：

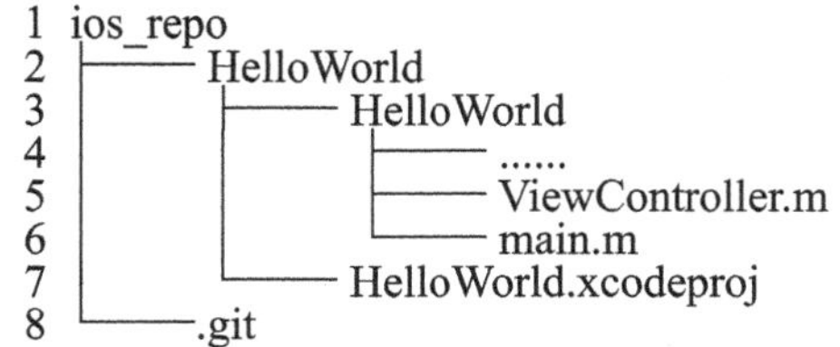

第一行的ios_repo是代码库的根目录，第二行的HelloWorld是工程目录。对于这样的结构，一个代码库可以放置多个工程，是一对多的关系。而Xcode生成的方式是代码库就是工程目录，它们是一对一的关系。

为了在Xcode中配置和使用Git，本节中我们重点介绍Xcode生成的代码库。它们是一对一的关系，我们在创建工程时不要忘记选中Create Git repository on复选框。如果我们创建的工程名字为HelloWorld，那么本地代码库的名字也就是HelloWorld。我们需要管理员在服务器上创建一个远程代码库HelloWorld，这个过程可以参考26.2.2节。

然后就可以在Xcode中修改这个工程了。修改并保存文件后，我们会看到在导航面板中文件的后面有一个M图标，如图26-10所示，这说明文件修改了但没有提交。

如果只是想提交选中的文件，可以点击鼠标右键，从弹出的快捷菜单中选择Source Control→Commit Selected Files...菜单项，其中Source Control菜单都是有关代码控制的。如果想提交全部的修改文件，可以使用Source Control→Commit...菜单项，此时弹出的对话框如图26-11所示。

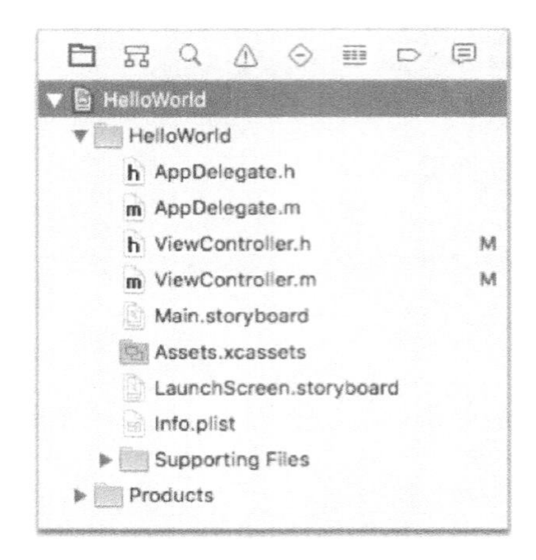

图26-10　文件修改了但没有提交的状态

图26-11　提交文件

其中有两个代码窗口，左边是本地未提交版本，右边是代码库中的版本，这里可以比较看看修改了哪些内容。在下面的输入框中添加注释，点击Commit按钮就可以提交了。

- 问题2

要解决问题2，可以选择Source Control→Push...菜单项，将本地代码库推送给远程服务器代码库，此时会出现如图26-12所示的对话框。点击Push按钮，就可以将本地代码的内容提交给远程服务器的代码库了。

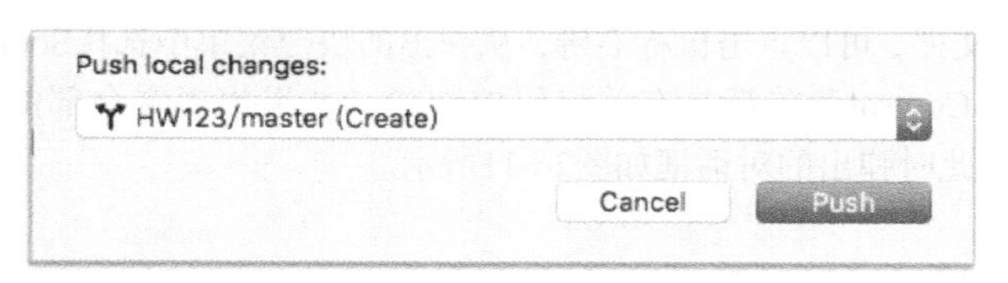

图26-12　推送给远程服务器

● 问题3

这个问题是从服务器代码库克隆到本地。可以选择Source Control→Check Out...菜单项，此时弹出的对话框如图26-13所示，选择其中要克隆的代码库，然后点击Next按钮进行认证，如果通过，就会提示我们选择本地的保存位置，接着点击Check Out就可以克隆了。

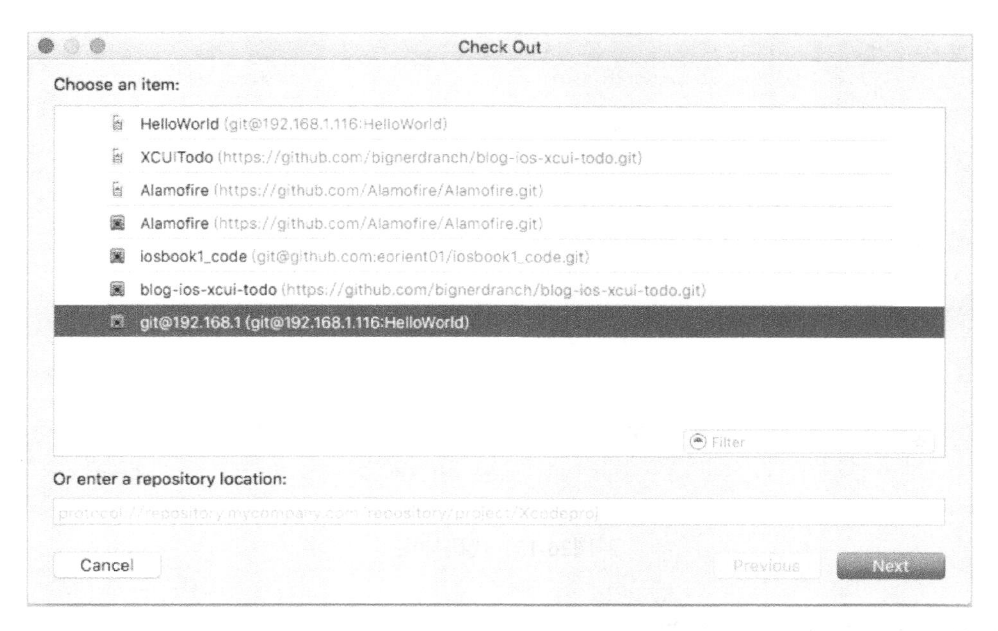

图26-13 克隆远程代码库

● 问题4

如果服务器代码有新的版本，要将其数据获取到本地，可以通过如下方法：选择Source Control→Pull...菜单项，此时弹出的对话框如图26-14所示，然后点击Pull按钮即可。

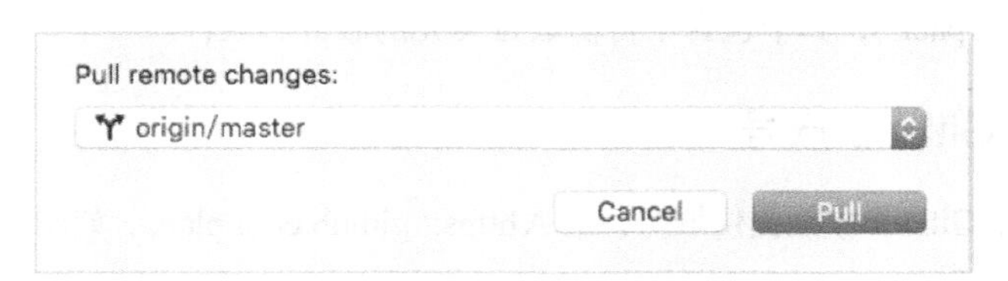

图26-14 获取远程代码库信息

如果这个过程中有冲突发生，会弹出如图26-15所示的对话框，从中可以看到它们的冲突点。

在图26-15中，最下面的4个按钮用于合并和编辑冲突点。如果没有冲突，Pull按钮就可以点击了，此时直接点击该按钮就可以了。

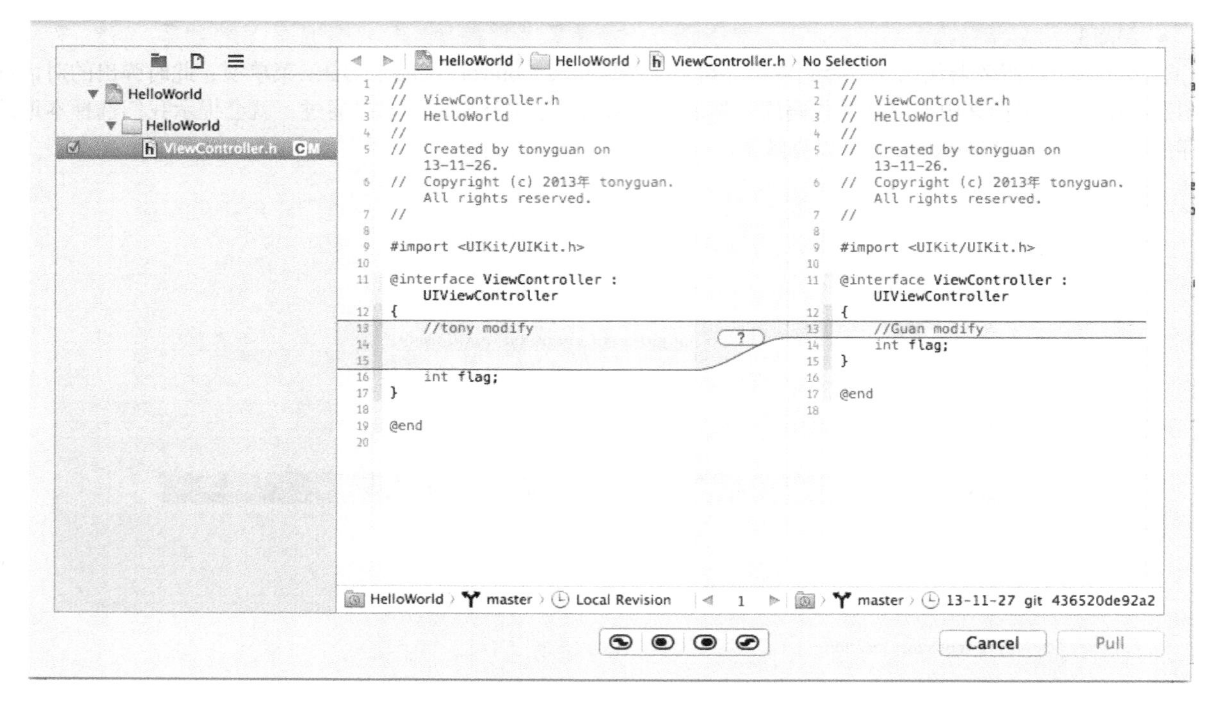

图26-15　代码冲突

26.3　GitHub 代码托管服务

GitHub（https://github.com）是全球最大的编程社区及代码托管网站，它可以提供基于Git版本控制系统的代码托管服务。如果我们的团队成员分散在不同的地方，使用GitHub代码托管服务是一个不错的选择。GitHub同时提供商业账户和免费账户：免费账户不能创建私有项目。针对学生，GitHub推出了免费的Student Developer Pack（https://education.github.com/pack），其中包含了价值数百美元的服务和软件。

26.3.1　创建和配置 GitHub 账号

只有用户注册账号了，GitHub才能提供服务。进入https://github.com/plans，如图26-16所示，从中可以创建免费账号、收费账号和收费组织。

这里我们创建免费账号。进入https://github.com/join页面创建免费账号，输入账号、邮箱和密码，验证通过就可以创建了。进入网址https://github.com/login可以登录，这里可以使用刚才创建的账号测试一下。登录成功后的页面如图26-17所示。

我们知道Git常用的协议是SSH协议，我们需要在本机上生成后，将公钥提供给GitHub网站。点击图26-17右上角的View profile and more图标，然后在弹出菜单中点击Settings菜单项，此时将进入如图26-18所示的设置页面。

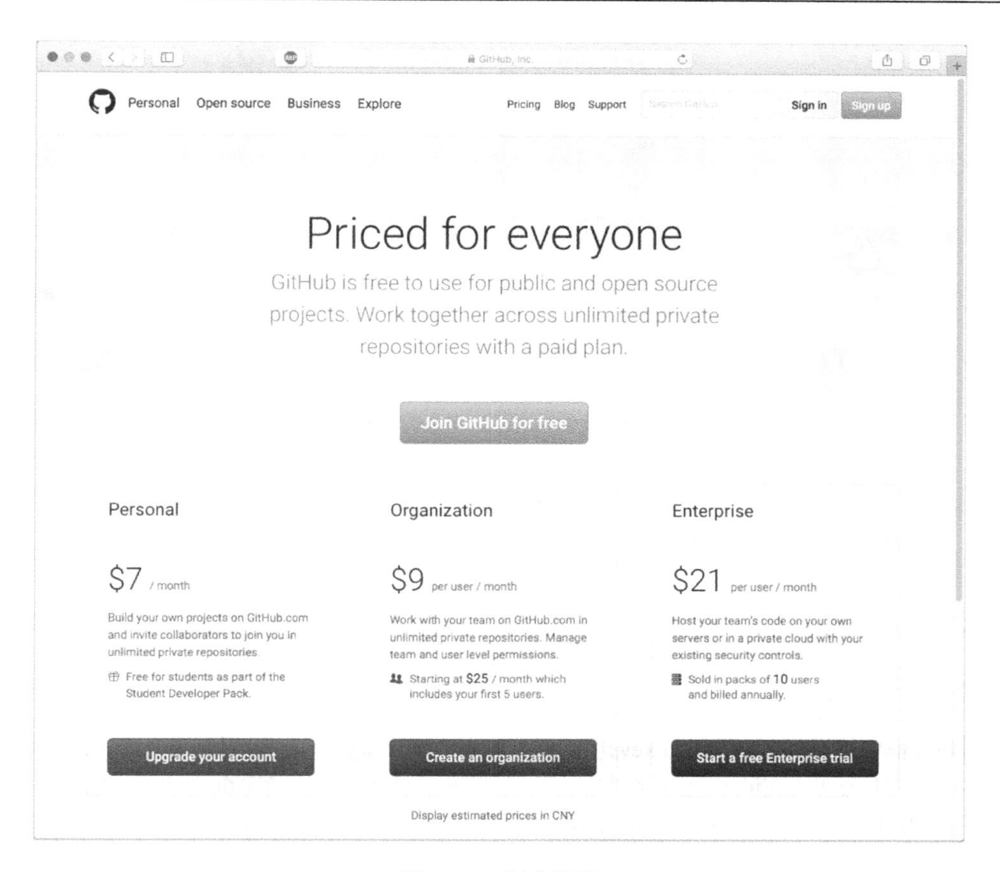

图26-16 创建账号

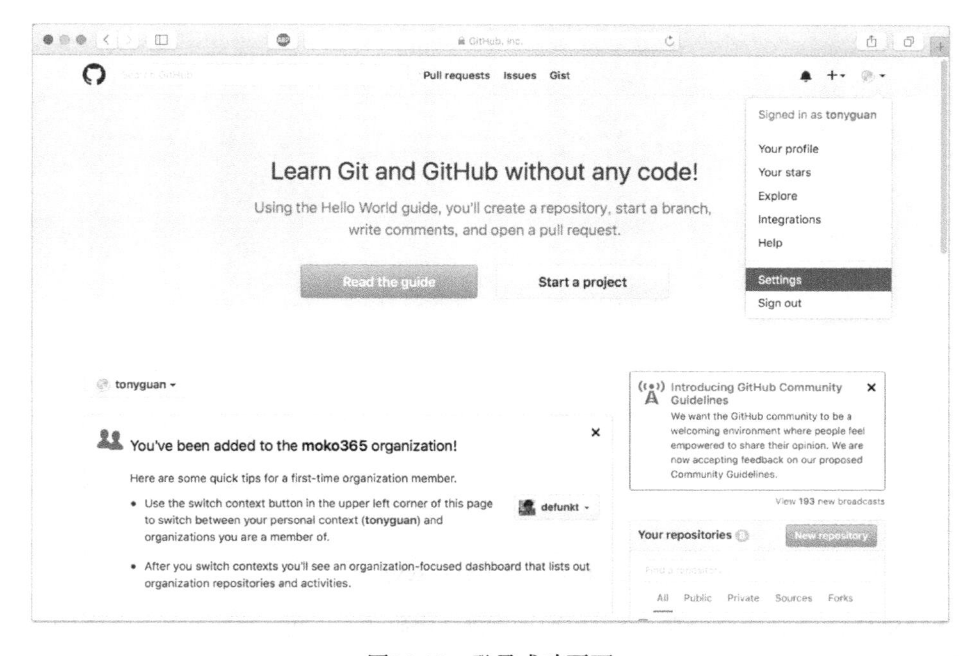

图26-17 登录成功页面

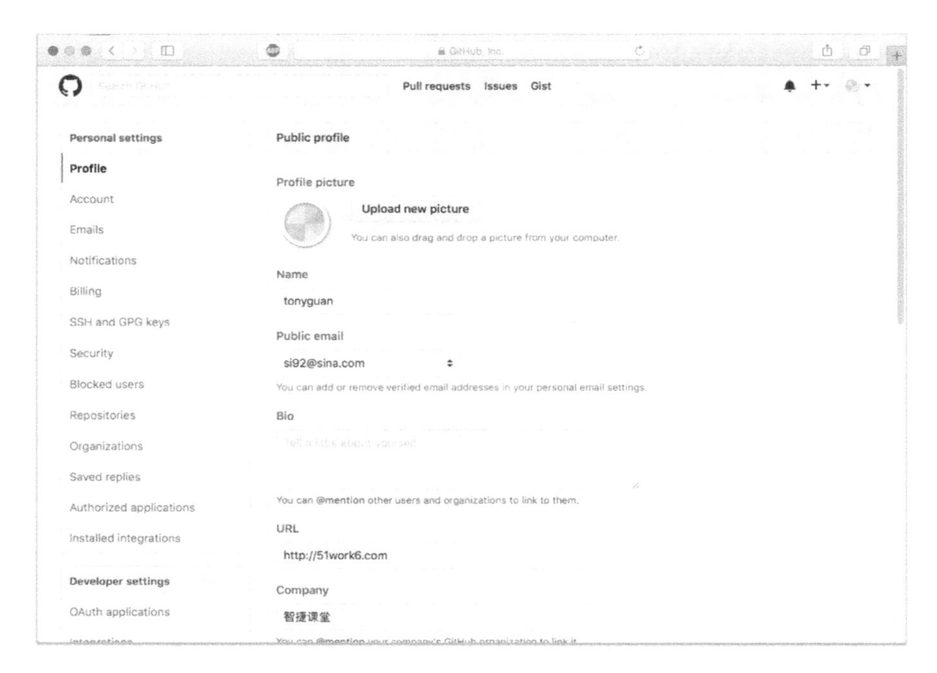

图26-18　设置页面

从左边的列表框中选择SSH and GPG keys项，再点击Add SSH key按钮，此时会进入如图26-19所示的页面，我们将本机生成的SSH公钥（在id_rsa.pub文件中）粘贴到Key文本框中，并在Title文本框中输入一个标题（这个可以随便命名）。另外，对于生成SSH key不熟悉的读者，可以点击generating SSH keys超链接查看帮助。

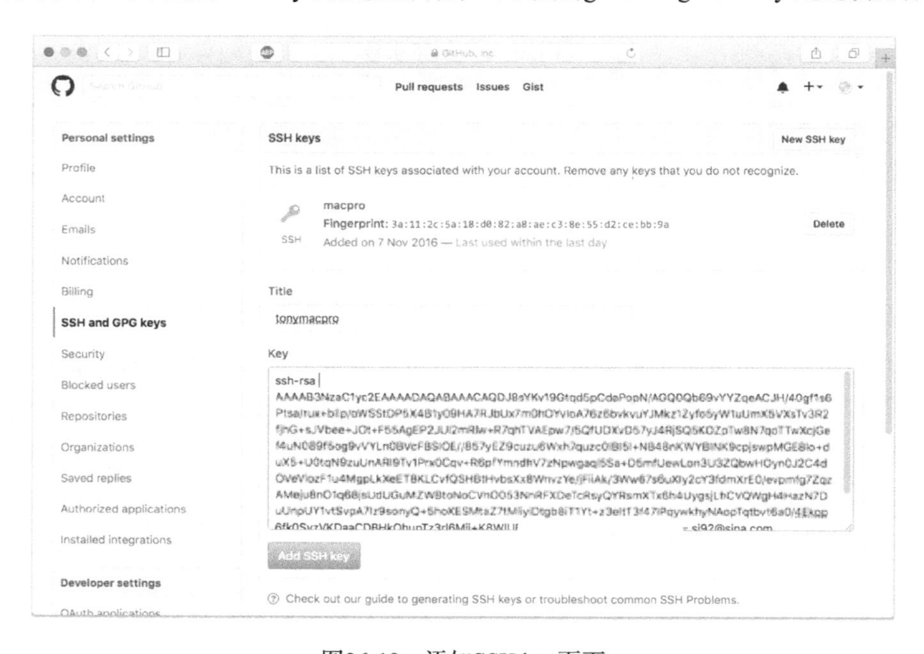

图26-19　添加SSH key页面

然后点击Add SSH key按钮提交内容，接着需要再次确认密码才能成功。这里可以提交多个key，用来管理同一用户在不同机器上的登录情况。

26.3.2 创建代码库

在登录成功页面（见图26-17）的右下角部分，或在设置页面中（见图26-18）点击左边列表中Repositories项，都可以看到当前用户下的代码库列表。这些代码库可以自己创建，也可以从别人那边派生（fork）过来。

如果要在GitHub中创建代码库，可以单击页面右上角的Create new...图标+，然后在弹出菜单中选择New repository菜单项，如图26-20所示。

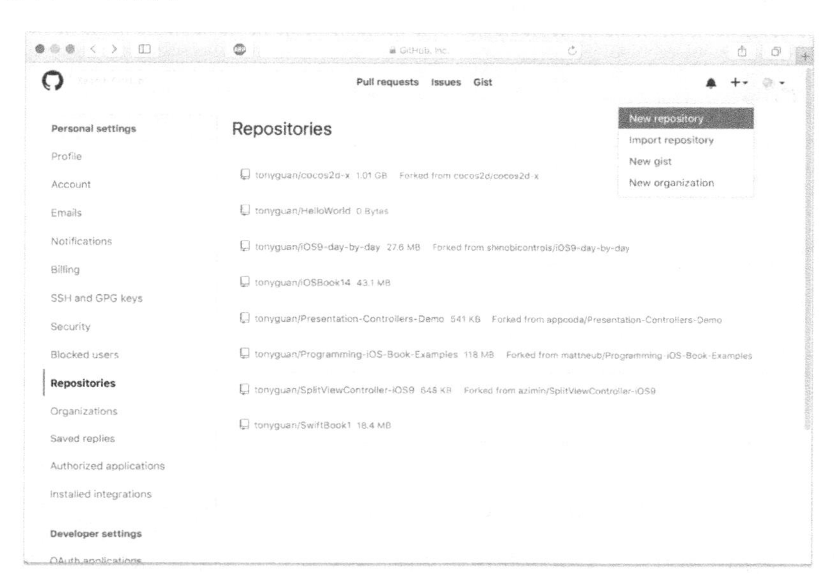

图26-20　创建代码库

此时将进入如图26-21所示的创建代码库页面，这里可以输入代码库的名字（Repository name）和描述信息（Description）。在这里，我们可以创建私有代码库（Private）或公有代码库（Public）。需要说明的是，只有付费账户才可以创建私有库。此外，我们还可以选中Initialize this repository with a README复选框创建一个README文件，用来说明这个代码库。

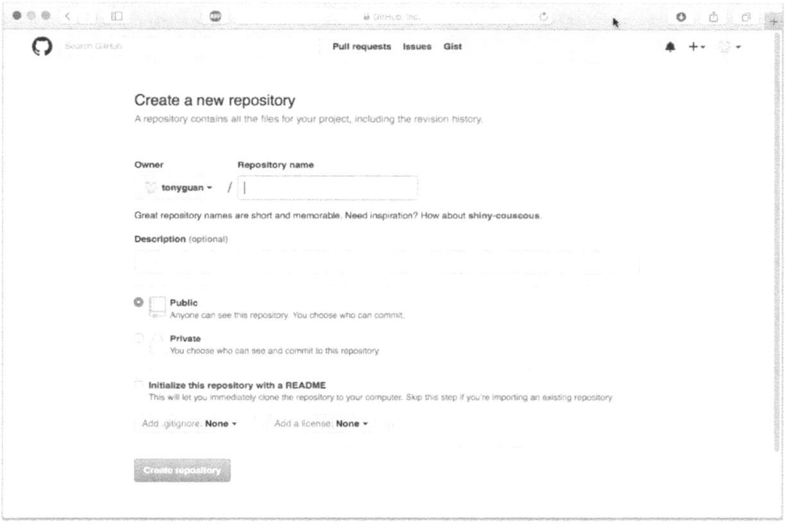

图26-21　创建代码库页面

此时点击Create repository按钮即可创建代码库，但此时代码库中还是空的，我们需要在本地电脑中推送一个Xcode项目到GitHub代码库。

由于GitHub上HelloWorld代码库是空的，我们需要将本地的HelloWorld代码库推送到GitHub服务器。这可以通过Source Control→Push...菜单项推送到GitHub服务器，其中选择Remote为origin/master，具体请参考26.2.6节。

推送成功后，可以在GitHub中查看。登录GitHub，进入tonyguan/HelloWorld代码库，如图26-22所示，如果在这里可以看到推送的HelloWorld工程了，就说明推送成功了。

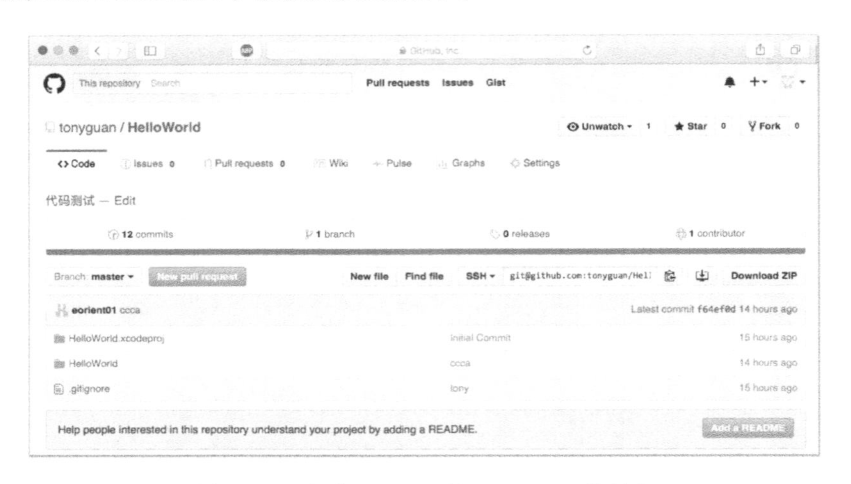

图26-22　查看GitHub上的HelloWorld代码库

GitHub上的代码一般都应该有一个README说明文件，我们可以点击Add a README按钮添加内容。

26.3.3　删除代码库

有的时候我们需要删除代码库，此时可以点击如图26-23所示的Settings标签，进入如图26-24所示的代码库维护页面，接着将页面滚动到下面，此时会看到Delete this repository按钮，点击该按钮，会弹出确认对话框，在对话框的文本框中输入要删除的代码库名，然后点击删除按钮完成操作。这个操作破坏性比较大，大家操作时一定要谨慎。

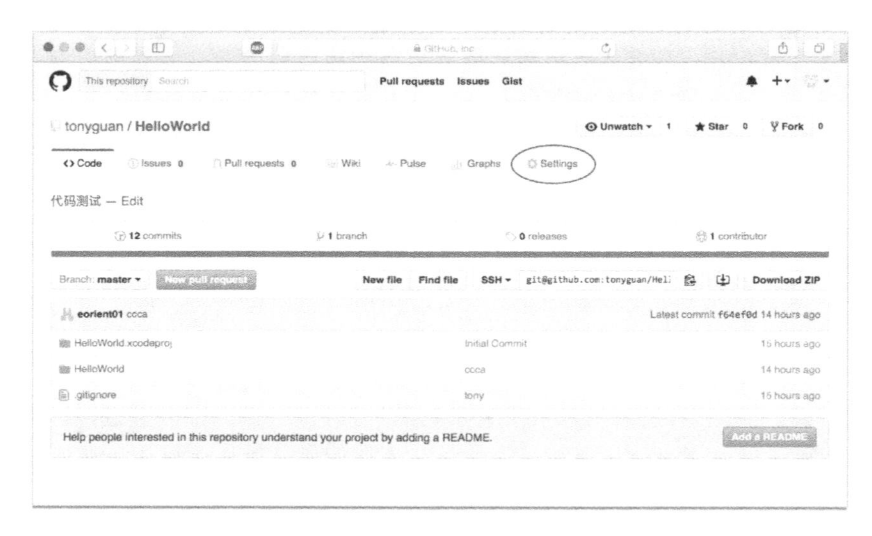

图26-23　进入Settings页面

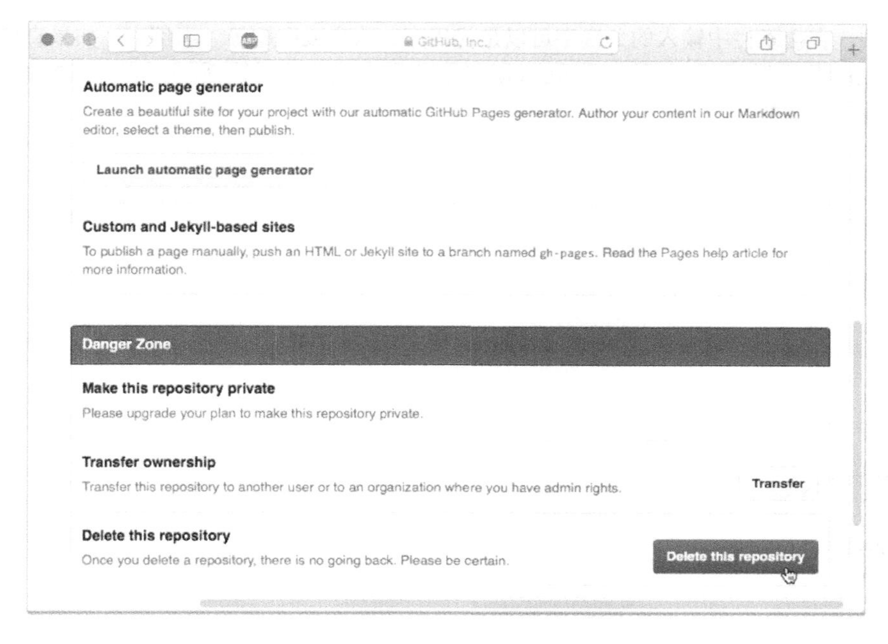

图26-24 删除代码库

26.3.4 派生代码库

获得代码库的最简单方式是从别人那里派生代码库。我们可以修改该代码库，然后提供给开发者。因此，派生与Git中的分支很像，可以把它理解为代码库级别的分支。

假设一个GitHub账户想zzr从tonyguan账户那里派生HelloWorld代码库，此时首先需要在GitHub中找到HelloWorld代码库。GitHub搜索功能在https://github.com/search页面中提供了，如图26-25所示。

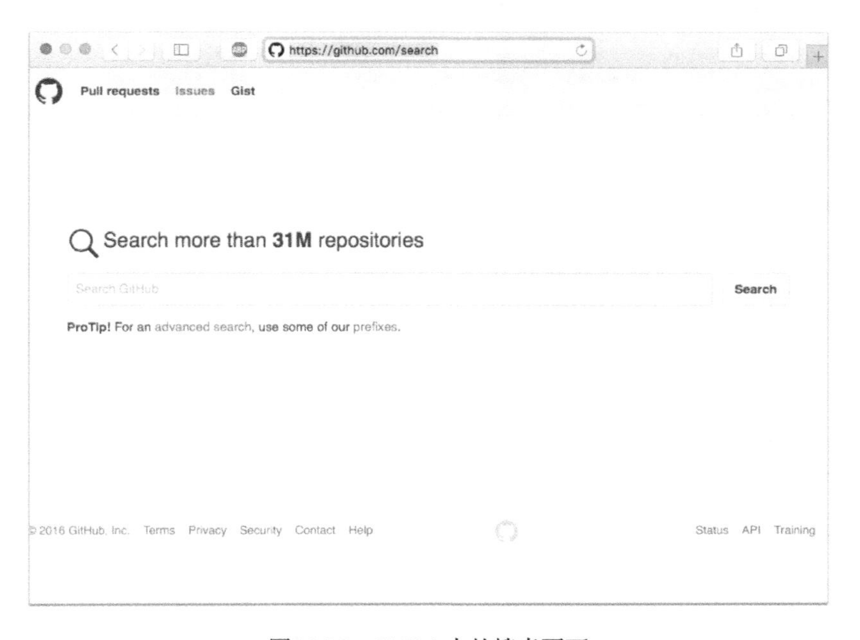

图26-25 GitHub中的搜索页面

在这个网页中，在搜索栏中输入的关键字可以是代码库、账户或问题等。本例中，如果直接查找HelloWorld代码库，结果会比较多。我们可以先搜索tonyguan账户，在搜索结果中再选择Users，如图26-26所示，其中第一条搜索结果就是我们需要的。

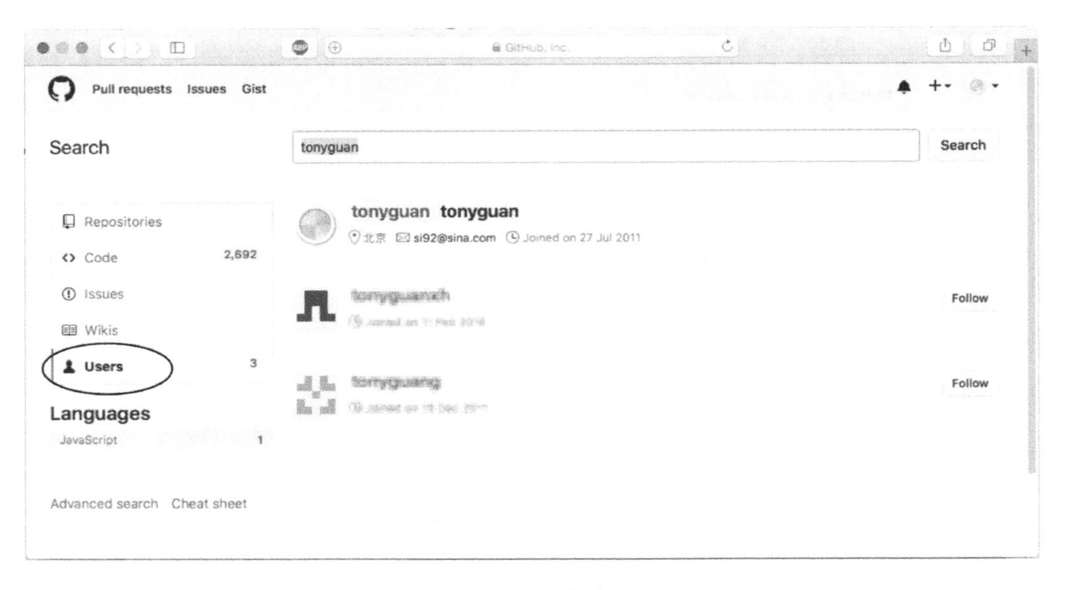

图26-26　搜索结果

点击tonyguan账户进入，然后点击Repositories标签，找到HelloWorld代码库，如图26-27所示。点击HelloWorld代码库，就可以看到该库的详细信息，如图26-28所示。

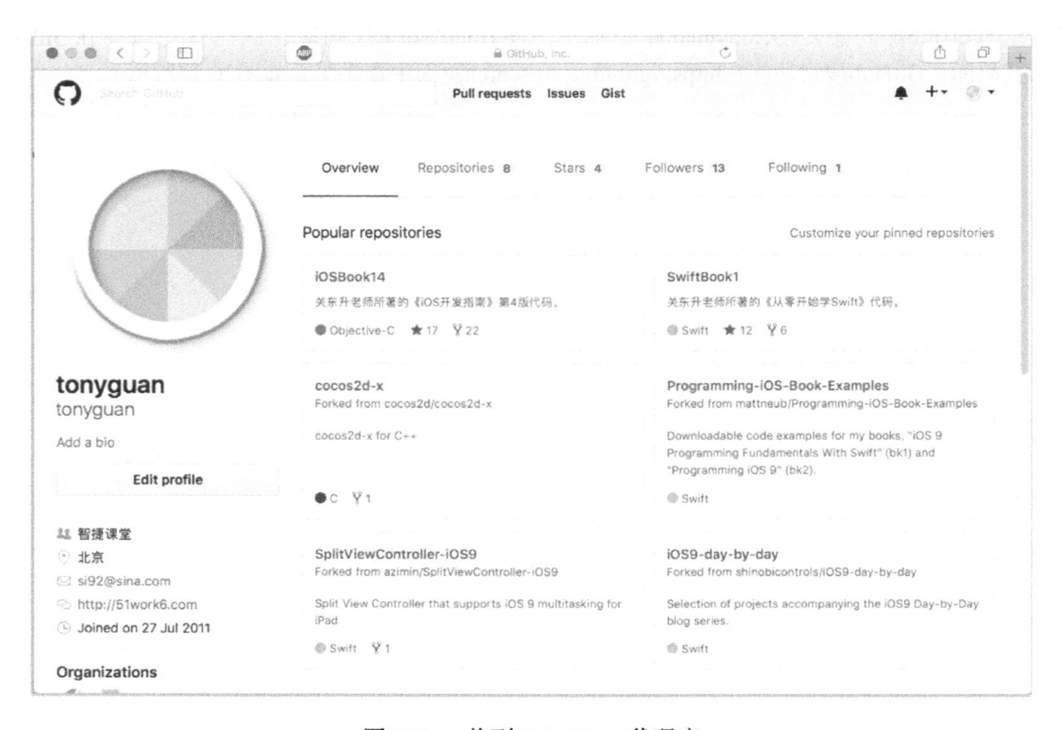

图26-27　找到HelloWorld代码库

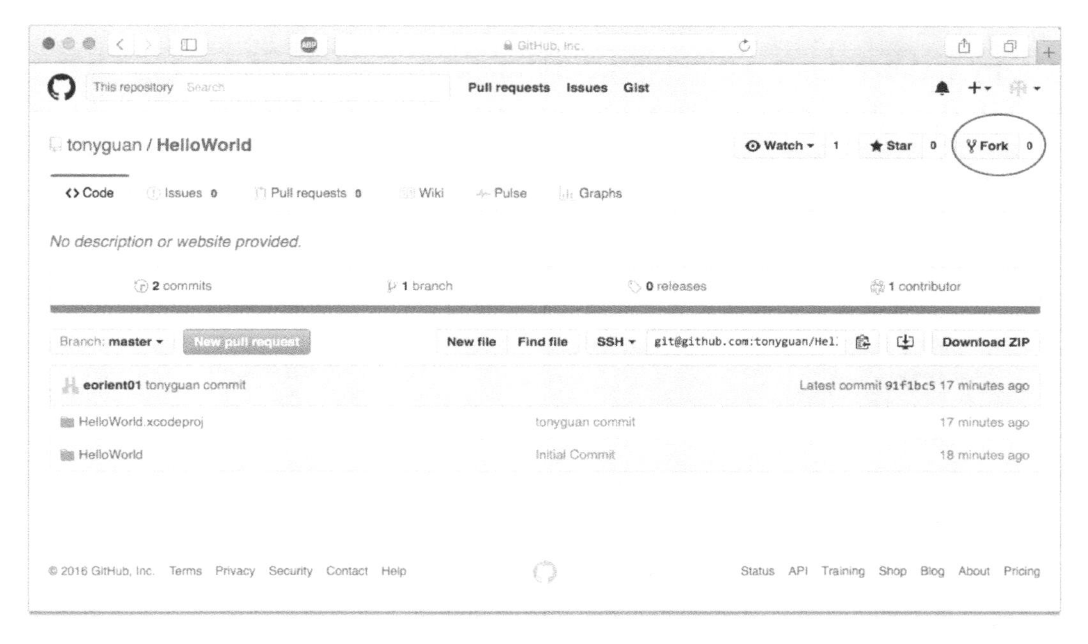

图26-28 tonyguan账户的HelloWorld代码库

点击右上角的Fork按钮，此时会弹出一个确认对话框，如图26-29所示，从中选择@zzr用户。

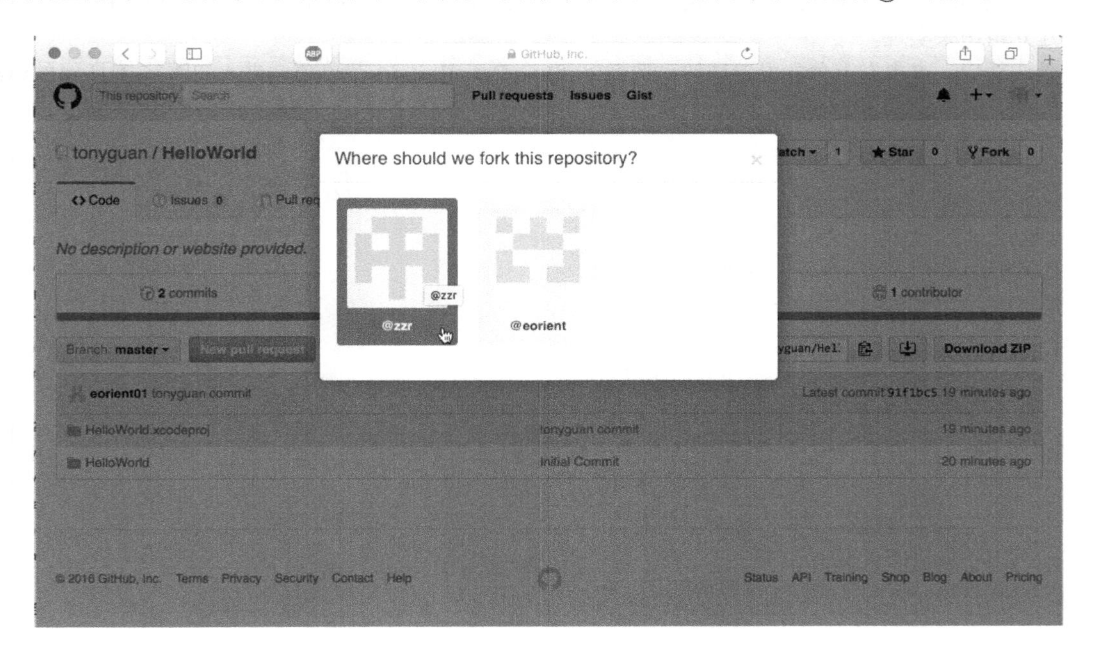

图26-29 选择用户

确认之后，就可以把Fork代码库派生到当前账户下面了，如图26-30所示。

此时，当前账号可以像使用自己创建的其他代码库那样使用这个库了。如果我们只是想参考别人的代码学习，这样派生过来后我们的工作就结束了。

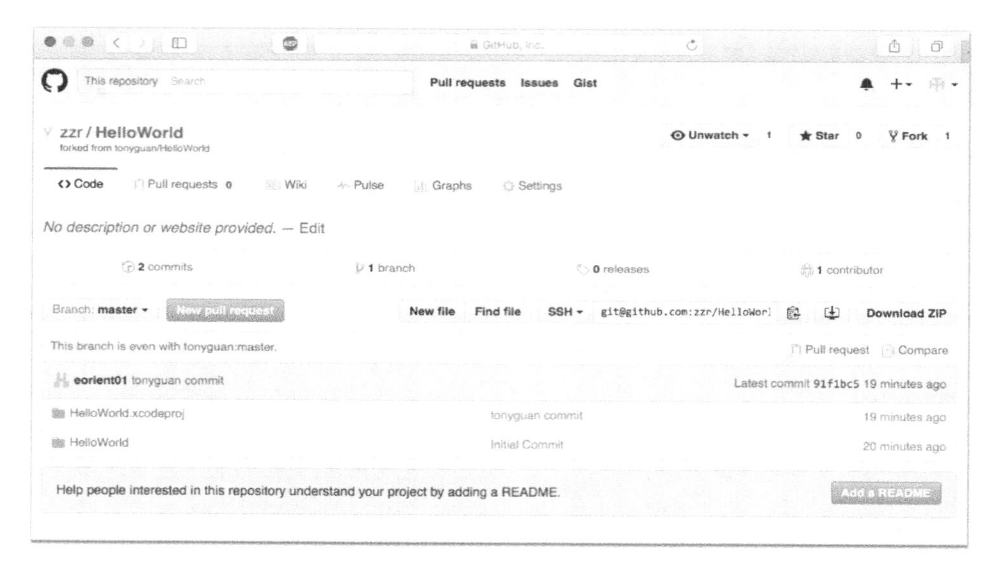

图26-30　派生代码库到当前账号

26.3.5　管理组织

在协同开发的时候，管理员会让其他人员一起管理代码库，包括响应推送请求、合并推送等，这可以通过在GitHub中建立组织来实现。

在登录成功后，点击右上角的设置按钮，选择Profile→Organizations菜单项，进入如图26-31所示的组织管理页面。

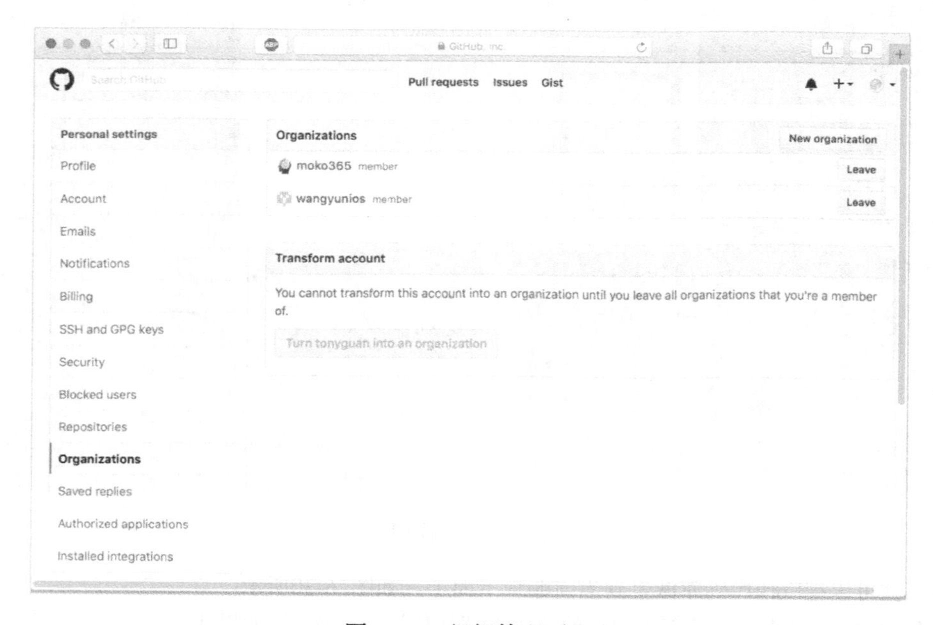

图26-31　组织管理页面

此时可以看到当前账号所属的组织，点击Leave按钮即可脱离该组织。点击New organization按钮，进入创建组织页面，如图26-32所示。

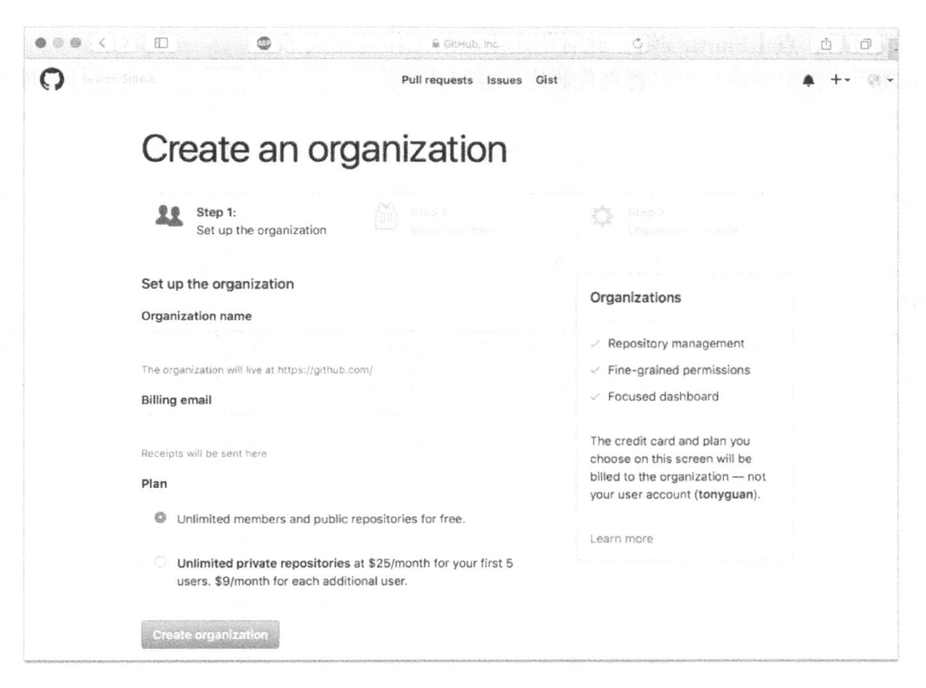

图26-32 创建组织页面

在图26-32中输入组织名和账单邮箱后，点击Create Organization按钮，即可进入添加组织成员页面。此时在添加成员文本框中输入GitHub账号，会出现下拉列表框，在其中选择要添加的成员，然后点击后面的+按钮邀请该成员进入组织，如图26-33所示。

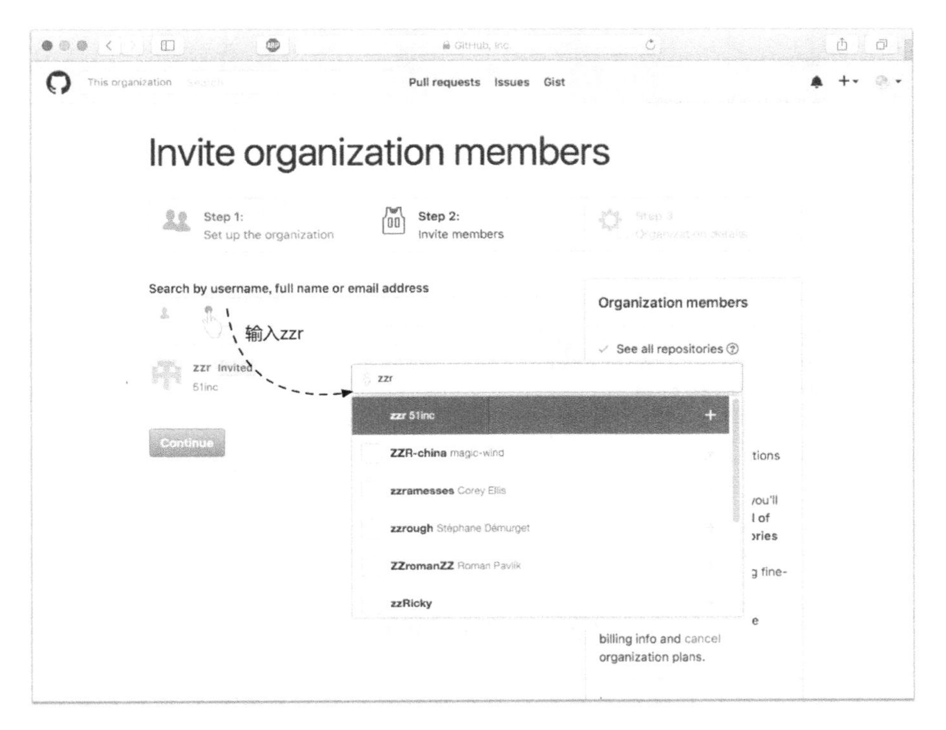

图26-33 添加组织成员页面

添加完组织成员后，点击Finish按钮，此时创建组织的工作就完成了。

组织创建成功后，组织的成员可以管理代码库。组织获得代码库有两种方法：一种是自己创建，这个过程可以参考一般账户的创建过程；另一种是由成员转移过来。对于后者，可以采用下面的方式来实现：首先要转移库的成员（例如：tonyguan）登录并进入到代码库，点击Settings标签，进入如图26-34所示的代码库设置页面，滚动屏幕到页面底部，接着点击Transfer按钮，此时弹出的对话框如图26-35所示，从中输入要转移的代码库和新的拥有者。

如果要转移给的组织是当前账户创建的，就可以在组织中看到这个库了，如图26-36所示。如果是其他的组织或成员，GitHub会有一个审核过程，不会马上看到。

组织中会有管理员，他可以管理代码库，对代码库重新命名、删除和转移代码库等。而普通组织成员不能够对代码库进行这些操作，这需要组织对该成员进行授权。但普通成员对代码库进行接受推送请求、合并推送等工作完全没有问题。

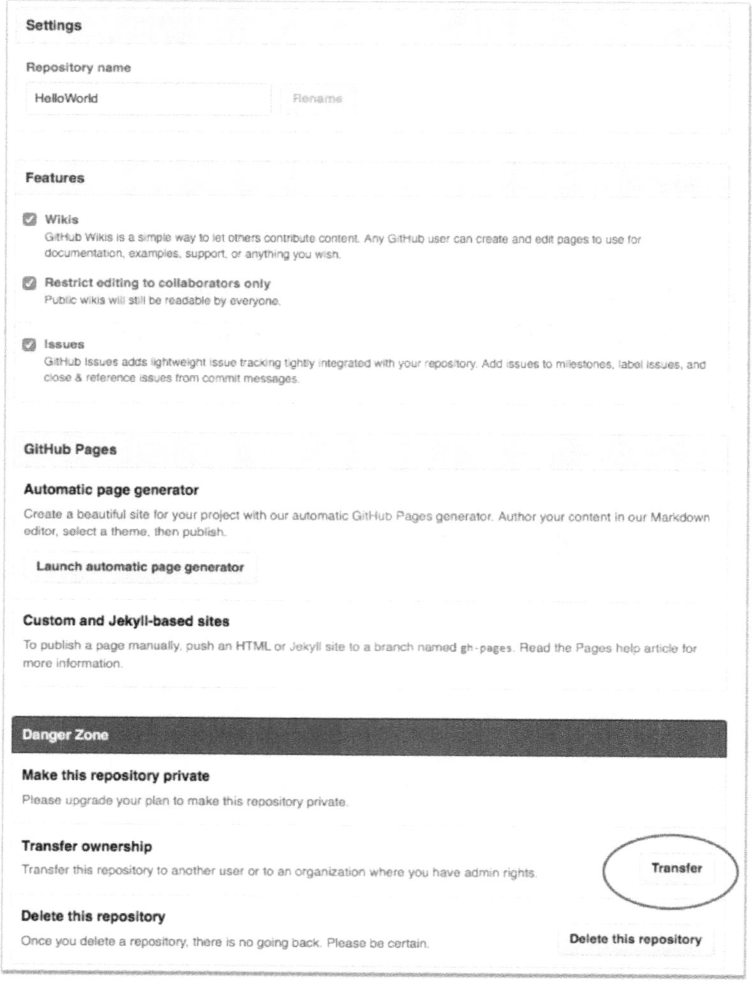

Settings

Repository name

HelloWorld

Rename

Features

Wikis

GitHub Wikis is a simple way to let others contribute content. Any GitHub user can create and edit pages to use for documentation, examples, support, or anything you wish.

Restrict editing to collaborators only

Public wikis will still be readable by everyone.

Issues

GitHub Issues adds lightweight issue tracking tightly integrated with your repository. Add issues to milestones, label issues, and close & reference issues from commit messages.

GitHub Pages

Automatic page generator

Create a beautiful site for your project with our automatic GitHub Pages generator. Author your content in our Markdown editor, select a theme, then publish.

Launch automatic page generator

Custom and Jekyll-based sites

To publish a page manually, push an HTML or Jekyll site to a branch named gh-pages. Read the Pages help article for more information.

Danger Zone

Make this repository private

Please upgrade your plan to make this repository private.

Transfer ownership

Transfer this repository to another user or to an organization where you have admin rights.

Transfer

Delete this repository

Once you delete a repository, there is no going back. Please be certain.

Delete this repository

图26-34　代码库设置页面

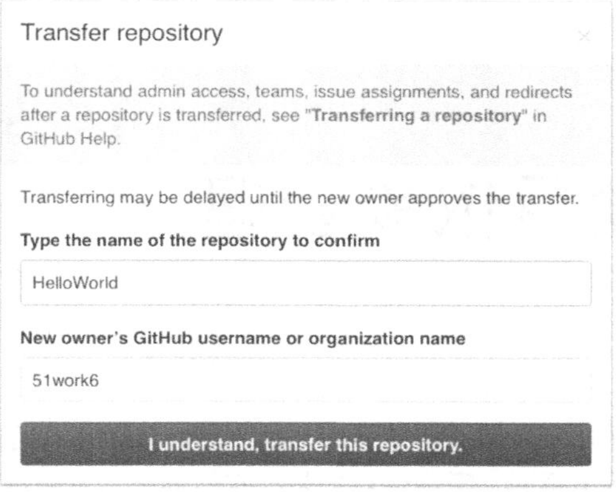

Transfer repository

To understand admin access, teams, issue assignments, and redirects after a repository is transferred, see "**Transferring a repository**" in GitHub Help.

Transferring may be delayed until the new owner approves the transfer.

Type the name of the repository to confirm

HelloWorld

New owner's GitHub username or organization name

51work6

I understand, transfer this repository.

图26-35　转移代码库给组织

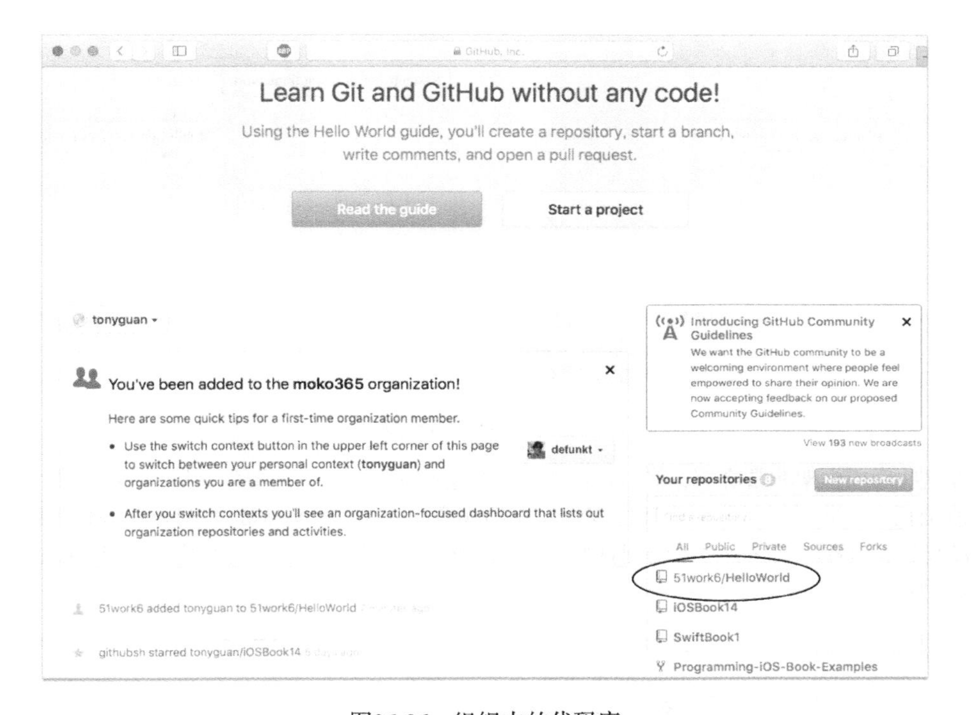

图26-36　组织中的代码库

26.4　小结

通过对本章的学习，我们了解了如何使用Git进行代码版本控制，其中包括Git服务器的搭建、Git常用命令和协同开发，还介绍了如何配置和使用Git工具。GitHub是一个优秀的Git开发社区，使用GitHub代码托管服务是一个不错的选择。

第 27 章 项目依赖管理

我们在开发过程中会用到（依赖于）其他第三方库，这些库或框架还有可能依赖于其他库或框架，这样就形成了一个如图27-1所示的复杂依赖关系网。需要注意的是，它们的版本也是有要求的。

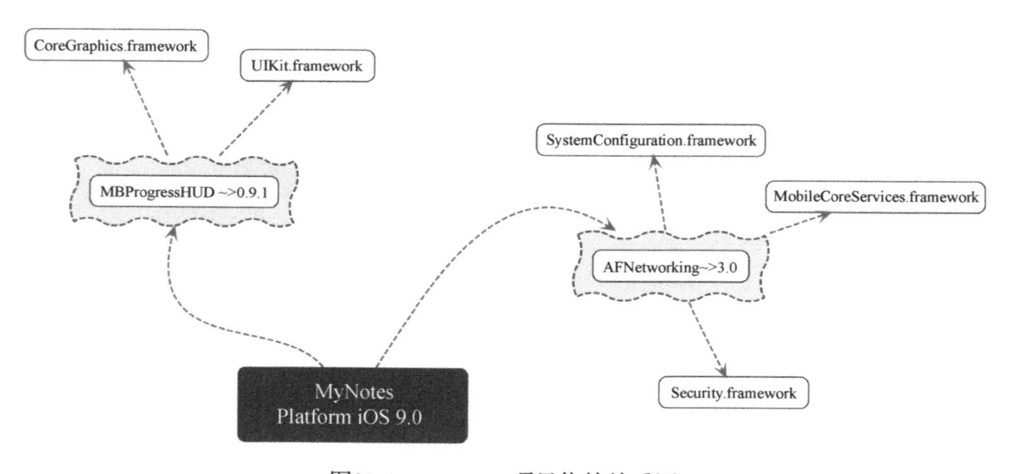

图27-1　MyNotes项目依赖关系网

手动管理这些库或框架非常麻烦，首先要找到并下载这些库或框架，还要注意版本的兼容性问题；然后把它们添加到当前工程中，此时还需要配置环境变量。在第16章介绍AFNetworking和Alamofire框架的使用时，我们采用的是手动配置，读者应该领教过手动配置的麻烦。因此，本章我们介绍如何采用CocoaPods和Carthage工具管理项目的依赖关系。

提示　为了简练，本章中将“第三方库或框架”简称为“第三方库”。

27.1　使用 CocoaPods 工具管理依赖

CocoaPods是最有影响力的macOS和iOS项目依赖管理工具，于2011年发布，经过多年的发展，它已经非常完善了。CocoaPods支持在项目中采用Objective-C或Swift语言。CocoaPods会将第三方库的源代码编译为静态链接库.a文件或者动态框架.framework文件的形式，并将它们添加到项目中，建立依赖关系。

27.1.1　安装 CocoaPods

由于CocoaPods（https://cocoapods.org/）工具是使用Ruby语言开发的，在macOS系统下默认安装了Ruby运行

环境，我们不需要另外安装了。CocoaPods的下载和安装可以使用Ruby包管理工具gem，此时需要在终端中执行如下指令：

```
$ sudo gem install cocoapods
```

提示 如果这个命令执行了很长时间但没有反应，或出现如下错误：

```
ERROR:  Could not find a valid gem 'cocoapods' (>= 0), here is why:
        Unable to download data from https://rubygems.org/ - Errno::ECONNRESET: Connection reset by peer - SSL_connect
(https://rubygems.org/latest_specs.4.8.gz)
```

这是由于gem工具无法访问默认的Ruby源（rubygems.org），此时我们可以把Ruby源换为国内的镜像源。在终端中执行如下指令：

```
$ sudo gem sources --remove https://rubygems.org/
$ sudo gem sources -a https://gems.ruby-china.org/
```

install指令成功执行之后，还需要进行设置。在终端中执行如下命令：

```
$ pod setup
```

setup指令会在本地创建~/.cocoapods/文件夹，并将CocoaPods的GitHub库（https://github.com/CocoaPods/Specs）下载到该文件夹中。这样一来，当配置依赖关系时，我们可以直接使用本地文件。由于下载内容很多（大约500 MB），setup执行时间会比较长。

27.1.2 搜索库

如何知道CocoaPods库中是否有我们所需要的第三方库呢？这可以通过网站或pod指令进行搜索。

1. 使用网站搜索

我们可以通过网站https://cocoapods.org（如图27-2所示）进行搜索，在SEARCH*中输入第三方库的关键字，例如输入“AF”关键字，搜索结果如图27-3所示，这里我们可以指定搜索的平台或者语言。

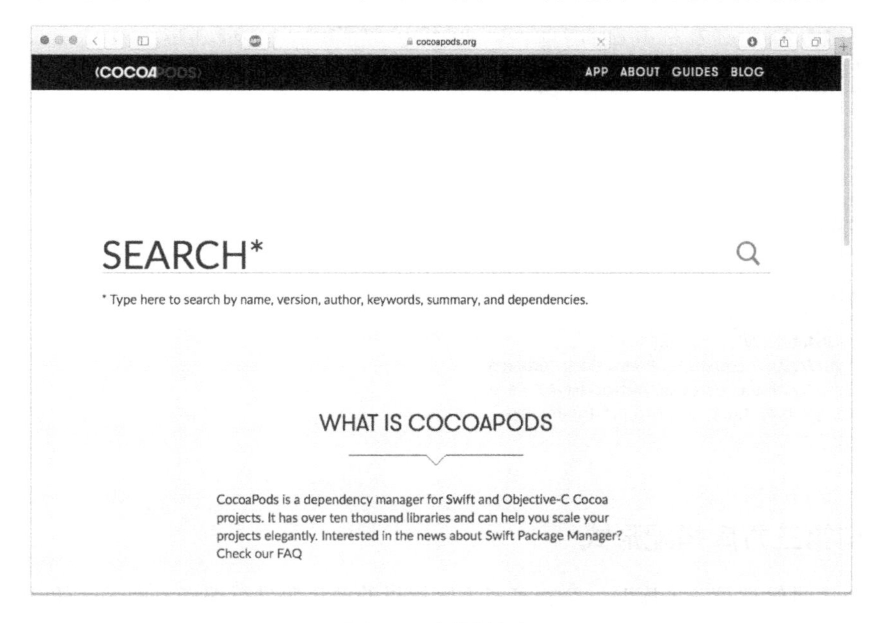

图27-2 网站搜索

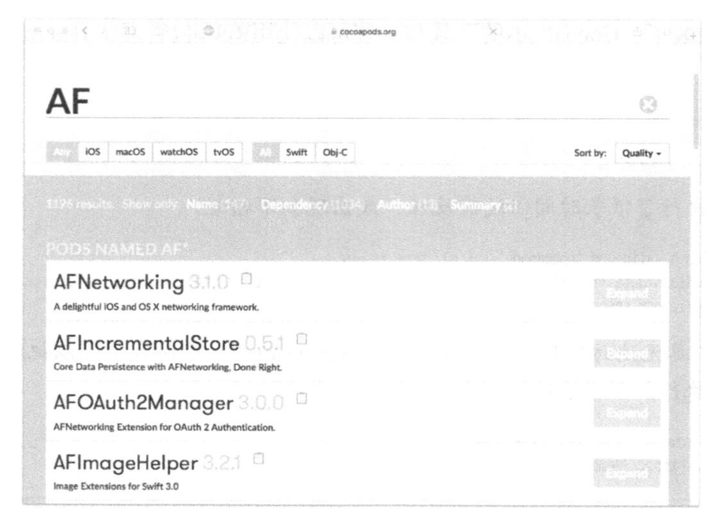

图27-3　网站搜索结果

2. 使用pod指令搜索

此外，我们可以使用pod search指令来搜索。在终端中搜索第三方库，例如搜索AFNetworking框架，相关指令如下：

```
$ pod search AFNetworking
```

搜索结果如下：

```
-> AFNetworking (3.1.0)
   A delightful iOS and macOS networking framework.
   pod 'AFNetworking', '~> 3.1.0'
   - Homepage: https://github.com/AFNetworking/AFNetworking
   - Source:   https://github.com/AFNetworking/AFNetworking.git
   - Versions: 3.1.0, 3.0.3, 3.0.2, 3.0.1, 3.0.0, 3.0.0-beta.3, 3.0.0-beta.2, 3.0.0-beta.1, 2.6.3, 2.6.2,
   2.6.1, 2.6.0, 2.5.4, 2.5.3, 2.5.2, 2.5.1, 2.5.0, 2.4.1, 2.4.0, 2.3.1, 2.3.0, 2.2.4, 2.2.3, 2.2.2,
   2.2.1, 2.2.0, 2.1.0, 2.0.3, 2.0.2, 2.0.1, 2.0.0, 2.0.0-RC3, 2.0.0-RC2, 2.0.0-RC1, 1.3.4, 1.3.3, 1.3.2,
   1.3.1, 1.3.0, 1.2.1, 1.2.0, 1.1.0, 1.0.1, 1.0, 1.0RC3, 1.0RC2, 1.0RC1, 0.10.1, 0.10.0, 0.9.2, 0.9.1,
   0.9.0, 0.7.0, 0.5.1 [master repo]
   - Subspecs:
     - AFNetworking/Serialization (3.1.0)
     - AFNetworking/Security (3.1.0)
     - AFNetworking/Reachability (3.1.0)
     - AFNetworking/NSURLSession (3.1.0)
     - AFNetworking/UIKit (3.1.0)


-> AFNetworking+AutoRetry (0.0.5)
   Auto Retries for AFNetworking requests
   pod 'AFNetworking+AutoRetry', '~> 0.0.5'
   - Homepage: https://github.com/shaioz/AFNetworking-AutoRetry
   - Source:   https://github.com/shaioz/AFNetworking-AutoRetry.git
   - Versions: 0.0.5, 0.0.4, 0.0.3, 0.0.2, 0.0.1 [master repo]

……
```

27.1.3　项目与第三方库搭配形式

使用CocosPods工具管理第三方库时，会将第三方库提供的源代码进行编译，而编译的结果有两种形式：静态链接库.a文件和动态框架.framework文件。

那么，选择静态链接库还是框架呢？这首先要看你的项目中调用第三方库的代码采用何种语言，另外还要看第三库采用何种语言。

提示 如果想真正实现零配置，那么项目中调用第三方库的代码采用何种语言（Swift或Objective-C），就应该选择采用哪种语言（Swift或Objective-C）的第三方库。但是有时候未必能找到合适语言的第三方库，此时如果要调用不同语言的第三方库，就会涉及Swift与Objective-C代码混合调用的问题了。

项目与第三方库的搭配形式如表27-1所述。

表27-1 项目与第三方库的搭配形式

第三方库 / 项目代码	Objective-C语言	Swift语言
Objective-C	支持.a和.framework	支持.framework，需要Xcode生成头文件（Xcode-generated header），文件命名为<框架名/框架名-Swift.h> 例如，在Objective-C中调用Swift语言的Alamofire框架，就需要引入头文件代码#import <Alamofire/Alamofire-Swift.h>
Swift	支持.a和.framework，.a需要引入桥接头文件	支持.framework

由于Swift不能编译为静态链接库，只能编译为框架，所有用Swift编写的第三方库只能编译为框架以被项目使用。如果项目中使用Objective-C调用Swift编写的第三方库，就需要使用Xcode生成头文件。如果项目中使用Swift调用Objective-C编写的第三方库，并且第三方库编译为静态链接库，就需要引入桥接头文件。

27.1.4 实例：静态链接库形式管理依赖

下面通过一个示例介绍一下CocosPods的具体用法。我们在16.4节的MyNotes工程中使用AFNetworking框架实现了网络通信，这一节重构该工程，使用CocosPods工具添加AFNetworking框架。

提示 在本书的配套代码中，本节MyNotes-Starter是配置之前的工程，我们可以在这个工程的基础上进行如下配置。而MyNotes-Final工程是使用CocosPods工具配置完成的。

首先，通过终端进入MyNotes工程目录，即*.xcodeproj文件所在的目录，执行如下指令来初始化：

```
$ pod init
```

该指令执行成功后，在当前目录下生成一个Podfile文件，该文件是一个文本文件，其内容如下：

```
# Uncomment this line to define a global platform for your project
# platform :ios, '10.0'                        ①
# Uncomment this line if you're using Swift
# use_frameworks!                              ②

target 'MyNotes' do                            ③

end                                            ④
```

其中的#号表示注释，第①行指定该工程的平台，第②行说明将第三方代码编译为框架.framework文件，否则将编译为静态链接库.a文件，第③行至第④行用于为工程中的MyNotes目标指定依赖关系，这里可以指定多个第三方库。如果想了解更多的Podfile文件配置，可以参考https://guides.cocoapods.org/syntax/podfile.html。

修改Podfile文件，具体如下：

```
# Uncomment this line to define a global platform for your project
platform :ios, '10.0'
# Uncomment this line if you're using Swift
# use_frameworks!

target 'MyNotes' do
    pod 'MBProgressHUD', '~> 1.0.0'
    pod 'AFNetworking', '~> 3.1.0'
end
```

其中指定了工程所依赖的两个第三方库MBProgressHUD和AFNetworking。MBProgressHUD和AFNetworking不能任意编写，它们是通过Pod search或网站搜索到的。依赖库后面的数值表示依赖的版本，它有几种表示方式，具体如下。

- ~> 3.1.0，表示大于等于3.1.0版本，小于下一个主版本，即小于4.0。
- >= 3.1.0，表示大于等于3.1.0版本。
- <= 3.1.0，表示小于等于3.1.0版本。
- >3.1.0，表示大于3.1.0版本。
- < 3.1.0，表示小于3.1.0版本。
- 3.1.0，表示等于3.1.0版本。

提示　MBProgressHUD提供了多种形式的活动指示器和进度条，其GitHub地址为https://github.com/jdg/MBProgressHUD.git。

Podfile文件修改完成后，就可以安装依赖关系了。这可以通过如下指令来执行：

```
$ pod install
Analyzing dependencies
Downloading dependencies
Installing AFNetworking (3.1.0)
Installing MBProgressHUD (1.0.0)
Generating Pods project
Integrating client project

[!] Please close any current Xcode sessions and use `MyNotes.xcworkspace` for this project from now on.
Sending stats
Pod installation complete! There are 2 dependencies from the Podfile and 2 total pods installed.
```

安装成功后，MyNotes工程的目录结构如下：

```
.
├── MyNotes
├── MyNotes.xcodeproj
├── MyNotes.xcworkspace
├── Podfile
├── Podfile.lock
└── Pods
```

MyNotes.xcworkspace、Podfile.lock和Pods是安装过程中创建的，其中Pods目录是工程所依赖库或框架源代码工程，Podfile.lock文件记录了已经安装的依赖库版本，用于团队协作开发时的依赖库版本控制。注意，如果通过Xcode启动工程，应该通过MyNotes.xcworkspace工作空间文件启动，之后看到的界面如图27-4所示，其中第三方库或框架源代码都放在Pods工程中，并且被编译为静态链接库.a文件，并配置到MyNotes工程中。

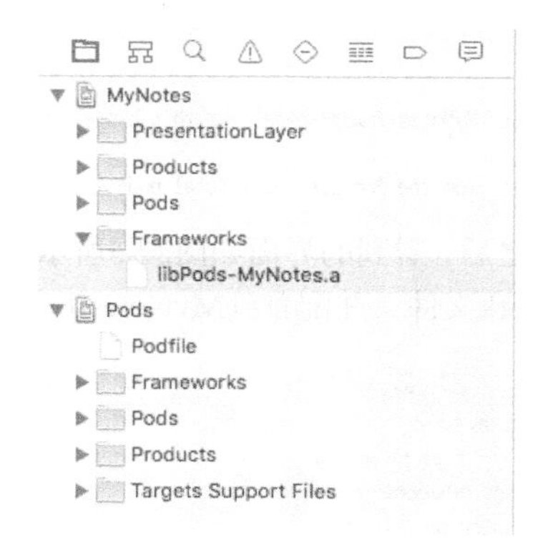

图27-4 启动Xcode工作空间

安装完成后，我们不需要再做任何配置，直接就可以运行MyNotes工程了。

注意

使用CocosPods工具配置完工程后，如果这些工程被复制或移动到其他目录，编译工程会发现有错误clang: error: linker command failed with exit code 1 (use -v to see invocation)，我们使用pod update指令更新即可。另外，无论Podfile.lock是否存在，pod update指令都会重新创建，从而更新到最新依赖库版本。如果存在Podfile.lock文件，pod install指令就不会重新创建，它会根据Podfile.lock所记录的依赖库版本进行安装。

27.1.5 实例：框架形式管理依赖

我们在上一节中演示了添加静态链接库的示例，这一节来介绍一个添加动态框架的示例。

16.5.2节的MyNotes工程使用了Alamofire框架实现网络通信，这一节重构该工程，使用CocosPods工具添加Alamofire框架。

首先，通过终端进入MyNotes工程目录，即*.xcodeproj文件所在的目录。生成Podfile文件并将其内容修改如下：

```
# Uncomment the next line to define a global platform for your project
# platform :ios, '9.0'

target 'MyNotes' do
    # Comment the next line if you're not using Swift and don't want to use dynamic frameworks
    use_frameworks!                         ①

    # Pods for MyNotes
    pod 'MBProgressHUD', '~> 1.0.0'         ②
    pod 'AFNetworking', '~> 3.1.0'          ③

end
```

第①行指示将第三方代码编译为框架.framework文件。第②行指定依赖MBProgressHUD库（注意，MBProgressHUD源代码是Objective-C代码）。第③行指定依赖Alamofire框架。

Podfile文件修改完后，就可以安装依赖关系了。这可以通过如下指令执行：

```
$ pod install
Analyzing dependencies
Downloading dependencies
Installing Alamofire (4.0.1)
Installing MBProgressHUD (1.0.0)
Generating Pods project
```

```
Integrating client project

[!] Please close any current Xcode sessions and use `MyNotes.xcworkspace` for this project from now on.
Sending stats
Pod installation complete! There are 2 dependencies from the Podfile and 2 total pods installed.
```

启动MyNotes.xcworkspace工作空间之后，看到的界面如图27-5所示，其中第三方库或框架源代码都放在Pods工程中，并且被编译为动态框架.framework文件，并配置到MyNotes工程中。

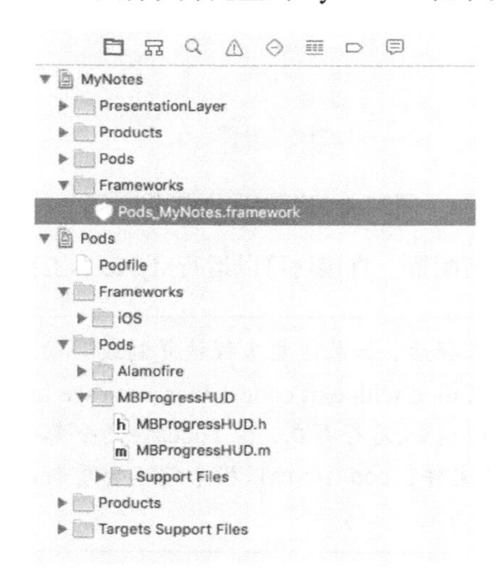

图27-5　启动Xcode工作空间

安装完成后，我们不需要再做任何配置，在程序中引入如下模块：

```
import Alamofire
import MBProgressHUD
```

27.2　使用 Carthage 工具管理依赖

CocoaPods已经是一个非常优秀的工具了，而我们还有另外一个选择——Carthage，它是一个轻量级的项目依赖管理工具。

Carthage主张“去中心化”和“非侵入性”。CocoaPods搭建了一个中心库（cocoapods.org），第三方库被收录到该中心库中，没有收录的第三方库是不能使用CocoaPods管理的，这就是所谓的“中心化”思想。而Carthage没有这样的中心库，第三方库基本上都是从GitHub或私有Git库中下载的，这就是“去中心化”。

另外，CocoaPods在下载第三方库后，会将其编译成静态链接库或动态框架文件，然后会修改Xcode项目属性配置依赖关系，这就是“侵入性”。而Carthage下载成功之后，会将第三方库编译为动态框架，由开发人员自己配置依赖关系，Carthage不会修改Xcode项目属性，这就是“非侵入性”。

提示　使用Carthage有两个限制：第三方库只能编译为框架，iOS项目必须是iOS 8以上的版本。

27.2.1　安装 Carthage

Carthage（https://github.com/Carthage/Carthage）工具可以通过两种方式安装：一种是在https://github.com/

Carthage/Carthage/releases网站下载已经编译好的安装包Carthage.pkg，然后进行安装；另一种是通过Homebrew工具安装，Homebrew是用来管理macOS系统下安装包的工具。

我们推荐使用Homebrew工具安装Carthage。首先，要确定是否安装了Homebrew工具，此时可以在终端中执行如下指令：

```
$ brew update
```

如果没有安装Homebrew工具，终端会提示brew命令无法执行，此时可以在终端中执行如下指令：

```
$ ruby -e "$(curl -fsSL https://raw.githubusercontent.com/Homebrew/install/master/install)"
```

命令执行成功之后，就可以通过如下指令在终端中执行：

```
$ brew install carthage
```

最后，提示安装完成。

27.2.2　项目与第三方库搭配形式

使用Carthage工具管理第三方库时，无论第三方库的源代码是Swift语言编写的还是Objective-C语言编写的，编译的结果都是动态框架。

项目与第三方库的搭配形式如表27-2所述。

表27-2　项目与第三方库的搭配形式

第三方库 / 项目代码	Objective-C语言	Swift语言
Objective-C	支持.framework	支持.framework。需要使用Xcode生成的头文件，文件命名为：<框架名/框架名-Swift.h> 例如：在Objective-C中调用Swift语言的Alamofire框架，需要引入头文件`#import <Alamofire/Alamofire-Swift.h>`
Swift	支持.framework	支持.framework

从表27-2中可见，如果项目中使用Objective-C调用Swift编写的第三方库，那么需要使用Xcode生成的头文件。

27.2.3　Cartfile 文件

与CocoaPods的Podfile类似，Carthage使用名为Cartfile的文件描述依赖库，该文件的内容示例如下：

```
github "Alamofire/Alamofire" >= 4.0                                        ①
github "https://github.com/AFNetworking/AFNetworking.git" ~> 3.0           ②
github "jdg/MBProgressHUD" "master"                                        ③
git "https://192.168.1.188/HelloWorld.git" "main"                          ④
```

从Carthage文件的内容可见依赖库的来源有Git和GitHub，Git是私有库。git或github是依赖库名或git地址，依赖库名见第①行的Alamofire/Alamofire，指定依赖库名一般是指GitHub上库的“用户名/项目名”，这个名字如果不能确定，需要到GitHub上查询确定。图27-6所示是GitHub上的Alamofire库，页面中左上角的“Alamofire/Alamofire”即是了。第②行的http://github.com/AFNetworking/AFNetworking.git是github上依赖库的地址。

git地址可见第④行，支持git所支持的所有协议，如图https、git和ssh等。我们知道git库可以创建多个分支，还可以在Carthage文件中指定分支名，其中第③行中的master以及第④行中的main都是分支名。有了分支，不需要指定版本号。

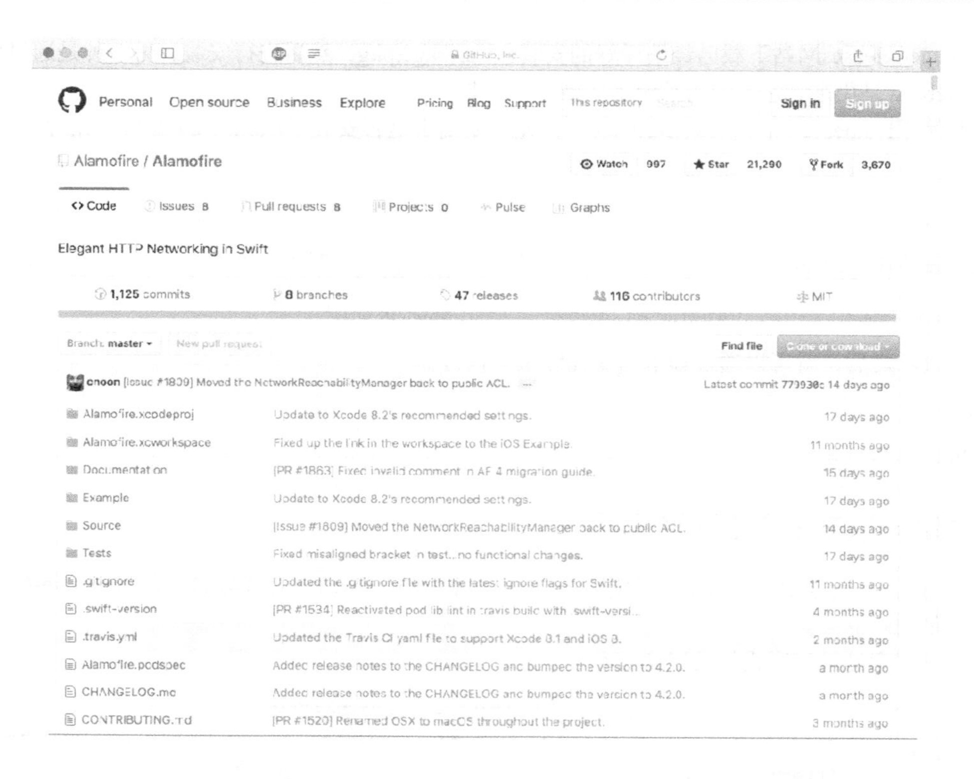

图27-6　GitHub上的库名

与Podfile类似，在Cartfile文件中可以指定依赖库版本，它主要有几种表示方式，具体如下：

- ~> 3.0，表示大于等于3.0版本，小于下一个主版本，即小于4.0；
- >= 3.0，表示大于等于3.0版本；
- == 3.0，表示等于3.0版本。

27.2.4　实例：重构 MyNotes 依赖关系

下面我们通过一个示例介绍一下Carthage的具体用法。我们在16.4节的MyNotes工程中使用AFNetworking框架实现了网络通信，这一节中重构该工程，使其使用Carthage工具添加AFNetworking框架。

首先，通过终端进入MyNotes工程目录下，即*.xcodeproj文件所在的目录，然后创建Cartfile文本，将其内容修改如下：

```
github "https://github.com/AFNetworking/AFNetworking.git" ~> 3.0
github "jdg/MBProgressHUD" "master"
```

然后就可以安装依赖关系了，这可以通过如下指令执行：

```
$ carthage update --platform iOS
*** Cloning MBProgressHUD
*** Cloning AFNetworking
*** Downloading AFNetworking.framework binary at "3.1.0"
*** Checking out AFNetworking at "3.1.0"
*** Checking out MBProgressHUD at "51291d888e06adde3dde3918a95241829e57a9d8"
*** xcodebuild output can be found in /var/folders/3k/jl0frl3j3xj0ppsdq3s_5jnm0000gn/T/carthage-xcodebuild.WLpTyv.log
*** Building scheme "AFNetworking iOS" in AFNetworking.xcworkspace
*** Building scheme "MBProgressHUD Framework" in MBProgressHUD.xcworkspace
```

参数platform iOS表示只考虑iOS平台。执行过程分为两个阶段，首先是将库下载到本地，然后再使用

`xcodebuild`命令将库编译为框架。

`update`命令执行成功后，MyNotes工程目录中会多出一个Carthage目录，该目录中又包含了两个目录 Build 和 Checkouts。Build目录中存放编译好的 framework文件，如图27-7所示，根据不同的平台它会有不同的目录，如 Mac、iOS、tvOS和watchOS。Checkouts目录是下载的第三方库源代码，如图27-8所示。

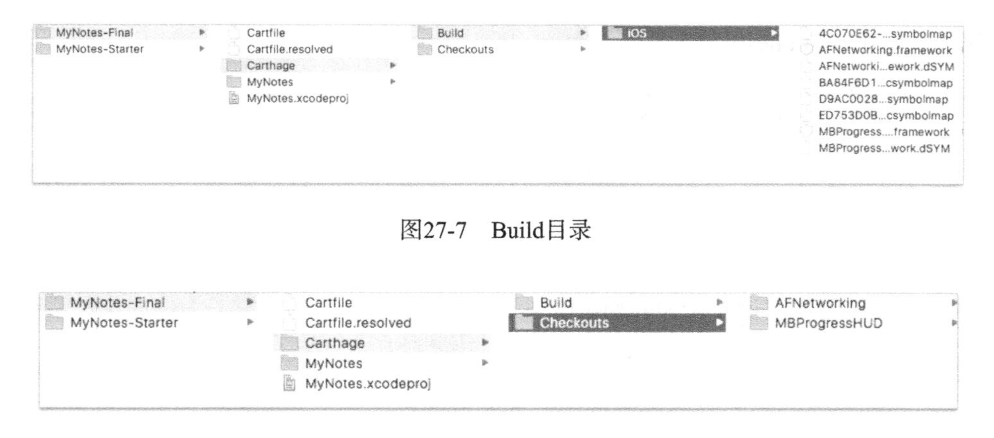

图27-7　Build目录

图27-8　Checkouts目录

然后，我们需要自己配置MyNotes工程与第三方库的依赖关系。在Xcode工程中打开TARGETS→MyNotes→General，如图27-9所示，Embedded Binaries或Linked Frameworks and Libraries中点击“+”按钮，在弹出的对话框中单击Add Other...按钮，找到Build目录中的framework文件，添加依赖框架。考虑到未来在设备中运行，我们最好在Embedded Binaries中添加依赖框架。

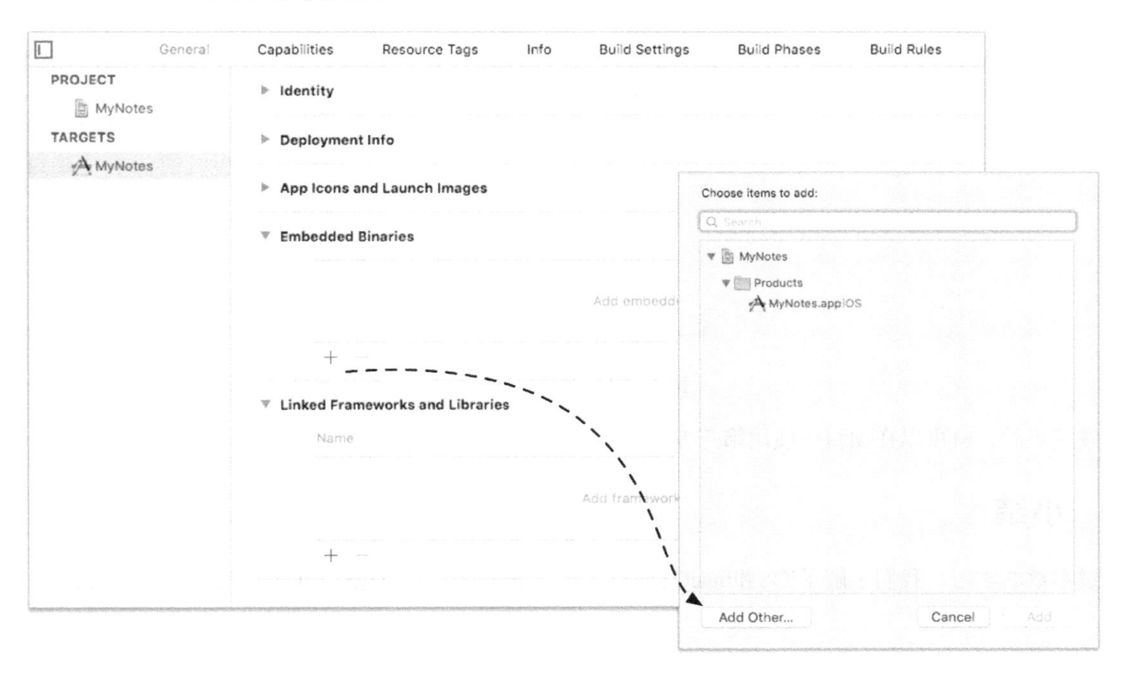

图27-9　配置第三方库的依赖关系

添加完第三方库之后，我们还需要设置运行脚本（Run Script）。所谓运行脚本，就是在编译时执行命令。Carthage提供的运行脚本可以将框架复制到运行环境目录下。

在Xcode工程中打开TARGETS→MyNotes→Build Phases，如果没有Run Script部分，可以通过菜单Editor→Add Build Phase→Add Run Script Build Phases菜单项添加。

如图27-10所示，在Run Script部分的脚本区域输入：

```
/usr/local/bin/carthage copy-frameworks
```

然后将路径添加到Input Files中：

```
$(SRCROOT)/Carthage/Build/iOS/AFNetworking.framework
$(SRCROOT)/Carthage/Build/iOS/MBProgressHUD.framework
```

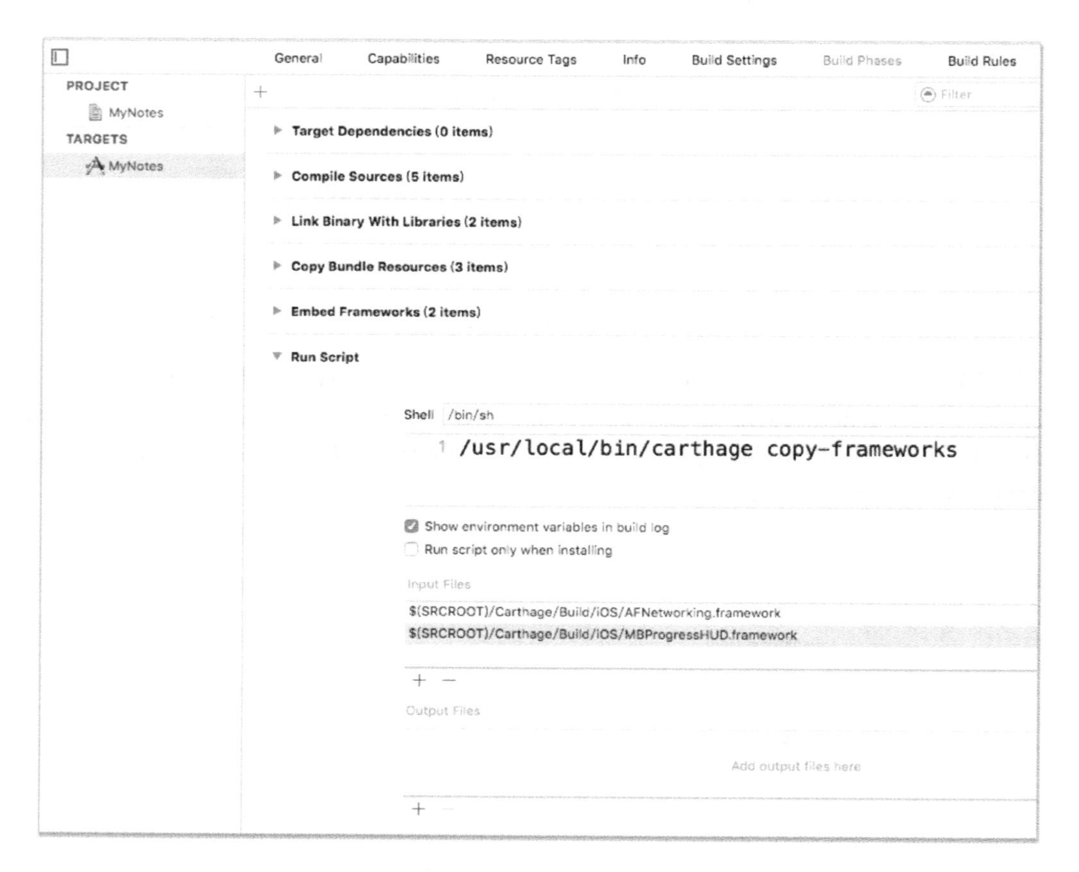

图27-10　配置Run Script

配置完成后，就可以在项目中使用第三方库了。

27.3　小结

通过本章的学习，我们了解了iOS和macOS项目依赖管理工具，其中包括：CocoaPods和Carthage。

第 28 章 把应用放到App Store上

当阅读到本章的时候，恭喜你已经学习了本书的大部分知识，完全可以开发一个完整的应用，并将其发布到App Store了。但是在真正把应用发布到App Store上之前，还有“最后一公里”要走，这就是28.1节所要介绍的内容。应用全部开发妥当后，就可以在App Store上发布了，发布过程有点复杂，本章会介绍这个流程。发布完成后，我们就等苹果公司审核了。苹果公司对于发布在App Store上的应用审核非常严格，我们需要避免应用不通过的情况。

28.1 收官

在应用的开发过程中，我们往往只关注程序本身和功能的实现，而不注重其“外表”，但是当应用要发布时，就需要为它添加启动界面和图标等。此外，我们还需要调整产品属性和进行编译操作等。

28.1.1 在 Xcode 中添加图标

用户第一眼看到的就是应用的图标。图标是应用的“着装”，“着装”应该大方得体，但图标设计已经超出了本书的讨论范围，这里我们只介绍iOS图标的设计规格以及如何把图标添加到应用中去。

iOS应用使用的图标（App Icon）分为设备上使用的图标（见图28-1）和App Store上使用的图标（见图28-2），它们之间只是尺寸大小不同。

图28-1　设备上使用的图标

图28-2　App Store上使用的图标

此外，iOS上使用的图标还有Spotlight搜索图标、设置图标、工具栏（或导航栏）图标、标签栏图标。这些图标在不同设备上的规格也不同，大家可以参考苹果的iOS Human Interface Guidelines（https://developer.apple.com/library/prerelease/ios/documentation/UserExperience/Conceptual/MobileHIG/IconMatrix.html#//apple_ref/doc/uid/TP40006556-CH27-SW1）。

这么多规格的图标真是让人很难记住，Xcode帮我们解决了这一问题。在Xcode中，有一个新的图标添加方法。如果创建的是通用（包括iPhone和iPad）的工程，选择Assets.xcassets→AppIcon，打开如图28-3所示的界面。在这个界面中，我们可以很直观地看到各种图标的规格，其中Notifications图标是通知栏所使用的图标，Spotlight图标是搜索栏，Settings是在设置应用中的图标，App图标是显示在桌面上的图标。另外，图标下面的数值代表图标的规格，pt单位是“点”。例如：图中2x iPhone App iOS 7-10 60 pt，60点在2x设备上图标规格为2 × 60 pt = 120像素，在3x设备上图标规格为3 × 60 pt = 180像素。这些图标的规格一定要严格按照图28-3的提示设计，而文件可以随意命名。

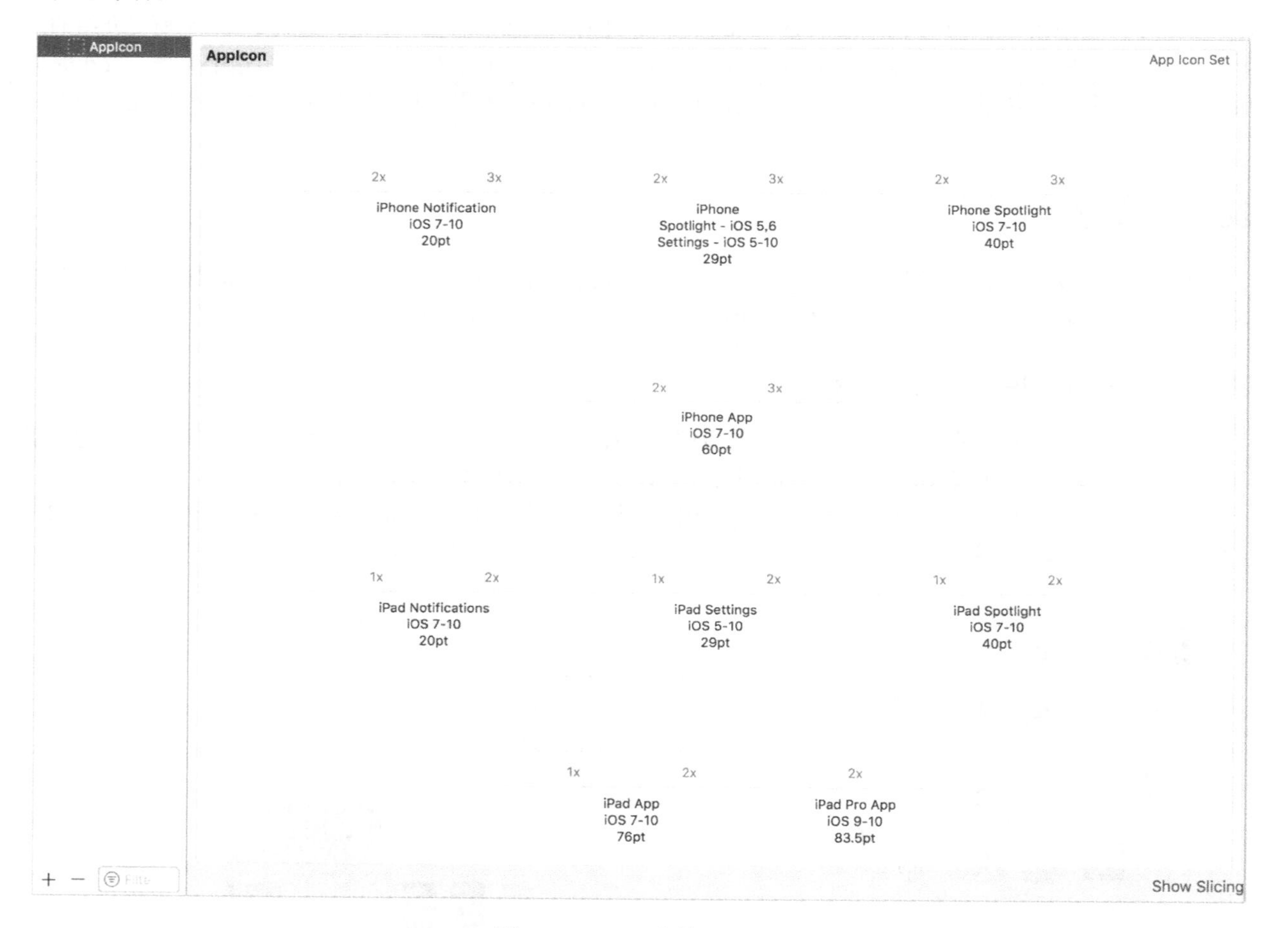

图28-3　Xcode中的AppIcon

添加这些图标的过程：在Finder里打开这些图标所在的文件夹，然后在Xcode中打开如图28-3所示的界面，从macOS操作系统的Finder中拖动图标到Xcode中对应的位置，如图28-4所示。

这些图标究竟放在什么地方呢？我们可以在Finder中打开Xcode工程目录，如图28-5所示，在Assets.xcassets→AppIcon.appiconset下可以看到刚才添加的几个图标。此外，还有一个Contents.json文件，这是个元数据描述文件。

全部拖曳完成后，就可以看到效果了。图28-6所示为IconDemo图标，其中图28-6a是应用图标、图28-6b是Spotlight搜索图标，图28-6c是设置图标。

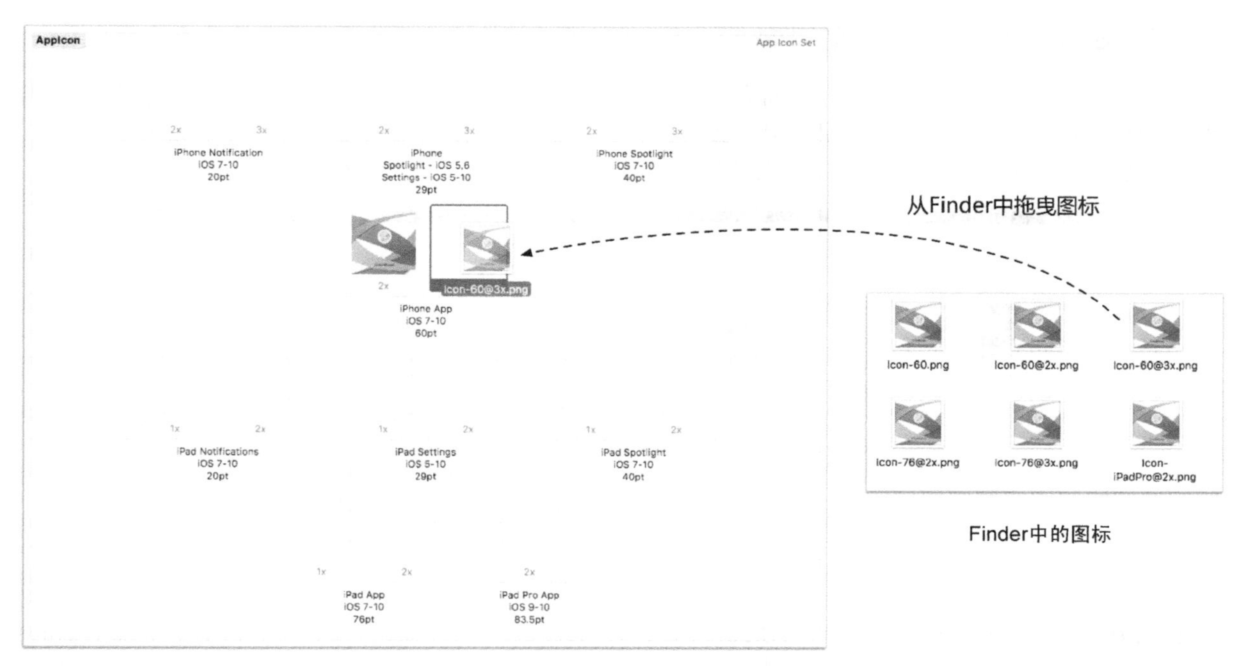

图28-4 拖动图标到Xcode

图28-5 图标所存放的目录

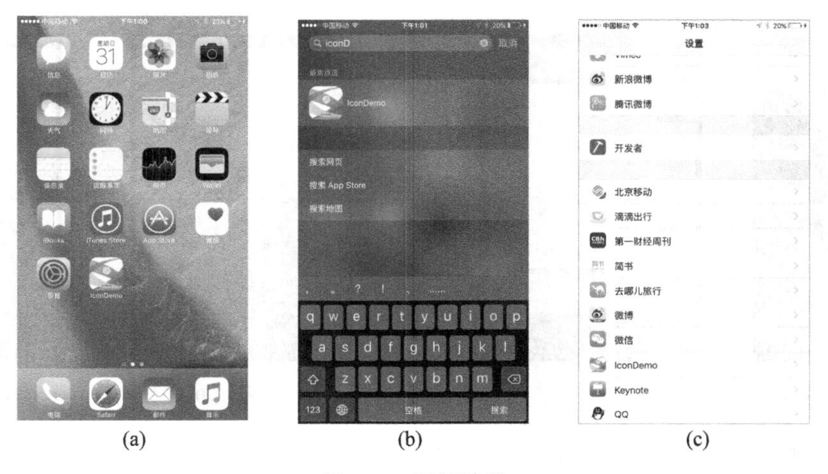

(a) (b) (c)

图28-6 运行效果

28.1.2　在 Xcode 中添加启动界面

启动界面是应用启动与进入到第一个屏幕之间的界面。没有启动界面的应用进入第一个屏幕之前是黑屏，这会影响用户体验。虽然这在开发阶段没有什么影响，但是在应用发布前，还是需要添加启动界面的。

添加启动界面有很多学问，下面我们先看看两个应用的启动过程，如图28-7和图28-8所示。

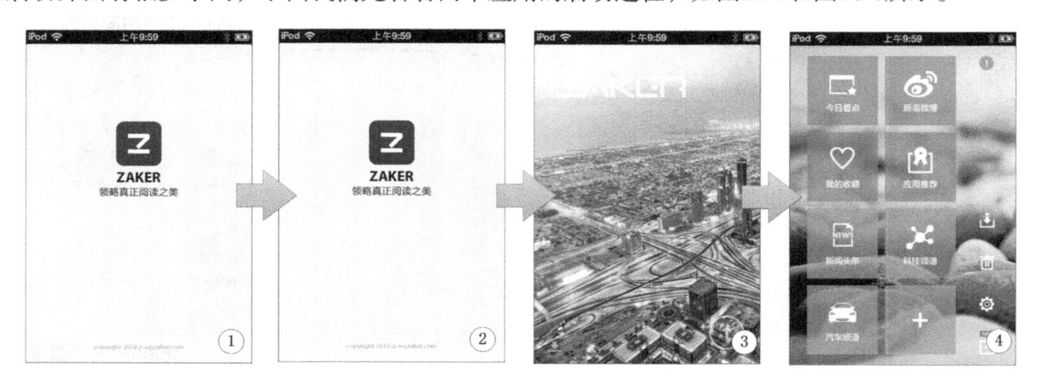

图28-7　启动界面场景1

你认为这两个启动界面哪个更好呢？图28-7所示的应用进入第一个屏幕（④界面）时，要经过3个界面的变化，其中启动界面放有大大的logo，并且设计者还故意让其延迟了几秒钟，让用户看清楚这个logo。这样做或许很酷，也能宣传自己，但事实上用户既然能够下载你的应用，肯定知道你是谁了！如果第一次看，用户可能感觉很新奇，但是如果每天看会是什么感觉呢？

提示　图28-8所示的界面①不是真的已经进入应用了，而是一张图片，它与应用的第一个屏幕非常相似，能够使用户感觉到很快进入了应用，让用户感觉不到等待，这才是以用户体验为中心的设计。

为了获得更好的用户体验，苹果对于iOS上的启动界面推荐采用图28-8所示的设计。在iOS中，苹果自带的应用都是这样设计的，比如iOS 5自带的“股市”应用界面（见图28-9）。

图28-9中的①号图片是启动界面，与第一个屏幕（②号图片）非常类似，就是将动态部分去掉了，它给用户的感觉是应用快速启动，几秒钟后，内容就会填充到屏幕中。

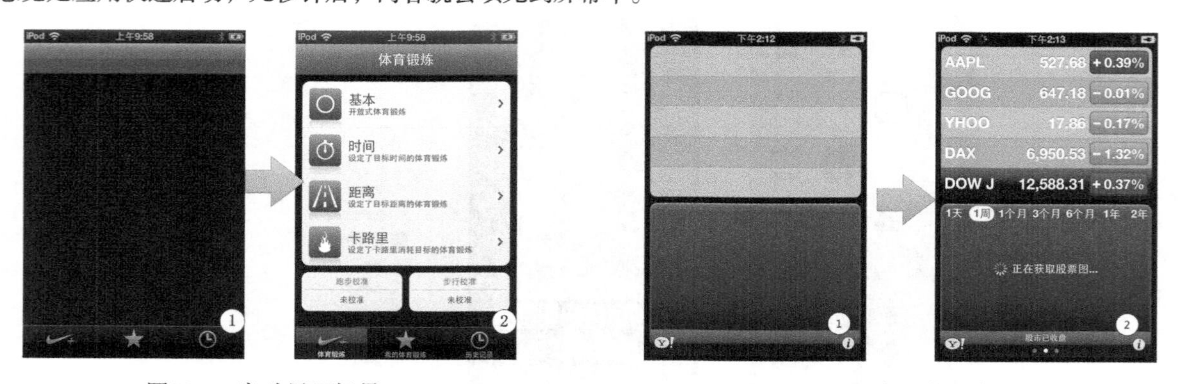

图28-8　启动界面场景2　　　　图28-9　“股市”应用启动界面

在启动界面的规格上，苹果也给出了相应的尺寸要求，参考8.1.2节中表8-2所述。在Xcode 7及其之后版本中，我们有两种方法添加启动界面：启动图片和启动文件。下面我们分别介绍一下。

1. 启动图片

启动图片方式是在Xcode 7之前采用的，就是整个启动界面是一张图片，图片的规格参考8.1.2节中的表8-2。在Xcode中添加启动界面与添加图标非常类似，打开由Xcode创建的通用（包括iPhone和iPad）工程，选择打开Assets.xcassets，如图28-10所示，右键点击+按钮，在弹出菜单中选择App Icons & Launch Images→New iOS Launch Image菜单项。

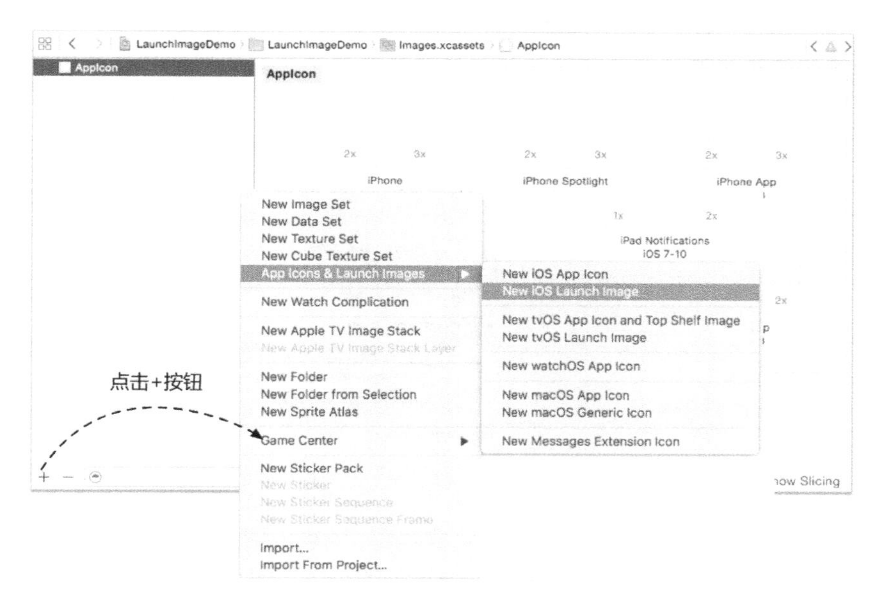

图28-10 在Xcode中添加Launch Image

创建Launch Image之后的界面如图28-11所示，从中可以很直观地看到21种规格的启动界面。

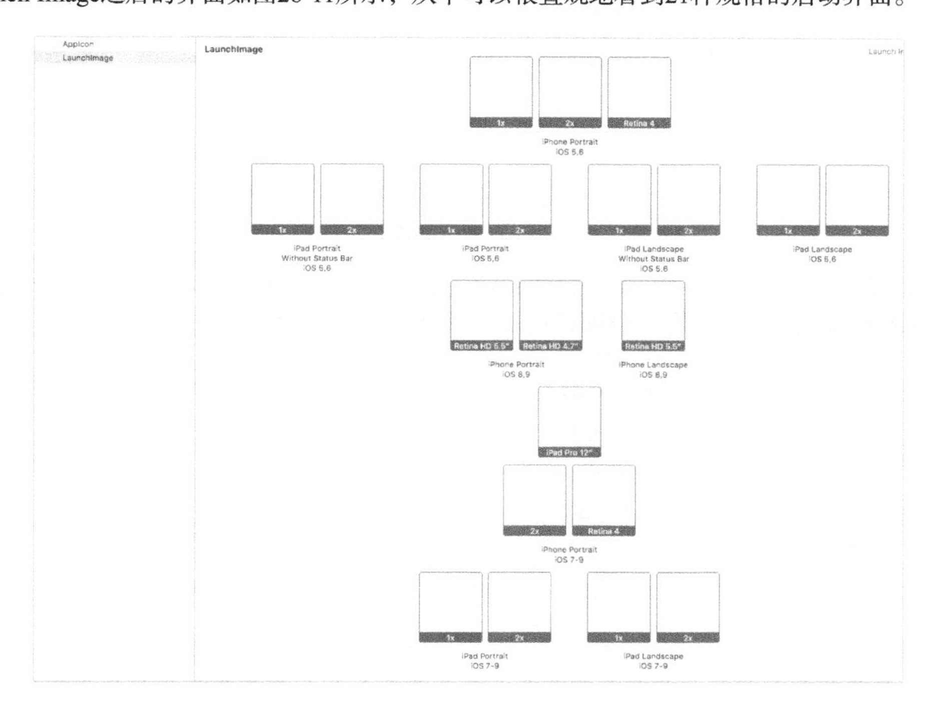

图28-11 Xcode中的Launch Image

请设计师为我们准备好启动图片，然后按照与上一节类似的方式，将这些图片拖曳到Xcode中。

然后选中TARGETS中LaunchImageDemo，如图28-12所示，找到App Icons and Launch Images设置项目，点击Launch Images Source后面的Use Asset Catalog按钮，此时会弹出如图28-13所示的对话框，点击该对话框中的Migrate按钮，关闭该对话框。

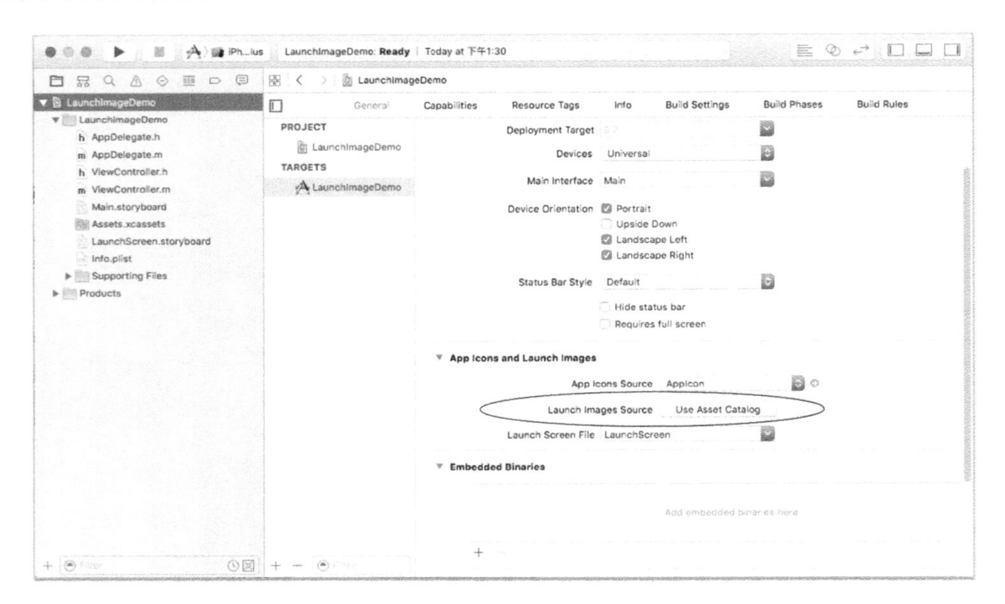

图28-12　设置启动界面

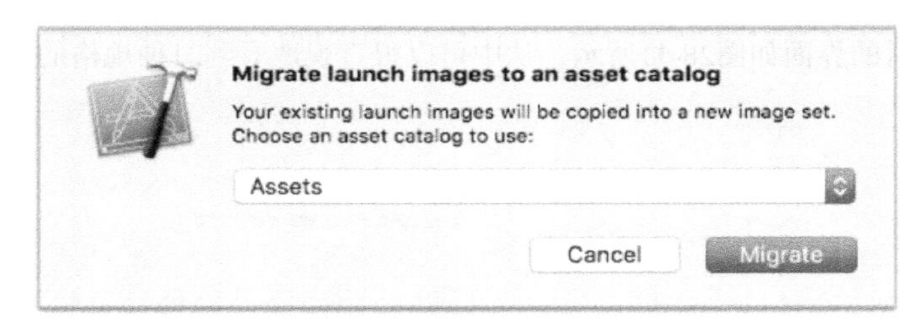

图28-13　设置对话框

此时会回到设置界面。如图28-14所示，选择Launch Images Source后面的下拉列表，从中选择LaunchImage，然后将Launch Screen File中的内容删除。设置完成后的界面如图28-15所示。

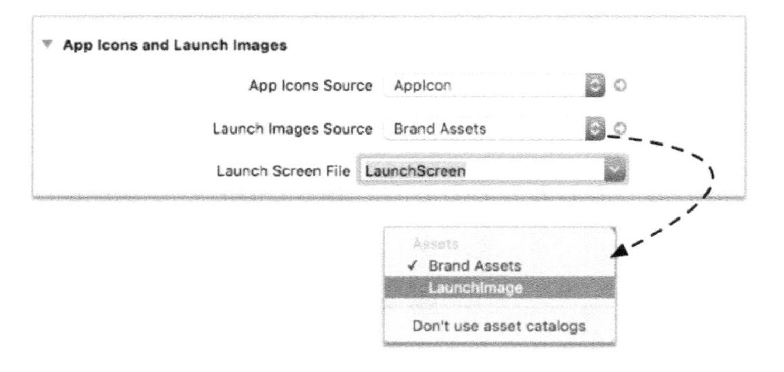

图28-14　选择LaunchImage

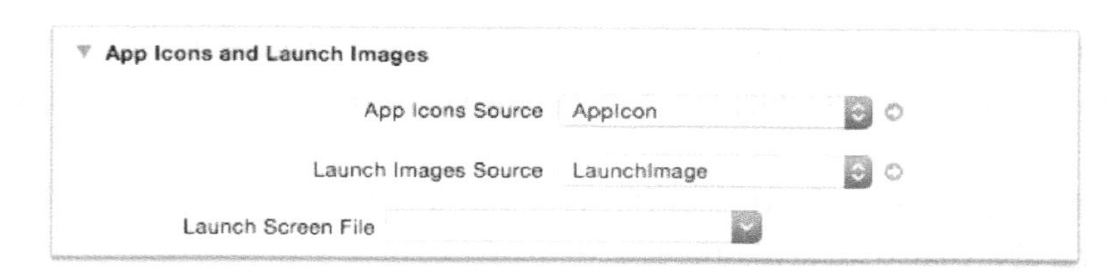

图28-15　设置完成的启动界面

设置完成之后，我们可以选择不同的模拟器测试一下。

2. 启动文件

启动文件是Xcode 6之后的新功能，其实原理很简单，就是可以创建基于故事板或XIB的启动界面文件，就像是我们在设计其他界面一样，这种方式要比启动图片方式灵活。由于默认情况下Xcode 8会生成一个LaunchScreen.storyboard文件，这里我们不需要设置。但是如果想使用其他故事板或XIB文件作为启动界面，则在图28-12中将Launch Screen File选择为需要的故事板或XIB文件即可。

由于这种界面设计方式与前面章节介绍的故事板没有区别，这里就不再赘述了。设置完成之后，大家可以选择不同的模拟器测试一下。

28.1.3　调整 Identity 和 Deployment Info 属性

在编程过程中，有些产品的属性并不影响我们开发。即便这些属性设置不正确，一般也不会有什么影响。但是在产品发布时，正确地设置这些属性就很重要了。如果设置不正确，就会影响产品的发布。这些产品属性主要是TARGETS中的 Identity和Deployment Info属性，如图28-16所示。

General　Capabilities　Resource Tags　Info　Build Settings　Build Phases

PROJECT
LostRoutes
TARGETS
LostRoutes

Identity
Bundle Identifier　com.work6.LostRoutes
Version　1.0
Build　1
Team　dongsheng guan (tonyguan58...

Deployment Info
Deployment Target　9.0
Devices　iPhone
Main Interface　Main
Device Orientation　Portrait
Upside Down
Landscape Left
Landscape Right
Status Bar Style　Default
Hide status bar
Requires full screen

图28-16　Identity和Deployment Info属性

在这些属性中，Identity部分主要包括Bundle Identifier（包标识符）、Version（发布版本）、Build（编译版本）和Team（开发者账号）。在Deployment Info部分，主要是Deployment Target（部署目标）。下面分别介绍它们的含义和重要性。

- **Bundle Identifier（包标识符）**。包标识符在开发过程中对我们似乎没有什么影响，但是在发布时非常重要。本例中我们设置的是com.51work6.LostRoutes。
- **Version（发布版本）**。这个版本号看起来无关紧要，但是在发布时，如果这里设定的版本号与iTunes Connect中设置的应用版本号不一致，在打包上传时就会失败。
- **Build（编译版本）**。它是编译时设定的版本号。
- **Team（开发者账号）**。测试时这里可以使用普通Apple ID，但如果是在App Store上发布，就必须是开发者账号Apple ID。
- **Deployment Target（部署目标）**。选择部署目标是开发应用之前就要考虑的问题，这关系到应用所能够支持的操作系统。如果考虑到支持老版本的操作系统（如低于5.0版本），则考虑到支持64位ARM CPU，至少需要iOS 7.0。本例中，我们设置为9.0。

28.2 为发布进行编译

从编写到发布应用会经历3个阶段：在模拟器上运行调试、在设备上运行调试和发布编译。其中完整的编译发布流程如图28-17所示，它需要经历创建开发者证书、创建App ID、创建描述文件和发布编译这4个阶段，本节就来详细介绍一下。

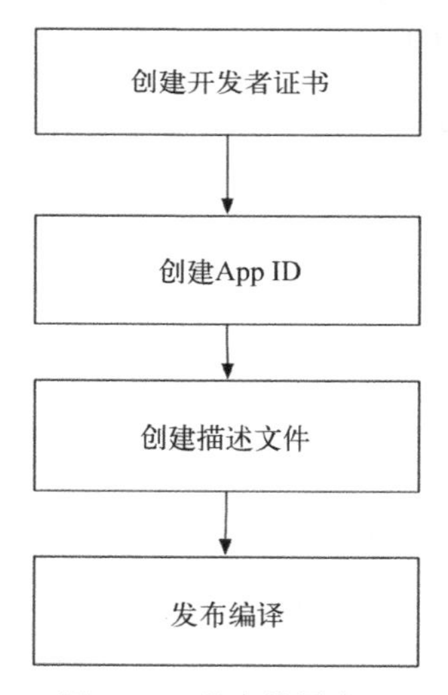

图28-17　发布编译流程

28.2.1 创建开发者证书

证书的管理可以在苹果开发者网站（网址为https://developer.apple.com）上进行。登录该网站时需要苹果开发者账号，登录成功后的界面如图28-18所示。

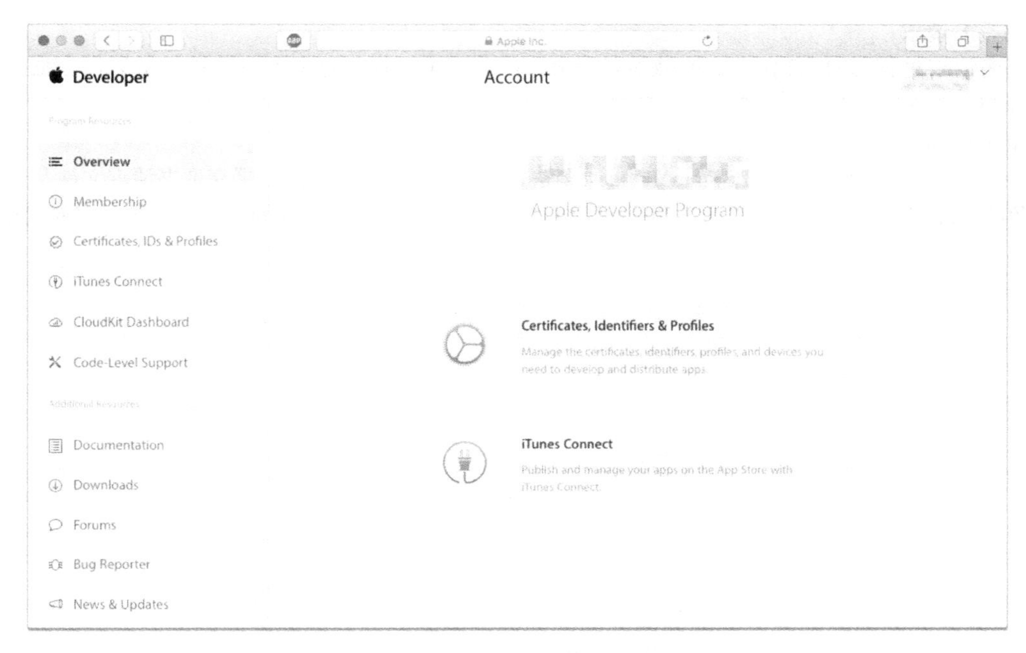

图28-18　成功登录苹果开发者网站

点击左边导航菜单“Certificates, IDs & Profiles”下的Certificates（证书）导航菜单，得到的证书管理界面如图28-19所示，这里可以下载和删除证书。

创建证书的过程分成以下两步：

(1) 使用钥匙串访问工具生成签名公钥；

(2) 提交请求到开发者网站并生成证书。

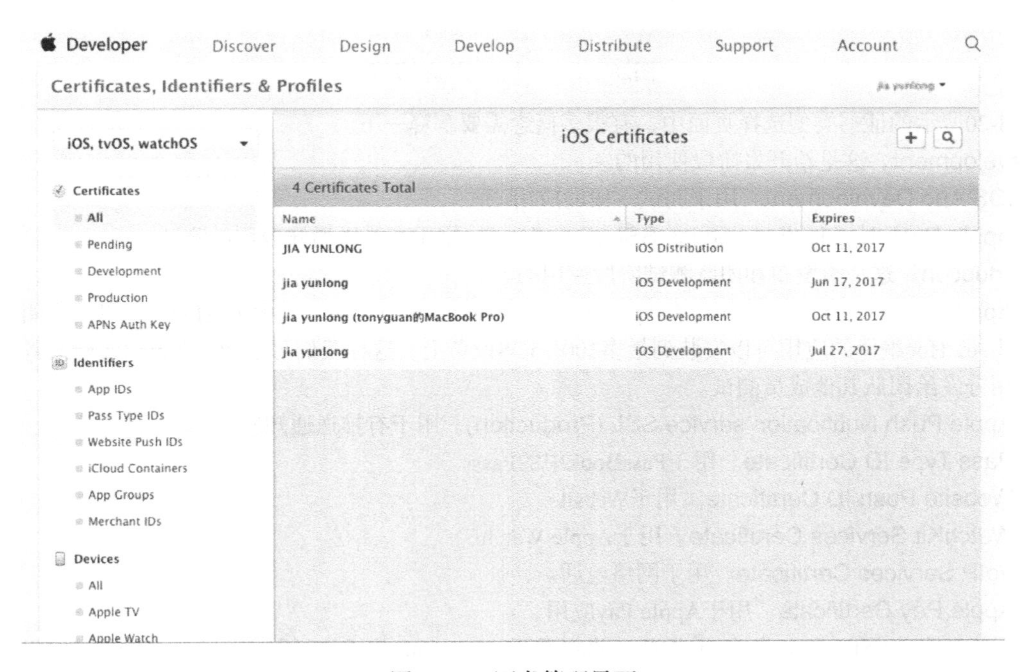

图28-19　证书管理界面

1. 证书签名请求

证书签名请求过程与21.3.2节“1. 证书签名请求”部分完全一样，这里不再赘述。最后保存证书请求文件CertificateSigningRequest. certSigningReques到桌面。

2. 提交请求到开发者网站并生成证书

生成CertificateSigningRequest.certSigningRequest文件后，重新回到开发者网站提交证书请求文件。点击图28-19页面右上角的添加按钮[+]，将打开如图28-20所示的证书类型选择页面，在这个页面中可以选择需要创建的证书。

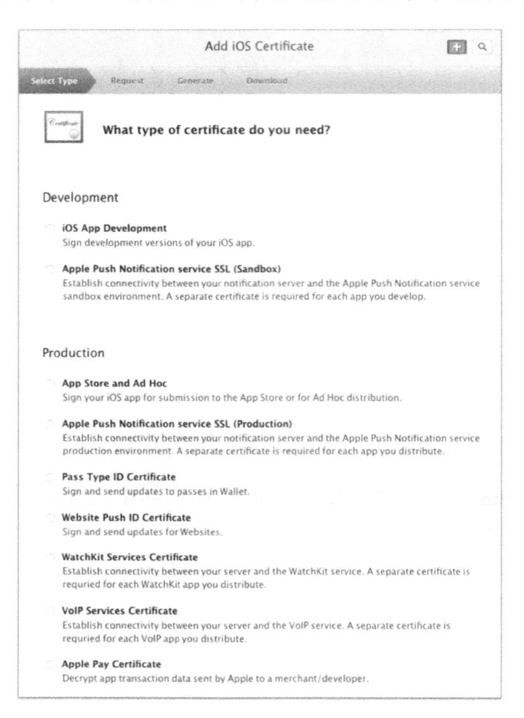

图28-20　证书类型选择页面

在图28-20所示的证书类型选择页面中，有很多概念需要解释一下。

- Development。这是给开发阶段使用的。
  - iOS App Development。用于测试一般的应用。
  - Apple Push Notification service SSL (Sandbox)。用于有推送通知应用的测试。
- Production。这是给发布和团队测试阶段使用的。
  - App Store and Ad Hoc。用于在App Store或Ad Hoc上发布应用，其中Ad Hoc也可以用于团队测试。采用Ad Hoc编译的应用可以安装到最多100个iOS设备上，这样我们可以通过E-mail或网站将要测试的应用分发给团队其他成员测试。
  - Apple Push Notification service SSL (Production)。用于有推送通知应用的发布。
  - Pass Type ID Certificate。用于PassBook中的Pass。
  - Website Push ID Certificate。用于Website。
  - WatchKit Services Certificate。用于Apple Watch。
  - VoIP Services Certificate。用于网络电话。
  - Apple Pay Certificate。用于Apple Pay应用。

我们应该选择App Store and Ad Hoc类型，此时下面的Continue按钮就可用了。点击Continue按钮，进入证书签名请求文件介绍页面，再点击证书请求文件，进入如图28-21所示的证书签名请求文件的上传页面。在页面的左下角找到Choose File按钮，选取桌面上的CertificateSigningRequest.certSigningRequest文件，然后点击Generate按钮就可以生成证书了，生成后的页面如图28-22所示。

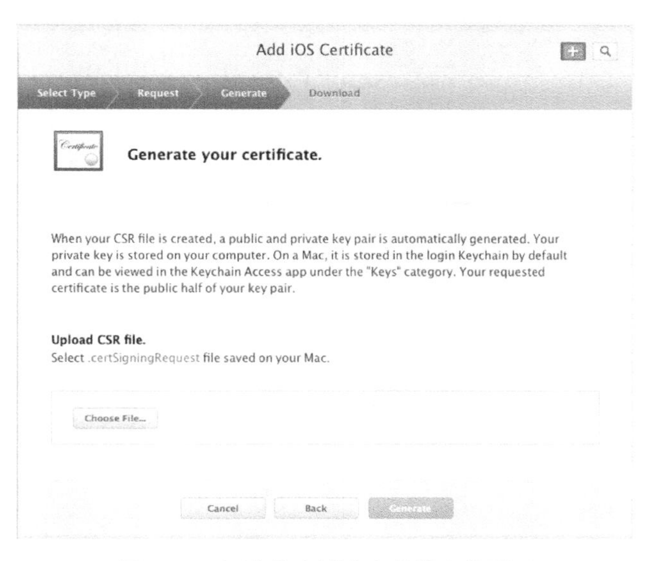

图28-21 证书签名请求文件的上传页面

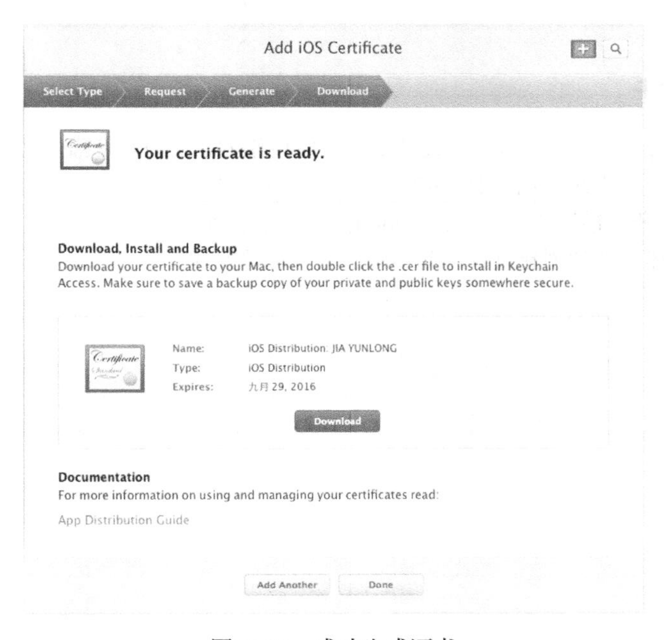

图28-22 成功生成证书

在这个页面下，我们可以下载证书文件用于发布编译。

28.2.2 创建 App ID

应用还需要有App ID，创建App ID的过程与21.3.2节中的“2. 创建App ID”部分基本一致，只是不同的应用会选择不同的服务，如图28-23所示，其他过程不再赘述。

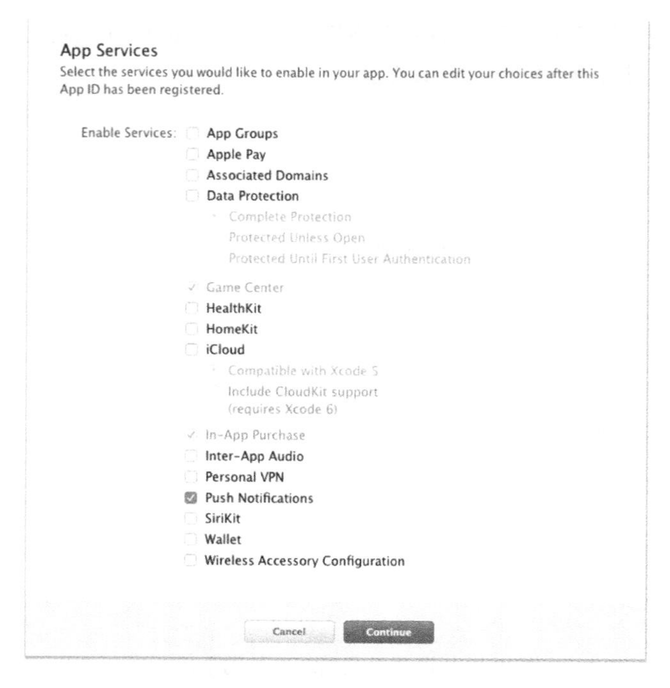

图28-23　选择服务

28.2.3　创建描述文件

描述文件（Provisioning Profile）是应用在设备上编译时使用的，分为开发描述文件和发布描述文件，分别用于开发和发布。管理描述文件的页面如图28-24所示，通过左边的Provisioning Profiles导航菜单进入，其中Development标签用于管理开发描述文件，Distribution标签用于管理发布描述文件。本节简要介绍一下创建开发描述文件的过程，发布描述文件的创建过程与此类似，这里不再赘述。

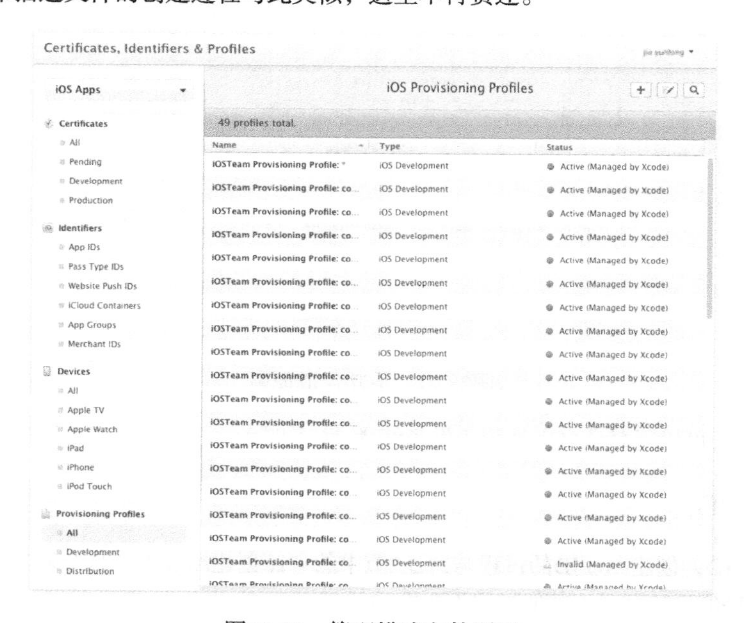

图28-24　管理描述文件页面

在图28-24所示的页面中，点击页面右上角的添加按钮 + ，进入创建描述文件选择页面（如图28-25所示）。这里我们需要选择Distribution中的App Store类型，接着点击下面的Continue按钮，进入如图28-26所示的页面，并在这个页面中选择前面创建好的App ID。

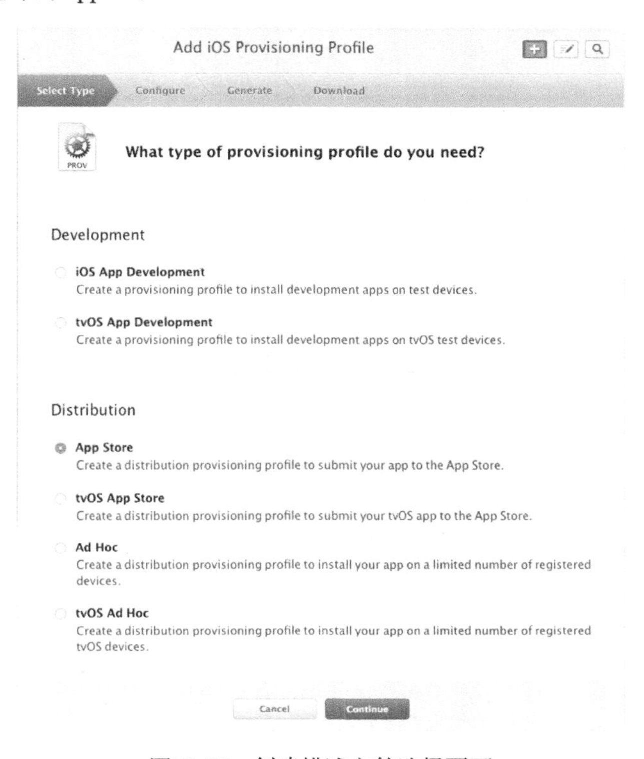

图28-25 创建描述文件选择页面

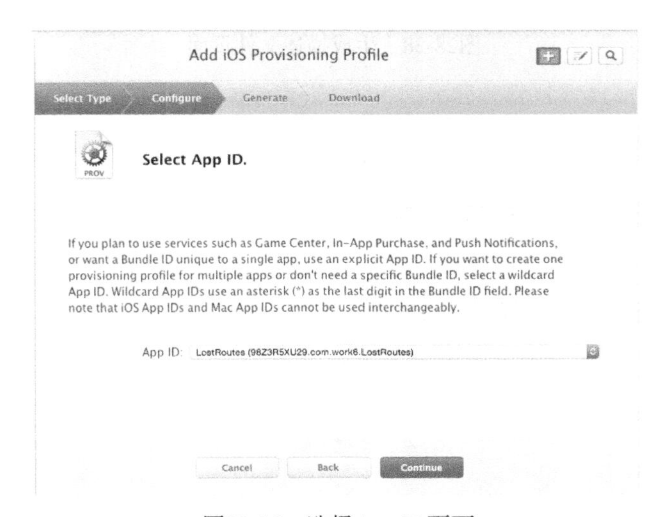

图28-26 选择App ID页面

接着点击下面的Continue按钮，进入如图28-27所示的页面，从中选择前面创建好的证书。然后点击Continue按钮，进入如图28-28所示的生成描述文件页面，在这个页面中输入描述文件名，然后点击下面的Generate按钮创建描述文件。创建完成后将进入如图28-29所示的页面，在这个页面中可以下载这个描述文件到本地。

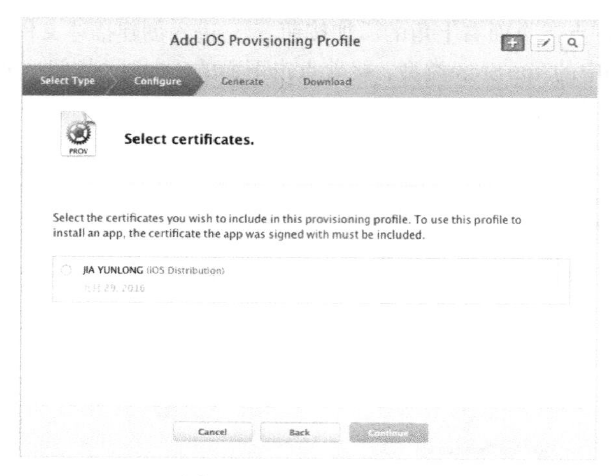

图28-27　选择证书页面

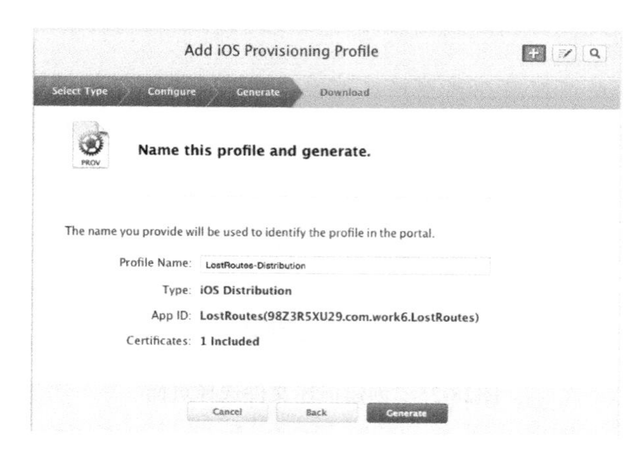

图28-28　生成描述文件页面

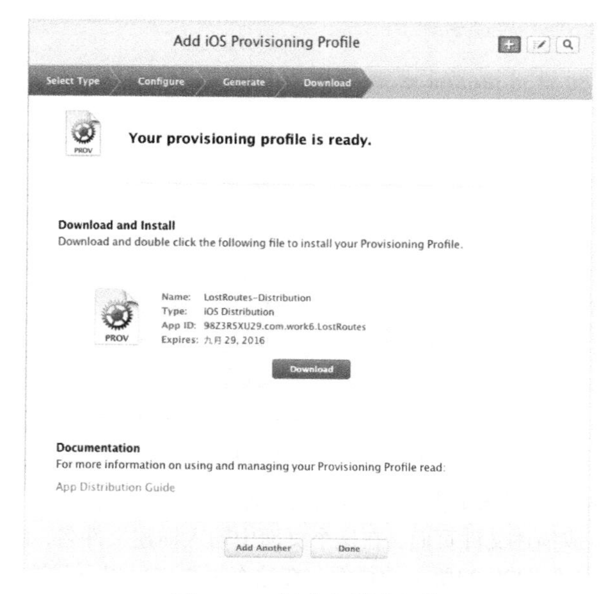

图28-29　创建完描述文件

28.2.4 发布编译

在Xcode中打开工程，选择要编译的目标，如图28-30所示，找到General→Signing。默认情况下Automatically manage signing是选中的，这个选项用于让Xcode自动管理目标数字签名，这种自动管理并不太适合最后的发布编译。因此，我们不要选中Automatically manage signing，去掉后如图28-31所示，会发现两个数字签名选项：Signing (Debug)和Signing (Release)。我们可以在Provisioning Profile中选中刚刚创建的数字证书文件，正确选择后如图28-32所示。

图28-30　目标数字签名

图28-31　不要选中Automatically manage signing

图28-32　选择数字证书文件

配置完成之后，可以进行编译了。如果编译结果有错误或警告，必须解决，忽略警告往往也会导致发布失败。

28.3　发布上架

应用程序编译通过后，就可以发布上架了。发布应用在iTunes Connect中完成，发布完成后等待审核，审核通过后就可以在App Store上架销售了，详细的发布流程如图28-33所示。

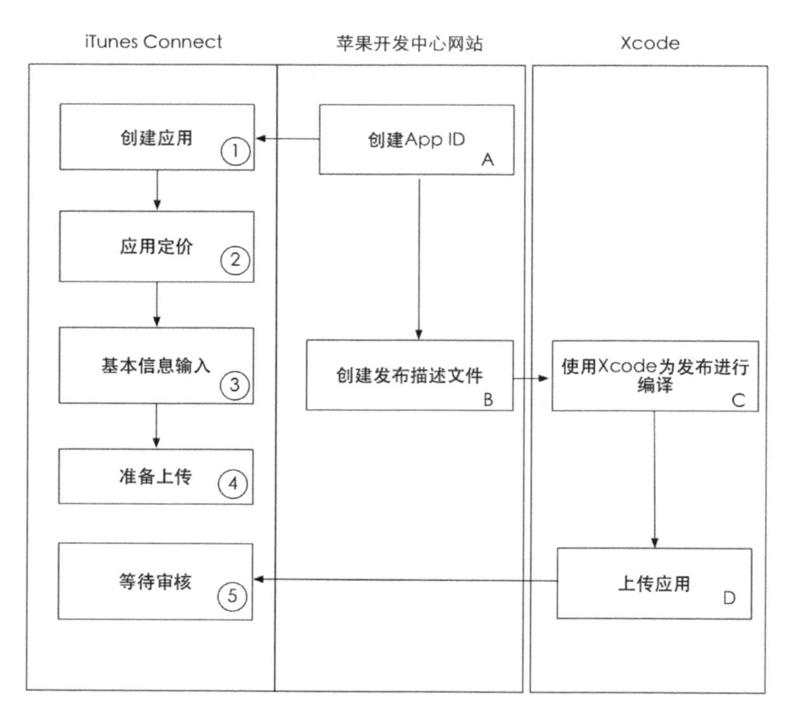

图28-33　发布流程图

其中步骤A和步骤B是在苹果开发中心网站中完成的，这在本章前面已经介绍过了。这里我们介绍其他几个流程，其中主要的流程是在iTunes Connect中完成的，而编译和上传应用是在Xcode中完成的。此外，上传应用还可以使用Application Loader工具完成。

28.3.1 创建应用

通过网址https://itunesconnect.apple.com打开iTunes Connect登录页面，使用苹果开发账号登录，登录成功后的iTunes Connect页面如图28-34所示。

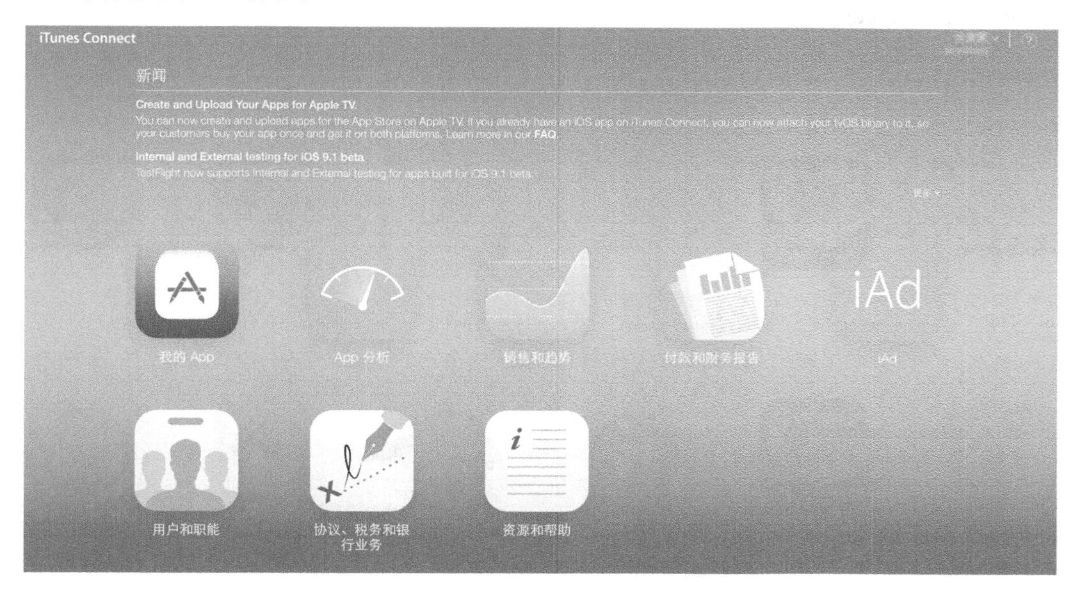

图28-34 iTunes Connect登录成功

点击“我的App”图标，进入应用管理页面，如图28-35所示，在这里可以管理我们的应用，其中显示审核中的、未通过的以及已经上架的所有应用。

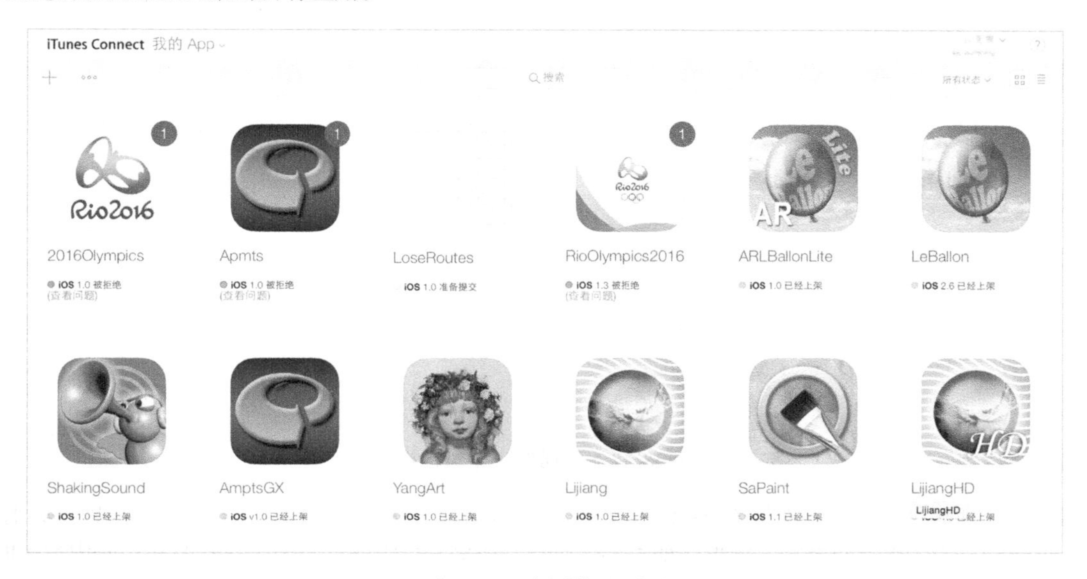

图28-35 应用管理页面

点击左上角的+按钮，如图28-36所示，在下拉菜单中选择“新App”，此时会弹出如图28-37所示的添加新应用对话框，在这里可以输入应用的如下信息。

- **平台**：选择iOS。
- **名称**：为LostRoutes，这个名称是显示到App Store上面的名字，是不能重复的。
- **主要语言**：选择Simplified Chinese（简体中文）。
- **套装ID**：即Bundle ID，它可以从开发者网站创建的App ID中选择。
- **SKU**：应用程序编号，具有唯一性，因此建议使用公司的“域名反写+应用名”，这里我们输入的是com.work6.LostRoutes。

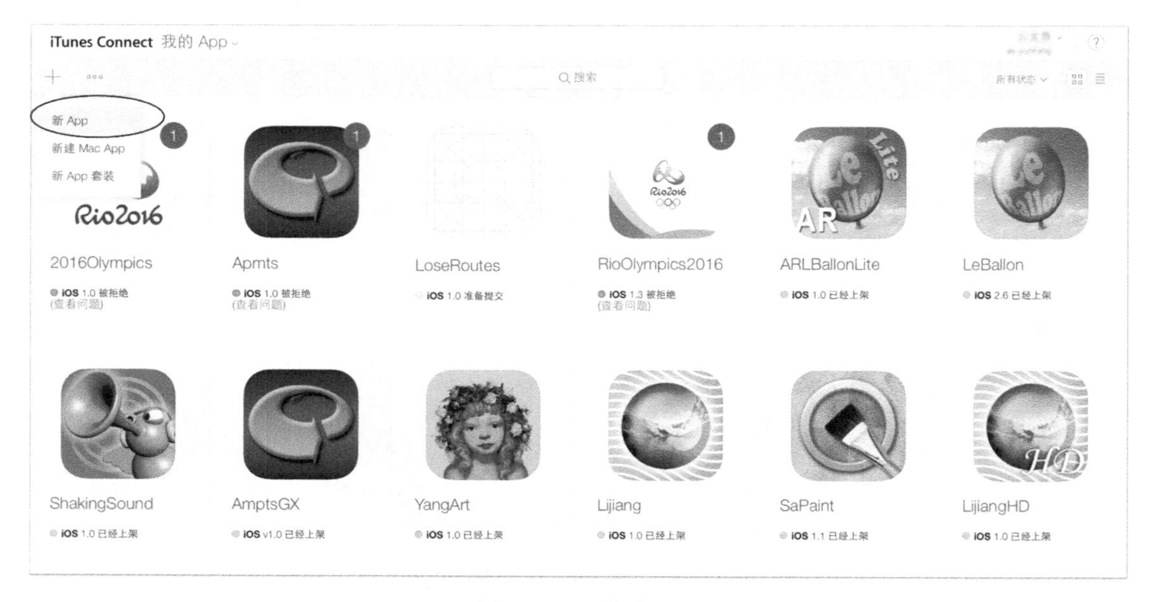

图28-36　添加应用

图28-37　添加新应用对话框

接着点击“创建”按钮，进入如图28-38所示的信息输入页面，这里我们可以设置游戏的类别。类别是应用的分类，也就是应用会被发布到哪个频道，如果选择游戏，还要进行细分，因为游戏是App Store中最多的应用，所以分得比较细。这两个分类选项可以根据自己的应用进行填写，要求不是特别严格。

图28-38　新应用页面

28.3.2　应用定价

应用定价或许是我们最关心的了。在图28-38所示的页面中，点击左边的导航菜单“APP STORE信息”→“价格与定价”，会出现如图28-39所示的页面，选择价格标签，会出现价格等级选择，这里我们选择“CNY0(免费)”。理论上，你可以定到很高的价格，是否能够卖得出就要看市场反馈了。这个定价很灵活，以后也可以修改。

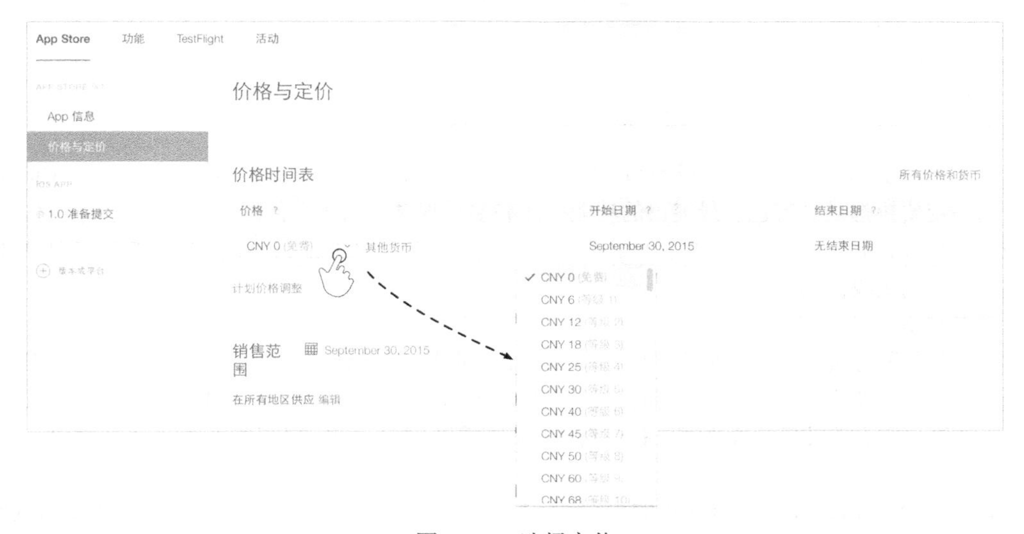

图28-39　选择定价

定价完成后，点击页面右上角的“存储”按钮保存。

28.3.3　基本信息输入

在图28-35中点击左上角的+按钮，出现下拉菜单，从中选择“iOS APP”→“1.0准备提交”，出现基本信息输入页面，其中包含更加详细的部分，包括应用基本数据信息、审核信息、版本发布，以及上传应用图标、截图和介绍视频，下面我们分别介绍一下。

1. 上传应用预览视频和截图

为了介绍、推广和宣传应用，App Store允许我们为应用上传相关的视频和截图。除了3.5英寸的设备只能上传截图外，其他设备可以上传视频或截图，每一种设备所能上传的视频和截图总数不能超过5个。我们可以准备好视频和截图，然后把它们拖曳到图28-40中的选择文件区域，如果视频或截图的大小规格没有问题，就可以上传了。

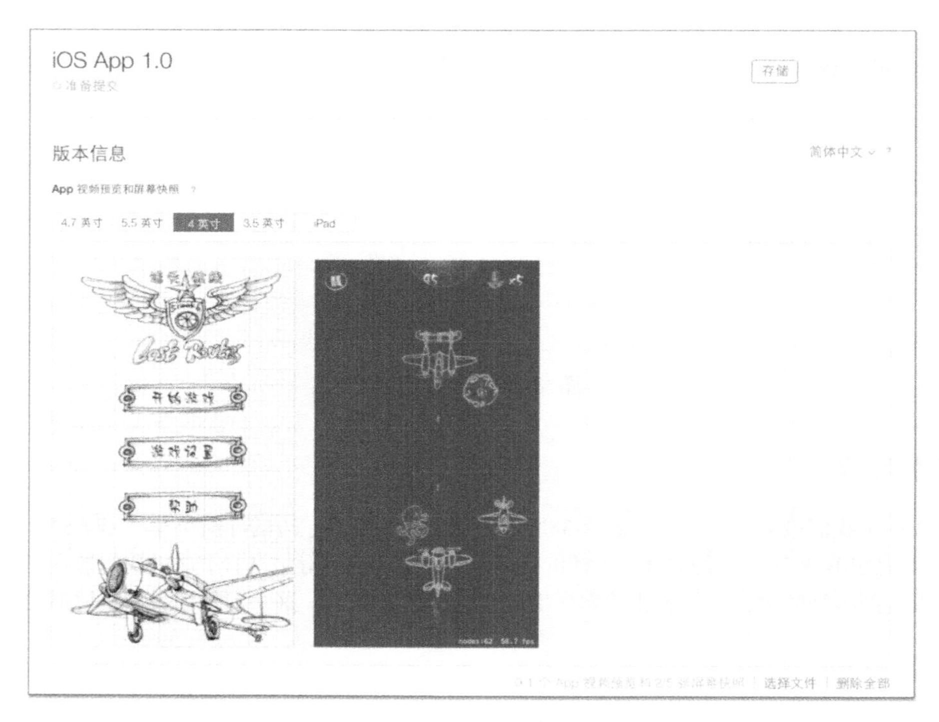

图28-40　上传应用预览视频和截图

截图可以让用户了解我们的应用，视频和截图往往比文字描述更形象、更具有说服力。应用可能有很多情景和功能，我们一定要挑选最具特色、最突出的功能截图和演示视频。由于上传的截图和视频不能超过5个，一定要把最好的截图和视频放到前面，因为后面的截图和视频需要向后滑动才能出现，这样才能吸引用户对我们的应用产生兴趣，考虑购买我们的应用。

2. 基本信息输入

App 视频预览和屏幕快照的下面是应用的基本信息输入页面，如图28-41所示，“描述”对应用很重要，将出现在App Store的应用介绍中。用户购买应用时，主要通过这段文字来了解我们的应用到底是做什么的，有什么用。因此，请一定认真、用心地准备这段文字，描述清楚应用的所有功能，体现出应用的特点、特色等，从而吸引用户来购买。

“关键词”是在App Store上查询该应用的关键词。“技术支持网址”里面需要填写应用技术支持的网址，“营销网址”里面填写应用营销的网址，主要是针对应用做进一步介绍。由于“描述”的文字和图片数是有限制的，可能不会把应用介绍得很详尽，所以我们可以自己创建一个网页，更详细地介绍应用。

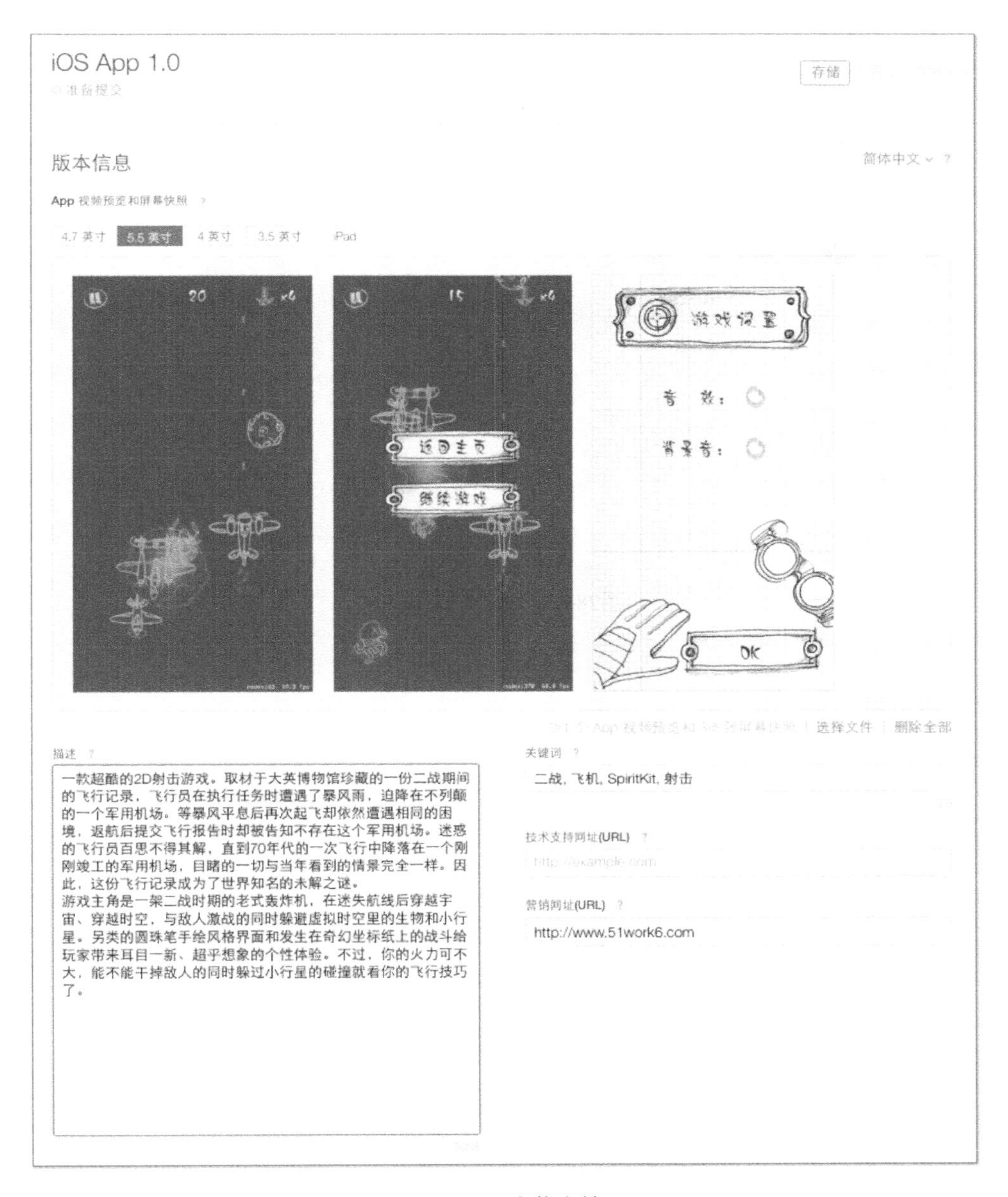

图28-41　基本信息输入

3. 综合信息输入

将图28-41的页面向下滚动，则出现如图28-42所示的界面，我们可以在该界面中输入App程序图标、版本号、分级和版权声明等信息。

这里的应用程序图标是在App Store上显示的，它要求1024 × 1024像素的PNG或JPEG格式图片。关于图标设计，我们一定要下点工夫，图标能够给用户带来第一印象，所以一定要用心去设计。

点击“分级”旁边的“编辑”链接，会弹出如图28-43所示的评级界面。这是根据应用中含有色情、暴力等内容的程度进行评级的。不同的等级标识使用该应用的年龄段。同时，也会有一些国家根据这个评级高低来限制是否在本国销售。在这个选项中，开发者应该按应用的实际情况来填写，如果与所描述的内容不符，苹果会拒绝审核通过。

App 综合信息

App 图标 ?

版本 ?

1.0

分级 编辑

4+ 岁

其他分级

版权 ?

商务代表联系信息 ?

在韩国 App Store 中显示的商务代表联系信息。

jia yunlong

姓氏　名字

楼号、单元号、房间号（可不填）

电话号码　电子邮件

App 地区范围文件 ?

选择文件（可不填）

图28-42　综合信息输入

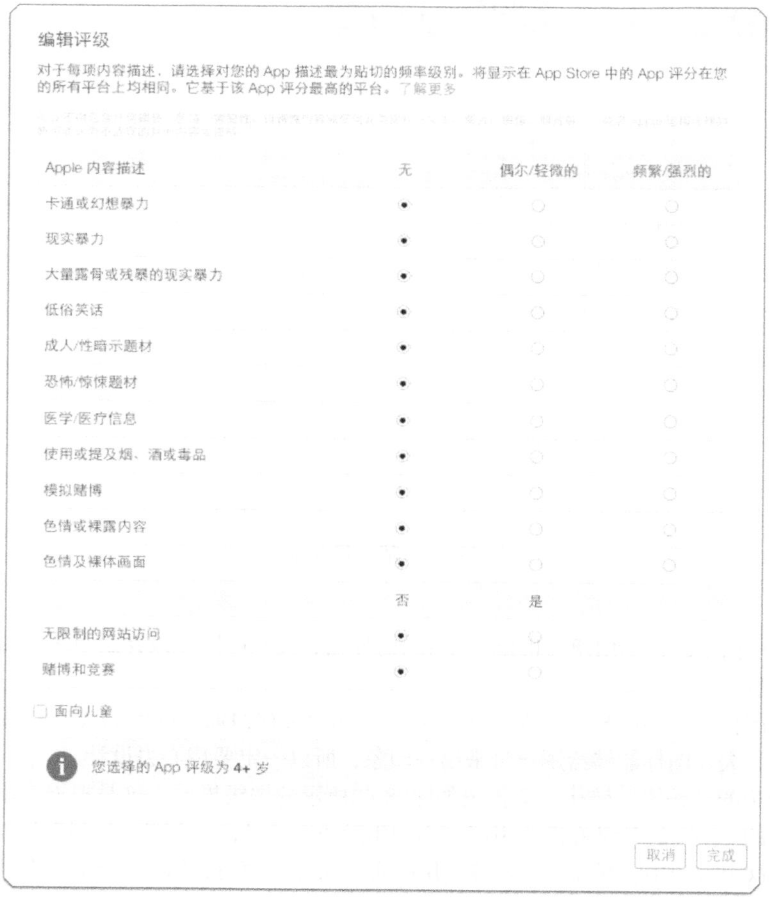

编辑评级

对于每项内容描述，请选择对您的 App 描述最为贴切的频率级别。将显示在 App Store 中的 App 评分在您的所有平台上均相同。它基于该 App 评分最高的平台。了解更多

Apple 内容描述	无	偶尔/轻微的	频繁/强烈的
卡通或幻想暴力	•	○	○
现实暴力	•	○	○
大量露骨或残暴的现实暴力	•	○	○
低俗笑话	•	○	○
成人/性暗示题材	•	○	○
恐怖/惊悚题材	•	○	○
医学/医疗信息	•	○	○
使用或提及烟、酒或毒品	•	○	○
模拟赌博	•	○	○
色情或裸露内容	•	○	○
色情及裸体画面	•	○	○
	否	是	
无限制的网站访问	•	○	
赌博和竞赛	•	○	

□ 面向儿童

您选择的 App 评级为 **4+** 岁

取消　完成

图28-43　评级界面

4. 应用审核信息

将图28-42的页面向下滚动，则出现如图28-44所示的界面，该界面是应用审核信息输入界面，这里的信息主要是给苹果审核团队的工作人员看的。在“联系信息”中填写开发者团队中负责与苹果审核小组联系的人员的信息，包括姓名、邮箱和电话号码。

图28-44 应用审核信息界面

在“备注”中，请填写应用细节和一些特别的功能，帮助审核人员快速了解该应用。在“演示账户”中，请填写应用中的测试账号和密码，提供给审核人员测试，以便于更加顺畅地通过审核。

5. 版本发布

将图28-44的页面向下滚动，则出现如图28-45所示的版本发布界面。在该界面中，我们可以设置在应用审核通过后自动发布还是手动发布。

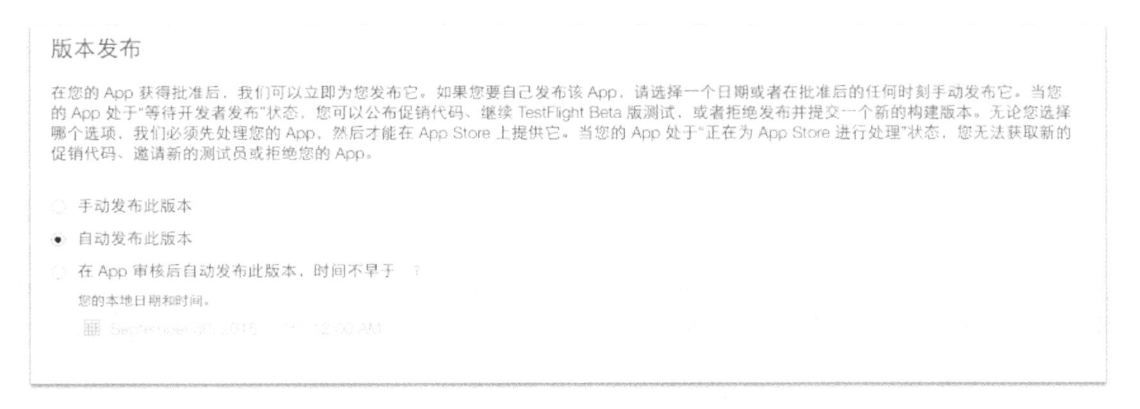

图28-45 版本发布界面

确认这些输入信息无误后，请点击页面上部的“存储”按钮保存输入。

28.3.4 上传应用

基本信息输入完成后，在点击图28-40中的“提交以供审核”按钮前，需要先上传应用。上传应用可以使用Xcode或Application Loader工具，或者两者结合使用。为了简单，我们推荐只使用Xcode工具。

首先，在Xcode中选择菜单项Product→Archive为应用归档。归档结束后，会打开如图28-46所示的界面。选中刚刚归档的项目，然后点击右边的Validate...按钮进行验证，这时候会弹出验证确认对话框，如图28-47所示。然后点击Validate按钮开始验证，验证结束后会有一个是否成功的提示。这里需要说明的是，在上传之前进行验证是非常有必要的。

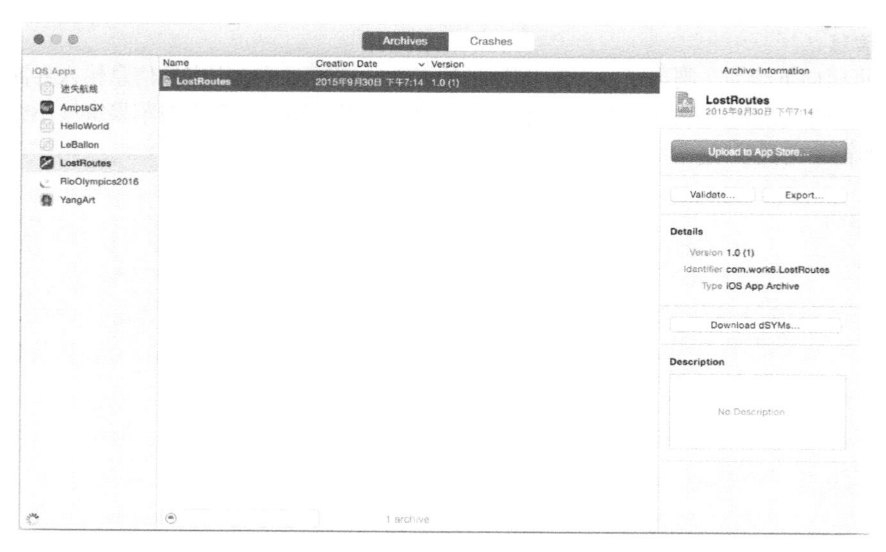

图28-46　归档界面

图28-47　验证程序包

在图28-46所示的界面中点击Upload to App Store...按钮，会弹出如图28-48所示的对话框，点击Submit按钮开始上传程序包，上传结束之后会有一个关于是否成功的提示。

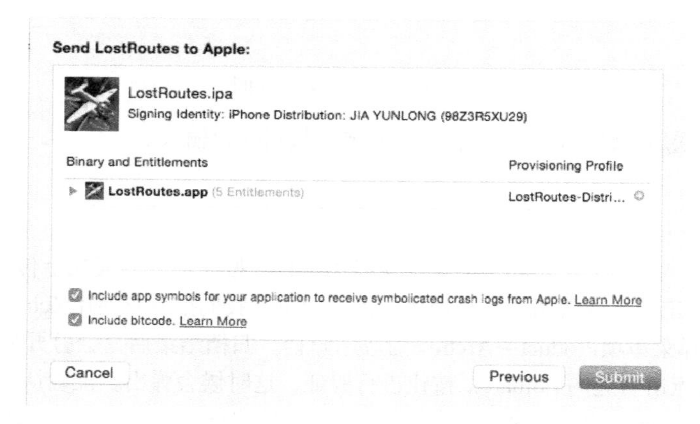

图28-48　提交程序包

应用程序上传之后的状态，可以在iTunes Connect中查看。登录iTunes Connect，打开应用，我们会看到已经上传的应用，如图28-49所示。点击应用，进入应用管理页面，在左边导航菜单中选择“iOS APP”→“1.0准备提交”页面，在“构建版本”部分点击+按钮，如图28-50所示，此时会弹出对话框，从中选择要上传的应用程序包，然后点击“完成”按钮即可。

图28-49 上传之后的应用

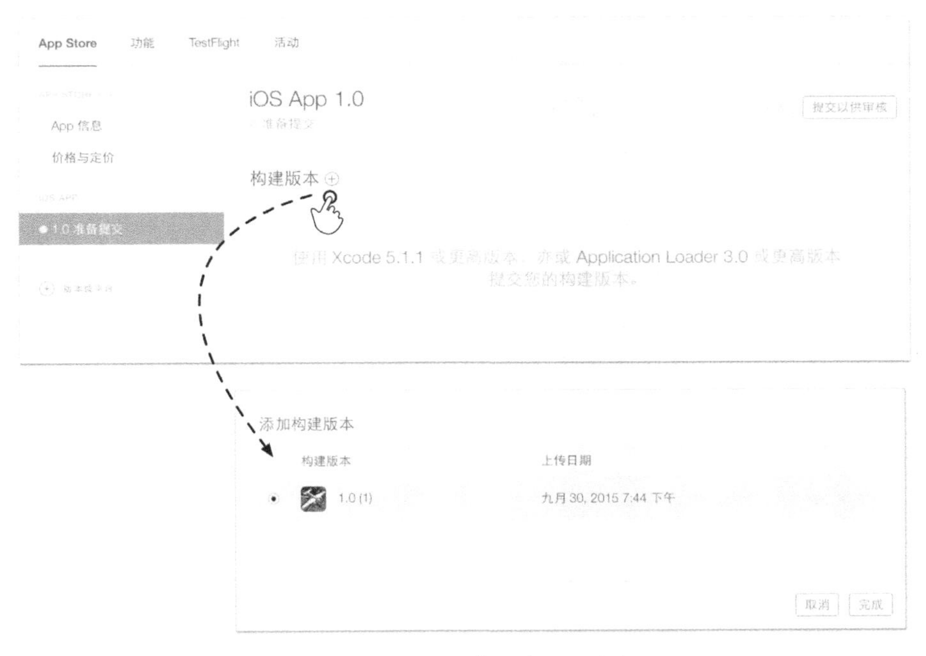

图28-50 选择上传的应用程序包

28.3.5 提交审核

选择上传的应用程序包后，如果确认无误，我们就可以点击“提交以供审核”按钮了。

在提交审核之前，iTunes Connect还需要我们最后确认几件事情，如图28-51所示，包括出口合规信息、内容版权和广告标识符。

出口合规信息是询问程序代码中是否有加密算法，美国出口法律规定禁止任何加密的软件流向国外，这里我们选择“否”即可。内容版权是询问应用中是否包含、显示或者会访问第三方内容。广告标识符是询问应用中是否使用了广告标识符（IDFA）。这些内容一定要实事求是地填写，特别是广告标识符，如果选择“是”，它还会有更加详细的询问，如图28-52所示。

取消
提交
出口合规信息
您的 App 是否设计使用了加密或含有整合加密功能？（即使您的 App 只使用了 iOS 或 OS X 提供的加密功能，也请选择“是”。）
是
否
内容版权
您的 App 是否包含、显示或者会访问第三方内容？
是
否
广告标识符
此 App 是否使用广告标识符 (IDFA)?
广告标识符 (IDFA) 是每台 iOS 设备的唯一 ID，是投放定向广告的唯一方法。用户可以选择在其 iOS 设备上限制广告定位。
是
否
对于广告标识符的 (IDFA) 的使用，请务必选择正确的答案。如果您的 App 包含 IDFA 而您选择了“否”，此二进制文件将永久遭绝，您必须提交另一个二进制文件。

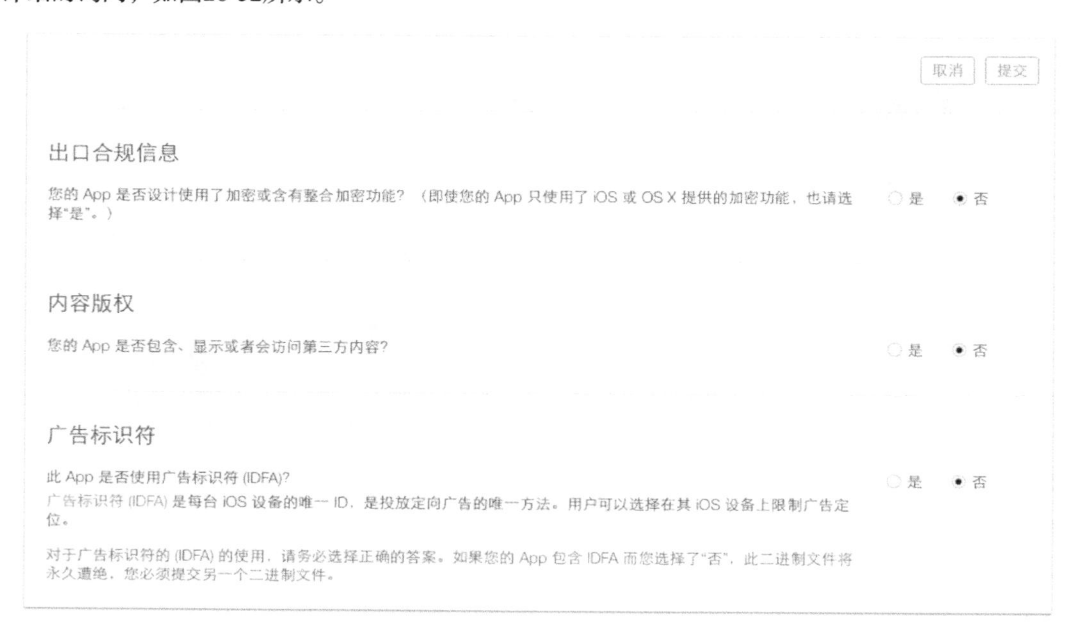

图28-51　最后确认的事情

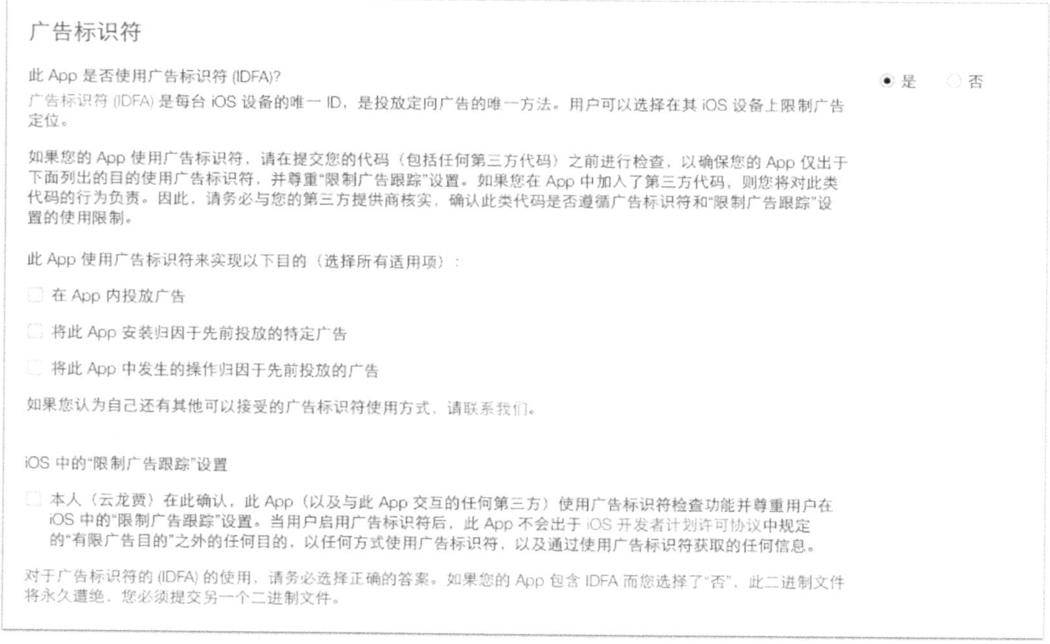

广告标识符
此 App 是否使用广告标识符 (IDFA)?
广告标识符 (IDFA) 是每台 iOS 设备的唯一 ID，是投放定向广告的唯一方法。用户可以选择在其 iOS 设备上限制广告定位。
是
否
如果您的 App 使用广告标识符，请在提交您的代码（包括任何第三方代码）之前进行检查，以确保您的 App 仅出于下面列出的目的使用广告标识符，并尊重“限制广告跟踪”设置。如果您在 App 中加入了第三方代码，则您将对此类代码的行为负责。因此，请务必与您的第三方提供商核实，确认此类代码是否遵循广告标识符和“限制广告跟踪”设置的使用限制。
此 App 使用广告标识符来实现以下目的（选择所有适用项）：
在 App 内投放广告
将此 App 安装归因于先前投放的特定广告
将此 App 中发生的操作归因于先前投放的广告
如果您认为自己还有其他可以接受的广告标识符使用方式，请联系我们。
iOS 中的“限制广告跟踪”设置
本人（云龙费）在此确认，此 App（以及与此 App 交互的任何第三方）使用广告标识符检查功能并尊重用户在 iOS 中的“限制广告跟踪”设置。当用户启用广告标识符后，此 App 不会出于 iOS 开发者计划许可协议中规定的“有限广告目的”之外的任何目的，以任何方式使用广告标识符，以及通过使用广告标识符获取的任何信息。
对于广告标识符的 (IDFA) 的使用，请务必选择正确的答案。如果您的 App 包含 IDFA 而您选择了“否”，此二进制文件将永久遭绝，您必须提交另一个二进制文件。

图28-52　广告标识符确认

完成这些工作后，点击“提交”按钮提交应用，然后的应用状态就变成了“正在等待审核”。如果没有任何问题，接下来就是等待了。因为每天都有很多程序要发布到App Store中，所以等待审核也要排队。如果状态变为“已经上架”，就说明审核通过并且可以销售了。

28.4 常见审核不通过的原因

App Store的审核是出了名的严格，相信大家也都略有耳闻。苹果官方提供了一份详细的审核指南，包括20多大项、100多小项的拒绝上线条款，并且条款在不断增加中。此外，还有一些模棱两可的条例，所以稍有“闪失”，应用就有可能被拒绝。但是有一点比较好，那就是每次拒绝时，苹果会给出拒绝的理由，并指出你违反了审核指南的哪一条，开发者可以根据评审小组给的回复修改应用重新提交。下面我们讨论一下常见的被拒原因。

28.4.1 功能问题

在发布应用之前，我们一定要对产品进行认真的测试，如果在审核中出现了程序崩溃或者程序错误，无疑会被审核小组拒绝。如果我们想发布一个演示版的程序，通过它给客户演示这也是不会被通过的。应用的功能与描述不相符，或者应用中含有欺诈虚假的功能，那么应用将被拒绝。比如在应用中有某个按钮，但是点击这个按钮没有反应或者不能点击，这样的程序是不会通过的。

苹果不允许访问私有API，有浏览器的网络程序必须使用iOS WebKit框架和WebKit JavaScript。还有几点让人比较头痛的规则，那就是如果App没有什么显著的功能或者没有长久娱乐价值，也会被拒绝。如果你的应用在市场中已经存在了，在相关产品比较多的时候也可能被拒绝。

28.4.2 用户界面问题

苹果审核指南规定开发者的应用必须遵守苹果《iOS用户界面指导原则》中解释的所有条款和条件，如果违反了这些设计原则，就会被拒绝上线，所以开发者在设计和开发产品之前，一定要认真阅读《iOS用户界面指导原则》。这些原则中也渗透着苹果产品的一些理念，不仅是为了避免程序被拒绝而看，而且还可让开发者们设计出更好的App。苹果不允许开发者更改自身按键的功能（包括声音按键以及静音按键），如果开发者使用了这些按键并利用它们做一些别的功能，应用将会被拒绝。

28.4.3 商业问题

在要发布的应用中，首先不能侵犯苹果公司的商标及版权。简单地说，在应用中不能出现苹果的图标，不能使用苹果公司现在产品的类似名字为应用命名，涉及iPhone、iPad、iTunes等相关或者相近的名字都是不可以的。苹果认为这会误导用户，认为该应用是来自苹果公司的产品。误导用户认为该应用是受到苹果的肯定与认可的，也一样不行。

私自使用受保护的第三方材料（商标、版权、商业机密和其他私有内容），需要提供版权认可。如果你的应用涉及第三方版权的信息，开发者们要仔细考虑了。由于有些开发者对版权法律意识比较淡薄，总会忽视这一点，然而这一点是非常致命的。苹果处理这种被起诉的侵权应用，最轻的处罚是下架应用，有时需要将开发者账户里的钱转到起诉者账户。再严重的就是，起诉者将你告上法庭，除了自己账户中的钱被扣除外，还要另赔付起诉者相关费用。

28.4.4 不当内容

若包含一些不合适、不和谐的内容，苹果当然不会允许应用上架，比如具有诽谤、人身攻击的应用，含有暴力倾向的应用，低俗、令人反感、厌恶的应用，赤裸裸的色情应用等。含有赌博性质的应用必须明确表示苹果不是发起者，也没有以任何方式参与活动。

28.4.5　其他问题

关于宗教、文化或种族群体的应用，或评论包含诽谤性、攻击性或自私性内容的应用不会被通过，使用第三方支付的应用会被拒绝，模仿iPod界面的应用将会被拒绝，怂恿用户造成设备损坏的应用会被拒绝。这里有个小故事，有一款应用，其功能是比比谁将设备扔得高，最后算积分，这个应用始终没能上架，因为在测试应用的时候就摔坏了两部手机。此外，未获得用户同意便向用户发送推送通知，要求用户共享个人信息的应用都会被拒绝。

28.5　小结

通过对本章的学习，我们了解了如何在App Store上发布应用，以及发布之前需要处理哪些问题。发布者需要了解应用的发布流程，更应该熟悉应用审核不通过的一些常见原因，从而在开发时注意，以免等到审核时被拒绝，耽误了应用的上架时间。

第 29 章

iOS开发项目实战——2020东京奥运会应用开发及App Store发布

这是本书的最后一章，也是本书的画龙点睛之笔。我想通过一个实际的应用使读者能够将本书前面讲过的知识点串联起来，了解iOS应用开发的一般流程，了解当下最为流行的开发方法学——敏捷开发。在开发过程中，我们会发现敏捷方法非常适用于iOS应用的开发。

29.1 应用分析与设计

本节中，我们从计划开发这个应用开始进行分析和设计，其中设计过程包括原型设计、数据库设计和架构设计。

29.1.1 应用概述

2016年里约奥运会结束时，我和智捷课堂就在想开发一个介绍体育比赛的应用。比赛类应用是有时效性的，用户只会在比赛前使用，比赛结束后就没人使用了，而且比赛项目和日程表等信息会有一些变化，但是鉴于对体育的热爱，我们还是决定开发下一届奥林匹克运动会的应用。

针对将在2020年东京举办的奥运会，我们智捷课堂决定开发一款应用，并将源代码发布在我们的网站www.51work6.com。

我们首先介绍一下应用的需求。根据我们现在了解到的资料，能够在应用中提供的信息有举办城市、会徽、开幕时间、比赛项目和比赛日程等部分信息。因此，我们整理了这个应用提供的一些功能：2020东京奥运会的一些基本信息、比赛项目、倒计时和比赛日程等。

29.1.2 需求分析

根据上面的功能描述，确定需求如下：

- 2020东京奥运会基本信息
- 比赛项目
- 倒计时
- 比赛日程
- 关于我们

这里我们采用用例分析方法描述用例图，如图29-1所示。

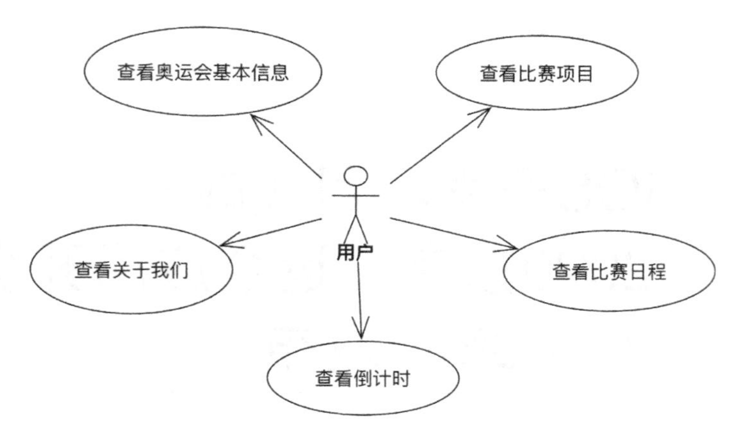

图29-1　2020东京奥运会应用用例图

29.1.3　原型设计

原型设计草图对于应用设计人员、开发人员、测试人员、UI设计人员以及用户都非常重要，该案例的原型如图29-2所示。

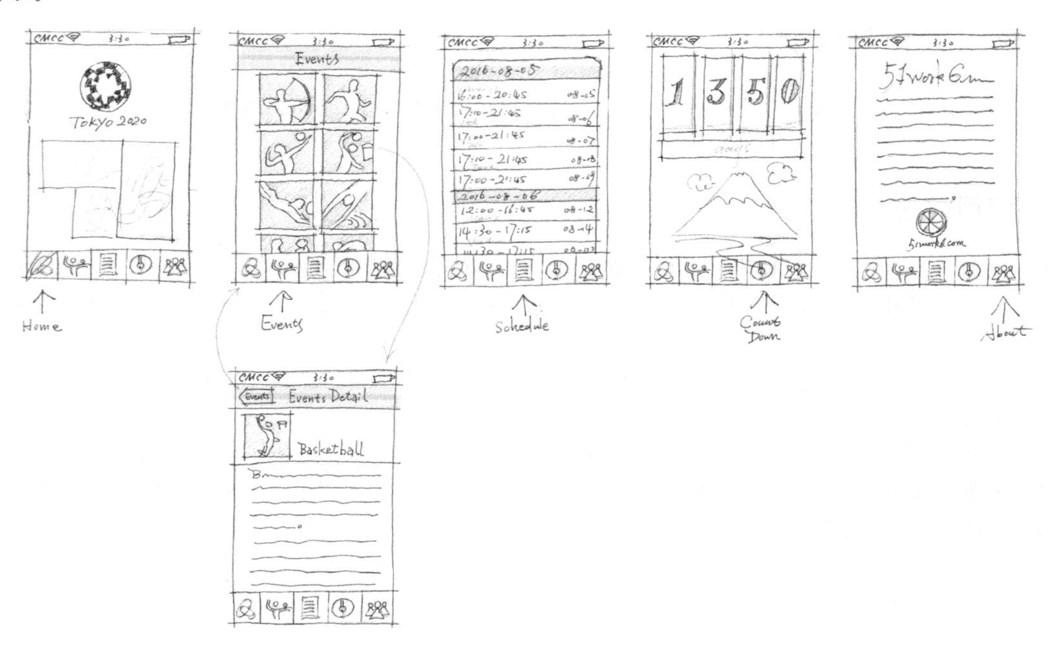

图29-2　原型设计图

29.1.4　数据库设计

从用例图可以了解到，这个应用只是查看信息，没有信息插入、修改和删除等功能，那么这些信息放在哪里了呢？原本考虑放在一个文本文件中，但是文本文件的结构不好，管理起来也不方便，于是考虑到使用数据库，我们可以在第一次启动时将数据导入iOS设备中的SQLite本地数据库。

在数据库概念设计阶段，我们首先需要找出应用中的实体，然后确定实体的属性以及实体关系。在数据库物理设计阶段，实体将演变成为表，实体属性演变成为字段。

应用中的实体有2020奥运会基本信息、比赛项目、倒计时信息和比赛日程，仔细分析后发现只有比赛项目和比赛日程，这是因为2020奥运会基本信息需要放到数据库中，倒计时信息是动态变化的，不需要放到数据库中。比赛项目和比赛日程是一对多的关系。最后的数据库物理数据模型如图29-3所示。

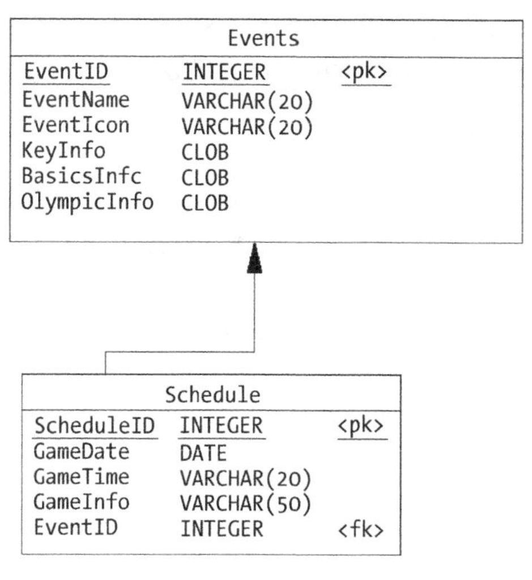

图29-3　数据库物理数据模型

该模型中的名词解释如下所示。

- **Events**：比赛项目。
- **EventID**：比赛项目编号，为主键，它是自增长整数类型。
- **EventName**：比赛项目名。
- **EventIcon**：比赛项目的图标名。
- **KeyInfo**：比赛项目的关键信息。由于包含的字符很长，其数据类型采用大文本类型（CLOB）。
- **BasicsInfo**：比赛项目基本信息，其数据类型采用大文本类型。
- **OlympicInfo**：比赛项目奥运会历史信息，其数据类型采用大文本类型。
- **Schedule**：比赛日程表。
- **ScheduleID**：比赛日程ID，为主键，它是一个没有实际意义的自增长整型字段。由于其他字段都不太适合做主键，所以采用这种设计。
- **GameDate**：比赛日期，日期类型。
- **GameTime**：比赛时间，是在某个比赛日中的比赛时间段。
- **GameInfo**：比赛描述，如男子单打和女子单打等。

两个表是通过EventID（比赛项目编号）关联在一起的。很多数据库建模工具都可以从物理数据模型生成DDL语句[1]，使用这些语句可以很方便地创建和维护数据库中的表结构。

29.2　任务1：创建应用工程

在开发项目之前，应该由一个人搭建开发环境，然后把环境复制给其他人使用。任务1是由老关创建好工作空间，将源代码提交到GitHub中，然后再由其他成员克隆到本地。在项目开发过程中，要求严格遵守并使用GitHub进行源代码的版本控制。

在Objective-C和Swift版本中，我们都采用基于同一个工作空间的分层架构设计，即Objective-C为WFOOO模

① 数据定义语句，用于创建、删除和修改数据库对象，包括DROP、CREATE、ALTER、GRANT、REVOKE和TRUNCATE等语句。

式，Swift为WFSSS模式。

29.2.1　迭代 1.1：创建工程

使用Xcode创建一个工程TokyoOlympics，使用的模板为Simple View Application，在Language中选择Swift或Objective-C，在Devices中选择Universal，然后参考第12章创建数据持久化逻辑组件。

29.2.2　迭代 1.2：发布到 GitHub

完成了这些工作并编译通过后，需要将这个空白的工程分发给小组的其他成员，他们使用GitHub发布应用的第一个版本。注意，其他成员就不要自己再创建工程，只需要从GitHub上克隆代码，然后在自己的工程中添加自己的内容就可以了。

我们需要使用GitHub账号登录，并创建一个名为TokyoOlympics的代码库，如图29-4所示。

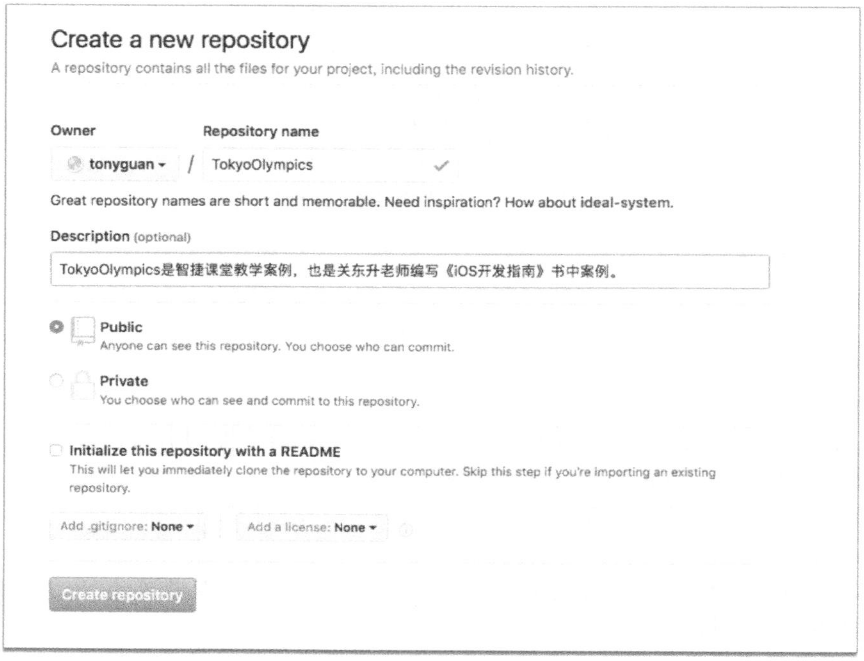

图29-4　创建代码库

创建完成后，需要将本地代码上传到GitHub服务器，具体步骤可参考26.2.5节。其他的成员使用`$git clone git@github.com:tonyguan/TokyoOlympics.git`克隆到本地。

29.3　任务 2：数据库与数据持久化逻辑组件开发

该任务负责人应该对数据库、SQL语句和SQLite API非常熟悉，可以不需要熟悉UI相关的技术。

29.3.1　迭代 2.1：编写数据库 DDL 脚本

我们要按照图29-3所示的数据库ER模型来编写数据库SQL DDL①脚本。当然，也可以通过一些工具生成DDL脚本，然后把这个脚本放在数据库中执行就可以了。下面是编写的DDL脚本：

① SQL中的数据定义语言，用于定义和管理数据库中的所有对象，如建表`create table`等。

```
/*==============================================================*/
/* DBMS name:      SQLite3 DB                                   */
/* Created on:     2016/12/21 15:59:37                          */
/*==============================================================*/

drop table if exists Events;

drop table if exists Schedule;

/*==============================================================*/
/* Table: Events                                                */
/*==============================================================*/
create table Events
(
    EventID             INTEGER   primary key autoincrement     not null,
    EventName           VARCHAR(20),
    EventIcon           VARCHAR(20),
    KeyInfo             CLOB,
    BasicsInfo          CLOB,
    OlympicInfo         CLOB
);

/*==============================================================*/
/* Table: Schedule                                              */
/*==============================================================*/
create table Schedule
(
    ScheduleID          INTEGER   primary key autoincrement     not null,
    GameDate            DATE                  not null,
    GameTime            VARCHAR(20)           not null,
    GameInfo            VARCHAR(50),
    EventID            INTEGER,
    constraint FK_SCHEDULE_REFERENCE_EVENTS foreign key (EventID) references Events (EventID)
);
```

29.3.2 迭代2.2：插入初始数据到数据库

在应用中没有插入、删除和修改功能，那么原始的数据如何插入到数据库中呢？我们可以在应用启动并且建表完成后，使用insert语句插入到数据库中。我们在脚本文件中预先写好insert语句，类似于下面的语句：

```
insert into Events(EventName,EventIcon,KeyInfo,BasicsInfo,OlympicInfo)
    values ('Athletics','athletics.gif','Athletics is…','There are four
    main strands…','The ancient Olympic …');
insert into Schedule (GameDate,GameTime,GameInfo,EventID) values
    ('2020-08-05','16:00 - 20:45','Women''s',29);
```

使用上面的语句，我们可以一起编写脚本文件create_load.sql，具体可参考工程中DBFile目录下的create_load.sql文件。

29.3.3 迭代2.3：数据库版本控制

程序代码负责执行DLL脚本来创建数据库并插入数据到数据库中，但是并不是每次运行都进行建表和插入数据的操作，这里通过一个数据库版本号来管理什么时候创建和插入。

我们在数据库中建立一个表DBVersionInfo，它有一个字段version_number，用来记录当前数据库的版本号。同时，我们也在应用程序资源文件DBConfig.plist中保存了版本号。如果数据库中的版本号与资源文件DBConfig.plist中的版本号一致，则不会进行建表和插入数据的操作；如果不一致，就执行创建表和插入数据的操作，并修改DBVersionInfo表的version_number字段，使之加一。

DBConfig.plist的内容如下：

```
<?xml version="1.0" encoding="UTF-8"?>
```

```
<!DOCTYPE plist PUBLIC "-//Apple//DTD PLIST 1.0//EN" "http://
    www.apple.com/DTDs/PropertyList-1.0.dtd">
<plist version="1.0">
<dict>
    <key>DB_VERSION</key>
    <integer>8</integer>
</dict>
</plist>
```

其中DB_VERSION项是版本号。如果发布新版本，我们只需要修改DBConfig.plist文件中的DB_VERSION项，让它加一即可。DBConfig.plist相当于配置文件。

29.3.4　迭代 2.4：配置数据持久化逻辑组件

配置数据持久化逻辑组件比较麻烦，我们将该配置工作分成如下几个任务。

1. 迭代2.4.1：添加DDL脚本文件和DBConfig.plist文件到工程

前面的任务已经创建了数据持久化逻辑组件所必需的DDL脚本文件和DBConfig.plist文件，这些文件应该添加到工程中作为应用程序包的一部分。这两个文件被放到DBFile文件夹中，我们需要拖曳它们到工程中。添加之后的结果如图29-5所示。

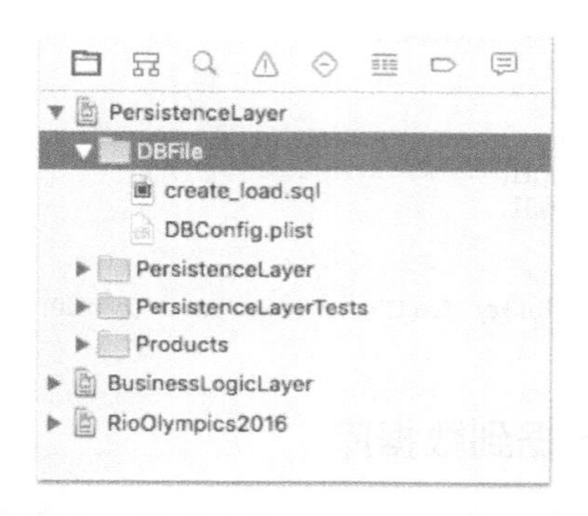

图29-5　添加DDL脚本文件和DBConfig.plist文件到工程

2. 迭代2.4.2：为工程添加SQLite3库

要使用SQLite3数据库，需要为工程添加SQLite3库（libsqlite3.0.tbd或libsqlite3.tbd），具体步骤参考14.4.2节。

3. 迭代2.4.3：为Swift版本配置环境

在Swift中添加SQLite3库比较麻烦，需要有桥接头文件，具体创建和添加步骤详见14.4.3节，这里不再赘述。

29.3.5　迭代 2.5：编写实体类

实体是应用中的人、事、物，在数据库设计的时候演变成为“表”，而在面向对象分析和设计时，实体演变成为“实体类”。因此，实体类与数据库的表有共同的渊源。图29-6是应用的实体类图，它看起来很像如图29-3所示的数据库ER模型。

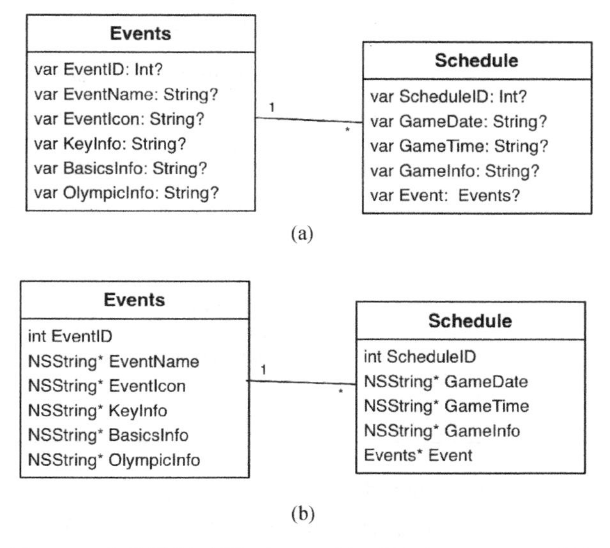

图29-6 实体类图（图a为Swift版，图b为Objective-C版）

在实体类图中，实体比赛项目（Events）和比赛日程（Schedule）是一对多的关联关系。实体类Events的代码如下：

```
//Events.swift文件
import Foundation

public class Events {
    //编号
    public var EventID: Int?
    //项目名
    public var EventName: String?
    //项目图标
    public var EventIcon: String?
    //项目关键信息
    public var KeyInfo: String?
    //项目基本信息
    public var BasicsInfo: String?
    //项目奥运会历史信息
    public var OlympicInfo: String?
}
```

```
//Events.h文件
#import <Foundation/Foundation.h>

//比赛项目实体类
@interface Events : NSObject

//编号
@property(nonatomic, assign) int EventID;
//项目名
@property(nonatomic, strong) NSString* EventName;
//项目图标
@property(nonatomic, strong) NSString* EventIcon;
//项目关键信息
@property(nonatomic, strong) NSString* KeyInfo;
//项目基本信息
@property(nonatomic, strong) NSString* BasicsInfo;
//项目奥运会历史信息
@property(nonatomic, strong) NSString* OlympicInfo;

@end

//Events.m文件
#import "Events.h"

@implementation Events

@end
```

实体类Schedule的代码如下：

```
//Schedule.swift文件
import Foundation

//比赛日程表实体类

public class Schedule {
    //编号
    public var ScheduleID: Int?
```

```
//Schedule.h文件
#import <Foundation/Foundation.h>
#import "Events.h"

//比赛日程表实体类
@interface Schedule : NSObject

//编号
```

```
    //比赛日期
    public var GameDate: String?
    //比赛时间
    public var GameTime: String?
    //比赛描述
    public var GameInfo: String?
    //比赛项目
    public var Event: Events?

}
```

```
@property(nonatomic, assign) int ScheduleID;
//比赛日期
@property(nonatomic, strong) NSString* GameDate;
//比赛时间
@property(nonatomic, strong) NSString* GameTime;
//比赛描述
@property(nonatomic, strong) NSString* GameInfo;
//比赛项目
@property(nonatomic, strong) Events* Event;

@end

//Schedule.m文件
#import "Schedule.h"
@implementation Schedule

@end
```

提示　在对象模型中，关联关系是通过在Schedule中定义属性Event关联在一起的。但是在Events实体类定义中，并没有与Schedule关联的属性，这说明这种关联关系是单向的。

29.3.6　迭代2.6：编写DAO类

编写DAO类是我们主要的工作。图29-7是应用的DAO类的类图。

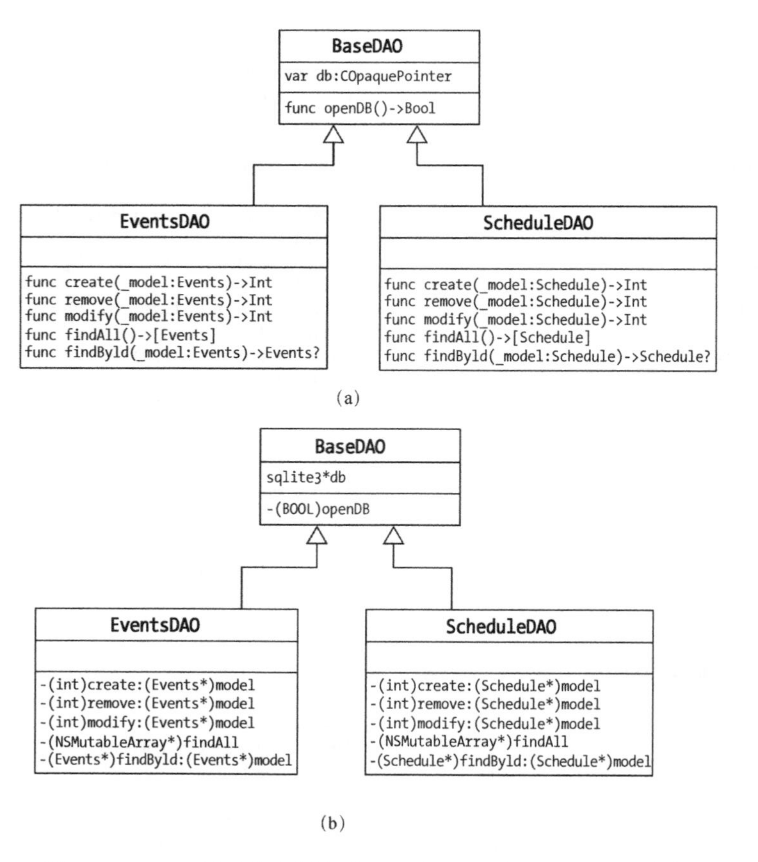

图29-7　DAO类的类图（图a为Swift版，图b为Objective-C版）

BaseDAO是DAO基类，它的成员变量db在Objective-C版中为sqlite3指针类型，而在Swift版中对应的类型是COpaquePointer。

EventsDAO和ScheduleDAO类都继承自BaseDAO类，采用单例设计模式设计。每一个DAO类都有5个方法，它们分别用于实现数据的插入、删除、修改、按照主键查询和查询所有数据。

BaseDAO类的代码如下：

```
//BaseDAO.swift代码

public class BaseDAO: NSObject {

    internal var db:COpaquePointer = nil

    override init() {
        //初始化数据库
        DBHelper.initDB()
    }

    //打开SQLite数据库，如果返回值为true，表示打开成功；如果为false，
    //表示打开失败
    internal func openDB()->Bool {

        //数据文件全路径
        let dbFilePath = DBHelper.applicationDocumentsDirectoryFile
        ➥(DB_FILE_NAME)!

        print("DbFilePath = \(String.fromCString(dbFilePath))")

        if sqlite3_open(dbFilePath, &db) != SQLITE_OK {
            sqlite3_close(db)
            print("数据库打开失败。")
            return false
        }
        return true
    }

    //获得字段数据
    internal func getColumnValue(index:CInt, stmt:OpaquePointer)->String? { ①

        if let ptr = UnsafeRawPointer.init(sqlite3_column_text(stmt, index)) {
            let uptr = ptr.bindMemory(to:CChar.self, capacity:0)
            let txt = String(validatingUTF8:uptr)
            return txt
        }
        return nil
    }

}
```

```
//BaseDAO.h代码
#import "sqlite3.h"
#import "DBHelper.h"

@interface BaseDAO : NSObject {
    sqlite3 *db;
}
//打开SQLite数据库，如果返回值为true，表示打开成功；如果为false，表示打开失败
- (BOOL)openDB;
@end

//BaseDAO.m代码
#import "BaseDAO.h"

@implementation BaseDAO

- (id)init {
    self = [super init];
    if (self) {
        //初始化数据库
        [DBHelper initDB];
    }
    return self;
}

- (BOOL)openDB {

    const char* dbFilePath = [DBHelper
    ➥applicationDocumentsDirectoryFile:DB_FILE_NAME];

    NSLog(@"DbFilePath = %s", dbFilePath);

    if (sqlite3_open(dbFilePath, &db) != SQLITE_OK) {
        sqlite3_close(db);
        NSLog(@"数据库打开失败。");
        return FALSE;
    }
    return TRUE;
}

@end
```

这里定义了SQLite成员变量db，这样子类EventsDAO和ScheduleDAO中就不用再定义了。openDB是打开数据库的方法，由于DAO在子类中多次使用，因此我们在父类BaseDAO中定义了该方法。

在Swift代码第①行定义的getColumnValue方法获得字段数据，这些内容14.4节介绍过，这里不再赘述。

下面我们看看EventsDAO中的关键代码，其中插入数据的代码如下：

```
//EventsDAO.swift文件
//插入方法
public func create(_ model: Events) -> Int {

    if self.openDB() {
        let sql = "INSERT INTO Events (EventName,
        ➥EventIcon,KeyInfo,BasicsInfo,OlympicInfo) VALUES (?,?,?,?,?)"
        let cSql = sql.cString(using: String.Encoding.utf8)
```

```
//EventsDAO.m文件
//插入Events方法
- (int)create:(Events *)model {

    if ([self openDB]) {

        NSString *sqlStr = @"INSERT INTO Events (EventName,
        ➥EventIcon,KeyInfo,BasicsInfo,OlympicInfo) VALUES (?,?,?,?,?)";
```

```swift
        var statement: OpaquePointer? = nil
        //预处理过程
        if sqlite3_prepare_v2(db, cSql, -1, &statement, nil) == SQLITE_OK {

            let cEventName = model.EventName!.cString(using: String.Encoding.
            ➥utf8)
            let cEventIcon = model.EventIcon!.cString(using: String.Encoding.
            ➥utf8)
            let cKeyInfo = model.KeyInfo!.cString(using: String.Encoding.utf8)
            let cBasicsInfo = model.BasicsInfo!.cString(using: String.
            ➥Encoding.utf8)
            let cOlympicInfo = model.OlympicInfo!.cString(using: String.
            ➥Encoding.utf8)

            //绑定参数开始
            sqlite3_bind_text(statement, 1, cEventName, -1, nil)
            sqlite3_bind_text(statement, 2, cEventIcon, -1, nil)
            sqlite3_bind_text(statement, 3, cKeyInfo, -1, nil)
            sqlite3_bind_text(statement, 4, cBasicsInfo, -1, nil)
            sqlite3_bind_text(statement, 5, cOlympicInfo , -1, nil)

            //执行插入
            if (sqlite3_step(statement) != SQLITE_DONE) {
                sqlite3_finalize(statement)
                sqlite3_close(db)
                assert(false, "插入数据失败。")
            }
        }
        sqlite3_finalize(statement)
        sqlite3_close(db)
    }
    return 0
}
```

```objc
        sqlite3_stmt *statement;
        //预处理过程
        if (sqlite3_prepare_v2(db, [sqlStr UTF8String], -1, &statement, NULL)
        ➥== SQLITE_OK) {

            const char* cEventName = [model.EventName UTF8String];
            const char* cEventIcon = [model.EventName UTF8String];
            const char* cKeyInfo = [model.KeyInfo UTF8String];
            const char* cBasicsInfo = [model.BasicsInfo UTF8String];
            const char* cOlympicInfo = [model.OlympicInfo UTF8String];

            //绑定参数开始
            sqlite3_bind_text(statement, 1, cEventName, -1, NULL);
            sqlite3_bind_text(statement, 2, cEventIcon, -1, NULL);
            sqlite3_bind_text(statement, 3, cKeyInfo, -1, NULL);
            sqlite3_bind_text(statement, 4, cBasicsInfo, -1, NULL);
            sqlite3_bind_text(statement, 5, cOlympicInfo, -1, NULL);

            //执行插入
            if (sqlite3_step(statement) != SQLITE_DONE) {
                sqlite3_finalize(statement);
                sqlite3_close(db);
                NSAssert(NO, @"插入数据失败。");
            }
        }
        sqlite3_finalize(statement);
        sqlite3_close(db);
    }

    return 0;
}
```

删除数据的代码如下：

```swift
//EventsDAO.swift文件
//删除方法
public func remove(model: Events) -> Int {

    if self.openDB() {

        //先删除从表（比赛日程表）相关数据
        let sqlScheduleStr = NSString(format: "DELETE
        ➥from Schedule where EventID=%i", model.EventID!)                ①
        let cSqlScheduleStr = sqlScheduleStr.cString(using:
        ➥String.Encoding.utf8)

        //开启事务、立刻提交之前事务
        sqlite3_exec(db, "BEGIN IMMEDIATE TRANSACTION", nil, nil, nil)    ②

        if sqlite3_exec(db, cSqlScheduleStr, nil, nil,  nil) != SQLITE_OK {  ③
            //回滚事务
            sqlite3_exec(db, "ROLLBACK TRANSACTION", nil, nil, nil)       ④
            assert(false, "删除数据失败。")
        }

        //先删除主表（比赛项目）数据
        let sqlEventsStr = String(format: "DELETE
        ➥from Events where EventID =%i", model.EventID!)                 ⑤
        let cSqlEventsStr = sqlEventsStr.cString(using: String.Encoding.utf8)

        if sqlite3_exec(db, cSqlEventsStr, nil, nil, nil) != SQLITE_OK {  ⑥
            //回滚事务
            sqlite3_exec(db, "ROLLBACK TRANSACTION", nil, nil, nil)       ⑦
```

```objc
//EventsDAO.m文件
//删除方法
- (int)remove:(Events *)model {

    if ([self openDB]) {

        //先删除从表（比赛日程表）相关数据
        NSString *sqlScheduleStr = [[NSString alloc] initWithFormat:@"DELETE
        ➥from Schedule where EventID=%i", model.EventID];                 ①

        //开启事务、立刻提交之前事务
        sqlite3_exec(db, "BEGIN IMMEDIATE TRANSACTION", NULL, NULL, NULL); ②

        char* err;

        if (sqlite3_exec(db, [sqlScheduleStr UTF8String], NULL, NULL, &err) !=
        ➥SQLITE_OK) {                                                      ③
            //回滚事务
            sqlite3_exec(db, "ROLLBACK TRANSACTION", NULL, NULL, NULL);    ④
            NSAssert(FALSE, @"删除数据失败。");
        }

        //先删除主表（比赛项目）数据
        NSString *sqlEventsStr = [[NSString alloc] initWithFormat:@"DELETE
        ➥from Events where EventID =%i;", model.EventID];                 ⑤

        if (sqlite3_exec(db, [sqlEventsStr UTF8String], NULL, NULL, NULL) !=
            SQLITE_OK) {                                                    ⑥
            //回滚事务
```

```
            assert(false, "删除数据失败。")
        }
        //提交事务
        sqlite3_exec(db, "COMMIT TRANSACTION", nil, nil, nil)           ⑧

        sqlite3_close(db)
    }
    return 0
}
```

```
            sqlite3_exec(db, "ROLLBACK TRANSACTION", NULL, NULL, NULL);     ⑦
            NSAssert(FALSE, @"删除数据失败。");
        }
        //提交事务
        sqlite3_exec(db, "COMMIT TRANSACTION", NULL, NULL, NULL);          ⑧

        sqlite3_close(db);
    }
    return 0;
}
```

这个删除方法非常特殊。由于比赛项目与比赛日程表之间有"主从"关系，当删除主表中的数据时，也要删除从表中的数据。比如，我们在比赛项目表中删除了赛马比赛项目，那么在比赛日程表中当然也不能有赛马项目的比赛日程了，因此需要先删除从表（比赛日程表）中的相关数据，再删除主表（比赛项目表）中的数据。第①行代码是删除从表中数据的SQL语句，第③行使用sqlite3_exec函数执行SQL语句，第⑤行代码是删除主表中数据的SQL语句，第⑥行使用sqlite3_exec函数执行SQL语句。

这两条语句（第③行和第⑤行代码）的执行应该放在一个事务[①]里，它们具有原子性，要么全部成功，要么全部失败，这是事务的原子性。第②行代码立刻提交之前的事务，并开启一个新事务。在执行删除操作失败的情况下，第④行代码和第⑦行代码要回滚（撤销）事务。在成功删除的情况下，需要提交（确定）事务，如第⑧行代码所示。

查询所有数据的方法findAll的代码如下：

```
//EventsDAO.swift文件
//查询所有数据的方法
public func findAll() -> [Events] {

    var listData = [Events]()

    if self.openDB() {

        let sql = "SELECT EventName, EventIcon,KeyInfo,
➥BasicsInfo,OlympicInfo,EventID FROM Events"
        let cSql = sql.cString(using: String.Encoding.utf8)

        var statement: OpaquePointer? = nil

        //预处理过程
        if sqlite3_prepare_v2(db, cSql, -1, &statement, nil) == SQLITE_OK {

            //执行
            while sqlite3_step(statement) == SQLITE_ROW {

                let events = Events()

                if let strEeventName = getColumnValue(index:0, stmt:statement!) {
                    events.EventName = strEeventName
                }
                if let strEventIcon = getColumnValue(index:1, stmt:statement!) {
                    events.EventIcon = strEventIcon
                }
                if let strKeyInfo = getColumnValue(index:2, stmt:statement!) {
                    events.KeyInfo = strKeyInfo
                }
                if let strBasicsInfo = getColumnValue(index:3, stmt:statement!) {
                    events.BasicsInfo = strBasicsInfo
                }
                if let strOlympicInfo = getColumnValue(index:4, stmt:statement!) {
```

```
//EventsDAO.m文件
//查询所有数据的方法
- (NSMutableArray *)findAll {

    NSMutableArray *listData = [[NSMutableArray alloc] init];

    if ([self openDB]) {

        NSString *qsql = @"SELECT EventName, EventIcon,
➥KeyInfo,BasicsInfo,OlympicInfo,EventID FROM Events";

        sqlite3_stmt *statement;
        //预处理过程
        if (sqlite3_prepare_v2(db, [qsql UTF8String], -1, &statement, NULL) ==
➥SQLITE_OK) {

            //执行
            while (sqlite3_step(statement) == SQLITE_ROW) {

                Events *events = [[Events alloc] init];

                char *cEventName = (char *) sqlite3_column_text(statement, 0);
                events.EventName = [[NSString alloc] initWithUTF8String:
➥cEventName];

                char *cEventIcon = (char *) sqlite3_column_text(statement, 1);
                events.EventIcon = [[NSString alloc] initWithUTF8String:
➥cEventIcon];

                char *cKeyInfo = (char *) sqlite3_column_text(statement, 2);
                events.KeyInfo = [[NSString alloc] initWithUTF8String:
➥cKeyInfo];

                char *cBasicsInfo = (char *) sqlite3_column_text(statement, 3);
                events.BasicsInfo = [[NSString alloc] initWithUTF8String:
```

① 数据库中的事务（transaction）是访问和更新数据库多个操作的执行单元。事务具有4个特性：原子性、一致性、隔离性、持久性，这4个特性通常称为事务的ACID特性。

```
                events.OlympicInfo = strOlympicInfo
            }
            events.EventID = Int(sqlite3_column_int(statement, 5))

            listData.append(events)
        }
    }
    sqlite3_finalize(statement)
    sqlite3_close(db)
  }
  return listData
}
```

```
            ➥cBasicsInfo];

            char *cOlympicInfo = (char *) sqlite3_column_text(statement,
            ➥4);
            events.OlympicInfo = [[NSString alloc] initWithUTF8String:
            ➥cOlympicInfo];

            events.EventID = sqlite3_column_int(statement, 5);

            [listData addObject:events];

        }
    }
    sqlite3_finalize(statement);
    sqlite3_close(db);

  }
  return listData;
}
```

findById:和modify:就不再介绍了。ScheduleDAO类与EventsDAO类也非常相似，读者可以下载我们的源代码查看。

29.3.7　迭代 2.7：数据库帮助类 DBHelper

DBHelper可以进行数据库初始化等操作，下面我们看看该类的代码：

```
//DBHelper.swift文件
import Foundation


let DB_FILE_NAME = "app.db"

public class DBHelper {

    static var db: COpaquePointer = nil

    //获得沙箱Document目录下的全路径
    static func applicationDocumentsDirectoryFile(_ fileName: String) ->
    ➥[CChar]? {                                                          ①

        let documentDirectory = NSSearchPathForDirectoriesInDomains
        ➥(.documentDirectory,
        ➥.userDomainMask, true)
        let path = (documentDirectory[0] as AnyObject)
        ➥.appendingPathComponent(DB_FILE_NAME) as String

        let cpath = path.cString(using: String.Encoding.utf8)

        return cpath
    }

    //初始化并加载数据
    public static func initDB() {

        let frameworkBundle = Bundle(for: DBHelper.self)

        let configTablePath = frameworkBundle.path(forResource: "DBConfig",
        ➥ofType: "plist")                                                 ②

        let configTable = NSDictionary(contentsOfFile: configTablePath!)  ③

        //从配置文件获得数据库版本号
        var dbConfigVersion = configTable?["DB_VERSION"] as? NSNumber
```

```
//DBHelper.h文件
#import <Foundation/Foundation.h>
#import "sqlite3.h"

#define DB_FILE_NAME @"app.db"

static sqlite3 *db;

@interface DBHelper : NSObject

//获得沙箱Document目录下的全路径
+ (const char *)applicationDocumentsDirectoryFile:(NSString *)fileName;

//初始化并加载数据
+ (void)initDB;

//从数据库获得当前数据库的版本号
+ (int)dbVersionNubmer;

@end


//DBHelper.m文件
#import "DBHelper.h"

@implementation DBHelper

+ (const char *)applicationDocumentsDirectoryFile:(NSString *)fileName {    ①

    NSString *documentDirectory = [NSSearchPathForDirectoriesInDomains
    ➥(NSDocumentDirectory, NSUserDomainMask, TRUE) lastObject];
    NSString *path = [documentDirectory stringByAppendingPathComponent:
    ➥fileName];

    const char *cpath = [path UTF8String];

    return cpath;
```

```
    if (dbConfigVersion == nil) {
        dbConfigVersion = 0
    }
    //从数据库DBVersionInfo表记录返回的数据库版本号
    let versionNubmer = DBHelper.dbVersionNubmer()

    //版本号不一致
    if dbConfigVersion?.int32Value != versionNubmer {
        let dbFilePath = DBHelper.applicationDocumentsDirectoryFile
        ➥(DB_FILE_NAME)
        if sqlite3_open(dbFilePath!, &db) == SQLITE_OK {
            //加载数据到业务表中
            print("数据库升级...")
            let createtablePath = frameworkBundle.path(forResource:
            ➥"create_load", ofType: "sql")
            let sql = try? NSString(contentsOfFile: createtablePath!,
            ➥encoding: String.Encoding.utf8.rawValue)                 ④
            let cSql = sql?.cString(using: String.Encoding.utf8.rawValue)
            sqlite3_exec(db, cSql!, nil, nil, nil)

            //把当前版本号写回到文件中
            let usql = NSString(format: "update
            ➥DBVersionInfo set version_number = %i", dbConfigVersion!.
            ➥intValue)                                                ⑤
            let cusql = usql.cString(using: String.Encoding.utf8.rawValue)

            sqlite3_exec(db, cusql, nil, nil, nil)

        } else {
            print("数据库打开失败。")
        }
        sqlite3_close(db)
    }

}

public static func dbVersionNubmer() -> Int32 {

    var versionNubmer: Int32 = -1

    let dbFilePath = DBHelper.applicationDocumentsDirectoryFile
    ➥(DB_FILE_NAME)

    if sqlite3_open(dbFilePath!, &db) == SQLITE_OK {
        let sql = "create table if not exists DBVersionInfo ( version_number
        ➥int )"
        let cSql = sql.cString(using: String.Encoding.utf8)

        sqlite3_exec(db, cSql!, nil, nil, nil)

        let qsql = "select version_number from DBVersionInfo"
        let cqsql = qsql.cString(using: String.Encoding.utf8)

        var statement: OpaquePointer? = nil
        //预处理过程
        if sqlite3_prepare_v2(db, cqsql!, -1, &statement, nil) == SQLITE_OK {
            //执行查询
            if (sqlite3_step(statement) == SQLITE_ROW) {
                //有数据情况
                print("有数据情况")
                versionNubmer = Int32(sqlite3_column_int(statement, 0))
            } else {
                print("无数据情况")
                let insertSql = "insert into DBVersionInfo (version_number)
                ➥values(-1)"
                let cInsertSql = insertSql.cString(using: String.Encoding.
                ➥utf8)
                sqlite3_exec(db, cInsertSql!, nil, nil, nil)
```

```
}

//初始化并加载数据
+ (void)initDB {

    NSBundle *frameworkBundle = [NSBundle bundleForClass:[DBHelper class]];

    NSString *configTablePath = [frameworkBundle
    ➥pathForResource:@"DBConfig" ofType:@"plist"];                    ②

    NSDictionary *configTable = [[NSDictionary alloc] initWithContentsOfFile:
    ➥configTablePath];                                                ③
    //从配置文件中获得数据库的版本号
    NSNumber *dbConfigVersion = configTable[@"DB_VERSION"];
    if (dbConfigVersion == nil) {
        dbConfigVersion = 0;
    }
    //从数据库的DBVersionInfo表记录返回的数据库版本号
    int versionNubmer = [DBHelper dbVersionNubmer];

    //版本号不一致
    if ([dbConfigVersion intValue] != versionNubmer) {
        const char *dbFilePath = [DBHelper applicationDocumentsDirectoryFile:
        ➥DB_FILE_NAME];
        if (sqlite3_open(dbFilePath, &db) == SQLITE_OK) {
            //加载数据到业务表中
            NSLog(@"数据库升级...");
            NSString *createtablePath = [frameworkBundle
            ➥pathForResource:@"create_load" ofType:@"sql"];
            NSString *sql = [[NSString alloc] initWithContentsOfFile:
            ➥createtablePath encoding:NSUTF8StringEncoding error:nil];    ④

            sqlite3_exec(db, [sql UTF8String], NULL, NULL, NULL);

            //把当前版本号写回到文件中
            NSString *usql = [[NSString alloc] initWithFormat:@"update
            ➥DBVersionInfo set version_number = %i", [dbConfigVersion
            ➥intValue]];                                                  ⑤
            sqlite3_exec(db, [usql UTF8String], NULL, NULL, NULL);
        } else {
            NSLog(@"数据库打开失败。");
        }
        sqlite3_close(db);
    }
}

+ (int)dbVersionNubmer {

    int versionNubmer = -1;

    const char *dbFilePath = [DBHelper applicationDocumentsDirectoryFile:
    ➥DB_FILE_NAME];

    if (sqlite3_open(dbFilePath, &db) == SQLITE_OK) {
        NSString *sql = @"create table if not exists DBVersionInfo (
        ➥version_number int )";
        sqlite3_exec(db, [sql UTF8String], NULL, NULL, NULL);

        NSString *qsql = @"select version_number from DBVersionInfo";
        const char *cqsql = [qsql UTF8String];

        sqlite3_stmt *statement;
        //预处理过程
        if (sqlite3_prepare_v2(db, cqsql, -1, &statement, NULL) == SQLITE_OK)
            {
            //执行查询
            if (sqlite3_step(statement) == SQLITE_ROW) { //有数据情况
                NSLog(@"有数据情况");
```

```
            }
        }
        sqlite3_finalize(statement)
        sqlite3_close(db)
    } else {
        sqlite3_close(db)
    }
    return versionNubmer
  }

}
```

```
                versionNubmer = sqlite3_column_int(statement, 0);
            } else {//无数据情况，插入数据
                NSLog(@"无数据情况");
                NSString *insertSql = @"insert into DBVersionInfo (
                ➥version_number) values(-1)";
                const char *cInsertSql = [insertSql UTF8String];
                sqlite3_exec(db, cInsertSql, NULL, NULL, NULL);
            }
        }

        sqlite3_finalize(statement);
        sqlite3_close(db);
    } else {
        sqlite3_close(db);
    }

    return versionNubmer;
}

@end
```

DBHelper中定义了静态变量db（它是sqlite3指针类型）以及3个静态方法，其中applicationDocuments-DirectoryFile:方法用于获得数据库文件的全路径，initDB用来初始化数据库，dbVersionNubmer用来返回当前数据库的版本号。

上述代码中，第①行用来定义applicationDocumentsDirectoryFile:方法，在Swift版本中其返回值是[CChar]?类型，其中[CChar]相当于C语言中的const char *类型，便于给SQLite API使用。而在Objective-C版本中，返回值类型是const char *，因此在调用SQLite API时，需要使用UTF8String方法进行转换。

在initDB方法中，比较数据表DBVersionInfo的version_number字段内容与DBConfig.plist的DB_VERSION数据项内容是否一致：如果一致，则执行DLL脚本，创建数据库并插入数据到数据库。

第④行代码用于从create_load.sql脚本文件中读取SQL语句，从而实现建表和插入数据的操作，程序执行到这里的时候是最耗时的。这个文件中的字符量很大，我们进行数据库版本的管理是有必要的。

ScheduleDAO的测试类ScheduleDAOTests与EventsDAOTests非常类似，这里就不再介绍了。

29.3.8　迭代 2.8：发布到 GitHub

整个任务需要将代码提交给GitHub服务器，此时在终端执行下面的命令即可：

```
$ git add .
$ git commit -m 'jia commit'
$ git remote add hw git@github.com:tonyguan/TokyoOlympics.git
$ git push hw master
```

由于我们的任务是整个团队多人完成的，每个人的任务是独立的，相互之间没有交叉，因此如果不出现纰漏的话，应该不会出现版本冲突。关于版本冲突的相关内容，可参考26.2.5节。

29.4　任务 3：表示逻辑组件开发

从客观上讲，表示逻辑组件开发的工作量很大，不仅有很多细节工作，而且还要做两套表示逻辑，分别针对iPhone和iPad。该任务由对UIKit非常熟悉的成员负责完成。

表示逻辑组件实现的方式有两种：一种是在同一个工程中包含两套UI，就是故事板（在Xcode 6及其之后版本中，可以使用一个故事板文件实现）或者XIB文件有两套，发布的时候是一个应用；另一种是对于两个工程（iPhone是一个工程，iPad是另外一个工程）会有两个不同的应用。这里我们采用故事板方式。

表示逻辑相关的内容如图29-8所示。

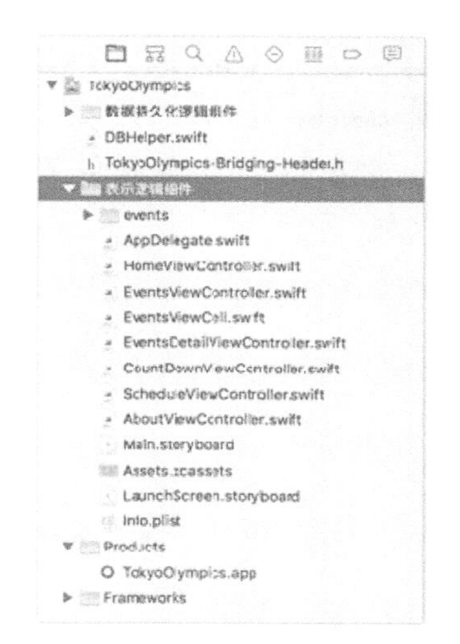

图29-8 表示逻辑相关的内容

表29-1简要介绍了表示逻辑的主要组件。

表29-1 表示逻辑的主要组件

组 件 名	说 明
AppDelegate	应用程序委托对象
HomeViewController	首页模块视图控制器类
EventsViewController	比赛项目模块视图控制器类
EventsViewCell	比赛项目模块视图中使用的集合视图中的单元格类
EventsDetailViewController	比赛项目详细模块视图控制器类
ScheduleViewController	比赛日程表模块视图控制器类
CountDownViewController	倒计时模块视图控制器类
AboutViewController	关于我们模块视图控制器类
Main.storyboard	iPhone和iPad版的故事板文件

29.4.1 迭代3.1：使用资源目录管理图片和图标资源

我们会在表示逻辑中用到很多图片和图标，除了应用图标和启动界面外，还有标签栏、工具栏等使用的图标。在本例中，我们还在Home等界面中用到了背景图片。这些图标和图片资源可以采用Xcode提供的资源目录管理。资源目录相关的技术在8.3节介绍过。本节我们先介绍使用资源目录管理标签栏图标和背景图片，而应用图标和启动界面的管理将在29.5节介绍。

下面我们先添加背景图片集，考虑应用的屏幕适配包括：iPhone和iPad视网膜显示屏。因此，我们需要为Home、Count Down和About三个模块分别准备多种不同规格的背景图片。然后在Xcode中打开Assets.xcassets目录，参考8.3节添加图片集Home-bg、CountDown-bg和About-bg，接着从Finder中拖曳不同规格的图片到设计界面。图29-9是设置图片集中About-bg的具体界面，Home-bg和CountDown-bg图片集的设置可以参考About-bg。

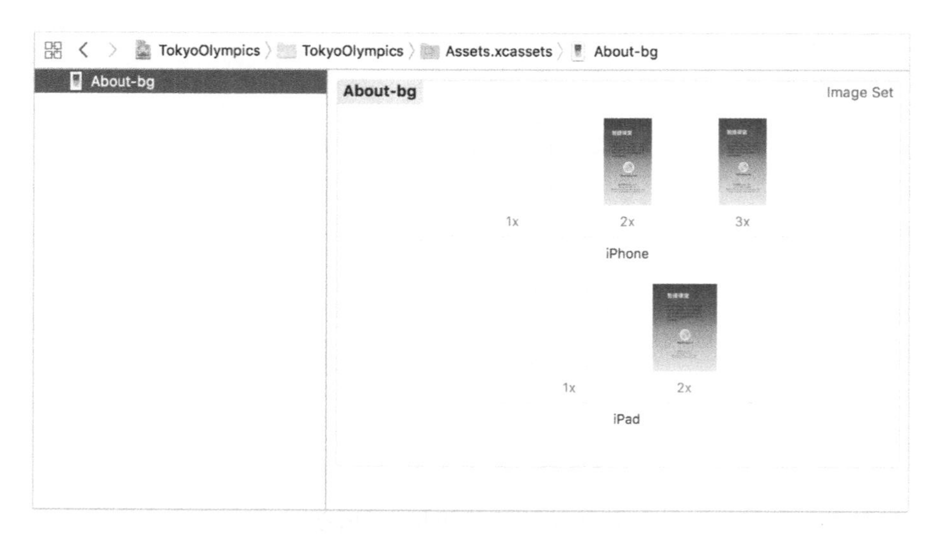

图29-9　About-bg图片集

下面我们再为标签栏图标添加图片集：Home-TabBarIcon、Events-TabBarIcon、Schedule-TabBarIcon、CountDown-TabBarIcon和About-TabBarIcon。图29-10是设置图片集中About-TabBarIcon的具体界面，其他标签栏图标的图片集可以据此设置。

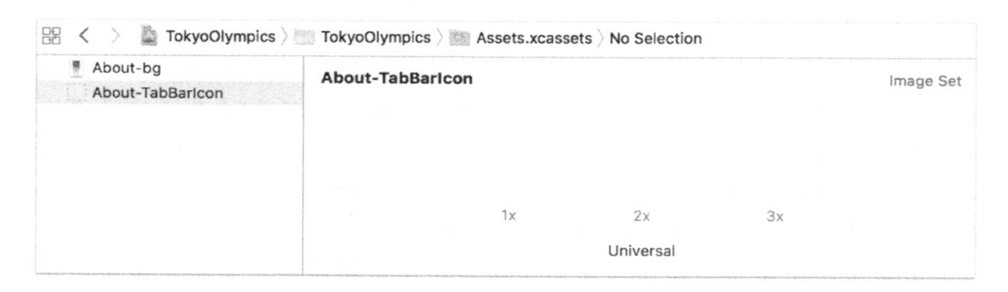

图29-10　About-TabBarIcon图片集

29.4.2　迭代 3.2：根据原型设计初步设计故事板

我们在设计应用原型的时候曾经绘制了原型设计图（见图29-2），它对于表示逻辑开发非常重要。通过原型设计图，我们可以了解界面中有哪些UI元素以及应用模块中界面之间的导航关系。所以，原型设计图可以帮助我们绘制故事板。苹果公司推出故事板技术的一个目的，也是便于开发人员查看界面之间的跳转关系。

但是随着业务复杂程度的增加，故事板会变得越来越大，这也是故事板的一个缺点。在Xcode 8中，苹果改进了Auto Layout和Size Class技术，使得屏幕适配更加简单。我们需要在一个故事板文件中分别设计iPhone和iPad应用界面。参考8.2节，分别设计iPhone和iPad应用界面，并选择iPhone竖屏，设置Size Class的值为wCompact | hRegular，选择iPad竖屏，设置Size Class的值为wRegular | hRegular。如图29-11所示是iPhone的设计界面。

从图29-11可见，每个单页面没法看清楚，笔者加上标号，通过标号给读者解释一下。

①：应用的根视图控制器，类型是UITabBarController。

②：首页的视图控制器，类型是UIViewController。

③：比赛项目模块的根视图控制器，类型是UINavigationController。

④：比赛项目模块的比赛项目视图控制器，类型是UIViewController。

⑤：比赛日程表模块的根视图控制器，类型是UINavigationController。

⑥：比赛日程表模块的视图控制器，类型是UITableViewController。
⑦：倒计时模块的视图控制器，类型是UIViewController。
⑧：关于我们模块的视图控制器，类型是UIViewController。

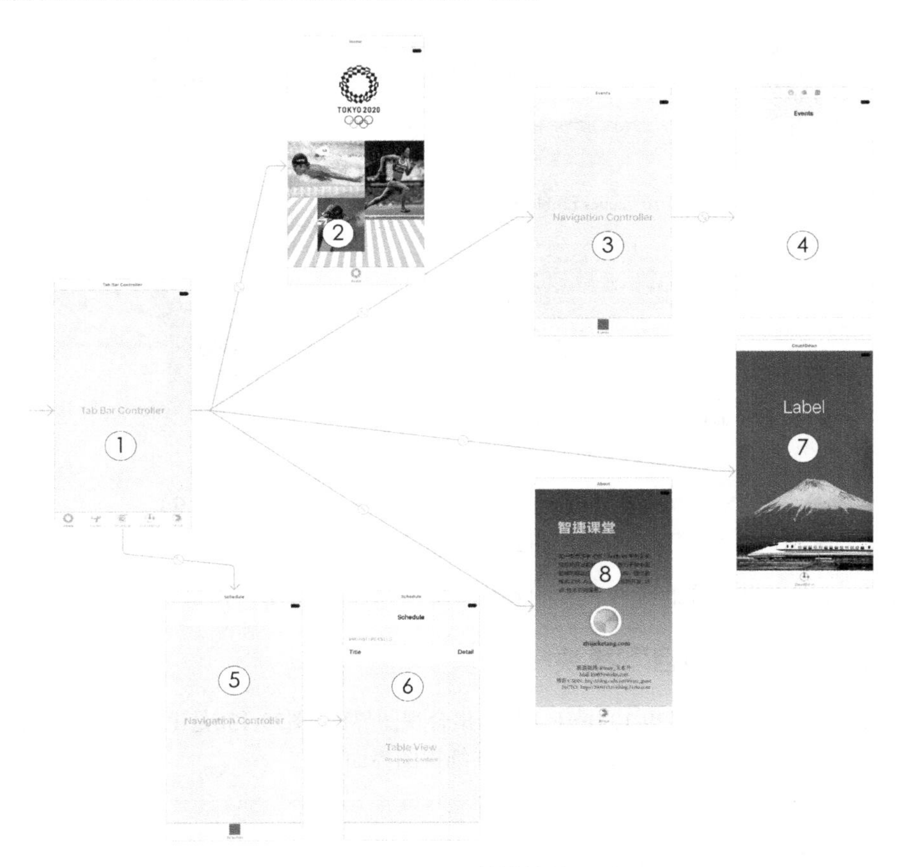

图29-11　iPhone的设计界面

29.4.3　迭代3.3："首页"模块

"首页"模块只有一个界面，其中图29-12a是iPad版的"首页"界面，图29-12b是iPhone版的"首页"界面。视图控制器是HomeViewController类，它继承自UIViewController类。我们按照图29-12所示的界面UI元素在iPad和iPhone故事板中进行设计。

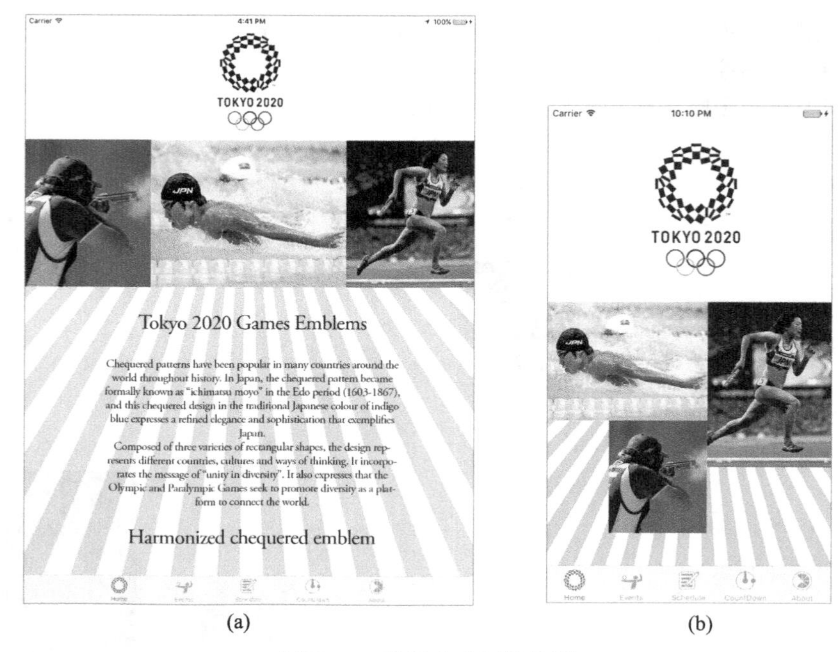

(a)　　　　　　　　(b)

图29-12　奥运会应用的首页

完成UI设计之后，在工程中添加HomeViewController类。首页只是展示一张图片，很简单，我们不用添加代码。

29.4.4　迭代 3.4："比赛项目"模块

"比赛项目"模块有两个界面，图29-13a是iPad版的集合视图界面，图29-13b是iPad版的详细视图界面。图29-14a是iPhone版的集合视图界面，图29-14b是iPhone版的详细视图界面。

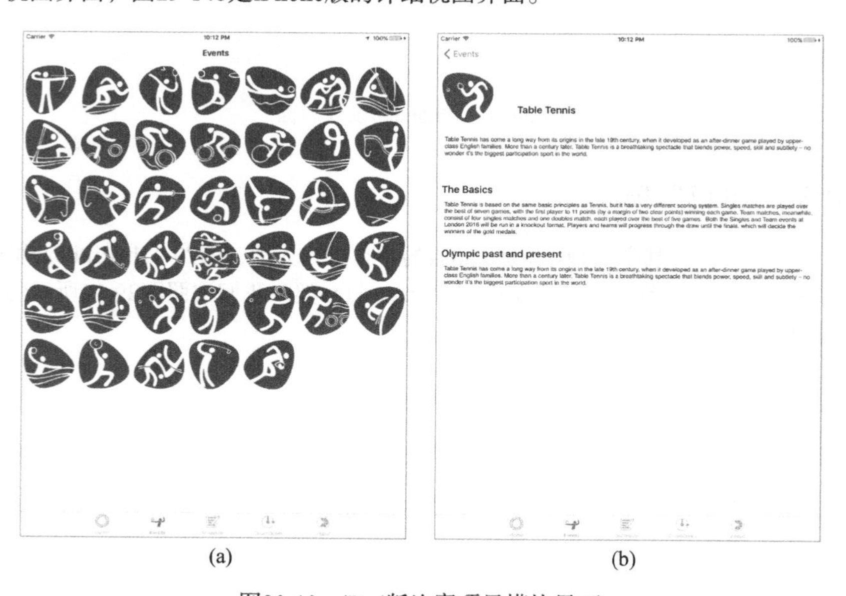

(a)　　　　　　　　(b)

图29-13　iPad版比赛项目模块界面

实现图29-13和图29-14所示的界面，我们推荐代码实现构建界面，图29-13所示的集合视图界面类似于5.3节~5.4节的实例。

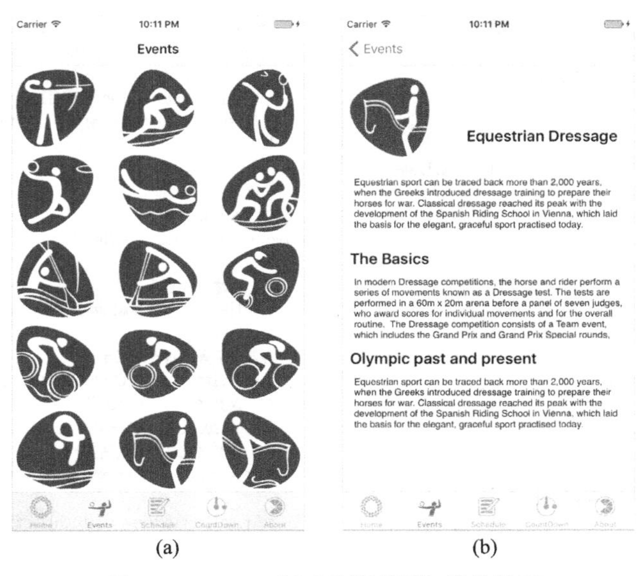

图29-14 iPhone版“比赛项目”模块界面

下面我们再看看代码部分。在该模块中EventsViewController类没有直接继承UICollectionViewController类，而是实现UICollectionViewDataSource和UICollectionViewDelegate两个协议，EventsDetailViewController类继承于UIViewController类。EventsViewController文件中类、属性定义以及初始化等相关的代码如下：

```
//EventsViewController.swift文件
import UIKit

private let reuseIdentifier = "cellIdentifier"

class EventsViewController: UIViewController, UICollectionViewDataSource,
➥UICollectionViewDelegate  {

    //一行中列数
    ///如果是iPhone设备，列数为2
    var COL_COUNT = 2

    var events : [Events]!

    var collectionView: UICollectionView!

    override func viewDidLoad() {
        super.viewDidLoad()

        self.setupCollectionView()

        if (self.events == nil || self.events.count == 0) {
            let dao = EventsDao.sharedfInstance
            let bl = EventsBL()
            //获取全部数据
            self.events = dao.findAll()
            self.collectionView.reloadData()
        }

    }

    func setupCollectionView() {

        //1.创建流式布局
        let layout = UICollectionViewFlowLayout()
        //2.设置每个单元格的尺寸
```

```
//EventsViewController.h文件
#import <UIKit/UIKit.h>

@interface EventsViewController : UIViewController

@end

//EventsViewController.m文件
#import "EventsViewController.h"
#import "EventsViewCell.h"
#import "EventsDetailViewController.h"

#import "Events.h"
#import "EventsDAO.h"


@interface EventsViewController ()  <UICollectionViewDataSource,
➥UICollectionViewDelegate>
{
    //一行中列数
    NSUInteger COL_COUNT;
}

@property (strong, nonatomic) NSArray * events;

@property (strong, nonatomic) UICollectionView* collectionView;

@end

@implementation EventsViewController

- (void)viewDidLoad {
    [super viewDidLoad];

    [self setupCollectionView];
```

```swift
        layout.itemSize = CGSize(width: 101, height: 101)
        //3.设置整个collectionView的内边距
        layout.sectionInset = UIEdgeInsetsMake(5, 5, 5, 5)
        //4.设置单元格之间的间距
        layout.minimumInteritemSpacing = 1

        self.collectionView = UICollectionView(frame: self.view.frame,
➥collectionViewLayout: layout)

        //设置可重用单元格标识与单元格类型
        self.collectionView.register(EventsViewCell.self,
➥forCellWithReuseIdentifier: "cellIdentifier" )

        self.collectionView.backgroundColor = UIColor.white

        self.collectionView.delegate = self
        self.collectionView.dataSource = self

        self.view.addSubview(self.collectionView)

        let screenSize  = UIScreen.main.bounds.size
        //计算一行中列数
        COL_COUNT = Int(screenSize.width / 106)                    ①

    }
    ......
}
```

```objc
    if (self.events == nil || [self.events count] == 0) {
        EventsDAO *dao = [EventsDAO sharedInstance];
        //获取全部数据
        self.events = [dao findAll];
        [self.collectionView reloadData];
    }
}

//设置集合视图布局
- (void) setupCollectionView {

    //1.创建流式布局
    UICollectionViewFlowLayout *layout = [[UICollectionViewFlowLayout alloc]
    ➥init];
    //2.设置每个格子的尺寸
    layout.itemSize = CGSizeMake(101, 101);
    //3.设置整个collectionView的内边距
    layout.sectionInset = UIEdgeInsetsMake(5, 5, 5, 5);
    //4.设置单元格之间的间距
    layout.minimumInteritemSpacing = 1;

    self.collectionView = [[UICollectionView alloc]
    ➥initWithFrame:self.view.frame collectionViewLayout:layout];

    //设置可重用单元格标识与单元格类型
    [self.collectionView registerClass:[EventsViewCell class]
    ➥forCellWithReuseIdentifier:@"cellIdentifier" ];

    self.collectionView.backgroundColor = [UIColor whiteColor];

    self.collectionView.delegate = self;
    self.collectionView.dataSource = self;

    [self.view addSubview:self.collectionView];

    CGSize screenSize  = [UIScreen mainScreen].bounds.size;
    //计算一行中列数
    COL_COUNT = screenSize.width / 106;                              ①
}
......
@end
```

在setupCollectionView方法中代码①行是根据当前设备屏幕宽度计算一行有几列。

下面我们再看看数据源UICollectionViewDataSource的实现方法，主要代码如下：

```swift
func numberOfSections(in collectionView: UICollectionView) -> Int {

    let num = self.events.count % COL_COUNT                           ①
    if (num == 0) { //偶数
        return self.events.count / COL_COUNT
    } else {        //奇数
        return self.events.count / COL_COUNT + 1
    }
}

func collectionView(_ collectionView: UICollectionView,
➥numberOfItemsInSection section: Int) -> Int {                      ②
    return COL_COUNT
}

func collectionView(_ collectionView: UICollectionView,
➥cellForItemAt indexPath: IndexPath) -> UICollectionViewCell {      ③

    let cell = collectionView.dequeueReusableCell(withReuseIdentifier
    ➥"cellIdentifier", for: indexPath) as! EventsViewCell

    //计算events集合下标索引
```

```objc
- (NSInteger)numberOfSectionsInCollectionView:(UICollectionView *)
➥collectionView {

    unsigned long num = [self.events count] % COL_COUNT;              ①
    if (num == 0) { //偶数
        return [self.events count] / COL_COUNT;
    } else {        //奇数
        return [self.events count] / COL_COUNT + 1;
    }
}


- (NSInteger)collectionView:(UICollectionView *)collectionView
➥numberOfItemsInSection:(NSInteger)section {
    return COL_COUNT;
}

- (UICollectionViewCell *)collectionView:(UICollectionView *)collectionView
cellForItemAtIndexPath:(NSIndexPath *)indexPath {

    EventsViewCell *cell = [collectionView
    ➥dequeueReusableCellWithReuseIdentifier: @"cellIdentifier"
```

```swift
    let idx = indexPath.section * COL_COUNT + indexPath.row

    if idx < self.events.count {
        let event = self.events[idx]
        cell.imageView.image = UIImage(named : event.EventIcon!)
    } else { //防止下标越界
        //下标越界 清除图标
        cell.imageView.image = nil
    }

    if (self.events.count <= idx) {//防止下标越界
        return cell;
    }

    return cell
}
```

```objc
➥forIndexPath:indexPath];
    //计算events集合下标索引
    NSInteger idx = indexPath.section * COL_COUNT + indexPath.row;

    if (idx < self.events.count) {
        Events *event = self.events[idx];
        cell.imageView.image = [UIImage imageNamed:event.EventIcon];
    } else { //防止下标越界
        //下标越界 清除图标
        cell.imageView.image = nil;
    }

    return cell;
}
```

numberOfSectionsInCollectionView:方法用来指定集合视图中节的个数，即有多少行，其中代码第①行是计算总比赛项目数（self.events长度）和每行列数（COL_COUNT）的余数，如果这个余数等于0，则整除，除数就是节的个数；如果不等于0，则没有整除，节数=除数+1，因为还需要一行显示内容。

当用户点击单元格时，界面会跳转到详细界面，这是通过实现集合视图的委托协议实现的，代码如下：

```swift
//MARK: --UICollectionViewDelegate
func collectionView(_ collectionView: UICollectionView,
➥didSelectItemAt indexPath: IndexPath) {

    //计算events集合下标索引
    let idx = indexPath.section * COL_COUNT + indexPath.row

    let event = self.events[idx]
    print("select event name : \(event.EventName)")

    let detailViewController = EventsDetailViewController()
    detailViewController.event = event

    self.navigationController?.pushViewController(detailViewController,
    ➥animated: true)                                                    ①

}
```

```objc
#pragma mark - UICollectionViewDelegate
- (void)collectionView:(UICollectionView *)collectionView
➥didSelectItemAtIndexPath:(NSIndexPath *)indexPath {

    //计算events集合下标索引
    NSInteger idx = indexPath.section * COL_COUNT + indexPath.row;
    Events *event = self.events[idx];

    NSLog(@"select event name : %@", event.EventName);

    EventsDetailViewController *detailViewController =
    ➥[[EventsDetailViewController alloc] init];
    detailViewController.event = event;
    [self.navigationController pushViewController:detailViewController
                                         animated:YES];                  ①
}
```

collectionView:didSelectItemAtIndexPath:方法在选择单元格时触发，然后通过代码第①行navigation-Controller的pushViewController:animated:方法进入详细界面。

EventsDetailViewController中的界面也是通过代码实现的，不需要故事板文件，其中主要的代码如下：

```swift
//EventsDetailViewController.swift文件
class EventsDetailViewController: UIViewController {

    var event: Events!

    var imgEventIcon: UIImageView!
    var lblEventName: UILabel!
    var txtViewKeyInfo: UITextView!
    var txtViewBasicsInfo: UITextView!
    var txtViewOlympicInfo: UITextView!

    override func viewDidLoad() {
        super.viewDidLoad()

        let screen = UIScreen.main.bounds

        self.view.backgroundColor = UIColor.white

        ///1.添加ImageView
        self.imgEventIcon = UIImageView(frame: CGRect(x: 10, y: 80, width: 102,
```

```objc
//EventsDetailViewController.h文件
#import <UIKit/UIKit.h>
#import <PersistenceLayer/Events.h>

@interface EventsDetailViewController : UIViewController

@property(nonatomic,strong) Events *event;

@end

//EventsDetailViewController.m文件
#import "EventsDetailViewController.h"

@interface EventsDetailViewController ()

@property (nonatomic) UILabel *lblEventName;
@property (nonatomic) UIImageView *imgEventIcon;
@property (nonatomic) UITextView *txtViewKeyInfo;
@property (nonatomic) UITextView *txtViewBasicsInfo;
@property (nonatomic) UITextView *txtViewOlympicInfo;
```

```
➥height: 102))
        self.imgEventIcon.image = UIImage(named : self.event.EventIcon!)
        self.view.addSubview(self.imgEventIcon)
        ......
    }
    ......
}
```

```
@end

@implementation EventsDetailViewController

- (void)viewDidLoad {
    [super viewDidLoad];

    CGRect screen = [[UIScreen mainScreen] bounds];

    self.view.backgroundColor = [UIColor whiteColor];

    ///1.添加ImageView
    self.imgEventIcon = [[UIImageView alloc] initWithFrame: CGRectMake(10, 80,
➥102, 102)];
    self.imgEventIcon.image = [UIImage imageNamed:self.event.EventIcon];
    [self.view addSubview:self.imgEventIcon];
    ......
}
......

@end
```

event属性是上一个界面传递过来的比赛项目数据。在viewDidLoad方法中，我们需要初始化各个控件的内容。

在集合视图中需要自定义单元格。EventsViewCell类是我们自定义的单元格类，其代码如下：

```
//EventsViewCell.swift文件
import UIKit

class EventsViewCell: UICollectionViewCell {

    var imageView: UIImageView!

    override init(frame: CGRect) {
        super.init(frame: frame)

        //单元格的宽度
        let cellWidth: CGFloat = self.frame.size.width

        ///1.添加ImageView
        let imageViewWidth: CGFloat = 100
        let imageViewHeight: CGFloat = 100
        let imageViewTopView: CGFloat = 0

        self.imageView = UIImageView(frame:
➥CGRect(x: (cellWidth - imageViewWidth) / 2,
➥y: imageViewTopView, width: imageViewWidth, height: imageViewHeight))

        self.addSubview(self.imageView)

    }

    required init?(coder aDecoder: NSCoder) {
        fatalError("init(coder:) has not been implemented")
    }
}
```

```
//EventsViewCell.h文件
#import <UIKit/UIKit.h>
@interface EventsViewCell : UICollectionViewCell

@property (strong, nonatomic) UIImageView *imageView;

@end

//EventsViewCell.h文件
#import "EventsViewCell.h"

@implementation EventsViewCell

- (id)initWithFrame:(CGRect)frame {

    self = [super initWithFrame:frame];
    if (self) {

        //单元格的宽度
        CGFloat cellWidth = self.frame.size.width;

        ///1.添加ImageView
        CGFloat imageViewWidth = 100;
        CGFloat imageViewHeight = 100;
        CGFloat imageViewTopView = 0;

        self.imageView = [[UIImageView alloc] initWithFrame:
➥CGRectMake((cellWidth - imageViewWidth) / 2,
➥imageViewTopView, imageViewWidth, imageViewHeight)];

        [self addSubview:self.imageView];
    }
    return self;
}

@end
```

其中自定义单元格类EventsViewCell继承自UICollectionViewCell类，单元格视图都是通过代码添加的。

29.4.5　迭代 3.5："比赛日程"模块

"比赛日程"模块有一个界面，其中图29-15a是iPad版的界面，图29-15b是iPhone版的界面。

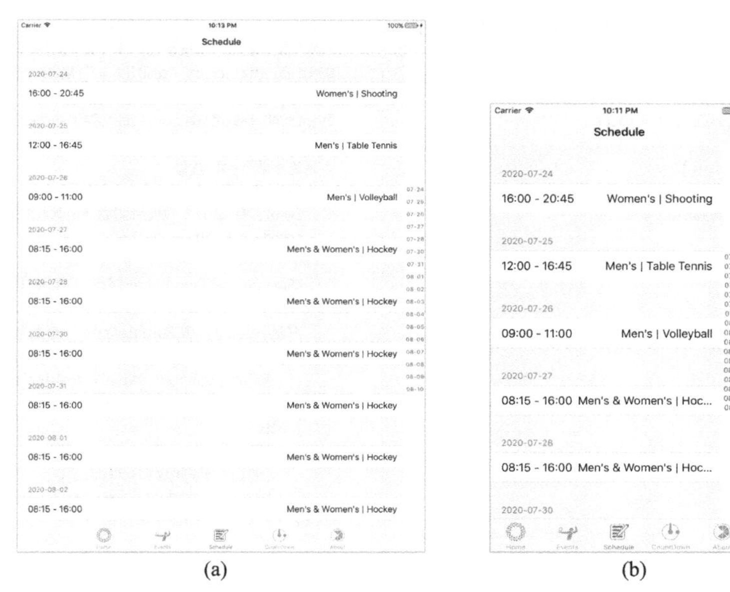

(a)　　　　(b)

图29-15　“比赛日程”界面

下面我们看看代码部分。在该模块中，ScheduleViewController类继承自UITableViewController类。ScheduleViewController中类定义、属性声明以及初始化等相关的代码如下：

```
//ScheduleViewController.swift文件
import UIKit

class ScheduleViewController: UITableViewController {

    //表视图使用的数据
    var data: [String : [Schedule]]!
    //比赛日期列表
    var arrayGameDateList: [String]!

    override func viewDidLoad() {
        super.viewDidLoad()

        if self.data == nil || self.data.count == 0 {
            self.data = self.readData()
            let keys  = self.data.keys
            //对key进行排序
            self.arrayGameDateList = keys.sorted(by: <)
        }
    }
    //查询所有数据的方法
    public func readData() -> [String : [Schedule]] {

        let scheduleDAO = ScheduleDAO.sharedInstance
        let schedules  = scheduleDAO.findAll()                          ①
        var resDict = [String : [Schedule]]()                           ②

        let eventsDAO = EventsDAO.sharedInstance

        //延迟加载Events数据
        for schedule in schedules {                                     ③

            let event = eventsDAO.findById(schedule.Event!)             ④
            schedule.Event = event                                      ⑤

            let allkey = resDict.keys
```

```
//ScheduleViewController.h文件
#import <UIKit/UIKit.h>

@interface ScheduleViewController : UITableViewController

@end
//ScheduleViewController.m文件
#import "ScheduleViewController.h"
#import "EventsDetailViewController.h"

#import "Schedule.h"
#import "ScheduleDAO.h"
#import "EventsDAO.h"

@interface ScheduleViewController ()
//表视图使用的数据
@property (strong, nonatomic) NSDictionary * data;
//比赛日期列表
@property (strong, nonatomic) NSArray * arrayGameDateList;
@end

@implementation ScheduleViewController

- (void)viewDidLoad {
    [super viewDidLoad];

    if (self.data == nil || [self.data count] == 0) {
        self.data = [self readData];
        NSArray* keys = [self.data allKeys];
        //对key进行排序
        self.arrayGameDateList = [keys
        ➥sortedArrayUsingSelector:@selector(compare:)];
    }
}
//查询所有数据的方法
-(NSMutableDictionary*) readData {
    ScheduleDAO *scheduleDAO = [ScheduleDAO sharedInstance];
```

```
        //把数组结构 ([Schedule]) 转化为字典结构 ([String : [Schedule]])
        if allkey.contains(schedule.GameDate!) {                          ⑥
            var value = resDict[schedule.GameDate!]                       ⑦
            value?.append(schedule)
            resDict[schedule.GameDate!] = value
        } else {
            var value = [Schedule]()                                      ⑧
            value.append(schedule)
            resDict[schedule.GameDate!] = value                           ⑨
        }
    }
    return resDict
}
  ......
}
```

```
    NSMutableArray* schedules  = [scheduleDAO findAll];                            ①
    NSMutableDictionary *resDict = [[NSMutableDictionary alloc] init];             ②

    EventsDAO *eventsDAO = [EventsDAO sharedInstance];

    //延迟加载Events数据
    for (Schedule *schedule in schedules) {                                        ③
        Events *event = [eventsDAO findById:schedule.Event];                       ④
        schedule.Event = event;                                                    ⑤

        NSArray  *allkey = [resDict allKeys];

        //把数组结构 (NSMutableArray) 转化为字典结构 (NSMutableDictionary)
        if([allkey containsObject:schedule.GameDate]) {                            ⑥
            NSMutableArray* value = resDict[schedule.GameDate];                    ⑦
            [value addObject:schedule];
        } else {
            NSMutableArray* value = [[NSMutableArray alloc] init];                 ⑧
            [value addObject:schedule];
            resDict[schedule.GameDate] = value;                                    ⑨
        }
    }
    return resDict;
}
......

@end
```

data属性保存从数据库返回的所有数据，arrayGameDateList属性是从data属性中提取出日期的集合。

初始化查询处理在viewDidLoad方法中进行，其原理与比赛项目模块一样。从数据持久化逻辑组件返回的数据是放到字典类型的data中，然后再从data中取出所有的键，这些键事实上就是所有的比赛日期集合。在赋值给arrayGameDateList属性前，要进行排序，因为字典结构中的数据是无序的。注意Objective-C与Swift对于集合排序使用了不同的方法。

readData方法比较复杂。为了提高查询效率，第①行查询出来的比赛日程数据并不包含比赛项目信息，而是在获取比赛项目信息数据时再进行加载。而我们的这个业务需要比赛日程和比赛项目信息，因此第③行循环遍历schedules集合，重新使用第④行查询比赛项目对象，然后再把比赛信息对象赋值给比赛日程的Event属性（如第⑤行所示）。

在readData方法中，除了延迟加载，还有一个问题需要解决：把数组结构数据转化为字典结构数据。从数据持久化逻辑组件返回的数据是数组结构的，返回的部分数据如表29-2所述。

表29-2　数组数据结构

ScheduleID	GameDate	GameTime	EventID
1	2020-08-05	16:00 - 20:45|Women's	29
2	2020-08-05	17:10 - 21:45|Women's	37
3	2020-08-05	17:00 - 21:45|Women's	36
4	2020-08-06	12:00 - 16:45|Men's	31
5	2020-08-06	14:30 - 17:15|Men's	30
6	2020-08-06	7:00 - 21:45|Men's	32
7	2020-08-07	9:00 - 21:45|Men's	35
8	2020-08-07	19:45 - 21:45|Men's	33

从表29-2中可见，GameDate（比赛日期）字段的很多记录是重复的，1、2、3记录的是2020年8月5日的3场比赛，类似地还有4、5、6和7、8。从存储空间、访问效率和方便访问等几个方面考虑，我们需要的数据结构应该是字典结构，如表29-3所述。

表29-3 字典数据结构

ScheduleID	GameDate	GameTime	EventID
1		16:00 - 20:45|Women's	28
2	2020-08-05	17:10 - 21:45|Women's	37
3		17:00 - 21:45|Women's	36
4		12:00 - 16:45|Men's	31
5	2020-08-06	14:30 - 17:15|Men's	30
6		7:00 - 21:45|Men's	32
7	2020-08-07	9:00 - 21:45|Men's	35
8		19:45 - 21:45|Men's	33

如表29-3所述，返回的数据放在字典中，它的键是GameDate，值是一个数组。例如，表29-3中的GameDate（2020-08-05）键，对应的值GameTime（16:00-20:45|Women's、17:10-21:45|Women's和17:00-21:45|Women's）为数组。

第⑥行用于判断字典中是否已经有某个GameDate键，如果有，则通过第⑦行取出数组，并把当前的比赛日程对象schedule添加到该数组中。如果没有，则通过第⑧行创建一个新的数组，然后再把当前的比赛日程对象schedule添加到数组中。第⑨行将GameDate和GateTime键值对数据放到字典中。

注意

在readData方法中把数组结构转化为字典结构的过程中，Swift版使用的数组和字典都是Swift原生类型，是值类型，Objective-C版本使用的数组和字典是Foundation框架提供的，是引用数据类型。所以在Swift版中，无论第⑥行为true还是false，都需要执行第⑨行，实现将GameDate和GateTime键值对数据放到字典中的操作。而Objective-C版本中，第⑥行为true的情况则不需要。因为第⑦行从字典中取出一个数组value，value数组对于Swift版本是值类型，修改的是它的副本，修改之后必须重新添加到字典中；而对于Objective-C版本，value数组是引用类型，修改的是引用所指的对象，也就是字典中的数组。

ScheduleViewController表视图数据源的实现方法如下：

```
//MARK: -- 表视图数据源
override func numberOfSections(in tableView: UITableView) -> Int {
    let keys = self.data.keys
    return keys.count
}

override func tableView(_ tableView: UITableView, numberOfRowsInSection
➥section: Int) -> Int {
    //比赛日期
    let strGameDate = self.arrayGameDateList[section]
    //比赛日期下的比赛日程表
    let schedules = self.data[strGameDate]!
    return schedules.count
}

override func tableView(_ tableView: UITableView, titleForHeaderInSection
➥section: Int) -> String? {
    //比赛日期
    let strGameDate = self.arrayGameDateList[section]
    return strGameDate
}

override func tableView(_ tableView: UITableView, cellForRowAt indexPath:
➥IndexPath) -> UITableViewCell {

    let cell = tableView.dequeueReusableCell(withIdentifier: "cellIdentifier",
    ➥for: indexPath)
```

```
#pragma mark -- 表视图数据源

- (NSInteger)numberOfSectionsInTableView:(UITableView *)tableView {
    NSArray* keys = [self.data allKeys];
    return [keys count];
}

- (NSInteger)tableView:(UITableView *)tableView
➥numberOfRowsInSection:(NSInteger)section {
    //比赛日期
    NSString* strGameDate = self.arrayGameDateList[section];
    //比赛日期下的比赛日程表
    NSArray *schedules = self.data[strGameDate];
    return [schedules count];
}

- (NSString *)tableView:(UITableView *)tableView titleForHeaderInSection:
➥(NSInteger)section {
    //比赛日期
    NSString* strGameDate = self.arrayGameDateList[section];
    return strGameDate;
}

- (UITableViewCell *)tableView:(UITableView *)tableView
➥cellForRowAtIndexPath:(NSIndexPath *)indexPath {

    UITableViewCell *cell = [tableView dequeueReusableCellWithIdentifier:
```

```swift
    //比赛日期
    let strGameDate = self.arrayGameDateList[indexPath.section]
    //比赛日期下的比赛日程表
    let schedules = self.data[strGameDate]!
    let schedule = schedules[indexPath.row]

    let subtitle = String(format: "%@ | %@", schedule.GameInfo!,
    ➥schedule.Event!.EventName!)
    cell.textLabel?.text = schedule.GameTime
    cell.detailTextLabel?.text = subtitle

    return cell
}

override func sectionIndexTitles(for tableView: UITableView) -> [String]? {    ①
    var listTitles = [String]()
    //2016-08-09 -> 08-09
    for item in self.arrayGameDateList {
        let title = (item as NSString).substring(from: 5)
        listTitles.append(title)
    }
    return listTitles
}
```

```objc
    ➥@"cellIdentifier" forIndexPath:indexPath];

    //比赛日期
    NSString* strGameDate = self.arrayGameDateList[indexPath.section];
    //比赛日期下的比赛日程表
    NSArray *schedules = self.data[strGameDate];

    Schedule *schedule = schedules[indexPath.row];
    NSString* subtitle = [[NSString alloc] initWithFormat:@"%@ | %@",
    ➥schedule.GameInfo, schedule.Event.EventName];

    cell.textLabel.text = schedule.GameTime;
    cell.detailTextLabel.text = subtitle;

    return cell;
}

-(NSArray *) sectionIndexTitlesForTableView: (UITableView *) tableView {    ①
    NSMutableArray *listTitles = [[NSMutableArray alloc] init];
    //2020-08-09 -> 08-09
    for (NSString *item in self.arrayGameDateList) {
        NSString *title = [item substringFromIndex:5];
        [listTitles addObject:title];
    }
    return listTitles;
}
```

在第①行代码中，sectionIndexTitlesForTableView:用于为表视图添加索引，返回一个索引的数组集合。但是如果我们直接返回arrayGameDateList属性，显示表视图索引是2020-08-09，这样的索引标题太长了，没有必要显示2020这个年份了。因此，我们在第②行代码中进行遍历，将集合截取为08-09的格式。

29.4.6 迭代 3.6："倒计时"模块

"倒计时"模块只有一个界面，其中图29-16a是iPad版的倒计时界面，图29-16b是iPhone版的"倒计时"界面。视图控制器是CountDownViewController类，它继承自UIViewController类。

按照图29-16所示的界面UI元素在iPad和iPhone的故事板中进行设计。

(a)

(b)

图29-16 "倒计时"界面

完成UI设计之后，在工程中添加CountDownViewController类。CountDownViewController的代码如下：

```swift
class CountDownViewController: UIViewController {

    //显示倒计时
    @IBOutlet weak var lblCountDownPhone: UILabel!
    @IBOutlet weak var lblCountDownPad: UILabel!

    override func viewDidLoad() {
        super.viewDidLoad()

        //创建DateComponents对象
        var comps = DateComponents()
        //设置DateComponents对象的日期属性
        comps.day = 24
        //设置DateComponents对象的月属性
        comps.month = 7
        //设置DateComponents对象的年属性
        comps.year = 2020

        //创建日历对象
        let calender = NSCalendar(calendarIdentifier:NSCalendar.Identifier.
        ➥gregorian)

        //获得2020-7-24的Date日期对象
        let destinationDate = calender!.date(from: comps)

        let date = Date()
        //获得当前日期到2020-7-24的DateComponents对象
        let components = calender!.components(.day, from: date ,
        ➥to:destinationDate!, options:.wrapComponents)
        //获得当前日期到2020-7-24相差的天数
        let days = components.day

        let strDays = String(format:"%li", days!)
        //设置iPhone中的标签
        self.lblCountDownPhone.text = strDays
        //设置iPad中的标签
        self.lblCountDownPad.text = strDays

    }
}
```

```objectivec
#import "CountDownViewController.h"

@interface CountDownViewController ()

@property (weak, nonatomic) IBOutlet UILabel *lblCountDownPhone;
@property (weak, nonatomic) IBOutlet UILabel *lblCountDownPad;

@end

@implementation CountDownViewController

- (void)viewDidLoad {
    [super viewDidLoad];

    //创建NSDateComponents对象
    NSDateComponents *comps = [[NSDateComponents alloc] init];
    //设置NSDateComponents对象的日期属性
    [comps setDay:24];
    //设置NSDateComponents对象的月属性
    [comps setMonth:7];
    //设置NSDateComponents对象的年属性
    [comps setYear:2020];
    //创建日历对象
    NSCalendar *calender = [[NSCalendar alloc]
    ➥initWithCalendarIdentifier:NSCalendarIdentifierGregorian];
    //获得2020-7-24的NSDate日期对象
    NSDate *destinationDate = [calender dateFromComponents:comps];
    NSDate *date = [NSDate date];
    //获得当前日期到2020-7-24的NSDateComponents对象
    NSDateComponents *components = [calender components:NSCalendarUnitDay
    ➥fromDate:date toDate:destinationDate
    ➥options:NSCalendarWrapComponents];
    //获得当前日期到2020-7-24相差的天数
    NSInteger days = [components day];

    NSString *strDays = [NSString stringWithFormat:@"%li",(long)days];
    //设置iPhone中的标签
    self.lblCountDownPhone.text = strDays;
    //设置iPad中的标签
    self.lblCountDownPad.text = strDays;
}

@end
```

上述获得2020-7-24日倒计时代码，我们在20.2.2节介绍过，这里不再赘述。

29.4.7 迭代3.7："关于我们"模块

"关于我们"模块只有一个界面，其中图29-17a是iPad版的"关于我们"界面，图29-17b是iPhone版的"关于我们"界面。视图控制器是AboutViewController类，它继承自UIViewController类。

(a)

(b)

图29-17　“关于我们”界面

按照图29-17所示的界面UI元素在iPad和iPhone故事板中进行设计。完成UI设计之后，在工程中添加AboutViewController类。

29.4.8　迭代 3.8：发布到 GitHub

整个任务完成后，需要将代码提交给GitHub服务器，此时在终端执行下面的命令即可：

```
$ git add .
$ git commit -m 'li commit'
$ git remote add hw git@github.com:tonyguan/TokyoOlympics.git
$ git push hw master
```

如果此时出现版本冲突，请参考26.2.5节查看并解决。

29.5　任务 4：收工

程序编写到现在，基本上已经完成了，但还有最后一点工作需要完成，这些工作包括图标和启动界面的添加、测试性能、提交代码到GitHub，以及在App Store上发布应用。

29.5.1　迭代 4.1：添加图标

图标是非常重要的，我们的UI设计师绘制好了一个图标，但是这个应用有iPad版本和iPhone版本，并且同时在一个工程中。因此，我们需要准备多种规格的图标，具体的规格和添加过程可以参考28.1.1节。

29.5.2　迭代 4.2：设计和添加启动界面

我们在28.1.2节介绍过添加启动界面的两种方式：启动图片和启动文件。本例中我们采用启动文件方式，需要打开LaunchScreen.storyboard文件重新设计界面，如图29-18所示，在其中添加一个图标。

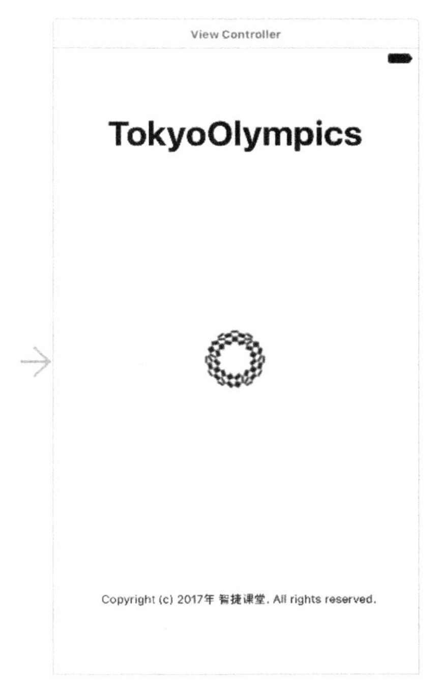

图29-18　启动界面

为了添加启动界面中的图标，我们需要添加图片集。在Xcode中打开Assets.xcassets文件夹，添加Icon图片集，如图29-19所示，添加3种图片。

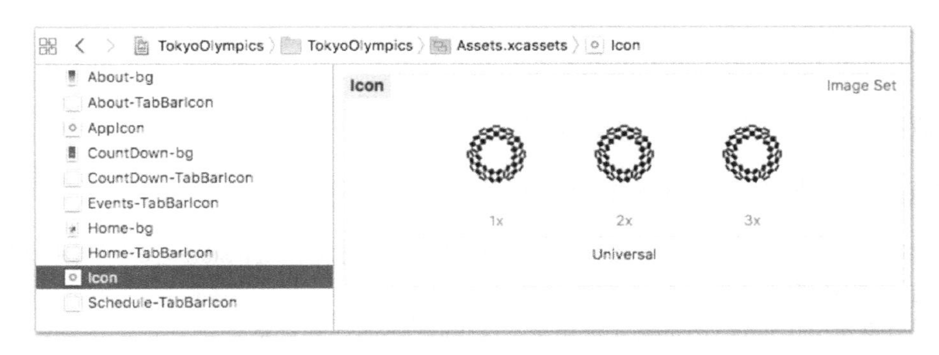

图29-19　添加Icon图片集

29.5.3　迭代 4.3：性能测试与改善

到现在为止，奥运会应用的主要功能基本实现了，此时可以在设备上测试一下。在性能、UI布局和用户体验等诸多方面，设备可能与模拟器不同，而且设备之间也有比较大的不同。如果有条件，凡是支持iOS 10所有型号的设备都应该拿来测试一下，这主要考虑用户设备的多样化。

我们对这个应用使用几个设备进行了测试，实际测试的结果是发现了问题。当第一次进入比赛项目或比赛日程表界面时，界面比较“卡”。经过分析发现，第一次查询使用数据库时，是由DAO对象创建数据库并添加初始化数据的，这个时间比较长。解决方法是把这些数据库的初始化放到子线程中处理，这个初始化只进行一次，除非数据库升级。因此，我们可以考虑在应用程序委托对象AppDelegate中进行初始化，主要是在启动界面加载时启动一个子线程，开始初始化数据库。在AppDelegate中添加的代码如下：

```
//AppDelegate.swift文件
......
func application(_ application: UIApplication,
➥didFinishLaunchingWithOptions launchOptions:
➥[UIApplicationLaunchOptionsKey: Any]?) -> Bool {

    DispatchQueue.global(qos: .utility).async {
        //加载并初始化数据库，提供效率，改善用户体验
        DBHelper.initDB()                                          ①
    }

    return true
}
```

```
//AppDelegate.m文件

#import <PersistenceLayer/DBHelper.h>
......

@implementation AppDelegate

- (BOOL)application:(UIApplication *)application
➥didFinishLaunchingWithOptions:(NSDictionary
  *)launchOptions {

    dispatch_queue_t queue = dispatch_get_global_queue(
    ➥DISPATCH_QUEUE_PRIORITY_DEFAULT, 0);
    dispatch_async(queue, ^ {
        //加载并初始化数据库，提高效率，改善用户体验
        [DBHelper initDB];                                         ①
    });
    return TRUE;
}
......

@end
```

这里我们使用GCD方式异步启动子线程，第①行代码用于初始化数据库。在这个方法中，我们会判断数据是否需要更新，如果需要，才去更新。

29.5.4　迭代 4.4：发布到 GitHub

整个任务完成后，应将代码提交给GitHub服务器，此时在终端执行下面的命令即可：

```
$ git add .
$ git commit -m 'guan commit'
$ git remote add hw git@github.com:tonyguan/TokyoOlympics.git
$ git push hw master
```

如果出现版本冲突，请参考26.2.5节查看并解决。

29.5.5　迭代 4.5：在 App Store 上发布应用

我们还需要进行一些其他的测试（如综合测试和用户测试等）才能在App Store上发布应用，但这些与本书关系不是很大，这里就不再介绍了。

现在，我们可以在App Store上发布这个应用了，发布流程可参考第28章。

29.6　小结

本章介绍了完整的iOS应用分析与设计、编程、测试和发布过程，整个过程采用的是敏捷开发方法。敏捷开发方法非常适合于iOS开发，希望广大读者能够认真学习。

欢迎加入

图灵社区 iTuring.cn

——最前沿的IT类电子书发售平台

电子出版的时代已经来临。在许多出版界同行还在犹豫彷徨的时候，图灵社区已经采取实际行动拥抱这个出版业巨变。作为国内第一家发售电子图书的IT类出版商，图灵社区目前为读者提供两种DRM-free的阅读体验：在线阅读和PDF。

相比纸质书，电子书具有许多明显的优势。它不仅发布快，更新容易，而且尽可能采用了彩色图片（即使有的书纸质版是黑白印刷的）。读者还可以方便地进行搜索、剪贴、复制和打印。

图灵社区进一步把传统出版流程与电子书出版业务紧密结合，目前已实现作译者网上交稿、编辑网上审稿、按章发布的电子出版模式。这种新的出版模式，我们称之为“敏捷出版”，它可以让读者以较快的速度了解到国外最新技术图书的内容，弥补以往翻译版技术书“出版即过时”的缺憾。同时，敏捷出版使得作、译、编、读的交流更为方便，可以提前消灭书稿中的错误，最大程度地保证图书出版的质量。

优惠提示：现在购买电子书，读者将获赠书款20%的社区银子，可用于兑换纸质样书。

——最方便的开放出版平台

图灵社区向读者开放在线写作功能，协助你实现自出版和开源出版的梦想。利用“合集”功能，你就能联合二三好友共同创作一部技术参考书，以免费或收费的形式提供给读者。（收费形式须经过图灵社区立项评审。）这极大地降低了出版的门槛。只要你有写作的意愿，图灵社区就能帮助你实现这个梦想。成熟的书稿，有机会入选出版计划，同时出版纸质书。

图灵社区引进出版的外文图书，都将在立项后马上在社区公布。如果你有意翻译哪本图书，欢迎你来社区申请。只要你通过试译的考验，即可签约成为图灵的译者。当然，要想成功地完成一本书的翻译工作，是需要有坚强的毅力的。

——最直接的读者交流平台

在图灵社区，你可以十分方便地写作文章、提交勘误、发表评论，以各种方式与作译者、编辑人员和其他读者进行交流互动。提交勘误还能够获赠社区银子。

你可以积极参与社区经常开展的访谈、乐译、评选等多种活动，赢取积分和银子，积累个人声望。